Principles of Power Electronics

Second Edition

Substantially expanded and updated, the new edition of this classic textbook provides unrivaled coverage of the fundamentals of power electronics.

It includes:

- Comprehensive and up-to-date coverage of foundational concepts in circuits, magnetics, devices, dynamic models, and control, establishing a strong conceptual framework for further study.
- Extensive discussion of contemporary practical considerations, enhanced by real-world examples, preparing readers for any design scenario from low-power dc/dc converters to multi-megawatt ac machine drives.
- New topics including SiC and GaN wide-bandgap materials, superjunction MOSFET and IGBT devices, advanced magnetics design, multi-level and switched-capacitor converters, RF converter circuits, and EMI.
- Over 300 new and revised end-of-chapter problems, designed to enhance and expand understanding of the material, with solutions for instructors.

Unique in its breadth and depth, and providing a range of flexible teaching pathways for instructors at multiple levels, this is the definitive guide to power electronics for graduate and senior undergraduate students in electrical engineering, and practicing electrical engineers.

John G. Kassakian is Professor of Electrical Engineering Emeritus at the Massachusetts Institute of Technology. He is the founding President of the IEEE Power Electronics Society, a Fellow of the IEEE, a member of the US National Academy of Engineering, and has taught, conducted research, and consulted in power electronics for over 45 years.

David J. Perreault is Ford Professor of Engineering at the Massachusetts Institute of Technology, with over 25 years of experience in power electronics research and teaching. He is a Fellow of the IEEE, and a member of the US National Academy of Engineering.

George C. Verghese is Henry Ellis Warren Professor of Electrical and Biomedical Engineering at the Massachusetts Institute of Technology, with over 40 years of research and teaching experience. He is an MIT MacVicar Faculty Fellow for outstanding contributions to undergraduate education, and a Fellow of the IEEE.

Martin F. Schlecht is the founder of SynQor, a supplier of high-performance power conversion solutions, and prior to that was Professor of Electrical Engineering at the Massachusetts Institute of Technology for 15 years. He has over 40 years of research, teaching, and industrial practice in power electronics.

"*Principles of Power Electronics* was a landmark in power electronics pedagogy when it was first published more than three decades ago. It is thrilling to see how these distinguished authors have not only thoroughly brought the book's core contents up to date, but also expanded its coverage to include several new topics that are increasingly important for students to learn as they prepare to enter this exciting field. Bravo!"

Thomas Jahns, *University Wisconsin – Madison*

"The second edition of the *Principles of Power Electronics* makes this classic book even more valuable. The book teaches power electronics from the ground up, providing the formal framework to learn its fundamentals and many advanced topics. This highly accessible book is an excellent text for a foundational course in power electronics. A must-have for both beginners and experienced practitioners."

Khurram Afridi, *Cornell University*

"The new edition of *Principles of Power Electronics* is a must for anyone in the field of power electronics, from the student learner to a working professional. The coverage is comprehensive, with detailed explanations backed up by well-chosen worked examples and illustrative problems. The treatment of magnetics, analysis, and design is particularly strong and a welcome addition."

Gerard Hurley, *University of Galway*

Principles of Power Electronics

SECOND EDITION

John G. Kassakian
Massachusetts Institute of Technology

David J. Perreault
Massachusetts Institute of Technology

George C. Verghese
Massachusetts Institute of Technology

Martin F. Schlecht
SynQor Inc.

CAMBRIDGE
UNIVERSITY PRESS

CAMBRIDGE
UNIVERSITY PRESS

Shaftesbury Road, Cambridge CB2 8EA, United Kingdom

One Liberty Plaza, 20th Floor, New York, NY 10006, USA

477 Williamstown Road, Port Melbourne, VIC 3207, Australia

314–321, 3rd Floor, Plot 3, Splendor Forum, Jasola District Centre, New Delhi – 110025, India

103 Penang Road, #05-06/07, Visioncrest Commercial, Singapore 238467

Cambridge University Press is part of Cambridge University Press & Assessment,
a department of the University of Cambridge.

We share the University's mission to contribute to society through the pursuit of
education, learning and research at the highest international levels of excellence.

www.cambridge.org
Information on this title: www.cambridge.org/highereducation/isbn/9781316519516

DOI: 10.1017/9781009023894

First published by Pearson College Div. 1991
Second edition published by Cambridge University Press & Assessment 2024 (version 2, February 2024)

Printed in Great Britain by CPI Group (UK) Ltd, Croydon CR0 4YY, February 2024

A catalogue record for this publication is available from the British Library.

Library of Congress Cataloging-in-Publication Data

Names: Kassakian, John G., author. | Perreault, David J., author. |
Verghese, George C., author. | Schlecht, Martin F., author.
Title: Principles of power electronics / John G. Kassakian,
David J. Perreault, George C. Verghese, Martin F. Schlecht.
Description: Second edition. | New York : Cambridge University Press, 2023.
| Includes bibliographical references and index.
Identifiers: LCCN 2023003028 (print) | LCCN 2023003029 (ebook) |
ISBN 9781316519516 (hardback) | ISBN 9781009023894 (epub)
Subjects: LCSH: Power electronics.
Classification: LCC TK7881.15 .K37 2023 (print) | LCC TK7881.15 (ebook) |
DDC 621.31/7–dc23/eng/20230126
LC record available at https://lccn.loc.gov/2023003028
LC ebook record available at https://lccn.loc.gov/2023003029

ISBN 978-1-316-51951-6 Hardback

Additional resources for this publication at www.cambridge.org/Kassakian_et_al

To our students,
who have been our best teachers,

and to Daniel Perreault Nakajima,
in memoriam.

Contents

Preface *page* xv
Acknowledgments xviii

1 Introduction 1
1.1 Power Electronic Circuits 1
1.2 Power Semiconductor Switches 2
1.3 Transformers 5
1.4 Nomenclature 7
1.5 Bibliographies 8
1.6 Problems 8

Part I Form and Function

2 Form and Function: An Overview 11
2.1 Functions of a Power Circuit 11
2.2 AC/DC Converters 13
2.3 DC/DC Converters 18
2.4 AC/AC Converters 20
2.5 Influence of Switch Implementation 22
Problems 24

3 Introduction to Rectifiers 27
3.1 Power in Electrical Networks 27
3.2 Single-Phase Half-Wave Rectifier 29
3.3 AC-Side Reactance and Current Commutation 35
3.4 Measures and Effects of Distortion 38
3.5 Bridge Rectifiers 44
Notes and Bibliography 49
Problems 50

4 Phase-Controlled Converters 57
4.1 Single-Phase Configurations 58
4.2 Phase Control with AC-Side Reactance 64
4.3 Inversion Limits 67

Notes and Bibliography 71
Problems 72

5 Pulse-Width-Modulated DC/DC Converters 76
5.1 The DC/DC Converter Topology 77
5.2 The Canonical Switching Cell 82
5.3 Direct Converter 82
5.4 Indirect Converter 89
5.5 Other PWM DC/DC Converters 92
5.6 Choice of Capacitor and Inductor Values 95
5.7 Semiconductor Device Stresses 103
5.8 Three-Level Flying-Capacitor Converter 104
5.9 Converter Operation with Discontinuous Conduction 106
Notes and Bibliography 108
Problems 109

6 Switched-Capacitor and Related Converters 116
6.1 Switched-Capacitor DC/DC Converter 116
6.2 Two-State Switched-Capacitor Converter 118
6.3 Switch Implementation 121
6.4 Other Switched-Capacitor DC/DC Converters 123
6.5 Other Kinds of Switched-Capacitor Converters 131
Notes and Bibliography 132
Problems 133

7 Isolated Pulse-Width-Modulated DC/DC Converters 136
7.1 Single-Ended Isolated Forward Converter 136
7.2 Double-Ended Isolated Forward Converter 145
7.3 Dual Active-Bridge Converter 150
7.4 Flyback Converter 153
7.5 Other PWM Isolated Converters 154
7.6 Effects of Transformer Leakage Inductance 155
7.7 Converters with Multiple Outputs 158
Notes and Bibliography 158
Problems 159

8 Single-Phase Switched-Mode DC/AC Converters 167
8.1 Basic Variable-Frequency Bridge Converter 167
8.2 Harmonic Reduction 172
8.3 Pulse-Width-Modulated DC/AC Converters 179
8.4 Current Control of Inverters 187

8.5 Multi-level Converters 190
8.6 Transformer-Isolated DC/AC Converters 192
8.7 Other DC/AC Converter Topologies 194
8.8 Power Balance in Single-Phase DC/AC Converters 196
8.9 Switched-Mode Rectifiers and Power Factor Correction 197

Notes and Bibliography 202

Problems 203

9 Polyphase Sources and Converters 207

9.1 Polyphase Sources 207
9.2 Three-Phase Sources 209
9.3 Introduction to Polyphase Rectifier Circuits 213
9.4 Phase-Controlled Three-Phase Converters 215
9.5 Commutation in Polyphase Rectifiers 216
9.6 Three-Phase Inverters 221
9.7 Space-Vector Representation and Modulation for Three-Phase Systems 229
9.8 Multi-Level Inverters 237

Notes and Bibliography 242

Problems 243

10 Resonant Converters 249

10.1 Review of Second-Order System Behavior 250
10.2 Quality Factor 254
10.3 Resonant Converter Analysis 256
10.4 Soft Switching of Resonant Converters 260
10.5 Resonant DC/DC Converters 264
10.6 Radio-Frequency Converters 271

Notes and Bibliography 284

Problems 285

11 AC/AC Converters 289

11.1 Introduction to AC/AC Converters 289
11.2 Energy Storage Requirements in a DC-Link Converter 290
11.3 The Naturally Commutated Cycloconverter 294
11.4 An Isolated High-Frequency-Link Cycloconverter 299
11.5 Solid-State Transformer 302
11.6 Matrix Converter 305

Notes and Bibliography 306

Problems 306

Part II Dynamic Models and Control

12 Dynamic Models and Control: An Overview 313
12.1 Dynamic Behavior of Power Converters 314
12.2 Dynamic Models 322
12.3 Averaged-Circuit Models 323
12.4 Linearized Models 329
12.5 Feedback Control 332
Notes and Bibliography 342
Problems 343

13 Averaged-Circuit and State-Space Models 346
13.1 Averaged-Circuit Models 346
13.2 Generalizing Circuit Averaging to the Fundamental Component 350
13.3 Continuous-Time State-Space Models 354
13.4 Discrete-Time or Sampled-Data Models 365
13.5 Generalized State-Space Models 368
13.6 Models for Controllers and Interconnected Systems 372
Notes and Bibliography 373
Problems 374

14 Linear Models and Feedback Control 378
14.1 Linearization 378
14.2 Linearizing an Averaged-Circuit Model 379
14.3 Linearizing Continuous-Time State-Space Models 383
14.4 Analysis of Continuous-Time LTI Models 385
14.5 Piecewise LTI Models 394
14.6 Linearizing Discrete-Time Generalized State-Space Models 395
14.7 Analysis of Discrete-Time LTI Models 399
14.8 Feedback Control Design 405
14.9 Multi-Loop Control 414
14.10 State Feedback 422
14.11 Digital Control 425
Notes and Bibliography 428
Problems 429

Part III Components and Devices

15 Components and Devices: An Overview 437
15.1 Practical Semiconductor Switches 438
15.2 Practical Energy Storage Elements 438
15.3 Semiconductor Devices 439

15.4 Capacitors 452
15.5 Inductors and Transformers 455
Problems 457

16 Review of Semiconductor Devices 461
16.1 Elementary Physics of Semiconductors 461
16.2 Simple Analysis of a Diode 472
16.3 Bipolar Junction Transistor 478
16.4 MOSFET 483
16.5 The Safe Operating Area 486
Notes and Bibliography 487
Problems 488

17 Power Semiconductor Devices 494
17.1 Bipolar Diode 494
17.2 Switch Transitions in a pin Diode 504
17.3 The Schottky Barrier Diode 507
17.4 Power BJT 511
17.5 Power MOSFET 517
17.6 Superjunction MOSFET 523
17.7 IGBT 524
17.8 Silicon Carbide and Gallium Nitride Devices 528
17.9 Thyristor 530
17.10 Datasheet 538
Notes and Bibliography 541
Problems 542

18 Introduction to Magnetics 545
18.1 Inductor 546
18.2 Saturation, Hysteresis, and Residual Flux 555
18.3 Losses In Magnetic Components 557
18.4 Transformers 566
18.5 Magnetic Material Properties 575
Notes and Bibliography 576
Problems 577

19 Magnetic Component Modeling 582
19.1 Mathematical Representation: Inductance Matrix 582
19.2 Circuit Representations for Two-Port Transformers 584
19.3 Determining Transformer Parameters 587
19.4 Multi-Winding Model 588

19.5 Extending the Magnetic Circuit Model Concept 597
19.6 Physical Electric Circuit Models for Magnetic Systems 603
Notes and Bibliography 607
Problems 608

20 Introduction to Magnetics Design 612
20.1 Filter Inductor Design and Core Factor 613
20.2 Thermal Constraints in Magnetics Design 618
20.3 Inductor Energy Storage Limits 621
20.4 AC Magnetics Sizing and Core Area Product 624
20.5 Performance Factor for Magnetic Materials 634
20.6 An Iterative Inductor Design Algorithm 636
Notes and Bibliography 638
Problems 640

21 Magnetics Loss Analysis and Design 643
21.1 Magnetic Diffusion 643
21.2 Winding Loss Calculation 655
21.3 Core Loss 675
Notes and Bibliography 684
Problems 686

Part IV Practical Considerations

22 Practical Considerations: An Overview 693
22.1 Gate and Base Drives 693
22.2 Thyristor Commutation Circuits 695
22.3 Snubbers, Clamps, and Soft Switching 697
22.4 Thermal Considerations 700
Problems 701

23 Gate and Base Drives 704
23.1 MOSFET and IGBT Gate Drives 705
23.2 Bipolar Transistor Base Drives 713
23.3 Thyristor Gate Drives 716
Notes and Bibliography 719
Problems 719

24 Snubber Circuits, Clamps, and Soft Switching 724
24.1 Turn-Off Snubber 725
24.2 Turn-On Snubber 730

24.3 Combined Turn-On/Turn-Off Snubber 733
24.4 Alternative Placements of the Snubber Circuit 735
24.5 Dissipation in Snubber Circuits 736
24.6 Soft Switching 739
Notes and Bibliography 748
Problems 750

25 Thermal Modeling and Heat Sinking 753
25.1 Static Thermal Models 754
25.2 Thermal Interfaces 758
25.3 Transient Thermal Models 762
Notes and Bibliography 767
Problems 768

26 Electromagnetic Interference and Filtering 771
26.1 EMI Specifications and Measurement 771
26.2 Filter Design 773
26.3 Common-Mode and Differential-Mode Variables 781
26.4 Parasitics and Circuit Layout 787
Notes and Bibliography 789
Problems 790

Index 793

Preface

The field of power electronics has advanced substantially since the initial publication of *Principles of Power Electronics* in 1991. New semiconductor devices, magnetic materials, fabrication technologies, and new modeling and control techniques have all combined to create an increasingly diverse universe of applications in which power electronics is embedded. Many component advances and the demands of new applications have pushed power converter switching frequencies into the hundreds of megahertz, more than an order of magnitude higher than what was practical at the time of publication of the first edition. And, simultaneously, the number of power electronics courses being taught worldwide has experienced a manifold increase. At a time when the efficient and socially responsible generation and use of energy are increasingly critical concerns globally, the importance of power electronics cannot be overstated.

As with the first edition, this second edition of *Principles of Power Electronics* is not intended as a reference book. It is a textbook specifically designed to *teach* the discipline of power electronics. Although the coverage is broad, we develop topics in sufficient depth to expose the *fundamental* principles, concepts, techniques, methods, and circuits necessary to understand and design power electronic systems for applications as diverse as a 100 mW switching converter operating at 100 MHz, a 25 MW motor drive, or a 1 GW high-voltage dc transmission terminal. All power electronics shares a common base, and we have tried to make this fact clear.

Principles of Power Electronics is divided into four parts, and each part has undergone significant rethinking, revision, and updating for this edition, as outlined below. Each begins with an overview chapter that establishes context for the remaining chapters of the part. These overviews are substantial enough to stand independently, and are intended to do so for certain teaching purposes.

Part I, "Form and Function," is the book's backbone. There we present the relationship between the form, or topology, of a power circuit and the functions it performs. The common features of circuits that perform the basic electrical energy conversion functions – ac/dc, dc/dc, dc/ac, and ac/ac – are introduced in this part. The deeper purpose of Part I, however, is to present ways of thinking about power electronic circuits, visualizing their behavior, and understanding their relationships with one another, so as to enable extension to new situations, and serve as the basis for synthesis as well as analysis. There is new material in this part on dc/dc converter topologies, multi-level converters and the use of flying capacitors, switched-capacitor converters, polyphase sources and converters, the concept of space-vector modulation for three-phase inverters, resonant converters (including RF converter designs), and soft switching.

Part II, "Dynamic Models and Control," considers the unique problems of modeling and controlling power electronic systems. We present analytical approaches to modeling and understanding their dynamic behavior, and show how to use these in designing and evaluating practical feedback control schemes. The emphasis is on fundamental formulations that apply across a range of power electronic systems, as illustrated by extensive examples in these chapters. The structuring of the material in this part is substantially revised relative to the first edition, for improved accessibility. The development of averaged models has also been extended considerably beyond dc/dc converters, with generalizations to track the dynamics of the fundamental (and harmonics) of converter waveforms. On the other hand, we have condensed the treatment of material that is now standard fare in undergraduate electrical engineering classes (and well supported by other textbooks), for example the analysis and feedback control of continuous-time, linear, time-invariant (LTI) systems described in the frequency domain or via LTI state-space models. Because of its role in stability evaluation of power electronic systems and its importance in the design of fully digital control systems, we retain our introduction to the topic of sampled-data modeling and control.

Part III, "Components," examines the behavior and characterization of the elements from which power electronic circuits are constructed. A review of semiconductor device physics precedes a discussion of specific device behaviors. The first edition's detailed development of the physics of specific devices has been replaced by a more phenomenological treatment. Also developed in this part is a discussion of the benefits and challenges of using the new wide-bandgap materials, SiC and GaN. The presentation of magnetic components has been expanded in this edition from a single chapter to four. This additional material addresses the challenge of designing magnetic components for the high switching frequencies now made possible by new MOSFET structures and GaN devices. We spend considerable time describing the behavior of magnetic materials, and the design and construction of inductors and transformers used at these high frequencies.

Part IV, "Practical Considerations," treats a variety of important additional topics that must be considered in the design of any practical system. Among these issues are gate and base drives, electromagnetic interference and filtering, snubbers, clamps and soft switching, and thermal modeling and heat sinking. New to this part is a discussion of circuit fabrication technologies necessary for very-high-frequency operation.

Unlike many power electronics texts which are designed for a single course, the scope of *Principles of Power Electronics* encompasses a *curriculum* of several sequential courses. A course in power electronics might use this book in one of several ways. Chapters 1 through 8, 9 (through Section 9.6), 10 (through Section 10.5), and the overview chapters in Parts II through IV would serve well as the basis for an advanced undergraduate or first graduate subject. Chapters 23 (Gate and Base Drives) and 18 (Introduction to Magnetics) might also be included. A more advanced graduate course might skim Part I and address Part II in detail. Other advanced courses may be tailored to need by selecting various chapters from Parts II through IV. Each chapter in Parts I, III, and IV is relatively self-contained. Selections from Part II can be made in at least two ways. Chapter 12 (Dynamic Models and Control: An Overview) and Chapter 13 (Averaged-Circuit and State-Space Models) may be used together in a course that emphasizes dynamic modeling of power electronic systems. An advanced graduate course that is particularly concerned with control could include Chapter 14 (Linear Models and Feedback Control), and if addressing machine control, could add Section 9.7 (Space-Vector Representation and Modulation for Three-Phase Systems).

We use examples extensively in this book to illustrate concepts or techniques introduced in the text, and also to introduce ways of thinking about problems, methods of analysis, and the use of approximations. The examples also form the basis for many of the end-of-chapter problems, and the creative instructor can use them to generate additional exercises, problems, or examples. We designed the end-of-chapter problems to stimulate thinking about the material presented in the chapter. That is, they are not intended as routine exercises to drill students in the use of particular equations in the text. Often, we introduce new circuits, concepts, or ways of approaching problems by using previous discussions in the text as a basis for considering the new material. We also present practical variations of circuits discussed in the text.

The notes and bibliography at the end of each chapter point you to selected papers in the research literature, and to books that underlie, complement, or extend the chapter material. These bibliographies, however, are not exhaustive.

We hope that instructors find this book to be a valuable teaching resource, and that students find it provides a challenging but enlightening learning experience.

Acknowledgments

It is no exaggeration to say that this second edition of *Principles of Power Electronics* would not exist if it weren't for the commitment, energy, care, and good-humored responsiveness of Sandeep Kaler at Toronto Metropolitan (formerly Ryerson) University. Despite the demands of the final stages of a doctoral program, he single-handedly worked to generate almost all the figures in this edition – recreating them from the first edition as needed, and preparing many new ones, then adjusting them wherever and whenever necessary. He also provided helpful feedback on various sections of the text. We are immensely grateful.

We are similarly indebted to Mike Ranjram on the faculty at Arizona State University who helped develop some portions of the text, contributing especially to the newer material in Part II, and revisiting various elements in this part from the first edition – checking the analysis, running new simulations, generating associated figures. He also vetted most of the end-of-chapter problems in the book while preparing a solutions manual for instructors.

We have had the benefit of careful reading and feedback from many colleagues (among them, former students, and their former students!) with whom we shared various draft chapters as we developed this second edition. They include Khurram Afridi, Arijit Banerjee, Richard Blanchard, Jessica Boles, Vahe Caliskan, Minjie Chen, Jesus del Alamo, Malik Elbuluk, Alex Hanson, Gerard Hurley, Jeffrey Lang, John H. Lienhard II, David Otten, Mohammad Qasim, Colm O'Rourke, Seth Sanders, Kenji Sato, Aleksandar (Alex) Stanković, Charles Sullivan, Joseph Thottuvelil, Amirnaser Yazdani, Xin Zan, and Yuhao Zhang. (We regret any inadvertent omissions there might be in this listing.) This brief mention cannot do justice to their generous and expert efforts – and the book was notably improved by their involvement.

Generations of students and teaching assistants in our introductory and advanced power electronics classes have worked with the material here, using the first edition of the book, and then drafts of this edition. We have counted on their being critical, fearless, and constructive in their feedback, and have never been disappointed.

The LaTeXwizardry of Amy Hendrickson, founder of TeXnology Inc, was invaluable throughout our work on this book. Her rapid and effective responses to our many requests helped shape the styling of the book and got us unstuck from typesetting quandaries countless times.

We are grateful to Montreal-based designer Rachel Paul for helping us reimagine EPSEL's "shazam" on its way to lighting up the cover of this book.

Nicola Chapman, Elizabeth Horne, Richard Hutchinson, Chloe Mcloughlin, Sarah Strange, and Hemalatha Subramanian of the editorial and production staff at Cambridge University Press have been most responsive and accommodating through the entire project, and we are very grateful.

Emanuel (Manny) Landsman has consistently – and in many different ways – supported our endeavors and those of MIT's Laboratory for Electromagnetic and Electronic Systems. Funding from the Landsman Charitable Trust was especially helpful in launching our effort here.

Wilma Kassakian, Heidi Nakajima, and Ann Kailath – doubtless the better and wiser halves of John, David, and George – have been patient and encouraging with our involvement in this multi-year labor, and have helped us keep perspective. They will be almost as happy as us to see this book wrapped up and on its way.

All four of us have had the good fortune to be shaped in fundamental ways, as students and professors, by MIT's stellar and lively Department of Electrical Engineering and Computer Science. Being part of this community for a combined total of more than 15 decades has been such a privilege, and we have gained more there than we could ever repay. Perhaps with this book we can pay some portion of our debt forward.

1 Introduction

In this chapter we describe power electronics and present a brief introduction to semiconductor switching devices and magnetic components. An introduction to these circuit elements is necessary because we use them in Part I, although we do not discuss them in detail until Part III. We also introduce in this chapter nomenclature that we use throughout the book.

1.1 Power Electronic Circuits

The dominant application of electronics today is to process information. The computer industry is the biggest user of semiconductor devices, and consumer electronics, including cameras and cell phones, is second. While all these applications require power (from a wall plug or a battery), their primary function is to process information; for instance, to take the digital optical signal produced by a cell phone camera and transform it into a photographic image. Power electronic circuits, on the other hand, are principally concerned with processing energy. They convert electrical energy from the form supplied by a source to the form required by a load. For example, the part of a computer that takes the ac mains voltage and changes it to the low-voltage dc required by the logic chips is a power electronic circuit (often abbreviated as *power circuit*). In many applications, the conversion process concludes with mechanical motion. In these cases the power circuit converts electric energy to the form required by the electromechanical transducer, such as a dc or ac motor.

Efficiency is an important concern in any energy processing system, for the difference between the energy into the system and the energy out is usually converted to heat. Although the cost of energy is sometimes a consideration, the most unpleasant consequence of generating heat is that it must be removed from the system. This consideration alone largely dictates the size of power electronic apparatus. A power circuit must, therefore, be designed to operate as efficiently as possible. The efficiency of very large systems exceeds 99%. High efficiency is achieved by using power semiconductor devices (where their voltage is nearly zero when they are on, and their current is nearly zero when they are off) to minimize their dissipation.† The only other components in the basic power circuit are inductors, capacitors, and transformers, so the ideal power circuit is lossless.

† Exceptions, such as linear voltage regulators, are so few that we do not consider them explicitly in this book.

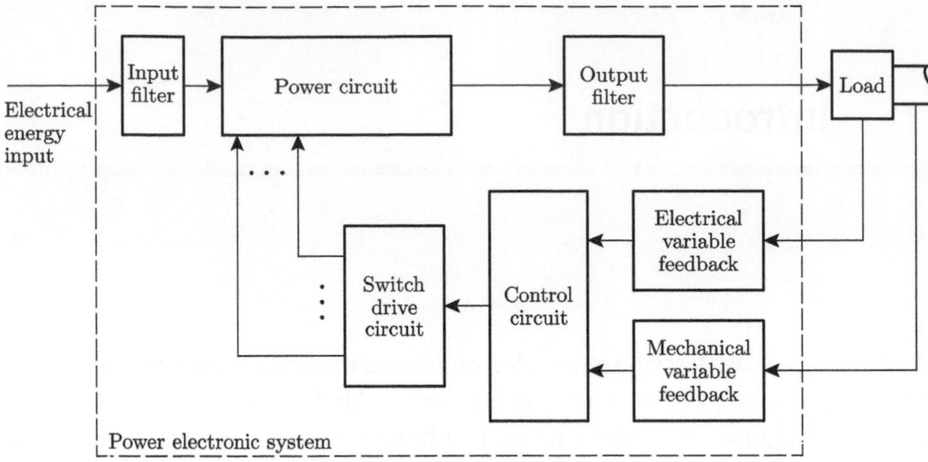

Figure 1.1 A block diagram of a typical power electronic system.

A power electronic system typically consists of much more than a power circuit, as shown in Fig. 1.1. Switching creates waveforms with harmonics that may be undesirable because they interfere with proper operation of the load or other equipment, so filters are often employed at the inputs and outputs of the power circuit. The system load, which may be electrical or electromechanical, is controlled via the feedback of electrical and/or electromechanical variables to a control circuit. This control circuit processes the feedback signals and drives the switches in the power circuit according to the demands of these signals. The system also includes mechanical elements, such as heatsinks and structures to support the physically large components of the power circuit.

In our circuit drawings, we use a capacitor symbol with a curved plate to differentiate it from the symbol for a contactor. We also find it a more elegant symbol than a pair of parallel plates. For electrolytic capacitors, which are polarized, the curved plate represents the negative terminal. We do not worry about this in our diagrams as the type of capacitor is irrelevant.

1.2 Power Semiconductor Switches

The basic semiconductor devices used as switches in power electronic circuits are the bipolar and Schottky diodes, the bipolar junction transistor (BJT), the metal-oxide-semiconductor field-effect transistor (MOSFET), the insulated-gate bipolar transistor (IGBT), and a class of latching bipolar devices known as thyristors, the most common of which is the silicon controlled rectifier (SCR). Their circuit symbols and operating regions in the v–i plane are shown in Fig. 1.2. We discuss these and other hybrid devices in detail in Part III.

What follows is a brief description of the salient operating characteristics of each device shown in Fig. 1.2. This information allows us to present the basic operation of power electronic circuits without first mastering Part III.

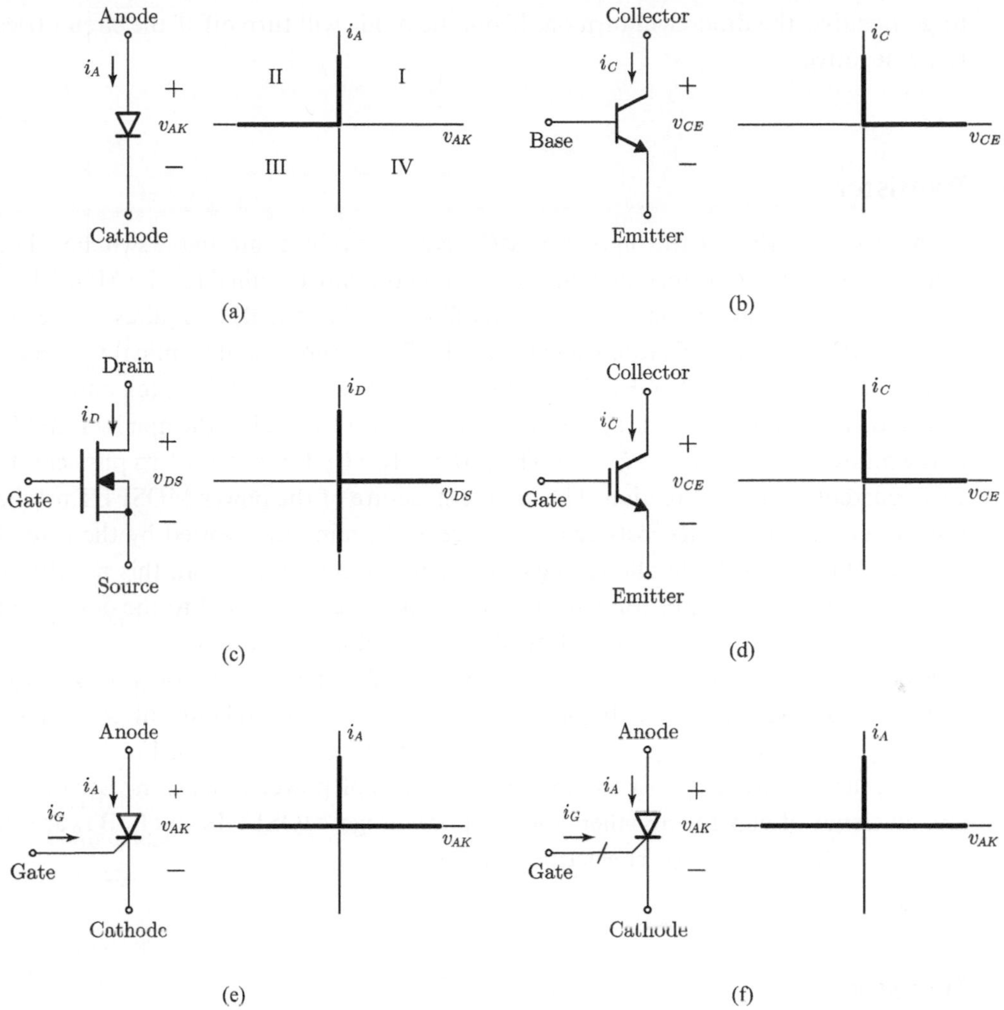

Figure 1.2 Circuit symbols and operating regions for power semiconductor devices: (a) diode; (b) (npn) bipolar junction transistor (BJT); (c) (n-channel) power metal-oxide-semiconductor field-effect transistor (MOSFET); (d) insulated gate bipolar transistor (IGBT); (e) silicon controlled rectifier (SCR); (f) gate turn-off thyristor (GTO).

1.2.1 Diode

The diode, whose symbol and variable definitions are shown in Fig. 1.2(a), is an uncontrollable semiconductor switch. It is uncontrollable because whether it is on or off is determined by the voltages and currents in the network, not by any action we can take. When on, its anode current, i_A, is positive. When off, its anode–cathode voltage, v_{AK}, is negative.† The diode switches in response to the behavior of its terminal variables. If it is off and the circuit causes v_{AK} to try

† The use of "*K*" instead of "*C*" reflects the Greek origin of the word cathode, or *kathodos*, meaning "way down," that is, the negative terminal.

to go positive, the diode will turn on. If on, the diode will turn off if the circuit tries to force i_A to go negative.

1.2.2 Transistor

Transistors, whether of the bipolar or MOS type, are fully controllable switches. They possess a third terminal (the *base* terminal for the BJT and the *gate* terminal for the MOSFET and IGBT) from which we can turn the device on and off. Control of the BJT requires a base current, while the MOSFET and IGBT require a gate voltage. The symbols and terminal variables for the npn BJT, n-channel power MOSFET, and IGBT are shown in Fig. 1.2(b), (c), and (d) respectively.

The BJT and IGBT can carry current in only one direction. For the npn BJT and IGBT shown in the figure, this direction is $i_C > 0$. The power MOSFET, because of its physical structure, can carry current in either direction. The unique structure of the power MOSFET results in a diode, known as the *body diode*, between its source and drain, represented by the arrowhead in the symbol. While this diode allows negative drain current conduction, this negative current can be supported with a lower forward drop by applying a gate signal to the device to turn on the channel and use it instead of the body diode to conduct the current.

When off, all three transistors can support only one polarity of voltage, which, for the transistors shown, are $v_{CE} > 0$ and $v_{DS} > 0$. These voltage and current polarities are reversed for the pnp BJT and the p-channel MOSFET. For reasons discussed in Part III, npn and n-channel devices are the most commonly used types of power transistors. Because of their ease and efficiency of control, and their low on-state voltage, MOSFETs and IGBTs are the dominant form of switch used in power electronic circuits.

1.2.3 Thyristor

The only members of the thyristor family that we describe in this introduction are the SCR and the gate turn-off thyrisor (GTO), whose circuit symbols are shown in Fig. 1.2(e) and (f).

The SCR is a switch that in some ways can be thought of as a “semi-controllable” diode. If no signal is applied to the gate, the device will remain off, independent of the polarity of v_{AK}. To turn the SCR on, a brief pulse of current, i_G, is applied to the gate terminal during a time when $v_{AK} > 0$. This initiates a regenerative turn-on process that quickly latches the SCR in the on state, in which $v_{AK} \approx 0$ and the gate no longer has any control over the device. When in this on state, the SCR can conduct only positive i_A. It turns off when i_A tries to go negative. So once on, the SCR behaves as a diode. In summary, the SCR is a diode whose turn-on can be inhibited by not applying a gate pulse.

The GTO is an SCR which has been constructed to enable a negative gate current to turn it off in the presence of $i_{AK} > 0$. The turn-off gain, defined as the ratio of I_{AK} to the negative gate current required to turn the device off, is typically in the low single digits. Two types of GTOs are available: asymmetrical GTOs, which cannot block a high reverse voltage, and symmetrical GTOs, which can.

1.3 Transformers

Transformers are a prominent feature of power electronic circuits. We treat them extensively in Part III, but the following introduction to their behavior permits us to use them as circuit elements in Parts I and II.

Transformers are employed to provide electric isolation and the step-up or step-down of ac voltages and currents. The *ideal transformer* shown in Fig. 1.3(a) has two windings of N_1 and N_2 turns. Dots indicate the orientation of the windings. If a voltage is applied to one winding so that the dot is positive, the dotted ends of all the other windings (only one in this case) are also positive. If its terminal variables are defined relative to the dots, as shown in Fig. 1.3(a), the ideal transformer has the following terminal relationships:

$$\frac{v_1}{v_2} = \frac{N_1}{N_2}, \tag{1.1}$$

$$\frac{i_1}{i_2} = -\frac{N_2}{N_1}. \tag{1.2}$$

From (1.1) and (1.2), we deduce that $v_1 i_1 = -v_2 i_2$; that is, the instantaneous power into one port is equal to the instantaneous power out of the other. The ideal transformer neither dissipates nor stores energy. A straightforward application of the above relations also shows that if an impedance of value Z_1 is connected to terminals 1–1′, an impedance of value $Z_2 = (N_2/N_1)^2 Z_1$ is measured at terminals 2–2′.

A transformer is ideal if it obeys (1.1) and (1.2), but no practical transformer is ideal. In most transformers, the principal departures from ideal result in some voltage and current being "lost" in the transformation, so terminal variables are not precisely related by (1.1) and (1.2). A model that represents these effects is shown in Fig. 1.3(b). Some of the terminal current i_1' is shunted through the *magnetizing inductance* L_μ and is called the *magnetizing current*. This is the current required to establish the necessary flux in the transformer core. So whereas i_1 and i_2 are still related by (1.1) and (1.2), the real terminal currents i_1' and i_2' are not. Similarly, the real terminal voltages v_1' and v_2' differ from v_1 and v_2 by the drops across $L_{\ell 1}$ and $L_{\ell 2}$, which are called *leakage*

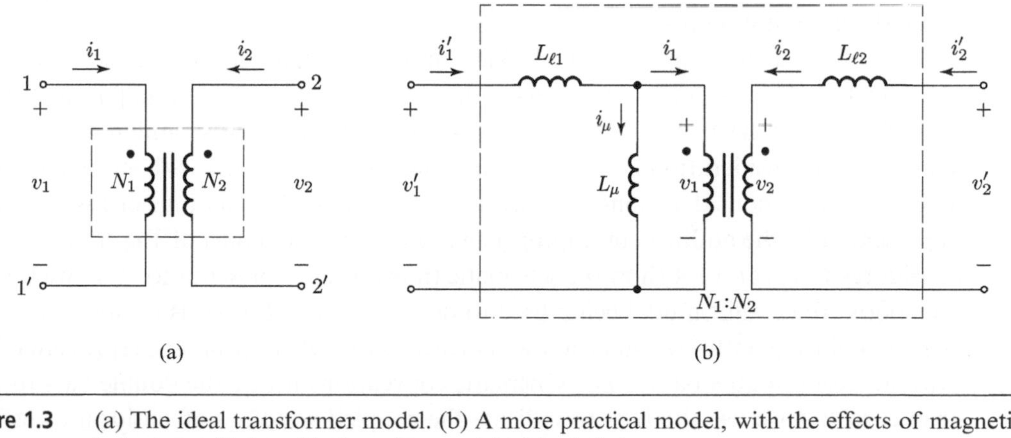

Figure 1.3 (a) The ideal transformer model. (b) A more practical model, with the effects of magnetizing inductance (L_μ) and leakage ($L_{\ell 1}$ and $L_{\ell 2}$) included.

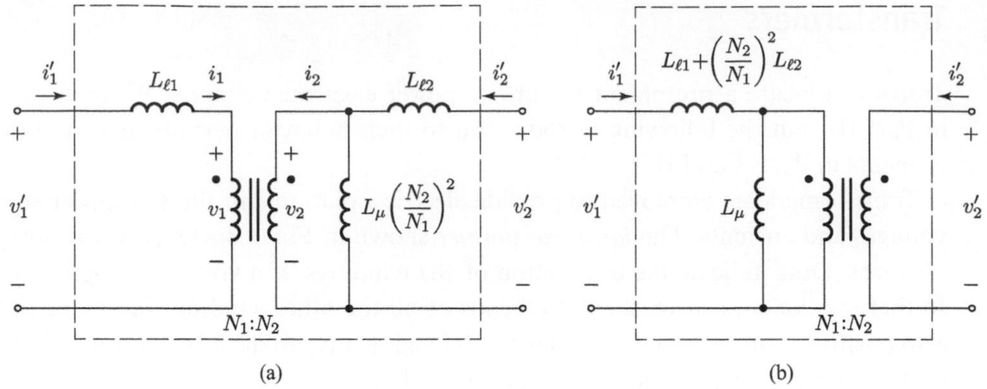

Figure 1.4 (a) The model of Fig. 1.3(b), with the magnetizing inductance placed on the N_2 side of the ideal transformer. (b) The model of Fig. 1.3(b), simplified by reflecting $L_{\ell 2}$ through the ideal transformer and combining it with $L_{\ell 1}$.

inductances. These represent flux linking one winding of the transformer but not the other. In Chapter 18 we describe the physical origins of these effects in much greater detail.

Figure 1.3(b) shows L_μ across the winding N_1. We can, however, *reflect* it through the ideal transformer so that it appears across the N_2 winding, as shown in Fig. 1.4(a). Sometimes we do this because the result is analytically more convenient to use. Although two leakage inductances, one for each winding, are shown in Fig. 1.3(b), they are often combined by reflecting one through the ideal transformer. If the voltage drop across this inductor is small relative to the voltage across L_μ, then L_μ can be moved inside this reflected inductance without introducing much of an error, and the two leakage inductances can be combined. The resulting approximate model is shown in Fig. 1.4(b).

Another useful model transformation is to reflect the entire circuit on one side of the ideal transformer to the other side. A transformation of this kind is shown in Fig. 1.5. There, the N_2 side circuit, C_o and R_o, has been "brought through" the ideal transformer. Of course, the isolation function is lost in the transformation, which makes the technique inappropriate for the analysis of some circuits.

We can calculate or measure the leakage and magnetizing inductances of transformers, and we sometimes construct transformers to have specific values for these parameters. And, even though we have been discussing only two-winding transformers, similar but somewhat more complicated considerations apply to the modeling of transformers with more than two windings. Other practical considerations, such as the resistance of the windings or losses in the core, are represented by the addition of appropriate elements to the model of Fig. 1.3(b).

Figures 1.3(b) and 1.4 show the schematic transformer representation that we use throughout this book: the circuit model being used to describe a transformer is enclosed in a dashed box. The model frequently has an ideal transformer as one of its elements, represented by windings with adjacent double bars. Some schematic conventions utilize the double bars to represent an iron core, but we use the bars to indicate the coupled windings of an ideal transformer when

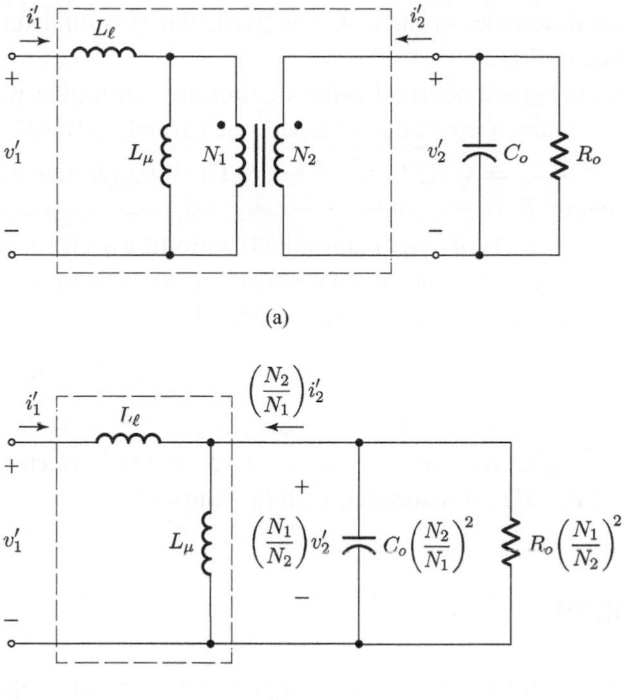

Figure 1.5 (a) A transformer with an RC load on the N_2 side. (b) The circuit of (a) with all the N_2 side components reflected to the N_1 side so that the ideal transformer can be eliminated from the transformer model.

it appears inside a dashed box. This convention avoids ambiguity and schematic clutter when more than two windings are involved.

1.4 Nomenclature

Because we discuss several different kinds of variables, we establish some notation and definitions now to permit a quick entry into the subject, and to avoid confusion later.

- Variables that may be time dependent are represented by lowercase names, such as v_1. When necessary for clarification, the time dependence is explicitly indicated, as in $v_1(t)$.
- Variables that are held constant are represented by uppercase names, such as V_1 or V_{dc}.
- The *average value* or *dc component* of a variable that varies periodically in time is denoted by angle brackets around the variable, for example $\langle v_o(t)\rangle = V_o$. Since this average value is a constant, it is represented by an uppercase name.
- The *local average* is defined for a possibly non-periodic waveform, say $x(t)$. It is the average over a time window of fixed length that moves with t; the window is usually chosen to be short

relative to the timescales of interest. The particular version defined in Chapter 12 is denoted by an overbar, so $\overline{x}(t)$.

- Perturbations around a constant value are indicated by a tilde, for instance $\widetilde{v}_C = v_C - V_C$.
- The *root mean square* (rms) value of a periodic variable $v(t)$ with period T is denoted by V_{rms} and defined as $V_{\text{rms}} = \sqrt{(1/T)\int_0^T v^2(\tau)\,d\tau}$. The integral may be taken over any contiguous interval of length T.
- Harmonic components of a (non-sinusoidal) periodic waveform are indicated by an additional subscript representing the harmonic number, for example $v_a = v_{a_1} + v_{a_3} + v_{a_5} + \cdots$.
- A sinusoidal function $v(t)$ can be written in the form

$$v(t) = V\cos(\omega t + \phi) = \text{Re}\left(Ve^{j\phi}e^{j\omega t}\right) = \text{Re}\big(\widehat{V}e^{j\omega t}\big),$$

where "Re" denotes the real part of the complex number that follows it. The *complex amplitude* of $v(t)$ is the complex number $\widehat{V} = Ve^{j\phi}$, and (particularly when plotted in the complex plane) is also termed the *phasor* associated with the sinusoid.

1.5 Bibliographies

We include an annotated bibliography at the end of most chapters. It provides sources of additional information on topics that you might want to pursue further.

1.6 Problems

Each chapter includes end-of-chapter problems that are designed to introduce variations or extensions of the chapter's material, or to provide students the opportunity to examine more closely concepts or circuits introduced in the chapter. They are not simply exercises in applying formulas. Further practice and confidence in applying and extending the material in a chapter can be obtained by filling in details of the derivations and examples, or by developing parallel results for circuits or converters in the same family or category as those explicitly treated in the chapter.

Part I

Form and Function

2 Form and Function: An Overview

Power electronic circuits change the character of electrical energy: from dc or ac to ac or dc, from one voltage level or frequency to another, or in some other way. We refer to such circuits generically as *converters*, *static converters* (because they contain no moving parts), *power processors*, or *power conditioners*. The part of the system that actually manipulates the flow of energy is the *power circuit*. It is the scaffold for the system's other components, such as the control circuit or the thermal management parts.

The power circuit has a basic topology to which we add other circuit elements that perform ancillary functions, such as protection against transient overvoltages, and filtering to eliminate electromagnetic interference. Although these other elements are important, they do not affect the primary function of the power circuit. Their purpose is to modify certain aspects of the power circuit's behavior, such as the rates of rise of currents or voltages during switch transitions. The study of a power circuit with all these additional elements can quickly become dominated by particulars rather than fundamentals. We therefore concern ourselves only with the basic forms of power circuits in Part I of this book. We describe methods of building on these basic structures to transform them into practical power circuits, and eventually into systems, in Parts II, III, and IV.

In Part I we show how a desired conversion function influences the form, or topology, of the power circuit. We also use these forms to illustrate the analytic tools and ways of thinking that you should apply when studying a power circuit. In most cases we keep both the function and the form simple. Where we present a more advanced topology, our goal is to show the connection between it and its simpler form – and the benefits gained from the added complexity.

2.1 Functions of a Power Circuit

Before we can specify the form of a power circuit, we must define its function. In general, its function is to alter the characteristics of electrical energy provided by one external system to those required by another. For instance, the power supply for a computer must convert the sinusoidal mains energy (60 Hz, 110 V rms in the United States) to a 5 V dc waveform. Another example is a power circuit for driving a variable-speed ac motor, which might draw power from a battery and deliver a sinusoidal current waveform to the motor. A dominant concern in almost every application is for power conversion at *high efficiency* to minimize power loss.

The types of functions that a power circuit can perform are limited only by the characteristics of electric power that are to be altered. As already mentioned, transformations from ac to dc (rectification) or from dc to ac (inversion) are two possible functions. Interfacing two systems with waveforms that are similarly shaped, but have different amplitudes, is another. If both waveforms are constant in time, we call the power circuit a *dc/dc converter*, and if both are alternating, we call the circuit an *ac/ac converter*. In the latter case, we might want to change the frequency or phase as well as the amplitude. The converter family also includes the ac/dc and dc/ac converters.

It is important that you not think of a power circuit's function as fixed for all time. The value of such a circuit is not just its ability to alter the form of electric energy, but also its ability to do so in response to a control signal. For instance, we can make the output of a computer power supply remain at 5 V even though the amplitude of the utility waveform changes by more than $\pm 20\%$. In some applications, such as light dimmers, the entire function of the power circuit is to provide this controllability.

A power circuit provides an interface between two other systems external to it, and therefore imposes relationships between the voltage and current waveforms at one port and those at the other. Exactly what waveforms exist at these ports depends not only on these relationships, but also on how the two external systems respond to being related in this manner. Thus, it is important always to describe the operation of a power circuit in the context of the external systems to which it is connected.

In certain situations, one external system – such as a voltage source – will dictate a waveform at one port, independent of the power circuit or the other external system. In these cases, we treat the waveform as an input to be processed by the power circuit to create an output. The external system at the output then defines the variables at its port, and we work backward through the power circuit to determine the shape of the covariable waveform at the input. Because of their simplicity, we often use these cases when presenting the initial topologies in the following chapters.

The direction of power usually determines whether ports serve as input or output. However, many power circuits are capable of processing bidirectional power, and identifying their ports as input or output leads to ambiguity. This ambiguity is aggravated by the fact that the types of semiconductor devices used to construct the circuit also constrain the direction of power. Thus, we could construct two identical topologies to process power in opposite directions. This is an important issue, which we address frequently, for it emphasizes that two visually identical circuits can behave differently, and conversely.

The basic form of a power circuit stems primarily from the need to provide efficient energy conversion. This need precludes the use of a transistor operating as a linear amplifier, regardless of its designed power level. For almost all energy conversion applications, such operation dissipates too much energy relative to the amount it processes. Similarly, using a resistor in conjunction with an energy storage element is not a practical way to make a low- or high-pass filter in a power circuit (unless the resistor is already present as part of the load at the output). All elements in the basic power circuit, at least in their ideal form, must be lossless. This requirement leaves us with three kinds of components with which to build power circuits: switches (semiconductor devices that are either fully on or off); energy storage elements (inductors and capacitors); and transformers.

2.2 AC/DC Converters

Without regard for the direction of energy flow, ac/dc converters comprise the broadest class of power electronic circuits. They are present in every piece of line-operated electronic equipment – from televisions to computers. They are also used extensively in industrial controls and processes, such as variable-speed motor drives, induction heating, plating, and the electrolytic production of chemicals.

Because of its symmetry, the basic converter topology is capable of bidirectional power; that is, the same topology can convert ac to dc, or dc to ac. For this reason we do not attach any significance to the order of the "ac" and "dc" in the name "ac/dc converter."

2.2.1 Basic Topology and Energy Flow

Our first example of a power circuit is one that creates a dc voltage from an ac voltage source. In this case, the power circuit produces a waveform that has an average value (the dc voltage) from one that does not (the ac source). Using switches configured in the topology of Fig. 2.1(a) produces the desired waveform. Because one of the two external networks connected to the converter is a resistor, there is no ambiguity about input or output ports: energy must flow from the source to the resistor. When the ac voltage is positive, closing the two switches marked P

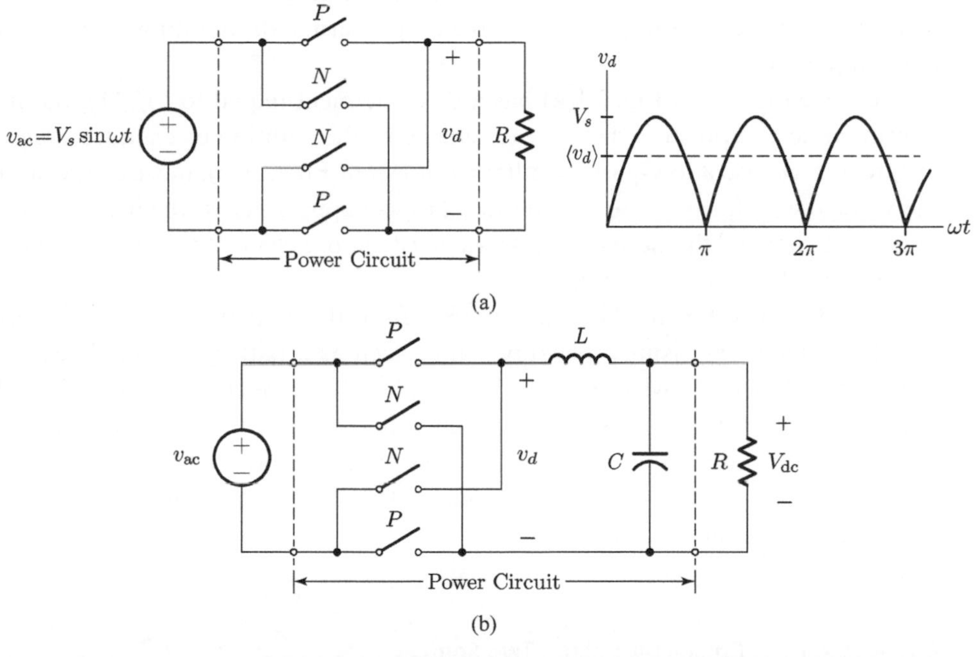

Figure 2.1 (a) A power circuit – consisting only of switches – that converts an ac voltage, v_{ac}, to one containing a dc component, v_d. (b) The ac/dc converter of (a) with the addition of filter elements L and C to remove the unwanted ac components from v_d, producing V_{dc}.

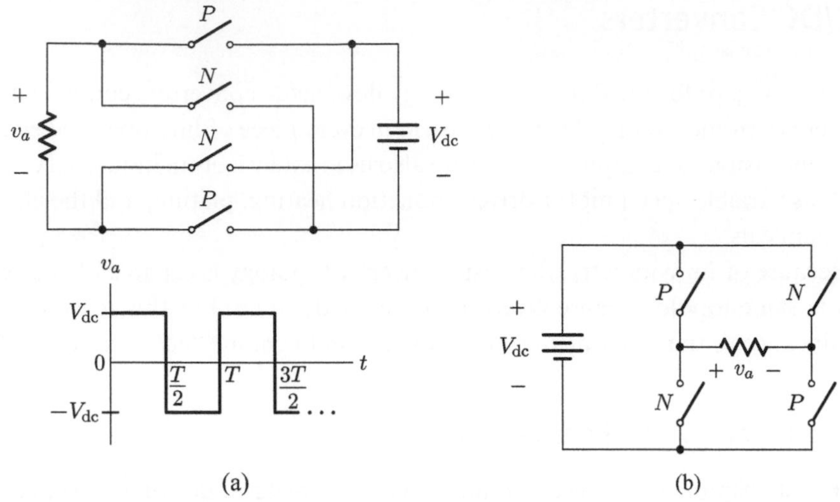

Figure 2.2 (a) The bridge converter of Fig. 2.1(a) connected and controlled to provide inversion. (b) The conventional way of drawing a bridge circuit.

and leaving the two switches marked N open connects the input voltage to the output in the positive sense. When the ac voltage is negative, reversing the states of the switches reverses the connection of the ac voltage source to the output terminals, resulting in an output voltage that is again positive.

The waveform v_d in Fig. 2.1(a) has a dc component (equal to $2V_s/\pi$), but it also contains unwanted ac components that we can remove by the addition of energy storage elements. For instance, if we use a low-pass LC filter as shown in Fig. 2.1(b), most of the ac components in the voltage waveform created by the switches will appear across the inductor instead of at the output. As part of the design process, we must choose element values large enough to achieve the level of attenuation desired.

An ac/dc converter in which energy flows from the ac network to the dc network is called a *rectifier*. However, using an energy source such as a battery for the dc external system, as Fig. 2.2(a) shows, allows energy to flow in the other direction. The circuit is then called an *inverter*. Note that the same power circuit provides both functions. In practice the external networks and switch implementation and control determine the function. The topology of this connection of four switches is called a *bridge*. It is used extensively in power electronics and is usually drawn as shown in Fig. 2.2(b).

Example 2.1 A Converter Linking Two Sources

If the inductor and/or the capacitor in the low-pass filter in the topology of Fig. 2.1(b) is large enough, the output voltage will be constant at $V_{dc} = 2V_s/\pi$. We now replace the capacitor and resistor with a voltage source of this value, as shown in Fig. 2.3(a), without changing the operation of the circuit. The result is a converter whose ports are connected to sources capable of supplying energy. What is the direction of power?

The answer is not at all obvious, because we cannot determine i_d without knowledge of the operating history of the switches, and the direction of i_d determines the direction of power. What we can do is write an expression for i_d in terms of v_d (which is explicitly determined by the switches) and V_{dc}:

$$i_d = \frac{1}{L}\int_{-\infty}^{t} (v_d - V_{dc})\, dt. \tag{2.1}$$

Figure 2.3(a) shows the voltage $v_d - V_{dc} = v_L$ for the case where the switches are controlled so that conduction alternates between the P and N switches at zero-crossings of v_{ac}. The average value of this voltage calculated over an interval π is

$$\langle v_L \rangle = \frac{1}{\pi}\int_{0}^{\pi} \left(V_s \sin \omega t - \frac{2V_s}{\pi} \right) d(\omega t) = 0, \tag{2.2}$$

making no net change in i_d. This condition is known as operation in the *periodic steady state*, because the circuit is in the same state at the beginning and end of each switching period. If L were very large ($L \approx \infty$), i_d would not vary, even during the interval π. If the switches were controlled in this manner for all time, there would never be an average value of voltage across L. and i_d would be forever zero.

However, we control the opening and closing of switches. Shifting the switching instants by an angle α from the zero-crossings of v_{ac}, as shown in Fig. 2.3(b), creates a nonzero average voltage across L. This voltage causes the current to change, decreasing (going negative) in this case. When the current reaches the value corresponding to the desired power, the switching times are changed to the zero-crossings of v_{ac} (Fig. 2.3(a). The average value of v_L is now once again zero and i_d remains constant.

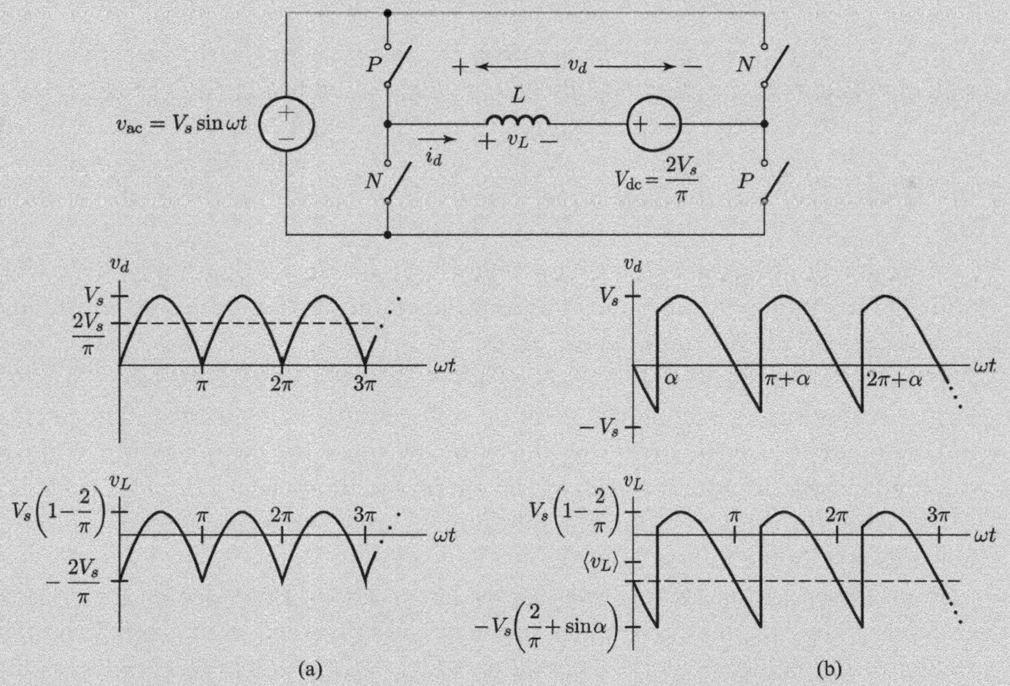

Figure 2.3 (a) The converter of Fig. 2.1(b) with the load resistor and filter capacitor replaced by a voltage source equal to the average value of the rectified voltage v_d. (b) Waveforms resulting when the switches are controlled to give an average (negative) value to v_L.

Note that because we have chosen a dc source voltage equal to the maximum possible value of the dc component of v_d, we cannot control the switches to give a positive average value to v_L. Thus we restrict the circuit to energy flow from the dc network to the ac network. A smaller value of V_{dc} would permit flow in either direction. This circuit, controlled as described, is one example of a class of circuits called *phase-controlled converters*. We consider them in detail in Chapter 4.

2.2.2 Filtering

The use of basic power electronic converter topologies, such as the ac/dc converter topology of Fig. 2.1(a), frequently results in deviation from the desired waveform by one or more of the port variables. In these cases we must modify the topology by adding filters to remove the unwanted components from the port variables. Figure 2.1(b) shows one way of doing this for the dc side of the ac/dc converter. Let's now consider this issue more generally.

We can obtain a simpler alternative to the filter of Fig. 2.1(b) by removing the capacitor and making the inductor very large. The resulting filter has a single pole at $1/\tau = R/L$. Placing this pole at a frequency that is very low compared to the switching frequency yields an inductor (and resistor) current that is nearly constant at some value I_{dc}. But the current in the ac source, i_a, is now a square wave instead of a sinusoid, which is undesirable for reasons we discuss in Chapter 3. We must employ another filter to eliminate all but the fundamental component from i_a.

Figure 2.4(a) shows the ac/dc converter topology of Fig. 2.1(b) with the filter on the dc side modified as in the preceding discussion, and a second-order filter consisting of L_a and C_a inserted on the ac side. The alternating action of the P and N switches creates the square-wave current i_a by alternately reversing the direction of I_{dc} as it is reflected through the switches to the dc side. The characteristics of the ac network connected to the converter strongly influence the form of the ac filter. In this case the network is simply the source, v_{ac}, which ideally has an incremental impedance of zero at any frequency. Therefore a shunt filter alone will not work, and the filter topology must present an impedance (ωL_a in this case) in series with the source at all but (ideally) the fundamental frequency.

In practice the low-pass ac filter does not work very well. The reason is that the first, and largest, undesirable harmonic of i_a is the third. Because it is so close to the fundamental, the filter pole cannot be placed at a frequency that strongly attenuates the third without also influencing the fundamental. Figure 2.4(b) shows an alternative filter circuit. It uses a series trap, L_3 and C_3, to shunt the third harmonic from the output, and the low-pass filter, L_5 and C_5, to remove harmonics from the fifth and above.

From these examples you can see that the introduction of filters complicates the basic power circuit topology required to perform the conversion function. It is equally important that you recognize the influence of the external networks on the form and effectiveness of the filter circuits. When these filters are part of the power circuit, you can determine the performance of the converter only in the context of an application that specifies the characteristics of the external networks.

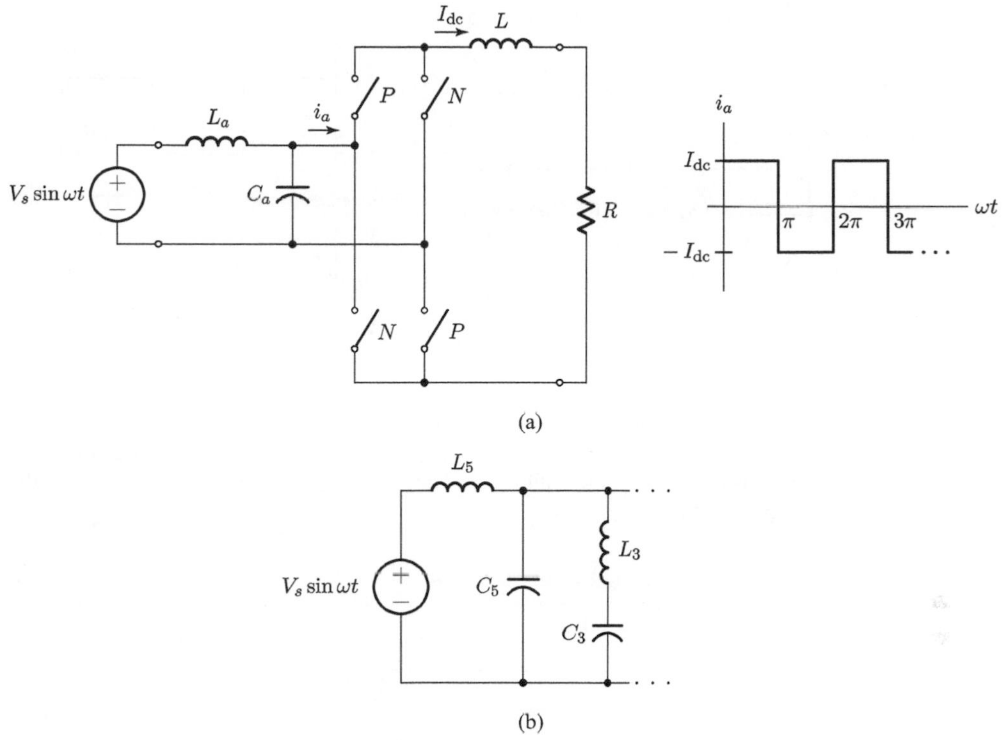

Figure 2.4 (a) The ac/dc converter topology of Fig. 2.1(a), with a first-order low-pass filter (RL) on the dc side and a second-order low-pass filter ($L_a C_a$) on the ac side. (b) An alternative and more effective ac-side filter.

Example 2.2 A Resonant Converter

In Example 2.1 we discussed a way of controlling power by varying the phase angle between the zero-crossings of the ac waveform and the switching times. When the ac port of the converter incorporates a resonant filter, we can sometimes use an alternative control technique based on the strong variation with frequency of the magnitude of the filter's transfer function.

Figure 2.5(a) is a dc/ac converter with a series resonant filter on the ac side. This topology, consisting of a split source and only two switches, is known as a *half bridge*. Figure 2.5(b) shows the waveforms v_a and v_{ac} when the filter is tuned to the fundamental of v_a. That is, the switching frequency ω_s is equal to the resonant frequency $\omega_o = 1/\sqrt{LC}$. If the Q of the series RLC circuit is high ($R \ll \omega_o L$), the filter provides good selectivity and v_{ac} is nearly sinusoidal. At switching frequencies other than $\omega_s = \omega_o$, we can determine the amplitude of the output from the magnitude of the admittance $Y(j\omega_s)$:

$$|V_{ac}| = RY(j\omega_s)V_{a_1} = R\left|\frac{-j\omega_s C}{\omega_s^2 LC - j\omega_s RC - 1}\right|\left(\frac{4}{\pi}V_{dc}\right), \tag{2.3}$$

where V_{a_1} denotes the amplitude of the fundamental (ω_s) component of v_a.

The magnitude of $Y(j\omega)$ is shown in Fig. 2.6. Switching at a frequency higher than ω_o allows the filter to still do a good job of removing harmonics from v_a. The sharp attenuation provided by the filter away from resonance reduces the amplitude of the almost sinusoidal output voltage. Thus, by varying the switching frequency, we can control the power delivered to the load resistor.

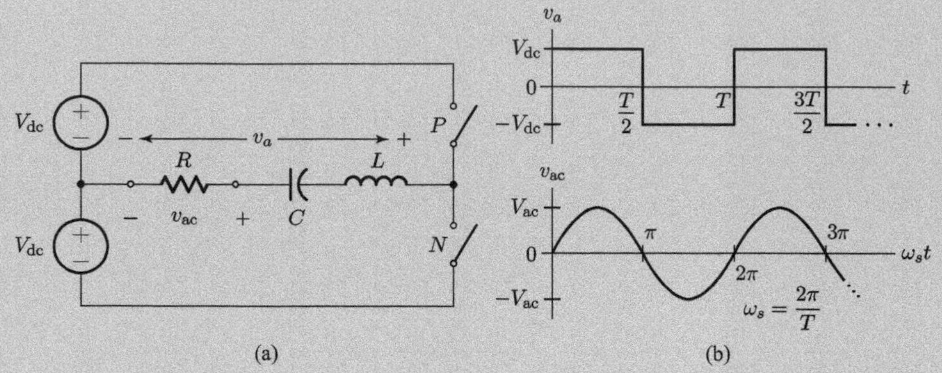

Figure 2.5 (a) A half-bridge dc/ac converter topology employing a series-tuned filter on the ac side. (b) The relationship between the unfiltered voltage, v_a, and the output voltage, v_{ac}, when the filter is tuned to the switching frequency ω_s.

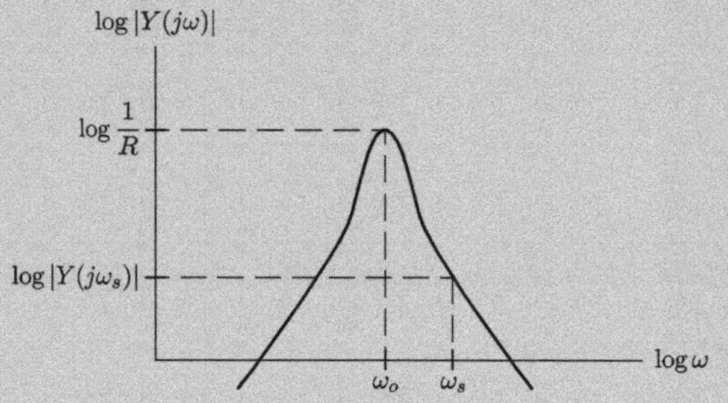

Figure 2.6 Admittance as a function of frequency for the *RLC* circuit on the ac side of Fig. 2.5(a).

The topology of Fig. 2.5(a) is known as a *series-resonant converter*. It is one member of a family of dc/ac converters called *resonant converters*, treated in more detail in Chapter 10. Designers use them principally to obtain clean, high-frequency ac waveforms, often for induction heating applications. Note that the technique of power control that we just discussed results in a varying output frequency, so any application of this type of control must not require a precise output frequency.

2.3 DC/DC Converters

Used extensively in power supplies for electronic equipment, dc/dc converters control the flow of energy between two dc systems. The dc/dc converter takes the dc output of the ac/dc converter and transforms it to the different dc voltages required by the electronics, 5 V and ±15 V, for example. These converters are also used in battery-powered equipment and to control the speed of dc motors in many traction applications, such as battery-powered forklifts or trains operating from a dc third rail or catenary. In relatively high-power applications, such as traction, the dc/dc converter is known as a *chopper*.

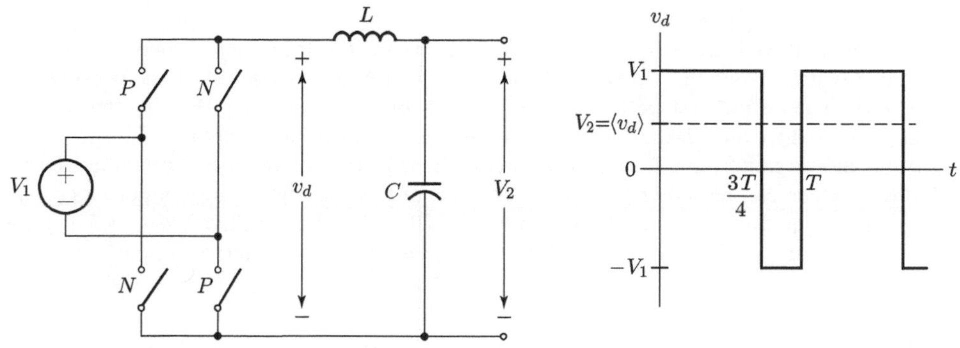

Figure 2.7 Basic dc/dc converter topology. The ratio V_2/V_1 can vary between ± 1.

2.3.1 Basic Topology

The basic dc/dc converter topology is shown in Fig. 2.7. The striking feature of this circuit is its similarity to the ac/dc converter topology shown in Fig. 2.1(b). In fact, the only difference is the control of switches. In the dc/dc converter, the switches are controlled to produce a voltage v_d that contains a nonzero dc component. This component is then extracted by the low-pass *LC* filter to produce the output dc voltage V_2.

As before, we alternately operate the *P* and *N* switches. Instead of leaving each set closed for exactly half the cycle, as we did in the ac/dc converter, here we control them to have asymmetric on-times, so that v_d contains a dc component. For example, if we leave the *P* switches on for three quarters of the switching period T, the voltage v_d shown in Fig. 2.7 results. It has an average value of $V_1/2$. In this circuit we can obtain an average voltage between V_1 and $-V_1$ by adjusting the relative conduction periods of the two sets of switches. The fraction of its switching period during which a switch is on is known as the *duty ratio D* of the switch. Here, the duty ratio of the *P* switches is 0.75, and that of the *N* switches is 0.25.

Example 2.3 A Simplified DC/DC Converter Topology

We can simplify the topology of Fig. 2.7 if we require an output voltage of only one polarity, for example, $0 < V_2 < V_1$. The circuit of Fig. 2.8, which contains only two switches, is sufficient to do the job. As we turn these switches on and off sequentially, the voltage waveform they create steps between V_1 and 0.

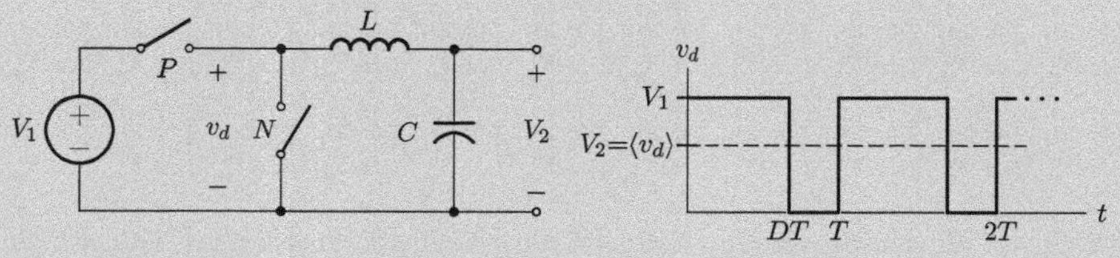

Figure 2.8 A simplified dc/dc converter topology limited to an output voltage of $0 < V_2 < V_1$.

Assume that we leave switch P on for time DT, where T is the switching period and D is the duty ratio of the P switch. We then turn switch P off and switch N on for the rest of the period, that is, for a time $(1 - D)T$. The output voltage V_2, the average value of v_d, is then DV_1. We can alter this average value by changing the duty ratio D, but the lower limit of the output voltage is zero.

What did we gain from this simplification? One benefit is the reduction in the number of switches required. Another, which can be shown through a harmonic analysis, is that the ac components of v_d in this circuit are smaller than those in v_d for the bridge circuit of Fig. 2.7. We can therefore use smaller filter elements to achieve the same level of ripple in the waveforms presented to the external system.

Because the frequency of the desired output (dc) of the converter of Fig. 2.7 is much less than the switching frequency $(1/T)$, this is a member of a class of converters known as *high-frequency switching converters*. We discuss high-frequency dc/dc converters in Chapter 5. As you will see in Chapter 8, another member of this class is the high-frequency dc/ac converter.

We can also achieve the transformation from dc to dc in quite a different way, using a dc/ac converter to create an ac waveform from the dc at one port, and then an ac/dc converter to transform this ac waveform back to dc at the other port. At first glance this approach appears to be wasteful because it requires so many switches and filter elements. It does, however, give us a point in the circuit where the waveforms are ac. As shown in Fig. 2.9, we can install a transformer at this point to provide electrical isolation and to make use of its turns ratio when the difference between the two external voltages is large. We discuss this type of converter, called an *isolated high-frequency converter*, in Chapter 7.

2.4 AC/AC Converters

An ac/ac converter converts an ac waveform of one amplitude, phase, and/or frequency to another ac waveform with different parameters. One of three distinct topological approaches can be used, depending on system requirements. The first and simplest can be used to change the amplitude parameters of an ac waveform (the rms value, for instance). It is known as an *ac controller*, and functions by simply taking symmetric "bites" out of the input waveform. The second can be used if the output frequency is much lower than the input source frequency, allowing the use of a *cycloconverter*, which approximates the desired output waveform by

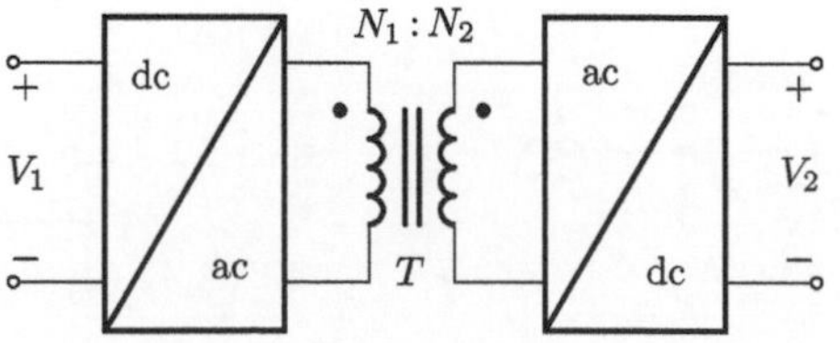

Figure 2.9 A dc/dc converter consisting of two cascaded ac/dc converters. This configuration permits the use of an isolating transformer T, as shown.

synthesizing it from pieces of the input waveform. The third approach consists of two ac/dc converters with their dc ports connected. The result is known as a *dc-link converter*.

2.4.1 AC Controller

Figure 2.10 illustrates the operation of an ac controller as a light dimmer. The switch is opened at every zero-crossing and reclosed at some later point in the half-cycle. This circuit does not control the fundamental frequency of the output waveform. However, it can control the rms value of the output or the amplitude of its fundamental component by varying the turn-on point, α, of the switch. To keep the harmonic components of the waveform created by the switch from reaching the external systems, we can again add filters to both the input and output ports of the power circuit. Although the waveform of Fig. 2.10 shows the "bite," or *notch*, starting at the zero-crossing, we can place the notch anywhere in the waveform so long as the switch implementation permits it.

2.4.2 Cycloconverter

Figure 2.11 illustrates the basic operation of the cycloconverter. The topology of this circuit is identical to that of the ac/dc converter of Fig. 2.4 without the ac-side filter. The difference between the two circuits lies in switch implementation and control. For v_2 to be positive, the P switches are closed when v_1 is positive, and the N switches conduct when the source is negative. But if the P switches are closed when v_1 is negative – and the N switches when v_1 is positive – the filtered output voltage v_2 will be negative. Therefore, by phase controlling the switches as described in Example 2.1, we can obtain any value for v_2 between $2V_1/\pi$ and $-2V_1/\pi$. If we vary the controlling phase angle α sinusoidally at a frequency ω_2 that is very low compared to the source frequency ω_1, and choose the inductor so that

$$\frac{1}{\omega_1} \ll \frac{L}{R} \ll \frac{1}{\omega_2}, \tag{2.4}$$

the low-pass filter does a good job of filtering out the source frequency ripple. However, the filter provides little attenuation at the modulating frequency ω_2, and v_2 is then an ac voltage at a frequency ω_2.

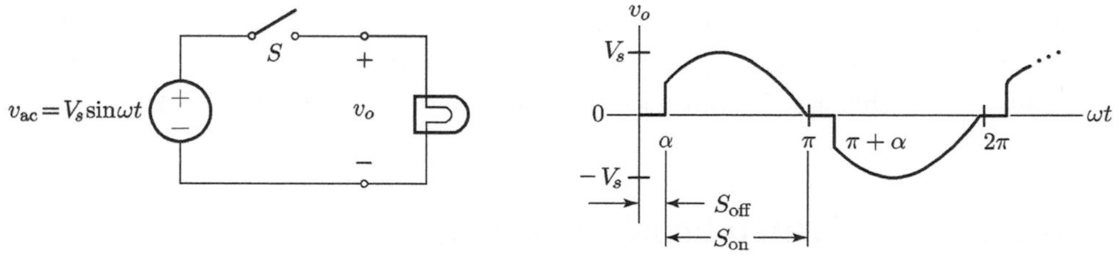

Figure 2.10 An ac controller utilized as an incandescent light dimmer.

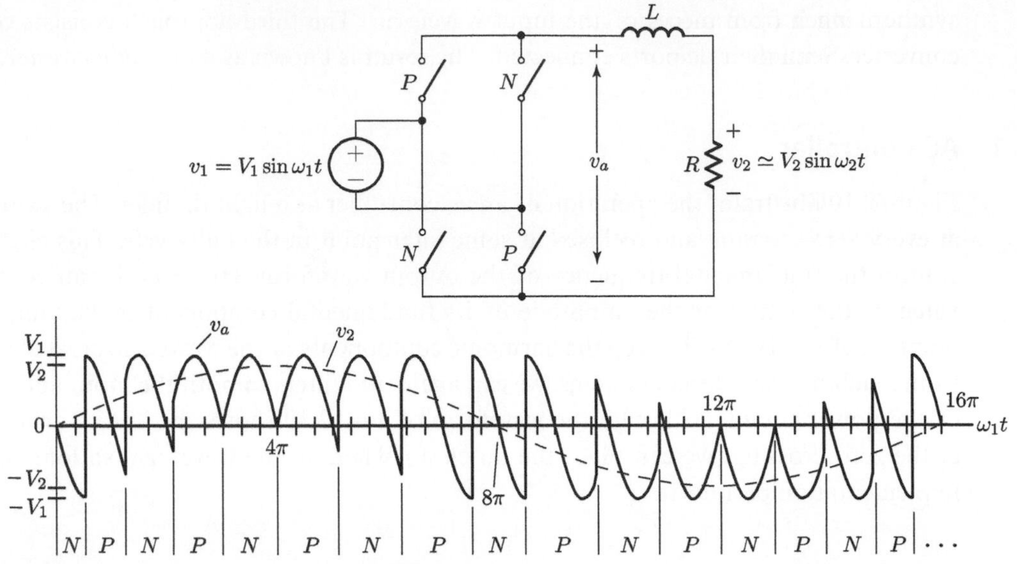

Figure 2.11 A bridge circuit operated as a cycloconverter.

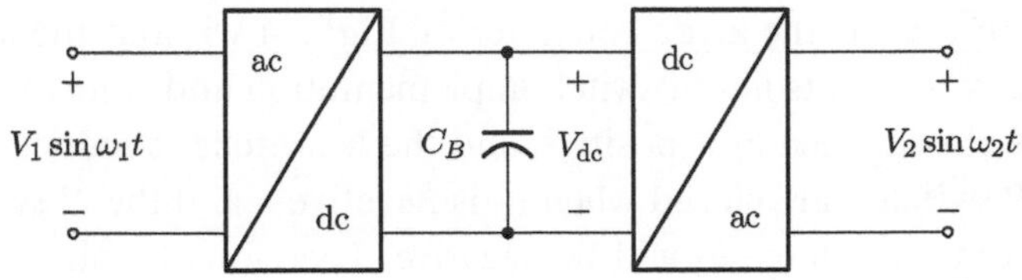

Figure 2.12 An ac/ac converter topology utilizing a dc link.

2.4.3 DC-Link Converter

Figure 2.12 illustrates the dc-link converter topology. The first ac/dc converter creates a dc waveform from the input ac. The second converts the dc energy back to ac with the desired parameters at the output. The part of the circuit where the energy exists in dc form is the dc link, or *dc bus*. The energy storage elements located at this bus can provide more than filtering. They can also store any momentary mismatch in energy between the input and output power. This function is called *load-balancing energy storage*. In Fig. 2.12 this function is provided by the dc bus capacitor C_B. For some applications the energy stored in the dc link is made large enough to support the continued operation of the system in the event of a power failure. A converter configured in this way is known as an *uninterruptible power supply* (UPS).

2.5 Influence of Switch Implementation

So far in this chapter we have discussed the operation of power circuit topologies with ideal switches. We can open or close these switches at will. When closed, they can carry current in either direction; when open they can support a voltage of either polarity. But there is no ideal switch in practice. Semiconductor switches share only some of the characteristics of ideal switches. For example, the bipolar transistor can carry current in only one direction when it is on; when it is off, it can support only one voltage polarity. And we cannot exert any control over the turn-on or turn-off of a diode. The limitations of semiconductor switches have an enormous influence on the performance of the topologies we have presented in this chapter. Let's now consider the implications of various switch implementations, using some examples.

Example 2.4 Switch Implementation in a DC/DC Converter

A dc/dc converter circuit similar to that of Fig. 2.8, with $V_1 > 0$, is shown in Fig. 2.13. The P switch has been created with an n-channel MOSFET, and both ports are connected to dc sources through filters. In what direction is energy flowing? What semiconductor device can we use for the N switch?

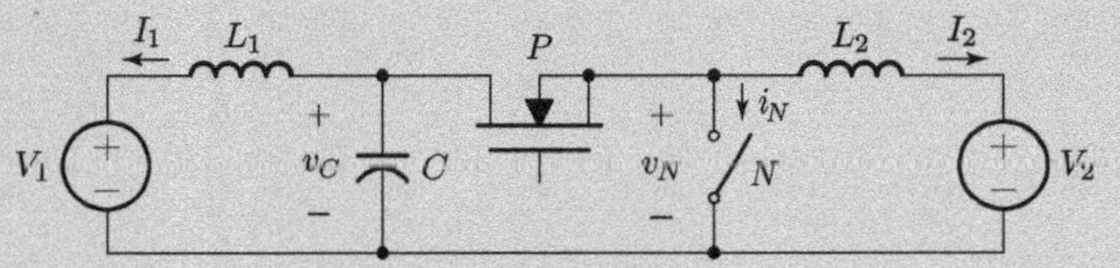

Figure 2.13 A dc/dc converter linking two dc sources. The filters are designed to make the source currents dc.

The n-channel MOSFET will permit current to flow in either direction. Therefore, depending on what device is used for the N switch, the circuit can transfer energy in either direction. If we restrict energy transfer to be from V_1 to V_2, we can determine the requirements on the N switch by looking at its voltage and current. When P is open, N must be closed, so $i_N = -I_2 < 0$ during conduction. When P is on, N must be open and $v_N = v_C$. If the filter $L_1 C$ is designed properly, the capacitor voltage v_C will be approximately constant and equal to V_1. (L_1 cannot support an average voltage.) Therefore $v_N > 0$ when N is open. This v_N–i_N characteristic is that of a diode. But we cannot control when a diode turns on and off. Can a diode be used here? The answer is yes, because the diode will be forced off (reverse biased) when the controlled switch P turns on. The current I_2, which must be continuous, will force the diode on when P turns off.

Example 2.5 The AC/DC Converter Using Diodes

Figure 2.14 shows the ac/dc converter of Fig. 2.1(a) with diodes serving as the switches. A diode can carry only positive current (anode to cathode), will turn on if its anode–cathode voltage attempts to go positive, and will turn off if the current tries to go negative. Applying these constraints to Fig. 2.14(a), you can see that both i_d and v_d must be positive. Similarly, Fig. 2.14(b), in which the diodes are reversed, produces negative values of i_d and v_d.

The current constraint imposed by the diode directions is easy to see, but the restricted polarity of v_d is a bit less obvious. A straightforward way of convincing yourself of this constraint is to assume that the P switches are on in Fig. 2.14(a), and the source $V_s \sin \omega t$ is crossing zero from positive to negative. What is happening to the voltage across the N diodes? Writing Kirchhoff's voltage law around the loop consisting of a P diode (which is on and has no voltage across it), an N diode (which is off), and the source shows that the anode–cathode voltage of the N diode is $-v_{ac}$. This anode–cathode voltage is negative (and consistent with the N diode's being off) while $V_s \sin \omega t$ is positive, but becomes positive when $V_s \sin \omega t$ goes negative. Therefore the N diodes will turn on when $V_s \sin \omega t$ passes through zero from positive to negative. Then, because the current will reverse at this time also, the *P* diodes will turn off. The use of diodes forces switching at the zero-crossings of the source $V_s \sin \omega t$ and does not allow any control of the output voltage.

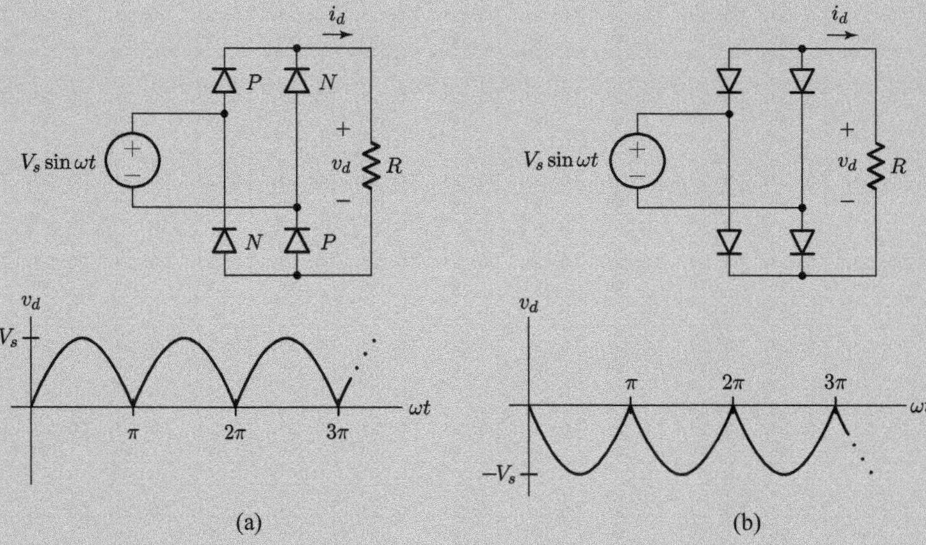

Figure 2.14 The ac/dc converter of Fig. 2.1(a) with diodes used as the switches: (a) diode directions giving a positive v_d; (b) diode directions giving a negative v_d.

PROBLEMS

2.1 If the inductor in the converter of Fig. 2.3(a) is very large, i_d can be considered constant at a value I_{dc} determined by the history of the switches. The inductor and dc voltage source can be replaced by a current source, as shown in Fig. 2.15. Sketch the ac source current i_a. Superimpose this sketch on the waveform of the source voltage $V_s \sin \omega t$ to show their relationship in time. Determine the average power delivered to the current source in terms of I_{dc} and V_s.

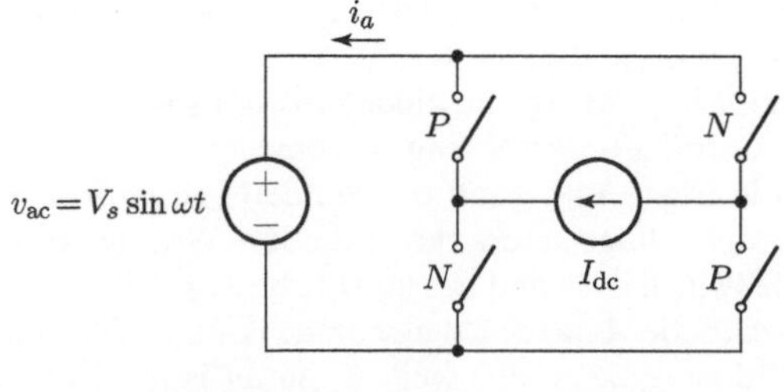

Figure 2.15 The converter circuit of Fig. 2.3(a) with its dc voltage source and inductor (assumed large) replaced by an equivalent current source.

2.2 Determine the current in the voltage source, v_{ac}, of Fig. 2.1(a). What is the average power delivered by this source?

2.3 Determine the current in the battery, V_{dc}, of Fig. 2.2. What is the power delivered to the resistor in this system?

2.4 The bridge converter of Fig. 2.3(a) has an ac source of value $v_{ac} = 170 \sin \omega t$ V and a dc source of value $V_{dc} = 75$ V. What value of α results in periodic steady-state operation?

2.5 A resistor replaces the dc source in the phase-controlled converter of Fig. 2.3(a). Determine and plot the average value of the voltage across this resistor as α is varied between 0 and π.

2.6 Determine the output voltage V_2 of the dc/dc bridge converter of Fig. 2.7 in terms of V_1 and the duty ratio D of the P switches.

2.7 A series resonant converter using the topology of Fig. 2.5(a) is constructed with the following element values:

$$L = 159\,\mu\text{H}, \qquad C = 0.25\,\mu\text{F}, \qquad R = 5\,\Omega, \qquad V_{dc} = 100\,\text{V}.$$

(a) What is the power delivered to the 5 Ω load resistor if $\omega_s = \omega_o$? Make an intelligent approximation of the effect of damping. What is the amplitude of the third harmonic in v_{ac}? Express your answer as a percentage of the fundamental of v_{ac}.

(b) The switching frequency is now adjusted so that $\omega_s = 3\omega_o$. Sketch v_a and v_{ac} on the same axes. What is the power delivered to the load? What is the amplitude of the third harmonic?

2.8 The filter and load of a dc/dc converter are as shown in Fig. 2.16. The input dc voltage source is 100 V, and the required dc load voltage (the average value of v_o) is 50 V. The size of L must be chosen such that the peak–peak ripple on the output current i_o does not exceed 0.5 A. These requirements can be met with the switch topology of either Fig. 2.7 or Fig. 2.8. Sketch the voltage v_d produced in this application by each topology. For each topology calculate the value of L required to meet the ripple specification. (*Hint:* Because the ripple is small, a close approximation to the exact answer can be obtained by assuming v_o to be constant.) Assume $T = 1$ ms.

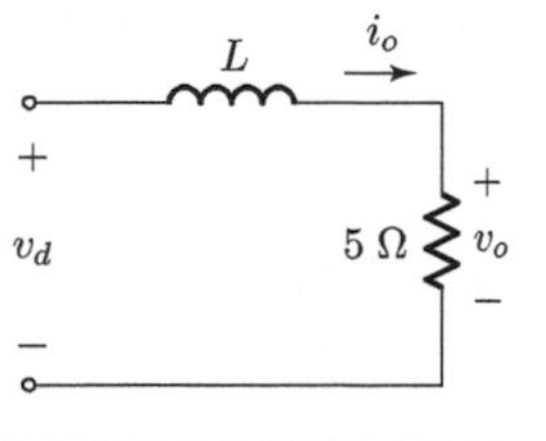

Figure 2.16 The filter (L) and load (R) of a dc/dc converter. The required value of L is determined in Problem 2.8.

2.9 Determine and plot the rms output voltage as a function of α for the ac controller of Fig. 2.10.

2.10 Can the bridge inverter of Fig. 2.1 be constructed using diodes for the switches?

2.11 Two dc sources are connected through a dc/dc converter, as shown in Fig. 2.17. In its periodic steady state the converter delivers 100 W from the 10 V source, V_2, to the 25 V source, V_1.

(a) What is the lowest frequency at which the switches can operate to produce a peak–peak ripple of no more than 60 mA in the current i_2?

(b) When this circuit is started, the switches must be controlled so that i_2 builds up to its periodic steady state dc value (10 A). Determine a control strategy for the switches (their on-state durations as a function of time) that will minimize the duration of this start-up transient.

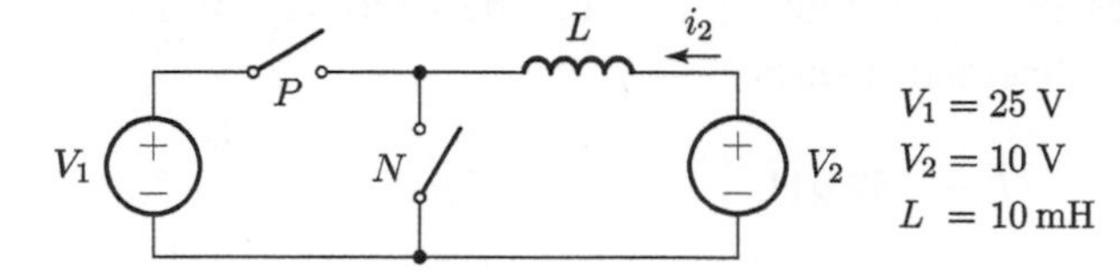

Figure 2.17 A dc/dc converter designed to transfer energy from V_2 to V_1.

2.12 How should the diode used to implement the switch N in Fig. 2.13 be connected in the circuit? What is the conversion ratio V_2/V_1 in terms of the duty ratio D of the P switch?

2.13 Determine a switch implementation (both N and P) for the topology of Fig. 2.13 that would permit the transfer of energy from V_2 to V_1. What is the conversion ratio V_1/V_2 in terms of the duty ratio D of the N switch?

2.14 Figure 2.18 shows a dc/ac converter designed to connect a battery to a resistive load which requires a sinusoidal voltage at a frequency of 50 Hz. The switches are operated to make v_a a square-wave voltage. Determine the amplitudes of the fundamental and third harmonic components of the load voltage v_{ac}.

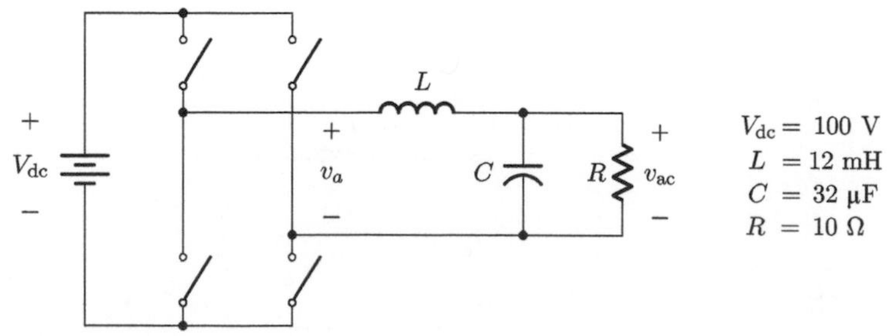

Figure 2.18 A dc/ac converter containing a second-order low-pass filter on the ac side.

3 Introduction to Rectifiers

A rectifier converts ac to dc. A basic rectifier circuit produces dc in the electrical engineering sense, that is, unipolar current flow. It does not produce dc in the mathematical sense, that is, a waveform that is constant in time and whose spectrum consists of a single zero-frequency component. A rectifier's output contains considerable ac content. These ac components result in fluctuations, called *ripple*, about the average value of the dc output. Eliminating this ripple and obtaining an approximation to "pure" dc requires insertion of a filtering process after the basic rectification function.

Ever since George Westinghouse emerged the victor over Thomas Edison in the great ac/dc battle of the twentieth century, rectifiers have become the backbone of power electronic circuits. Although they come in a large number of different configurations (for example, single phase, three phase, half wave, and full wave), rectifier circuits possess fundamentally similar principles of operation. Our purpose in this chapter is to introduce you to a number of different concepts, using two simple single-phase rectifier circuits as vehicles. These behave in a qualitatively transparent manner and thus are well suited to this purpose. We reserve the discussion of more complex rectifier circuits for Chapters 4 and 9.

Because power electronic circuits include switches, the constraints on circuit behavior almost always vary with time. Thus we can describe their behaviors most easily in the time domain. Another purpose of this chapter, then, is to reintroduce you to the techniques of time domain analysis. Most electrical engineers either forget or never fully understand these techniques in their preoccupation with frequency domain methods. We also introduce the analytic method of *assumed states*, which is particularly appropriate to the analysis of power electronic circuits. Our discussion is based on the assumption that the diodes in the rectifier circuits are ideal; that is, they have no forward drop when they are on or reverse leakage current when they are off.

Throughout this book, we refer to the *average value of variables*. We use angle brackets to denote the average value of a quantity, say, $f(t)$, as $\langle f(t) \rangle$. We use this notation to specify average values of periodic functions. We later extend the notion of "average" to the "local average" of non-periodic functions, but with different notation.

3.1 Power in Electrical Networks

The ideal power electronic circuit provides a lossless transformation of the form of electric power. The reason is that the circuit is made up of (ideally) lossless components – inductors,

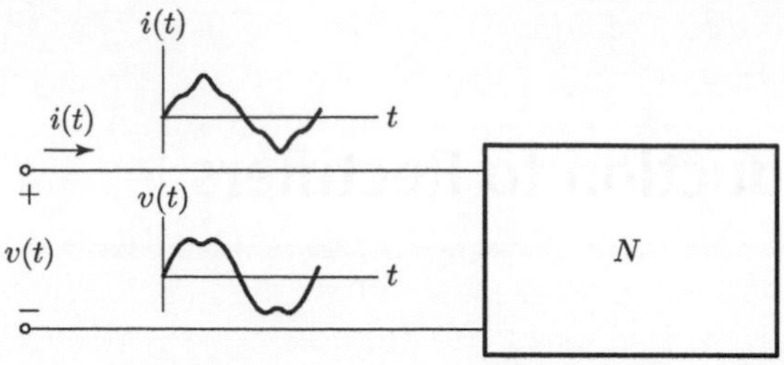

Figure 3.1 Time varying nonlinear network, N, having periodic terminal variables v and i.

capacitors, transformers, and switches. Average power in and average power out of lossless transformations are equal. We state this condition in terms of *average* power rather than instantaneous power in order to accommodate the possibility of energy storage, as in inductors and capacitors, within the transformation. Keeping your eye on the power is often insightful in the analysis of multi-port power electronic networks (such as rectifiers), which – because they are nonlinear and vary with time – do not yield easily to conventional analysis.

Consider the network N (which perhaps varies with time and/or is nonlinear) shown in Fig. 3.1. Assume that the terminal variables $i(t)$ and $v(t)$ are periodic with period $T = 2\pi/\omega$, but not necessarily sinusoidal, in which case they may be expressed in terms of their Fourier series,

$$i(t) = I_0 + \sum_{n=1}^{\infty} I_n \sin(n\omega t + \theta_n), \tag{3.1}$$

where I_0 is the average or dc value, and $I_n = \sqrt{a_n^2 + b_n^2}$ for $n > 0$, with

$$a_n = \frac{2}{T}\int_0^T i(t)\sin(n\omega\tau)\,d\tau, \quad b_n = \frac{2}{T}\int_0^T i(t)\cos(n\omega\tau)\,d\tau, \tag{3.2}$$

and with similar expressions for $v(t)$. (The integrals can be taken over any contiguous interval of length T.) The amplitude of the fundamental is I_1, and of the nth harmonic is I_n. Also, $\theta_n = \tan^{-1}\left(\frac{b_n}{a_n}\right)$.

The time-average power $\langle p(t)\rangle$ at the terminals of N is

$$\langle p(t)\rangle = \frac{1}{T}\int_0^T vi\,dt. \tag{3.3}$$

Because the product of sinusoidal variables of different frequencies integrated over a common period is zero (the variables are said to be *orthogonal*), only components of v and i that are of the same frequency contribute to the average power at the terminal pair. If either v or i consists of a single-frequency component (possibly dc), only the corresponding component in the covariable contributes to the integral in (3.3).

For a lossless network containing no energy storage (one containing only switches and ideal transformers, for instance), the instantaneous input and output powers must be equal. This fact helps explain the separate roles of the switches and the energy storage elements in a power circuit. For example, it shows that a rectifier system that converts 60 Hz ac to pure dc cannot do so with switches only. Because the input voltage is a 60 Hz sinusoid, there must be a 60 Hz component of current to produce a nonzero average input power (to match the average output power). But this 60 Hz component of input current, when multiplied by the 60 Hz input voltage, produces a 120 Hz component of instantaneous input power. As there is no 120 Hz component of output power, the circuit must contain energy storage in addition to switches. Stated simply, there is no magic you can do with switches to eliminate the need for energy storage at this low frequency (120 Hz) – a need that translates into physically large capacitors or inductors.

3.2 Single-Phase Half-Wave Rectifier

The half-wave rectifier is the simplest type of rectifier circuit, consisting of only a single diode. Unfortunately, its output voltage contains ripple at the ac input frequency, which makes filtering more difficult than for other circuits having ripple frequencies that are multiples of the input frequency. Moreover, its ac source current contains a dc component, which can make the circuit impractical to use with an input transformer, as you will see in Chapter 20. When vacuum tubes with their large forward drops were used as rectifying diodes, the half-wave circuit was frequently employed instead of alternative circuits in which two or more conducting diodes appeared in series. Today, however, the low forward drop of semiconductor diodes makes it unnecessary to accept the disadvantages of this circuit in most cases. The exceptions are in low-voltage applications, such as power supplies for computers and electrochemical processes (plating, for instance), or in very cost-sensitive applications, such as battery chargers for portable power tools. Nevertheless, its simplicity makes the half-wave circuit a good way to introduce basic rectifier concepts.

3.2.1 Half-Wave Rectifier with Resistive Load

The purely resistive load of the half-wave rectifier circuit shown in Fig. 3.2 constrains the rectified voltage v_d to be a positive multiple of the current i_d at all times. The diode cannot conduct current in the reverse direction, so it must be off when the source voltage is negative. If we instead assume that it is on under this condition, we find an inconsistency between the current direction and the diode state; that is, we find the diode current to be negative. Achieving consistency between a diode's assumed state and its associated voltage and current is the essence of the

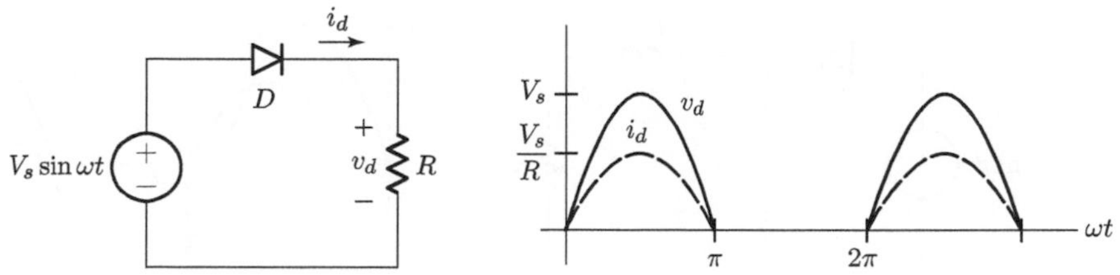

Figure 3.2 Half-wave rectifier with resistive load.

method of assumed states. Although this example is simple, more complicated arrangements of switches and passive components can present a greater challenge.

To apply the method of assumed states to an arbitrary network, we assume that each diode in the network is in a particular state, either on or off. We then solve the network for the currents in the diodes that are on and the voltages across those that are off. If an on-diode's current is negative or an off-diode's voltage is positive, there is an inconsistency between our assumptions about the diode states and the terminal characteristics of the diodes. In this case, the assumed states are not consistent, and we try another combination of on and off diodes. Once we obtain a consistent set of states, we determine the diode voltages and currents as functions of time and test them to see which diode first presents an inconsistency. We then assume that this diode (or diodes) changes state at this point, and solve the circuit for the new set of states. By progressing in this way, we eventually return to our original consistent set of states, in the process having analyzed the circuit for a complete period.

Because we generally want the dc component, that is, the average value, of the rectifier output, let's calculate this component for the simple rectifier of Fig. 3.2:

$$\langle v_d \rangle = \frac{1}{2\pi} \int_0^{2\pi} v_d(\omega t)\, d(\omega t) = \frac{V_s}{\pi}. \tag{3.4}$$

The function of most rectifiers is to produce an output voltage with a low ripple content. When low ripple is important, a low-pass filter is placed between the rectifier and the load. However, the presence of a filter often changes the operating characteristics of the rectifier, and therefore the output of the filter is not always the average value of v_d determined before the filter is added. For example, putting a low-pass filter in series with the load resistor R in Fig. 3.2 changes the circuit's operation, so that $\langle v_d \rangle$ is no longer equal to V_s/π, as we show next.

3.2.2 Half-Wave Rectifier with Inductive Load

Figure 3.3 shows a half-wave rectifier containing a simple low-pass filter consisting of an inductor L in series with the load resistor R. The influence of L on v_d is clear when we compare the waveforms of Fig. 3.2 with those of Fig. 3.3. With the inductive load, v_d goes negative, and the ac line current (which is the same as i_d) is out of phase with the line voltage. Although "phase" is an inexact concept when used to refer to distorted waveforms, what we generally mean is the phase of the fundamental component of the waveform, which we discuss in Section 3.4.

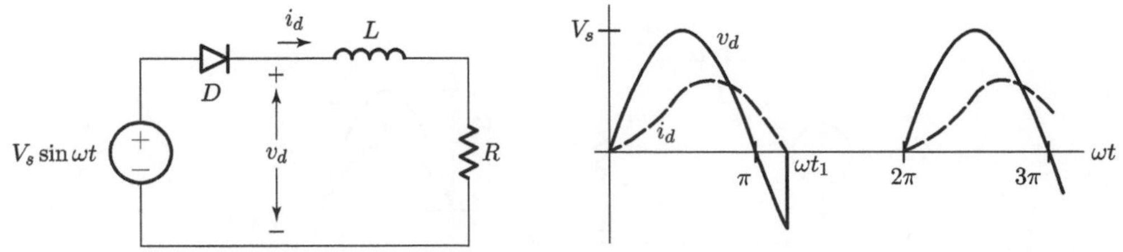

Figure 3.3 Half-wave rectifier with inductive load.

We can again use the method of assumed states to derive the waveforms of Fig. 3.3. If we assume that the diode is on, i_d must be positive. We will now show that with the circuit operating in the cyclic steady state, the diode must turn off for part of the cycle. First, let's assume that it does not. Then $v_d = V_s \sin \omega t$, which results in $\langle v_d \rangle = 0$. Because the average voltage across L must be zero and the average voltage across the load resistor equals $\langle v_d \rangle$, which is also zero, the average voltage across R must also be zero. Thus, current through the resistor will either be zero or have both positive and negative excursions. Zero current is not possible for finite values of L, and the diode precludes negative current. Therefore, 360° conduction of the diode is inconsistent with the constraints imposed by both the circuit and the diode, and the load current must be zero for part of each cycle. Hence, this operating condition is known as *discontinuous conduction*.

During periods when the diode is off, i_d is identically zero and v_d must be too, so the source voltage cannot be positive. The time at which the diode turns on must then be the time when the source voltage crosses zero, from negative to positive. If the diode did not turn on at this time, it would be supporting a forward voltage. The problem now is to determine when the diode turns off.

Figure 3.4 shows an equivalent circuit for the calculation of i_d, where the switch is closed when the source voltage crosses zero from negative to positive. The resulting response, i_d, consists of two parts: one has the same form as the excitation and is variously known as the *driven*, *forced*, or *particular* response; the other is characterized by the eigenvalues, or natural frequencies, of the circuit and is known as the *natural* or *homogeneous* response. For the circuit of Fig. 3.4, the network equations in terms of i_d are

$$\frac{di_d}{dt} + \frac{R}{L} i_d = \frac{V_s}{L} \sin \omega t, \qquad t > 0, \tag{3.5}$$

with initial condition $i_d(0) = 0$. Solving (3.5), we obtain i_d for $0 < t < t_1$:

$$i_d = \frac{V_s}{Z} \sin(\omega t - \phi) + Ae^{-Rt/L}, \tag{3.6}$$

where

$$Z = \sqrt{R^2 + (\omega L)^2}, \quad \phi = \tan^{-1}\left(\frac{\omega L}{R}\right), \quad A = \frac{V_s}{Z} \sin \phi = \frac{V_s \omega L}{Z^2}. \tag{3.7}$$

The time t_1 in Fig. 3.3 is the first zero-crossing of i_d, or the time at which the diode turns off. We can find the value of t_1 from (3.6) by iteratively solving the transcendental equation $i_d(t_1) = 0$.

An interesting but unfortunate characteristic of this simple rectifier is that $\langle v_d \rangle$, the voltage in which we are interested, is a function of the load R, that is, this rectifier exhibits *load regulation*:

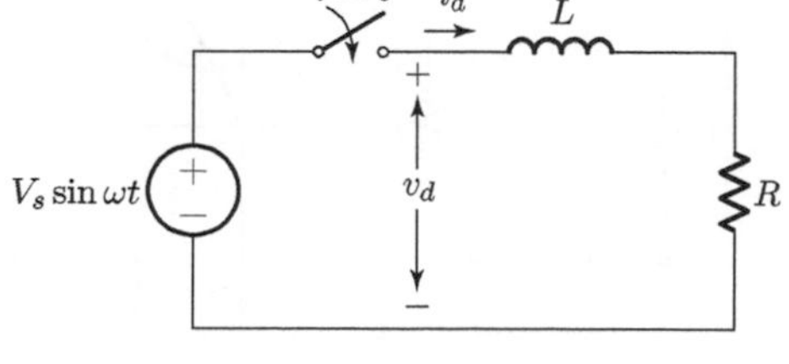

Figure 3.4 Equivalent circuit for the calculation of i_d in Fig. 3.3.

the load (through its current) affects, or regulates, the output voltage. This may be confusing at first, because when we say that something is regulated, we usually imply that it is held constant. When we use the term *regulation* in the context of rectifiers, however, the implicit meaning is load regulation. The source of regulation in this rectifier circuit is the fact that t_1 depends on the time constant $\tau = L/R$, a function of R (the load) for constant L.

The behavior of this circuit is insensitive to the location of L relative to the diode; that is, it can be on either the ac or the dc side. Thus, as a practical ac source invariably contains inductance, especially if the source includes a transformer, some degree of regulation is difficult to avoid.

3.2.3 Half-Wave Rectifier with Freewheeling Diode

The half-wave rectifier with low-pass filter shown in Fig. 3.5 is a more practical circuit than the one shown in Fig. 3.3. The addition of D_2 now permits the load current i_d to be continuous, and prevents v_d from going negative. Using the method of assumed states, we can prove that a further constraint on circuit operation is that D_1 and D_2 cannot be on simultaneously. When D_1 is off, D_2 allows the energy in the circuit to maintain continuity by providing a path through which the inductor current can "free wheel." For this reason D_2 is known as a *freewheel* (or *freewheeling*) diode. Diodes can perform this function in other power circuits, in which case they are called *bypass*, *flyback*, or *catch* diodes.

If the inductor is large enough, i_d never decays to zero (an operating condition known as *continuous conduction*, where $L/R \gg \pi/\omega$). We can then further show that D_1 and D_2 cannot be off simultaneously because there must always be a path for i_d. With this condition, the circuit has two topological states: in one, D_1 is on and D_2 is off; in the other, D_1 is off and D_2 is on.

The behavior of i_d shown in Fig. 3.5 is the result of solving the circuit equations for each of the two topological states of the network. For the state D_1 on and D_2 off, the circuit equation in terms of i_d is still (3.5). For the freewheeling state in which D_1 is off and D_2 is on, i_d simply decays exponentially with time constant $\tau = L/R$.

From the waveform of v_d in Fig. 3.5, we see that $\langle v_d \rangle$ is still given by (3.4). Note that (3.4) gives the maximum possible average value of v_d, independent of either L or R. This rectifier circuit, then, exhibits *no regulation* (unless the inductor contains significant resistance). The result is

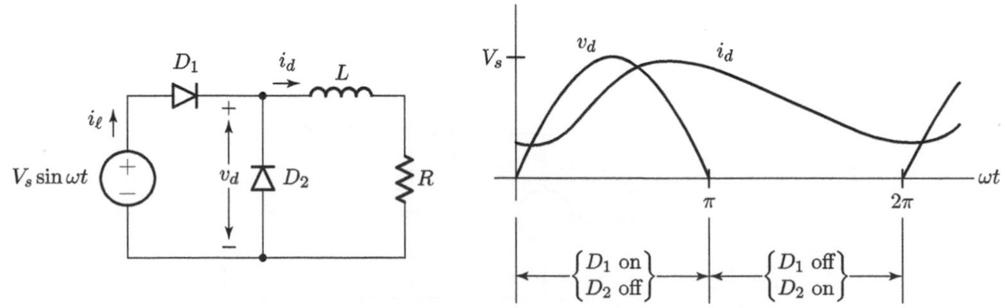

Figure 3.5 Half-wave rectifier with a freewheeling diode.

that we can make L as large as necessary to achieve the desired degree of filtering. As L goes to infinity, i_d becomes constant, and the voltage across R is only the dc component of v_d.

3.2.4 Periodic Steady State

The waveform of i_d in Fig. 3.5 illustrates the *periodic steady state*. It contains both the particular and homogeneous solutions to the differential equations that describe the network. In this sense it is different from the "steady state" solution that we associate with the driven response of a linear, time-invariant (LTI) network to a sinusoid. Furthermore, it is composed of two distinct analytic expressions, one valid for $0 < \omega t < \pi$ and the other for $\pi < \omega t < 2\pi$. The first is (3.6), repeated here for the appropriate interval:

$$i_d = \frac{V_s}{Z}\sin(\omega t - \phi) + Ae^{-Rt/L}, \qquad 0 < \omega t < \pi. \tag{3.8}$$

The second is simply the natural response of the LR load circuit, or

$$i_d = Be^{-Rt/L}, \qquad \pi < \omega t < 2\pi. \tag{3.9}$$

Two boundary conditions are required in order to determine the coefficients A and B. The two boundaries are $\omega t = 0$ and $\omega t = \pi$, the points at which the diodes switch. However, we must exploit periodicity in setting up these boundary conditions. The first of them is

$$i_d(0+) = i_d(2\pi-). \tag{3.10}$$

Again, this condition simply represents the constraint imposed by the required continuity of state variables or energy. The second boundary condition is

$$i_d(\pi-) = i_d(\pi+). \tag{3.11}$$

We do not solve these equations for A and B here because the solutions are just straightforward algebra. Of use for sketching the waveform of i_d is that its derivative is continuous at the boundaries. Convince yourself that a discontinuity in the derivative would require a step in v_d (which does not exist). The ability to quickly determine such details of waveforms is of considerable use in many circumstances where a qualitative description of a circuit's behavior is desired.

3.2.5 Circuit Replacement by Equivalent Source

The circuit of Fig. 3.5 presents an opportunity to demonstrate another very useful technique of circuit analysis called *circuit replacement by equivalent source*. Under the condition that i_d is never zero, the voltage v_d is defined irrespective of the details of the load circuit. We can therefore determine the detailed behavior of the dc, or load, side of the circuit by replacing the ac source and diodes with a voltage source having a waveform v_d, as shown in Fig. 3.6. Such circuit replacement by an equivalent source is an especially useful technique for numerical simulations, because the complexity of coding the replaced circuit is avoided.

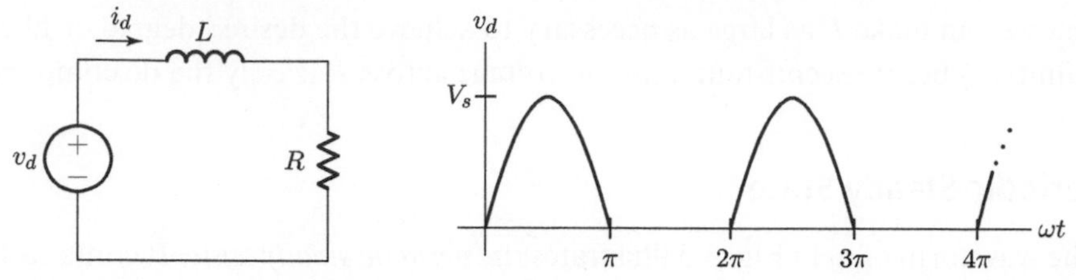

Figure 3.6 The technique of circuit replacement by equivalent source illustrated by replacing the source and diodes of Fig. 3.5 with the equivalent source v_d.

Example 3.1 A Half-Wave Rectifier with a Capacitive Filter

In many applications, especially those in which cost is a major consideration, a capacitive filter is used rather than the inductive filter just discussed. Figure 3.7 shows a half-wave rectifier circuit containing such a filter and its output voltage v_d. We determine both $v_d(t)$ and $\langle v_d \rangle$ for this circuit.

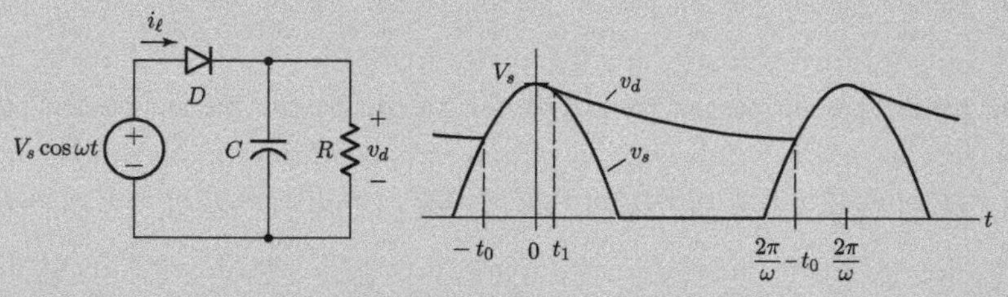

Figure 3.7 A simple half-wave rectifier with a capacitive filter.

It is clear that the capacitor can only be charged – and only to positive values of v_d – when the diode is conducting, and during this time v_d equals the source voltage. When the diode is off, that is, when v_d exceeds the source voltage, the capacitor discharges exponentially with time constant $\tau = RC$ through the resistor. We thus expect v_d to be positive throughout, and to have the general form shown in Fig. 3.7, where $T = 2\pi/\omega$, the period of the source.

While the diode is conducting, the current through the diode is

$$i_D(t) = C\frac{d}{dt}(V_s \cos \omega t) + \frac{1}{R}V_s \cos \omega t. \tag{3.12}$$

When (3.12) reaches 0 at time $t = t_1$, satisfying the expression $-\omega\tau \sin \omega t_1 + \cos \omega t_1 = 0$, the diode turns off. Thus ωt_1 will be in the interval $(0, \pi/2)$, where $\cos \omega t_1 < 0$ and $\sin \omega t_1 > 0$; the larger τ is, the closer ωt_1 will be to $\pi/2$. An explicit expression for t_1 is

$$t_1 = \frac{1}{\omega} \cot^{-1}(\omega\tau). \tag{3.13}$$

The subsequent decay of v_d is now given by

$$v_d(t) = (V_s \cos \omega t_1)\, e^{-(t-t_1)/\tau}. \tag{3.14}$$

The time $(2\pi/\omega) - t_o$ at which the diode turns on again is found by setting the value of v_d at the end of the previous discharge period $(-t_o)$ equal to the preceding expression evaluated at the end of the current discharge cycle $(2\pi/\omega) - t_o$:

$$(V_s \cos \omega t_1)\, e^{-\left(\frac{2\pi}{\omega}-t_o-t_1\right)/\tau} = V_s \cos(-\omega t_0). \tag{3.15}$$

To determine t_o, we solve the nonlinear equation (3.13) for t_1 and use this value in (3.15) to solve the resulting nonlinear equation for t_o. The solution of such nonlinear equations requires iterative methods, typically starting with an initial guess and then refining it (e.g., using the Newton–Raphson method).

We now assume $V_s = 24$ V, solve (3.15) for t_o by iteration, and calculate $\langle v_d \rangle$ and the ripple. In most applications of rectifiers employing capacitive filters, τ is large enough that t_1 is close to 0. Assuming $\omega = 377$ (60 Hz) and $\tau = 20$ ms, (3.13) yields $t_1 = 0.35$ ms, that is, very close to $t = 0$, so (3.15) becomes

$$e^{-\left(\frac{2\pi}{\omega}-t_o\right)/\tau} = V_s \cos(-\omega t_0), \tag{3.16}$$

$$t_o = \frac{1}{\omega}\cos^{-1}\left\{e^{-\left(\frac{2\pi}{\omega}-\frac{\pi}{4\omega}\right)}\right\}. \tag{3.17}$$

Our initial guess for t_o is $t_{o_1} = \pi/4\omega$. We plug this value into into the right-hand side of (3.17), which gives a new value of t_{o_2}:

$$t_{o_2} = \frac{1}{\omega}\cos^{-1}\left(e^{-0.296}\right) = \cos^{-1}(0.744) = 1.94\,\text{ms}.$$

We now use this new value for t_o in the right-hand side of (3.17) and proceed to determine t_{o_3}. After two more iterations, t_{o_4} and t_{o_5} converge to $t_o = 2.78$ ms. Using this value of t_o we calculate the average value of v_d,

$$\langle v_d \rangle = \frac{\omega V_s}{2\pi}\left(\int_{-t_o}^{0} \cos \omega t\, dt + \int_0^{\frac{2\pi}{\omega}-t_o} V_s e^{\frac{-t}{\tau}}\right) = 0.74 V_s,$$

which is greater than that of the rectifier with inductive filter ($0.32V_s$). However, a rectifier with capacitive filter experiences load regulation as a function of τ.

3.3 AC-Side Reactance and Current Commutation

We mentioned in the previous section that the behavior of the half-wave rectifier shown in Fig. 3.3 is insensitive to the location of L. The same is not true of the circuit of Fig. 3.5. In particular, the presence of inductance on the ac side as well as on the dc side, as shown in Fig. 3.8, creates a third topological state of the network, where both diodes are on simultaneously. This state is known as the *commutation* state because, as we will show, the load current is being transferred, or *commutated*, from one diode to the other during this state. The inductance responsible for the existence of this state is known as the *commutating inductance*, L_c, with corresponding *commutating reactance*, $X_c = \omega L_c$.

Throughout the following discussion we assume that $L_d/R \gg \pi/\omega$, so that, for all practical purposes, the load current is constant at I_d. Once you understand the behavior of the circuit under this condition, you will see clearly how to include the effects of ripple or discontinuous conduction. The assumption of constant load current during the commutation state is often employed in the analysis of rectifier circuits, as in most cases $L_d \gg L_c$.

3.3.1 Commutation Processes and Equivalent Circuits

If we assume that the circuit of Fig. 3.8 is in the state with D_1 on and D_2 off, and note that there is no drop across L_c because the current through it is constant, we see that v_d is forced to equal the source voltage, $V_s \sin \omega t$. This state becomes inconsistent with the diode voltages when the source voltage goes negative. Two states are now possible, only one of which can be consistent with the circuit constraints. Either D_2 turns on and D_1 turns off, as in the circuit of Fig. 3.5, or D_2 turns on and D_1 remains on. The latter state must occur next, because the former would require a step change in the energy stored in L_c. The time during which both D_1 and D_2 are on is known as the *commutation period* and has a duration u in electrical degrees.

Commutation is the process of moving a current from one branch of a circuit to another. In the case of Fig. 3.8, the current I_d is being moved from the D_1 branch to the D_2 branch. The process is analogous to that performed by the commutator and brushes in a dc motor during the interval in which two bars of the commutator are shorted by the brush, and the armature current is being transferred from one bar to the other.

The equivalent circuit of the rectifier during the commutation period is shown in Fig. 3.9, from which we can calculate i_ℓ (or $i_{D_2} = I_d - i_\ell$), whose waveform is shown in Fig. 3.8(b):

$$
\begin{aligned}
i_\ell(\omega t) &= i_\ell(\pi) + \int_\pi^{\omega t} \frac{V_s}{\omega L_c} \sin \omega t \, d(\omega t) \\
&= I_d - \frac{V_s}{\omega L_c}(1 + \cos \omega t), \qquad \pi < \omega t < \pi + u.
\end{aligned}
\tag{3.18}
$$

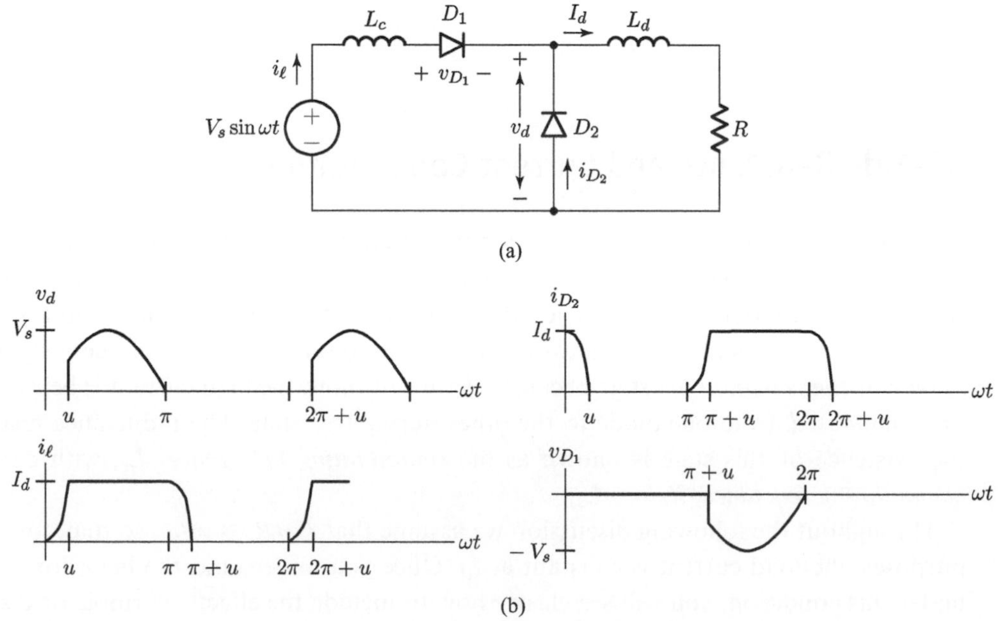

Figure 3.8 (a) A half-wave rectifier with commutating inductance L_c. (b) The branch variables.

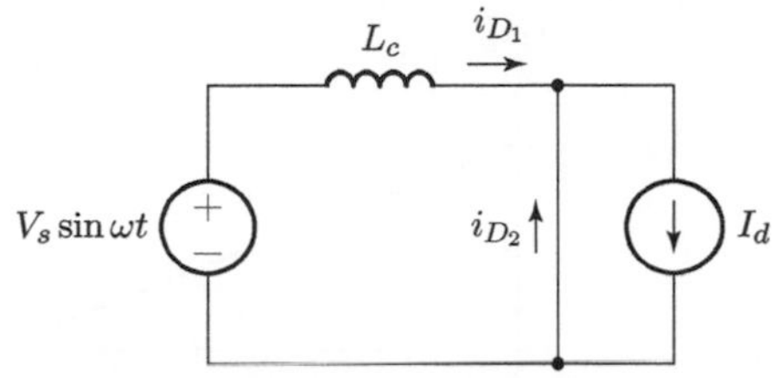

Figure 3.9 The equivalent circuit for Fig. 3.8(a) during the commutation interval, $\pi < \omega t < (\pi + u)$.

We determine the duration u of the commutation period from the condition $i_\ell(\pi + u) = 0$; that is,

$$u = \cos^{-1}\left(1 - \frac{\omega L_c I_d}{V_s}\right) = \cos^{-1}\left(1 - \frac{X_c I_d}{V_s}\right). \tag{3.19}$$

The commutating inductance L_c is the inductance across which appears the voltage forcing the commutating branch current up or down. The voltage doing the forcing is called the *commutating voltage*, which in this case equals $V_s \sin(\omega t)$.

The diode states for which Fig. 3.9 is drawn are consistent with their branch variables until $i_{D_1} = 0$ at $\omega t = \pi + u$. The next consistent state is D_1 off, D_2 on. The voltage across D_1 is now the source voltage, so this state persists until $\omega t = 2\pi$, when the source changes polarity. The resulting process of the current commutating from the D_2 branch back to the D_1 branch is identical to that just described, except that the commutating voltage is reversed. The same electrical angle u as before is required for the current in L_c to build up from zero to I_d. Note that during this commutation period, $v_d = 0$ instead of $V_s \sin \omega t$. Figure 3.8(b) shows the diode branch variables during normal operation.

3.3.2 Effects of Commutation

The presence of commutating inductance has two effects on the terminal characteristics of the rectifier of Fig. 3.8(a). First, it causes load regulation of the output voltage; second, it changes the waveform of the ac source current. We can determine the regulation characteristic by calculating $\langle v_d \rangle$ as a function of u and substituting (3.19), which contains the load current explicitly:

$$\langle v_d \rangle = \frac{V_s}{2\pi} \int_u^\pi \sin\theta \, d\theta = \frac{V_s}{2\pi}(1 + \cos u) = \frac{V_s}{\pi}\left(1 - \frac{X_c I_d}{2V_s}\right). \tag{3.20}$$

This result is plotted in Fig. 3.10. You can see that the commutating reactance $X_c = \omega L_c$ has the same effect on the average output voltage as a resistor of value $X_c/2\pi$ in series with the load. (Note that in other respects there is a considerable difference between X_c and a resistance. For example, X_c is lossless.) The term $X_c I_d / V_s$ will show up again later when we discuss more complicated rectifier circuits. It is a normalized parameter known as the *reactance factor*, and as (3.20) shows, it is directly related to the degree of regulation exhibited by a rectifier circuit.

You can also see the effect of commutating inductance on the source current by comparing the waveform of i_ℓ in Fig. 3.5 (which is equal to i_d for $0 < \omega t < \pi$ and 0 for $\pi < \omega t < 2\pi$)

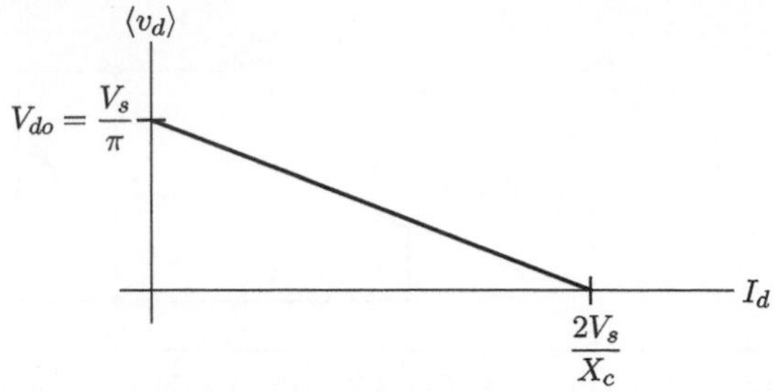

Figure 3.10 Load regulation curve for the half-wave rectifier with commutating reactance of Fig. 3.8.

to i_ℓ in Fig. 3.8(b). If we assume L/R in the circuit of Fig. 3.5 to be very large so $i_d \approx I_d$, as we did in the circuit of Fig. 3.8(a), then i_ℓ for the circuit of Fig. 3.5 is a rectangular pulse with a duration of half a cycle and an amplitude I_d. When we add commutating inductance to obtain the circuit of Fig. 3.8(a), the pulse duration is extended by u and the pulse shape becomes approximately trapezoidal. For the same peak value, the source current of the rectifier with commutating inductance has a higher rms value than the source current in the rectifier circuit of Fig. 3.5. This result means that the *power factor*, a measure we discuss next, is reduced when commutating inductance is added to the circuit.

3.4 Measures and Effects of Distortion

All the rectifier circuits discussed so far have in common a distorted line (source) current waveform. One practical effect is that the source must have a volt-ampere (VA) rating that is higher than it would need to be if it were supplying the same average power to a linear resistor. The VA *rating* of a source is the product of its maximum deliverable rms voltage and current. For example, a conventional 110 V_{rms} outlet in an American home generally has a maximum current rating of 15 A_{rms}, as determined by wire size, fixture capability, and fuse or circuit-breaker rating. Therefore this outlet (source) has a rating of 1650 VA.† A sinusoidal, 60 Hz voltage source delivers average power only through the 60 Hz component of its current, as shown by (3.3). Thus, if the rms value of its total current exceeds the rms value of its fundamental, the outlet cannot deliver its rated power, although it may be delivering its rated rms current. We also speak of the VA *product* at a terminal pair, which means the product of V_{rms} and I_{rms} being supplied to the load connected to the terminals. This product is called the *apparent power* and is given the symbol S.

The *power factor*, k_p, of a terminal pair is the ratio of the average power to the apparent power S at the terminals. The factor embodies the effects of both distortion and phase shift between voltage and current. A measure of the distortion – caused by the undesirable frequency components in a waveform – is given by the *total harmonic distortion* (THD).

† In practice, a wall receptacle is loaded to only 80% of its maximum VA rating, or 1320 VA in this case.

3.4.1 Power Factor

We define the power factor k_p of a two-terminal network as the ratio of the average power measured at the terminals to the product of the rms values of the terminal voltage and current; that is,

$$k_p = \frac{\langle p(t) \rangle}{V_{\text{rms}} I_{\text{rms}}} = \frac{\langle p(t) \rangle}{S}. \tag{3.21}$$

The importance of this measure is that it reflects how effectively available power is being used. A source supplying an average power less than the apparent power $V_{\text{rms}} I_{\text{rms}}$ at its terminals is not operating at its full capability at this voltage and current.

Example 3.2 Power Factor of an AC Controller

Light dimming is one application of a class of circuits known as *ac controllers*. These circuits are also used to control the speed of small appliances and ac-powered hand-tools. You will learn how they work in Chapter 4. We assume that the ac controller of Fig. 3.11 is feeding a resistive load R that results in the current waveform shown. What is the power factor k_p at the ac source?

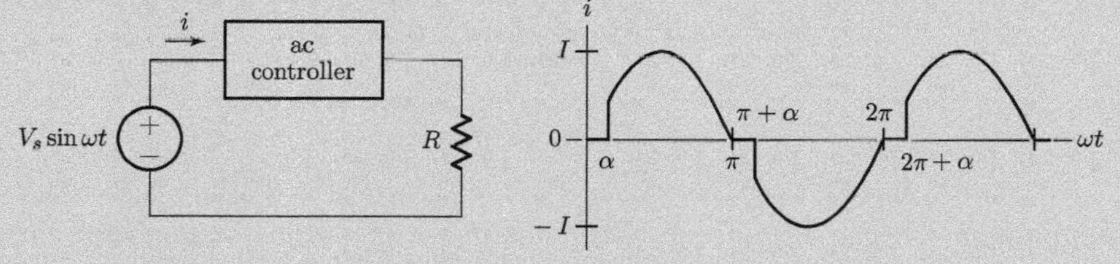

Figure 3.11 An ac controller supplying a resistive load.

We calculate k_p as follows:

$$\langle p(t) \rangle = \frac{V_s I}{\pi} \int_{\alpha}^{\pi} \sin^2 \omega t \, d(\omega t) = \frac{V_s I}{2} \left[\left(1 - \frac{\alpha}{\pi}\right) + \frac{1}{2\pi} \sin 2\alpha \right], \tag{3.22}$$

$$V_{\text{rms}} = \frac{V_s}{\sqrt{2}}, \tag{3.23}$$

$$\begin{aligned} I_{\text{rms}} &= \sqrt{\frac{1}{\pi} \int_{\alpha}^{\pi} I^2 \sin^2 \omega t \, d(\omega t)} \\ &= \frac{I}{\sqrt{2}} \sqrt{\left(1 - \frac{\alpha}{\pi}\right) + \frac{1}{2\pi} \sin 2\alpha}, \end{aligned} \tag{3.24}$$

$$S = \frac{V_s I}{2} \sqrt{\left(1 - \frac{\alpha}{\pi}\right) + \frac{1}{2\pi} \sin 2\alpha}, \tag{3.25}$$

$$k_p = \frac{\langle p(t) \rangle}{S} = \sqrt{\left(1 - \frac{\alpha}{\pi}\right) + \frac{1}{2\pi} \sin 2\alpha}. \tag{3.26}$$

This power factor is plotted as a function of α in Fig. 3.12. From this plot you can see that if $\alpha = \pi/2$, the source can supply only 71% of the power it could supply at $\alpha = 0$, for the same value of S.

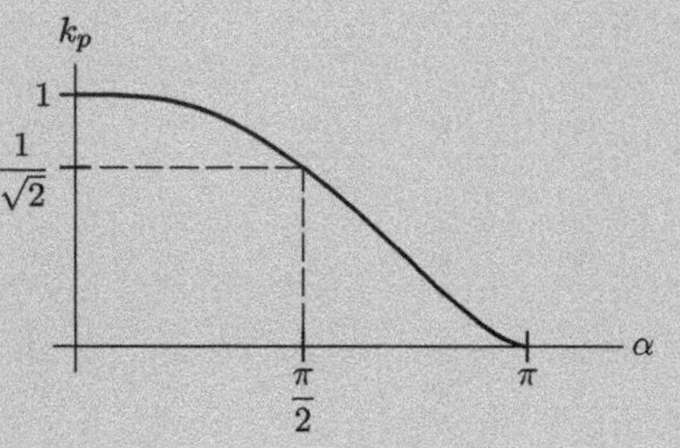

Figure 3.12 Power factor k_p as a function of α for the ac controller of Fig. 3.11.

3.4.2 Real and Reactive Power for Sinusoidal Variables

For sinusoidal voltage and current waveforms of the same frequency, the power factor is the cosine of the phase angle between them. We can see this by calculating the average power delivered by the source voltage v_s and the current i_s shown in Fig. 3.13. This average power, which we also call the *real power* P, is

$$\begin{aligned} P = \langle p(t) \rangle &= \frac{1}{2\pi} \int_0^{2\pi} v_s i_s \, d(\omega t) = \frac{V_s I_s}{2} \cos\theta \\ &= V_{srms} I_{srms} \cos\theta = S \cos\theta = k_p S. \end{aligned} \tag{3.27}$$

We call the angle θ the *power factor angle*. Electricians and utility engineers commonly use this parameter because their concern is primarily with sinusoidal voltages and currents. If $\theta > 0$ (an inductive load), the current lags the voltage in time, and the result is a *lagging power factor*. A *leading power factor* results for $\theta < 0$ (a capacitive load).

In addition to the real power P, we *define* a quantity called *reactive power* Q as

$$Q \equiv V_{srms} I_{srms} \sin\theta = S \sin\theta. \tag{3.28}$$

With this definition,

$$|P + jQ| = V_{srms} I_{srms} = S. \tag{3.29}$$

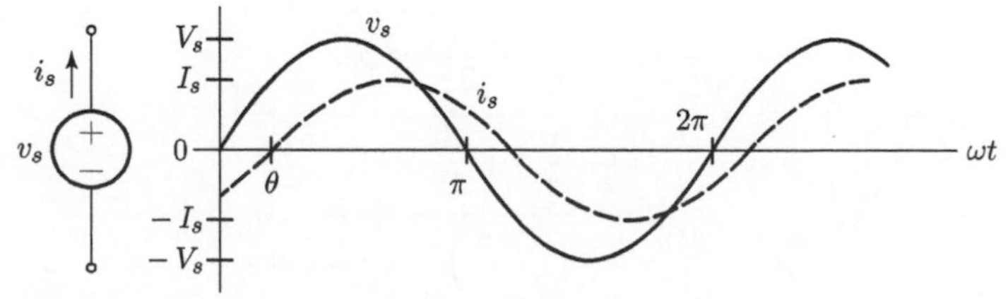

Figure 3.13 Voltage and current waveforms at the terminals of a source. The angle θ is called the power factor angle.

The usefulness of Q is that it tells us how to *compensate* a load using, for instance, reactive elements (inductors and capacitors) to make the resulting power factor unity. For example, the reactive power being delivered by the source in Fig. 3.13 is positive and given by (3.28). A capacitor placed across the source would draw negative reactive power Q_C because $\theta_C = -\pi/2$. If we choose the value of this capacitor so the net reactive power Q' delivered by the source is 0, then only real power would be delivered by the source. For this, we require

$$Q' = Q_C + Q = -V_{srms}^2 C\omega + V_{srms} I_{srms} \sin\theta = 0, \tag{3.30}$$

from which we can deduce the required capacitance to be

$$C = \frac{I_{srms} \sin\theta}{\omega V_{srms}}. \tag{3.31}$$

Now, (3.29) will tell us what the new source current I_s' is if P remains unchanged from its value given by (3.27):

$$|P + jQ'| = P = V_{srms} I'_{srms} = V_{srms} I_{srms} \cos\theta, \tag{3.32}$$

$$I'_{srms} = I_{srms} \cos\theta < I_{srms}. \tag{3.33}$$

The compensated source current I'_{srms} is thus reduced from its uncompensated value I_{srms}, permitting the source to deliver more real power at a given current.

3.4.3 Power Factor of Distorted Waveforms

Because of the switching occurring in power electronic circuits, the voltage and current waveforms at a port are seldom both sinusoidal. For example, none of the rectifier circuits we have presented so far produces a sinusoidal line current, although their line voltages are sinusoidal. The concept of power factor angle is therefore not particularly useful when we are dealing with power electronic circuits.

The following derivation of k_p explicitly and separately accounts for both waveform distortion and the phase displacement of equal-frequency components of the covariables. Thus, we express the power factor k_p as the product of two terms, one representing the effect of distortion and the other the effect of displacement:

$$k_p = \frac{\langle p \rangle}{S} = k_d k_\theta. \tag{3.34}$$

In this expression, k_θ is the *displacement factor* and k_d is the *distortion factor*, both defined below. The case we consider here is that of distortion in only one of the variables; we assume that the other is sinusoidal, an assumption that is often well justified.

To derive such an expression for the power factor, we assume a port, such as that of network N in Fig. 3.1, to have periodic voltage and current waveforms that we can describe analytically as

$$v(t) = V_s \sin\omega t, \qquad \omega \neq 0, \tag{3.35}$$

$$i(t) = I_o + I_1 \sin(\omega t + \theta_1) + \sum_{n>1} I_n \sin(n\omega t + \theta_n). \tag{3.36}$$

In terms of these variables, the average power is

$$\langle p \rangle = \frac{1}{T}\int_0^T vi\,dt = \frac{V_s I_1}{2}\cos\theta_1 = V_{s\text{rms}} I_{1\text{rms}} \cos\theta_1, \tag{3.37}$$

where the term $\cos\theta_1$ is the displacement factor k_θ, and $I_{1\text{rms}}$ is the rms value of the fundamental component of i. By factoring out I_{rms}, we can find the distortion factor k_d:

$$\langle p \rangle = V_{s\text{rms}} I_{\text{rms}} \frac{I_{1\text{rms}}}{I_{\text{rms}}} \cos\theta = S k_d k_\theta, \tag{3.38}$$

$$k_d = \frac{I_{1\text{rms}}}{I_{\text{rms}}}. \tag{3.39}$$

3.4.4 Total Harmonic Distortion

One measure of distortion in waveforms is a quantity known as the *total harmonic distortion*, or THD, which we define as

$$\text{THD} = \sqrt{\frac{\sum_{n\neq 1} I_{n\text{rms}}^2}{I_{1\text{rms}}^2}}, \tag{3.40}$$

where $I_{n\text{rms}}$ is the rms value of the nth harmonic component of the line current. In practical terms, the THD is the square root of the ratio of the following quantities: (i) the power that would be dissipated in a resistor by the distortion components of a waveform; (ii) the power that would be dissipated by the fundamental component alone.

Example 3.3 An Example THD Calculation

Let us calculate the THD of the line current i_ℓ in the half-wave rectifier with freewheeling diode of Fig. 3.5. We assume that $L/R = \infty$, so the line current comprises periodic rectangular pulses of amplitude I_d, as shown in Fig. 3.14.

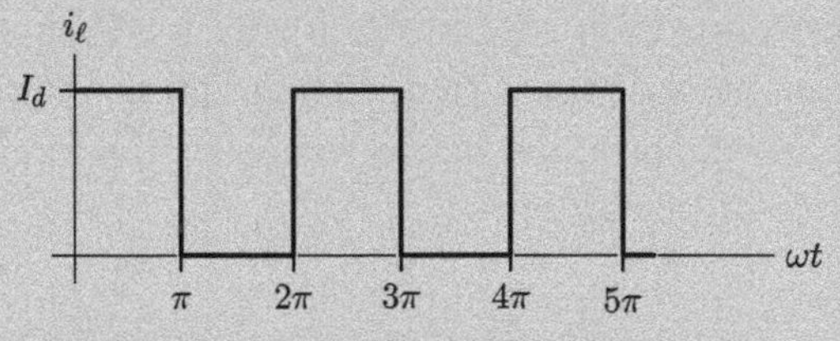

Figure 3.14 Waveform of the line current i_ℓ for the rectifier of Fig. 3.5, assuming that $L/R = \infty$.

In this case we may express the THD in terms of the rms amplitudes of the components as

$$\text{THD} = \sqrt{\frac{I_{\ell\text{rms}}^2 - I_{1\text{rms}}^2}{I_{1\text{rms}}^2}}, \tag{3.41}$$

where $I_{1\text{rms}}$ is the rms value of the fundamental component of i_ℓ, namely $I_1/\sqrt{2}$, and I_1 is computed using (3.1) and (3.2). Since i_ℓ is constant in the first half of the period and 0 for the remainder – see Fig. 3.14 – the integral for b_1 in (3.2) is 0, and we have $I_1 = a_1$ so $I_{\text{rms}} = a_1/\sqrt{2}$:

$$I_{1\text{rms}} = \frac{1}{\sqrt{2}\pi} \int_0^{2\pi} i_\ell \sin \omega t \, d(\omega t)$$
$$= \frac{1}{\sqrt{2}\pi} \int_0^{\pi} I_d \sin \omega t \, d(\omega t) = \frac{\sqrt{2} I_d}{\pi}. \tag{3.42}$$

The square of the rms value of the total line current is

$$I_{\ell \text{rms}}^2 = \frac{1}{2\pi} \int_0^{2\pi} i_\ell^2 \, d(\omega t) = \frac{I_d^2}{2}. \tag{3.43}$$

We can now calculate the THD for this waveform:

$$\text{THD} = \sqrt{\frac{I_d^2/2 - 2I_d^2/\pi^2}{2I_d^2/\pi^2}} = 121\%. \tag{3.44}$$

You will encounter this particular waveform often, so remembering its THD will be useful. In addition, this THD provides a benchmark for comparison of the THDs of other waveforms.

In terms of the THD of the distorted waveform as defined in (3.40), we can express the distortion factor k_d as

$$k_d = \sqrt{\frac{1}{1 + (\text{THD})^2}}. \tag{3.45}$$

Example 3.4 Using Power Relationships to Compute THD

By exploiting the power relationships discussed in Section 3.1, we can calculate the THD of the line current for the rectifier of Fig. 3.2 (which is equal to i_d in this case) without explicitly calculating its Fourier components. We do so by first equating the average powers on the ac and dc sides of the rectifier to determine I_1, the amplitude of the fundamental component of the line current:

$$\langle p_{\text{dc}} \rangle = \frac{\omega}{2\pi} \int_0^{2\pi/\omega} v_d i_d \, dt = \frac{V_s^2}{2\pi R} \int_0^{\pi} \sin^2 \omega t \, d(\omega t) - \frac{V_s^2}{4R} \tag{3.46}$$

and

$$\langle p_{\text{ac}} \rangle = \frac{V_s I_1}{2} = \langle p_{\text{dc}} \rangle = \frac{V_s^2}{4R}. \tag{3.47}$$

Hence:

$$I_1 = \frac{V_s}{2R}, \qquad I_{1\text{rms}} = \frac{V_s}{2\sqrt{2}R}. \tag{3.48}$$

We may now calculate the THD in terms of the rms values:

$$I_{1\text{rms}}^2 + \sum_{n \neq 1} I_{n\text{rms}}^2 = I_{\text{rms}}^2 = \frac{1}{2\pi} \int_0^{2\pi} i_d^2 \, d(\omega t) = \frac{V_s^2}{4R^2}, \tag{3.49}$$

giving

$$\sum_{n \neq 1} I_{n\text{rms}}^2 = \left[\frac{V_s}{2R}\right]^2 - \left[\frac{V_s}{2\sqrt{2}R}\right]^2 = \left[\frac{V_s}{2\sqrt{2}R}\right]^2. \tag{3.50}$$

Thus:

$$\text{THD} = \sqrt{\frac{\sum_{n\neq 1} I_{n\text{rms}}^2}{I_{1\text{rms}}^2}} = 100\%. \tag{3.51}$$

3.5 Bridge Rectifiers

Although easily understood, and an effective vehicle for understanding basic rectifier operation, the half-wave rectifier is seldom used today. The reasons are that the half-wave circuit has a high ratio of ac source VA rating to dc power (that is, the ac source operates at a low power factor), produces a dc component in the ac source current (which causes problems for transformers in the source network), and requires a large amount of dc filtering to achieve a specified ripple. As you will see when we present alternative circuits, the singular advantage of the half-wave circuit is that only a single diode drop is present between its ac and dc sides. When there was no practical alternative to vacuum diodes this advantage made the half-wave rectifier a popular circuit, because vacuum diode drops range from tens to hundreds of volts, and represent appreciable loss and thermal problems. Semiconductor diodes have effectively neutralized this advantage of the half-wave circuit, except in low-voltage, high-power applications, the most notable being in the electrochemical industry (for example, low-potential electrolytic cells for the production of chlorine), or in applications where ripple is not a problem, such as battery charging.

The disadvantages of the half-wave circuit are largely overcome by circuits that function as *full-wave* rectifiers. In these circuits the diodes connect the dc load to the ac source during both the positive and negative half-cycles of the source. The result is a bilateral source current having no dc component, and an increase in the fundamental dc-side ripple frequency, which reduces the dc filtering requirements. The most common of these circuits is the *bridge rectifier*. An alternative, the *full-wave center-tapped* or *half-bridge* rectifier analyzed in Problem 3.10, is also used extensively.

In this section we introduce the concept of full-wave rectification by an appropriate interconnection of two half-wave circuits.

3.5.1 Single-Phase Full-Wave Bridge Rectifier

The full-wave bridge rectifier is the workhorse of single-phase rectifier circuits. Its ac source current contains no dc component, and the fundamental frequency of the ripple on the dc voltage is twice that of the half-wave circuit. For the same ac source voltage, the full-wave rectifier produces an average output voltage twice that of the half-wave circuit with freewheeling diode, even though the voltage and current ratings of the diodes in the two circuits are the same. A disadvantage of all bridge circuits is that the input and output ports have no common terminal. In a rectifier this means that the ac source and the dc load cannot share the same ground.

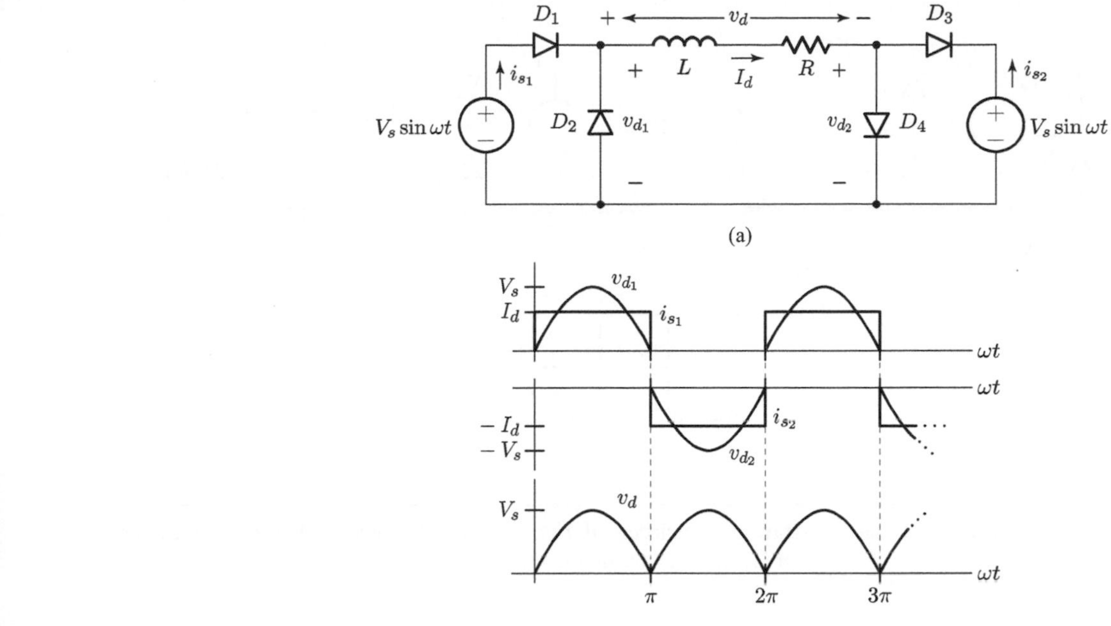

Figure 3.15 (a) The single-phase full-wave bridge circuit as a connection of two half-wave rectifiers with freewheeling diodes (half bridges). (b) Waveforms of variables in the circuit of (a).

The half-wave rectifier with freewheeling diode of Fig. 3.5 serves as the basic building block for the bridge circuit, and in this context that circuit is sometimes called a *half bridge*. So long as i_d is continuous (nonzero), v_d has the waveform shown there.

A circuit similar to that of Fig. 3.5 in every respect except the directions of the diodes, which are reversed, produces a load current i_d and voltage v_d of the same form as those shown in Fig. 3.5, but of opposite polarity. If we now connect the load between these two half-wave rectifiers, its voltage will be the difference between the output voltages of the two circuits, or $v_d = v_{d_1} - v_{d_2}$, as shown in Fig. 3.15(a).

The waveform v_d in Fig. 3.15(b) is known as a full-wave rectified voltage because both the positive and negative half-cycles of the ac source waveform are present on the dc side of the circuit. Because the two ac sources in Fig. 3.15(a) are identical, the anode of D_1 and cathode of D_3 are at the same potential and may be connected. The circuit thus requires only a single ac source. Two different ways of drawing the result are shown in Fig. 3.16(a). We call this connection of diodes a *bridge* because the circuits connected to its input and output (the ac source and the *RL* load) *bridge* one another topologically. Although the source currents i_{s_1} and i_{s_2} of Fig. 3.15 contain dc components, the current i_s in the equivalent source in Fig. 3.16 does not. The reason is that the source current in the bridge is the sum of the two source currents in Fig. 3.15, which have dc components of equal magnitude but opposite polarity.

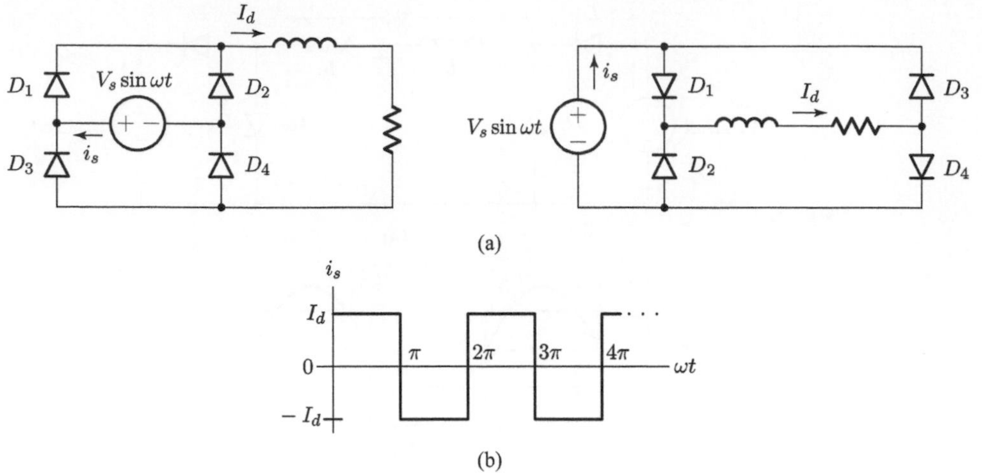

Figure 3.16 (a) Equivalent ways of drawing the circuit of Fig. 3.15 as a bridge. (b) The resulting source current, $i_s = i_{s_1} + i_{s_2}$, where i_{s_1} and i_{s_2} are defined in Fig. 3.15.

3.5.2 Output Voltage and Power Factor of the Single-Phase Bridge

All the operating characteristics of the single-phase bridge rectifier may be inferred from those of the half-wave rectifier with freewheeling diode. First, since the two half bridges in Fig. 3.15(a) can be viewed as independent voltage sources, v_{d_1} and v_{d_2}, the average output voltage of the bridge is twice that of the half-wave circuit with freewheeling diode, which is given by (3.4). That is,

$$\langle v_d \rangle = \langle v_{d_1} \rangle - \langle v_{d_2} \rangle = \frac{2V_s}{\pi} = 0.64V_s. \tag{3.52}$$

For the same dc currents, the rms value of the ac line current for the bridge of Fig. 3.16 is equal to only $\sqrt{2}$ times the rms value of the line current for one of the half bridges of Fig. 3.15(a). The power factor k_p of the ac source in the bridge circuit is

$$k_p = \frac{\langle p \rangle}{S} = \frac{(0.64V_s)(I_d)}{(V_s/\sqrt{2})(I_d)} = 0.91. \tag{3.53}$$

For comparative purposes, the power factor of the half-wave circuit with freewheeling diode is 0.64.

Example 3.5 Half-Bridge Full-Wave Rectifier

A half-bridge circuit can be used to create a full-wave rectified voltage if it is supplied by a center-tapped voltage source, as shown in the *center-tapped rectifier* of Fig. 3.17(a). What are $\langle v_d \rangle$ and k_p at the sources?

An assumed-state analysis shows that D_1 and D_2 cannot be on or off at the same time. Therefore, the top and bottom halves of the circuit behave as two independent half-wave rectifiers. The result is the waveforms shown in Fig. 3.17(b). Although v_d is a full-wave rectified voltage, each source carries a dc component of current, resulting in a source power factor of

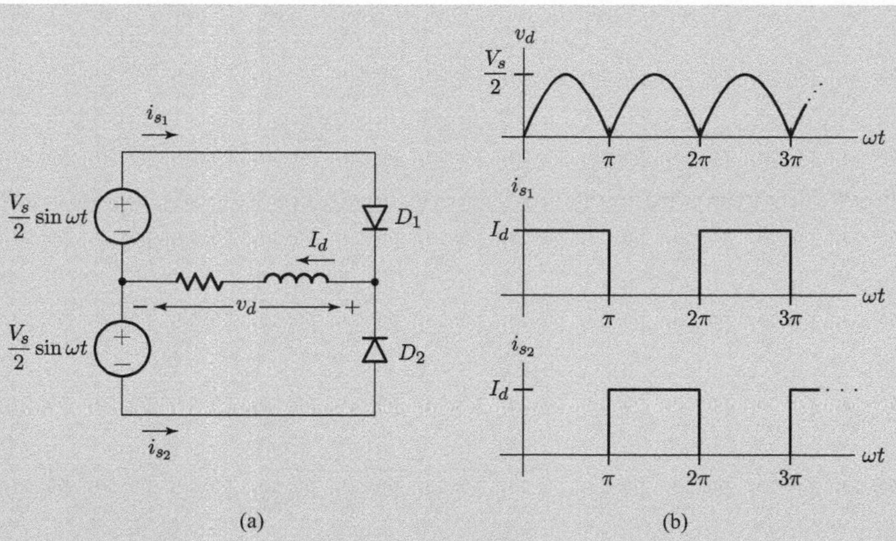

Figure 3.17 (a) Half-bridge rectifier circuit with center-tapped source. (b) Waveforms of variables in the circuit of (a).

$$k_p = \frac{\langle v_d \rangle \langle i_{s1} \rangle}{(V_s/2\sqrt{2})(I_d/\sqrt{2})} = \frac{(V_s/\pi)(I_d/2)}{V_s I_d/4} = \frac{2}{\pi} = 0.64. \tag{3.54}$$

If the two sources in the half bridge are created by center tapping the secondary of a transformer, the power factor at the primary terminals will be $k_p = 0.91$ because the primary current will have no dc component. However, the current rating (and corresponding size of wire) chosen for the secondary winding will have to be based on the poorer power factor of 0.64.

3.5.3 Commutation and Regulation in the Single-Phase Bridge

We can derive the commutation process in a bridge rectifier with ac-side reactance from the circuit of Fig. 3.18(a), which shows the two half bridges, each with commutating inductance $2L_c$. The current source I_d represents the load. Commutation of the two half bridges is simultaneous, but oppositely directed.

That is, while the load current is commutating from D_1 to D_2, it is also commutating from D_4 to D_3. Thus, all four diodes are on during the commutation process, $v_{x_1} = v_{x_2}$, and the derivatives of i_{s_1} and i_{s_2} are equal. During periods when no commutation is taking place, the voltage drops across the commutating inductances are zero because of the constant-load-current constraint. At all times, then, the anode of D_1 and cathode of D_3 are at the same potential and may be connected. The resulting circuit is shown in Fig. 3.18(b), which is simply the bridge circuit with a commutating inductance equal to one half that in the half bridges of Fig. 3.18(a).

The operations of the two circuits in Fig. 3.18 are the same. Thus the commutation period u for the bridge with commutating inductance L_c is equal to that for a half bridge having commutating inductance $2L_c$. We obtain the expression from (3.19):

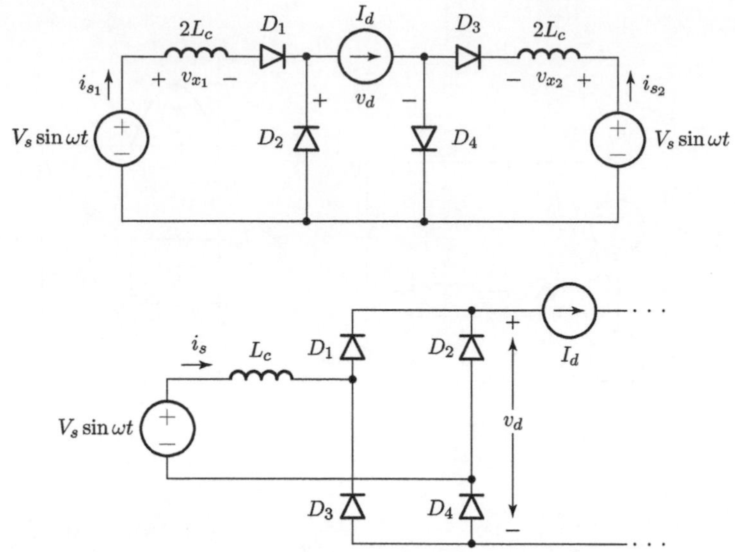

Figure 3.18 (a) Full-wave rectifier consisting of two half bridges with commutating inductance. (b) Full-wave bridge equivalent to (a).

$$u = \cos^{-1}\left(1 - \frac{2X_c I_d}{V_s}\right). \tag{3.55}$$

The commutation periods are equal because the current in the inductance in the full-bridge circuit swings between I_d and $-I_d$ during commutation ($\Delta i_s = 2I_d$), whereas it varies only between I_d and 0 in the half-wave circuit.

The average dc voltage as a function of I_d (the load regulation characteristic) is twice that for a half-wave rectifier with freewheeling diode having a commutating reactance of $2X_c$. Applying this factor of two to (3.20), we obtain

$$\langle v_d \rangle = \frac{V_s}{\pi}(1 + \cos u) = \frac{2V_s}{\pi}\left(1 - \frac{X_c I_d}{V_s}\right) = V_{do}\left(1 - \frac{X_c I_d}{V_s}\right), \tag{3.56}$$

where V_{do} is the name we use to designate the maximum possible value of the output of a rectifier circuit.

The waveforms of the source current i_s and the output voltage v_d in the circuit of Fig. 3.18(b) are shown in Fig. 3.19. The regulation characteristics of the half-wave and full-wave single-phase rectifiers are compared in Fig. 3.20. Normalized to V_{do}, the bridge circuit produces twice the regulation of the half-wave circuit. As Δi_s and X_c always appear as a product in the expression for $\langle v_d \rangle$, the effect of $\Delta i_s = 2I_d$ in the bridge is the same as that of doubling the commutating reactance in the half bridge. However, the regulation advantage of the half-wave circuit is rarely important enough to overcome its ac-side disadvantages, namely poor power factor and the presence of a dc component in the current.

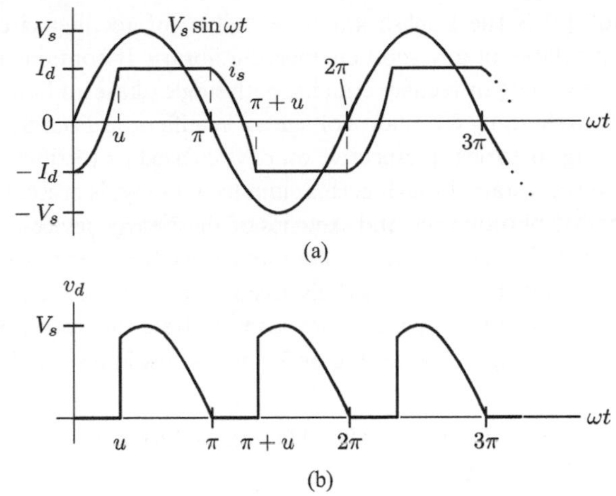

Figure 3.19 Waveforms for the bridge rectifier of Fig. 3.18(b): (a) source voltage $V_s \sin \omega t$ and source current i_s; (b) rectified output voltage v_d.

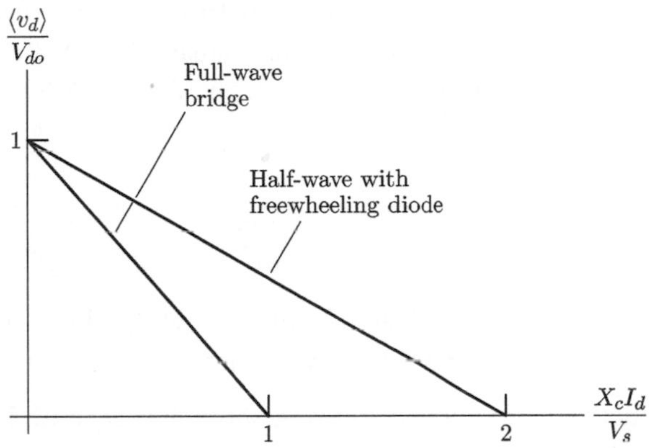

Figure 3.20 Normalized load regulation curves for the single-phase half-wave and bridge rectifiers.

Notes and Bibliography

Rectifiers are the oldest electronic circuit, and their analyses and configurations have changed little over the years. Although the references below may appear dated, their only fault is that some may be difficult to obtain.

A good discussion of the clash of personalities and professional opinions between Edison and Westinghouse can be found in [1]. Although it was these two men who debated the ac/dc issue, if they had not, others would have. The battle also raged in other countries with similar results. As recently as 1950, a section of New York City was supplied with dc instead of ac, and as late as 1969 dc was in some of the student dormitories of Boston University.

Single-phase rectifier circuits are treated extensively in [2]. Included are a variety of loads and filters, including capacitive filters and parallel LR loads. This book is extraordinarily comprehensive, but will prove an inconvenient source of information if you do not read German.

Schaefer's book [3] is the English language "bible" of rectifier circuits. Although out of print, the book should be available in any good engineering library. It contains tables, diagrams, and graphs that characterize a wide variety of rectifier circuits, both single phase and three phase. It also includes a chapter discussing capacitive loading (constant voltage) of rectifiers, as does Schade's paper [4].

A very interesting historical perspective on devices used as rectifier switches (vacuum tube, gas filled, mercury pool, and solid state through germanium technology) is presented in Chapters 2–5 of [5]. The text contains a number of photographs and sketches of these early devices.

Emanuel's paper [6] is an excellent and comprehensive treatment of power factor in systems producing non-sinusoidal variables. The six detailed discussions at the end of the paper and Emanuel's response add considerably to the value of this paper. They also illustrate the degree of confusion and controversy that still exists about the concept of power factor for non-sinusoidal variables.

1. T. S. Reynolds and T. Bernstein, "The Damnable Alternating Current," *Proc. IEEE*, vol. 64, no. 9, pp. 1339–1343, Sept. 1976.
2. Th. Wasserab, *Schaltaungslehre Der Stromrichtertechnik*, Berlin: Springer, 1962.
3. J. Schaefer, *Rectifier Circuits: Theory and Design*, New York: Wiley, 1965.
4. O. H. Schade, "Analysis of Rectifier Operation," *Proc. IRE*, vol. 31, pp. 341–361, 1943.
5. F. G. Spreadbury, *Electronic Rectification*, New Jersey: Van Nostrand, 1962.
6. A. E. Emanuel, "Powers in Non-sinusoidal Situations – A Review of Definitions and Physical Meaning," *IEEE Trans. Power Delivery*, vol. 5, pp. 1377–1389, July 1990.

PROBLEMS

3.1 By using the method of assumed states, show that in the circuit of Fig. 3.3 the diode off state is consistent with its branch variables after t_1.

3.2 Use the method of assumed states to show that D_1 and D_2 cannot be on simultaneously in the rectifier circuit of Fig. 3.5.

3.3 The circuit of Fig. 3.21 is representative of a class of ancillary circuits called *snubbers*, which we discuss in Chapter 24. Its purpose is to limit the rate of rise of the voltage v_Q.

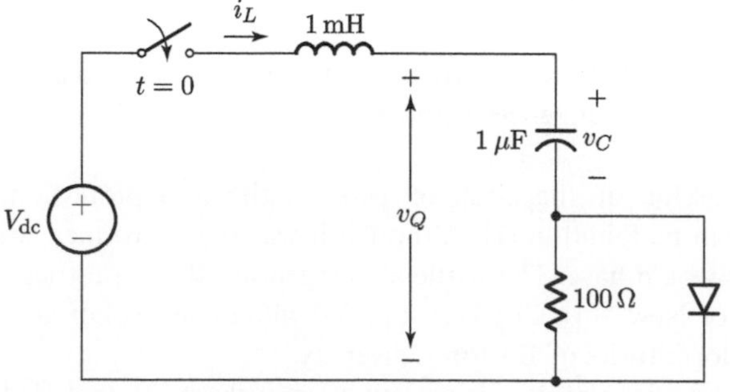

Figure 3.21 The snubber circuit analyzed in Problem 3.3.

Determine, sketch, and dimension i_L, v_C, and v_Q. What purpose do the diode and resistor serve?

3.4 For most engineering purposes, it is not necessary to calculate exactly ωt_1 in Example 3.1. Determine the error that would result in the calculation of $\langle v_d \rangle$ if we approximated ωt_1 as zero, that is,

$$e^{-(3\pi/2+\omega t_1)/\omega RC} \approx e^{-3\pi/2\omega RC}. \tag{3.57}$$

Would this be an appropriate approximation to make in calculating the ripple in v_d?

3.5 Construct a regulation curve for the circuit of Fig. 3.8 if we replace L_c with a resistor of value $R_c = X_c$.

3.6 The circuit in Fig. 3.22 varies with time because L_2 is switched in and out of the circuit by the diode D. The circuit is initially at rest, and at $t = 0$ the switch S is closed. Calculate and plot i_ℓ, i_1, i_2, and v_C for $0 < t < 50$ μs.

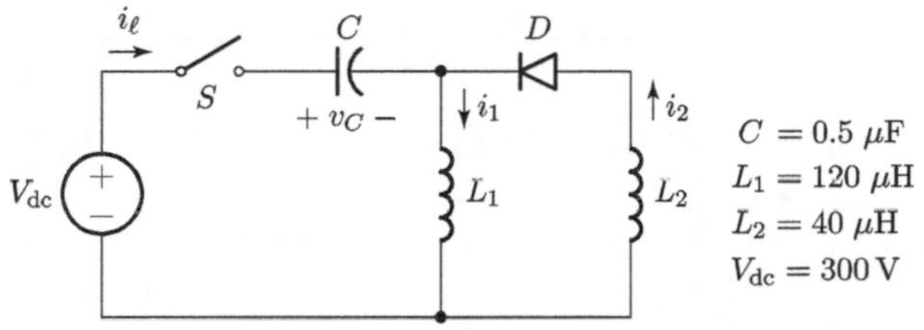

Figure 3.22 The second-order switched circuit analyzed in Problem 3.6.

3.7 Determine the power factor of the half-wave rectifier with freewheeling diode of Fig. 3.5. Assume that $L/R \gg 2\pi/\omega$, so $i_d \approx I_d$.

3.8 Determine the THD of the line current for the rectifier with commutating inductance shown in Fig. 3.8(a). What is the power factor of this circuit? How does it compare to the power factor if $L_c = 0$? Approximate the line current as a trapezoid whose rising and falling edges have duration u.

3.9 Sketch the ac source current i_ℓ in the half-wave rectifier with capacitive filter of Example 3.1. What is the power factor of this circuit?

3.10 Figure 3.23 is a schematic drawing of a full-wave, center-tapped rectifier. The two ac sources are usually created by a transformer with a center-tapped secondary winding.

(a) Assume that $L_c = 0$ and sketch $v_d(t)$.

(b) Determine and sketch i_{ℓ_1} and i_{ℓ_2} for $0 < \omega t < 2\pi$.

(c) Determine and sketch the load regulation curve for this circuit and compare it quantitatively to that in Fig. 3.10 for the half-wave rectifier of Fig. 3.8.

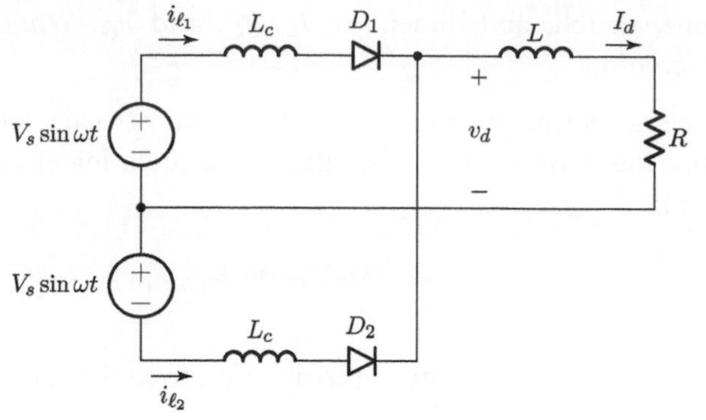

Figure 3.23 The full-wave, center-tapped rectifier of Problem 3.10.

3.11 The circuit of Fig. 3.24 is often used to create a dual voltage supply, ±15 V, for instance. Sketch and dimension v_{d_1} and v_{d_2}. (*Hint:* try to recognize the independence of the two outputs.)

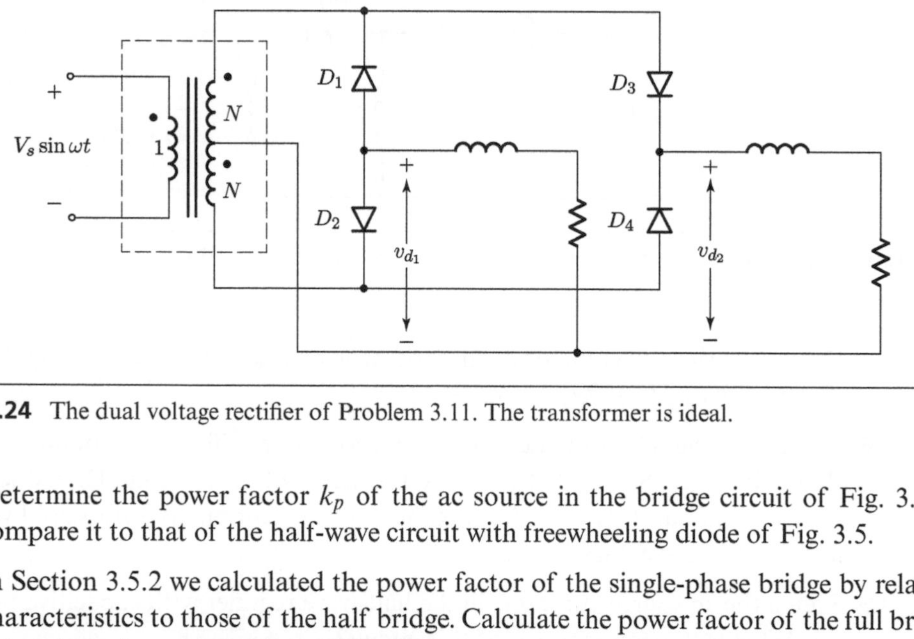

Figure 3.24 The dual voltage rectifier of Problem 3.11. The transformer is ideal.

3.12 Determine the power factor k_p of the ac source in the bridge circuit of Fig. 3.16 and compare it to that of the half-wave circuit with freewheeling diode of Fig. 3.5.

3.13 In Section 3.5.2 we calculated the power factor of the single-phase bridge by relating its characteristics to those of the half bridge. Calculate the power factor of the full bridge by using the ac input variables of Fig. 3.16.

3.14 In some applications a rectifier is supplied from an ac current source instead of a voltage source. If the frequency is high enough, the junction capacitance of the rectifiers has an appreciable effect on the dc output current. Consider the circuit of Fig. 3.25, in which the diode junction capacitance C_j has been modeled as constant and in parallel with an ideal diode. Determine the load regulation characteristic for this circuit,

$$\langle i_d \rangle = f\left(\frac{V_d}{X_j I_s}\right), \tag{3.58}$$

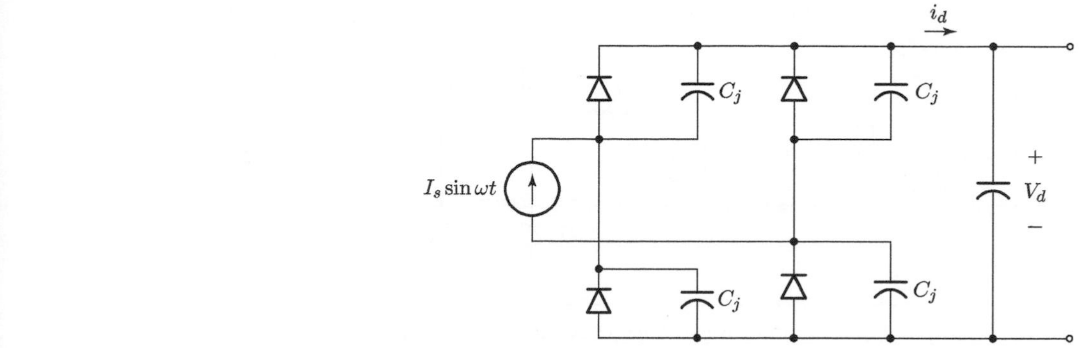

Figure 3.25 The current-fed bridge rectifier of Problem 3.14, including the diode junction capacitances, C_j, which have an important effect at high frequencies.

where X_j is the reactance of the junction capacitance, $X_j = 1/(\omega C_j)$. The junction capacitance of a 20 A, 100 V Schottky diode is approximately 200 pF. At what frequency would you expect the effects of C_j to become important?

3.15 Figure 3.26 shows a half-wave rectifier driven by a sinusoidal current source supplying a capacitively filtered output. (Such a configuration is sometimes found in resonant dc/dc converters.) Determine the power factor seen by the current source, assuming that the diodes act ideally and the capacitance C_f is large enough that the output voltage has small ripple ($v_D \approx V_D$). What are the displacement factor and distortion factor presented by this rectifier?

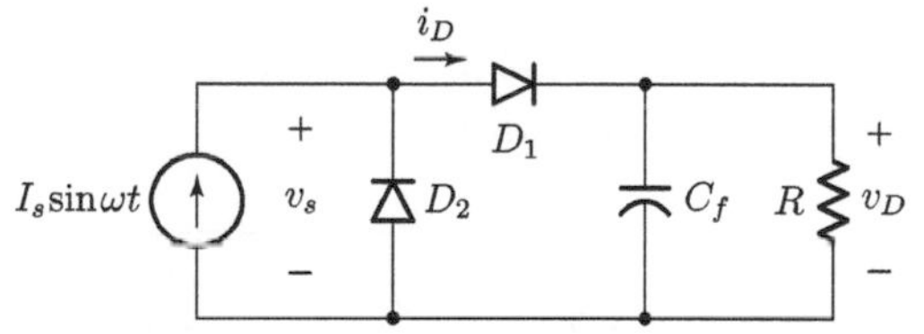

Figure 3.26 A half-wave rectifier driven from a current source analyzed in Problem 3.15.

3.16 Figure 3.27 shows a square-wave source voltage of magnitude V_s driving a rectifier via a small inductance L_c. The diodes can be treated as ideal, and the inductor L is sufficiently large that the ripple in current i_D may be neglected ($i_D \approx I_D$). Such a condition can occur on the secondary side of an isolated dc/dc converter, for example, where $v(t)$ is the voltage induced on the transformer secondary and the inductance L_c is the transformer leakage inductance.

(a) Sketch and dimension the voltage v_x and current i_c (under periodic steady-state conditions) for the case where $L_c = 0$. You should denote the values and intervals that define the waveform shape in terms of circuit parameters V_s, T, and R.

(b) What is the dc output voltage v_D in this case?

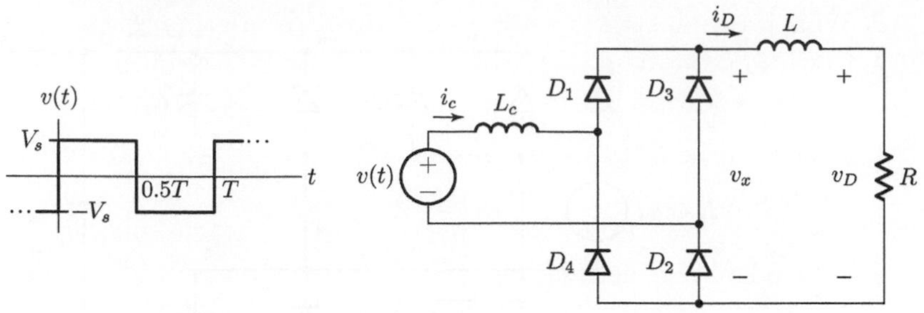

Figure 3.27 A rectifier driven by a square-wave voltage source considered in Problem 3.16.

(c) The inductance L_c can significantly affect the operation of the rectifier. Sketch and dimension the voltage v_x and current i_c (under periodic steady-state conditions) for the case where L_c is nonzero. You should denote the values and intervals that define the waveform shape in terms of circuit parameters V_s and T, and the output current i_D.

(d) Label the sketch of part (c) to clearly illustrate when each diode conducts, and indicate the duration of the commutation interval (during which i_c changes).

(e) Again referring to the system of part (c), express the average output voltage v_D in terms of V_s, T, and L_c, and the output current i_D.

(f) Make a sketch of how the output voltage v_D varies with load (that is, plot a load regulation curve).

3.17 Consider the rectifier system shown in Fig. 3.28, which is known as a "current doubler" rectifier. It has the following parameters: $V_s = 170$ V, $\omega = 377$. Also, assume that $L_{D1}/R \gg 2\pi/\omega$, so $i_{D1} \approx I_{D1}$, and $L_{D2}/R \gg 2\pi/\omega$, so $i_{D2} \approx I_{D2}$. Note that, by symmetry, we expect $I_{D1} = I_{D2}$ in periodic steady state.

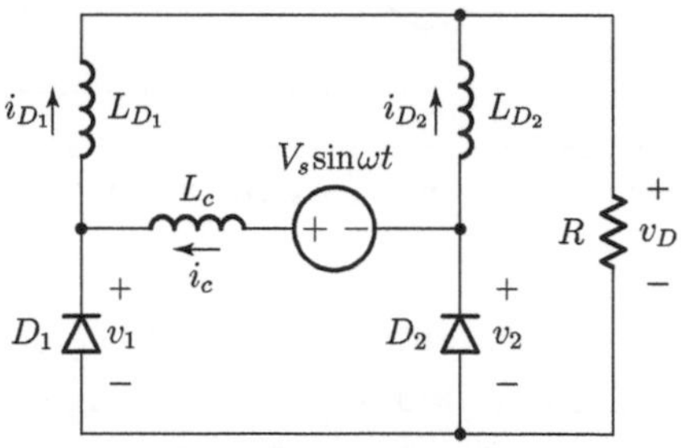

Figure 3.28 A current doubler rectifier considered in Problem 3.17.

(a) Plot $v_1(t)$ and $v_2(t)$ in the case where the ac-side inductance $L_c = 0$, and label the duration when each of the two diodes conducts.

(b) Compute the power factor k_p, the distortion factor k_d, and the displacement factor k_θ for the system (at the terminals of the voltage source), in the case where the ac-side inductance $L_c = 0$.

(c) What is the maximum power that could be drawn from the ac source in this case ($L_c = 0$) without exceeding a 15 A rms limit on the ac line current? How does this compare to the amount of power that could be drawn with a resistor (instead of the rectifier)?

(d) Now suppose the ac-side inductance is $L_c = 3$ mH (a rather large value). Ignoring line current limits, what is the maximum average power this rectifier can draw (over all possible values of load resistance)? (Hint: one way to find the answer is to look at the system from the dc side and apply the famous *maximum power transfer theorem*.)

3.18 Some types of rectifiers can only draw current from the grid over a limited portion of the line cycle. Consider the rectifier system and associated waveforms shown in Fig 3.29. For a line voltage $v_l(t) = V_s \sin(\omega t)$, the rectifier line current $i_l(t)$ is a stepped square wave having maximum value I_D and a conduction duration specified by an angle β.

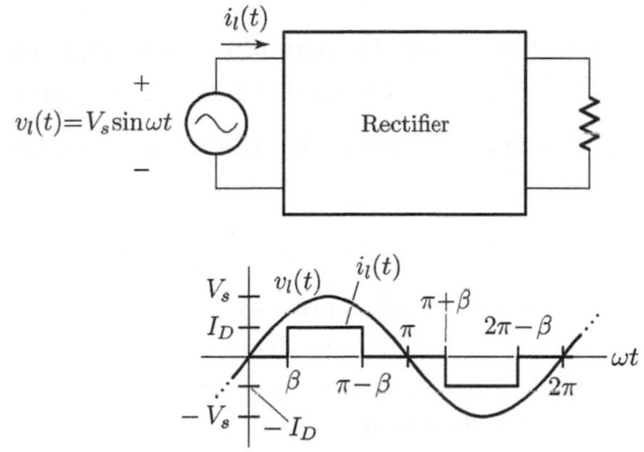

Figure 3.29 A rectifier that only draws current over a limited portion of the line cycle and its associated ac waveforms analyzed in Problem 3.18.

(a) What is the power factor of the rectifier at its ac port (v_l, i_l) as a function of the angle β $(0 \leq \beta < \pi/2)$?

(b) What is the distortion factor k_D and displacement factor k_θ for this rectifier?

3.19 Consider the high-frequency rectifier of Fig. 3.30. The capacitor C_f is sufficiently large that $RC_f \gg 1/\omega$, and assume the diodes are ideal. Ignore the capacitor C_p for parts (a) and (b).

(a) Sketch and dimension the input voltage v_X (under periodic steady-state conditions) for $R = 10\,\Omega$, $I = 2$ A, and $\omega = 1\times10^6$ rad/s. (You should calculate and specify numerical values that define the waveform shape for the specified operating condition.)

(b) What is the power factor (at the rectifier input) in this case?

(c) The parasitic junction capacitance of the diodes can significantly affect the operation of the rectifier. For our purposes here, we model the effects of junction capacitance with a small capacitor C_p in parallel with the current source. Sketch and dimension

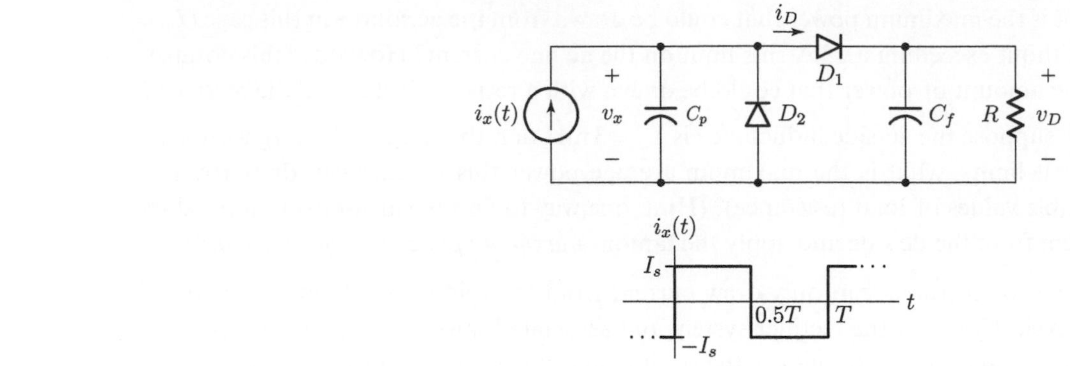

Figure 3.30 The high-frequency rectifier driven by a square-wave current source of Problem 3.19.

the input voltage v_X (under periodic steady-state conditions) for operation with a nonzero capacitance C_p. Dimension the plot in terms of V_d, I_s, C_p, and T.

(d) Label the sketch of part (c) to clearly illustrate when each diode conducts.

(e) Again referring to the system of part (c), express the duration of the commutation interval (during which v_X changes sign) as a function of V_d, I_s, C_p, and T.

3.20 Figure 3.31 is a rectifier circuit known as a *voltage doubler*. It is often used to provide for dual voltage operation – for instance, from both the 110 V residential service in the United States and the 220 V service in Europe – thereby avoiding the expense of manufacturing separate products for the two markets.

(a) If the switch S is open, draw the equivalent circuit and calculate V_{dc}.

(b) Repeat (a) if S is closed.

(c) For what voltage V_{dc} must the equipment powered by this supply be designed?

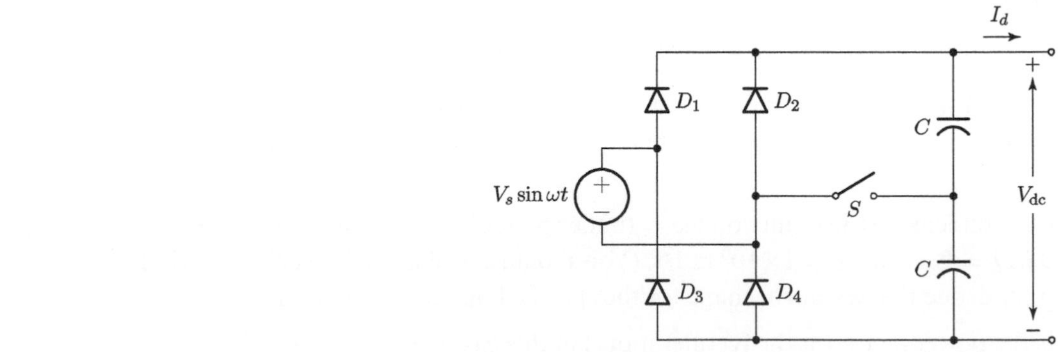

Figure 3.31 The voltage doubler circuit of Problem 3.20.

4 Phase-Controlled Converters

The rectifier circuits discussed so far are uncontrollable; that is, their output voltage is a function of system parameters and cannot be adjusted in response to parametric changes, such as variations in load (I_d) or ac voltage (V_s). You saw this characteristic in Fig. 3.10 for the single-phase circuit with freewheeling diode. However, we can make rectifier circuits controllable if we replace the diodes with a device known as a *silicon controlled rectifier* (SCR). This device is one member of a family of controllable switches known collectively as *thyristors*. Think of the SCR as a diode that will not conduct when forward biased until an appropriate control signal is applied to a third terminal, called the *gate*.

A rectifier circuit containing one or more SCR devices for controlling the dc output voltage is known as a *phase-controlled rectifier*, or simply a *controlled rectifier*. In addition to providing voltage control, phase-controlled rectifiers can be designed to permit power flow from the dc side to the ac side (as long as there is a source of energy on the dc side), opposite to the direction of power flow in a diode rectifier circuit. This process is called *inversion*, and a circuit operating in this way is called a *phase-controlled inverter*. In this chapter we discuss the behavior of both the phase-controlled rectifier and the phase-controlled inverter. These circuits differ primarily in the way they are controlled, so we refer to them as *phase-controlled converters* when no operating mode is specified.

We discuss the physics and detailed behavior of SCRs in Chapter 17. In this chapter you need to become familiar with only three characteristics of the SCR:

- Turning on an SCR requires applying a positive signal to its gate while its anode–cathode voltage is positive.
- Once on, an SCR remains on independent of the presence or absence of a signal at its gate, until its anode current goes to zero. That is, once on, an SCR behaves like a diode.
- When the anode current goes to zero, a small amount of time, known as the *turn-off time* t_q, must elapse before a positive anode–cathode voltage can be reapplied to the SCR without its turning on.

Phase-controlled rectifiers have a broad range of applications. They are frequently used as preregulators to create a dc supply from which one or more dc/dc converters operate. In the electrochemical industry they are used to control the power to electrolytic processes, such as electroplating and chlorine production. SCRs can be manufactured with very high current and voltage ratings. The ratings of commercially available devices range from about 250 mA and 50 V to 4 kA and 6 kV, and are used to build the converters for high-voltage dc transmission systems. The speed of dc motors and certain types of induction motors can be controlled using a simple phase-controlled rectifier. And the *ac controller*, a derivative of the phase-controlled rectifier circuit, is the basis for consumer products such as light dimmers. The ac controller generally contains a member of the thyristor family called a *triode ac switch* (TRIAC), which is functionally equivalent to a pair of SCRs connected in anti-parallel; that is, the TRIAC is a thyristor capable of controlled conduction in either direction.

4.1 Single-Phase Configurations

Single-phase controlled converters are generally used at power levels below about 10 kW. Some applications are dc motor drives, lighting controls, battery chargers, and preregulators for ac motor drives and switching power supplies. Besides the importance of their applications, these single-phase circuits demonstrate all the basic phase-controlled converter concepts – such as commutation, regulation, inversion, and non-unity power factor – and therefore serve as good vehicles for introducing phase control.

4.1.1 Half-Wave Controlled Rectifier with Resistive Load

If a phase-controlled converter is supplying a resistive load, we know that it is operating as a rectifier, because there is no source of energy on the dc side to permit inversion. We first consider the circuit of Fig. 3.2 with the diode replaced by an SCR, as shown in Fig. 4.1. Note the circuit symbol for an SCR, which is a diode symbol with a third terminal, called the gate, connected to its cathode. We do not show the circuit that provides the gate signal, because in this chapter we are concerned only with the operation of the power circuit. (We generally consider the gate circuit to be part of the control circuitry, and discuss it in Chapter 23.) Here, only the gate circuit's ultimate function at the gate terminal is important, and it is represented by either the presence or the absence of a signal.

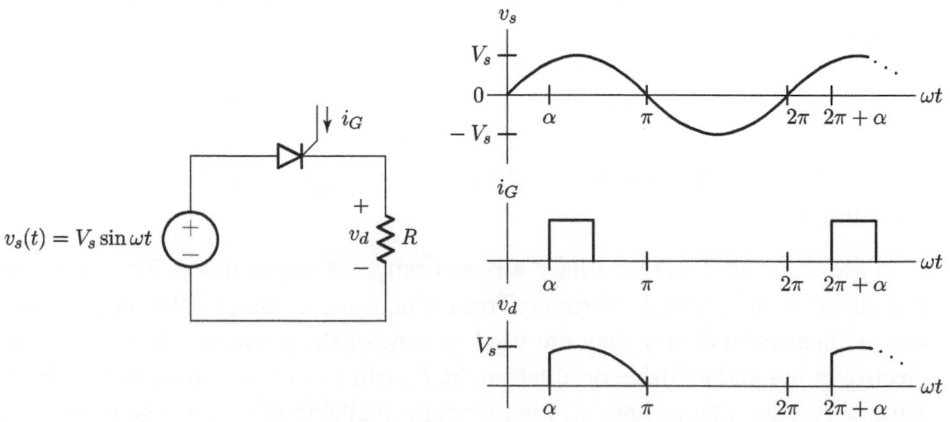

Figure 4.1 Half-wave phase-controlled rectifier with resistive load and its associated waveforms.

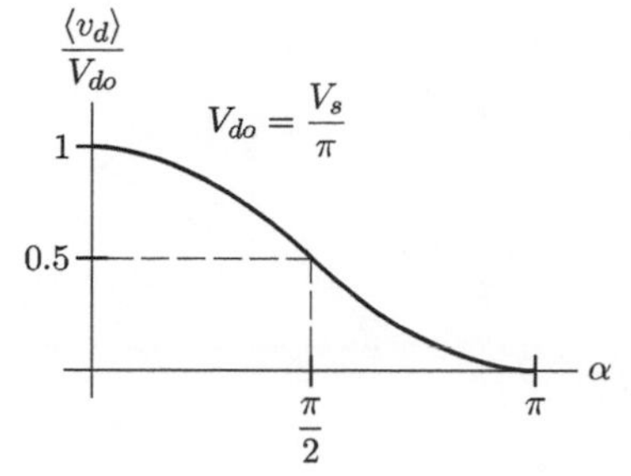

Figure 4.2 Control characteristic (4.1) for the rectifier of Fig. 4.1.

If no gate signal is ever applied to the SCR of Fig. 4.1, the output voltage of the rectifier is zero, because the SCR never turns on. Proper operation of this circuit requires application of the gate signal at some time during the period of normal diode conduction, that is, when the anode–cathode voltage of the SCR is positive. Figure 4.1 shows the gate signal being applied at an electrical angle α between 0 and π, which is during the interval in which a diode would conduct if it replaced the SCR. This signal is a current pulse (whose width is not critical) applied once every cycle of the line voltage waveform. The SCR is a regenerative device, that is, once it is on, it remains on – even after the gate pulse ends – until its anode current goes to zero (which in this circuit occurs at the zero-crossing of the ac input voltage). The angle α is variously called the *firing angle*, the *angle of retard*, or the *delay angle*. The retard or delay is measured relative to the angle at which the device would have turned on if it were a diode. The resulting output voltage v_d is also shown in Fig. 4.1, and from it we can infer the effect of gate control. The important characteristic of this voltage is its dc component,

$$\langle v_d \rangle = \frac{V_s}{2\pi}(1 + \cos\alpha) = \frac{V_{do}}{2}(1 + \cos\alpha), \tag{4.1}$$

where V_{do}, the maximum possible value of $\langle v_d \rangle$, is also the output of an equivalent diode rectifier. The voltage $\langle v_d \rangle$ as a function of α is known as the *control characteristic* of the rectifier and is shown in Fig. 4.2.

Example 4.1 Linearizing the Phase-Control Characteristic

The problem with the control characteristic shown in Fig. 4.2 is that its incremental gain, $d\langle v_d \rangle / d\alpha$, approaches zero near full voltage or zero voltage. This is not a desirable feature in a system's transfer function, if we want to close a feedback loop around it.

Figure 4.3 illustrates a common technique for linearizing the control characteristic, that is, making its gain constant over the entire operating range. A scaled version of the ac input voltage, v_1, is first integrated, then given a dc offset to produce v_2. This voltage is then compared to the new controlling variable V_α to produce a pulse v_3, commencing at $\omega t = \alpha$. The input voltage, v_1, has an amplitude of 1 V, the dc offset is 1 V, and $0 < V_\alpha < 1$ V. The output of the comparator v_3 is high when $V_\alpha > v_2$, producing a pulse starting at $\omega t = \alpha$. In terms of V_α, the pulse starts when $V_\alpha = 1 + \cos\alpha$, so

$$\alpha = \cos^{-1}(V_\alpha - 1). \tag{4.2}$$

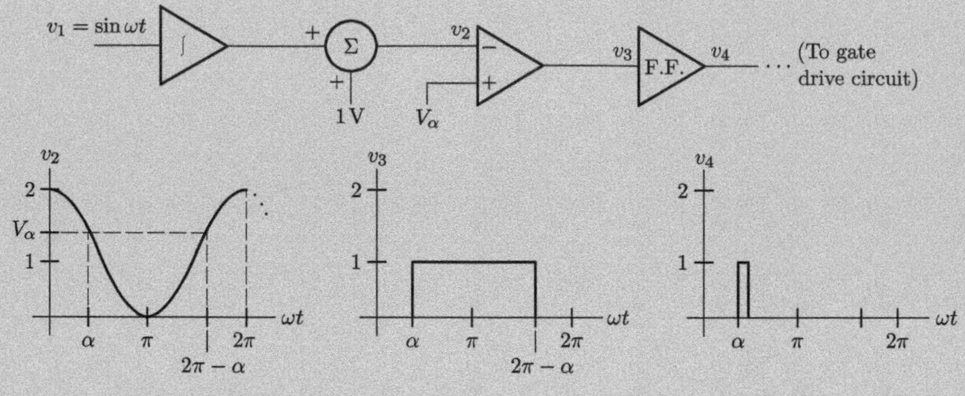

Figure 4.3 A functional block diagram of a circuit for linearizing the control characteristic of a phase-controlled rectifier.

Substituting this value of α into (4.1), we obtain the new control characteristic having constant gain:

$$\langle v_d \rangle = \frac{V_s}{2\pi}(V_\alpha), \qquad \frac{dv_d}{dV_\alpha} = \frac{V_s}{2\pi}. \tag{4.3}$$

The duration of the pulse v_3 is $2\pi - 2\alpha$, which can be too long, depending on the rectifier circuit. We can use a one-shot flip-flop, triggered on the rising edge of v_3, to create a shorter pulse v_4 starting at α, which is also shown in Fig. 4.3.

4.1.2 Full-Wave Phase-Controlled Bridge Rectifier

The phase-controlled rectifier circuit of Fig. 4.1 is not very interesting in terms of practical applications. A much more useful circuit is the phase-controlled bridge shown in Fig. 4.4. In this circuit, one diagonal pair of thyristors conducts for some part of each half cycle. We assume in the following discussion that we can model the load as a current source of value I_d.

The behavior of the output voltage waveform v_d can be understood by referring to the diode bridge rectifier, in which one pair of diodes is turned off by the action of the second pair turning on. In the phase-controlled rectifier, if the second pair does not turn on, the load current must continue to flow in the first pair, keeping it on. This means that Q_1 and Q_2 continue to conduct after the source voltage changes sign and becomes negative. They turn off when Q_3 and Q_4 receive a gate signal and turn on. Because there is no commutating reactance in the circuit, the load current commutates instantaneously to Q_3 and Q_4.

The average value of v_d, calculated from the waveform of v_d in Fig. 4.4, is

$$\langle v_d \rangle = \frac{1}{\pi}\int_{\alpha}^{\pi+\alpha} V_s \sin \omega t \, d(\omega t) = \frac{2V_s}{\pi}\cos\alpha = V_{do}\cos\alpha. \tag{4.4}$$

This voltage as a function of α is plotted in Fig. 4.5(a). The most interesting characteristic of the relationship is that, for $\pi/2 < \alpha < \pi$, the average value of the output voltage is negative.

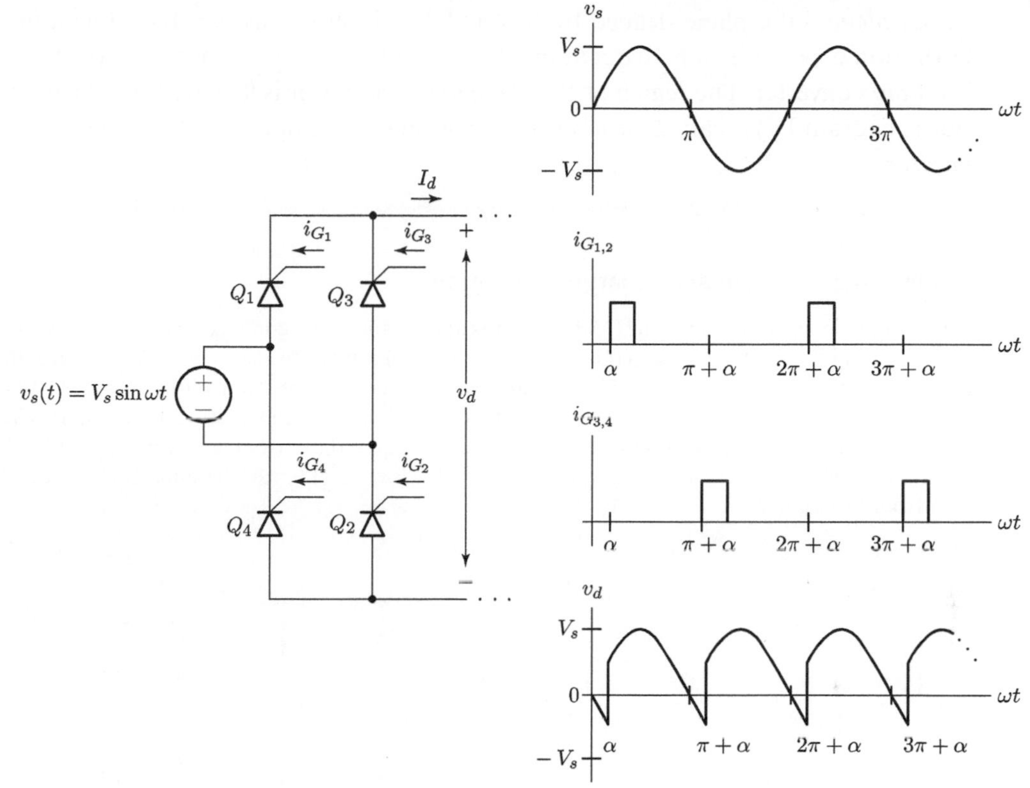

Figure 4.4 Phase-controlled bridge rectifier with no ac-side (commutating) reactance.

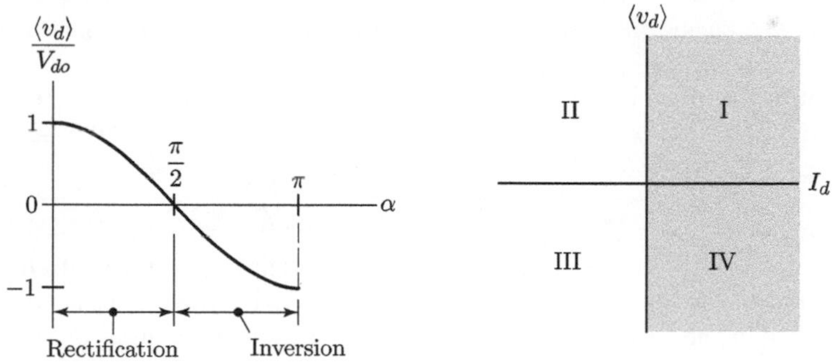

(a) (b)

Figure 4.5 (a) Control characteristic, and (b) quadrants of operation, for the bridge rectifier of Fig. 4.4.

Therefore, over this range of α, power is flowing from the dc side of the circuit to the ac side. As described in the introduction to this chapter, operation with this direction of power flow is called *inversion*.

The circuit of Fig. 4.4 can operate only if the load current I_d is positive; otherwise the thyristors cannot conduct. The region of possible operation for this circuit, represented in the *output plane* – the plane defined by $\langle v_d \rangle$ and I_d – is shown as the shaded area in Fig. 4.5(b). Operation is possible only in quadrants I and IV of this plane; hence the circuit is called a *two-quadrant* converter. The region of the plane in which power is flowing from the dc side to the ac side (quadrant IV) is called the *inversion region*, and when operating here, the circuit is called an *inverter*.

Example 4.2 A Magnet Discharge Application

Both magnetic resonance imaging (MRI) systems and magnetically confined fusion reactors require precise control of high magnetic fields. In this example we use a phase-controlled converter to control the magnetic field of a large electromagnet, such as might be used in these applications. Figure 4.6 shows the magnet modeled as the series combination of an inductance of value 0.5 H and a resistance of 2.5 Ω. The 60 Hz line voltage has a peak value of 2000 V. Assuming that the required steady-state magnet current is 400 A, what is the value of α that results in this value of current? How quickly can the magnet current be brought to zero from this value?

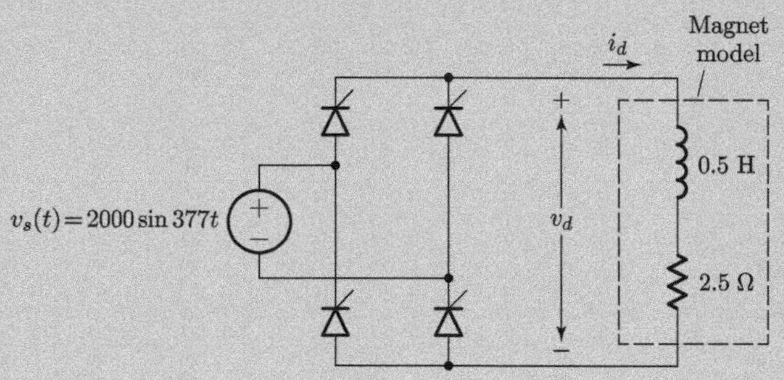

Figure 4.6 A phase-controlled converter being used to control the magnetic field of a high-field magnet.

The time constant of the load is $\tau = L/R = 0.2\,\text{s} \gg 1/60\,\text{Hz}$, so i_d has very little ripple. The inductor cannot support an average voltage in the steady state, so $\langle v_d \rangle$ appears across R, and $\langle i_d \rangle$ is

$$\langle i_d \rangle = \frac{\langle v_d \rangle}{R} = \frac{2(2000)}{\pi R} \cos \alpha. \tag{4.5}$$

Setting $i_d = 400$ A and solving for α, we get $\alpha = 38.2°$. This value of α results in $\langle v_d \rangle = 1000$ V.

The fastest way to bring the magnet current to zero is to remove the magnetic stored energy by operating the converter in the inversion region at the value of α that produces the maximum negative voltage. In this example the appropriate value of α is $\alpha = \pi$, which results in $\langle v_d \rangle = -1273$ V. (In Section 4.3.2 we show that an angle slightly less than $\alpha = \pi$ must be used.) We can now replace that part of the circuit consisting of the ac source and SCRs with an equivalent source whose value steps from $\langle v_d \rangle = 1000$ V to $\langle v_d \rangle = -1273$ V when we begin the magnet discharge. This equivalent circuit is shown in Fig. 4.7(a), and the resulting behavior of i_d is shown in Fig. 4.7(b). The current as a function of time is

$$i_d = -509 + 909e^{-t/\tau}.$$

Setting $\tau = 0.2$ s and $i_d = 0$, then solving for $t\,(= t_o)$, we find $t_o = 0.12$ s.

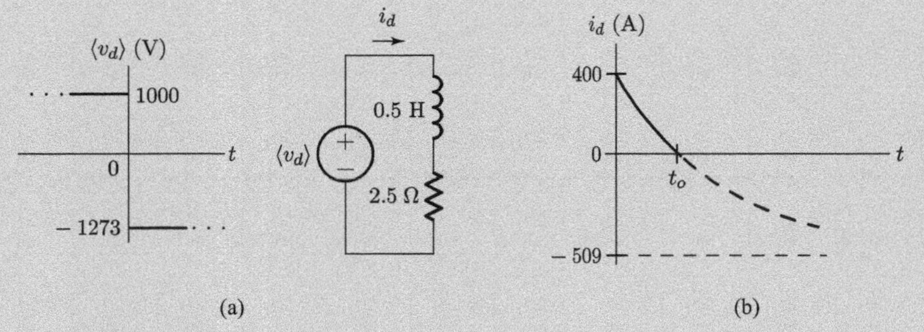

Figure 4.7 (a) Equivalent circuit for the calculation of i_d. The line and SCRs have been replaced by the source $\langle v_d \rangle$. (b) The behavior of i_d in the equivalent circuit as a function of time.

4.1.3 Converter Power Factor

An unfortunate consequence of phase control is that it degrades the power factor from what it would be for a diode rectifier. Consider the bridge of Fig. 4.8, which also shows the line voltage and current. We assume that the load is a constant current source of value I_d, no commutating reactance, and we determine the power factor as a function of α.

In Chapter 3 we discussed the power factor for distorted waveforms and showed how to write the power factor as the product of two factors: a displacement factor k_θ and a distortion factor k_d. Figure 4.8 represents the case in which the line voltage is sinusoidal, but the current contains distortion (it is a square wave). At this point we intuitively know what to expect. The square wave will retain its shape independent of α (although its amplitude might change). Therefore k_d will be constant. As α increases, the phase shift of the fundamental of the current waveform with respect to the voltage will increase, thus decreasing k_θ. Therefore we expect the power factor to decrease with increasing α.

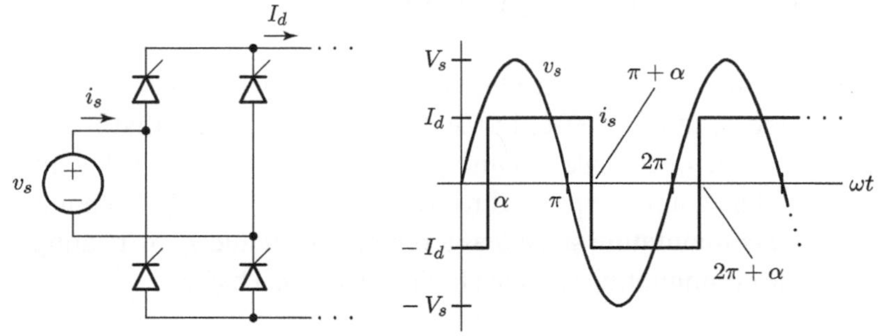

Figure 4.8 Phase-controlled bridge converter with a constant current load and no commutating reactance.

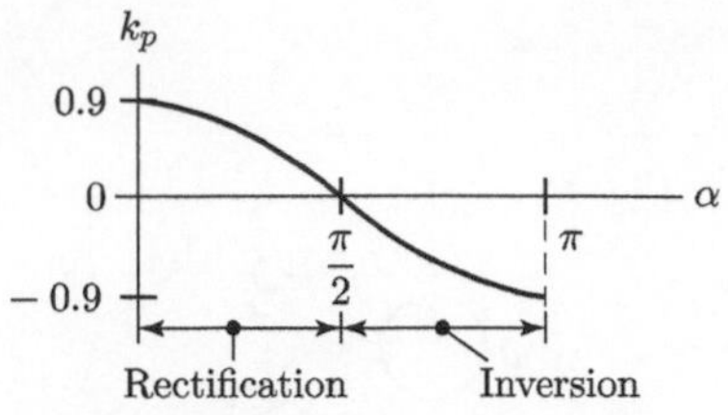

Figure 4.9 Power factor as a function of α for the bridge converter of Fig. 4.8.

We can write the power factor k_p as

$$k_p = k_d k_\theta = \frac{I_{1\text{rms}}}{I_{\text{rms}}} \cos\theta_1. \tag{4.6}$$

The fundamental component of a square wave has an amplitude equal to $4/\pi$ times the amplitude of the square wave. The fundamental is in phase with the square wave, which means that the fundamental component of the current lags the voltage by the firing angle α. For this case, then, k_p becomes

$$k_p = \frac{(4I_d/\sqrt{2}\pi)\cos\alpha}{I_d} = 0.9\cos\alpha. \tag{4.7}$$

This function is plotted in Fig. 4.9 for $0 < \alpha < \pi$, which includes inversion operation in quadrant IV.

The power factor of this circuit varies from 0.9 to −0.9, depending on the value of α. If the circuit contained commutating reactance, the edges of the current waveform would not be so abrupt, and k_d would be slightly larger. However, k_θ would be smaller, because the fundamental component of the current would be shifted to the right of α (see Problem 4.6).

4.2 Phase Control with AC-Side Reactance

Figure 4.10 shows a phase-controlled bridge rectifier with ac-side reactance. The effect of this reactance is identical to its effect in the diode rectifier, except that the commutation process is delayed by the electrical angle α. During the commutation interval u, all four thyristors are on and $v_d = 0$. The source appears directly across L_c and is the commutating voltage. We can determine the commutating angle u as a function of α and I_d by relating the current change in L_c $(2I_d)$ to the commutating voltage $(V_s \sin\omega t)$ and u. That is,

$$2I_d = \frac{1}{\omega L_c}\int_\alpha^{\alpha+u} V_s \sin\omega t \, d(\omega t). \tag{4.8}$$

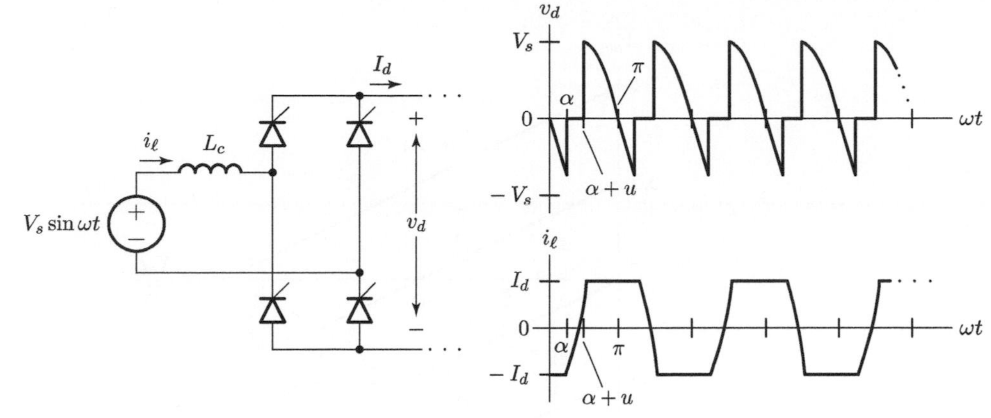

Figure 4.10 Phase-controlled bridge rectifier with ac-side (commutating) inductance L_c.

Solving this equation for u (and using $X_c = \omega L_c$) yields

$$u = \cos^{-1}\left(\cos\alpha - \frac{2X_c I_d}{V_s}\right) - \alpha. \tag{4.9}$$

We seldom need to determine the value of u explicitly, but (4.9) is useful in eliminating u from the expression for $\langle v_d \rangle$. We do so as follows:

$$\begin{aligned}\langle v_d \rangle &= \frac{V_s}{\pi}\left\{\int_0^{\alpha}(-\sin\omega t)\, d(\omega t) + \int_{\alpha+u}^{\pi} \sin\omega t\, d(\omega t)\right\}\\ &= \frac{V_s}{\pi}\{\cos\alpha + \cos(\alpha+u)\}.\end{aligned} \tag{4.10}$$

But, from (4.9), we know that

$$\cos(\alpha + u) = \cos\alpha - \frac{2X_c I_d}{V_s},$$

which, when substituted into (4.10), gives

$$\langle v_d \rangle = \frac{2V_s}{\pi}\left\{\cos\alpha - \frac{X_c I_d}{V_s}\right\} = V_{do}\left\{\cos\alpha - \frac{X_c I_d}{V_s}\right\}. \tag{4.11}$$

Note the similarity between this expression and that for the diode bridge, (3.56). Also note that the term $X_c I_d / V_s$, which is responsible for load regulation, is independent of the phase angle α. Figure 4.11 shows the family of regulation curves described by (4.11). The curves extend into quadrant IV, which represents inversion with $I_d > 0$ and $\langle v_d \rangle < 0$. in Section 4.3 we explain the diagonal boundary of the curves in this quadrant.

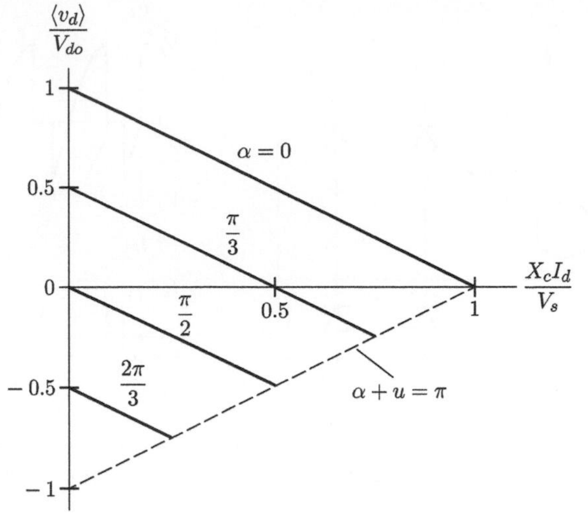

Figure 4.11 Family of regulation curves for the phase-controlled converter of Fig. 4.10.

Example 4.3 Phase Control of a Battery Charger

Figure 4.12(a) shows a phase-controlled bridge rectifier being used as a battery charger. A large inductor L smooths the charging current I_B. The battery is modeled as a voltage source, V_B (= 72 V). Note that the polarity of V_B does not permit steady-state inversion, although a transient excursion into inversion is possible until the inductor L is discharged. We are interested in determining the charging current I_B as a function of α.

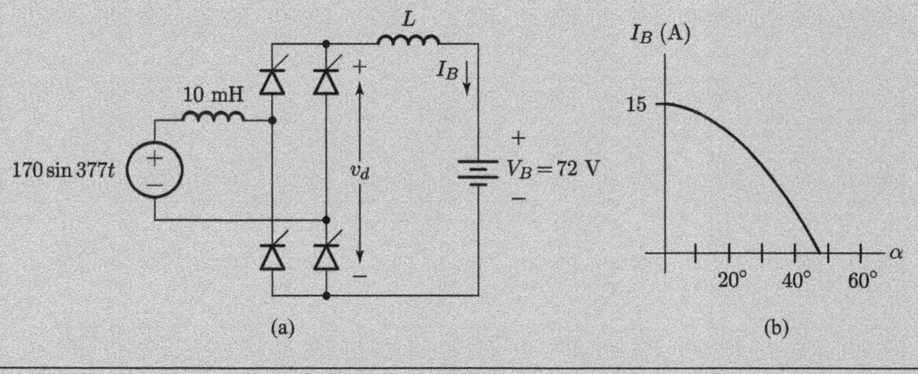

Figure 4.12 (a) Battery charger using a phase-controlled rectifier containing commutating reactance. (b) Relationship between charging current and firing angle.

The average value of v_d in the periodic steady state must equal the battery voltage, that is, $\langle v_d \rangle = 72$ V. The reason is that there can be no average voltage across L in the periodic steady state. Furthermore, this condition is independent of α. Using (4.11) to state this condition, $\langle v_d \rangle = 72$ V, in terms of I_B and α, we obtain

$$I_B = \frac{V_s}{X_c}\cos\alpha - \frac{\pi V_B}{2X_c} = (45\cos\alpha - 30)\,\text{A}, \tag{4.12}$$

which is plotted in Fig. 4.12(b). We can control the charging current between 0 and 15 A by varying α between approximately 48° and 0°. Note that the only reason that we can control this circuit is the presence of commutating reactance. As we change α, the current changes just enough to cause u given by (4.9) to assume the value necessary to ensure the condition $\langle v_d \rangle = 72$ V. If X_c equaled 0, we would be forced to set α so that $\alpha = \cos^{-1}(\pi V_B/2V_s) = 48.3°$, and the charging current would be indeterminate. In practice, however, there is always some resistance in both L and the battery, providing a degree of controllability; see Problem 4.12.

4.3 Inversion Limits

Inversion requires a source of energy on the dc side of the converter, as well as an ac voltage source on the ac side to commutate the thyristors. For the circuits we have been discussing, the dc source must be in quadrant IV of the output ($\langle v_d \rangle$ versus I_d) plane; that is, its current must be positive and its voltage negative, as these variables are defined in Fig. 4.4. Such a source might be a battery, a dc generator, or a solar photovoltaic array. A phase-controlled converter may also operate transiently in the inversion region to remove stored energy from the load, even though the load cannot provide steady-state power. Example 4.2 illustrated this mode of operation. In this section we explore more completely the behavior of phase-controlled converters operating in the inversion mode.

4.3.1 Commutation Failure

When a phase-controlled rectifier such as the bridge circuit shown in Fig. 4.13(a) operates in the inversion region, we must ensure that commutation is complete before the next zero-crossing of line voltage, at which point the commutating voltage changes polarity. If commutation were not complete by this time, the current in the SCR pair that is turning off would begin to increase, thus keeping the pair on. Figure 4.13(b) shows the commutation process during normal operation, when the load current properly commutates from one device pair to the other.

Figure 4.13(c) shows commutation failure caused by changing the firing angle from α_1 in the first half-cycle to α_2 in the second. For the selected value of α_2, the load current in the SCR pair turning off at the end of the second half-cycle is not given sufficient time to reach zero before the commutating voltage changes sign and the current begins to increase. Nothing can be done to recover from this failure until the fourth half-cycle, when the control circuit returns the firing angle to α_1. We assume that the load current I_d is constant during this entire transient, as it frequently is in practice.

Commutation failure is not necessarily catastrophic and may even be used as part of the gate control system. That is, we can set α for maximum inversion by sensing the onset of commutation failure and then backing α off slightly.

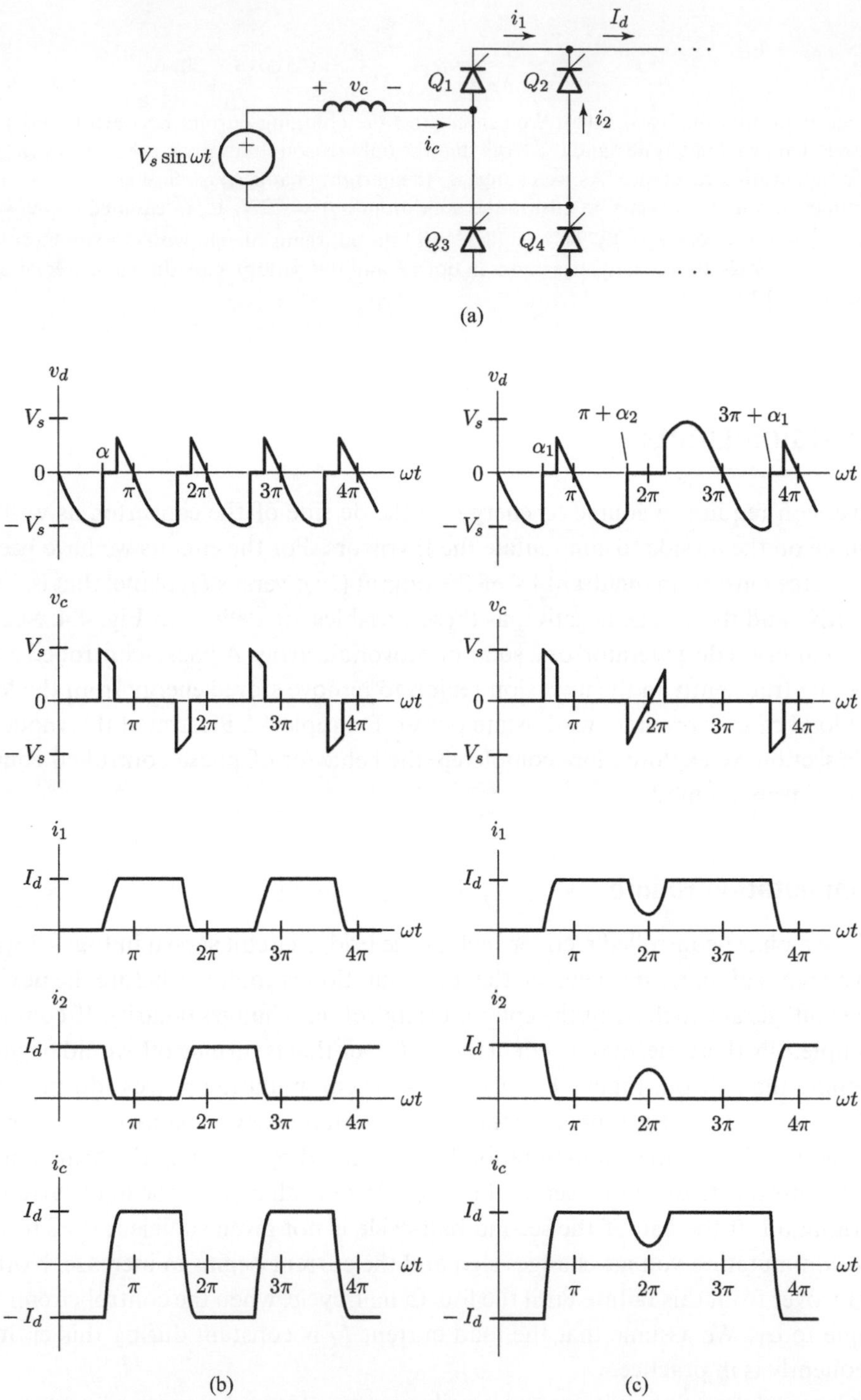

Figure 4.13 (a) A phase-controlled bridge converter with commutating inductance. (b) Voltages and currents during normal commutation for a constant firing angle α. (c) The same variables during commutation failure induced by moving the firing angle from α_1 to α_2 in the second half-cycle. Recovery is achieved by moving the firing angle back to α_1 during the fourth half-cycle.

Example 4.4 Commutation Failure

In this example we assume that the magnet power supply of Example 4.2 (shown in Fig. 4.6) experiences commutation failure while discharging the magnet. The cause might be control system noise that pushes α just beyond 180°. The current at the time of failure was 200 A. By how much does the magnet current increase if the converter recovers at the earliest possible time?

The voltage v_d is shown in Fig. 4.14(a). Commutation failure occurs at $\omega t = 0$, and recovery is not possible until $\omega t = 2\pi$. That is, the same pair of SCRs remain on for three half-cycles. The magnet current, which had been decreasing, increases during the positive half-cycle and reaches its maximum incremental excursion at $\omega t = \pi$.

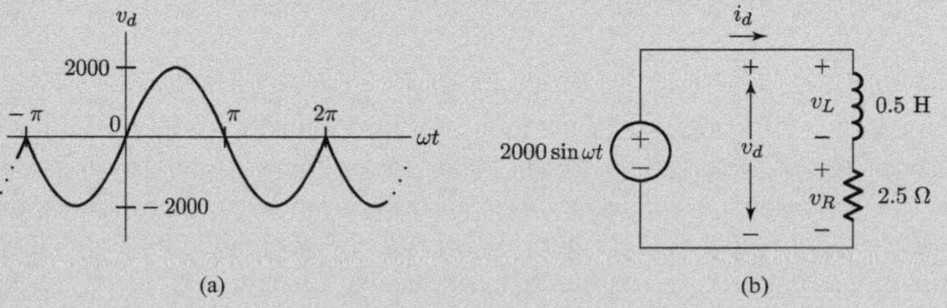

Figure 4.14 (a) The voltage v_d during commutation failure for the magnet power supply of Example 4.2. (b) Equivalent circuit of the converter for $-\pi < \omega t < 2\pi$.

The equivalent circuit for the converter during the three half-cycles in which the same pair of SCRs conduct is shown in Fig. 4.14(b). At $\omega t = 0$, $i_d = 200$ A, and we can calculate $i_d(\pi)$ by integrating the inductor voltage from 0 to π. Because the magnet time constant is quite long, we assume that the resistive drop v_R will not change much between $\omega t = 0$ and π; that is, it remains constant at close to 500 V. Based on this assumption, $i_d(\pi)$ is

$$\begin{aligned} i_d(\pi) &= 200 + \frac{1}{\omega L}\int_0^{\pi}(v_d - v_R)\,d(\omega t) \\ &\approx 200 + \frac{2000}{(377)(0.5)}\int_0^{\pi}\sin\omega t\,d(\omega t) - \frac{500\pi}{(377)(0.5)} = 212.9\text{ A}. \end{aligned} \tag{4.13}$$

Therefore, the magnet current has increased by 12.9 A before proper inverter operation is restored. The result is a maximum change in v_R of 32 V, which, compared to 500 V, is probably small enough to ignore – as we did.

4.3.2 Margin Angle

Commutation failure is a hazard when we are trying to obtain the largest possible negative voltage from a phase-controlled rectifier operating in the inversion mode. From Fig. 4.14 you can see that for a given load current and commutating voltage, a certain minimum area is necessary under the commutation voltage waveform to ensure complete commutation of the load current. Thus we cannot exceed a certain value of α, say α_{max}, without causing commutation failure.

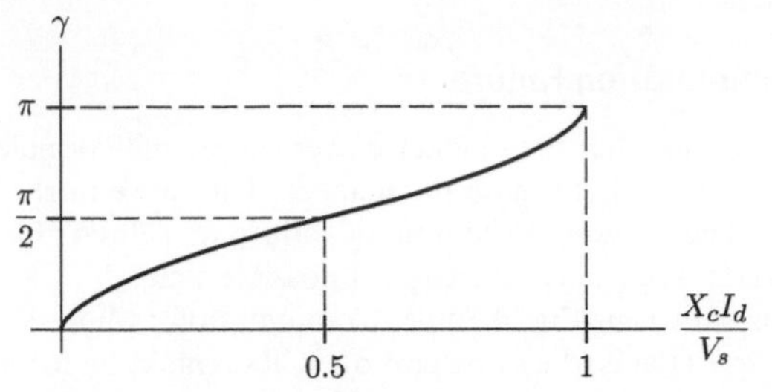

Figure 4.15 Margin angle γ as a function of the reactance factor $X_c I_d / V_s$ for the bridge converter of Fig. 4.6.

The angle $\pi - \alpha_{\max}$ is called the *margin angle*, γ.† It differs from $u_{\max}$ by the time required for the thyristors that are turning off to recover their ability to block (remain off) when a positive anode–cathode voltage is reapplied. This recovery time is given by the device manufacturer as the parameter t_q. The value of t_q can range from a few microseconds to several hundred microseconds, depending on thyristor size and the application for which it was designed. (We discuss this important parameter in more detail in Chapter 17.)

The margin angle for the single-phase bridge is equal to $u_{\max} + \omega t_q$, and $\alpha_{\max} = \pi - (u_{\max} + \omega t_q)$. The boundary $\alpha + u = \pi$ in Fig. 4.11 represents the margin angle constraint, assuming that $t_q = 0$. For the bridge converter, we can determine the margin angle as a function of the reactance factor $(X_c I_d / V_s)$ by substituting $u_{\max} = \gamma = \pi - \alpha_{\max}$ into (4.9):

$$\gamma = \cos^{-1}\left\{\cos(\pi - \gamma) - \frac{2X_c I_d}{V_s}\right\} + \gamma - \pi, \tag{4.14}$$

which implies that

$$\cos(\pi - \gamma) - \frac{2X_c I_d}{V_s} = -1$$

and

$$\gamma = \pi - \cos^{-1}\left\{\frac{2X_c I_d}{V_s} - 1\right\}. \tag{4.15}$$

We have plotted the margin angle γ as a function of the reactance factor and for $t_q = 0$ in Fig. 4.15.

Example 4.5 Inversion with Commutating Reactance

Let's reconsider the magnet discharge calculation of Example 4.2 and introduce 1 mH of commutating reactance. The result is the circuit of Fig. 4.16. What is the maximum negative magnet voltage at the onset of inversion?

† The term "margin angle" sometimes refers to the quantity $\pi - \alpha_{\text{design}}$, where α_{design} includes a safety factor to ensure against failure under any design condition, such as low line voltage or overloads. We do not use the term in this ambiguous sense.

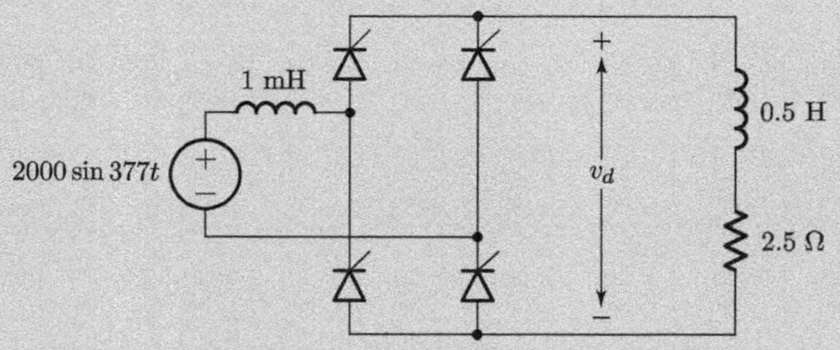

Figure 4.16 Phase-controlled magnet power supply containing commutating reactance.

We know the reactance factor of this converter at the onset of inversion, so we can obtain the margin angle from (4.15). First, we calculate the reactance factor:

$$\frac{X_c I_d}{V_s} = \frac{(377)(10^{-3})(400)}{2000} = 0.075, \tag{4.16}$$

which, from (4.15), gives $\gamma = 31.9°$. Thus $\alpha_{max} = 180° - \gamma = 148.1°$, and the inverting voltage is

$$\langle v_d \rangle = \frac{V}{\pi}[\cos\alpha_{max} + \cos(\alpha_{max} + \gamma)] = -1177\,\text{V}. \tag{4.17}$$

Note that as the magnet discharges and I_d decreases, the reactance factor also decreases and α_{max} increases, permitting the inverting voltage to increase.

Margin angle considerations are especially important in three-phase, high-power phase-controlled rectifier systems, which are discussed in Chapter 9. The reason is that the large thyristors used in these systems have large values of t_q, and the higher-order modes of three-phase rectifiers aggravate the commutation problem by extending u.

Notes and Bibliography

The GE SCR Manual [1] is the bible of the SCR. Included are principles of operation, device parameters and specifications, circuit applications, gate-drive circuits, and techniques for accommodating rate effects. Though out of print, it is a valuable reference, and still available through used-book sellers.

Written by some of the pioneers in the development of the SCR, [2] contains a wealth of information about the device and its applications. Chapter 8 of [3] discusses phase control circuits, and contains a very extensive chart summarizing possible circuits and their characteristics. The authors also present alternatives to the single-phase bridge circuit, in addition to the hybrid bridge you are asked to analyze in Problem 4.8. Various gate trigger circuits and their characteristics are also presented.

Rissik's work [4] is one of the earliest and most comprehensive on the operation of rectifiers. Although the rectifying switch is an arc tube, the analyses are still relevant to modern rectifier circuits. The mathematical developments are detailed and cover not only single- and polyphase-controlled and uncontrolled circuits, but also the cycloconverter.

1. D. R. Grafham, F. G. Golden, and A. P. Connolly, *GE SCR Manual*, 6th ed., New Jersey: Prentice-Hall, 1982.
2. F. E. Gentry, F. W. Gutzwiler, N. Holonyak, and E. E. Von Zastrow, *Semiconductor Controlled Rectifiers: Principles and Applications of p–n–p–n Devices*, New Jersey: Prentice-Hall, 1964.
3. B. D. Bedford and R. G. Hoft, *Principles of Inverter Circuits*, New York: Wiley, 1964.
4. H. Rissik, *The Fundamental Theory of Arc Converters*, London: Chapman & Hall, 1939.

PROBLEMS

4.1 Derive a phase-controlled converter circuit capable of four-quadrant operation and specify the necessary gate drives. (Such a circuit is known as a *dual converter*.)

4.2 What fraction of the energy removed from the magnet in Example 4.2 returns to the source? Does the ratio of energy recovered (returned to the source) to energy dissipated in R depend on the rate at which energy is removed from the 0.5 H inductor?

4.3 How would the circuit of Fig. 4.6 behave if i_d were not zero and the gate signals to all four SCRs were removed? Sketch $i_d(t)$ and $v_d(t)$ under this condition.

4.4 Derive and plot $\langle v_d \rangle = f(\alpha)$ for the phase-controlled rectifier with freewheeling diode shown in Fig. 4.17. Is this circuit capable of inversion?

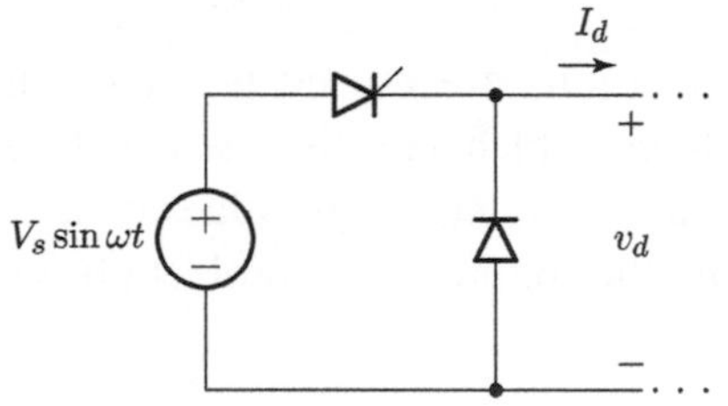

Figure 4.17 Half-wave phase-controlled rectifier with freewheeling diode analyzed in Problem 4.4.

4.5 An SCR replaces the diode in the rectifier of Fig. 4.17 as shown in Fig. 4.18. Determine and plot the control characteristic for this converter.

4.6 Repeat the power-factor calculation of Section 4.1.3 for the bridge with ac-side reactance in Fig. 4.10. Because of the sinusoidal rising and falling edges of the line current waveform, a computer will help you get an exact solution, but you can obtain a reasonably close approximation by assuming that the waveform is trapezoidal.

4.7 An SCR with a specified t_q of 150 μs is used in the 400 Hz converter of Fig. 4.19. What is the maximum magnitude of the voltage, $|\langle v_d \rangle|$, at which this converter can invert?

4.8 In certain applications, a controlled bridge rectifier consisting of two SCRs and two diodes is adequate. Such a *hybrid bridge* circuit is shown in Fig. 4.20. Determine and plot the dc voltage $\langle v_d \rangle$ as a function of α for $0 < \alpha < \pi$. Why is this circuit incapable of inversion?

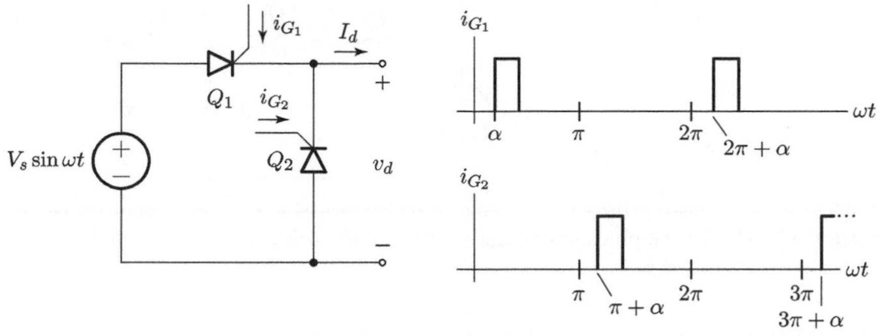

Figure 4.18 Phase-controlled converter whose control characteristic is determined in Problem 4.5.

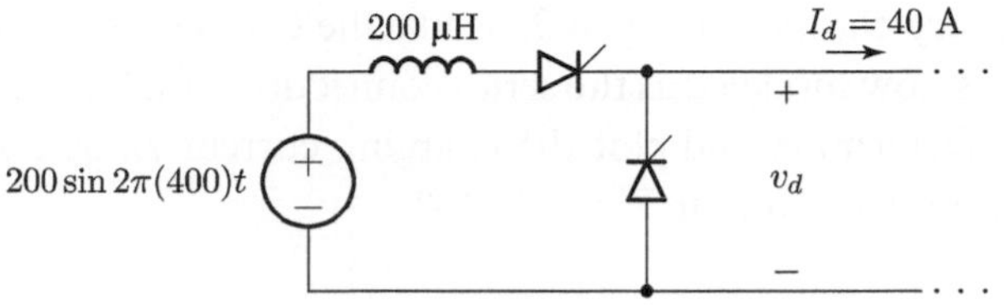

Figure 4.19 The 400 Hz converter analyzed in Problem 4.7. The SCRs have a specified t_q of 150 μs.

Compare the power factor of this hybrid bridge to that of the four-SCR bridge of Fig. 4.4.

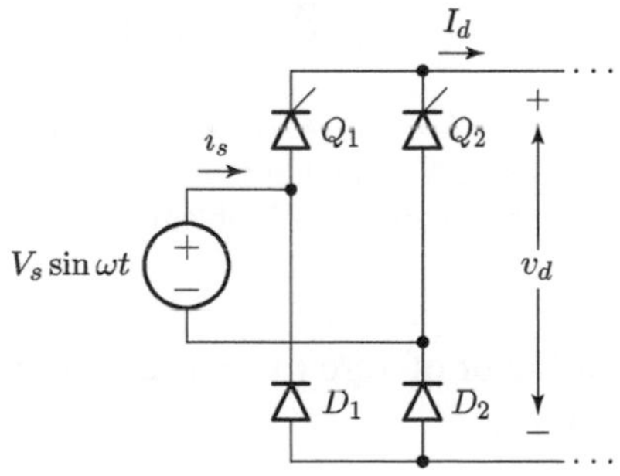

Figure 4.20 The hybrid phase-controlled bridge rectifier circuit discussed in Problem 4.8.

4.9 Determine (approximately) the minimum time required to discharge the magnet of Example 4.2.

4.10 The circuit of Fig. 4.21 is a half-wave phase-controlled converter. Determine and plot a family of regulation curves for this circuit. How do the regulation characteristics of this circuit compare to those of Fig. 4.11 for the bridge converter?

4.11 If the diodes in the half-bridge rectifier of Fig. 3.17 were instead SCRs, and their firing angle was 30°, draw the resulting waveforms for v_d, i_{S_1}, and i_{S_2}.

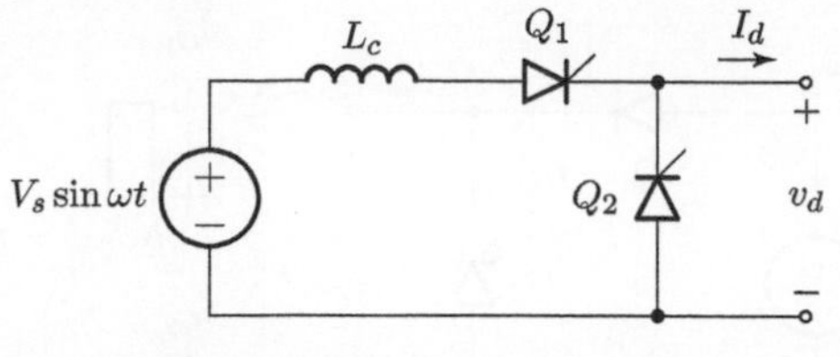

Figure 4.21 Half-wave phase-controlled converter, the subject of Problem 4.10.

4.12 Batteries, such as the lead–acid storage battery used in automobiles, always contain some resistance. For example, a 12 V battery with a short circuit (approximately the "cold-cranking") capacity of 240 A can be modeled as having an internal resistance of 50 mΩ. The battery charger of Fig. 4.22 is like the one discussed in Example 4.3, except that the battery is now modeled as the series connection of a 0.240 Ω resistance and a 72 V voltage source. Determine and plot the charging current I_B as a function of α. How does this curve differ from that in Fig. 4.12(b)?

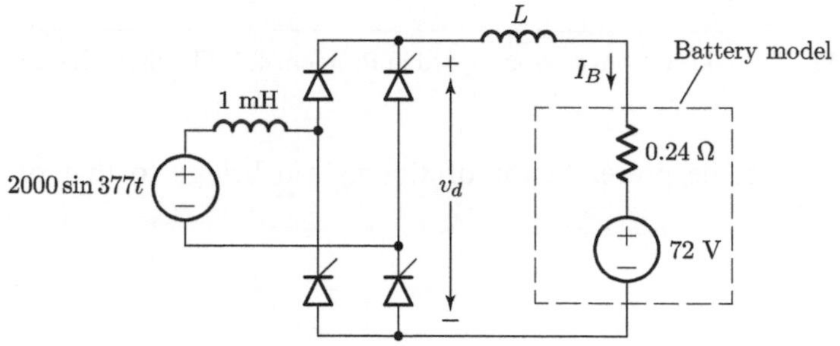

Figure 4.22 A battery charger with commutating reactance charging a battery containing internal resistance. This circuit is analyzed in Problem 4.12.

4.13 Reconcile a power factor of zero ($\alpha = \pi/2$) with what is happening on the dc side of the converter of Fig. 4.8.

4.14 Figure 4.23 shows a phase-controlled bridge converter with commutating resistance R_c. Determine the regulation curves for this converter and compare them with the curves in Fig. 4.11 for the converter with commutating inductance.

4.15 The ac controller of Fig. 4.24(a) is supplying a resistive load, perhaps a heater.

(a) Sketch $i_o(t)$ for this circuit, assuming that $\alpha \neq 0$.

(b) Determine and plot $v_{o\,\mathrm{rms}}$ as a function of α for this circuit.

In many applications (incandescent light dimming, for example), the large values of di_o/dt present in the circuit of Fig. 4.24(a) cause both radio frequency interference (RFI) and acoustic noise (lamp "buzz"). To reduce these effects, the manufacturer often places an inductor in series with the resistive load, as shown in Fig. 4.24(b).

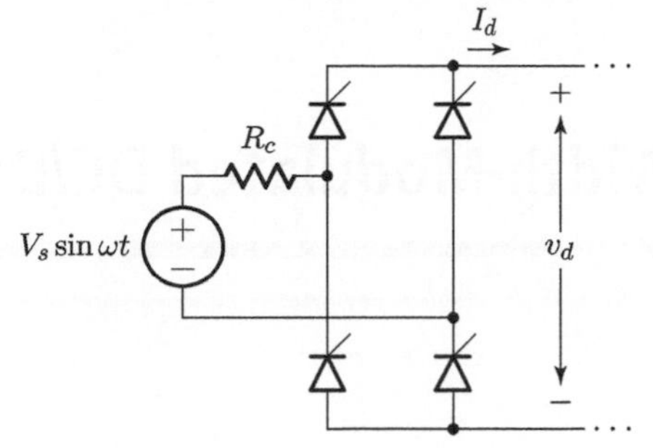

Figure 4.23 Phase-controlled bridge converter with commutating resistance R_c for Problem 4.14.

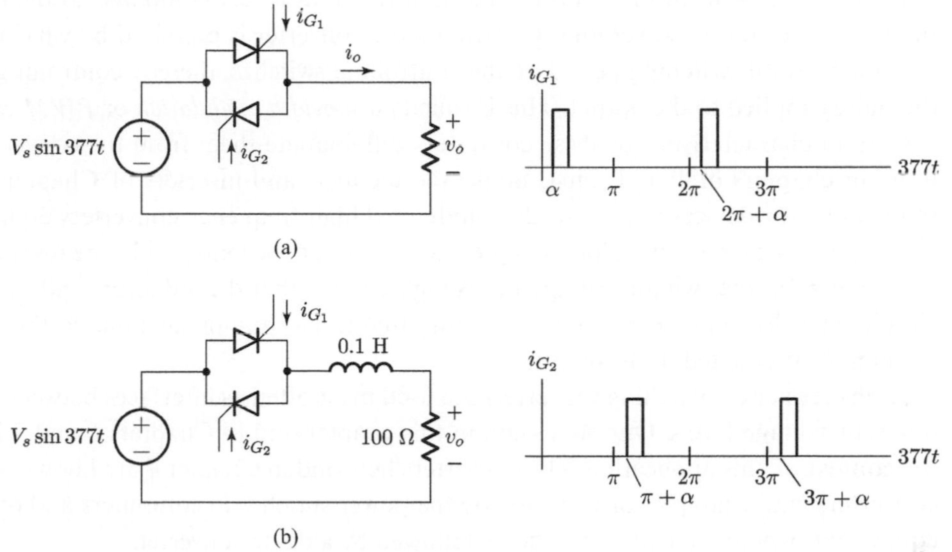

Figure 4.24 The ac controllers analyzed in Problem 4.15.

(c) For what range of α is the conduction of the controller discontinuous? Determine and carefully sketch i_o for some value of α within this range. Why are the RFI and acoustic noise both reduced?

(d) Determine and sketch i_o for α outside the range determined in (c).

4.16 Derive the relationship $\langle v_d \rangle = f(X_c I_d / V_s)$, which forms the inversion limit $\alpha + u = \pi$ in Fig. 4.11.

5 Pulse-Width-Modulated DC/DC Converters

Pulse-width-modulated switching converters are power circuits in which the semiconductor devices switch between on and off states at a rate that is fast compared to the frequencies of the input and output waveforms. Control of the converter is exercised by varying the ratio of on-time to total switching period of the controlling switches, thereby controlling the width of the pulses applied to the output. This is called *pulse-width modulation* or *PWM* control.

Certain characteristics of these converters differentiate them from the converters described in other chapters of Part I. First, unlike the rectifiers and inverters of Chapters 3 and 4, the semiconductor devices in pulse-width-modulated high-frequency converters do not usually use the reversal of external waveforms to turn themselves off. Second, unlike the resonant converters of Chapter 10, the switching frequency is high enough that the inductors and capacitors in the circuit act as low-pass filters to the switching-frequency components; that is, these components are greatly attenuated in the output.

High-frequency switching converters are used most often as interfaces between dc systems of different voltage levels. Our discussion in this chapter and in Chapters 6 and 7 is therefore in the context of this application. The converters here and in Chapter 7 are known as PWM dc/dc converters, and examples of their use are the power supplies in computers and other electronic equipment, where typically a rectifier is followed by a dc/dc converter.

High-frequency switching converters can also be used to interface between dc and ac systems. Although we look at the unique aspects of this application in Chapter 8, much of what we say about dc/dc converters in this chapter applies to dc/ac converters as well. The reason is that in the dc/ac application, the switching frequency is usually so much higher than the ac-output (or -input) frequency that we can consider the variables at the ac port to be constant over many switching periods. In other words, for times on the order of the switching period, we can view the dc/ac converter as a dc/dc converter.

In this chapter we present the structure and operation of high-frequency switching converters that use both inductors and capacitors to store energy. Chapter 6 discusses another class of dc/dc converters that use only capacitors. Rather than just categorizing a large number of circuits, we develop the fundamental concepts and topological relationships on which all such circuits are based. Again, you will see that the specification of a topology does not simultaneously specify which port is the input and which is the output. Only when we specify the implementation and control of the switches do the ports obtain identities as input or output.

5.1 The DC/DC Converter Topology

The simplest form of a switching dc/dc converter is that of Fig. 5.1. The switch opens and closes at a frequency $1/T$, with the ratio of the on-time to the period defined as D, the *duty ratio*. The resulting load voltage v_2 is a *chopped* version of the input – a series of pulses having an amplitude of V_1 and an average, or dc, value of DV_1. But this dc value comes with a substantial amount of ripple, which is present not only in the load voltage v_2 but also in the source current i_1. Few applications can make use of this dc component in the presence of so much distortion. The high frequencies contained in the ripple can cause both conducted and radiated interference with other apparatus, such as computers or communications equipment. Moreover, some loads, such as many electronic circuits, function properly only if operated from a dc power supply with little ripple. Therefore, a dc/dc converter generally has terminal voltages and currents that deviate only minutely and instantaneously from their dc values. This restriction means that we have to modify the elementary topology of Fig. 5.1. We develop these modifications by first exploring the topological constraint imposed by the requirement of nearly constant terminal variables.

The circuit in Fig. 5.2(a) shows a converter connecting two systems whose terminal voltages and currents are dc with the values shown. Whatever is in the box has to produce terminal variables that are free from pulsations or ripple. The difference between input and output voltages, $V_i - V_o = 50$ V, must drop across an element connected between the ports, and the box must contain a shunt element to provide a path for the difference between input and output currents, $I_o - I_i = 5$ A. This minimal connection of elements is shown in Fig. 5.2(b). Note that the power absorbed by the series element is equal to the power supplied by the shunt element, so the box is lossless – but the energy must be transferred from the shunt to the series element. It is difficult to imagine a way to implement this simple two-element topology.

A circuit element that can support both a nonzero average voltage and a nonzero average current without dissipating energy is a switch. Furthermore, we can control the average value of its variables by varying the ratio of its on to off times. The problem with using switches as the shunt and series elements, however, is that instantaneous values of the input current and output voltage will differ from their average values. We can, however, extract their average values by using low-pass filters. By defining the converter to include these filters, we create an interface that produces the desired terminal variables.

Figure 5.3 shows switches as the shunt and series elements in our example converter, connecting the dc input voltage, $V_i = 200$ V, to the dc output current, $I_o = 20$ A. The switches

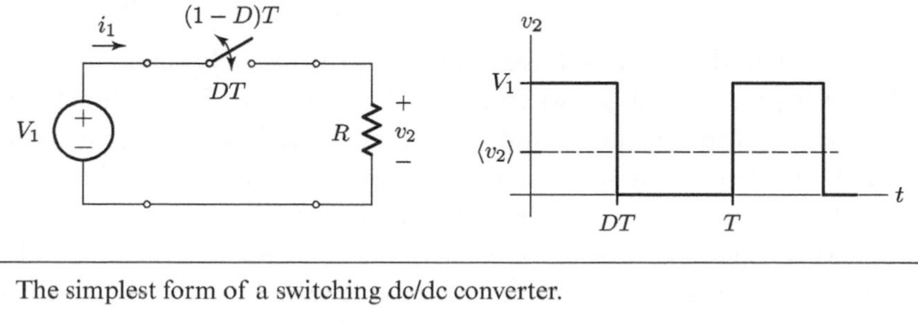

Figure 5.1 The simplest form of a switching dc/dc converter.

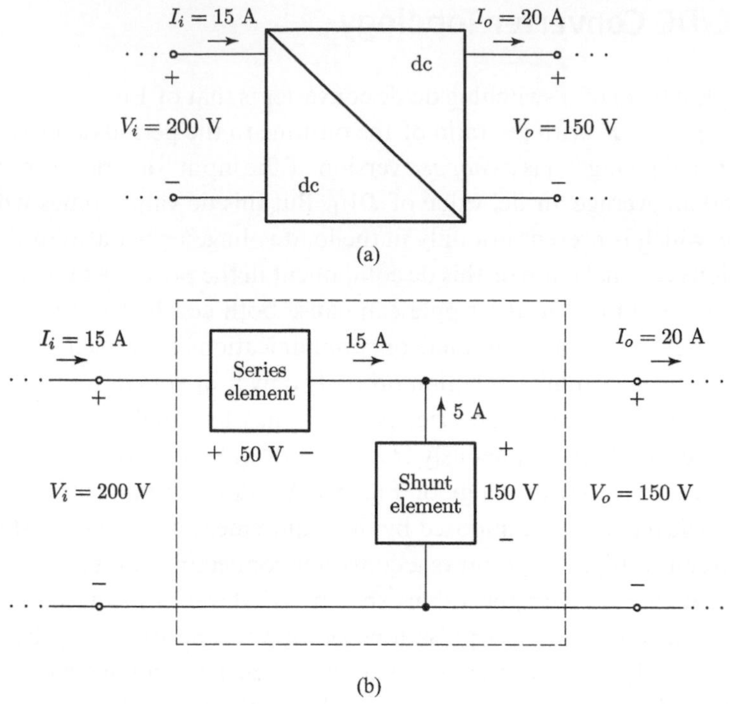

Figure 5.2 (a) A converter connecting two dc systems. (b) The minimal necessary topology for performing the conversion function of (a).

operate at a constant switching frequency $1/T$, and are controlled to be complementary. That is, when one is closed, the other is open (so that a path is always provided for the load current and the input voltage is never shorted). The waveforms of v_{S_1} and v_{S_2} show that switches S_1 and S_2 are on for times DT and $(1-D)T$, respectively, giving average values of

$$\langle v_{S_1} \rangle = (1-D)V_i \quad \text{and} \quad \langle v_{S_2} \rangle = DV_i.$$

Because $\langle v_o \rangle = \langle v_{S_2} \rangle = 150$ V, we must control the switches so that $D = 0.75$. The average voltage across the series switch is 50 V, or the difference between the average input and output voltages that we said was necessary when discussing Fig. 5.2. Similarly, we can show that the average current in the shunt switch is the difference between the average input and output currents, as required, or

$$\langle i_{S_2} \rangle = \langle i_i \rangle - I_o = -(1-D)I_o = -5\,\text{A}. \tag{5.1}$$

We assume that the input voltage contains no ripple. However, the input current contains substantial ripple caused by switching, as shown in Fig. 5.3. Although the output current is ripple-free, the output voltage is not. To obtain the desired ripple-free input current and output voltage, we must insert low-pass filters at the input and output. In their simplest form, these filters consist of a shunt capacitor at the input and a series inductor at the output. By making these elements very large, small values of external network input impedance and output conductance will result in acceptably small ripples in all the terminal variables. The resulting high-frequency

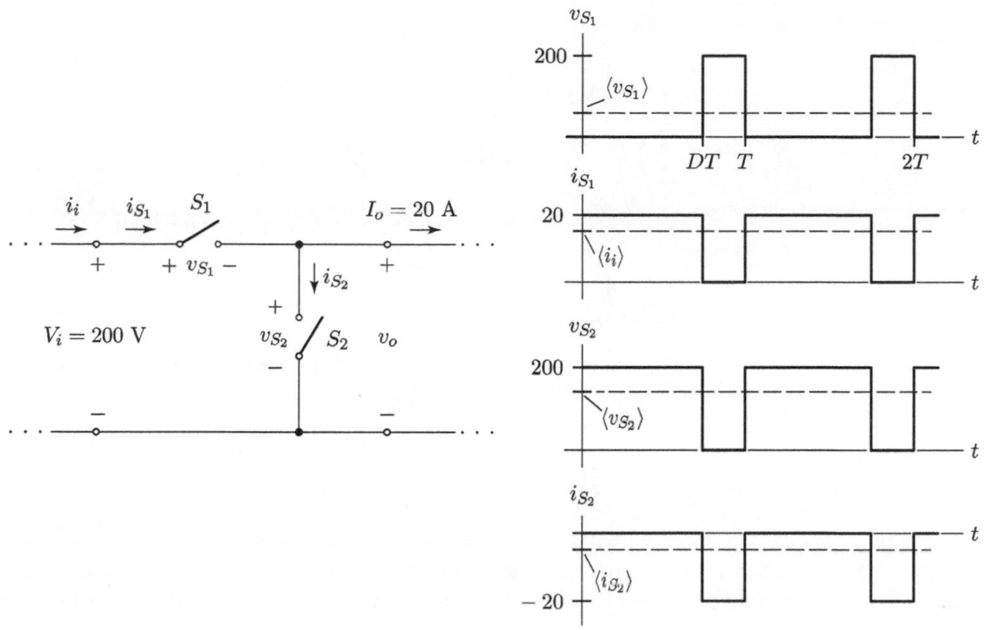

Figure 5.3 Two dc systems connected by a lossless interface, using switches as the shunt and series elements.

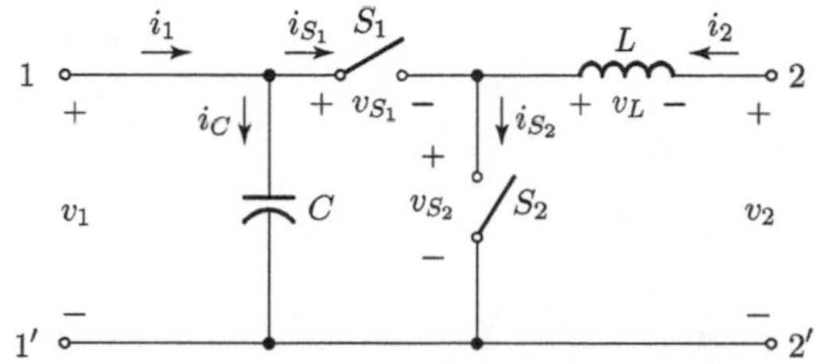

Figure 5.4 Simplest topology for a high-frequency dc/dc converter with low-pass filtering at its ports.

dc/dc converter topology is shown in Fig. 5.4. We have not identified input and output ports, because energy can flow in either direction, depending on how we control the switches.

Example 5.1 Input Filter Effectiveness

Figure 5.5 shows a converter whose external input circuit is modeled by a voltage source V_i behind a resistor R, which might represent wiring resistance or the internal resistance of a battery. We assume that the converter inductor L is large enough to make the output current constant at I_o. What is the peak–peak amplitude of the ripple on the input current i_i?

If the switches are operating at a fixed duty ratio, D, we can replace the circuit to the right of the capacitor by an equivalent source i_e, as shown in Fig. 5.5(b). Furthermore, being concerned only with the ripple component of i_i, we can replace i_e with its ac component i_e'. The circuit that we analyze is shown in Fig. 5.6(a), where i_i' is the ac component of the input current, and the voltage source has been modeled as an incremental short.

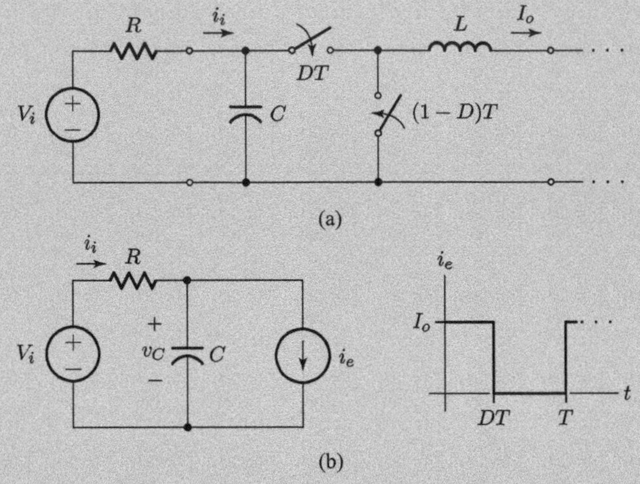

Figure 5.5 (a) Switching converter with a low-pass filter to eliminate ripple from the current flowing through the input circuit. (b) The switches, inductor, and external output network replaced by an equivalent source, i_e.

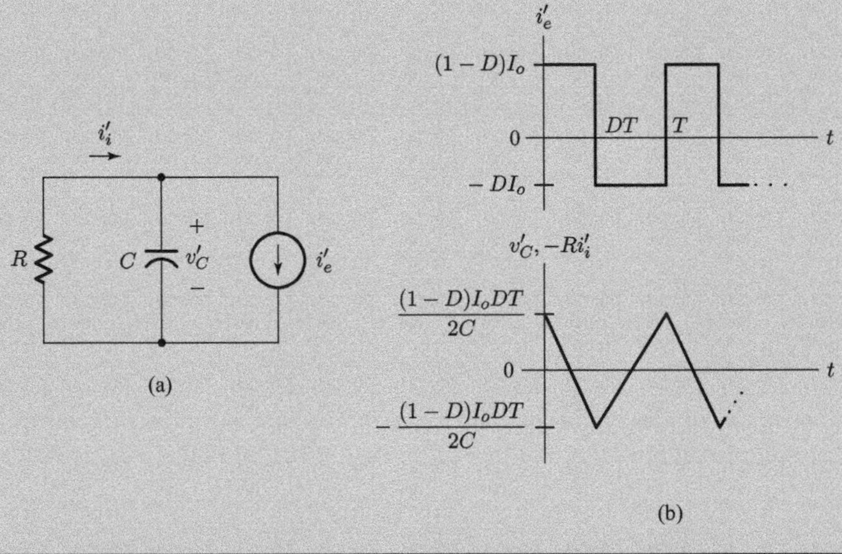

Figure 5.6 (a) Equivalent circuit for the calculation of the ripple component of the input current in the circuit of Fig. 5.4. (b) Branch variables for the circuit of (a), assuming that $RC \gg T$.

If the low-pass filter is effective, that is, $\tau = RC \gg T$, or $R \gg 1/\omega C$, then almost all the ripple current will pass through C. Therefore the capacitor voltage ripple is triangular with a peak–peak amplitude of

$$\Delta v_C = \frac{(1 - D)I_o}{C} DT, \tag{5.2}$$

as shown in Fig. 5.6(b). The peak–peak amplitude of the input ripple current will thus be

$$\Delta i_i = \frac{\Delta v_C}{R} = \frac{(1-D)I_o DT}{RC}. \tag{5.3}$$

From (5.3) we see that if $C \to \infty$ and R is nonzero, then $\Delta i_i \to 0$.

In many cases, we add additional energy storage elements to the external systems to enhance filtering. For instance, if the external network connected to port 1–1′ in Fig. 5.4 has a very low ac impedance, C will have to be very large. However, if we increase the impedance of the external network by placing an inductor in series with it, C can be much smaller. Similarly, if the external impedance at port 2–2′ is very high, L will have to be large unless we place a capacitor in parallel with the port to reduce its impedance. These additional elements on each side of the converter make the filters second-order. Third-, fourth-, and even-higher-order filters are used if the need to attenuate the ac components is great enough. These elements do not change the basic function of the original capacitor and inductor, so we do not include them in the topologies of this chapter.

5.1.1 Average Value Analysis

The average values of currents and voltages in a circuit obey Kirchhoff's current and voltage laws (KVL and KCL), providing a straightforward analysis technique for determining the relationship among average variables in converters. To illustrate, consider Fig. 5.4. Writing KVL for the average voltages around the loop formed by v_1, S_1, v_L, and v_2 gives

$$-\langle v_1\rangle + \langle v_{S_1}\rangle + \langle v_L\rangle + \langle v_2\rangle = 0.$$

Recognizing that in a periodic steady state the average value of the voltage across an inductor is zero,

$$-\langle v_1\rangle + \langle v_{S_1}\rangle + \langle v_2\rangle = 0.$$

Referring to Fig. 5.3, with $\langle v_1\rangle = V_1$ and $\langle v_2\rangle = V_2$,

$$\langle v_{S_1}\rangle = (1-D)V_1$$

so

$$\begin{aligned} -V_1 + (1-D)V_1 + V_2 &= 0 \\ V_2 &= DV_1. \end{aligned}$$

A similar analysis can be done for the currents, recognizing that the average value of the current through a capacitor in a periodic steady state is zero. Since the basic circuit has three nodes, two KCL equations are required. We choose the nodes between S_1 and C, and S_1 and S_2:

$$\begin{aligned} \langle i_1\rangle - \langle i_{S_1}\rangle - \langle i_C\rangle &= 0, \\ \langle i_{S_1}\rangle - \langle i_{S_2}\rangle + \langle i_2\rangle &= 0, \\ \langle i_1\rangle = I_1 &= \langle i_{S_1}\rangle, \\ \langle i_2\rangle = I_2 &= \langle i_{S_2}\rangle - I_1, \\ \langle i_{S_2}\rangle &= I_2(1-D), \\ I_2 &= -\frac{1}{D}I_1. \end{aligned}$$

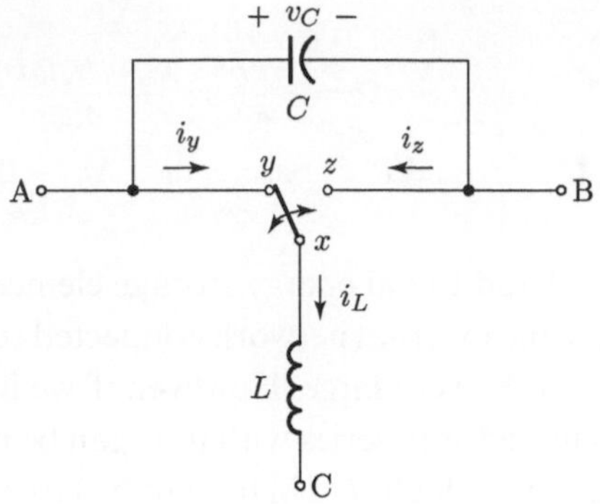

Figure 5.7 The canonical switching cell, the basic building block of many high-frequency dc/dc converters.

5.2 The Canonical Switching Cell

We refer to the minimal practical dc/dc converter topology developed in Section 5.1 and shown in Fig. 5.4 as the *canonical switching cell.*† We redraw it symmetrically in Fig. 5.7, using a single-pole double-throw switch, which satisfies the condition that the switches be neither on nor off simultaneously. The canonical cell is the basic building block for many high-frequency switching converters. The distinctions between these converters arise mainly from the way in which the external systems are connected to the cell. These connections determine both the input/output conversion ratios and the levels of current and voltage stress imposed on the cell's components.

We may connect the three-terminal canonical cell of Fig. 5.7 between two dc systems in one of three ways. The connections differ with respect to the cell node that we make common to both ports. Because of the symmetry of the cell, however, the use of node A or B as a common node results in indistinguishable topologies. Thus there are only two unique connections to consider. If node A or B is common, the resulting topology is that of Fig. 5.8(a), which is identical to Fig. 5.4. Figure 5.8(b) shows the topology that results from the use of node C as the common terminal.

In the connection of Fig. 5.8(a), there is a direct dc path between the input and output ports when switch S_{xy} is conducting. For this reason we call this a *direct converter* connection. In the connection of Fig. 5.8(b), there is no dc path between the ports in any switch state. Therefore we call this an *indirect converter* connection. The use of the names *direct* and *indirect* highlights the essential difference among various specific high-frequency dc/dc converter circuits. The presence of transformers or additional filter elements does not negate this distinction.

5.3 Direct Converter

We now consider the direct converter of Fig. 5.8(a) in detail. We are particularly interested in the dc conversion ratio and the implementation of the switches. Note that we cannot define the

† E. Landsman, "A Unifying Derivation of Switching DC–DC Converter Topologies," *Power Electronics Specialists Conference*, IEEE, 1979, pp. 239–243.

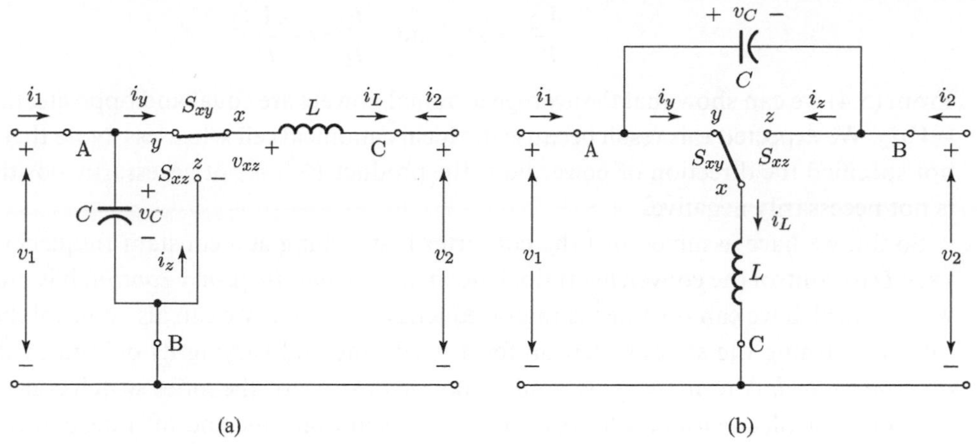

Figure 5.8 The two unique connections of the canonical cell between two dc systems: (a) a direct converter; (b) an indirect converter.

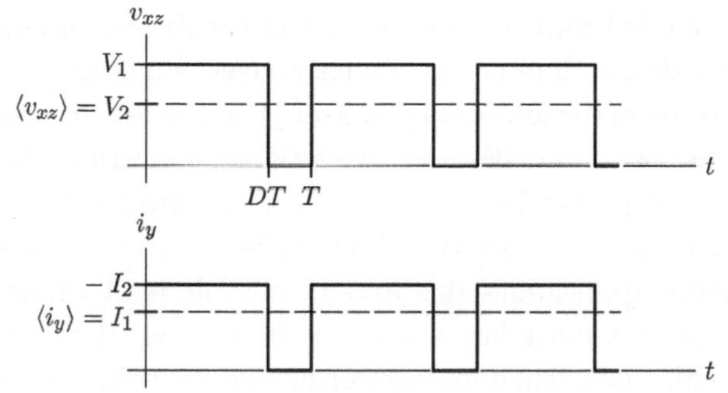

Figure 5.9 Waveforms of the switch variables v_{xz} and i_y for the direct converter of Fig. 5.8(a).

ports as input or output until we specify the semiconductor devices to be used as the switches, and possibly the external networks and controls as well.

5.3.1 DC Conversion Ratio of the Direct Converter

In the discussion that follows we assume that L and C are large enough to eliminate switching-frequency components from the terminal variables v_1, i_1 and v_2, i_2. Therefore we specify the terminal variables using uppercase letters to indicate dc quantities. For example, $i_1(t) = \langle i_1(t) \rangle = I_1$.

The values of the conversion ratios, V_2/V_1 and $-I_1/I_2$, depend on the duty ratio of the switches, and we can determine them as above using average KVL and KCL relations, or from the waveforms of v_{xz} and i_y shown in Fig. 5.9. Here we express the conversion ratios in terms of the duty ratio D of the series switch. As the inductor cannot support an average voltage, $V_2 = \langle v_{xz} \rangle = DV_1$. And as the capacitor cannot carry an average current, $I_1 = \langle i_y \rangle = D(-I_2)$. Solving these constraints for the conversion ratios gives

$$\frac{V_2}{V_1} = D \quad \text{and} \quad \frac{I_2}{I_1} = -\frac{1}{D}. \tag{5.4}$$

From (5.4) we can show that the average terminal powers are equal and opposite, that is, $V_1 I_1 = -V_2 I_2$. We expected this result because the ideal canonical cell is lossless. Note that we still have not specified the direction of power flow; the product $V_1 I_1$ is not necessarily positive, and $V_2 I_2$ is not necessarily negative.

So far, we have assumed that the converter is switching at a constant frequency, $1/T$, as we vary D to control the conversion ratio. Known as *constant frequency control*, it is only one of the ways by which we can operate the high-frequency converter. We can also control the conversion ratio by turning the series switch on for a fixed time and varying its off time, called *constant on-time control*. Alternatively, we could hold the off-time of the series switch constant and vary its on time, called *constant off-time control*. Constant on-time and off-time control both result in changes in the switching frequency as we vary the conversion ratio. In all three cases, the conversion ratio depends only on the percentage of the period the switches are in one state or the other, which represents a form of pulse-width modulation.

Because of the relationship between the two external voltages given by (5.4), the direct converter has come to be known as either a *buck* (or *down*) *converter* or a *boost* (or *up*) *converter*, depending on the direction of power; a buck converter has power sourced from the high-voltage side (V_1), a boost from the low-voltage side (V_2). These are the common names when we discuss these circuits, but they do not describe two separate converters. As we show in Section 5.3.2, the desired directions of power flow only affect how we implement the switches.

Note that these names are based on how the *voltages* of the two external systems are related. Although in many applications the voltage variable is of prime concern, in some situations the current variable is more important. In these cases, the circuit we called a voltage buck converter is actually a current boost converter, and the voltage boost converter is a current buck converter. Thus the popular names of these circuits can be misleading unless you understand the commonality of topology and function implied by the single (direct converter) connection of the canonical cell.

5.3.2 Implementation of Switches and Power Flow

To determine the direction of power in a dc/dc converter, we generally need to specify switch implementations, external networks, and the controls.

When we replace the generalized switches we have used so far with semiconductor devices, we also constrain the directions of their current and voltage, depending on the device. The networks external to the converter terminals may or may not be able to source or sink energy. And in a situation where both the switches and networks allow bidirectional power flow, how the switches are controlled determines in which direction it flows. Here we treat the simplest case in which the switch implementation is sufficient to determine the direction of power flow.

The buck converter To illustrate how switch type can determine the direction of power flow, let's consider the direct converter of Fig. 5.8(a) operating in the buck mode. That is, power is flowing from the higher-voltage side (V_1) to the lower-voltage side (V_2). In this case, $V_1 > 0$,

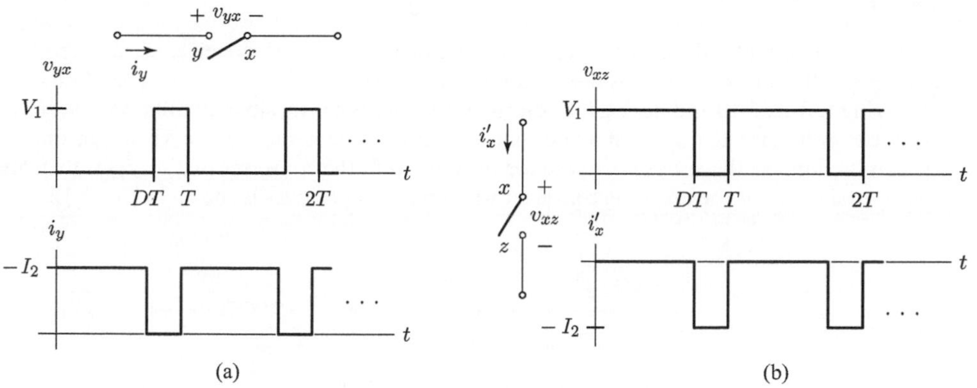

Figure 5.10 Voltage and current waveforms for the series and shunt switches in the direct converter of Fig. 5.8(a) when it is operating in the buck mode: (a) series switch; (b) shunt switch.

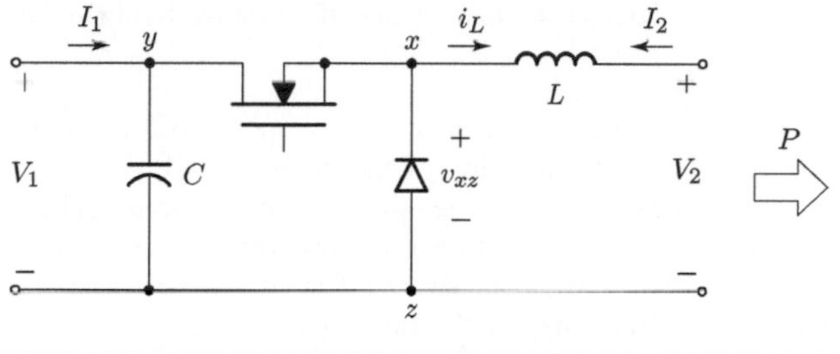

Figure 5.11 Switch implementation for the direct buck converter.

$I_1 > 0$, $V_2 > 0$, $I_2 < 0$; the switch voltages and currents are shown in Fig. 5.10, where the shunt switch component of i_x is designated as i'_x ($i_x = -i_z$ in Fig. 5.8(a)).

The current and voltage of the series switch, i_y and v_{yx}, are both positive, conditions satisfied by a transistor. The current and voltage of the shunt switch, i_x and v_{xz}, are of opposite polarities, as are those of a diode. It may not be immediately apparent that an uncontrollable device is suitable for the shunt switch. However, the diode is forced on and off by the series switch. When the switch closes, V_1 reverse biases the diode, and when it opens, the continuity of current in L forces the diode to conduct. The function and operation of the shunt switch is identical to that of the freewheeling diode in the half-wave rectifier of Fig. 3.5.

The buck converter – with the series and shunt switches replaced by a transistor and diode, respectively – is shown in Fig. 5.11. Although the integral diode in the MOSFET (known as the *body diode*) would permit reverse current in the device, the shunt diode, being uncontrollable, provides no means for reversing this current.

Example 5.2 A Buck Converter with Common Positives

At times we need to connect two dc systems so that their positive terminals are common. In this example, we design a buck converter for such an application.

Again, we consider the general direct converter of Fig. 5.8(a). The buck constraint requires that $v_1 i_1 > 0$ and $v_2 i_2 < 0$. The common positive requirement implies $v_1 < 0$ and $v_2 < 0$. Therefore $i_1 < 0$ and $i_2 > 0$. Referring to Fig. 5.10 and noting the polarity for the terminal variables in this application, you can see that the series switch is still a transistor and that the shunt switch is still a diode. However, since their directions of conduction are the reverse of those for the devices in the converter of Fig. 5.11, they are connected in the opposite sense. The resulting converter with common positives is shown in Fig. 5.12.

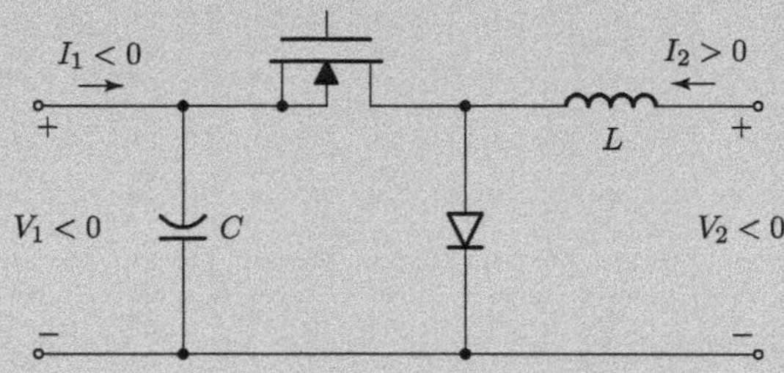

Figure 5.12 A buck converter configured so that the positive terminals of the input and output ports are common.

We can also derive the converter of Fig. 5.12 by simply transforming the circuit of Fig. 5.11. We do so by noting that we can change the position of the inductor and the series switch to the lower branches in their respective loops without changing the operation of the circuit. The resulting circuit is shown in Fig. 5.13. Now, by simply flipping the circuit over, we have the result shown in Fig. 5.12. An advantage of the circuit of Fig. 5.13 is that the gate of the MOSFET is referenced to the negative terminal of the source V_1, eliminating the need for level shifting of the gate drive.

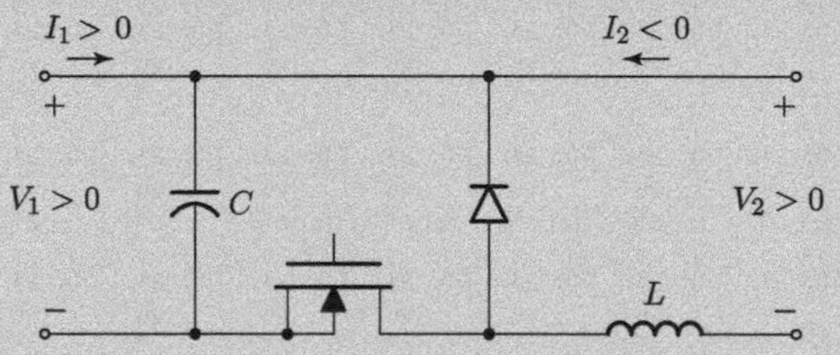

Figure 5.13 A buck converter with common positives derived by changing the positions of the inductor and series switch in the circuit of Fig. 5.12.

The boost converter We now consider the switch implementation necessary to create a boost converter. We choose the switches so that the power in the direct converter of Fig. 5.8(a) flows from the lower-voltage side (V_2) to the higher-voltage side (V_1). We do so by simply reversing the polarities of either the terminal currents or the terminal voltages, relative to the buck converter we just discussed.

If we choose to reverse the polarity of the terminal currents, leaving positive voltages V_1 and V_2, the switch current polarities of Fig. 5.10 are reversed. In this case, the shunt switch becomes a transistor and the series switch can be a diode. The result is the direct boost converter shown in Fig. 5.14. Note that the controllable switch is now the shunt switch instead of the series switch.

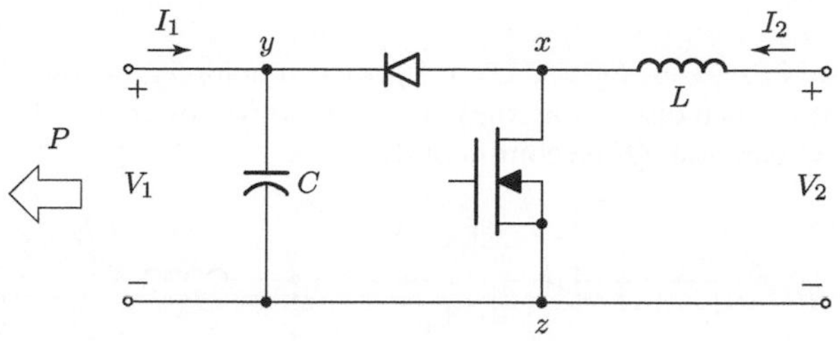

Figure 5.14 Switch implementation for the direct boost converter.

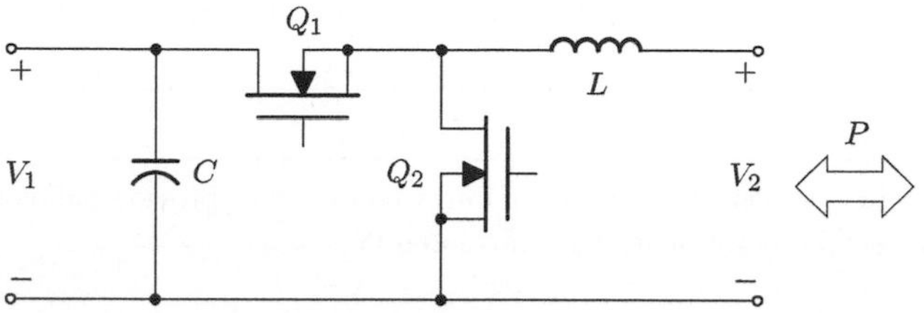

Figure 5.15 A direct converter configured with MOSFETs to provide bilateral terminal currents.

Direct converter with bilateral energy flow Bilateral power can be implemented in the direct converter by replacing the diode with another MOSFET, as shown in Fig. 5.15. For operation in the buck mode, that is, power flowing from V_1 to V_2, Q_1 controls D, and and the body diode of Q_2 can function as the shunt element. However, when on, a power MOSFET can conduct negative current, which makes it possible to synchronously control Q_2 so its current is conducted through the channel, bypassing the body diode. The benefit of this is that the source-drain drop is lower than it would be if the body diode were conducting, providing improved efficiency. For power flowing from V_2 to V_1, that is, operation in the boost mode, Q_2 controls D and Q_1 is controlled synchronously. Using power MOSFETs in this manner is the basis of *synchronous rectifiers*.

Example 5.3 Switch Implementation for Bilateral Terminal Voltages

Some applications of dc/dc converters require power to flow in either direction. Two examples are a dc motor that is actively braked, and a magnetic resonance imaging (MRI) machine magnet, which must create a precisely controlled time-varying magnetic field. In both cases the terminal voltages change sign but the current polarities remain unchanged (due to the inability to create a step change in a magnetic field absent a voltage impulse) in order to remove or introduce energy into the magnetic field system. That is, with reference to Fig. 4.5(b), the converters operate in quadrants I and IV as seen from their source terminals (V_1, I_1). How do we implement the switches in a direct converter designed for such applications?

We again consider the direct converter switch variables of Fig. 5.10, noting that if the terminal currents do not change sign, the switches will still only be required to conduct in one direction. However, because the direction of power is reversed by inverting the terminal voltages, the switches must now be capable of blocking both voltage polarities. A switch made up of a diode in series with a transistor has the required characteristic, and the resulting bilateral direct converter is shown in Fig. 5.16. Note that we must now control both switches explicitly and make sure that they are not off simultaneously. However, when energy flows from left to right, that is, when we are supplying energy to the magnetic system and operating in quadrant I ($V_1 > 0, I_1 > 0, V_2 > 0, I_2 < 0$), we can hold Q_2 on continuously, because D_2 will do the switching. Similarly, when discharging the magnet or motor, power flows from right to left ($V_1 < 0, I_1 > 0, V_2 < 0, I_2 < 0$) we can hold Q_1 on continuously.

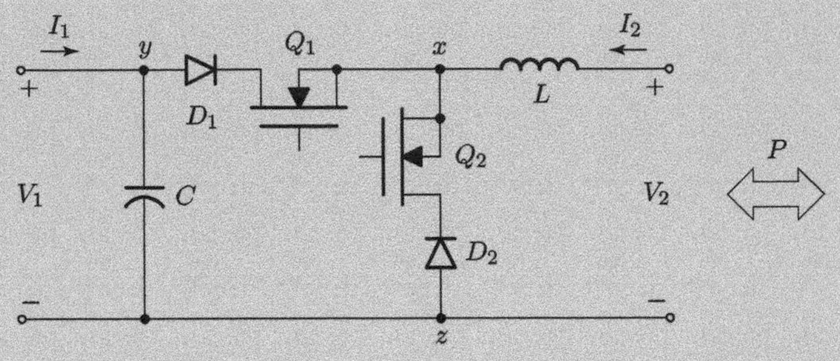

Figure 5.16 Switch implementation in the direct converter to provide bidirectional power flow with voltage reversal (operation in quadrants I and IV).

Note that the diodes in the circuit of Fig. 5.16 are on the drain sides of Q_1 and Q_2. Therefore we can connect the sources of Q_1 and Q_2. This approach simplifies the level-shifting and gate drive circuit by referencing the drives for both transistors to the same point.

5.3.3 About the Duty Ratio

In Section 5.1 we defined the duty ratio for the direct converter as the fraction of the switching period during which the series switch is on. Actually, common practice is to define the duty ratio as the fraction of the period during which the *controllable* switch (the transistor) conducts, regardless of whether it is in the series or shunt position. When we use this definition for D, (5.4), written for the boost converter of Fig. 5.14, becomes

$$\frac{V_2}{V_1} = 1 - D \quad \text{and} \quad \frac{I_2}{I_1} = -\frac{1}{1-D}. \tag{5.5}$$

We sometimes refer to the quantity $1 - D$ as D'. We also commonly express the conversion ratio for a dc/dc converter in the form of the output over the input. Therefore we can write the voltage conversion ratio for the boost converter as

$$\frac{V_{\text{out}}}{V_{\text{in}}} = \frac{1}{1-D} = \frac{1}{D'}. \tag{5.6}$$

Finally, the conventional conversion ratio expressions are ambiguous when power can flow in either direction, because we cannot label either of the external systems as the input or the output, and both switches are controllable. In this case we normally use the terms and expressions that are appropriate for the direction power is flowing at the time.

5.4 Indirect Converter

The indirect converter connection of the canonical cell is shown in Fig. 5.8(b). As you will see in this section, the indirect converter exhibits operating characteristics that are very different from those of the direct converter just discussed.

5.4.1 The DC Conversion Ratio of the Indirect Converter

An important distinction between direct and indirect connections of the canonical cell is the relative polarity of their actual (measured) terminal voltages and currents, as defined in Fig. 5.8. The voltages are of the same polarity in the direct connection, but they are of opposite polarities in the indirect connection. In contrast, the terminal currents are of opposite signs in the direct converter, but are of the same sign in the indirect converter.

We can derive the polarity relationships between the terminal variables of the indirect converter of Fig. 5.8(b) by considering the inductor voltage and the capacitor current. The voltage across the inductor is V_1 when switch S_{xy} is closed and V_2 when S_{xz} is closed, so one of these two voltages must be positive and the other negative to satisfy the condition that the average inductor voltage be zero. Similarly, the capacitor current is $-I_2$ when S_{xy} is closed and I_1 when S_{xz} is closed, so both I_1 and I_2 must be of the same polarity to satisfy the condition that the average capacitor current be zero.

If we leave switch S_{xz} on for a long time, both V_2 and I_1 become zero, regardless of the values of V_1 and I_2. Therefore both the ratios V_2/V_1 and I_1/I_2 become zero. However, if we leave switch S_{xy} on for a long time, these ratios become minus and plus infinity, respectively. As we inferred for the direct converter, when the two switches are alternating, these ratios will be somewhere between their two extremes. For the indirect converter, the magnitude of the conversion ratio ranges from zero to infinity, compared with zero to one or one to infinity for the direct converter.

The simplest way to determine the exact dependence of the conversion ratios on the duty ratio is to set the average inductor voltage to zero by balancing its positive and negative volt–time integrals. If we define D to be the fraction of the switching period T during which S_{xy} is closed, the average inductor voltage is zero when

$$V_1 DT = -V_2(1-D)T,$$

which gives a voltage conversion ratio of

$$\frac{V_2}{V_1} = -\frac{D}{1-D}. \tag{5.7}$$

Similarly, we can determine the current conversion ratio by setting the average capacitor current to zero, that is,

$$I_2 DT = I_1(1-D)T,$$

which gives a current conversion ratio of

$$\frac{I_2}{I_1} = \frac{1-D}{D}. \tag{5.8}$$

We can also obtain (5.8) from (5.7) by setting the net power into the converter to zero.

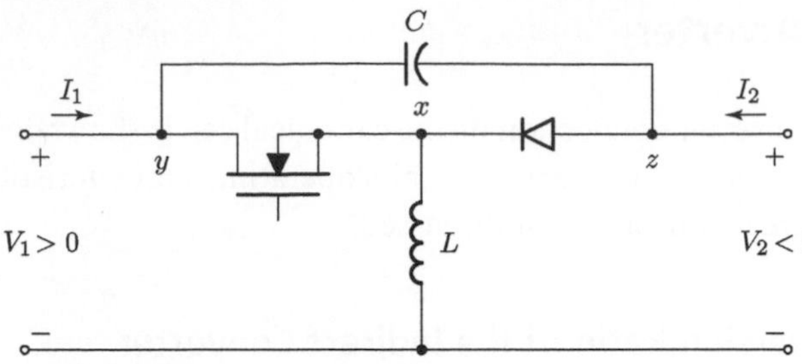

Figure 5.17 Switch implementation for the indirect converter with power flowing from left to right and $V_1 > 0$ (implying I_1 and I_2 are also > 0).

We commonly refer to the indirect converter as a *buck/boost* converter, because the output voltage (or respectively current) can be higher or lower than the input, for either direction of power.

5.4.2 Implementation of Switches

If we consider the switch voltages and currents as we did for the direct converter, we can determine appropriate semiconductor devices to use for the indirect converter under different conditions. Again, we must specify the direction(s) of power flow and the polarity of one of the terminal voltages or currents. For power flowing left to right in the circuit of Fig. 5.8(b) (there is no higher or lower voltage side in the indirect converter) and $V_1 > 0$, switch S_{yx} conducts positive current and blocks positive voltage (note that $V_2 < 0$ for the assumed conditions on power and V_1). Therefore this switch can be a transistor. Switch S_{zx}, on the other hand, conducts positive current but blocks negative voltage, which are conditions met by a diode. The resulting indirect converter circuit is shown in Fig. 5.17.

Example 5.4 Indirect Converter with Power Flowing Right to Left

One way to make an indirect converter with power flowing from right to left in the circuit of Fig. 5.8(b) is to take the converter of Fig. 5.17 and flip it about the inductor. In this case, the voltage at the left terminal pair becomes negative. If the requirement were that this voltage be positive, we would invert the voltage and current for both switches and turn the diode and transistor around, as shown in Fig. 5.18.

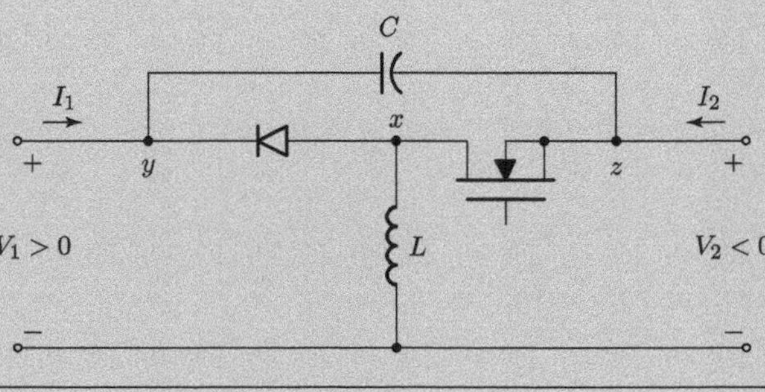

Figure 5.18 Indirect converter switch implementation for power flowing from right to left and $V_1 > 0$.

5.4.3 A Variation of the Basic Indirect Converter Topology

There is an important variation of the basic indirect converter topology. This variation leaves the conversion ratios given by (5.7) and (5.8) unchanged, and the voltage and current waveforms at the semiconductor devices are unaffected. The only thing that we change is deployment of the capacitor to filter the high-frequency switch currents.

The capacitor and inductor in the indirect converter of Figs. 5.17 and 5.18 perform their high-frequency filtering functions in the following ways. The ac components of the switch currents must circulate through C to prevent their appearing at the terminals. But note that because we can model L as an open circuit at the switching frequency, the series connection of the external networks is in parallel with C. Therefore the high-frequency impedance of C must be much smaller than the sum of the high-frequency impedances of the external networks. Similarly, the ac components of the switch voltages must appear across L if they are not to appear at the terminals. In this case, we model the capacitor as a short circuit at the switching frequency, which results in the parallel connection of the two external networks appearing in series with L. Thus the high-frequency impedance of L must be much larger than the parallel combination of the high-frequency impedances of the external networks. (We derive these conditions in more detail in Section 5.6.)

The implications of these conditions are: if the capacitor is to be of reasonable size, at least one of the external networks must have a high ac impedance; if the inductor is to be small, at least one of the external networks needs to have a low ac impedance. In the ideal case, one of the external systems has a high ac impedance and the other a low one.

Unfortunately, both external systems often have a large impedance, particularly when the switching frequency is high, and long (inductive) connection wires are used. Under these conditions the inductor may be unreasonably large. However, if we modify the topology by placing an additional capacitor across one of the two ports, as shown in Fig. 5.19(a), we make one of the external systems appear to have a low ac impedance. Note that the purpose of this capacitor C_1 is to allow most of the ac switch voltage to drop across a reasonably sized inductor. This capacitor does not help the original capacitor C_{12} filter the ac switch current. This switch current continues to flow through C_{12}, because one of the two external systems still has a large ac impedance.

At the switching frequency, the three nodes of the canonical cell are shorted together by the two capacitors of Fig. 5.19(a). We can achieve the same result with the arrangement of capacitors shown in Fig. 5.19(b). If we treat the inductor as an open circuit at the switching frequency, we can see that both capacitors now help filter the ac switch currents. Each carries the full ac switch current if its ac impedance is small compared to the impedance of its corresponding external system. However, across each appears a dc voltage equal only to that at its port, rather than the sum of the terminal voltages, which is what the capacitor in the indirect topology of Fig. 5.16 must withstand. Each of the two capacitors is therefore smaller than the original. In fact, under certain assumptions, we can show the sum of these two capacitors' peak energies to be equal to that of the original capacitor (see Section 5.6.5). Using the names *up/down* or *buck/boost* to identify a converter circuit usually implies the circuit variation of Fig. 5.19(b). In practice the configuration of Fig. 5.19(b) is preferred to that of Fig. 5.18(a) because it eliminates capacitive coupling between the two converter ports.

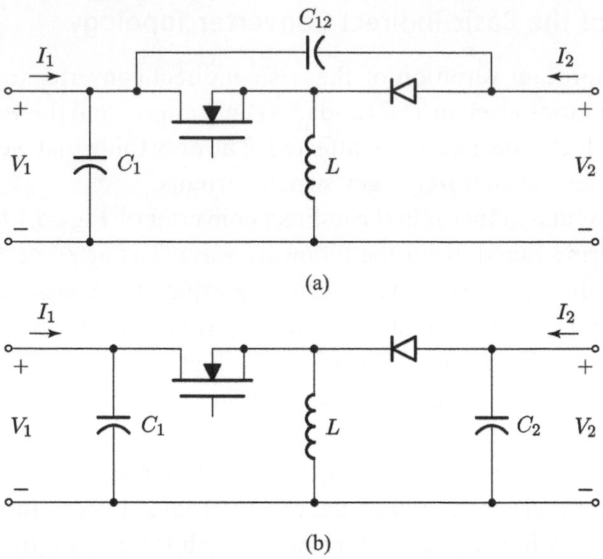

Figure 5.19 Variations of the basic indirect converter topology to provide ac terminal voltage filtering for two external networks having high ac impedances. (a) Addition of capacitor C_1 to the basic indirect topology. (b) Alternative placement for the capacitors in (a). This circuit is commonly referred to as the *buck/boost* converter.

5.5 Other PWM DC/DC Converters

There are numerous other PWM dc/dc converter topologies and variants. These can be arrived at through circuit transformations on canonical cell converters (e.g., through topological duality, addition and rearrangement of components, and addition of circuit blocks); as realizations of the polarity of other, more complex, canonical cells; through intuition; and even by exhaustive search. In this section we introduce two of the more useful and widely applied such topologies. We illustrate both techniques that can be used to synthesize a topology suitable for a particular goal, and methods that can be used to analyze any such topology.

5.5.1 Ćuk Converter

Consider further the canonical cell indirect converter of Fig. 5.17. When both external systems have a low ac impedance, an inconveniently large capacitor must be used to remove ac ripple from the terminal currents. We can, however, increase the impedance of one of the external networks by placing an inductance in series with it, such as L_1 in Fig. 5.20(a). This approach results in a return to the ideal situation in which an indirect converter interfaces one high- and one low-impedance dc network. Although we can now reduce the size of the cell capacitor, this additional inductor has no influence on the size of the original cell inductor L_{12}. The parallel impedance of the two external systems is still low because at least one of them has a low ac impedance, so most of the ac switch voltage will still appear across L_{12}. So far we have simply traded off capacitor size for an additional inductor.

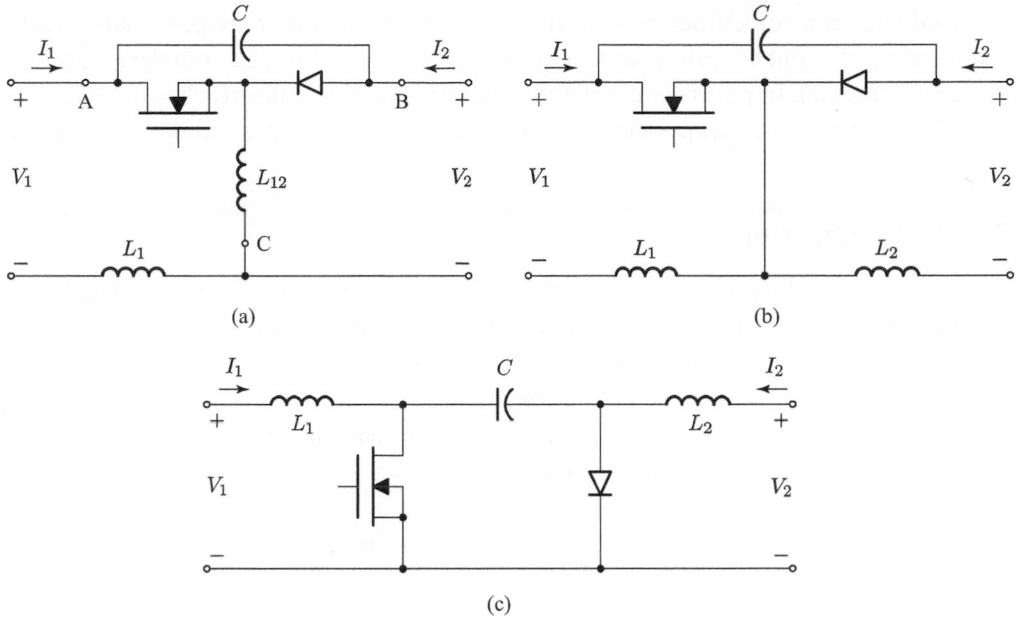

Figure 5.20 Indirect converter topology variations designed to provide ac terminal current filtering for two external networks having low ac impedances. (a) Addition of inductor L_1 to the basic indirect converter topology. (b) Alternative placement for the inductor in (a). (c) The circuit of (b) redrawn in the form commonly known as the Ćuk converter.

Of the three branches stemming from node C in Fig. 5.20(a), two have inductors that, if they have been properly sized to eliminate ac voltage ripple from the converter terminals, we can model as open circuits at the switching frequency. Therefore the ac current flowing through the third branch is also zero. We can achieve the same result if we arrange the two inductors as L_1 and L_2, shown in Fig. 5.20(b). Now both inductors contribute to filtering the ac switch voltage. Each must have an ac impedance that is large compared to the impedance of its corresponding external system. But each carries only the terminal current rather than the sum of the terminal currents that L_{12} must carry. Therefore each inductor is smaller than L_{12}. For the same ripple currents, the sum of the peak stored energies in these two inductors is equal to the peak energy in L_{12}.

Figure 5.20(b) shows the inductors connected in series with the negative terminals of the converter; however, positioning them in series with the positive terminals, as shown in Fig. 5.20(c), is more practical. This arrangement changes nothing functionally, but it results in a circuit node common to the input, output, and source of the transistor, which is both convenient and often required. This variant of the indirect converter is commonly known as the *Ćuk converter*. It is worth noting that one can also arrive at the Ćuk converter by applying topological duality techniques to the indirect converter of Fig. 5.19(b).

The Ćuk converter has some attractive features, starting with the desirable filtering properties described above. In addition, it employs a switch referenced to circuit common, which can greatly simplify level shifting, gate drive, and sensing. Moreover, the two inductors of the Ćuk converter can sometimes be wound on the same core and/or coupled magnetically to improve filtering of input or output waveforms (at the expense of design complexity and manufacturing requirements). In part because of these features, the Ćuk converter has found use despite having a higher magnetic component count as compared to the conventional buck/boost converter.

5.5.2 SEPIC Converter

Here we introduce another widely used indirect converter topology, the "Single-Ended Primary Inductor Converter" known as the SEPIC. It is essentially a cascade of a boost and buck/boost converter using common switching elements. If we take the direct boost converter of Fig. 5.14 and insert a voltage source equal to the input source value V_i between the switch and the diode, as shown in Fig. 5.21, the conversion ratio can be calculated by setting the average value of the inductor voltage v_{L_1} to zero. The result is

$$V_o = V_i\left(\frac{D}{1-D}\right). \tag{5.9}$$

That is, we now have the conversion ratio characteristics of a buck/boost converter but without voltage inversion. We would like to replace the inserted source with a capacitor, but as the circuit is presently configured, the diode current, which has a dc component, passes through the source. Simply replacing the source with a capacitor is prevented by the presence of this dc current. However, if we can provide an appropriate shunt component between the capacitor and the diode to carry the diode's dc current component, a capacitor would function as did the source. We note that the average voltage between the source-diode node and ground, $\langle v_x \rangle$, is zero, so this shunt element must carry a dc current without producing a dc voltage. These two requirements are satisfied by an inductor. The resulting circuit is shown in Fig. 5.22.

We now calculate the conversion ratio of our new circuit to see if our replacements for the source do what we expected. Recognizing that the average inductor voltages in periodic steady state must be zero, we get $v_C = V_i$. Since the average voltage across L_2 over the cycle must be zero in periodic steady state,

$$DV_i + (1-D)(-V_o) = 0, \tag{5.10}$$

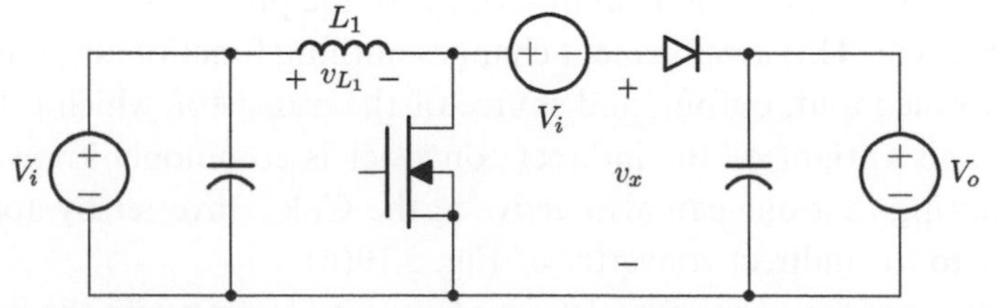

Figure 5.21 The boost converter of Fig. 5.14 with a voltage source of value V_i inserted between the diode and the switch.

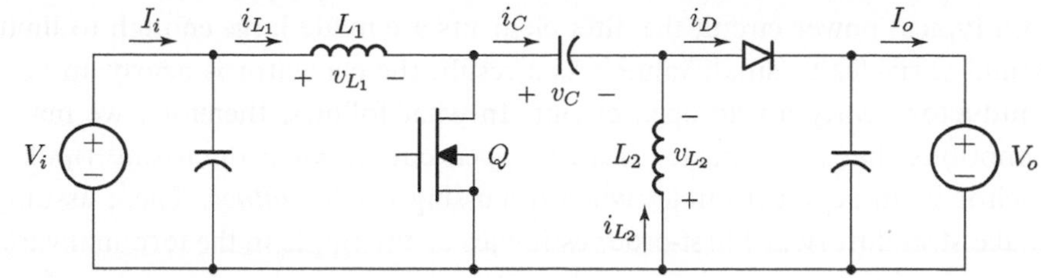

Figure 5.22 The SEPIC converter.

which yields the result sought above:

$$V_o = \frac{D}{(1-D)} V_i. \tag{5.11}$$

The SEPIC converter, which we developed from the direct connection of the canonical cell, is an indirect converter. Its primary advantages over the buck/boost converter of Fig. 5.19(b) are that the input and output voltages are not inverted, and that the MOSFET is referenced to ground. Its disadvantages are that since all the energy is transferred through the coupling capacitor, this capacitor is quite large and must be rated to carry high currents. Also, the circuit's four energy storage elements make control of transient deviations from periodic operation more of a challenge. (As we'll see in Part II, the transfer functions needed for control design will have four poles.)

There are, of course, other means to achieve a non-inverting buck/boost function, such as by cascading a buck and a boost stage to form a *four-switch buck/boost* converter, as explored in Problem 5.14. Such an approach requires more switches (not all ground referenced) than a SEPIC converter, but requires only a single magnetic component. Part of the art of power electronics design is selecting a topology (from among myriad possibilities) that best meets the design requirements in a particular application. It is thus important to be able to both analyze and synthesize designs to meet a particular goal.

5.6 Choice of Capacitor and Inductor Values

So far in our discussion of dc/dc converters, we have assumed that the high-frequency filter elements (e.g., L and C of the canonical cell) are "large enough" to reduce the switching-frequency ripple in the terminal variables to an acceptable level. We now consider the specification of these components in more detail, assuming a second-order filter, that is, a single capacitor and inductor. This subject is important because the physical size of a converter is strongly influenced by the size of its energy storage elements. Furthermore, comparisons among different converter circuits are often based on ripple amplitude for a given amount of energy storage. In this section we discuss the source of ripple in terminal currents and voltages, develop models for analyzing ripple for various converter topologies, and specify the relative sizes of L and C for direct and indirect converters.

In a typical power circuit the filter elements are made large enough to limit ac ripple in the terminal variables to small values. As a result, the capacitor is nearly an ac short circuit and the inductor nearly an ac open circuit. In what follows, therefore, we make the simplifying assumptions that the inductor is an ac open circuit when discussing ripple *current*, and the capacitor is an ac short circuit when discussing ripple *voltage*. These assumptions permit us to make straightforward first-order estimates of the ripple in the terminal variables.

5.6.1 Ripple-Frequency Model for the Direct Converter

We first consider the source and distribution of ripple current in the direct converter of Fig. 5.8(a), repeated here as Fig. 5.23(a) for convenience. With L assumed to be an open circuit at the ripple frequencies, there is no ripple in i_2, and the ac components of i_y and i_z must be equal and opposite. We designate the ac component of i_y as i'_y. For the purpose of calculating the ripple component of i_1, denoted by i'_1, we now replace the circuit to the right of C in Fig. 5.23(a) with an equivalent current source of value i'_y. The resulting ripple-current model is shown in Fig. 5.23(b). Note that the ripple current i'_1 depends on how i'_y divides between C and the impedance of the external network, Z_1.

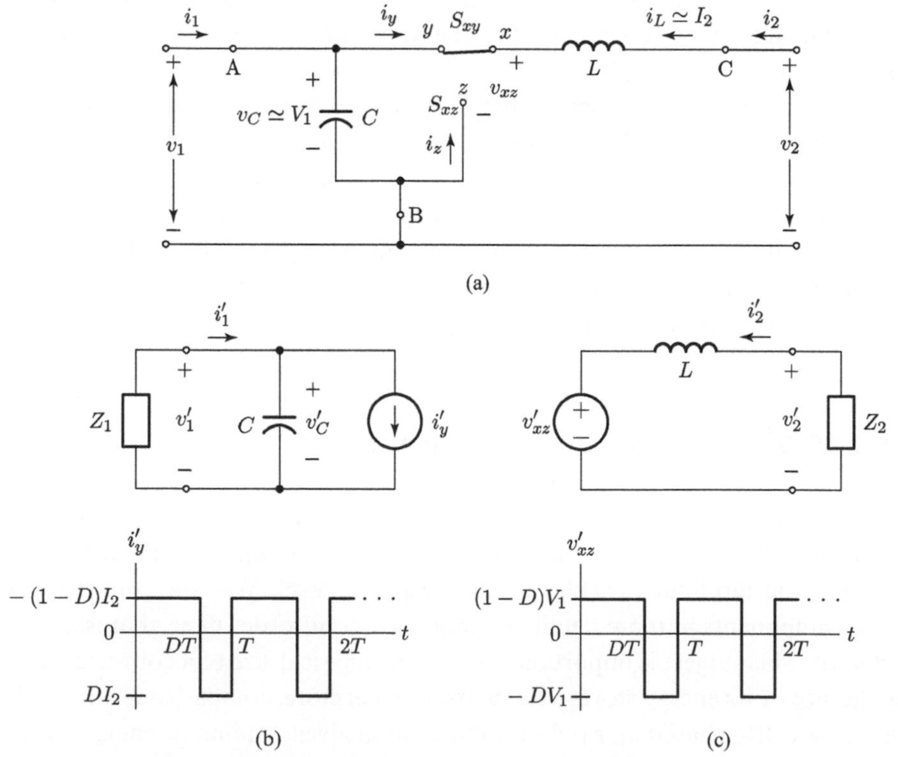

Figure 5.23 (a) The direct converter connection of the canonical cell. (b) First-order ripple-current model. (c) First-order ripple-voltage model for this converter.

We now develop a model that we can use to calculate the ripple voltage at port 2 of the direct converter. Based on the assumption that the capacitor is an ac short circuit at the switching frequency, the ac component of the voltage v_{xz}, which we denote by v'_{xz}, is as shown in Fig. 5.23(c). We now replace the circuit to the left of L in Fig. 5.23(a) by an equivalent voltage source of value v'_{xz}. The resulting ripple-voltage model is shown in Fig. 5.23(c). We calculate the first-order voltage ripple at port 2 by dividing v'_{xz} between the impedance of L and that of the external network Z_2.

5.6.2 Ripple-Frequency Model for the Indirect Converter

For the indirect converter of Fig. 5.8(b), repeated here as Fig. 5.24(a), the interactions of the capacitor and the inductor with the impedances of the external systems are more complicated than they are for the direct converter. In the direct converter, the high-frequency impedance

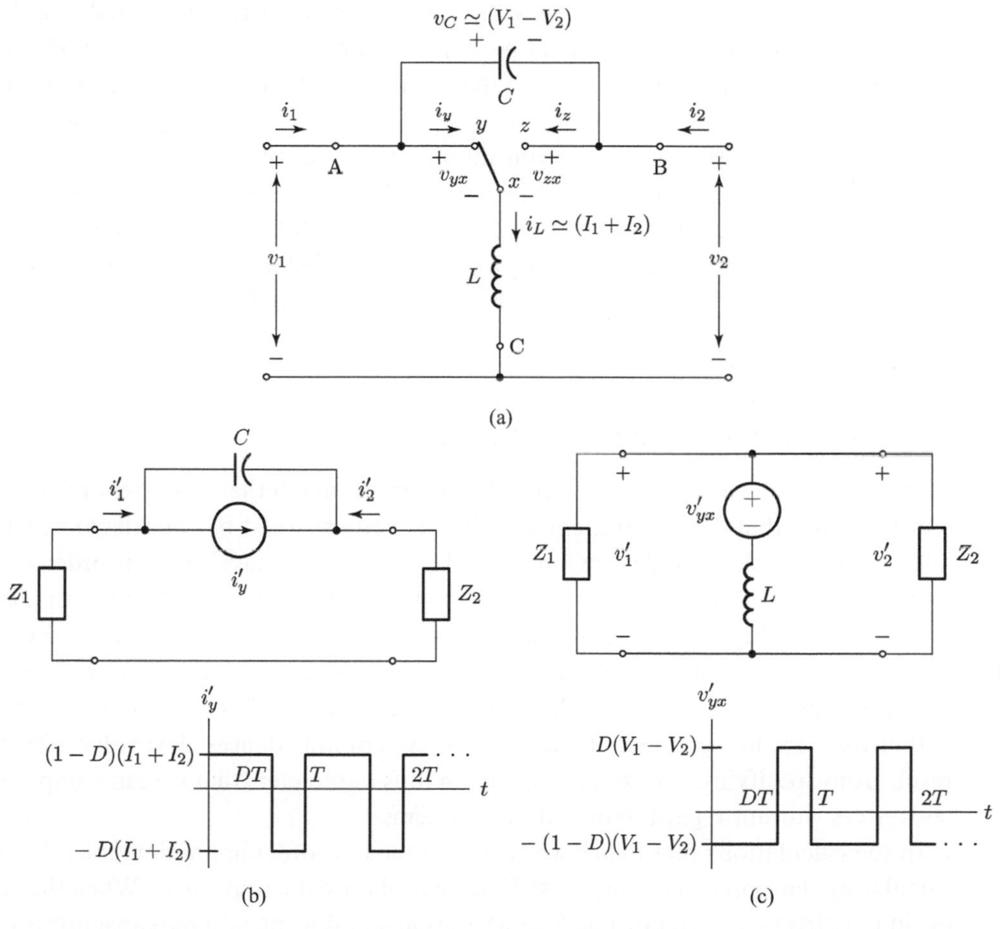

Figure 5.24 (a) The indirect converter connection of the canonical cell. (b) First-order ripple-current model. (c) First-order ripple-voltage model.

of only one external system influences the filtering effectiveness of the L or C in the canonical cell. For the indirect converter, we must simultaneously consider the impedance of both external systems to determine the ripple in the terminal variables.

We first consider the effect of capacitor value on the terminal ripple currents. We assume that the inductor is large enough for its ripple current to be negligible, permitting us to treat it as an infinite impedance (open circuit) at the switching frequency. Again, $i'_y = -i'_z$. The result is the switching-frequency ripple-current model shown in Fig. 5.24(b). We derive the waveform of i'_y by recognizing that the inductor current is $I_1 + I_2$ and is switched back and forth between nodes y and z. The ac current i'_y has two paths through which it can flow. One is through the capacitor, and the other is through the *series* connection of the two external system impedances. Therefore the high-frequency impedance of the capacitor must be small compared to the sum of the external impedances, Z_1+Z_2, if we are to keep the switching-frequency current from flowing through the external network. The ripple current that does show up at the converter terminals has the same amplitude at both ports.

Now we consider the requirements on the inductor, assuming that the capacitor is large enough that its voltage ripple is negligible and the capacitor can be treated as a short circuit in the switching frequency model. This assumption forces $v'_{yx} = v'_{zx}$, putting nodes y and z at the same potential in the ac model. We can therefore connect these nodes, and in turn connect them to the inductor through a source of value v'_{yx}, creating the voltage-ripple model of Fig. 5.24(c). (In deriving the waveform of v'_{yx}, you should remember that if $V_1 > 0$, then $V_2 < 0$ for the indirect converter.) The source v'_{yx} is in series with the inductor and the *parallel* combination of the two external network impedances. Therefore the high-frequency impedance of the inductor must be large compared to that of Z_1 and Z_2 in parallel if the switching-frequency voltage is not to appear at the converter terminals. To the extent that it does appear, the ripple voltage is the same at both ports.

5.6.3 Minimum *L* and *C* for the Direct Converter

Up to this point we have assumed that L was large enough that its current could be considered constant, and that C was large enough that its voltage could be considered constant. We now address the problem of determining how large L and C have to be in order to justify these approximations. Any ripple in i_L or v_C shows up directly as first-order ripple in i_2 or v_1, respectively, in Fig. 5.23(a). It would have a second-order effect on i'_1 or v'_2 by adding ripple to the flat portions of the waveforms of the ripple sources i'_y and v'_{xz} in Fig. 5.23(b) and (c). Another reason for calculating the first-order capacitor voltage and inductor current ripples is that they are independent of the external system impedances. The values for L and C that result from specifying a maximum ripple are thus parameters that we can compare for different converters without regard to the external systems.

In the calculations that follow we again use the first-order approximations that $i_1 \approx I_1$ when calculating the ripple in v_1, and $v_2 \approx V_2$ for calculating the ripple in i_2. When the series switch is off in the direct converter of Fig. 5.23(a), the current flowing into the capacitor is I_1, and during this interval (of length $(1-D)T$) the capacitor voltage v_1 will increase, giving a peak–peak ripple amplitude of

$$\Delta v_1 = \frac{1}{C}\int_0^{(1-D)T} I_1\, dt = \frac{I_1(1-D)T}{C}. \tag{5.12}$$

During this same interval, the voltage across the inductor equals V_2, and the inductor current $i_L = -i_2$ decreases by an amount equal to its peak–peak ripple amplitude:

$$-\Delta i_L = \Delta i_2 = \frac{1}{L}\int_0^{(1-D)T} V_2\, dt = \frac{V_2(1-D)T}{L}. \tag{5.13}$$

With specified limits on the terminal current and voltage ripple amplitudes Δi_2 and Δv_1, we can use (5.12) and (5.13) to determine minimum values of L and C:

$$C \geq \frac{I_1(1-D)T}{\Delta v_1}, \tag{5.14}$$

$$L \geq \frac{V_2(1-D)T}{\Delta i_2}. \tag{5.15}$$

After we have determined values for L and C from these expressions, we can determine the maximum energy stored in each element, which is useful because the stored energy is an indication of the element's size and cost. To express the stored energy in a useful way, we first define the *ripple ratios* $\mathcal{R}_C$ and $\mathcal{R}_L$ for the capacitor voltage and the inductor current, respectively. These are the ratios of the peak ripple amplitude to the dc (average) value of the capacitor voltage and inductor current, or

$$\mathcal{R}_C = \frac{\Delta v_1/2}{V_1}, \tag{5.16}$$

$$\mathcal{R}_L = \frac{\Delta i_2/2}{I_2}. \tag{5.17}$$

By combining (5.14) with (5.16), and (5.15) with (5.17), we can find the minimum peak stored energy in the capacitor and inductor:

$$E_C = \frac{1}{2}CV_{1p}^2 = \left(\frac{1-D}{4}\right)\left(\frac{P_o}{f_s}\right)\frac{(1+\mathcal{R}_C)^2}{\mathcal{R}_c}, \tag{5.18}$$

$$E_L = \frac{1}{2}LI_{2p}^2 = \left(\frac{1-D}{4}\right)\left(\frac{P_o}{f_s}\right)\frac{(1+\mathcal{R}_L)^2}{\mathcal{R}_L}. \tag{5.19}$$

In these expressions, f_s is the switching frequency $1/T$; V_{1p} and I_{2p} are the peak values of the capacitor voltage and inductor current, respectively; and P_o is the average power of the converter, that is, $P_o = I_1V_1 = -I_2V_2$. The value P_o/f_s is the amount of energy transferred from input to output during one cycle of the switching frequency. As we decrease the power or increase the switching frequency, the energy storage requirement goes down. Equations (5.18) and (5.19) show that, for a fixed ripple ratio, as the value of $1-D$ gets larger, the peak energy storage gets proportionally larger. Therefore, as the difference between the input and output voltage or current becomes greater, the energy storage requirement increases proportionally.

Example 5.5 Specifying *L* and *C* Values for a Direct Converter

The circuit of Fig. 5.25(a) shows a direct (buck) converter between a source of 101 V having an internal resistance of 0.5 Ω, a resistive load of 2 Ω, and a load voltage of 20 V. What values of L and C will ensure that the input and output currents have ripples with peak-to-peak amplitudes of no more than 5% of their dc values if $f_s = 500\,\text{kHz}$? If these minimum values of L and C are used in the circuit, what are the peak-to-peak terminal ripple-voltage amplitudes?

We can determine the average input-port variables $\langle v_1 \rangle$ and $\langle i_1 \rangle$ by equating the powers at the input and output ports of the converter:

$$\begin{aligned}
\langle v_1 \rangle \langle i_1 \rangle &= \frac{\langle v_2 \rangle^2}{2} = 200\,\text{W}, \\
\langle v_1 \rangle &= 101 - (0.5)\langle i_1 \rangle, \\
\langle i_1 \rangle &= 2\,\text{A}, \\
\langle v_1 \rangle &= 100\,\text{V}.
\end{aligned}$$

The duty ratio $D = 20/100 = 0.2$ and $T = 2\,\mu\text{s}$.

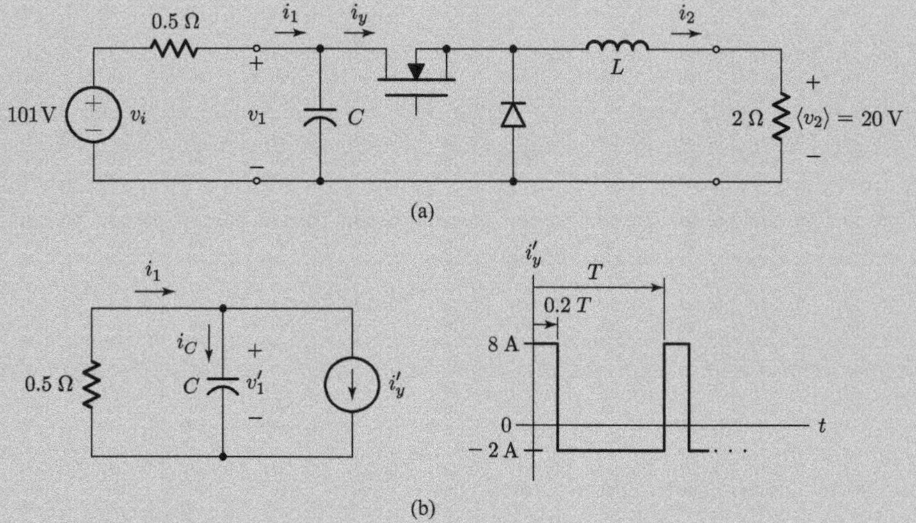

Figure 5.25 (a) Buck converter circuit. (b) Input ripple-current model for the circuit in (a).

We can easily calculate the value of L from (5.15):

$$L \geq \frac{20(0.8)\left(2 \times 10^{-6}\right)}{0.05(10)} = 64\,\mu\text{H}.$$

To determine the value of C, we must first use the input ripple-current model of Fig. 5.25(b) to calculate i_1'. Because $i_1' \ll i_y'$ with the proper values of L and C, we can make the first-order approximation that $i_C \approx -i_y'$. Therefore the peak-to-peak ripple voltage on C, $\Delta v_1'$, is

$$|\Delta v_1'| \approx \frac{1}{C} \int_0^{DT} i_y'\,dt = \frac{(0.2)(8)\left(2 \times 10^{-6}\right)}{C} = \frac{6.4 \times 10^{-7}}{C}, \tag{5.20}$$

$$\Delta i_1' = \frac{\Delta v_1'}{0.5} \leq .05\,\langle i_1 \rangle = 0.1\,\text{A},$$

$$C = \frac{6.4 \times 10^{-7}}{0.1} = 64\,\mu\text{F}.$$

We can now calculate the first-order ripple in the terminal voltages. Because the load is resistive, the output voltage ripple as a percentage of V_2 is the same as the ripple-current specification. That is,

$$\frac{\Delta v_2'}{V_2} = \frac{\Delta i_2'}{I_2} = 5\%.$$

The input voltage ripple given by (5.20) is

$$\frac{\Delta v_1'}{V_1} = \frac{0.05}{100} = 0.05\%.$$

In practice one would include a capacitor at the output to reduce the size of the inductor and to provide the converter with better dynamic performance.

5.6.4 Minimum *L* and *C* for the Indirect Converter

We use the same approach to determine the capacitor and inductor values for the indirect converter of Fig. 5.8(b). The resulting values of L and C are

$$L \geq \frac{V_1 DT}{\Delta i_L}, \tag{5.21}$$

$$C \geq \frac{I_1(1-D)T}{\Delta v_C}. \tag{5.22}$$

In the indirect converter the average capacitor voltage is the sum of $|V_1|$ and $|V_2|$, and the average inductor current is the sum of $|I_1|$ and $|I_2|$. Expressed in terms of these terminal variables, the ripple ratios for the capacitor and inductor are:

$$\mathcal{R}_C = \frac{\Delta v_C/2}{|V_1| + |V_2|}, \tag{5.23}$$

$$\mathcal{R}_L = \frac{\Delta i_L/2}{|I_1| + |I_2|}. \tag{5.24}$$

We can again express the peak stored energy in the capacitor and inductor in terms of the ripple ratios:

$$E_C = \left(\frac{1}{4}\right)\left(\frac{P_o}{f_s}\right)\frac{(1+\mathcal{R}_C)^2}{\mathcal{R}_C}, \tag{5.25}$$

$$E_L = \left(\frac{1}{4}\right)\left(\frac{P_o}{f_s}\right)\frac{(1+\mathcal{R}_L)^2}{\mathcal{R}_L}. \tag{5.26}$$

Note that, unlike for the direct converter, these storage requirements do not depend on the duty ratio. We can again minimize them by choosing a ripple ratio equal to unity.

The peak energy storage requirements of the indirect converter are typically larger than they are for the direct converter. Only when the direct converter operates at a duty ratio near zero are they equal. For example, at an operating point of $D = 0.5$ for the direct converter, the indirect converter requires twice as much storage. Indirect conversion also imposes increased semiconductor stresses in a converter, as detailed below.

5.6.5 Calculations for the Buck/Boost and Ćuk Converters

The buck/boost and Ćuk variations of the indirect converter use two capacitors and two inductors, respectively. For these converters we compare the total capacitive or inductive peak energy storage requirements to the corresponding values calculated for the indirect converter. The requirement placed on the inductor in the buck/boost variant is the same as that for the inductor in the basic indirect converter. The requirements on the capacitors in the Ćuk variant and the basic circuit are identical. Let's consider the buck/boost circuit of Fig. 5.19(b).

The terminal voltage ripples in the indirect converter are related by the relative sizes of the external impedances connected to the two ports of the converter, as discussed in Section 5.6.2. In the analysis that follows, we assume that the ripple amplitudes at the two ports are proportional to their dc voltages. That is, the ripple ratio at each port is the same as the ripple ratio of the capacitor in the basic indirect converter against which we are comparing the stored energy requirements of the buck/boost variant. As a result, the two capacitors in the buck/boost converter each have ripple ratios equal to that of the single capacitor in the basic indirect converter, or $\mathcal{R}_C$.

In the buck/boost circuit of Fig. 5.19(b), the capacitor at the left-hand port, C_1, is charged in the same way as the capacitor in the direct converter. It therefore stores the same peak energy, which is given by (5.18):

$$E_{C_1} = \left(\frac{1-D}{4}\right)\left(\frac{P_o}{f_s}\right)\frac{(1+\mathcal{R}_C)^2}{\mathcal{R}_C}. \tag{5.27}$$

The ripple voltage on C_2 is

$$\Delta v_{C_2} = \frac{DTI_2}{C_2}, \tag{5.28}$$

giving a minimum capacitance of

$$C_2 \geq \frac{DTI_2}{2V_2\mathcal{R}_C}. \tag{5.29}$$

The resulting peak energy stored in C_2 is

$$E_{C_2} = \left(\frac{D}{4}\right)\left(\frac{P_o}{f_s}\right)\frac{(1+\mathcal{R}_C)^2}{\mathcal{R}_C}. \tag{5.30}$$

The sum of E_{C_1} and E_{C_2} is exactly equal to the peak energy stored in the single capacitor of the basic indirect converter and given by (5.25).

A similar analysis done on the Ćuk variant of the indirect converter shown in Fig. 5.19(c) gives the same results. That is, the total energy storage requirement is identical to that for the basic indirect converter if we postulate the same ripple requirements at the ports – current ripple in this case.

5.7 Semiconductor Device Stresses

Semiconductor switches must be rated to carry the peak current I_p and withstand the peak voltage V_p presented to them by the power circuit. These ratings not only affect the cost of a switch, but they also affect various device performance parameters, such as storage time, current gain, and switching speed. Therefore they are meaningful parameters for comparing the attributes of different circuits. The product of the two peak stresses $V_p I_p$ is frequently useful for making comparisons among the various topologies. We call this product the *switch-stress parameter*.

For both the direct and indirect converters of Fig. 5.8, the peak voltage at the switches is equal to the maximum capacitor voltage. Similarly, the peak current carried by either switch is equal to the maximum inductor current. In the case of the buck/boost variant of the indirect converter of Fig. 5.19(b), the peak switch voltage is the sum of the two peak capacitor voltages. For the Ćuk variant of Fig. 5.20(c), the peak current is the sum of the two peak inductor currents.

For the direct converter circuit of Fig. 5.8(a), the stress parameter for both switches is

$$V_p I_p = V_1(1+\mathcal{R}_C)I_2(1+\mathcal{R}_L) = P_o\frac{(1+\mathcal{R}_C)(1+\mathcal{R}_L)}{D}. \tag{5.31}$$

For the indirect converter of Fig. 5.8(b), the stress parameter is

$$\begin{aligned} V_p I_p &= (|V_1|+|V_2|)(1+\mathcal{R}_C)(|I_1|+|I_2|)(1+\mathcal{R}_L) \\ &= P_o\frac{(1+\mathcal{R}_C)(1+\mathcal{R}_L)}{D(1-D)}. \end{aligned} \tag{5.32}$$

Note that for both converters, the stress parameter is smallest when the conversion ratio is unity, that is, when $D=1$ in the direct converter and when $D=0.5$ in the indirect converter.

We call the switch-stress parameters normalized to the power P_o the *switch-stress factors*. Figure 5.26 shows them for both direct (buck or boost) and indirect converters under the condition of zero ripple ratios. We need to plot only the stresses for output-to-input voltage ratios between zero and one because of circuit symmetry. We obtain the switch-stress factors for ratios between one and ∞ by exchanging the input and output roles of the ports. For example, the factors for a conversion ratio of 2.5 are the same as those for a conversion ratio of 0.4 ($= 1/2.5$). When the conversion ratio is unity, the stresses on the switches in the indirect converter are four times greater than they are for the direct converter. At a 50% conversion ratio, the difference is more than 2. Even at a 25% ratio the difference is greater than 1.5. Clearly, we would not choose to use an indirect converter unless we needed either a negative output voltage or *both* buck and boost conversion ratios from the same converter. The relative energy storage requirements for the two converters, as derived in Sections 5.6.3 and 5.6.4, also support this decision.

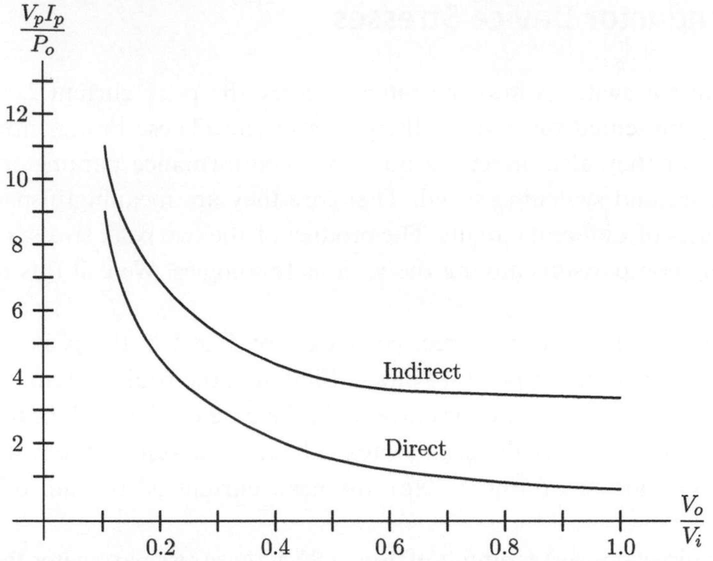

Figure 5.26 Switch-stress factors for direct and indirect converters.

5.8 Three-Level Flying-Capacitor Converter

If large variation of the conversion ratio is necessary, both the energy storage requirements and switch-stress factors can be reduced by varying the converter input voltage to avoid extreme duty ratios. One circuit technique for accomplishing this is through the use of *multi-level* power conversion. An example of this is the *flying-capacitor* multi-level (FCML) converter, a three-level version of which is shown in Fig. 5.27(a). The FCML itself is just one of a whole variety of multi-level converters that attain improved performance and reduced individual device stresses at the cost of increased complexity. Multi-level conversion will be further treated in Chapters 8 and 9, as it is often employed in the context of dc/ac inverters.

The circuit of Fig. 5.27(a) can have four possible switch states $S_{k,\ell}$, where the subscripts indicate which switches are closed. These states yield four different voltages v_d:

$$\begin{aligned} &\text{state 1} \quad S_{1,2}: v_d = V_i, \\ &\text{state 2} \quad S_{3,4}: v_d = 0, \\ &\text{state 3} \quad S_{1,3}: v_d = V_i - V_C, \\ &\text{state 4} \quad S_{2,4}: v_d = V_C. \end{aligned}$$

If V_C is controlled to be $V_\text{i}/2$, states 3 and 4 produce $V_\text{i}/2$ and the filter input voltage can be V_dc, $V_\text{i}/2$, or 0. The converter can thus be operated to produce four different waveforms for v_d, as shown in Fig. 5.27(b).

Switching between states 1 and 2 produces a conventional buck converter. However, for $V_o \leq V_i/2$, considerable performance improvement can be obtained by employing the flying capacitor to reduce the voltage at the filter input to $V_i/2$ and switching among states 1, 2, 3, and 4 as

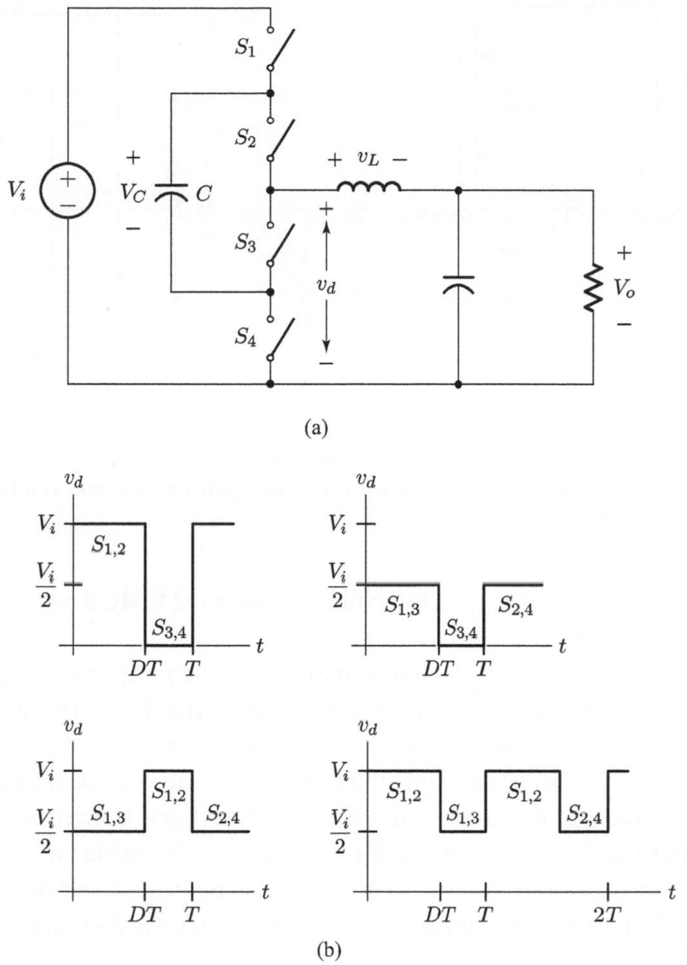

Figure 5.27 (a) A flying-capacitor buck converter producing an output voltage V_o. (b) The three possible voltages v_d if $V_C = V_i/2$.

shown in the middle waveform in Fig. 5.27(b). Alternating between states 1,3 and 2,4 assures C is charged and discharged equally to maintain $V_C = V_i/2$. In this operating mode the filter sees twice the switching frequency of the individual switches and $\Delta v_L = V_i/2$. Compared to a conventional buck converter with an output voltage $\leq V_i/2$, the three-level flying-capacitor converter produces twice the ripple frequency and half the inductor ripple voltage. The result is a smaller inductor for the same output ripple current. And although there are four switches, each changes state only twice per period T, as do the switches in a conventional buck. However, each switch sees $V_i/2$, not V_i, so switching losses are reduced by half per switch, giving total switching losses comparable to those of the conventional converter. If V_C can be precharged to $V_i/2$ without subjecting the devices to V_i at start-up, the switches can be rated for $V_i/2$ instead of V_i, further reducing losses, as lower-voltage devices with lower $R_{ds(on)}$ can be used. Because of these advantages and despite their complexity, flying-capacitor multi-level converters have

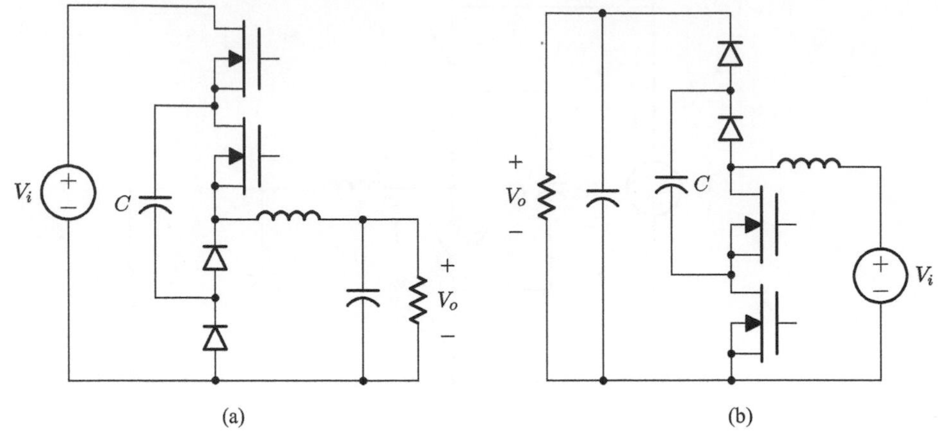

Figure 5.28 The circuit of Fig. 5.27 with switch implementations for (a) a buck converter and (b) a boost converter.

found uses in applications from low-voltage integrated CMOS power converters to high-voltage industrial machine drives.

Theoretically, V_C is maintained at its preset voltage by the charging symmetry of states 3 and 4, but any variations in switch timings or device rise and fall times will cause V_C to drift from its setpoint value. Therefore, some means to maintain $V_C = V_i/2$ must be provided. Feeding back V_C to control the relative on-times of states 3 and 4 is one technique frequently employed.

Although the circuit shown in Fig. 5.27 is a buck converter, depending on the switch implementations it can also be configured as a boost converter, as shown in Fig. 5.28.

Flying capacitors can also be used to produce multiple levels in dc/ac inverters, as discussed in Chapter 8. In addition they are the foundation of *switched-capacitor converters* presented in Chapter 6.

5.9 Converter Operation with Discontinuous Conduction

So far we have implicitly assumed that the peak inductor current ripple was smaller than the dc component of the inductor current. The total current therefore was always positive, and the diode was forced to be on when the transistor was off. However, if the dc component of the current is smaller than the peak ripple, the total current will fall to zero during the time when the diode is on, that is, during the period $1 - D$. The diode will then turn off, and the inductor current will remain zero until the transistor is turned on again. When this sequence occurs, we say that the converter is operating in the *discontinuous conduction* mode (DCM).

Figure 5.29 shows the current and voltage waveforms of the direct converter of Fig. 5.11 operating in DCM. In drawing these waveforms, we assumed an additional capacitor across the output terminal pair, so that v_2 is constant at V_2. Note that when the inductor current is zero, the voltage across the inductor is also zero, and the diode is reverse biased by the voltage $v_{xz} = V_2$. The average output voltage (which in this case is V_2) is still equal to $\langle v_{xz} \rangle$,

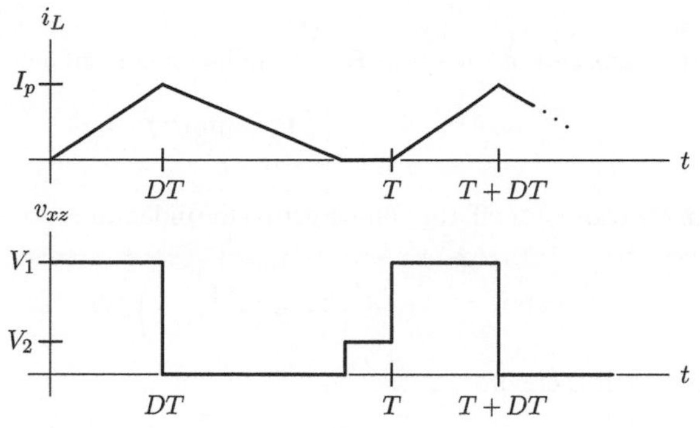

Figure 5.29 Waveforms of i_L and v_{xz} for the buck converter of Fig. 5.11 operating in the discontinuous conduction mode.

but its value in this mode of operation is higher than (5.4) predicts for the continuous conduction mode.

In practice, owing to switch and other capacitances, the voltage at the switch node x in Fig. 5.11 can ring with the inductor during the portion of the cycle when both switches are off. This can be a practical concern as regards electromagnetic interference (EMI), and is considered a disadvantage of DCM operation.

For the direct converter operating in continuous conduction, the voltage across the inductor is set at all times by the state of the switches. The average of this voltage must be zero, so the duty ratio specifies, by (5.4) and (5.5), the input/output conversion ratios. In discontinuous conduction, however, the inductor voltage averaged over a switching cycle is always zero, independent of D. The reason is that discontinuous conduction, by definition, ensures this condition. Therefore the duty ratio does not directly control the voltage conversion ratio. Instead, given V_1 and V_2, the duty ratio controls the average current in the inductor, and hence the dc components of i_1 and i_2, or I_1 and I_2. That is, in discontinuous conduction the current and voltage conversion ratios are functions of the duty ratio, but the functional dependencies are *not* those of (5.4). We must specify the external systems in order to determine these conversion ratios.

Example 5.6 Direct Converter in Discontinuous Conduction

The converter of Fig. 5.30 is operating in DCM and supplying a constant voltage V_2 to a resistive load in parallel with a large capacitor. What is the value of V_2?

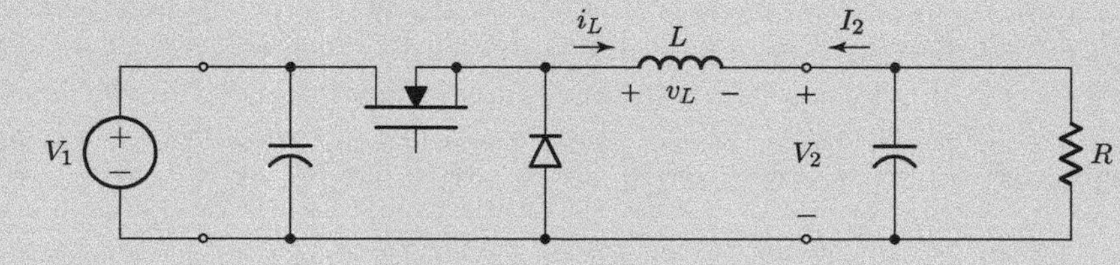

Figure 5.30 A buck converter analyzed under the condition of discontinuous conduction.

When we turn the transistor on for a time DT, the inductor current increases linearly from zero to a peak value of

$$i_{Lp} = \frac{(V_1 - V_2)DT}{L}. \tag{5.33}$$

When we then turn the transistor off, the voltage across the inductor is $-V_2$, and the time required to reduce its current to zero is

$$\Delta t = \frac{Li_{Lp}}{V_2} = \left(\frac{V_1}{V_2} - 1\right) DT. \tag{5.34}$$

The average value of i_L is therefore

$$\langle i_L \rangle = \frac{i_{Lp}(DT + \Delta t)}{2T} = -\langle i_2 \rangle = \frac{V_2}{R}. \tag{5.35}$$

When we combine (5.33), (5.34), and (5.35), we obtain

$$V_2^2 + V_2\left(\frac{V_1 RD^2 T}{2L}\right) - \left(\frac{V_1^2 RD^2 T}{2L}\right) = 0, \tag{5.36}$$

which we can solve for V_2.

Designers often avoid operation in DCM, at least under nominal conditions. One reason is that the peak rms current stresses on the semiconductor devices are high compared to their values in a converter operating at the same power level in the continuous conduction mode. If the load varies over a wide range, however, the converter may enter the discontinuous mode at the low-power end. When this happens, the response of the converter to changes in the duty ratio is altered. In this case, we must make sure that the control circuit continues to work properly.

Positive aspects of DCM operation include the fast transient response resulting from the small inductor value and high ripple, simplified converter dynamics due to the inductor current returning to zero each cycle, and the very wide power range that can be spanned with a small inductor.

Notes and Bibliography

The evolution of the canonical cell is described in [1]. Landsman's approach is insightful and clear, and the paper represents the first time the fundamental similarity among various dc/dc converter topologies is identified; subsequent papers have identified and applied other, more complex, canonical cells to realize dc/dc converters. An excellent coverage of topologies and all sorts of variations on them can be found in [2], which reflects Severns' extensive practical experience with dc/dc converters. Mitchell's book [3] does a more concise job of treating circuits and is more mathematical than [2]. It also presents a good discussion of control of dc/dc converters. The Ćuk converter was first presented in [4]. The flying-capacitor multilevel converter was first presented in [5]; for this reason, the FCML is sometimes called a "Meynard converter."

1. E. Landsman, "A Unifying Derivation of Switching dc–dc Converter Topologies," in *IEEE Power Electronics Specialists Conference (PESC)*, pp. 239–243, 1979.

2. R. P. Severns and G. Bloom, *Modern DC-to-DC Switchmode Power Converter Circuits*, New York: Van Nostrand, 1985.
3. D. M. Mitchell, *DC–DC Switching Regulator Analysis*, New York: McGraw-Hill, 1988.
4. R. D. Middlebrook and S. Ćuk, "A New Optimum Topology Switching dc-to-dc Converter," in *IEEE Power Electronics Specialists Conference (PESC)*, pp. 160–179, 1977.
5. T. A. Meynard and H. Foch, "Multi-Level Conversion: High-Voltage Choppers and Voltage-Source Inverters," *IEEE Power Electronics Specialists Conference (PESC)*, pp. 397–403, 1992.

PROBLEMS

5.1 Proceeding along the lines of Example 5.1, determine the amplitude of the ripple on the voltage v_o in Fig. 5.31. Assume that the input capacitor C is large enough that the input voltage is constant at value V_i and that the Norton equivalent resistance R_N of the external load network is small.

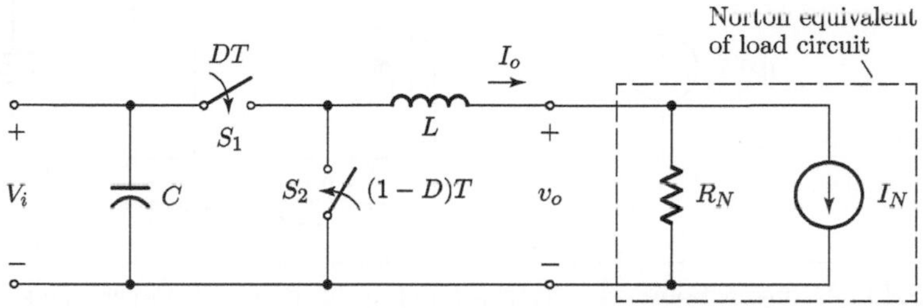

Figure 5.31 The dc/dc converter whose output voltage ripple is calculated in Problem 5.1.

5.2 Figure 5.32 shows a buck converter supplying 5 V to a load of 0.1 Ω from a 25 V source having an internal resistance of 0.2 Ω. Determine the duty ratio D at which the converter is operating.

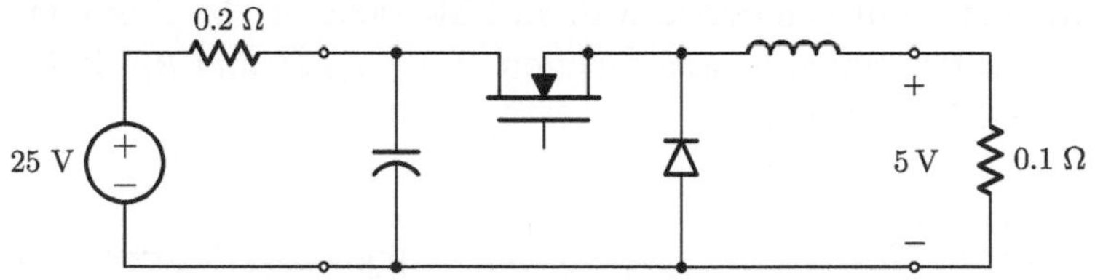

Figure 5.32 The buck converter whose operating duty ratio is calculated in Problem 5.2.

5.3 With reference to Fig. 5.21 it was stated that $\langle v_X \rangle = 0$. Show that this is true.

5.4 Derive a switch implementation for the direct boost converter so that the input and output circuits share a common positive terminal.

5.5 Reconsider Example 5.3 and develop a bilateral direct converter in which the terminal currents reverse direction but the terminal voltage polarities remain unchanged.

5.6 Determine a switch implementation for a bilateral indirect converter in which the terminal currents change sign.

5.7 Derive (5.8) from (5.7) by setting the net power into the converter to zero.

5.8 A *four-quadrant* converter can operate with its terminal variables I_t and V_t in any of the four quadrants of the I_t–V_t plane. Note that this implies bilateral power flow, and that it makes no difference which terminal pair you consider in a two-pair network. Derive the switch implementation for a four-quadrant indirect converter.

5.9 The direct boost converter shown in Fig. 5.33 serves as a battery charger. The control circuit provides constant-current charging at a switching frequency of 400 kHz. The current i_L is continuous. Determine the minimum value of L that will give a peak-to-peak ripple of less than 100 mA in i_L. If $I = 20$ A, what is the average value of i_L?

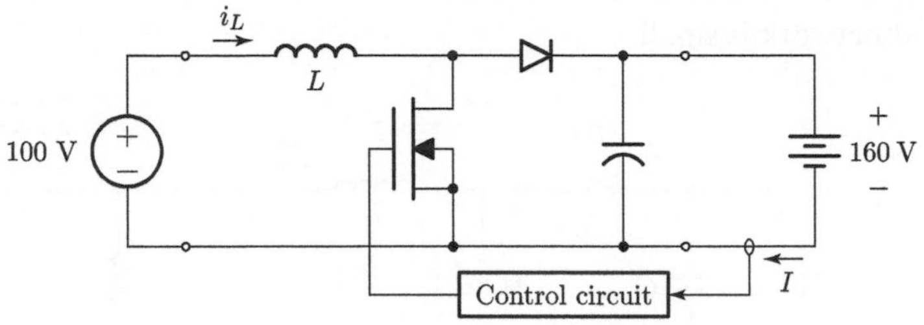

Figure 5.33 A direct boost converter serving as a battery charger. This circuit is the subject of Problem 5.9.

5.10 The *equivalent series resistance* (ESR) of capacitors in dc/dc converters is often an important parameter. The ESR not only contributes to converter loss, but it also sometimes affects the converter's control characteristics.

An indirect converter with a second-order output filter is shown in Fig. 5.34. The output filter capacitor is modeled with an ESR value of R_C. Assuming that all Ls and Cs are infinite, determine V_2 as a function of V_1, I_1, D, and R_C. If $I_1 = 10$ A, $D = 0.25$, and $R_L = 0.5\,\Omega$, what is V_2?

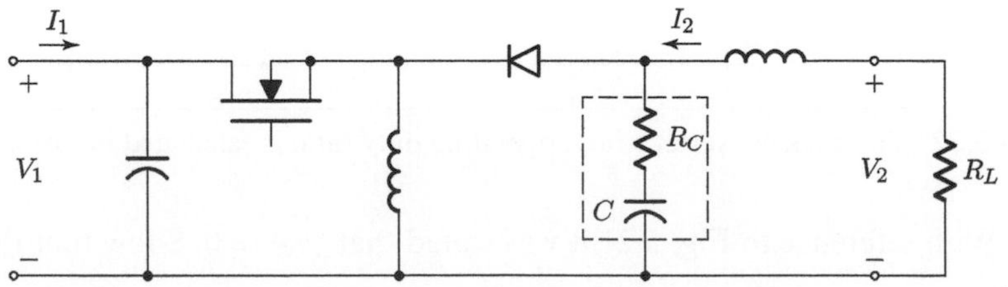

Figure 5.34 The indirect converter of Problem 5.10. The output capacitor C has an ESR value of R_C.

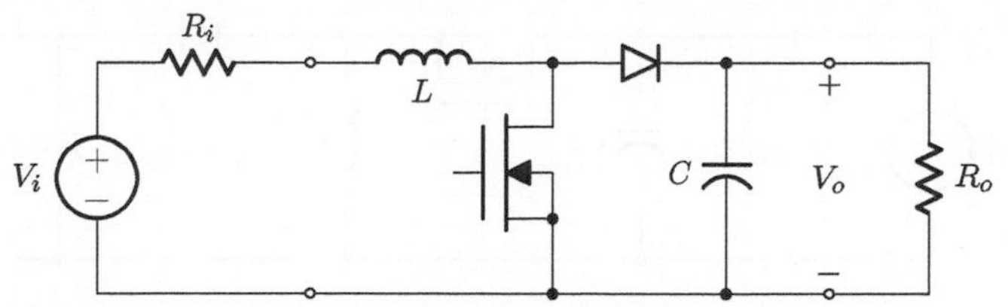

Figure 5.35 The direct boost converter analyzed in Problem 5.11.

5.11 A direct boost converter connects two external systems, as shown in Fig. 5.35. Assume that L and C are large enough that ripple is not a concern. As a function of R_o/R_i and the duty ratio D, find expressions for:

(a) V_o/V_i;

(b) the efficiency of the system, $\eta = P_o/P_i$, where P_i is the power from the source V_i and P_o is the power delivered to R_o;

(c) the duty ratio at which the output voltage is maximized.

5.12 Compare the energy storage requirements of a direct converter operating at $D = 0.5$ to those of an indirect converter if their ripple ratios are the same. Check your answer with the statement at the end of Section 5.6.4.

5.13 Show that the total stored-energy requirement of the Ćuk variant of the indirect converter of Fig. 5.20(c) is identical to that of the basic indirect converter of Fig. 5.17 if their terminal ripple current ratios are the same.

5.14 A 100 V regulated dc supply must be designed using an unregulated 50–200 V source. Two possible ways of designing this supply are shown in Fig. 5.36. Figure 5.36(a) is the buck/boost variant of the indirect converter. Figure 5.36(b) is a cascade connection of a direct buck and a direct boost converter. This converter is often referred to as a *four-switch buck/boost converter* when MOSFETs are used in place of the diodes. Compare these alternative designs in the following manner.

(a) Ignoring the ripple in all capacitor voltages and inductor currents, express the transistor switch-stress parameter in terms of the output power P_o and the voltage conversion ratio V_o/V_i for the buck/boost converter of Fig. 5.36(a).

(b) Repeat (a) for the circuit of Fig. 5.36(b), but express the transistor stresses in terms of P_o, V_i/V_m, and V_o/V_m.

(c) Find the value of V_m that minimizes the sum of the switch-stress parameters for the circuit of Fig. 5.36(b). Interpret this result in terms of the duty ratios at which the two parts of the cascade operate as V_i varies from 50 V to 200 V.

(d) Compare the switch-stress parameters for the circuit of Fig. 5.36(a) with the optimized sum of the parameters found in (c). Which is lower?

(e) Discuss other issues that would affect your choice of topology for this application.

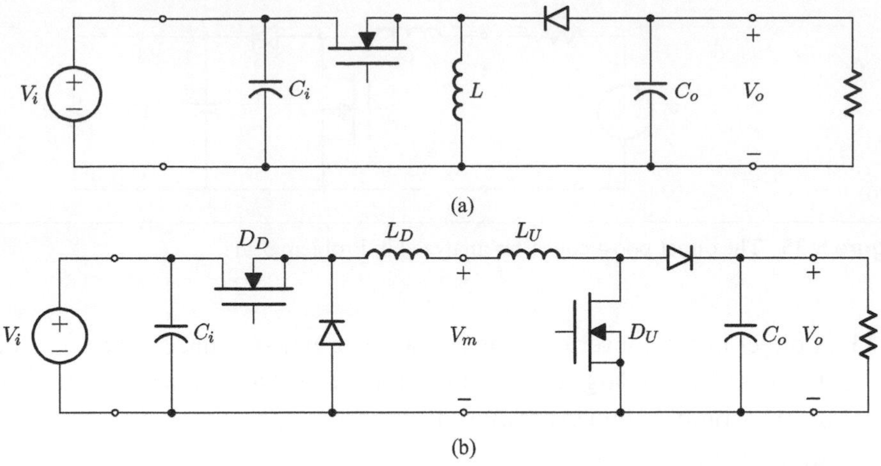

Figure 5.36 Power supply alternatives compared in Problem 5.14. (a) A single buck/boost converter. (b) A cascade of two direct converters, a buck on the left and a boost on the right.

5.15 The converter of Fig. 5.30 has the following parameters:

$$V_1 = 25\,\text{V}, \qquad L = 1\,\mu\text{H}, \qquad R = 2\,\Omega, \qquad f = 300\,\text{kHz}, \qquad D = 0.2.$$

(a) Determine and sketch the inductor current i_L.

(b) What is V_2?

(c) At what value of D does the transition from discontinuous to continuous conduction mode take place?

5.16 Determine the energy storage requirement for the SEPIC converter and compare it to that for the indirect converter.

5.17 The topological dual of the SEPIC converter is known as the *Zeta* converter and is shown in Fig 5.37. Assume that it operates in continuous conduction mode, and that the capacitors and inductors have small ripple.

(a) Find the steady-state voltage V_{C_1} on capacitor C_1.

(b) What is the steady-state voltage conversion ratio V_2/V_1 of this topology as a function of the switch duty ratio D?

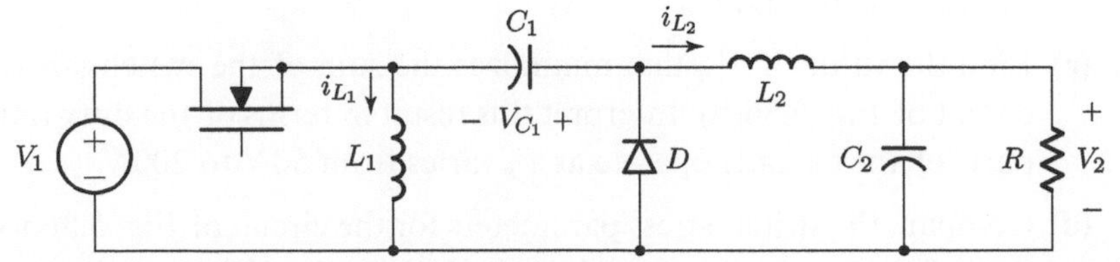

Figure 5.37 The Zeta converter, the topological dual of the SEPIC converter, analyzed in Problem 5.17.

5.18 Figure 5.38 shows a unique configuration of the canonical cell, sometimes referred to as the *Landsman converter*.‡ Assume that the converter operates in continuous conduction mode with small ripple in the inductor currents and capacitor voltages.

(a) What polarity (positive, negative, or both) of the output voltage V_o is possible in this converter in periodic steady-state operation? Please justify your answer.

(b) What polarity (positive, negative, or both) of the input voltage V_g is possible in this converter in periodic steady-state operation? Justify your answer.

(c) Find the average current $\langle i_m \rangle$ in inductor L_m as a function of the average currents i_p and/or i_s for periodic steady-state operation.

(d) What is the periodic-steady-state current conversion ratio i_s/i_p of this topology as a function of the switch duty ratio D? (Note the designated current directions.)

(e) What is the periodic steady-stage voltage conversion ratio V_o/V_g of this topology as a function of the switch duty ratio D?

(f) Find an expression for the peak off-state voltage across the transistor as a function of the input voltage V_g and the duty ratio.

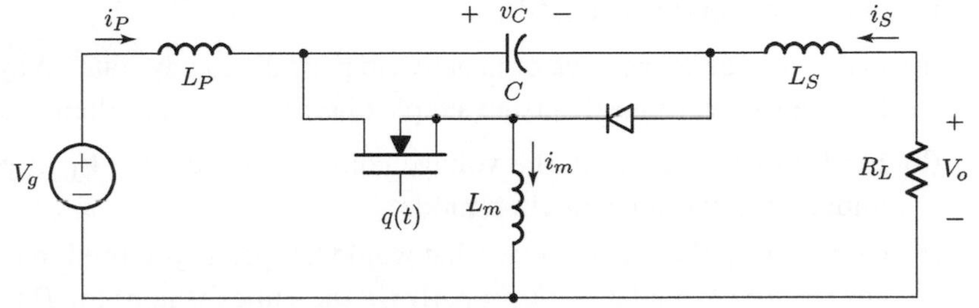

Figure 5.38 The Landsman converter analyzed in Problem 5.18

5.19 Figure 5.39 shows a boost (up) converter supplying 12 V to a load of 5 Ω from a 5 V source having an internal resistance of 0.2 Ω. Determine the duty ratio D at which the converter operates. (Neglect semiconductor device drops in your calculations.)

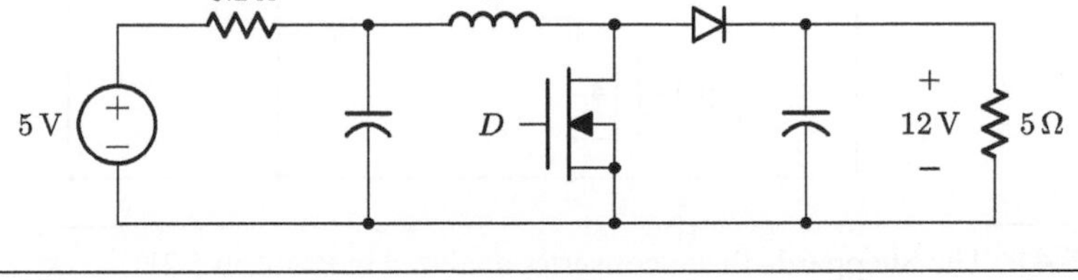

Figure 5.39 The boost converter of Problem 5.19 operating from a source with output resistance.

‡ Named after Dr. Emanuel Landsman, this figure is an annotated version of the one appearing in Dr. Landsman's original 1979 paper referred to earlier. The paper is about unifying aspects of dc/dc converter topologies and not about this topology per se.

5.20 Derive the current conversion ratio I_2/I_1 for the converter in Fig. 5.40. Which direction(s) can power flow in this converter? This circuit was discussed in Example 5.3.

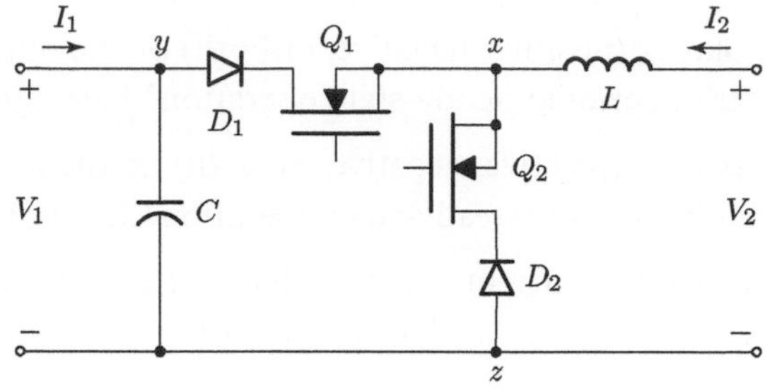

Figure 5.40 The dc/dc converter of Problem 5.20, which is configured with switches capable of blocking bidirectional voltages but carrying only unidirectional current.

5.21 Figure 5.41 shows a version of the Sheppard–Taylor converter topology. In analyzing this circuit assume that V_1 is positive, $D < 0.5$, that the inductors and capacitors are all large enough to produce small ripples in the current and voltage, and that the converter operates in continuous conduction mode.

(a) Is the voltage V_2 positive or negative in periodic steady state? Why? (Note: you should be able to determine this from simple observation rather than detailed analysis.)

(b) Find an expression for the voltage conversion ratio V_2/V_1 in periodic steady state under continuous conduction mode.

(c) Considering the result above, what would happen if you tried to operate this converter at $D = 0.5$? Would it work properly (in the same fashion) for $D > 0.5$?

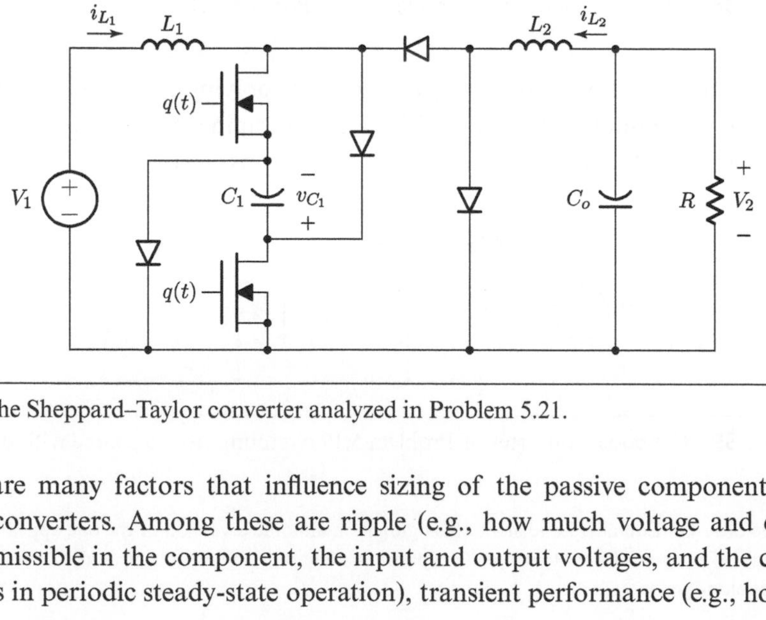

Figure 5.41 The Sheppard–Taylor converter analyzed in Problem 5.21.

5.22 There are many factors that influence sizing of the passive components in practical power converters. Among these are ripple (e.g., how much voltage and current ripple are permissible in the component, the input and output voltages, and the corresponding currents in periodic steady-state operation), transient performance (e.g., how much peak

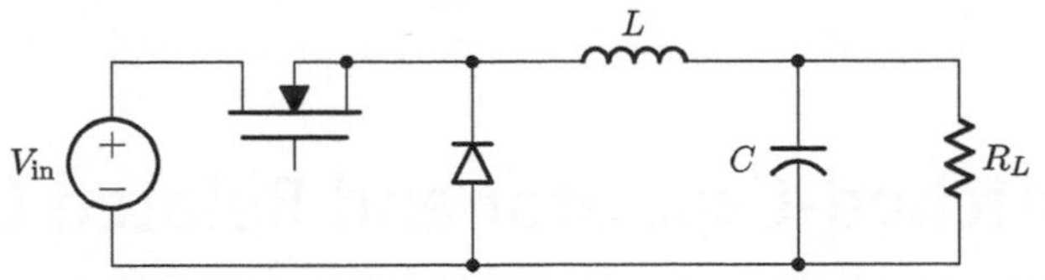

Figure 5.42 A buck converter supplying a resistive load for which you are to determine values of L and C in Problem 5.22.

deviation away from steady state will occur during a transient condition, such as when the load resistance changes), and the desire to limit the size and cost of the passive components (inductors and capacitors).

This problem concerns the selection of passive component values for the buck converter of Fig. 5.42. This converter operates with a switching frequency of $f_s = 250$ kHz from an input voltage of $V_{in} = 48$ V at a constant duty ratio of $D = 0.5$. The load resistance R_L can vary over the range $0.5\,\Omega < R_L < 1\,\Omega$.

(a) A decision has been made to design the system such that the capacitor receives less than 4 A (rms) of ripple current in periodic steady-state operation. It is also desirable to keep the inductor value reasonably small to save cost and space. Select an appropriate inductor value and calculate the rms capacitor current in periodic steady-state operation.

(b) A decision has been made that the output ripple voltage must be less than 1.2 V peak-to-peak in periodic steady-state operation, but that the capacitor should be kept reasonably small to save cost and space. Select an appropriate capacitor value to meet this requirement. Calculate an approximation for the expected peak-to-peak ripple voltage on the capacitor.

(c) Simulate your design and verify that the requirements in parts (a) and (b) are met.

(d) A transient specification is now added. The output voltage should remain between 16 and 32 V during a transient when the load steps between 0.5 and 1 Ω (either direction). Simulate this transient. Does your design comply with this new requirement? If not, propose a second set of L and C values that meet all of the above requirements.

6 Switched-Capacitor and Related Converters

Converter circuits in the class known as *switched-capacitor converters* (SCCs) comprise only switches and capacitors. They operate on the principle of charge transfer, wherein capacitors are charged in one switching state, then reconfigured in a second state to deliver charge to the output. Depending on the circuit topology and how the switches are configured and controlled, the converter function can be either step-up or step-down.

6.1 Switched-Capacitor DC/DC Converter

There are several unique characteristics of switched capacitor converters:

- The current conversion ratio between output and input is always rational, and determined by the configuration of switches and capacitors and the switching sequence.
- The voltage conversion ratio is generally not rational, and depends upon losses in the network. The ideal no-load voltage conversion ratio is the inverse of the current conversion ratio, so is rational; this no-load voltage conversion ratio is often referred to as the conversion ratio of the circuit.
- Because energy storage is solely electrostatic, that is, the circuit is free of inductors, these converters can achieve very high power densities and lend themselves to integrated circuit fabrication technology.
- Very high conversion ratios can be achieved, but at the expense of a large number of switches and capacitors.

Unlike switched-mode converters employing inductors to transfer energy from input to output, where the inductive stored energy can be brought to zero (discontinuous conduction mode) without efficiency concerns, the capacitive stored energies in a switched-capacitor converter must usually have small ripple. That is, the capacitor voltages cannot sustain wide swings without significantly compromising efficiency if the converter is connected to stiff voltage sources and loads. This can be seen by considering the circuit of Fig. 6.1, where a capacitor with an initial voltage V_O is charged through a small resistance ϵ by a switched voltage source V_S. The energy delivered by the source is $\Delta E_s = V_S \Delta q$, where Δq is the charge delivered to the capacitor as it charges from its initial voltage V_O to its final value V_S. Now,

$$\Delta q = C(V_s - V_O),$$

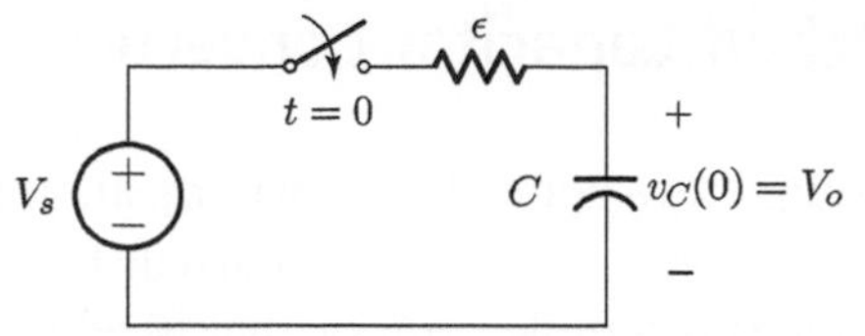

Figure 6.1 A capacitor being charged by a switched voltage source.

so

$$\Delta E_s = C(V_s^2 - V_s V_o).$$

The energy, ΔE_c, delivered to the capacitor is

$$\Delta E_c = \frac{1}{2} C(V_s^2 - V_o^2). \tag{6.1}$$

The charging efficiency as a function of V_o is therefore

$$\eta = \frac{\Delta E_C}{\Delta E_S} = \frac{1}{2} \frac{(V_S^2 - V_O^2)}{(V_S^2 - V_S V_O)} = \frac{1}{2}\left(1 + \frac{V_O}{V_S}\right) \tag{6.2}$$

and does not depend on the value of ϵ. Defining $\Delta V = V_s - V_o$,

$$\eta = 1 - \frac{\Delta V}{2V_S}. \tag{6.3}$$

Note that η ranges from 50%, if V_o is zero, to near 100% as ΔV approaches zero.

The capacitors in an SCC alternately charge and discharge over a switching cycle to transfer energy from the source to the load, while providing voltage and current transformation. In a two-state converter, this is accomplished by using the switches to create two alternating topological states of the network. The efficiency of conversion depends upon the voltage ripple on the capacitors over a switching cycle. The capacitor voltage ripple depends upon the capacitor values, switching frequency, and load, with larger capacitors, higher switching frequencies, and smaller load currents tending to reduce capacitor charging loss.

The steady-state current conversion ratio in an SCC is always a rational number, owing to the requirement to maintain charge balance on the capacitors over a cycle. The current conversion ratio is determined by the circuit topology and switching pattern, and for a given topology may only be selected from among one or more discrete values by choosing the switching pattern. Capacitor charging losses affect the voltage conversion ratio. As capacitor charging loss is a function of switching frequency, the voltage conversion ratio can be somewhat controlled by adjusting the frequency, but at the expense of efficiency. A typical application of a switched-capacitor converter is where a fixed conversion ratio (close to the ideal no-loss conversion ratio) is desired. Where regulation is needed, a common approach is to modify the switching pattern in order to adjust the output in (approximately) discrete steps. A subsequent stage, for example a linear or duty-ratio-controllable switching converter, can then be used for fine control.

There are numerous types of switched-capacitor converters, each having their own characteristics. For a given type of SCC, achieving a higher conversion ratio requires a larger number of capacitors and switches.

6.2 Two-State Switched-Capacitor Converter

Figure 6.2 shows a simple two-state SCC providing an approximate 2 : 1 step down in voltage from V_i to V_o (and an exact 1 : 2 step up in current from I_i to I_o). The switching of the odd- and even-labeled switches is complementary, that is, when switches 1 and 3 are closed (state 1), switches 2 and 4 are open (state 2), and vice versa. The output capacitor C_o is assumed large enough that V_o is constant over a switching period. The capacitor C functions as the energy transfer element between input and output.

6.2.1 The Slow-Switching Limit

Let us first consider operation where the switching frequency is low enough that v_C reaches its steady-state value by the end of each switch state. This operating regime is known as the *slow-switching limit* (SSL). In in SSL, the voltage v_C takes on steady-state values V_{C1} and V_{C2} at the end of states 1 and 2, respectively:

$$V_{C1} = V_i - V_o, \qquad V_{C2} = V_o. \tag{6.4}$$

During state 1, v_C charges by an amount Δv_C, and discharges by this same amount during state 2. From (6.4), $\Delta v_C = V_{C1} - V_{C2} = V_i - 2V_o$, from which we can determine the charge transferred in each portion of the switching cycle:

$$\begin{aligned} \Delta q_{C1} &= C\Delta v_C = C(V_i - 2V_o), & \Delta q_{C2} &= -C(V_i - 2V_o); \\ \Delta q_{i1} &= C\Delta v_C = C(V_i - 2V_o), & \Delta q_{i2} &= 0; \\ \Delta q_{o1} &= C(V_i - 2V_o), & \Delta q_{o2} &= C(V_i - 2V_o); \end{aligned}$$

where Δq_C is the charge into the + terminal of C, Δq_i is the charge out of the source, and Δq_o is the charge into the load. The total charge transfers at the input and output are the sums of the transfers in each state. When multiplied by the switching frequency f_s they yield the input and output currents:

$$I_i = f_s \Delta q_i = f_s C(V_i - 2V_o), \tag{6.5}$$

$$I_o = f_s \Delta q_{o1} + \Delta q_{o2} = 2f_s C(V_i - 2V_o) = 2I_i. \tag{6.6}$$

The current ratio I_o/I_i is precisely two. Note that this result is independent of any parasitic series resistances of the switches or capacitor that we might have included in the circuit. This result

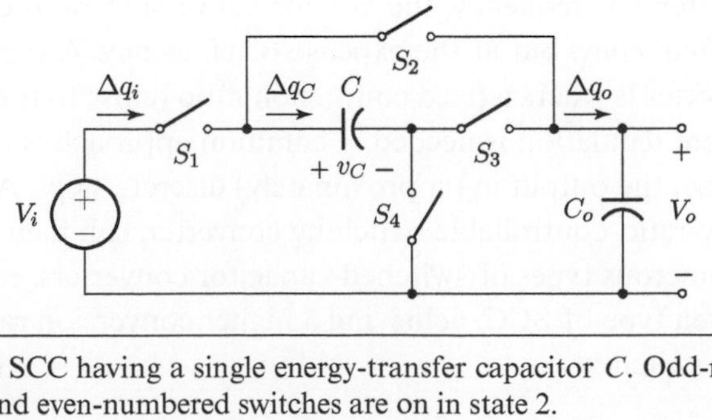

Figure 6.2 A two-state SCC having a single energy-transfer capacitor C. Odd-numbered switches are on in state 1, and even-numbered switches are on in state 2.

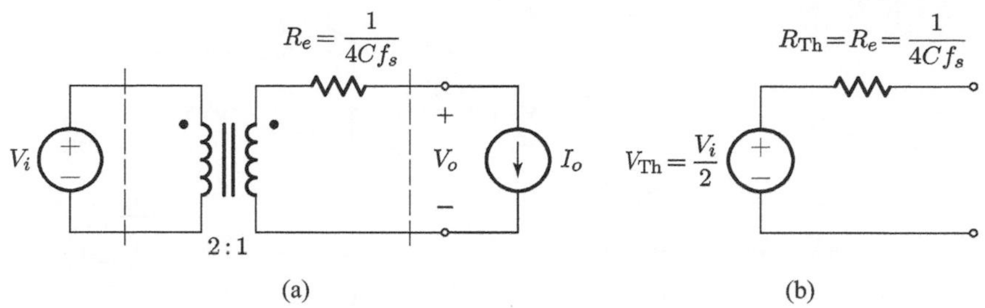

Figure 6.3 (a) A two-port model of the SCC of Fig. 6.2 operating in the SSL regime. (b) A Thevénin equivalent circuit for the SCC as seen from its output port, with Thevénin voltage $V_{Th} = V_i/2$ and Thevénin resistance $R_{Th} = 1/(4Cf_s)$.

can also be intuited by considering that the net charge variation in C over one cycle must be zero for periodic steady-state operation. During state 1, Δq_C is supplied by the source and also delivered to the output. During state 2, the source is disconnected and the current in C reverses, again supplying Δq_C to the output. Thus, during a cycle of time T, the output receives twice the charge supplied by the source, that is, $I_o = 2I_i$.

The output voltage V_o can now be obtained from (6.6):

$$V_o = \frac{V_i}{2} - \left(\frac{1}{4Cf_s}\right) I_o. \tag{6.7}$$

This result, which is valid for periodic steady-state operation of the SCC, can be represented by the two-port equivalent circuit shown in Fig. 6.3(a). The transformer captures the fixed current conversion ratio and the ideal (no-load) voltage conversion ratio, while the resistor R_e represents the effects of capacitor charging loss on voltage conversion ratio and efficiency. Specifically, R_e captures the losses inherent in charging and discharging the energy transfer capacitor from stiff dc voltages. Its value in SSL is not dependent on actual circuit resistances, which we have not specified here, but does depend on the capacitance and switching frequency. Seen from its output port, the SCC can be represented by a Thévenin equivalent circuit where $V_{Th} = V_i/2$ and $R_{Th} = R_e = 1/(4Cf_s)$, as shown in Fig. 6.3(b).

Note that the analysis above is particular to the slow-switching limit of a switched capacitor converter, that is, for frequencies and duty ratios where the voltages on the energy transfer capacitors reach steady state in each switching state. What this implies is that the duration of each state is long compared to the time constants of that state. The time constants for a state arise from the capacitor(s) being charged through the circuit's parasitic resistances (e.g., switch on-state resistances and capacitor ESR). Operation in the SSL regime allows us to ignore the specific values of such resistances. Likewise, when in the SSL, the voltage conversion ratio does not depend on the fraction of time spent in each switch state.

6.2.2 The Fast-Switching Limit

We now consider operation in which the switching frequency is high enough that the duration of the switching states are much shorter than the circuit time constants of those states. This is known as the *fast-switching limit* (FSL). In the FSL the capacitor does not have time to

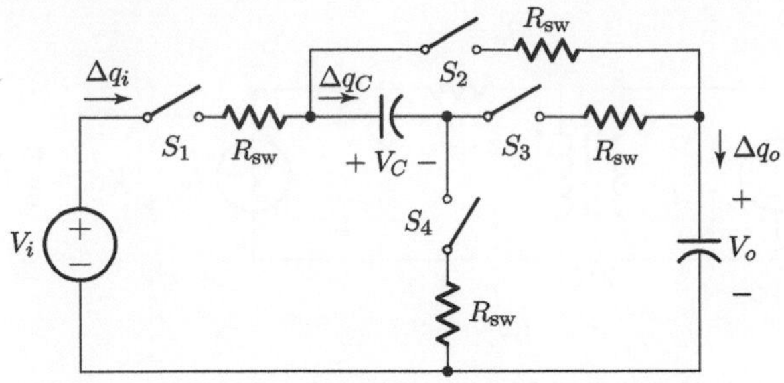

Figure 6.4 The SCC of Fig. 6.2 with switch resistances, R_{SW}, added to allow analysis of the fast-switching limit.

appreciably charge or discharge, so v_C can be treated as approximately constant ($v_C \approx V_C$). To analyze operation in the FSL, we must explicitly include parasitic resistance in the circuit, since under this assumption I_i and I_o are undefined without resistance in the circuit. The SCC circuit of Fig. 6.2 is shown for the FSL case in Fig. 6.4.

We assume a switch duty ratio of D = 0.5 and calculate the various charge transfers:

$$\Delta q_{C1} = \frac{V_i - V_o - V_C}{2R_{SW}}(0.5T), \qquad \Delta q_{C2} = \frac{V_o - V_C}{2R_{SW}}(0.5T); \tag{6.8}$$

$$\Delta q_{i1} = \Delta q_{C1}, \qquad \Delta q_{i2} = 0; \tag{6.9}$$

$$\Delta q_{o1} = \Delta q_{C1}, \qquad \Delta q_{o2} = -\Delta q_{C2}. \tag{6.10}$$

Using the fact that $\Delta q_C = \Delta q_{C1} + \Delta q_{C2} = 0$ over a switching period, we can calculate V_C, I_i, I_o, and V_o:

$$V_C = \frac{V_i}{2}, \tag{6.11}$$

$$I_i = (\Delta q_{i1} + \Delta q_{i2})f_s = \frac{1}{2}\left(\frac{1}{4R_{SW}}\right)(V_i - 2V_o), \tag{6.12}$$

$$I_o = (\Delta q_{o1} + \Delta q_{o2})f_s = \left(\frac{1}{4R_{SW}}\right)(V_i - 2V_o) = 2I_i, \tag{6.13}$$

$$V_o = \frac{V_i}{2} - 2R_{SW}I_o. \tag{6.14}$$

Again we see that the current conversion ratio is rational but the voltage conversion ratio is a function of the output current. This result leads to the equivalent circuit of Fig. 6.5. This is similar to the model for SSL in Fig. 6.3(a), but now R_e is a function of the circuit parasitic resistances.

In the FSL the SCC exhibits a constant output resistance, $R_e = 2R_{SW}$, while in the SSL regime the resistance is inversely proportional to the switching frequency $R_e = 1/(4Cf_s)$. More generally, R_e has a switching-frequency-dependent value that transitions between the asymptotes defined by the SSL and FSL, as illustrated in Fig. 6.6. We might characterize this transition by a break frequency at which the SSL and FSL output resistance asymptotes are equal. Operation at frequencies far above this break frequency is of little benefit as it incurs increased

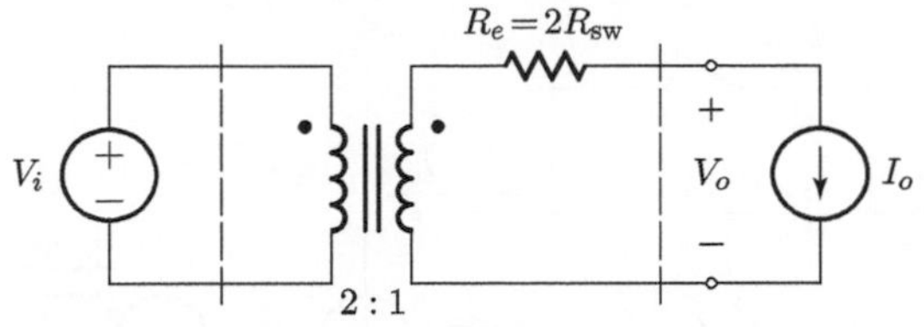

Figure 6.5 The equivalent circuit of the two-state SCC of Fig. 6.2 operating in the FSL regime.

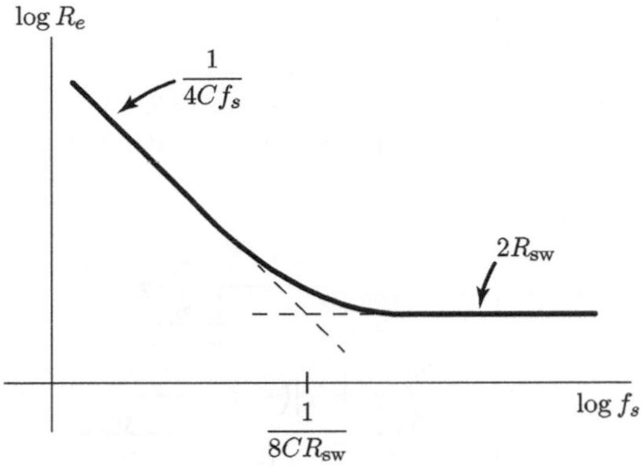

Figure 6.6 Equivalent output resistance R_e of the SCC of Fig. 6.2 as a function of switching frequency. The duty ratio is 0.5 in the FSL regime.

frequency-dependent parasitic losses without significantly improving output resistance or capacitor charging loss.

While R_e for operation in the SSL is independent of the fraction of time spent in each switch state, this is not true for the FSL. For the circuit of Fig. 6.2 in the FSL regime and having a duty ratio D in state 1, the output voltage and equivalent output resistance can be shown to be

$$V_o = \frac{V_i}{2} - \frac{R_{SW}}{2(D - D^2)} I_o, \tag{6.15}$$

$$R_e = \frac{R_{SW}}{2(D - D^2)}. \tag{6.16}$$

6.3 Switch Implementation

The circuit of Fig. 6.2 is redrawn in Fig. 6.7(a), where the switch currents and voltages have been identified for down-conversion from a positive input voltage. In state 1, $i_1 \geq 0$, $i_3 \geq 0$, $v_2 > 0$, and $v_4 < 0$. In state 2, $v_1 > 0$, $v_3 < 0$, $i_2 \geq 0$, $i_4 \geq 0$. This is true for both the SSL and FSL. When the blocking voltage and current are of the same polarity, a transistor is an appropriate switch, and when they are of opposite polarities, a diode can suffice. These facts lead to one possible switch implementation where S_1 and S_2 are transistors and S_3 and S_4 are diodes, as shown in Fig. 6.7(b).

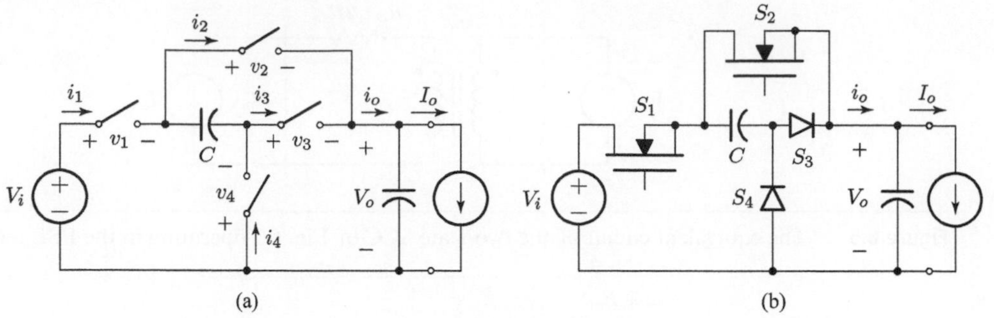

Figure 6.7 Switch implementation in the two-state SCC of Fig. 6.2. (a) Definitions of switch variables. (b) An example switch implementation for step-down conversion from a positive-valued input voltage V_i.

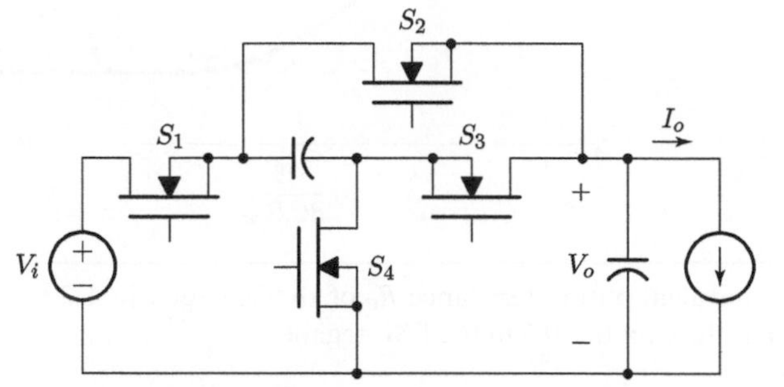

Figure 6.8 The SCC of Fig. 6.7(b) with diodes replaced by MOSFETs. This circuit is capable of bidirectional power flow between its two ports.

A disadvantage of the simple switch implementation shown in Fig. 6.7(b) is that the diode drops can have a considerable practical impact on voltage conversion ratio and loss, especially in low-voltage systems. This often motivates the use of active switches in place of diodes. As mentioned in Chapters 1 and 5 and discussed in more detail in Chapter 15, because of its body diode, a power MOSFET can conduct in both directions. Moreover, when on, it can conduct in reverse through its channel, bypassing the body diode, which can result in a lower on-state voltage. Lateral MOSFETs in integrated circuits can likewise conduct bidirectionally. This bidirectional current capability makes it possible for the diodes in the circuit of Fig. 6.7(b) to be replaced by MOSFETs acting as "synchronous rectifiers," as shown in Fig. 6.8. This is particularly advantageous for low-voltage converters, where the additional cost and complexity of the substitution is compensated for by the lower on-state loss and voltage drop of the MOSFET compared with the diode.

The circuit of Fig. 6.7(b) is capable of bidirectional power flow, stepping down from V_i to V_o (power flowing left to right, with $I_o > 0$) or stepping up from V_o to V_i (power flowing right to left, with $I_o < 0$). The two-port steady-state models of Figs. 6.3 and 6.5 are valid in SSL and FSL, respectively, regardless of power flow direction (i.e., regardless of the sign of I_o). Analysis of the bidirectional power flow capability of this circuit is the subject of Problem 6.2(d).

6.4 Other Switched-Capacitor DC/DC Converters

There are numerous types of switched-capacitor dc/dc converters. Examples of several popular SCC topologies are provided in Fig. 6.9 in their voltage step-up forms. The step-down forms of these topologies may be found by exchanging V_i and V_o. Different conversion ratios may be realized by expanding or contracting the illustrated structures (i.e., using different numbers of switches and energy-transfer capacitors).

Each of the topologies of Fig. 6.9 has its own advantages and limitations. For example, the Fibonacci topology provides the largest conversion ratio for a given number of capacitors, but requires diverse voltage ratings for its switches and capacitors, which is often impractical. The ladder topology requires the same voltage ratings of all of its switches and capacitors, but requires more capacitors than a Fibonacci converter to achieve a high conversion ratio. Some

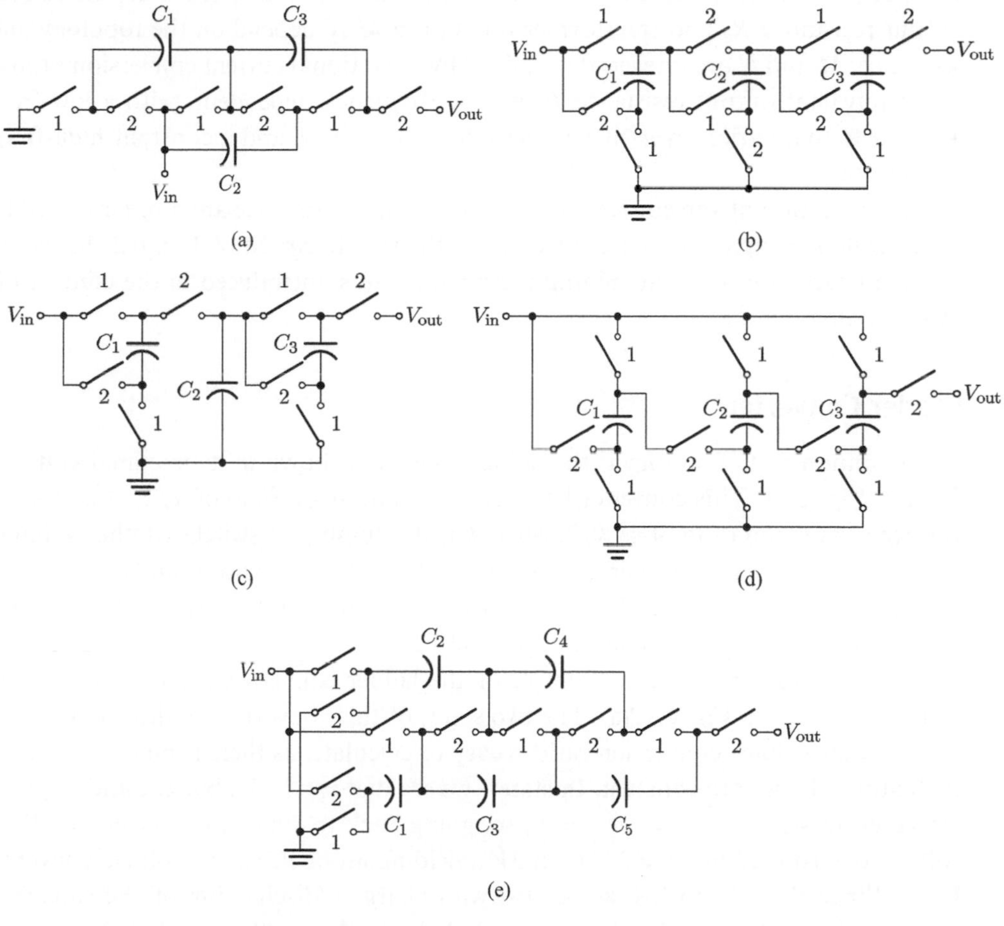

Figure 6.9 Examples of five common switched-capacitor dc/dc converter topologies (step-up form): (a) ladder; (b) Fibonacci; (c) doubler; (d) series-parallel; (e) Dickson.

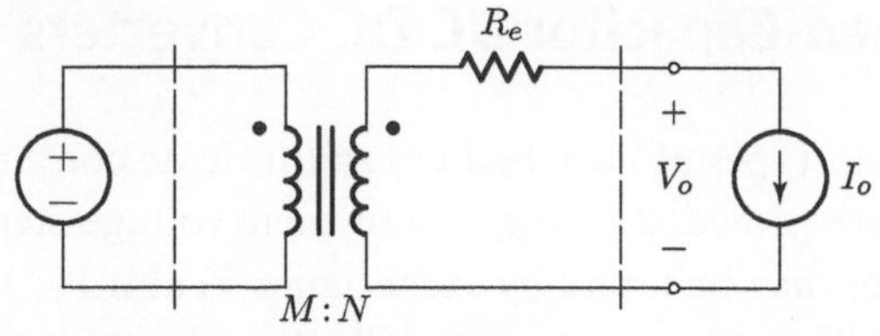

Figure 6.10 A general two-port steady-state model for a SCC.

topologies make good use of switches or of capacitors, or perform well in SSL or FSL, but none achieve all of these goals.†

All of the switched-capacitor converters of Fig. 6.9 share the general characteristics of the simple converter of Fig. 6.2. For example, they exhibit SSL and FSL operating regimes and can be represented with models having the form of Figs. 6.3 and 6.5, albeit with different transformer turns ratios and different expressions for resistor R_e. Figure 6.10 shows a general two-port model that represents the steady-state behavior of a switched-capacitor converter. The output resistance R_e and transformer turns ratio M:N depend on the topology and switching sequence. M and N are integers determined by the rational current conversion ratio of the SCC. The equivalent output resistance R_e will be frequency dependent, with a low-frequency SSL asymptote that varies inversely with switching frequency, and a constant high-frequency FSL asymptote.

Determination of the conversion ratio and output resistance are more involved for a higher-order switched capacitor converter than for the simple circuit of Fig. 6.2. In the following we provide procedures for determining these parameters, introduced in the context of a popular SCC topology.

6.4.1 Ladder Converter

One common SCC topology is the ladder converter, shown in its maximal voltage step-down form in Fig. 6.11. This converter has a current conversion ratio of $k/2 : 1$ and an open-circuit voltage conversion ratio of $1 : k/2$, where k is the number of switches. Other voltage step-down conversion ratios can be obtained by replacing C_o with a capacitor C and taking the output from a different point in the right-hand capacitor stack. The step-up form of the topology is formed by switching V_i with C_o and its voltage label V_o.

We now consider a specific example of the ladder converter, the $1 : 3$ step-up form of the converter shown in Fig. 6.12(a). The two states of this converter are shown in Fig. 6.12(b). The open-circuit voltage conversion ratio is easy to calculate, as there is no charge transfer and the capacitor voltages are constant. In state 1 we see that $v_{C_1} = V_i$, but v_{C_2} and v_{C_3} are undefined. However, in state 2, $v_{C_2} = v_{C_1} = V_i$ so, going back to state 1, $v_{C_3} = v_{C_2} = V_i$. The output voltage determined in state 2 is then $3V_i$, yielding an open-circuit voltage conversion ratio of $1 : 3$. Since there is no loss associated with charging/discharging of the capacitors (as their voltages are constant) and we have not included other losses, the current conversion ratio is $3 : 1$.

† A good comparison of these aspects for the topologies in Fig. 6.9 is provided in [1].

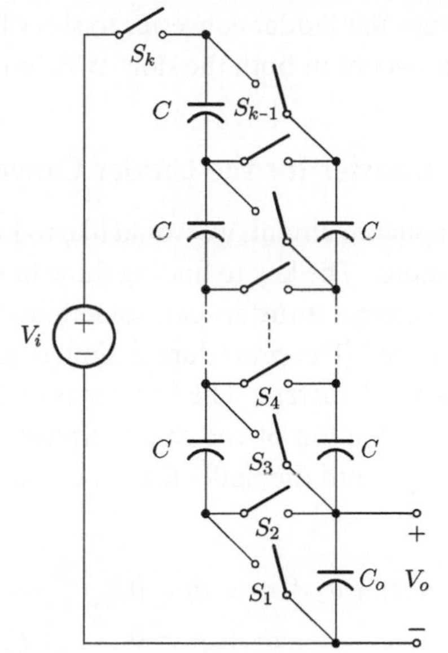

Figure 6.11 A ladder converter of k switches and an open-circuit voltage conversion ratio of $k/2:1$

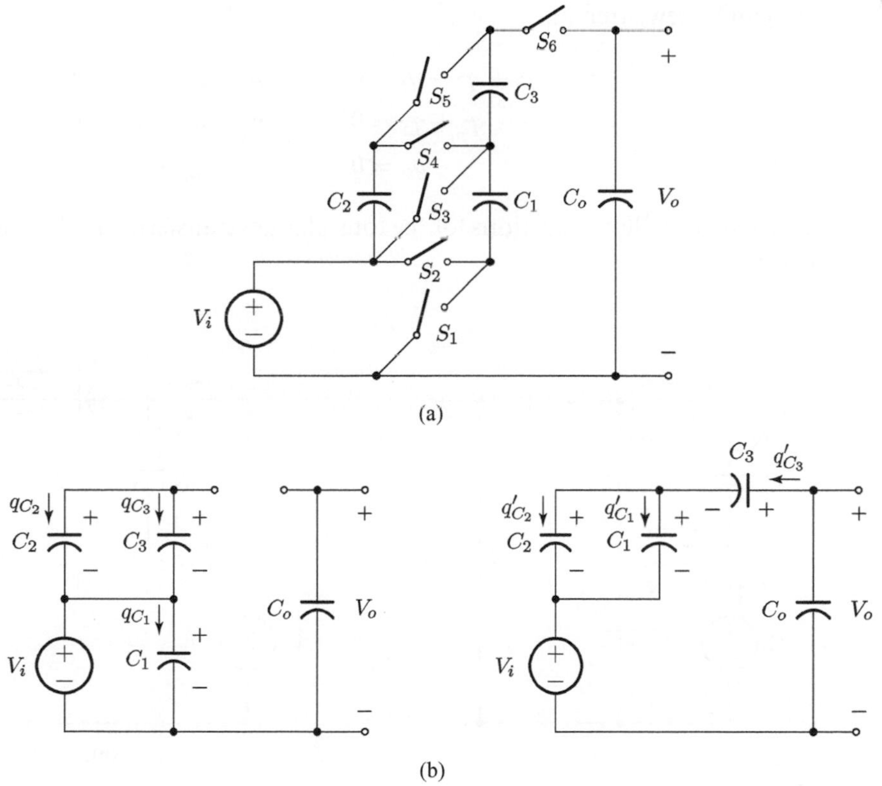

Figure 6.12 (a) A 1 : 3 ladder converter. (b) The state 1 and state 2 circuits.

In the following, we use this ladder converter to show how one can determine the behavior of a switched-capacitor converter in both the slow-switching and fast-switching limits.

6.4.2 Calculating Charge Transfer for the Ladder Converter

To model a switched-capacitor circuit, we would like to know its current conversion ratio and its equivalent output resistance. The key to finding these in general is knowing the charge transfers in each state. While the charge transfers can sometimes be determined by inspection, a formal calculation is often required. The procedure is similar to writing KCL equations, but in terms of charge transfer instead of current. The two states of the 1 : 3 ladder converter are shown in Fig. 6.13, including the definition of the charge transfers associated with the elements in each state. Summing the charges into the nodes for the two circuits yields the following equations:

$$-q_1 + q_2 + q_3 - q_i = 0, \qquad -q_1' - q_2' + q_3' = 0; \tag{6.17}$$

$$q_2 + q_3 = 0, \qquad q_1' + q_2' - q_i' = 0; \tag{6.18}$$

$$q_o = 0, \qquad q_o' + q_i' = 0. \tag{6.19}$$

Since over a cycle the net charge into a capacitor is zero, we know that $q_k = -q_k'$ and these equations can be rewritten as:

$$-q_1 + q_2 + q_3 - q_i = 0, \qquad q_1 + q_2 - q_3 = 0; \tag{6.20}$$

$$q_2 + q_3 = 0, \qquad -q_1 - q_2 - q_i' = 0; \tag{6.21}$$

$$q_o = 0, \qquad q_o' + q_i' = 0. \tag{6.22}$$

We also have two ancillary equations for the total charges transferred to the input and the output over a cycle:

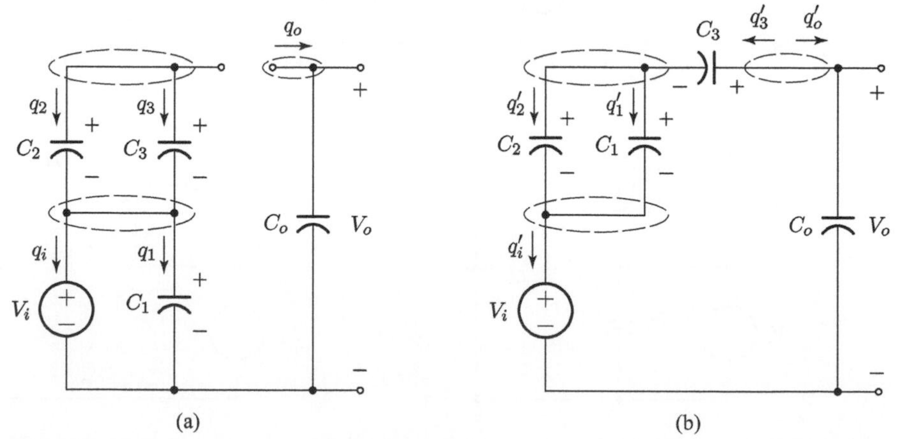

Figure 6.13 The two circuit states of the 1 : 3 ladder converter of Fig. 6.12, showing the definition of the charge variables and the nodes for writing the node equations.

$$Q_i = q_i + q_i',$$
$$Q_o = q_o + q_o'.$$

We will express the charge transfers for each capacitor and the input in terms of the total charge transferred to the output Q_o. To do so, we can reduce the six equations above to a set of four:

$$-q_1 + q_2 + q_3 - q_i = 0, \tag{6.23}$$
$$q_1 + q_2 - q_3 = 0, \tag{6.24}$$
$$q_2 + q_3 = 0, \tag{6.25}$$
$$-q_1 - q_2 = -Q_o, \tag{6.26}$$

where we have recognized that $q_i' = -Q_o$. This set of equations can be written in matrix form:

$$\begin{bmatrix} -1 & 1 & 1 & -1 \\ 1 & 1 & -1 & 0 \\ 0 & 1 & 1 & 0 \\ -1 & -1 & 0 & 0 \end{bmatrix} \begin{bmatrix} q_1 \\ q_2 \\ q_3 \\ q_i \end{bmatrix} = \begin{bmatrix} 0 \\ 0 \\ 0 \\ -Q_o \end{bmatrix}.$$

Solution of the equations yields the results

$$q_1 = 2Q_o, \tag{6.27}$$
$$q_2 = -Q_o, \tag{6.28}$$
$$q_3 = Q_o, \tag{6.29}$$
$$q_i = -2Q_o, \tag{6.30}$$

which in turn gives us $Q_i = q_i + q_i' = -3Q_o$. This approach thus provides the individual capacitor charge flows in terms of the charge delivered to the output. It also directly gives us the current conversion ratio $(-Q_o/Q_i)$ of the SCC. Note that – as expected – the current conversion ratio is independent of the relationship between the input and output voltages.

6.4.3 The Slow-Switching Limit for the Ladder Converter

The current conversion ratio of an SCC is fixed by the charge-balance requirement on its capacitors and is the same in both the FSL and SSL, 3 : 1 in our example design. This gives us the transformer ratio $N : M$ in the equivalent circuit model of Fig. 6.10 (here, 1 : 3). To determine the output voltage we need to calculate the equivalent output resistance R_e resulting from the capacitor charging/discharging losses. To do this for the slow-switching limit, we set $V_i = 0$ and determine the output current I_o with a fixed voltage V_o applied to the output. This is essentially the "test source" method for finding the Thévenin equivalent resistance of a system.

Figure 6.14 shows the circuit in states 1 and 2 with the input voltage set to zero and a fixed voltage V_o at the output. To derive the desired result, we employ Tellegen's theorem, which requires that the vi product summed over all elements in a circuit is 0, where the set of v's across the elements and the set of i's through the elements can be determined from different

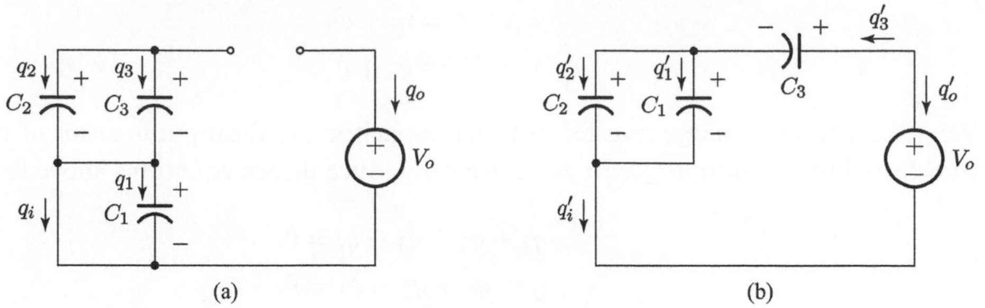

Figure 6.14 The circuits corresponding to (a) state 1 and (b) state 2, with $V_i = 0$.

experiments at different times.‡ Recognizing that charge is linearly related to current, we can express this constraint in each state as:

$$\text{State 1} \quad V_o q_o + \sum_{k=1}^{3} v_{C_k} q_k = 0 + q_{C_1} v_{C_1} + q_{C_2} v_{C_2} + q_{C_3} v_{C_3} = 0, \tag{6.31}$$

$$\text{State 2} \quad V_o q_o' + q_{C_1}' v_{C_1}' + q_{C_2}' v_{C_2}' + q_{C_3}' v_{C_3}' = 0, \tag{6.32}$$

where v_{C_k} represents the voltage on capacitor k at the end of state 1 and v_{C_k}' represents the voltage on capacitor k at the end of state 2. Adding these two equations, defining the voltage ripple on capacitor k as $\Delta v_{C_k} = v_{C_k} - v_{C_k}'$, and recognizing that $q_C = -q_C'$, we get an equation in terms of Δv_{C_k}:

$$q_o' V_o + q_{C_1} \Delta v_{C_1} + q_{C_2} \Delta v_{C_2} + q_{C_3} \Delta v_{C_3} = 0.$$

But $\Delta v_{C_k} = q_{C_k}/C_k$ and $Q_o = q_o + q_o' = q_o'$, so we can rewrite this as

$$-V_o Q_o = \frac{q_{C_1}^2}{C_1} + \frac{q_{C_2}^2}{C_2} + \frac{q_{C_3}^2}{C_3}. \tag{6.33}$$

Since $I_o = f_s Q_o$, the Thévenin output resistance $R_e = -V_o/I_o$ for the SSL case is

$$R_{e\text{SSL}} = \frac{1}{f_s}\left[\left(\frac{1}{C_1}\right)\left(\frac{q_{C_1}}{Q_o}\right)^2 + \left(\frac{1}{C_2}\right)\left(\frac{q_{C_2}}{Q_o}\right)^2 + \left(\frac{1}{C_3}\right)\left(\frac{q_{C_3}}{Q_o}\right)^2\right]. \tag{6.34}$$

The quantity q_{C_n}/Q_o is known as the *charge multiplier*, a_{C_n}, for capacitor C_n. With this notation,

$$R_{e\text{SSL}} = \frac{1}{f_s}\left(\frac{a_{C_1}^2}{C_1} + \frac{a_{C_2}^2}{C_2} + \frac{a_{C_3}^2}{C_3}\right). \tag{6.35}$$

From the previous subsection, the charge multipliers for this circuit may be found as

$$a_{C_1} = +2, \qquad a_{C_2} = -1, \qquad a_{C_3} = +1, \tag{6.36}$$

‡ Tellegen's theorem is quite broad, as it derives purely from constraints related to the graph of a network. It states that across all circuits having a given graph, any vector of branch voltages satisfying KVL must be orthogonal to any vector of branch currents satisfying KCL.

such that for $C_1 = C_2 = C_3 = C$ we get $R_{e\mathrm{SSL}} = 6/(f_s C)$.

For the general case of a switched-capacitor converter in SSL, one finds the charge transfers, calculates the charge multipliers for the capacitors $a_{C_k} = q_{C_k}/Q_o$, and finds the output resistance R_e in the slow-switching limit as

$$R_{e\mathrm{SSL}} = \frac{1}{f_s}\left(\sum_k \frac{a_{C_k}^2}{C_k}\right). \tag{6.37}$$

Note that the charge multipliers depend only on the topology and switching patterns of the network. Also, since $q_{C_k} = C_k \Delta v_{C_k}$, increasing C_k reduces Δv_{C_k} and R_e but a_{C_k} remains unchanged.

6.4.4 The Fast-Switching Limit for the Ladder Converter

In the FSL, circuit loss and output resistance are functions of the circuit resistances, assumed here to be only switch resistances (though other resistances could be included). The charge transfer relationships and capacitor charge multipliers are still valid in the fast-switching limit. We can use this information to find the currents through the switches, and thereby calculate loss and output resistance in the FSL.

Figure 6.15(a) shows the 1 : 3 ladder converter of Fig. 6.12(a) including switch voltages v_{S_j} and currents i_{S_j}, where we have defined the polarities such that the switch voltages are positive during the switch off states and the switch currents flow into the positive terminals of the switch voltages. Figure 6.15(b) and (c) show the two states of the converter, in which the switch on-state resistances have been included and we have illustrated the net charge (q_{S_j}) flowing through each switch during its on-state period. The net switch charge transfers may be expressed as linear combinations of the capacitor charge transfers:

$$\begin{aligned} q_{S_1} &= q_{C_1}, & q_{S_2} &= -q'_{C_1} = q_{C_1}, \\ q_{S_3} &= q_{C_3} - q_{C_1}, & q_{S_4} &= -q'_{C_2} = q_{C_2}, \\ q_{S_5} &= q_{C_2}, & q_{S_6} &= q'_{C_3} = -q_{C_3}. \end{aligned}$$

Assuming the circuit is operating at a duty ratio of 50% and the current through the switch is constant during its on-time, the average on-state loss of the jth switch is $(R_{S_j} i_{S_j}^2)/2$. Since each switch conducts for half the cycle, the switch current may be expressed in terms of switch charge flow as $i_{S_j} = 2q_{S_j} f_s$. The average power lost in switch j is then $\langle P_{S_j}\rangle = R_{S_j}(2q_{S_j} f_s)^2/2$. The total power loss is the sum of the switch losses.

Loss and output resistance may be conveniently expressed in terms of *switch charge multipliers* $a_{S_j} = q_{S_j}/Q_o$ which reflect the charges delivered through each of the switches during their on-state periods normalized to the total output charge over a cycle. Since we have previously determined the capacitor charge multipliers, the switch charge multipliers are easily determined:

$$\begin{aligned} a_{S_1} &= 2, & a_{S_2} &= 2, \\ a_{S_3} &= -1, & a_{S_4} &= -1, \\ a_{S_5} &= -1, & a_{S_6} &= -1. \end{aligned}$$

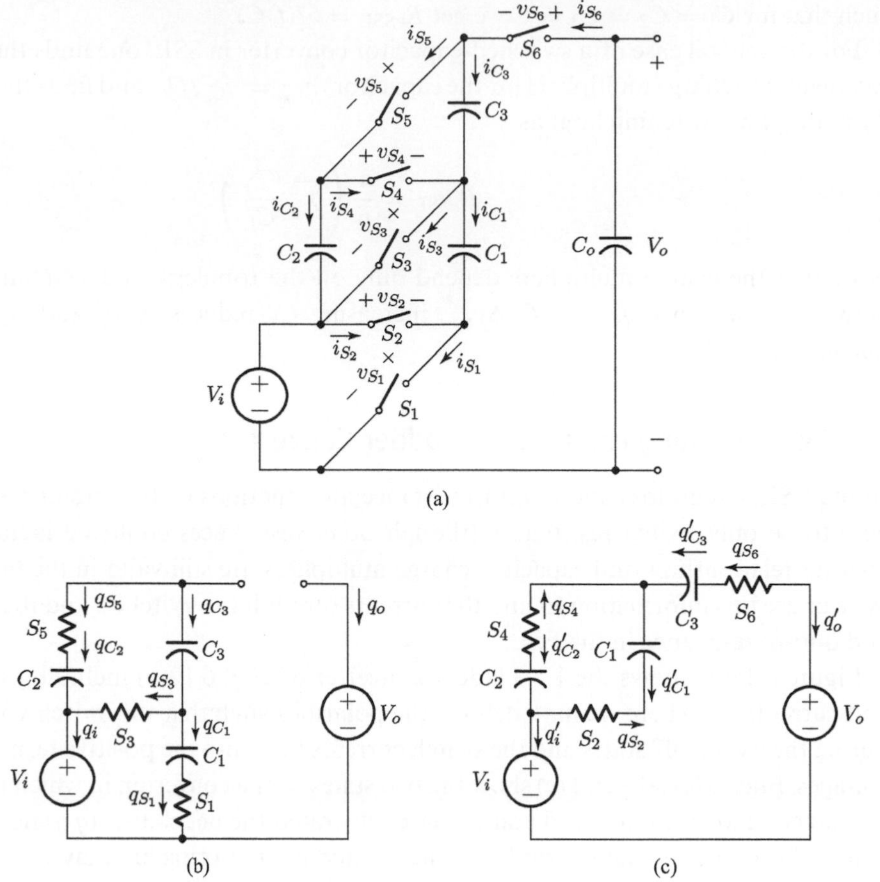

Figure 6.15 (a) The ladder converter of Fig. 6.12(a). (b) State 1, and (c) state 2, of the converter, including the switch on-state resistance necessary to determine losses in the fast-switching limit.

This results in $\langle P_{S_j} \rangle = R_{S_j}(2Q_o a_{S_j} f_s)^2/2$ and a total power loss of

$$\langle P \rangle = \frac{1}{2}(2Q_o f_s)^2 \sum_j R_{S_j} \left(a_{S_j}\right)^2 = 2I_o^2 \sum_j R_{S_j} \left(a_{S_j}\right)^2. \tag{6.38}$$

It follows that the equivalent output resistance for a switched-capacitor converter operating in the FSL may be expressed as

$$R_{e\text{FSL}} = 2 \sum_j R_{S_j} \left(a_{S_j}\right)^2. \tag{6.39}$$

For our 1 : 3 ladder converter with $R_{S_j} = R_{\text{sw}}$ for all switches, this yields

$$R_{e\text{FSL}} = 2R_{\text{sw}} \sum_{j=1}^{6} \left(a_{S_j}\right)^2 = 24R_{\text{sw}}. \tag{6.40}$$

This in turn yields an output voltage in FSL of

$$V_o = 3V_i - R_{eFSL}I_o. \tag{6.41}$$

Note that a switch charge multiplier a_{S_j} being negative means that if switch j blocks a positive voltage during its off state, it will carry a negative current during its on state for a positive output current I_o. The switch can consequently be implemented with a diode if only unidirectional power flow is required and the diode drop and loss are acceptable.

6.5 Other Kinds of Switched-Capacitor Converters

While we have focused on switched-capacitor dc/dc converters, other conversion functions such as inverters (dc/ac converters) can also be realized with switched-capacitor techniques. This can be done, for example, by changing switching patterns and/or output connections of dc/dc structures to dynamically vary the conversion ratio. Moreover, some kinds of multi-level converters (such as the flying-capacitor multi-level converter described in Chapters 5 and 8) are closely related to switched-capacitor converters. Switched-capacitor techniques are also used in rectifiers. Diode-based switched-capacitor rectifiers providing voltage step up ("voltage multiplier" rectifiers) are often used to provide voltage gain in high-voltage power supplies, for example. Perhaps the most famous and widely used such voltage multiplier rectifier is the Cockcroft–Walton circuit described below.

Switched-capacitor designs can also be augmented with magnetics and/or hybridized with magnetics-based converters to achieve some of the benefits of both switched-capacitor and magnetic power conversion. We outline this below and provide further references at the end of the chapter.

6.5.1 The Cockcroft–Walton Voltage Multiplier

Figure 6.16 shows a rectifier circuit known as a two-stage Cockcroft–Walton (C–W) multiplier. Its function is to provide a dc output voltage that is ideally four times the peak value of the input ac voltage. With more stages, it can provide a very high multiplication factor. Cockcroft and Walton used such a circuit to create the very high voltage (800 kV) that they used in

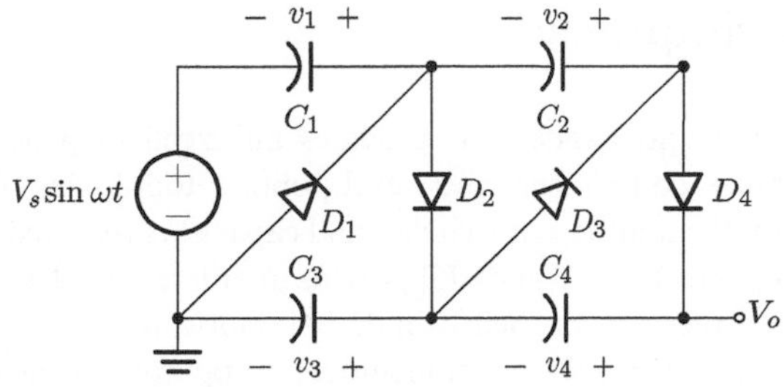

Figure 6.16 A two-stage Cockcroft–Walton, or Greinacher, multiplier.

their particle accelerator, earning them the 1951 Nobel Prize "for their pioneer work on the transmutation of atomic nuclei by artificially accelerated atomic particles." Although the circuit is commonly referred to as the Cockcroft–Walton multiplier, it was invented previously by Heinrich Greinacher, with a voltage-doubler version published in 1914 and higher-order design in 1921 (both in German). The circuit is therefore sometimes called the Greinacher multiplier.

The circuit operation is straightforward, but not intuitive. Consider the first negative half-cycle of the ac source. The diode D_1 will conduct, charging C_1 to V_s. During the next positive half-cycle, D_1 will be off and D_2 conducts to discharge C_1 and charge C_3. During the next (negative) half-cycle, D_2 will be off while D_1 and D_3 conduct. One can use the method of assumed states described in Section 3.2.1 to complete the charging sequence. The fully-charged "open circuit" state has all the diodes off (otherwise capacitors would be charging or discharging), and takes many cycles to achieve. When in this fully charged state, $v_1 = V_s$, $v_2 = v_3 = v_4 = 2V_s$, and $V_o = 4V_s$. In general, an open-circuit output voltage $V_o = 2NV_s$ is provided, where N is the number of stages, equal to the number of capacitors stacked between ground and the output. Of course, as with switched-capacitor dc/dc converters, the output voltage under load depends upon the load current, drooping down from the open-circuit voltage.

The C–W multiplier can produce very high voltages (large ac-to-dc voltage gain), but with limited current delivery capability, as the capacitors must retain their charge to avoid voltage droop. (That is, the equivalent output resistance of the C–W multiplier is high for a large number of stages.) Voltage multiplier rectifiers such as the C–W multiplier are often applied in high-voltage-output power supplies.

6.5.2 Hybrid Magnetic/Switched-Capacitor Converters

An SCC can be made efficient over a wide load range, especially if the switching frequency is varied with load. However, to achieve high efficiency, its voltage conversion ratio is generally constrained to a narrow range. In the SSL, regulation of the output voltage is possible by varying f_s, as (6.7) suggests, but such control is inefficient. To overcome this limitation a hybrid converter can be used, in which a switched-capacitor stage is cascaded or otherwise interconnected with a magnetics-based dc/dc converter, such that the SC stage provides (possibly variable) transformation and the magnetics-based stage provides regulation capability.

Notes and Bibliography

The paper by Seeman and Sanders [1] provides an excellent general analysis of switched-capacitor converters along with a comparison of several popular topologies. The classic paper by Makowski [2] introduces bounds on the numbers of switches and capacitors required to achieve a given conversion ratio.

In their paper, Cockcroft and Walton [3] provide an interesting description of the development of their multiplier, which used rectifiers of their own design, fabricated in long glass tubes, evacuated, and sealed with plasticine (modeling clay). The paper includes photographs of their 800 kV apparatus, which was used for the experiments that won them the Nobel Prize. At the other end of the scale, in [4] Dickson introduces his eponymous SCC topology in an integrated circuit, an approach that is still widely used in on-chip converters. Of course, the underlying techniques can be used at many scales and can also be combined in various forms, such as in the hybrid Cockcroft–Walton/Dickson voltage multiplier of [5].

Lei and Pilawa-Podgurski [6] analyze both resonant switching and soft charging of switched-capacitor converters, which are techniques in which inductors are added to conventional switched-capacitor structures to reduce capacitor charge/discharge loss and/or realize high-efficiency voltage regulation. Examples of hybrid switched-capacitor/magnetic converters are presented in [7] and [8].

1. M. D. Seeman and S. R. Sanders, "Analysis and Optimization of Switched-Capacitor DC–DC Converters," *IEEE Trans. Power Electron.*, vol. 23, pp. 841–851, Mar. 2008.
2. M. S. Makowski, "Realizability Conditions and Bounds on Synthesis of Switched-Capacitor DC–DC Voltage Multiplier Circuits," *IEEE Trans. Circ. and Sys. I*, vol. 44, no. 8, pp. 684–691, Aug. 1997.
3. J. D. Cockcroft and E. T. S. Walton, "Experiments with High Velocity Positive Ions. (I) Further Developments in the Method of Obtaining High Velocity Positive Ions," *Proc. R. Soc. Lond. A*, vol. 136, pp. 619–630, 1932.
4. J. F. Dickson, "On-Chip High-Voltage Generation in MNOS Integrated Circuits Using an Improved Voltage Multiplier Technique," *IEEE J. Sol. State Circ.* vol. SC-11, no. 3, pp. 374–378, Jun. 1976.
5. S. Park, J. Yang, and J. Rivas-Davila,"Hybrid Cockcroft–Walton/Dickson Multiplier for High Voltage Generation," *IEEE Trans. Power Electron.* vol. 35, no. 3, pp. 2714–2723, Mar. 2020.
6. Y. Lei and R. C. N. Pilawa-Podgurski, "A General Method for Analyzing Resonant and Soft-Charging Operation of Switched-Capacitor Converters," *IEEE Trans. Power Electron.*, vol. 30, pp. 5650–5664, Oct. 2015.
7. R. C. N. Pilawa-Podgurski and D. J. Perreault, "Merged Two-Stage Power Converter with Soft Charging Switched-Capacitor Stage in 180 nm CMOS," *IEEE J. Sol. State Circ.*, vol. 47, pp. 1557–1567, Jul. 2012.
8. C. Schaef and J. T. Stauth, "A Highly Integrated Series-Parallel Switched Capacitor Converter with 12 V Input and Quasi-Resonant Voltage-Mode Regulation," *IEEE J. Emerg. Sel. Top. Power Electron.*, vol. 6, pp. 456–464, Jun. 2018.

PROBLEMS

6.1 Add a resistor of arbitrary value in series with the capacitor in the circuit of Fig. 6.1 and show that the charging efficiency given by (6.3) is unchanged.

6.2 We want to operate the converter of Fig. 6.7(a) as a step-up converter, that is, with power flowing from right to left.

(a) Redraw the circuit showing the direction of the switch variables for step-up conversion.

(b) Redraw the circuit again with a switch implementation appropriate for step-up conversion, using diodes where possible.

(c) Draw the equivalent circuit model for this step-up converter, as shown in Fig. 6.3 for the step-down version.

(d) Can the circuit of Fig. 6.8 provide bilateral conversion, that is, step-down conversion from left to right and step-up conversion from right to left? Why, or why not?

6.3 Figure 6.17 shows a switched-capacitor dc/dc converter circuit. The circuit modulates with 50% duty ratio between switches 1 on and 2 on at a frequency f. The switches have an on-state resistance R_{on}. Capacitors C_1 to C_3 have capacitance C and may charge and discharge significantly, but the voltage ripple on capacitor C_{big} can be ignored.

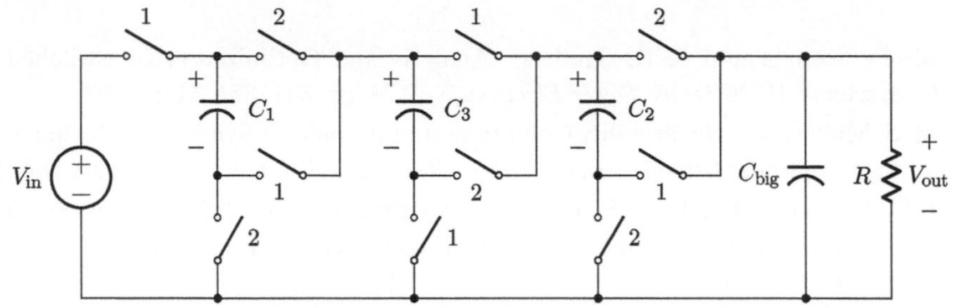

Figure 6.17 The switched-capacitor dc/dc converter circuit analyzed in Problem 6.3.

(a) Calculate the open-circuit voltage conversion ratio V_{out}/V_{in} as $R \to \infty$.

(b) What is the output resistance of this converter in the slow-switching limit? (That is, find the output resistance when the switching frequency is low enough that capacitors C_1 to C_3 are fully charged in each portion of the switching cycle.)

(c) What is the output resistance of this converter in the fast-switching limit? (That is, the output resistance when the switching frequency is high enough that the capacitors remain fully charged during the switching cycle.) Neglect device gating losses associated with turning the switches on and off, and losses due to capacitor ESR, but include the switch conduction losses during capacitor charge and discharge.

(d) For a circuit with $V_{in} = 25\,\text{V}$, $C = 1\,\mu\text{F}$, $R_{on} = 0.01\,\Omega$, and $f = 500\,\text{kHz}$, what are the output voltage and efficiency for load currents of 1 mA, 10 mA, 100 mA, and 1 A?

6.4 Figure 6.18 shows a circuit in which a nonlinear capacitor C is charged from a voltage source V_{in}. The switch is closed at $t = 0$, and the capacitor voltage before the switch closes is $v_x(0^-) = 0$. The capacitance is a function of voltage such that $C = f(v_x) = K(v_x)^{-(1/2)}$. (This function crudely approximates the capacitance of a reversed-biased p–n junction with abrupt grading, such as for a diode or MOSFET in the off state.)

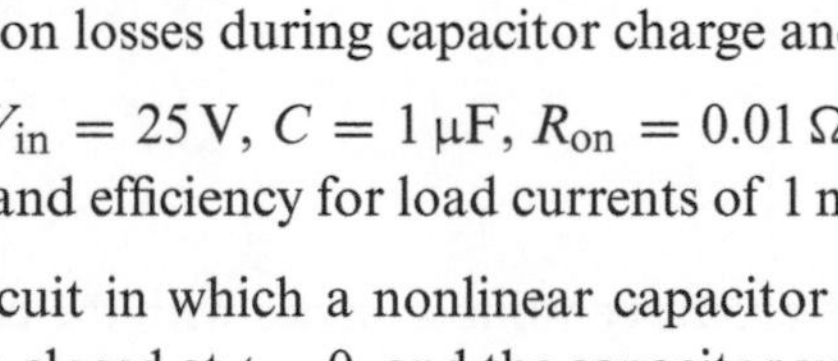

Figure 6.18 The circuit analyzed in Problem 6.4 in which a nonlinear capacitor is charged from a constant-voltage source.

(a) Find the amount of energy delivered by the source, the energy delivered to the capacitor, and the energy dissipated in the resistor in the charging process.

(b) Determine if/how the answer changes for the condition $v_x(0-) = V_{\text{in}}/2$.

(c) Determine if/how the answer changes for the case where the resistance is nonlinear: $R = r(i) = Bi^2$.

(d) Find an expression giving the value of an equivalent linear capacitance C_E that stores the same amount of energy as the nonlinear capacitor C when charged from 0 V to a positive voltage V_B. The value of C_E must (of course) be a function of V_B.

(e) Find an expression giving the value of an equivalent linear capacitance C_Q that accepts the same amount of charge as the nonlinear capacitor C when charged from 0 V to a positive voltage V_B. The value of C_Q is of course a function of V_B.

6.5 Consider the simple switched-capacitor converter of Fig. 6.2.

(a) Find the capacitor charge multiplier for the energy transfer capacitor, and use this to compute $R_{e\text{SSL}}$ for this circuit according to (6.37).

(b) Find the switch charge multipliers for the switches, and use this to compute $R_{e\text{FSL}}$ for this circuit according to (6.39), assuming all switches have resistance R_{sw}.

6.6 Consider a 3:1 step-down ladder converter having the structure of Fig. 6.11 with $k = 6$.

(a) Find the parameters of the model of Fig. 6.10 for this converter in the SSL, considering all energy-transfer capacitors to have value C.

(b) Find the parameters of the model for this converter in the FSL, assuming all switches have resistance R_{sw}.

(c) Compare these models to those found for the step-up converter of Fig. 6.12 in the FSL and SSL. How are the models for the step-up and step-down converters related?

7 Isolated Pulse-Width-Modulated DC/DC Converters

We add transformers to the topology of a high-frequency converter for three reasons: to provide electrical isolation between two (or more) external systems; to reduce the component stresses that result when the input/output conversion ratio is far from unity; and to create multiple related outputs in a simple manner. (We showed the relationship between switch-stress factor and the conversion ratio in Fig. 5.26.) There are many ways in which we can include the transformer in the topology of a dc/dc converter; we present and discuss some of them in this chapter. The overview of transformers presented in Chapter 1 should be sufficient background for understanding the material in this chapter. However, Section 18.4 provides additional background if desired.

Recall that a transformer winding cannot have a dc voltage across it, because the magnetizing inductor is a short circuit at zero frequency. We therefore need to create, from the dc voltages of the external systems, an ac voltage with no average value. Thus we usually – but not necessarily – have the switches that produce this ac voltage also control the input/output conversion ratio.

There are essentially two types of high-frequency transformers in isolated converters. The first is exemplified by that in the *forward converter*, an isolated converter based on the buck topology. The second is exemplified by the transformer in the *flyback* converter based on the buck/boost topology. In the forward converter, the algebraic sum of the instantaneous power over all windings is ideally zero. That is, the transformer is not required to store significant energy. Although some energy is stored in the transformer's magnetizing inductance, we minimize this energy by making the inductance large. In the flyback converter, however, the transformer is required to store energy. During one part of the switching cycle, the primary winding takes energy from the input system and stores it in the magnetizing inductance. During the second part of the cycle, a second winding removes this energy and delivers it to the load. These two ways of using transformers underpin a wide range of isolated dc/dc converters.

7.1 Single-Ended Isolated Forward Converter

Figure 7.1(a) shows one version of an isolated buck converter. The transformer is modeled as an ideal transformer with a shunt inductor L_μ representing its magnetizing inductance. (For now we are going to ignore the transformer's leakage inductance.) We call this converter a *single-ended* isolated forward converter – "single-ended" because power flows through the transformer for only one polarity of the primary voltage. The diode D_3 and the voltage source V_c form a *clamping circuit* to provide a path for the magnetizing current i_μ when the transistor turns off.

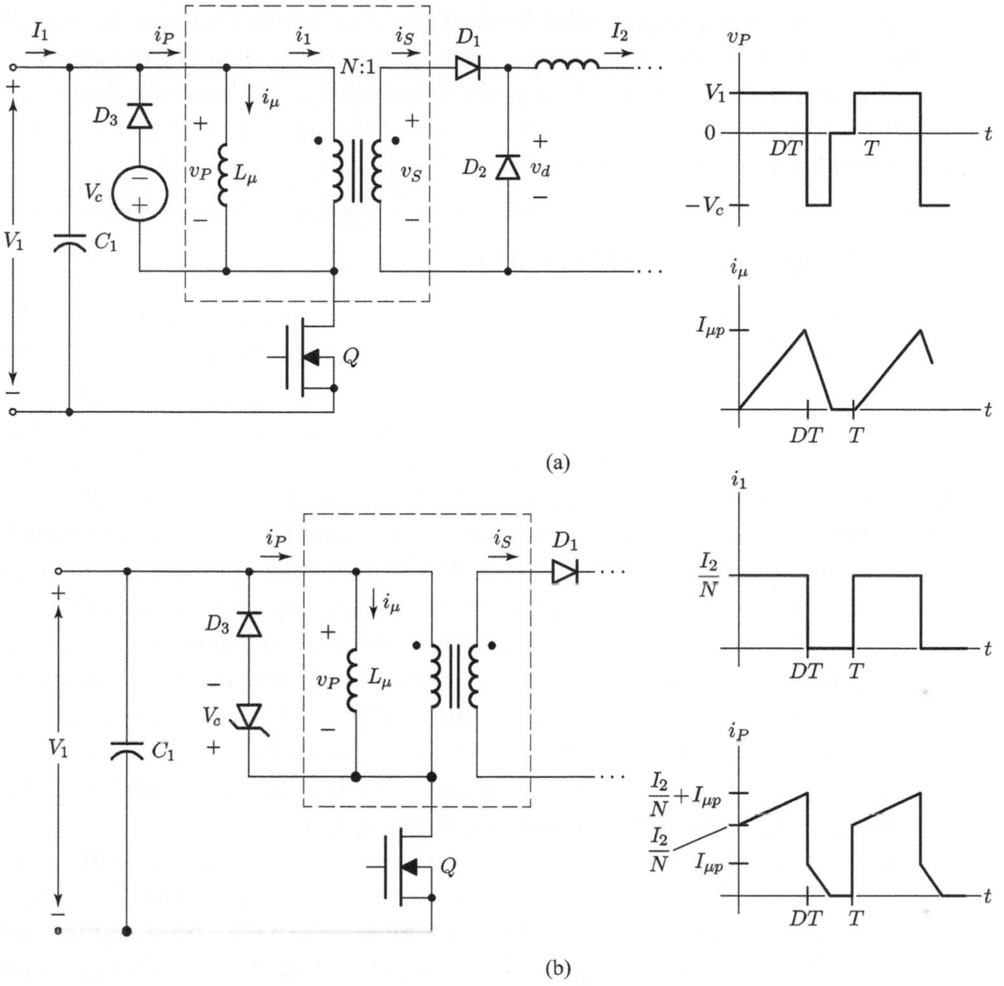

Figure 7.1 Single-ended isolated buck converter with clamp. (a) Clamp voltage created by a voltage source V_C. (b) Clamp voltage produced by a Zener diode.

When the transistor is on, the transformer primary voltage v_P is equal to the input voltage V_1. The secondary voltage v_S is related to v_P by the turns ratio, and its polarity is positive. Therefore the output rectifier diode D_1 is on, the freewheeling diode D_2 is reverse biased, and $v_d = V_1/N$. The output current I_2 is reflected to the primary circuit as $i_1 = I_2/N$. Note that if we ignore the magnetizing current, and if $N = 1$, operation of the single-ended isolated forward converter is indistinguishable from that of the non-isolated buck converter at this point. Furthermore, a non-unity turns ratio simply provides a step-up or step-down of the voltage, and inversely for the current.

Continuing to ignore the magnetizing inductance for now, the transformer's primary and secondary currents i_P and i_S are zero when the transistor is off. The freewheeling diode D_2 will therefore turn on to carry the output current, so $v_d = 0$. This condition corresponds to that part of the cycle when the shunt switch S_{xz} in the buck converter of Fig. 5.8(a) is conducting.

A practical isolated single-ended forward converter has two features that complicate both its performance and its design. The first is the need to provide for the consequences of nonzero stored energy in the transformer magnetizing branch L_μ. The second is the fact that processing this magnetizing energy results in switch stresses that exceed those of the buck converter. We now consider these two features in detail.

7.1.1 Magnetizing Current and Clamping

If the transformer magnetizing current i_μ in Fig. 7.1 is nonzero when the transistor Q is on, we can no longer ignore it when Q turns off. We must provide the current with a path, in order to prevent it from being discontinuous and generating impulsive voltages. Furthermore, while i_μ is flowing during the time Q is off, v_P must be negative so that $di_\mu/dt < 0$ and $i_\mu(T) = i_\mu(0)$. This is the periodic steady-state condition, and it guarantees that $i_\mu(T)$ is not increasing every cycle. Stated another way, the magnetic flux B in the core must be returned to its starting value at the end of every cycle, referred to as *resetting* the core. Therefore we must provide a means for reversing the polarity of v_P when Q is off. We do so with the diode D_3 and the clamp voltage V_c.

You can best understand the operation of the clamp by first assuming that the magnetizing current is zero when Q turns on. Because v_P is positive and constant at the value V_1 when Q is on, i_μ increases linearly, to the current carried by Q but not by D_1. When Q turns off, I_2 immediately commutates to D_2, and D_1 turns off. The ideal transformer currents i_S and i_1 step to zero at this time, but i_μ must be continuous. It cannot flow into the primary winding of the ideal transformer because the corresponding secondary current would have to flow the wrong way through D_1. Instead, i_μ forces D_3 on, clamping v_P at $-V_c$.

The magnetizing current i_μ must be zero by the start of the next cycle. If i_μ were not zero, it would continue to increase each cycle until the transformer saturated. If i_μ is to return to zero by the start of the next cycle, v_P must be negative when Q is off, and at a value sufficient to force i_μ to zero before Q turns on again. Specifically, the time integral of v_P (generally referred to as the *primary volt-seconds*) when Q is off must at least equal the negative of the primary volt-seconds when Q is on, so $(1-D)V_c \geq DV_1$. Equality here is the same as $\langle v_P \rangle = 0$, and occurs if V_c is precisely the value necessary to balance the positive and negative primary volt-seconds, returning the transformer flux at the end of each cycle to its starting value of zero. However, a transient or an imperfection in control may cause this starting value to be nonzero. In such a case, the peak flux in the transformer will be higher than necessary, and we may still have to worry about saturation. To guarantee that $i_\mu(T) = 0$, we usually make the clamp voltage slightly greater than the critical value necessary for balancing the volt-seconds. Diode D_3 will prevent i_μ from actually going negative, so the minimum flux level in the transformer is $B_{\min} = 0$. When D_3 turns off, v_P steps to zero, where it stays until the transistor turns on again. The waveforms to the right in Fig. 7.1 illustrate the preceding discussion.

The duty ratio is often a control variable, so we must choose V_c to guarantee that i_μ will be reset to zero in the worst case. For example, the larger the duty ratio in the converter of Fig. 7.1, the higher will be the positive volt-seconds seen by the transformer and the larger V_c must be. Therefore we must choose V_c on the basis of the expected maximum duty ratio. For duty ratios below this maximum value, i_μ will simply return to zero more quickly than necessary, and a greater amount of the transistor's off time will be spent with $v_P = 0$.

There are many different ways of implementing the clamp function, only some of which we discuss. In practice, we frequently use a breakover device – such as a Zener diode as shown in Fig. 7.1(b), or a metal-oxide varistor – as a replacement for the clamp voltage source of Fig. 7.1(a) if the energy involved is within the ratings of the breakover device. However, the principles involved are independent of the specifics of implementation. For this reason, we use easily understood clamp circuits to illustrate the important issues of the clamping function.

Example 7.1 Choosing a Clamp Voltage

Assume that the circuit of Fig. 7.1(a) has an input voltage $V_1 = 50\,\text{V}$ and that the transformer has a turns ratio $N = 2$. If the transistor's duty ratio is D, the output voltage will be

$$V_2 = D\left(\frac{V_1}{N}\right) = D(25)\,\text{V}. \tag{7.1}$$

If the requirement on V_2 is $0 < V_2 < 20\,\text{V}$, we need a maximum duty ratio of 80%. We can find the minimum necessary value of V_c by equating the positive and negative volt-seconds on the primary:

$$V_1 D_{\max} T = (V_c)(1 - D_{\max})T, \tag{7.2}$$

or

$$V_c = V_1 \frac{D_{\max}}{1 - D_{\max}} = (50)(4) = 200\,\text{V}. \tag{7.3}$$

The clamp voltage should be somewhat greater than this minimum value. How much greater depends on how sure we are that the maximum duty ratio will not exceed 80%, or how carefully we control transients that can cause flux offsets. Note that we must choose a transistor rated to withstand $V_1 + V_c$, which is high relative to other voltages in the circuit.

The energy stored in L_μ, namely $L_\mu I_\mu^2/2$, is absorbed by the clamping circuit every time it operates, that is, once a cycle. This energy is often difficult and uneconomical to recover from the source V_c, and hence it represents a loss. If the application is a very high-power one, we might justifiably use an auxiliary dc/dc converter to recover this energy by transferring it to either the source or the load.

Example 7.2 Using a Capacitor to Produce V_c

Figure 7.2 illustrates the use of a capacitor to provide the clamping voltage V_c. The advantage of this circuit is that V_c will automatically assume the value necessary to assure core reset. The disadvantage is that a large capacitor is required.

We assume that C_c is large enough that its voltage is constant, and that

$$V_1 = 100\,\text{V}, \qquad T = 10\,\mu\text{s}, \qquad L_\mu = 10\,\text{mH}.$$

We determine the necessary value of R for operation at $D = 0.5$ and then calculate the resulting value of V_c when $D = 0.8$. For $D = 0.5$ we need $V_c = 100\,\text{V}$ to reset the core. The peak magnetizing current (assuming the core is reset each cycle) is

$$I_{\mu p} = \frac{DTV_1}{L_\mu} = 50\,\text{mA}.$$

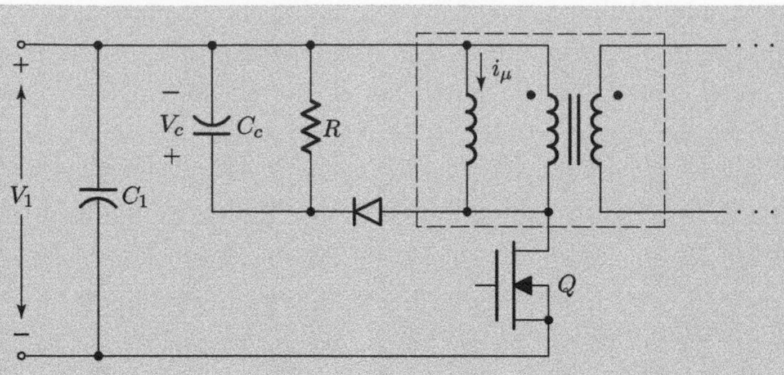

Figure 7.2 A capacitor alternative to providing the source V_c in Fig. 7.1.

The resulting energy stored in L_μ is thus 12.5×10^{-6} J. This energy must be dissipated in R during time T (since R is always connected to C_c). Therefore

$$\frac{V_c^2 T}{R} = 12.5 \times 10^{-6}\,\mathrm{J}$$
$$R = 8\,\mathrm{k}\Omega.$$

Repeating this calculation for $D = 0.8$, we get $I_{\mu p} = 80$ mA, the energy stored in $L_\mu = 32 \times 10^{-6}$ J and $V_c = 160$ V. That is, V_c scales linearly with the duty ratio.

If R deviates from 8 kΩ, the value calculated here, V_c will assume whatever value is necessary to reset the core.

7.1.2 Transformer-Coupled Clamp

We show another way to provide the clamp function in Fig. 7.3, where a third winding of N_T turns has been added to the transformer. This third winding, called a *tertiary* or *clamp* winding, permits i_μ to circulate through the primary and clamp windings when Q turns off. The current i_T is therefore

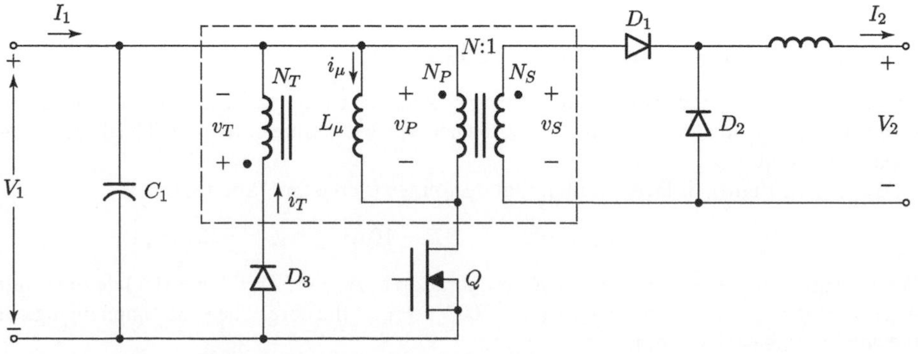

Figure 7.3 Clamp function provided by a tertiary transformer winding.

$$i_T = \frac{N_P}{N_T} i_\mu.$$

Note that when Q turns off, i_μ will flow out of the dot on N_P, causing i_T to flow into the dot on N_T, forcing D_3 on. Because $v_P = v_T(N_P/N_T)$ and $v_T = -V_1$ while D_3 is on, L_μ discharges, as desired. However, i_T cannot go negative, and therefore we can still reduce the transformer flux only to zero. The advantage of this approach is that the magnetizing energy returns directly to the source, instead of to a separate clamp circuit.

We can adjust the effective clamp voltage (the voltage across L_μ when Q is off) by changing the ratio N_P/N_T. For a maximum duty ratio of 0.5, N_P/N_T must have a minimum value of 1. For a maximum duty ratio of 0.75, N_P/N_T must have a minimum value of 3. Note that this turns ratio affects the off-state voltage of the transistor. For the 1:1 ratio, the transistor's off-state voltage is $2V_1$, and for the 3:1 ratio it is $4V_1$.

A problem with the circuit of Fig. 7.3 is that any leakage inductance between the primary and tertiary windings keeps the magnetizing current from commutating immediately to the tertiary winding when the transistor turns off. An additional clamp, or a transient-suppressing circuit called a *snubber*, must be placed across the transistor to keep its voltage from rising too much. (We discuss snubber circuits in detail in Chapter 24.) The energy associated with this leakage inductance is typically small compared to that of the magnetizing inductance, so the dissipation in this clamp or snubber is not a problem unless the circuit is switching at a very high frequency.

Instead of connecting the clamp winding across V_1, we could connect it across V_2, transferring the magnetizing energy to the output rather than back to the input. Such a circuit is sometimes called a *flyforward* converter. The result is a slight improvement in the circuit's efficiency. Owing to safety isolation specifications that often call for a minimum spacing between source and load windings, however, this approach makes a tight coupling of the primary and tertiary windings difficult. The greater dissipation caused by the larger leakage inductance usually offsets the expected gain in efficiency.

7.1.3 Isolated Hybrid Bridge

A technique frequently used to maintain current continuity in inductive loads switched by transistors is to place a diode around the load, as shown in Fig. 7.4(a). The problem with this technique for our present purposes, though, is that the discharging voltage is only a diode drop plus the drop across the resistance of the winding. However, the circuit of Fig. 7.4(a) suggests another possibility.

When Q_1 is off, the bottom terminal of the load is connected to the positive rail through D_1. Thus, if we could connect the top terminal of the load to the negative rail, we could apply $-V_1$ across the inductor to discharge it. This is accomplished by adding Q_2 and D_2, as shown in Fig. 7.4(b). The transistors Q_1 and Q_2 turn on and off simultaneously. Note that again, because the diodes do not permit i_μ to reverse, the transformer core resets to a minimum flux value of only $B_{\min} = 0$. Because only two of the four bridge switches are controllable, this circuit is still only single ended, and is called a *single-ended isolated hybrid bridge*.

Although this converter has twice as many primary-side switches as that of Fig. 7.3, they all have a smaller required voltage rating. And as the magnetizing current does not need to

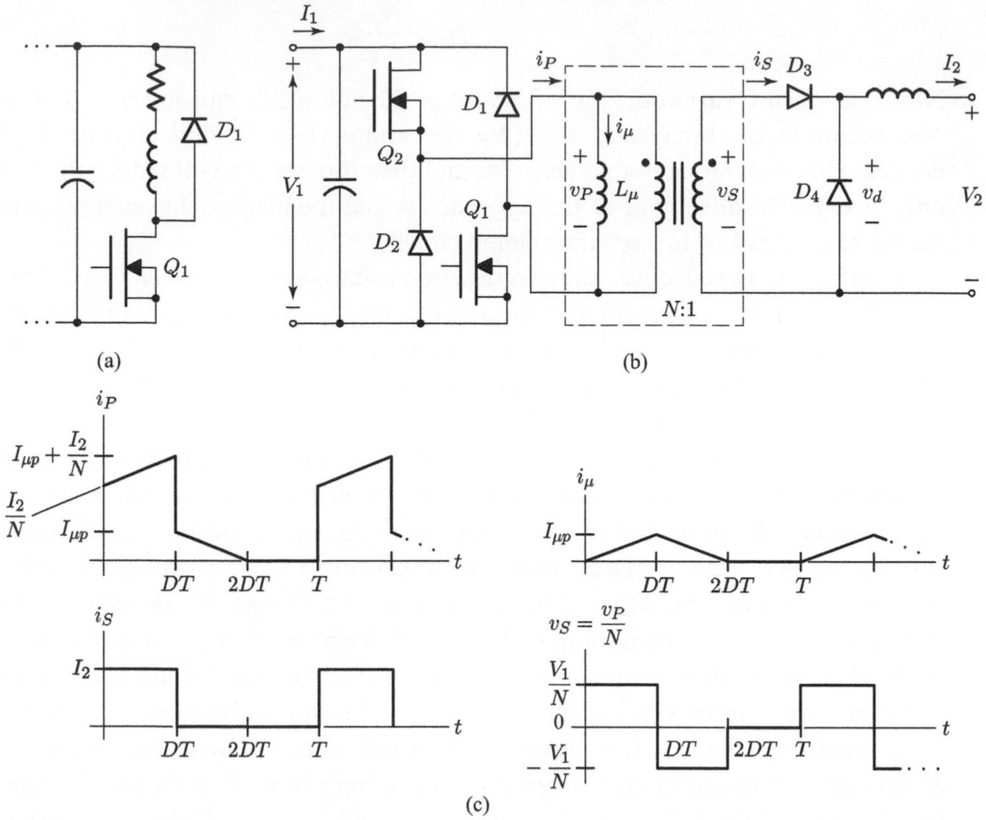

Figure 7.4 (a) Conventional application of a clamp diode to an inductive load. (b) Extension of the circuit of (a) to provide an L_μ discharge voltage equal to $-V_1$. (c) Waveforms for the circuit of (b).

commutate from the primary winding to a clamp winding, we avoid the leakage inductance in the circuit of Fig. 7.3. We do need to drive two transistors whose sources are at different potentials, however, and the circuit is limited to a maximum duty ratio of 50%. We can infer this duty-ratio limit from the current waveforms shown in Fig. 7.4(c). These waveforms are based on the assumption that the transformer leakage inductance is zero, which is not true in a practical transformer: a commutation period results, during which both D_3 and D_4 are on, as we discuss in Section 7.6.

7.1.4 Switch Stresses in the Single-Ended Forward Converter

One drawback to the single-ended forward converter is that power flows through the transformer only during the period DT. As the circuit is idle for the period $(1-D)T$, the voltage and current stresses on the transformer, switches, and filter elements are higher than they would be if the circuit were transferring energy from input to output continuously. You can see the reason for these effects by considering the voltage and current stresses on the transistor in the circuit of Fig. 7.1.

If the maximum duty ratio of the converter in Fig. 7.1 is 50%, the minimum clamp voltage is V_1, and therefore the minimum peak voltage across the transistor is $2V_1$. If P_o is the average power flowing through the converter, the primary-side current, which is the current carried by Q (ignoring i_μ), is rectangular with a 50% duty ratio and an amplitude of $2P_o/V_1$. The stress parameter of the transistor is therefore

$$V_{Qp}I_{Qp} = (2V_1)\left(\frac{2P_o}{V_1}\right) = 4P_o, \tag{7.4}$$

showing that it operates with a stress factor of 4 in this circuit. Similarly, the two output diodes carry the output current and withstand twice the output voltage, so their stress factors are 2.

In general, if the maximum duty ratio is $D_{\max}$, the clamp voltage must be at least $V_1/(1 - D_{\max})$, and the current carried by the transistor is $P_o/(D_{\max}V_1)$. The stress parameter of the transistor is therefore

$$V_{Qp}I_{Qp} = P_o\frac{1}{D_{\max}(1 - D_{\max})}. \tag{7.5}$$

This parameter is minimal when $D_{\max} = 0.5$. Even at the minimum value of its stress parameter, the switch must be rated at a stress factor of 4. Therefore, if we do not need isolation, we would not replace a conventional, non-isolated direct converter with one containing a transformer and a single transistor unless the output/input conversion ratio V_2/V_1 was less than 0.25 (see Fig. 5.26).

7.1.5 Active-Clamp Forward Converter

The active-clamp forward converter shown in Fig. 7.5(a) takes a somewhat different approach to resetting the core than previously shown. A second transistor Q_c and a capacitor C_c are used here to losslessly reset the core flux, ideally to the negative of the peak positive core flux. This yields converter characteristics that fall in between those of the single-ended converters described so far and the double-ended converters introduced in the next section.

In this converter, the clamp switch Q_c is turned on in a complementary fashion to the main switch Q. As shown in Fig. 7.5, when the main switch is off, the clamp switch Q_c connects capacitor C_c across the primary and imposes a voltage $v_P = -V_c$ across the magnetizing inductance. To identify the periodic steady-state value of voltage V_c, we recognize that capacitor C_c will be charged up over time by the current i_c (equal to the magnetizing current i_μ during the on time of the clamp switch) until the voltage V_c is sufficient to make the average of i_c (and hence the average of i_μ) zero. From a volt-seconds balance on the magnetizing inductance, we can identify this steady-state voltage to be

$$V_c = \frac{D}{1 - D}V_1, \tag{7.6}$$

which leads to a peak off-state voltage, for both the main and auxiliary switches, of

$$V_{pk} = \frac{1}{1 - D}V_1, \tag{7.7}$$

which is the same switch stress as with a conventional clamp. A key difference from a conventional clamp, though, is that the transformer flux and magnetizing current are now bipolar.

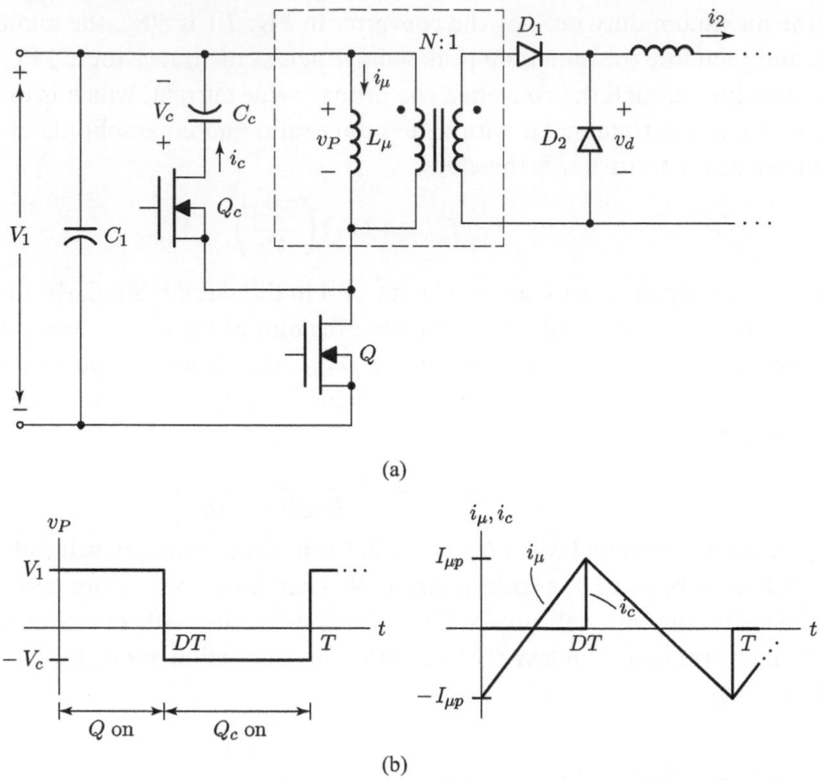

Figure 7.5 The active-clamp forward converter and its associated waveforms.

Essentially, C_c is used to losslessly reverse the polarity of the magnetizing current during the off state of the main switch. In periodic steady-state operation, the magnetizing current starts from a negative value $-I_{\mu p}$ when the main switch Q turns on, traverses linearly to a positive value $I_{\mu p}$ when Q turns off, and traverses (approximately) linearly back to $-I_{\mu p}$ during the off-state period of Q.

The active clamp technique has a number of benefits. First, it recycles the energy stored in the magnetizing inductance, rather than dissipating it. Second, because the transformer flux swing is now bipolar, one can achieve better core utilization, because the core flux density can potentially swing between $-B_s$ and $+B_s$ rather than 0 and $+B_s$. Even in cases where the core flux swing is limited by core loss rather than saturation, there can be a benefit, because core loss owing to a given flux swing often increases with a dc flux offset in practice, as we describe in Chapter 21. Third, the bipolar nature of the magnetizing current enables it to be used to provide zero-voltage soft switching of both Q and Q_c, reducing switching loss (switching loss and soft switching are treated in Chapter 24).

One limitation of the active clamp approach is the need for a second switch and a "flying" switch driver. Another limitation is that the clamp voltage must go through a transient readjustment whenever the input voltage or duty cycle change, and performance of the converter can suffer during this transient period. Even so, the active clamp approach is attractive when the additional cost of the auxiliary switch and driver can be justified.

7.2 Double-Ended Isolated Forward Converter

Even though the secondary voltage in the circuit of Fig. 7.4(b) is a symmetric ac waveform, diodes D_1 and D_2 prevent i_P from going negative, which rules out simply using a full-wave rectifier circuit on the secondary voltage to create the desired dc output. We are thus stuck with the ripple in v_d being at the switching frequency. *Double-ended* converters, by contrast, permit full-wave rectification and a doubling of the fundamental ripple frequency – and also result in better utilization of transformer core material. Their disadvantage is that they are more complicated than their single-ended counterparts.

7.2.1 Double-Ended Bridge Converter

If we replace the diodes and transistors in the hybrid bridge circuit of Fig. 7.4(b) with bilaterally conducting switches (e.g., power MOSFETs), we can use a full-wave rectifier on the output. This approach has two advantages. First, we can make the output inductor smaller, because the ripple frequency would be twice that of the single-ended circuits. Second, the magnetizing current i_μ can change sign, thereby permitting the core material to be better used, because the flux can now vary between $\pm B_s$ instead of between 0 and $+B_s$, where B_s is the saturation flux density of the material. Such a double-ended converter is shown in Fig. 7.6(a).

The waveform of v_d for the double-ended converter is shown in Fig. 7.6(b). Here, we define T as the period of the full-wave rectified voltage v_d. If $D = 1$, each of the primary-side

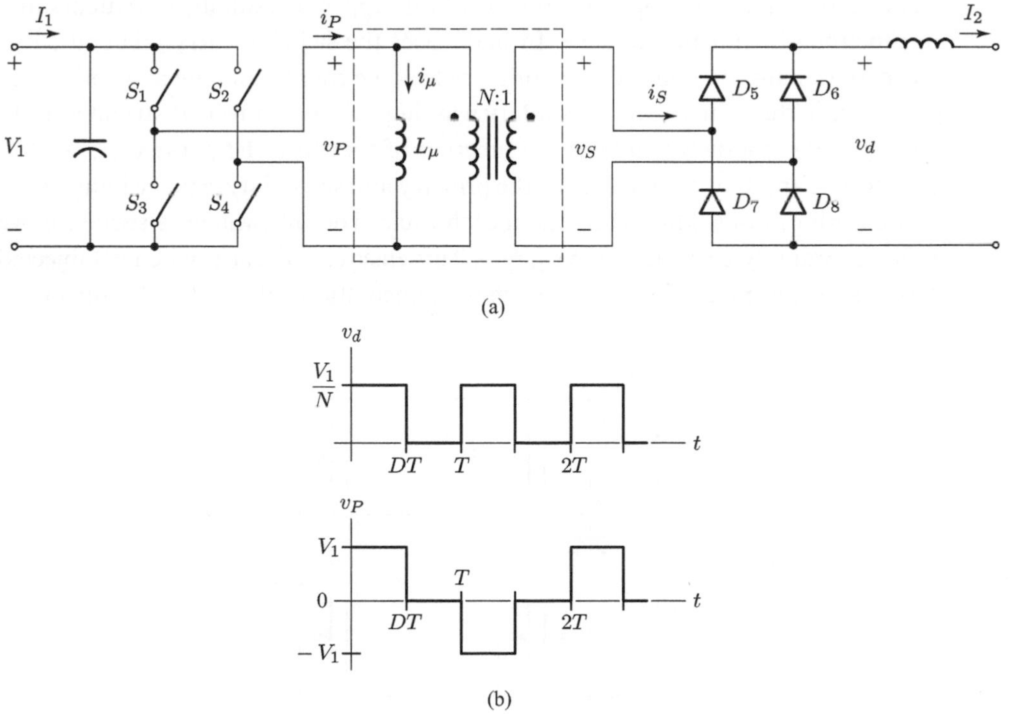

Figure 7.6 (a) Double-ended converter with primary-side switches arranged in a bridge. (b) Rectifier output voltage v_d and primary voltage v_P.

switches would be on 50% of the time, with $\langle v_d \rangle = V_1/N$. We can create the waveform of v_d by several different primary-side switching sequences. They differ in the path provided for i_μ during the time between DT and T, and their applicability is constrained by the magnitude of i_μ relative to the load current reflected to the primary, I_2/N. They also result in different switch implementations. In every case, however, the core resets during the next interval, DT, when v_P reverses. The only function that the switches must perform during time $(1 - D)T$ is to maintain the continuity of i_μ. We do not have to apply a resetting voltage to the transformer during this time because the core resets during the next period DT when v_P reverses. We can maintain the continuity of i_μ by using any one of three switch combinations. First, we can turn off all the switches, forcing the magnetizing current to flow in the secondary winding, so $i_S = -Ni_\mu$. The second combination is S_1 and S_2 on, and the third combination is S_3 and S_4 on.

The first option, with all the primary-side switches off, is possible only if $Ni_\mu < I_2$, so that at least three secondary diodes can be on simultaneously to allow $i_S \neq I_2$. For instance, if $0 < Ni_\mu < I_2$, the diodes D_5, D_6, and D_7 (or, alternatively, D_6, D_7, and D_8) would have to be on, with I_2 circulating through D_5 and D_7, and $i_S = -Ni_\mu$ circulating in the negative direction through D_5 and in the positive direction through D_6. If $Ni_\mu > I_2$, then D_5 would be conducting a net negative current, which would be inconsistent with its on state. (Note that the state of D_8 during this time is ambiguous but makes no difference to the result.) The advantage of this option is that the primary switches carry only a unipolar current. Its disadvantage is that the continuity of energy stored in any transformer leakage inductance (which is not included in the figure) cannot be maintained. This is therefore not a practical switching sequence, and illustrates the problem of erroneous design if inadequate models are used (in this case omitting leakage inductance).

The second and third options, in which both upper or both lower switches are conducting, are functionally indistinguishable. In both cases the switches carry bilateral current, a path is provided to maintain continuity of primary leakage energy, and the magnetizing current i_μ will split between the primary and secondary windings. Exactly how i_μ distributes itself between the primary and secondary windings is a function of second-order parameters, such as switch and diode drops, and the relative sizes of the primary and secondary leakage inductances. In practice we use both options, alternating them each cycle. The full switching sequence over two cycles might accordingly be $S_{1,4}$, $S_{4,3}$, $S_{3,2}$, $S_{2,1}$. This ensures that each switch is subjected to the same current stress. Figure 7.7 shows a switch implementation suitable for this operating mode.

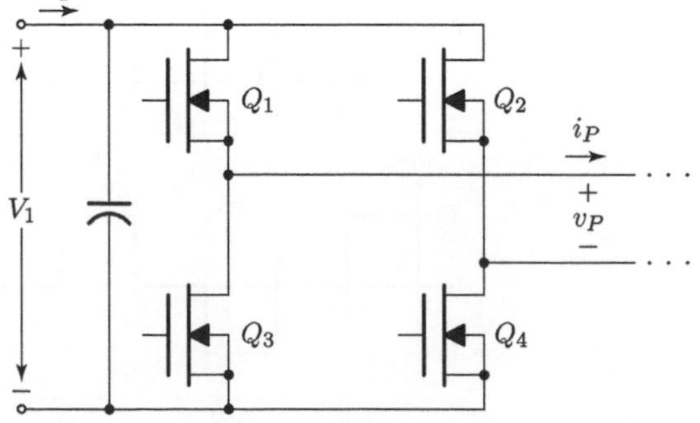

Figure 7.7 Switch implementation for the double-ended converter of Fig. 7.6(a).

Example 7.3 Transformer Primary Current in the Double-Ended Full-Bridge Converter

Let's determine the primary current i_P in the double-ended converter of Fig. 7.6. We use the switching sequence $S_{1,4}$, $S_{4,3}$, $S_{3,2}$, $S_{2,1}$, and we assume that i_μ circulates only on the primary side during the period $(1-D)T$, when S_1 and S_2 or S_3 and S_4 are on. (Recall that T is the period of v_d, as shown in Fig. 7.6b.) During this time, i_μ remains constant at $I_{\mu p}$ because $v_P = 0$. A possible path for I_2 while $v_P = 0$ is the freewheeling path created if D_5 and D_7 or D_6 and D_8 are on.

At the start of the period, $t = 0$, we assume that $i_\mu(0) = -I_{\mu p}$. During the interval DT, switches S_1 and S_4, and diodes D_5 and D_8, are conducting, and the primary current is

$$i_P = \frac{1}{N}I_2 - I_{\mu p} + \frac{1}{L_\mu}\int_0^t V_1 dt. \tag{7.8}$$

At $t = DT$, $i_\mu = +I_{\mu p}$ (we assume operation in the periodic steady state), S_1 turns off, and S_3 turns on, maintaining a path for i_μ. With $v_P = 0$ during the remainder of the period, for a duration $(1-D)T$, $i_\mu = I_{\mu p}$ until S_4 turns off and S_2 turns on at $t = T$. The resulting waveforms of i_μ, i_P, and i_S/N are shown in Fig. 7.8(a).

Although we assumed that I_2 freewheeled through D_5 and D_7, or D_6 and D_8, during $(1-D)T$, the possibility exists that D_5 and D_8 remain on and that D_6 and D_7 remain off during this time, causing an

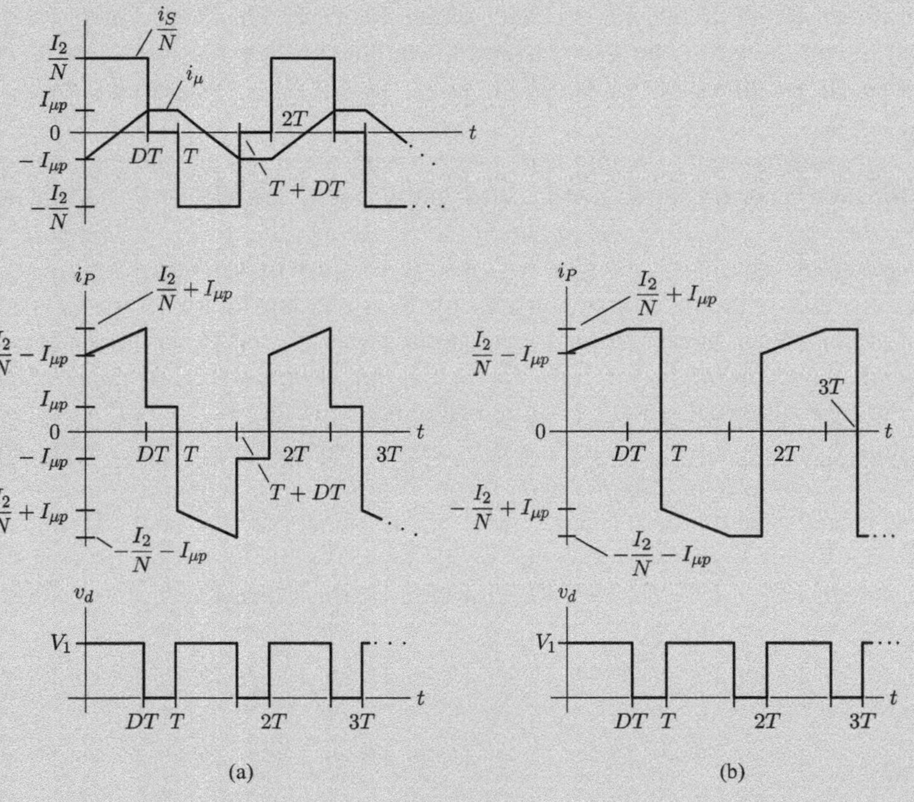

Figure 7.8 Waveforms for the circuit of Fig. 7.6. (a) I_2 freewheeling during $(1-D)T$. (b) D_5 and D_8, or D_6 and D_7, remaining off during $(1-D)T$.

additional current I_2/N to circulate through S_3 and S_4. In this case, i_P is as shown in Fig. 7.8(b). The ambiguity present in this example disappears if the transformer model includes some leakage, for then i_S could not change instantaneously at $t = DT$, and D_5 and D_8 would remain on.

7.2.2 Double-Ended Isolated Half-Bridge Converter

For the bridge converter of Fig. 7.7, the fact that the sources of the transistors are at three different potentials complicates the gate-drive circuits. We can use a center-tapped primary winding, as in the circuit shown in Fig. 7.9(a), to solve this problem. Now the emitters of both transistors connect to the same point, the negative rail of the input supply in this case. We also call this converter a *push–pull* converter because of the alternating action of the two transistors. Although we show a full-wave center-tapped, or half-bridge, rectifier on the secondary, a full-bridge rectifier circuit could be used instead.

We control the output voltage in the same way as in the full-bridge converter, but we can create the zero-voltage interval $(1 - D)T$ only by turning off both primary-side switches. If we ignore the magnetizing current, and the transformer has no leakage inductance, then both primary winding currents are zero during this time. The transformer forces $i_1' = i_2'$, splitting I_2 evenly between the two secondary windings, which are now shorted because both D_3 and D_4 are on. Under this condition, all the winding voltages must be zero, and therefore $v_d = 0$.

We now consider the effects of magnetizing current by modeling L_μ as an inductor across the lower primary winding of the ideal four-winding transformer. (We could model L_μ as being across any of the four windings (with a change in value to L_μ/N^2 when it is placed across a secondary winding), or even across the two primary or secondary windings; in the latter case the circuit symmetry is more obvious.) If S_1 is on, the voltage across L_μ is equal to $-V_1$, and i_μ ramps down linearly from its peak positive value $I_{\mu p}$. Because the switches are controlled to make v_P a symmetric waveform, $i_\mu = -I_{\mu p}$ at the end of the S_1 conduction period. During the S_1 conduction period, v_P is negative and D_3 is on, so $i_1' = I_2$. As $i_2 = -i_\mu$ during this time, $i_1 = i_\mu - I_2/N$, because the ideal transformer requires that the sum of all the Ni products of the windings be zero. When S_1 turns off at DT, i_1 becomes zero, and both $i_\mu(= -I_{\mu p})$ and I_2 must be continuous. The only way for this to happen is for both D_3 and D_4 to be on. The result is that I_2 splits between the two diodes so that the difference between the two secondary currents is $i_1' - i_2' = Ni_\mu = -NI_{\mu p}$. Combining this expression with the constraint $i_1' + i_2' = I_2$ yields

$$i_1' = (I_2 - NI_{\mu p})/2 \tag{7.9}$$

and

$$i_2' = (I_2 + NI_{\mu p})/2. \tag{7.10}$$

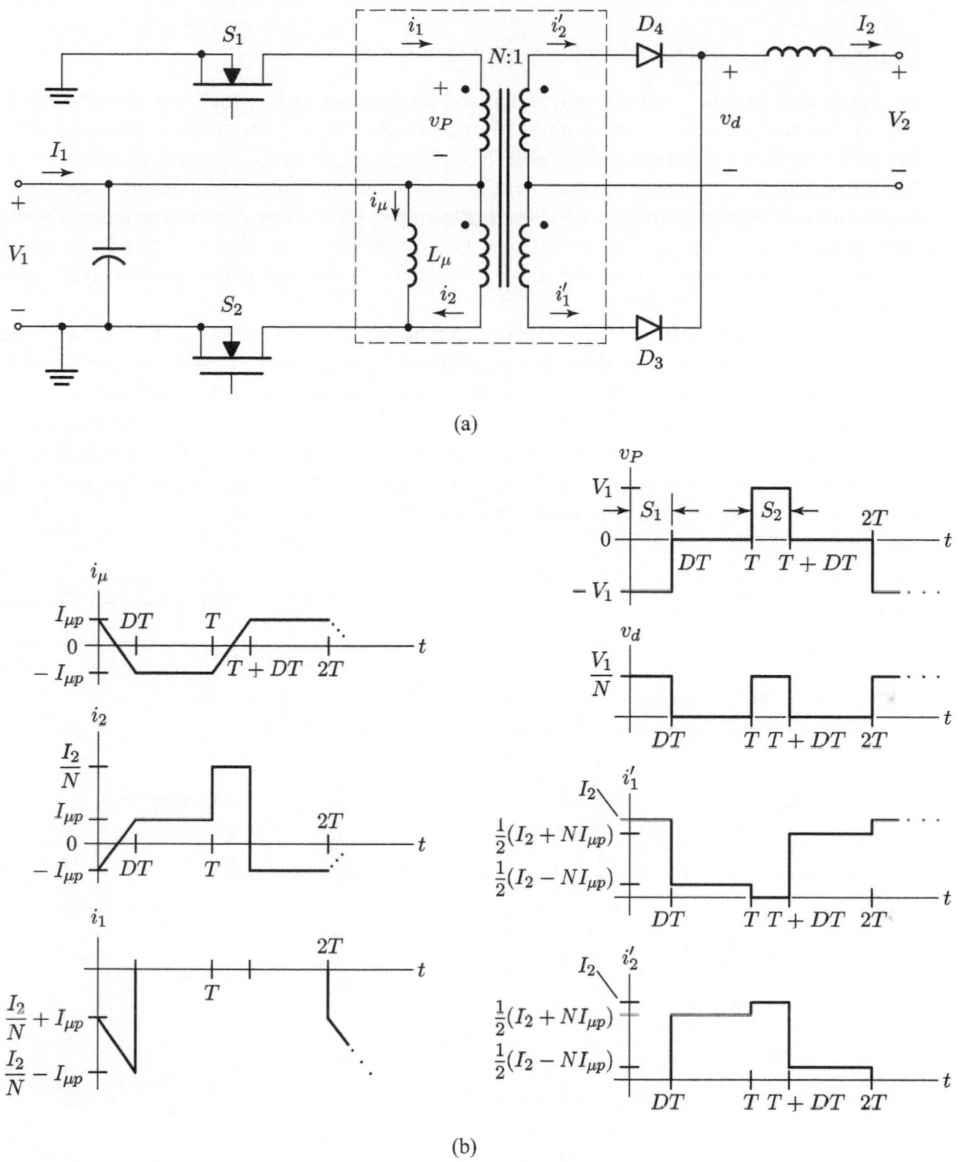

Figure 7.9 (a) A transformer-coupled half-bridge, or push–pull, converter. (b) Waveforms of branch variables.

Continuing this analysis results in the waveforms of Fig. 7.9(b), which were drawn for $I_{\mu p} < I_2/N$. In this case, the switch currents are unilateral, and the body diodes of S_1 and S_2 do not conduct. However, if $I_{\mu p} > I_2/N$, the switch currents are bilateral and the body diodes (or other means to provide bilateral switch conduction) are necessary. Note that $i_\mu(t) + i_2(t) = -i_1(t - T)$; that is, the switch currents are identical but shifted in time by T. When we recognize that L_μ is part of the transformer, we see that the circuit is symmetrical.

Example 7.4 A Current-Fed Push–Pull Converter

So far in this chapter we have only discussed power circuits in which the power flows from the voltage source side to the current source side of the canonical cell. In Chapter 6 we showed that power can flow in the other direction if we control the switches differently. To do so, in practical terms, we have to rearrange the transistors and diodes. For example, Fig. 7.10 shows the push–pull converter of Fig. 7.9(a) configured to provide for power from V_2 to V_1. In this example, we assume L_μ to be infinite. This circuit is called a *current-fed push–pull converter*. As with the boost converter on which it is based, the output voltage V_1 of this converter can range from a minimum value of V_2/N (the input voltage transformed by the turns ratio) to infinity.

The operation of a current-fed converter is slightly different from the operation of a voltage-fed converter. When we turn on only one of the switches, the input current flows through the transformer to the load, and the *load voltage* determines the voltage across the transformer. To obtain an interval of duration $(1 - D)T$ when power does not flow to the load, we turn both S_1 and S_2 *on*. The input current then splits evenly between the two primary windings, just as the load current did in the secondary windings of the voltage-fed converter when both switches were off, and the transformer voltage is forced to be zero. Branch variables for the current-fed converter are shown in Fig. 7.10(b).

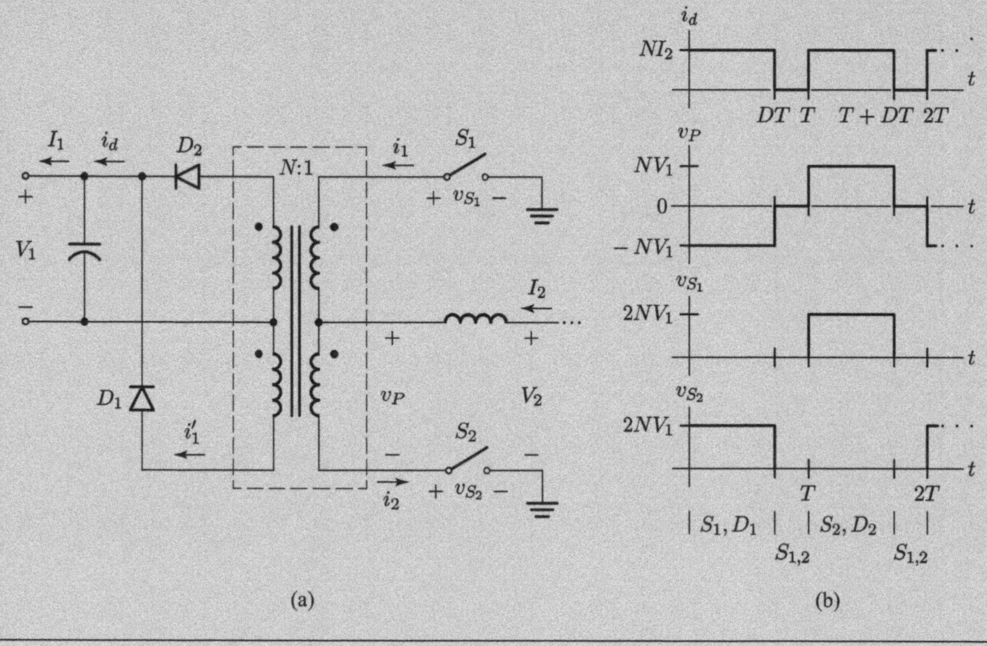

Figure 7.10 (a) Current-fed half-bridge converter of Fig. 7.9 configured to provide power from V_2 to V_1 ($L_\mu = \infty$). (b) Branch variables in the circuit of (a).

7.3 Dual Active-Bridge Converter

An interesting variant of the double-ended converter of Fig. 7.6 has the diode bridge replaced by an active bridge, as shown in Fig. 7.11. This circuit is known as a *dual active-bridge* (DAB) isolated dc/dc converter. If the two dc sources are approximately equal when scaled by the

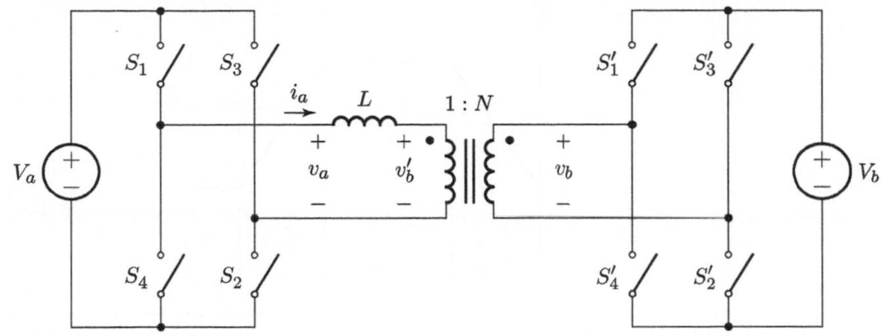

Figure 7.11 Dual active-bridge converter.

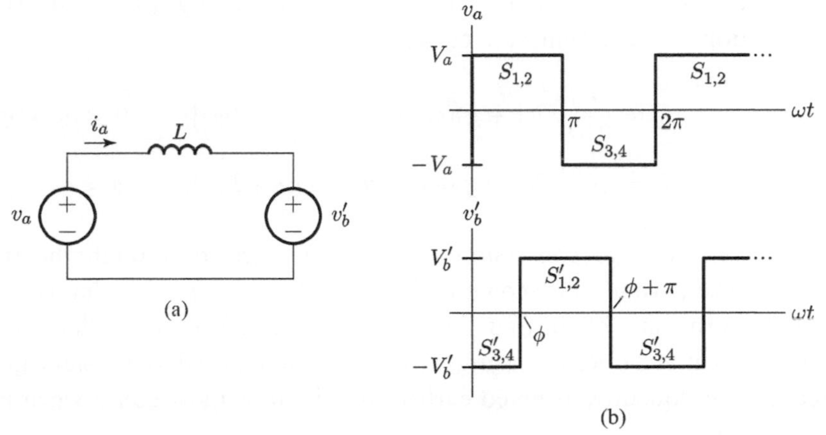

Figure 7.12 (a) Dual active-bridge converter with the output reflected to the input through the transformer. (b) The output voltage waveforms of the bridges.

transformer turns ratio, the circuit provides trapezoidal current waveforms in the transformer, yielding low conduction losses. Moreover, the circuit can provide zero-voltage switching, that is, each device turning on when the voltage across it is very low, for low loss and/or high-frequency operation. (The topic of zero-voltage or zero-current switching is discussed in Chapter 25.) Control is typically exercised through the phase shift ϕ between the square-wave outputs of the two bridges. The resulting voltage difference between the bridges controls the voltage across the inductor and therefore the current through it.

With proper design of the transformer, the inductor L in Fig. 7.11 can be realized as the transformer leakage inductance. If we reflect the voltage v_b to the primary side of the transformer as v'_b, the result is the circuit of Fig. 7.12(a), where $v'_b = v_b/N$. The waveforms for v_a and v'_b, where v'_b is phase-shifted by an angle ϕ, are shown in Fig. 7.12(b).

We now determine the current i_a and the average power delivered from V_a to V_b.

$$i_a = I(0) + \frac{v_a - v'_b}{\omega L}(\omega t) = I(0) + \frac{V_a + V'_b}{\omega L}(\omega t), \qquad 0 < \omega t < \phi, \tag{7.11}$$

$$i_a = I(\phi) + \frac{v_a - v'_b}{\omega L}(\omega t - \phi) = I(\phi) + \frac{V_a - V'_b}{\omega L}(\omega t - \phi), \quad \phi < \omega t < \pi. \tag{7.12}$$

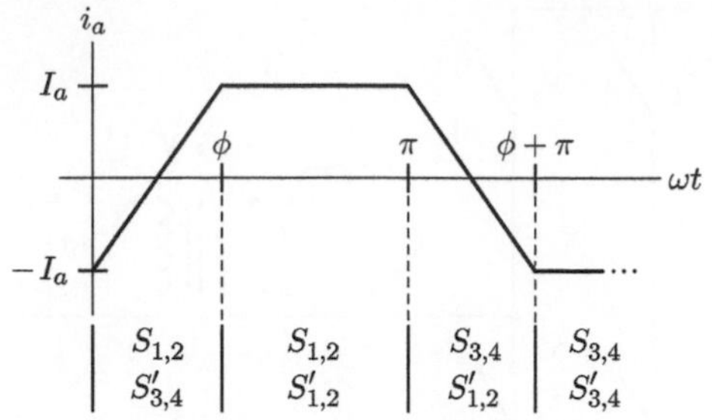

Figure 7.13 The DAB primary current, i_a, for $g \approx 1$, showing the switching sequences.

Solving (7.11) for $I(\phi)$, substituting it into (7.12), then setting $I(0) = -I(\pi)$ (the periodic steady-state condition), we determine the current i_a to be

$$i_a = \frac{V_a}{2\omega L}[2(1+g)(\omega t) - \pi(1-g) - 2g\phi], \qquad 0 < \omega t < \phi, \tag{7.13}$$

$$i_a = \frac{V_a}{2\omega L}[2(1-g)(\omega t) - \pi(1-g) + 2g\phi], \qquad \phi < \omega t < \pi, \tag{7.14}$$

where g is the voltage conversion ratio (gain) reflected through the transformer, that is, $V_b/(NV_a)$. The current i_a is shown in Fig. 7.13 assuming $g = 1$; that is, V_b/V_a is determined only by the turns ratio of the transformer. As g deviates from 1, the flat portions of i_a develop a positive or negative slope, increasing the rms current required to transfer a given power, reducing efficiency. Consequently, as noted earlier, this circuit is most useful when the required voltage ratio is close to N.

Because the waveform is trapezoidal, the MOSFET body diodes (or the channels, if gate voltage is maintained) initially conduct during transitions between $\pm V_a$. The transistor therefore turns on at zero voltage when the current commutates from the diodes to the transistors, ideally resulting in zero turn-on loss.

We can now calculate the average power, $P = \langle v_a i_a \rangle$, again assuming $g \approx 1$:

$$\begin{aligned}\langle p(t)\rangle = P &= \frac{1}{\pi}\int_0^{\pi} v_a(t) i_a(t)\, d(\omega t) \\ &= \frac{V_a^2}{\pi\omega L}\left[\int_0^{\phi}(2\omega t - \phi)\, d(\omega t) + \int_{\phi}^{\pi}\phi\, d(\omega t)\right] = \frac{V_a^2}{\omega L}\left(\phi - \frac{\phi^2}{\pi}\right).\end{aligned} \tag{7.15}$$

This result can be generalized to the cases where (i) $V_b \neq NV_a$ and (ii) $\phi < 0$. The result is almost identical to (7.15):

$$P = \frac{V_a V_b}{N\omega L}\phi\left(1 - \frac{|\phi|}{\pi}\right), \qquad -\pi < \phi < \pi. \tag{7.16}$$

Unlike classic dc/dc converters where the duty ratio, D, is the variable controlling the output voltage, the DAB output voltage is indirectly controlled by varying ϕ, and thus the power P, in response to the fed-back voltage V_b.

7.4 Flyback Converter

We can easily transform the buck/boost converter into a circuit that provides electrical isolation, because the voltage across its filter inductor has no dc component. For example, to provide isolation in the buck/boost circuit of Fig. 5.19(b), we substitute a transformer having a magnetizing inductance of value $L_\mu = L$ for the filter inductor L in the circuit, as shown in Fig. 7.14(a). When Q turns on, D turns off, and energy flows into L_μ. When Q turns off, D turns on, and the stored magnetizing energy flows to the load through the secondary winding. Cathode-ray tube television receivers had a converter of this type with $N \ll 1$ to generate a very high voltage that drove the retrace of the beam in the picture tube. What happens visually is that the spot on the screen "flies back" to start another horizontal line. Thus the isolated buck/boost converter is commonly known as a *flyback converter*. Because the input and output need not share the same common node, the switch Q in Fig. 7.14(a) is typically repositioned so that its control port is referenced more conveniently to the negative of the input supply, as shown in Fig. 7.14(c).

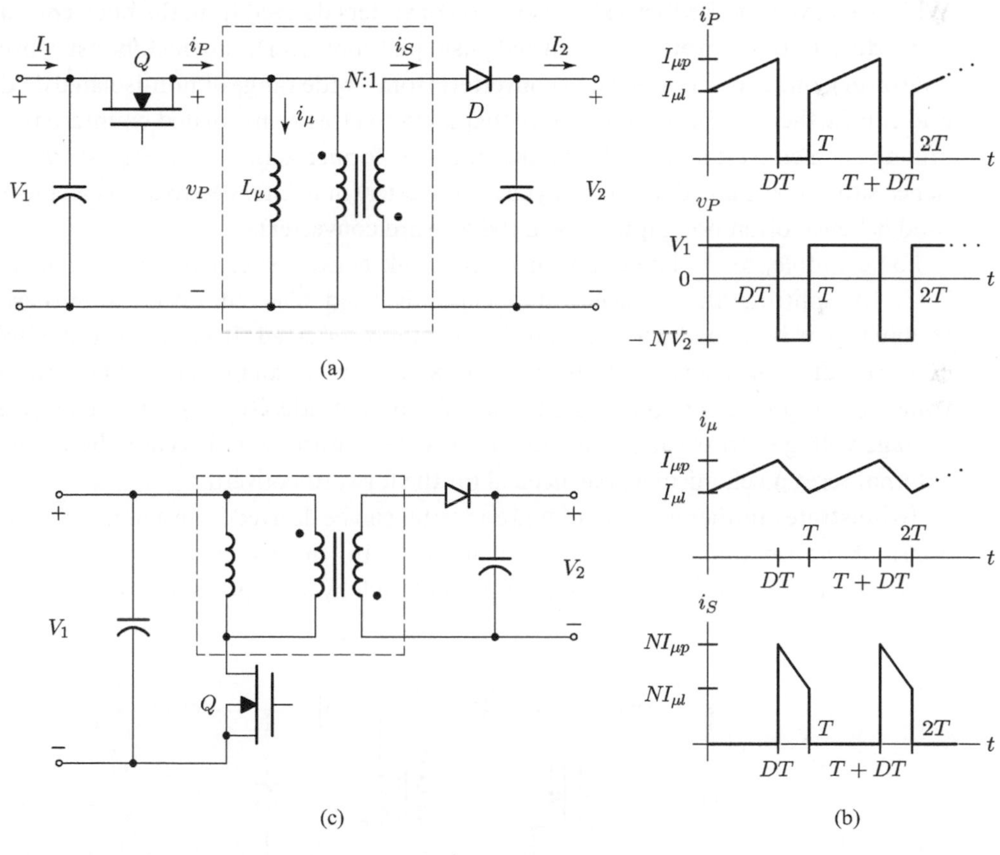

Figure 7.14 (a) The flyback converter, an isolated version of the buck/boost converter of Fig. 5.19(b). (b) Branch variables for the converter of (a). (c) The circuit of (a) with the transistor moved to provide for a ground-referenced gate drive.

If it were not for leakage inductance, the operation of the flyback converter would be identical to that of the non-isolated buck/boost converter, except for the step-up or step-down created by the transformer. Primary-side leakage stores energy that, in the circuit of Fig. 7.14(a), has no place to go when the transistor turns off. We usually place an *RC* circuit, called a snubber, across the transistor to solve this problem (see Chapter 25).

Unlike the transformers used to provide isolation in the direct converter topologies discussed previously – which are transformers designed for minimum magnetizing current – the magnetizing inductance of the transformer used in a flyback converter must have a specified value in order to store energy. Such transformers are known as *energy-storage transformers*, and in the case of the flyback converter, they are called *flyback transformers*. The transformer core is typically gapped to store the required energy, just as the inductor core in a non-isolated buck/boost converter would be. The purpose of the gap is discussed in Chapter 18.

7.5 Other PWM Isolated Converters

While we have thus far focused on isolated converters derived from the buck converter (forward converter), boost converter (current-fed push–pull converter), and buck/boost converter (flyback converter), one can derive isolated converters from a wide range of non-isolated dc/dc converters, and can further synthesize topologies that have no clear non-isolated counterpart. Distinctions among such converters include whether the transformer acts as an energy storage element (and hence should be gapped, as in the flyback converter) or is intended to act as an ideal transformer (and is hence often not gapped, as in the forward converter).

For example, an isolated version of the Ćuk converter can be realized as shown in Fig. 7.15. By splitting the capacitor in the non-isolated Ćuk converter, a zero-average-voltage terminal pair is created where a transformer can be inserted. In contrast to the isolated SEPIC converter discussed next, the transformer does not replace an inductor, but operates like an ideal transformer, providing voltage gain and isolation but (ideally) storing no energy. Because zero average voltage across the transformer is provided naturally in this case, there is no need for an external reset mechanism as was needed for the forward converter.

To illustrate another way an isolated converter can be derived from a non-isolated counterpart, consider the isolated SEPIC converter of Fig. 7.16. In this circuit, one of the two inductors in the non-isolated SEPIC converter is replaced by a gapped-core transformer, such that

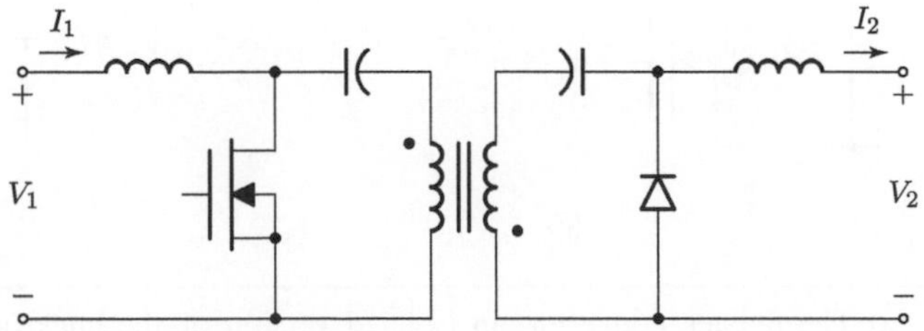

Figure 7.15 The isolated Ćuk converter.

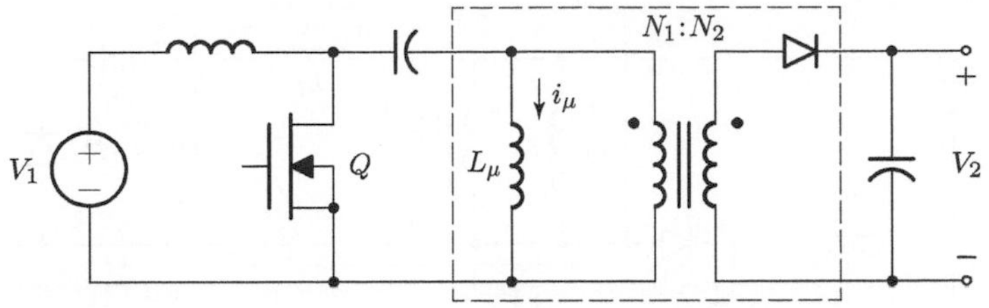

Figure 7.16 The isolated SEPIC converter.

the transformer magnetizing inductance takes the place of the original inductor. We thereby obtain electrical isolation and an additional factor in converter voltage gain of N_2/N_1 without disrupting the underlying operation of the converter. An isolated *Zeta converter* (or "inverse SEPIC") can likewise be created through inductor substitution, as explored in a problem at the end of the chapter.

7.6 Effects of Transformer Leakage Inductance

So far in this chapter we have not considered the detailed effects of transformer leakage inductance. In this section we describe how the primary to secondary winding leakage inductance affects the operation of isolated converters.

7.6.1 Leakage Effects in the Single-Ended Converter

In the single-ended converter without leakage shown in Fig. 7.1, the load current immediately commutates from D_2 to D_1 when Q turns on, and the voltage v_d makes a step change from zero to V_1/N. Figure 7.17(a) shows the same converter with leakage modeled as an inductor L_ℓ in series with the secondary winding. Now when the transistor turns on, the current in D_1 cannot make a step change. Instead, there is a commutation period of duration τ_{u_1} during which both diodes are on and the commutating voltage is $v_x = V_1/N$. The current in L_ℓ then increases linearly with slope V_1/NL_ℓ until it reaches the full load current I_2, as shown in the waveforms of Fig. 7.17(b). The time τ_{u_1} is

$$\tau_{u_1} = \frac{NL_\ell I_2}{V_1}. \tag{7.17}$$

Only at τ_{u_1} does D_2 turn off and allow the voltage v_d to make its step change to V_1/N.

Because $v_d = 0$ instead of V_1/N during the commutation period, the output voltage $V_2 = \langle v_d \rangle$ is less than DV_1/N, its value without leakage. Therefore, to maintain a desired output voltage over a specific range of input voltage, we must either decrease the turns ratio N or increase the maximum duty ratio. For the same values of V_2 and I_2, decreasing N increases the peak and rms values of the transistor current, and the reverse voltage at the diodes. Increasing the maximum duty ratio increases the transistor's required voltage rating.

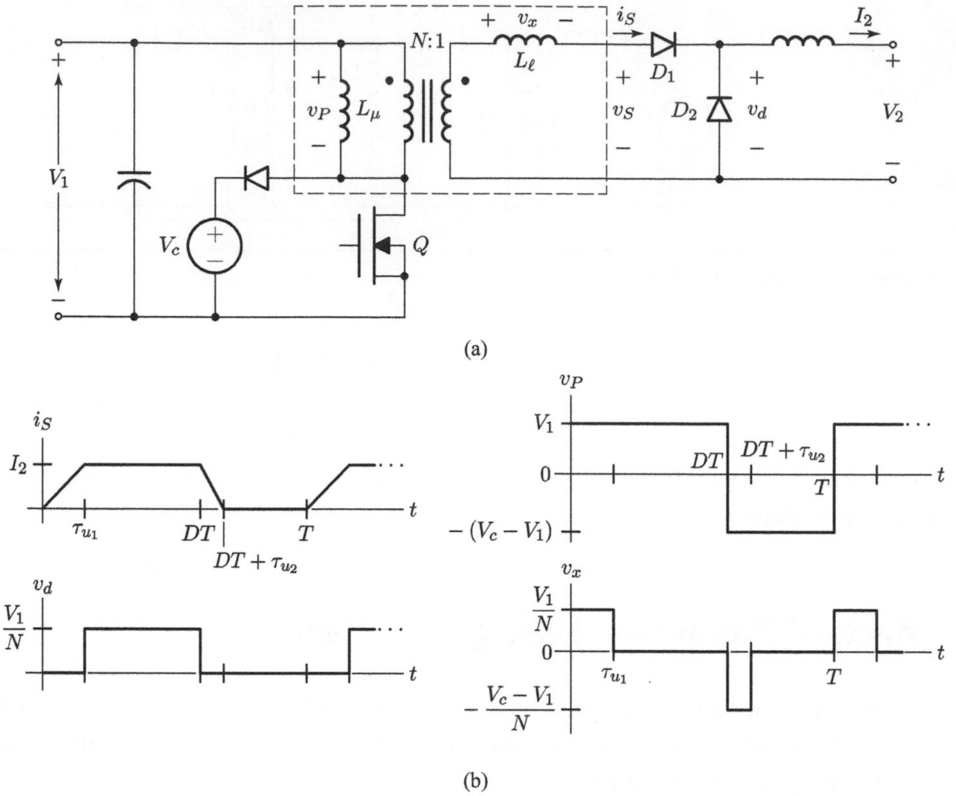

Figure 7.17 (a) A single-ended forward converter with leakage inductance. (b) The behavior of i_S, v_d, v_P, and the commutating voltage v_x in the converter of (a).

Turning the transistor off transfers the energy stored in L_ℓ to the clamp voltage source V_c in the following manner. The transformer primary voltage v_P immediately steps from $+V_1$ to $-(V_c - V_1)$, initiating commutation of I_2 from D_1 to D_2. The commutating voltage is now $-(V_c - V_1)/N$, so this commutation period, τ_{u_2}, is different from τ_{u_1}. During this time, the direction of energy flow is from the secondary winding to the primary winding, as the polarities of v_P and i_S show.

Example 7.5 Determining *D* for a Single-Ended Forward Converter with Leakage Inductance

The single-ended forward converter of Fig. 7.17(a) has the following parameters:

$$V_1 = 50\,\text{V}, \quad V_2 = 25\,\text{V}, \quad I_2 = 8\,\text{A}, \quad N = 1, \quad L_\ell = 2.5\,\mu\text{H}, \quad L_\mu = \infty, \quad f_s = 200\,\text{kHz}.$$

What must D be to give the specified 5 V output?

With leakage inductance, we need to increase the duty ratio to account for the commutation interval τ_{u_1}. From (7.17), we determine this time to be

$$\tau_{u_1} = \frac{(1)(2.5 \times 10^{-6})(8)}{50} = 0.4\,\mu\text{s}. \tag{7.18}$$

This time corresponds to 8% of the switching period. Therefore, $D = 58\%$. The clamp voltage must be high enough to satisfy the condition that the core be reset at this duty ratio:

$$V_1 D < (V_c - V_1)(1 - D). \quad (7.19)$$

From this condition, we determine that $V_c > 120$ V.

If this converter had no leakage, the duty ratio would be 50%, and V_c could be as low as 100 V. Because the transistor voltage rating must be at least as high as V_c, another result of leakage in this circuit is to increase the transistor voltage rating.

The peak stored energy E_ℓ in L_ℓ is

$$E_\ell = \frac{1}{2} L_\ell I_2^2 = 80\ \mu\text{J}. \quad (7.20)$$

However, because the input source V_1 is also supplying energy while L_ℓ is discharging, more than E_ℓ is absorbed by the clamp. We leave the calculation of this energy to an end-of-chapter problem, but the result in this case is that $E_c = 137\ \mu$J, which is approximately 70% greater than E_ℓ. If this energy were dissipated every cycle, the power lost from leakage inductance would be 27.5 W, or about 14% of the output power. If instead we use the more efficient placement of the discharge resistor discussed in Example 7.2, we lose only E_ℓ per cycle, or 8% of the output power. Even so, the efficiency demands on power converter circuits today would motivate reduction of this loss, either through redesigning the transformer for lower leakage inductance or using a circuit variant capable of fully recovering clamp energy.

Note from Example 7.5 that the energy lost from leakage inductance in a single-ended forward converter is significant. For this reason, we want to return the clamp energy to the input voltage source, such as by using the methods shown in Figs. 7.3 and 7.4. Double-ended circuits automatically return the clamp energy to the source.

The presence of leakage also increases the required voltage rating of the transistor in the flyback converter – but for a different reason. Figure 7.18 shows a flyback converter from which the ideal part of the transformer has been removed, which does not change its operation in any

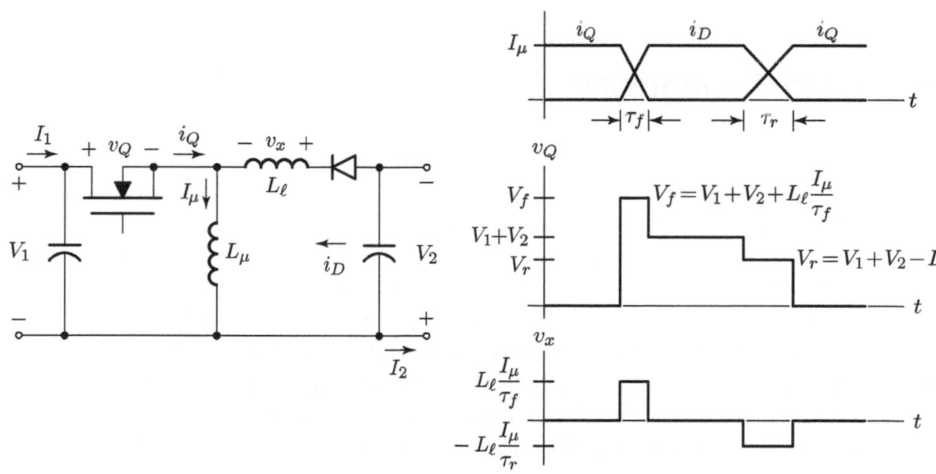

Figure 7.18 A simplified flyback converter analyzed to show the effect of leakage inductance. The magnetizing current is assumed constant at I_μ.

way that is significant to this discussion. For simplicity we also assume that L_μ is large enough that i_μ is constant at I_μ. If we now turn off Q so that i_Q falls linearly to zero in time τ_f, then during this fall time $v_x = L_\ell(I_\mu/\tau_f)$, and the transistor voltage is

$$v_Q = V_1 + V_2 + L_\ell\left(\frac{I_\mu}{\tau_f}\right). \tag{7.21}$$

This voltage is greater than the transistor voltage in the absence of leakage (namely $V_1 + V_2$) by the amount necessary to force the current to commutate from Q to L_ℓ. If we were to clamp the transistor voltage at $V_1 + V_2$ while it was turning off, the current would never commutate to L_ℓ. Therefore, leakage increases the required voltage rating of the transistor by an amount determined by the desired switching speed. This characteristic often makes the flyback topology less desirable than the forward converter topology, particularly at power levels high enough to justify the cost of the additional parts for the forward converter.

7.7 Converters with Multiple Outputs

Isolated converters supplying power to electronic equipment, such as computers, must usually provide several outputs at different voltages. Rather than make a separate supply for each output, it is often better to add additional secondary windings to the transformer, each with its own rectifier and output filter. The problem created by this approach is that, invariably, each output requires a slightly different value of duty ratio D, because of different load regulation characteristics of the different secondary windings and rectifiers.

One approach to solving this problem is to control the highest power output directly with the duty ratio and to adjust the turns ratios of the other outputs to ensure that they will always be slightly high. We can then use a linear regulator† on each of these lower power outputs to give the desired voltage.

Notes and Bibliography

Some of the practical problems of using transformers in dc/dc converters are described nicely in Section 4.4 of [1]. This classic textbook presents a wide range of isolated converter topologies. There are numerous variants and combinations of the circuits described in this chapter; for example, the "flyforward" converter – a hybrid between the forward converter and flyback converter – is described in [2]. The active-clamp technique is widely employed commercially and was the subject of substantial patent litigation. To the authors' knowledge, [3] is the earliest paper describing the use of this technique in a forward converter. Introduction of isolation and multiple outputs in single-switch PWM converter topologies is exemplified by the isolated Ćuk converter, described in [4]. The dual active-bridge converter was introduced in [5] along with a number of variants. When isolation is not required but the voltage transformation benefits of a transformer are desired, tapped-inductor topologies are sometimes employed, as described in [6] and explored in the problems.

† A variable resistance created by a transistor operating in its active gain region.

1. R. Severns and G. Bloom, *Modern DC-to-DC Switchmode Power Converter Circuits*, New York: Van Nostrand, 1985.
2. J. N. Park and T. R. Zaloum, "A Dual Mode Forward/Flyback Converter," *IEEE Power Electronics Specialists Conference (PESC)*, pp. 3–13, June 14–17, 1982.
3. B. Carsten, "High Power SMPS Require Intrinsic Reliability," *Proc. Third Annual International Power Conversion Conference (PCI)*, pp. 118–133, Sept. 14–17, 1981.
4. R. D. Middlebrook and S. Ćuk, "Isolation and Multiple Outputs of a New Optimum Topology Switching DC-to-DC Converter," *IEEE Power Electronics Specialists Conference (PESC)*, pp. 256–264, 1978.
5. R. W. A. A. De Doncker, D. M. Divan, and M. H. Kheraluwala, "A Three-Phase Soft-Switched High-Power-Density DC/DC Converter for High-Power Applications," *IEEE Trans. Industry Applications*, Vol. 27, pp. 63–73, Jan./Feb. 1991.
6. D. A. Grant, Y. Darroman, and J. Suter, "Synthesis of Tapped-Inductor Switched-Mode Converters," *IEEE Trans. Power Electronics*, Vol. 22, pp. 1964–1969, Sept. 2007.

PROBLEMS

7.1 Reconsider Example 7.1, but assume now that the output voltage is held constant at 5 V, while the input voltage varies between 50 and 100 V, that is, $50 < V_1 < 100$ V. What is the minimum value for V_c?

7.2 We stated in Section 7.2.1 that the option of opening all the switches in the double-ended bridge converter of Fig. 7.6(a) during the period $(1 - D)T$ created a problem if the transformer contained leakage. Describe this problem, and argue that it makes no difference whether the leakage is on the primary or secondary side of the transformer.

7.3 Repeat Example 7.3 under the condition that $I_{\mu p} > I_2/N$.

7.4 Determine and sketch the waveforms of i_μ, i_P, and i_S in Fig. 7.4(b) when leakage inductance is present between the primary and secondary windings of the transformer. Assume that all the leakage is on the secondary side, as shown in Fig. 7.19.

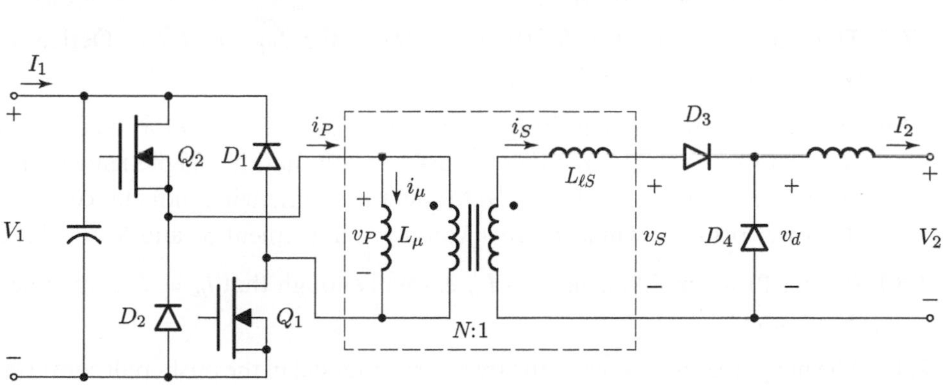

Figure 7.19 The single-ended converter of Fig. 7.4(b) with the addition of secondary leakage inductance $L_{\ell S}$. The effect of this leakage is the subject of Problem 7.4.

7.5 Repeat Example 7.3 under the assumption that the transformer has secondary-side leakage $L_{\ell S}$, as shown in Fig. 7.20(a). How does the behavior of the circuit differ if the leakage is on the primary side, that is, $L_{\ell P} = N^2 L_{\ell S}$, as shown in Fig. 7.20(b)?

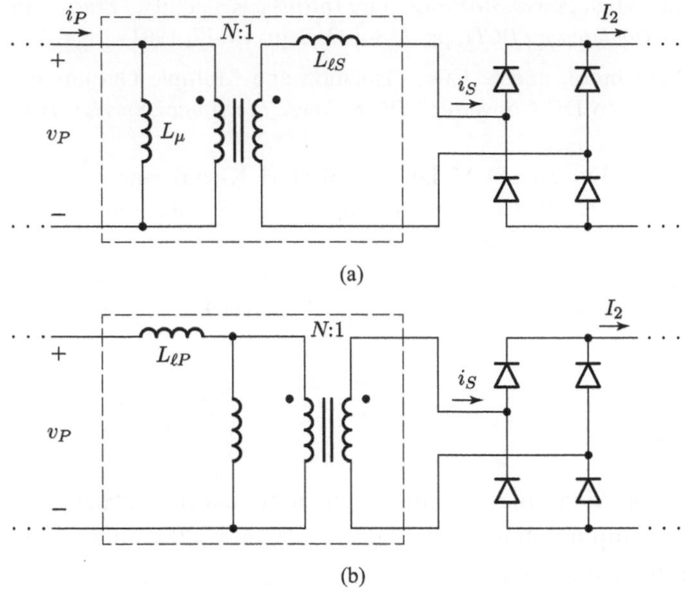

Figure 7.20 The double-ended converter of Fig. 7.6 with the addition of transformer leakage inductance. (a) All the leakage referred to the secondary winding. (b) All the leakage referred to the primary winding. This circuit is the subject of Problem 7.5.

7.6 Explain why the switch implementation shown in Fig. 7.7 will not permit the first switching option (i.e., all switches off) during the period $(1 - D)T$.

7.7 Compare the switch voltage stresses of the bridge converter of Fig. 7.7 and the push–pull converter of Fig. 7.9.

7.8 The waveforms of Fig. 7.9(b) were drawn for $I_{\mu p} < I_2/N$. Derive and draw these waveforms for the case $I_{\mu p} > I_2/N$.

7.9 Determine and sketch the branch variables i_1, i_μ, i_d, and v_P for the current-fed push–pull converter of Fig. 7.10(a) under the assumption that the magnetizing current is no longer negligible but that $0 < I_{\mu p} < I_2$. Model the magnetizing inductance by placing it across the lower primary winding. How would you implement S_1 and S_2 for this circuit?

7.10 Repeat Problem 7.9, assuming L_μ is small enough that $i_\mu = I_2$ sometime in the interval $0 < t < DT$.

7.11 The magnetizing branch of the transformer model in the push–pull converter of Fig. 7.9(a) was arbitrarily placed across the lower primary winding in the transformer model. However, the magnetizing branch may be connected in several places in the model. Figure 7.21 shows an alternative placement, where L'_μ is across the entire secondary winding.

Show, by deriving and plotting the real transformer's terminal currents i_1, i_2, i'_1, and i'_2, that the behavior of this circuit is identical to that of Fig. 7.9. How is L'_μ related to L_μ in Fig. 7.9(a)?

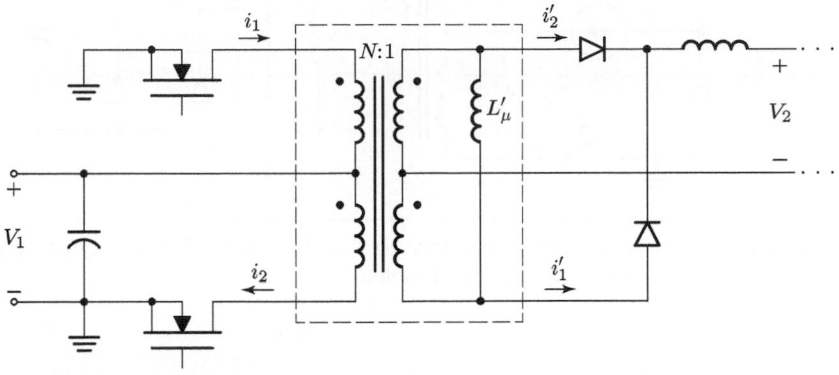

Figure 7.21 Push–pull converter of Fig. 7.9(a) redrawn with the magnetizing inductance L'_μ across the secondary winding. This circuit is analyzed in Problem 7.11.

7.12 In many respects, the circuit of Fig. 7.10(a) is the dual of the circuit of Fig. 7.9(a). With this in mind, we might expect that the switch implementations, for nonzero magnetizing currents, would be duals of each other. As Problems 7.9 and 7.10 showed, this is not so. Where does the duality between these two circuits break down?

7.13 How do the device stresses in the isolated flyback converter of Fig. 7.14(a) compare with those in the non-isolated version having the same input voltage and the same output voltage?

7.14 Determine and sketch the branch variables shown in Fig. 7.14(b), assuming secondary-side leakage.

7.15 When the transistor turns off in the circuit of Fig. 7.17(a), the energy absorbed by the clamp source V_c is greater than the energy stored in the transformer leakage L_ℓ. Determine and plot the energy absorbed by the clamp as a function of the ratio V_1/V_c. Interpret the result for $V_c = V_1$. What is the disadvantage of making V_c very large?

7.16 The center-tapped converter topology of Fig. 7.9(a) can also be used to provide power flow from right to left, that is, from V_2 to V_1. How would you implement and control the switches in this case?

7.17 Determine the conversion ratio V_1/V_2 as a function of D for the current-fed half-bridge circuit of Fig. 7.10(a).

7.18 Figure 7.22 shows the current-fed push–pull converter of Example 7.4 with the magnetizing inductance L_μ now included. Determine the behavior of this circuit for the case $I_{\mu p} < NI_2$. Sketch i_1, i_2, i_μ, and i_d. What is the maximum possible value of i_μ? Does the bridge rectifier change the behavior of this circuit from that of Fig. 7.10(a)? How would you implement S_1 and S_2 in the circuit of Fig. 7.22?

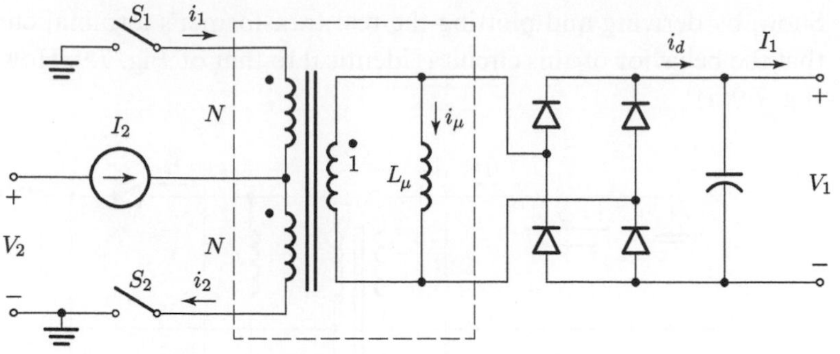

Figure 7.22 The current-fed push–pull converter of Example 7.4 with magnetizing inductance included. This circuit is the subject of Problem 7.18.

7.19 We stated in Example 7.2 that the ratio of E_R to E_c is equal to the ratio of the resistor voltage to the clamp voltage. Show that this statement is true.

7.20 What is the stress factor for the transistors in the hybrid-bridge converter of Fig. 7.4(b)? (Neglect magnetizing current.) Based only on minimizing the switch-stress factor, at what conversion ratio V_2/V_1 would you choose this isolated circuit over the non-isolated direct converter?

7.21 Show that the switch currents in the circuit of Fig. 7.9(a) are symmetrical.

7.22 Consider the actively clamped forward converter of Figure 7.5 discussed in Section 7.1.5.

(a) Sketch the shape of the magnetizing inductance current in periodic steady-state operation for $V_1 = 48$ V, $D = 0.4$, $T = 5\,\mu$s, and $L_{\mu 1} = 50\,\mu$H (magnetizing inductance referred to the primary side). Does this transformer have a unipolar or bipolar flux swing?

(b) To minimize the energy stored in the transformer magnetizing inductance, should the transformer be wound on a gapped core or an ungapped core? (Recall that the peak stored energy is $0.5 L_\mu I^2_{\mu(\text{peak})}$). This topology has been widely used in dc/dc converters for telecommunications applications, and was the subject of substantial litigation.

7.23 Consider the isolated Zeta converter of Fig. 7.23. As with the flyback converter, transformer magnetizing inductance is important, and the transformer acts as an energy storage element. The transformer magnetizing inductance is illustrated as an explicit circuit element in the figure. You may assume the converter operates in heavy continuous conduction mode.

(a) What is the input-to-output conversion ratio with a transformer turns ratio of N_1:N_2?

(b) If the nominal input voltage is 12 V, and the desired output voltage is 100 V, what is the optimum transformer turns ratio to minimize the device stresses? At what nominal duty ratio would the converter operate in this case?

(c) How do the device stresses of this case compare to that for a non-isolated indirect converter with the same input and output voltage magnitudes (cf. Section 5.7).

(d) Would it be better to use a gapped core or ungapped core for the transformer? Why?

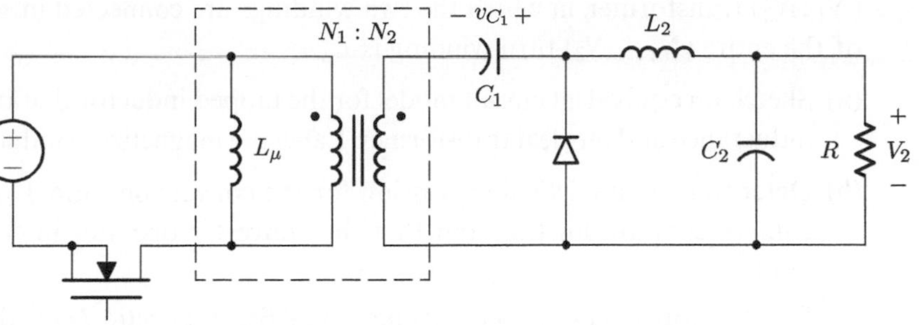

Figure 7.23 The isolated Zeta converter of Problem 7.23. The magnetizing inductance of the transformer is shown as an explicit circuit element.

7.24 Consider the isolated SEPIC converter of Fig. 7.24. (As with the flyback converter, transformer magnetizing inductance is important, and the transformer acts as an energy storage element. The transformer magnetizing inductance is illustrated as an explicit circuit element in the figure.) You may assume the converter operates in heavy continuous conduction mode.

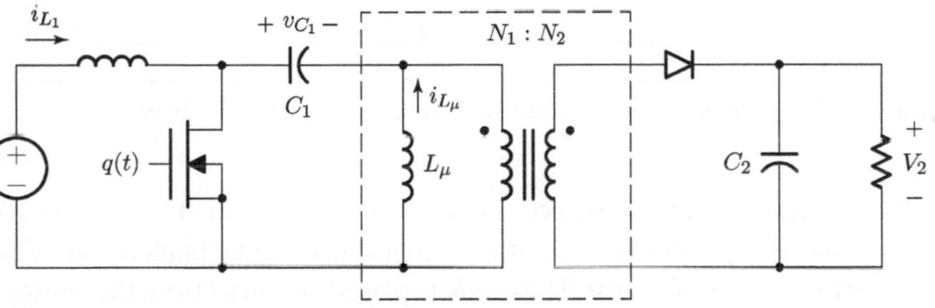

Figure 7.24 The isolated SEPIC converter of Problem 7.24, including the magnetizing inductance of the transformer.

(a) What is the input-to-output conversion ratio with a transformer turns ratio of $1{:}N$?

(b) If the nominal input voltage is 12 V, and the desired output voltage is 60 V, what is the optimum transformer turns ratio to minimize the device stresses? At what nominal duty ratio would the converter operate in this case?

(c) How do the device stresses of this case compare to that for a non-isolated indirect converter with the same input and output voltage magnitudes (cf. Section 5.7).

(d) Would it be better to build the transformer with a gapped core or an ungapped core? Why?

7.25 Figure 7.25 shows a tapped-inductor boost converter. This approach is sometimes used to achieve larger conversion ratios as compared to a conventional boost design. The inductor winding contains a total of $N_1 + N_2$ turns. The switch is placed N_1 turns from the source side of the inductor, as shown. The tapped inductor can be viewed as a two-winding ($N_1 : N_2$) transformer, in which the two windings are connected in series. The inductance of the entire ($N_1 + N_2$)-turn winding is L.

(a) Sketch an equivalent circuit model for the tapped inductor that includes a magnetizing inductance and an ideal transformer. Label the magnetizing inductance and turns ratio.

(b) Determine an analytical expression for the conversion ratio V_2/V_1 assuming that all components are lossless, and that the converter operates in continuous conduction mode.

(c) Plot the conversion ratio as a function of the duty ratio D for the case $N_1 = N_2$, and compare to the conventional non-tapped case ($N_2 = 0$).

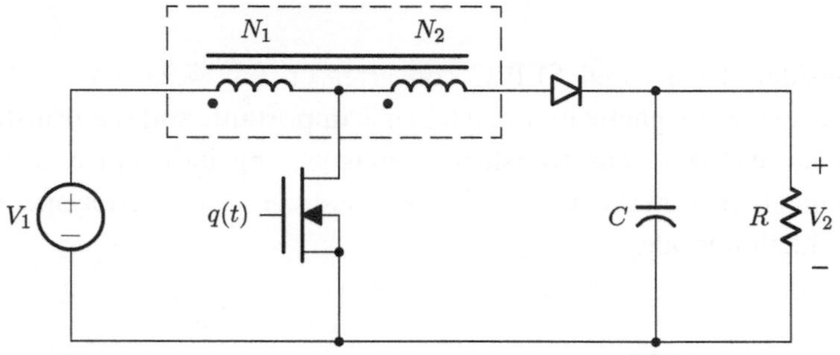

Figure 7.25 A tapped-inductor boost converter, the subject of Problem 7.25.

7.26 Figure 7.26 shows a tapped-inductor buck converter. This approach is sometimes used to achieve larger conversion ratios than a conventional buck design. The inductor contains a total of $N_1 + N_2$ turns. The diode is placed N_1 turns from the source side of the inductor, as shown. The tapped inductor can be viewed as a two-winding ($N_1 : N_2$) transformer, in which the two windings are connected in series. The inductance of the entire (N_1+N_2)-turn winding is L.

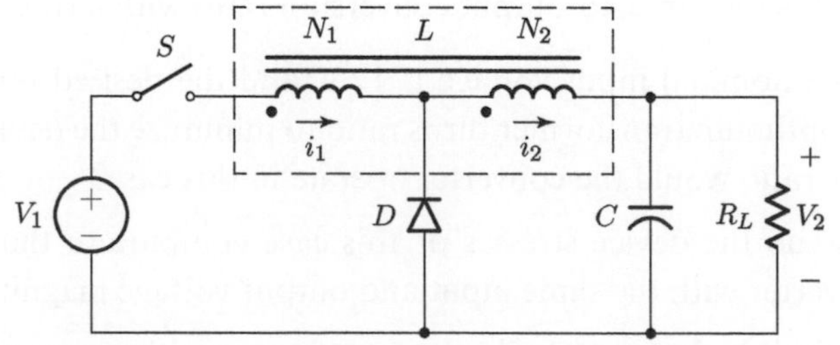

Figure 7.26 The tapped-inductor buck converter analyzed in Problem 7.26.

(a) Sketch an equivalent circuit model for the tapped inductor that includes a magnetizing inductance and an ideal transformer. Label the magnetizing inductance and turns ratio.

(b) Find an analytical expression for the conversion ratio V_2/V_1 assuming that all the components are lossless, and that the converter operates in continuous conduction mode.

(c) Plot the conversion ratio as a function of the duty ratio D for the case $N_1 = N_2$, and compare to the conventional non-tapped case ($N_1 = 0$).

7.27 Figure 7.27 shows a version of an "asymmetric push–pull" isolated dc/dc converter. The two switches in the half-bridge operate in a complementary fashion, such that the top switch has a duty ratio D and the bottom switch has a duty ratio $1 - D$. The transformer is "series wound" (wound on a core with one magnetic flux path) with N_p primary turns and two secondaries each with N_s turns. (This configuration is sometimes referred to as having a "center-tapped" secondary.) Assume the transformer magnetizing inductance is infinite and its leakage inductance is zero. Also assume that the converter operates in continuous conduction mode with negligibly small ripple in the inductor current i_L and the capacitor voltages v_{C_1} and v_2.

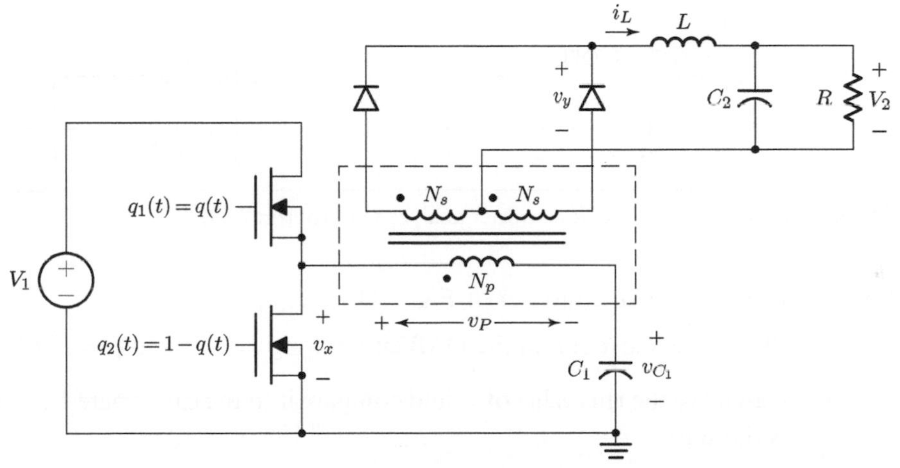

Figure 7.27 One version of the "asymmetric push–pull" dc/dc converter analyzed in Problem 7.27.

(a) In periodic steady state, what is the dc voltage v_{C_1} on capacitor C_1?

(b) Plot and dimension v_X and v_Y in terms of V_1, D, N_P, N_S, and the switching period T.

(c) What is the periodic steady-state voltage conversion ratio V_2/V_1 of this topology as a function of the switch duty ratio D?

(d) What are the off-state voltages on the switches and diodes in terms of V_1 and D? (Note that the two diodes see different voltages in the off state at a given duty ratio. Indicate both off-state voltages.)

7.28 Show that (7.16) is correct for the conditions where $g \neq 1$ and $-\pi < \phi < \pi$.

7.29 Figure 7.28 shows an isolated dc/dc converter and the switching pattern for its two switches. For all parts of this question assume that C_1, C_2, L_1, and L_2 are large so the converter operates in continuous conduction mode and ripple can be neglected. Also assume that the transformer is ideal.

(a) Find the peak currents carried by switches S_1 and S_2 in terms of currents I_1 and I_2.

(b) Find the periodic steady-state relationship between voltages V_1 and V_2 for this converter in terms of the duty ratio.

(c) Given that $V_1 > 0$, show the simplest switch implementation to allow power to flow from V_2 to V_1 in periodic steady state.

(d) Referring to Chapter 5, what type of converter is this?

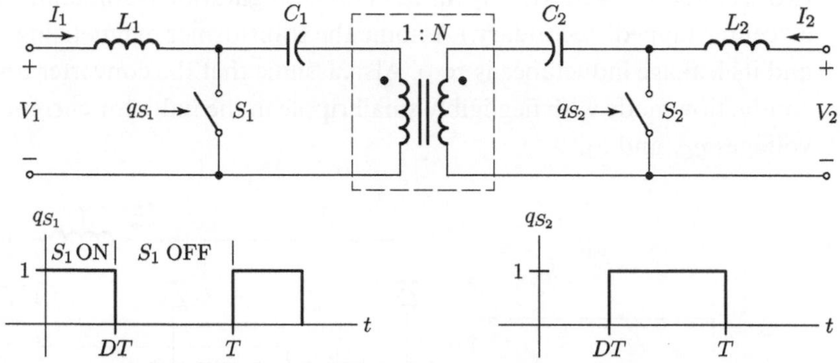

Figure 7.28 The isolated dc/dc converter analyzed in Problem 7.29.

7.30 Consider the DAB converter of Fig. 7.11.

(a) Sketch the current i_a in the DAB circuit for $g = 1.25$ and $\phi = \pi/6$.

(b) Calculate the rms value of i_a and compare it to the case where $V_a = V_b$ and the power is the same.

8 Single-Phase Switched-Mode DC/AC Converters

The phase-controlled dc/ac converter introduced in Chapter 4 requires that the external ac system be a voltage source, typically the ac utility line. This condition is necessary because the phase-controlled converter uses the reversal of the ac voltage to drive the commutation process. Therefore the ac frequency in these circuits is constrained to be that of the ac source. In this chapter we remove the restriction that the ac system be a voltage source, realizing that by doing so we must use means other than line commutation to turn devices off. These circuits, which we call *switched-mode dc/ac converters*, use fully controllable devices such as MOSFETs and IGBTs, and switch at frequencies that are higher (often much higher) than the ac-side frequency.

A common application of variable-frequency dc/ac converters is to drive ac motors at varying speeds. These *variable-speed drives* are used to control the speed of electrically driven vehicles; to vary the speed of pumps and compressors so that they operate at maximum efficiency under varying loads; to control converter speeds; to control and coordinate the speed of sequential rollers in manufacturing operations such as those found in steel, paper, and textile mills; and to control the speed and positioning of machine tools. Other applications of variable-frequency converters include uninterruptible power supplies (which use batteries to provide standby ac power), frequency changers, mobile power supplies, and systems that match ac loads to alternative energy sources that produce dc, such as photovoltaic arrays.

In the discussion of dc/dc converters in Chapter 5, we emphasized that inputs and outputs remain undefined until we specify the switches and external networks. The same is true of dc/ac converters. Until we specify the types of switches to be used, the networks connected to the converter ports, and sometimes the converter controls, we cannot tell whether power is flowing from the dc side to the ac side (inversion), or vice versa (rectification).

8.1 Basic Variable-Frequency Bridge Converter

The basic bridge converter circuit is shown in Fig. 8.1(a). It is nothing more than a bridge connection of switches controlled so as to periodically reverse the polarity of the voltage applied across the ac system. Here, the ac system is a resistive load, R (so the circuit is an inverter), and in Fig. 8.1(b) the ac voltage shown is a square wave. The switches are controlled so that either switches S_1 and S_4 or S_2 and S_3 are on; that is, the source is always connected to the load. The frequency is controlled by the switching rate. An inverter circuit such as this, in which the switches create an ac *voltage* from a dc voltage source, is called a *voltage-source inverter*. Its dual, the *current-source inverter*, creates an ac current from a dc current source.

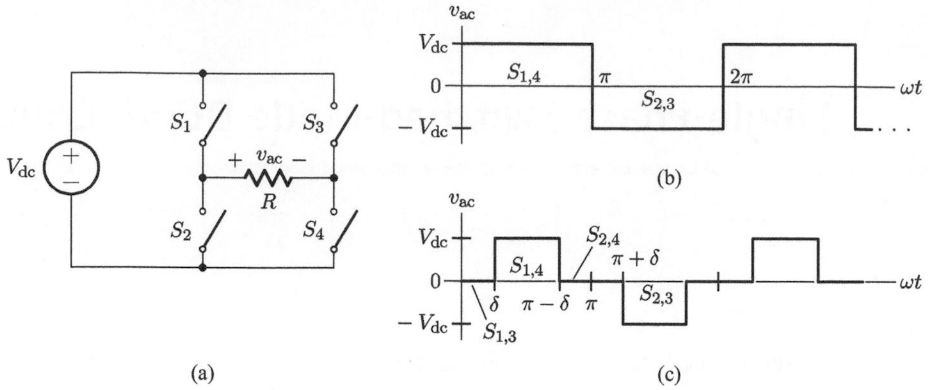

Figure 8.1 (a) A bridge inverter driving a resistive load. (b) The ac voltage when diagonal switches open and close simultaneously. (c) The ac voltage when the switches are controlled to provide a variable amplitude ac voltage.

We can control parameters of the ac voltage (its rms value or the amplitude of its fundamental component, for instance) by varying the dc-port voltage. This requires a complicated dc system that might, for instance, use a dc/dc converter. This approach is relatively straightforward and we do not discuss it further. An alternative technique is to use a third switch state during which $v_{ac} = 0$ to create the waveform of Fig. 8.1(c). In the third switch state, switches S_1 and S_3 or S_2 and S_4 close for a time $2\delta/\omega$, shorting the ac system. A bridge converter capable of providing a zero-voltage state at its output is known as a *tri-state inverter*, with the states 1, 0, and −1 denoting the amplitude of the output relative to the dc input voltage. Which parameters of the ac voltage we control depends on the specific requirements of the load. Here, we have modeled the load as a resistor, so we might want to control the rms value of the output voltage. As a function of our controlling variable δ, $V_{ac\,rms}$ is

$$V_{ac\,rms} = \sqrt{\frac{1}{\pi}\int_{\delta}^{\pi-\delta} V_{dc}^2 \, d(\omega t)} = V_{dc}\sqrt{1 - \frac{2\delta}{\pi}}. \qquad (8.1)$$

In general, the ac loads for inverters are not as simple as the resistor of Fig. 8.1(a). Almost invariably the power factor of the load is not unity, and in many cases average power is transferred at only one frequency, generally the fundamental. For instance, an ac rotating machine (a machine without a commutator, either mechanical or electronic) accepts or supplies average power only at that electrical frequency corresponding to the speed of its rotating flux field, which, except for induction machines, is its mechanical speed.† Although the converters discussed in this chapter have the ability to produce variable-frequency ac, they are often connected to ac sources (or sinks) of constant frequency, such as a utility line. In these applications, the need for a high-power-factor and low-distortion interface to the line makes the simpler phase-controlled rectifier/inverter unsuitable, because the THD of its line current is high, and its power factor is

† The relationships between power and frequency in multi-port systems are described by the *Manley–Rowe relations*. See P. Penfield, *Frequency–Power Formulas*, MIT Press, 1960.

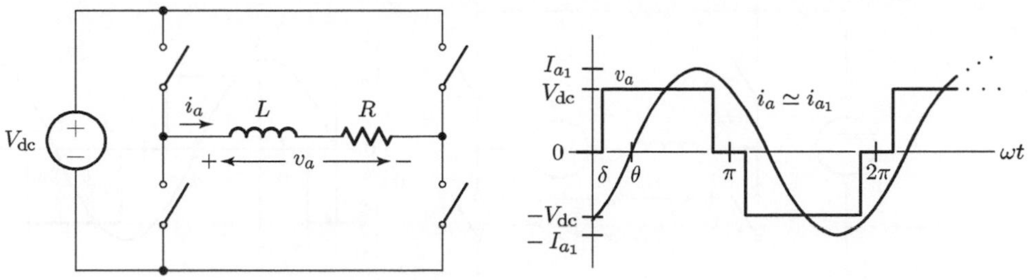

Figure 8.2 A variable-frequency bridge converter with a non-unity-power-factor load.

low at large values of α. Instead, we can use a variable-frequency converter operated as described in Section 8.2.

8.1.1 Bridge Converters with Non-Unity-Power-Factor Loads

Figure 8.2 shows a voltage-source inverter supplying a reactive load. If $L/R > \pi/\omega$, the third harmonic component of i_a is on the order of 10% of its fundamental, allowing us to approximate i_a as

$$i_a(t) \approx I_{a_1} \sin(\omega t - \theta), \tag{8.2}$$

where

$$\theta = \tan^{-1}\left(\frac{\omega L}{R}\right), \qquad I_{a_1} = \frac{V_{a_1}}{\sqrt{(\omega L)^2 + R^2}}, \tag{8.3}$$

$$V_{a_1} = \frac{2V_{dc}}{\pi}\int_{\delta}^{\pi-\delta} \sin(\omega t)\, d(\omega t) = \frac{4V_{dc}}{\pi}\cos\delta. \tag{8.4}$$

The average power P delivered to R is correspondingly

$$P = I_{a_1 \text{rms}} V_{a_1 \text{rms}} \cos\theta = \frac{I_{a_1} V_{a_1}}{2}\cos\theta. \tag{8.5}$$

Substituting (8.3) and (8.4) into (8.5) yields

$$P = \frac{8V_{dc}^2}{\pi^2\sqrt{(\omega L)^2 + R^2}}\cos^2\delta\cos\theta. \tag{8.6}$$

The conclusion we draw from this analysis is that the functional dependence of power on δ, as given by (8.6) for a reactive load, is different than it is for a resistive load. If we were to consider additional harmonic contributions to the power, the relationship would be more complicated, but would still depend on δ and the fundamental and harmonic displacement power factors. For the passive load of Fig. 8.2, the power factor is fixed, and we can control the power only by varying δ or V_{dc}.

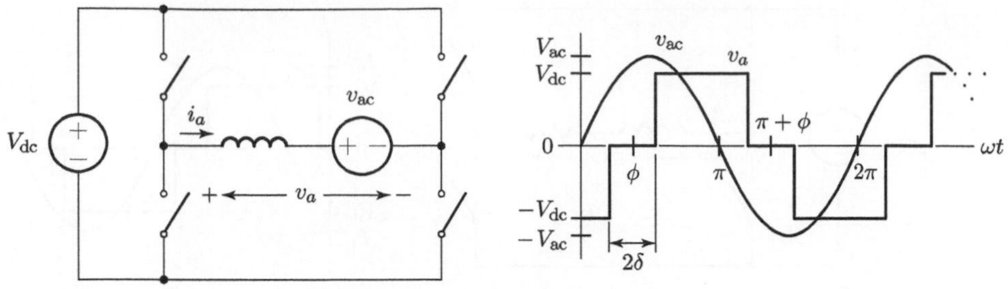

Figure 8.3 A full-bridge inverter feeding an ac load with a source of electromotive force (EMF).

8.1.2 Power Control for a Load Containing an AC Voltage Source

If we replaced the resistive load in Fig. 8.2 with an ac voltage source, we could use θ in addition to δ as a control variable. Figure 8.3 shows a full-bridge inverter connecting a dc voltage source, V_{dc}, to a load modeled by an inductance in series with a sinusoidal voltage source, v_{ac}. This might represent, for instance, a single-phase synchronous motor under certain operating conditions (in which case the ac source is a model for the *back-EMF* of the motor), or a utility grid (in which case the dc source might represent a photovoltaic array). The inverter produces a tri-state output as shown in Fig. 8.3. As we can now control the angle ϕ between v_a and v_{ac}, we can also control the phase angle between the fundamental component of i_a and v_{ac}, giving us another handle on power.

Example 8.1 Control of an Inverter Feeding an AC Voltage Source

Our problem is to specify values of δ and ϕ for the inverter of Fig. 8.3 so that the power delivered to v_{ac} is 10 kW. The circuit parameters are

$$v_{ac} = 400\sin(377t)\,\text{V}, \qquad V_{dc} = 350\,\text{V}, \qquad L = 10\,\text{mH}.$$

No unique combination of δ and ϕ yields a power of 10 kW, unless we add a requirement on the performance of the circuit. The additional constraint we choose is that the ac source operate at unity displacement factor; that is, v_{ac} is in phase with i_{a_1}.

The complex amplitude of the fundamental component of the current i_a is

$$\widehat{I}_{a_1} = \frac{\widehat{V}_{a_1} - \widehat{V}_{ac}}{j\omega L}. \tag{8.7}$$

As v_{ac} and i_{a_1} are in phase, we can express the average power as

$$\langle p(t)\rangle = P = \frac{I_{a_1}V_{ac}}{2} = 10^4\,\text{W}, \tag{8.8}$$

which gives $I_{a_1} = 50$ A. Choosing $\widehat{V}_{ac}$ as our reference for angular measurement (thus $\widehat{V}_{ac} = V_{ac}$), we can express $\widehat{V}_{a_1}$ in terms of ϕ and δ and use (8.4) and (8.7) to determine these angles:

$$\frac{((4V_{dc}/\pi)\cos\delta)e^{-j\phi} - V_{ac}}{j\omega L} = 50\,\text{A}. \tag{8.9}$$

By equating the real and imaginary parts on the two sides of (8.9), we obtain

$$\phi = -25.2^\circ, \qquad \delta = 7.1^\circ.$$

8.1.3 Current-Source Inverter

The inductor in Fig. 8.3 serves as a buffer between two voltage sources, v_a (created by the inverter) and v_{ac}. This inductor absorbs the instantaneous difference between these voltages, and its value depends on the magnitude and duration of the difference. In many cases its actual value is very large – on the order of $V_a/\omega I_{a\max}$ – and degrades the power factor of the ac network. Alternatively, we can provide the necessary buffering by placing an inductor on the dc side of the bridge. If the inductor is large enough, we can model the dc voltage source and inductor as a current source. This buffering technique is sometimes used because the inductor can be made arbitrarily large without degrading the power factor of the ac network, and because it provides time to shut down the system (without high currents) in case one of the switches fails. It does, however, degrade the dynamic performance of the converter, because more time is required for the load current to change in response to a control command. A current-source converter of this type is shown in Fig. 8.4, and we now analyze its performance.

The circuit diagram of Fig. 8.4 contains insufficient information for us to know whether the function performed is inversion or rectification. Only when we specify the switch control can we determine the direction of power. Let's assume that it flows from the dc side to the ac side (inversion). The waveforms of Fig. 8.4 reflect this assumption.

The average power P delivered to the ac source is

$$P = \frac{1}{T}\int_0^T v_{ac} i_a \, dt = \frac{V_{ac} I_{a_1}}{2} \cos\theta, \tag{8.10}$$

where $I_{a_1}\cos\theta$ is the amplitude of the fundamental component of i_a that is in phase with the ac source v_{ac}. Thus, for a fixed δ, we may control the average power delivered to the load by varying θ through switch timing. But because the amplitudes of all the frequency components of i_a are dependent on δ, we may also control the power by varying δ. Calculating I_{a_1} in terms of δ gives

$$I_{a_1} = \frac{2I_{dc}}{\pi}\int_\delta^{\pi-\delta} \sin(\omega t)\, d(\omega t) = \frac{4I_{dc}}{\pi}\cos\delta, \tag{8.11}$$

so the average power is

$$P = \frac{2V_{ac}I_{dc}}{\pi}\cos\delta\cos\theta. \tag{8.12}$$

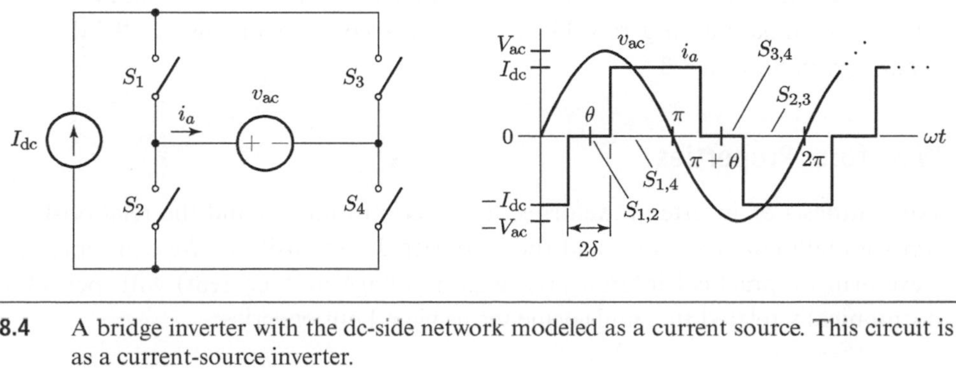

Figure 8.4 A bridge inverter with the dc-side network modeled as a current source. This circuit is known as a current-source inverter.

In certain situations, controlling power by varying δ rather than θ may be more desirable, because θ control generally requires bidirectional switches. You can see the need for switches with bidirectional blocking capability in the current-source inverter by studying the waveforms of Fig 8.4. Note that the voltage across the open switches is v_{ac}, which changes sign during the S_1, S_4 and S_2, S_3 conduction periods if δ is zero and θ varies. Thus, each switch blocks both voltage polarities. Similarly, we can show that the switches in the voltage-source converter must carry bidirectional current. If θ is zero, δ control does not require bidirectional switches.

Another disadvantage of θ control is that it reduces power by decreasing the power factor at the load. For instance, if we decrease the output power to zero in the circuit of Fig. 8.4 by setting $\theta = \pi/2$, we still have an output current with a peak of I_{dc}, and many of the inverter losses remain unchanged from their full-load ($\theta = 0$) values.

8.2 Harmonic Reduction

Reducing the harmonic content of the ac port voltage or current is of great importance in many applications. Harmonics not only reduce the power factor of the ac port, but can also interfere with the proper operation of the converter or other equipment by appearing as noise in control circuits, or can excite mechanical resonances in electromechanical loads, causing them to emit acoustic noise and/or causing additional loss.

We have already seen in Chapter 5 that a low-pass filter consisting of a shunt capacitor and a series inductor was required to prevent the ac components caused by switching from appearing at the terminals of a dc/dc converter. However, a low-pass filter is not nearly as effective when applied to dc/ac converters. The reason is that while the ratio of the switching frequency to the input or output frequency is infinite for a dc/dc converter, it is finite for a dc/ac converter. For a dc/dc converter, therefore, the amount of ripple attenuation is limited only by the physical size of the inductor or capacitor or, perhaps, by the desired control bandwidth. For the dc/ac converter, on the other hand, the size and effectiveness of the filter elements are determined by factors such as the tolerable attenuation or phase shift in the fundamental. As a result, we often control the switches in a dc/ac converter to achieve *active harmonic reduction*. In this section we discuss two ways of doing this. The first is *harmonic elimination*, in which we control the switches to eliminate certain harmonics. The second is *harmonic cancellation*, in which we add the outputs of two or more converters so as to cancel certain harmonics.

An alternative approach to harmonic reduction is to move the harmonics to frequencies high enough to make filtering possible with smaller components, using PWM, which we discuss separately in Section 8.3.

8.2.1 Waveform Properties

For synthesis of inverter waveforms, it is useful to understand the relationship between their time-domain characteristics and their Fourier decomposition. We can represent any periodic waveform of practical interest $f(t)$ (e.g., a voltage or a current) with period T in terms of harmonically related sine and cosine terms via a Fourier series:

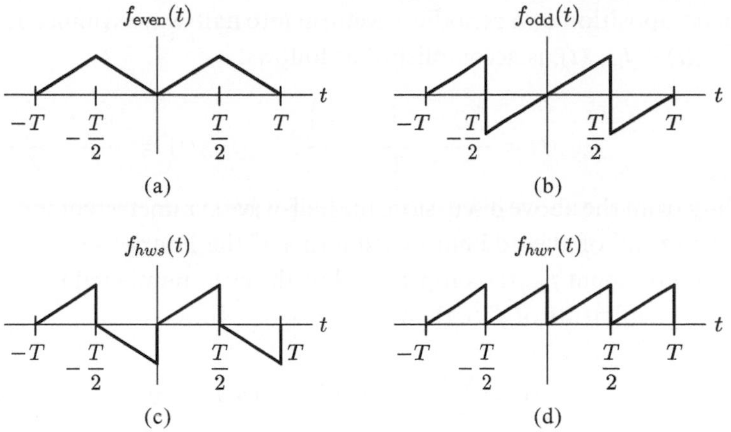

Figure 8.5 (a) An even waveform. (b) An odd waveform. (c) A half-wave symmetric waveform. (d) A half-wave repeating waveform.

$$f(t) = \frac{b_0}{2} + \sum_{n=1}^{\infty} a_n \sin(n\omega t) + b_n \cos(n\omega t), \tag{8.13}$$

where $\omega = 2\pi/T$ and the Fourier coefficients are

$$a_n = \frac{2}{T}\int_0^T f(t)\sin(n\omega t)\,dt, \qquad b_n = \frac{2}{T}\int_0^T f(t)\cos(n\omega t)\,dt. \tag{8.14}$$

Some waveforms have special symmetry characteristics that are reflected in their Fourier decomposition. For example, an *even* waveform has the characteristic that $f(t) = f(-t)$, while an *odd* waveform has the characteristic that $f(t) = -f(-t)$. Example even and odd waveforms are shown in Fig. 8.5(a) and (b). Because cosine is even and sine is odd, an even function will have only b_n (cosine) terms in its Fourier series, while an odd function will have only a_n (sine) terms.

We can decompose any waveform into even and odd components, $f(t) = f_E(t) + f_O(t)$, where

$$f_E(t) = \frac{f(t) + f(-t)}{2}, \qquad f_O(t) = \frac{f(t) - f(-t)}{2}. \tag{8.15}$$

It can also be valuable to understand the implications of waveform symmetry within a cycle. Any periodic waveform can be decomposed into *half-wave symmetric* and *half-wave repeating* components, that is, $f(t) = f_{hws}(t) + f_{hwr}(t)$. A half-wave symmetric waveform has the characteristic that $f_{hws}(t) = -f_{hws}(t - T/2)$, where T is the waveform period. That is, the waveform may be broken into two half-cycles such that the second half-cycle is a mirror across the horizontal axis of the first half-cycle. A half-wave repeating waveform has the characteristic that $f_{hwr}(t) = f_{hwr}(t - T/2)$, such that the second half-cycle of the waveform is simply a repetition of the first half-cycle. Examples of half-wave symmetric and half-wave repeating waveforms are shown in Fig. 8.5(c) and (d).

The decomposition of a periodic waveform into half-wave symmetric and half-wave repeating, $f(t) = f_{hws}(t) + f_{hwr}(t)$, is accomplished as follows:

$$f_{hws}(t) = \frac{f(t) - f\left(t - \frac{T}{2}\right)}{2}, \qquad f_{hwr}(t) = \frac{f(t) + f\left(t - \frac{T}{2}\right)}{2}. \tag{8.16}$$

Expanding from the above discussion, the half-wave symmetric component $f_{hws}(t)$ of a waveform $f(t)$ comprises all of the odd-numbered terms of the Fourier series of $f(t)$, while the half-wave repeating component $f_{hwr}(t)$ comprises all of the even-numbered terms. That is, if a_n and b_n are the Fourier coefficients of $f(t)$, then

$$\begin{aligned} f_{hws}(t) &= \sum_{k=1}^{\infty} a_{2k-1} \sin(2k-1)\omega t + b_{2k-1} \cos(2k-1)\omega t, \\ f_{hwr}(t) &= \frac{b_0}{2} + \sum_{k=1}^{\infty} a_{2k} \sin(2k)\omega t + b_{2k} \cos(2k)\omega t. \end{aligned} \tag{8.17}$$

The Fourier or spectral content of a waveform is closely tied to its symmetry. A half-wave symmetric waveform contains only a fundamental component and odd harmonics of the fundamental frequency. This can be seen from the integrals (8.14) defining the Fourier coefficients. Likewise, a half-wave repeating waveform can contain only a dc component and even harmonic components of the defined fundamental frequency $1/T$ (because its period is actually $T/2$).

Symmetry is important to consider when synthesizing periodic inverter waveforms, especially when trying to approximate a sinusoid. By synthesizing a half-wave symmetric waveform, we automatically null all even-order harmonics, including the second harmonic (which can be difficult to attenuate by filtering because of its close proximity to the fundamental). It is for this reason that half-wave symmetric patterns are usually selected for inverter waveform synthesis, as has been done in the examples shown thus far. The relationship between symmetry and harmonic content is sometimes applied in the other direction as well. In some radio-frequency applications, for example, notch filters are used to null one or more even harmonics in order to promote half-wave symmetry in the resulting waveform.

A useful property of the decomposition into the even and odd components in (8.15), or half-wave symmetric and half-wave repeating components (8.16), is that in each case the decomposed signal types are *orthogonal*. Two waveforms $x(t)$ and $y(t)$ are orthogonal on an interval $[0, T]$ if

$$\int_0^T x(t)y(t)\,dt = 0.$$

Because sine and cosine waves are orthogonal on any interval that is an integer multiple of T, and sinusoids at different frequencies are orthogonal on a commensurate interval, any even waveform is orthogonal to any odd waveform. Likewise, any half-wave symmetric waveform is orthogonal to any half-wave repeating waveform. One consequence of this is that average power transfer requires voltages and current components of the same decomposition type. For example, at a port having a half-wave symmetric voltage, one can only deliver average power via the half-wave symmetric component of the port current. Similarly, no average power transfer at a port can be obtained with an even voltage and an odd current. These waveform decompositions thus enable information about average power transfer to be inferred directly from time-domain properties of the voltage and current waveforms.

8.2.2 Harmonic Elimination

An important potential benefit of controlling δ in the converter of Fig. 8.3 is that the amplitude of the third harmonic of v_a may be controlled. In fact, by appropriate choice of δ we can completely eliminate the third harmonic. Because the synthesized waveform is half-wave symmetric, there will be no even harmonics. The size of passive filters is generally determined by the lowest frequency to be eliminated. Thus, elimination of the third harmonic by control of δ has a major beneficial effect on the size of the ac-side filter components, because the lowest harmonic present in v_a will then be the fifth harmonic.

If the converter of Fig. 8.3 is controlled so that $\theta = 0$ (which is not necessary to the result, but makes the math a bit simpler), the third-harmonic amplitude in v_a is

$$V_{a_3} = \frac{2V_{\mathrm{dc}}}{\pi} \int_{\delta}^{\pi-\delta} \sin 3\omega t \, d(\omega t) = \frac{4V_{\mathrm{dc}}}{3\pi} \cos 3\delta. \tag{8.18}$$

The third harmonic is therefore eliminated if we control the switches to obtain $\delta = \pi/6$. In fact, all harmonics of order $3n$ are thereby eliminated, and the waveform is said to be free of *triple-n* harmonics. Of course, if we fix δ, we no longer use this parameter to control v_{a_1} or power.

So far we have assumed that the 0 state of our tri-state inverter occurs as an intermediate step between the 1 and -1 states. This is not necessary, nor is it necessary that there be only two 0 states per cycle. We can obtain more sophisticated harmonic control by creating 0-state regions, called *notches*, during the 1 and -1 states of the ac waveform.

Example 8.2 Simultaneous Elimination of Third and Fifth Harmonics

We shall use a graphical approach in this example to show that the introduction of appropriate notches in the v_a waveform of Fig. 8.2 allows elimination of both the third and fifth harmonics.

First consider the graphic representation in Fig. 8.6(a) of the integrand in (8.18), drawn for the case $\delta = 30°$. We focus on only a half-cycle of the fundamental, because v_a is half-wave symmetric. In any waveform manipulation we undertake, we use the same manipulations in each half-cycle, to preserve half-wave symmetry and avoid introducing even-harmonic components in the waveform. The portion of the third-harmonic sinusoid that contributes to the integral in (8.18) is shown shaded. The positive lightly shaded area is equal to the negative heavily shaded area, so the net area (that is, the integral) is 0, which is the result given by (8.18) for $\delta = 30°$.

Figure 8.6(b) similarly shows the portion of the fifth-harmonic sinusoid that contributes to the integral that determines the amplitude of the fifth harmonic in v_a, again with the positive area shaded lightly and the negative area shaded heavily. The net area, which is the result of the integration, is negative. So the fifth-harmonic amplitude is nonzero.

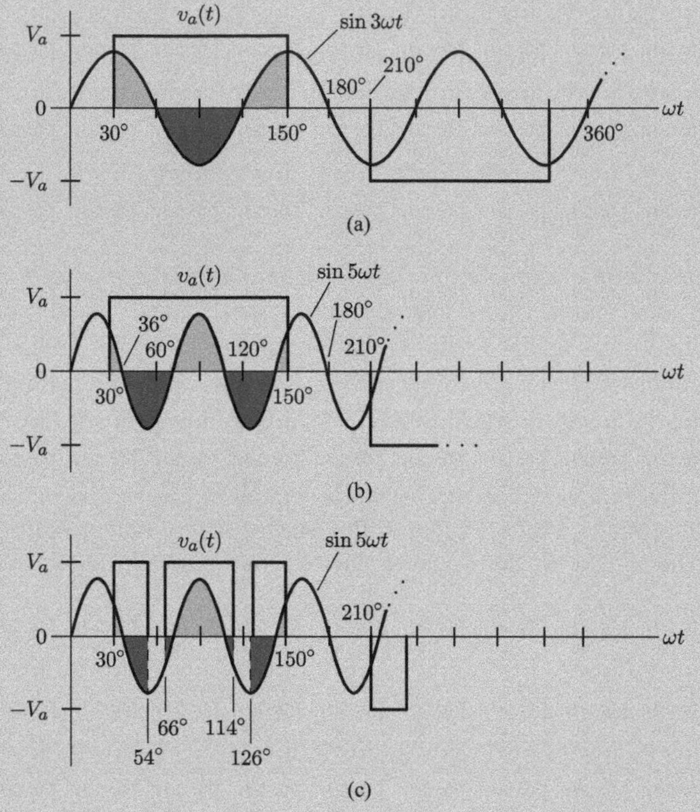

Figure 8.6 (a) Graphical representation of the integrand of (8.3). The positive and negative areas are equal, canceling the third harmonic of the square wave. (b) The third-harmonic-free waveform of (a) superimposed on a fifth-harmonic sine wave. (c) The square wave notched so that it is free from both third and fifth harmonics. A fifth-harmonic sine wave illustrates that the product of it and $v_a(t)$ has zero net area.

We now introduce notches into v_a, as shown in Fig. 8.6(c). The notches must be positioned so as to maintain the cancellation of the third harmonic, while providing the additional degree of freedom needed to cancel the fifth harmonic. In terms of the graphical construction, we must position the notches so as to eliminate equal amounts of the positive and negative areas of Fig. 8.6(a). Notches placed symmetrically about 60° and 120°, as shown in Fig. 8.6(c), meet this criterion.

Placing notches in these positions has the effect of zeroing out part of the integrand that determines the fifth-harmonic magnitude. We need to make these notches wide enough to offset the area that resulted in a negative fifth harmonic. Visualizing notches of varying width around the 60° and 120° points, we conclude that symmetrically placed notches of width 12° will result in a net area of zero. The small positive and negative triangular segments cancel each other, and the two negative quarter-cycles sum to cancel the single positive half-cycle. The resulting v_a, which is free of both the third and fifth harmonics, is shown in Fig. 8.6(c).

Through this development and associated examples, we have illustrated that by introducing notches (or, equivalently, synthesizing additional positive pulses in each half-cycle) we can eliminate one or more harmonics. As we seek to eliminate more and more harmonics, the switch timing required becomes increasingly critical, and switching losses increase. Additionally, at high multiples of the fundamental frequency, switch turn-on and turn-off times make it difficult to precisely set the notch locations. These can represent significant practical constraints. It is also important to realize that as we eliminate low-order harmonics through more pulses per half-cycle, the high-frequency harmonic content of the waveform increases. As high-frequency components are easier to filter out, however, this is often an acceptable trade-off.

8.2.3 Harmonic Cancellation

We can derive an alternative harmonic reduction technique by recognizing that we can create the tri-state waveform of v_a in Fig. 8.6(a) by adding two square waves of amplitude $V_{dc}/2$, shifted by 60° with respect to each other. Two voltage-source square-wave inverters configured this way are shown in Fig. 8.7. If the switching of one circuit is delayed by 60° with respect to the other, then v_a will have no third harmonic. We have already considered this cancellation in

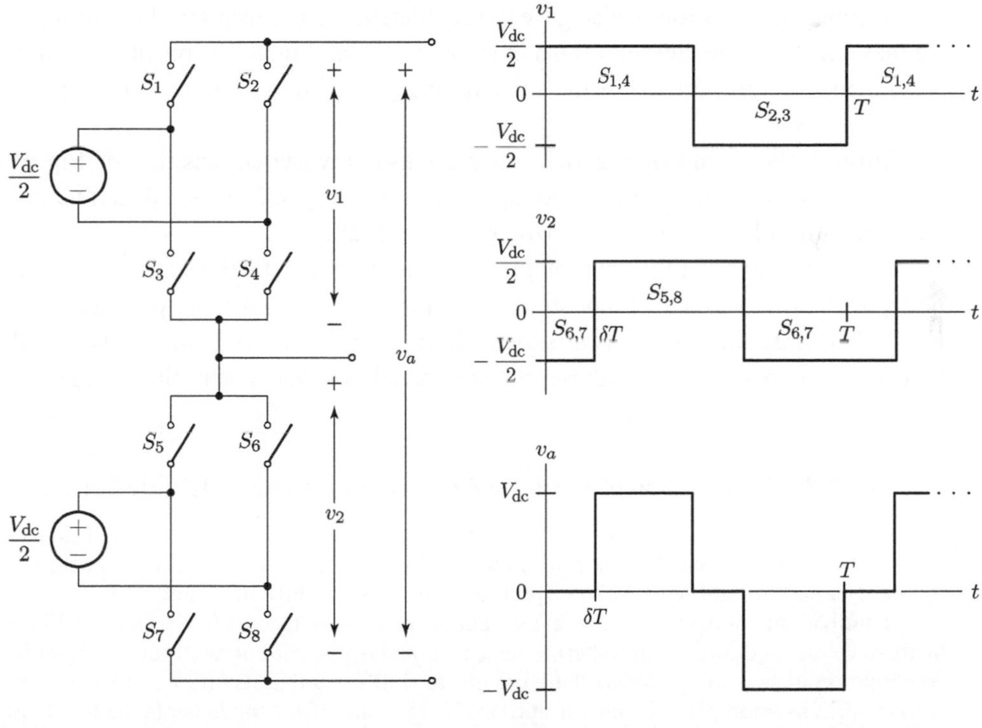

Figure 8.7 Harmonic cancellation achieved by adding the out-of-phase voltages of two similar bridge converters.

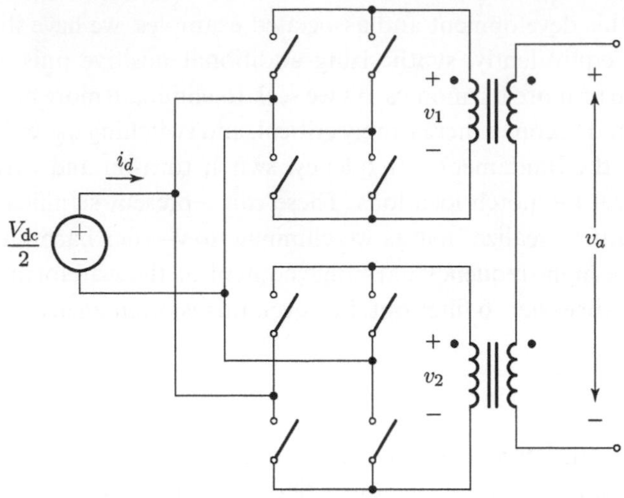

Figure 8.8 Two harmonic-canceling bridge converters with a common dc-side voltage and transformer-coupled ac-side voltages.

the graphical integration, which gave the coefficients of the terms in the Fourier series. However, we may interpret the lack of a third harmonic in v_a of Fig. 8.7 as resulting from the addition of two harmonic components of the same frequency (third) and amplitude, but differing in phase by $3 \times 60° = 180°$.

Although the circuit of Fig. 8.7 uses twice as many switches as that of Fig. 8.3, each of the switches is rated at only half the voltage of those of Fig. 8.3. This reduced voltage stress can be an important advantage if the ac voltage is very high.

Implementation of the circuit of Fig. 8.7 is not very practical because the two dc sources must be separate; that is, they cannot share a common terminal. A more practical circuit is shown in Fig. 8.8. The dc sides of the two bridges share a common terminal, but the ac sides are isolated by transformers whose secondaries are connected in series to sum the ac-side voltages.

Example 8.3 Cancellation of Third and Fifth Harmonics Using Multiple Converters

We can easily adapt the circuit of Fig. 8.8 to the cancellation of both the third and fifth harmonics of the ac output voltage. We do so by controlling the individual bridge switches to cancel the third harmonic from v_1 and v_2, and then shifting v_2 with respect to v_1 to cancel the fifth harmonic.

The individual voltages v_1 and v_2 are similar to the third-harmonic-free voltage of Fig. 8.6(a). Neither of these voltages contains a third-harmonic component, so we cannot reintroduce it by creating any linear combination of v_1 and v_2. We are therefore free to shift v_2 so that its fifth harmonic is 180° out of phase with the fifth harmonic in v_1. This is a shift of $180°/5 = 36°$ of the fundamental period. Now when we add v_1 to the phase-shifted v_2, the fifth-harmonic terms will cancel. The appropriate v_1 and v_2 and their sum are shown in Fig. 8.9.

We accomplish the phase shift of v_2 simply by delaying the switching sequence of the lower bridge by 36° relative to the upper bridge. The switching patterns of both bridges are identical. The resulting ac voltage waveform is called a *stepped waveform*, whereas the ac voltage of Fig. 8.6(c) is called a *notched waveform*.

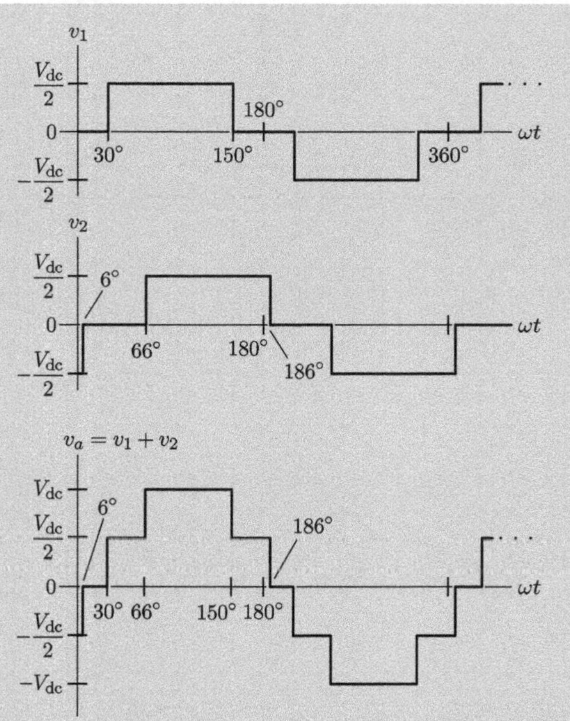

Figure 8.9 Waveforms for the circuit of Fig. 8.8, showing how the third and fifth harmonics can be simultaneously eliminated from the ac voltage.

The generation of a stepped waveform, such as v_a in Fig. 8.9, requires considerably more circuit complexity than the creation of a notched waveform, but the stepped waveform has a much lower THD. For instance, although the notched waveform of Fig. 8.6(c) and the stepped waveform of Fig. 8.9 are both free from third- and fifth-harmonic components, the notched waveform has a THD of 63%, compared to 17.5% for the stepped waveform. The notches in Fig. 8.6(c) eliminate only the fifth harmonic, while the addition of the two phase-shifted voltages to create the stepped waveform of Fig. 8.9 cancels more than just the fifth harmonic.

8.3 Pulse-Width-Modulated DC/AC Converters

In the discussion of dc/dc converters in Chapter 5, we pointed out that the output voltage of a buck converter could vary from zero to the input voltage V_i, depending upon the duty ratio D. If we were to vary D slowly, relative to the switching frequency, we could synthesize a waveform whose "average" value varied with time and was given by $d(t)V_i$, where $d(t)$ denotes the prevailing duty ratio at time t. The averaging time must be long relative to the switching period but short relative to the rate of change of $d(t)$. We refer to the result of this averaging process on some

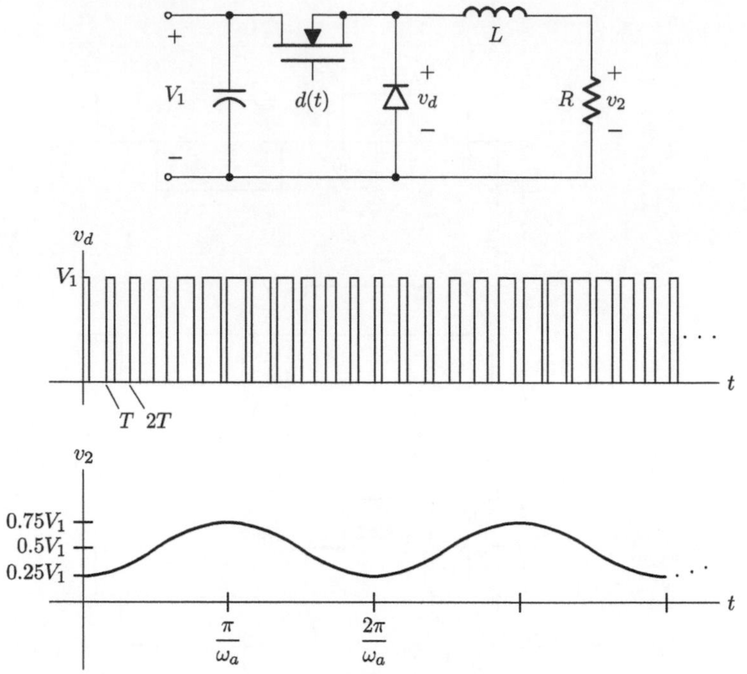

Figure 8.10 A buck converter modulated to produce a load voltage having a dc value and a sinusoid of frequency ω_a much lower than the switching frequency.

quantity $x(t)$ as the *local average* of $x(t)$, denoted by an overbar, $\overline{x}(t)$. (Such local averaging is discussed in more detail in Chapter 12.)

An example of a buck converter with a modulated $d(t)$ is shown in Fig. 8.10. Here, $d(t) = 0.5 + 0.25 \sin \omega_a t$, and $T \ll L/R \ll 2\pi/\omega_a$. The voltage v_d is a PWM waveform with a dc component, a fundamental component of frequency ω_a, and additional, unwanted components at and above the switching frequency, $1/T$. The load voltage v_2 is the local average $\overline{v}_d$ resulting from low-pass filtering v_d through L and R. The local average contains a dc component of $V_1/2$.

In this section we explore the use of this high-frequency PWM, or *waveshaping*, technique in the control and construction of variable-frequency dc/ac converters. The advantage that high-frequency PWM techniques have over the relatively low-frequency switching techniques already discussed is that the undesirable harmonics in the output are at a much higher frequency and thus easier to filter.

8.3.1 Waveshaping and Unfolding

If we modulate $d(t)$ in Fig. 8.10 to produce a waveform $v_2 = |V_2 \sin \omega_a t|$, that is, the same waveform produced by a full-wave rectifier, we can then use a bridge of switches to *unfold* this waveform across the load resistor. A buck converter functioning in this way is shown in Fig. 8.11. The bridge transistors are switched at the cusps of v_2. The time-dependent duty ratio necessary to create v_2 is $d(t) = k|\sin \omega_a t|$, where k is a constant between zero and one and is known as the *depth of modulation* or the *modulation index*. The amplitude of the resulting sinusoidal load voltage is $V_2 = kV_1$. This circuit, although shown generating a sinusoidal output, actually

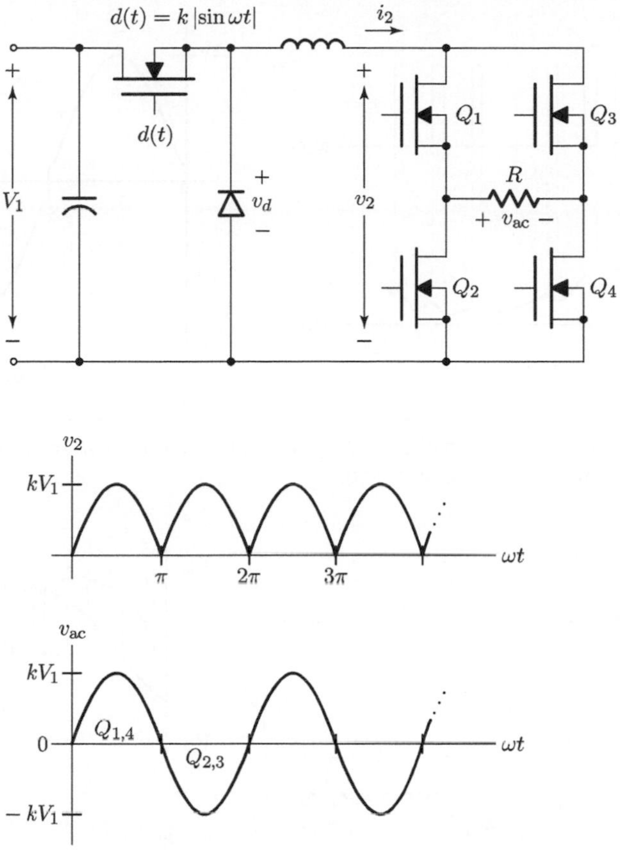

Figure 8.11 A waveshaping buck converter and an unfolding bridge controlled to produce a sinusoidal load voltage.

is a more general *switching power amplifier*. If the reference waveform $f(t)$ is arbitrary with a bandwidth less than R/L (a Mozart sonata, for instance) normalized so that $|f(t)| < 1$, we could amplify this signal by setting $d(t) = |f(t)|$ and controlling the bridge switches according to the polarity of $f(t)$.

A problem with the circuit of Fig. 8.11 is that, contrary to the way the waveform appears, the wave-shaped voltage v_2 does not approach zero sinusoidally. Instead, the current i_2 and voltage v_2 approach zero exponentially with time constant L/R. Figure 8.12 shows the v_d and v_2 of Fig. 8.11 expanded around $\omega_a t = \pi$ to illustrate the problem. The fact that v_2 never quite reaches zero at times $\omega t = n\pi$ results in a step at these times in the unfolded waveform v_{ac}, as shown in Fig. 8.12(b). This *crossover distortion* results in the generation of harmonics of the output frequency ω_a, which, if they are large enough, can compromise the advantage of using high-frequency PWM.

Reversing the voltage across the inductor L in Fig. 8.11 could force the current i_2 to zero, as we desire. We can do so by placing L inside the bridge, as shown in Fig. 8.13. The reference voltage v_{ref}, the voltage v_{ac}, and the current i_a (scaled) are also shown in the figure. Because the effective load now has an inductive component, the current lags the voltage. Because of this current lag, i_2 is negative for short periods. Therefore we must configure the waveshaping buck converter for

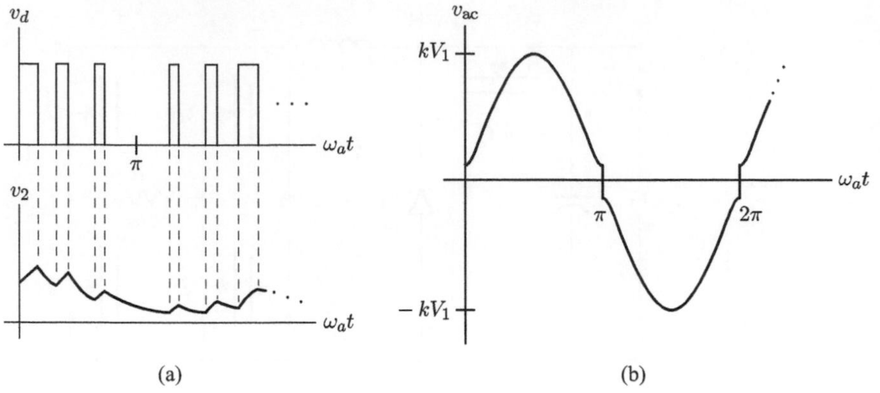

Figure 8.12 (a) An expansion of v_d and v_2 of Fig. 8.11 around $\omega_a t = n\pi$. (b) The output voltage v_{ac}, which exhibits crossover distortion.

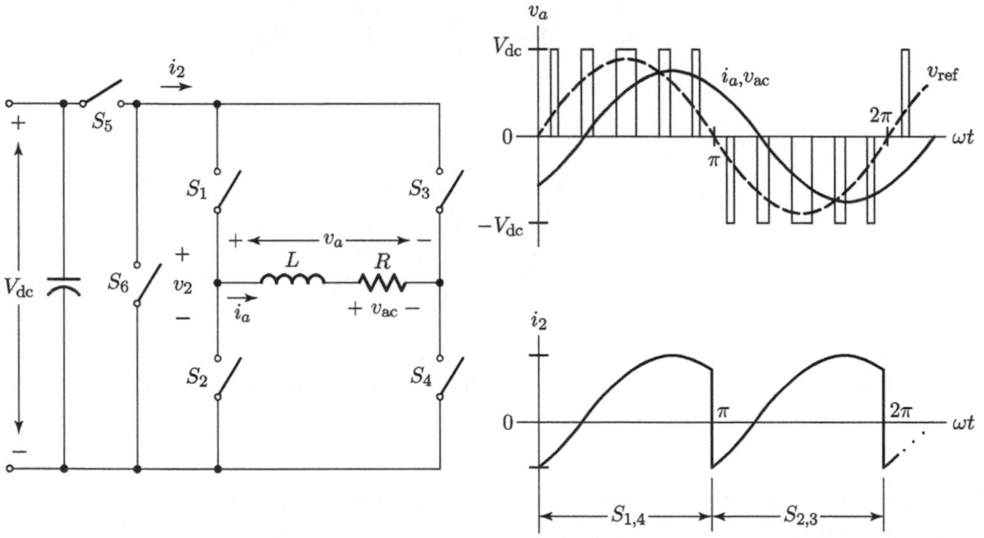

Figure 8.13 A waveshaping buck converter and an unfolding bridge with the inductor placed inside the bridge to reduce crossover distortion.

two-quadrant operation: positive v_2 and positive or negative i_2. The bridge switches also must carry bipolar current. In this configuration, the switches S_5 and S_6 are not necessary, as their function can be assumed by one of the bridge legs, as illustrated in Fig. 8.14 and discussed next.

8.3.2 High-Frequency Bridge Converter

A practical high-frequency bridge converter is shown in Fig. 8.14, along with a sketch of v_a showing the switching sequence. When Q_1 and Q_4 are on, $v_a = V_{dc}$; when Q_1 and Q_3, or Q_2 and Q_4, are on, $v_a = 0$; and when Q_2 and Q_3 are on, $v_a = -V_{dc}$. Moreover, careful consideration of this switching sequence shows that when $v_a \geq 0$, Q_4 can remain on and v_a can be created by

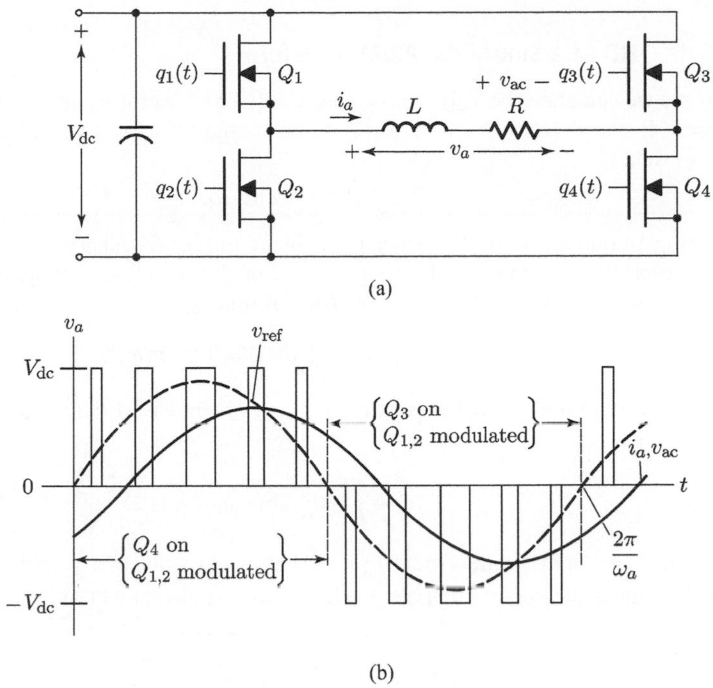

Figure 8.14 (a) A PWM bridge converter, showing a practical switch implementation. (b) The reference voltage v_{ref}, output voltage v_{ac}, and current i_a.

switching Q_1 and Q_2 on and off in a complementary fashion. Similarly, when $v_a \leq 0$, Q_3 can remain on while Q_1 and Q_2 alternate. That is, Q_1 and Q_2 are switching at the high frequency of the carrier $1/T$, while Q_3 and Q_4 unfold the modulated carrier by switching at the much lower bridge frequency ω_a. We now have a circuit performing the same function as that of Fig. 8.13, with four switches instead of six, and yet only two of them being high-frequency switches.

The low-frequency variation of v_a, $v_{ac} = \bar{v}_a(t)$, is the load voltage we desire. Although in this case we have again assumed $\bar{v}_a$ to be a sinusoid, it can have any arbitrary time variation so long as the variation is slow enough to permit proper filtering. We choose the value of L to filter the high switching frequency from the load current, that is,

$$\frac{2\pi}{\omega_a} \gg \frac{L}{R} \gg T. \tag{8.19}$$

In addition to the desired low-frequency fundamental, v_a contains harmonics centered around the high switching frequency $1/T$. The advantage of PWM over the harmonic reduction schemes presented in Section 8.2 is that the harmonic distortion is moved to higher frequencies, making filtering easier. The THD of the PWM waveform, however, is greater than that of a square wave, as we show in Example 8.4. However, the important harmonic issue is almost always the spectrum of the load current, which can be limited most easily in the case of a high-frequency PWM voltage waveform. Furthermore, controlling the amplitude of the fundamental in a notched or stepped waveform, while simultaneously trying to control harmonics, is difficult.

Example 8.4 THD of a Sinusoidal PWM Waveform

In this example we calculate the THD of a sinusoidal PWM waveform, such as that of v_a in Fig. 8.14(b), and compare it to the THD of a square wave. We create the PWM waveform by varying $d(t)$ sinusoidally with $k = 1$, so that:

$$d(t) = |\sin \omega_a t|. \tag{8.20}$$

If the switching frequency is much greater than the ac output frequency, we can assume that $\sin \omega_a t$ is constant over each switching period. If the amplitude of the dc voltage is V_{dc}, the height of each pulse is V_{dc}, and the squared rms value of a pulse occurring at time t_o is

$$v^2_{a(\text{rms})}(t_o) = V^2_{\text{dc}} d(t_o) = V^2_{\text{dc}} \sin \omega_a t_o. \tag{8.21}$$

We can now determine the squared rms value, $V^2_{a(\text{rms})}$, of the PWM pulse train by averaging (8.21) over half a cycle of v_a:

$$V^2_{a(\text{rms})} = \frac{1}{\pi} \int_0^{\pi} V^2_{\text{dc}} \sin \omega_a t \, d(\omega_a t) = \frac{2V^2_{\text{dc}}}{\pi}. \tag{8.22}$$

Because we assumed that $d(t)$ varies from 0 to 1, v_{ac} has a peak value of V_{dc} and an rms value of $V_{\text{ac(rms)}} = V_{\text{dc}}/\sqrt{2}$. We can now calculate the THD using the definition given by (3.41):

$$\text{THD} = \sqrt{\frac{V^2_{\text{a(rms)}} - V^2_{\text{ac(rms)}}}{V^2_{\text{ac(rms)}}}} = 52\%. \tag{8.23}$$

To put this THD value into perspective, we compare it to that for a square wave of fundamental frequency ω_a. If the square wave has an amplitude V, its rms value is also V, and its fundamental amplitude is $4V/\pi$ with an rms value of $2\sqrt{2}\,V/\pi$. The THD for this square wave is

$$\text{THD} = \sqrt{\frac{V^2 - (2\sqrt{2}V/\pi)^2}{(2\sqrt{2}V/\pi)^2}} = 48\%. \tag{8.24}$$

Thus, $v_a(t)$ and a square wave have essentially the same THD, although the significant harmonics of the PWM waveform are at much higher frequencies than those of the square wave. If we supplied the load through a simple inductive filter, as shown in Fig. 8.14, the required value of the inductor to obtain a given THD for the current would be much smaller for the PWM converter than for a square-wave converter.

8.3.3 Generation of *d*(*t*) for the PWM Inverter

There are many ways we can generate the PWM waveform of the bridge converter of Fig. 8.14. One method of generating a sinusoidally varying $d(t)$ from an analog reference waveform is to use the *sine–triangle intercept* technique, one example of what is termed *naturally sampled PWM*. Its essential features are shown in Fig. 8.15. A full-wave-rectified sine wave v_S with peak amplitude k, and a unipolar triangle wave v_T (known as the *carrier*) are fed into a comparator whose output $g(t)$ is high precisely when the value of the sine wave exceeds that of the triangle wave. The duration of each output pulse is therefore weighted by the value of the sine wave at that time; that is, $d(t) \approx k|\sin \omega t|$, which is what we desire. These pulses then drive the high-frequency transistors in the circuit of Fig. 8.14(a), alternating between Q_1 and Q_2 depending upon the sign of the reference signal.

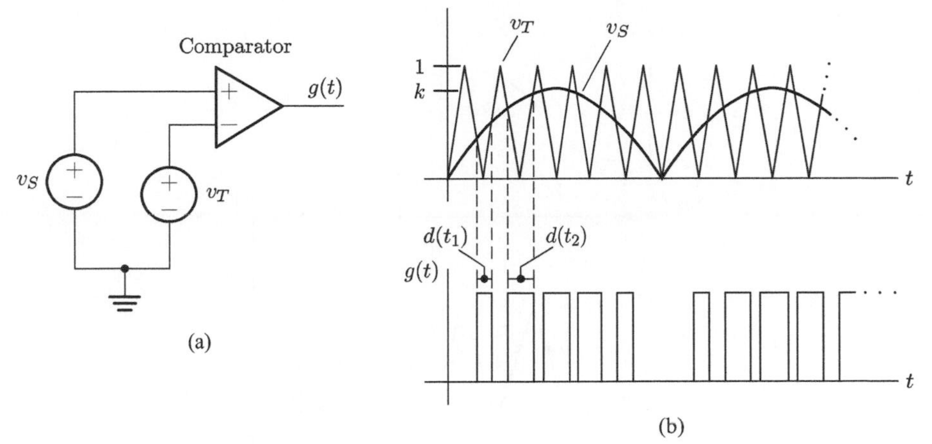

Figure 8.15 The sine–triangle intercept technique for generating a sinusoidally weighted pulse train $g(t)$. (a) A circuit to produce the pulse train $g(t)$. (b) The waveforms of v_S, v_T, and $g(t)$.

For a sinusoidal reference signal $v_S(t)$, we control the amplitude of the fundamental by varying the modulation index, k. This, in turn, varies the absolute widths of the pulses comprising $g(t)$, while retaining their relative widths. However, changing the amplitude of the fundamental by varying k changes the THD of v_a.

Example 8.5 PWM Modulation of a Full Bridge Inverter

In the full-bridge inverter of Fig. 8.14, it is often preferable to use the same high-frequency switch for all four switches, and to modulate the switches such that each switch sees the same switching frequency and loss. As is the case for the inverter of Fig. 8.14, we want to synthesize an output waveform that matches the reference waveform, at least as far as its low-frequency content is concerned. To do so, we use the capability of the full bridge to develop a three-level output (i.e., in which $v_a(t)$ can take on each of $+V_{dc}$, 0, and $-V_{dc}$), and to take advantage of the fact that each side of the bridge operates at a high switching frequency.

Figure 8.16 shows a pulse-width modulator that accomplishes these goals. As with the modulator of Fig. 8.15, the modulator of Fig. 8.16 operates through comparison of a reference waveform to a triangular carrier waveform. As mentioned at the beginning of this section, this approach of generating PWM waveforms by comparison of a reference waveform to a carrier waveform is broadly referred to as naturally sampled PWM, and in the case where the carrier is a triangle wave and the reference is a sine wave, is often called sine–triangle PWM. Many other carrier waveforms are possible (e.g., leading- or trailing-ramp sawtooth waveforms, yielding leading- or trailing-edge PWM modulation within a cycle). A balanced triangle (as in Figs. 8.16 and 8.17) is often preferable because it provides "double-edged modulation," with PWM pulses that are symmetric around the peaks of the triangular carrier, yielding improved high-frequency content in the output waveform.

To generate switching signals for the two sides of the bridge, the triangular carrier is compared to both the reference waveform and the negative of the reference waveform, with each comparison providing PWM signals for one side of the bridge. The PWM comparisons, the left and right half-bridge outputs, and the resulting full-bridge output voltage waveform are shown in Fig. 8.17.

Through this method, the comparison and switching of one side of the full bridge is delayed by half of a switching cycle from the other side of the full bridge. Consequently, the full-bridge output waveform achieves pulse-width modulation at twice the switching frequency f_{sw} of the individual half bridges. Essentially, this approach provides sampling of the reference waveform at twice the carrier frequency, and cancels ripple components such that the output waveform has a pulse repetition rate at twice the carrier frequency, making it easier to filter.

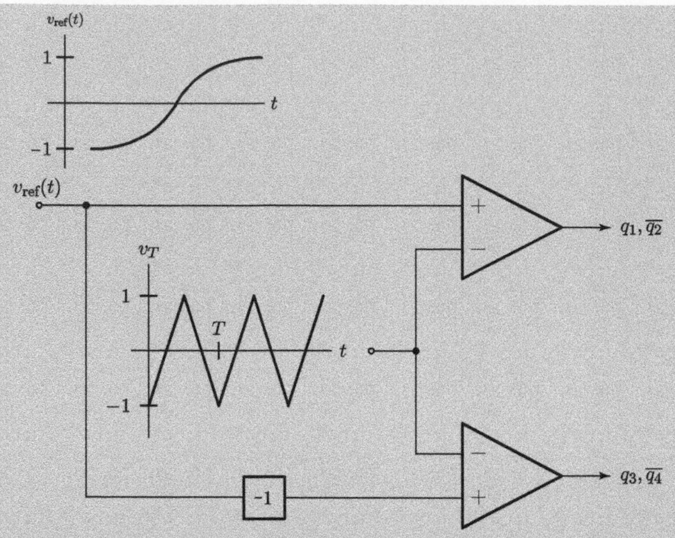

Figure 8.16 A pulse-width modulator for a bridge consisting of four high-frequency switches.

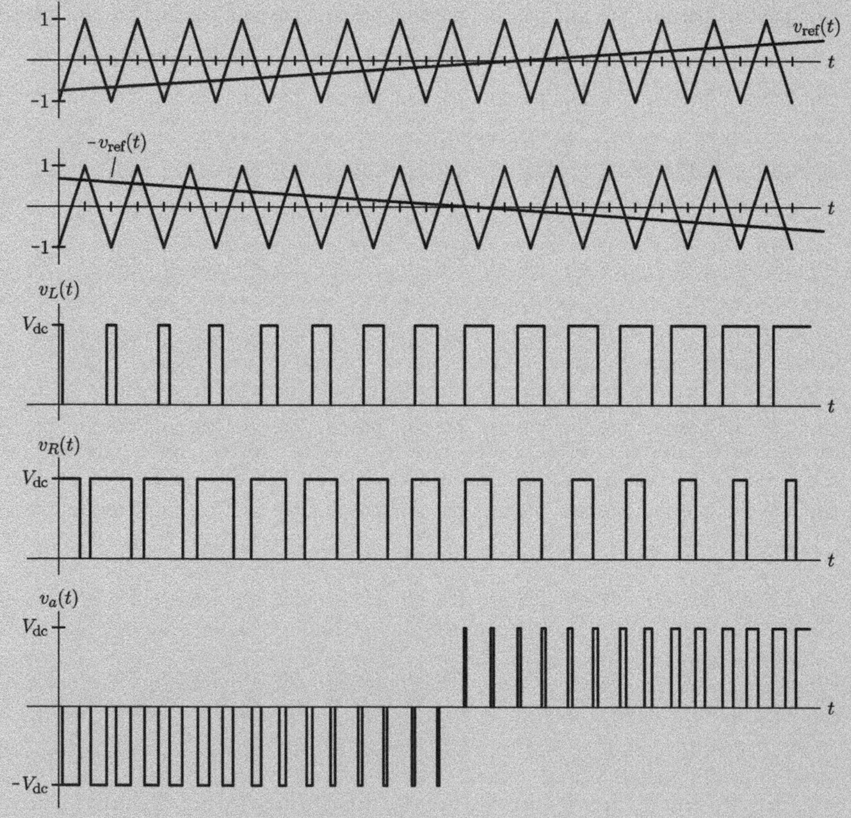

Figure 8.17 (a) The inputs to the two comparators of Fig. 8.16. (b) The outputs of the left and right legs of the bridge of Fig. 8.14. (c) The output voltage v_a.

8.4 Current Control of Inverters

While many applications use an inverter to synthesize an output *voltage* waveform, in others it is preferable to control inverter output *current*. For example, in a grid-tied inverter (e.g., to deliver energy from a photovoltaic system into the grid), the ultimate output voltage is determined by the grid itself, and one regulates real and reactive power by controlling the inverter current waveform. Likewise, in a motor drive, torque is directly related to current, so it is often preferred to regulate the inverter output current waveform directly. Even when one does regulate output voltage, an inner current control loop is sometimes included, as it can help to constrain the instantaneous output current of the inverter and can be used to facilitate current sharing among paralleled inverters supplying a common load.

Consider the single-phase inverter of Fig. 8.18, driving a time-varying voltage source $v_{ac}(t)$ through an inductor L. This load could represent an ac motor, or could represent the grid voltage being driven via a filter inductor, for example. To achieve current control, the inductor current i_L is measured, and is used to either directly or indirectly determine the switching of the four bridge switches.

Two general classes of current control techniques for inverters are carrier-based methods and hysteretic-based methods. Carrier-based current control systems are closely related to the PWM voltage control methods described in Section 8.3.3, using the same kind of carrier/reference comparison mechanism. Consider the carrier-based PWM control structure of Fig. 8.19. The measured inverter output current $i_L(t)$ is subtracted from the desired inverter output current $i_{ref}(t)$ to provide a current error signal $i_{err}(t)$. This current error is fed to a high-gain compensator that generates a reference voltage for the inverter bridge output $v_{ref}(t)$, which is the input to a conventional carrier-based PWM modulator, such as that shown in Example 8.5. So long as the compensator gain is sufficiently high, the inverter has sufficient voltage capability to drive the desired current, and the closed-loop dynamics are stable, this approach can synthesize the desired output current waveform. This current control method produces a fixed switching frequency but current ripple that varies with operating point. It also provides only indirect control of the output current via the compensator.

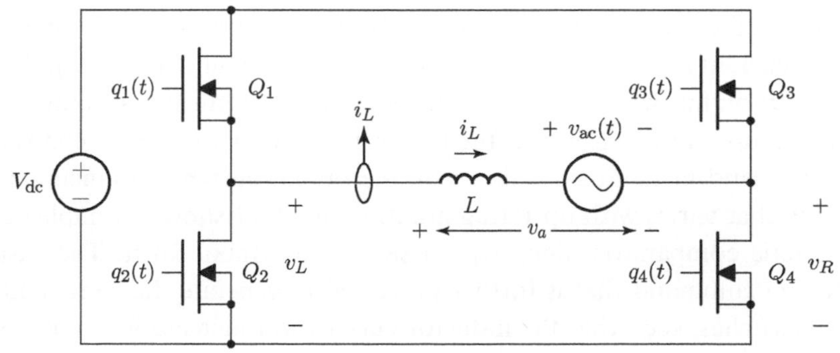

Figure 8.18 A single-phase full-bridge inverter with an inductive filter and time-varying voltage-source load. The output current is sensed for the purposes of control.

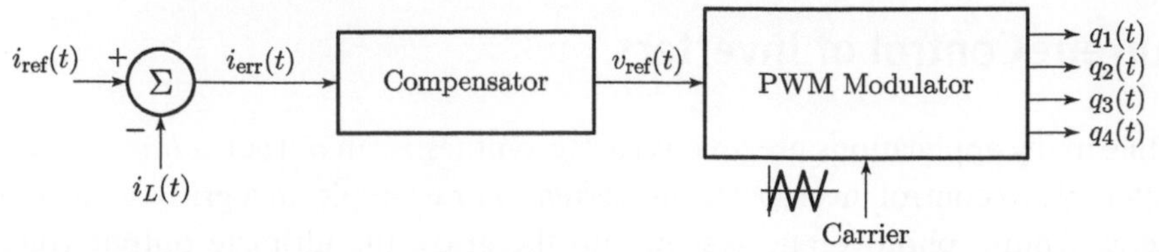

Figure 8.19 A current controller for a full-bridge inverter using a compensator and carrier-based PWM modulator.

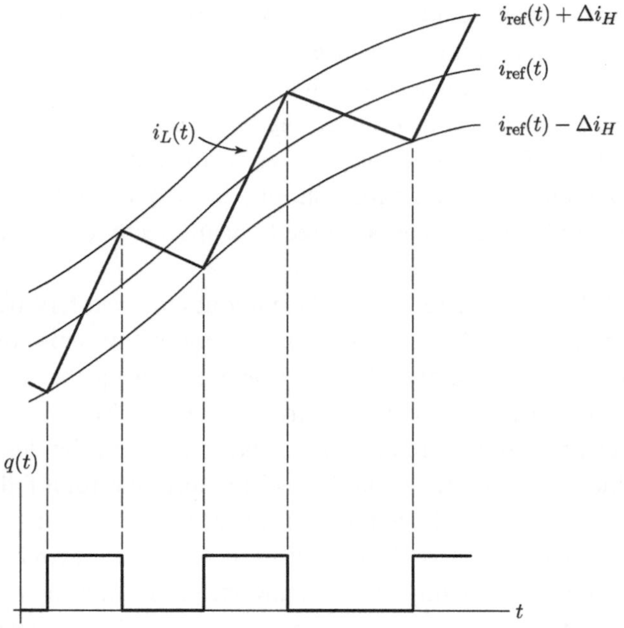

Figure 8.20 An illustration of the waveforms for hysteretic current control. An inverter reference current i_{ref} is provided with a hysteresis band of $\pm\Delta i_H$.

Hysteretic current control operates on a different principle. The inverter is switched in such a manner that the inverter output current $i_L(t)$ continuously stays within an amount $\pm\Delta i_H$ of a reference current $i_{ref}(t)$, as illustrated in Fig. 8.20. The boundaries $\pm\Delta i_H$ define a *hysteresis band* about $i_{ref}(t)$ within which we maintain the current, producing an inductor current waveform that ramps up and down within the hysteresis band. This current control approach provides fixed ripple and instantaneous limits on the output current, producing an inverter switching frequency that varies with operating point. Figure 8.21 shows an implementation of one type of hysteretic comparator, along with a simplified symbol for it. The variable $q(t)$ represents a switching command that is further processed to generate the commands $q_1(t)$–$q_4(t)$ for the inverter switches, such that the inductor current will increase when $q(t)$ is high and decrease when $q(t)$ is low.

Hysteretic current controllers are differentiated in part by how one drives the inverter switches to maintain output current within the hysteresis band. One way to drive the inverter switches is

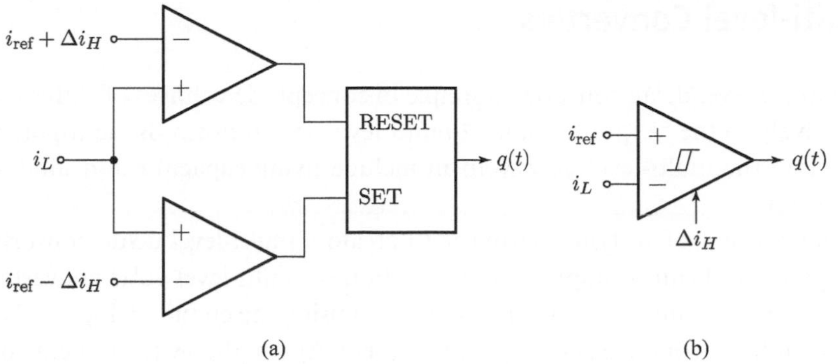

Figure 8.21 (a) One implementation of a hysteretic comparator based on two comparators and an SR latch. The hysteretic comparator generates the switching signal $q(t)$ in Fig. 8.20 on which inverter switching decisions are based. (b) A symbol sometimes used for a hysteretic comparator.

to make $q_1(t) = q_4(t) = q(t)$ and $q_2(t) = q_3(t) = \bar{q}(t)$, where the overbar here denotes the complement (not the average). So long as $|v_{ac}(t)| < V_{dc}$, this hysteresis controller will switch the inverter such that the inductor current enters and remains within the hysteresis band. While this technique is simple, it only produces the output states $v_a = \pm V_{dc}$ and does not take advantage of the inverter zero state, $v_a = 0$. Consequently, in one of the switch states, the voltage is higher than necessary, and the inductor current is quickly driven across the hysteresis band, resulting in a higher switching frequency than necessary to achieve the specified ripple amplitude.

One alternative hysteretic current control strategy is to switch only the left half-bridge with this hysteresis control law ($q_1(t) = q(t)$ and $q_2(t) = \bar{q}(t)$) and switch the right half-bridge based on the polarity of $v_{ac}(t)$, with q_3 high when $v_{ac}(t) < 0$ and q_4 high when $v_{ac}(t) > 0$. This uses the left half-bridge to provide hysteretic current control while using the right half-bridge as an unfolder, and results in similar current ripple performance, but with a reduced switching frequency. While we do not describe the details here, the same effect can be accomplished without knowledge of the output voltage $v_{ac}(t)$ by switching the right half-bridge based on inverter output current, using a second, larger hysteresis band. Even more sophisticated techniques switch both sides of the bridge at high frequency (e.g., by using a state machine to determine the switching patterns). In all of these approaches to hysteretic current control, we can maintain the inverter output current within a small band about a reference current (provided the dc bus voltage is large enough to enforce this). However, because the switching durations depend upon the rate of change of inductor current, the switching frequency varies with V_{dc} and $v_{ac}(t)$.

In summary, carrier-based current control methods provide a well-defined (fixed) switching frequency for the inverter switches, but lead to variable switching ripple amplitude in the ac output current waveform as the output voltage $v_{ac}(t)$ varies. Moreover, they do not provide direct instantaneous control over the inverter output current, but only indirect control via the compensator. Hysteretic-based methods, by contrast, suffer from variable switching frequency of the inverter switches, but provide fixed output current switching ripple and direct instantaneous control over inverter output current. Each of these classes of current-control techniques sees widespread use in practice.

8.5 Multi-level Converters

The multi-level dc/ac converter is simple in concept: dc voltages of differing values are switched selectively to the output, creating a multi-level ac waveform. Basic topologies that can be used to create the multi-level ac waveform include flying-capacitor and diode-clamped multi-level converters.

The concept of a flying capacitor to create a multi-level dc/dc converter was developed in Chapter 5 and can be applied to the creation of multi-level dc/ac converters. As shown in Fig. 8.22, we can create a three-level ac waveform using the circuit of Fig. 5.27 as the leg of a bridge in which the second leg is simply an unfolder.‡ Again, the switching sequence must be such that over a cycle the net capacitor charge is zero, and this sequence is shown in the figure. In this single-phase case, the duration of the three states can be varied to control the harmonic content

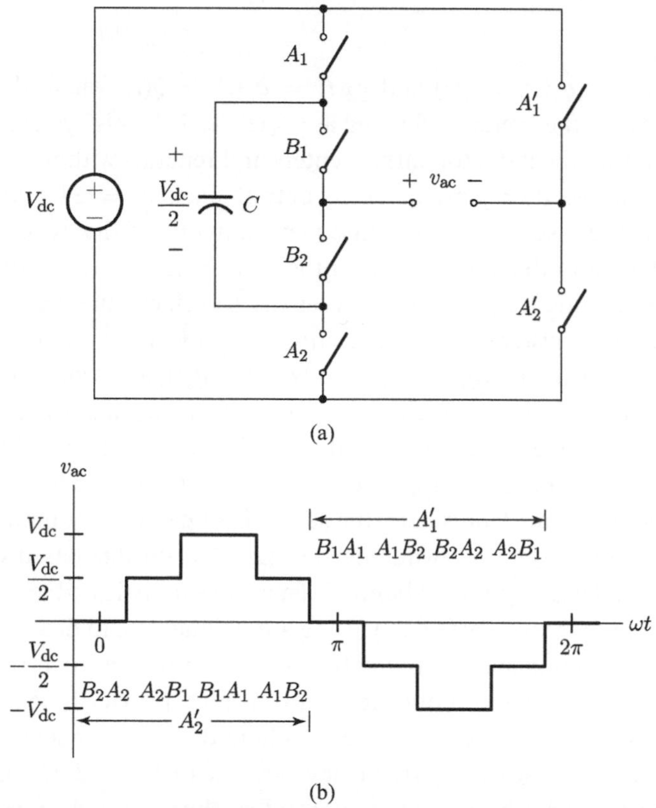

Figure 8.22 (a) A full bridge three-level dc/ac converter based on the flying-capacitor circuit of Fig. 5.27. (b) The ac output voltage showing the switching sequence required to balance the capacitor charges.

‡ In the nomenclature of multi-level converters, the number of levels is related to the number of dc potentials that can be applied at a single unipolar output port. So, for example, a full-bridge inverter is considered a two-level inverter, in that a given half-bridge output has two levels, V_{dc} and 0.

of the waveform. In addition, the V_d and $V_d/2$ states can be pulse-width modulated to further control harmonic content.

The use of flying-capacitors can be extended to an arbitrary number of levels. For a bridge leg with n levels, $n-1$ capacitors are required. The kth capacitor is controlled to have a voltage kV_d/n. Example 8.6 illustrates the structure of a four-level bridge leg.

Example 8.6 A Four-Level Flying-Capacitor Bridge Leg

Figure 8.23 shows a four-level bridge leg using flying capacitors. When combined with a second unfolding bridge leg, a bilateral four-level ac waveform results.

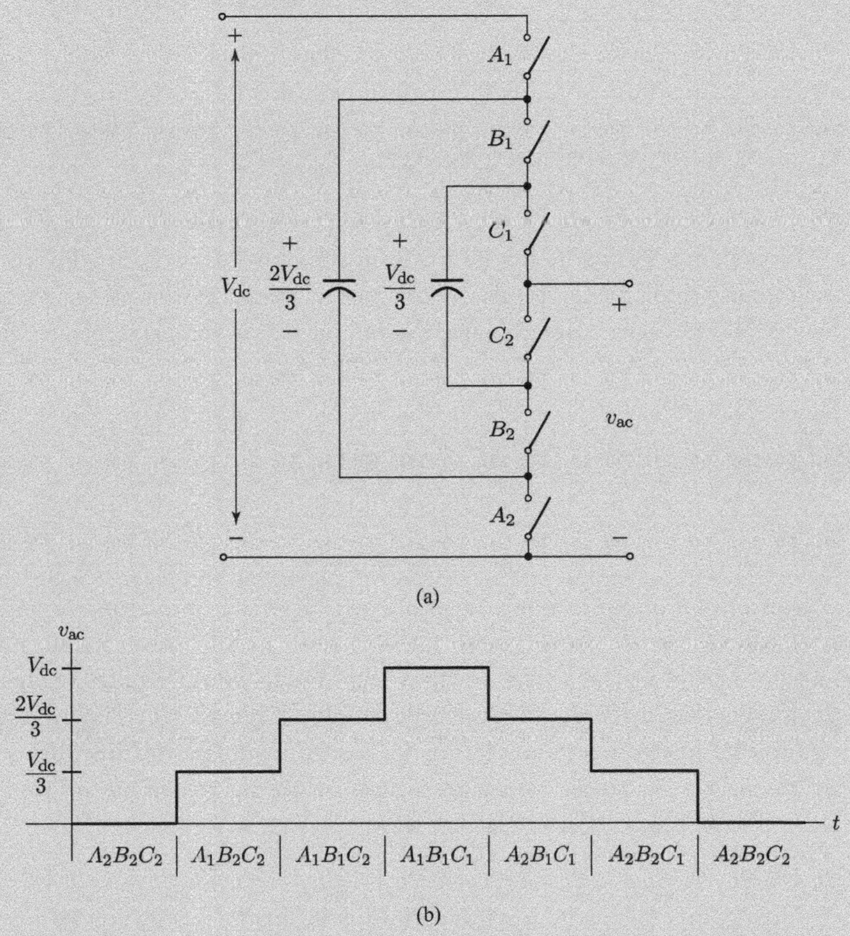

Figure 8.23 A four-level flying capacitor bridge leg with its half-wave waveform showing the switching sequence necessary to balance the capacitor voltages.

Figure 8.24 illustrates the basic diode-clamped three-level bridge leg. Like the flying-capacitor circuit, it can be used to create a three-level full-bridge converter. However, if the split dc source is implemented with capacitors, care must be taken that no current is drawn from the neutral

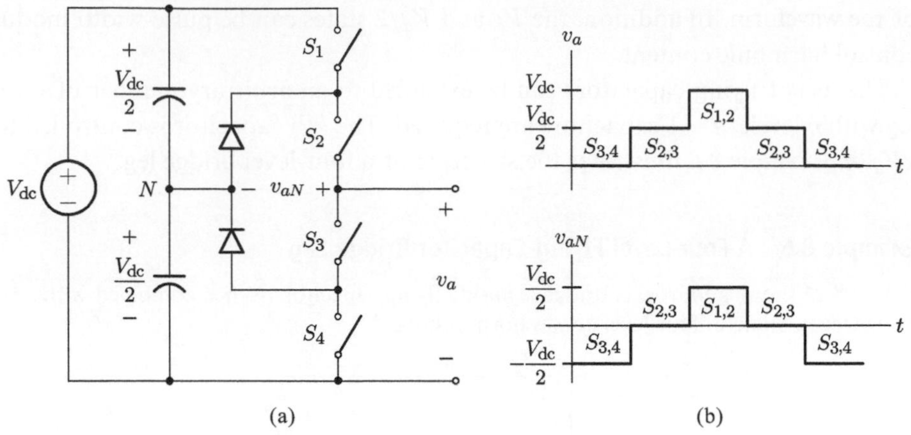

Figure 8.24 (a) A three-level diode-clamped bridge leg. (b) The line-to-ground voltage v_a and the corresponding line-to-neutral voltage v_{aN}.

point. The switches S_2 and S_3 can be switched simultaneously to simplify control. This circuit is a single-phase implementation of one method of creating what is known as a *neutral-point-clamped (NPC) inverter*, covered in more detail for three-phase systems in Chapter 9.

8.6 Transformer-Isolated DC/AC Converters

In principle, we can couple any dc/ac converter to a load through a properly designed transformer. For instance, we can couple both the PWM bridge converter of Fig. 8.14 and the square-wave converter of Fig. 8.1 through a transformer by replacing the resistor with the transformer primary and loading the secondary with the resistor. In both cases the transformer must have sufficient core cross-sectional area to carry the fundamental frequency magnetic flux without saturating. (We discuss this requirement in detail in Chapter 18.) The important constraint is that the product of the core cross section A_c, the maximum permissible flux density B_s, and the number of turns on one of the windings N must exceed the maximum value of the integral with respect to time of the voltage across that winding. That is,

$$NA_cB_s \geq \int v_{ac}(t)\, dt. \tag{8.25}$$

The maximum value of the volt–time integral is about the same for the output voltage of either the square-wave or PWM converter. The PWM converter, however, generates a voltage v_a containing components of much higher frequencies than those produced by the square-wave converter. Consequently, we must use a magnetic material with better high-frequency characteristics in the transformer for the PWM circuit, unless these high frequencies are filtered before the transformer.

8.6.1 High-Frequency Transformer Isolation

The PWM approaches that we have discussed so far result in a component of v_a at the low frequency ω_a. By utilizing a modulating process that moves all components of the modulated waveform to frequencies in the vicinity of the switching frequency, we can use a much smaller transformer to provide the requisite isolation. Unfortunately, a demodulating circuit (instead of a simple low-pass filter) is now required to recover the low-frequency ac waveform.

A modulation technique that moves the lowest-frequency component of the transformer voltage to the switching frequency is shown in Fig. 8.25(a). We control switches S_1–S_4 to create pulses of alternating polarity across the transformer primary. The result is a waveform v_a' whose maximum volt–time integral value is approximately the area of the largest pulse. Contrast this result with the sum of the areas of all the pulses between $\omega_a t = 0$ and π for the PWM inverter of Fig. 8.14. Therefore, the transformer linking the modulator and demodulator in the converter of Fig. 8.25 can be considerably smaller than one in a converter like that of Fig. 8.14. Although (8.25) shows that we can reduce the transformer size (implicit in A_c) in direct proportion to the

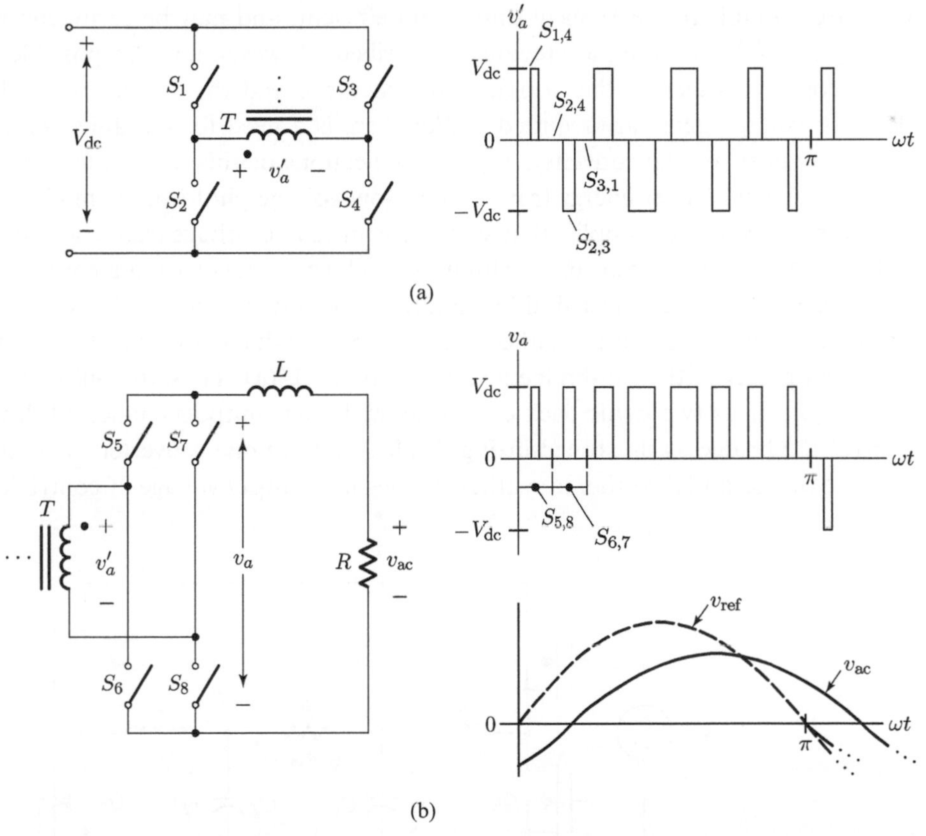

Figure 8.25 A method of PWM modulation that translates all frequency components of v_a' to the region above the switching frequency, permitting the use of a small transformer. (a) The modulator and transformer primary voltage. (b) The demodulator and its waveforms on the transformer (1:1) secondary.

reduction in the volt–time integral, the size reduction is less than this in practice. The reason is the higher-frequency waveform (that of Fig. 8.25a) generally requires the use of a magnetic core material with a lower value of B_s.

Figure 8.25(b) is one version of a demodulator that we can use to recover the desired low-frequency waveform from v_a'. The circuit shown is similar to the unfolder of Fig. 8.11, except that it unfolds each pulse instead of each half-cycle of the desired output waveform. The voltage v_a of Fig. 8.25(b) is simply the PWM waveform of Fig. 8.14, which we can low-pass filter to give the desired voltage v_{ac}.

The price we pay for the relatively small transformer in the circuit of Fig. 8.25 is greater control complexity and the need for all the switches in the circuit to operate at the switching frequency.

8.7 Other DC/AC Converter Topologies

The full-bridge inverter of Fig. 8.14(a) is by far the most common single-phase PWM dc/ac converter circuit. It is extremely simple and efficient, and may be easily controlled through a variety of PWM schemes, as previously described. However, it is also possible to realize PWM dc/ac converters using other topological structures, and this is sometimes done when large dc-to-ac voltage gains are required and/or there is a need for galvanic isolation between the dc and ac ports of the converter. Typical applications of this sort include photovoltaic *micro-inverters* (which deliver energy from a single low-voltage photovoltaic module to the grid) and uninterruptible power supplies that synthesize an ac line voltage from a low-voltage battery.

Figure 8.26 shows a pair of synchronous buck converters having a common dc input and a single resistive load connected differentially between their outputs. If we remove the optional output capacitors C_1 and C_2, and combine the buck inductors L_1 and L_2 into a single inductor L, we get precisely the full-bridge inverter of Fig. 8.14(a). Thus, the full-bridge inverter might be thought of as two dc/dc buck converters with their outputs connected differentially. If we control the bridge as illustrated in Fig 8.14(b), we use one converter to synthesize a positive output voltage, and the other to synthesize a negative output voltage. If controlled as in Example

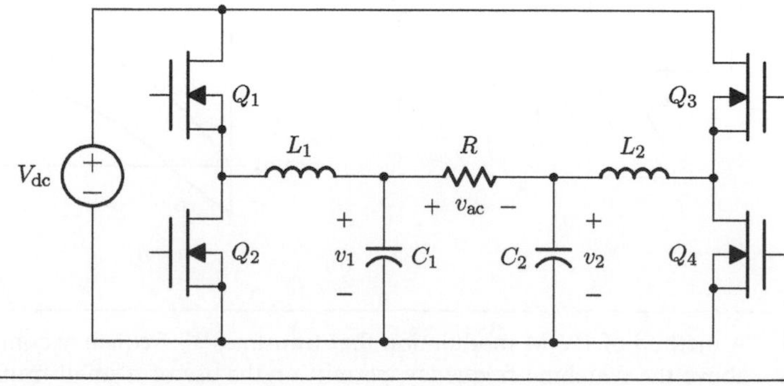

Figure 8.26 A dc/ac converter formed from a pair of synchronous buck converters with a common input, and the load connected between their differential outputs.

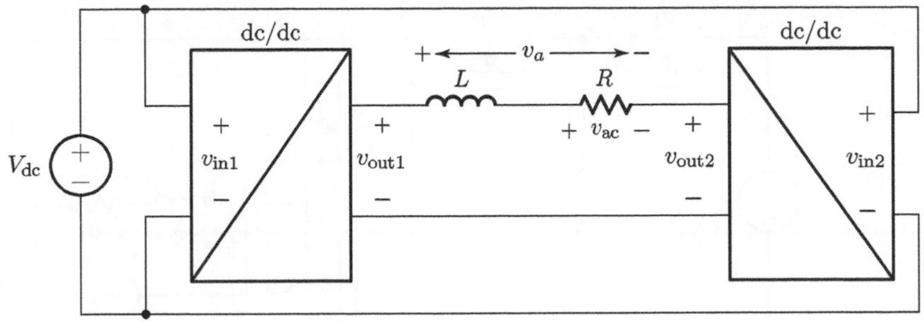

Figure 8.27 A dc/ac converter implemented as a pair of dc/dc converters with a common input and a single load connected differentially between their outputs.

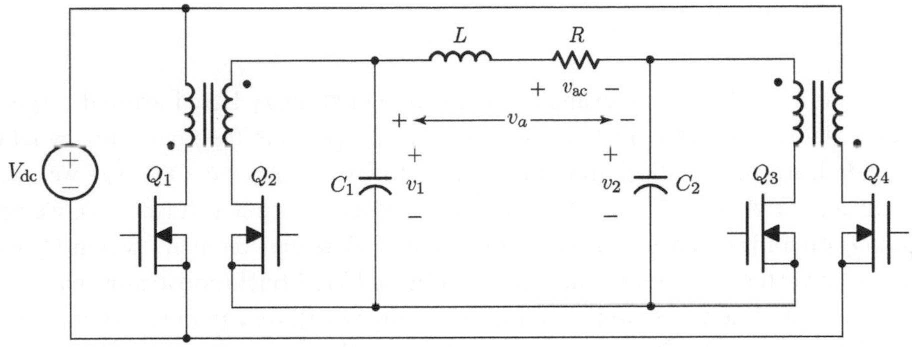

Figure 8.28 A full-bridge dc/ac converter implemented as a pair of synchronous flyback converters with a common input and a single load connected differentially between their outputs.

8.5, each buck converter synthesizes a complementary ac signal plus a dc offset, with the dc components canceled and ac components reinforced in the differential output $v_{ac}(t)$.

While the full-bridge inverter might be thought of as a dc/ac converter based upon synchronous buck converters, we could instead use other dc/dc converter topologies, as suggested by Fig. 8.27. One requirement on the underlying dc/dc converters used in such a system is that they be able to provide the necessary bilateral energy flow and voltage polarities. One example of such a design based on back-to-back synchronous flyback converters is shown in Fig. 8.28. Such a circuit might find use in a low-power application where a large voltage gain and electrical isolation between the dc source and the ac output of the system are required.

Another approach to building PWM dc/ac converters is suggested by the modulator/demodulator design of Fig. 8.25, which uses a full-bridge inverter to provide PWM inversion to high-frequency ac, and another full bridge to provide down-conversion from the high-frequency ac waveform to low-frequency ac. We might think of the demodulation (or down-conversion) as a combination of rectification and unfolding. Almost any magnetics-based PWM dc/dc converter incorporates an inverter (or inverting switch) and a rectifier (or rectifying switch). So, a variety of PWM dc/dc converters can be used as the core for a dc/ac converter, taking care that the rectifier portion of the converter also provides unfolding.

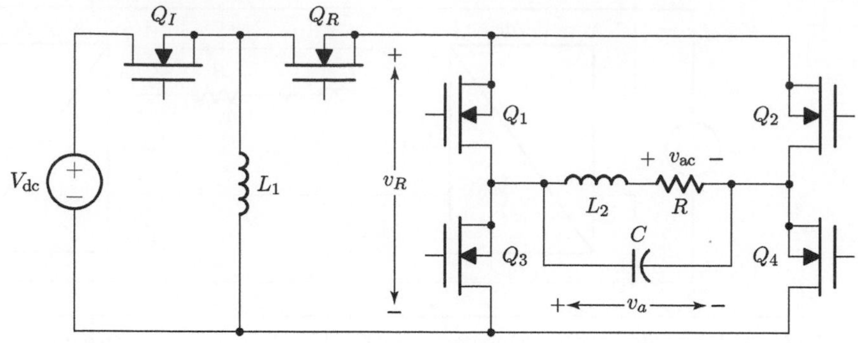

Figure 8.29 A dc/ac converter based upon a synchronous buck/boost converter, but with a rectifier circuit capable of both rectification and unfolding to low-frequency ac from the high-frequency PWM waveform.

Figure 8.29 shows one example of this design strategy based around a synchronous buck-/boost dc/dc converter. In this converter, switch Q_R acts together with switches Q_1 to Q_4 to provide both rectification and unfolding to low-frequency ac. One way we might control this dc/ac converter is as follows. For positive unfiltered output voltage $v_a(t)$, we hold switches Q_2 and Q_3 on and switches Q_1 and Q_4 off, and modulate the "rectifier" switch Q_R complementarily to the "inverter" switch Q_I, such that we obtain PWM buck/boost conversion between V_{dc} and $v_R = -v_a$. For negative unfiltered output voltage $v_a(t)$, we can hold switches Q_1 and Q_4 on and switches Q_2 and Q_3 off, and again modulate switch Q_R complementarily to the switch Q_I, such that we get buck/boost conversion between V_{dc} and $v_R = v_a$. In this topology, which allows bidirectional power flow between V_{dc} and v_{ac}, the high-frequency switches Q_I and Q_R handle PWM conversion while the low-frequency switches Q_1–Q_4 manage unfolding, much as in the buck-converter-based design of Fig. 8.11. However, with the use of switches with bidirectional blocking capability, we could instead integrate the rectifying and unfolding functions together into one bridge structure.

8.8 Power Balance in Single-Phase DC/AC Converters

Figure 8.30 illustrates a generalized single-phase dc/ac converter that may generally have energy flow in either direction, though the waveforms are illustrated with flow from ac to dc. If the converter operates with pure dc currents and voltages at one port and sinusoidal ac with unity power factor at the other, there is an instantaneous power mismatch between ports. The power at the dc port, $P_{dc} = V_{dc} I_{dc}$, is constant, but that at the ac port, $p_{ac} = i_{ac} v_{ac}$, is varying sinusoidally around the average (dc) power at twice the ac frequency:

$$p_{ac}(t) = I_{ac} \sin(\omega t) V_{ac} \sin(\omega t) = \frac{1}{2} I_{ac} V_{ac} [1 - \cos(2\omega t)]. \tag{8.26}$$

The instantaneous difference between the ac and dc powers must be provided by alternately supplying and absorbing energy within the converter, and thus the converter must contain energy

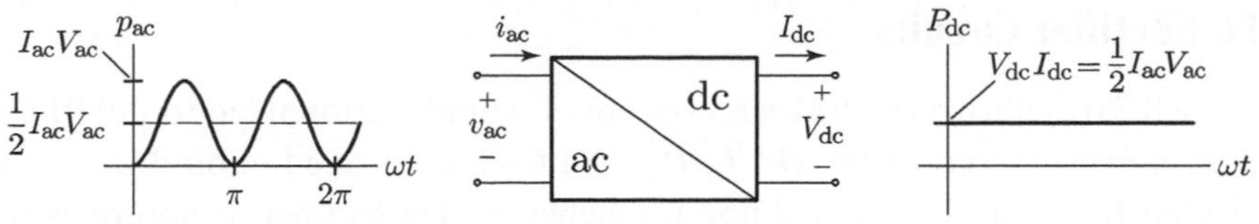

Figure 8.30 A generalized ac/dc converter showing $p(t)$ at both ports.

storage. This twice-line-frequency energy storage requirement is considerably greater than what we have discussed for reducing ripple at the terminals of a dc/dc converter. A capacitor is generally used for this purpose and is known as the *bulk storage capacitor*, denoted by C_B in what follows.

Defining the average power being processed by the converter in Fig. 8.30 as $P_{av} = \frac{1}{2}V_{ac}I_{ac}$, consider the amount of energy storage required for the purposes of buffering the "twice-line-frequency" energy. The peak swing in stored energy is the integral of the difference in powers at the dc and ac ports over one quarter of a cycle (e.g., from $\omega t = -\pi/4$ to $\omega t = \pi/4$):

$$E_c = \int_{-\pi/(4\omega)}^{\pi/(4\omega)} P_{av}\cos(2\omega t)\, d(t) = \frac{P_{av}}{\omega}. \tag{8.27}$$

Thus, to operate at unity power factor at the ac port and provide constant power at the dc port, the converter must buffer an amount of energy proportional to the average power divided by the line frequency. This energy storage is considerable (especially because of the low value of the line frequency), and often represents a large portion of converter volume in applications such as single-phase photovoltaic converters and ac/dc power supplies. This storage requirement represents a particular design challenge in that it cannot be changed through techniques such as increases in switching frequency. It can only be reduced by allowing reductions in power factor at the ac port (e.g., inclusion of harmonic currents) or by allowing a degree of power ripple at the dc port, each of which is sometimes done to reduce the energy storage requirement of the converter. This energy storage requirement is also one of the reasons why three-phase systems are preferred at high power, as three-phase converters do not have the same buffering requirement.

8.9 Switched-Mode Rectifiers and Power Factor Correction

So far in this chapter we have been concerned with the quality of the converter's ac output, with a focus on conversion of power from dc to ac. However, we can use the same techniques to convert power from ac to dc (e.g., from the ac grid) with high waveform quality. Depending on the application, there can be strict limits imposed on the power factor (both its displacement and harmonic components) of the ac line current drawn by a rectifier, as dictated by standards of the International Electrotechnic Commission (IEC) and others. In many rectifier applications these standards cannot be met by simply correcting the displacement component of power factor. In these cases the power factor of the ac line input can be corrected to near unity by using a *switched-mode rectifier*, also referred to as a *power factor correction (PFC)* rectifier.

8.9.1 PFC Rectifier Circuits

Figure 8.31(a) illustrates the basic concept of a unidirectional (ac-to-dc) PFC rectifier. The basic boost converter comprising Q_1, L, D_5, and C_B is controlled to produce a current i_L that, when reflected to the ac-side of the line-frequency bridge rectifier, is approximately sinusoidal and in phase with the ac line voltage, as illustrated in Fig. 8.31(b). The fact that a simple diode rectifier can provide unfolding of the boost converter input current to provide an ac line current is among the simplifying characteristics of a grid-connected ac/dc rectifier, as compared to a general-purpose dc/ac converter that can provide bidirectional power flow.

Use of the boost topology as the dc/dc stage of PFC rectifiers is extremely common. With an inductor at its input, a boost converter has a smooth input current waveform, and a boost converter offers the ability to draw current across the wide input voltage range (effectively down to zero volts), while providing a fixed output voltage (often selected in the vicinity of 400 V, considering single-phase line voltages across the world). A high output voltage is also conducive to storing the twice-line-frequency energy at relatively low cost and volume. As a direct converter topology, it also offers relatively low device stress.

Many other dc/dc converter stages can be used together with a line-frequency rectifier to provide power factor correction. A SEPIC converter, for example, can provide smoothed current and wide input voltage capability, and can be used to achieve a low output voltage, but at the expense of increased component stresses. A buck converter, on the other hand,

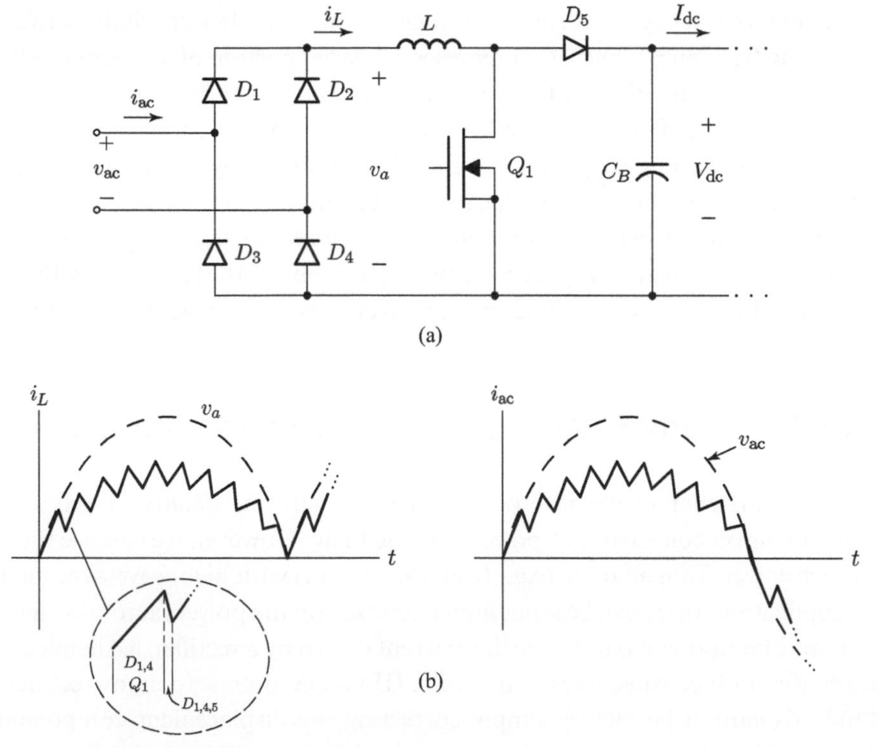

Figure 8.31 (a) The basic PFC rectifier. (b) The inductor and line currents i_L and i_{ac}.

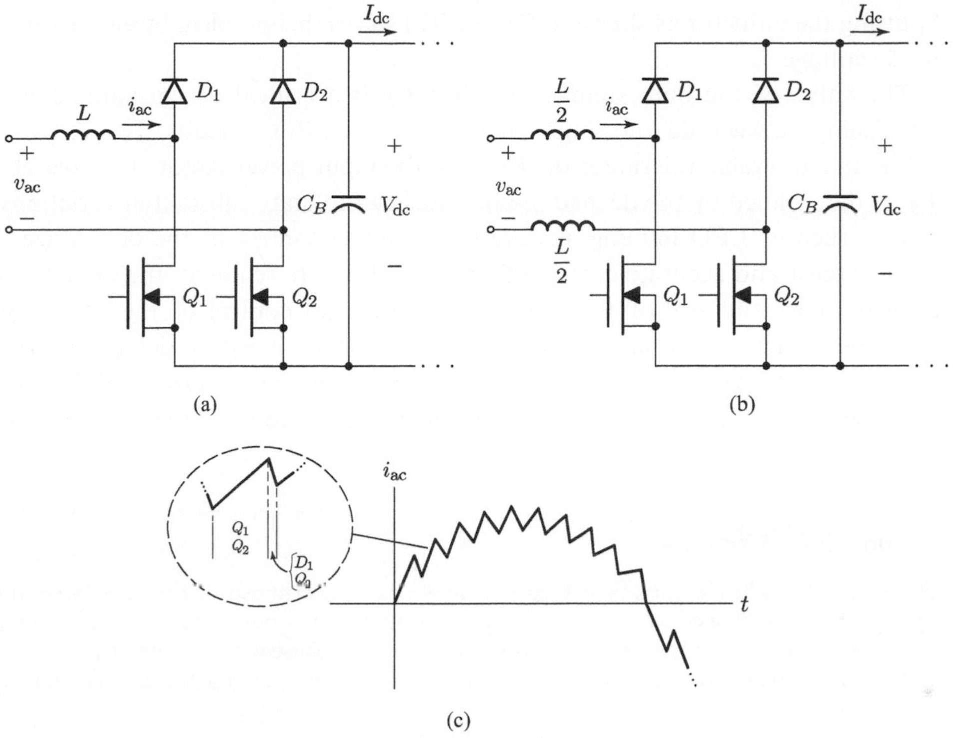

Figure 8.32 (a) A "bridgeless" PFC rectifier with a single ac line inductor. (b) The rectifier of (a) with the inductor split between the + and − ac lines to eliminate a common-mode voltage at the output. (c) The line current i_{ac}.

has low device stresses and enables a low output voltage, but can only operate when the instantaneous input voltage is higher than the specified output voltage, resulting in power factor limitations. In some cases, the power factor correction power supply is also designed to provide isolation and step down to a low-voltage output, yielding a "single-stage" power factor corrected power supply. Flyback converters and their variants are used extremely widely in such applications.

An alternate PFC circuit is shown in Fig. 8.32(a), where the inductor is placed on the ac side of the bridge and the lower bridge-leg diodes are replaced by MOSFETs. For obscure reasons this circuit is referred to as a *bridgeless* PFC converter (or sometimes as a *semi-bridge* switched-mode rectifier), though it clearly contains a bridge. This design partly combines the devices used for line frequency rectification with those used for high-frequency PWM conversion. The MOSFETs are controlled to again produce a current i_{ac} in phase with v_{ac}. During the positive half cycle of v_{ac}, D_1 and the body diode of Q_2 alternate conduction with D_1 and Q_1. The duty ratio of the two conducting paths is controlled to provide the appropriate shaping of the input current i_{ac}.

With a single ac-side inductor, as shown in Fig. 8.32(a), the terminals of V_{dc} will experience a common-mode voltage fluctuation with respect to circuit ground, caused by the Ldi/dt voltage introduced in series with the ac line. The result is noise that can interfere with control circuits.

Splitting the inductor as shown in Fig. 8.32(b) solves this problem by eliminating the common-mode voltage.

The value of the bulk storage capacitor C_B is a critical design parameter for converters interfacing between dc and single-phase ac, such as PFC circuits. Two criteria determine its value: the allowable tolerances on V_{dc}, and the input power factor. The tolerable variation in V_{dc} is determined by the dc load. Some loads can tolerate substantial variations in V_{dc}, while others, such as LED lighting, require more precise control of the dc voltage. Where a high power factor and accurate control of the dc voltage are required, the capacitor can dominate both the cost and size of the converter. Where tight control of the dc voltage is required, the bulk capacitor can be reduced in size by following it with a dc/dc converter. In this case the circuits of Figs. 8.31 and 8.32 are known as *front-end converters*. While these two circuits represent the basic structures of PFC rectifiers, there are a number of variations on them in practical use.

Example 8.7 Sizing C_B

The value of C_B in the circuits of Figs. 8.31 and 8.32 is determined by the acceptable ripple amplitude on V_{dc}. Assuming the converter is operating at constant output power P_{dc}, C_B must source and sink the difference between $p_{ac}(t)$ and P_{dc}. If i_{ac} is sinusoidal and in phase with v_{ac}, then I_{ac} is proportional to V_{ac} and we can express it as $i_{ac}(t) = kV_{ac}\sin(\omega t)$. The instantaneous power at the ac port can then be written as

$$p_{ac}(t) = \frac{1}{2}kV_{ac}^2(1 - \cos 2\omega t).$$

The power to C_B, plotted in Fig. 8.33, is then

$$p_{C_B}(t) = p_{ac} - P_{dc} = \frac{1}{2}kV_{ac}^2\cos(2\omega t).$$

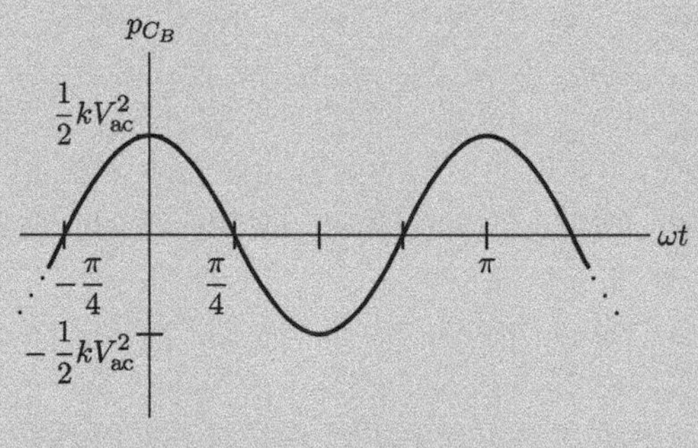

Figure 8.33 The power to C_B which is sinusoidal at twice the line frequency.

If P_{dc} is constant, that is, Δv_{C_B} is constant, (8.27) states that the peak swing in stored energy that C_B is required to buffer is P_{dc}/ω. To approach this condition requires C_B to be of extremely large value, particularly at high powers. We therefore specify an acceptable nonzero value for Δv_{C_B} to determine a practical value for C_B.

The median stored energy in C_B is $E_{C_B} = (1/2)C_B V_{dc}^2$. If we specify a ripple Δv_{C_B}, the corresponding change in E_{C_B}, ΔE_{C_B}, is $\Delta E_{C_B} = C_B V_{dc}\Delta v_{C_B}$. If Δv_{C_B} is small, then (8.27) remains approximately true, and

$$\Delta E_{C_B} = C_B V_{\mathrm{dc}} \Delta v_{C_B} \approx \frac{P_{\mathrm{dc}}}{\omega}.$$

The ripple ratio, $\mathcal{R}$, of the dc voltage is

$$\mathcal{R} = \frac{\Delta E_{\mathrm{dc}}}{2V_{\mathrm{dc}}} = \frac{P_{\mathrm{dc}}}{\omega C_B V_{\mathrm{dc}}^2}.$$

So, for a given $\mathcal{R}$, we can determine the required C_B:

$$C_B = \frac{2P_{\mathrm{dc}}}{\mathcal{R}\omega V_{\mathrm{dc}}^2}.$$

For $P_{\mathrm{dc}} = 1\,\mathrm{kW}$, $\omega = 377$ (60 Hz), $V_{\mathrm{dc}} = 200\,\mathrm{V}$, and $\mathcal{R} = 5\%$,

$$C_B = \frac{2(10^3)}{(0.05)(377)(4 \times 10^4)} = 2.7\,\mathrm{mF}.$$

Note that if the ripple ratio is large, then our assumption of constant output power is compromised and C_B will be a value different from that calculated here.

8.9.2 Control of PFC Rectifiers

Unlike diode rectifiers, the PFC rectifier must be actively controlled. There are a number of approaches to controlling the ac line current. We discuss here the general approach to control, using a current controller to determine when to switch the controlling transistors. The basic structure of such a controller with reference to the circuit of Fig. 8.31 is shown in Fig. 8.34, in which the output voltage is determined by controlling the input current to be both sinusoidal and of the value necessary to produce the desired V_{dc}.

The output V_{dc} is scaled and fed back to the controller input as v_{ref}, creating an error voltage that is processed by the compensator to produce g_i, which – when multiplied by the scaled and rectified v_{ac} – produces the required reference sinusoidal current, i_{ref}. This reference current is then fed into a current controller, which could take several forms, as discussed in Section 8.4. Figure 8.35 illustrates a hysteretic controller that could occupy the position of the current controller block.

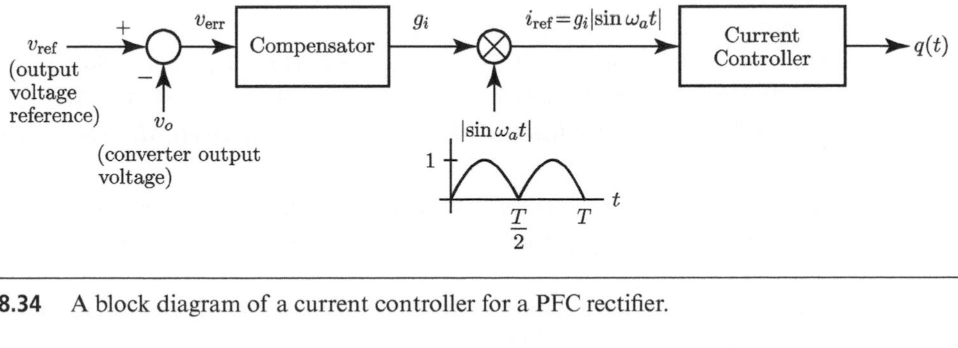

Figure 8.34 A block diagram of a current controller for a PFC rectifier.

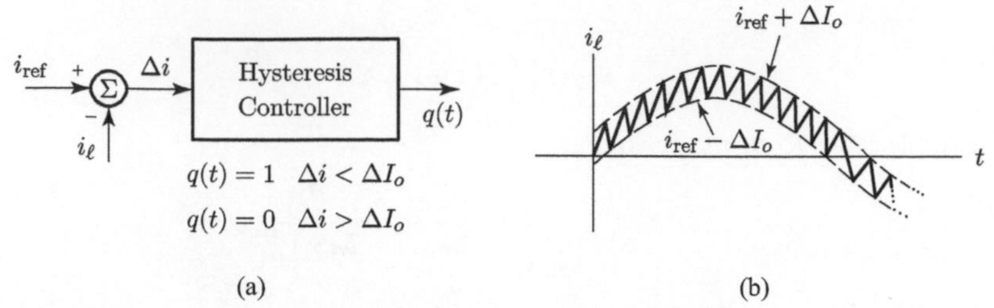

Figure 8.35 A hysteretic current controller that could occupy the "current controller" block in Fig. 8.34.

Notes and Bibliography

Harmonic elimination through proper choice of PWM patterns is the subject of [1], [2], and [3]. In [1], Patel and Hoft derive patterns to eliminate five harmonics from the ac waveform, and in [2] they consider the result of varying these patterns to control voltage as well as harmonics. In [3], Chiasson *et al.* expand upon this work and derive alternative switching patterns that in addition to eliminating low-order harmonics, reduce the increase in higher-order harmonics.

The origins of the inverter are described in [4]; the word *inverter* likely dates to the classic paper by Prince in 1925 [5]. A wealth of information about carrier-based pulse-width modulation techniques for inverters can be found in [6]. This reference provides tremendous detail about different available methods in both single-phase and polyphase inverters, including analysis of the frequency content resulting from various techniques. Description of hysteresis-based current control of single-phase inverters, including a comparison of various techniques, can be found in [7]. Current control methods in polyphase inverters, including both hysteretic and carrier-based methods, are reviewed in [8].

A study of the twice-line-frequency energy storage requirements of PFC converters, and means to minimize them within regulatory specifications, are well documented in [9].

Dc/ac converters based around a variety of dc/dc converter topologies are overviewed in [10] and [11]. These dc/ac converters are used, for example, as "microinverters," sometimes called "module integrated converters," to convert energy from individual low-voltage solar panels (or modules) to the ac grid.

1. H. S. Patel and R. G. Hoft, "Generalized Techniques of Harmonic Elimination and Voltage Control in Thyristor Inverters: Part I – Harmonic Elimination Techniques," *IEEE Trans. Industry Applications* vol. 9, pp. 310–317, May/June 1973.
2. H. S. Patel and R. G. Hoft, "Generalized Techniques of Harmonic Elimination and Voltage Control in Thyristor Inverters: Part II – Voltage Control Techniques," *IEEE Trans. Industry Applications*, vol. 10, pp. 666–673, Sept./Oct. 1974.
3. J. N. Chiasson, L. M. Tolbert, S. J. McKenzie, and Z. Du, "A Complete Solution to the Harmonic Elimination Problem," *IEEE Trans. Power Electronics*, vol. 19, pp. 491–499, Mar. 1974.
4. E. L. Owen, "Origins of the Inverter," *IEEE Industry Applications Magazine*, pp. 64–66, Jan./Feb. 1996.
5. D. C. Prince, "The Inverter," *General Electric Review*, vol. 28, pp. 676–681, Oct. 1925.
6. D. G. Holmes and T. A. Lipo, *Pulse Width Modulation for Power Converters: Principles and Practice*, New York: IEEE Press, 2003.
7. P. A. Dahono, "New Hysteresis Current Controller for Single-Phase Full-Bridge Inverters," *IET Power Electronics*, vol. 2, pp. 585–594, 2009.

8. M. P. Kazmierkowski and L. Malenesi, "Current Control Techniques for Three-Phase Voltage-Source PWM Converters: A Survey," *IEEE Trans. Industrial Electronics*, vol. 45, pp. 691–703, Oct. 1998.
9. A. J. Hanson, A. Martin, and D. J. Perreault, "Energy and Size Reduction of Grid-Interfaced Energy Buffers Through Line Waveform Control," *IEEE Trans. Power Electronics*, vol. 34, pp. 11442–11453, Nov. 2019.
10. Q. Li and P. Wolfs, "A Review of the Single Phase Photovoltaic Module Integrated Converter Topologies With Three Different DC Link Configurations," *IEEE Trans. Power Electronics*, vol. 23, pp. 1320–1333, May 2008.
11. D. Meneses, F. Blaabjerg, O. Garcia, and J. A. Cobos, "Review and Comparison of Step-Up Transformerless Topologies for Photovoltaic AC-Module Application," *IEEE Trans. Power Electronics*, vol. 28, pp. 2649–2663, June 2013.

PROBLEMS

8.1 Determine the average power delivered to the load in Fig. 8.2 if the third-harmonic component of i_a is not neglected. (*Hint:* Do your analysis in the frequency domain.)

8.2 In Example 8.1 we determined values of ϕ and δ that would result in 10 kW being delivered to the source v_{ac} at unity displacement factor. Redo that example under the constraint that the *inverter* has a unity displacement factor at the fundamental frequency. That is, i_{a_1} and v_{a_1} are in phase. Which of these two control schemes results in the lowest value of rms load current?

8.3 Determine and sketch the load current i_a in Example 8.1. (*Hint:* Use circuit replacement by equivalent source to represent v_a and then use superposition of the responses to v_a and v_{ac}.) Can switches permitting only unilateral current flow be used in this circuit? Explain.

8.4 Consider the inverter and switching sequence shown in Fig. 8.1. Determine the value of δ that would result in elimination of the seventh harmonic of the load voltage. Calculate the third and fifth harmonic amplitudes for this case and compare them to the corresponding amplitudes for a true square wave (that is, a square wave having no 0 state).

8.5 The waveforms of Figs. 8.6(c) and 8.9 were constructed to be free of both the third and fifth harmonics. Compare these waveforms with respect to their THDs.

8.6 Example 8.2 showed how both the third and fifth harmonics could be eliminated by appropriately placed notches. However, doing so eliminates the possibility of controlling the fundamental of v_a. But we can use the notches in the waveform of Fig. 8.6(c) to control the fundamental of v_a if we are not required to eliminate the fifth harmonic. Determine the dependence of the fundamental on the width of the two notches in v_a of Fig. 8.6(c).

8.7 For an appropriately-modified definition of THD to normalize to the dc component, calculate the THD of the dc source current i_d in Fig. 8.8 if the converters are controlled as described in Example 8.3. Assume that the converter has a resistive load at its ac terminals.

8.8 Example 8.4 analyzes the THD of a sine–triangle PWM waveform having a depth of modulation $k = 1$. Show that decreasing the depth of modulation k below 1 increases the THD of the PWM waveform.

8.9 Example 8.3 showed how multiple converters could be used to eliminate the third and fifth harmonics from the ac voltage. How would you eliminate three harmonics (e.g., third, fifth, and seventh) using two bridge converters? Draw the resulting ac voltage waveform.

8.10 The low-pass filter used in the PWM inverter of Fig. 8.25 introduces a phase shift between the fundamental of v_a and v_{ac}. Express this phase shift in terms of L and R.

8.11 The resistive load in the PWM inverter of Fig. 8.25 is replaced by a sinusoidal voltage source having the same polarity as that shown for v_{ac}. How would the eight switches be implemented and controlled so that power flows from this ac source to the dc side of the circuit?

8.12 Figure 8.36 shows a half-bridge inverter driving a load consisting of a sinusoidal voltage source v_s in series with an inductor L. How should the switches be controlled to maximize the power *delivered to* the load? Express this maximum power in terms of the parameters of the circuit. How could the switches be implemented for this condition?

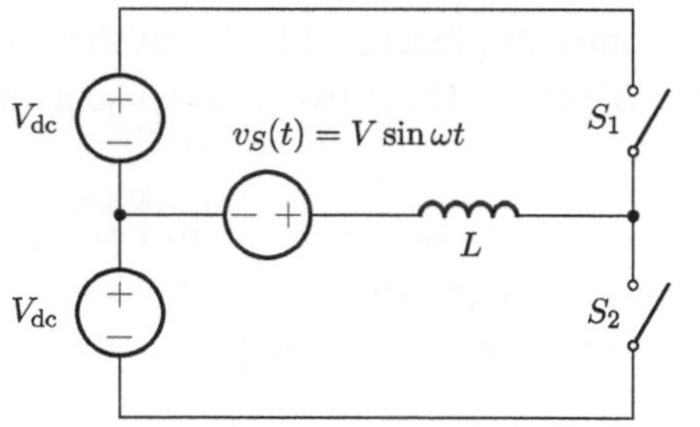

Figure 8.36 The half-bridge inverter discussed in Problem 8.12.

8.13 Is there any operating condition for the current-source inverter of Fig. 8.4 under which the switches are required to carry bilateral current?

8.14 Propose an implementation of the switches in the inverter of Example 8.1.

8.15 In Section 8.2 we discussed the elimination of harmonics from the inverter's ac terminal variables. It is almost always equally important to consider harmonics generated in the dc variables.

(a) Calculate, sketch, and dimension the dc terminal current for the inverter of Fig. 8.3 under the operating conditions determined in Example 8.1.

(b) One possible filter circuit designed to eliminate the ac components from the dc input current is shown in Fig. 8.37. Using intelligent approximations, determine values for L and C so that the peak-to-peak ripple current at the dc source is 5% of the dc current, and the peak-to-peak ripple voltage at the inverter input terminals is 5% of V_{dc}.

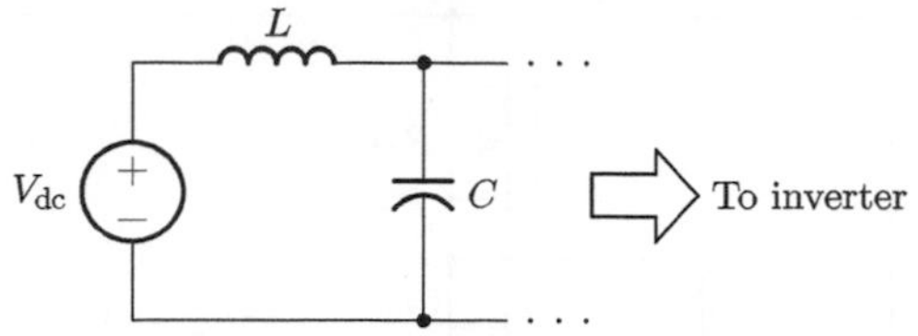

Figure 8.37 Input filter for the inverter of Fig. 8.3. Values for L and C are determined in Problem 8.15.

8.16 Reconsider the filter designed in Problem 8.15. Sketch and dimension the inverter input voltage if all four switches of the inverter are suddenly opened simultaneously. Does this behavior create a problem? If so, how would you solve it?

8.17 In the text we stated that more than just the third and fifth harmonics had been canceled in the waveform of v_a in Fig. 8.9. What are the other harmonics that had been canceled?

8.18 Assuming that a center-tapped secondary winding is permissible in the high-frequency transformer of Fig. 8.25, design a simpler demodulator circuit to replace the four-switch bridge shown.

8.19 Compare the fundamental components of v_a of Fig. 8.6(a), which contains no third harmonic, and v_a of Fig. 8.6(c), which contains neither the third nor the fifth harmonic. Assume that V_a is the same for both.

8.20 Compare the THD of a sinusoidal PWM waveform, as calculated in Example 8.4, with the THD of v_a in Fig. 8.6(c), which is notched to eliminate both the third and fifth harmonics.

8.21 Figure 8.38 shows the schematic of a full-bridge three-level FCML (see Chapter 5) inverter. In this topology, the voltages on the flying capacitors are controlled to be approximately half of the input voltage (by balancing the switching patterns appropriately).

(a) Propose a switching pattern for the devices in the flying capacitor inverter that takes advantage of the multi-level capabilities of the inverter and results in an (unfiltered) output voltage $v_x(t)$ having no third, fifth, or even harmonics. You may assume that the voltages on the flying capacitors are at one half the input voltage. Note that each switch is switched oppositely to its corresponding "primed" switch. Consequently, the independent switching functions are q_{A1}, q_{A2}, q_{B1}, and q_{B2}. How many times does each active device switch on and off per ac output cycle?

(b) Which has higher harmonic content (as described by THD): the unfiltered waveform $v_x(t)$ you have determined, or that of a conventional inverter using the harmonic elimination scheme of Fig. 8.6(c)?

8.22 Figure 8.39 shows a simplified model of a current-source inverter. The circuit takes a dc input current and generates an ac output current into the load/filter combination. The transformer is wound with a 1:1 turns ratio; you may assume it is ideal.

(a) Considering the case $I_{dc} > 0$, propose a switch implementation for switches S_1 to S_4.

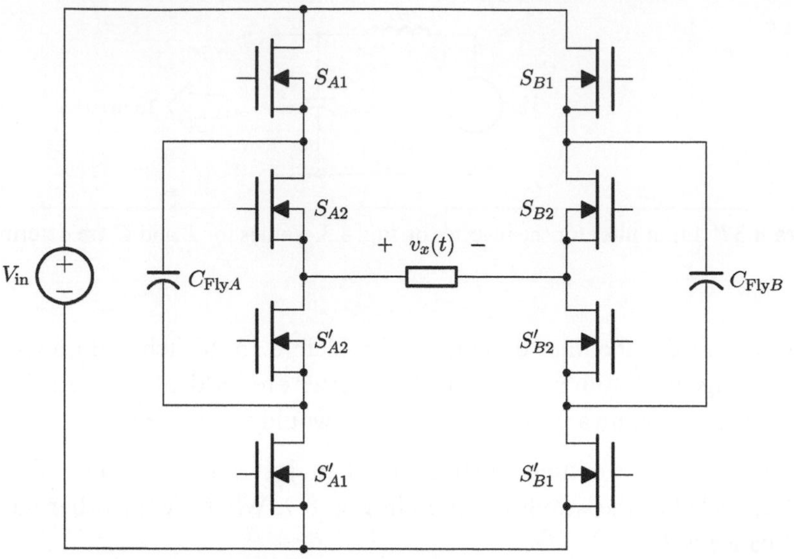

Figure 8.38 Structure of a full-bridge three-level FCML inverter.

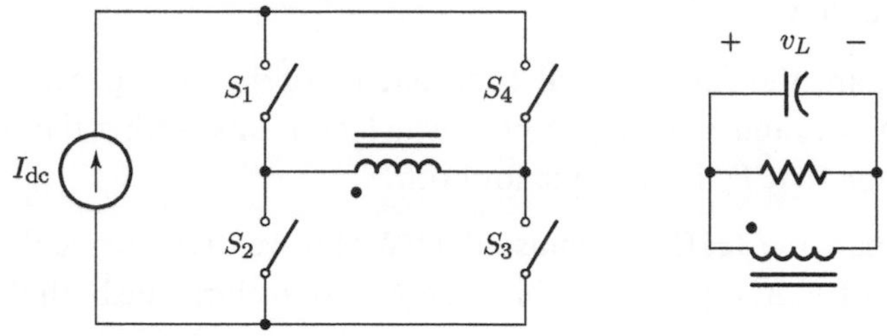

Figure 8.39 A simplified model for a current-source inverter. The current source is usually realized as a large inductor in series with a source voltage.

(b) The switches S_1–S_4 are controlled such that they each switch on and off once per ac output cycle. Determine a switching pattern for S_1–S_4 such that the ac output current (and voltage) have no third-harmonic content. Sketch the switching functions $q_1(t)$–$q_4(t)$ (for S_1 to S_4) over $0 < \omega t < 2\pi$, and sketch the resulting current waveform into the load/filter combination.

(c) Illustrate how a second current-source inverter might be connected to implement a harmonic cancellation scheme.

9 Polyphase Sources and Converters

In this chapter we introduce polyphase systems, their advantages relative to single-phase systems, and how they are represented. We then use the single-phase bridge circuit as a building block to create polyphase rectifiers and inverters. We also discuss the operating characteristics of these circuits from both the ac and dc sides.

9.1 Polyphase Sources

A polyphase ac source provides several sinusoidal voltages simultaneously, with identical frequencies and with phases shifted from one another by a predetermined electrical angle. Such sources, particularly those consisting of three voltages phase-shifted by 120° (known as *three-phase* sources), provide functional advantages to many high-power systems such as rectifiers, inverters, generators, and power transmission lines. These systems benefit by both reduced complexity and improved performance relative to systems supplied by single-phase sources. The first polyphase systems originated in the late 1800s and were two-phase systems with phase voltages 90° out of phase, that is, with sine and cosine phase voltages.†

A central advantage of a *balanced* polyphase ac system, that is, one whose sources all have the same amplitude (and frequency), and are equally separated in phase, is that the total power being supplied by the sources to a balanced linear load is constant. This is not true for a single-phase source, which supplies power pulsating at twice the source frequency, for example

$$p(t) = \frac{V_s^2}{2R}(1 - \cos(2\omega t))$$

for a sine-wave source of amplitude V_s supplying a load resistor R. This pulsating power characteristic can be problematic for many kinds of loads. A good example of the benefits of constant power is the powering of electric motors, where a single-phase motor will produce torque pulsations while a three-phase motor will produce a constant torque.

9.1.1 Phasor Representation of Sinusoids

A sinusoidal voltage having amplitude V_x and phase ϕ_x can be represented as the real part, Re{ }, of the product of a complex vector, $\widehat{V}_x$, and the complex time-dependent exponential

† Important early two-phase systems included the ac generation system provided by Westinghouse for powering the 1893 Columbian Exposition in Chicago and the generation system installed by Westinghouse at Niagara Falls in 1895 [1].

$e^{j\omega t}$. The complex vector is known as the complex amplitude of the sinusoid and contains both the magnitude and phase of the voltage:

$$v_x(t) = V_x \cos(\omega t + \phi_x) = \mathrm{Re}\{\widehat{V}_x e^{j\omega t}\}, \qquad \widehat{V}_x = V_x e^{j\phi_x}, \tag{9.1}$$

where Re{·} takes the real part of a complex number. When plotted in the complex plane, $\widehat{V}_x$ is known as a *phasor* and the plot is known as a *phasor diagram*.

Consider the two-phase system illustrated in Fig. 9.1(a). The two sinusoidal voltage sources are 90° out of phase and can be expressed as

$$\begin{aligned} v_a(t) &= \mathrm{Re}\{V_a e^{j\omega t}\} = \mathrm{Re}\{\widehat{V}_a e^{j\omega t}\}, \\ v_b(t) &= \mathrm{Re}\{V_b e^{j(\omega t - \frac{\pi}{2})}\} = \mathrm{Re}\{\widehat{V}_b e^{j\omega t}\}, \end{aligned} \tag{9.2}$$

where $\widehat{V}_a = V_a$ and $\widehat{V}_b = V_b e^{j(-\pi/2)}$. Figure 9.1(b) is the phasor diagram for the two sources in (a). Since it is the real part (Re) of the phasor that represents its associated voltage, the projection of the phasor onto the real axis gives its value. Also indicated in the figure is the rotation of the phasors with time. As the phasor rotates, its projection on the real axis varies sinusoidally with time. Note that the phases of the voltages referenced to one another are constant, and unrelated to the angle of the phasors with respect to the real axis.

9.1.2 Constant Power and Rotating Fields

The two-phase source in Fig. 9.1(a) is coupled by a "two-phase, three-wire" interconnection to a balanced load of resistors of equal value. Only three wires are required because the return wire is common between the two phases. Calculating the instantaneous power delivered by the two sources to the two loads shows that it is constant at $p(t) = V_s^2/R$. The oscillating terms cancel owing to the identity $\sin^2(\theta) + \cos^2(\theta) = 1$. This two-phase source can thus provide constant power to a balanced two-phase load, despite using purely ac voltages and currents to deliver the power. Were the two sources to represent the outputs of a generator with windings in quadrature, for example, the loading on the generator would be constant, eliminating torque pulsations that would occur for a single winding producing a single-phase output. Similar benefits, including reductions in filtering and energy storage requirements, accrue for many types of loads.

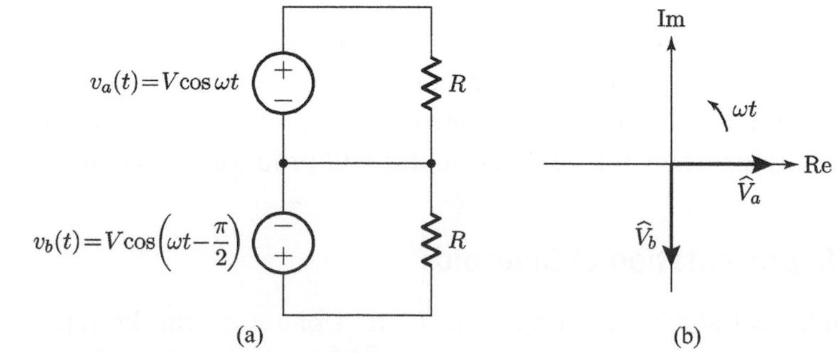

Figure 9.1 (a) A two-phase, three-wire power system with sine and cosine sources, and a balanced resistive load. (b) A phasor diagram showing the relative phases of the two source voltages.

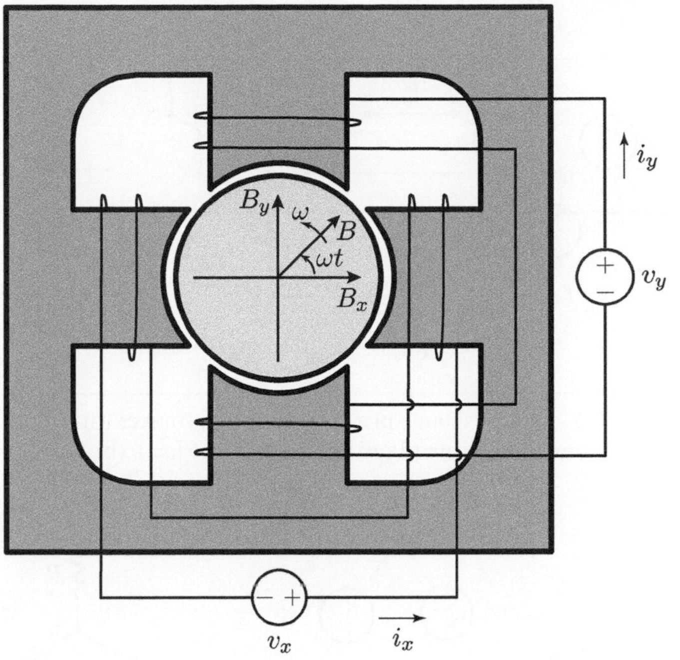

Figure 9.2 Conceptual illustration of a motor armature with two spatially orthogonal windings driven by two temporally orthogonal voltage sources.

A second advantage of polyphase systems relates to their ability to generate rotating magnetic fields for ac electric machines. Consider the conceptual illustration of an electric machine shown in Fig. 9.2, in which the armature comprises two windings – one x-directed and one y-directed. If the two windings, which are orthogonal *in space*, are driven by ac voltage sources that are orthogonal *in time* (that is, 90° out of phase), then the induced magnetic flux density will have the form $\widehat{B} = B_0 \cos(\omega t)\widehat{x} + B_0 \sin(\omega t)\widehat{y}$. Such a magnetic flux density has a constant amplitude B_0 and forms an angle ωt with respect to the x axis. Thus, the flux rotates in space with angular velocity ω. Such a rotating field is valuable for driving a variety of ac electric machines, providing for self-starting motors.

9.2 Three-Phase Sources

A three-phase source consists of three sinusoidal voltages having equal amplitudes, and phases shifted by one third of a cycle ($2\pi/3$ radians or 120° at the fundamental frequency) with respect to one another:

$$\begin{aligned} v_{an} &= V_s \cos \omega t = \text{Re}\{\widehat{V}_{an} e^{j\omega t}\}, \\ v_{bn} &= V_s \cos\left(\omega t - \frac{2\pi}{3}\right) = \text{Re}\{\widehat{V}_{bn} e^{j\omega t}\}, \\ v_{cn} &= V_s \cos\left(\omega t + \frac{2\pi}{3}\right) = \text{Re}\{\widehat{V}_{cn} e^{j\omega t}\}. \end{aligned} \tag{9.3}$$

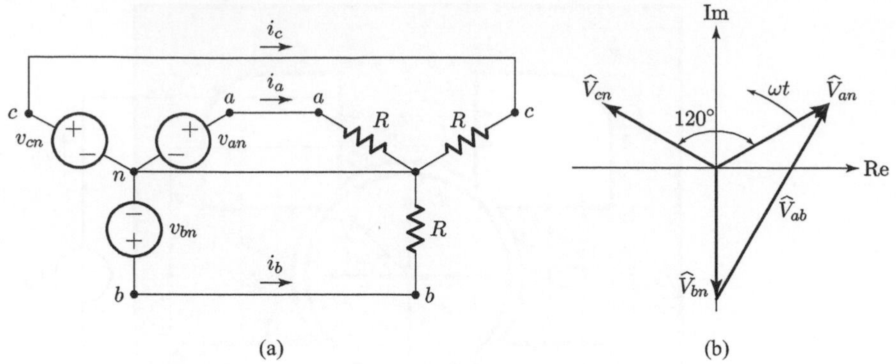

Figure 9.3 (a) A Y-connected three-phase set of voltage sources with neutral (n) connected by four wires to a Y-connected balanced resistive three-phase load. (b) A phasor diagram of the load voltages.

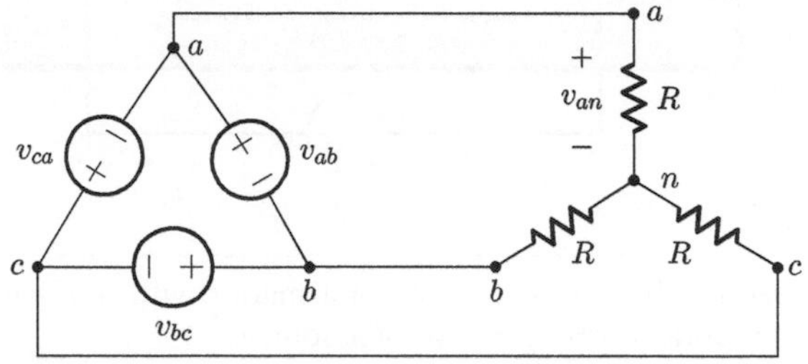

Figure 9.4 (a) A Δ-connected three-phase set of voltage sources feeding a Y-connected load. This connection does not provide a neutral point.

It can be constructed in two ways. One connection is shown in Fig. 9.3(a), where the three sources are referenced to a common point called *neutral*, or the *Y-point*, the latter because in the connection as drawn, the three sources form a "Y." Depending on the system, the neutral point may or may not be connected to electrical ground.

The voltages shown in this connection are called *line-to-neutral* voltages. We identify such voltages by a subscript indicating their phase. For instance, v_{an} is the voltage of phase a with respect to neutral. Figure 9.3(b) is the phasor diagram for the three phase voltages. To find the instantaneous value of a voltage $v_{xn}(t)$, one may simply rotate the phasor $\widehat{V}_{xn}$ counterclockwise in the complex plane by an angle ωt and take its projection onto the real axis. The phasor vectors $\widehat{V}_{an}$, $\widehat{V}_{bn}$, and $\widehat{V}_{cn}$ of Fig. 9.3(b) clearly illustrate the equal magnitudes and 120° relative phase displacements of the three-phase voltage set. They may be thought of as rotating together counterclockwise at an angular velocity ω in the complex plane to synthesize the instantaneous voltages $v_{an}(t)$, $v_{bn}(t)$, and $v_{cn}(t)$ as their individual projections onto the real axis.

An alternative connection is shown in Fig. 9.4. The sources in this configuration do not have a common node but are arranged in a *delta* (Δ) connection. These sources create *line-to-line* voltages, and we likewise designate them by a subscript – for example, v_{ab}. Fig. 9.3b illustrates how the line–line phasor $\widehat{V}_{ab}$ can be geometrically constructed from the line-to-neutral phasors $\widehat{V}_{an}$ and $\widehat{V}_{bn}$. The line-to-line voltages form a balanced three-phase set that is a factor of $\sqrt{3}$

larger than the line-to-neutral voltages, and lead them by $\pi/6$ radians or 30°. Consequently, for the same load power, load resistors R' connected in delta would have a resistance that is a factor of three larger than the resistors R connected in Y. Thus, by selecting how three-phase sources and loads are constructed and connected, one can provide a degree of matching of voltage to load. For example, a three-phase 208 V_{rms} line–line voltage provides 120 V_{rms} line–neutral. Both are standard voltages (with a tolerance of $+10\%/-15\%$) in the US, and can be provided by a single three-phase service.

The Y and Δ three-phase sources produce the same line–line voltages, that is, $v_{ab} = v_{an} - v_{bn}$. If $v_{an} = V_s \sin \omega t$ and $v_{bn} = V_s \sin(\omega t - 2\pi/3)$, then

$$v_{ab} = V_s \sin \omega t - V_s \sin\left(\omega t - \frac{2\pi}{3}\right) = \sqrt{3} V_s \sin\left(\omega t + \frac{\pi}{6}\right). \tag{9.4}$$

Although the Y- and Δ-connected sources produce the same line-to-line voltages, they are not always interchangeable. The Δ source has three interconnections, while the Y source can have four (a, b, c, n). Therefore, if a load requires the line–neutral voltage, it cannot be supplied by a Δ-connected source.

Like a two-phase source, a three-phase source also provides constant power to a balanced load, and a rotating magnetic field. It also has a number of further benefits, discussed below, that have made it ubiquitous.

Owing to the large number of connections and waveforms in three-phase systems, various means to ease their representation have evolved. For example, the phase voltages are named such that the ac line-to-neutral voltages peak in order of their names (e.g., v_{an}, v_{bn}, v_{cn}, v_{an}, v_{bn}, v_{cn}, ...). Specific color codings are also used for different wires in a three-phase system, with the details specified by code requirements. Likewise, the way in which the voltage sources are drawn in Fig. 9.3(a) intentionally evokes their relative orientations in the phasor diagram of Fig. 9.3(b), though this is not universally done. As we will see later in the chapter, specific naming schemes are also often selected for the switches in polyphase converters to denote conduction order.

One key advantage of three-phase systems over single- and two-phase systems is that, as is clearly visible in the phasor diagram of Fig. 9.3(b), the three-phase line-to-neutral voltages sum to zero. For a *balanced load*, that is, one in which the loading on each phase is identical, the sum of the three phase currents $i_a + i_b + i_c$ is also zero, and the neutral wire carries no current. Effectively, for each phase the other two phases carry the current return. The result of this is that for a nearly balanced load, the neutral wire has little or no loss, and, subject to electrical code constraints, can be reduced in size or even omitted, with significant savings in size, cost, and loss as compared to three independent single-phase systems. This zero-sum of currents and – in the case of magnetic components, magnetic fluxes – can be applied to advantage in the design of three-phase equipment, though the practical benefits depend heavily on the application.

The construction of the phasor $\widehat{V}_{ab}$ in Fig. 9.3 suggests a further way in which three-phase systems can be leveraged to create sources of higher phase number. By using transformer connections to symmetrically sum scaled elements of individual phase voltages, one can ideally construct new three-phase sets having *any* phase shift with respect to the original. For example, two three-phase transformers, one with a Δ-connected primary and the other Y-connected – each with a Y-connected secondary – will produce two three-phase sets of voltages 30° out of

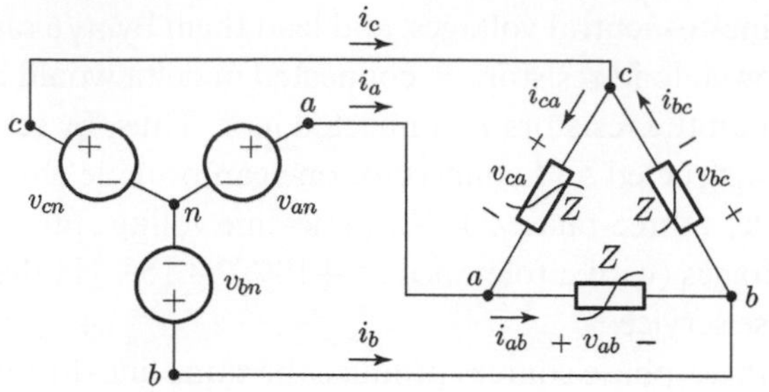

Figure 9.5 A Y-connected three-phase source feeding a balanced delta-connected load which may be nonlinear.

phase with each other. This is valuable in applications where one desires subsystems that operate with known phase shifts (e.g., for purposes of canceling harmonic currents or voltages). The 12-pulse rectifiers described in Example 9.1 use this technique, for example.

A further, and extremely important, benefit of balanced three-phase systems is that they provide natural cancellation of all triple-n harmonics. For example, consider the three-phase system of Fig. 9.5, in which a Y-connected source supplies a balanced Δ-connected load. The load network may be nonlinear and drawing non-sinusoidal currents in response to the sinusoidal line-to-line voltages. However, the load is balanced, with each load responding identically to a given impressed voltage pattern: each load draws the same current waveform for a given applied voltage waveform, with a time-shifted voltage producing an equally time-shifted current. Owing to harmonic cancellation, the source phase currents will have no component at the third harmonic or any multiple of the third, regardless of the harmonic components of the individual load currents.

Yet another benefit of three-phase systems is that one can interface between dc and three-phase waveforms with a reduced number of devices, compared to doing so with three independent single-phase full-bridge converters. This benefit and its constraints are detailed in Section 9.6.

9.2.1 Calculating Power in Three-Phase Systems

The power supplied to the Y-connected three-phase load of Fig. 9.3 is simple to calculate as we know the line–neutral voltages. The power is the sum of the powers in each phase. For a resistive load the average power is

$$P_Y = \frac{1}{2}\sum_{x=a}^{c} V_{xn} I_x = \frac{3}{2} V_{\ell n} I_\ell,$$

where ℓ denotes "line." In the case of a Δ-connected load as in Fig. 9.5, however, the line currents do not correspond directly to the currents in each of the Δ-connected load resistors. The phases of the line–line currents in the Δ-connected load are related to those of the line currents in the same fashion as the phases of the line–line voltages and line–neutral voltages are related in a

Y-connection. That is, the current i_{ab} leads the line current i_a by 30°. The magnitude relationship, however, is reversed, with $I_{ab} = I_a/\sqrt{3}$. The average power to a Δ-connected load is therefore

$$P_\Delta = \frac{1}{2}(3I_{\ell\ell}V_{\ell\ell}) = \frac{1}{2}\left(\sqrt{3}I_\ell V_{\ell\ell}\right).$$

9.3 Introduction to Polyphase Rectifier Circuits

We can most easily describe polyphase rectifier circuits as diode OR gates, in which the output assumes the value of the highest input. Figure 9.6 shows a four-input OR gate that is also a four-phase half-wave rectifier circuit. So long as one of the inputs is greater than zero, a diode will be on. The source voltages shown have arbitrary phase relationships with one another. In practical polyphase rectifier circuits, the ac sources are generally symmetric in phase, but the basic operation of these circuits does not depend on this condition.

We can now obtain full-wave, three-phase, polyphase rectification by connecting two half-wave three-phase circuits in series with the load, similar to what we did in evolving the single phase full-wave bridge circuit. Such a connection is shown in Fig. 9.7(a), where a pair of three-phase half-wave circuits is used. The sources are shown as two conventional three-phase sets. The three voltages in each set are equal in amplitude and frequency, but are phase-displaced from each other by 120°. We designate them as the *a*, *b*, or *c* phase voltage, as shown in Fig. 9.7(b).

The individual half-wave voltages v_{d_1} and v_{d_2}, and their difference v_d, are shown in Fig. 9.7(c). The presence of six pulses in v_d for every cycle of line voltage is the reason we call the three-phase full-wave circuit a *six-pulse rectifier*. We call the three-phase half-wave circuit a *three-pulse rectifier* for similar reasons.

We can redraw the circuit of Fig. 9.7(a) in a more conventional way to illustrate its bridging topology if we recognize that nodes at the same voltage can be connected, thereby eliminating three sources. Figure 9.8(a) shows the resulting three-phase bridge rectifier circuit, along with one of the phase voltages, v_a, and its relationship to its line current i_a. The relationships between the other phase voltages and their associated currents are identical, apart from being shifted by $\pm 2\pi/3$ from the waveforms shown. The six-pulse circuit is the basic building block for constructing higher pulse-number rectifiers. For instance, we may construct a 12-pulse circuit by connecting two 6-pulse circuits in series, as we show in Example 9.1; or in parallel, as you will do in Problem 9.10.

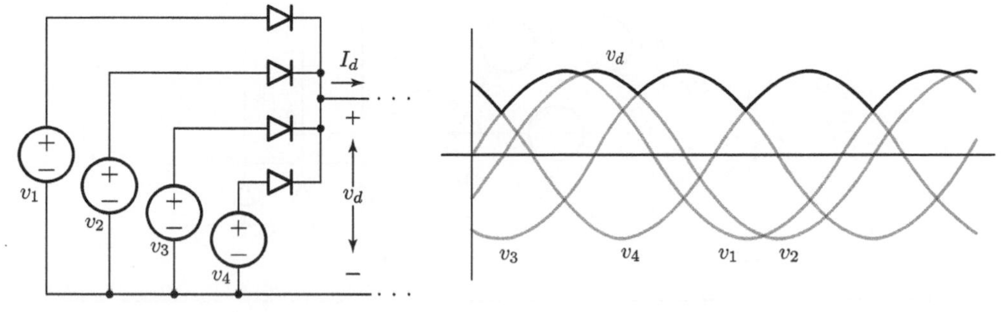

Figure 9.6 A four-phase half-wave rectifier viewed as a four-input OR gate.

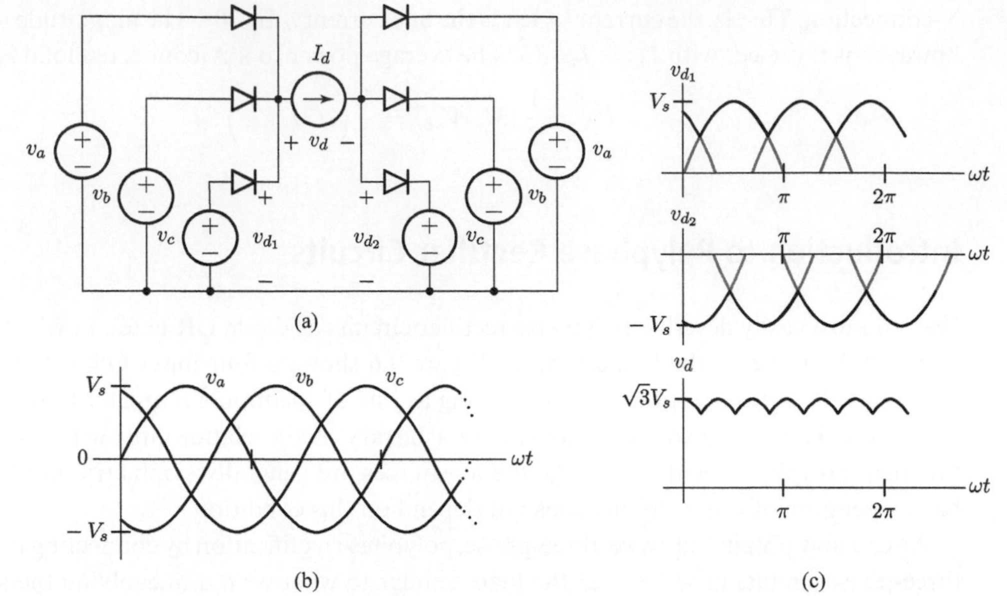

Figure 9.7 A three-phase, full-wave rectifier constructed from a pair of three-phase, half-wave (three-pulse) circuits. (a) Interconnection of the three-pulse circuits. (b) The source voltages. (c) The individual half-wave voltages and their difference, v_d.

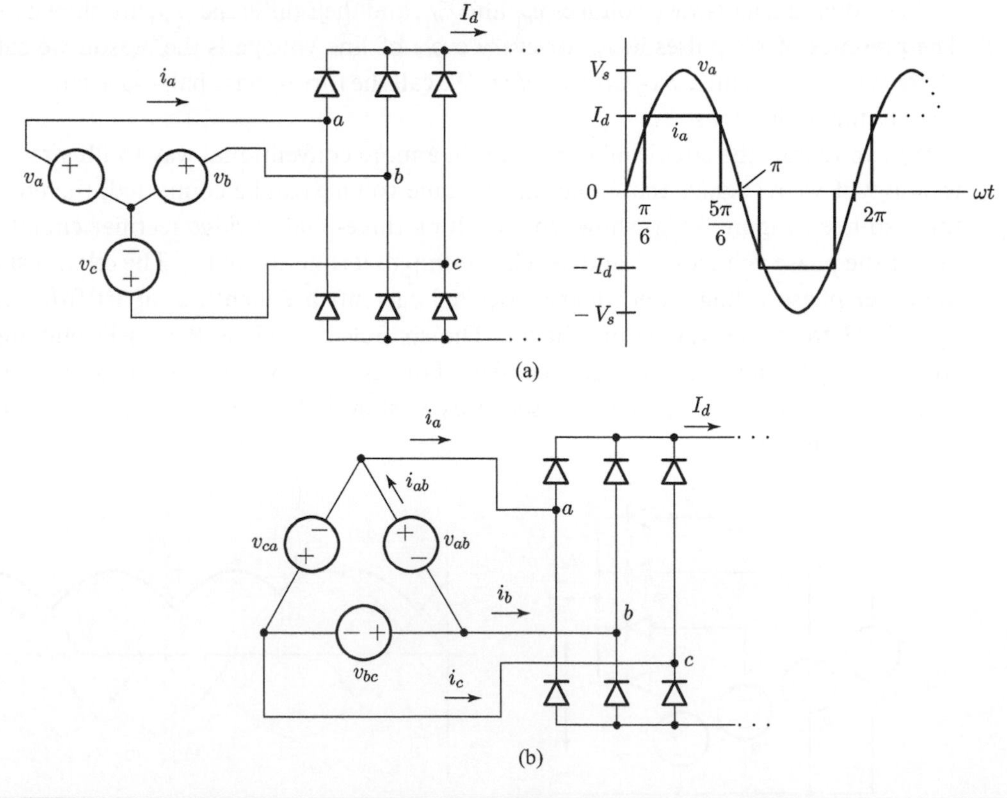

Figure 9.8 (a) The circuit of Fig. 9.7 redrawn as a three-phase bridge (six-pulse). The three-phase source is Y-connected. (b) The circuit of (a) with the sources Δ-connected.

Example 9.1 A 12-Pulse Rectifier Circuit

A series connection of two 6-pulse bridges results in a 12-pulse rectifier only if the ripples of the two bridges are phase-shifted relative to each other. Otherwise the ripple waveforms of the two bridges are simply congruent, resulting in six-pulse performance. Shifting the three-phase ac sources supplying the two bridges by $\pi/6$ with respect to one another results in 12 symmetric pulses. This shift usually involves special three-phase transformer connections called Y/Y and Δ/Y connections.

Two six-pulse bridges connected in series and supplied by phase-shifted ac sources are shown in Fig. 9.9(a). The Y/Y and Δ/Y boxes represent the necessary transformer connections. The resulting line-to-line voltages v_{ab} and $v_{a'b'}$ illustrate the 30° phase shift created by the transformer connections. Because each of the six-pulse bridges operates independently, the output voltage v_d is the sum of v_{d_1} and v_{d_2}. The waveforms of v_{d_1}, v_{d_2}, and v_d in Fig. 9.9(b) show how the 12-pulse output occurs, with $\langle v_d \rangle = 1.93\ V_s$.

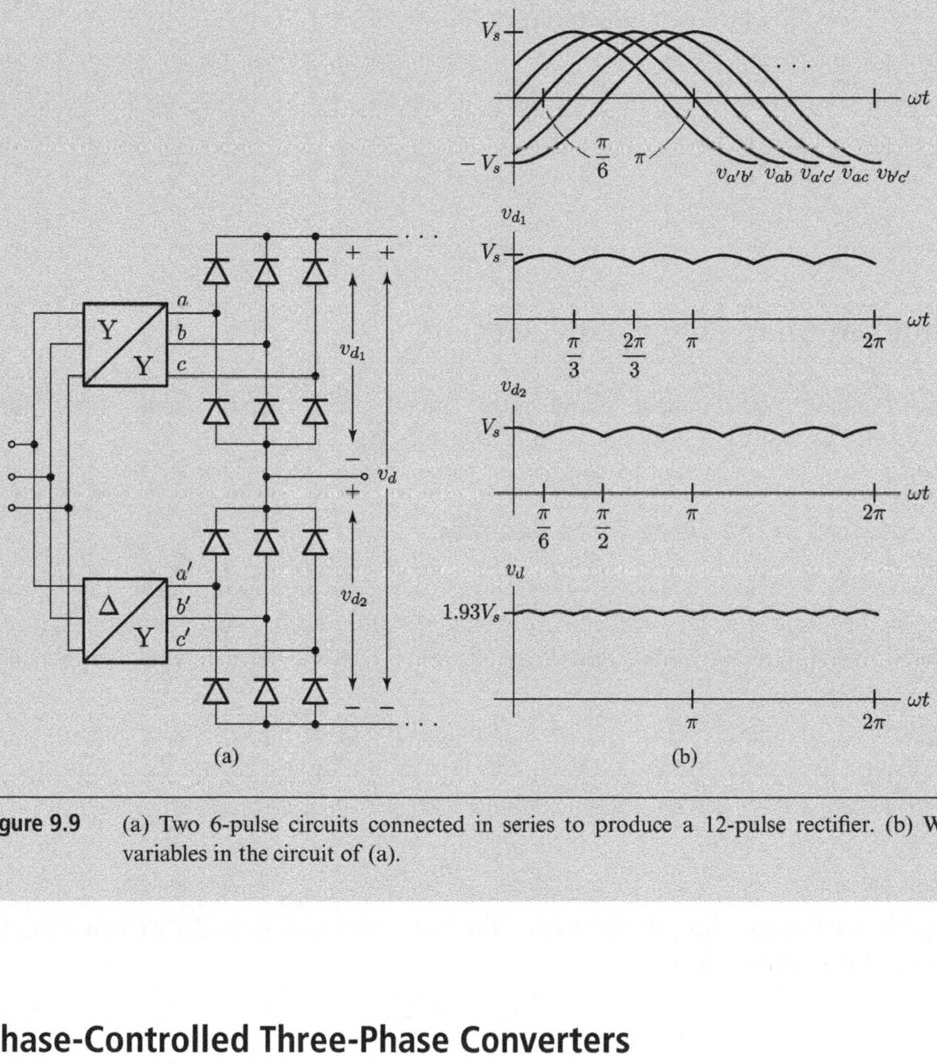

Figure 9.9 (a) Two 6-pulse circuits connected in series to produce a 12-pulse rectifier. (b) Waveforms of variables in the circuit of (a).

9.4 Phase-Controlled Three-Phase Converters

A phase-controlled three-phase bridge converter is shown in Fig. 9.10(a). Its operation is similar to the three-phase rectifier of Fig. 9.8(a), except that commutation is delayed by an angle α

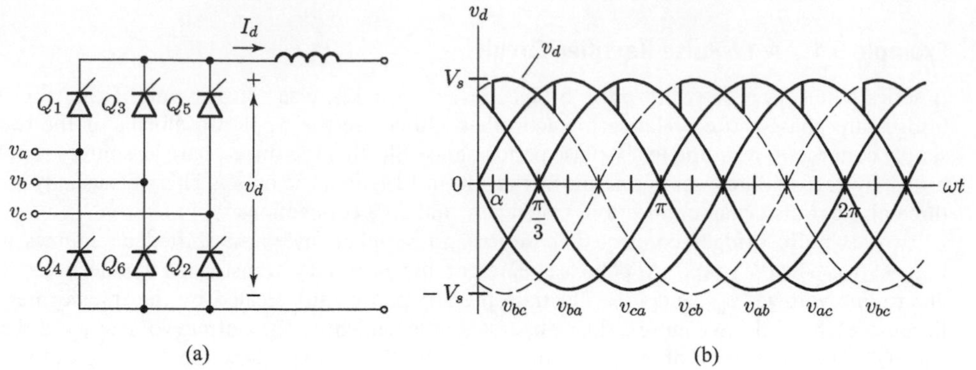

Figure 9.10 (a) A phase-controlled three-phase bridge converter. (b) The line-to-line voltages and the voltage v_d resulting with a firing angle α.

from when commutation occurs between diodes in Fig. 9.8(b). Calculating $\langle v_d \rangle$ in Fig. 9.10(b) gives us

$$\langle v_d \rangle = \frac{3V_s}{\pi} \cos \alpha.$$

9.5 Commutation in Polyphase Rectifiers

We first consider commutation in the simple three-pulse half-wave circuit of Fig. 9.11(a) and then address the commutation behavior of the six-pulse circuit.

9.5.1 Commutation in the Three-Pulse Rectifier

If we assume a constant dc current I_d, commutation between diodes in the three-pulse rectifier begins at the same time that the current would instantaneously commutate if $L_c = 0$. Except for the value of the commutating voltage, the process is identical to that for the single-phase half-bridge circuit.

Figure 9.11(b) illustrates the equivalent circuit of the three-pulse rectifier during commutation from phase c to phase a. Both diodes D_1 and D_3 are conducting during the commutation period u. For this interval we can express the voltage v_d in terms of either v_a or v_c:

$$v_d = v_a - L_c \frac{di_a}{dt}, \quad \text{or} \quad v_d = v_c - L_c \frac{di_c}{dt}.$$

As di_a/dt must equal $-di_c/dt$ because of the constraint $i_a + i_c = I_d$, we can solve these two equations for v_d during u:

$$v_d = \frac{v_a + v_c}{2}. \tag{9.5}$$

Therefore v_d is simply the average of v_a and v_c during u. We can again determine the commutation period u by equating the volt–time integral for the sum of the two commutating inductances to the dc current I_d:

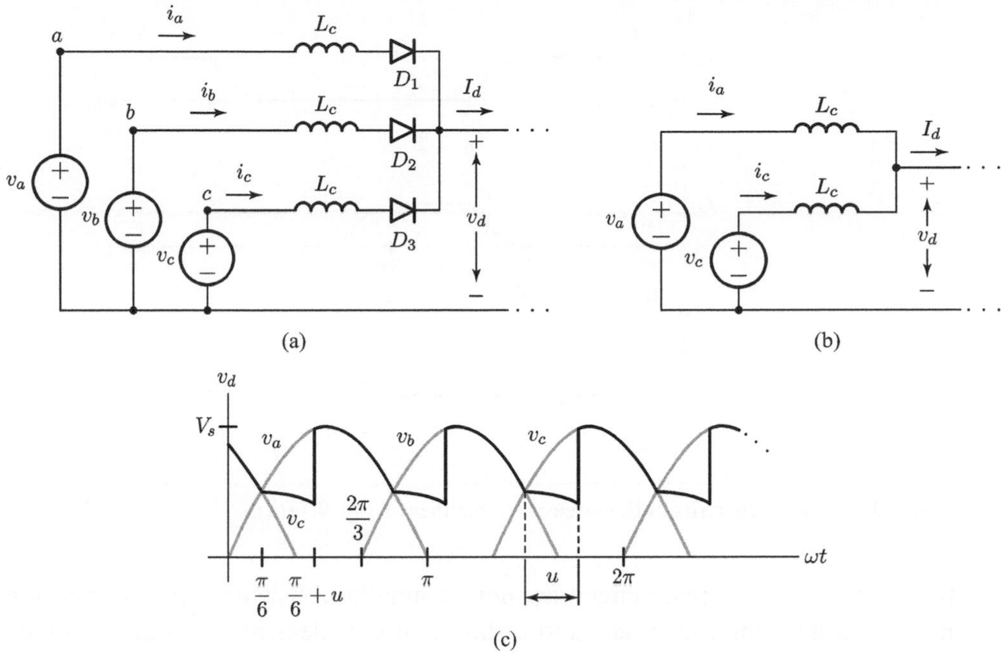

Figure 9.11 Commutation between phases in the three-pulse rectifier circuit of (a). (b) The equivalent circuit during commutation from phase c to phase a. (c) The voltage v_d of (a), showing the effect of the commutation period u.

$$
\begin{aligned}
I_d &= \frac{1}{2\omega L_c}\int_{\pi/6}^{(\pi/6)+u} V_s\left[\sin\omega t - \sin\left(\omega t + \frac{2\pi}{3}\right)\right] d(\omega t) \\
&= \frac{1}{2X_c}\int_0^u \sqrt{3}V_s \sin\omega t \, d(\omega t) = \frac{\sqrt{3}V_s}{2X_c}(1-\cos u)
\end{aligned}
\tag{9.6}
$$

Solving for u yields

$$
u = \cos^{-1}\left(1 - \frac{2X_c I_d}{\sqrt{3}V_s}\right). \tag{9.7}
$$

The effect of u on v_d is shown in Fig. 9.11(c). The resulting line currents – i_a, i_b, and i_c – are shown in Fig. 9.12. Note that, as with the single-phase half-wave rectifier with freewheeling diode, the line currents contain a dc component. Also, because line-current pulse width in the absence of commutating reactance is shorter for the three-pulse circuit ($2\pi/3$ compared to π), the ratio of the rms to fundamental values of the line current is higher for the three-pulse circuit, and a lower power factor results (0.48 versus 0.64).

9.5.2 Commutation in the Six-Pulse Bridge Rectifier

We cannot simply derive the behavior of the six-pulse circuit as the sum of the behaviors of two three-pulse circuits, as we did in the case of the single-phase bridge. The reason is that all

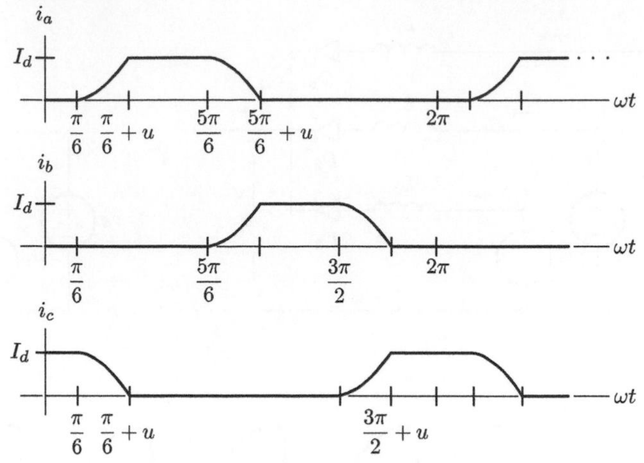

Figure 9.12 Line currents in the three-pulse rectifier of Fig. 9.11(a).

the diodes in the six-pulse circuit are not on simultaneously during u. This condition prevents us from connecting the anodes and cathodes of complementary diodes, as we did in deriving the single-phase bridge circuit of Fig. 3.17(a). We must derive the commutation behavior of the six-pulse circuit explicitly.

A six-pulse bridge with commutating reactance and a Δ-connected source is shown in Fig. 9.13(a). The diodes are numbered in their conducting sequence. We begin our analysis by assuming that only D_1 and D_2 are conducting. The line currents are

$$i_a = I_d, \qquad i_b = 0, \qquad i_c = -I_d.$$

During this time there are no voltage drops across the commutating reactances, and the behavior of the circuit is identical to that of Fig. 9.7 or Fig. 9.8. The voltages across the other (off) diodes are

$$v_3 = v_{b'a'} = v_{ba}, \qquad v_4 = v_5 = v_{c'a'} = v_{ca}, \qquad v_6 = v_{c'b'} = v_{cb}.$$

These voltages are shown in Fig. 9.13(c), from which we can conclude that D_3, D_4, D_5, and D_6 can be simultaneously reverse biased only for $2\pi/3 < \omega t < \pi$. Because v_3 is the first diode voltage to go positive, D_3 turns on at $\omega t = \pi$, and $v_{a'b'}$ goes to zero. Commutation of the current from D_1 to D_3 now commences at $\omega t = \pi$.

The equivalent circuit during commutation is shown in Fig. 9.14. Because $i_a + i_b = I_d$, $di_b/dt = -di_a/dt$. Therefore $v_{x_b} = -v_{x_a}$, $v_{ab} + 2v_{x_b} = 0$, and

$$v_{x_b} = -\frac{v_{ab}}{2}. \tag{9.8}$$

We can now determine u:

$$\begin{aligned} I_d &= \frac{1}{X_c}\int_{\pi}^{\pi+u} v_{x_b}\, d(\omega t) = \frac{1}{X_c}\int_{\pi}^{\pi+u} -\frac{v_{ab}}{2}\, d(\omega t) \\ &= \frac{1}{X_c}\int_{\pi}^{\pi+u} -\frac{V_s}{2}\sin\omega t\, d(\omega t) = \frac{V_s}{2X_c}(1-\cos u) \end{aligned}, \tag{9.9}$$

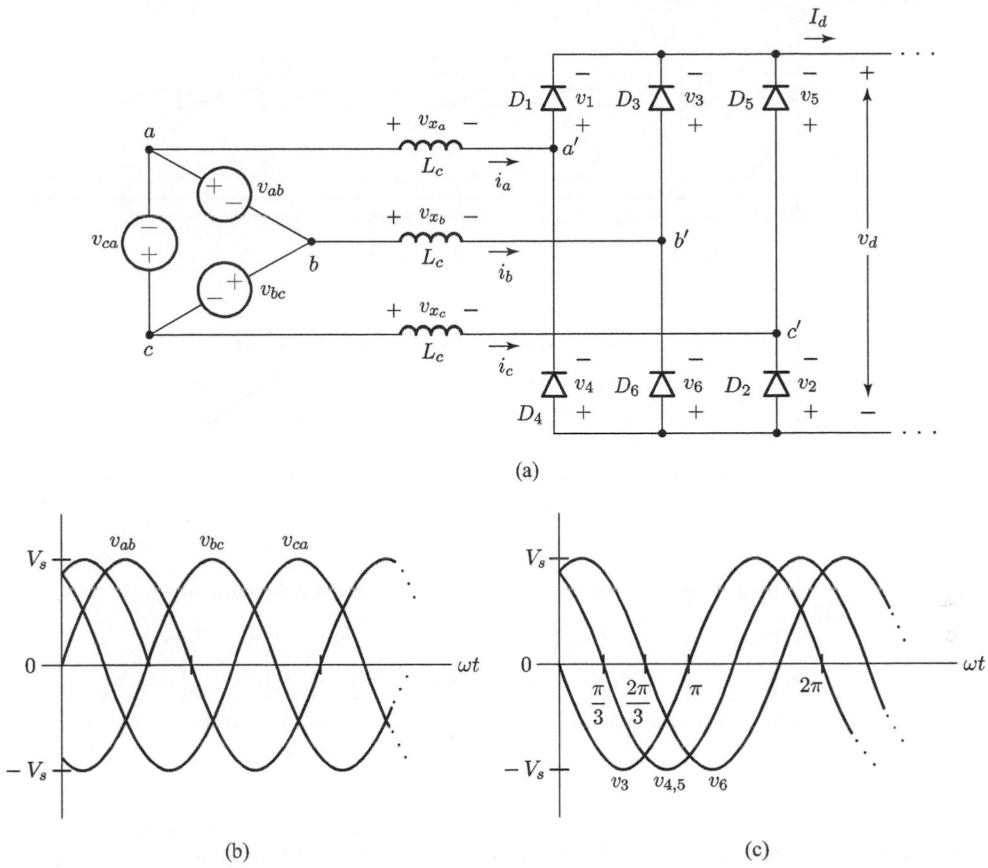

Figure 9.13 (a) A six-pulse bridge with commutating reactance and a Δ-connected three-phase source. (b) The three-phase source voltages. (c) Diode voltages, assuming that D_1 and D_2 are on.

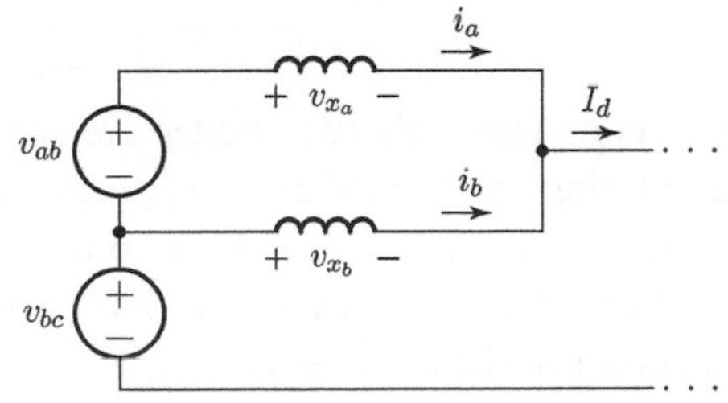

Figure 9.14 Equivalent circuit of the rectifier of Fig. 9.13(a) during commutation from D_1 to D_3.

$$u = \cos^{-1}\left(1 - \frac{2X_c I_d}{V_s}\right). \tag{9.10}$$

During the commutation interval u, the voltage v_d is equal to the average of v_{bc} and v_{ac}. We can show this condition by expressing v_d as $-v_{ca} - v_{x_a}$ and as $v_{bc} - v_{x_b}$. We then use $v_{ab} + v_{bc} + v_{ca} = 0$ and (9.8) to eliminate v_{x_a} and v_{x_b} and obtain

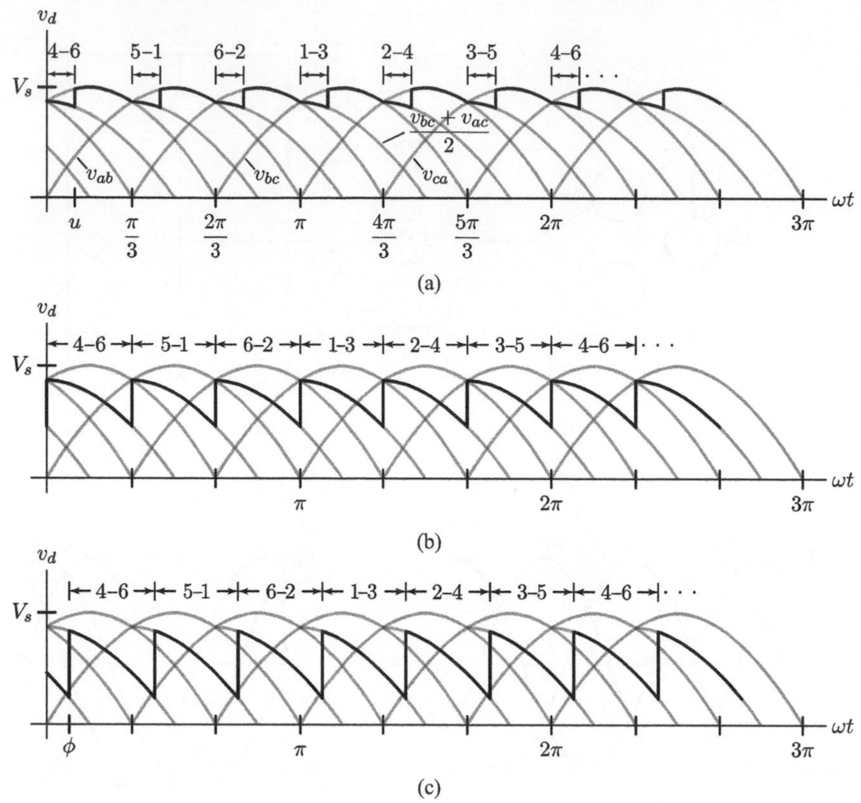

Figure 9.15 Output voltage v_d for the rectifier of Fig. 9.14(a) (the diodes undergoing commutation are identified by their numbers). (a) Mode I ($u < \pi/3$). (b) Boundary between modes I and II ($u = \pi/3$). (c) Mode II ($u = \pi/3$).

$$v_d = \frac{v_{bc} - v_{ca}}{2} = \frac{v_{bc} + v_{ac}}{2}. \tag{9.11}$$

Once commutation is complete, only two diodes are conducting (D_3 and D_2), there are no drops across the commutating reactances, and v_d assumes the value of the line-to-line source voltage connected to the output by the conducting diodes. If we continue this analysis for a complete cycle, the waveform for v_d shown in Fig. 9.15(a) results. From this waveform we can calculate the regulating characteristic $\langle v_d \rangle = f(X_c I_d / V_s)$:

$$\langle v_d \rangle = \frac{3V_s}{2\pi}(1 + \cos u) = \frac{3V_s}{\pi}\left(1 - \frac{X_c I_d}{V_s}\right) = V_{do}\left(1 - \frac{X_c I_d}{V_s}\right). \tag{9.12}$$

The analysis leading to (9.12) is valid only if $u < \pi/3$, because it is based on the assumption that commutation of the current between two diodes commences at the angles $\omega t = n\pi/3$ (Fig. 9.15a). We call operation in this region *mode I*. However, increasing the reactance factor $X_c I_d / V_s$ beyond the point where $u = \pi/3$ delays commutation beyond the angles $\omega t = n\pi/3$, preventing the next pair of diodes from starting to commutate at $n\pi/3$. This result becomes evident if you consider the voltages across the off-state diodes. While three diodes are conducting

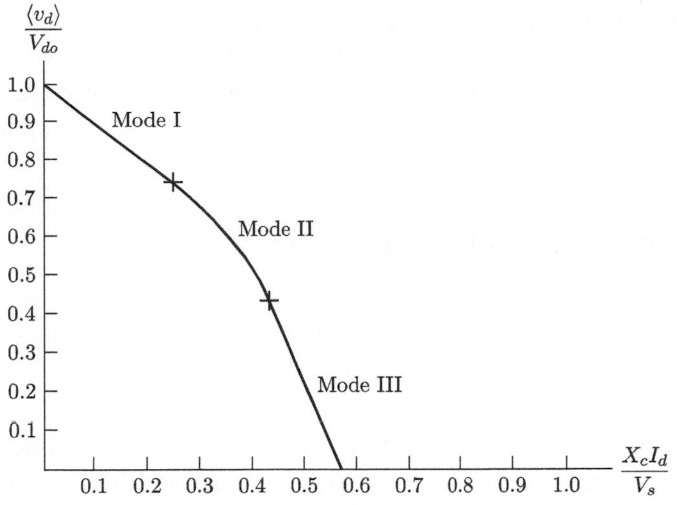

Figure 9.16 Regulation curve for the rectifier of Fig. 9.13(a) operating in modes I, II, and III.

during commutation – for instance D_1, D_3, and D_2 – the voltage across all other diodes is the same and equal to $-v_d$. Therefore, with three diodes conducting and $v_d > 0$, a fourth diode cannot turn on. When $u = \pi/3$, Fig. 9.15(b) shows that v_d is still greater than zero. An increase in the reactance factor $X_c I_d / V_s$ now results in the commutation interval "sliding" to the right by an angle ϕ, as shown in Fig. 9.15(c). Because we are assuming that a periodic steady state exists, requiring six identical pulses per cycle, u remains constant at $\pi/3$ for this mode, which we call *mode II*. Although we will not derive it here, the regulating characteristic for mode II is

$$\langle v_d \rangle = \frac{V_{do}\sqrt{3}}{2}\sqrt{1 - \left(\frac{2X_c I_d}{V_s}\right)^2}. \tag{9.13}$$

Mode II lasts until $\phi = \pi/6$, at which point $v_d = 0$ at the end of the commutation interval. A fourth diode now turns on if we further increase $X_c I_d / V_s$, initiating *mode III*, in which there are periods where $v_d = 0$. Analysis of the rectifier operating in this mode is straightforward, but tedious, so we do not present it here.

The complete regulating characteristic for the rectifier, including mode III, is shown in Fig. 9.16. Note that the regulation becomes more severe as $X_c I_d / V_s$ is increased and a new mode is entered.

9.6 Three-Phase Inverters

Many applications, such as the control of rotating machinery, require three-phase ac sources of varying frequency and/or amplitude. We could in principle use three single-phase inverters to generate a three-phase set of waveforms from a dc source, assuming adequate isolation were provided among the individual units. One could do so by controlling the switches of each inverter to produce an output phase-shifted by 1/3 of a cycle (or $2\pi/3$ radians of the fundamental) with respect to the outputs of the other two inverters. However, considering direct connection of full-bridge single-phase inverters, six of the twelve switches become redundant, as described below. A three-phase inverter circuit thus requires only six switches, configured as a three-phase bridge.

9.6.1 Evolution of the Three-Phase Bridge Circuit

Figure 9.17(a) illustrates the connection of three inverters to create a symmetrical three-phase output. Each block might be implemented as a single-phase full-bridge inverter such as that shown in Fig. 8.1. Such a configuration is sometimes used to drive electric machines with *open-ended* windings, in which the individual phase windings of the machine are not electrically connected. However, most three-phase loads – including most large electric machines – are not configured as three separate loads isolated from one another; rather, they are connected in a Y or Δ configuration. In the Y connection, one output terminal of each inverter is attached to the common point of the Y. In the Δ connection, the inverter outputs are chained together in a loop. In either case, because all three inverters share a common dc bus, we cannot control their switches arbitrarily, but must instead coordinate them to prevent shorting the dc source.

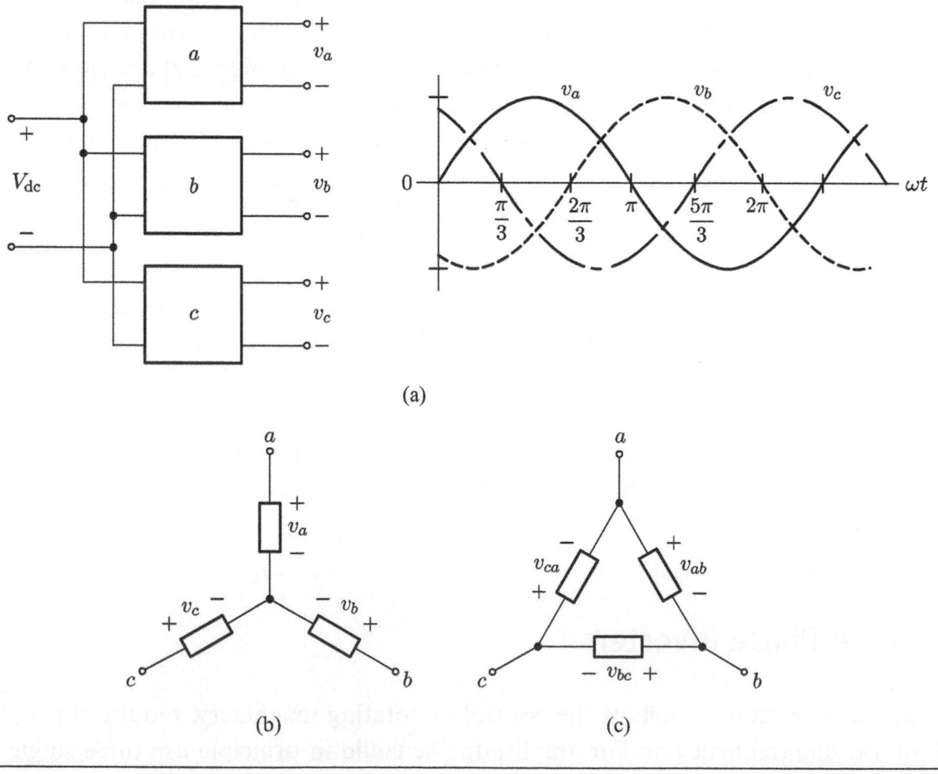

Figure 9.17 (a) A three-phase inverter consisting of three single-phase circuits operated from the same dc bus. (b) A Y-connected three-phase load. (c) A Δ-connected three-phase load.

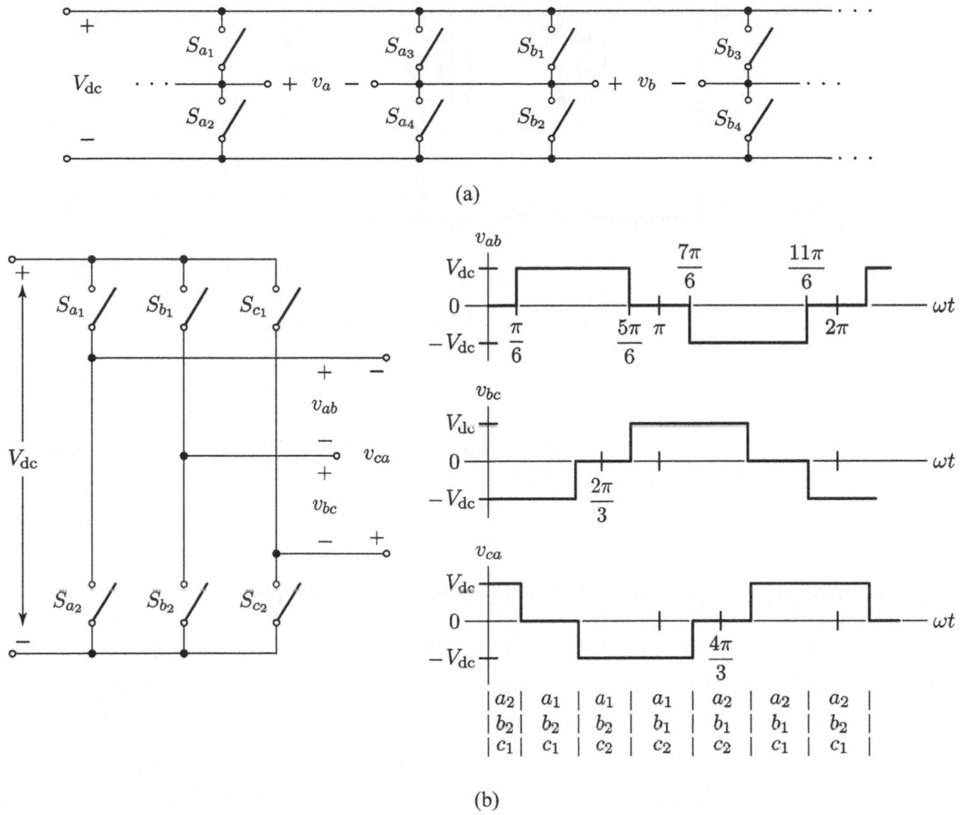

Figure 9.18 Simplification of the converter circuit for the case of a Δ-connected load. (a) Connection between two of the independent converters. (b) The simplified six-switch three-phase converter and its phase-to-phase waveforms.

If we consider the Δ-connected load and the connection of two of the converter bridges, the circuit of Fig. 9.18(a) results. Here, it is clear that S_{a_3} and S_{b_1} are redundant, as are S_{a_4} and S_{b_2}. Note that this prevents the independent operation of the redundant switches and thus the arbitrary control of the individual single-phase converters. If the third converter were added to the figure, the switches S_{a_1}, S_{a_2}, S_{b_3}, and S_{b_4} would each have a redundant counterpart. Consequently, we can eliminate six switches from the individual converter version, resulting in the three-phase bridge converter circuit of Fig. 9.18(b), shown with its six-step waveforms. Figure 9.19 shows this circuit with the six switches implemented using power MOSFETs, numbered according to their conduction sequence in *six-step* switching mode, which refers to the six different switch states occurring each cycle. (The three-phase load is illustrated as Y-connected, though it could also be Δ-connected.) A center-tap on the dc bus, having potential v_r, is illustrated as a voltage reference for modeling the system behavior. When used to convert power from dc to ac, this three-phase bridge converter is also sometimes referred to as a *three-phase voltage source inverter*.

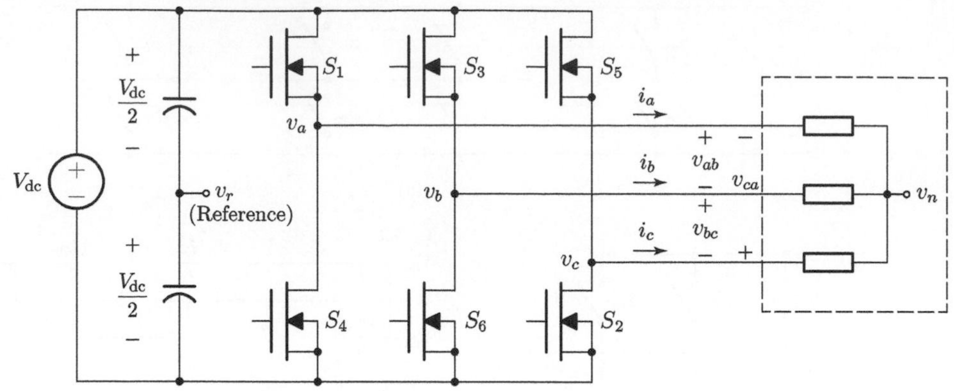

Figure 9.19 A three-phase bridge inverter using power MOSFETs with intrinsic diodes. The potential v_r is used as a reference in modeling the system behavior.

9.6.2 Switching States of the Three-Phase Bridge Inverter

The three-phase bridge inverter of Fig. 9.19 comprises three half-bridge legs (one for each of the phases *a, b, c*). The devices are often numbered as illustrated, conveying their conduction order in square-wave or six-step operation, as is also done for rectifiers. For convenience in analyzing circuit waveforms, we use the midpoint of the dc bus (having potential v_r) as a voltage reference node. The load on the inverter could be connected either Y or Δ, but we illustrate it as a Y connection with an internal (unconnected) neutral point.

Considering inverter states in which one switch in each half-bridge is always on (for continuous conduction at the load), there are $2^3 = 8$ switch state possibilities for the three-phase bridge. Each state may be assigned a vector designation and associated number corresponding to whether the bottom ("0") or top ("1") switch in each half-bridge is on, as shown in Table 9.1. For each state, we can directly calculate the resulting output line-to-reference voltages (v_{ar}, v_{br}, and v_{cr}) and line-to-line voltages (v_{ab}, v_{bc}, and v_{ca}). For a *balanced* Y-connected three-phase load with a floating neutral point, we can further calculate the line-to-neutral voltages (v_{an}, v_{bn}, and v_{cn}) and the neutral-to-reference output voltage v_{nr}. (By a "balanced" load, we mean a load that acts to divide the applied voltages just as would a set of three equal impedances connected in Y to a floating neutral point.)

Figure 9.20 shows the waveforms for the inverter of Fig. 9.19 for the simple case of six-step operation, where the inverter cycles sequentially through six of the eight switch states shown in Table 9.1. In particular, the inverter cycles through states X_1 to X_6 in Table 9.1, generating outputs having relatively modest harmonic content, while only requiring each device in the three-phase bridge to turn on and off once per ac output cycle. Since the line-to-line voltages are formed from differences of identical waveforms shifted by $T/3$, they do not contain any triple-*n* harmonics.

For a balanced three-phase load of three identical impedances connected to a (floating) neutral point, the neutral-point voltage v_{nr} is the *average* of the three phase voltages. If v_{ar}, v_{br},

Table 9.1 Available voltage vectors (normalized to V_{dc}) for the converter of Fig. 9.19.

Vector {a b c}	Sw. on	v_{ar}	v_{br}	v_{cr}	v_{ab}	v_{bc}	v_{ca}
$X_0 = \{000\}$	4, 6, 2	−1/2	−1/2	−1/2	0	0	0
$X_1 = \{100\}$	1, 6, 2	+1/2	−1/2	−1/2	+1	0	−1
$X_2 = \{110\}$	1, 3, 2	+1/2	+1/2	−1/2	0	+1	−1
$X_3 = \{010\}$	4, 3, 2	−1/2	+1/2	−1/2	−1	+1	0
$X_4 = \{011\}$	4, 3, 5	−1/2	+1/2	+1/2	−1	0	+1
$X_5 = \{001\}$	4, 6, 5	−1/2	−1/2	+1/2	0	−1	+1
$X_6 = \{101\}$	1, 6, 5	+1/2	−1/2	+1/2	+1	−1	0
$X_7 = \{111\}$	1, 3, 5	+1/2	+1/2	+1/2	0	0	0

and v_{cr} are identical but shifted by $T/3$, they have identical triple-n harmonic amplitude and phase, and thus the neutral-point voltage v_{nr} will also have the same triple-n harmonic content but no fundamental component. Consequently, the line-to-neutral voltages ($v_{xn} = v_{xr} - v_{nr}$ for $x \in a, b, c$) will have no triple-n content. For the special case of six-step operation, the neutral voltage v_{nr} becomes a square wave at a frequency three times the fundamental, and the line-to-neutral voltages take on six-step patterns shifted from one another by $T/3$, as seen in Fig. 9.20.

The remaining switching states (X_0 and X_7) shown for the three-phase bridge in Table 9.1 are *zero-output* states, which effectively short circuit the terminals of the three-phase load together. These states provide a means to apply zero-state voltages to the load, such as for PWM control of the output voltage, eliminating harmonic voltage components in the output, etc. Even when applying these zero states, so long as the line-to-line voltages are formed by the individual bridge legs switching in a balanced three-phase switching pattern (i.e., with identical waveforms shifted by $T/3$), they *cannot* contain any triple-n harmonic content. For a balanced Y-connected load with floating neutral, the line-to-neutral voltages likewise cannot contain triple-n harmonic components. For this reason, harmonic elimination or reduction techniques in three-phase inverters often focus on mitigating other harmonic components (e.g., fifth, seventh, etc.). *Space vector modulation* is a pulse-width modulation technique that utilizes these zero states and is specifically suited to three-phase systems. It is the topic of Section 9.7.

Example 9.2 A Three-Phase Inverter with a Center-Tapped DC Voltage Source

While the bridge inverter connected as in Fig. 9.19 is ubiquitous, there are both other ways to apply the bridge inverter and other three-phase inverter circuits. One example is a three-phase inverter feeding a Y-connected load from a center-tapped dc source, as shown in Fig. 9.21. In this circuit a separate half-bridge inverter supplies each phase, making independent control of the phase voltages possible. The conduction angle of each bridge leg may therefore exceed 120°, as shown in the waveforms of Fig 9.20, where each switch is conducting for 180°. However, the circuit is capable only of bi-state operation; the zero-volt state is not available.

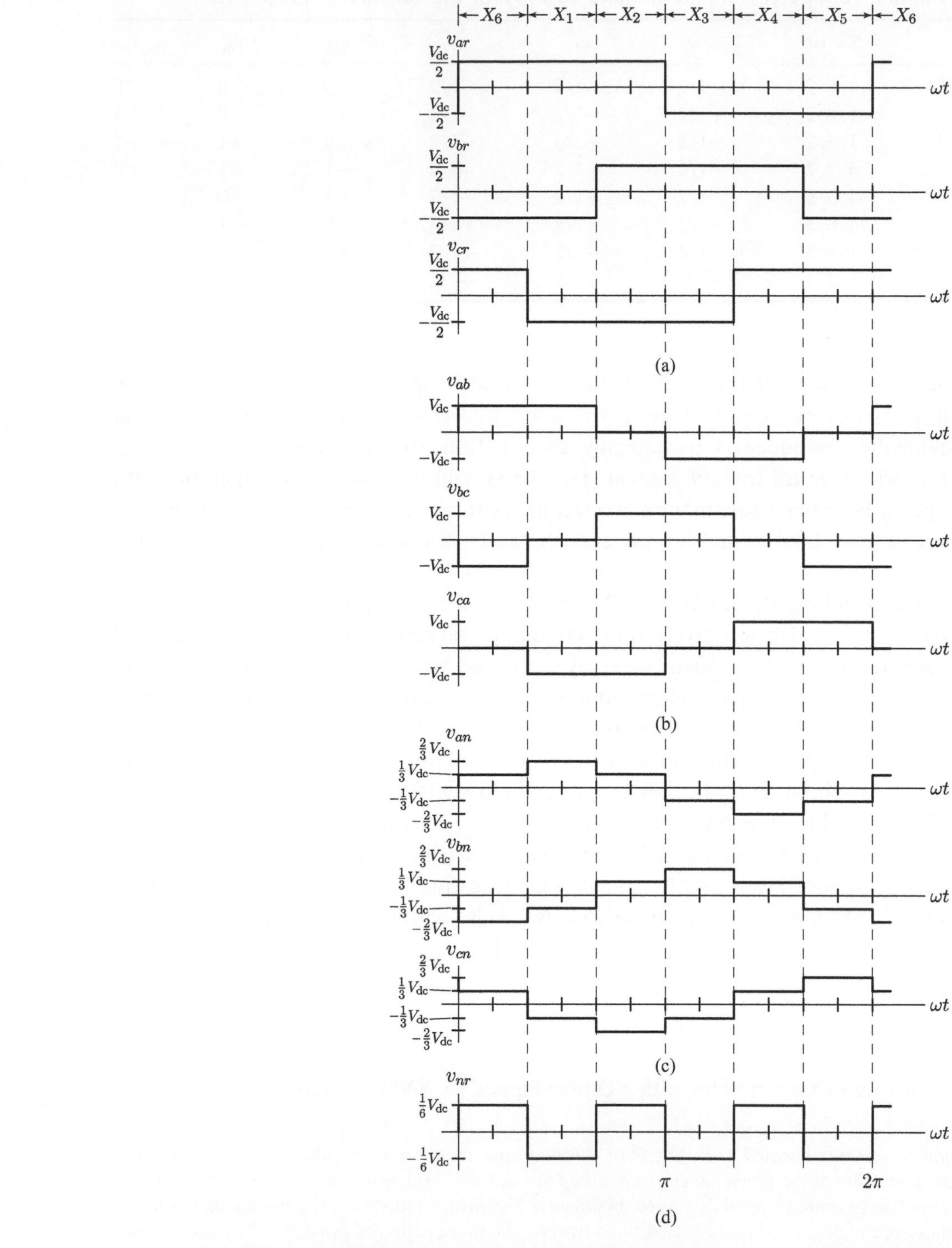

Figure 9.20 Waveforms of the three-phase bridge inverter of Fig. 9.19 driving a balanced Y-connected load in six-step switching mode.

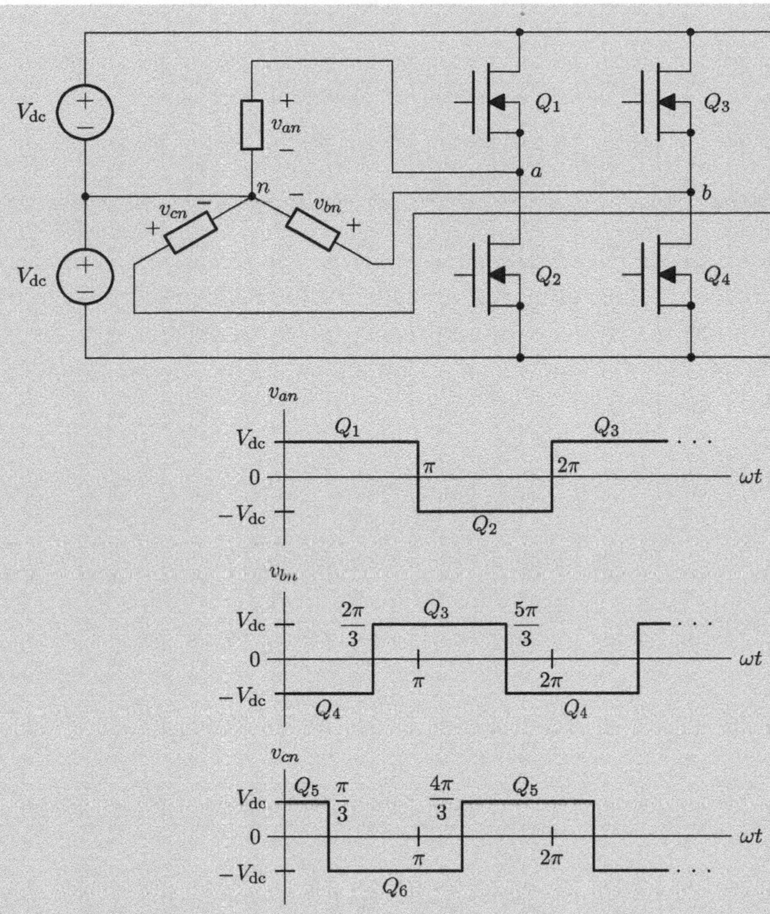

Figure 9.21 A three-phase inverter with a center-tapped dc source connected to the neutral point of the three-phase load, and its phase-to-neutral waveforms.

The benefits of this circuit are that the center point of the Y-load can be grounded to the source; better use is made of the switches because they conduct for 180° instead of 120°; and the circuit ensures that the Y-connected load has a balanced three-phase voltage, even if the load itself is not balanced. The disadvantages are that the circuit cannot produce a zero-volt state, and the load *must* be Y-connected, with an accessible center point. With no center-point connection, the circuits of Figs. 9.19 and 9.21 are identical.

9.6.3 Pulse-Width Modulation of the Three-Phase Bridge Inverter

Consider the three-phase inverter of Fig. 9.19, but in which the balanced three-phase set of outputs having angular frequency ω is synthesized using PWM at an angular switching frequency of $\omega_s \gg \omega$. The switching-period duty ratios d_1–d_6 of switches S_1–S_6 are set as follows:

$$d_1 = 1 - d_4 = \frac{1}{2} + \frac{m}{2}\sin\omega t, \tag{9.14}$$

$$d_3 = 1 - d_6 = \frac{1}{2} + \frac{m}{2}\sin\left(\omega t - \frac{2\pi}{3}\right), \tag{9.15}$$

$$d_5 = 1 - d_2 = \frac{1}{2} + \frac{m}{2}\sin\left(\omega t + \frac{2\pi}{3}\right), \tag{9.16}$$

where m is known as the *modulation index* or *depth of modulation.* Setting duty ratios appropriately can be accomplished using sine–triangle PWM techniques similar to those shown for single-phase inverters in Chapter 8, or using space-vector modulation as in Section 9.7.

For $m \leq 1$ this leads to the following local average values (i.e., over a switching period) of the line-to-reference voltages:

$$\begin{aligned}
\overline{v}_{ar} &= m\frac{V_{dc}}{2}\sin\omega t,\\
\overline{v}_{br} &= m\frac{V_{dc}}{2}\sin\left(\omega t - \frac{2\pi}{3}\right),\\
\overline{v}_{cr} &= m\frac{V_{dc}}{2}\sin\left(\omega t + \frac{2\pi}{3}\right),
\end{aligned} \tag{9.17}$$

and these associated local-average values of the line-to-line voltages:

$$\begin{aligned}
\overline{v}_{ab} &= m\frac{\sqrt{3}V_{dc}}{2}\sin\left(\omega t + \frac{\pi}{6}\right),\\
\overline{v}_{bc} &= m\frac{\sqrt{3}V_{dc}}{2}\sin\left(\omega t - \frac{\pi}{2}\right),\\
\overline{v}_{ca} &= m\frac{\sqrt{3}V_{dc}}{2}\sin\left(\omega t + \frac{5\pi}{6}\right).
\end{aligned} \tag{9.18}$$

By selecting the value of m, we can control the amplitude of the three-phase sinuisoidal line-to-line voltages synthesized at frequency ω to have any value between 0 and $(\sqrt{3}V_{dc})/2$. Attempting to set m larger than unity with this scheme – or "overmodulation" – results in distortion (i.e., low-frequency harmonic content), as the duty ratio of a switch is limited to be between zero and one. In the limit of large m, saturation in duty ratios will lead to the six-step switching pattern shown in Fig. 9.20.

It is possible to extend the modulation index m slightly beyond unity (up to a value of $(2/\sqrt{3}) \approx 1.15$) without incurring low-frequency distortion. This can be accomplished using space-vector modulation or through a technique called *third-harmonic injection.* In third-harmonic injection, each phase leg synthesizes a third-harmonic component in addition to the fundamental component, such that m can be made slightly larger than unity without saturating the duty ratios. As the third-harmonic component is eliminated in the line-to-line waveforms, no distortion in the output results. This technique is explored in Problem 9.18.

9.7 Space-Vector Representation and Modulation for Three-Phase Systems

In this section we describe methods used to model three-phase ac systems, and then examine a class of pulse-width modulation techniques for three-phase converters known as *space-vector modulation* (SVM) that takes advantage of these methods. As described earlier, balanced three-phase systems offer substantial advantages in hardware cost and performance over both single-phase and other polyphase systems, and so have become ubiquitous in high-power applications. The modeling techniques described in this chapter provide a way to represent balanced three-phase quantities compactly and eliminate the dependencies among variables. Space-vector modulation takes advantage of these techniques to provide better means for pulse-width modulation and control of three-phase converters. SVM techniques allow a more straightforward representation of the modulation process, are well suited to digital implementation, and make it easier to lower loss and shape spectral content.

By their nature, three-phase systems have numerous variables of interest. For example, in the three-phase inverter and Y-connected load of Fig. 9.19, there are three voltages applied to the load and three currents in response. This system can be modeled directly by representing the variables of interest for each phase. However, because of the connection of the system, there are dependencies among the variables. For example, the phase currents are not independent variables (because they must sum to zero at the neutral point, by KCL), nor are the applied line-to-line voltages (because they form a loop, so must sum to zero, by KVL). Even though some quantities are not interdependent (such as the applied phase-to-reference voltages in Fig. 9.19), there are constraints in how power is transferred in the system owing to the quantities that are interdependent. Consequently, one might imagine that there is some way to represent the system that usefully captures the dependencies, and enables their elimination when appropriate. In this section we introduce representations, coordinate systems, and transformations developed to better represent three-phase systems.

9.7.1 Space Vectors

A set of three phase variables (x_a, x_b, x_c) can be expressed as the vector

$$\widehat{x}_{abc} = \begin{bmatrix} x_a \\ x_b \\ x_c \end{bmatrix} = x_a \widehat{a} + x_b \widehat{b} + x_c \widehat{c}, \tag{9.19}$$

where

$$\widehat{a} = \begin{bmatrix} 1 \\ 0 \\ 0 \end{bmatrix}, \quad \widehat{b} = \begin{bmatrix} 0 \\ 1 \\ 0 \end{bmatrix}, \quad \widehat{c} = \begin{bmatrix} 0 \\ 0 \\ 1 \end{bmatrix}. \tag{9.20}$$

This representation of a three-phase set of variables is commonly known as a *space vector*, and enables any three-phase quantity to be visualized as a vector in space, see Fig. 9.22. As $x_a(t)$, $x_b(t)$, and $x_c(t)$ change over time, the space vector $\widehat{x}_{abc}(t)$ points to different locations in space as t varies.

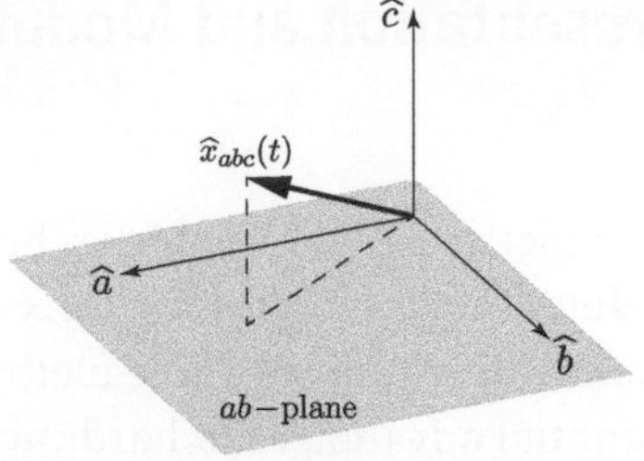

Figure 9.22 A space vector representing a three-phase quantity (x_a, x_b, x_c). Each phase variable is associated with an independent dimension in Cartesian coordinates $\widehat{a}, \widehat{b}, \widehat{c}$, enabling any three-phase set of variables to be represented as a vector in three-dimensional space.

9.7.2 (α, β, γ) Coordinates

For a *balanced* three-phase set, we have the additional constraint that the instantaneous phase variables must sum to zero:

$$x_a + x_b + x_c = 0. \tag{9.21}$$

Even in unbalanced situations, such a constraint often applies to certain variable sets in a three-phase system, owing to KCL or KVL. For example, the phase currents ($i_a(t)$, $i_b(t)$, and $i_c(t)$) and line-to-line voltages ($v_{ab}(t)$, $v_{bc}(t)$, and $v_{ca}(t)$) in the system of Fig. 9.19 are each bound by such a constraint. This requirement places a significant restriction on the possible locations for the space vector of such three-phase sets, confining the space vector to a specific plane in three-space. In particular, the plane in $\widehat{a}, \widehat{b}, \widehat{c}$ coordinates represented by (9.21) passes through the origin, that is, the point $[\,0\;0\;0\,]^{\mathrm{T}}$ (where the superscript T denotes the transpose) and has a unit normal vector

$$\widehat{\gamma} = \frac{1}{\sqrt{3}}\begin{bmatrix}1\\1\\1\end{bmatrix}, \tag{9.22}$$

since $\widehat{\gamma}^{\mathrm{T}}\widehat{x}_{abc} = 0$ if and only if (9.21) is true. For a more general $\widehat{x}_{abc}$, the quantity $\widehat{\gamma}^{\mathrm{T}}\widehat{x}_{abc}$ is proportional to the *common-mode* component of the three-phase set, which is defined as the average of x_a, x_b, and x_c.

We refer to the plane defined by (9.21) as the "α–β plane." As this plane includes the origin, it also comprises a subspace of three-space, meaning that any linear combination of vectors in the plane will also be in the plane. We can represent the space vector of any balanced three-phase set as a vector in this plane. It is useful to represent this two-dimensional subspace with its own orthonormal basis (that is, with two orthogonal unit vectors that can be used to define where in this subspace a given balanced three-phase quantity lies). The two perpendicular unit vectors that define this α–β plane are denoted $\widehat{\alpha}$ and $\widehat{\beta}$, such that $\widehat{\alpha}$, $\widehat{\beta}$, and $\widehat{\gamma}$ form a right-handed coordinate system.

By convention, the $\widehat{\alpha}$ axis is defined by the projection of the $\widehat{a}$ axis onto the α–β plane; we then define the $\widehat{\beta}$ axis such that $\widehat{\alpha} \times \widehat{\beta} = \widehat{\gamma}$. (We can find $\widehat{\alpha}$ by subtracting the projection of $\widehat{a}$ onto $\widehat{\gamma}$ from $\widehat{a}$ and normalizing the result to unit length. We can then find $\widehat{\beta} = \widehat{\gamma} \times \widehat{\alpha}$.) This results in the orthonormal basis

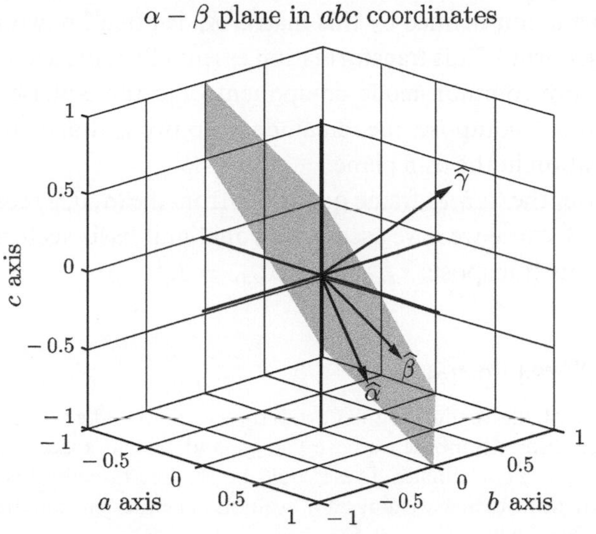

Figure 9.23 The unit vectors $\widehat{\alpha}$, $\widehat{\beta}$, and $\widehat{\gamma}$, and the α–β plane, shown in (a, b, c) coordinates.

$$\widehat{\alpha} = \begin{bmatrix} \frac{2}{\sqrt{6}} \\ \frac{-1}{\sqrt{6}} \\ \frac{-1}{\sqrt{6}} \end{bmatrix}, \quad \widehat{\beta} = \begin{bmatrix} 0 \\ \frac{1}{\sqrt{2}} \\ \frac{-1}{\sqrt{2}} \end{bmatrix}, \quad \widehat{\gamma} = \begin{bmatrix} \frac{1}{\sqrt{3}} \\ \frac{1}{\sqrt{3}} \\ \frac{1}{\sqrt{3}} \end{bmatrix} \tag{9.23}$$

that can be used to represent any space vector. Figure 9.23 illustrates the $\widehat{\alpha}$, $\widehat{\beta}$, and $\widehat{\gamma}$ axes and α–β plane as defined by (9.23).

Any common-mode, or average, component of a three-phase quantity must exist only within the one-dimensional subspace defined by the $\widehat{\gamma}$ axis (and not within the α–β plane). Any balanced three-phase quantity (or balanced component of a three-phase quantity) similarly must exist in the subspace defined by the $\widehat{\alpha}$, $\widehat{\beta}$ axes.

9.7.3 The Clarke Transform

The $\widehat{\alpha}$, $\widehat{\beta}$, $\widehat{\gamma}$ axes themselves represent a Cartesian coordinate system, and we can express space vectors in terms of their (α, β, γ) coordinates. We can find the representation of any vector in the (a, b, c) frame in terms of (α, β, γ) coordinates by taking the projections of the (a, b, c) vector onto the α, β, γ axes. Defining the components x_α, x_β, x_γ as scaling constants of the projections of $\widehat{x}_{abc}$ onto $\widehat{\alpha}$, $\widehat{\beta}$, and $\widehat{\gamma}$ respectively, we get

$$\widehat{x}_{\alpha\beta\gamma} = \begin{bmatrix} x_\alpha \\ x_\beta \\ x_\gamma \end{bmatrix} = \begin{bmatrix} \widehat{\alpha}^{\mathrm{T}}\widehat{x}_{abc} \\ \widehat{\beta}^{\mathrm{T}}\widehat{x}_{abc} \\ \widehat{\gamma}^{\mathrm{T}}\widehat{x}_{abc} \end{bmatrix} = \begin{bmatrix} \widehat{\alpha}^{\mathrm{T}} \\ \widehat{\beta}^{\mathrm{T}} \\ \widehat{\gamma}^{\mathrm{T}} \end{bmatrix} \widehat{x}_{abc} = \begin{bmatrix} \frac{2}{\sqrt{6}} & \frac{-1}{\sqrt{6}} & \frac{-1}{\sqrt{6}} \\ 0 & \frac{1}{\sqrt{2}} & \frac{-1}{\sqrt{2}} \\ \frac{1}{\sqrt{3}} & \frac{1}{\sqrt{3}} & \frac{1}{\sqrt{3}} \end{bmatrix} \begin{bmatrix} x_a \\ x_b \\ x_c \end{bmatrix}, \tag{9.24}$$

or $\widehat{x}_{\alpha\beta\gamma} = K_c \widehat{x}_{abc}$.

The transformation defined by this matrix K_c is often known as the "α–β–γ" transform or the Clarke transform.‡ This transformation essentially represents a rotation of coordinates that separates out any common-mode component of a three-phase vector quantity (into the x_γ component), and decomposes the remainder into two orthogonal components (x_α and x_β) that identify its location in the α–β plane.

We can recover the (a, b, c)-frame quantities from the (α, β, γ) components using the inverse of the K_c matrix. Because we have used orthogonal unit basis vectors, K_c is an orthogonal matrix, so its inverse is its transpose: $\widehat{x}_{abc} = K_c^{-1}\widehat{x}_{\alpha\beta\gamma} = K_c^{\mathrm{T}}\widehat{x}_{\alpha\beta\gamma}$.

Example 9.3 Power Invariance

A useful attribute of the version of the Clarke transform described here is that it preserves power in the different representations. Suppose we have voltage and current space vectors $\widehat{v}_{abc}(t)$, $\widehat{i}_{abc}(t)$ at some set of terminals in (a, b, c) coordinates. These could be the line-to-neutral voltages and phase currents of a three-phase set of transformer windings, for example. The total instantaneous power into the three-phase windings would then be

$$p_{abc}(t) = \widehat{v}_{abc}^{\mathrm{T}}\,\widehat{i}_{abc}. \tag{9.25}$$

Now consider the voltage $\widehat{v}_{\alpha\beta\gamma}(t)$, current $\widehat{i}_{\alpha\beta\gamma}(t)$, and power calculated in (α, β, γ) coordinates:

$$p_{\alpha\beta\gamma}(t) = \widehat{v}_{\alpha\beta\gamma}^{\mathrm{T}}\,\widehat{i}_{\alpha\beta\gamma}. \tag{9.26}$$

Rewriting the expression in (α, β, γ) coordinates in terms of the Clarke transform, and recognizing that $K_c^{\mathrm{T}} K_c = I$ (the identity matrix), we get

$$p_{\alpha\beta\gamma}(t) = \widehat{v}_{\alpha\beta\gamma}^{\mathrm{T}}\,\widehat{i}_{\alpha\beta\gamma} = (K_c\widehat{v}_{abc})^{\mathrm{T}}(K_c\widehat{i}_{\alpha\beta\gamma}) = \widehat{v}_{abc}^{\mathrm{T}} K_c^{\mathrm{T}} K_c\widehat{i}_{abc} = p_{abc}(t). \tag{9.27}$$

As might be desired, we calculate the same instantaneous power regardless of which way we represent our voltages and currents.

Of course, in cases where we need not concern ourselves with common-mode signals (e.g., if we are only dealing with balanced three-phase sources and loads), we can focus only on the α and β components, dropping the bottom row of K_c and the rightmost column of K_c^{-1}:

$$\begin{bmatrix} x_\alpha \\ x_\beta \end{bmatrix} = \begin{bmatrix} \frac{2}{\sqrt{6}} & \frac{-1}{\sqrt{6}} & \frac{-1}{\sqrt{6}} \\ 0 & \frac{1}{\sqrt{2}} & \frac{-1}{\sqrt{2}} \end{bmatrix} \begin{bmatrix} x_a \\ x_b \\ x_c \end{bmatrix}, \quad \begin{bmatrix} x_a \\ x_b \\ x_c \end{bmatrix} = \begin{bmatrix} \frac{2}{\sqrt{6}} & 0 \\ \frac{-1}{\sqrt{6}} & \frac{1}{\sqrt{2}} \\ \frac{-1}{\sqrt{6}} & \frac{-1}{\sqrt{2}} \end{bmatrix} \begin{bmatrix} x_\alpha \\ x_\beta \end{bmatrix}. \tag{9.28}$$

The α, β representation provides substantial simplification in this case: instead of managing three dependent phase variables, we need only handle two independent variables representing

‡ Named for Edith Clarke, a pioneer in ac circuit analysis who developed the first version of it. Among many accomplishments in a trailblazing career, in 1919 she became the first woman to receive a Master's degree in electrical engineering from MIT, and in 1947 she was the first female professor of electrical engineering in the US, joining the faculty of the University of Texas at Austin.

orthogonal components. The benefits of this in describing three-phase pulse-width modulation will become apparent in the next subsection.

Example 9.4 Clarke Transform and Phasors

To illustrate the application of the Clarke transform, we consider the following balanced set of three-phase sinusoidal voltages:

$$\begin{aligned} v_a(t) = V_s \cos(\omega t) &= \mathrm{Re}\{V_s e^{j0} e^{j\omega t}\} = \mathrm{Re}\{\widehat{V}_A e^{j\omega t}\}, \\ v_b(t) = V_s \cos\left(\omega t - \frac{2\pi}{3}\right) &= \mathrm{Re}\{V_s e^{-j\frac{2\pi}{3}} e^{j\omega t}\} = \mathrm{Re}\{\widehat{V}_B e^{j\omega t}\}, \\ v_c(t) = V_s \cos\left(\omega t + \frac{2\pi}{3}\right) &= \mathrm{Re}\{V_s e^{j\frac{2\pi}{3}} e^{j\omega t}\} = \mathrm{Re}\{\widehat{V}_C e^{j\omega t}\}, \end{aligned} \tag{9.29}$$

where Re{·} denotes the real part, and $\widehat{V}_A, \widehat{V}_B, \widehat{V}_C$ are the *phasors* associated with the voltages $v_a(t), v_b(t), v_c(t)$ respectively. Applying (9.28) to this, we get

$$v_\alpha(t) = \sqrt{\frac{3}{2}} V_s \cos(\omega t), \qquad v_\beta(t) = \sqrt{\frac{3}{2}} V_s \sin(\omega t), \qquad v_\gamma(t) = 0. \tag{9.30}$$

Thus, at $t=0$ the space vector has length $\sqrt{\frac{3}{2}}\, V_s$ in the α–β plane, directly along the α axis. This vector stays at constant length over time, rotating counterclockwise in the α–β plane (about the $\widehat{\gamma}$ axis) with angular velocity ω, forming an angle ωt with the α axis at time t. This trajectory is illustrated in Fig. 9.24.

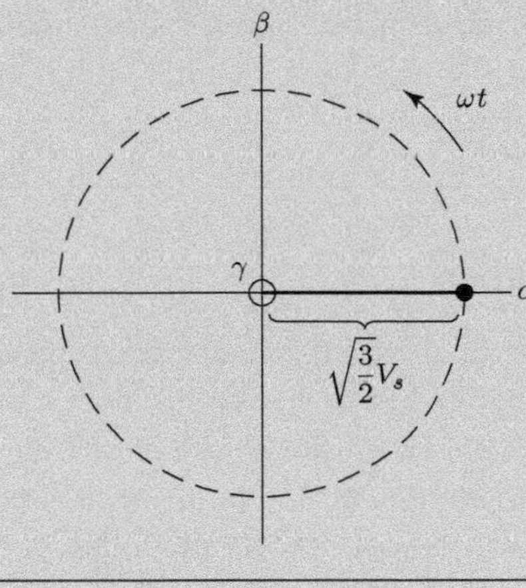

Figure 9.24 The trajectory for the space vector of the three-phase sinusoidal set of signals (9.29). The space vector sweeps out a circle in the α–β plane.

The sinusoidal three-phase set in (9.29) can alternatively be represented in the phasor domain, as illustrated in Fig. 9.25(a). The voltages $v_a(t), v_b(t), v_c(t)$ can be obtained by computing the real parts of the corresponding phasors after rotating them counterclockwise through the angle ωt. Based on (9.30), the α–β representation can also be described in terms of phasors, as shown in Fig. 9.25(b).

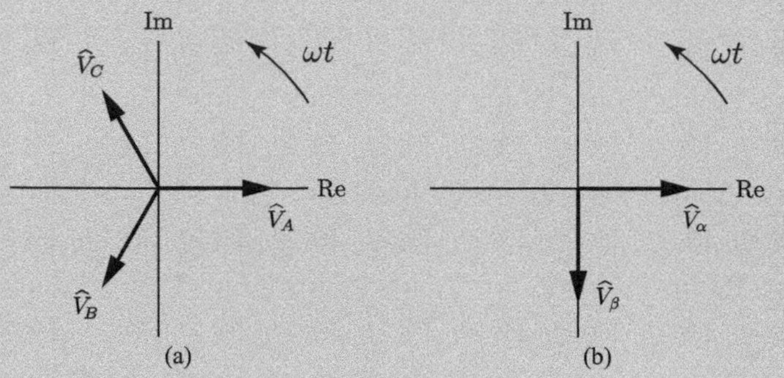

Figure 9.25 (a) The graphical phasor representation of (9.29). (b) The phasor representation of the same three-phase set in an α–β representation.

The transform scalings introduced here are somewhat different from that in Clarke's original paper [8], though the general concepts are similar. Unfortunately, in addition to scaling considerations, there is not widespread agreement on axis definitions and naming, so care is required when reading the literature. The (α, β, γ) coordinates are also closely related to the so-called "direct-quadrature-zero," or *d-q-0*, coordinates. The main difference is that the *d-q-0* representation includes a further rotational transformation to map space vectors into a *rotating* reference frame that rotates about the $\widehat{\gamma}$ axis; this transformation from *a-b-c* to *d-q-0* – known as the the Park transform – can be convenient in applications such as field-oriented control of electric machines. In the preceding example, the phasor representation in Fig. 9.25(b) is effectively a *d-q-0* representation.

9.7.4 Space-Vector Modulation

Pulse-width modulation of three-phase inverters can be realized through a wide variety of means, many of which are direct extensions of techniques used in the single-phase case. However, space vectors and the (α, β, γ) coordinate system provide a superior means to characterize the outputs of a three-phase inverter and to implement three-phase PWM. The resulting class of PWM techniques is known as space-vector modulation, and is well matched to digital control implementations, making it easier to achieve high performance.

Consider the three-phase inverter of Fig. 9.19 connected to a balanced load. Because the load acts like three identical impedances connected to a floating neutral point, there can be no common-mode component to the phase currents. Looked at in (α, β, γ) coordinates, the space vector representing the phase currents can have no γ component. The same may be said of the line-to-neutral voltages synthesized by the inverter. As a result, there can be no γ-component contribution to power flow between the inverter and load. For this reason, one can ignore γ components entirely in controlling the inverter, and focus only on the α and β components of space vectors.

Table 9.2 The different voltage vectors that can be synthesized by the three-phase bridge converter of Fig. 9.19 in (*a,b,c*) coordinates and their resulting α, β components (γ components neglected). Voltages are relative to the bridge midpoint voltage v_r and normalized to the dc-bus voltage V_{dc}.

Vector {a b c}	Sw. on	v_{ar}	v_{br}	v_{cr}	v_α	v_β	$\sqrt{v_\alpha^2 + v_\beta^2}$	$\tan^{-1}\left(\frac{v_\beta}{v_\alpha}\right)$
$X_0 = \{000\}$	4, 6, 2	−1/2	−1/2	−1/2	0	0	0	—
$X_1 = \{100\}$	1, 6, 2	+1/2	−1/2	−1/2	$\frac{2}{\sqrt{6}}$	0	$\frac{2}{\sqrt{6}}$	0°
$X_2 = \{110\}$	1, 3, 2	+1/2	+1/2	−1/2	$\frac{1}{\sqrt{6}}$	$\frac{\sqrt{3}}{\sqrt{6}}$	$\frac{2}{\sqrt{6}}$	60°
$X_3 = \{010\}$	4, 3, 2	−1/2	+1/2	−1/2	$\frac{-1}{\sqrt{6}}$	$\frac{\sqrt{3}}{\sqrt{6}}$	$\frac{2}{\sqrt{6}}$	120°
$X_4 = \{011\}$	4, 3, 5	−1/2	+1/2	+1/2	$\frac{-2}{\sqrt{6}}$	0	$\frac{2}{\sqrt{6}}$	180°
$X_5 = \{001\}$	4, 6, 5	−1/2	−1/2	+1/2	$\frac{-1}{\sqrt{6}}$	$\frac{-\sqrt{3}}{\sqrt{6}}$	$\frac{2}{\sqrt{6}}$	240°
$X_6 = \{101\}$	1, 6, 5	+1/2	−1/2	+1/2	$\frac{1}{\sqrt{6}}$	$\frac{-\sqrt{3}}{\sqrt{6}}$	$\frac{2}{\sqrt{6}}$	300°
$X_7 = \{111\}$	1, 3, 5	+1/2	+1/2	+1/2	0	0	0	—

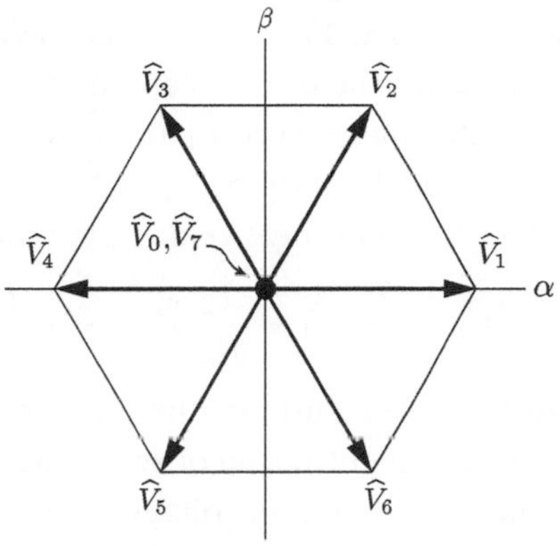

Figure 9.26 Projections of the phase-voltage space vectors of Table 9.2 onto the α–β plane. Six of these vectors are of equal length, while the other two have zero length in this plane.

There are eight possible switch configurations for the three-phase inverter, assuming one switch of each half-bridge is always on at a given time. Table 9.2 (which extends Table 9.1) shows the instantaneous phase voltages that the inverter can synthesize (with respect to the dc-bus midpoint voltage v_r), and the resulting α, β components of the associated phase voltage space vectors, all normalized to the dc-bus voltage. These eight inverter states X_0 to X_7 thus correspond to eight possible inverter output vectors in the α–β plane. As illustrated in Fig. 9.26, six of these inverter output vectors have the same length, pointing in different directions in the α–β plane at intervals of $\pi/3$. The remaining two output vectors are of zero length, as they originate from γ-directed space vectors.

The vectors in Fig. 9.26 represent the different outputs that the three-phase bridge can produce to synthesize a desired output waveform or to control power or current in a balanced three-phase

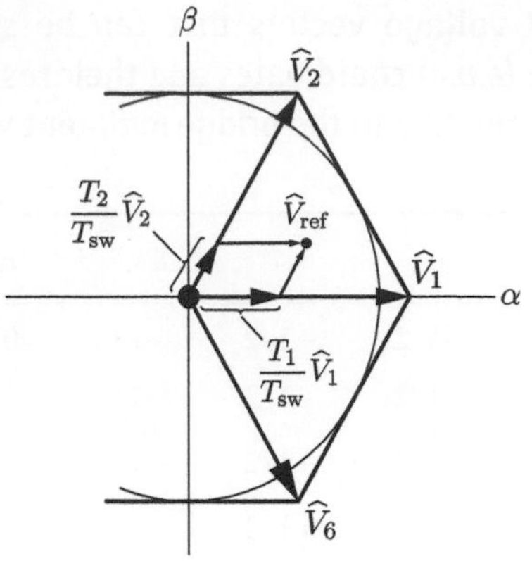

Figure 9.27 Taking the average of adjacent voltage vectors and one or more zero vectors over a switching period T_{sw} to create a local average voltage vector $\hat{V}_{ref}$.

load. By modulating between these voltage vectors (e.g., over some switching period T_{sw}), one can synthesize any local time-average voltage in the α–β plane that falls within the hexagon surrounding the voltage vector set. For example, consider the triangular sector delineated by $\hat{V}_1$, $\hat{V}_2$, and the line connecting them. One can get an average voltage anywhere on the line connecting $\hat{V}_1$ and $\hat{V}_2$ by switching between $\hat{V}_1$ and $\hat{V}_2$ with an appropriate duty ratio. Moreover, as illustrated in Fig. 9.27, by modulating between $\hat{V}_1$, $\hat{V}_2$, and one or both of the zero vectors $\hat{V}_0$ or $\hat{V}_7$ one can obtain any local average value inside the sector:

$$\overline{v}_{ref} = \langle \hat{V} \rangle = \frac{T_1}{T_{sw}}\hat{V}_1 + \frac{T_2}{T_{sw}}\hat{V}_2 + \frac{1 - T_1 - T_2}{T_{sw}}\{\hat{V}_0 \text{ or } \hat{V}_7\}. \tag{9.31}$$

This SVM technique lets us conveniently synthesize a desired (switching-cycle-average) voltage vector in the α–β plane, or, equivalently, a set of switching-cycle-average *a, b, c* phase voltages. We can use SVM to control the inverter output voltages and/or to get a desired response (e.g., current or flux) in the load.

Note that SVM generally dictates the fraction of a switching period spent in adjacent nonzero states, but not which zero state (or states) are used to complete the switching cycle or the relative position of the various time periods within a switching cycle. These details influence aspects such as the spectral content of the output waveforms and the number of switching transitions required (and associated switching loss). Thus, we may really think of SVM as a *class* of PWM techniques that employ space vectors as a means to organize the modulation process. Also note that extensions to the basic SVM approach can be formulated for multi-level inverters, for inverters that do not maintain balanced operation among the phases (e.g., that actively modulate the load neutral point), or that allow discontinuous conduction of phases.

Example 9.5 Space-Vector Modulation for Sinusoidal Outputs

As described above, by adjusting the modulation time between adjacent output vectors, one can generate a desired voltage output $\overline{v}_{ref}(t)$ in the α–β plane, and hence at the *a, b, c* phase outputs of the inverter. An interesting question is: What is the largest three-phase sinusoidal set of voltages that can be synthesized without incurring distortion or harmonics in the output?

A sinusoidal reference voltage becomes a circle in the α–β plane. Consequently, the largest sinusoidal reference that can be synthesized without distortion is delimited by the circle inscribed *inside* the hexagon defining the achievable outputs, as illustrated in Fig. 9.27. This circle has magnitude

$$|V_{r,\sin}| = \frac{1}{2}|\widehat{V}_1 + \widehat{V}_2| = \frac{1}{\sqrt{2}}V_{dc}, \tag{9.32}$$

and corresponds to an amplitude of the a, b, c components of

$$|V_{ar}| = |V_{br}| = |V_{cr}| = \frac{\sqrt{2}}{\sqrt{3}}\frac{1}{\sqrt{2}}V_{dc} = \frac{1}{\sqrt{3}}V_{dc} \tag{9.33}$$

or a line-to-line output amplitude of

$$|V_{ab}| = |V_{bc}| = |V_{ca}| = V_{dc}. \tag{9.34}$$

Thus, we can generate inverter output sine waves without low-frequency distortion (but still containing switching frequency ripple) so long as the line-to-line output voltage does not exceed the dc-bus voltage.

Note that if instead of SVM PWM synthesis we simply used whichever space vector $\widehat{V}_1, \ldots, \widehat{V}_6$ was closest to the sinusoidal reference voltage $\widehat{v}_{ref}(t)$, we would get exactly the six-step waveform described for three-phase inverters in Section 9.6.2 (with the attendant harmonics and distortion six-step switching entails).

9.8 Multi-Level Inverters

The converter circuit of Fig. 9.19 is capable of selecting from among two line-to-reference voltage levels at each of the three half-bridge outputs: $+V_{dc}/2$ and $-V_{dc}/2$. For this reason, the three-phase bridge of Fig. 9.19 is known as a *two-level* inverter. These available voltage levels yield three possible line-to-line output voltage levels: V_{dc}, 0, and $-V_{dc}$. However, as shown in conjunction with single-phase inverters in Chapter 8, the availability of additional voltage levels can enable higher-quality (lower THD) waveforms to be synthesized. This has motivated the development of multi-level inverters that can provide more than two levels at each output.

An m-level three-phase inverter provides for selection from among m distinct available voltage levels at each phase output. This enables $2m - 1$ possible voltage levels to be synthesized in the line-to-line outputs of the three-phase system. Thus, a three-level inverter enables three possible levels of each phase voltage and five possible states for the line-to-line voltages.

Multi-level inverters have become widespread in high-power applications, including drives, photovoltaic systems, converters for high-voltage dc, etc. Their popularity stems from both their improved waveform quality and from the ability to use devices having voltage ratings less than the full converter voltage. To illustrate the possibilities with multi-level inverters, we consider the neutral-point-clamped (NPC) converter, a three-level inverter that was the first multi-level inverter to gain widespread attention. Other approaches to multi-level conversion are described in Chapters 5 and 8, and in the Notes and Bibliography section at the end of this chapter.

9.8.1 Neutral-Point-Clamped Inverter

Figure 9.28(a) illustrates the bridge-leg building block of the NPC inverter. If switches S_1 and S_3 are on, then $v_{ar} = +V_{dc}/2$, while if switches S_2 and S_4 are on, then $v_{ar} = -V_{dc}/2$. While these states can be synthesized by a two-level inverter, we also have the option of turning on combinations of switches S_3, S_4, S_5, and S_6 to obtain $v_{ar} = 0$. Figure 9.28(b) shows an example waveform for v_{ar}, illustrating the ability for this circuit to create a zero-voltage state, which the conventional two-switch bridge leg cannot do.

The circuit of Fig. 9.28(a) is illustrated with two dc voltages of value $V_{dc}/2$. More typically, the system is implemented with a single dc-bus voltage of value V_{dc} that is split with a pair of large capacitors to form the center point or *neutral point*. This approach works if the inverter is controlled to maintain the center-point voltage at half the dc-bus voltage (that is, with no dc current into the center-point node) or if some auxiliary balancing circuit is provided.

In the traditional version of the NPC inverter, S_1–S_4 are actively controlled devices, while diodes are used for S_5 and S_6, as illustrated in Fig. 9.29. This version of the circuit, known as a *diode-clamped multi-level inverter*, thus requires only four controlled switches for an NPC bridge leg, with the switches supporting unidirectional off-state voltage and bilateral on-state current. In the active neutral-point-clamped (ANPC) inverter, switches S_5 and S_6 are also actively controlled unidirectional-blocking and bilateral-carrying switches. While both versions can synthesize the same output voltage waveforms, the ANPC version can support different switching schemes for doing so. Moreover, in the ANPC inverter, so long as S_6 is held on when S_1 and S_3 are on, and S_5 is held on when S_2 and S_4 are on, each of the switches need

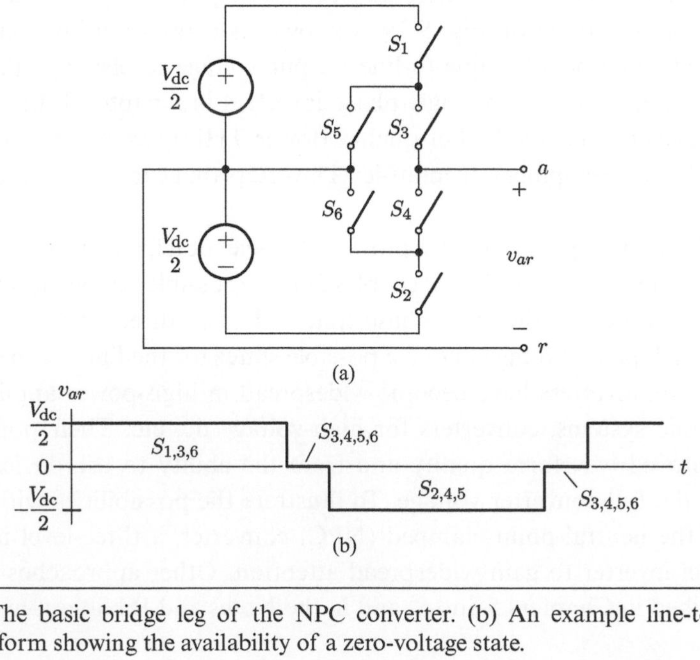

Figure 9.28 (a) The basic bridge leg of the NPC converter. (b) An example line-to-reference voltage waveform showing the availability of a zero-voltage state.

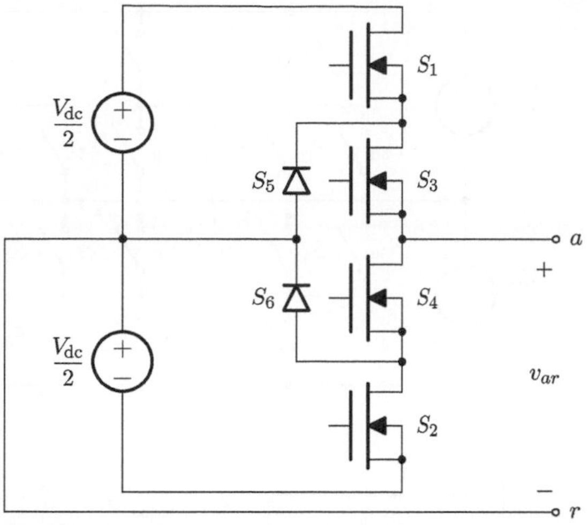

Figure 9.29 A practical implementation of the NPC inverter using diodes for clamping.

only be rated (ideally) for a blocking voltage of $V_{dc}/2$, with some practical derating for safety and dynamic variations in the center-point voltage. The traditional diode-clamped version also permits voltage derating of the devices, but does not actively control the voltage sharing between S_1 and S_3 or S_2 and S_4 when they are off.

The basic three-phase NPC circuit and illustrative waveforms for *stairstep* operation are shown in Figure 9.30. Because each leg can output three possible levels, there are five states of the line–line voltage: V_{dc}, $V_{dc}/2, 0, -V_{dc}/2$, and $-V_{dc}$. The three-phase NPC inverter may be employed in much the same way as a conventional three-phase bridge inverter, albeit with greater modulation flexibility and improved waveform quality. PWM can be employed to improve the THD and reduce harmonics in the output, but with pulse depths limited to line-to-reference voltages of $\pm V_{dc}/2$.

Because the NPC inverter offers an additional level compared to the conventional three-phase bridge inverter, it affords considerably greater flexibility in tailoring the switching patterns to reduce THD and harmonics. Figure 9.31 shows line-to-reference voltage waveforms that eliminate both the fifth and seventh harmonics for both a conventional bridge and the NPC converter. The amplitudes of the fundamentals are similar for the two waveforms: 0.594 V_{dc} for the conventional bridge waveform and 0.585 V_{dc} for the NPC waveform. Because triple-n harmonics are canceled at the load and there are no even harmonics owing to half-wave symmetry, the lowest-order remaining harmonics of interest are the 11th, 13th, and 17th. For the phase-to-reference waveform of the conventional bridge, these harmonics are 20.3%, 27.1%, and 17.1% of the fundamental, respectively. By contrast, for the NPC waveform they are much lower, at 5.05%, 12.5%, and 5.3% of the fundamental. The improvements carry over to the line-to-line waveforms, which are shown in Fig. 9.32. The line-to-line voltage for the three-phase bridge has a THD of approximately 51%, while for the NPC line-to-line voltage waveform it is only 22%. The NPC inverter can thus offer much better output waveform quality. Moreover,

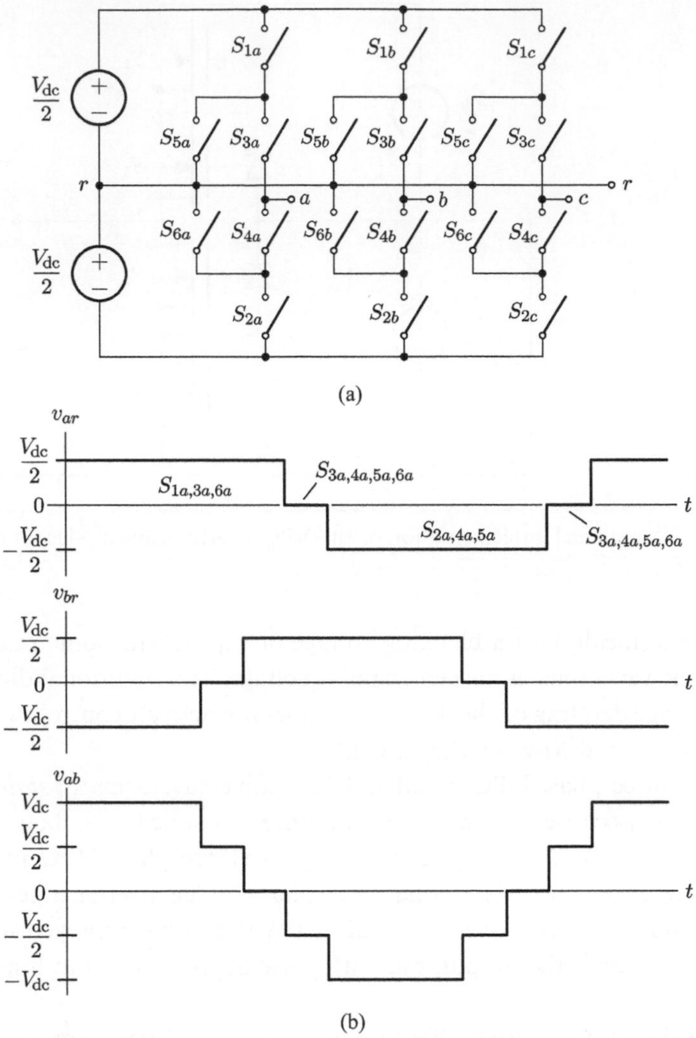

Figure 9.30 (a) The three-phase NPC circuit. (b) Line-to-reference and line-to-line voltage waveforms for stairstep operation.

this higher waveform quality is achieved with smaller-voltage switching transitions using lower-voltage devices.

The NPC inverter can be seen as achieving higher performance than the conventional three-phase bridge at the expense of a higher degree of complexity. For high-power systems, this increase in complexity is often warranted. Still higher performance at greater degrees of complexity can be achieved with multi-level inverters having higher numbers of levels. The diode-clamped family of multi-level inverters may be thought of as a generalization of the NPC inverter to more levels.

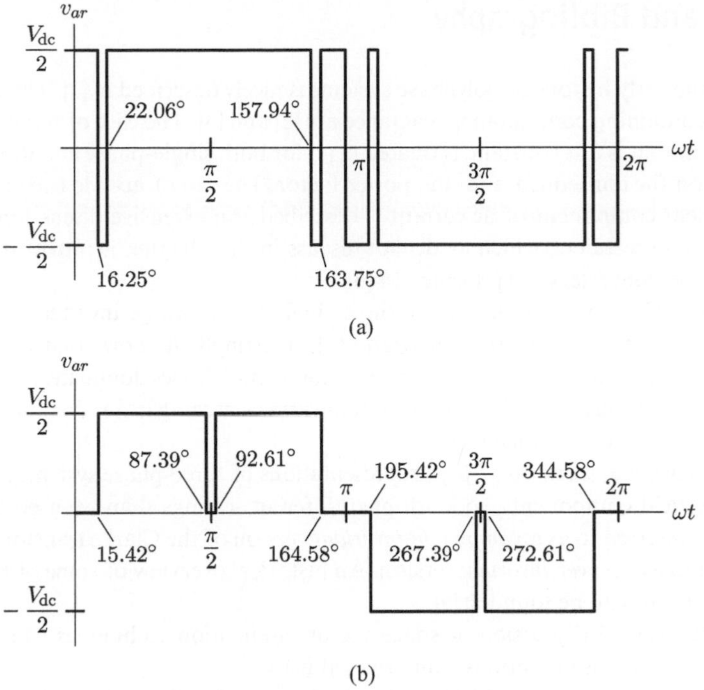

Figure 9.31 Line-to-reference voltage waveforms having fifth and seventh harmonics eliminated. (a) A two-level waveform suitable for the conventional three-phase bridge inverter. (b) A three-level waveform suitable for the three-phase NPC inverter.

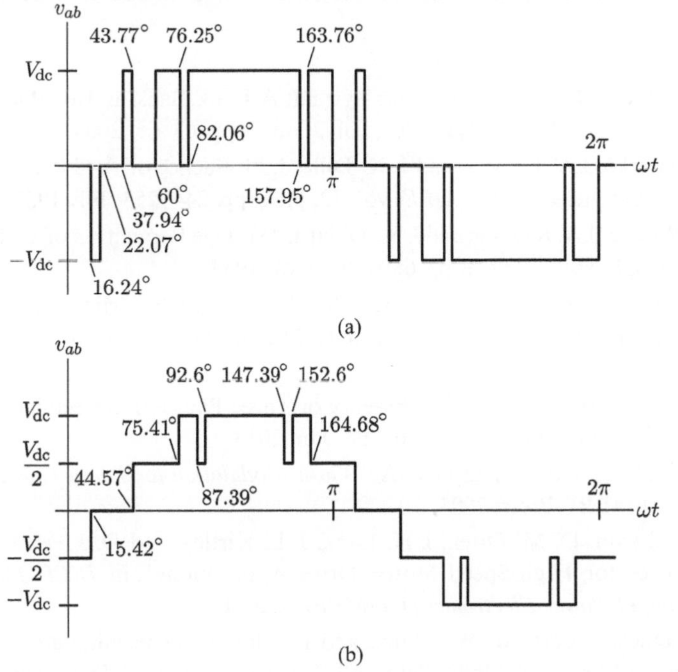

Figure 9.32 Line-to-line voltage waveforms for the line-to-reference voltage waveforms of Fig. 9.31. (a) A three-level waveform suitable for the three-phase bridge inverter. (b) A five-level waveform suitable for the three-phase NPC inverter.

Notes and Bibliography

Some of the early history of polyphase systems is nicely described in [1]. The classic papers on multi-pulse rectifiers containing commutating reactance are [2] and [3]. The case of finite dc inductance, and therefore a dc current that is not constant, is treated in [4] for both single-phase and three-phase rectifiers. Attention is placed on the consequence for the power factor. The use of ac-side capacitance to compensate for the displacement component of ac current is described. An extensive discussion of power-factor correction in three-phase rectifiers, which we do not discuss in this chapter, is provided in [5]. PWM techniques for inverters and converters are presented in [6].

Reference [7] compares the use of three single-phase bridge inverters to the three-phase inverter in applications that allow either configuration. Interestingly, it shows that while the three-phase bridge is preferable in applications for which converter conduction losses dominate, use of three single-phase bridges can be advantageous in applications where converter switching losses dominate, such as when driving high-speed or low-inductance machines.

Clarke's original work on (α, β, γ) representations of three-phase systems [8] was based on an extension of symmetrical components, so it adopted different scalings than we used here. Her original version is sometimes referred to as an *amplitude-invariant* version of the Clarke transform, while that developed here is referred to as a *power-invariant* version. An historical overview of some of these coordinate systems and transformations can be found in [9].

A more detailed discussion of space-vector modulation techniques can be found in [6], which uses slightly different axis definitions than are used here.

Multi-level inverters are widely used at high power levels, especially in three-phase applications. The neutral-point-clamped inverter was the earliest multi-level inverter to gain widespread attention. It was first described in [10], which provides principles of operation, harmonic analyses, and experimental results. Other types of multi-level converters also exist, as surveyed in [11], [12], and [13]. Some of these approaches are also outlined in Chapters 5 and 8. A review of programmed PWM techniques for multi-level inverters is presented in [14].

1. T. J. Blalock "The First Polyphase System: A Look Back at Two-Phase Power for ac Distribution," *IEEE Power and Energy Magazine*, vol. 2, no. 2, pp. 63–66, 2004.
2. R. L. Witzke, J. V. Kresser, and J. K. Dillard, "Influence of A-C Reactance on Voltage Regulation of 6-Phase Rectifiers," *Trans. AIEE*, vol. 72, pt. 1, pp. 244–253, July 1953.
3. R. L. Witzke, J. V. Kresser, and J. K. Dillard, "Voltage Regulation of 12-Phase Double-Way Rectifiers," *Trans. AIEE*, vol. 72, pt. 1, pp. 689–697, Nov. 1953.
4. A. W. Kelley and W. F. Yadusky, "Rectifier Design for Minimum Line Current Harmonics and Maximum Power Factor," in *Fourth Annual IEEE Applied Power Electronics Conference and Exposition (APEC)*, pp. 13–22, 1989.
5. J. W. Kolar and T. Friedli, "The Essence of Three-Phase PFC Rectifier Systems – Part I," *IEEE Trans. Power Electronics* vol. 28, pp. 176–198, Jan. 2013.
6. D. G. Holmes and T. A.Lipo, *Pulse Width Modulation for Power Converters: Principles and Practice*, New York: IEEE Press, 2003.
7. M. M. Qasim, D. M. Otten, J. H. Lang, J. L. Kirtley, and D. J. Perreault, "Comparison of Inverter Topologies for High-Speed Motor Drive Applications," in *IEEE 22nd Workshop on Control and Modeling of Power Electronics (COMPEL)*, 2021.
8. W. C. Duesterhoeft, M. W. Schulz, and E. Clarke, "Determination of Instantaneous Currents and Voltage by Means of Alpha, Beta and Zero Components," *Transactions of the AIEE*, vol. 70, no. 2, pp. 1248–1255, 1955.

9. C. J. O'Rourke, M. M. Qasim, M. R. Overlin, and J. L. Kirtley, "A Geometric Interpretation of Reference Frames and Transformations: dq0, Clarke and Park," *IEEE Trans. Energy Conversion*, vol. 34, pp. 2070–2083, Dec. 2019.
10. A. Nabae, I. Takaashji, and H. Akagi, "A New Neutral-Point Clamped PWM Inverter," *IEEE Trans. Industry Applications*, vol. 17, pp. 518–523, Sept./Oct. 1981.
11. H. Akagi, "Multilevel Converters: Fundamental Circuits and Systems," *Proc. IEEE*, vol. 105, pp. 2048–2065, Nov. 2017.
12. J. Rodriguez, J.-S. Lai, and F. Z. Peng, "Multilevel Inverters: A Survey of Topologies, Controls and Applications," *IEEE Trans. Industrial Electronics*, vol. 49, pp. 724–738, Aug. 2002.
13. J. Rodriguez, L. G. Franquelo, S. Kouro, and J. I. Leon, "Multilevel Converters: An Enabling Technology for High-Power Applications," *Proc. IEEE*, vol. 97, pp. 1786–1817, Nov. 2009.
14. M. S. A. Dahidah, G. Konstantinou, and V. G. Agelidis, "A Review of Multilevel Selective Harmonic Elimination PWM: Formulations, Solving Algorithms, Implementations and Applications," *IEEE Trans. Power Electronics*, vol. 30, pp. 4091–4106, Aug. 2015.

PROBLEMS

9.1 Commercial buildings are often supplied with three-phase 208 V_{rms} line-to-line electrical service. Lighting circuits are then connected between line and neutral, giving $V_{\ell n} = V_{\ell\ell}/\sqrt{3} = 120\,V_{rms}$. If light dimmers, which can be thought of as ac controllers, are connected to the lights, the result is the equivalent circuit of Fig. 9.33. Assume that the lamps are incandescent and can be modeled as resistors, that each phase has the same number of lamps, and that the dimmers are adjusted to give the same light levels on all circuits (α is the same for all dimmers). Sketch the neutral current i_n and determine the fundamental frequency of this current.

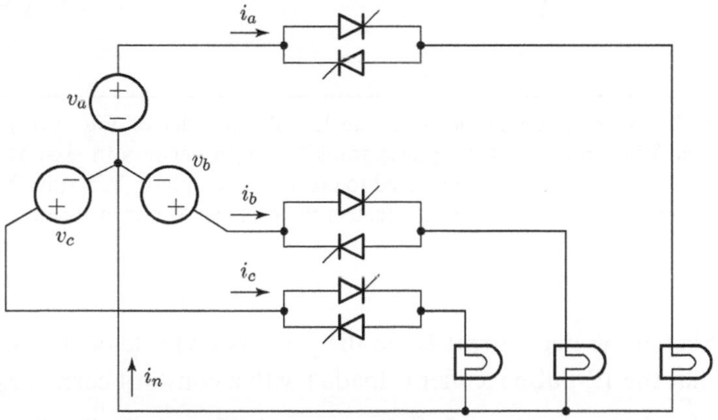

Figure 9.33 The three-phase lighting control circuit for Problem 9.1.

9.2 One result of the phase-shifting transformer connections used in the 12-pulse rectifier of Fig. 9.9(a) is that the net primary line current has a lower THD than that of a 6-pulse circuit. Figure 9.34 shows the transformer connections necessary to generate the 6-phase source for the 12-pulse rectifier of Fig. 9.9(a).

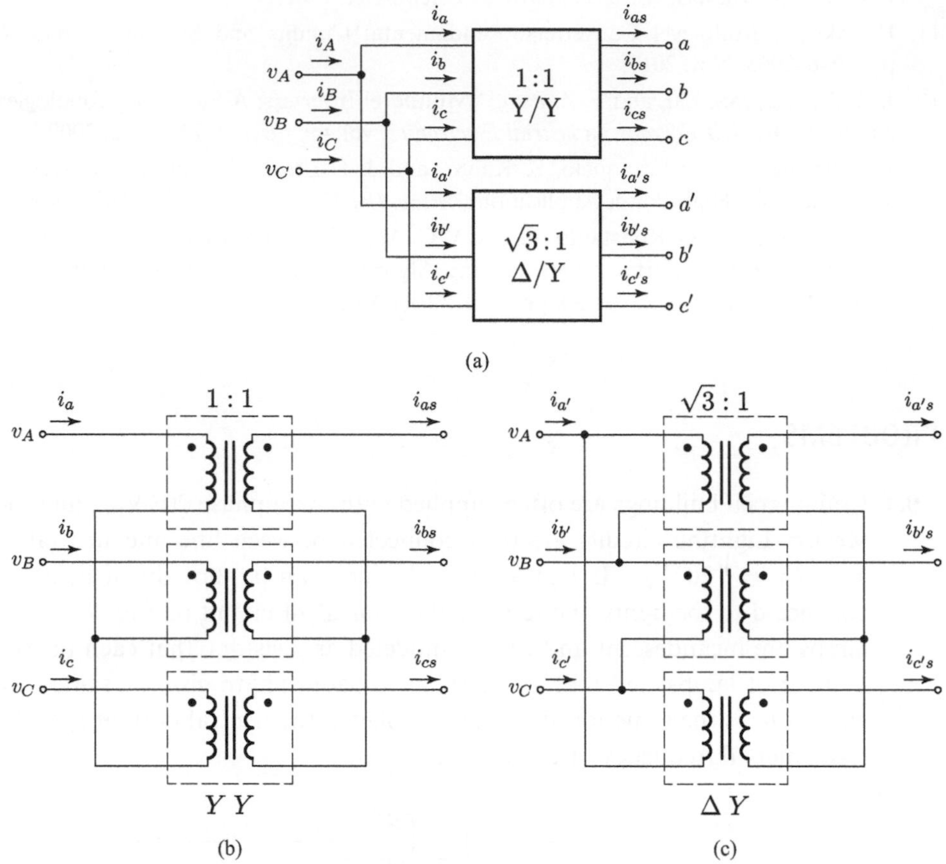

Figure 9.34 Transformer connections for the 12-pulse rectifier of Fig. 9.9(a). (a) The interconnection of a Y/Y and Δ/Y three-phase transformer to generate a six-phase source. (b) Three single-phase transformers connected to create a three-phase Y/Y transformer. (c) Three single-phase transformers connected to create a three-phase Δ/Y transformer.

(a) Sketch the line current $i_{a'}$ on the primary (Δ) side of the Δ/Y transformer, assuming that the 12-pulse rectifier is loaded with a constant current I_d.

(b) Sketch the primary-side line current i_A. What are its first two nonzero harmonics?

(c) Compare the first two nonzero line current harmonics of this 12-pulse circuit to those of the 6-pulse circuit, as determined in Problem 9.8.

(d) Calculate the THD of the 12-pulse rectifier line current and compare it to that of the 6-pulse rectifier.

9.3 If we wanted to operate the two series-connected six-pulse bridges in Example 9.1 as a six-pulse rectifier, that is, without the phase shift between the ac sources created by the Y/Y and Δ/Y transformer connections, could we operate the circuit without transformers?

9.4 What is $\langle v_d \rangle$ for the 12-pulse circuit of Fig. 9.9?

9.5 The two 6-pulse waveforms in Fig. 9.9, v_{d_1} and v_{d_2}, have the 6th harmonic of the ac frequency as their first nonzero component, whereas the 12-pulse waveform, v_d, has the 12th harmonic as its first nonzero component. Show why the 6th harmonic is missing from v_d.

9.6 Calculate the power factor of the three-phase bridge of Fig. 9.8(a) and compare it with that of a single-phase bridge.

9.7 Commutating inductance is added to the phases of the three-phase converter of Fig. 9.10.

(a) Sketch v_d for $\alpha = 30°$ and $u = 15°$. What is $\langle v_d \rangle$?

(b) Repeat (a) for $\alpha = 100°$ and $u = 20°$. In this case (inversion), is $|\langle v_d \rangle|$ less than or greater than $|\langle v_d \rangle|$ for $u = 0$, that is, for a converter with no commutating reactance?

9.8 An important advantage of high-pulse-number rectifiers is that as the pulse number goes up so does the order of the first nonzero harmonic in the line current. The higher frequencies are easier to filter from the line current. What are the first two nonzero harmonics in the line current of the six-pulse rectifier of Fig. 9.8(a)?

9.9 For a balanced three-phase set of Y-connected voltage sources (equal frequencies, amplitudes, and phase displacements), derive the equivalent Δ-connected set.

9.10 An alternative to the 12-pulse circuit of Fig. 9.11(a) is shown in Fig. 9.35, where the two 6-pulse bridges are connected in parallel instead of series.

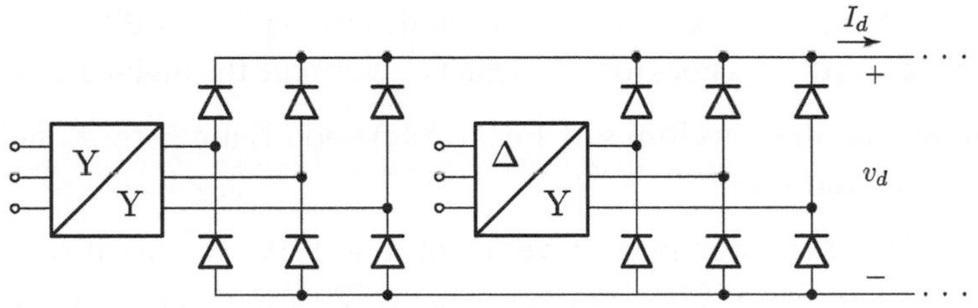

Figure 9.35 A 12-pulse rectifier consisting of two 6-pulse circuits connected in parallel.

(a) Sketch v_d for this circuit and calculate $\langle v_d \rangle$.

(b) Compare $\langle v_d \rangle$ for this circuit with $\langle v_d \rangle$ for the 12-pulse circuit of Example 9.1.

(c) What is the rms current in the diodes in this circuit, and how does it compare to that for the diodes in Fig. 9.11(a)? (The use of an *interphase transformer* to reduce the current ratings of the diodes in Fig. 9.35 is discussed in Chapter 18, Problem 18.6.)

9.11 Plot v_{ab} and i_a for the six-pulse rectifier with a Δ-connected source shown in Fig. 9.8(b), and compare it to the plot in Fig. 9.8(a). What are the power factors of these two circuits? Remember that the circuits are functionally indistinguishable.

9.12 Determine the source current i_{ab} in Fig. 9.8(b) and plot it relative to v_{ab}. What is the power factor at this source? (Assume there are no circulating current components in the Δ. That is, there are no current components present in the legs of the Δ that are not also present in the line currents i_a, i_b, and i_c.)

9.13 How is the operation of the six-pulse circuit of Fig. 9.8(a) affected if two large capacitors of equal value are connected as shown in Fig. 9.36, with their center point connected to the neutral of the Y-connected source? What is V_d?

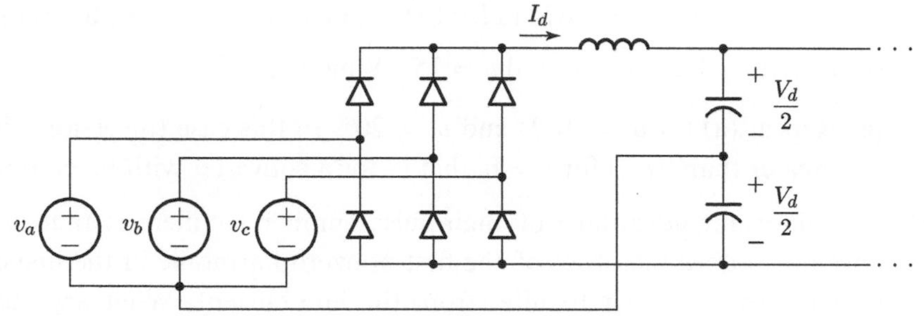

Figure 9.36 A six-pulse rectifier with dc center-point connection.

9.14 The mode III region of the regulation curve for a six-pulse rectifier is presented in Fig. 9.16 without a supporting derivation. Show that the termination of this region, the point where $\langle v_d \rangle = 0$, occurs at $X_c I_d / V_s = 1/\sqrt{3}$. (The solution to this problem is not complicated. Assume four diodes are on all the time, so the line currents are sinusoidal and I_d circulates through a pair of series diodes. Then determine the condition under which the current in one of the diodes goes to zero. That is, determine the high-current boundary of mode III.)

9.15 Show that the waveforms of Fig. 9.20(c) and (d) are correct, and fill in Table 9.1 for v_{an}, v_{bn}, v_{cn}, and v_{nr}.

9.16 Consider the three-phase inverter of Fig. 9.19. A typical dc-bus voltage for a circuit constructed with 1200 V devices is $V_{dc} = 720$ V (60% of the nominal rated device voltage). For this bus voltage and the six-step switching mode shown in Fig. 9.20, what is the fundamental amplitude of the line-to-neutral voltage? What is the fundamental amplitude of the line-to-line voltage? Can a larger three-phase set of line-to-line voltages be created with this inverter?

9.17 Determine the line–neutral waveforms implicit in Fig. 9.31 when the inverter drives a Y-connected load such as shown in Fig. 9.17(b).

9.18 Section 9.6.3 considers PWM modulation of the three-phase inverter of Fig. 9.19 at variable modulation index m. Modulation using the duty ratio scheme of (9.16) results in distortion in the local average voltage patterns for $m > 1$. However, a slight increase in modulation index and resulting output voltage can be achieved without distorting the line-to-line voltages using the following duty ratio formulation:

$$\begin{aligned} d_1 = 1 - d_4 &= \frac{1}{2} + \frac{m}{2}\sin(\omega t) + \frac{m}{12}\sin(3\omega t), \\ d_3 = 1 - d_6 &= \frac{1}{2} + \frac{m}{2}\sin\left(\omega t - \frac{2\pi}{3}\right) + \frac{m}{12}\sin(3\omega t), \\ d_5 = 1 - d_2 &= \frac{1}{2} + \frac{m}{2}\sin\left(\omega t + \frac{2\pi}{3}\right) + \frac{m}{12}\sin(3\omega t). \end{aligned} \tag{9.35}$$

(a) Show that the revised duty ratio selection in (9.35) yields the same local-average line-to-line output voltages as in (9.18).

(b) We can find an allowable limit on the modulation index m to avoid saturating duty ratios in (9.35) using d_1 only, as the other duty ratio expressions are time-shifted and/or complementary versions of d_1. First, show that the locations of extremae of d_1 across ωt must satisfy the relationship $2\cos(\omega t) + \cos(3\omega t) = 0$. From this relationship, d_1 has extremae at $\omega t = 60°, 90°, 120°, 240°, 270°$, and $300°$.

(c) Show that a value $m = 2/\sqrt{3}$ results in d_1 having a maximum of 1 at $\omega t = 60°$ and $120°$, and a minimum of 0 at $240°$ and $300°$. So the duty ratio scheme of (9.35) can be used without distorting the line-to-line output voltage for values of m up to $2/\sqrt{3}$.

9.19 A three-phase bridge converter, like that in Fig 9.19, is used to drive a three-phase Y-connected motor. It has a dc-bus voltage $V_{dc} = 600$ V and uses sine–triangle PWM at a modulation index of $m = 1$. The motor draws real (average) power $P = 75$ kW at a power factor of $k_p = 0.9$, and provides sufficient filtering for its currents to be treated as sinusoidal. What is the rms current drawn by each phase of the motor?

9.20 A balanced Y-connected load is supplied by the inverter of Fig. 9.19. The switches are controlled so that their conduction angles are all 120°. The three phase-to-reference voltages, v_{ar}, v_{br}, v_{cr}, are plotted in Fig. 9.37.

(a) The three phase-to-reference voltages form a three-phase set. Plot their α, β, and γ components, $v_{lr,\alpha}$, $v_{lr,\beta}$, $v_{lr,\gamma}$, over the cycle.

(b) Plot the three phase-to-neutral voltages, v_{an}, v_{bn}, v_{cn}. (*Hint:* Find and plot the neutral-to-reference voltage v_{nr}, which can be found by voltage division at the load.)

(c) The three phase-to-neutral voltages also represent a three-phase set. Plot the α, β, and γ components of the phase-to-neutral voltages, $v_{ln,\alpha}$, $v_{ln,\beta}$, $v_{ln,\gamma}$, over the cycle.

9.21 As explored in Section 9.6.3 and Problem 9.18, the three-phase bridge of Fig. 9.19 can synthesize sinusoidal local-average line-to-neutral voltages with amplitudes up to $V_{dc}/\sqrt{3}$ without low-frequency distortion using a technique known as third-harmonic injection. Show that this same output voltage range can be realized directly with space-vector modulation.

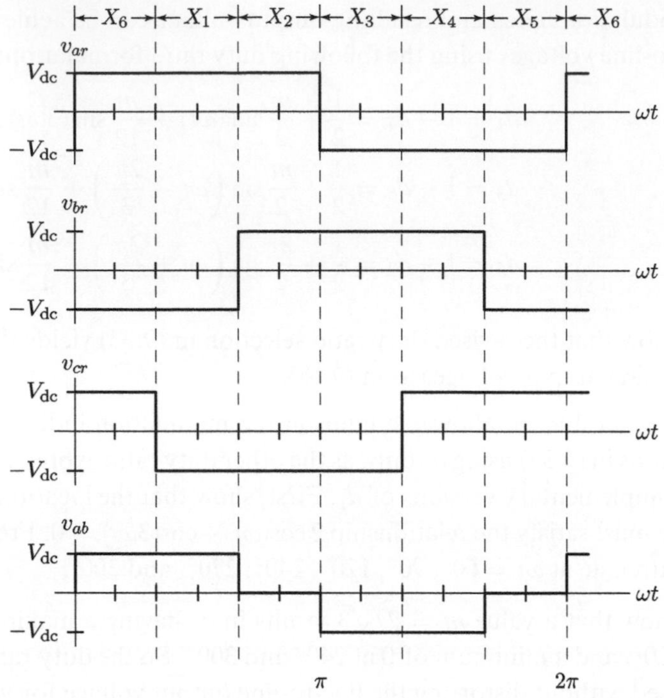

Figure 9.37 Phase-to-reference voltages of the three-phase bridge of Fig. 9.19 operating in six-step switching mode.

10 Resonant Converters

There are many power electronics applications in which one converts power between dc and high-frequency sinusoidal ac. In applications such as induction heating, plasma generation, fluorescent lamp ballasts, electro-surgical tools, radio-frequency (RF) welding, and RF communications, the ac waveform is the converter output. In other applications, such as wireless power transfer and resonant dc/dc converters, a high-frequency sinusoid is an intermediate waveform in the overall conversion process.

Processing of energy using high-frequency sinusoids permits effective use of frequency-selective resonant networks (e.g., *LC* resonant tanks) and components (e.g., piezoelectric resonators and transformers). Power converters designed for such applications are thus referred to as resonant converters. In this chapter we consider topologies, designs, and control methods that are suitable for such applications.

Resonant converters, like the line-frequency phase-controlled converters of Chapter 4, connect a dc system to an ac system. They also share the feature that the switching frequency typically equals the fundamental ac frequency. Unlike line-frequency phase-controlled converters, however, resonant converters usually operate at extremely high frequencies, such that device switching loss is a major concern. Moreover, they often control energy flow by varying their switching frequency, thereby taking advantage of the frequency selectivity of resonant networks. Other common control techniques include phase-shift (or "outphasing") control and on–off (or "burst") control.

Although there are many kinds of resonant inverters and converters, they typically share a number of common features. First, the switches in a resonant inverter are used to create an ac waveform (often a square wave) from a dc source. A resonant tank (e.g., comprising inductors and capacitors) is used to remove unwanted harmonic components from this waveform and also to provide voltage or current scaling. As the difference in frequency between the fundamental component and the lowest harmonic (e.g., the third harmonic for a square wave) is small, a resonant circuit tuned near the switching frequency is used to remove harmonics from the fundamental, rather than using a simple low-pass filter. Hence the name *resonant converter* for this class of converters. The tuned filter can be very selective, providing a sharp gain-versus-frequency characteristic. This selectivity also provides one of the means by which we can control output voltage and power.

Second, a resonant converter's semiconductor devices can have significantly lower per-cycle switching losses than those in a conventional PWM dc/dc or dc/ac converter. The energy lost when a device Q switches on or off is

$$E_{\mathrm{loss}} = \int_0^{t_{\mathrm{on,off}}} v_Q i_Q \, dt, \tag{10.1}$$

where $t_{\mathrm{on,off}}$ is the time it takes the semiconductor device to turn on or off, that is, the rise or fall time of its current and/or voltage. We can design a resonant converter to have one of the switch variables remain near zero during this time, resulting in low switching losses. This technique – known as *soft switching* – is explored in detail in Chapter 25. While not all resonant converter designs have this feature, those that do can operate at higher switching frequencies than otherwise possible, especially when *zero-voltage switching* is employed. This advantage is often the basis for making a decision to use a resonant converter topology. Unfortunately, in return for the lower switching losses, the passive components and semiconductor devices are often subjected to higher currents and/or voltages than in a non-resonant topology operating at the same power level.

In this chapter we provide an introduction to resonant converters. We begin by reviewing the time- and frequency-domain characteristics of second-order resonant circuits, and then explore how these characteristics can be leveraged in the design of inverters and dc/dc converters. Rather than focusing on specific topologies, we develop analysis and design approaches that are common to many such converters. Included in this development are approaches for "radio-frequency" switched-mode converters that are suitable for extremely high-frequency operation (e.g., tens of MHz and above).

10.1 Review of Second-Order System Behavior

In each of its topological states, a typical resonant converter comprises a second-order system, though some designs have third- or higher-order behavior. Even in such higher-order cases one can often approximate the behavior as that of a second-order system. It is thus useful to briefly review the behavior of a second-order system in both the time and frequency domains.

10.1.1 Time-Domain Response

We analyze the switched *RLC* network shown in Fig. 10.1 to determine $v_C(t)$, assuming initial conditions $i_a(0) = 0$ and $v_C(0) = V_{Co}$.

Applying KVL around the circuit for $t > 0$, we can write the differential equation for v_C as follows, dropping the time argument t for notational simplicity:

$$\frac{d^2 v_C}{dt^2} + \frac{R}{L}\frac{dv_C}{dt} + \frac{1}{LC} v_C = \frac{V_{\mathrm{dc}}}{LC}. \tag{10.2}$$

Defining the *damping factor* $\alpha = R/2L$ and *undamped natural frequency* $\omega_o = 1/\sqrt{LC}$, we can rewrite (10.2) as

$$\frac{d^2 v_C}{dt^2} + 2\alpha \frac{dv_C}{dt} + \omega_o^2 v_C = \omega_o^2 V_{\mathrm{dc}}. \tag{10.3}$$

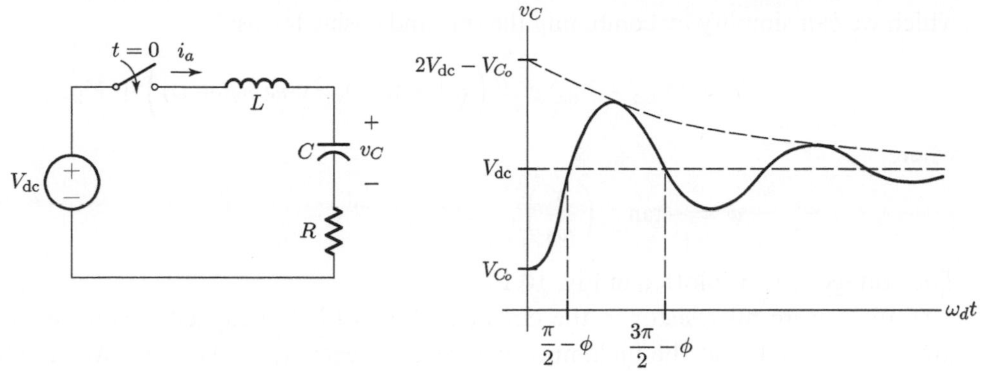

Figure 10.1 A switched *RLC* network and its response illustrated for the underdamped case in which the natural frequencies are complex.

This equation has natural frequencies, s_1 and s_2, at

$$s_{1,2} = -\alpha \pm \sqrt{\alpha^2 - \omega_o^2}. \tag{10.4}$$

If $\omega_o > \alpha$ then $s_{1,2}$ are complex, and we can write them in terms of the *damped resonant frequency* of the circuit ω_d:

$$s_{1,2} = -\alpha \pm j\omega_d, \tag{10.5}$$

$$\omega_d = \sqrt{\omega_o^2 - \alpha^2}. \tag{10.6}$$

This case, which is of primary interest in resonant converters, is known as the underdamped case. We can now write the general solution to (10.3) for the underdamped case as

$$v_C(t) = e^{-\alpha t}(A \sin \omega_d t + B \cos \omega_d t) + V_{dc}, \tag{10.7}$$

where V_{dc} is a particular solution, and is the value of $v_C(t)$ after a long time. The initial conditions on $v_C(t)$ are

$$v_C(0) = V_{C_o}, \tag{10.8}$$

$$\left.\frac{dv_C(t)}{dt}\right|_{t=0} = 0. \tag{10.9}$$

Applying these conditions to (10.7), we can determine A and B:

$$A = \frac{\alpha(V_{Co} - V_{dc})}{\omega_d}, \qquad B = V_{Co} - V_{dc}. \tag{10.10}$$

The specific solution to our problem then is

$$v_C = (V_{Co} - V_{dc})\, e^{-\alpha t}\left(\frac{\alpha}{\omega_d} \sin \omega_d t + \cos \omega_d t\right) + V_{dc}, \tag{10.11}$$

which we can simplify by combining the sine and cosine terms:

$$v_C = (V_{Co} - V_{\mathrm{dc}})\, e^{-\alpha t} \left(\sqrt{1 + (\alpha/\omega_d)^2} \cos(\omega_d t + \phi)\right) + V_{\mathrm{dc}}, \tag{10.12}$$

where

$$\phi = -\tan^{-1}\left(\frac{\alpha}{\omega_d}\right). \tag{10.13}$$

The voltage $v_C(t)$ is plotted in Fig. 10.1.

If there were no resistor in the circuit of Fig. 10.1, the capacitor voltage would oscillate forever around V_{dc} at the undamped natural frequency $\omega_o = 1/\sqrt{LC}$. We are always careful to distinguish between undamped and damped natural frequencies by using the symbols ω_o and ω_d, respectively.

10.1.2 Frequency-Domain Response

The admittance of the series RLC circuit of Fig. 10.2(a), which is driven by a sinusoidal source of amplitude V_a, is

$$Y(s) = \frac{1}{sL + 1/sC + R} = \frac{sC}{s^2LC + sRC + 1}. \tag{10.14}$$

The magnitude of $Y(j\omega)$ is graphed on a log–log plot as a function of frequency ω in Fig. 10.2(b). We can rewrite (10.14) in terms of α and ω_o as

$$Y(s) = \left(\frac{1}{R}\right) \frac{2\alpha s}{s^2 + 2\alpha s + \omega_o^2}. \tag{10.15}$$

At resonance, $s = j\omega_o$, the impedances of the inductor and capacitor cancel, the admittance takes its maximum magnitude, and is a pure conductance $1/R$.

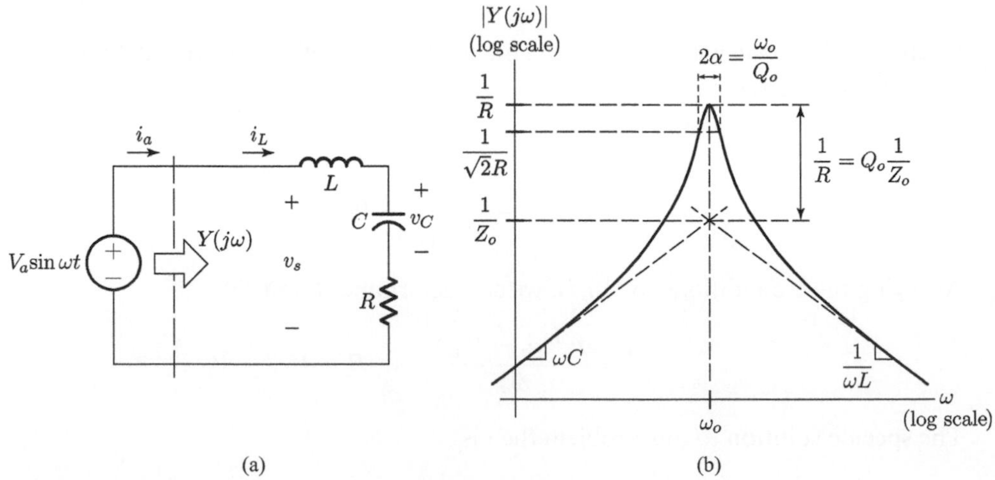

Figure 10.2 (a) A series resonant circuit. (b) The magnitude of its admittance $Y(j\omega)$.

Another way the admittance can be expressed is in terms of the *quality factor* Q_o of the network, where

$$Q_o = \frac{\omega_o}{2\alpha}, \tag{10.16}$$

or, expressed in terms of the components for a series-resonant circuit,

$$Q_o = \frac{\omega_o L}{R} = \frac{1}{\omega_o CR}. \tag{10.17}$$

The higher the value of Q_o, the less damped the network is and the sharper the $|Y(j\omega)|$ plot becomes. In terms of Q_o, (10.15) becomes

$$Y(s) = \left(\frac{1}{R}\right) \frac{(1/Q_o)(s/\omega_o)}{(s/\omega_o)^2 + (1/Q_o)(s/\omega_o) + 1}. \tag{10.18}$$

There is a direct relationship between Q_o and both the amount and bandwidth of resonant peaking in $|Y(j\omega)|$. First, consider the amount of peaking that occurs. At the resonant frequency ω_o, the magnitudes of the impedances presented by the capacitor and inductor are equal, and define the *characteristic impedance* of the circuit Z_o:

$$Z_o = \omega_o L = \frac{1}{\omega_o C} = \sqrt{\frac{L}{C}}. \tag{10.19}$$

The capacitive and inductive (low- and high-frequency) asymptotes of the admittance curve intersect at the resonant frequency ω_0 and an admittance value $Y_o = \frac{1}{Z_o}$. Considering (10.17) and (10.19), we can see that the ratio of the admittance at resonance to that of the intersection of the asymptotes is exactly

$$\frac{1/R}{1/Z_o} = \frac{Z_o}{R} = Q_o. \tag{10.20}$$

Thus, the smaller R is relative to Z_o, the higher the Q_o and the more the admittance peaks at resonance (by a factor Q_o). Conversely, as R becomes larger than Z_o, $Q_o < 1$ and there is no peaking in the admittance curve, with $Q_o < \frac{1}{2}$ representing an overdamped system.

Consider the bandwidth over which significant resonant peaking occurs in Fig. 10.2(b). In particular, consider the frequency range over which the power delivered to R is at least half that delivered when being driven at ω_o. This *half-power bandwidth* is the frequency range over which the magnitude of the admittance is at least a factor of $1/\sqrt{2}$ (or -3 dB) times its magnitude at ω_o. The -3 dB points where the admittance magnitude is exactly $1/\sqrt{2}$ times its peak value are located at

$$\omega_{1,2} = \sqrt{\frac{1}{LC} + \left(\frac{R}{2L}\right)^2} \mp \frac{R}{2L} = \sqrt{\omega_o^2 + \alpha^2} \mp \alpha. \tag{10.21}$$

Note that the half-power frequency range is *geometrically* centered at ω_o, but not linearly centered there, though the difference is small at high Q_o. From (10.21) we can find the half-power bandwidth as

$$\omega_2 - \omega_1 = \frac{R}{L} = 2\alpha. \tag{10.22}$$

The damping coefficient α thus determines the half-power bandwidth. Normalized to the geometric center frequency ω_o, this becomes

$$\frac{\omega_2 - \omega_1}{\omega_o} = \frac{2\alpha}{\omega_o} = \frac{1}{Q_o}.$$

Thus, the *fractional* half-power bandwidth is $1/Q_o$. Consequently, at higher Q_o we get both more peaking and higher frequency selectivity of the admittance curve.

A further observation about the behavior of resonant circuits is the effect of Q_o on the magnitude of one or the other of the capacitor voltage or inductor current. At resonance, the voltage across the resistor in Fig. 10.2 is equal to the source voltage. The capacitor voltage, however, can be substantially larger than the source voltage, since

$$\frac{V_C}{V_a} = \left|\frac{\omega_o^2}{s^2 + 2\alpha s + \omega_o^2}\right|_{s=j\omega_o} = \frac{\omega_o}{2\alpha} = Q_o, \tag{10.23}$$

where V_C is the amplitude of the sinusoidal voltage v_C. This result means that if $Q_o = 10$ and the source has an amplitude of 100 V, the capacitor would see a peak voltage of Q_o times the source voltage at resonance, or 1000 V! Such an excessive voltage (or current, in the case of a parallel resonant circuit) can be a significant concern when using resonant converters, though it can sometimes be put to good use.

It is also worth noting that there is a direct relationship between the amplitudes V_C and I_L respectively of capacitor voltage and inductor current when driven at resonance (or in the natural response of an undamped resonant circuit). Considering (10.19), it is easily seen that in the circuit of Fig. 10.2,

$$\left.\frac{V_C}{I_L}\right|_{\omega_o} = \frac{1}{\omega_o C} = Z_o. \tag{10.24}$$

The characteristic impedance Z_o thus relates the amplitude of the capacitor voltage to that of the inductor current at resonance. Z_o can be helpfully thought of as a relative measure of voltage and current amplitudes in a second-order circuit at resonance.

While the parameters L, C, and R characterize the behavior of a given second-order resonant circuit, it is often useful to translate these figures into more intuitive measures: ω_o, Z_o, and Q_o. The specific conversions vary among circuits (for example, in a parallel resonant circuit $Q_o = R/Z_o$), but the general interpretations of ω_o, Z_o, and Q_o carry over.

10.2 Quality Factor

The concept of quality factor Q is broader than suggested by the previous section. Indeed, one can define quality factor for any physical system under sinusoidal ac excitation. The fundamental definition of quality factor Q for such a system is

$$Q = 2\pi \frac{\text{Peak energy stored}}{\text{Energy lost per cycle}}. \tag{10.25}$$

Quality factor is thus a measure of how much peak ac energy one stores in a system relative to how much is being lost each cycle under sinusoidal ac excitation. The energies stored and lost are only those owing to the ac excitation. Energy lost may represent energy dissipated as heat (e.g., owing to resistance in a circuit), but may also be defined to include energy delivered into a load (leading to the concept of a "loaded quality factor"). Given this definition, we can expect the quality factor of a system to be a function of the excitation frequency. Q is invariant to drive amplitude in a linear system, but can become a function of drive amplitude in a nonlinear system.

Example 10.1 Quality Factor of an LR Circuit

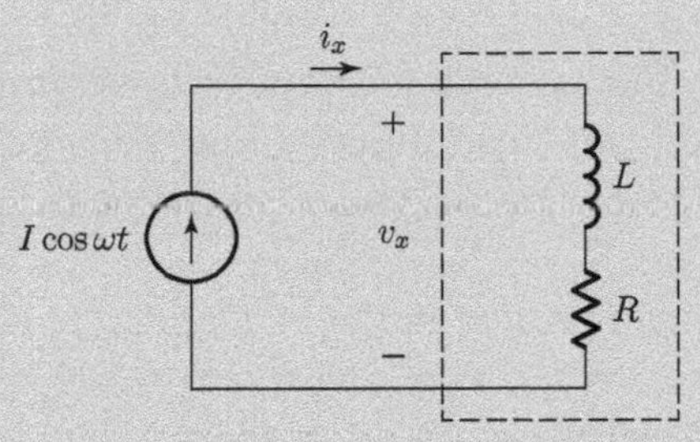

Figure 10.3 A series LR circuit driven by a sinusoidal current source.

Consider a series LR circuit driven by a sinusoidal current source as shown in Fig. 10.3. The quality factor of the LR network in the dashed box may be calculated as follows.

The peak energy stored is

$$E_{\text{stored}} = \frac{1}{2} L I^2, \tag{10.26}$$

while the energy dissipated over a period $T = 2\pi/\omega$ is

$$E_{\text{diss}} = \frac{1}{2} I^2 R T = \frac{\pi I^2 R}{\omega}, \tag{10.27}$$

which, when applied in (10.25), yields

$$Q = \frac{\omega L}{R}. \tag{10.28}$$

The Q of a series LR circuit is thus proportional to the drive frequency and the inductance, and inversely proportional to the resistance. That is, the ratio of energy stored to that lost over an ac cycle increases with higher frequency and inductance, and decreases with higher resistance.

Because practical inductors are lossy, they are often modeled as an ideal inductor in series with an *equivalent series resistance*, or ESR, similar to the network in the dashed box in Fig. 10.3. Datasheets for inductors often indicate inductor Q for some specified set of operating conditions. Equation (10.28) can be used to determine the ESR of a practical inductor from an indicated quality factor, where R is the ESR. (In fact, the concept of Q was originally developed in the early 1900s as a figure of merit for practical inductors – or "coils" in the parlance of the time – with the ratio of reactance to resistance of a coil defining Q, just as in (10.28).) It may also be appreciated that – at least for frequencies where inductance and ESR remain constant – one can expect the achievable Q of an inductor to increase with frequency (so one gets better performance at higher frequencies).

In discussing second-order circuits in Section 10.1.2, we referred to the quality factor Q_o "of the circuit." While it is often not stated, what is meant by this is the Q of the circuit *when driven at the resonant frequency* ω_o. The expression for Q_o given in (10.16) thus really means $Q_o = Q|_{\omega=\omega_o}$. At any other frequency, the quality factor at which the circuit operates – as defined by (10.25) – will actually be different.

10.3 Resonant Converter Analysis

A basic resonant inverter consists of a switching network (comprising switches and possibly passive components) interposed between a dc source and a resonant filter network (e.g., an LC "tank"), and connected to a load (e.g., a resistor). The switching network synthesizes an ac waveform from the dc source, which is filtered by the resonant network to extract a single frequency (usually the fundamental) from the ac waveform; the resulting sinusoidal waveform is delivered to the load. This basic structure may be augmented with additional elements for voltage transformation, isolation, and interconnection (e.g., transformers, matching networks, transmission lines, etc.)

This basic structure can be seen in the half-bridge series-resonant inverter of Fig. 10.4. A dc source voltage V_{dc} is split by two large bypass/blocking capacitors C_{big}. These capacitors are sufficiently large that they each hold an approximately constant voltage $V_{dc}/2$ and do not participate significantly in circuit resonance. The blocking capacitors and half-bridge pair of switches S_1 and S_2 form a switching network that can create a square-wave voltage v_a across the LC series-resonant filter network and load resistor R to drive a sinusoidal current i_a through R.

Figure 10.5(a) shows a model for the inverter of Fig. 10.4 in which the dc source and switching network are replaced by an equivalent square-wave voltage source v_a. We could use this equivalent source model to calculate the current i_a and power transfer to R in detail. However, to simplify analysis we can make an approximation known as the *fundamental harmonic approximation* (FHA) in which we assume that the resonant tank provides sufficient filtering that the current i_a is purely sinusoidal. The FHA is often made to simplify the first-pass analysis of resonant converters, with the quantity that is assumed sinusoidal depending upon the

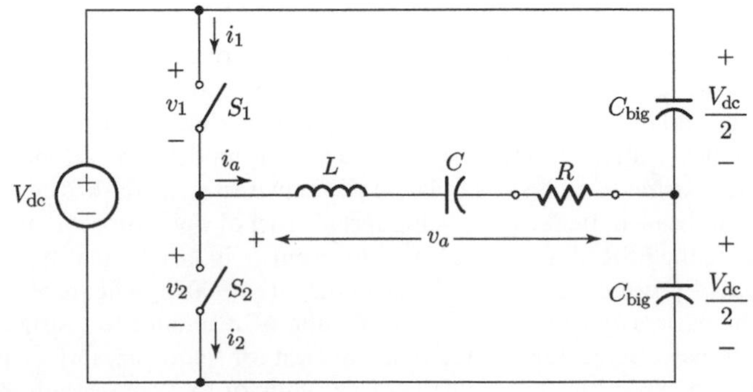

Figure 10.4 A half-bridge series-resonant inverter.

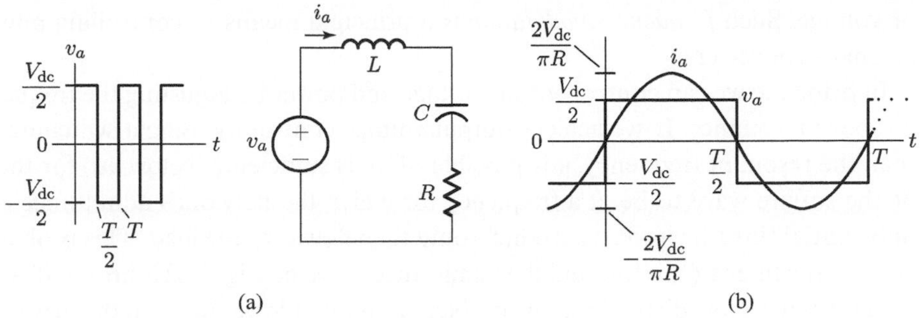

Figure 10.5 (a) A model for the series-resonant inverter of Fig. 10.4 with the dc source and switching network replaced by the equivalent source v_a. (b) The load current i_a drawn for $\omega_s = 2\pi/T = \omega_o$.

type of resonant network, and the accuracy of the results depending upon the quality factor Q_o of the network and how close the switching frequency is to resonance. With this approximation, we only need to use the fundamental of v_a to calculate i_a. Likewise, under the FHA we only need the fundamental of v_a to analyze power delivery, since one can only transfer real power with voltages and currents having the same frequency. Analyzing the inverter of Fig. 10.4 under the FHA, we can replace the square-wave source $v_a(t)$ in Fig. 10.5(a) with its fundamental component $v_{a_1}(t) = (2V_{dc}/\pi)\sin(\omega_s t)$. This simplified model matches the second-order system of Fig. 10.2(a).

In keeping with the FHA, the series resonant LC filter of Fig. 10.5(a) placed in series with the load R makes the load current i_a nearly sinusoidal. At the resonant frequency ω_o, the impedance of the inductor and the capacitor exactly cancel, and the admittance of the RLC network equals $1/R$. Therefore, when the resonant converter is switched at $\omega_s = \omega_o$, the full fundamental component of the square wave, which has amplitude $V_{a_1} = 2V_{dc}/\pi$, appears across the load resistor. If the Q_o of the filter is high, the harmonic content of i_a is low, and i_a is nearly sinusoidal. Thus:

$$i_a \approx |Y(j\omega_o)|\, V_{a_1} \sin\omega_o t = \left(\frac{1}{R}\right)\left(\frac{2V_{dc}}{\pi}\right)\sin\omega_o t. \tag{10.29}$$

This current, whose amplitude is inversely proportional to R, and its relationship to v_a are shown in Fig. 10.5(b). Note that the fundamental of v_a and of i_a are in phase.

10.3.1 Control of the Output Waveform

It is not necessary that $\omega_s = \omega_o$ for the LC network to adequately filter harmonics from the current i_a. Focusing on the fundamental frequency, if ω_s is slightly higher or lower than ω_o, the LC filter looks like a small inductor or a large capacitor, respectively. The additional impedance it provides reduces the magnitude of voltage across the load resistor. If the filter Q_o is high, we can achieve a large change in the output power by shifting ω_s only a small amount. This shifting of the drive frequency, therefore, becomes a means by which we can control the output power or voltage. Such *frequency modulation* is a principal means of controlling power flow in many resonant converters.

In principle we can control output voltage and power by adjusting the frequency either below or above resonance. If we achieve output voltage control by using a switching frequency lower than the resonant frequency, it is possible (if ω_s is sufficiently below ω_o) for the third harmonic of the square wave to be at a frequency for which the network's admittance is relatively high. Substantial third-harmonic currents would then flow into the load. This problem is particularly acute if the filter's Q_o is low and the admittance curve of Fig. 10.2 is broad. If we want an output waveform with low distortion, this problem imposes a lower limit on the drive frequency and/or the Q_o of the filter.

However, if the switching frequency is higher than ω_o, the harmonic components of the square wave are always filtered to some extent because $|Y(n\omega_s)| < |Y(\omega_s)|$ for $\omega_s > \omega_o$. For this reason, among others, above-resonance operation is often preferred for series-resonant inverters. This mode of control also has its limitations, though, because device switching losses become worse at higher switching frequencies. Consequently, some additional form of control (e.g., burst-mode control) is often used to achieve extremely low output power levels.

Example 10.2 Load Current Determination When Switching Off-Resonance

If the resonant converter of Fig. 10.4 is driven above resonance by a factor $\omega_s/\omega_o = \beta > 1$, then at ω_s the combined impedance of the inductor and capacitor is

$$Z(j\omega_s) = \frac{(j\omega_s)^2LC+1}{j\omega_s C} = \frac{-(\beta^2-1)}{j\omega_s C} = j\omega_s L\left(\frac{\beta^2-1}{\beta^2}\right) = j\omega_s L_e, \tag{10.30}$$

where

$$L_e = L\left(1-\frac{1}{\beta^2}\right). \tag{10.31}$$

The LC network is an equivalent inductor of value L_e at the fundamental component of v_a. Therefore, at the fundamental, v_a appears across an equivalent series LR network. Figure 10.6(a) shows the resulting equivalent circuit, which we can use to determine the fundamental component i_{a_1} of i_a. We assume that the higher harmonics are filtered well enough that $i_a \approx i_{a_1}$. Under this assumption, the current i_a has amplitude I_{a_1} and lags v_a by an angle θ. That is,

$$i_a \approx I_{a_1}\sin(\omega_s t-\theta), \tag{10.32}$$

$$I_{a_1} = \frac{2V_{\text{dc}}}{\pi R}\left(\frac{1}{\sqrt{\omega_s^2L_e^2/R^2+1}}\right), \tag{10.33}$$

$$\theta = \tan^{-1}\frac{\omega_s L_e}{R}. \tag{10.34}$$

Note that the amplitude of i_a is less than its amplitude when $\omega_s = \omega_o$, as given by (10.29).

When the converter is driven below resonance by a factor of $\omega_s/\omega_o = \gamma < 1$, the LC circuit looks like an equivalent capacitor of value C_e. Thus,

$$Z(j\omega_s) = \frac{(j\omega_s)^2LC+1}{j\omega_s C} = \frac{(1-\gamma^2)}{j\omega_s C} = \frac{1}{j\omega_s C_e} \tag{10.35}$$

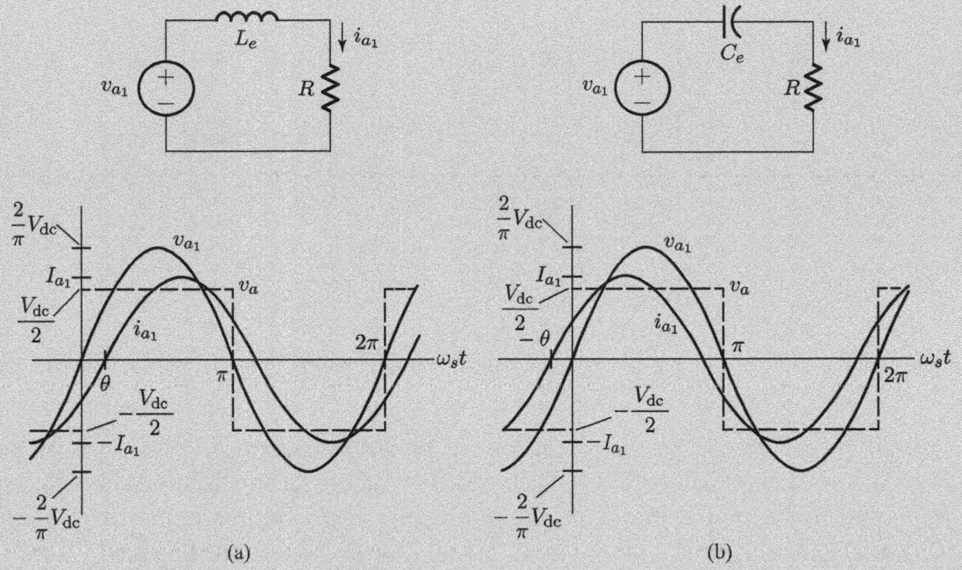

Figure 10.6 Behavior of the circuit of Fig. 10.4 for $\omega_s \neq \omega_o$. (a) Equivalent circuit and waveforms for $\omega_s > \omega_o$. (b) Equivalent circuit and waveforms for $\omega_s < \omega_o$.

and

$$C_e = \frac{C}{1-\gamma^2}. \tag{10.36}$$

The fundamental of v_a therefore sees a series RC load. If ω_s is not too much lower than ω_o, the harmonic content of i_a is low, and again we assume that $i_a \approx i_{a_1}$. Thus:

$$i_a \approx I_{a_1}\sin(\omega_s t + \theta), \tag{10.37}$$

$$I_{a_1} = \frac{2V_{dc}}{\pi R}\left(\frac{1}{\sqrt{1/(\omega_s C_e R)^2 + 1}}\right), \tag{10.38}$$

$$\theta = \tan^{-1}\frac{1}{\omega_s R C_e}. \tag{10.39}$$

The equivalent circuit and waveforms for $\omega_s < \omega_o$ are shown in Fig. 10.6(b).

10.3.2 Series-Resonant Inverter: Design Considerations

The switches in the series-resonant inverter of Fig. 10.4 must carry current in both directions if, to obtain control, ω_s is to deviate from ω_o. At ω_o, the load current i_a is in phase with v_a, so it is positive when S_1 is on and negative when S_2 is on. Therefore both switches carry a current that is always positive. At any other frequency, however, i_a is either leading or lagging v_a, as shown in (10.6), so both switches must carry bilateral current. One common switch implementation is to use power MOSFETs as the switching devices, as shown in Fig. 10.7.

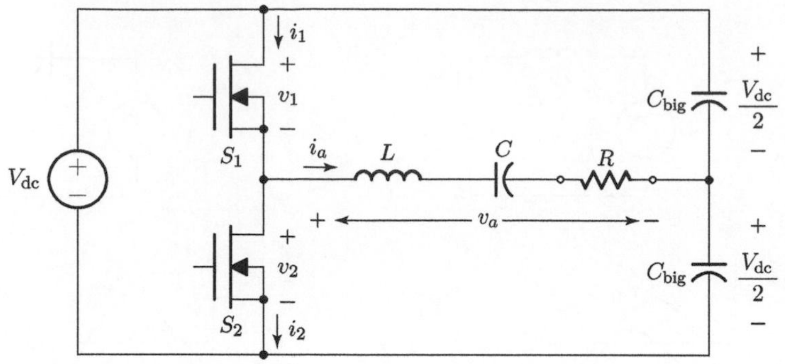

Figure 10.7 A half-bridge series-resonant inverter implemented with power MOSFETs.

To attenuate the harmonic components of the square wave in a series-resonant inverter and to achieve a wide range of output voltage control with only a small change in switching frequency, we would like to make the Q_o of the filter large. However, the higher the Q_o, the higher the peak energy storage requirements of the inductor and capacitor become. To see this, consider the expression for Q_o in (10.20). We increase Q_o at constant R by increasing the characteristic impedance Z_o, which requires increasing L and decreasing C – as per (10.19) – while keeping their product constant (to maintain constant ω_o). For a given peak current I_p at ω_o, the energy storage required of L and C can be expressed as

$$E_L = \frac{1}{2} L I_p^2, \tag{10.40}$$

$$E_C = \frac{1}{2} C \left(\frac{I_p}{\omega_o C} \right)^2 = \frac{1}{2\,\omega_o^2\, C} I_p^2 = E_L. \tag{10.41}$$

Thus, the required energy storage of the filter at ω_o is proportional to L and inversely proportional to C, and thus proportional to Z_o and Q_o. Another way of looking at this relationship is to consider Q_o (Q at ω_o) in terms of the general definition of Q in (10.25): the energy lost per cycle is proportional to the load power, which is specified by the application, so the only way to increase Q is to increase the peak energy stored.

There are two negative consequences of increasing Q_o. First, the larger energy storage requirements of L and C tend to make the filter components larger and more expensive. Moreover, increasing L and decreasing C tends to result in higher equivalent series resistance (parasitic resistance) of the filter components, which will in turn give higher loss. There is thus a fine balance between making Q_o large enough to provide adequate harmonic filtering and narrow-band power control while keeping Q_o small enough to limit the size and loss of the filter components themselves.

10.4 Soft Switching of Resonant Converters

As described in the chapter introduction, a major benefit of using a resonant converter is that one can achieve soft switching – zero-voltage switching (ZVS) and/or zero-current switching

(ZCS) – of the devices. Not all resonant converter implementations provide soft switching, but those that do can achieve especially high efficiency. Chapter 24 contains a general treatment of soft switching, but we illustrate its use in resonant converters here.

The series-resonant inverter implementation of Fig. 10.7 can achieve ZVS turn-on and turn-off of each device when operating above resonance, so long as we provide an appropriate amount of capacitance across the transistors and a suitable "dead time" during which both switches S_1 and S_2 are off. Figure 10.8(a) shows a simplified model of the inverter in which we treat each transistor as comprising an ideal switch in anti-parallel with a diode and in parallel with a capacitance C_{sw}. The capacitance C_{sw} can model the transistor output capacitance C_{oss} and may also include additional discrete "snubber" capacitance to reduce turn-off loss.

As shown in Fig. 10.6, when the switching frequency ω_s is higher than the resonant frequency ω_o, the inverter is inductively loaded so the load current lags v_a. Therefore each switch/diode element starts its conduction period with a negative current and ends with a positive current. Figure 10.8(b) shows the switching sequence that occurs above resonance, including dead times when both transistors are turned off, while Fig. 10.8(c) shows the associated waveforms. During phase 1, S_1 carries positive current, and turns off while its current is still positive. Assuming S_1 can be turned off in a short time, the capacitances C_{sw} hold the voltage across S_1 low during its turn-off transition, providing low-loss "zero-voltage" turn-off. Because i_a is positive, it charges the capacitance across S_1 and discharges the capacitance across S_2 during the dead time when both switches S_1 and S_2 are held off (phase 2). Once the voltage across S_2 reaches zero, the diode D_2 turns on, beginning phase 3. Switch S_2 can then be turned on with zero voltage across it (ZVS turn-on) and starts to conduct forward current when i_a becomes negative, initiating phase 4. S_2 continues to carry forward current ($i_a < 0$) when it turns off to end phase 4, again with ZVS owing to the presence of the capacitances C_{sw}. As i_a is negative, it charges the capacitance across S_2 and discharges the capacitance across S_1 until diode D_1 turns on, beginning phase 6. S_1 can then be turned on with zero voltage, with phase 1 starting once i_a reverses direction to become positive.

Considering the switching sequence shown in Fig. 10.8, a benefit of above-resonance operation is that the presence of the capacitances C_{sw} across the switches enables low turn-off switching loss of each device, while the direction of i_a charges/discharges these capacitances during the dead-time periods such that each device can also be turned on with low loss. ZVS turn-on is particularly valuable for high-frequency operation as there is no loss from discharging the intrinsic device capacitance at turn-on. Of course, the behavior indicated in Fig. 10.8 assumes that (i) the devices can be turned off quickly relative to the capacitor charge/discharge times, and that (ii) the circuit operates sufficiently above resonance, at high enough quality factor, and with enough dead time that zero-voltage turn-on of the devices can be achieved.

Below-resonance operation of the circuit in Fig. 10.7 results in each of the switches being gated off when carrying negative current, such that the device body diodes pick up the current at turn-off. This enables low-loss turn-off of the switches. For below-resonance operation, one can even use SCRs with anti-parallel diodes in place of MOSFETs, because each SCR will naturally turn off when the load current changes direction. However, a limitation of below-resonance operation in a series-resonant inverter is that the transistors do not turn on with soft switching, so can exhibit significant turn-on loss. There are other circuit variants (e.g., the "Mapham" inverter)

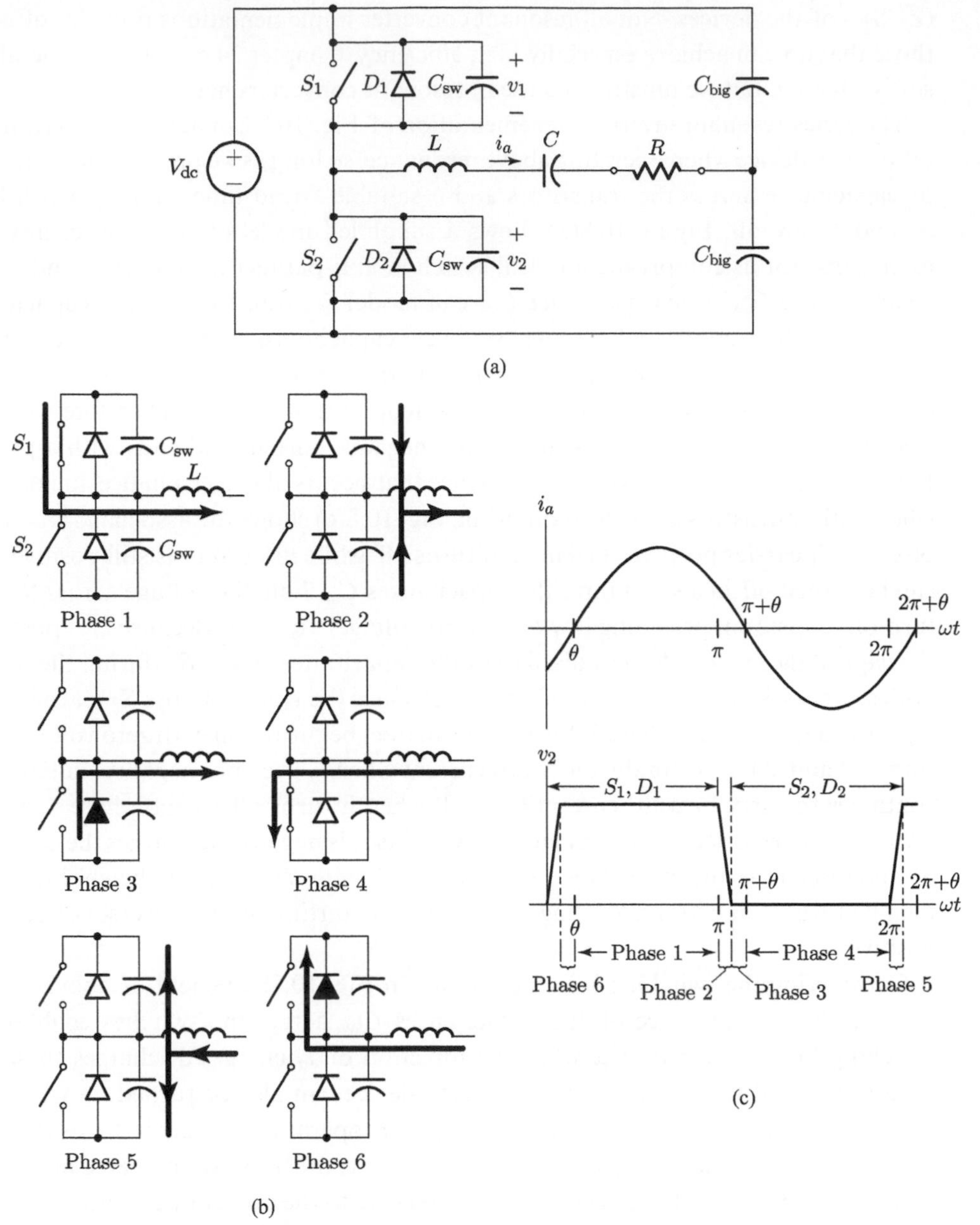

Figure 10.8 (a) A half-bridge series-resonant inverter including switch capacitances C_{sw}. (b) Switching sequence for above-resonance operation, including dead time between the switches. (c) Waveforms for the switching sequence (b).

that can achieve both ZCS turn-off and ZCS turn-on of the switches. However, even in designs providing full ZCS soft switching, any internal device capacitance is discharged with loss at switch turn-on. Consequently, above-resonance operation is strongly preferred for operating at high frequencies.

Example 10.3 Selecting Values of L and C for a Series-Resonant Inverter

Consider the design of the series-resonant inverter of Fig. 10.4 to deliver power with sinusoidal waveforms into a 25 Ω resistive load. We would like to deliver power of $P_1 = 800\,\text{W}$ at a frequency of $f_{s1} = 1.5\,\text{MHz}$, and to be able to reduce the power down to a value of $P_2 = 200\,\text{W}$ without exceeding the frequency $f_{s2} = 1.75\,\text{MHz}$. Additional specifications are that the converter operate from a dc input voltage $V_{\text{dc}} = 400\,\text{V}$, and that it operate with ZVS over this power range.

To accomplish this, we operate above resonance and design the resonant tank such that the current delivered to the resistor can be approximated as sinusoidal (i.e., that the fundamental harmonic approximation holds). For this situation, we can model the system as in Fig. 10.6(a) and apply (10.30)–(10.34). Given a load resistor R and a desired average power P, we require a fundamental peak ac current through the resistor of

$$|I_{a_1}| = \sqrt{\frac{2P}{R}}.$$

From this, we find that at $P_1 = 800\,\text{W}$ we have $|I_{a_1}| = 8\,\text{A}$, while at $P_2 = 200\,\text{W}$ we have $|I_{a_1}| = 4\,\text{A}$. Solving (10.33) for the equivalent inductance L_e we get

$$L_e = \frac{\sqrt{4V_{\text{dc}}^2 - \pi^2 R^2 I_{a_1}^2}}{\pi I_{a_1} \omega_s}.$$

Applying this at our first operating point P_1, f_{s1} we get $L_{e1} = 2.09\,\mu\text{H}$, and at our second operating point P_2, f_{s2} we get $L_{e2} = 5.32\,\mu\text{H}$. With a bit of algebra, we can find an expression for the resonant frequency ω_o that allows us to achieve both equivalent inductance L_{e1} at $\omega_{s1} = 2\pi f_{s1}$ and equivalent inductance L_{e2} at $\omega_{s2} = 2\pi f_{s2}$:

$$\omega_o = \omega_{s1} \sqrt{\frac{(L_{e2}/L_{e1}) - 1}{(L_{e2}/L_{e1}) - (\omega_{s1}/\omega_{s2})^2}}. \tag{10.42}$$

From (10.31) we can get the desired inductance,

$$L = \frac{L_{e1}}{1 - (\omega_o/\omega_{s1})^2}, \tag{10.43}$$

and can calculate the desired capacitance as

$$C = \frac{1}{\omega_o^2 L}. \tag{10.44}$$

Applying our system parameters in (10.42)–(10.44) we get $\omega_o = 8.71 \times 10^6$ rad/s (1.39 MHz), $L = 14.3\,\mu\text{H}$, and $C = 923\,\text{pF}$. Given these values, we can calculate the tank characteristic impedance $Z_o = 124\,\Omega$ and quality factor $Q_o = 4.97$. Given the high tank quality factor and that the converter operates in the region of resonant peaking (as evidenced by the fact that $L_{e2} \ll L$), we can expect that the load waveforms will be nearly sinusoidal.

The inverter operates with inductive loading over the specified operating range, giving the opportunity for ZVS. We can find the load current phase θ with respect to the voltage waveform (i.e., as in Figs. 10.6(a) and 10.8(c)) using (10.34). At $\omega_{s1}, \theta = 38°$, while at $\omega_{s2}, \theta = 67°$. We can only achieve ZVS if the load current is capable of completely charging/discharging the switch capacitances C_{sw} in the interval $0 < \omega t < \theta$, as illustrated in Fig. 10.8. The total charge Q_T that the load current can transfer in this interval is

$$Q_T = \int_0^{\frac{\theta}{\omega_s}} I_{a_1} \sin(\omega t)\,dt = \frac{I_{a_1}}{\omega_s}[1 - \cos(\theta)]. \tag{10.45}$$

To achieve soft switching, we require $Q_T \geq 2C_{\text{sw}} V_{\text{dc}}$, which gives us a maximum allowable switch capacitance C_{sw} for soft switching of

$$C_{\text{sw}} \leq \frac{I_{a_1}}{2\omega_s V_{\text{dc}}}[1 - \cos(\theta)]. \tag{10.46}$$

At ω_{s1} (the limiting case) this gives us $C_{\text{sw}} \leq 228$ pF. As switches are available having the required voltage and current ratings and smaller switch capacitance, the converter can be implemented with soft switching.

10.5 Resonant DC/DC Converters

A resonant dc/dc converter comprises a resonant inverter – including its resonant tank – and a rectifier, and may include additional transformation and/or isolation stages (such as a transformer or matching network) between them. Frequency control is often used to regulate the converter output by taking advantage of the frequency-dependent gain characteristics of the resonant tank.

10.5.1 Rectifier Analysis

Consider how we might model the rectifier in a resonant dc/dc converter. If at least one of the rectifier input voltage or current is sinusoidal (e.g., owing to the filtering of the resonant tank), we can use the fundamental harmonic approximation to model it as an equivalent impedance at the drive frequency ω_s. This is sometimes referred to as a *describing function* model. In many cases the rectifier simply acts as an equivalent resistance loading the inverter and resonant tank, with the resistor value depending upon the operating point.

Figure 10.9(a) shows a full-bridge rectifier driven by a sinusoidal current source, which might represent the current from a resonant inverter. Given that the current driving the rectifier is sinusoidal, average power can only be delivered to the rectifier via the fundamental component $v_{\text{rect},1}$ of the rectifier ac input voltage v_{rect} (i.e., the ac voltage component at the same frequency as the current). Figure 10.9(b) shows the voltage waveform v_{rect} at the rectifier input and its fundamental component $v_{\text{rect},1}$. For an ideal full-bridge rectifier with a sine-wave drive current $I\sin(\omega_s t)$ and negligible output voltage ripple ($v_d \approx V_d$), we can express the fundamental ac-side voltage as $v_{\text{rect},1} = (4/\pi)V_d \sin(\omega_s t)$. Considering the relationship between the rectifier ac-side current and fundamental voltage, we might model the rectifier as a resistor of value

$$R_{\text{ac(eff)}} = \frac{4}{\pi}\frac{V_d}{I}. \tag{10.47}$$

The average power P delivered to the rectifier with these waveforms can be expressed as $P = (2/\pi)V_d I$. Solving this for I and substituting it into (10.47) gives us an expression for the effective ac resistance presented by the rectifier in terms of the rectifier output voltage and power:

$$R_{\text{ac(eff)}} = \frac{8}{\pi^2}\frac{V_d^2}{P}. \tag{10.48}$$

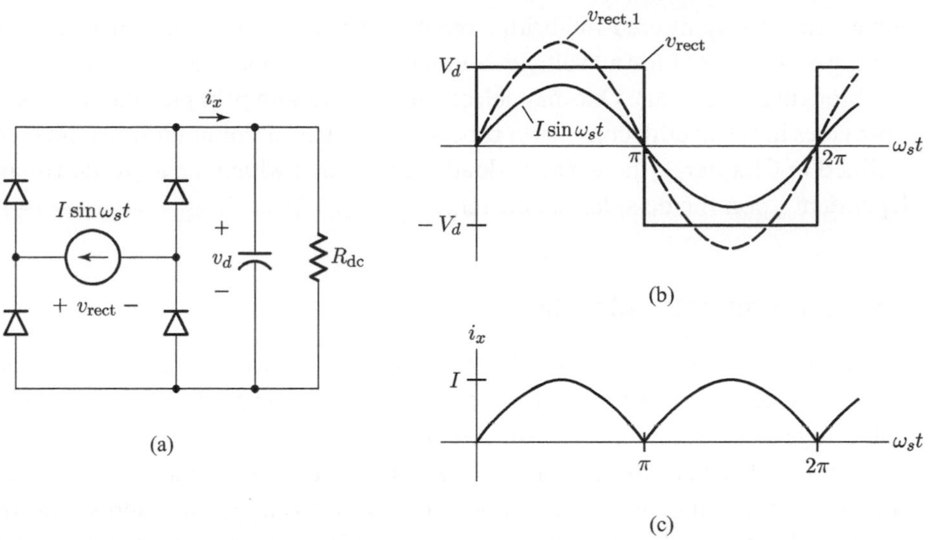

Figure 10.9 (a) A capacitively filtered full-bridge rectifier with a sinusoidal current drive. (b) The ac-side voltage waveform and its fundamental component. (c) The waveform at the filter capacitor and load of the rectified current.

The effective resistance loading the resonant inverter thus depends on both the rectifier dc output voltage and the delivered power. The form of this equivalent resistance, $k(V_d^2/P)$, is common to many types of rectifiers, with the leading constant k depending upon rectifier topology.

For the special case where the rectifier is loaded with a resistor R_{dc} as illustrated in Fig. 10.9(a), we can develop a further expression for the effective resistance in terms of R_{dc}. For the full-bridge rectifier, the rectified current waveform i_x shown in Fig. 10.9(b) has an average value of $\langle i_x \rangle = (2/\pi)I$, which gives $V_d = (2/\pi)IR_{dc}$. Substituting this into (10.47), we get another way to represent the ac equivalent resistance of the full-bridge rectifier:

$$R_{ac(eff)} = \frac{8}{\pi^2} R_{dc}. \tag{10.49}$$

Many other rectifiers provide a similar form (that is, $k' R_{dc}$ for some constant k'). If we include non-idealities such as diode capacitance, we often obtain an effective ac impedance having a reactive component.

Modeling the rectifier in a resonant dc/dc converter with a resistance or impedance as outlined above can be very helpful for describing its operating characteristics. Considering the forms of the equations above, we can expect that the behavior of such a converter will depend upon both the output power and the dc output voltage at which it operates.

Another way to think about the rectifier in a resonant dc/dc converter is in terms of its ac-to-dc voltage gain. In particular, we often consider the rectifier gain G_r to be the ratio of its dc output voltage to the amplitude, $V_{rect,1}$, of the fundamental component of its ac input voltage:

$$G_r = \frac{V_d}{V_{rect,1}}. \tag{10.50}$$

For a capacitively filtered full-bridge rectifier, we get $G_r = \pi/4$. Other rectifiers give different gains, e.g., $G_r = \pi/2$ for a "voltage doubler" rectifier (that is, a half-bridge rectifier with an ac-side blocking capacitor). One may select a rectifier topology in part based upon the voltage gain it provides in the overall conversion process. Of course, if commutation effects come into play as outlined in Chapter 3, there may be load regulation in which the ac-to-dc voltage gain becomes dependent upon the dc-side load current.

10.5.2 Resonant Converter Modeling

An example of a series-resonant dc/dc converter is shown in Fig. 10.10(a), comprising a half-bridge inverter, a series-resonant tank, a transformer, and a full-bridge rectifier. Together, the half-bridge inverter and the series-resonant tank form a series resonant inverter like that in Fig. 10.7, except that here the capacitor C serves as both the resonant capacitor and the blocking capacitor preventing the dc component of v_{inv} from appearing across the transformer. The inverter is loaded by the equivalent ac resistance of the rectifier reflected through the transformer. Figure 10.10(b) segregates the functional stages of the circuit of (a) into the inverter, the resonant tank, the transformation stage, and the rectifier. (By "inverter" we mean only the dc-to-ac switching network portion of the circuit, separate from the resonant tank.) This block diagram representation of the inverter permits us to define the gains, G, of the processed fundamental voltages for the individual blocks.

It is useful to have a model for a resonant dc/dc converter that describes its steady-state transfer characteristic from dc input to dc output. To develop such a model, we define a voltage gain for each stage, where the voltage gains for the inverter and rectifier stages include appropriate

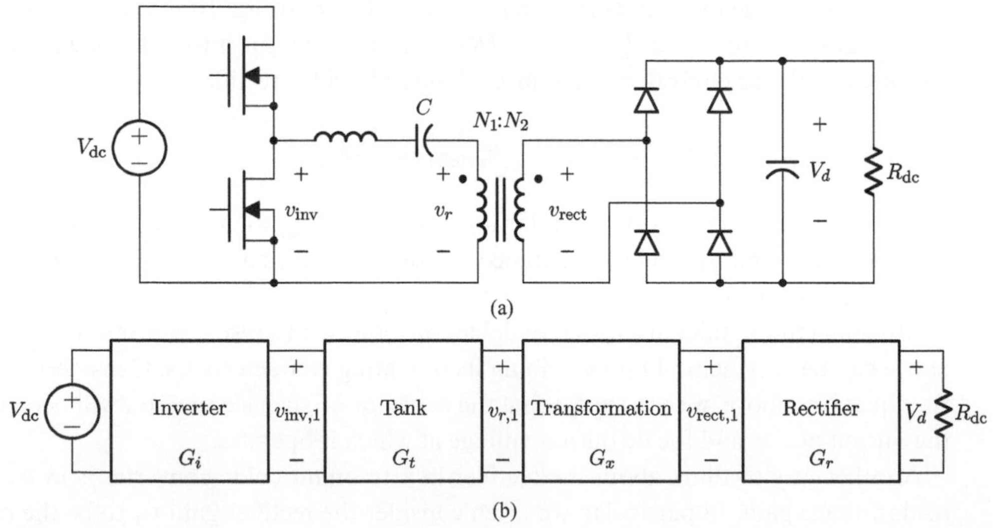

Figure 10.10 (a) A series-resonant dc/dc converter. (b) A block diagram representing the functional sections of the converter of (a), showing the fundamental terminal variables, the ratio of whose amplitudes, $V_{x,1}$, defines the block gains, G.

conversions between the dc and fundamental ac components. The voltage gain of the inverter stage is thus represented as $G_i = V_{\text{inv},1}/V_{\text{dc}}$, the voltage gain of the tank as $G_t = V_{r,1}/V_{\text{inv},1}$, the voltage gain of the transformation stage as $G_x = V_{\text{rect},1}/V_{r,1}$, and the gain of the rectifier as $G_r = V_d/V_{\text{rect},1}$. We can then model the overall transfer characteristic of the system of Fig. 10.10(a) as

$$\frac{V_d}{V_{\text{dc}}} = G_i \cdot G_t \cdot G_x \cdot G_r. \tag{10.51}$$

In this representation, the gains of the inverter, transformation stage, and rectifier are typically simply constants determined by design selection. For example, in the series-resonant converter of Fig. 10.10(b) we have $G_i = \frac{2}{\pi}$, $G_x = \frac{N_2}{N_1}$, and $G_r = \frac{\pi}{4}$. The gain G_t provided by the resonant tank, however, is a function of switching frequency and load, and provides a means by which the output may be controlled. For the series-resonant tank, we can derive the following expression for the voltage gain:

$$G_t = \frac{1}{\sqrt{1 + Q_o^2\left(\frac{\omega_s}{\omega_o} - \frac{\omega_o}{\omega_s}\right)^2}}, \tag{10.52}$$

where Q_o is the loaded quality factor of the tank as defined by (10.17), with R being the equivalent ac resistance loading the tank. Thus, for the circuit of Fig. 10.10(b), we have

$$Q_o = \frac{\pi^2 N_2^2\, \omega_o L}{8\, N_1^2\, R_{\text{dc}}}. \tag{10.53}$$

By adjusting the switching frequency ω_s, and hence G_t, we can control the steady-state output voltage V_d to a desired value over a range of input voltage V_{dc} and load R_{dc}.

Example 10.4 Design of a Series-Resonant DC/DC Converter

Consider the design of a series-resonant dc/dc converter to operate from an input voltage V_{dc} varying between 300 V and 400 V and provide an output voltage $V_d = 100$ V over a load power range of 80 W to 400 W (i.e., a dc load resistance range of $25\,\Omega \le R_{\text{dc}} \le 125\,\Omega$). We will use the topology of Fig. 10.10(a), operating above resonance for soft switching at frequencies in the high hundreds of kHz to low MHz range. Consequently we choose a tank resonant frequency of $\omega_o = 2\pi \cdot 600\,\text{kHz} = 3.770 \times 10^6$ rad/s.

The transformer turns ratio is selected based on the required voltage conversion range. At the minimum dc input voltage $V_{\text{dc(min)}}$, the series-resonant inverter can generate a maximum fundamental ac voltage (at resonance) of $|v_{r,1}| = (2/\pi)V_{\text{dc(min)}}$. To achieve the desired dc output voltage, we must drive a fundamental ac voltage at the input of the rectifier of $V_{\text{rect},1} = (4/\pi)V_d$. This requires a transformer turns ratio of $N_2/N_1 > 2V_d/V_{\text{dc(min)}} = 2/3$. To provide adequate drive voltage for operation above resonance, a transformer turns ratio of 1:1 is selected. With this turns ratio, the tank gain should be controlled between $G_t = 1/2$ at $V_{\text{dc(max)}}$ and $G_t = 2/3$ at $V_{\text{dc(min)}}$ to achieve the desired steady-state output voltage, according to (10.51).

The resonant tank components are selected to provide an acceptable resonant tank loaded quality factor Q_o. Too low a quality factor can lead to poor tank frequency selectivity and unacceptable waveforms, while too high a quality factor results in excess component stress and energy storage. Q_o is a function of the effective ac resistance loading the tank, and is thus a function of the dc resistance loading the converter, as seen in (10.53).

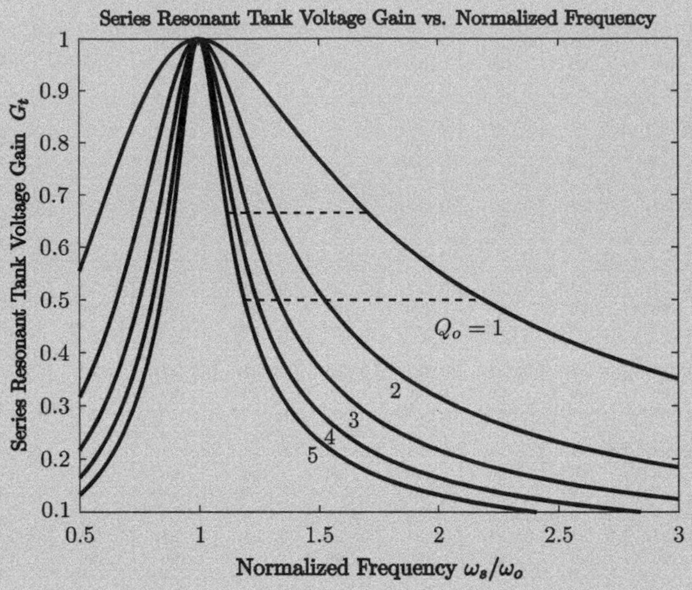

Figure 10.11 Series-resonant tank gain G_t vs. normalized frequency ω_s/ω_o, parametric in the loaded quality factor Q_o of the tank.

Expressed in terms of ac equivalent resistance, with a 1:1 transformer turns ratio the resonant inverter sees a load of $R_{ac} = (8/\pi^2)R_{dc}$, which falls in the range $R_{ac(min)} = 20.26\,\Omega$ to $R_{ac(max)} = 101.32\,\Omega$. The minimum tank quality factor $Q_{o,\min}$ occurs at light load (i.e., maximum load resistance): $Q_{o,\min} = (\omega_o L)/R_{ac(max)}$. For an acceptable value of $Q_{o,\min} = 1$, we obtain $L = (Q_{o,\min} R_{ac(max)})/\omega_o = 26.87\,\mu H$ and $C = 1/(\omega_o^2 L) = 2.618\,nF$. For this component selection, at heavy load (minimum load resistance) we get a maximum circuit quality factor $Q_{o,\max} = (\omega_o L)/R_{ac(min)} = 5$, which represents a reasonable range over which to operate.

The operating range of the converter can be seen in Fig. 10.11, which shows the series-resonant tank gain G_t in (10.52) plotted against normalized frequency ω_s/ω_o and parameterized by different values of tank loaded quality factor Q_o. The proposed converter design operates over values of G_t between 1/2 and 2/3 (determined by the dc/dc voltage conversion ratio) and over values of Q_o between 1 and 5 (determined by the output power), such that its operating region is delineated by the dashed horizontal lines and the $Q_o = 1$ and $Q_o = 5$ curves. Based on (10.52), the converter operates over a normalized frequency range from $\omega_s/\omega_o = 1.118$ (at minimum input voltage, maximum load power) to $\omega_s/\omega_o = 2.189$ (at maximum input voltage, minimum load power), or from 670.8 kHz to 1.313 MHz.

10.5.3 Other Resonant Tanks

As can be seen in the previous example, resonant tank design for a dc/dc converter is closely tied to the dc/dc conversion ratio (relating to the needed range of tank gain G_t) and to output power (which affects the tank quality factor). The tank design directly influences aspects such as component ratings, switching frequency range, and the ability to maintain soft switching.

The use of a series-resonant tank as in Example 10.4 has a number of advantages. Among these is the fact that as power decreases (R_{dc} increases), the currents in the converter switches and resonant components also decrease, which helps maintain high efficiency across load power. At the same time, the use of a series-resonant tank imposes some design constraints. One of these relates to voltage gain. As seen in Fig. 10.11, the series-resonant tank can provide a maximum tank gain $G_t = 1$ as part of the overall voltage conversion gain in (10.51). While voltage conversion gain can be achieved through other factors (inverter topology, transformer ratio, rectifier topology, etc.), it can sometimes be useful to have a tank gain greater than unity, especially if a large overall voltage step up is desired. Even more important, however, is the effect of load resistance variations on tank quality factor. As load power decreases in a series-resonant dc/dc converter, the ac resistance loading the series-resonant tank increases. This causes tank Q_o to decrease, which reduces frequency selectivity (impacting waveform quality and soft switching) and can result in unreasonably high switching frequencies at very light loads. (Conversely, if tank quality factor Q_o is selected to have a reasonable minimum for a very light-load condition, one can get unreasonably high Q_o under heavy-load conditions, resulting in excessive component size and loss.) It is thus challenging to design a series-resonant dc/dc converter using only frequency control for a very wide range of loads.

An alternative approach yielding complementary properties is to use a parallel-resonant tank. One approach to realizing a parallel-resonant inverter is to create a square-wave current and drive it through a parallel *RLC* network, resulting in a sinusoidal voltage across the network. However, a more common approach is to use a voltage-source drive (e.g., from a half-bridge inverter) in series with the resonant inductor of a parallel tank as shown in Fig. 10.12(a). Figure 10.12(b) shows a fundamental-frequency ac model for this voltage-source parallel-resonant inverter; it can be seen that the undriven network is a parallel *RLC* network.

For the parallel-resonant network we have the same expression for ω_o as a series-resonant network but the inverse expression for the tank quality factor Q_o:

$$Q_o = \frac{R}{Z_o} = \frac{R}{\sqrt{L/C}}. \tag{10.54}$$

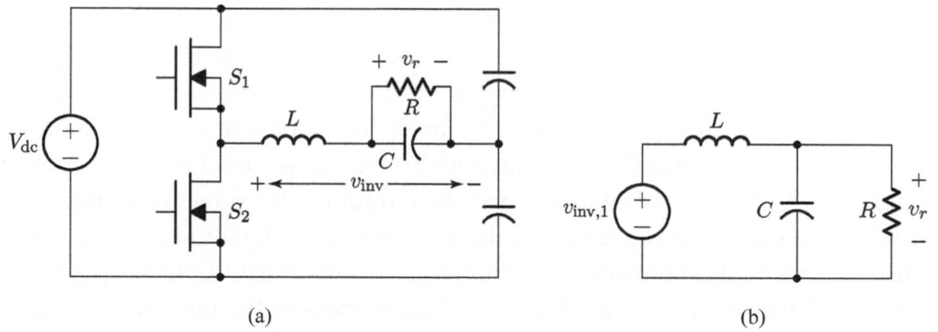

Figure 10.12 (a) A half-bridge parallel-resonant inverter implemented with power MOSFETs. (b) A fundamental-frequency model for the inverter in (a).

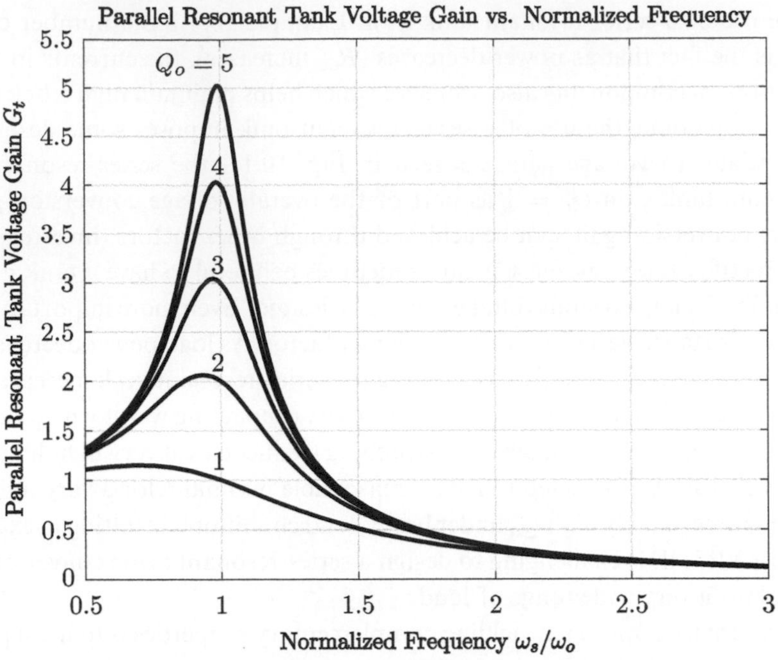

Figure 10.13 Parallel-resonant tank gain G_t vs. normalized frequency ω_s/ω_o, parameterized by different values of the loaded quality factor Q_o of the tank.

Consequently, Q_o for a parallel-resonant tank *increases* with increasing load resistance. Considering the fundamental-frequency model for the parallel-resonant inverter in Fig. 10.12(b), the resonant tank voltage gain $G_t = V_r/V_{\text{inv},1}$ may be found as

$$G_t = \frac{1}{\sqrt{\frac{1}{Q_o^2}\left(\frac{\omega_s}{\omega_o}\right)^2 + \left(1 - \left(\frac{\omega_s}{\omega_o}\right)^2\right)^2}}. \tag{10.55}$$

Figure 10.13 plots the parallel-resonant tank voltage gain against normalized frequency ω_s/ω_o, parameterized in tank quality factor Q_o. As compared to the series-resonant inverter, we can see that the parallel-resonant tank can give values of $G_t > 1$, with $G_t = Q_o$ at $\omega_s/\omega_o = 1$. (For inductive loading of the inverter to achieve ZVS soft switching, we operate at normalized frequencies $\omega_s/\omega_o > 1$.) The voltage gain of a parallel-resonant tank is often useful when high ac voltages need to be synthesized (e.g., for striking an arc in a cold-cathode fluorescent lamp).

To implement a parallel-resonant dc/dc converter, we use a rectifier topology that accepts a sinusoidal voltage at its input (such as the output of the inductively filtered bridge rectifier of Fig. 3.16). Such a parallel-resonant dc/dc converter has interesting properties. First, we can use the resonant tank to provide greater-than-unity voltage gain G_t (e.g., to generate a high-voltage output). Likewise, as the ac effective resistance loading the tank increases (e.g., at light load), the resonant tank Q_o *increases* as shown by (10.54). Consequently, a parallel-resonant dc/dc converter can operate with good waveforms even down to no load. At the same time, such a network imposes disadvantages. For example, unlike the series-resonant converter, the resonant tank currents in a parallel-resonant converter do not back off proportionally to load current, so

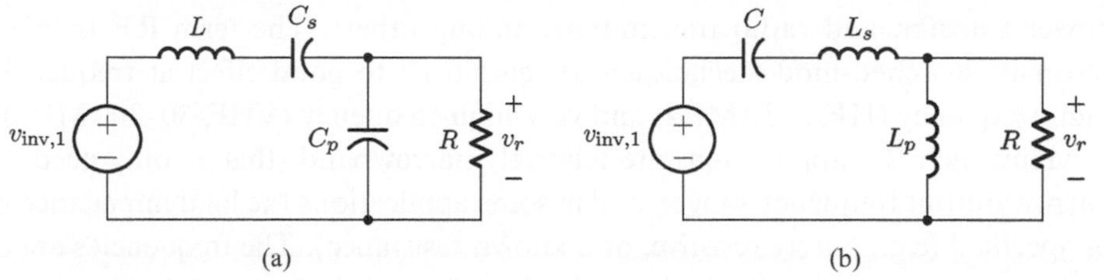

Figure 10.14 (a) Fundamental-frequency model of a series-parallel, or LCC, resonant inverter. (b) Fundamental-frequency model of an LLC resonant inverter.

it is difficult to maintain high efficiency across load. Also, because Q_o decreases at low effective ac resistance, problems such as low waveform quality and loss of soft switching can occur under unusually heavy load conditions. The use of a parallel-resonant tank thus offers complementary advantages and disadvantages to those of a series-resonant tank.

To bypass the limitations of series- and parallel-resonant tanks, higher-order tanks may be used. Fundamental-frequency models of two such tanks are shown in Fig. 10.14. The LCC, or "series-parallel," resonant tank of Fig. 10.14(a) shares some of the best features of each of the series- and parallel-resonant tanks at slightly higher complexity, with detailed characteristics that depend upon the relative values of C_s and C_p. Qualitatively, for frequencies and load resistance values for which the load resistance R is small compared to the impedance of capacitor C_p, the load is in series with C_s and the circuit operates somewhat like a series-resonant inverter. For frequencies and load resistance values for which the load resistance R is large compared to the impedance of C_p, the load is in parallel with C_p and the circuit acts more like a parallel-resonant inverter. Thus, series-parallel converters blend the features of pure series- and parallel-resonant converters.

The LLC resonant tank of Fig. 10.14(b) provides qualitatively similar advantages, including facilitating expanded gain and resistance ranges. For small values of R compared to the impedance of L_p, the converter acts somewhat like a series-resonant converter. For large values of R compared to the impedance of L_p, L_p serves to provide inductive current to enable soft switching of the inverter. Moreover, under this condition, inductive loading of the inverter can be maintained at frequencies *below* the series-resonant frequency $1/\sqrt{L_s C}$, where resonance between the net capacitive impedance of the series L_s–C branch and L_p can be used to provide voltage gain $G_t > 1$. It is also notable that the two magnetic elements L_s and L_p can be realized as leakage and magnetizing inductances of a transformer, yielding a simple physical structure.

While the series, parallel, series-parallel, and LLC resonant tanks are among the most common tank designs found in resonant power converters, there are many other variants (including still-higher-order designs) that may be selected based on the needs of a particular application.

10.6 Radio-Frequency Converters

There are many applications where one would like to use switched-mode techniques to efficiently generate RF power from a dc source. Such applications include RF plasma generation, wireless power transfer, and radio transmitters, among others. The term RF is loosely defined, but resonant switched-mode techniques are employed to good effect at frequencies including the high-frequency (HF, 3–30 MHz) and very-high-frequency (VHF, 30–300 MHz) ranges.

Many such RF applications are relatively narrowband (that is, only need to operate over a narrow output frequency range), and in some applications the load impedance is well controlled or specified (e.g., purely resistive, or a known resistance). The frequencies are often sufficiently high that one must explicitly factor in the effect of device capacitances in circuit operation, and it becomes difficult to drive "flying" devices (that is, devices having control ports that are referenced to rapidly changing potentials).

Here we describe resonant power converters and associated techniques that are well suited to the needs of these applications. Because the circuits and techniques used evolved in both the power electronics and RF communities, there is a mixture of terminology in this space. For example, one might refer to the same circuit as a resonant inverter or as a switched-mode RF power amplifier (and we will use the terms interchangeably). As we will see, the approaches used are closely related to those shown earlier in the chapter.

10.6.1 ZVS Class D Inverters

ZVS half-bridge resonant inverters are widely used in the HF frequency range. In the RF community, such a design would be referred to as a ZVS class D power amplifier. Power amplifier class refers to how the power devices are operated to synthesize an output and/or to other characteristics of the design; classes A through C are not switched mode, but class D indicates that the transistors are used as switches, with the switches seeing quasi-square-wave voltages and carrying currents that are segments of the output waveform. The series- and parallel-resonant half-bridge inverters of Figs. 10.8 and 10.12 would thus both be characterized as class D amplifiers. A more general diagram of a half-bridge ZVS class D inverter that encompasses these designs is shown in Fig. 10.15(a).

Realizing zero-voltage switching in a class D inverter becomes more challenging as frequency is increased. The effects of a given switch capacitance C_{sw} become more pronounced at higher operating frequencies: a greater percentage of the switching cycle is required for dead time between switch on-state durations, and the range of impedances loading the half-bridge that enable soft switching is reduced.

Figure 10.15(b) shows the range of loading impedances $Z_a(j\omega) = R_a + jX_a$ for which ZVS is achievable in the inverter of Fig. 10.15(a), assuming that the loading network maintains a sinusoidal current i_a. The R' axis in the plot is the effective loading resistance R_a normalized to the reactance of the total switch capacitance $1/(2\omega_s C_{\text{sw}})$, while the X' axis is the effective loading reactance X_a normalized to the reactance of the total switch capacitance. We can show that the region of this normalized impedance plane in which ZVS can be maintained is defined by the X' axis and a cycloid curve

$$R' = \frac{1 - \cos(2\theta)}{2\pi}, \qquad X' = \frac{2\theta - \sin(2\theta)}{2\pi} \tag{10.56}$$

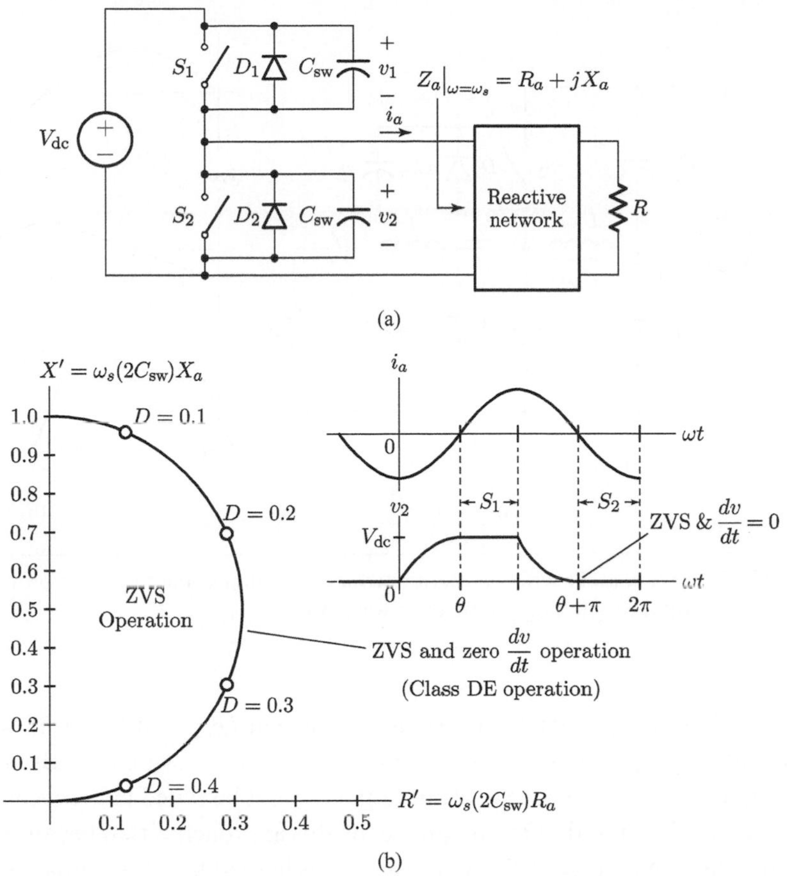

Figure 10.15 (a) A class D inverter driving an impedance $Z_a = R_a + jX_a$. (b) The region in the load–impedance plane in which ZVS soft switching is achieved. The waveforms on the boundary of the ZVS region, known as class DE operation, are also shown.

for $0 \leq \theta \leq \pi$. ZVS can be achieved for normalized loading impedances falling within the delineated region so long as appropriate switch dead times (and associated duty ratios D) are employed and the loading network maintains i_a sinusoidal. For loading impedances inside this region, the half-bridge inverter follows the switching sequence illustrated in Fig. 10.8(b) and (c). On the boundary of this ZVS region, the loading current i_a provides only just enough charge to transition the bus voltage for ZVS. Under this condition, the diode conduction phases 3 and 6 in Fig. 10.8(b) vanish, and the transistors turn on at both zero voltage and with zero dv/dt. The waveforms for this case, which is known as class DE operation, are illustrated in Fig. 10.15(b).

Notice that the load network in Fig. 10.15 is tuned inductively; that is, the load seen by the inverter has an inductive reactive component at the switching frequency. While sinusoidal currents from inductive tuning of the load network are often used to provide ZVS soft switching as described above, it is sometimes preferred to provide other means to enable soft switching. One approach for doing this is shown in Fig. 10.16 in the context of an inverter with a series-resonant output tank. The half-bridge in Fig. 10.16 imposes a trapezoidal-wave voltage across inductor

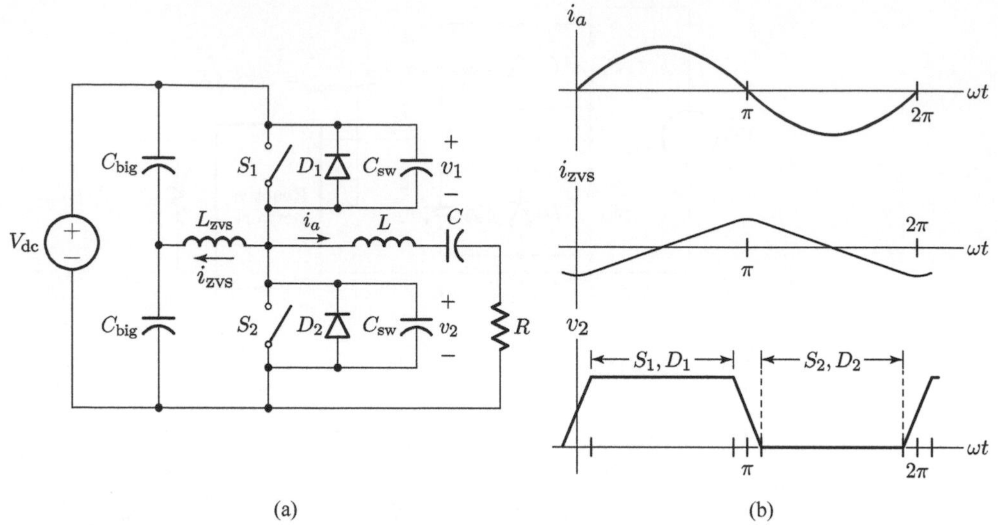

Figure 10.16 (a) A class D series-resonant inverter with an inductor L_{ZVS} to provide current i_{ZVS} for zero-voltage switching. (b) Inverter waveforms.

L_{ZVS}, yielding an approximately triangular current i_{ZVS} that has peak values at the switching times, with appropriate directions to enable zero-voltage switching. Because L_{ZVS} provides the necessary current for ZVS, the output tank need not be tuned inductively (e.g., may be tuned on resonance, as illustrated). One advantage of this approach is that because the current for ZVS is generated independently of the load, variations in load have less impact on ZVS of the inverter. Moreover, because the ZVS current is triangular with peaks at the switching times, the rms switch currents are often lower than when using a sinusoidal loading current to implement ZVS. These advantages are accomplished at the cost of requiring an additional magnetic component in the design. It should be noted that similar effects can be realized by incorporating L_{ZVS} (or a similar element) as part of the loading network.

10.6.2 Class E and Related Inverters

One challenge with half-bridge-based inverter designs is the need to drive a high-side switch (e.g., S_1 in Fig. 10.16). This becomes challenging as frequency increases because most suitable switches have their control ports referenced to the flying node (e.g., as in the N-channel MOSFET-based inverter of Fig. 10.7). It can be difficult to find level shifters and gate drivers that can function with the high dv/dt and short on-state pulses that are required. Moreover, a half-bridge-based circuit requires driving two switches at different reference potentials with carefully controlled timing and dead times between them, making the drive challenge even more difficult. Limiting factors on frequency for a half-bridge design thus include common-mode transient immunity (CMTI, explained in Section 23.1.5), minimum pulse width, pulse-width distortion, and maximum delay skew of the level-shifting and driving circuitry.

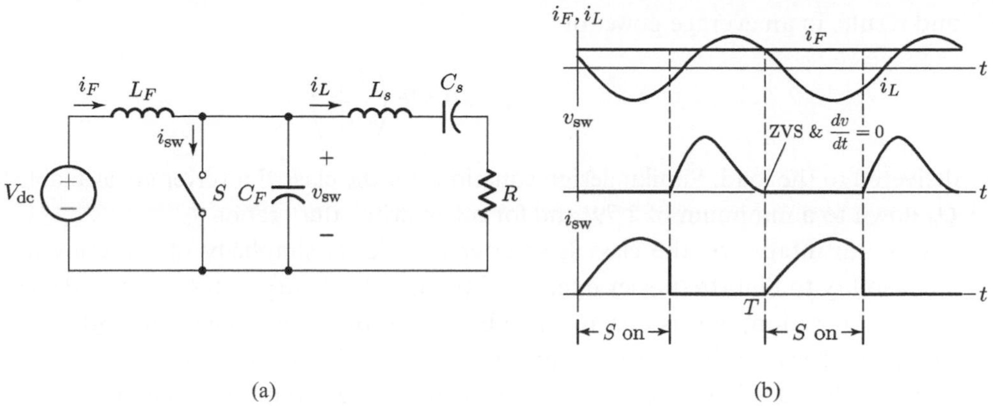

Figure 10.17 (a) A single-ended class E inverter. (b) Class E inverter waveforms.

At sufficiently high frequencies, it becomes preferable to use designs that do not have high-side switches and are less sensitive to switch timing. The most well known of these designs is the single-ended class E inverter shown in Fig. 10.17(a). This inverter effectively replaces the high-side switch in a half-bridge resonant inverter with a network comprising inductor L_F and capacitor C_F. In the traditional implementation, L_F is a choke inductor that supplies nearly constant current from the input and maintains the average voltage across the switch at V_{dc}. Capacitor C_F (which incorporates the intrinsic switch capacitance) ensures that switch S turns off with ZVS. Furthermore, its value is selected in conjunction with those of the inductively tuned load network (L_s, C_s, R) to shape the switch voltage v_{sw} during the switch off-time. In particular, the values are selected to provide a resonant pulse in the switch voltage that enables the switch to turn back on with both ZVS and zero dv/dt, as shown in Fig. 10.17(b). The output tank network is driven by the ac component of v_{sw}, yielding an approximately sinusoidal load current i_L. We thus have a ZVS resonant inverter that only requires a single ground-referenced switch!

Considered one way, the switch in the class E inverter acts to generate resonant pulses that are filtered by the tank network to deliver power to the load at the switching frequency. Considered another way, the switch *absorbs* average power via the dc components of its voltage and current waveforms and *outputs* average power at RF via the fundamental-frequency components of its waveforms. Seen this way, the switch may be thought of as an (ideally lossless) frequency changer.

The set of switching conditions embodied in Fig. 10.17(b) – ZVS turn-off and both ZVS and zero dv/dt turn-on of the switch – has come to be known as class E switching. To obtain class E switching, careful selection of the resonant elements is necessary. For a high loaded quality factor Q_o of the series-resonant tank ($Q_o \geq 10$), an angular switching frequency ω_s, and a switch duty ratio of $D = 0.5$, it can be shown that the following component selection provides the desired waveforms:

$$L_s = \frac{Q_o R}{\omega_s} \qquad C_s = \frac{1}{\omega_s(Q_o - 1.1525)R} \qquad C_F = \frac{1}{5.446\,\omega_s R} \tag{10.57}$$

and results in an average power of

$$P = 0.576\left(\frac{V_{\rm dc}^2}{R}\right) \tag{10.58}$$

delivered to the load. Similar design equations for the class E inverter are available for values of Q_o down to a minimum of 1.79, and for other switch duty ratios.

Key advantages of the class E inverter include its simplicity of structure and drive, and amenability to high-frequency operation. It naturally incorporates parasitic device capacitance into its operation, and is easy to drive because it only has a single ground-referenced switch. Moreover, because the transistor turns on with both ZVS and zero dv/dt (and hence with zero current), the class E inverter is very forgiving of timing variations in the gate drive waveform and of slow gate drive transitions. These characteristics become particularly valuable for operation at high frequencies.

Limitations of the class E inverter include relatively high transistor stresses as compared to a class D inverter. The transistor off-state voltage ideally reaches a peak value of approximately 3.6 $V_{\rm dc}$, and with device capacitance nonlinearity can sometimes exceed 4 $V_{\rm dc}$. Likewise, the peak switch current is approximately 1.7 $V_{\rm dc}/R$. Moreover, the traditional implementation of the class E inverter is very sensitive to load impedance, and will quickly lose ZVS with variations in load resistance or reactance. Consequently, one would typically only use a class E inverter in situations where a half-bridge-based inverter is not suitable.

Various design approaches have been developed to overcome the limitations of the basic class E inverter. One such limitation is the sensitivity of ZVS to variations in load resistance R. To reduce load sensitivity, one can implement a "variable-load" or "load-modulation" class E inverter in which the input-side network L_F, C_F is itself a resonant network that acts to maintain ZVS. This is accomplished by tuning the load network (L_s, C_s, R) to be resonant at the switching frequency, and by tuning the input-side network to resonate at a frequency $f_{\rm F} = 1/(2\pi\sqrt{L_F C_F})$ somewhat above $f_{\rm sw}$. L_F and C_F are further selected to provide a specified characteristic impedance $Z_F = \sqrt{L_F/C_F}$. The best design values depend upon the desired load resistance range, with $f_{\rm F}$ between 1.3 and 1.5 × $f_{\rm sw}$, and with Z_F between 0.2 and 1.5 × $R_{\rm min}$, where $R_{\rm min}$ is the minimum value of R. (Within this range, Z_F is usually made as large as possible while preserving ZVS to reduce resonant network losses.) In this variant of the class E inverter, one maintains ZVS but loses zero dv/dt at turn-on (and hence "true" class E switching), with the transistor voltage more closely approximating a half-sine wave during its off state. Nonetheless, so long as the switch has a body diode (or equivalent characteristic) to "catch" the ringdown of $v_{\rm sw}$, this is usually acceptable. For an inverter tuned this way, we get the following approximate relation between load resistance and power:

$$P = 1.32\frac{V_{\rm dc}^2}{R}. \tag{10.59}$$

A second limitation of the basic class E inverter is its extremely high device voltage stress. This can be improved by using higher-order tuning to shape $v_{\rm sw}$ to approximate a trapezoid with a peak transistor voltage stress near 2 $V_{\rm dc}$. The traditional class E $v_{\rm sw}$ waveform of Fig. 10.17 dominantly comprises dc, the fundamental, and second-harmonic components. A pure

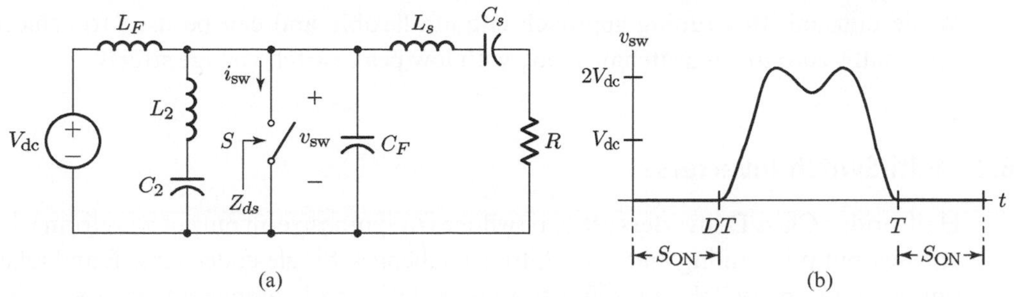

Figure 10.18 (a) A single-ended class Φ_2 (or EF_2) inverter. (b) The voltage waveform v_{sw} showing approximate half-wave symmetry and a peak switch voltage of just above 2 V_{dc}.

trapezoidal waveform would have an ac component that is half-wave symmetric (varying about an average value V_{dc}), and hence would contain no even-harmonic content, as described in Section 8.2.1. By using a switch duty ratio somewhat below 0.5 and tuning out one or more even-harmonic components in v_{sw} while retaining appropriate amounts of odd harmonics, v_{sw} can be made to approximate a trapezoid with a peak value slightly above 2 V_{dc}. This general technique is referred to as class EF operation or – if the input network is used in the tuning – class Φ operation. In the RF literature, class F operation indicates the use of harmonic tuning of the load network to "square up" the transistor voltage, but does not imply switched-mode operation.

The class Φ_2 (or EF_2) inverter of Fig. 10.18(a) illustrates this general approach. The subscript 2 is adopted because the L_2, C_2 network is tuned to twice the switching frequency in order to null the second-harmonic component in v_{sw}. The resulting switch voltage waveform is dominated by dc, fundamental, and third-harmonic components, and thus has approximate half-wave symmetry as shown in Fig. 10.18(b). With appropriate tuning, the fundamental and third-harmonic components combine in a way that yields a maximum value of v_{sw} that is just over 2 V_{dc}.

There are various approaches by which the resonant network in 10.18(a) can be tuned to enable ZVS and provide low switch voltage stress. One effective (but underspecified) approach involves designing the network based on the impedance Z_{ds} seen looking into the network port at which the switch is connected. (See Fig. 10.18(a); the output capacitance of the switch is included as part of C_F.) The network is tuned to provide the following characteristics:

- The impedance Z_{ds} is tuned to be inductive at the fundamental switching frequency f_s (typically $+30^\circ < \angle Z_{ds}|_{\omega_s} < +60^\circ$), in order to ensure ZVS.
- The impedance at the second harmonic $2f_s$ is tuned to be zero by selection of L_2, C_2 to help impose approximate half-wave symmetry in v_{sw}.
- The impedance at the third harmonic $3f_s$ is tuned to be capacitive (i.e., $-90^\circ < \angle Z_{ds} < 0$) with a magnitude that is typically 4–8 dB below that at the fundamental, with a value selected to achieve low peak voltage stress.
- The output power is set by selecting the impedance of the L_s, C_s, R branch such that appropriate power is delivered to the load when driven by the quasi-trapezoidal voltage v_{sw}.

While unusual, this tuning approach is quite flexible and can be used to achieve ZVS and, optionally, zero dv/dt switching along with low peak switch voltage stress.

10.6.3 Multi-Switch Inverters

Half-bridge Class D inverters use two switches to synthesize an output waveform with low device stresses, but pose timing and level-shifting challenges. Single-ended class E and related inverters only require one switch, but suffer from high device and component stresses. In between these approaches are inverters that use multiple ground-referenced switches. While more complex and timing sensitive than single-switch designs, they can attain better efficiency and waveform quality without the level-shifting challenges of half-bridge designs. Most common among such inverters are double-ended or "push–pull" designs in which two inverter halves operate 180° out of phase, with the load network driven differentially between them.

Example 10.5 A Push–Pull Inverter

One example of a push–pull inverter is shown in Fig. 10.19(a). Its operating waveforms are shown in Fig. 10.19(b). To understand its operation, first assume that the input inductors L_F are infinite (so carry constant currents $i_{F1} = i_{F2} = I_F$) and that the switch capacitances are negligible ($C_{\text{sw}} = 0$). If switches S_1 and S_2 are switched in a complementary fashion at 50% duty ratio with a switching frequency

$$f_{\text{sw}} = \frac{1}{T} = \frac{1}{2\pi}\frac{1}{\sqrt{L_P C_P}}, \tag{10.60}$$

then a square-wave current of amplitude I_F will be driven through the parallel L_p, C_p, R tank at its resonant frequency. The filtering action of the tank results in a sinusoidal voltage $v_o(t)$ in phase with the driving square-wave current, and switch voltages v_{sw1} and v_{sw2} that are half-sine sections. We thus obtain a sinusoidal output voltage across the load resistance R using ground-referenced switches that turn on and off with ZVS.

Because the switches in Fig. 10.19(a) are connected to the dc source V_{dc} via inductors L_F, the average switch voltages must be V_{dc}. Based on this, it is easily shown that the peak switch voltages v_{pk} and the output ac voltage amplitude V_o must be

$$v_{pk} = V_o = \pi V_{\text{dc}}, \tag{10.61}$$

resulting in an average output power of

$$P = \frac{\pi^2}{2}\frac{V_{\text{dc}}^2}{R}. \tag{10.62}$$

It can be further deduced (e.g., based on conservation of energy) that the current from the dc source (equal to the switch on-state currents) is

$$2I_F = \frac{\pi^2}{2}\frac{V_{\text{dc}}}{R}. \tag{10.63}$$

This circuit is basically a current-fed version of a parallel-resonant inverter operated at resonance. It is often referred to as an "inverse class D inverter" or a "current-mode class D inverter" because the transistors ideally see off-state voltages that are half-sine-wave sections and on-state currents that are square waves. A desirable trait of this circuit is that the half-sine switch voltages provide ZVS across a wide range of load resistances R.

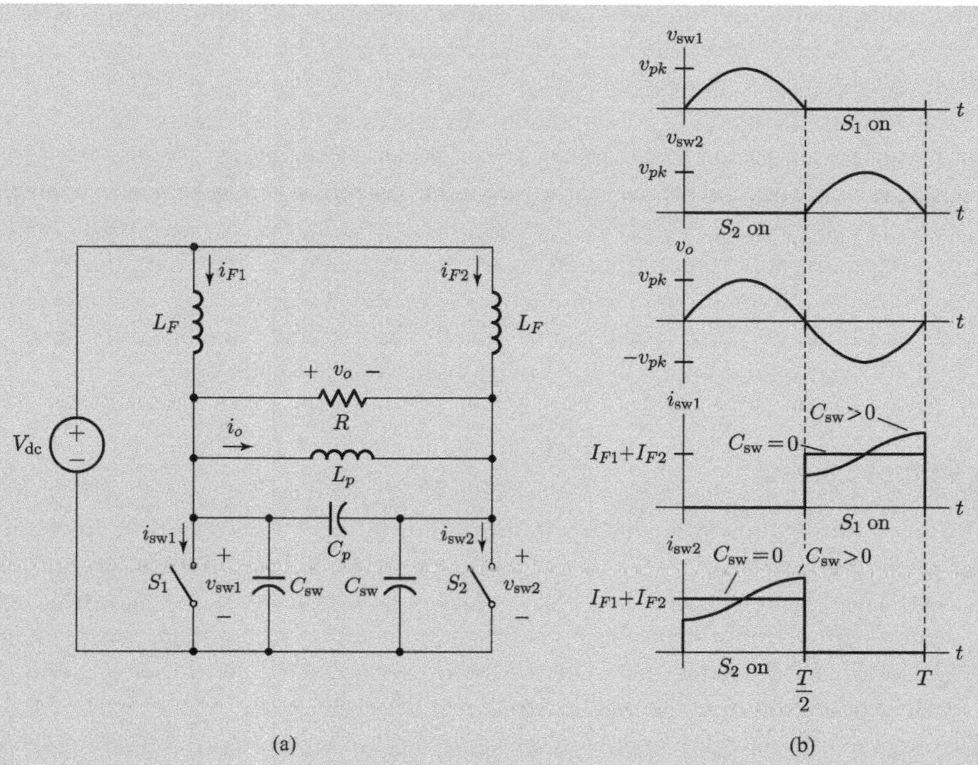

Figure 10.19 (a) A push–pull RF inverter with a parallel-resonant tank. This circuit is often referred to as an "inverse class D" or "current-mode class D" inverter. (b) Inverter waveforms.

In many practical cases, we must allow for the impact of non-negligible switch capacitances C_{sw}. C_{sw} of the off-state transistor is effectively in parallel with C_p. Consequently, we can absorb C_{sw} as part of the parallel-resonant tank, and operate the circuit at the net resonant frequency,

$$f_{sw} = \frac{1}{T} = \frac{1}{2\pi} \frac{1}{\sqrt{L_p(C_p + C_{sw})}}. \tag{10.64}$$

A side effect of the switch capacitances C_{sw} alternately acting as part of the resonant tank is that part of the parallel-resonant current travels through the on-state switch, changing its current waveform from a square wave to a square wave plus a half-sine section as shown in Fig. 10.19(b). The fact that the switch capacitances can be neatly absorbed as part of the parallel resonant tank is a significant advantage of this circuit. It is also possible to incorporate the inductors L_F as part of the resonant circuit.

One consideration with this and other push–pull circuits is that the load resistor R is connected differentially between the two inverter halves. If the ultimate load must be ground referenced then an additional transformer or other mechanism is needed to provide the necessary differential to single-ended conversion. (In the RF world, such a device is called a balun, so named for providing a "balanced to unbalanced" conversion.) If a transformer is used for this purpose, the transformer magnetizing inductance can be absorbed as part of L_p.

One advantage of push–pull inverters such as shown in Example 10.5 is how the two out-of-phase circuit halves interact to support output waveform quality. Because the load network is connected differentially between the inverter halves, any even-harmonic voltage or current components arising from the individual inverter halves will naturally be canceled in the load network. A push–pull design can thus often use lower tank quality factors than are acceptable with a single-ended design.

A further advantage of push–pull designs relates to the inclusion of higher-order harmonic tuning techniques such as employed in the class EF_2 inverter. While the details of such techniques are beyond the scope of this text, the basic concept may be readily appreciated: tuning networks connected differentially between the circuit halves only respond to the fundamental and odd-harmonic components, owing to the cancellation of even harmonics between the two circuit halves. By contrast, tuning networks connected from each of the halves to ground will respond to all frequency components. This can be useful because one often seeks to tune even and odd harmonics differently (e.g., to impose waveform symmetries).

10.6.4 Output Control

As described in the introduction, frequency control is widely used in resonant converters. However, it is less common in RF applications, because the output frequency content is often constrained by the application. For example, applications using the industrial, scientific, and medical (ISM) frequency bands often face tight frequency limits: the allowed bandwidth for the 13.56 MHz ISM band is only 14 kHz. Moreover, many RF power topologies only achieve their desired operation over relatively narrow frequency ranges (e.g., the inverter of Example 10.5). Consequently, many RF applications require some other means of controlling the RF output.

The RF output amplitude of a switched-mode inverter is ideally proportional to its dc supply voltage V_{dc}, while its output power is ideally proportional to the square of V_{dc} (e.g., as seen in (10.61) and (10.62)). A common technique for controlling the output amplitude or power of an RF inverter is thus to use a dc/dc converter to dynamically adjust its dc supply voltage, as illustrated in Fig. 10.21. Versions of this control technique are variously referred to as "envelope tracking," "drain modulation," "adaptive-power tracking," and "envelope elimination and restoration." This technique can be highly effective, and has the advantage that both voltage-related and current-related losses in the inverter reduce as inverter output power is reduced. Drawbacks of this technique center around the size, loss, and bandwidth limitations of the dc/dc converter. Energy in such a design must be processed twice – once through a dc/dc converter, which must handle wide output voltage and power ranges, and once through the inverter itself. Moreover, the achievable bandwidth (speed) with which the RF output voltage can be adjusted is often limited by the dynamic capabilities of the dc/dc converter. Lastly, device capacitance nonlinearities can sometimes impose a minimum supply voltage for which an inverter will work well, limiting the range over which its output can be controlled in this way.

Phase-shift control, or "outphasing," is also widely used in RF applications. In this control method, the outputs of two inverters are combined to drive the load such that their phase difference modulates the overall output. For example, Fig. 10.21(a) shows a system with a common-mode combiner in which the system output voltage v_{OUT} is the average of the two inverter outputs v_1 and v_2: $v_{OUT} = (v_1 + v_2)/2$. Figure 10.21(b) shows a phasor diagram representing this relationship, with control angle Δ defining the difference in phase 2Δ between the two inverter outputs. If $v_1 = V\cos(\omega_s t + \Delta) = \mathrm{Re}\{Ve^{j\Delta}e^{j\omega_s t}\}$ and $v_2 = V\cos(\omega_s t - \Delta) =$

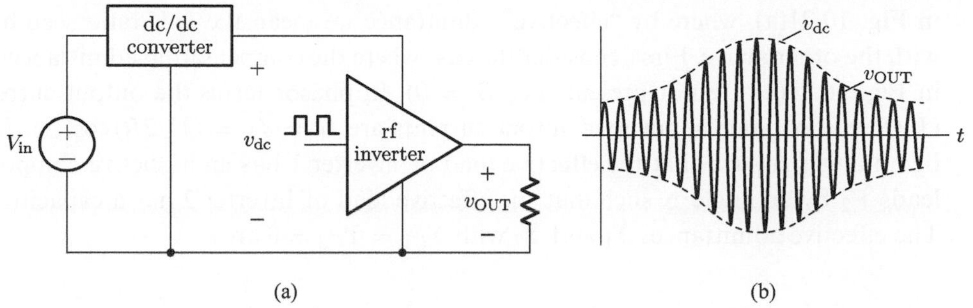

Figure 10.20 (a) An inverter with a dc/dc converter providing dynamic adjustment of the source voltage v_{dc}. (b) Qualitative waveforms of the system of (a).

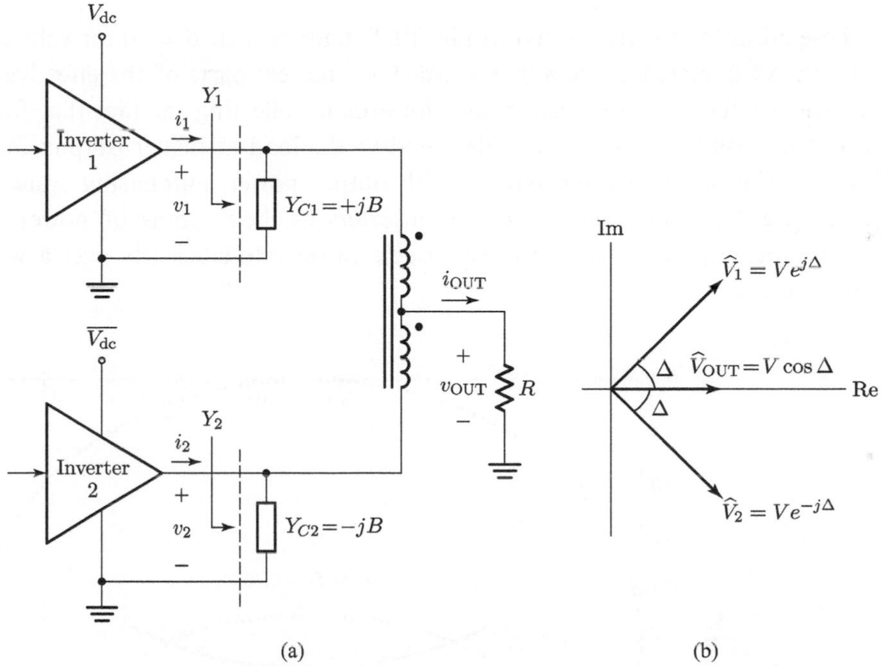

Figure 10.21 (a) An inverter system designed for outphasing control, comprising two inverters and a transformer-based power combiner. The Chireix compensating elements Y_{C1}, Y_{C2} improve the loading of the inverters. (b) A phasor diagram showing the voltage $\widehat{V}_{OUT}$ as a weighted vector sum of the output phasors $\widehat{V}_1$ and $\widehat{V}_2$.

$\text{Re}\{Ve^{-j\Delta}e^{j\omega_s t}\}$, we can express their phasors as $\widehat{V}_1 = Ve^{j\Delta}$ and $\widehat{V}_2 = Ve^{-j\Delta}$. The output voltage phasor is the weighted vector sum of the inverter phasors, $\widehat{V}_{OUT} = (\widehat{V}_1 + \widehat{V}_2)/2 = V\cos(\Delta)$, giving an output voltage $v_{OUT} = V\cos(\Delta)\cos(\omega_s t)$. The RF voltage driving the load is thus controlled by the phase difference between the inverter outputs, defined by control angle Δ.

An important consideration in outphasing is how the loading of the inverters varies with outphasing angle Δ. Consider the effective load admittances Y_1 and Y_2 seen by inverters 1 and 2 in Fig. 10.21(a), where by "effective" admittance we mean the $\widehat{I}/\widehat{V}$ ratio seen by one inverter with the other active. First, consider the case where the compensating admittances Y_{C1} and Y_{C2} in Fig. 10.21(a) are not present (i.e., $B = 0$). In phasor terms the output current is $\widehat{I}_{\text{OUT}} = (V/R)\cos(\Delta)$ and the inverter output currents are $\widehat{I}_1 = \widehat{I}_2 = (V/2R)\cos(\Delta)$. $\widehat{I}_1$ thus lags $\widehat{V}_1$ by an angle Δ such that the effective load of inverter 1 has an inductive component, while $\widehat{I}_2$ leads $\widehat{V}_2$ by an angle Δ such that the effective load of inverter 2 has a capacitive component. The effective admittances Y_1 and Y_2 with $Y_{C1} = Y_{C2} = 0$ are

$$\begin{aligned} Y_1 &= \frac{1}{2R}\cos(\Delta)e^{-j\Delta} = \frac{1}{2R}\cos^2\Delta - j\frac{1}{2R}\cos(\Delta)\sin(\Delta), \\ Y_2 &= \frac{1}{2R}\cos(\Delta)e^{j\Delta} = \frac{1}{2R}\cos^2\Delta + j\frac{1}{2R}\cos(\Delta)\sin(\Delta). \end{aligned} \tag{10.65}$$

These admittances are plotted in Fig. 10.22 (curves with $B = 0$) for values of Δ between 0° and 90°. As Δ increases from 0° towards 90°, the real parts of the effective admittances (the effective conductances) reduce from 1 towards 0, reflecting the fact that for a given inverter RF output voltage, less power is delivered to the load at higher outphasing angles. Because of this variation in inverter loading with output power, outphasing systems are sometimes said to use "load modulation" of the inverters as their means of power control. Inverters for such an outphasing system must be able to operate efficiently over a wide range of load conductances.

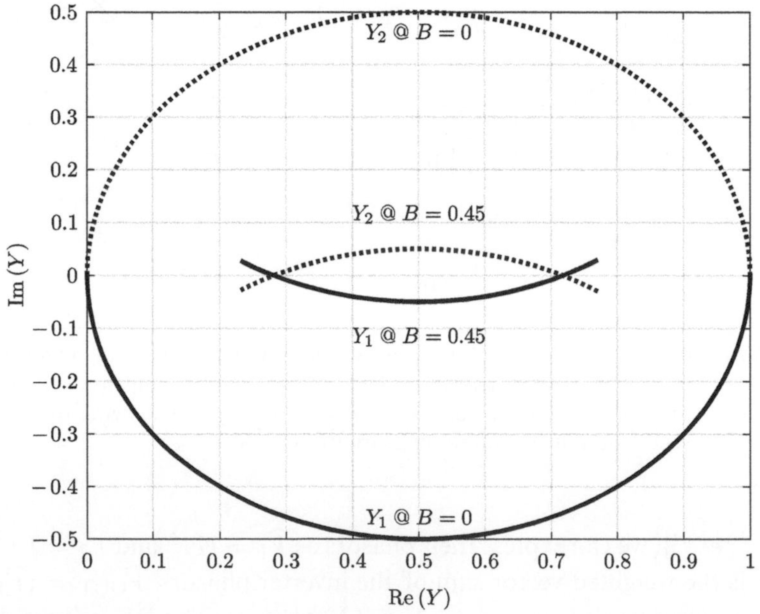

Figure 10.22 Effective admittances Y_1, Y_2 seen by the inverters in Fig. 10.21 for $R = 0.5$. The effective admittances are shown with Chireix compensating elements $Y_{C1} = Y_{C2} = 0$ and an example with elements of value $Y_{C1} = -Y_{C2} = 0.45j$.

It can be seen that the imaginary parts of the effective admittances (the effective susceptances) also vary considerably (and non-monotonically) with Δ, with the two inverters having opposite susceptive loading components. The effective susceptances and their variations are quite large (on the same scale as the admittance variations), which can pose a problem for many RF inverters. Susceptive compensating admittances Y_{C1} and Y_{C2} are thus sometimes introduced to help reduce the susceptive components of loading. As illustrated in Fig. 10.21(a), the leading-phase inverter is often loaded with an additional capacitive shunt admittance $Y_{C1} = jB$, while the lagging-phase inverter is loaded with an additional inductive shunt admittance $Y_{C1} = -jB$. With these compensating elements, we see modified values of the effective loading admittances:

$$\begin{aligned} Y_1 &= \frac{1}{2R}\cos(\Delta)e^{-j\Delta} + Y_{C1} = \frac{1}{2R}\cos^2\Delta - j\left(\frac{1}{2R}\cos(\Delta)\sin(\Delta) - B\right), \\ Y_2 &= \frac{1}{2R}\cos(\Delta)e^{j\Delta} + Y_{C2} = \frac{1}{2R}\cos^2\Delta + j\left(\frac{1}{2R}\cos(\Delta)\sin(\Delta) - B\right). \end{aligned} \tag{10.66}$$

As can be seen in Fig. 10.22 (curves with a representative value $B = 0.45$), the effective admittances remain much closer to being purely conductive over some range of outphasing values Δ, making them more suitable for many inverter designs. A power combiner providing such compensation is sometimes known as a "Chireix combiner" after its inventor.† Alternatively, to facilitate soft ZVS of the inverters, susceptances Y_{C1} and Y_{C2} are sometimes chosen asymmetrically such that both inverters always see an inductive loading component.

An advantage of outphasing is that it can achieve very high control bandwidth, as inverter phase can be varied very quickly. Moreover, current loading – and current-related losses – of the inverters reduce as power is reduced, which helps preserve efficiency at low power. However, the inverter voltage amplitudes do not change with output power, so that voltage-related losses do not reduce with output power. It is also worth noting that there are many variations of the outphasing concept, including the use of more than two inverters to enhance performance ("multi-way" outphasing) and use of different means for combining inverter outputs.

A third control method that can be used in some applications is on/off, or "burst-mode," control. In this approach the inverter is turned on and off at some rate far below the switching frequency ω_s, with its fraction of on-time used to control the average power delivered to the load. This technique can be effective in applications where the power pulsations owing to the on/off bursts are adequately filtered by the load (e.g., using the thermal mass of the load in an induction heating application). It has the advantage that the inverter always operates at full power (or a single load point), and efficiency can stay relatively constant as power is reduced. Challenges to maintaining high efficiency include addressing losses and transient responses that occur during startup and shutdown of the converter for on/off pulses.

† In his seminal 1935 paper introducing outphasing [16], Chireix (pronounced "she wrecks") described this technique in the context of improving the power factor seen at the RF amplifier outputs.

Notes and Bibliography

An interesting history of quality factor Q and its many applications is presented in [1]. A whole range of resonant inverters, rectifiers, and dc/dc converters is analyzed in [2] and [3]. Analysis of the soft-switching range of half-bridge class D / class DE inverters is provided in [4]. Concise treatment of some popular half-bridge resonant dc/dc converters may be found in [5] and [6].

The classic paper by the Sokals introducing the single-ended class E inverter is [7]. An intuitive description of the class E inverter may be found in [8], while [9] treats its design across quality factor and duty ratio. Higher harmonic tuning techniques for single-switch inverters are described in [10]–[13]; the impedance-based design procedure for the Φ_2 inverter is described in [12].

The design of single-switch inverters for variable load ranges is detailed in [14], while design techniques for push–pull RF converters, including the use of common- and differential-mode connections of tuning networks to improve performance, are presented in [10] and [15].

The classic 1935 paper by Henri Chireix [16] introducing outphasing is exceptionally clear. Interestingly, it also references the use of envelope tracking for high-efficiency RF modulation, indicating that this basic concept is even older. A modern review of outphasing techniques is presented in [17], while multi-way outphasing is described in [18] and related works.

1. E. I. Green, "The Story of Q," *American Scientist*, vol. 43, pp. 584–594, Oct. 1955.
2. M. K. Kazimierczuk and D. Czarkowski, *Resonant Power Converters*, 2nd ed., Hoboken, NJ: Wiley, 2011.
3. A. Grebennikov, N. O. Sokal, and M. J. Franco, *Switchmode RF and Microwave Power Amplifiers*, 2nd ed., New York: Academic, 2012.
4. D. C. Hamill, "Impedance Plane Analysis of Class DE Amplifier," *Electronics Lett.*, vol. 30, no. 23, pp. 1905–1906, 1994.
5. R. L. Steigerwald, "A Comparison of Half-Bridge Resonant Converter Topologies," *IEEE Trans. Power Electron.*, vol. 3, no. 2, pp. 174–182, Apr. 1988.
6. H. Huang, "Designing an LLC Resonant Half-Bridge Power Converter," *2010 Texas Insturments Power Supply Design Seminar*, TI Literature Number SLUP263, 2010.
7. N. Sokal and A. Sokal, "Class E – A New Class of High-Efficiency Tuned Single-Ended Switching Power Amplifiers," *IEEE J. Solid-State Circ.*, vol. SC-10, no. 3, pp. 168–176, Jun. 1975.
8. N. Sokal, "Class E RF Power Amplifiers," *QEX*, pp. 9–20, Jan./Feb. 2001.
9. M. Kazimierczuk and K. Puczko, "Exact Analysis of Class-E Tuned Power Amplifier at any Q and Switch Duty Cycle," *IEEE Trans. CAS*, vol. 34, pp. 149–159, Feb. 1987.
10. S. D. Kee, I. Aoki, A. Hajimiri, and D. Rutledge, "The Class E/F Family of ZVS Switching Amplifiers," *IEEE Trans. Microwave Theory and Tech.*, vol. 51, pp. 1677–1690, Jun. 2003.
11. Z. Kaczmarczyk, "High-Efficiency Class E, EF2, and E/F3 Inverters," *IEEE Trans. Ind. Electron.*, vol. 53, no. 5, pp. 1584–1593, Oct. 2006.
12. J. M. Rivas, Y. Han, O. Leitermann, A. D. Sagneri, and D. J. Perreault, "A High-Frequency Resonant Inverter Topology with Low Voltage Stress," *IEEE Trans. Power Electron.*, vol. 23, no. 4, pp. 1759–1771, July 2008.
13. S. Aldhaher, D. C. Yates, and P. D. Mitcheson, "Modeling and Analysis of Class EF and Class E/F Inverters with Series-Tuned Resonant Networks," *IEEE Trans. Power Electron.*, vol. 31, no. 5, pp. 3415–3430, May 2016.
14. L. Roslaniec, A. S. Jurkov, A. Al Bastami, and D. J. Perreault "Design of Single-Switch Inverters for Variable Resistance / Load Modulation Operation," *IEEE Trans. Power Electron.*, vol. 30, no. 6, pp. 3200–3214, Jun. 2015.

15. L. Gu, G. Zulauf, Z. Zhang, S. Chakraborty, and J. Rivas-Davila, "Push–Pull Class Φ_2 RF Power Amplifier," *IEEE Trans. Power Electron.*, vol. 35, no. 10, pp. 10515–10531, Oct. 2020.
16. H. Chireix, "High Power Outphasing Modulation," *Proc. IRE*, vol. 23, no. 11, pp. 1370–1392, Nov. 1935.
17. T. Barton, "Not Just a Phase: Outphasing Power Amplifiers," *IEEE Microwave Mag.*, pp. 18–31, Feb. 2016.
18. A. S. Jurkov, L. Roslaniec, and D. J. Perreault, "Lossless Multi-Way Power Combining and Outphasing for High-Frequency Resonant Inverters," *IEEE Trans. Power Electron.*, vol. 29, no. 4, pp. 1894–1908, Apr. 2014.

PROBLEMS

10.1 Determine and plot the current i_a in the switched *RLC* network of Fig. 10.1. If the switch opens at the first zero-crossing of i_a, how much energy is trapped in C?

10.2 Figure 10.23 shows two parallel *RLC* networks. The network of Fig. 10.23(a) is driven by a sinusoidal voltage source. The network of Fig. 10.23(b) is excited by a current source. Calculate the inductor current i_L in (a) and the capacitor current i_C in (b) when the circuits are driven at resonance, that is, at $\omega = \omega_o$. Define α and Q_o for these circuits and express your answers in terms of α, Q_o, and the characteristic impedance of the tank, $Z_o = \sqrt{L/C}$.

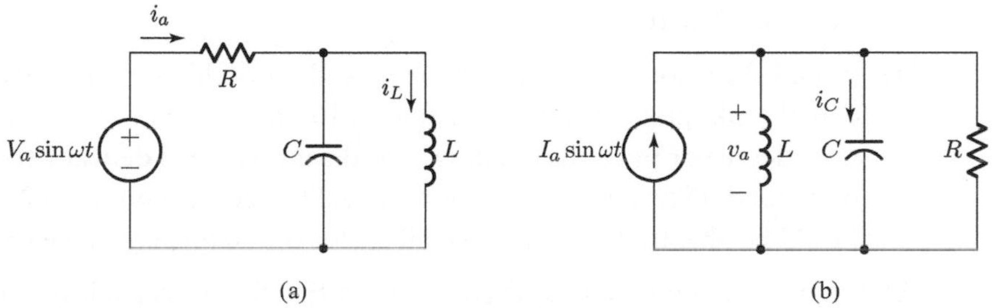

Figure 10.23 Two parallel-resonant networks with different input drives. (a) A voltage-driven parallel-resonant circuit. (b) A current-driven parallel-resonant circuit.

10.3 Consider a parallel-resonant circuit driven by a current source as in Fig. 10.23(b).

(a) Find the driving-point impedance seen by the current source, and express it in a form similar to that used in (10.15) for the admittance of the series-resonant circuit. Define α, ω_o, Q_o, and Z_o for this circuit in terms of the circuit parameters L, R, and C.

(b) Show that the expression for Q_o given by (10.16) applies to this parallel-resonant circuit and is equivalent to the general definition of Q given by (10.25) when the circuit is driven at the resonant frequency ω_o.

10.4 Here we revisit the discussion of practical inductors in Example 10.1. For a practical inductor in which the inductance and ESR are constant across frequency, one can expect to get better performance (higher Q) from the inductor at higher frequencies. Often, however, skin and proximity effects can make ESR increase and inductance decrease with frequency. Consider the case of a coreless solenoid inductor in a frequency regime where the inductance L is approximately constant but the ESR R_{ac} varies with angular frequency ω as

$$R_{ac} \approx \begin{cases} R_{dc}, & \omega < \omega_1, \\ R_{dc}\sqrt{\dfrac{\omega}{\omega_1}}, & \omega \geq \omega_1. \end{cases} \tag{10.67}$$

What is the expected quality factor of the inductor as a function of frequency ω? (Note that the result is a reasonable model for the quality factor of a single-layer coreless solenoid inductor at frequencies below those for which capacitive effects become important.)

10.5 Revisiting the system of Example 10.3, suppose we want instead to have power $P_1 =$ 800 W delivered at $f_{s1} = 3$ MHz, and power $P_2 = 200$ W at $f_{s2} = 3.5$ MHz. What values of L and C should be chosen for the resonant inverter in this case? What is the allowed upper value of C_{sw} for such a design? How do the new component values compare to those of the original design?

10.6 Revisiting Example 10.3, derive (10.42).

10.7 Design a series-resonant half-bridge inverter as in Fig. 10.4. Assume that the inverter should be designed to operate at a frequency of at least 1 MHz from $V_{dc} = 200$ V into a resistive load $R = 10\,\Omega$.

(a) Specify the resonant tank components and a switching frequency range such that the loaded tank quality factor Q_o is at least 10, the inverter switches always operate with some degree of inductive loading, and the inverter can deliver between 70 and 250 W to the load. (The inverter may be designed to deliver more than 250 W at 1 MHz, but should be able to deliver at least 250 W for the frequency range specified.)

(b) What is the minimum required voltage rating for the capacitor in this design?

10.8 Figure 10.24 shows a "voltage doubler" rectifier driven from a sinusoidal current source (e.g., representing a resonant inverter) via a transformer. Note that you may consider the blocking capacitor C_B to be large such that voltage V_B is approximately constant at a voltage such that the average of voltage v_x is zero. You may also consider the diodes D_1 and D_2 to be ideal.

(a) Plot voltage v_x and current i_x, and dimension them in terms of the circuit parameters given.

(b) We would like to find an equivalent circuit model for this rectifier. Given that the rectified output voltage is constant at a voltage V_d, find an effective ac-side resistance for the rectifier R_{ac} at port X as a function of the dc-side voltage V_d and the ac current magnitude into the port I_X. (You may use the fundamental harmonic approximation in modeling the effective resistance.) The resulting value R_{ac} should correctly model the power transfer from the ac waveforms to the dc output of the rectifier.

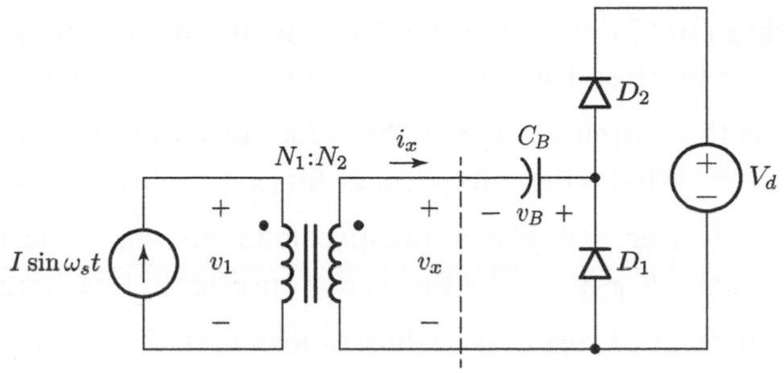

Figure 10.24 A voltage doubler rectifier driven from a sinusoidal current source via a transformer.

(c) Assuming an effective resistance R_{ac} of 1 Ω at port X (for specified operating voltage and current levels), what is the effective resistance R_A seen by the current source at the input port 1? Justify your answer.

10.9 Parallel-resonant and LCC dc/dc converters are implemented with rectifier topologies that accept a sinusoidal ac input voltage, such as the inductively filtered bridge rectifier of Fig. 3.16.

(a) Find an effective ac-side resistance R_{ac} to model this rectifier in terms of the rectifier dc-side load resistance. You may use the fundamental harmonic approximation in modeling the effective resistance, and assume that the filter inductor is sufficiently large that there is negligible ripple in the load current I_d. The resulting value R_{ac} should correctly model the average power transfer from the ac waveforms to the dc output of the rectifier.

(b) Find the ac-to-dc voltage gain G_r for this rectifier. G_r is defined to be the ratio of its dc output voltage (i.e., the dc voltage across the rectifier load resistor) to the magnitude of its ac input voltage V_s.

10.10 Derive the expression (10.52) for the voltage gain G_t of a series-resonant network. This is the magnitude of the voltage across the resistive load divided by the magnitude of the voltage driving the series-resonant network.

10.11 Using (10.57) and (10.58), select circuit component values L_s, C_s, and C_F and dc-bus voltage V_{dc} for a class E inverter operating at a switching frequency of $f_s = 30$ MHz that delivers $P = 200$ W into a load resistance $R = 1\,\Omega$. Design the inverter to have a loaded tank quality factor of $Q_o = 15$. The input inductor can be selected to have a value of $L_F = 100L_s$.

10.12 Derive the expressions (10.61) and (10.63) for the peak switch voltages and switch currents of the current-mode class D inverter of Example 10.5.

10.13 Design an envelope-tracking inverter system as in Fig. 10.20 in which the dc/dc converter is a synchronous buck converter and the RF inverter is a class E inverter. The system should provide a 10 MHz RF output into a 50 Ω load resistor, with the RF output controllable over a power range of 10 W to 160 W.

(a) Using (10.57) and (10.58), select circuit component values L_s, C_s, and C_F for the class E inverter. Design the inverter to have a loaded tank quality factor of $Q_o = 10$.

(b) v_{dc} is the output voltage of the dc/dc converter and the input voltage of the class E inverter. What is the voltage range for v_{dc} to deliver the desired range of output power?

(c) The inverter acts like a resistive load on the dc/dc converter. What equivalent resistance R_{eff} is seen by the dc/dc converter in this design?

(d) What range of steady-state duty ratios will the buck converter run over if it is supplied from an input voltage $V_{in} = 150\,V$?

10.14 Consider an outphasing inverter system like that in Fig. 10.21.

(a) Write an expression for the power factor seen at the outputs of each of the two inverters in terms of Δ, R, and B.

(b) Find an expression for the outphasing angles Δ at which unity power factor is achieved in terms of Δ, R, and B. That is, at what outphasing angles do the inverters see a purely resistive effective load admittance?

(c) Find the outphasing angles at which $\mathrm{Im}(Y) = 0$ for the example in Fig. 10.22 (i.e., for $R = 0.5$ and $B = 0.45$).

11 AC/AC Converters

AC/AC converters take power from one ac system and deliver it to another with waveforms of the same or different amplitude, frequency, or phase. The ac systems can be single phase or polyphase, and reactive power can exist at the input, output, or both, depending on how we configure the converter. The simplest and most common ac/ac converter is the transformer.

A major application of ac/ac converters is variable-speed motor drives. These devices range in complexity from simple ac controllers found on products such as variable-speed electric drills or kitchen appliances, to highly sophisticated four-quadrant pulse-width modulated drives used for traction or roller-mill applications. The most common approaches to designing ac/ac converters use a dc link between the two ac systems, or provide direct conversion. The first are called *dc-link converters*. Those that fall into the second category are *cycloconverters* or a variant of the isolated dual-active-bridge converter discussed in Chapter 7.

11.1 Introduction to AC/AC Converters

In the dc-link approach, we first rectify the input ac waveform to obtain a dc waveform, which we then invert to produce the output ac waveform. A capacitor or inductor placed between the two converter stages stores the energy necessary to supply the instantaneous difference between the input and output powers. This intermediate dc stage is known as a *dc link* or *dc bus*. We can control the ac/dc and dc/ac converters independently, providing frequency conversion, so long as the *average* energy flows of the two are equal. And as usual, the converters can be configured and controlled to provide bidirectional power. We can implement the input and output stages of the dc-link converter with a phase-controlled, high-frequency dc/ac converter, or with a resonant converter, depending on the application. We have already studied these basic topologies, so we focus our discussion of dc-link converters on the requirements of the dc energy storage element.

The cycloconverter avoids an intermediate dc bus by converting the input ac directly into the desired output waveforms. Figure 11.1 illustrates the basic principle of cycloconversion. The input voltage source v_i is sinusoidal with a higher frequency than that of the output (a factor of three in this case). We control the switches in the cycloconverter to synthesize a waveform having a fundamental equal to the desired output frequency, which is a sub-multiple of the input. We then use filters to eliminate higher-frequency components. Although the design shown here uses a single-phase input source, a polyphase source is more common, and produces a more flexible converter.

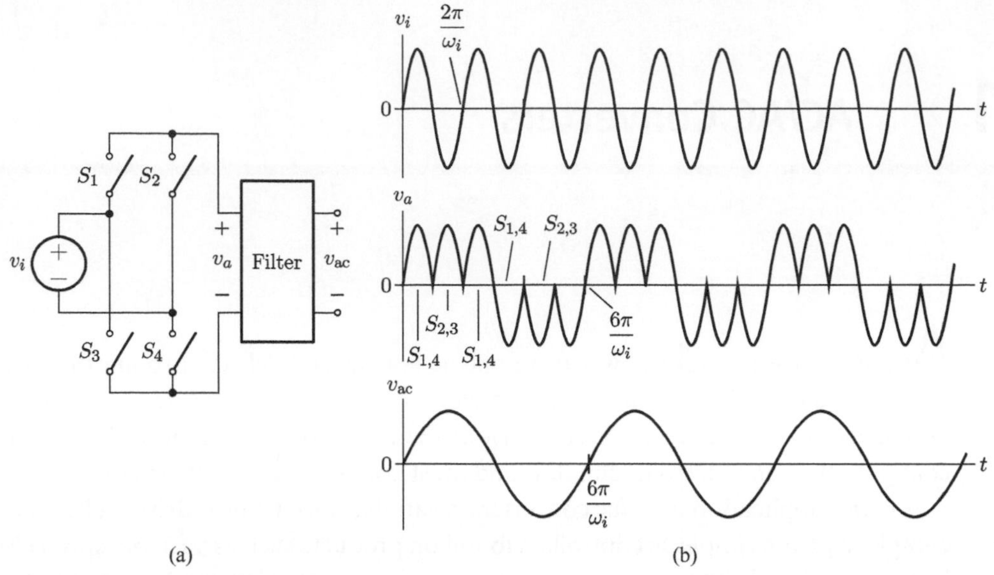

Figure 11.1 An illustration of the principle of operation of a cycloconverter. (a) An elementary cycloconverter circuit. (b) Waveforms of the ac source voltage v_i, the unfiltered output voltage v_a, and the filtered output voltage v_{ac}.

Applications requiring very high power, such as traction or steel-mill roll drives, historically required the use of SCRs because of the high voltages and currents at which these drives operate. An important advantage of the cycloconverter for these high-power applications is that it can use line commutation to turn off the SCRs. When operated this way, the cycloconverter is called a *naturally commutated cycloconverter*. The IGBT in a dc-link converter has displaced the SCR-based cycloconverter in many of these applications.

An increasingly important class of ac/ac converters is the *solid-state transformer* (SST). It is a system employing a high-frequency link between two much-lower-frequency ac systems, enabling voltage conversion to be done with a high-frequency transformer. The result is a device that is smaller and lighter than an iron-core transformer. Typically the two ac systems are of the same frequency, and the vision is that these solid-state transformers will replace conventional transformers in electric utility distribution systems. However, the SST can certainly support different frequencies at its input and output.

11.2 Energy Storage Requirements in a DC-Link Converter

In general, the instantaneous power flowing into a dc-link converter does not equal the instantaneous power flowing out of it. For instance, assume that the input and output waveforms have negligible distortion – and are therefore sinusoids – but have different amplitude, frequency, and phase, as shown in Fig. 11.2. The power waveforms at the two ports then each have a dc

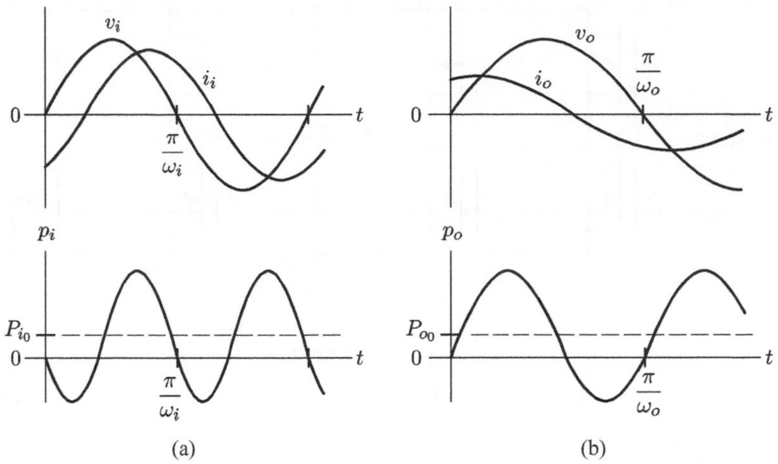

Figure 11.2 Waveforms at the input and output ports of a dc-link converter (neglecting harmonic distortion). (a) Input voltage and current and input power $p_i(t)$. (b) Output voltage and current, and output power $p_o(t)$.

component and an ac component (at the second harmonic of the respective port's fundamental frequency). These instantaneous input and output powers, p_i and p_o, are also shown in Fig. 11.2.

If the converter is 100% efficient, the average input power equals the average output power. Therefore the dc components of the power waveforms shown in Fig. 11.2(a) and (b) must be the same. However, as the input and output can be at different frequencies, the ac components of the two power waveforms need not be the same. The difference between the instantaneous input and output powers must be absorbed or delivered by an energy storage element within the converter. For the dc-link converter, this *load-balancing* energy storage element is located at the dc bus.

The load-balancing energy storage element can be a capacitor, an inductor, or both. The choice depends on the topologies used for the input and output conversion stages. For those topologies that require the dc system to resemble a voltage source, we need a capacitor; for those that require the dc system to resemble a current source, we need an inductor. For instance, in Section 8.3.2 you learned how to create an ac voltage from a dc voltage by modulating the duty ratio of bridge switches operated at a high frequency. So long as the switches have anti-parallel diodes, we can make the average power flow in either direction. Hence we can use one of these converters for the input stage of our dc-link converter and another for the output stage. Both converters require a voltage source at their dc ports, so we place a large capacitor across the dc bus, as shown in Fig. 11.3(a).

We can also use a phase-controlled converter for the input and output stages of a dc-link converter. As you saw in Chapter 5, the phase-controlled converter needs a large inductor in series with its dc port to hold the dc current constant over a cycle of the ac waveform. These inductors serve as the load-balancing energy storage element of the dc-link converter for this configuration, as shown in Fig. 11.3(b). The use of phase-controlled converters imposes the additional constraint that both the input and output be connected to ac voltage sources (lines).

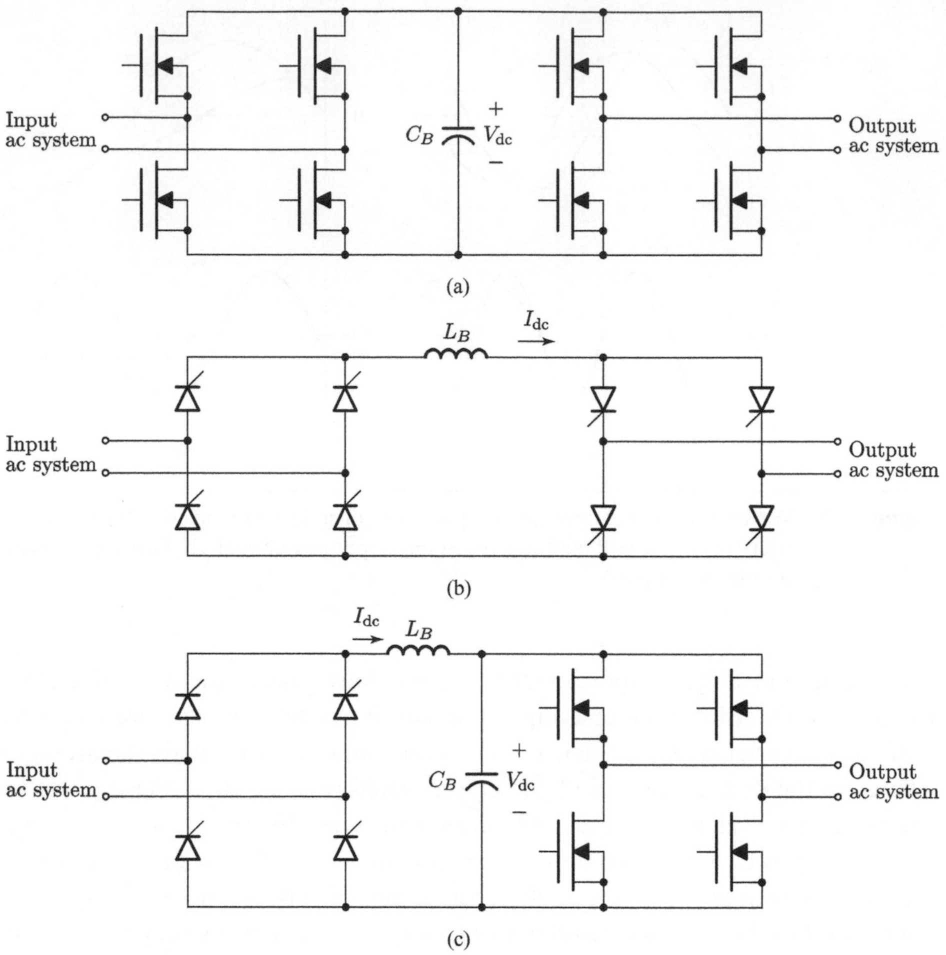

Figure 11.3 Possible choices of dc-bus energy storage elements. (a) A capacitor, when both dc/ac converters require a dc voltage source. (b) An inductor, when both converters require a dc current source. (c) Both an inductor and a capacitor, when one dc voltage source and one dc current source are needed.

This is precisely the method used for high-voltage dc transmission, in which the dc bus is the dc transmission line.

A typical implementation of a dc-link converter would use a phase-controlled rectifier to take power from the utility, and a voltage-source dc/ac converter to deliver it to an ac load. The former requires a current source at the dc bus, the latter a voltage source. In this case, we must achieve the load-balancing energy storage with two components: a series inductor and a shunt capacitor, as shown in Fig. 11.3(c).

The size of a load-balancing energy storage element depends on the amount of ac energy it must absorb and the level of ripple that its corresponding energy storage (or state) variable can tolerate. For instance, to determine the value of C_B in Fig. 11.3(a), we need to first find the difference between input and output power waveforms. The integral of this difference is the energy that flows in and out of C_B. We can then relate the peak-to-peak energy amplitude, ΔE_C, to the peak-to-peak ripple in the capacitor voltage, Δv_{dc}. If $\langle v_{dc} \rangle = V_{dc}$, then

$$\Delta E_C = \left(\frac{1}{2}\right) C_B \left[\left(V_{dc} + \frac{\Delta v_{dc}}{2} \right)^2 - \left(V_{dc} - \frac{\Delta v_{dc}}{2} \right)^2 \right], \tag{11.1}$$

from which we can determine the value of C_B.

Example 11.1 Sizing the DC-Bus Capacitor

We want to determine the size of C_B in the dc-link converter of Fig. 11.3(a) such that $\Delta v_{dc} < 0.1 V_{dc}$, that is, the dc bus ripple is less than 10% of V_{dc}. We first determine the maximum value of ΔE_C created by the difference between the input and output powers, and then use (11.1) to determine C_B. We assume that the input and output waveforms have negligible distortion.

The input power waveform will have a dc component and a second-harmonic component, as shown in Fig. 11.2(a). If the input voltage and current are in phase, the amplitude of the second harmonic P_{i_2} equals the amplitude of the dc component P_{i_0}. If they are out of phase, meaning that there is reactive power flow, $P_{i_2} > P_{i_0}$. In general,

$$P_{i_2} = \frac{P_{i_0}}{\cos \theta_i}, \tag{11.2}$$

where θ_i is the phase between the input voltage and the current waveforms. The output power waveform has a similar relationship between its dc and second-harmonic components.

The two dc components P_{i_0} and P_{o_0} are equal, and the difference between the two ac components flows into C_B. If the frequencies of the two power waveforms are ω_i and ω_o, these components are

$$p_{i_2}(t) = P_{i_2} \cos 2\omega_i t = \frac{P_{i_0}}{\cos \theta_i} \cos 2\omega_i t, \tag{11.3}$$

$$p_{o_2}(t) = P_{o_2} \cos 2\omega_o t = \frac{P_{o_0}}{\cos \theta_o} \cos 2\omega_o t. \tag{11.4}$$

As $P_{i_0} = P_{o_0}$, the difference between (11.3) and (11.4) is

$$p_{i_2}(t) - p_{o_2}(t) = \frac{P_{i_0}}{\cos \theta_i} \left(\cos 2\omega_i t - \beta \cos 2\omega_o t \right), \tag{11.5}$$

where $\beta = \cos \theta_i / \cos \theta_o$.

The dc-bus capacitor C_B has a total energy $E_C(t)$ composed of a dc component E_{dc} and an ac component $E_{ac}(t)$. We can find the ac component by integrating (11.5):

$$E_{ac}(t) = \int \left[p_{i_2}(t) - p_{o_2}(t) \right] dt = \frac{P_{i_0}}{\cos \theta_i} \left(\frac{\sin 2\omega_i t}{2\omega_i} - \beta \frac{\sin 2\omega_o t}{2\omega_o} \right). \tag{11.6}$$

The positive peak of this function occurs when $\sin 2\omega_i t = 1$ and $\sin 2\omega_o t = -1$; its negative peak occurs when $\sin 2\omega_i t = -1$ and $\sin 2\omega_o t = 1$. Therefore, the peak-to-peak variation in E_C is

$$\Delta E_C = \frac{P_{i_0}}{\cos \theta_i} \left(\frac{1}{\omega_i} + \frac{\beta}{\omega_o} \right). \tag{11.7}$$

Equating ΔE_C to the change in energy of a capacitor whose voltage changes from 0.95 V_{dc} to 1.05 V_{dc} (10%), we can find C_B:

$$C_B = \frac{\Delta E_C}{(1/2)\left[(1.05\, V_{dc})^2 - (0.95\, V_{dc})^2 \right]}. \tag{11.8}$$

If we use the parameters $P_{i_0} = 10\,\text{kW}$, $\omega_i = 2\pi \times 60$, $\omega_o = 2\pi \times 400$, $\theta_i = \pi/6$, $\theta_o = 0$, and $V_{dc} = 200$ V, then $\beta = 0.866$, $\Delta E_C = 34.6$ J, and $C_B = 8650\ \mu$F.

11.3 The Naturally Commutated Cycloconverter

The cycloconverter is typically used to provide ac/ac conversion at very high power levels, typically in excess of 100 kW. A thyristor can meet the switch voltage and current ratings needed at these power levels, and naturally commutated cycloconverter circuits provide a means of using SCRs without requiring forced commutation. Although we can design cycloconverters for lower power levels using fully controlled switches, such as transistors, the number of required switches and the complexity of their control usually makes a dc-link converter a more attractive choice. We therefore confine ourselves here to the naturally commutated cycloconverter utilizing SCRs.

Generally speaking, we control the switches in a cycloconverter to connect the output terminals of the converter to the input voltage source whose value is closest to the desired output voltage at the time. The switches that make these connections must usually support bipolar voltage and carry bidirectional current, which is one of the reasons why cycloconverter circuits are frequently quite complex. We can consider the ac controller described in Chapter 2 as the simplest form of cycloconverter, but we can use that only when frequency conversion is not required.

11.3.1 Principles of Operation

In Section 9.4 we discussed the operation of the six-pulse phase-controlled converter shown in Fig. 11.4(a). For fixed α, the output voltage of this converter is

$$v_o = V_d = \langle v_d \rangle = \frac{3V_{\ell\ell}}{\pi}\cos\alpha = V_{do}\cos\alpha, \tag{11.9}$$

where $V_{\ell\ell}$ is the amplitude of the line-to-line voltage. Note that as α increases from 0 to π, the output voltage varies from V_{do} to $-V_{do}$. If we modulate α slowly compared to the input frequency ω_i, the short-time average of $v_d(t)$ yields an output voltage $\overline{v}_d(t)$ that varies with time, $v_o(t) = \overline{v}_d(t)$. This technique of synthesizing a waveform that varies with time by varying α is similar to the waveshaping technique discussed in Section 8.3. There, we used a duty ratio $d(t)$ that varied with time to modulate the output of a dc/dc converter.

The wave-shaped output of a modulated dc/dc converter is directly proportional to $d(t)$. Shaping the output of a phase-controlled converter, however, is complicated by the nonlinear relationship between $\alpha(t)$ and $\overline{v}_d(t)$ given by (11.9). For example, suppose we want $\overline{v}_d(t) = v_o$ to be sinusoidal, or $v_o = V_o \sin\omega_o t$, where $\omega_o \ll \omega_i$ and $V_o < V_{do}$. The required $\alpha(t)$ must be such that $v_o = V_{do}\cos\alpha(t) = V_o \sin\omega_o t$, which gives

$$\alpha(t) = \cos^{-1}\left(\frac{V_o}{V_{do}}\sin\omega_o t\right). \tag{11.10}$$

Figure 11.4(b) shows the waveforms that result from this $\alpha(t)$. The waveform we are trying to synthesize is v_o', and if $\omega_o \ll \omega_i$ so that the inductor effectively removes the harmonics from v_d, then $v_o \approx v_o'$. However, as ω_o and ω_i get closer, the output filter will affect both the phase and amplitude of the fundamental of v_d.

In a practical control circuit, the function for α given in (11.10) is generally implemented indirectly by comparing the line-to-line input voltages with the desired output waveform v_o' and

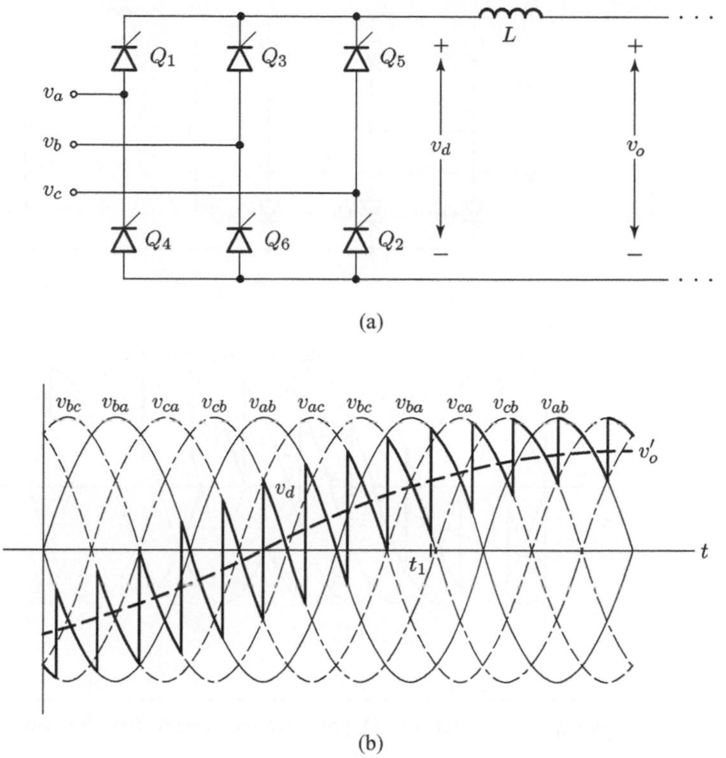

Figure 11.4 (a) A six-pulse phase-controlled two-quadrant (I and II) converter. (b) Output voltage when the circuit of (a) is controlled to be a naturally commutated cycloconverter. The voltage v_o' is the desired waveform and, if L is properly sized, $v_o \approx v_o'$.

turning on the next SCR when the voltage v_d that would result is closer to v_o' than is the present value of v_d. For instance, just before t_1 in Fig. 11.4(b), v_{bc} is connected to the output through Q_3 and Q_2. The next commutation event is from Q_2 to Q_4, connecting v_{ba} to the output. At t_1, v_{bc} and v_{ba} are equidistant from v_o', but v_{bc} is diverging while v_{ba} is converging on v_o'. At this time, then, Q_4 is triggered and v_d assumes the value of v_{ba}. The next event is initiated based on a comparison of $v_{ca} - v_o'$ with $v_o' - v_{ba}$, and so on.

The problem with the ac/ac converter we have developed so far is that, although the output voltage can be either positive or negative, the output current can only be positive. To obtain bilateral load current, we need to place in parallel with this "positive" converter a "negative" converter, created by inverting its SCRs relative to those in the positive converter, so that they can carry negative load current. A negative converter is shown in Fig. 11.5(a); its output voltage waveform is shown in Fig. 11.5(b). Note that, because the SCRs are inverted, the output voltage takes a negative step at commutation. The combined circuit, known as a *four-quadrant naturally commutated cycloconverter*, is shown in Fig. 11.6(a). The waveforms of Fig. 11.6(b) illustrate how the two converters produce an ac output voltage for the output current shown. Note that the direction of the step change in v_d at commutation indicates which converter is operating at any point in the cycle. At t_o, operation shifts from the negative to the positive converter.

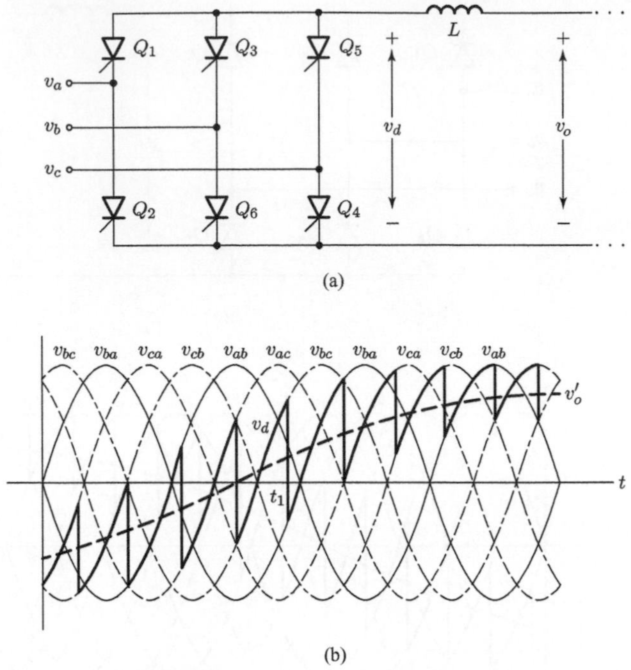

Figure 11.5 (a) A two-quadrant (III and IV) negative converter. (b) The voltage v_d of the converter in (a) and the desired output waveform v_o'.

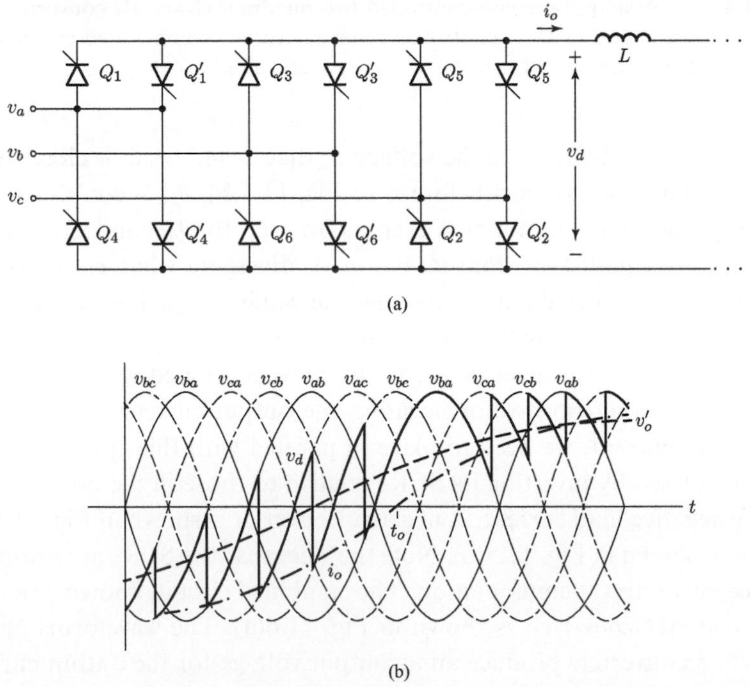

Figure 11.6 (a) A naturally commutated four-quadrant cycloconverter made by combining a positive and a negative phase-controlled converter. (b) Output voltage and current for the cycloconverter of (a).

Example 11.2 A Control Circuit For the Naturally Commutated Cycloconverter

Figure 11.7 is a block diagram of a circuit suitable for controlling the six naturally commutated switches (each comprised of Q and Q') in the cycloconverter of Fig. 11.6. The finite-state machine has six states, one allocated to each of the six switches (12 SCRs). The output of analog multiplexer MUX_1, v_{d_1}, is a synthetic, signal-level version of the converter voltage v_d. The second analog multiplexer MUX_2 has an output v_{d_2} that is one commutation event (60°) ahead of v_d and v_{d_1}.

The essence of this control scheme is that commutation of the converter begins when v_{d_1} and v_{d_2} are equidistant from, but on opposite sides of, v_o'. At this point, the algebraic sum of $v_{d_1} - v_o'$ and $v_{d_2} - v_o'$ is zero. The voltages v_1 and v_2 are the differences between the multiplexer outputs and v_o'. For example, just before t_1 in Fig. 11.4(b), $v_{d_1} = v_d = v_{bc}$, and $v_{d_2} = v_{ba}$. At t_1, $v_1 < 0$ and $v_2 = -v_1$. At this time the zero-crossing detector clocks the finite-state machine and the commutation event at t_1 takes place, as well as the switching of the multiplexers. This description is based on the assumption that the load current i_o is positive and therefore that the positive converter is functioning. In this case the zero-crossing is from positive to negative. If load current were negative, the zero-crossing would be oppositely directed, from negative to positive, as you can see from Fig. 11.5(b). The polarity of the load current is thus an input to the zero-crossing detector, determining which zero-crossing direction is valid.

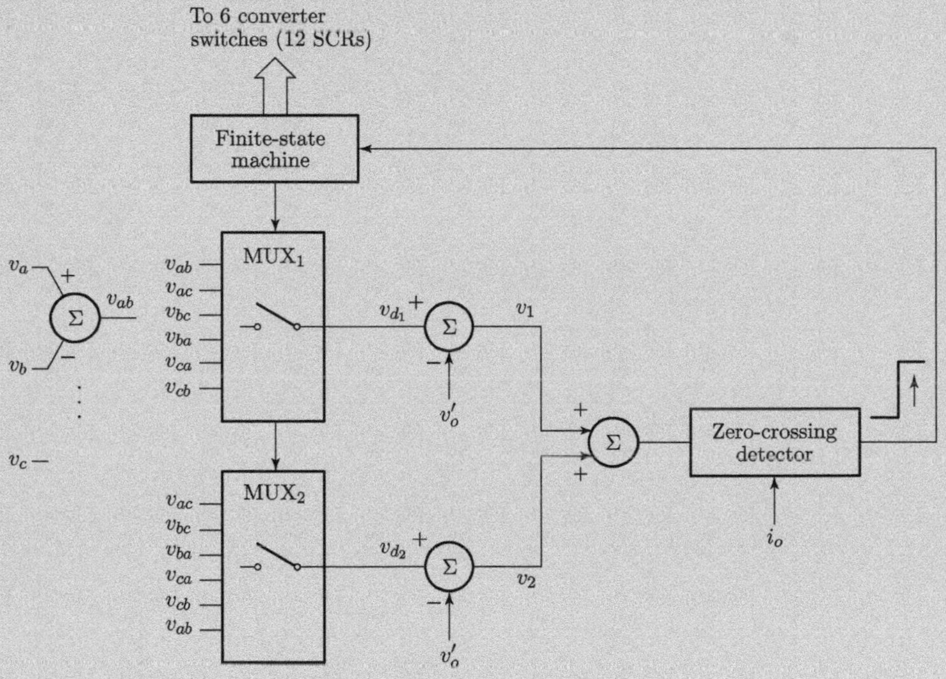

Figure 11.7 A circuit for controlling the cycloconverter of Fig. 11.6.

11.3.2 Cycloconverters with Polyphase Outputs

Most high-power applications of cycloconverters require polyphase output waveforms. For instance, we may need to drive a large three-phase machine at varying speeds from a three-phase fixed-frequency service. We could duplicate the cycloconverter of Fig. 11.6 to create the three output waveforms required, but could not directly connect the outputs of the individual converters in a delta or Y configuration. The reason is that the output terminals of each converter are directly connected to the input voltages by the SCRs that are on, and if two such terminals were connected, the input voltage sources would be shorted. Therefore, if we use independent cycloconverters for each phase, we must incorporate into the circuit input or output isolation transformers to allow cycloconverter interconnection at the load.

Another way to create a three-phase cycloconverter, which avoids the need for isolation transformers and halves the number of switches required, is shown in Fig. 11.8(a). This approach

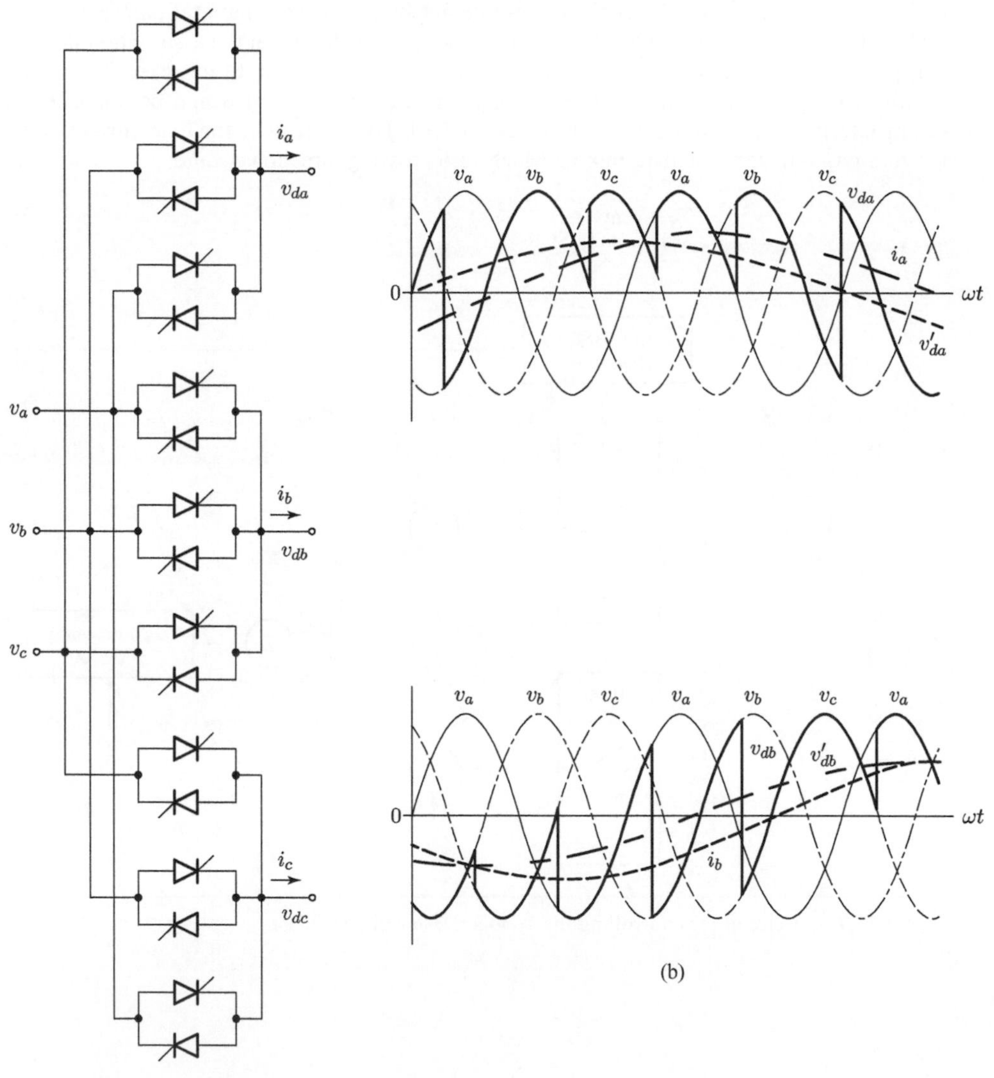

Figure 11.8 (a) A three-phase in, three-phase out cycloconverter using three-pulse bridges. (b) Waveforms at the *a* and *b* phases in the cycloconverter of (a).

uses three three-pulse phase-controlled converters to create the three output voltages. Each converter has six SCRs, three to carry positive load current and three to carry negative load current. The input voltage sources are indicated as line-to-neutral sources to make it easier for you to see how the output voltage waveforms are created. Figure 11.8(b) shows the line-to-neutral output voltages v_{da} and v_{db}. Note that as each three-pulse converter has only three input voltages to choose from (the negative input voltages are not available), commutations occur half as often as with the six-pulse circuit. The deviations of v_{da}, v_{db}, and v_{dc} from v'_{da}, v'_{db}, and v'_{dc} are therefore larger than the deviation of v_d from v'_o in Fig. 11.5(b). Other than that, v_{da}, v_{db}, and v_{dc} have the same features as v_d, including positive step changes when the load current is positive and negative step changes when the load current is negative.

11.4 An Isolated High-Frequency-Link Cycloconverter

The cycloconverters discussed so far are naturally commutated and require bidirectional voltage-blocking switches. The SCR provides this capability and is used when the frequencies of concern are within their capability, typically around 400 Hz. One can also implement cycloconverters that need not rely on natural commutation, and which can operate at substantially higher frequencies. This can be accomplished by using "back-to-back" transistors or other four-quadrant switch configurations, and in some cases such designs can be operated with ac waveforms in the hundreds of kHz and above.

There are still other cycloconverter configurations that avoid the necessity for four-quadrant switches. An example of such a circuit is shown in Fig. 11.9. It is a cycloconverter that converts a high-frequency input current to a low-frequency output feeding a sinusoidal voltage source. One application of such a circuit is to interface a solar panel to the ac grid. The origin of the high-frequency current source i_s is not shown, but can take the form of a resonant dc/ac converter, as discussed in Chapter 10.

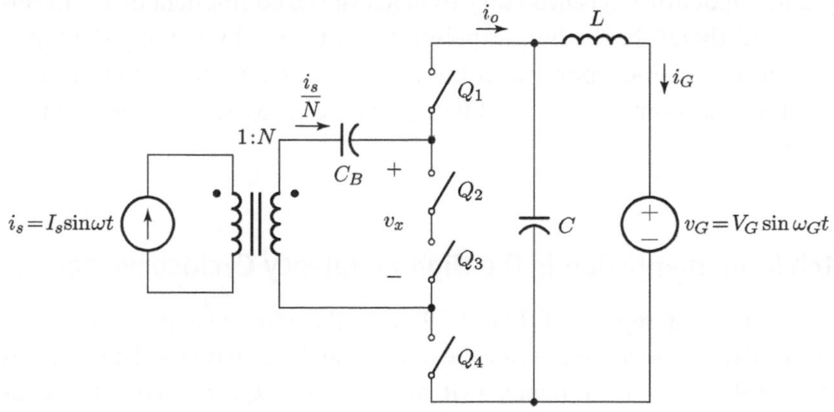

Figure 11.9 An isolated high-frequency-link cycloconverter.

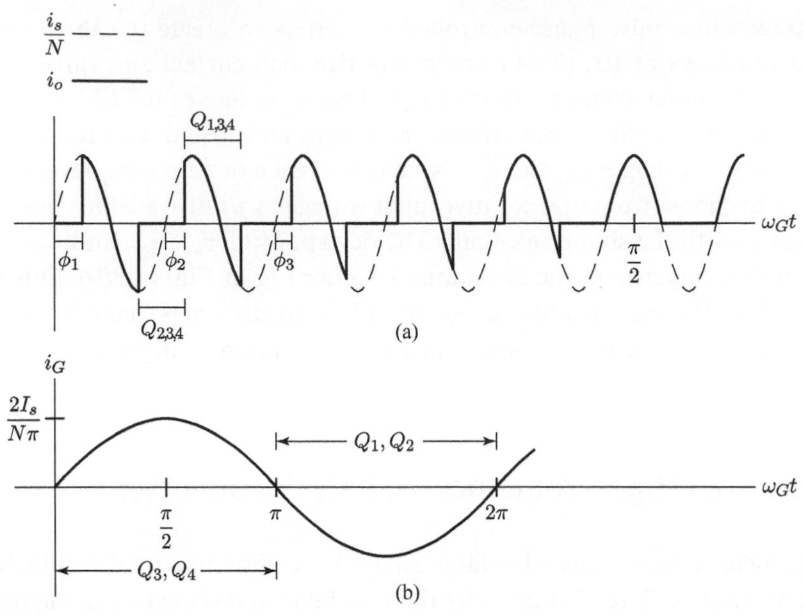

Figure 11.10 Current waveforms in the cycloconverter of Fig. 11.9. (a) The source current i_s reflected through the transformer, and the current i_o for $0 < \omega_G t < \pi/2$. (b) The output current i_G showing the on-switches in the positive and negative half-cycles.

In operation, the switches Q_3 and Q_4 are on during the positive half-cycle of the grid voltage v_G, and Q_1 and Q_2 are complementarily switched to take 180° sections out of the secondary current i_S/N, as shown in Fig. 11.10. These sections are phase-shifted each cycle of i_S so their average values follow a sinusoid at the grid frequency. During the negative half-cycle of v_G the switch roles are reversed: Q_1 and Q_2 are on, and Q_3 and Q_4 are switched to extract sections of i_S/N. The grid-side LC filter passes the average of these pieces as i_G. The result is a sinusoidal current i_G at grid frequency and in phase with v_G, that is, the power factor at the grid interface is unity. The capacitor C_B is necessary to block the dc component of v_x. In this circuit, the output current, and therefore the output power, is controlled by varying the input current amplitude I_S through means incorporated in the current-source implementation, though one could also control the cycloconverter to modulate power (e.g., by selecting less than 180° sections of the waveform).

11.4.1 Switch Implementation in the High-Frequency Cycloconverter

Figure 11.11 is a repeat of Fig. 11.9 with the reference polarity for Q_1 shown, which is representative of the references for the remaining three switches. The switches have four states, as shown in Table 11.1. To minimize switching, Q_3 and Q_4 remain on during the positive half-cycle of the grid voltage, and Q_1 and Q_2 during the negative half-cycle.

Table 11.1 The states and voltage polarities for the switches in Fig. 11.11.

State		Q_1	Q_2	Q_3	Q_4
1	$v_G > 0, i_o > 0$	On ($v = 0$)	Off ($v > 0$)	On ($v = 0$)	On ($v = 0$)
2	$v_G > 0, i_o = 0$	Off ($v > 0$)	On ($v = 0$)	On ($v = 0$)	On ($v = 0$)
3	$v_G < 0, i_o < 0$	On ($v = 0$)	On ($v = 0$)	Off ($v < 0$)	On ($v = 0$)
4	$v_G < 0, i_o = 0$	On ($v = 0$)	On ($v = 0$)	On ($v = 0$)	Off ($v < 0$)

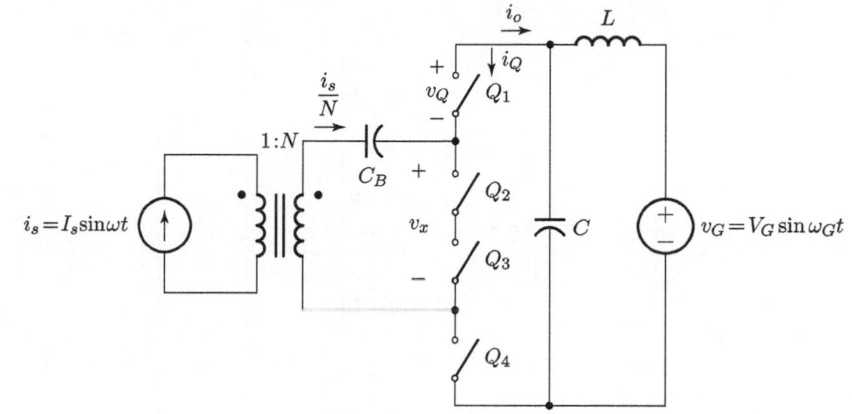

Figure 11.11 The isolated high-frequency link cycloconverter of Fig. 11.9 defining the polarities of the switch variables.

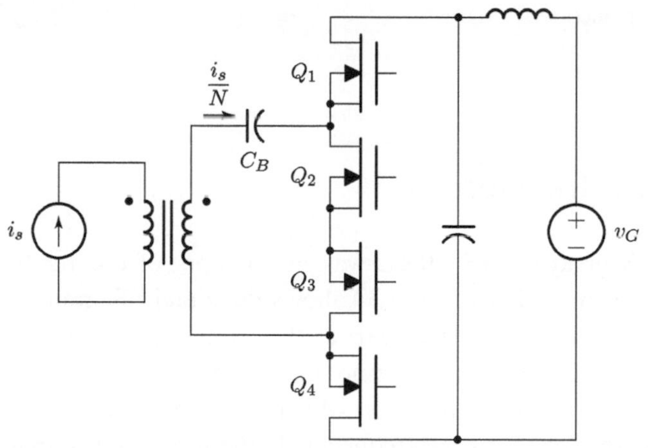

Figure 11.12 A switch implementation for the isolated high-frequency-link cycloconverter of Fig. 11.9.

Since MOSFETs can conduct in either direction but can block only a $v_{DS} > 0$, we can use them as the switches with Q_3 and Q_4 inverted, as shown in Fig. 11.12. Providing gate drives for the four MOSFETs is somewhat complex, as their sources are floating and do not have a common reference. The design of gate drives is covered in Chapter 23.

Example 11.3 Reducing Switch Resistive Loss in the High-Frequency-Link Cycloconverter

Figure 11.13 shows the cycloconverter of Fig. 11.12 with the addition of two high-frequency, low-loss bypass capacitors C_1 and C_2 across the output, with their midpoint connected between Q_2 and Q_3. Consider the positive half-cycle of v_G. The bottom transistors Q_3 and Q_4 turn on at the positive zero-crossing of v_G, at which time the voltage of both bypass capacitors is zero. During the remainder of the half-cycle, v_{C_2} remains zero and Q_3 and Q_4 are effectively in parallel. Their combined on-state drop is therefore $R_{DS}/2$. The total resistive drop during operation is thus $3R_{DS}/2$, compared with $2R_{DS}$ without the bypass capacitors.

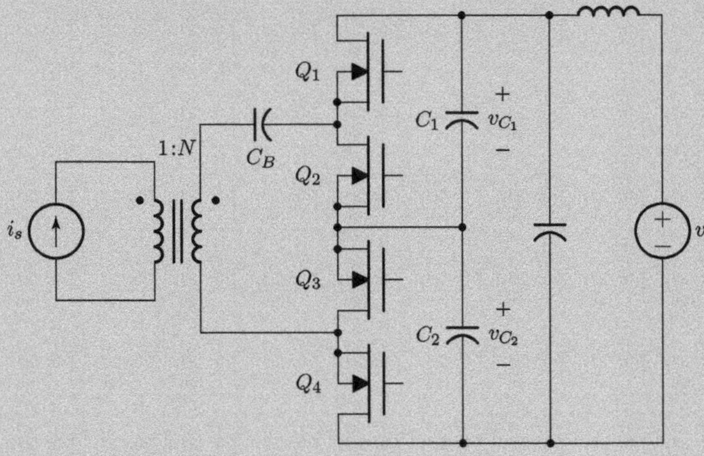

Figure 11.13 Splitting the output capacitor in the isolated high-frequency-link cycloconverter of Fig. 11.12 to reduce device loss.

11.5 Solid-State Transformer

William McMurray of GE CR&D was, in 1968, one of the first to propose a design capable of functioning as an SST. Figure 11.14 shows the circuit diagram from his patent. The left half-bridge creates a high-frequency carrier that is amplitude modulated (AM) by the input (ac) waveform. The right half-bridge demodulates the AM signal, after being converted in amplitude by the high-frequency transformer, which is much smaller than a conventional transformer designed for the lower output frequency. If the input and output are dc, the circuit is simply an isolated dc/dc converter. This figure provides the essence of a solid-state transformer.

As noted earlier, replacing electric utility system transformers, particularly in the distribution network, is one application motivating the development of SSTs. Others include traction and electric vehicle charging systems. The challenge of SSTs is their cost and lower efficiency relative to conventional transformers. However, the power- and voltage-control capabilities that the SST provides may compensate for the cost and efficiency loss in appropriate applications. For example, although a distribution transformer may be more efficient at load, there are periods when the load is very light. During these periods their copper loss is very small but their core loss

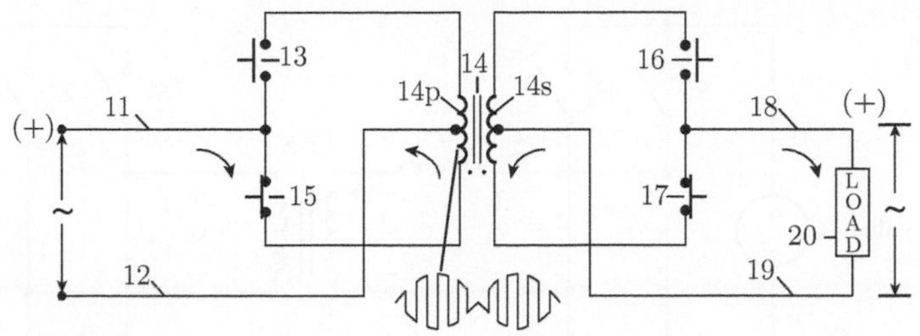

Figure 11.14 The circuit concept proposed by McMurray, which illustrates the essence of a solid-state transformer. (US Patent no. 3517300, June 23, 1970)

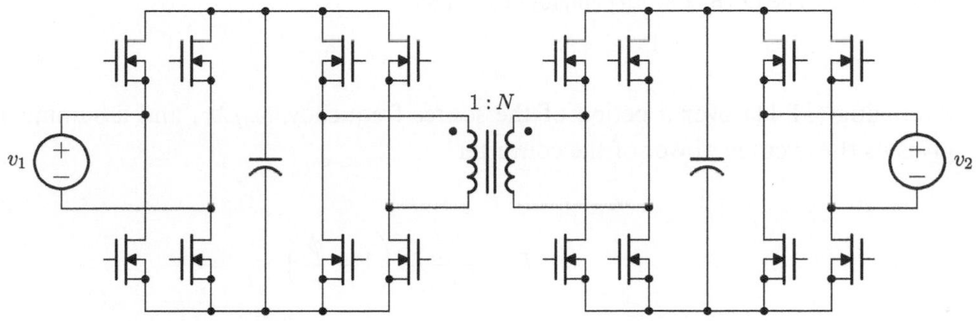

Figure 11.15 The cascaded approach to an SST.

remains unchanged, as it is a function of the voltage, not the current. An SST with appropriate controls can reduce this "idling" loss.

11.5.1 Cascaded SST

Conceptually, an SST can be created by taking any of the isolated dc/dc converters discussed in Chapter 5 and adding rectifier/inverter stages to the input and output. Figure 11.15 shows a "brute force" design based on this concept. It illustrates the large number of switches and stages required. Though straightforward, the cascaded SST incorporates two dc buses and is accompanied by complex control and low efficiency relative to a conventional transformer.

11.5.2 DAB as an SST

Figure 11.16 shows the DAB converter discussed in Chapter 7, with ac sources replacing their dc counterparts. The bridges are again operated at a high frequency ω_s relative to the sources, that is, $\omega_s \gg \omega_o$, with $V_2 = NV_1$ and phase-shifted by an angle ϕ of the switching period. During a switching period the ac waveforms can be assumed constant, so, using (7.15), the local average power is

$$\overline{p} = \frac{\overline{v}_a^2}{\omega_s L}\phi\left(1 - \frac{\phi}{\pi}\right), \qquad 0 < \phi < \pi. \tag{11.11}$$

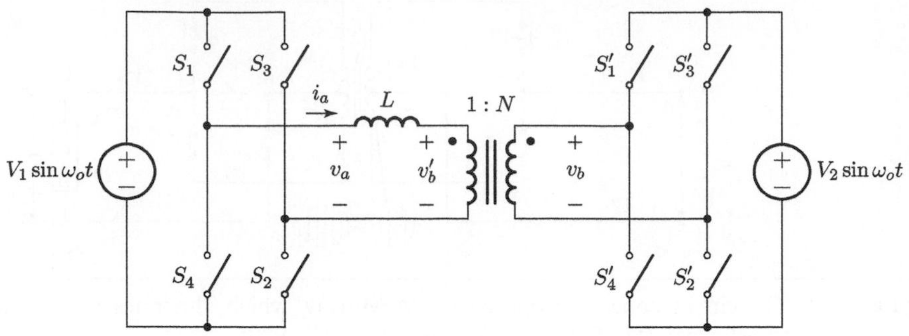

Figure 11.16 The DAB converter employed as an SST.

Averaging (11.11) over a period of the source frequency, $\omega_o/2\pi$, and assuming $V_1 = V_2/N$, gives us the average power of the converter:

$$P = \frac{V_1^2}{2\omega_s L}\phi\left(1 - \frac{\phi}{\pi}\right). \tag{11.12}$$

As with the DAB discussed in Chapter 7, the DAB as an SST is controlled by feeding back one of the output variables to establish ϕ. In the case of the SST linking two sources, one would feed back the output current. In an application where the load is other than a voltage source, either the current or the voltage is fed back, depending on whether the desire is to control the output power or voltage.

Example 11.4 A DAB With an RC Load

Figure 11.17 shows a DAB SST delivering 500 W to an RC load with $N = 1$, so $V_1 = V_2 = 100$ V. If $\omega_s = 10^6$, what value of L should we specify?

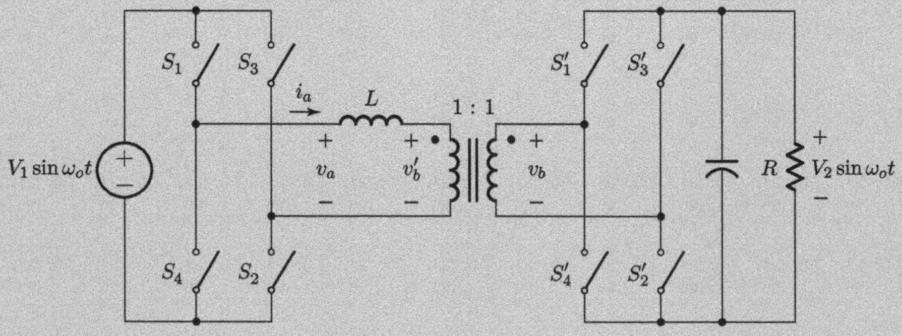

Figure 11.17 A DAB as an SST with a resistive load.

The average power delivered to the load is

$$P = 500\,\mathrm{W} = \frac{1}{\pi}\int_0^{\pi} \overline{p}(t)\, d(\omega_o t) = \frac{V_a^2}{2\omega_s L}\left(\phi - \frac{\phi^2}{\pi}\right) = \frac{V^2}{2R}, \tag{11.13}$$

$$\phi = \frac{\pi}{2} \pm \frac{1}{2}\sqrt{\pi^2 - \frac{4\omega_s L\pi}{R}}. \tag{11.14}$$

Requiring the radicand to be positive shows we must have $L < 7.85\,\mu\mathrm{H}$ for the parameters of the example.

Choosing $L = 5\,\mu\mathrm{H}$ results in $\phi = 0.63\,\mathrm{rad}$. If we use negative feedback to control ϕ, then ϕ must be less than $\pi/2$. So this value of L provides us with room to maintain 100 V at the load for variations in R.

11.6 Matrix Converter

Figure 11.18 illustrates the essence of what is known as a *matrix converter*. Subject to certain constraints, the nine switches forming the matrix can connect any of the input sources to any of the outputs. Analogous to the cycloconverter, the output waveform is constructed of pieces of the three input waveforms to create an output having a frequency and amplitude different from those of the input.

While the concept is easy to visualize, implementation of the matrix converter is complicated. The switches must be bilateral in both voltage and current (which is not difficult to achieve), and gate drive must be supplied to 18 switches for a three-phase to three-phase circuit if back-to-back unilateral switches are used. The commutation among the switches must be controlled carefully to prevent short circuiting the input or output. Also, depending on the nature of the conversion, large harmonic content may be present at the input and output, requiring aggressive filtering.

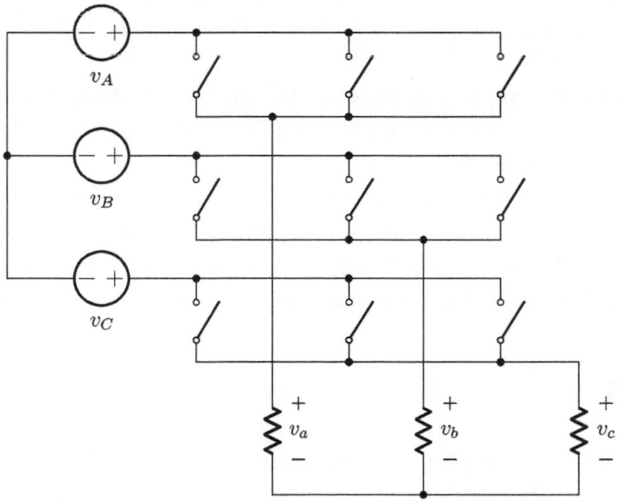

Figure 11.18 The essential elements of a matrix converter.

Most difficult is the design of a modulation scheme that creates the desired output from sections of the input. Possible modulation schemes have been the subject of many academic papers, but practical application of the converter is rare because of the complexity of implementation. With the advent of more powerful microcontrollers and modulation algorithms, the future may see the matrix converter achieve greater utility.

Notes and Bibliography

An exhaustive and detailed treatment of cycloconverters can be found in [1]. The treatment tends to be more mathematical and theoretical than practical, but there is a good chapter on control of cycloconverters that includes a number of block diagrams of control system implementations. The isolated high-frequency-link cycloconverter and its control are presented in [2].

An ac/ac converter of a type sometimes referred to as a high-frequency-link converter is described in numerous papers by Fransisc Schwarz and J. Ben Klaassens. A good introduction to the concept is given in [3]. A more detailed application of the high-frequency-link converter is to be found in [4].

A discussion of the benefits of the SST in an electric utility distribution system is presented in [5]. The application of the dual active-bridge converter as an SST is discussed in [6]. In addition to presenting the benefits of an SST, the paper summarizes high-frequency transformer technology and compares several SST topologies. A good introduction to matrix converters and the challenge of their control is presented in [7].

1. L. Gyugyi and B. Pelly, *Static Power Frequency Changers*, Chichester: Wiley, 1976.
2. A. Trubitsyn, B. J. Pierquet, A. K. Hayman, G. E. Gamache, C. R. Sullivan, and D. J. Perreault, "High-Efficiency Inverter for Photovoltaic Applications," *IEEE Energy Conversion Congress and Exposition (ECCE)*, pp. 2803–2810, Sept. 2010.
3. F. C. Schwarz, "A Double-sided Cycloconverter," *IEEE Power Electronics Specialists Conference (PESC)*, San Diego, CA, 1979, pp. 437–447.
4. J. B. Klaassens, "DC-AC Series-Resonant Converter System with High Internal Frequency Generating Multiphase AC Waveforms for Multikilowatt Power Levels," *IEEE Trans. Power Electronics*, vol. 2, pp. 247–256, July 1987.
5. X. She, A. Q. Huang, and R. Burgos, "Review of Solid-State Transformer Technologies and Their Application in Power Distribution Systems," *IEEE J. Emerging and Selected Topics in Power Electronics*, vol. 1, pp. 186–198, Sept. 2013.
6. H. Qin and J. W. Kimball, "Solid-State Transformer Architecture Using AC–AC Dual-Active-Bridge Converter," *IEEE Trans. Industrial Electronics*, vol. 60, no. 9, pp. 3720–3730, Sept. 2013.
7. P. W. Wheeler, J. Rodriguez, J. C. Clare, L Empringham, and A. Weinstein., "Matrix Converters: A Technology Review," *IEEE Trans. Industrial Electronics*, vol. 49, pp. 276–288, Apr. 2002.

PROBLEMS

11.1 The simple single-phase cycloconverter of Fig. 11.1 is designed to drive a resistive load through a filter consisting of an inductor, as shown in Fig. 11.19. Design a suitable implementation for the switches.

11.2 The cycloconverter of Fig. 11.19 is designed to produce an output v_{ac} with a fundamental frequency of $\omega_i/3$, as shown in Fig. 11.1. If $R/L = \omega_i/3$, what are the amplitudes of the fundamental and first nonzero harmonic of v_{ac} relative to the amplitude of the input voltage V_i? Sketch v_{ac}, considering only these two components.

11.3 Sketch the input current i_i for the cycloconverter of Fig. 11.19 with the values of L and R as specified in Problem 11.2. What is the power factor of the converter?

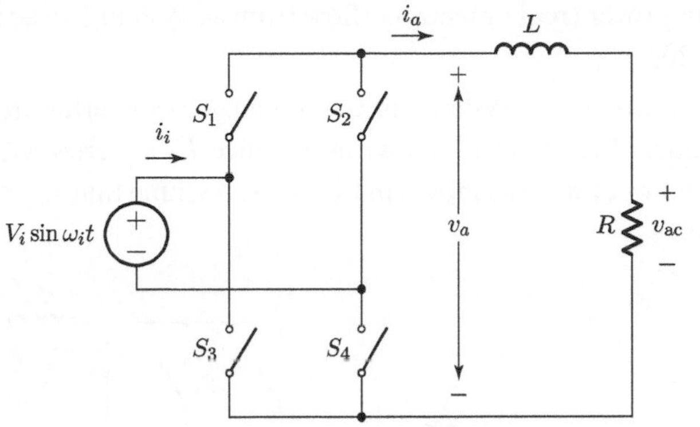

Figure 11.19 The cycloconverter of Fig. 11.1 with a simple low-pass filter. This circuit is the subject of Problems 11.1, 11.2, and 11.3.

11.4 A high-voltage dc (HVDC) transmission system consists of two phase-controlled converters connected by a dc link, as shown in Fig. 11.20 for a simplified single-phase ac system. It is often used to tie together ac utility systems that have different frequencies or are out of phase relative to each other. One of the advantages of an HVDC intertie between two ac systems is that the intertie can be used to control real and reactive power independently, enhancing the stability of the overall system. In answering the following questions, assume that the ac filters remove all but the fundamental components of the line current and do not affect the fundamental at all.

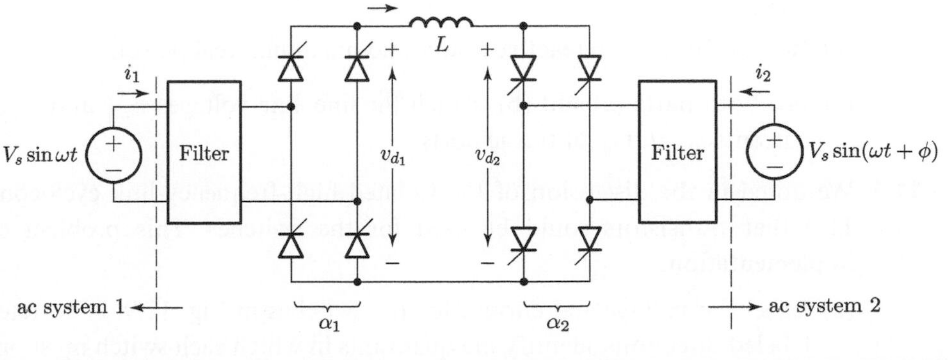

Figure 11.20 The HVDC intertie analyzed in Problem 11.4.

(a) How is the link current I_{dc} controlled?

(b) Sketch v_{d_1} and v_{d_2} if α_1 and α_2 are controlled so that the real power flow is zero, but the reactive power flowing into system 2 is maximized for a given I_{dc}, and the power factor for ac system 2 is leading – that is, the source of this reactive power acts like a capacitor.

(c) Repeat (b) for zero reactive power and maximum real power.

(d) Can power (real or reactive) flow from ac system 2 to ac system 1 in the circuit of Fig. 11.20?

11.5 The ac sources in a system containing a cycloconverter are seldom without some series impedance. Figure 11.21 shows inductance L_c in series with the source $V_i \sin \omega_i t$ of the converter operating as shown in Fig. 11.1. Assume that $L_c \ll L$. Make qualitative sketches of i_i and v_a.

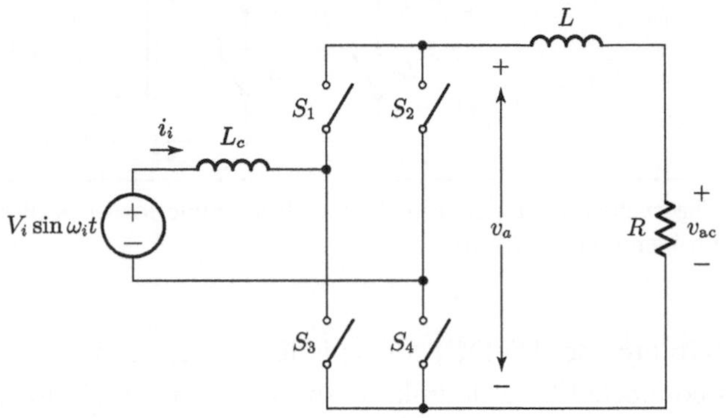

Figure 11.21 The single-phase cycloconverter with commutating inductance L_c of Problem 11.5.

11.6 Figure 11.22 shows an HVDC intertie between two three-phase ac systems.

(a) Sketch v_{d_1} and v_{d_2} if α_1 and α_2 are controlled so that the real power flow is zero, but the reactive power flowing into system 2 is maximized for a given I_{dc}, and the power factor for ac system 2 is lagging – that is, the source of this reactive power acts like an inductor.

(b) Repeat (b) for zero reactive power and maximum real power.

(c) For both parts (a) and (b), sketch the line–line voltages v_{ab_1} and v_{ab_2}, and the line currents i_{a_1} and i_{a_2} at the ac ports.

11.7 We noted in the discussion of the isolated high-frequency-link cycloconverter of Fig. 11.9 that transistors could be used for the switches. This problem concerns their implementation.

(a) Label the voltage and current for the switches in Fig. 11.9. With reference to your labeled directions, identify the quadrants in which each switch must operate.

(b) Provide a switch implementation for this converter.

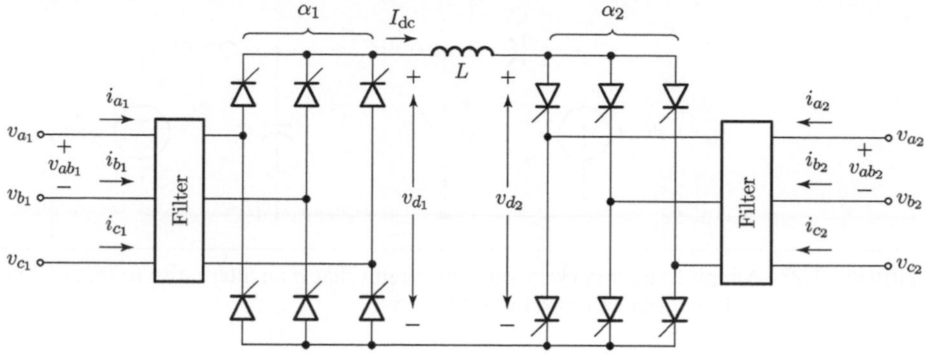

$$v_{a_1} = V \sin \omega t \qquad v_{a_2} = V \sin(\omega t + \phi)$$

$$v_{b_1} = V \sin\left(\omega t - \frac{2\pi}{3}\right) \qquad v_{b_2} = V \sin\left(\omega t + \phi - \frac{2\pi}{3}\right)$$

$$v_{c_1} = V \sin\left(\omega t + \frac{2\pi}{3}\right) \qquad v_{c_2} = V \sin\left(\omega t + \phi + \frac{2\pi}{3}\right)$$

Figure 11.22 A dc-link converter connecting two three-phase ac networks. This dc intertie is considered in Problem 11.6.

11.8 Select a region of v_G in the cycloconverter of Fig. 11.9 and plot v_x for several cycles of i_S. Label the y axis with the values of v_x.

11.9 Consider Example 11.4.

(a) Show that the statement "If we use negative feedback to control ϕ, ϕ must be less than $\pi/2$." is true.

(b) Given the value of L we have chosen in the example (5 μH), what is the range of R that will still allow the DAB SST to maintain 100 V at the load?

(c) If the converter were interfacing a source (such as a solar panel) to the grid, so that the configuration was as shown in Fig. 11.16, with $V_a = V_b = 100$ V, what is the range of power that could be supplied to the grid?

11.10 Figure 11.23 is an alternate possibility for the cycloconverter circuit of Fig. 11.9. The transformer has been omitted for simplicity but does not affect circuit operation. The switches operate at the input source frequency, ω_s.

(a) Select a region of v_G and sketch and dimension the voltage v_1 for several cycles of i_s.

(b) What is the function of C_1?

(c) Provide an implementation of the switches S_1 and S_2.

(d) Considering only resistive losses in the switches, how does the efficiency of this circuit compare with that of Fig. 11.12?

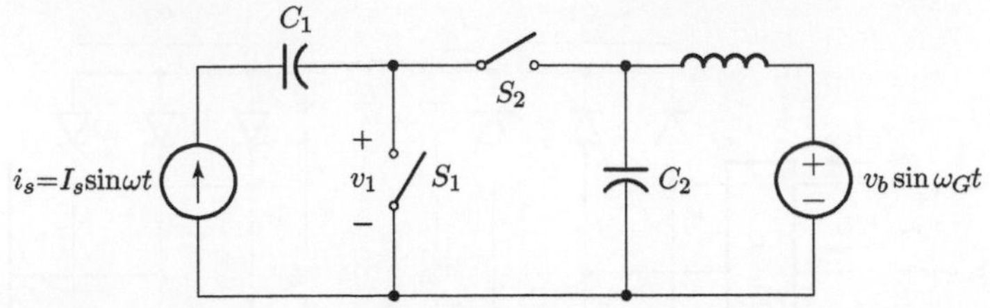

Figure 11.23 A high-frequency cycloconverter circuit that is an alternative to that of Fig. 11.9. This circuit is the subject of Problem 11.10.

11.11 Determine what devices you would use and their connections for the eight switches in the DAB SST of Fig 11.17. Draw the resulting circuit.

Part II

Dynamic Models and Control

12 Dynamic Models and Control: An Overview

In Part I we examined the form and function of the major families of power electronic converters. Our goal was to show how the intended power conversion function is achieved in each case by appropriate configuration of the circuit components and by proper operation of the switches. Throughout those earlier chapters, our concern was with *nominal* operating conditions, that is, the ideal operating conditions in which a converter is designed to perform its primary conversion function. As nominal operation in most power electronic circuits involves a *periodic steady state*, we focused on situations in which circuit operation and behavior are the same from cycle to cycle.

Now we need to deal with the consequences of the inevitable disturbances or errors that cause circuit operation to deviate from nominal. These disturbances include variations and uncertainties in source, load, and circuit parameters, perturbations in switching times, startup and shutdown, or component failure. We refer to the resulting deviations from nominal behavior as the *dynamic behavior* of the circuit. If such deviations have a negligible effect on converter operation, the user may be content to let the converter run without corrective action. This is rarely the case, however.

Departures from desired nominal conditions have to be counteracted through properly designed controls. We show examples of circuits that do not recover, or whose recovery is incomplete or too slow, without such controls. A controller or "compensator" must provide the user with a simple and convenient means of specifying the desired nominal operating condition, and must automatically regulate the circuit at this operating condition by appropriately adjusting the times at which switches are turned on and off.

Our aim here in Part II is to build the basis for analyzing the dynamic behavior of power circuits, and for designing and implementing controls that regulate these dynamics, ensuring operation close to the desired condition, despite disturbances or errors. Our focus will be on analysis and control design using appropriate models of the dynamic behavior. The dynamic models will allow us to anticipate behavior under diverse operating conditions, to generate candidate controller structures and parameters, plan simulation studies, understand experimental results, recognize which regimes of operation call for further investigation, and so on. Such an approach is especially critical in power electronics because of the effort and cost involved even in breadboarding, and the expense (not to mention the distress and, sometimes, smoke!) associated with component or circuit failure. Analytical studies must, of course, be combined with engineering experience and intuition, experimentation, and other ingredients of the design process in order to be successful.

12.1 Dynamic Behavior of Power Converters

We aim in this section to provide examples of the dynamic behavior associated with deviations from – and recovery to – nominal operation, for both uncontrolled and controlled power converters. As a prelude to that, we first review the notions and notation required to describe waveforms in nominal periodic operation, and then introduce some extensions that allow us to describe deviations from such nominal behavior.

12.1.1 Nominal Periodic Operation

Ideal or nominal operation for a vast majority of power converters involves a periodic steady state, with a period T fixed by the characteristics of a periodic switching scheme or a periodic driving source. Very often – as in the case of dc/dc converters – our interest is in the *average* values of voltages and currents in the converter, rather than their instantaneous values, provided the ripple or harmonics are sufficiently small.

For a periodically varying quantity $x(t)$ with period T we have been using the notation $\langle x(t)\rangle$ to denote the constant *average* value, so

$$\langle x(t)\rangle = \frac{1}{T}\int_{\text{interval } T} x(\tau)\, d\tau. \tag{12.1}$$

The range of integration here is any contiguous interval of length T; all such intervals give the same result.

In other cases our interest is in the value of the response at the *fundamental frequency* $1/T$ Hz or $\omega_s = 2\pi/T$ rad/s that is established by the nominally periodic *switching* or *source* frequency (hence our choice of s for the subscript on ω_s). The fundamental-frequency component or ω_s-*component* of a periodic real-valued signal $x(t)$ of period T may be extracted from its Fourier series, which can be written as in (3.1) or (8.13), or more conveniently for present purposes as

$$x(t) = \sum_{k=-\infty}^{\infty} \langle x\rangle_k\, e^{j2\pi kt/T} = \sum_{k=-\infty}^{\infty} \langle x\rangle_k\, e^{jk\omega_s t}, \tag{12.2}$$

where the kth Fourier coefficient is

$$\langle x\rangle_k = \frac{1}{T}\int_{\text{interval } T} x(\tau)e^{-j2\pi k\tau/T}\, d\tau = \frac{1}{T}\int_{\text{interval } T} x(\tau)e^{-jk\omega_s\tau}\, d\tau. \tag{12.3}$$

The range of integration here can again be any contiguous interval of length T, as all such intervals give the same result. The equivalence with (3.1) and (8.13) follows from the Euler identity: $e^{j\theta} = \cos\theta + j\sin\theta$.

Note from (12.3) that $\langle x\rangle_0$ is the same as the average value $\langle x(t)\rangle$. From (12.2) we see that the fundamental-frequency component of $x(t)$, or its ω_s-component, is

$$\langle x\rangle_{-1}\, e^{-j\omega_s t} + \langle x\rangle_1\, e^{j\omega_s t}. \tag{12.4}$$

The *amplitude* of this ω_s-component is easily calculated to be $2\,|\langle x\rangle_1|$ (taking account of the fact that $\langle x\rangle_{-1}$ and $\langle x\rangle_1$ are complex conjugates of each other when $x(t)$ is real).

12.1.2 Analyzing Deviations from Nominal Operation

We now introduce the notions of a *local average* and a *local fundamental*, which are helpful in studying the time evolution of deviations from nominal periodic operation of power converters.

Local Average As we begin to study deviations of a waveform $x(t)$ from behavior that is nominally periodic with period T, what is often of interest is some form of local average of $x(t)$, representing the low-frequency content of the variable of interest – whether a converter voltage or current, or a signal in an associated control system. The *centered* local average over an interval T, denoted by $\check{x}(t)$, is defined as

$$\check{x}(t) = \frac{1}{T}\int_{t-\frac{T}{2}}^{t+\frac{T}{2}} x(\tau)\, d\tau. \tag{12.5}$$

As an example, suppose

$$x(t) = At + p(t) \tag{12.6}$$

over some interval of time, where A is a constant and $p(t)$ is a periodic waveform of period T; then the centered local average over this interval is easily computed to be

$$\check{x}(t) = At + \langle p(t)\rangle = At + \langle p\rangle_0\,. \tag{12.7}$$

An alternative choice of averaging window is one of duration T that trails the instant of interest t rather than being centered on it. We shall denote this *trailing* local average by $\overline{x}(t)$ and define it as

$$\overline{x}(t) = \frac{1}{T}\int_{t-T}^{t} x(\tau)\, d\tau. \tag{12.8}$$

Hence,

$$\overline{x}(t) = \check{x}\left(t - \frac{T}{2}\right), \tag{12.9}$$

with a corresponding relationship in the transform domain – so the Fourier transform of $\overline{x}(t)$ is $e^{-j\omega T/2}$ times the transform of $\check{x}(t)$. Problem 12.1 explores the frequency–domain relationships in more detail, and makes clear that both the above local averages are particular lowpass-filtered versions of the original waveform. The advantage of the trailing window is that the corresponding local average can be computed or approximated in a *causal* fashion, as would be desired if the low-frequency content of a switching signal is (causally) extracted to drive a real-time feedback control system. From now on we shall assume a trailing window unless otherwise specified, and refer to $\overline{x}(t)$ simply as the *local average*.

Note that $\overline{x}(t)$ is a more smoothly varying function of t than $x(t)$ is; in particular, it will be a continuous function of time even if $x(t)$ has discontinuities (as long as $x(t)$ has no impulses). When and if the quantity $x(t)$ settles to a periodic steady state of period T, the local average will equal the associated average, with $\overline{x}(t) = \langle x(t)\rangle = \langle x\rangle_0$, regardless of the value of t.

Local ω_s-Component Of similar interest when examining deviations from nominally periodic behavior is a quantity that can describe variations in the fundamental or ω_s-component of a real signal $x(t)$. For this we shall use a variable inspired by (12.3) for $k = 1$, namely

$$\widehat{x}(t) = \frac{1}{T}\int_{t-T}^{t} x(\tau)e^{-j\omega_s\tau}\,d\tau, \tag{12.10}$$

where $\omega_s = 2\pi/T$. We again use a trailing rather than a centered window, in order to have a causally computable (or approximable) quantity. We shall refer to $\widehat{x}(t)$ as the *local fundamental* or *local ω_s-component*, though strictly speaking the local fundamental would be a sum analogous to (12.4), specifically

$$\widehat{x}^*(t)e^{-j\omega_s t} + \widehat{x}(t)e^{j\omega_s t}, \tag{12.11}$$

where * denotes complex conjugation. The amplitude of this sum is given by $2|\widehat{x}(t)|$. In this case too, if the quantity $x(t)$ settles to a periodic steady state of period T, then the local ω_s-component will – regardless of the value of t – equal the first Fourier coefficient $\langle x\rangle_1$, obtained by setting $k = 1$ in (12.3). (Given this, an alternative notation for $\widehat{x}(t)$ would be $\langle x\rangle_1(t)$, but the former choice is simpler).

We shall be using the local average in our discussions in this chapter, while more detailed examination of the local fundamental is left to Chapter 13. Note also that the constructs used to study the local average and local fundamental can be directly extended to the case of local harmonics, that is, $k \geq 2$ in (12.3).

12.1.3 Open-Loop and Feedforward Control

The block diagram in Fig. 12.1 represents the typical configuration of a controlled power converter. Each connection between the blocks represents one or more actions, measurements, or information flows affecting the block at which the arrowhead terminates – and originating in the block at the other end of the connection. Control of a power circuit entails specifying the desired nominal operating condition and then regulating the circuit so that it stays close to the nominal in the face of disturbances, measurement noise, and modeling errors that cause its operation to deviate dynamically from the desired nominal.

In simple *open-loop* control, the controller is not given any information about the system during operation, although the open-loop controller may be constructed on the basis of prior

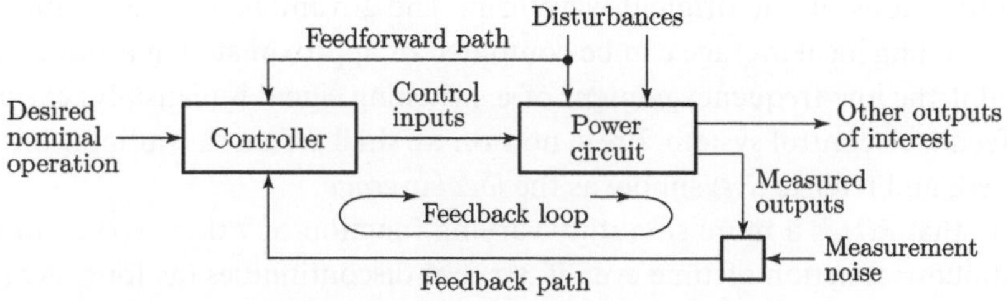

Figure 12.1 Typical configuration of a controlled power converter.

information or models. The feedback path in Fig. 12.1 is therefore missing in open-loop control. In open-loop control with *feedforward*, the controller uses measurements of some of the disturbances affecting the system, but still has no information about the current behavior of the system itself. Using feedforward, the controller can attempt to cancel the anticipated effects of measured disturbances. Feedforward alone is usually insufficient, however, to obtain satisfactory performance in power electronic systems, as the next example illustrates.

Example 12.1 Open-Loop Control with Feedforward for a Buck/Boost Converter

Let us consider a buck/boost dc/dc converter whose power circuit is built according to the circuit schematic in Fig. 12.2 and is operated at a frequency of 50 kHz, or with a switching period of $T = 20\,\mu\text{s}$. Let $R = 2\,\Omega$, $C = 220\,\mu\text{F}$, and $L = 0.25\,\text{mH}$. We discussed such circuits in Chapter 5, though typically operating at significantly higher frequencies there. The principles developed in the context of this example carry over to those settings.

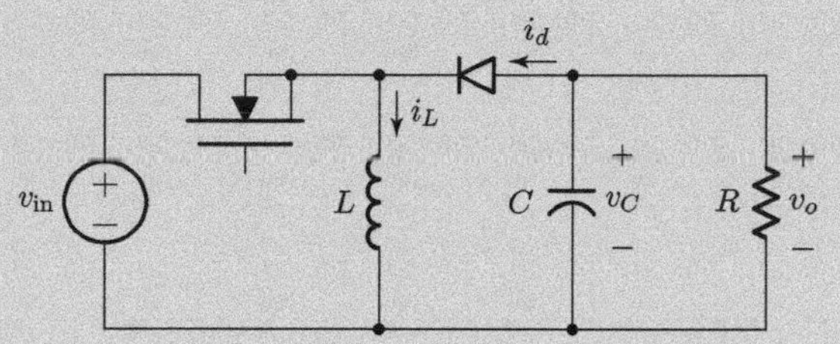

Figure 12.2 Circuit schematic of the power stage of a buck/boost converter.

We would like to maintain the average output voltage $\langle v_o(t)\rangle$ within 5% of the nominal or reference value of $V_{ref} = -9$ V, despite step changes in the input voltage v_{in} from a nominal dc value of $V_{in} = 12$ V down to values as low as 8 V. For the purposes of this example, we assume that there are no other non-idealities or uncertainties in the circuit and, in particular, that the transistor and diode function as ideal switches. We shall see that even if the system behaves according to this idealized model, the circuit response can be unsatisfactory.

Recall from Chapter 5 that if the transistor is turned on periodically and operated with a duty ratio D, and if the converter is operating in continuous conduction mode with a constant input voltage $v_{in} = V_{in}$, then $\langle v_o(t)\rangle = -V_{in}D/D'$ to a good approximation, where $D' = 1 - D$. The duty ratio must therefore be set at a nominal value of $D = V_{ref}/(V_{ref} - V_{in}) = 0.43$ in order to obtain the desired operation under nominal conditions.

Figure 12.3(a) shows the response of this idealized circuit to a step in the input voltage from 12 V to 8 V. The waveform before time t_0 corresponds to operation in a cyclic steady state with an input voltage of 12 V and duty ratio $D = 0.43$. Not unexpectedly, the ideal circuit produces the correct average output voltage under nominal conditions – before t_0. At t_0 the input voltage drops to 8 V. The circuit undergoes significant oscillatory transients and settles to an average value of -6 V at the output, rather than the desired -9 V. This is because information about the change in the input voltage has has not been "fed forward" to the controller, which maintains the duty ratio at the value $D = 0.43$. We can completely explain the steady-state behavior using the static model that we already have, but the explanation of the oscillatory transition from the prior steady state to the final one must await the modeling results in Chapter 13.

One natural idea for obtaining better responses to changes in the input voltage v_{in} is to use feedforward. If D is varied rather than being fixed by the nominal value V_{in}, so that $-v_{in}D/D'$ is held constant at V_{ref}, then v_o will attain the desired average value despite variations in input voltage. This control approach requires that we select $D = V_{ref}/(V_{ref} - v_{in})$. The resulting response to a step change in the input voltage is shown in Fig. 12.3(b). The ideal circuit with feedforward now settles down to the correct average output voltage, despite the step in input voltage. The peak excursions are considerably less, though still outside the allowed 5%. The transients, however, take as long to die out as they did without feedforward.

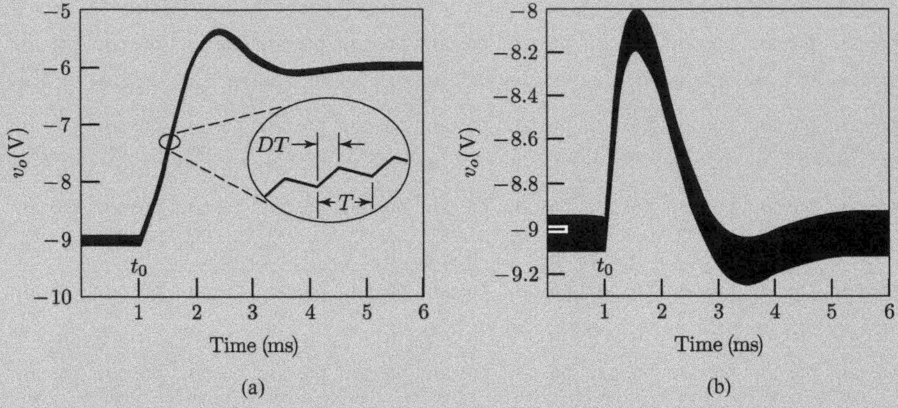

Figure 12.3 (a) Response of ideal buck/boost converter circuit to a step from 12 V to 8 V in the input voltage v_{in}. (b) Response to the same step after incorporating feedforward.

We will be able to explain why the transients in Example 12.1 appear as they do, after we obtain dynamic models for the buck/boost converter. We shall return to the converter in this example several times, in this chapter as well as in Chapters 13 and 14, to illustrate various aspects of dynamic modeling and control design.

Feedforward in Example 12.1 compensated for the steady-state effects of disturbances in the input voltage, but did not modify the dynamics of the transient. With non-ideal components, even the steady state would not be accurately restored using feedforward alone. More than feedforward is needed to obtain significantly better behavior.

12.1.4 The Need for Feedback Control

A controller can do better by also using real-time measurements of the system's present behavior to assess the departure from the desired behavior and to choose control actions aimed at restoring the system rapidly and safely to nominal operation. This strategy is the essence of *closed-loop* or *feedback control*. The following example illustrates the idea. (When the model that underlies the choice of control actions is itself updated on the basis of measurements, we refer to the controller as an adaptive or learning controller.)

Example 12.2 Closing the Loop on a Controlled-Rectifier Drive for a DC Motor

The circuit schematic in Fig. 12.4(a) represents a separately excited dc motor, driven by a phase-controlled bridge rectifier supplied from a sinusoidal voltage source v_{ac} of amplitude V. (We discussed such controlled rectifiers in Chapter 4.) The voltage applied to the motor is denoted by v_d and the armature current by i_d. The waveforms expected on the basis of the analysis of nominal operation in Chapter 4 are displayed in Fig. 12.4(b), with the assumptions that the armature current remains positive throughout each cycle and

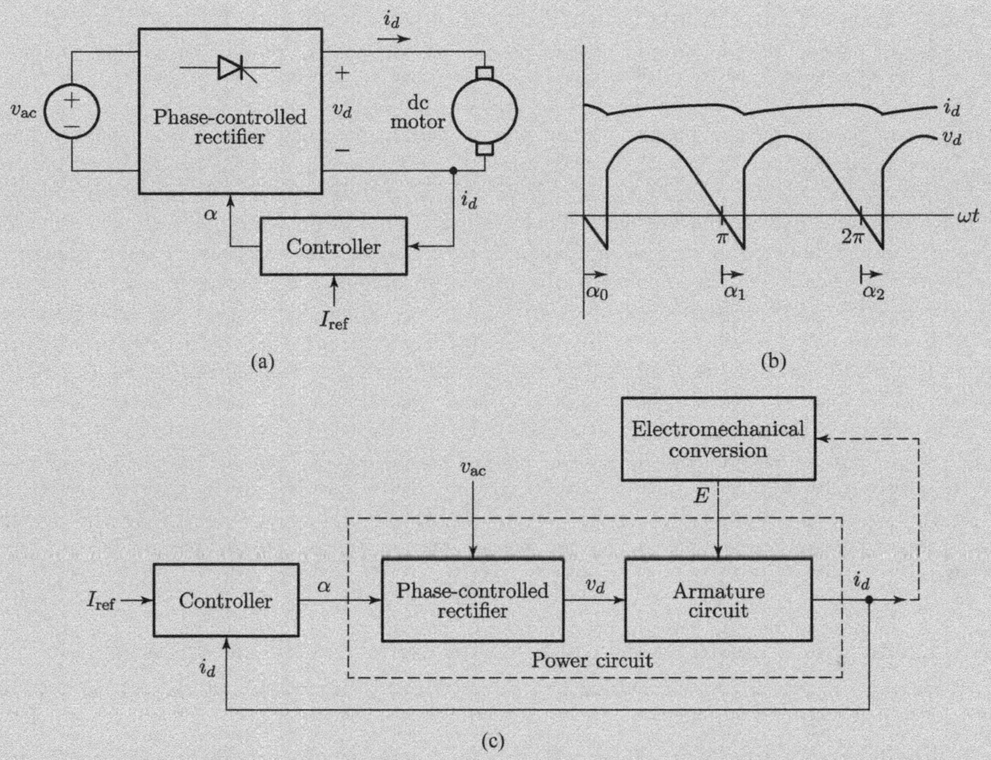

Figure 12.4 (a) Schematic diagram of a phase-controlled rectifier drive. (b) Waveforms of applied voltage v_d and current i_d, with a rectifier firing angle of α_k in the kth cycle. (c) Control block diagram of the drive.

that there is no commutating reactance. Figure 12.4(b) shows the role of the firing angle of the rectifier, denoted by α_k for the kth cycle, in specifying the voltage waveform.

Nominal operation of the drive system corresponds to a periodic or cyclic steady state in which the firing angle is maintained at a constant value $\alpha_k = \alpha$ and the armature current varies periodically at twice the frequency of v_{ac}. We know from Chapter 4 that the average voltage across the motor under this condition is $\langle v_d(t) \rangle = (2V/\pi)\cos\alpha$. Our objective is to use closed-loop control of the firing angle to regulate the average armature current $\langle i_d(t) \rangle$ at any specified reference value in some range. This reference value, I_{ref}, is specified by a higher-level speed or torque controller for the motor. The average current determines the average torque applied to the motor shaft. If the mechanical time constant of the motor is considerably larger than the period of the source, the speed of the motor displays very little ripple in the steady state, and depends essentially on the average current.

To represent the system entirely by a control block diagram – for comparison with the general control configuration presented in Fig. 12.1 – we can transform Fig. 12.4(a) to Fig. 12.4(c). The presence of commutating reactance would introduce a feedback connection (not shown here) from i_d to the phase-controlled rectifier in this block diagram, indicating that the applied voltage v_d would then depend on i_d.

Figure 12.4(c) represents the system at a finer level of modeling detail than in Fig. 12.4(a), with the dc motor model split into subsystems corresponding to the armature circuit and the electromechanical conversion process. The connections to and from the electromechanical subsystem are shown in dashed lines because the back-emf E is most commonly treated as a slowly varying external disturbance acting on the armature circuit, when designing the faster current-control loop. The rationale for this is that many cycles are needed for changes in the average current to cause changes in motor speed and hence in back-emf, so E may be treated as essentially constant during transients in the local average of the current.

One fairly satisfactory and widely used control solution for such systems is based on a *proportional–integral (PI)* controller. Its heart is the proportional part, whose action is consistent with simple reasoning. This part changes the firing angle from its nominal value by an amount *proportional* to the error $\bar{i}_d(t) - I_{\text{ref}}$. Note here that we are using the *local average* of $i_d(t)$, as defined in (12.8), as we are aiming to regulate deviations of $\bar{i}_d(t)$ from the constant value $\langle i_d(t) \rangle$ that it takes in the periodic steady state. This local average can in principle be computed or approximated by the controller in real time from measurement of $i_d(t)$. The firing angle is increased when the error is positive, because the average voltage, and hence the average current, are thereby decreased, which decreases the error. When the error is negative, the proportional part of the controller does the opposite, increasing the firing angle in proportion to the error magnitude.

The *integral* part of the controller acts on the integral of the error, and works on a slower timescale to correct for steady-state errors induced by parameter uncertainties and constant disturbances in the model. The integral of the error settles to a constant only when the error itself settles to zero.

It is also possible to use a proportional–integral–*derivative* (PID) controller instead, which can speed up the response by adding to the PI controller a contribution that depends on the derivative of the error or the derivative of the output (in which case it is called *rate feedback*). The PI, PID, and related controller structures, such as lag, lead, and lead–lag compensators, are ubiquitous in control.

Example 12.2 illustrates the form that the component blocks and signals in Fig. 12.1 may take in an application. We return to the system in this example later in this chapter, and again in Chapter 14.

Elementary reasoning in Example 12.2 could have led us to at least the proportional part of the PI controller. With some tuning, a proportional controller alone could permit stable operation over a range of operating points (though probably with unacceptably high steady-state error). In this respect, the controlled-rectifier drive is a benign system, because a reasonable controller can be derived from a model that is not much more sophisticated than the one in Example 12.2.

12.1.5 The Need for Dynamic Models

One can run into trouble, however, by following only common sense or using overly simplified analytical models when designing a feedback controller. A controller based only on an understanding of static operating characteristics, ignoring dynamic effects, can fail badly. The following example illustrates this.

Example 12.3 Problems with Proportional Feedback Control of a Buck/Boost Converter

The desire for better dynamic performance than that obtained with open-loop and feedforward control of the buck/boost converter in Example 12.1 leads us to consider a feedback control solution. Given that our interest is in regulating the average output voltage, we must now measure the deviation of the average output voltage from the desired value of $V_{\text{ref}} = -9$ V, and use the discrepancy $\bar{v}_o(t) - V_{\text{ref}}$, which we shall denote by $\tilde{v}_o(t)$, to adjust the duty ratio from the nominal value D to $D+\tilde{d}$. Just as we did with the armature current in the preceding example, we are using the *local average* $\bar{v}_o(t)$ rather than $\langle v_o(t) \rangle$ because we intend to regulate deviations from nominal periodic operation. As noted earlier, this local average can in principle be computed or approximated by the controller in real time, from measurement of $v_O(t)$. The correction $\widetilde{d}$ will depend on the polarity and magnitude of the voltage deviation $\widetilde{v}_O(t)$.

Examination of the (inverting) steady-state characteristic of the converter, $\langle v_O(t)\rangle = -v_{\text{in}}D/D'$, suggests that when the error $\widetilde{v}_O(t) = \overline{v}_O(t) - V_{\text{ref}} = \overline{v}_O(t) + 9$ is negative (indicating that $\overline{v}_O(t)$ is too negative), we should decrease the duty ratio. Similarly, we should increase the duty ratio when the error is positive. The *proportional* feedback control system represented in the block diagram in Fig. 12.5(a) is one implementation of this control law: it generates duty ratio perturbations that are proportional to the measured deviations of $\overline{v}_O$. The constant feedback gain or compensator gain h has to be negative to provide the corrective action suggested by examination of the steady-state characteristic. The block diagram also indicates that the earlier feedforward compensation may be implemented along with the feedback.

Figure 12.5(b) presents the response of the ideal circuit to the same input-voltage step as in Example 12.1, for increasingly negative values of the proportional gain, and with the previous feedforward in place. (The waveforms in this figure have no switching-frequency ripple because they have been computed using an averaged-circuit model of the sort discussed later in this chapter and also in Chapter 13. Such models generate close approximations of the local averages of the corresponding switched waveforms.)

The response of the circuit in this example clearly becomes *more* oscillatory as h is decreased, and the circuit actually goes *unstable* before h becomes very negative. On the other hand, the counterintuitive choice of a *positive* h does not immediately lead to the disaster that might be expected on the basis of the steady-state characteristic. As indicated by the waveforms in Fig. 12.5(c), the response can be stable for some interval of positive h. Evidently our reasoning based on the steady-state characteristic has led us quite astray. We shall see in Chapter 14 how to do better.

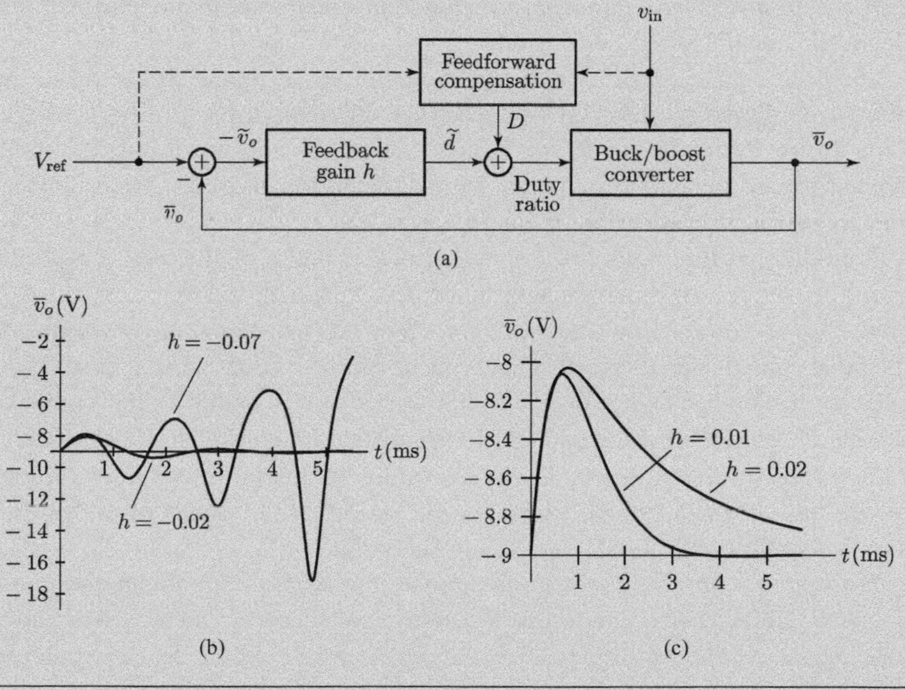

Figure 12.5 (a) Block diagram of a control system using proportional feedback with feedforward for a buck/-boost converter. (b) Step response of the ideal circuit for increasingly negative gains. (c) Response for increasingly positive gains.

Example 12.3 shows that a model based on steady-state characteristics may be inadequate for feedback controller design. This motivates us to develop models that incorporate dynamic effects. In fact, a clearer picture of what is happening with the feedback-controlled buck/boost converter in Fig. 12.5 emerges after we obtain a dynamic model for the response of the output voltage to variations in the duty ratio (see especially Examples 14.1 and 14.2).

Much of our attention in Part II centers on the feedback loop in Fig. 12.1. Feedback can speed up the response of a system to commanded changes, improve the recovery from unanticipated disturbances, and make performance less sensitive to system variations. However, a system that is stable and insensitive to perturbations in open loop could become sensitive, slow to recover, or even completely unstable in closed loop, if the controller reacts inappropriately to the feedback signals. The controller's actions could aggravate deviations from nominal operation instead of rapidly restoring nominal operation. To obtain the advantages of feedback control, we have to design and implement the feedback loop carefully.

12.2 Dynamic Models

Part I was largely devoted to nominal or ideal operation in steady state, but one has to go beyond static or steady-state models to develop controllers for power circuits. A major objective of Part II is to introduce a variety of fundamental approaches to modeling the *dynamics* of power electronic circuits and their controllers.

The use of an appropriate dynamic model or set of models is central to the control design process. Different models may be needed for different stages or aspects of the control design. Even for a specific stage of the design, there may be several possible models, differing in explicitness, complexity, accuracy, domain of definition, flexibility, tractability, and so on. Trade-offs are involved in selecting among these models.

A model that functions well for numerical simulation of the open-loop behavior of a power circuit may not necessarily be helpful for designing a closed-loop controller. A more complex model for a power circuit may predict observed behavior more accurately but may be less tractable analytically, less useful in generating candidate control designs, and perhaps less able to yield controllers that can withstand circuit variations. However, a simple model may miss crucial aspects of system behavior, and lead to unsatisfactory controllers. In practice, one should work with several models, checking the analytical and numerical predictions of one versus those of others, and also against experimental observations. The insights from this process are used to iteratively refine the models and the control design.

Most power converters are well modeled as interconnections of ideal switches and *linear, time-invariant* (LTI) circuit components. The analysis of such a circuit in each switch configuration (or topological state) is as simple as the analysis of an LTI circuit. Thus, we can use various special techniques – such as impedance methods – in any particular switch configuration. The challenge for an *analytical* understanding of overall behavior is in piecing together the solutions from successive configurations, especially when the times of transition are controlled as functions of the circuit's behavior. Switching generally causes the overall model to become time-variant, and if the switch operation is driven by circuit variables in addition to external action, then the overall model becomes nonlinear as well. On the other hand, the wide availability of excellent circuit simulators allows efficient *numerical* solution of power converter behavior in both transient and steady-state operation, even when nonlinear components are present.

We begin our discussion of modeling approaches by introducing the idea of an *averaged-circuit model*. Such models are especially valuable in describing the behavior of dc/dc converters, but extend usefully to certain other families of power circuits. The topic is continued and considerably extended in Chapter 13.

We also need to model the dynamics of the controller along with that of the power circuit in order to study the controlled, closed-loop system, but circuit methods are often not appropriate or convenient for controller modeling. In Chapter 13 we introduce *state-space models*, which embrace a much wider variety of modeling possibilities for converters and their controllers, and which are also amenable to averaging.

The most typical route to usable LTI models in power electronics is by the process of *linearization*, to describe *small deviations around a constant steady*. Such linearization is illustrated later in this chapter using examples of averaged-circuit models, and is treated more generally in Chapter 14. The design of feedback control to regulate a power converter in the vicinity of its nominal operating condition is largely done on the basis of LTI models (usually obtained via linearization), and we outline the key issues and tradeoffs for this in Chapter 14.

12.3 Averaged-Circuit Models

For many power electronic circuits, our interest is in the average values of voltages and currents rather than in their instantaneous values, provided the ripple is sufficiently small. In the context of departures from nominal operation for such converters, and feedback control to regulate them back to the desired condition, our focus turns to the *local* average. The local average is also of primary interest in the case of power converters such as PWM inverters for ac motor drives, of the kind discussed in Chapter 8. The objective in the latter case is regulation of the local average of the output current around a sinusoidal reference whose frequency is much lower than the switching frequency.

It is fortunately the case that for many classes of power converters, the constraints governing the local average values of circuit variables can be represented using an *averaged circuit* whose topology reflects that of the underlying switched circuit. This permits us to analyze the averaged variables using circuit ideas and tools that are very familiar to electrical engineers. Averaged models have traditionally been derived mainly for high-frequency switching dc/dc converters, usually through an averaging process applied to state-space models of these converters. Here we begin with a more fundamental approach, which starts directly from the circuit diagram of the power circuit. Chapter 13 shows that the averaging idea, and specifically circuit averaging, can be applied more widely, to develop useful dynamic models for other categories of power converters, for instance resonant converters, where the focus is on the local fundamental component.

12.3.1 Averaging a Circuit

The voltages across, and currents through, the elements in a circuit respectively satisfy Kirchhoff's voltage and current laws (KVL and KCL), as well as the equations defining each of the components in the circuit. We therefore examine the local averages associated with KVL, KCL, and the component equations.

Averaging KVL and KCL Since KVL and KCL impose LTI constraints at each instant on the circuit voltages and currents respectively, their form is not modified if we compute the local average of each of these equations. For example, if one KVL equation for a circuit is $v_1(t) + v_3(t) - v_7(t) = 0$, then taking the local average of both sides, we get $\overline{v}_1(t) + \overline{v}_3(t) - \overline{v}_7(t) = 0$; similarly for KCL equations.

We thus find that the locally averaged variables satisfy the same KVL and KCL constraints as the underlying instantaneous variables. Hence, replacing the instantaneous voltages and currents in the circuit by the corresponding local averages will yield a circuit satisfying Kirchhoff's laws for that circuit topology. Also, thanks to Tellegen's theorem, which states that the familiar power conservation equation $\sum v_k(t)i_k(t) = 0$ extends to any set of paired quantities that respectively satisfy KVL and KCL on the circuit topology, we can write $\sum \overline{v}_k(t)\overline{i}_k(t) = 0$ (though this does not begin to exhaust the implications of the theorem). The sum is over all the components in the circuit, with a consistent and conformable sign convention for voltages v and currents i.

Averaging an LTI Component Now consider averaging the constraint equation imposed by an LTI component in the circuit on its instantaneous terminal voltage and current. It is easily seen that the averaged terminal quantities are constrained in the same way as the instantaneous quantities. This is clear in the case of an LTI resistor, where taking the local average of the equation $v_R(t) = Ri_R(t)$ yields

$$\overline{v}_R(t) = R\overline{i}_R(t). \tag{12.12}$$

The case of an LTI inductor or capacitor is more subtle, but follows from the fact that the derivative of the local average equals the local average of the derivative, that is,

$$\frac{d}{dt}\overline{x}(t) = \overline{\frac{dx(t)}{dt}} = \overline{\dot{x}}(t), \tag{12.13}$$

where $\dot{x}(t)$ denotes $dx(t)/dt$; each side of this equation evaluates to $[x(t) - x(t-T)]/T$, as can be directly verified. Invoking this result to average the defining equation of an LTI inductor, namely $v_L(t) = L(di_L(t)/dt)$, we obtain

$$\overline{v}_L(t) = L\frac{d\overline{i}_L(t)}{dt}. \tag{12.14}$$

In other words, the local averages of the voltage across and current through an LTI inductor are constrained in the same way as the corresponding instantaneous variables. Similarly, from the defining equation $i_C(t) = C(dv_C(t)/dt)$ for an LTI capacitor, we deduce the averaged equation

$$\overline{i}_C(t) = C\frac{d\overline{v}_C(t)}{dt}. \tag{12.15}$$

Circuit Averaging The preceding observations allow us to construct an averaged version of a given circuit from the original circuit by simply replacing all instantaneous voltages and currents in the circuit by their local averages, and keeping all LTI components unchanged. Nonlinear or time-varying components in the original circuit, however, do not map into the same components in the averaged circuit. For example, switches in the original circuit become elements in the averaged circuit that simultaneously have a nonzero (average) voltage and nonzero (average) current – and so are no longer switches. We defer treatment of switch averaging to Chapter 13, apart from the special case of switching functions described in the next subsection.

Despite the fact that only LTI components of the original circuit are preserved intact, the transformation to the averaged circuit can often be very useful. Parts of the averaged circuit, if not all of it, may be amenable to LTI analysis. We may be able to usefully employ impedance methods or superposition or Thévenin/Norton equivalents on the LTI parts. Even when the exact averaged circuit is difficult or impossible to analyze, some approximations may yield useful insights. The following two examples illustrate the process of circuit averaging, and the understanding that can be gained from it.

Example 12.4 Averaged Circuit for a Controlled-Rectifier Drive

Figure 12.6(a) models the armature circuit of the controlled-rectifier drive of Example 12.2. Here, R and L represent the armature resistance and inductance, respectively, and E denotes the back-EMF of the motor. The waveform of the applied voltage v_d was shown in Fig. 12.4(b). The variable we want to control is the average armature current. More specifically in the setting of departures from nominal operation, we want to control the *local average* current, $\overline{i}_d(t)$. The natural choice of the averaging interval T for this system is the period of v_d (which is one half the period of the sinusoidal source). With this choice, $\overline{v}_d(t)$ and $\overline{i}_d(t)$ in the steady state are both constant, and respectively equal to $\langle v_d(t)\rangle = (2V/\pi)\cos\alpha$ and $\langle i_d(t)\rangle = (\langle v_d(t)\rangle - E)/R$, since the average inductor voltage must be zero in steady state.

Figure 12.6(b) shows the result of averaging the circuit in Fig. 12.6(a). The averaged controlled rectifier is represented simply as a voltage source, because $\overline{v}_d$ here is completely defined by the control variable, namely, the firing angle α.

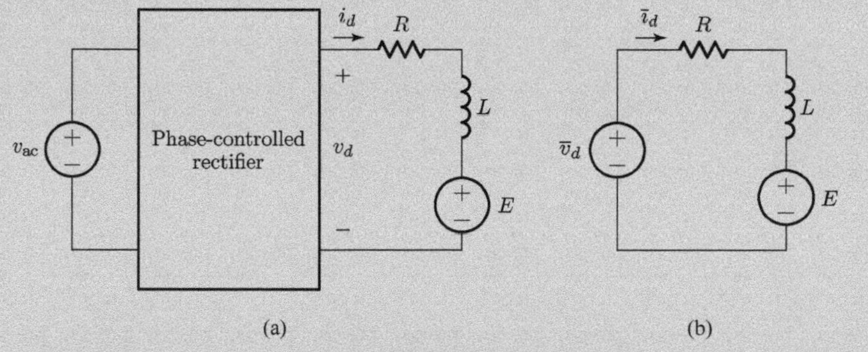

Figure 12.6 (a) Instantaneous and (b) averaged circuits for a controlled rectifier drive.

We can illustrate the usefulness of the averaged circuit by considering the open-loop step response from one steady state to another. A step change in the firing angle α from one cycle to the next leads to a transition in $\overline{v}_d(t)$ in a single cycle, from its steady-state value before the step to its steady-state value after the step. The averaged circuit shows that, following this first cycle, the local average $\overline{i}_d(t)$ approaches its new steady state exponentially, with a time constant of L/R. The first cycle is usually a small part of the transient. A typical value for L/R may be 40 ms, so the transient would last about 120 ms, whereas T for a 60 Hz supply is only 8.33 ms. (With a six-pulse converter derived from a three-phase supply, T would be just 2.78 ms.)

By working with the averaged circuit, we have been able to easily predict the overall form of the step response at a sufficient level of detail for many purposes. Examples 12.9 and 12.10 illustrate how we can use the averaged-circuit model as the basis for designing a simple feedback controller for the average armature current.

An exact analysis of average behavior becomes far more complicated if either of the initial assumptions in Example 12.2, namely, continuous conduction and no commutating reactance, is violated. The reason is that $\overline{v}_d(t)$ then depends on the instantaneous values of waveforms in the original circuit and can no longer be represented as a voltage source in the averaged circuit. Nevertheless, an approximate analysis of average behavior is often still possible and useful.

We might have intuitively anticipated the overall features of the step response in Example 12.4, but the analysis gives our intuition a firm foundation. The next example uses circuit averaging in a very different setting.

Example 12.5 Averaged Circuit for a Buck/Boost Converter in Discontinuous Conduction

We consider again the buck/boost converter used in Example 12.1, but now assume the resistance R of the load is high enough for the converter to be operating in discontinuous conduction mode. The corresponding inductor- and diode-current waveforms are shown in Fig. 12.7(a). We will construct an averaged model with the switching period T as the averaging interval.

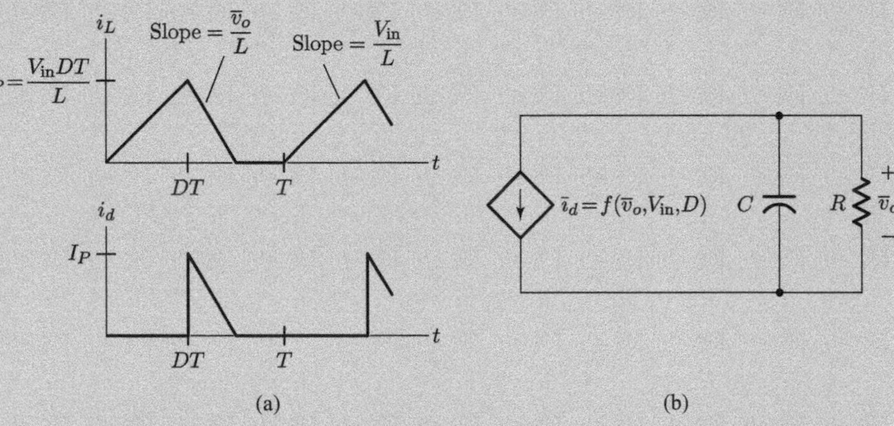

Figure 12.7 (a) Inductor- and diode-current waveforms for the buck/boost converter of Example 12.1, but operating in discontinuous conduction mode. (b) Output portion of the averaged circuit.

Provided the input voltage is constant over a switching period, the inductor current rises linearly from 0 to $I_P = V_{in}DT/L$ while the transistor is conducting. The current decay to zero is shown in Fig. 12.7(a) as being linear, but it is actually a portion of the ringing waveform of the parallel RLC network formed when the transistor is turned off and the diode begins to conduct. Nevertheless, this portion of the ringing waveform is very well approximated by a linear segment, provided we assume that the output voltage does not vary significantly over a single cycle. More specifically, if the output voltage is well approximated by its local average (that is, if the ripple is small) and the average does not change significantly over a cycle, the inductor and diode currents decay essentially linearly, with a slope of $\overline{v}_o/L$.

With this linear approximation, we can easily use Fig. 12.7(a) to compute the average current flowing through the diode over one switching period:

$$\overline{i}_d(t) = -\frac{V_{\text{in}}^2 TD^2}{2L\overline{v}_o(t)} = f(\overline{v}_o(t), V_{\text{in}}, D). \tag{12.16}$$

(Recall that v_o is negative, so $-\overline{v}_o$, i_d, and $\overline{i}_d$ are positive.) Assuming $\overline{v}_o$, V_{in}, and D are all slowly varying, (12.16) holds closely for all t, not just when the averaging interval is aligned with a switching cycle. Note the nonlinear dependence on $\overline{v}_o$, V_{in}, and D. We can now draw the output portion of the resulting averaged circuit as in Fig. 12.7(b), with the current through the diode represented by a voltage-controlled current source. We can use this nonlinear circuit to study the dynamics of $\overline{v}_o$ when V_{in} and D are constant or slowly varying.

The control variables in Examples 12.4 and 12.5 are respectively the firing angle α and the duty ratio D. These typically change from cycle to cycle, taking values α_k and d_k, respectively, in the kth cycle, so are most naturally thought of as discrete-time signals (with "time" being the cycle number). However, in analysis and control design, working with models that involve both continuous-time and discrete-time quantities is awkward. Sampled-data models get around this obstacle by using samples of the continuous-time waveforms in order to work entirely with discrete-time sequences. We discuss such models further in Chapter 13. In the context of averaged models, the opposite strategy is more natural, namely, representing the effects of discrete-time sequences such as α_k and d_k by continuous-time quantities. We shall see this in the subsection below, where the duty-ratio sequence d_k is replaced by a continuous duty-ratio signal $d(t)$.

12.3.2 Averaging a Switching Function

Figure 12.8(a) represents such circuits as buck converters (Chapter 5) and full-bridge inverters (Chapter 8), in which a controlled switching network is interposed between a dc voltage source V_{in} and an LTI load, represented in the figure by its Norton equivalent (though a Thévenin equivalent would have served as well). The voltage at the output of the switching network is $q(t)V_{\text{in}}$, where $q(t)$ is a *switching function* that constitutes the modulation of the source voltage.

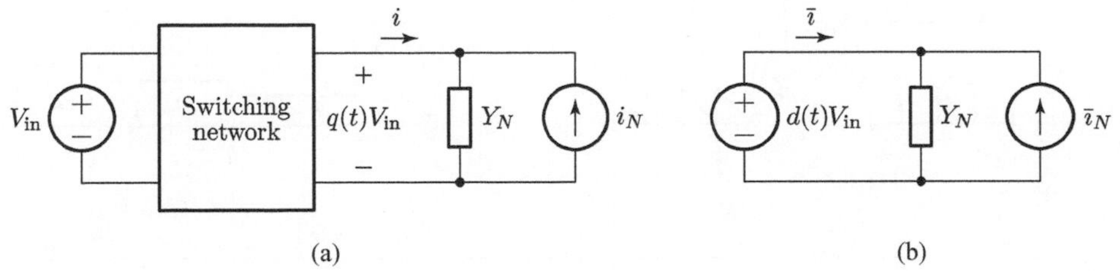

Figure 12.8 Averaging a modulated dc source to obtain a duty-ratio-controlled dc or ac source. (a) Switched circuit. (b) Averaged circuit with $d(t) = \overline{q}(t)$.

Thus, $q(t)$ is usually a waveform that switches in a controlled way between a finite set of values: 1 and 0 for buck converters, or 1, 0, and -1 for a full-bridge inverter such as that in Fig. 8.2,

Since the input voltage is fixed at V_{in}, circuit averaging directly yields the averaged circuit in Fig. 12.8(b), where $d(t)$ denotes the local average $\overline{q}(t)$ of the switching function, and may be called the *continuous duty ratio*. If $q(t)$ is periodic with period T (the averaging interval), then $d(t)$ is constant at some value D.

If instead of having a fixed voltage V_{in} behind the switching network, we had a voltage that depended on other circuit variables, the circuit averaging would not have been this straightforward. We defer the treatment of circuit averaging in more subtle cases, such as for the buck/boost converter, to Chapter 13.

The use of a switching function $q(t)$ and its average $\overline{q}(t) = d(t)$ is natural and convenient in analysis and control design for many other types of power converters. In most of these cases, the controller manipulates quantities closely related to $d(t)$ in order to control the average values of circuit waveforms.

Example 12.6 Generating a Switching Function

The case in which a switch is turned on when $q(t) = 1$ and turned off when $q(t) = 0$ is often encountered. The switching function $q(t)$ may be generated as the output of the latch in the circuit shown schematically in Fig. 12.9(a). The clock sets the latch output to 1 every T seconds, defining the beginning of a cycle. The output of the comparator is low initially but switches later in the cycle to its high value, resetting the latch to 0.

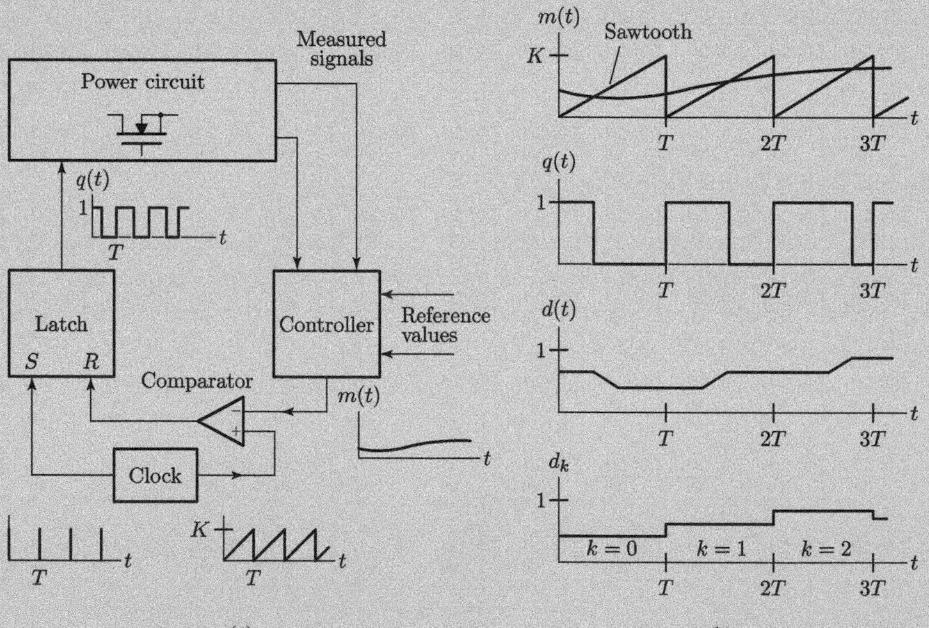

Figure 12.9 (a) Generating a switching function $q(t)$, with duty ratio determined by a modulating function $m(t)$. (b) Relationships among $m(t)$, $q(t)$, $d(t) = \overline{q}(t)$, and d_k.

The sawtooth waveform that is applied to the positive input of the comparator is synchronized to the clock, starting at 0 every T seconds and ramping up linearly to K. A modulating signal, $m(t)$, is fed to the negative input of the comparator and normally satisfies $0 \leq m(t) \leq K$. It follows that the output of the comparator is low at the start of each cycle and switches to its high value when the ramp crosses $m(t)$. The duty ratio d_k in the kth cycle therefore equals the ratio $m(t)/K$ at the instant t that the sawtooth ramp in that cycle first crosses $m(t)$. Thus, the modulating signal $m(t)$ controls the duty ratio.

Even though $m(t)$ controls the duty ratio, it is not the case that the duty ratio faithfully follows variations in $m(t)$. In fact, we can obtain a constant duty ratio either by keeping $m(t)$ constant or by varying it at precisely the switching frequency. These two possibilities in turn suggest two ways to obtain slow changes in the duty ratio: by varying $m(t)$ slowly or by varying it at frequencies in the vicinity of the switching frequency. In the latter case, the duty ratio varies at the beat frequency.

Both methods are used in practice. In fact, for so-called current-mode control of high-frequency switching dc/dc converters, both methods are used simultaneously. One part of $m(t)$ is derived from the average output voltage, and varies at low frequencies; the other part is derived from the instantaneous inductor or switch current and therefore has components both at low frequencies and near the switching frequency. The periodic sawtooth waveform in the case of current-mode control is termed a stabilizing ramp (or compensation ramp), for reasons that we explain in Chapter 14. We assume in what follows that $m(t)$ varies slowly.

If $m(t)$ does not change significantly over the course of a cycle, so that significant variations in $m(t)$ occur at substantially less than one half the switching frequency, then $m(t)/K$ at any time closely approximates the prevailing duty ratio. The control circuit varies $m(t)$ around its nominal value to vary the duty ratio in accordance with feedback signals that measure how far the power circuit is from nominal operation. Evidently, the closed loop "bandwidth" of such a system, or the speed with which the controlled system returns to its nominal operation, is substantially less than one half the switching frequency under these conditions.

Figure 12.9(b) shows how $m(t)$, $q(t)$, the continuous duty ratio $d(t) = \overline{q}(t)$, and the duty ratio d_k in the kth cycle are interrelated. A helpful check on your understanding of the trailing local average operation is to verify that the $d(t)$ shown here is indeed the trailing local average of the indicated $q(t)$. Note that $d(t)$ also closely approximates the prevailing duty ratio d_k, provided the change in duty ratio from cycle to cycle is sufficiently small. Although we can choose to constrain $m(t)$ to have only slow variations, $d(t)$ is intrinsically constrained to vary slowly. The fastest possible variation in d_k occurs when the duty ratio alternates between high and low values in successive cycles. The corresponding period of $d(t)$ is $2T$. Thus, $d(t)$ can never have a fundamental frequency higher than one half the switching frequency.

Note that variations in K could also be used to vary the duty ratio. Feedforward control that compensates for supply voltage variations is commonly implemented this way. With the buck/boost converter circuit in Examples 12.1 and 12.3, for instance, the required feedforward can be achieved by making K proportional to $V_{\text{in}} - V_{\text{ref}}$ (recall that V_{ref} is negative). For buck converters and other circuits of the form shown in Fig. 12.8, we make K proportional to V_{in}.

We have covered considerable ground without looking in detail at averaging the variables associated with a *switch* embedded in a circuit. We shall treat this topic in considerably more detail in Chapter 13, in the course of extending our treatment of averaged-circuit models.

12.4 Linearized Models

The static characteristics of power electronic circuits often depend nonlinearly on the driving sources and the control variables. Their dynamic characteristics, whether in open loop or under feedback control, are even more likely to be nonlinear. Assessing stability and designing or evaluating controllers using nonlinear models are usually difficult. The most common, systematic and generally successful approach to these tasks is *linearization*. It yields *linear* models that approximately describe small deviations or perturbations from nominal operation of a system; such models are therefore also termed *small-signal* models.

Linear models are far easier to analyze than nonlinear ones, because one can superpose the responses to simple inputs in order to obtain the response to a superposition of the simple inputs. Even more useful are linear models that are time invariant, that is, LTI models, because they can be fully characterized by their response to sinusoidal or exponential inputs. Linearization carried out to describe small deviations around a *constant* nominal operating condition yields LTI models, and this is our focus. Unless otherwise specified, we shall use the term linearization in what follows to denote this particular case.

LTI models obtained through linearization around a constant operating condition are crucial to evaluating the stability of the operating condition. Asymptotic stability of the linearized model ensures the nominal operating condition is stable for small perturbations at least. An initial goal for control design is therefore stabilization of the linearized model. This is much easier than direct stabilization of the nonlinear model.

We outline the basis of linearization in this section by showing how to linearize two averaged models described in earlier examples. Chapter 14 extends the idea to general state-space models. The application of the resulting linear models in stability evaluation and control design is also illustrated in Chapter 14.

12.4.1 Linearization

The place to start when linearizing a dynamic model is with a nominal solution, usually a steady-state solution. Even the steady state in a dynamic model for a power circuit will typically involve periodically varying rather than constant quantities. For example, steady state in the instantaneous model of a buck/boost converter involves a periodic switching function and periodic waveforms.

One way to get a constant steady state rather than a periodic steady state in power circuit models is by working with a discrete-time model that involves circuit variables sampled once per cycle; the reason for this becomes clearer in Chapter 13. Averaged models can also yield constant steady states. Steady state in the averaged-circuit model given in Fig. 12.8(b), when we set the continuous duty ratio $d(t) = \overline{q}(t)$ to a constant value D, corresponds to the averaged variables taking constant values. However, with the high-frequency PWM inverters described in Chapter 8, even an averaged model can have a steady state that varies periodically rather than remaining constant, if the inverter is generating a fixed-frequency output.

We can represent small deviations from the nominal by expanding all the nonlinear terms of the model into Taylor series around the nominal values. Keeping only first-order terms yields a *linear* model that approximately governs small deviations; this is the linearized model. The parameters of the linearized model depend on the nominal operating condition, because the Taylor series coefficients depend on the nominal condition. If the original nonlinear model is time invariant and if the nominal solution corresponds to constant values of the variables, the linearized model will be LTI.

12.4.2 Linearizing a Circuit

For nonlinear models in circuit form, we can further simplify the process of linearization. Using simple operations on the nonlinear circuit, we can obtain the linearized model itself in circuit form. The procedure and justification are analogous to those we used in Section 12.3 to obtain an averaged-circuit model from an instantaneous-circuit model. The arguments should also be familiar from small-signal analysis of transistor amplifier circuits, for example.

We begin by replacing every voltage in the nonlinear circuit by its *deviation* from the nominal. This step results in voltage deviations that satisfy Kirchhoff's voltage law equations on the given circuit topology. The reason KVL is satisfied is that we obtain the deviations by taking the difference between two sets of voltages – the perturbed set and the nominal – that each satisfy the same linear equations. Similarly, we replace every current in the nonlinear circuit by its deviation from the nominal to obtain current deviations that satisfy Kirchhoff's current law on the given circuit topology.

The final step is to replace every nonlinear component in the circuit with its linearized version. (The linear components do not need replacement, because they impose the same constraints on the deviations as they do on the original variables.) The linearization of a nonlinear component is obtained by Taylor expansion of its characterizing equation, up to first-order terms in the deviations from nominal values. The linearized component imposes linear constraints that approximately govern small deviations of the variables at the component's terminals, and small deviations of any control variables that govern the component. The result of all these manipulations is a *linear circuit* that governs small deviations from the nominal.

The nonlinear averaged circuits that we described earlier are time invariant, and the nominal solution of interest is usually the constant steady state. The linearization in these cases is an LTI circuit. In the case of averaged-circuit models for the PWM inverters in Chapter 8, the nominal steady state may be periodically varying, and the corresponding linearization is then a periodically varying circuit.

Example 12.7 Linearizing a Buck Converter Circuit

A buck converter can be represented as in Fig. 12.8(a), with a switching network interposed between a dc source and an LTI circuit represented in the figure by its Norton equivalent. The corresponding averaged model is shown in Fig. 12.8(b), and was derived under the assumption of a constant V_{in}. If the duty ratio, input voltage, and average Norton current are held constant at respective values D, V_{in}, and I_N, then the averaged circuit has a constant steady state. Suppose, however, that $d(t)$ and $\overline{i}_N(t)$ are perturbed around their nominal values, so

$$d(t) = D + \widetilde{d}(t), \qquad \overline{i}_N(t) = I_N + \widetilde{i}_N(t). \tag{12.17}$$

We use the superscript $\widetilde{\ }$ to denote a perturbation from the nominal. The linearized version of Fig. 12.8(b) then has the same structure, except that the source on the left is now $\widetilde{d}(t)V_{\text{in}}$ instead of $d(t)V_{\text{in}}$, the Norton source is now $\widetilde{i}_N(t)$ instead of $\overline{i}_N(t)$, and the current from the switching network is now $\widetilde{i}(t)$ instead of $\overline{i}(t)$. In this case of constant V_{in}, the components in the averaged circuit are linear in the variables that are being perturbed, so the perturbations do not have to be small for the higher-order terms to be negligible (there are none!).

Example 12.8 Linearized Circuit for a Buck/Boost Converter in Discontinuous Conduction

We describe a less straightforward case of circuit linearization next. A nonlinear averaged-circuit model for the buck/boost converter in discontinuous conduction was obtained in Example 12.5, Fig. 12.7(b). Our derivation of the model there allows the duty ratio and input voltage to be slowly varying rather than constant. We can therefore rewrite (12.16) with $d(t)$ instead of D and $\overline{v}_{\text{in}}$ instead of V_{in}. We reserve D and V_{in} to denote the constant nominal values.

The nominal steady-state solution corresponds to constant values of the averaged quantities. In particular, $\overline{i}_d(t) = I_d$ and $\overline{v}_o(t) = V_o$. The capacitor in the averaged circuit acts as an open circuit in the steady state, so we see from Fig. 12.7(b) that

$$V_o = -RI_d = -Rf(V_o, V_{\text{in}}, D) = R\frac{V_{\text{in}}^2 TD^2}{2LV_o}. \tag{12.18}$$

Rearranging and taking the square root yields $V_o = -V_{\text{in}}D\sqrt{RT/2L}$. (With our sign convention, we need the negative square root.)

Now let the duty ratio be perturbed from D to $d(t) = D + \widetilde{d}(t)$, but assume for simplicity that the input voltage is fixed at V_{in}. Correspondingly, let $\overline{i}_d(t) = I_d + \widetilde{i}_d(t)$ and $\overline{v}_o(t) = V_o + \widetilde{v}_o(t)$. The superscript $\widetilde{}$ again denotes a perturbation from the nominal. Linearizing the current source in the averaged circuit by a Taylor series expansion up to linear terms, we find

$$\begin{aligned} \widetilde{i}_d(t) &\approx \frac{\partial f}{\partial \overline{v}_o}\widetilde{v}_o(t) + \frac{\partial f}{\partial d}\widetilde{d}(t) \\ &= \left[\frac{V_{\text{in}}^2 TD^2}{2LV_o^2}\right]\widetilde{v}_o(t) - \left[\frac{V_{\text{in}}^2 TD}{LV_o}\right]\widetilde{d}(t). \end{aligned} \tag{12.19}$$

The partial derivatives in (12.19) are evaluated at the nominal solution. Using (12.18) to simplify (12.19), we have

$$\widetilde{i}_d(t) \approx (1/R)\widetilde{v}_o(t) + (V_{\text{in}}\sqrt{2T/RL})\,\widetilde{d}(t). \tag{12.20}$$

The resulting linearized averaged circuit is shown in Fig. 12.10.

The circuit directly shows, among other things, that if the duty ratio is held constant, that is, if $\widetilde{d}(t) \equiv 0$, then any (small) transient in the average output voltage decays exponentially with time constant $RC/2$, a fact that may not have been immediately obvious otherwise.

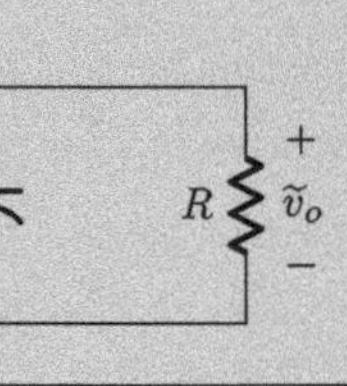

Figure 12.10 Linearized averaged circuit for a buck/boost converter in discontinuous conduction.

The topic of linearization is treated in more detail in Chapter 14, for averaged circuits and for state-space models.

12.5 Feedback Control

It is generally hard to assess stability and to design or evaluate feedback control schemes directly with time-variant and/or nonlinear models. In this section we focus on LTI models, such as those we obtain by linearization at a constant operating point. With LTI models, we can choose from a wide range of systematic analysis and design approaches to feedback control. Controllers designed using LTI models are usually LTI as well.

Controllers derived from linearized models cannot be guaranteed to provide satisfactory operation for large deviations from nominal operation, even if we expect them to function well for small perturbations. Also, the control design has to account for the fact that the linearized model will vary with the operating condition. Nevertheless, the majority of power circuit controllers are in fact designed on the basis of linearized models. A controller that performs well on the linearized model is likely to keep the circuit operating near the nominal and to make the operation relatively insensitive to small disturbances and errors. We can therefore reasonably expect that the nonlinearities have only secondary effects, except when there are major disturbances. We usually deal with nonlinear effects through refinements, modifications, or complements to a core LTI controller, evaluating these through simulation studies if they are not amenable to analysis.

For these reasons, we confine ourselves in this overview to discussing feedback control design issues in the context of "classical" control with continuous-time LTI models. This allows us to use simple transfer function computations but still provides ample opportunity to appreciate the potential benefits and pitfalls of feedback control. Further consideration of control design using such models, as well as state-space models and discrete-time models, is left to Chapter 14.

12.5.1 The Classical LTI Control Configuration

Considerable insight into what feedback can accomplish – and an understanding of what the dangers are – may be obtained by considering the interconnection of single-input, single-output, continuous-time LTI subsystems in Fig. 12.11. The simple feedback configuration here is of the same form as Fig. 12.1, and lies at the heart of classical control design. The lessons you learn from this configuration are directly applicable to controlling the small-signal dynamic behavior of a wide range of power circuits.

Control approaches rooted in the state-space methods developed since the early 1960s are used to tackle much more general problems, with interconnected multi-input, multi-output systems. In this text we can do no more than hint at these extensions. For the simple configuration in Fig. 12.11, however, the design approaches of classical control have withstood the test of time and provide a demanding benchmark.

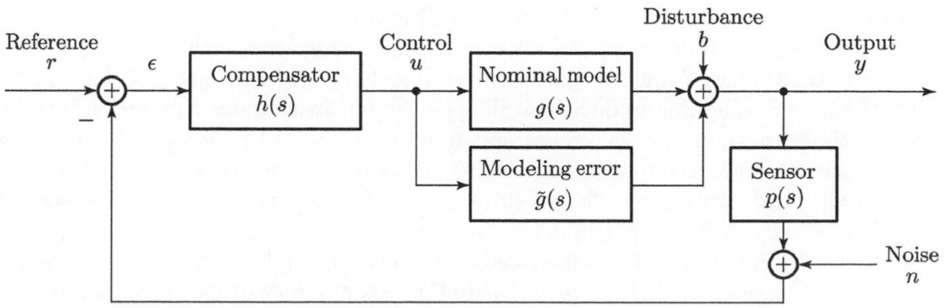

Figure 12.11 Block diagram of the classical LTI feedback configuration.

The quantity marked in each block of Fig. 12.11 is the transfer function that relates the output of the block to its input – in the Laplace transform domain (or frequency or impedance domain). Each block represents a *causal* subsystem; while non-causal subsystems are relevant to many offline signal processing or filtering applications, the concern in control design is with causal systems. Except where otherwise stated, we assume that all the transfer functions we work with are ratios of polynomials in s, or *rational* functions of s, as would be the case with *lumped* (as opposed to distributed) systems. On occasion we must step outside this category of transfer functions, as when representing time delays; a time delay of η between an input and an output corresponds to the non-rational transfer function $e^{-\eta s}$.

To keep our notation streamlined, we denote time-domain and transform-domain representations of a signal such as u in Fig. 12.11 by $u(t)$ and $u(s)$, respectively, whenever necessary to make the domain explicit. However, this convention is an abuse of notation, because $u(s)$ is *not* obtained through replacement of t by s in the expression for $u(t)$. (An unambiguous notation would, for example, be $U(s)$ for the transform, but such symbols are awkward to combine with our notation for steady-state values, averages, perturbations, and so on.) The context makes clear which domain we are working in. We always write transfer functions such as $g(s)$ with their arguments.

Figure 12.11 represents the controlled system or *plant* as having a transfer function $g(s)+\widetilde{g}(s)$, which relates the output y to the control input u. Here, $g(s)$ constitutes the *nominal* model of the plant, which is the basis for the control design, and $\widetilde{g}(s)$ represents errors in the nominal model. These errors could, for example, reflect uncertainties regarding the load and other circuit parameters, or simplifications, approximations, and compromises made during modeling.

The disturbances affecting the plant are represented in terms of their effects at the plant output, by means of the signal b. For simplicity, we assume that none of the disturbances can be measured, so no feedforward is possible. The feedback signal consists of a measurement of the system output, y, passed through a sensor whose transfer function is $p(s)$, and corrupted by measurement noise, n. For simplicity, we shall omit further consideration of sensor dynamics, so $p(s) = 1$. The feedback signal is compared with the external signal, r, which represents the reference (or desired or commanded) value for y; we would ideally like to have $r = y$. The result of the comparison is the noise-corrupted error signal, ϵ. The control signal to the plant, namely u, is produced by an LTI controller or compensator with transfer function $h(s)$ that acts on ϵ.

Example 12.9 Feedback Control of Armature Current in a Controlled-Rectifier Drive

The control structure shown in Fig. 12.12 is based on a proportional–integral (PI) control design for the controlled-rectifier drive treated in Examples 12.2 and 12.4.

The plant transfer function $1/(sL+R)$, which relates the averaged signals in the armature circuit, is simply the input admittance of the averaged circuit in Fig. 12.6(b) of Example12.4. We could have modeled the "disturbance" E as adding in directly at the plant input. Instead, we represent it by adding an equivalent signal at the plant output, to make the configurations in Figs. 12.11 and 12.12 look similar.

The local average current $\overline{i}_d(t)$ (or a good approximation to it) is compared with I_{ref}. Based on the discrepancy, the controller specifies the firing angle α for the phase-controlled rectifier. The output voltage waveform of the rectifier is thereby determined and hence so is its average value $\overline{v}_d(t)$.

We know from Chapter 4 that the steady-state relationship between the firing angle α and the average output voltage is $\langle v_d(t)\rangle = (2V/\pi)\cos\alpha$, where V is the amplitude of the sinusoidal voltage source. However,

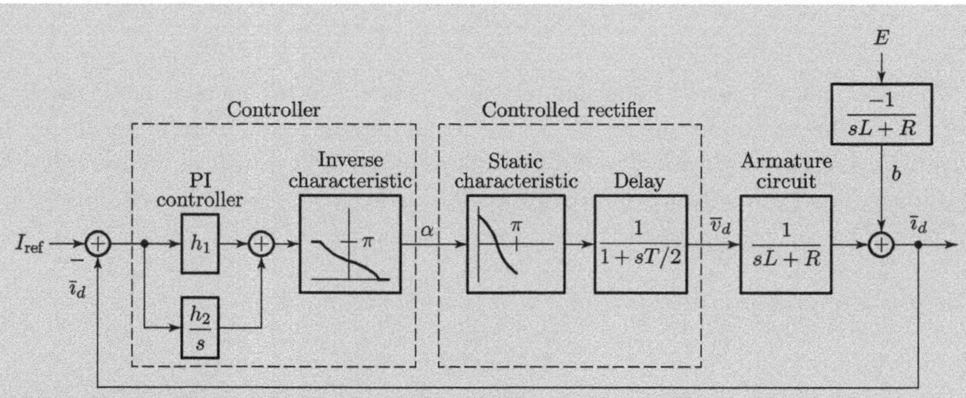

Figure 12.12 Feedback control for a phase-controlled-rectifier dc motor drive.

any transients in the average voltage, as reflected in variations in the local average $\overline{v}_d(t)$, depend in a more complicated way on variations in the firing angle of the rectifier. For instance, we have already noted, when discussing the open-loop response to a step in firing angle in Example 12.4, that it takes one cycle of duration T for the average output voltage to make the transition from its initial to its final value.

These facts lead to an approximate nominal model that is often used to represent the dynamics of the phase-controlled rectifier. The model comprises a cascade of the converter's steady-state characteristic and a dynamic block to represent a (mean) time delay of $T/2$ in the converter. This approximation of converter dynamics is reasonable when the signals in the system do not vary significantly during intervals of length T. We have used the rational transfer function $1/(1+sT/2)$ to approximate the non-rational transfer function $e^{-sT/2}$ of the time delay. Such models are often constructed in situations where sampling effects need to be approximately represented in a continuous-time model. Sometimes, however, the approximation is too crude to permit a satisfactory analysis of the system, and there is then little choice but to go to a sampled-data model of the sort described in Chapter 13.

The basic purpose of the controller is to vary $\overline{v}_d(t)$ in accordance with the error between the desired and actual average armature currents. To obtain a linear relationship from the error signal to the average voltage, the controller may be constructed as a cascade of a dynamic LTI section and a static nonlinearity that represents the *inverse* of the steady-state characteristic of the controlled rectifier. This approach, in the same spirit as Example 4.1, permits us to analyze and design the overall system using LTI models. The LTI portion of the controller shown in the block diagram is a PI controller, with h_1 representing the proportional gain and h_2 the integral gain. This controller is easily constructed by means of an operational amplifier circuit.

We can directly compute the transfer functions from the driving signals r, b, and n to the output y for the nominal system ($\widetilde{g}(s) = 0$) in Fig. 12.11, using the relationships represented in it. We state the result in terms of the nominal *loop transfer function* $\ell(s)$, which is the product of the transfer functions around the feedback loop, namely, $\ell(s) = p(s)g(s)h(s)$ for the general case but $g(s)h(s)$ under our assumption here that $p(s) = 1$. The desired result then is

$$y = \frac{1}{1+\ell(s)}b + \frac{\ell(s)}{1+\ell(s)}(r-n), \quad \text{with} \quad \ell(s) = g(s)h(s). \tag{12.21}$$

We can write (12.21) still more simply by defining what are known as the *sensitivity function* $\mu(s)$ and *complementary sensitivity function* $\mu'(s) = 1 - \mu(s)$:

$$\mu(s) = \frac{1}{1+\ell(s)}, \qquad \mu'(s) = \frac{\ell(s)}{1+\ell(s)}. \tag{12.22}$$

Note that $\mu'(s)$ is just the transfer function from r to y, so we also call it the *system transfer function*. One reason for giving $\mu(s)$ its particular name is that the fractional change in the system transfer function for a small fractional change in the loop transfer function is given by

$$\frac{d\mu'(s)}{\mu'(s)} = \mu(s)\frac{d\ell(s)}{\ell(s)}, \tag{12.23}$$

as you can verify quite simply. Now we can write (12.21) as

$$y = \mu(s)b + \mu'(s)(r-n). \tag{12.24}$$

The above expressions allow us to study three overlapping issues that are critical to assessing feedback control systems such as those in Figs. 12.11 and 12.12. These issues are *nominal stability*, *nominal performance*, and the *robustness* of both stability and performance to the presence of modeling errors. The nominal case corresponds to $\widetilde{g}(s) = 0$, and robustness refers to the preservation of stability and performance when $\widetilde{g}(s) \neq 0$.

12.5.2 Nominal Stability

Our assumption is that the given transfer functions $g(s)$ and $h(s)$ for the individual subsystems embody all that is relevant about the behavior of these subsystems. However, interconnecting them in the feedback loop provides multiple points at which to excite the system and to observe the response, and thus many possible transfer functions that could conceivably be examined.

The important notion is that of *asymptotic* or *internal* stability of the nominal system, which requires a decaying response to arbitrary initial conditions in the system. This goes beyond simply requiring a bounded output for all bounded inputs acting on the system when it starts from rest, that is, with zero initial conditions. It turns out that a necessary and sufficient condition for such internal stability in the case where $g(s)$ and $h(s)$ are *rational* functions of s is (i) that the transfer function $\mu(s)$ have all its poles strictly in the left half of the complex plane, *and* (ii) that any poles of $g(s)$ and $h(s)$ that are *not* strictly in the left half-plane are also poles of $\ell(s)$, that is, are not canceled by zeros of $h(s)$ and $g(s)$ respectively. (The second condition is sometimes overlooked in statements of conditions for nominal stability.) These conditions turn out to also guarantee that bounded inputs will always give rise to bounded outputs.

Poles of $g(s)$ or $h(s)$ that are canceled by zeros of $h(s)$ or $g(s)$ are termed *hidden poles* of the nominal system. They are poles of the plant or compensator that do not appear as poles of the loop transfer function and are therefore unaffected by the feedback. Hence, an equivalent statement of condition (ii) is that there are no unstable hidden poles. The choice of $h(s)$ is up to us, so satisfying (ii) is easy. We assume from now on that we have done so. The stability condition therefore reduces to condition (i), namely, that the sensitivity function (or the system transfer function) is stable.

The pole locations provide important information about the degree of stability in that they determine the speed of transients. The poles of $\mu(s)$, along with any hidden poles, constitute the system's *natural frequencies*. The number of natural frequencies is termed the *order* of the system. If there are no hidden poles, the denominator of $\mu(s)$ is the *characteristic polynomial* of the system, whose roots are the natural frequencies or *characteristic roots*. We will study these notions in more detail in Chapter 14, in the context of LTI state-space models.

A natural frequency at $\lambda_1 = \sigma_1 + j\omega_1$ contributes a term of the form $c_1 e^{\lambda_1 t} = c_1 e^{\sigma_1 t} e^{j\omega_1 t}$ to the system response. This term decays to zero when $\sigma_1 < 0$, that is, when λ_1 is in the left half-plane. The more negative σ_1 is, the faster the decay or damping of this term will be. The larger the magnitude of ω_1 is, the more oscillatory this term will be.

Classical control has various rules that permit rapid graphic determination of how the characteristic roots or natural frequencies move in the complex plane in response to variations in the parameters of the controller or nominal plant model. Such a *characteristic root locus* can be a significant aid to understanding how system behavior is affected by parameter variations, and to synthesizing candidate control structures.

With this basic understanding of nominal stability, and with the linearized averaged model of the buck/boost converter that we develop in Example 14.1, we will be able to make sense, in Example 14.2, of the rather non-intuitive behavior of the buck/boost converter under proportional feedback control in Example 12.3.

12.5.3 Nominal Performance

Once stability of the nominal closed-loop system is ensured, other aspects of performance can be pursued. Evaluation of nominal performance is considerably more involved, however. Some desirable performance features are best stated and evaluated in the time domain; others are best adapted to frequency-domain specification and treatment. Necessary trade-offs can be difficult to resolve.

For time-domain performance, our first interest is in the transient response determined by the natural frequencies. The more negative the real part of a natural frequency λ_1 is, the faster the associated exponential term $c_1 e^{\lambda_1 t}$ decays. A quick response is often desirable. However, the locations of the natural frequencies alone do not tell us about such things as the peak values of system variables during a transient, which may be critical in evaluating performance, especially in the context of power circuits. Time-domain simulations of system behavior are valuable, even with linear models, in evaluating performance measures such as peak overshoot in the output and in internal variables.

Various connections between pole–zero locations and transient behavior are known for second-order systems. For example, fast responses with small overshoots are typically associated with complex poles whose real and imaginary parts are approximately equal; the presence of a zero tends to increase overshoots. To take advantage of such results, higher-order systems are often controlled so as to make the dominant behavior essentially second order. This is done by placing a pair of closed-loop poles substantially closer to the imaginary axis than any of the other poles.

Performance specifications that are conveniently stated in the frequency domain can be studied using (12.24). The nominal performance of the feedback control system in Fig. 12.11 is primarily measured by how closely the output y matches the reference signal r in the nominal system, despite the disturbance b and measurement noise n. Although r is commonly a constant or slowly varying value, the inputs b and n are best thought of as time-varying but bounded signals, with their power distributed over certain frequency ranges. Hence, (12.24) suggests that the frequency response $\mu(j\omega)$ of the sensitivity function, or equivalently the loop gain $\ell(j\omega)$, plays a critical role.

Note first, for later comparison, that in the open-loop system the disturbance b appears without attenuation, because $y = g(s)u + b$. Even if $g(s)$ were stable and u could be chosen to obtain $g(s)u = r$, we would have $y - r = b$. Measurement noise is not an issue in the open-loop case, because measurements are not used.

For the closed-loop system, (12.24) shows that we can make $y \approx r$, more or less uniformly over frequency, if:

- $|\mu(j\omega)| \approx 0$, or, equivalently, $|\ell(j\omega)| \gg 1$, at frequencies ω where r and b have significant power compared to n; and
- $|\mu(j\omega)| \approx 1$, or, equivalently, $|\ell(j\omega)| \ll 1$, at frequencies ω where n has significant power compared to r and b.

In frequency ranges for which r and b have power comparable to n, good performance may not be attainable, at least with the configuration in Fig. 12.11.

The loop-gain magnitude of a physical system naturally falls off to low values as the frequency increases. Therefore the best situation for control generally occurs when the reference signal r and disturbance b are low-frequency signals, and the measurement noise n is confined to higher frequencies. We would then try and choose the compensator transfer function $h(s)$ to obtain a large loop-gain magnitude at low frequencies and have the magnitude fall off (or "roll" off) at higher frequencies. Of course, this "shaping" of the loop gain is subject to the constraint that closed-loop stability is maintained. The Nyquist criterion of classical control is especially useful because it relates stability to the loop gain; we mention it in Chapter 14, but omit a more detailed treatment.

The classical approach to control design tries to deal with all of these considerations. Related compensator design calculations are best carried out using *Bode plots* of the loop gain, which comprise a plot of its magnitude $|\ell(j\omega)|$ versus frequency on a log–log scale, and a plot of its phase $\angle\ell(j\omega)$ versus frequency on a linear–log scale. The Bode magnitude plot traditionally uses units of *decibels* (dB). The loop-gain magnitude in dB is, by definition, $20\log_{10}|\ell(j\omega)|$.

The frequency at which the loop gain drops to a magnitude of 1 (or 0 dB) is termed the *unity-gain crossover frequency*, ω_c. Examination of (12.22) shows that the system frequency response at ω_c, namely $\mu'(j\omega_c)$, has magnitude $\geq 1/2$. This magnitude is still comparable to the low-frequency value of approximately 1, so the system can still respond significantly to inputs at frequency ω_c. Hence ω_c provides some measure of the bandwidth of the system. (Strictly speaking, the bandwidth is defined as the frequency at which the system frequency response drops to $1/\sqrt{2}$ of its low-frequency value.)

Although we have summarized the traditional setup for control design, you should be alert to features of a problem that demand a different treatment. For example, many of the rules of thumb commonly used in control design apply only when behavior is dominated by a single pair of complex poles, and many do not apply when the plant has right-half-plane poles or zeros.

It should now be evident that feedback control can provide better performance than open-loop control. However, it should also be clear that a poor choice of compensator could increase the sensitivity to disturbances and measurement noise and make the performance worse than that of the open-loop system. The constraints on the loop transfer function imposed by disturbances, noise, and stability conditions may prevent a compensator from performing well for any choice of its parameters. In this case, an alternative compensator structure has to be sought.

Example 12.10 Nominal Closed-Loop Performance of a Controlled-Rectifier Drive

The loop transfer function for the controlled-rectifier drive model in Example 12.9 is

$$\ell(s) = \frac{(h_1 s + h_2)}{s(1 + sT/2)(sL + R)}. \tag{12.25}$$

Let's assume that $T = (1/120)\,\mathrm{s} = 8.33\,\mathrm{ms}$, $R = 0.1\,\Omega$, and $L/R = 10T = 83.3\,\mathrm{ms}$. The magnitude and phase of $\ell(j\omega)$ are shown in the Bode plots of Fig. 12.13(a) for a candidate controller in which $h_1 = 0.35$ and $h_2 = 10.5$. The magnitude is high at low frequencies, thanks to the integrator, and crosses over beyond 50 rad/s. The Bode plots in Fig. 12.13(b) are for $\mu'(j\omega)$, which is the closed-loop system frequency response from I_{ref} to $\bar{\imath}_d$.

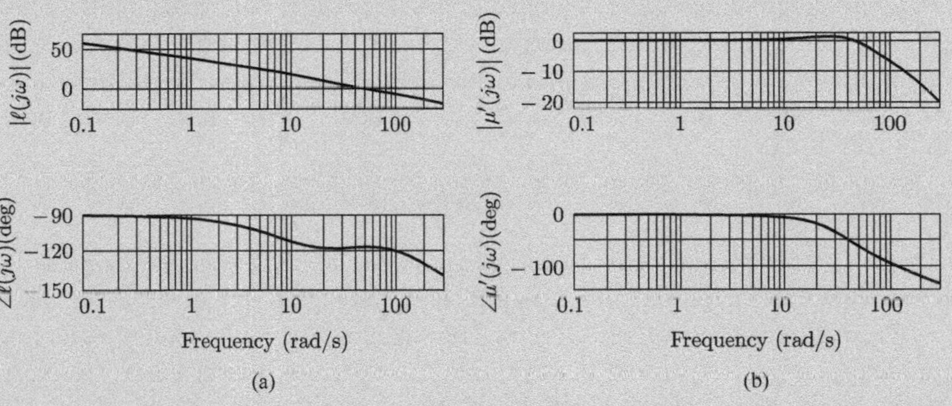

Figure 12.13 Bode plots of (a) loop gain and (b) system frequency response for a controlled-rectifier drive.

The closed-loop pole locations are -193 and $-29.5 \pm j26.4$. The complex pair is dominant and the real pole is a reflection of our representation of time delay in the converter. The real and imaginary parts of the dominant pair are approximately equal, so we expect good transient responses. The time constant associated with these responses is on the order of 34 ms (1/29.5), or significantly larger than the period T. The design is therefore consistent with our modeling assumptions for the converter.

The simulation in Fig. 12.14, using the model of Fig. 12.12, shows the closed-loop response of the averaged armature current following a step in I_{ref}. Note that $\bar{\imath}_d(t)$ in the steady state equals I_{ref}, despite the presence of the unknown constant disturbance representing the back-EMF E. The reason for the zero steady-state error can be traced to the integrator in the PI controller. The integrator introduces a pole at $s = 0$ in the loop transfer function, which causes the loop gain at $\omega = 0$ to be infinite, and thereby makes the system insensitive to constant disturbances. Another way to view this result is to note that in the constant steady state obtained when this stable LTI system is driven by constant disturbances, the input to the integrator must be zero in order for its output to be constant. Hence $\bar{\imath}_d = I_{\mathrm{ref}}$ in the steady state. More detailed simulations correlate well with predictions of the averaged model.

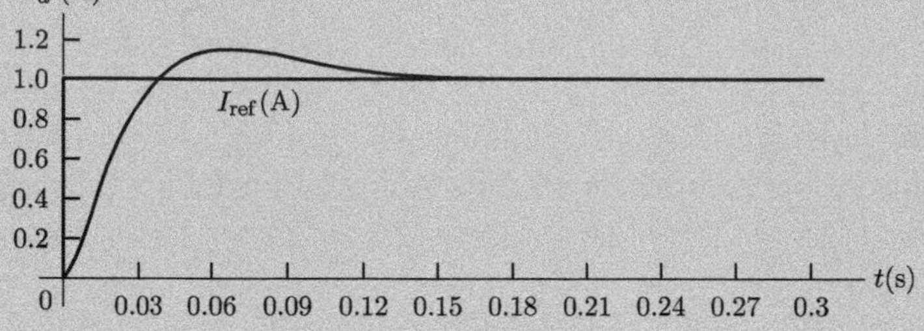

Figure 12.14 Closed-loop response of averaged armature current $\bar{\imath}_d$ to a step in the reference I_{ref}.

12.5.4 Robustness

Our robustness requirement is that stability and performance be preserved for some range of plant model uncertainty $\widetilde{g}(s)$. The constraints on the loop gain that are imposed by our *stability robustness* requirement are highlighted in the following result, which holds under the condition that the nominal loop transfer function $\ell(s)$ and the actual transfer function $\ell(s) + \widetilde{\ell}(s)$ have the same number of unstable poles (and no unstable *hidden* poles, as defined in Section 12.5.2). With this condition, the result is that stability of the actual system is guaranteed by stability of the nominal system if

$$|\widetilde{\ell}(j\omega)| < |1 + \ell(j\omega)|. \tag{12.26}$$

The proof of this result is based on the Nyquist criterion. The constraint implies that we are especially vulnerable to errors or uncertainties in the loop gain at frequencies for which $\ell(j\omega) \approx -1$.

When the variation of the loop transfer function from its nominal is the result entirely of modeling errors, $\widetilde{\ell}(s) = h(s)\widetilde{g}(s)$. Relating (12.26) specifically to this case, we obtain the following sufficient condition for robust stability: If the nominal plant transfer function $g(s)$ and the actual transfer function $g(s)+\widetilde{g}(s)$ have the same number of unstable poles, stability of the actual system follows from that of the nominal system if

$$\frac{|\widetilde{g}(j\omega)|}{|g(j\omega)|} < \frac{1}{|\mu'(j\omega)|}. \tag{12.27}$$

The constraint in (12.27) requires that the frequency response $\mu'(j\omega)$ of the nominal closed-loop system have small magnitude in frequency ranges for which the fractional error in the plant model can be large. For example, at frequencies for which we are completely uncertain about the phase of the plant's frequency response, the left-hand side of (12.27) is ≥ 2. Thus we need $|\mu'(j\omega)| < 1/2$, which requires that we be beyond the crossover frequency. Resonance peaks in the system frequency response $\mu'(j\omega)$ mark regions that are particularly sensitive to modeling errors. Translating (12.27) to conditions on the nominal loop gain, we see that $|\ell(j\omega)|$ can be large at frequencies for which the fractional modeling error is small – but must be small, and not close to -1, wherever the modeling error is large.

In classical control we commonly evaluate the distance of $\ell(j\omega)$ from -1 by means of a *phase margin*. This measures how much $\angle\ell(\omega_c)$ exceeds -180°, where ω_c is the crossover frequency, so $|\ell(j\omega_c)| = 1$. The phase margin tells us how much additional phase delay around the loop is needed to make $\ell(j\omega_c) = -1$. A phase margin of 45° is considered good if the modeling error around crossover is relatively small. For the candidate controller in Example 12.10, the phase margin is 60°. Also of interest is the *gain margin*, defined as $1/|\ell(j\omega_o)|$, where ω_o is the *phase crossover frequency*, at which the phase of $\ell(j\omega)$ becomes -180°. The gain margin is therefore the factor by which the loop gain must be increased to make $\ell(j\omega_o) = -1$.

A major potential advantage of feedback is that it can robustly stabilize an unstable or insufficiently stable system. However, we evidently cannot consider stability robustness in isolation from good nominal performance. Both objectives impose constraints on the loop gain, and these constraints could very well conflict. The best situation generally occurs when significant modeling errors are confined to higher frequencies, the reference signal r and disturbance b are low-frequency signals, and the measurement noise n is at higher frequencies. In this case, we can make the loop gain high at low frequencies and roll it off at high frequencies. In other cases, the design of a good LTI feedback controller can be much more difficult.

We tend to have good low-frequency models in power electronics (for instance, averaged models). The modeling error is then indeed confined to higher frequencies, but often not as high as the noise frequencies. Modeling error – rather than measurement noise – then determines how soon the loop gain magnitude must be made small. For example, in the controlled rectifier circuit of Example 12.9, we constructed the rectifier model under the assumption that signals varied slowly relative to the switching period, so the model becomes less accurate as we approach one half the switching frequency. This limitation might cause us to reduce the loop gain magnitude well before measurement noise (such as the switching-frequency ripple in the current measurement) becomes a concern. The candidate controller in Example 12.10 may not be sufficiently conservative in this respect.

It is important to recognize that we cannot shape the loop gain at will. A practical compensator has only a limited number of parameters that we can vary. Also, what we do in one frequency range has consequences in other ranges. As noted earlier, we want to keep $\mu(j\omega) \approx 0$, that is, have large loop gain, at frequencies for which good command following and disturbance rejection are needed. We want $\mu(j\omega) \approx 1$, that is, small loop gain, at frequencies for which sensor noise and modeling errors are large. We are also forced to have $\mu(j\omega) \approx 1$ for high frequencies because the loop gain of physical systems must fall off. It turns out then that to maintain stability of our closed-loop system we must pay for these requirements by having $|\mu(j\omega)| \gg 1$, or, equivalently, $|\mu'(j\omega)| \gg 1$, or, equivalently, $\ell(j\omega) \approx -1$ in the remaining frequency ranges. The system could thus become vulnerable to even small modeling errors in these other frequency ranges.

Ensuring *performance robustness* is more involved than ensuring stability robustness. The sensitivity relation in (12.23) suggests that, at least for small modeling errors, the closed-loop system frequency response does not change much at frequencies for which the sensitivity $\mu(j\omega)$ is small. We simply note here that, if nominal performance is satisfactory, actual performance is likely to be degraded but not destroyed, so long as the modeling errors satisfy (12.26) by an adequate margin.

We are also interested in studying the robustness of system stability and performance to *nonlinearities* in the model. One important nonlinearity in power electronics arises from the fact that control variables often have tight constraints. These constraints introduce saturation nonlinearities in the feedback loop, the effects of which we can sometimes approximate in an LTI analysis by assuming a reduced loop gain.

More generally, the effects of nonlinearities on a control system designed by LTI methods are studied through simulations on more detailed models than the ones used for the initial controller design. The simulations should examine a variety of expected operating conditions, including those ignored during the initial design. Such simulations help to validate the control design or to expose problems stemming from the various modeling assumptions and the design approach used. The simulations can form the basis for refining the initial controller.

Feedback control has the potential to turn an unstable open-loop system into a stable closed-loop system that performs significantly better than the open-loop system. It can also, unless care is taken, convert a stable open-loop system into a closed-loop system that performs poorly or becomes unstable. The design and implementation of feedback controls therefore require considerable care.

Notes and Bibliography

Averaging is a frequent theme in modeling and analysis of power electronic circuits, especially for high-frequency switching or PWM converters. An early reference is [1], and several others are listed and discussed in [2]. Tellegen's theorem is the subject of the monograph [3]. Our treatment of averaging and linearization continues in Chapters 13 and 14, including in the setting of state-space models. The comprehensive text [4] devotes substantial attention to all of this; see also [5]. The local ω_s-component is introduced in [6]. Dynamic modeling and control design for dc and ac machine drives are treated in [7], [8], and [9], for example. There are many textbooks that are appropriate for a first course on control, see for instance [10] and [11]. Our sketch in Section 12.5 of issues and objectives in control design is inspired by [11] and [12].

1. G. W. Wester and R. D. Middlebrook, "Low-Frequency Characterization of Switched DC-DC Converters," *IEEE Trans. Aerospace and Electronic Systems*, vol. 9, pp. 376–385, May 1973.
2. S. R. Sanders and G. C. Verghese, "Synthesis of Averaged Circuit Models for Switched Power Converters," *IEEE Trans. Circuits and Systems*, vol. 38, pp. 905–915, Aug. 1991.
3. P. Penfield Jr., R. Spence, and S. Duinker, *Tellegen's Theorem and Electrical Networks*, Cambridge, MA: MIT Press, 1970.
4. R. W. Erickson and D. Maksimović, *Fundamentals of Power Electronics*, 3rd ed., New York: Springer, 2020.
5. D. Maksimović, A. M. Stanković, V. J. Thottuvelil, and G. C. Verghese, "Modeling and Simulation of Power Electronic Converters," *Proc. IEEE*, vol. 89, pp. 898–912, June 2001.
6. S. R. Sanders, J. M. Noworolski, X. Z. Liu, and G. C. Verghese, "Generalized Averaging Method for Power Conversion Circuits," *IEEE Trans. Power Electronics*, vol. 6, pp. 251–259, April 1991.
7. W. Leonhard, *Control of Electrical Drives*, 3rd ed., New York: Springer, 2001.

8. N. Mohan, *Electric Machines and Drives: A First Course*, Chichester: Wiley, 2012.
9. P. C. Krause, O. Wasynczuk, S. D. Sudhoff, and S. D. Pekarek, *Analysis of Electric Machinery and Drive Systems*, 3rd ed., Chichester: Wiley-IEEE Press, 2013.
10. G. Franklin, J. Powell, and A. Emami-Naeini, *Feedback Control of Dynamic Systems*, 8th ed., London: Pearson, 2018.
11. K. J. Åström and R. M. Murray, *Feedback Systems: An Introduction for Scientists and Engineers*, 2nd ed., Princeton, NJ: Princeton University Press, 2021.
12. J. C. Doyle, B. A. Francis, and A. Tannenbaum, *Feedback Control Theory*, London: Macmillan, 1991.

PROBLEMS

12.1 **(a)** Determine the relation in the frequency domain between the Fourier transforms of: (i) a signal $x(t)$; (ii) its local average $\overline{x}(t)$ defined in (12.8); and (iii) its local ω_s-component $\widehat{x}(t)$ defined in (12.10).

(b) Use your results from (a) to make clear in what sense $\overline{x}(t)$ is a low-pass-filtered version of $x(t)$, and similarly describe how $\widehat{x}(t)$ relates to $x(t)$ in the frequency domain.

(c) Show that the derivative of the local average equals the local average of the derivative, and find an expression relating the derivative of the local ω_s-component to the local ω_s-component of the derivative.

(d) How is the derivative of the local average related to the local average of the derivative when the averaging interval T is varied as a function of time t, so it is $T(t)$ rather than a constant T?

12.2 We obtained an averaged circuit for a controlled-rectifier drive in Fig. 12.6(b), Example 12.4, under the assumptions of continuous conduction and no commutating reactance.

(a) Use the results of Section 4.2 to show that the effect of a commutating reactance X_c on the averaged circuit can be approximately accounted for by incorporating an additional resistor of value $2X_c/\pi$ in series with the voltage source. Why is this only an approximate result?

(b) Suppose that there is no commutating reactance but that the drive goes into discontinuous conduction. Find an approximate averaged circuit for this case, making clear why the circuit is only approximate. (A good place to start is with a sketch of the rectifier voltage v_d for discontinuous conduction.)

12.3 This problem deals with a buck/boost converter in discontinuous conduction.

(a) Extend the linearized model of the buck/boost converter in discontinuous conduction in Example 12.8 to include perturbations $\widetilde{v}_{\text{in}}$ in the source value. Show that the effect is to add a $\widetilde{v}_{\text{in}}$-dependent current source in parallel with the $\widetilde{d}$-dependent current source.

(b) Compute the transfer function from $\tilde{d}$ to $\tilde{v}_o$ in the linearized circuit of (a). Obtain Bode plots of the frequency response when $R = 300\ \Omega$ and the other parameter values are as in Example 12.1. (You should verify that the converter is indeed in discontinuous conduction for this R.)

(c) Draw a block diagram of a closed-loop PI control system that could be used to regulate the buck/boost converter in discontinuous conduction, maintaining its output voltage at -9 V. Find the choice of PI control gains that places the natural frequencies of the closed-loop system at $2p \pm j2p$, where p is the pole of the open-loop converter transfer function that you computed in (b). Obtain Bode plots of the loop gain for this closed-loop system.

(d) Show that the PI control system in (c) can be interpreted as replacing the $\tilde{d}$ current source in the linearized circuit by a parallel resistor–inductor combination. Explain from the circuit diagram what this replacement does to the dynamic behavior of $\tilde{v}_o$ and to the steady-state error in $\tilde{v}_o$ induced by parameter errors and constant disturbances, such as source deviations $\tilde{v}_{\text{in}} \neq 0$.

(e) Compute the transfer functions from $\tilde{v}_{\text{in}}$ to $\tilde{v}_o$ (the so-called *audio susceptibility*) for the open-loop and closed-loop systems in (b) and (c), respectively. Compare their frequency-domain characteristics using Bode plots.

(f) How would the controls in (c) perform if we actually had $R = 2\ \Omega$, with the converter in continuous conduction?

12.4 In *current-mode control* of high-frequency switching dc/dc converters, we specify the peak inductor current or switch current in each cycle, rather than specifying the duty ratio. This problem considers constant-frequency current-mode control, in which the switch is turned on every T seconds and turned off when its current reaches a specified threshold value, $\imath_{\text{th}}$. Develop averaged models and their linearizations for discontinuous-conduction operation of the buck/boost converter. Find the transfer function from small perturbations $\tilde{\imath}_{\text{th}}$ in the threshold to perturbations $\tilde{v}_o$ in the output for this converter. Also find the audio susceptibility transfer function, from $\tilde{v}_{\text{in}}$ to $\tilde{v}_o$. Can this converter become open-loop unstable for any discontinuous operating condition?

12.5 Derive (12.21), which relates the output y of the classical closed-loop control configuration in Fig. 12.11 to the driving signals r, b, and n. Derive the sensitivity result in (12.23). Verify the conditions for closed-loop stability given in Section 12.5.2. Show that the robustness condition (12.27) follows from (12.26).

12.6 This problem relates to the closed-loop control design for the controlled-rectifier drive in our examples.

(a) Determine how the characteristic roots vary for the closed-loop controlled rectifier system in Example 12.10, as the controller parameters vary over the following ranges: (i) $-3 \leq h_1 \leq 3$, with $h_2 = 0$; (ii) $-2 \leq h_2 \leq 60$, first with $h_1 = 0.35$ and then with $h_1 = 1$. A plot of these variations in the complex plane is referred to as a *root locus* diagram.

(b) Verify the step response shown in Fig. 12.14. Also construct a simulation of the full nonlinear model of the closed-loop drive and check if the step response of i_d is predicted well by our averaged analysis.

(c) Evaluate the closed-loop system obtained with $h_1 = 1$ and $h_2 = 60$. How does it compare with the system in Example 12.9?

12.7 Obtain an averaged-circuit model for the high-frequency PWM bridge inverter in Fig. 8.14 and use it to simulate the startup behavior of the system for different parameter values.

13 Averaged-Circuit and State-Space Models

The preceding chapter presented several examples of perturbations from nominal operation of power electronic systems. These examples motivated the need for dynamic modeling of power converters, in order to understand such perturbations and to design appropriate controllers. We introduced the idea of circuit averaging as a means of obtaining simple and informative dynamic models in circuit form. Linearization of such models around a constant nominal operating condition then gave rise to linear, time-invariant models to describe small deviations from nominal operation. These LTI models permit the systematic, analytically based design of feedback controllers to regulate a power converter in the neighborhood of a desired operating point.

In this chapter we significantly broaden the range of dynamic modeling approaches and tools available to us. The first part of the chapter extends circuit averaging ideas to broader classes of converters than covered in the previous chapter, including those for which the switch averaging is somewhat more subtle than seen so far. We also develop circuit averaging approaches for converters (such as resonant converters) in which the quantity of interest is not the local average value but rather the local fundamental.

The second part of this chapter introduces state-space models, for circuits and for more general systems (such as feedback controllers), and in both continuous time (CT) and discrete time (DT). Many power electronic converters can be modeled as circuits that are switched between LTI circuit structures, and CT state-space representations are particularly suited to modeling these. We also demonstrate that averaging can be carried out on such a switched CT state-space model of a power electronic circuit, rather than directly on the circuit itself, offering helpful flexibility and generality. DT state-space representations arise naturally in sampled-data models for power converters, as might be used to design digital controllers.

In Chapter 14 we shall return to the idea of linearization around an operating point, for both averaged-circuit and state-space models. The resulting LTI models can provide the basis for feedback control design, as also outlined in that chapter.

13.1 Averaged-Circuit Models

We defined the local averaging operation in (12.8), repeated here for convenience:

$$\overline{x}(t) = \frac{1}{T}\int_{t-T}^{t} x(\tau)\, d\tau. \tag{13.1}$$

It had the property (12.13), namely that the derivative of the local average was the local average of the derivative. We then invoked this definition and property to average a circuit, and also to average a switching function $q(t)$, resulting in the *continuous duty ratio* $d(t) = \overline{q}(t)$. However, we have not yet directly addressed how to average a switch in a power electronic circuit. It is clear that the average of a switch is no longer a switch in the averaged circuit, because the average voltage and average current on the switch are simultaneously nonzero. We address this task first, and then apply the result to obtaining averaged models of general dc/dc converters.

13.1.1 Averaging a Switch

Switch averaging is especially rewarding in the case of high-frequency switching or PWM converters, where local average values are of primary interest. In such circuits, we can approximately characterize the averaged switch entirely in terms of the (control-dependent) constraints that it imposes on the averaged circuit variables at its terminals. As a result, we can replace the switch in the instantaneous circuit with a corresponding component in the averaged circuit. The other components in the circuit are usually LTI, and are unchanged by averaging. Hence, circuit averaging for such a converter simply involves replacement of the switch by its averaged model.

It is natural to start with the canonical switching cell of Fig. 5.7, reproduced in Fig. 13.1(a) for convenience. The averaging interval is again chosen to be the switching period T. The assumptions that we make in obtaining an averaged model for a switch can be stated in different ways, but most usefully as follows:

- a *small-ripple* assumption, namely, that the voltage v_{yz} $(= v_C)$ and current i_x $(= i_L)$ at time t are well approximated by their local average values, $\overline{v}_{yz}(t)$ and $\overline{i}_x(t)$ respectively; and

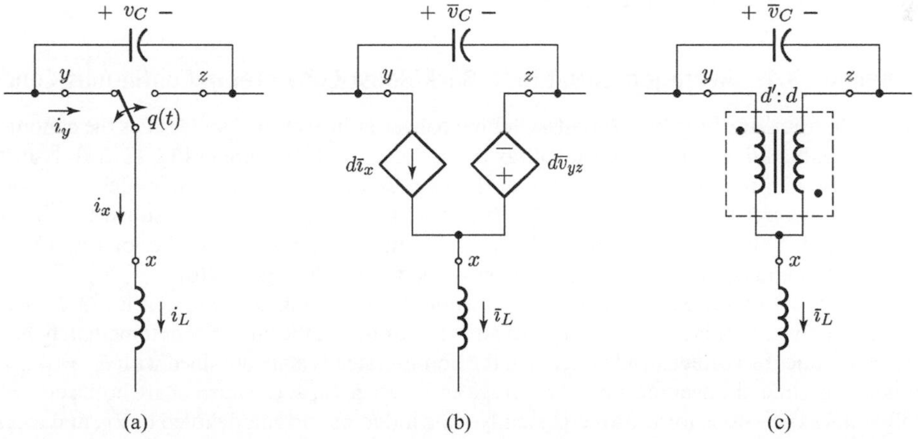

Figure 13.1 (a) The canonical switching cell for high-frequency switching converters (the time argument t is omitted on all variables other than $q(t)$, for simplicity). (b) Approximate averaged switching cell using controlled sources, for continuous conduction with duty ratio d. (c) Approximate averaged switching cell, using an ideal transformer; $d' = 1 - d$.

- a *slow-variation* assumption, namely, that these two local average values do not vary significantly over any interval of length T, that is, they vary substantially more slowly than half the switching frequency.

Both assumptions are generally satisfied in well-designed high-frequency switching converters operating in continuous conduction.

Suppose the switch in Fig. 13.1(a) is governed by a 0–1 switching function $q(t)$, such as that described in Section 12.3.2 (and specifically in Example 12.6), and is in position y with some duty ratio in each switching interval. As $i_y = qi_x$ (dropping the time argument t to simplify the notation), we can write $\bar{i}_y = \overline{qi}_x$. Our assumptions allow us to treat $i_x(\tau)$ as though it were approximately constant at $\bar{i}_x(t)$ over the averaging interval $t - T \le \tau \le t$, so

$$\bar{i}_y(t) = \overline{qi}_x(t) \approx \bar{q}(t)\bar{i}_x(t) = d(t)\bar{i}_x(t). \tag{13.2}$$

Similarly, $v_{xz} = qv_{yz}$, so

$$\bar{v}_{xz}(t) = \overline{qv}_{yz}(t) \approx \bar{q}(t)\bar{v}_{yz}(t) = d(t)\bar{v}_{yz}(t). \tag{13.3}$$

The relationships obtained in (13.2) and (13.3) constitute the desired approximate characterization of the averaged switch in terms of the control-dependent constraints imposed on the averaged circuit variables at its terminals. (Thanks to Kirchhoff's voltage and current laws, a three-terminal element is fully characterized by two constraint equations for its terminal variables.) Two circuit representations of this characterization are shown in Figs. 13.1(b) and (c). Although these two are equivalent, the representation in terms of controlled dependent sources is simpler for such tasks as linearization, which we consider later. The representation involving the ideal transformer is useful for situations in which the duty ratio is held constant, that is, with $d(t) = D$.

The high-frequency dc/dc converter topologies in Chapter 5 had the canonical switching cell at their core, with all other elements being LTI. Hence, approximate averaged circuits for all these converters are simply obtained by replacing the canonical switching cell by its approximate average. We see this in the next example.

Example 13.1 Averaged Circuit for a Buck/Boost Converter in Continuous Conduction

Our development of the buck/boost switching converter in Section 5.4 identifies the canonical cell structure within the buck/boost converter topology in Fig. 12.2, repeated here in Fig. 13.2(a). Note that in this case v_{yz} of Fig. 13.1 is actually $v_{\text{in}} - v_C$ rather than v_C. To obtain the averaged model, we replace the canonical cell by its averaged version from Fig. 13.1(b) or (c). Choosing the latter, we substitute the ideal transformer windings of d' and d turns, respectively, for the transistor and diode switches of Fig. 12.2. We also replace all instantaneous quantities by their averages. The result is in Fig. 13.2(b).

When the duty ratio $d(t)$ is constant at a value D, the averaged circuit of Fig. 13.2 is (passive and) LTI, so that analysis in this case is straightforward. For example, the circuit can immediately be used to read off average values of voltages and currents in the nominal steady state obtained with $\bar{v}_{\text{in}} = V_{\text{in}}$ and $d(t) = D$. In this steady state, the average inductor voltage and average capacitor current are both zero, which leads to the following expressions for the average steady-state inductor current, denoted by I_L, and average steady-state output voltage V_o (or capacitor voltage, V_C):

$$I_L = -\frac{V_o}{RD'}, \qquad V_o = -V_{\text{in}}\frac{D}{D'}. \tag{13.4}$$

These expressions are consistent with the results in Section 5.4.

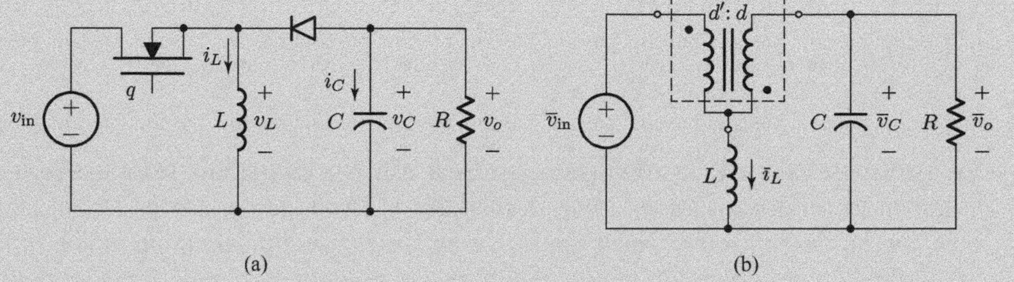

Figure 13.2 (a) Buck/boost converter circuit. (b) Averaged circuit for the buck/boost converter.

The averaged circuit also provides a basis for understanding the results in Example 12.1, where we examined the open-loop response of v_o ($= v_C$) to a step in the applied voltage v_{in}, with constant $d(t) = D$. Using the averaged circuit to compute the transfer function from the locally averaged input voltage to the locally averaged output voltage – referred to as the *audio susceptibility* transfer function of the circuit – we get

$$\frac{\overline{v}_o(s)}{\overline{v}_{in}(s)} = \frac{-D'D/LC}{s^2 + (1/RC)s + (D'^2/LC)}. \tag{13.5}$$

With a step change in $v_{in}(t)$, the *averaged* input $\overline{v}_{in}(t)$ actually takes one averaging interval to cross between its initial and final values. Also, the averaging approximations in (13.2) and (13.3) are poor during this interval. However, this interval T is much smaller than the time constants of the averaged model, so we can consider $\overline{v}_{in}(t)$ itself to be a step function, and can use the transfer function in (13.5) of the averaged model to compute the response to this step.

Rather than computing this step response in detail, it suffices for the purposes of this illustration to simply recall that the form of the response is largely determined by the *poles* of the transfer function (13.5). For the underdamped case, which is what the parameter values in Example 12.1 correspond to, the poles λ_1 and λ_2 are complex, and conjugate to each other:

$$\lambda_1 = \lambda_2^* = -\frac{1}{2RC} + j\omega_d, \qquad \omega_d = \sqrt{\frac{D'^2}{LC} - \frac{1}{4R^2C^2}}, \tag{13.6}$$

where * denotes complex conjugation. The transition from the steady state before the input step to the steady state after it is thus governed by a transient of the form

$$c_1 e^{\lambda_1 t} + c_1^* e^{\lambda_1^* t} = Ae^{-t/(2RC)} \sin(\omega_d t + \theta), \tag{13.7}$$

where the constants c_1, A, and θ are fixed (upon invoking the continuity of the inductor current and capacitor voltage) by the initial conditions prior to the step. The response is therefore an exponentially damped oscillation at the angular frequency ω_d – termed the *damped natural frequency* – with the damping governed by the time constant $2RC$. See Section 10.1 for an analysis of a very similar situation.

A detailed calculation based on (13.5) accounts very well – not only qualitatively but also quantitatively – for the average behavior of the open-loop step response seen in Fig. 12.3(a). The time constant $2RC$ is 880 μs or 44 switching cycles, and the period $2\pi/\omega_d$ is 2924 μs, or 146 cycles. A similar calculation can be carried out for the step response with feedforward, shown in Fig. 12.3(b), accounting for the fact that d has a step change as well, to a value determined by v_{in} after the step.

The averaged circuit in this example can be used as the basis for generating other types of analytical models, such as state-space models. The circuit can also be accepted directly by many standard circuit simulation packages. Note that the circuit depends nonlinearly on the control variable $d(t)$. The process of linearization that we examined in Section 12.4 (and describe further in Chapter 14) provides one way of dealing with the nonlinearity. Example 14.1 derives a transfer function that allows us to compute the response to small but otherwise arbitrary changes in d.

Sometimes the capacitor in the canonical cell has significant resistance, modeled as an equivalent series resistance (ESR). In this case, the voltage v_{yz} in Fig. 13.1(a) may no longer satisfy the assumption of small ripple that underlies the approximation in (13.3). It is possible to refine the approximation and obtain an averaged-switch model that accounts for ESR. Alternatively, the state-space approach to averaging described in Section 13.3.6 permits us to handle this situation systematically.

The switch averaging in Fig. 13.1 does not apply to *discontinuous* conduction (see Section 5.9). The switch of the canonical cell in this case takes a third position, contacting neither y nor z. Using the fact that the average voltage across the inductor is now approximately zero, we can still obtain an averaged model of the switch. The style of analysis demonstrated in Example 12.5 is usually simpler, however.

13.2 Generalizing Circuit Averaging to the Fundamental Component

In Chapter 12 we introduced the idea of a *local fundamental* or *local ω_s-component* of any waveform $x(t)$ whose nominal behavior was periodic with period $T = 2\pi/\omega_s$. We repeat the definition here:

$$\widehat{x}(t) = \frac{1}{T}\int_{t-T}^{t} x(\tau)e^{-j\omega_s\tau}\, d\tau. \tag{13.8}$$

Our motivation was to track the dynamic behavior of the component of a waveform at the frequency ω_s, where this could be the nominal switching frequency or the nominal fundamental frequency of the source. We have already seen in Section 12.3 – and continuing in Section 13.1.1 – how an averaged circuit can easily be constructed to permit tracking of the local average component of each variable in a power circuit, provided the circuit comprises LTI components interconnected by switches. It turns out that, following a similar route, this circuit averaging idea can be extended to the local ω_s-component, as described next.

The definition (13.8) shows that computing the ω_s-component is a linear operation, so computing this for each KVL and KCL equation associated with the underlying circuit results in the same equations written in terms of the ω_s-components. We can therefore replace the currents and voltages in the original circuit by their ω_s-components to obtain a set of quantities that satisfy the same respective KVL and KCL equations.

The next step is to understand the constraints imposed by LTI components in the original circuit on the local ω_s-components of their terminal voltages and currents. Any resistor in the original circuit will remain as a resistor of the same value in the ω_s-component circuit: if $v_R(t) = R\,i_R(t)$ is the relationship for all t between the voltage $v_R(t)$ across, and current $i_R(t)$ through, a resistor of value R in the original circuit, then applying the operation in (13.8) to both sides yields

$$\widehat{v}_R(t) = R\widehat{\imath}_R(t), \tag{13.9}$$

which is the same resistive relationship in the ω_s-component circuit.

The relationship for an inductor is slightly more elaborate, and hinges on the following expression for the time-derivative of the local ω_s-component $\widehat{x}(t)$:

$$\frac{d\widehat{x}(t)}{dt} = \widehat{\frac{dx(t)}{dt}} - j\omega_s\widehat{x}(t) = \widehat{\dot{x}}(t) - j\omega_s\widehat{x}(t), \tag{13.10}$$

where $\dot{x}(t)$ denotes $dx(t)/dt$. This equality can be verified by showing that each side evaluates to $[x(t)e^{-j\omega_s t} - x(t-T)e^{-j\omega_s(t-T)}]/T$. Note that if the local ω_s-component $\widehat{x}(t)$ is constant, then the derivative on the left of (13.10) is 0, so then $\widehat{\dot{x}}(t) = j\omega_s\widehat{x}(t)$. Thus, the local ω_s-component of the derivative is obtained through just multiplication by $j\omega_s$ in the case of constant $\widehat{x}(t)$. This is a familiar result from impedance analysis of circuits with sinusoidal drives of frequency ω_s, and also underlies the use of phasors in the analysis of power systems. The quantity $\widehat{x}(t)$ in the *non-constant* case is thus a generalization of the notion of a phasor, and is sometimes termed a *dynamic phasor*.

Accordingly, if $v_L(t)$ and $i_L(t)$ are respectively the voltage and current across and through the inductor, and therefore related by

$$v_L(t) = L\frac{di_L(t)}{dt},$$

then the identity in (13.10) shows that

$$\widehat{v}_L(t) = L\Big(\frac{d\widehat{\imath}_L(t)}{dt} + j\omega_s\widehat{\imath}_L(t)\Big). \tag{13.11}$$

This is equivalent to replacing the impedance sL of the inductor in the original circuit by an impedance $(s + j\omega_s)L$ in the ω_s-component circuit. Note that the subscript s on ω_s refers to "switching" or "source," and is unrelated to the Laplace transform variable s (which is the transform of the derivative operator) that enters the definition of impedance. Following a similar analysis for the capacitor, we find that the admittance sC of the capacitor in the original circuit needs to be replaced by the admittance $(s + j\omega_s)C$.

We illustrate the process of obtaining and analyzing the ω_s-component circuit in the following extended example.

Example 13.2 Switching-Frequency Dynamics of a Series-Resonant Converter

The underlying principle of a series-resonant converter – see Section 10.3 and Fig. 10.4 – is to drive a series-*RLC* circuit with a source that, in steady state, is periodically switched at some fundamental frequency ω_s rad/s. The periodic steady-state response will also have fundamental frequency ω_s. By adjusting the switching frequency ω_s to be closer to or further from the resonant frequency, we obtain a corresponding increase or decrease, respectively, in the amplitude of the current flowing through the resistor.

We would like to track the behavior of the ω_s-components of the voltages and currents in the circuit during transients of various kinds, for instance induced by changes in the switching frequency, or changes in the load. We can do this using an ω_s-averaged circuit, whose construction and analysis we outline next, following the route described above.

First consider the underlying series-*RLC* circuit. Setting the source voltage to 0 and then writing KVL around the loop in the impedance domain (alternatively, in the time domain and then taking the Laplace transform) yields

$$sL + \frac{1}{sC} + R = 0, \tag{13.12}$$

from which we get the characteristic polynomial governing the behavior of the *RLC* circuit:

$$s^2 + \frac{R}{L}s + \frac{1}{LC} = s^2 + 2\alpha s + \omega_o^2, \tag{13.13}$$

where $2\alpha = R/L$ s and $\omega_o = 1/\sqrt{LC}$ rad/s. The quantity ω_o is the resonant frequency of the circuit, at which the impedance is minimum. Also recall from (10.17) in Section 10.1 that the quality factor Q_o of the circuit is $\omega_o/(2\alpha) = 1/(\omega_o CR)$. Another commonly used parameter is the *damping factor* $\zeta = \alpha/\omega_o$. The roots of the characteristic polynomial for the (underdamped) case of $\zeta < 1$ are

$$-\alpha \pm j\omega_o\sqrt{1-\zeta^2} = -\alpha \pm j\omega_d, \tag{13.14}$$

where $\omega_d = \omega_o\sqrt{1-\zeta^2}$ is termed the *damped natural frequency* of the circuit. The settling of the circuit from some initial condition to a new steady state is marked by transient terms of the form

$$Ae^{-\alpha t}\sin(\omega_d t + \theta), \tag{13.15}$$

where A and θ are constants determined by the initial conditions.

We now construct the ω_s-averaged version of the series-*RLC* circuit, beginning by replacing all instantaneous voltages and currents by their local ω_s-components. We keep the resistor R unchanged, but replace the impedance sL of the inductor by $(s+j\omega_s)L$, and the admittance sC of the capacitor by $(s+j\omega_s)C$. With these substitutions, the characteristic polynomial describing transients of the ω_s-components in the circuit becomes

$$(s+j\omega_s)^2 + \frac{R}{L}(s+j\omega_s) + \frac{1}{LC} = (s+j\omega_s)^2 + 2\alpha(s+j\omega_s) + \omega_o^2, \tag{13.16}$$

with roots $-\alpha - j\omega_s \pm j\omega_d$.

To recover the real (as opposed to complex) underlying circuit variables, we need to track and combine complex conjugate components, as was done in the expression in (12.11). Hence, along with the component at frequency ω_s, we need to track the conjugate component at $-\omega_s$. The characteristic polynomial governing the latter component has roots at $-\alpha + j\omega_s \pm j\omega_d$.

From the analysis above, and comparing with (13.15), we expect the deviation of the full (real) ω_s-component from its steady state to be primarily governed by a transient of the form

$$Ae^{-\alpha t}\sin((\omega_d - \omega_s)t + \theta), \tag{13.17}$$

that is, a damped sinusoidal oscillation at the difference (or *beat*) frequency $\omega_d - \omega_s$. The component at the sum frequency $\omega_d + \omega_s$, which is much higher, may not be as visibly manifested in the response.

To proceed more quantitatively, consider a series-resonant converter that interfaces a 1 V dc input, via a half-bridge switching network as in Fig. 10.4, to a series-*RLC* circuit with a load resistance of $R = 1\,\Omega$, and with $L = 500\,\text{nH}$, $C = 50.66\,\text{nF}$. The associated resonant frequency is $\omega_o = 1\,\text{MHz}$. The inductor quality factor Q_o is designed to be 100 at resonance, corresponding to an effective series resistance of $r = 31.4\,\text{m}\Omega$ that adds to the load resistance R. The numerical illustrations below do not attempt to reflect the needs of a typical series-resonant converter application, but rather aim to show how the dynamics of such systems can be represented and evaluated generally.

Suppose the converter is operating in steady state at $f_s = 960\,\text{kHz}$, but that at $t = 0$ the load is short circuited, leaving only the small resistance r of the inductor to provide damping, thus generating a large transient response. A simulation of this transient is shown in Fig. 13.3. The rapidly oscillating quantity is the current at frequency ω_s through the circuit, essentially sinusoidal for $t < 0$, and quasi-sinusoidal for $t > 0$, but showing an impressive transient. The figure also highlights the more slowly varying *envelope* of the transient, which is essentially the behavior of the *magnitude* of the ω_s-component. We shall see in Example 13.6 how to generate this envelope using an appropriate ω_s-component state-space model.

To confirm numerically that the envelope is indeed governed by the expression (13.17), we evaluate the various parameters. Because of the short-circuited load, the series resistance drops at $t = 0$ from $R + r$ to just r. Thus, for $t > 0$, $\alpha = r/2L = 3.142 \times 10^4\,\mathrm{s}^{-1}$, which corresponds to a decay time constant of $1/\alpha = 31.8\,\mu\mathrm{s}$ in the expression (13.17). Also, since the resistance value is so low, the damped natural frequency for $t > 0$ is essentially the resonant frequency, so the beat frequency is 1 MHz − 960 kHz = 40 kHz, corresponding to a period of 25 µs. We therefore expect to see an envelope displaying a damped sinusoidal oscillation of period 25 µs, with the oscillation substantially reduced in amplitude in three time constants, that is, around 95 µs, and essentially gone in five time constants, that is, around 160 µs. This is indeed what is seen in Fig. 13.3. (There is no visible manifestation of any component of the response at the sum frequency, 1 MHz + 960 kHz = 1.96 MHz.)

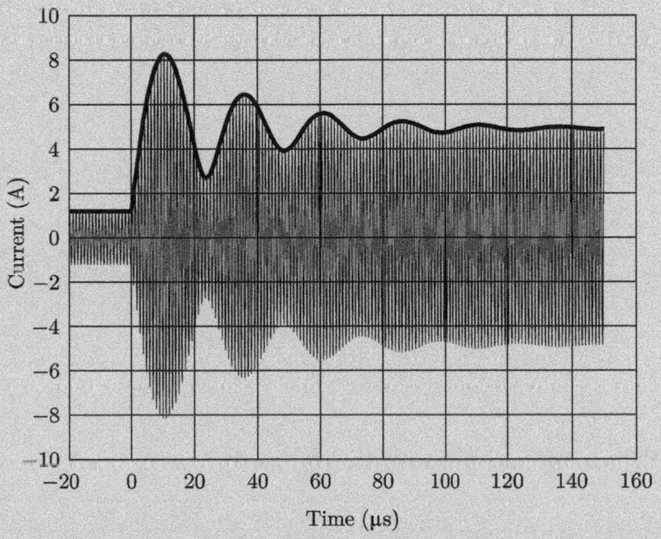

Figure 13.3 Transient response in our example converter when the load resistor is short circuited at $t = 0$ while operating at 960 kHz.

Now consider a quite different situation, with the load resistor intact, and with the converter initially switching at the resonant frequency of 1 MHz, but then dropped for $t > 0$ to a frequency of 800 kHz. Figure 13.4 shows a simulation of the transient associated with this experiment. The response is again quasi-sinusoidal at the frequency ω_s, with an envelope that shows a damped response settling to a constant value. As in the preceding figure, this envelope is the magnitude of the ω_s-component, and was generated by the ω_s-component state-space model described in Example 13.6. The steady-state envelope predicted by the model is slightly lower in magnitude than the measured response because 800 kHz is far enough away from resonance that higher-order harmonics (especially the third) are important. Slight distortions from sinusoidal can indeed be seen in the waveform for $t > 0$. The predicted steady-state value of the displayed envelope can be verified to be exactly the value of the fundamental of the steady-state response, as expected from our modeling assumptions.

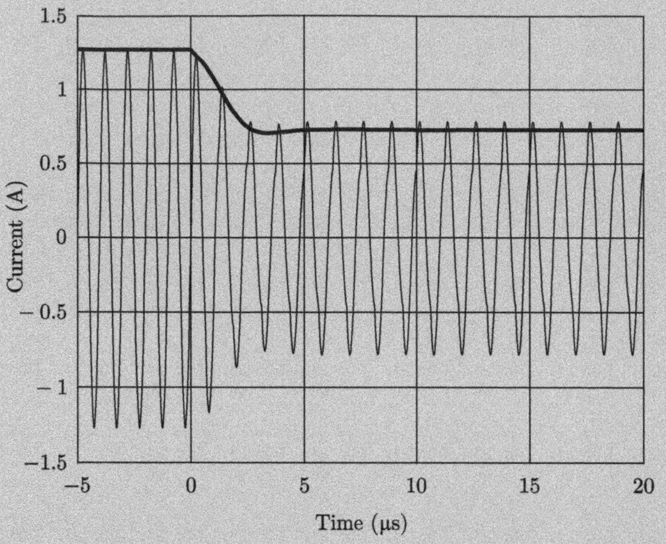

Figure 13.4 Transient response in our example converter for a step change in switching frequency at $t = 0$ from 1 MHz to 800 kHz.

With the operative parameter values for this case, $\alpha = (R+r)/2L = 1.031 \times 10^6$ Hz, the time constant is $1/\alpha = 0.97$ μs, and the associated damping factor is $\zeta = 0.164$. The damped natural frequency accordingly is $\omega_d = 6.198 \times 10^6$ rad/s $= 986.4$ kHz, so the beat frequency is 186.4 kHz, corresponding to a period of 5.36 μs. These numbers are very consistent with what is visible in Fig. 13.4. Since the period of the oscillation governing the envelope is slightly larger than five decay time constants, the transient dies within one period.

13.3 Continuous-Time State-Space Models

State-space models provide a very general and powerful basis for dynamic modeling. They include switched and averaged-circuit models as special cases but go considerably further. They are important in analyzing, simulating, and controlling both steady-state behavior and perturbations away from it. Our goal in this section is to introduce continuous-time state-space models, specifically in the context of circuits. We study discrete-time or sampled-data state-space models in the next section.

There are several reasons for the widespread use of state-space models in power electronics. Such models focus on those variables that are central to describing the dynamic evolution of a system. By aiming for a state-space model of a power circuit, we impose a valuable discipline and direction on the modeling process. The variables to highlight, relationships to examine, and route the analysis should take become clearer. The state-space approach also allows us to handle nonlinear and time-variant models in the same framework as LTI models. Furthermore, the formalism applies both to continuous-time and (with minor modifications, but the same underlying principles) to discrete-time or sampled-data systems. State-space models also provide a uniform and convenient starting point for such diverse tasks as steady-state computation, linearization, stability evaluation, control design, and simulation.

A state-space model of an LTI electrical circuit can be obtained systematically – and hence automatically by computer – from a circuit description. The resulting LTI description (as well as other LTI descriptions obtained from, say, linearization of a nonlinear model around a constant operating condition) can be analyzed by familiar techniques, including impedance or transform methods. This treatment can then be extended to piecewise LTI circuits, consisting of LTI elements and ideal switches.

No other framework has all these features and advantages. In addition, certain disadvantages sometimes associated with state-space models are usually not serious in the context of power electronic circuits. Specifically, a state-space model may not reflect the "sparsity" of the interconnections among components in the system. Although that is of great concern in large power systems or complicated analog circuits – with hundreds or thousands of variables to deal with – it is not of concern in a typical power electronic circuit, which has far fewer variables. Moreover, the generalized state-space models introduced later can be used to reflect sparsity, if this is a concern.

13.3.1 State Variables, Inputs, and Outputs

The key variables in a state-space model are the *state variables*, whose values together define the *state* of the system. State variables summarize *those aspects of the past that are relevant to the future*. They are the variables whose initial values are needed to determine future system behavior, after future inputs acting on the system are specified. They are thus typically associated with a system's memory mechanisms or energy storage mechanisms. Natural state variables in electrical circuits are the currents or flux linkages in inductors and the voltages or charges on capacitors. For a digital controller, the natural state variables would be the contents of its memory registers.

In addition to the state variables, other variables are of interest in a description of a dynamic system. The *inputs* to the system are external signals, such as the waveforms of voltage and current sources that drive a power circuit, and the signals that modulate the actions of controlled switches in it. Some of the inputs may be control variables that are under our command, whereas others may be disturbances that we have no control over. Specification of the inputs from the initial time onward, together with the initial values of the state variables, determines the future behavior of the state variables according to laws that govern their evolution. These governing laws are embodied in the *state-space model*.

The *outputs* of the system are either measured quantities or those whose values are of interest, even if not measured. State-space models are specified so as to allow the associated output values at any instant to be written as functions of the system state and the inputs at that instant. In the context of electrical circuits, typical outputs may be the voltages, currents, or dissipation associated with selected elements.

13.3.2 Mathematical Form

Suppose we have chosen to model a system with n state variables x_i, $i = 1, \ldots, n$, and with m inputs u_j, $j = 1, \ldots, m$. A general continuous-time state-space model of the system then takes the form of a set of coupled, nonlinear, time-variant, first-order differential equations:

$$\begin{aligned}
\frac{dx_1(t)}{dt} &= \dot{x}_1(t) = f_1(x_1(t),\ x_2(t), \ldots,\ x_n(t),\ u_1(t), \ldots,\ u_m(t),\ t),\\
\frac{dx_2(t)}{dt} &= \dot{x}_2(t) = f_2(x_1(t),\ x_2(t), \ldots,\ x_n(t),\ u_1(t), \ldots,\ u_m(t),\ t),\\
&\vdots\\
\frac{dx_n(t)}{dt} &= \dot{x}_n(t) = f_n(x_1(t),\ x_2(t), \ldots,\ x_n(t),\ u_1(t), \ldots,\ u_m(t),\ t).
\end{aligned} \tag{13.18}$$

They express the instantaneous rates of change of each of the n state variables as functions of the indicated arguments, namely, the instantaneous values of all the state variables and inputs, and the time argument t itself. We also refer to these equations as the *state evolution* equations. The model is said to be of nth *order*, because it involves n state variables. The functions $f_i(\cdot)$ can be fairly elaborate for even simple power circuits, as later examples show.

Also associated with the state-space model (13.18) are the output variables $y_\ell(t)$, $\ell = 1, \ldots, p$. We consider only models in which the outputs can be written in the form

$$y_\ell(t) = g_\ell(x_1(t), x_2(t), \ldots, x_n(t), u_1(t), \ldots, u_m(t), t), \tag{13.19}$$

so that the outputs are directly determined at any time by the state and inputs *at that time*. If an output variable is not of this form initially, we may be able to bring it to this form by appropriately defining additional state variables.

Figure 13.5 is a representation of the model in (13.18) and (13.19). The outputs of the integrators represent the state variables, and the interconnection constraints ensure that the system is governed by (13.18) and (13.19).

Linearity and Time Invariance The description in (13.18) is called *time invariant* if none of the functions $f_i(\cdot)$ explicitly involves t, so that

$$f_i(\cdot) = f_i(x_1(t), x_2(t), \ldots, x_n(t), u_1(t), \ldots, u_m(t)) \tag{13.20}$$

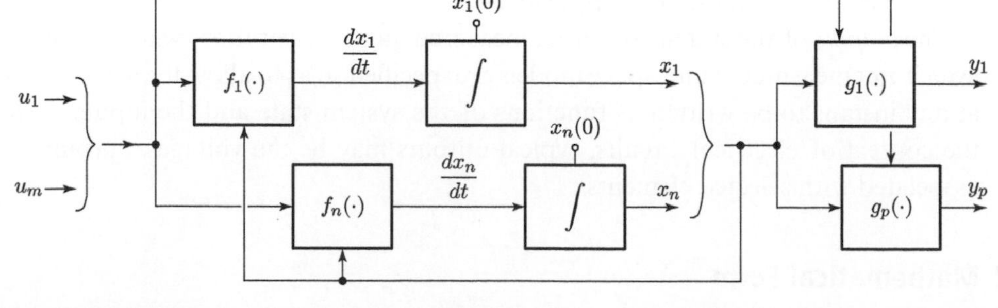

Figure 13.5 Representation of the continuous-time state-space model in (13.18) and (13.19).

for all i. In other words, the *manner* in which the arguments $x_1(t), x_2(t), \ldots, x_n(t)$ and $u_1(t), \ldots, u_m(t)$ are combined to determine $f_i(\cdot)$, and thereby to specify $\dot{x}_i(t)$, does not vary with time, even though $x_1(t), x_2(t), \ldots, x_n(t)$, $u_1(t), \ldots, u_m(t)$, and $f_i(\cdot)$ can all vary with time.

If all the $f_i(\cdot)$ in (13.18) are linear functions of the state variables and inputs, the model is termed *linear* and we have

$$\begin{aligned} f_i(\cdot) = a_{i1}(t)x_1(t) + a_{i2}(t)x_2(t) + \cdots + a_{in}(t)x_n(t) \\ + b_{i1}(t)u_1(t) + \cdots + b_{im}(t)u_m(t). \end{aligned} \tag{13.21}$$

If the various coefficients $a_{ij}(t)$ and $b_{ik}(t)$ vary *periodically* with a common period P, then the model is termed periodically varying, with period P.

A most important special case is a linear *and* time-invariant description, for which the coefficients $a_{ij}(t)$ and $b_{ik}(t)$ are constant. For such a *linear, time-invariant (LTI)* model, we have

$$\begin{aligned} f_i(\cdot) = a_{i1}x_1(t) + a_{i2}x_2(t) + \cdots + a_{in}x_n(t) \\ + b_{i1}u_1(t) + \cdots + b_{im}u_m(t). \end{aligned} \tag{13.22}$$

The output equations (13.19) can also be classified as linear and/or time invariant in the same manner. It is quite possible for the state evolution equations in (13.18) to be LTI while the output equations (13.19) are not, or conversely.

13.3.3 Notation

Having compact notation to represent systems of equations such as (13.18) and (13.19) is extremely useful. The language of vector and matrix algebra provides this.

We define the state vector $x(t)$, input vector $u(t)$, and output vector $y(t)$ by

$$x(t) = \begin{bmatrix} x_1(t) \\ x_2(t) \\ \vdots \\ x_n(t) \end{bmatrix}, \qquad u(t) = \begin{bmatrix} u_1(t) \\ u_2(t) \\ \vdots \\ u_m(t) \end{bmatrix}, \qquad y(t) = \begin{bmatrix} y_1(t) \\ y_2(t) \\ \vdots \\ y_p(t) \end{bmatrix}. \tag{13.23}$$

We shall *not* use boldface or any other notational device to denote vectors such as $x(t)$, relying instead on the context of a discussion for clarity. We restrict ourselves to a few letters for vectors, such as x, u, y, and a handful more that we introduce later. Thus you should soon begin to automatically interpret these symbols as vectors, without the aid (or burden!) of additional notational frills. We usually denote the components of a vector by subscripting the symbol used for the vector, as in x_i for the ith component of the vector x. The symbol dx/dt, or sometimes $\dot{x}(t)$, denotes the vector whose ordered entries are the derivatives dx_i/dt, $i = 1, \ldots, n$.

Now we can write (13.18) and (13.19) in far more compact notation:

$$\begin{aligned} \frac{dx(t)}{dt} = \dot{x}(t) = f(x(t), u(t), t), \\ y(t) = g(x(t), u(t), t), \end{aligned} \tag{13.24}$$

where the vector functions $f(\cdot)$ and $g(\cdot)$ are given by

$$f(\cdot) = \begin{bmatrix} f_1(\cdot) \\ f_2(\cdot) \\ \vdots \\ f_n(\cdot) \end{bmatrix}, \qquad g(\cdot) = \begin{bmatrix} g_1(\cdot) \\ g_2(\cdot) \\ \vdots \\ g_p(\cdot) \end{bmatrix}. \tag{13.25}$$

When the description in (13.18) is linear, the use of matrix notation allows us to write it in the following more explicit and convenient form:

$$\frac{dx(t)}{dt} = A(t)x(t) + B(t)u(t), \tag{13.26}$$

where $A(t)$ and $B(t)$ are $n \times n$ and $n \times m$ matrices whose entries are functions of t only:

$$A(t) = \begin{bmatrix} a_{11}(t) & \dots & a_{1n}(t) \\ \vdots & \ddots & \vdots \\ a_{n1}(t) & \dots & a_{nn}(t) \end{bmatrix}, \qquad B(t) = \begin{bmatrix} b_{11}(t) & \dots & b_{1m}(t) \\ \vdots & \ddots & \vdots \\ b_{n1}(t) & \dots & b_{nm}(t) \end{bmatrix}. \tag{13.27}$$

The $a_{ij}(t)$ and $b_{i\ell}(t)$ are the same as in (13.21). Although we generally denote matrices by capital letters, again we do *not* use any other special notational device, such as boldfacing, to distinguish them. As with vectors, we use only a handful of symbols for matrices, and you should soon begin to recognize them as matrices.

If the output equations (13.19) are also linear, they take the form

$$y(t) = E(t)x(t) + F(t)u(t), \tag{13.28}$$

where $E(t)$ and $F(t)$ are $p \times n$ and $p \times m$ matrices, respectively. For a continuous-time LTI model, the description is still of the form (13.26) and (13.28), except that now the coefficient matrices A, B, E, and F are *constant*:

$$\frac{dx(t)}{dt} = Ax(t) + Bu(t), \tag{13.29}$$

$$y(t) = Ex(t) + Fu(t). \tag{13.30}$$

We will use similarly condensed vector and matrix notation when we come to examining discrete-time and sampled-data models. The appropriate notational modifications for the discrete-time case will be quite apparent.

13.3.4 Properties of Solutions

We now examine two important properties of the solutions of continuous-time state-space models: the *state property* and the *continuity property*.

The State Property The significance of the special form of description in (13.18) is that, knowing the values $x_i(t_0)$, $i = 1, \dots, n$, of all the state variables at time t_0, and knowing the inputs $u_j(t)$ for t in some interval $t_0 \le t < t_f$, we can determine $x_i(t)$ for t in the interval $t_0 < t \le t_f$. For convenience, we refer to this property as the *state property* of the description (13.18).

The state property can be made plausible through examination of the representation of (13.18) in Fig. 13.5. This schematic representation can form the basis for *solutions* or *simulations* of the state-space model. A continuous-time, analog simulation based on Fig. 13.5 would require an appropriate physical realization of the integrators (typically using operational amplifiers), as well as some means of evaluating the functions $f_i(\cdot)$ and $g_\ell(\cdot)$. When we have set the initial conditions on the integrators, the system variables will evolve in response to the inputs. The state variables at any time will then be the outputs of the integrators at that time, which shows that the state property holds.

Alternatively, a numerical solution or simulation can be carried out. This method would involve using some numerical integration routine instead of the analog integrators in Fig. 13.5 to compute the state variable waveforms or "trajectories" at discrete points in time. The conceptually simplest such numerical integration scheme is the *forward Euler* method for stepping a solution forward in small time increments ϵ, for which – taking advantage of our compact notation – we would write (for the general case, the linear case, and the LTI case respectively, with the identity matrix denoted by I)

$$\begin{aligned} x(t+\epsilon) &\approx x(t) + \epsilon f(x(t), u(t), t) \\ &= [I + \epsilon A(t)]x(t) + \epsilon B(t)u(t) \\ &= (I + \epsilon A)x(t) + \epsilon Bu(t). \end{aligned} \tag{13.31}$$

The Continuity Property The state variables are continuous functions of time, as you can see by examining (13.18). If any state variables had discontinuities, the derivatives on the left-hand side of (13.18) would have impulses in them, which would have to be balanced by impulses in the inputs $u_j(t)$ on the right-hand side. Hence, in the absence of impulsive inputs, state variables are continuous. For piecewise LTI models, we can invoke this continuity to piece together the analytical solutions obtained in each LTI regime and thus obtain the overall solution.

The above state evolution and continuity properties are well illustrated by the following example.

Example 13.3 State Trajectories of a Resonant DC/DC Converter

The circuit in Fig. 13.6(a) represents a series-resonant dc/dc converter of the type discussed in Chapter 10 and in Example 13.2. We have omitted the input dc source and the switching circuit that modulates it, instead showing only the resulting square-wave voltage v_{in}, of fixed amplitude V_1 and controllable frequency f_s, that is applied to the resonant circuit. This voltage produces a nearly sinusoidal inductor current i_L, which is rectified by the diode bridge and passed through the load. By controlling the frequency of the square wave relative to the resonant frequency f_o of the LC circuit, we control the amplitude of the resonant current and thereby the average current delivered to the load. The load in our example is a dc voltage source of value V_2. The circuit response, even for fixed f_s, is complicated by the action of the bridge rectifier, which subtracts the load voltage from v_{in} when $i_L > 0$ and adds it to v_{in} when $i_L < 0$.

Natural state variables for this circuit are the inductor current and capacitor voltage. To be specific, take the nominal f_s to be above the resonant frequency. The current through the LC pair then lags the voltage across it in the steady state. We assume for simplicity that the transients we are modeling are not large enough to change this feature of the steady-state response, so that $i_L < 0$ at the instants that v_{in} goes from

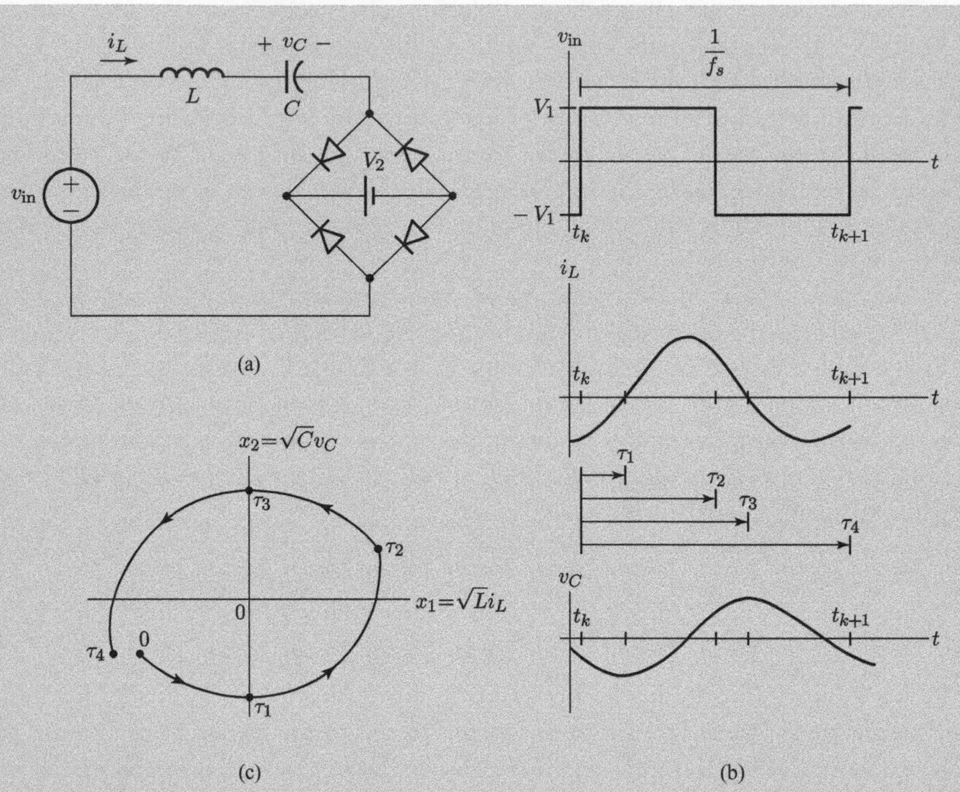

Figure 13.6 (a) Series-resonant dc/dc converter circuit. (b) Associated waveforms. (c) State-plane representation (with durations after time t_k marked on the trajectory).

$-V_1$ to $+V_1$ (and $i_L > 0$ when v_{in} goes from $+V_1$ to $-V_1$). We consider a cycle of operation beginning at such an instant.

Typical transient waveforms during the kth cycle of operation, which extends from $t = t_k$ to $t = t_{k+1}$, are shown in Fig. 13.6(b). Note that f_s and the indicated transition times τ_ℓ, $\ell = 1, 2, 3, 4$, can in general vary from cycle to cycle. Thus they should actually carry the argument k, but we drop the k for notational simplicity.

With these assumptions, the voltage applied to the resonant LC circuit at the start of every cycle is V_1+V_2 and remains constant at this value until i_L goes positive. During this interval, v_C is a section of a sinusoidal waveform of frequency f_o and mean value $V_1 + V_2$ (though the mean value of v_C over this interval is different). The corresponding i_L is also a section of a sinusoid (of the same frequency f_o and mean value zero, but the mean value of i_L over this interval is nonzero).

When i_L passes upward through zero at time τ_1 after the start of the kth cycle, v_C becomes a section of a different sinusoidal waveform (of frequency still f_o but mean value $V_1 - V_2$ now). The associated i_L is a section of a different sinusoid than before the transition (though still of frequency f_o and mean value zero).

At time τ_2 after the start of the kth cycle, the input voltage changes from V_1 to $-V_1$. The analysis during the second half of the cycle parallels that of the first half, except that all waveforms have the opposite polarity. At time τ_3, i_L passes downward through zero, and at time τ_4 the input voltage changes from $-V_1$ to V_1 again, signifying the start of the next cycle. During each cycle, therefore, the circuit goes through four configurations.

An illuminating companion to the waveform representation in Fig. 13.6(b) is the representation of state variable trajectories on the *state plane* (or "phase plane"), as shown in Fig. 13.6(c). The coordinates in such a representation are the state variables. The waveforms of i_L and v_C for the resonant converter would map into piecewise-elliptical trajectories on the state plane. When we use the scaled quantities $x_1 = \sqrt{L}\, i_L$ and $x_2 = \sqrt{C}\, v_C$ as state variables, instead of i_L and v_C, the trajectories turn out to be piecewise-*circular* arcs (centered on different points), as you can easily verify. This scaling simplifies the analysis, so we use scaled quantities as state variables for the rest of this example and in Example 13.9, where we continue the analysis. The state-plane diagram in Fig. 13.6(c) already reflects this choice.

The circular arcs are centered at points determined by the offsets about which the sinusoidal sections of i_L and v_C oscillate during the corresponding intervals. The center of each arc therefore has an x_1 coordinate of 0, and its x_2 coordinate takes one of the four values $\sqrt{C}(\pm V_1 \pm V_2)$. Because v_C increases when i_L is positive, the arcs are all traversed counterclockwise, with angular velocity $\omega_o = 2\pi f_o$. For clarity, only one cycle of operation is represented in Fig. 13.6(c). In the nominal periodic steady state, this trajectory would be a closed curve, because the state at the end of the cycle equals the state at the beginning.

State-plane representations often provide valuable insight into the dynamics of second-order circuits and can be a fruitful source of ideas for control design. We say more about this in Chapter 14.

13.3.5 State-Space Models for Electrical Circuits

Natural state variables for a circuit, as already noted, are the inductor currents or flux linkages and capacitor voltages or charges. The state-space description requires expressions for the derivatives of these quantities, expressed entirely as functions of themselves and of the sources and control inputs driving the circuit. Outputs must also be expressed in this form.

We limit ourselves to outlining the procedure for a circuit whose inductors and capacitors are all LTI. For such a circuit, if the current through an inductor L is $i_L(t)$, then the inductor voltage is $L\, di_L(t)/dt$; and, if the voltage across a capacitor C is $v_C(t)$, the capacitor current is $C dv_C(t)/dt$. These relationships suggest a way to express the state variable derivatives in the required form: determine the inductor voltages and capacitor currents at time t as functions of the inductor currents, capacitor voltages, source values, and control inputs at that same instant t. The state-space description follows readily, as illustrated in the next example. (When mutual inductance is present, the voltage across an inductor also depends on the derivatives of the currents in all the inductors coupled to it; this introduces additional steps in the derivation of the state-space model, but we omit the details.)

There are some exceptions to choosing all the inductor currents and capacitor voltages as state variables. If k capacitors form a loop that contains no other elements besides independent voltage sources, then Kirchhoff's voltage law shows that only $k-1$ of the capacitor voltages can have independent initial conditions, and hence only $k-1$ capacitor voltages should be chosen as state variables. A similar statement holds for the dual situation, in which a cutset of only inductors and current sources is constrained by Kirchhoff's current law to have currents that sum to zero.

The following example illustrates the above procedure for obtaining a state-space model, using a converter topology we have examined several times in this chapter and the previous one.

Example 13.4 State-Space Model for a Buck/Boost Converter

Consider the buck/boost converter shown in Fig. 13.7, which we used extensively as an example in Chapter 12. The only difference is that we have included an equivalent series resistance (ESR) of value R_C with the capacitor. We assume for simplicity that the converter is operated so that either the transistor or the diode is always conducting; this assumption rules out the discontinuous conduction mode, in which both devices are off during some part of each cycle. The switching function $q(t)$ represents the switch status, with $q(t) = 1$ when the transistor is on and $q(t) = 0$ when it is off, and $q'(t) = 1 - q(t)$.

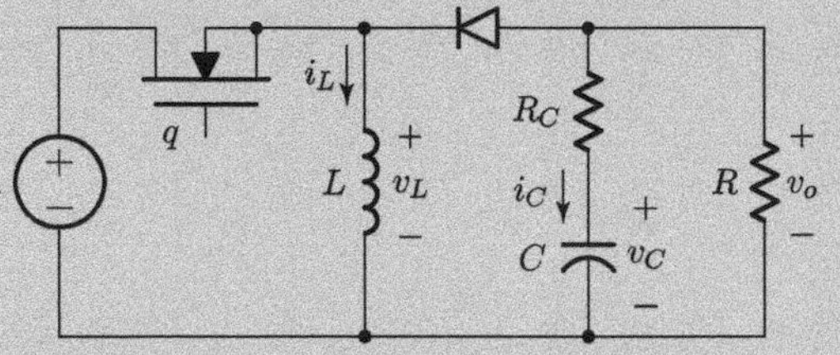

Figure 13.7 Buck/boost converter circuit.

To obtain a state-space model of the circuit, we focus on those variables whose initial values are essential to determining the future behavior of the system for a specified source value and switching pattern. The natural choice of state variables, as noted earlier, is the inductor current i_L and capacitor voltage v_C. The natural inputs are the source voltage $v_{\text{in}}(t)$ and the control signal $q(t)$.

When $q(t) = 1$,

$$\begin{aligned} v_L(t) &= L\frac{di_L(t)}{dt} = v_{\text{in}}(t), \\ i_C(t) &= C\frac{dv_C(t)}{dt} = \frac{-1}{R+R_C}v_C(t), \end{aligned} \tag{13.32}$$

and when $q(t) = 0$,

$$\begin{aligned} v_L(t) &= L\frac{di_L(t)}{dt} = \frac{R}{R+R_C}\big[-R_C i_L(t) + v_C(t)\big], \\ i_C(t) &= C\frac{dv_C(t)}{dt} = \frac{-1}{(R+R_C)}\big[R i_L(t) + v_C(t)\big]. \end{aligned} \tag{13.33}$$

Combining (13.32) and (13.33) by introducing $q(t)$, and rearranging them into the form (13.18), we obtain the desired state-space model:

$$\begin{aligned} \frac{di_L(t)}{dt} &= \frac{R}{L(R+R_C)}\big[-R_C q'(t) i_L(t) + q'(t) v_C(t)\big] + \frac{1}{L}q(t)v_{\text{in}}(t), \\ \frac{dv_C(t)}{dt} &= \frac{-1}{C(R+R_C)}\big[R q'(t) i_L(t) + v_C(t)\big]. \end{aligned} \tag{13.34}$$

We leave it to you to derive an output equation of the form (13.19), expressing the output voltage v_o as a function of the state variables and inputs.

Note that if $q(t)$ were fixed at either 0 or 1, the description in (13.34) would be LTI, justifying the label "piecewise LTI" for the model in (13.34). When $q(t)$ is considered as a control input, the description is nonlinear but time invariant. However, if $q(t)$ happened to be a prespecified function of time, then $q(t)$ would simply be a time-variant parameter of the system, not an input, and the corresponding state-space model would be linear and time-variant. Finally, in a feedback controlled system like that illustrated in Fig. 12.9, $q(t)$ is determined by the state variables as well as by a clocking waveform. The resulting description is nonlinear and time variant.

Although we have stated several times that inductor currents and capacitor voltages are the natural choices of state variables for electrical circuits, they are not the only ones. We could pick any other set of variables that provides equivalent information. For example, we could have chosen $Ki_L - v_C$ and $Ki_L + v_C$ for the circuit in the preceding example, where K is any nonzero constant. Another choice could have been the flux in the inductor and the charge on the capacitor, which are helpful choices when these components are nonlinear, because explicit expressions may be available for their time-derivatives.

13.3.6 State-Space Averaging

We can construct state-space models for the averaged circuits developed in Sections 12.3 and 13.1 by following the procedure just described. Alternatively, rather than constructing a state-space model for the averaged circuit, we can directly *average a state-space description of the switched circuit*. Such state-space averaging can sometimes be valuable. For example, when the capacitor in the buck/boost converter has significant ESR, as in the preceding example, the voltage across the series combination of R_C and C (in the notation of the preceding example) may have significant ripple. This condition violates the small-ripple assumption underlying the canonical cell averaging in Fig. 13.1. In such situations, state-space averaging can be easier and more systematic than an attempt at direct circuit averaging because the process of obtaining the state-space model of the circuit serves to organize some of the necessary computations, as the next example shows.

Example 13.5 State-Space Averaging for a Buck/Boost Converter

Averaging both sides of each expression in (13.34), with temporary suppression of the time argument to keep the notation simple, we find

$$\begin{aligned} \frac{d\overline{i}_L}{dt} &= \frac{R}{L(R+R_C)}\left(-R_C\overline{q'i_L} + \overline{q'v_C}\right) + \frac{1}{L}\overline{qv_{\text{in}}}, \\ \frac{d\overline{v}_C}{dt} &= \frac{-1}{C(R+R_C)}\left(R\overline{q'i_L} + \overline{v}_C\right). \end{aligned} \tag{13.35}$$

These two equations do not yet constitute a state-space model, because the right-hand sides are not expressed entirely as functions of the averaged state variables $\overline{i}_L$ and $\overline{v}_C$ and of the averaged inputs $\overline{q} = d$ and $\overline{v}_{\text{in}}$. However, assume that i_L is well approximated during any interval of length T by its local average over that interval, as are v_C and v_{in}. These assumptions are equivalent to those we made in Section 13.1.1 when averaging a switch. Then we can express the averages of the products $q'v_C$, $q'i_L$, and qv_{in} as the products of the averages of the individual terms, and approximate (13.35) by

$$\begin{aligned} \frac{d\overline{i}_L(t)}{dt} &\approx \frac{R}{L(R+R_C)}\left[-R_C d'(t)\overline{i}_L(t) + d'(t)\overline{v}_C(t)\right] + \frac{1}{L}d(t)\overline{v}_{\text{in}}, \\ \frac{d\overline{v}_C(t)}{dt} &\approx \frac{-1}{C(R+R_C)}\left[Rd'(t)\overline{i}_L(t) + \overline{v}_C(t)\right], \end{aligned} \tag{13.36}$$

which *is* in state-space form. This pair of equations is the desired state-space averaged model. The assumptions that permitted us to obtain this model are generally satisfied in well-designed switching converters. Note that the only difference between the averaged model and the switched model is that instantaneous quantities have been replaced by averages. Any output equations associated with the switched state-space model in (13.34) can be averaged similarly. A circuit that corresponds to the state-space averaged model in (13.36) can also be obtained, if desired.

When $R_C = 0$, the circuit of Fig. 13.7 is exactly the circuit of Fig. 12.2, which we averaged in Example 13.1. Not surprisingly, therefore, setting $R_C = 0$ in the preceding state-space averaged model turns out to yield a state-space description of the averaged circuit in Fig. 13.2 of Example 13.1.

13.3.7 State-Space Models for the ω_s-Component

In Section 13.2 we examined how to generalize circuit averaging to represent the ω_s-component of circuit variables. The process was illustrated in Example 13.2 for the case of a series-resonant *RLC* converter. In particular, Figs. 13.3 and 13.4 displayed envelopes for the current waveform in the converter. These envelopes show the time evolution of the magnitude of the ω_s-component of the converter current. We are now in a position to explain how to compute the envelopes. The key is to obtain a state-space model for the ω_s-components of the state variables governing the *RLC* circuit. The following example leads you through the key steps.

Example 13.6 State-Space Model for the ω_s-Component of a Series *RLC* Circuit

Consider a series-*RLC* circuit excited by a square-wave voltage source $v_{\text{in}}(t)$ that switches at a frequency ω_s between the values 0 and V_{in}, with a duty ratio of 0.5. (Such an applied voltage could be obtained by a half-bridge connecting a dc source of value V_{in} to the series-*RLC* combination.) We choose as state variables the inductor current and capacitor voltage, which we respectively denote here simply by $i(t)$ and $v(t)$. Choosing polarities such that positive $i(t)$ enters the positive terminal for $v(t)$, we directly obtain the following state-space model for the circuit:

$$\begin{aligned}\frac{dv(t)}{dt} &= \frac{1}{C}i(t),\\ \frac{di(t)}{dt} &= -\frac{1}{L}\big(v(t) + Ri(t) - v_{\text{in}}(t)\big).\end{aligned} \tag{13.37}$$

The first equation is simply the component characterization for the capacitor, while the second equation is a statement of KVL for the circuit, incorporating the component characterizations for the inductor and resistor.

We now compute the ω_s-component of each side of both the equations in (13.37), invoking (13.10), namely the relation

$$\widehat{\frac{dx(t)}{dt}} = \widehat{\dot{x}}(t) = \frac{d\widehat{x}(t)}{dt} + j\omega_s\,\widehat{x}(t), \tag{13.38}$$

to express the ω_s-component of the derivative of a variable in terms of the derivative of the ω_s-component. The result is the following state-space model:

$$\begin{aligned}\frac{d\widehat{v}(t)}{dt} &= -j\omega_s\widehat{v}(t) + \frac{1}{C}\widehat{\imath}(t),\\ \frac{d\widehat{\imath}(t)}{dt} &= -j\omega_s\widehat{\imath}(t) - \frac{1}{L}\big(\widehat{v}(t) + R\widehat{\imath}(t) - \widehat{v}_{\text{in}}(t)\big).\end{aligned} \tag{13.39}$$

This is a *complex-valued* second-order model, since the ω_s-components are complex and there are both real and imaginary parameters. Separating $\widehat{v}(t)$ and $\widehat{\imath}(t)$ into their real and imaginary components, and separately equating the real and imaginary parts of each equation in (13.39), results in a *real, fourth-order, LTI* state-space equation that can be solved conventionally for given initial conditions at a starting time $t = 0$ and input trajectory for $t \geq 0$. Note that the amplitude or envelope of the ω_s-component of the current $i(t)$ is given by $2|\widehat{\imath}(t)|$, as noted in the statements immediately below (12.4) and (12.11).

The initial conditions at $t = 0$ may correspond to a steady state existing prior to that time, as in the illustrations in Example 13.2. For the input term, suppose $v_{\text{in}}(t) = V_{\text{in}}$ for $0 < t < (T/2)$ and is 0 for $(T/2) < t < T$, repeating periodically with period T thereafter. Then $\widehat{v}_{\text{in}}(t)$ in (13.39) for all $t > T$ evaluates to

$$\widehat{v}_{\text{in}}(t) = \frac{1}{T}\int_{t-T}^{t} v_{\text{in}}(t)e^{-j\omega_s t}\,dt = \frac{V_{\text{in}}}{T}\int_{0}^{T/2} e^{-j\omega_s t}\,dt = j\frac{2V_{\text{in}}}{\pi}. \tag{13.40}$$

If T is much smaller than the duration of the transient behavior associated with (13.39)), then the precise value of $\widehat{v}_{\text{in}}(t)$ in the short interval $0 < t < T$ may not matter much, and it may suffice to approximate $\widehat{v}_{\text{in}}(t)$ as a step function that rises from 0 to the value in (13.40) at the starting time $t = 0$. This is what was done for the simulations that yielded the envelopes in Figs. 13.3 and 13.4 of Example 13.2.

13.4 Discrete-Time or Sampled-Data Models

The detailed behavior of a converter circuit's waveforms within one cycle of operation is certainly of interest. Often, however, we need a higher-level model that summarizes or condenses the *intra*-cycle details, and instead highlights the *inter*-cycle behavior. Averaged models are particularly helpful in this regard. We turn now to an alternative class of models that are very useful in describing the cycle-to-cycle behavior of power circuits, namely, discrete-time or sampled-data models in state-space form. These models can also form the basis for evaluating the local stability of cyclic steady-state operation of power converters, and are the starting point for digital control design, as we illustrate in Chapter 14.

A discrete-time state-space description most commonly comes from regular sampling of the state variables of a continuous-time description. Because the majority of power electronic circuits operate cyclically, working with models that involve quantities sampled once per cycle is especially natural.

Suppose that the underlying continuous-time description is given by (13.18) and is regularly sampled every T seconds. We often write $x_i[k]$ to denote $x_i(kT)$, where k is an integer that indexes the samples. To obtain useful sampled-data models, we need to assume that the inputs $u_j(t)$ in the continuous-time model (13.18) are determined by a finite set of variables in each sampling interval, with the precise form of dependence being allowed to vary from cycle to cycle. These *determining variables* are denoted by $p_1[k], \ldots, p_r[k]$ in the kth interval. The brackets around k serve to remind us that these quantities are discrete-time sequences.

If, for example, the input $u_1(t)$ in (13.18) changes only at sample times (remaining constant over the interval between successive sampling instants), we could use $p_1[k]$ to denote the value of $u_1(t)$ in the interval $kT \leq t < kT + T$. This situation is represented schematically in Fig. 13.8. As another example, suppose that the continuous-time input $u_2(t)$ in (13.18) during the kth sampling interval is a portion of a sinusoid, but with parameters that can vary from interval to interval. We could then represent $u_2(t)$ in the kth interval by $p_2[k]\sin(p_3[k]t + p_4[k])$.

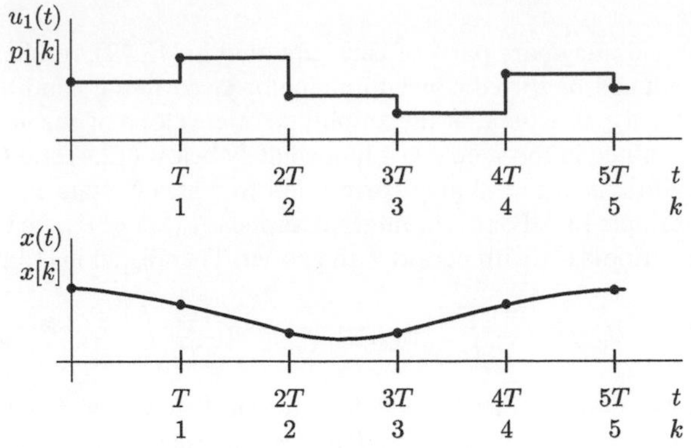

Figure 13.8 Sampling a continuous-time system whose inputs are piecewise constant.

The state and continuity properties of (13.18) show that specifying all the state variables $x_i[k]$ at time kT and all the inputs $u_j(t)$ over the intervening interval $kT \leq t < kT + T$ completely specifies all the state variables $x_i[k+1]$ at time $kT + T$. The evolution of the sampled state is therefore governed by some model of the form

$$\begin{aligned} x_1[k+1] &= \phi_1\big(x_1[k], x_2[k], \ldots, x_n[k], p_1[k], \ldots, p_r[k], k\big), \\ x_2[k+1] &= \phi_2\big(x_1[k], x_2[k], \ldots, x_n[k], p_1[k], \ldots, p_r[k], k\big), \\ &\vdots \\ x_n[k+1] &= \phi_n\big(x_1[k], x_2[k], \ldots, x_n[k], p_1[k], \ldots, p_r[k], k\big), \end{aligned} \tag{13.41}$$

which is what we term a discrete-time or sampled-data state-space model. The x_i are the state variables and the p_j are the inputs to the model.

The functions $\phi_i(\cdot)$ in (13.41) cannot in general be related in any simple, analytically expressible way to the underlying continuous-time model. Exceptions occur in some special but important cases. One such case is when the underlying continuous-time system is LTI or can be described by a succession of LTI models, which is the piecewise-LTI case treated in Chapter 14. Another case occurs when the sampling interval is very short compared to the times over which the right-hand sides of (13.18) vary significantly; in this case we obtain an approximate sampled-data model by using the same kind of approximation as in (13.31).

We can also obtain a discrete-time state-space model of the form (13.41) in ways that need not directly involve sampling an underlying continuous-time description. Whatever the origin of the discrete-time model, its key feature is that it expresses the state at the next discrete-time instant in terms of the state at the present instant and the inputs associated with the present instant. The description (13.41) therefore has a state property exactly like that of the continuous-time case (see Section 13.3.4): for a given sequence of inputs over some interval and a specified state at the beginning of the interval, the model allows us to compute the sequence of states, or the state trajectory, over the entire interval. This computation involves nothing more than iterative use of the state-space model.

The definitions of linearity and time invariance are the same as in the continuous-time case, with the functions $\phi_i(\cdot)$ of (13.41) used instead of the functions $f_i(\cdot)$ of (13.18). Also, each output associated with a discrete-time state-space model is given, as in the continuous-time case, by a possibly time-dependent function of the current state and inputs:

$$y_\ell[k] = \gamma_\ell\big(x_1[k], x_2[k], \ldots, x_n[k], p_1[k], \ldots, p_r[k], k\big). \tag{13.42}$$

A representation analogous to Fig. 13.5 for the discrete-time state-space model is easily obtained by using (one-step) *delay elements* instead of integrators, and making other obvious changes: replacing $x(t)$ with $x[k]$, dx/dt with $x[k+1]$, and $f_i(\cdot)$ and $g_\ell(\cdot)$ with $\phi_i(\cdot)$ and $\gamma_\ell(\cdot)$, respectively. A delay element is characterized by the property that the signal present at its input at discrete-time index k appears at its output at time $k+1$. Hence, when $x_i[k]$ is at the output of a delay element, its input must be $x_i[k+1]$.

As was done for the continuous-time case in Section 13.3.3, we can introduce convenient vector notation for the descriptions in (13.41) and (13.42), and write

$$x[k+1] = \phi\big(x[k], p[k], k\big), \qquad y[k] = \gamma\big(x[k], p[k], k\big) \tag{13.43}$$

for the state evolution equation and the output equation respectively. The definitions of the various vector quantities in the preceding two equations should be apparent by analogy with the development in Section 13.3.3.

Example 13.7 Approximate Sampled-Data Model for a Buck/Boost Converter

We begin by expressing the continuous-time state-space model (13.34) for the buck/boost converter of Example 13.4 in matrix form. Let $x_1 = i_L$ and $x_2 = v_C$. Then, for continuous conduction, and neglecting the capacitor ESR for simplicity (so $R_C = 0$), we have

$$\begin{aligned}\frac{dx(t)}{dt} &= \begin{bmatrix} 0 & q'(t)/L \\ -q'(t)/C & -1/RC \end{bmatrix} x(t) + \begin{bmatrix} q(t)/L \\ 0 \end{bmatrix} v_{\text{in}}(t) \\ &= A_{q(t)}x(t) + B_{q(t)}v_{\text{in}}(t).\end{aligned} \tag{13.44}$$

Everything in the rest of this example also applies to other high-frequency switching converters in continuous conduction, with the exception that the specific matrices $A_{q(t)}$ and $B_{q(t)}$ for other switching converters are different.

The switching period T in a well-designed buck/boost converter is much smaller than the time constants associated with the circuit in its different switch configurations, so the inductor current and capacitor voltage waveforms in each switch configuration are essentially straight-line segments. This allows approximations similar to (13.31), which – starting with the continuous-time model (13.44) – yields a sampled-data model for the buck/boost converter.

Suppose that the transistor is turned on every T seconds and turned off a time d_kT later in the kth cycle, so that d_k is the duty ratio in the kth cycle. (For consistency with our notation for discrete-time sequences, we should actually write $d[k]$ instead of d_k, but the latter notation is less cluttered and is unambiguous because we do not attach any other subscripts to d.) Assume also that the source voltage v_{in} takes the constant value $v_{\text{in}}[k]$ in the kth cycle. Setting $q(t) = 1$ in (13.44) to represent the interval when the transistor is on, we obtain an expression similar to (13.31):

$$x(kT + d_kT) \approx (I + d_kTA_1)x(kT) + d_kTB_1v_{\text{in}}[k]. \tag{13.45}$$

Now setting $q(t) = 0$ in (13.44) for the interval when the transistor is off, and letting $d'_k = 1 - d_k$, we get

$$x(kT + T) \approx (I + d'_kTA_0)x(kT + d_kT) + d'_kTB_0v_{\text{in}}[k]. \tag{13.46}$$

Using (13.45) to substitute for $x(kT + d_k T)$ in (13.46), and neglecting terms involving T^2, we get the desired approximate sampled-data model. (If we had wanted to track terms in T^2, we would have needed to retain such terms in the approximations (13.45) and (13.46) as well.)

To streamline the description of this model, we write $x[k] = x(kT)$ and use the natural symbols A_{d_k} and B_{d_k} for the "averages" of $A_{q(t)}$ and $B_{q(t)}$, respectively, taken with duty ratio d_k:

$$A_{d_k} = d_k A_1 + d'_k A_0, \qquad B_{d_k} = d_k B_1 + d'_k B_0. \tag{13.47}$$

We obtain these matrices simply by replacing $q(t)$ with d_k in the expressions for $A_{q(t)}$ and $B_{q(t)}$ in (13.44). The result of using (13.45) in (13.46) and dropping terms in T^2 is then

$$\begin{aligned} x[k+1] &\approx (I + TA_{d_k})x[k] + TB_{d_k} v_{\text{in}}[k] \\ &= \phi(x[k], v_{\text{in}}[k], d_k). \end{aligned} \tag{13.48}$$

This is a nonlinear, time-invariant, sampled-data model in state-space form, with inputs d_k and $v_{\text{in}}[k]$. We would obtain the same sampled-data model by starting with the averaged model instead of the switched model, and making approximations similar to (13.31).

Figure 13.9 represents the relationships among the state variable trajectories of the sampled-data model in (13.48), the underlying continuous-time switched model in (13.44), and the continuous-time averaged model in (13.36) of Example 13.5.

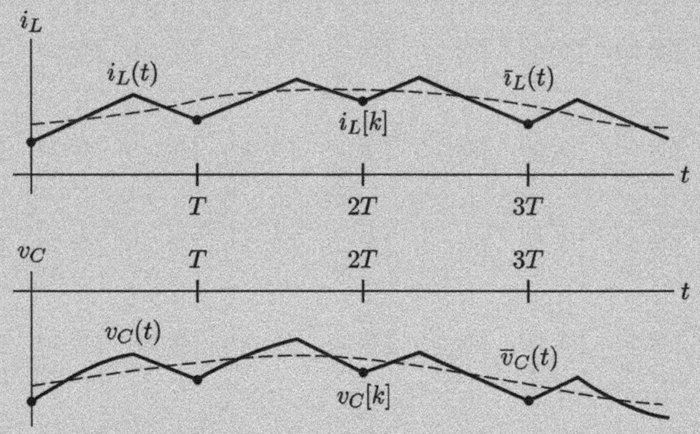

Figure 13.9 Relationships among the state variable trajectories of the sampled-data model (bullets); the underlying continuous-time switched model (solid line); and the continuous-time averaged model (dashed line).

13.5 Generalized State-Space Models

In Example 13.4, and again in the example above, we wrote a state-space model directly in terms of the variables that are allowed in such a description, namely the memory/energy-storage variables and the driving/control inputs. However, it is often convenient or necessary to introduce auxiliary variables during the derivation of a state-space model. Some auxiliary variables can be easily rewritten in terms of just the allowed variables. Others may require more work to solve for, or are only implicitly determined by some algebraic constraint equations. In such cases, it can be helpful or necessary to carry the constraints along with the description of state evolution. The combination of a state evolution description that contains auxiliary variables with associated algebraic constraints that determine these variables is what we term a *generalized state-space model*.

The most important case for us occurs in the context of sampled-data models for power circuits. The generalized state-space model in this case takes the form

$$\begin{aligned} x[k+1] &= \phi\big(x[k], p[k], w[k], k\big), \\ 0 &= \sigma\big(x[k], p[k], w[k], k\big). \end{aligned} \tag{13.49}$$

Here, $w[k]$ is a vector of auxiliary variables, and $\sigma(\cdot)$ is a vector of *constraint functions*. Given $x[k]$ and $p[k]$, we could in principle solve the second equation for $w[k]$ and substitute the result into the first equation to determine $x[k+1]$. However, as $\sigma(\cdot)$ is typically a nonlinear function of $w[k]$, solution of the second equation usually requires iterative numerical calculations, as well as some good initial guess of the solution. (There is an obvious continuous-time analog of (13.49), sometimes referred to as a differential/algebraic equation description, but we shall not have need for it.)

The auxiliary variables in $w[k]$ for power circuit models are primarily the time instants at which the topological state of the circuit changes in response to those switch transitions that are *not* directly controlled. (The times of directly controlled switch transitions appear as entries of the input vector $p[k]$.) Diodes turning on or off may correspond to such indirectly controlled transitions, if they occur at times that are functions of the state as well as of the control inputs. Transitions in fully controllable switches such as transistors also become indirectly controlled when their operation has been made state dependent through feedback control.

The following example illustrates how a generalized state-space model can arise.

Example 13.8 Generalized State-Space Model for a Buck/Boost Converter under Current-Mode Control

In (13.48) of Example 13.7, we obtained an approximate sampled-data model for a buck/boost converter, with the duty ratio as the control variable. In *current-mode control*, however, the controller specifies a peak switch current in each cycle, or equivalently a peak inductor current, rather than the duty ratio. The switch may be turned on regularly every T seconds, as in the implementations discussed earlier, but is turned off when the transistor current or inductor current reaches the specified upper threshold value i_{th}. This threshold value is now the primary control variable; the duty ratio becomes an indirectly determined auxiliary variable. (Several modifications are possible. For example, a *hysteretic* current-mode controller uses a lower threshold also on the inductor current, to determine the time at which the transistor is turned on again. We focus here on implementations in which the turn-on occurs at constant frequency.) We discuss current-mode control further in Chapter 14.

The constraint that determines d_k in terms of i_{th} can be written as

$$0 = i_L(kT + d_kT) - i_{\text{th}}(kT + d_kT). \tag{13.50}$$

The threshold i_{th} is usually chosen as the sum of two signals: a slowly varying signal i_P determined by the controller on the basis of the discrepancy between the actual and nominal average output voltages; and a regular sawtooth ramp of slope $-S$ at the switching frequency, termed a stabilizing ramp for reasons that we explain when examining the stability aspects of this model in Chapter 14. The resulting waveforms are shown in Fig. 13.10. Note the connection with Example 12.6, where we would choose $K = ST$ and $m = i_P - i_L$ in order to obtain current-mode control.

With the above choice of threshold, $i_{\text{th}}(kT + d_k T) = i_P[k] - Sd_k T$, where $i_P[k]$ is the value of i_P in the kth cycle. This expression can be substituted into (13.50), along with the approximate expression for $i_L(kT + d_k T)$ obtained from the first row of the matrix equation in (13.45). In the case of a buck/boost converter, the substitution yields

$$\begin{aligned} 0 &\approx i_L[k] + (d_k T v_{\text{in}}[k]/L) - i_P[k] + Sd_k T \\ &= \sigma\big(x[k], v_{\text{in}}[k], i_P[k], d_k\big). \end{aligned} \tag{13.51}$$

Together, (13.48) and (13.51) constitute a sampled-data model in generalized state-space form for a buck/boost converter under current-mode control. The appropriate expressions for other switching converters are obtained similarly.

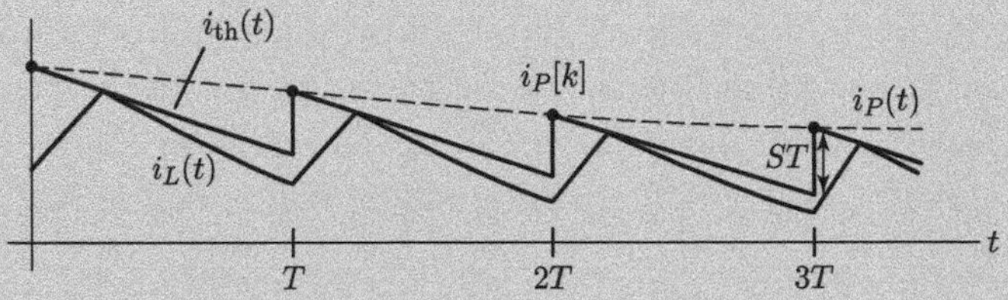

Figure 13.10 Inductor current waveform under current-mode control.

Example 13.8 is representative of the ways in which generalized state-space descriptions arise. In this particular case, solving explicitly for d_k from the constraint (13.51) actually is easy. Substituting the result into (13.48) yields an ordinary sampled-data state-space model. Nonetheless, it may be easier to work with the generalized form, even for this simple case. For example, we show in Chapter 14 that the task of linearization can be carried out completely on a generalized state-space model, and does not require reduction to an ordinary state-space model. In any case, with more complicated circuits or control laws and more accurate sampled-data models, carrying out such a reduction would be hard or impossible, and one would have to work with the generalized state-space model.

The following example shows a generalized state-space model arising in modeling a different sort of converter.

Example 13.9 Generalized State-Space Model for a Resonant DC/DC Converter

Choosing the normalized state variables $x_1 = \sqrt{L}\, i_L$ and $x_2 = \sqrt{C}\, v_C$ for the resonant converter circuit in Fig. 13.6(a), Example 13.3, we know that the state trajectory in each cycle comprises a succession of four circular arcs in the state plane. With the help of Fig. 13.11, we can easily verify that, if $x(t_\alpha)$ and $x(t_\beta)$ are the state vectors at two points on the same circular arc centered at x_c, and if $t_\beta > t_\alpha$, then

$$x(t_\beta) - x_c = \Theta(t_\beta - t_\alpha)\big[x(t_\alpha) - x_c\big], \tag{13.52}$$

where $\Theta(t_\beta - t_\alpha)$ is the "rotation matrix" defined by

$$\Theta(t_\beta - t_\alpha) = \begin{bmatrix} \cos(\omega_o(t_\beta - t_\alpha)) & -\sin(\omega_o(t_\beta - t_\alpha)) \\ \sin(\omega_o(t_\beta - t_\alpha)) & \cos(\omega_o(t_\beta - t_\alpha)) \end{bmatrix}. \tag{13.53}$$

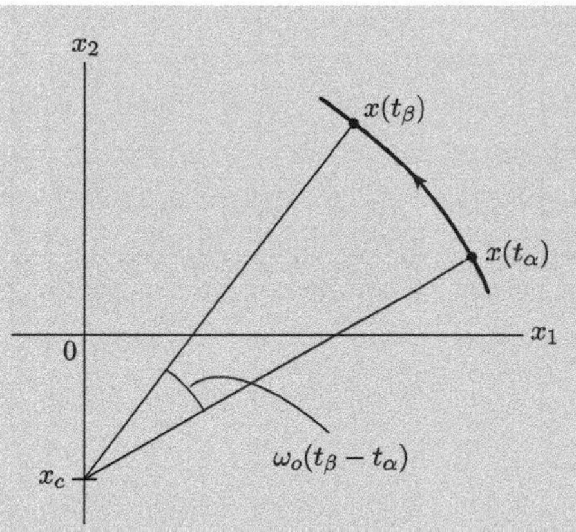

Figure 13.11 Relationship between two points on a circular trajectory in the state plane.

Using (13.53), we can translate the description of circuit behavior in Example 13.3 into the following succinct quantitative relationships for the first half of the kth cycle of operation (with the argument k again dropped from the times $\tau_1[k]$ and $\tau_2[k]$, for notational simplicity):

$$\begin{aligned} x(t_k + \tau_1) &= \Theta(\tau_1)\big[x(t_k) - x_{c_1}\big] + x_{c_1}, & x_{c_1} &= \begin{bmatrix} 0 \\ \sqrt{C}(V_1 + V_2) \end{bmatrix}, \\ x(t_k + \tau_2) &= \Theta(\tau_2 - \tau_1)\big[x(t_k + \tau_1) - x_{c_2}\big] + x_{c_2}, & x_{c_2} &= \begin{bmatrix} 0 \\ \sqrt{C}(V_1 - V_2) \end{bmatrix}. \end{aligned} \tag{13.54}$$

We now use the first expression in (13.54) to eliminate $x(t_k + \tau_1)$ from the second, then also use the fact that $\Theta(\tau_2 - \tau_1)\Theta(\tau_1) = \Theta(\tau_2)$, as rotation over an interval τ_1 followed by rotation for $\tau_2 - \tau_1$ is equivalent to rotation over an interval τ_2. The result is

$$x(t_k + \tau_2) = \Theta(\tau_2)x(t_k) - \Theta(\tau_2)x_{c1} + \Theta(\tau_2 - \tau_1)\big[x_{c_1} - x_{c_2}\big] + x_{c_2}. \tag{13.55}$$

The time τ_2 in (13.55) is a control variable, whereas τ_1 is an auxiliary variable. Recall from Fig. 13.6(b) that τ_1 is determined by the constraint $i_L(t_k + \tau_1) = 0$, or, equivalently, $x_1(t_k + \tau_1) = 0$. Using the first row of the first matrix equation in (13.54), we can express this constraint in the more detailed form

$$x_1(t_k)\cos(\omega_o\tau_1) - \big[x_2(t_k) - \sqrt{C}(V_1 + V_2)\big]\sin(\omega_o\tau_1) = 0. \tag{13.56}$$

Together, (13.55) and (13.56) constitute a sampled-data model in the generalized state-space form (13.49), describing the evolution of the state over the first half of the kth cycle.

The symmetry of the circuit and its operation allows us to easily write a similar description for the evolution over the second half, and to combine it with the model in (13.55) and (13.56) to obtain a generalized state-space model over the entire cycle. The control variables for this model are τ_2 and τ_4; in practice, the controller typically sets $\tau_4 = 2\tau_2$. The auxiliary variables are τ_1 and τ_3. With the obvious notation, namely, $x[k] = x(t_k)$ and so on, the result is exactly of the form (13.49).

Note that the constraint (13.56) in Example 13.9 can actually be solved for $\tau_1[k]$ in this case, expressing it as the arctangent of a quantity determined by the state $x(t_k)$ at the beginning of the cycle. A similar solution can be obtained for $\tau_3[k]$ in the second half-cycle. As with our model for current-mode control in Example 13.8, therefore, we can reduce the generalized state-space model to an ordinary state-space model. The result is messy, however, and not convenient for such tasks as linearization. We would thus be likely to retain the generalized form for many purposes. Furthermore, for slightly more involved circuits, no such reduction would be possible.

Examples 13.8 and 13.9 involve circuits with special features that we exploited to simplify the analysis. The buck/boost converter has (approximately) piecewise linear waveforms; the resonant converter has piecewise circular trajectories in the state plane. Nevertheless, as we show in Chapter 14, these examples actually contain the essential ingredients of dynamic modeling for much more general power electronic circuits. Using the *state-transition matrix* introduced in Chapter 14, we can conceptually deal with more complicated circuits in the same straightforward manner.

13.6 Models for Controllers and Interconnected Systems

Our focus so far has been on developing state-space models for power circuits. State-space models can also be used to represent controllers, as well as interconnections of controllers with power circuits. Most continuous-time controllers that are encountered in practice, and certainly all those considered in this book, have a continuous-time state-space description. This result holds even if the control design was carried out and implemented, as it often is, without any reference to state-space models or methods. Similarly, the discrete-time controllers of interest to us typically have a discrete-time state-space description.

Now consider an interconnection of a power circuit and a controller, as in Fig. 13.12. (We are representing any sensor dynamics as part of the controller.) A state-space description of the interconnection is simply obtained from the state-space models of the individual subsystems. We illustrate the procedure here for a continuous-time case, assuming time-invariant models. You can easily extend the procedure to other situations, including time-variant and discrete-time cases.

Assume that the power circuit is modeled by the state-space description

$$\dot{x} = f(x, u, v), \qquad y = g(x, v), \tag{13.57}$$

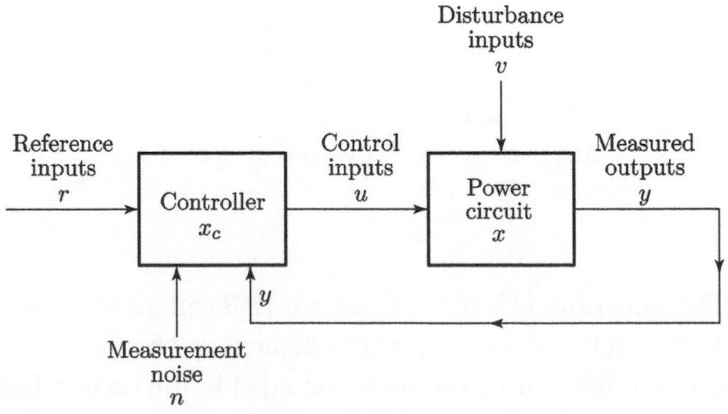

Figure 13.12 Interconnection of a power circuit and its controller.

where the vectors u and v represent control and disturbance inputs, respectively. Suppose the controller can be modeled by a state-space description of the form

$$\dot{x}_c = f_c(x_c, y, r, n), \qquad u = g_c(x_c, r), \tag{13.58}$$

where the vectors r and n represent reference signals and measurement noise, respectively. The output of the controller is the control vector u that is applied to the power circuit.

The state variables of the combined system are those of the two subsystems, taken together. We can now combine their descriptions to find a state-space model of the interconnected system. This step simply entails merging the state equations in (13.57) and (13.58) and making the substitutions necessary for the model to contain only those variables allowed in a state-space description, namely, the state variables and external inputs. The result is

$$\begin{bmatrix} \dot{x} \\ \dot{x}_c \end{bmatrix} = \begin{bmatrix} f\big(x, g_c(x_c, r), v\big) \\ f_c\big(x_c, g(x, v), r, n\big) \end{bmatrix}. \tag{13.59}$$

In more complicated situations, we may be able to obtain only a generalized state-space description rather than an ordinary description.

Notes and Bibliography

We pointed to a few references on circuit-averaged models at the end of Chapter 12, and they are relevant to the material covered in Sections 13.1 and 13.2 as well. An illustration of how the notion of a dynamic phasor can be applied is found in [1], which models a thyristor-controlled series capacitor of the sort used in high-power transmission grids.

For an introduction to state-space models, see, for example, [2].

The procedure in Section 13.3.5 for obtaining a state-space model for an electrical circuit is described in more detail in [3]. This imposing classic contains a great deal of useful material, including a treatment of numerical algorithms for solving state equations. A more accessible reference for computer methods in circuit analysis is [4].

The state-space averaging approach in Section 13.3.6 is comprehensively described in [5]; see also [6].

State-space modeling of the ω_s-component, as in Section 13.3.7, can be extended to model multiple frequencies simultaneously, as in [7] and [8].

For sampled-data modeling of power electronic systems, a more detailed treatment than in this chapter may be found in [9].

For more on state-plane analysis of resonant converters, which we encountered in Examples 13.3 and 13.9, see, for example, [10].

An early reference on current-mode control is [11], and associated modeling and control aspects are treated in [12]; see also [6]. We return to the topic in Chapter 14.

1. P. Mattavelli, G. C. Verghese, and A. M. Stankovic, "Phasor Dynamics of Thyristor-Controlled Series Capacitor Systems," *IEEE Trans. Power Systems*, vol. 12, pp. 1259–1267, Aug. 1997.
2. A. V. Oppenheim and G. C. Verghese, *Signals, Systems and Inference*, London: Pearson, 2017.
3. L. O. Chua and P.-M. Lin, *Computer-Aided Analysis of Electronic Circuits: Algorithms and Computational Techniques*, Hoboken, NJ: Prentice-Hall, 1975.

4. J. Vlach and K. Singhal, *Computer Methods for Circuit Analysis and Design*, 2nd ed., New York: Van Nostrand Reinhold, 1993.
5. R. D. Middlebrook and S. Ćuk, "A General Unified Approach to Modeling Switching Converter Power Stages," *Int. J. Electronics*, vol. 42, pp. 521–550, June 1977.
6. R. W. Erickson and D. Maksimović, *Fundamentals of Power Electronics*, 3rd ed., New York: Springer, 2020.
7. S. R. Sanders, J. M. Noworolski, X. Z. Liu, and G. C. Verghese, "Generalized Averaging Method for Power Conversion Circuits," *IEEE Trans. Power Electronics*, vol. 6, pp. 251–259, Apr. 1991.
8. V. A. Caliskan, G. C. Verghese, and A. M. Stanković, "Multi-frequency Averaging of DC/DC Converters," *IEEE Trans. Power Electronics*, vol. 14, pp. 124–133, Jan. 1999.
9. G. C. Verghese, M. E. Elbuluk, and J. G. Kassakian, "A General Approach to Sampled-Data Modeling for Power Electronic Circuits," *IEEE Trans. Power Electronics*, vol. PE-1, pp. 76–89, Apr. 1986.
10. R. Oruganti and F. C. Lee, "Resonant Power Processors, Part I – State Plane Analysis," *IEEE Trans. Industry Appl.*, vol. 21, pp. 1453–1460, Nov./Dec. 1985.
11. C. W. Deisch, "Simple Switching Control Method Changes Power Converter Into a Current Source," in *IEEE Power Electronics Specialists Conference (PESC)*, pp. 300–306, Syracuse, June 1978.
12. S. P. Hsu, A. Brown, L. Resnick, and R. D. Middlebrook, "Modeling and Analysis of Switching DC-to-DC Converters in Constant-Frequency Current-Programmed Mode," in *IEEE Power Electronics Specialists Conference (PESC)*, pp. 284–301, San Diego, June 1979.

PROBLEMS

13.1 The circuit of Fig. 13.13 is sometimes used to study the effect of an input filter on the dynamics of a controlled converter. The R and L represent, respectively, the combination of internal resistance and inductance of the source and externally added resistance and inductance in the filter. The block marked P is a nonlinear resistor that absorbs constant power P and is used as a simplified model for the controlled converter. This choice reflects the fact that a well-regulated, high-efficiency converter with a constant load absorbs essentially constant power from its input source. The purpose of the input filter is to reduce undesired current harmonics in the source.

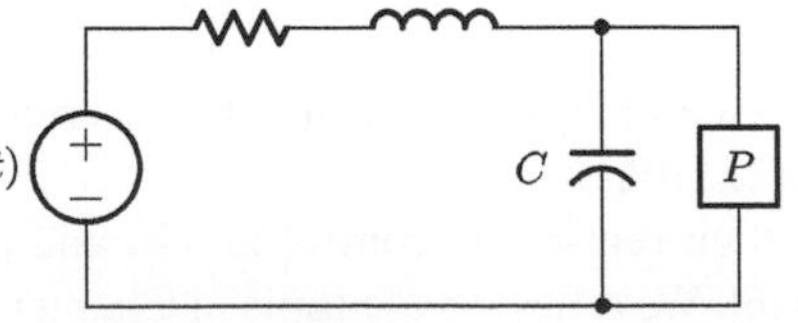

Figure 13.13 Model for a regulated converter with an input filter, analyzed in Problem 13.1.

(a) Obtain a state-space model of the circuit. Show that you can determine all the voltages and currents in the circuit at time t from knowledge of the state variables and source value at t.

(b) Use the state-space description in (a) to obtain a single second-order differential equation that describes the system. Explain how you would determine all voltages and currents in the circuit from knowledge of the solution of this equation.

(c) Suppose the capacitor has some equivalent series resistance R_C. Obtain a continuous-time generalized state-space description for the resulting circuit. Can you obtain an ordinary state-space model from it?

(d) Suppose that the inductor is modeled as nonlinear, with its flux linkage λ_L being some nonlinear function of its current i_L, say, $\lambda_L = \ell(i_L)$. Find a state-space model of the circuit in the figure (assuming no ESR).

13.2 In Example 13.4, we obtained a state-space model (13.34) for a buck/boost converter in continuous conduction. Complete the model by deriving an equation for the output voltage v_o as a function of the state variables and inputs. Then repeat the derivation of state and output equations for at least the following switching converters described in Chapters 5 and 7: buck, boost, flyback, and forward converters. Include the ESR of the capacitor in your model.

13.3 This problem deals with state-space averaged models for the buck/boost converter and the converters in Problem 13.2.

(a) Verify that the state-space averaged model in (13.36) for the buck/boost converter, with $R_C = 0$ (no ESR), describes the averaged circuit in Fig. 13.2 (Example 13.1).

(b) For $R_C \neq 0$ in the buck/boost converter, find a circuit representation of the state-space averaged description (13.36) by modifying the averaged circuit in Fig. 13.2. Also express the average of the output voltage v_o as a function of the state variables and inputs of the averaged model. You can do this from the averaged-circuit representation or by averaging your expression for v_o in Problem 13.2.

(c) Find state-space averaged models for the other converters in Problem 13.2. Repeat (a) and (b) for these converters, with the necessary modifications.

(d) Repeat your derivation of averaged models for all the preceding converters when the switches are no longer modeled as ideal. Specifically, instead of representing a conducting switch as a short circuit, model it as a (small) resistor in series with a (small) dc voltage source.

13.4 Figure 12.7(b) in Example 12.5 shows the averaged output circuit for a buck/boost converter in discontinuous conduction under duty-ratio control. Find state-space descriptions for the circuit, first using $\overline{v}_o$ as the state variable and then using $\overline{v}_o^2$ as the state variable. Can you see why using $\overline{v}_o^2$ may be preferable? How do your descriptions change if the circuit is operating under current-mode control?

13.5 This problem deals with the resonant converter in Examples 13.3 and 13.9.

(a) Obtain a state-space model for each of the four configurations encountered during a cycle of operation; use the scaled quantities $x_1 = \sqrt{L}\, i_L$ and $x_2 = \sqrt{C}\, v_C$ as state variables.

(b) Verify that the state-plane trajectories in each configuration are indeed circular arcs and that their centers are as claimed in Example 13.9.

(c) Use the simple geometry of the state-plane trajectories to obtain a closed-form solution for the steady-state radii of the circular arcs in (b), and draw the resulting diagram.

(d) Verify the properties of the "rotation matrix" that were used in Example 13.9.

(e) Even though the resonant *LC* pair actually sees square-wave voltages presented by the input and load sources, replace the square waves by their fundamental (switching frequency) components. Also, take the inductor current to be essentially a sinusoid at the switching frequency. Now determine the steady-state behavior of the resulting approximate circuit. Compare your result with that in (c).

(f) Verify that the analysis in (e) corresponds to a steady-state analysis of the ω_s-component, where ω_s is the switching frequency. Obtain a model for the non-steady-state behavior of the ω_s-component in this case.

13.6 Our analytical description of the operation of the resonant converter in Example 13.9 only went through one half-cycle of operation. Using the symmetry of the circuit and its operation, extend the description to the full cycle. Write the result in the standard generalized state-space form of (13.49).

13.7 Obtain an output equation of the form (13.42) for the sampled-data model in (13.48) of Example 13.7, with the output $y[k]$ being the *average* value of $v_o(t)$ in the kth cycle. (You can make approximations similar to those in Example 13.7.)

13.8 Obtain an approximate sampled-data model for a buck/boost converter in discontinuous conduction. Do so for (i) duty-ratio control and (ii) current-mode control.

13.9 Obtain approximate sampled-data models for the converters in Problem 13.2 for (a) continuous conduction and (b) discontinuous conduction. Again, do so for (i) duty-ratio control and (ii) current-mode control.

13.10 We made extensive use in Chapter 12 of a phase-controlled rectifier drive for a dc motor. Let $i_d[k]$ denote the armature current at the beginning of the kth cycle and α_k be the firing angle in this cycle. Find a sampled-data description for the open-loop drive.

13.11 Consider the buck/boost converter in Example 13.4. Suppose we decide to approximate each waveform $x(t)$ in the converter by the sum of its local average and its local first harmonic, see (12.11), and similarly for the switching function $q(t)$, in effect writing

$$x(t) = \widehat{x}^*(t)e^{-j\omega_s t} + \overline{x}(t) + \widehat{x}(t)e^{j\omega_s t}, \qquad q(t) = \widehat{q}^*(t)e^{-j\omega_s t} + \overline{q}(t) + \widehat{q}(t)e^{j\omega_s t}.$$

Recall that $\overline{q}(t) = d(t)$.

(a) Show that

$$\overline{qx} = \overline{q}\,\overline{x} + \widehat{q}^*\,\widehat{x} + \widehat{q}\,\widehat{x}^*,$$
$$\widehat{qx} = \overline{q}\,\widehat{x} + \widehat{q}\,\overline{x}$$

(where we have dropped the time argument t for notational simplicity).

(b) Now starting from (13.34), obtain a state-space model describing the dynamics of the system in terms of the appropriate local averages and local ω_s-components.

For particular numerical values of the components that would yield high ripple (making the above approximation worthwhile), explore how well the approximation captures the actual transient behavior of the system.

14 Linear Models and Feedback Control

In Chapter 12 we introduced the process of linearization for certain classes of nonlinear averaged-circuit models. This allowed us to obtain LTI circuit models for small perturbations of averaged values from their constant values in nominal, steady-state operating conditions. These LTI models then served as the basis for stability evaluation and control design in the examples considered in Chapter 12.

In Chapter 13 we extended the ideas of circuit averaging to derive an averaged model for the switch in high-frequency switched dc/dc converters, and also obtained averaged-circuit models for the local ω_s-component. The present chapter begins by showing how to linearize the averaged switch, and illustrates the application of this linearized model to analyzing the small-signal behavior of the associated converter.

Chapter 13 also expanded our modeling options to include state-space models. These can describe continuous-time signals, their averages, or their discrete-time samples in open- or closed-loop systems comprising power circuits and controllers. In the present chapter we show how to linearize models given in state-space form rather than circuit form. The chapter also develops some of the key concepts and results for the analysis of LTI state-space models. We then show applications of these results to analyzing *piecewise*-LTI models and evaluating the stability of cyclic nominal operation in power circuits. With this background, we are in a position to examine control design based on LTI models for power converters. Approaches to this, and illustrative examples, are presented in the last part of this chapter.

14.1 Linearization

Power circuit models are typically nonlinear, and therefore difficult to work with. The process of linearization introduced in Chapter 12 allows us to obtain linear models that approximately govern small deviations from some nominal solution. These linearized models or small-signal models are far more tractable than nonlinear models.

Linearized models are valuable for several tasks of analysis and design. In stability evaluation, we want to know whether a circuit can return to nominal operation after small perturbations away from it. Asymptotic stability of the linearized model implies that nominal operation will not be destroyed by sufficiently small disturbances; instability of the linearized model implies that the nominal operating condition cannot be maintained without further control action.

In controller design, we start by examining behavior under normal operating conditions, when the power circuit deviates only slightly from its nominal periodic steady-state operation. Models for small deviations are evidently central to this task. Linearized models are also useful in iterative numerical determination of steady-state behavior, because they allow us to predict the effects of small corrections to computed estimates at each stage of the iteration.

We showed in Chapter 12 that linearization of a circuit model or an averaged-circuit model is easy in principle: we replace the voltages and currents in the circuit model by their deviations from nominal, and replace each nonlinear component in the circuit by its linearized version. The next section shows this more explicitly for the case of averaged-circuit models for high-frequency switched dc/dc converters.

14.2 Linearizing an Averaged-Circuit Model

We have seen in Examples 12.7 and 12.8 respectively how a linearized averaged-circuit model can be obtained for a buck converter, or for a buck/boost converter in discontinuous conduction. In the case of other switched dc/dc converters, such as a buck/boost converter in continuous conduction, the key step in linearization of the corresponding averaged-circuit model is the linearization of the averaged-circuit model of the switch, Fig. 13.1(b) and (c).

14.2.1 Linearizing the Averaged Switch

Linearization of the averaged-switch model in Fig. 13.1(b) is easily carried out. Let us denote the nominal value of a quantity x by the corresponding uppercase letter X, and deviations from the nominal by $\widetilde{x}$, as in Example 12.8. Most commonly the nominal solution corresponds to a constant or periodic steady state, but this is not needed for the linearization below. We can write

$$d(t) = D + \widetilde{d}(t), \qquad d'(t) = D' - \widetilde{d}(t), \tag{14.1}$$

and similarly for the other variables.

We now expand the source terms in Fig. 13.1(b), namely, $d(t)\overline{i}_x(t)$ and $d(t)\overline{v}_{yz}(t)$, to first-order or linear terms in the perturbations. That is, we neglect terms that involve squares or products of the small perturbations. This linearized expansion yields the following first-order perturbations to these terms:

$$\begin{aligned} d(t)\overline{i}_x(t) - DI_x &\approx D\widetilde{i}_x(t) + I_x\widetilde{d}(t), \\ d(t)\overline{v}_{yz}(t) - DV_{yz} &\approx D\widetilde{v}_{yz}(t) + V_{yz}\widetilde{d}(t). \end{aligned} \tag{14.2}$$

The results of this calculation are represented in the linearized circuit in Fig. 14.1(a), with an equivalent representation in Fig. 14.1(b). The use of this linearized model of the averaged switch is illustrated in the following example.

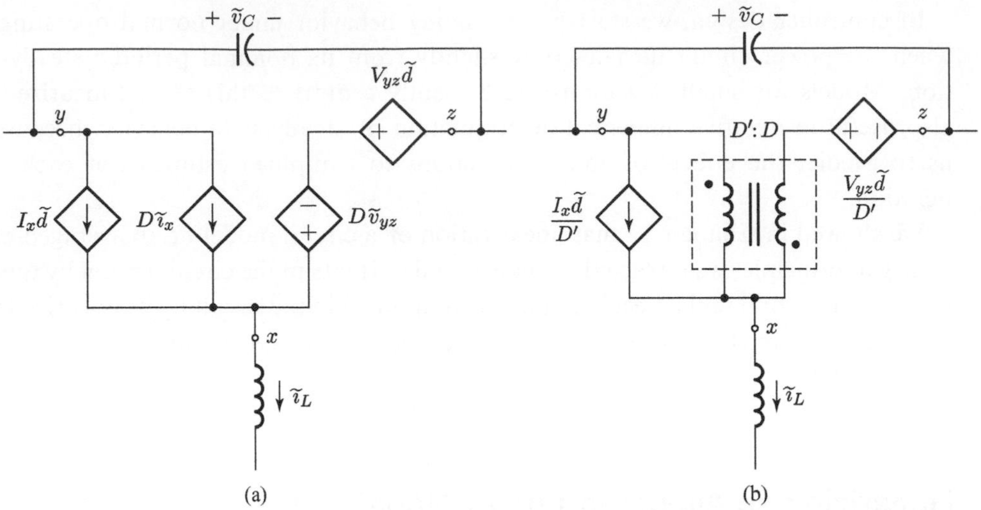

Figure 14.1 (a) Linearized averaged model of the canonical switching cell. (b) Alternative representation.

Example 14.1 Linearized Averaged-Circuit Model for a Buck/Boost Converter in Continuous Conduction

We return once more to the buck/boost converter of Examples 12.1, 12.3, and 13.1 to obtain a linearized averaged model that describes small perturbations from the nominal steady state in the continuous conduction mode.

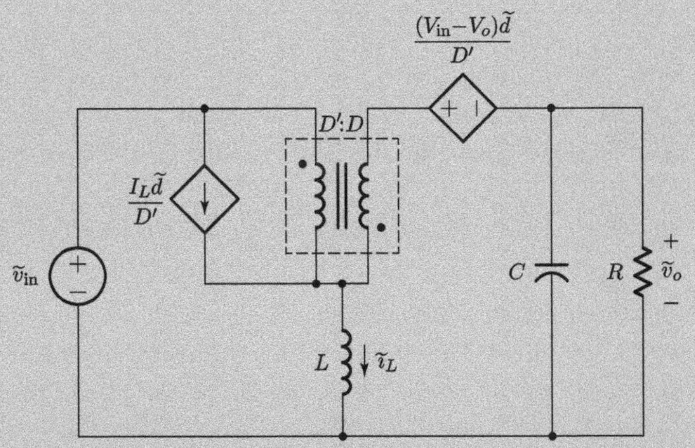

Figure 14.2 Linearized averaged model of a buck/boost converter.

The linearized version of Fig. 13.2 is obtained by replacing all voltages and currents by their perturbations from the nominal, and replacing the averaged canonical cell by its linearized version, using Fig. 14.1(b). All other converter elements are linear and are therefore preserved intact in the linearized circuit. The result is shown in Fig. 14.2, with $\tilde{v}_{\text{in}}$ representing deviations of the averaged input source from its nominal dc value. Note that, as in Fig. 13.2, we have placed the capacitor C in its more typical position across the converter output, unlike in the linearized canonical model of Fig. 14.1.

The linearized model corresponding to the constant nominal steady state is an LTI circuit. Hence, solving for the values of all circuit variables is straightforward (for example, using impedance methods). For control-design purposes, we must know the transfer function from perturbations $\tilde{d}$ in the duty cycle, which constitutes our control input, to perturbations $\tilde{v}_o$ in the output voltage. Setting $\tilde{v}_{\text{in}} = 0$, a straightforward computation with the circuit in Fig. 14.2 shows the transfer function to be

$$\frac{\tilde{v}_o(s)}{\tilde{d}(s)} = \left(\frac{I_L}{C}\right)\frac{s-(V_{\text{in}}/LI_L)}{s^2+(1/RC)s+(D'^2/LC)}. \tag{14.3}$$

(We have used the steady-state relationships in (13.4) to simplify the constant in the numerator.) This is the transfer function we referred to at the end of Example 13.1.

The transfer function computation in Example 14.1 illustrates the advantages of having an LTI circuit model for power circuit dynamics. Such a model allows us to apply the extensive array of concepts and techniques available for LTI circuits to the study of deviations from nominal operation in a power circuit. These concepts and techniques include superposition, impedance methods, Norton and Thévenin equivalents and their multi-port generalizations, and so on. For instance, in addition to the transfer function computed in Example 14.1, we could determine input and output impedances or admittances, as well as various other relevant transfer ratios. These transfer ratios are important in describing how dynamic behavior depends on characteristics of the source, the load, the circuit itself, and the control.

The transfer function (14.3) provides a good starting point for an explanation of why the proportional feedback scheme in Example 12.3 led to the behavior observed there. We pursue this explanation in the following example.

Example 14.2 Stability of a Buck/Boost Converter under Proportional Feedback

We now have the information needed to explain the results of our attempt in Example 12.3 to regulate the buck/boost converter by proportional feedback. The feedforward compensation in the block diagram of Fig. 12.5(a) has no dynamics associated with it and does not affect the discussion of stability. If we ignore the feedforward, the remainder of the diagram can be redrawn as in Fig. 14.3(a). This is in the standard form of Fig. 12.11.

The transfer function $g(s)$ from $\tilde{d}$ to $\tilde{v}_o$ is given by (14.3), and h is simply the feedback gain or compensator gain. The corresponding sensitivity function $\mu(s)$, as defined in Chapter 12, turns out to be

$$\mu(s) = \frac{s^2+(1/RC)s+(D'^2/LC)}{\left[s^2+(1/RC)s+(D'^2/LC)\right]+h(I_L/C)\left[s-(V_{\text{in}}/LI_L)\right]}. \tag{14.4}$$

The denominator has both its roots in the left half-plane if and only if the coefficient of s and the constant term are both positive. The right-half-plane zero of the transfer function $g(s)$ causes the stability condition to only hold for h satisfying

$$-\frac{1}{I_L R} < h < \frac{D'^2}{V_{\text{in}}}. \tag{14.5}$$

An alternative expression for the lower bound on h is $-D'^2/(V_{\text{in}}D)$.

The roots of the denominator in (14.4) are also the natural frequencies of the system. Figure 14.3(b) shows the locus of these natural frequencies, as h takes both negative and positive values. The damping decreases as h takes on increasingly negative values; the natural frequencies move closer to the imaginary axis and their imaginary parts increase. These variations correlate well with the time-domain responses shown in Fig. 12.5(b). For sufficiently small positive values of h, the natural frequencies move away from the imaginary axis and the damping increases. For more positive values of h, the roots become real, and one of them eventually crosses over to the right half-plane. These movements again correlate well with the time-domain responses in Fig. 12.5(c).

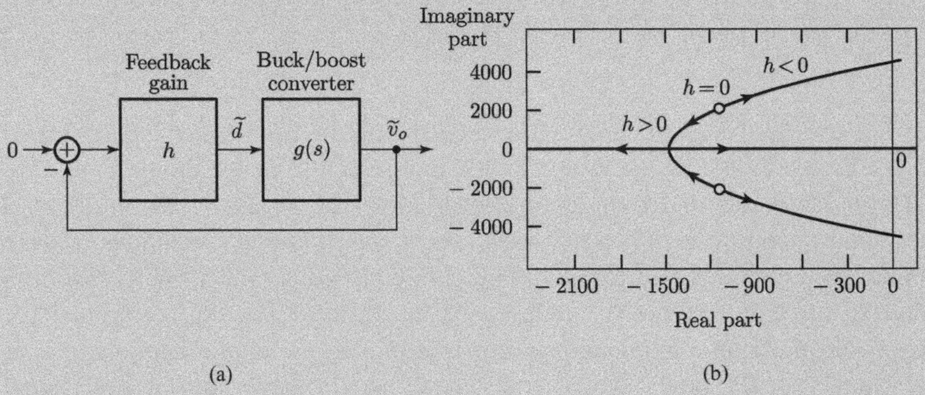

Figure 14.3 (a) Linearized model for a buck/boost converter with proportional feedback. (b) Locus of natural frequencies for $h < 0$ and $h > 0$.

Apart from serving as a basis for computations that support the tasks of stability analysis and control design, a circuit model can sometimes directly suggest control designs. For example, consider again the circuit in Fig. 12.10, which we obtained by linearizing the averaged model of the buck/boost converter in discontinuous conduction. Its damping would evidently be increased – and $\tilde{v}_o$ would decay to zero faster – if additional conductance were placed in parallel with the capacitor. Adding a physical conductance is clearly ruled out on grounds of efficiency, but it is possible to reproduce the effect of additional conductance entirely by means of control. If we use *proportional* feedback to enforce $\tilde{d}(t) = h\tilde{v}_o(t)$, where h is a constant, the effect is the same as that of a conductance of value $hV_{\text{in}}\sqrt{2T/RL}$ replacing the current source in Fig. 12.10. *Proportional–integral* control has the effect of replacing the current source with a parallel combination of a conductance and an inductance. This inductance causes the steady-state value of $\tilde{v}_o$ to be zero even in the presence of parameter errors and constant disturbances (such as constant deviations of the input voltage or converter inductance from their nominal values). Thus, using a circuit model to approach or interpret control design can often provide useful insights.

In the next section we describe the details of the linearization process for continuous-time *state-space* models. The discrete-time case is treated in Section 14.6 (for generalized state-space descriptions, which are more important for power circuits, but the results for discrete-time systems in ordinary state-space form follow easily).

14.3 Linearizing Continuous-Time State-Space Models

We now show how to linearize the continuous-time state-space model given in (13.24), and repeated here for convenience:

$$\begin{aligned} \frac{dx}{dt} &= f(x(t), u(t), t), \\ y(t) &= g(x(t), u(t), t). \end{aligned} \tag{14.6}$$

We shall use capital letters here to denote the nominal values of the variables in (14.6), so that in the nominal operating condition $x(t) = X(t)$, $u(t) = U(t)$, and $y(t) = Y(t)$. The linearization process is not predicated on these nominal values being constant or periodic; they can be any values that together satisfy (14.6). The most important cases for our purposes are when the nominal solutions are constant or periodic.

Let us now consider deviations from nominal, and mark these quantities with a tilde, as before, so that away from the nominal solution we have

$$x = X + \widetilde{x}, \qquad u = U + \widetilde{u}, \qquad y = Y + \widetilde{y}. \tag{14.7}$$

(We have dropped the time argument t for notational simplicity.) Substituting the relationships from (14.7) into (14.6) and expanding the resulting nonlinear terms in multi-variable Taylor series around the nominal solution, then using the fact that the nominal solution itself satisfies (14.6), we obtain the desired linearized model:

$$\begin{aligned} \frac{d\widetilde{x}}{dt} &\approx \left[\frac{\partial f}{\partial x}\right] \widetilde{x}(t) + \left[\frac{\partial f}{\partial u}\right] \widetilde{u}(t), \\ \widetilde{y}(t) &\approx \left[\frac{\partial g}{\partial x}\right] \widetilde{x}(t) + \left[\frac{\partial g}{\partial u}\right] \widetilde{u}(t). \end{aligned} \tag{14.8}$$

The notation in (14.8), although designed to be simple and suggestive, needs definition and interpretation. The idea of a partial derivative of a vector function such as $f(x, u, t)$ with respect to a vector argument such as x may not be familiar. The symbol $\partial f/\partial x$ denotes a *matrix* (termed the *Jacobian* matrix), whose entry in the ith row and jth column is the partial derivative of the ith component of $f(\cdot)$ with respect to the jth component of x, namely, $\partial f_i/\partial x_j$. A similar definition holds for the other partial derivatives in (14.8). In general, these partial derivatives are all functions of x, u, and t but must be evaluated at the nominal solution, with $x = X$ and $u = U$, when used in (14.8).

Note that the linearized model in (14.8) is itself in state-space form, with the deviations $\widetilde{x}$, $\widetilde{u}$, and $\widetilde{y}$ now respectively constituting the state variables, inputs, and outputs. We can easily represent the model by a block diagram of the form in Fig. 13.5. This block diagram of the linearized model is in fact a direct linearization of the one in Fig. 13.5. The major advantage of (14.8) over (14.6) is, of course, that (14.8) is linear, because the partial derivatives that make up the coefficient matrices are functions only of the known nominal quantities X and U, and do not depend on $\widetilde{x}$ or $\widetilde{u}$. We can write (14.8) in the form of (13.26)–(13.28), namely

$$\frac{d\widetilde{x}}{dt} \approx A(t)\widetilde{x}(t) + B(t)\widetilde{u}(t),$$
$$\widetilde{y}(t) \approx E(t)\widetilde{x}(t) + F(t)\widetilde{u}(t), \tag{14.9}$$

where the coefficient matrices are the partial-derivative matrices in (14.8).

If the original nonlinear model is time-invariant, so that $f(\cdot)$ and $g(\cdot)$ are not explicitly dependent on time, and if the nominal solution is constant, then all the partial derivatives that define the linearized model are constant. The linearized model in this case is therefore LTI. If the nominal solution is periodically varying, the linearized model will also be periodically varying rather than LTI, even if the underlying nonlinear model is time invariant.

The following example relates the switched and averaged models for the case of a buck/boost converter.

Example 14.3 Linearized Switched and Averaged Models for a Buck/Boost Converter

Let us return to the switched and averaged models of a buck/boost converter in (13.34) and (13.36) of Examples 13.4 and 13.5, respectively. These two models have the same structure; replacing all the instantaneous variables of the switched model by their local averages produces the averaged model, and vice versa. We therefore focus first on linearization of the *averaged* model and then make a few remarks about the switched case.

We can conveniently rewrite the averaged model (13.36) in the matrix notation introduced in (13.44) of Example 13.7. With capacitor ESR neglected, so $R_C = 0$, we write

$$\begin{aligned}\frac{dx(t)}{dt} &\approx \begin{bmatrix} 0 & d'(t)/L \\ -d'(t)/C & -1/RC \end{bmatrix} x(t) + \begin{bmatrix} d(t)/L \\ 0 \end{bmatrix} \overline{v}_{\text{in}}(t) \\ &= A_{d(t)}x(t) + B_{d(t)}\overline{v}_{\text{in}}(t) \\ &= f(x(t), \overline{v}_{\text{in}}(t), d(t)),\end{aligned} \tag{14.10}$$

where the components of $x(t)$ now denote averaged quantities: $x_1(t) = \overline{i}_L(t)$ and $x_2(t) = \overline{v}_C(t)$. If $R_C \neq 0$, we can still represent the averaged model in the form (14.10), except that the particular entries of $A_{d(t)}$ and $B_{d(t)}$ will be different. Furthermore, there are similar representations for other switching converters. We leave pursuit of these other cases to you.

With a constant duty ratio and a constant supply voltage, we have $d(t) = D$ and $\overline{v}_{\text{in}}(t) = V_{\text{in}}$. We find the corresponding constant nominal steady-state solution $x(t) = X$ by setting the derivative in (14.10) to zero and solving for X. This calculation yields

$$X = \begin{bmatrix} I_L \\ V_C \end{bmatrix} = -A_D^{-1} B_D V_{\text{in}} = \begin{bmatrix} D/RD'^2 \\ -D/D' \end{bmatrix} V_{\text{in}}. \tag{14.11}$$

These are the same formulas we obtained in (13.4) of Example 13.1 by direct analysis of the averaged circuit.

To compute the linearized model, we use (14.8), with $f(\cdot)$ as in (14.10) and with the superscript $\sim$ now denoting deviations of averaged quantities from their nominal values. The linearized model then takes the form

$$\begin{aligned}\frac{d\widetilde{x}}{dt} &\approx \frac{\partial f}{\partial x}\widetilde{x}(t) + \frac{\partial f}{\partial \overline{v}_{\text{in}}}\widetilde{v}_{\text{in}}(t) + \frac{\partial f}{\partial d}\widetilde{d}(t) \\ &= A_D\widetilde{x}(t) + B_D\widetilde{v}_{\text{in}}(t) + J\widetilde{d}(t),\end{aligned} \tag{14.12}$$

where the vector J is given by

$$J = (A_1 - A_0)X + (B_1 - B_0)V_{\text{in}}. \tag{14.13}$$

To derive (14.13), you should note, as we did in (13.47), that $A_d = dA_1 + d'A_0$ and $B_d = dB_1 + d'B_0$. We leave you to verify that the linearized state-space description in (14.12) actually describes the linearized circuit obtained in Fig. 14.2 of Example 14.1.

Linearization of the *switched* model (13.34) results in a description like (14.12), except that *instantaneous* quantities appear instead of averaged quantities. The nominal duty ratio D is replaced by the periodic nominal switching function $q(t) = Q(t)$; the nominal solution $X(t)$ in this case is periodically varying; $\tilde{d}(t)$ is replaced by $\tilde{q}(t)$; $\tilde{v}_{\text{in}}$ now denotes the deviation of the instantaneous rather than averaged supply voltage; and the entries of $\tilde{x}(t)$ now denote perturbations of the instantaneous waveforms from their periodic steady states. Figure 14.4 schematically represents the differences between the switched and averaged models. (The fact that $\tilde{q}$ takes only the values 0 and 1 introduces a subtle twist to the linearization argument; a "small" $\tilde{q}$ is one that has small *area* rather than small magnitude. We leave you to verify that the Taylor series linearization is still valid.)

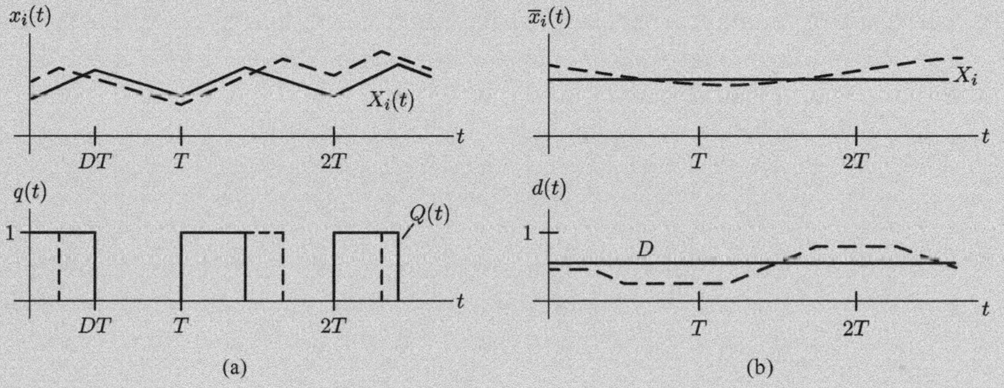

Figure 14.4 Representation of differences between the linearizations of switched and averaged models. (a) Switched model. (b) Averaged model.

14.4 Analysis of Continuous-Time LTI Models

We have shown how to develop and apply continuous-time LTI models in several ways in power electronics. We use such models to describe a fixed topological state of a power circuit model that is made up of LTI components and ideal switches. We obtain them when we linearize a nonlinear, time-invariant, continuous-time model at a constant nominal operating point, as with the averaged model in Example 14.1. We also use them to describe some of the most common controllers or compensators, such as the PI controller for the phase-controlled drive in Example 12.9.

In this section we develop some of the basic tools and results used to analyze continuous-time LTI models. The results are central to stability evaluation and control design. They also permit us, in Section 14.5, to describe the general solution of piecewise LTI models for power circuits, leading to useful sampled-data models.

Our starting point is an LTI model of the form (13.29) and (13.30), repeated here:

$$\frac{dx}{dt} = Ax(t) + Bu(t),$$
$$y(t) = Ex(t) + Fu(t). \tag{14.14}$$

The results in this section can, of course, also be applied to a linearized model of the form (14.9) when its coefficient matrices are constant.

14.4.1 Transform-Domain Solution

The easiest way to solve (14.14) for specific initial conditions and inputs is to use Laplace transforms to convert the differential equations into algebraic equations. This procedure corresponds to using impedance methods for an LTI circuit. As noted in Chapter 12, to avoid excessively cluttered notation, we shall denote the Laplace transform of a time function $h(t)$ simply by $h(s)$. This abuse of notation is not likely to cause confusion as long as we reserve the argument s for Laplace transforms and always display this argument. When more careful notation is called for, we write $\mathcal{L}\{h(t)\}$ instead of $h(s)$.

Recall the definition of the (one-sided) Laplace transform:

$$h(s) = \mathcal{L}\{h(t)\} = \int_0^\infty h(t)e^{-st}\,dt. \tag{14.15}$$

Define the Laplace transform of a vector function of time $x(t)$ to be its componentwise Laplace transform, so that the ith entry of $x(s)$ is $x_i(s)$. With this definition, we can easily show that the following key property of the Laplace transform carries over from the scalar case to the vector case:

$$\mathcal{L}\left\{\frac{dx(t)}{dt}\right\} = sx(s) - x(0), \tag{14.16}$$

where $x(0)$ is the value of the *time* function $x(t)$ at $t = 0$. This result is usually referred to as the differentiation theorem for Laplace transforms, and is the key to converting differential equations in the time domain to algebraic equations in the transform domain (or "frequency" domain). Many other results from the scalar case also have easy vector extensions. In particular, if A is a constant matrix, then $\mathcal{L}\{Ax(t)\} = Ax(s)$.

Taking transforms of the expressions in (14.14) and using the properties just discussed, we can write

$$sx(s) - x(0) = Ax(s) + Bu(s), \tag{14.17}$$

$$y(s) = Ex(s) + Fu(s). \tag{14.18}$$

Our aim is to solve for $x(s)$ from (14.17), given the initial condition $x(0)$ and the Laplace transform $u(s)$ of the input. We can then substitute the resulting expression for $x(s)$ into (14.18) to solve for $y(s)$.

Gathering terms involving $x(s)$ to one side of (14.17), we get

$$sx(s) - Ax(s) = (sI - A)x(s) = x(0) + Bu(s), \tag{14.19}$$

where I denotes the identity matrix. Solving (14.19) for $x(s)$ requires inversion of $(sI - A)$. The inverse exists if and only if the determinant of this matrix is not identically zero. For an $n \times n$ matrix A, $\det(sI - A)$ is always an nth-degree polynomial in s, of the form

$$\det(sI - A) = s^n + a_{n-1}s^{n-1} + \cdots + a_0 = a(s) \tag{14.20}$$

for some coefficients $\{a_j\}$ determined by A. This polynomial $a(s)$ is termed the *characteristic polynomial* of A or of the system (14.14), and its roots $\lambda_1, \ldots, \lambda_n$ are called the *characteristic roots*, or the *eigenvalues* of A. Because these roots have units of reciprocal time, they are also referred to as *characteristic frequencies*, or *natural frequencies*, of the system (14.14). The characteristic polynomial is bound to appear in any framework for analyzing an LTI system. In the state-space approach, the characteristic polynomial makes its appearance as $\det(sI - A)$.

As the characteristic polynomial is not identically zero, the inverse of $(sI - A)$ exists. The entries of the inverse are rationals in s, with the nth-degree characteristic polynomial $a(s)$ appearing as the denominator of every term, and with every numerator being of degree less than n. (If a numerator root happens to coincide with a characteristic root, this root cancels out of both the numerator and denominator, yielding a denominator of reduced degree. It can be shown, however, that every characteristic root will be represented in the denominator of at least one entry of $(sI - A)^{-1}$.)

The solution $x(s)$ of (14.19) follows on premultiplying that equation by $(sI - A)^{-1}$ to get

$$x(s) = (sI - A)^{-1}x(0) + (sI - A)^{-1}Bu(s). \tag{14.21}$$

The first term constitutes the response to initial conditions alone, in the absence of external inputs, and is called the *natural response* or *free response* or *zero-input response* of the system. The second term, similarly, is the *zero-state response*, which we also refer to as the *forced response*. The full solution is thus the superposition of these two components.

To complete the solution of the LTI system, we substitute for $x(s)$ in the output equation (14.18) to obtain

$$y(s) = E(sI - A)^{-1}[x(0) + Bu(s)] + Fu(s). \tag{14.22}$$

Transfer Functions With zero initial conditions, that is, $x(0) = 0$, we have

$$y(s) = [E(sI - A)^{-1}B + F]u(s) = G(s)u(s), \tag{14.23}$$

where $G(s)$ denotes the $p \times m$ matrix in brackets in (14.23). This matrix is the *transfer function* matrix of the system from the input u to the output y. Its entry in the ith row and jth column gives the transfer function from the jth component of the input to the ith component of the output.

The way in which $(sI - A)^{-1}$ enters the expression for $G(s)$ shows that each entry of the transfer function matrix will also be rational in s, with a denominator equal to the characteristic polynomial $a(s)$. Again, if a root of the numerator in some entry coincides with a characteristic root, the resulting pole–zero cancellation leads to a denominator of reduced degree in that entry. (Under typically satisfied "controllability" and "observability" conditions – beyond our scope to describe here – every characteristic root is represented in the denominator of at least one entry of $G(s)$.) The numerator degree equals the denominator degree if the corresponding entry of F is nonzero, and otherwise is less than the denominator degree. Also of interest for many computations is the transfer function matrix from u to x, which (14.21) shows to be $(sI - A)^{-1}B$.

Of particular significance in control design using LTI models is the *frequency response* of the system. This is simply the transfer function evaluated along the imaginary axis, that is, $G(s)$ evaluated for $s = j\omega$. In the next subsection we remind you of the role of the frequency response in describing input–output behavior.

Example 14.4 Transfer Functions for the Linearized Averaged Model of a Buck/Boost Converter

We return now to the linearized averaged model in (14.12) and examine it in the transform domain. Taking Laplace transforms and solving for $\widetilde{x}(s)$ yields

$$\widetilde{x}(s) = (sI - A_D)^{-1}[\widetilde{x}(0) + B_D\widetilde{v}_{\text{in}}(s) + J\widetilde{d}(s)]. \tag{14.24}$$

The transfer function matrix between, for example, $\widetilde{d}(s)$ and $\widetilde{x}(s)$ can be obtained by setting $\widetilde{x}(0) = 0$ and $\widetilde{v}_{\text{in}}(s) = 0$:

$$\widetilde{x}(s) = \begin{bmatrix} \widetilde{i}_L(s) \\ \widetilde{v}_C(s) \end{bmatrix} = (sI - A_D)^{-1}J\widetilde{d}(s). \tag{14.25}$$

For the case where the capacitor ESR is zero ($R_C = 0$),

$$\begin{aligned} \widetilde{x}(s) &= \begin{bmatrix} s & -D'/L \\ D'/C & s + (1/RC) \end{bmatrix}^{-1} \begin{bmatrix} (V_{\text{in}} - V_C)/L \\ I_L/C \end{bmatrix} \widetilde{d}(s) \\ &= \frac{1}{a_D(s)} \begin{bmatrix} s + (1/RC) & D'/L \\ -D'/C & s \end{bmatrix} \begin{bmatrix} (V_{\text{in}} - V_C)/L \\ I_L/C \end{bmatrix} \widetilde{d}(s), \end{aligned} \tag{14.26}$$

where

$$a_D(s) = s^2 + \frac{1}{RC}s + \frac{D'^2}{LC}. \tag{14.27}$$

The transfer function from the controlling input (duty-ratio perturbations $\widetilde{d}$) to the controlled output (perturbations $\widetilde{v}_o$ in the output voltage) is of special interest in the design of a feedback controller and may be derived from (14.25). When $R_C = 0$, $v_o = v_C$, so the desired transfer function can be read from (14.26). Using (14.11) to simplify the result, we find

$$\widetilde{v}_o(s) = \widetilde{v}_C(s) = \left(\frac{I_L}{C}\right) \frac{s - (V_{\text{in}}/LI_L)}{s^2 + (1/RC)s + (D'^2/LC)} \widetilde{d}(s), \tag{14.28}$$

which is exactly what we obtained from the linearized averaged circuit of Example 14.1. Figure 14.5(a) shows the locations of the poles and zero of this transfer function $g(s)$ for the parameter values in Example 12.1, and Fig. 14.5(b) shows Bode plots of the magnitude and phase of its frequency response. Although

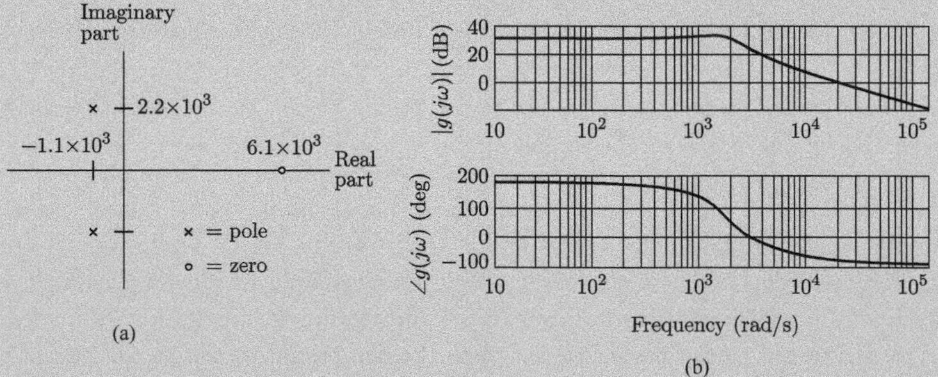

Figure 14.5 (a) Location of poles (**x**) and zero (**o**) of the transfer function $g(s)$ from perturbations in duty ratio to perturbations in output voltage. (b) Bode plots of frequency response corresponding to this transfer function.

the frequency response is shown up to one half the switching frequency ($\pi/T = 1.57 \times 10^5$ rad/s), you should keep in mind that the predictions of the averaged model become progressively less accurate as this frequency is approached.

Using (14.24) to compute the transfer function from $\widetilde{v}_{in}(s)$ to the output leads similarly to the transfer function (13.5) already obtained from the averaged circuit in Example 13.1. The polynomial appearing in the denominators of these transfer functions is indeed the characteristic polynomial of the matrix A_D in (14.12), as expected.

The preceding example – in particular, the concordance of the derived expressions with those obtained quite directly from the linearized averaged circuit in Example 14.1 – helps to build confidence in matrix computations and their results, if one is unused to working with matrices. This confidence becomes important when you confront situations in which there is no direct circuit-based route to the desired results.

14.4.2 Time-Domain Solution

To obtain the time-domain version of the transform-domain expression for $\widetilde{v}_C$ in Example 14.3, or more generally for the vectors x or y in (14.21), (14.22), or (14.23), we must compute inverse Laplace transforms. We can do so easily when each entry of the vector whose inverse transform we want is rational in s. We compute a partial fraction expansion of each entry to obtain a sum of elementary terms whose inverse transforms we can write by inspection.

Natural Response The transform-domain expression for the natural response, from (14.21), is $(sI - A)^{-1}x(0)$. We start with an example, to remind you of what is involved in computing the inverse transform of an expression such as this.

Example 14.5 Natural Response of the Linearized Averaged Model of a Buck/Boost Converter

Let us continue with the circuit of Example 14.4. Suppose that we want to determine the output response $\widetilde{v}_o(t) = \widetilde{v}_C(t)$, assuming an initial deviation $\widetilde{x}(0)$ from the nominal steady state but with $\widetilde{d} = 0$ and $\widetilde{v}_{in} = 0$. This formulation could be used, for example, to analyze how the circuit approaches the steady state corresponding to a new operating condition, if d and $\overline{v}_{in}$ have been set at their new values but $x(0)$ is still at a value determined by the old operating condition. The open-loop response of the circuit in Example 12.1 to a step in the source voltage could have been analyzed this way, rather than by the approach outlined in Example 13.1 (where we only presented the general form of the response).

From (14.24), we determine that

$$\widetilde{v}_C(s) = \frac{1}{a_D(s)}\left[\widetilde{v}_C(0)s - \widetilde{i}_L(0)\frac{D'}{C}\right]. \tag{14.29}$$

The characteristic polynomial $a_D(s)$ is given in (14.27). We can rewrite it as

$$a_D(s) = \left(s + \frac{1}{2RC}\right)^2 + \omega_D^2, \quad \text{where} \quad \omega_D = \sqrt{\frac{D'^2}{LC} - \frac{1}{4R^2C^2}}, \tag{14.30}$$

which identifies its roots as $\lambda_1 = -(1/2RC) + j\omega_D$, $\lambda_2 = \lambda_1^*$. Now, rewriting (14.29) as

$$\widetilde{v}_C(s) = \frac{1}{a_D(s)}\left[\widetilde{v}_C(0)\left(s + \frac{1}{2RC}\right) - \frac{1}{\omega_D}\left(\widetilde{\imath}_L(0)\frac{D'}{C} + \widetilde{v}_C(0)\frac{1}{2RC}\right)\omega_D\right] \tag{14.31}$$

allows us, using elementary Laplace transform properties and results, to conclude that

$$\widetilde{v}_C(t) = e^{-t/2RC}\left\{\widetilde{v}_C(0)\cos(\omega_D t) - \frac{1}{\omega_D}\left[\widetilde{\imath}_L(0)\frac{D'}{C} + \widetilde{v}_C(0)\frac{1}{2RC}\right]\sin(\omega_D t)\right\}. \tag{14.32}$$

The solution is thus a sinusoidal oscillation at a frequency determined by the imaginary part of the characteristic roots, but exponentially damped at a rate determined by the real part of the characteristic roots. The larger the imaginary part, the higher the frequency of oscillation will be; the more negative the real part, the greater the damping will be.

As suggested at the beginning of this example, we can use (14.32) to explain the waveforms in Fig. 12.3 of Example 12.1, where we obtained the open-loop response to a step in the source voltage. We fix the time origin to be the instant of the step and choose the initial conditions for the perturbed state variables in (14.32) to be the difference between their *actual* values before the step and their *nominal* steady-state values after the step. In other words, we set $\widetilde{x}(0) = x(0) - X_{\text{new}}$, where we use (14.10) to compute the new steady-state vector X_{new}, utilizing the values that apply after the step. Also, the values of D and ω_D in (14.32) must be those after the step.

Figure 14.6 shows the result of computing $\widetilde{v}_C(t)$ this way for the open-loop system with and without feedforward. In each case, we assumed that the circuit was in steady state prior to the step, so that $x(0) = X_{\text{old}}$. For example, in the case with feedforward, we have $\widetilde{v}_C(0) = 0$ and $\widetilde{\imath}_L(0) = I_{L,\text{old}} - I_{L,\text{new}} = (-V_{\text{ref}}/R)[(1/D'_{\text{old}}) - (1/D'_{\text{new}})]$, where D_{new} is the duty ratio needed to maintain the output at -9 V when the input drops to 8 V, so $D_{\text{new}} = 9/(9+8)$. Also, D'_{new} is used in (14.32) to compute ω_D. Note the correspondence with the waveforms in Fig. 12.3.

Since the perturbations in Fig. 14.6 are not small, you might expect discrepancies between the predictions of the nonlinear model and the linearized one. However, as seen in Examples 13.1 and 13.5, the averaged model is linear when the duty ratio is fixed, so the transients in this particular example actually coincide with those predicted by the nonlinear model.

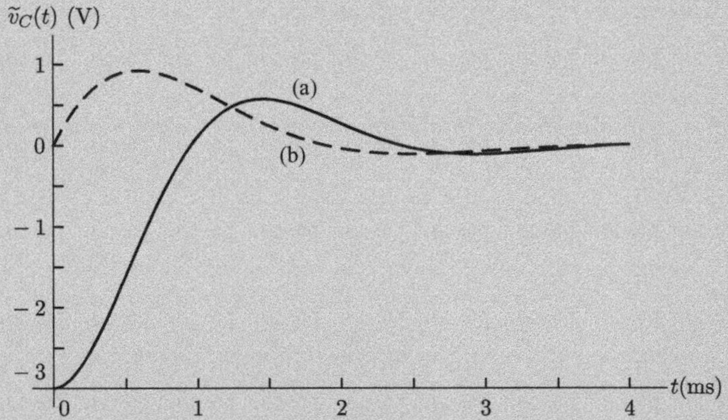

Figure 14.6 Output of the linearized model for two different initial conditions, representing the open-loop response $\widetilde{v}_C(t)$ of the buck/boost converter to a step in source voltage. (a) Without feedforward. (b) With feedforward.

We can generalize the approach used in Examples 14.4 and 14.5 to obtain a time-domain expression for the natural response of the state vector in any LTI system. As the transform-domain expression for the natural response is $(sI - A)^{-1}x(0)$, a useful first step in finding the corresponding time function is to compute the inverse Laplace transform of $(sI-A)^{-1}$, that is, to determine the $n \times n$ matrix whose entries are the inverse Laplace transforms of the corresponding entries of $(sI - A)^{-1}$.

Computation of the inverse transform of an entry of $(sI - A)^{-1}$ is as straightforward as the computations in Example 14.5. Because the denominator of each entry is $a(s) = \det(sI-A)$, inverse transformation yields a sum of exponentials and exponentially weighted sinusoids, respectively corresponding to the real and complex characteristic roots $\{\lambda_i\}$ of the system. The exponential $e^{\lambda_i t}$ decays or grows, depending on whether λ_i is in the left or right half of the complex plane, respectively.

The inverse transform of $(sI - A)^{-1}x(0)$, namely, the natural response, is therefore also a combination of these same exponentials and exponentially weighted sinusoids. The precise combination of terms in the natural response is determined by $x(0)$.

Asymptotic Stability It should be evident now that the natural response decays asymptotically to zero if and only if all the natural frequencies are strictly in the left half of the complex plane. We term the system *asymptotically stable* under this condition, or *exponentially stable*, because the asymptotic decay is bounded by some decaying exponential.

We turn now from the special cases we have been considering, and develop an expression for the general time-domain solution of an LTI system. We also use this solution to describe the general solution of piecewise LTI models in Section 14.5.

Matrix Exponential Because (14.21) shows that $(sI-A)^{-1}$ is the key to representing the general transform-domain solution, we would expect that the inverse transform of $(sI - A)^{-1}$ is the key to the time-domain solution. The inverse transform of the scalar $(s - a)^{-1}$ is the exponential e^{at}, so we refer to the inverse transform of the matrix $(sI - A)^{-1}$ as the *matrix exponential* and denote it by e^{At}:

$$e^{At} = \mathcal{L}^{-1}\{(sI - A)^{-1}\}. \tag{14.33}$$

In connection with our earlier determination of the natural response, we deduced that the entries of this matrix are made up of sums of (scalar) exponentials and exponentially weighted sinusoids, corresponding to the characteristic roots or eigenvalues $\{\lambda_i\}$ of A.

The matrix exponential has properties similar to the scalar exponential. For instance, we can write it as the following infinite series, which converges for all t:

$$e^{At} = I + At + \frac{1}{2!}A^2t^2 + \frac{1}{3!}A^3t^3 + \cdots . \tag{14.34}$$

For small enough t, the first two terms provide a good approximation:

$$e^{At} \approx I + At. \tag{14.35}$$

More precisely, for this to be a good approximation, we require $|\lambda_{max}t| \ll 1$, where λ_{max} is the eigenvalue of maximum absolute value among the eigenvalues of A.

The derivative of the matrix exponential, defined as the matrix whose entries are the derivatives of the corresponding entries of e^{At}, satisfies

$$\frac{de^{At}}{dt} = Ae^{At} = e^{At}A. \tag{14.36}$$

This result can be obtained by differentiating (14.34) term by term, an operation that is justified because the series is sufficiently well behaved. The main difference from the scalar case occurs when two matrix exponentials (of the same dimensions, of course) are multiplied together:

$$e^{A_1t}e^{A_2t} \neq e^{(A_1+A_2)t} \neq e^{A_2t}e^{A_1t}, \tag{14.37}$$

except when $A_1A_2 = A_2A_1$, in which case we have equalities in (14.37).

General Solution The matrix exponential allows us to write an expression for the general solution of an LTI system. Note first that we can now write the natural response in the time domain as

$$\mathcal{L}^{-1}\{(sI - A)^{-1}x(0)\} = e^{At}x(0). \tag{14.38}$$

The transform-domain representation of the forced response – the term $(sI - A)^{-1}Bu(s)$ in (14.21) – is the product of two transforms. Recall that multiplication of transforms corresponds to convolution of the associated time functions, in this case convolution of e^{At} with $Bu(t)$. Superposing the natural response and forced response leads us to the following expression for the complete time-domain equivalent of the frequency-domain solution in (14.21):

$$x(t) = e^{At}x(0) + \int_0^t e^{A(t-\xi)}Bu(\xi)\,d\xi. \tag{14.39}$$

The second term on the right is the forced response. The vector under the integral sign is integrated entry by entry. Also, as we are dealing with matrix products, the order of the factors must be respected. The forced response, like the natural response, contains exponentials at the natural frequencies of the system, but it also contains terms displaying the characteristics of the input. The choice of 0 as the time origin is arbitrary for a time-invariant system, so we can rewrite (14.39) as

$$x(t) = e^{A(t-t_0)}x(t_0) + \int_{t_0}^t e^{A(t-\xi)}Bu(\xi)\,d\xi. \tag{14.40}$$

In Section 13.3.4 we made plausible the state property of a state-space description but did not prove it. For LTI systems, however, the explicit analytical solution in (14.40) is the proof of the state property. The role of the matrix exponential in (14.40) leads to its other name, the *state-transition matrix* of the system.

We have already encountered matrix exponentials acting as state-transition matrices. The "rotation matrix" $\Theta(t - t_0)$ used to analyze the resonant converter of Example 13.9 is the matrix exponential associated with the state-space description of resonant converter dynamics. Also, the matrices $(I + d_kTA_1)$ and $(I + d'_kTA_0)$ that were used to analyze the switched buck/boost converter in Example 13.7 are the appropriate matrix exponentials for that converter model, approximated according to (14.35).

BIBO Stability A notion of stability that is useful when inputs are present is that of *bounded input, bounded output* (BIBO) stability, where "output" broadly means any response of interest. For BIBO stability, the response to any bounded input (i.e., one whose magnitude remains under a fixed finite bound for all time) must also be bounded, *with the system initially at rest*, that is, with $x(0) = 0$. A necessary and sufficient condition is that the transfer function matrix between the input and output has the poles of all its entries strictly in the left half-plane.

We can show from (14.39) that asymptotic stability of the system, which guarantees exponential decay of the zero-input response for arbitrary initial conditions, is sufficient to guarantee BIBO stability. The reason is that the entries of the matrix exponential all decay exponentially under this condition, so the integral that gives the forced response is bounded for any bounded input. Whenever we refer to stability of an LTI system from now on, we mean asymptotic or exponential stability. BIBO stability without asymptotic stability is not enough in practice, because initial conditions cannot be guaranteed to be zero.

Frequency Response The frequency response $G(j\omega)$ of a stable LTI system with transfer function $G(s)$ determines how the system responds to sinusoidal inputs or, more generally, to combinations of sinusoids. This response is important for two reasons. First, reference signals, persistent disturbances, and measurement noise can often be represented as combinations of sinusoids (though we may require Fourier transforms or power spectral densities to determine the combinations). Second, sinusoidal signals in LTI systems represent the boundary between stability and instability, so the frequency response of the loop transfer function – the *loop gain* – is sufficient to determine whether a closed-loop system is stable or unstable.

Suppose that the input to a single-input, single-output (SISO, or scalar) system with transfer function $G(s)$ is a sinusoid at a frequency ω_0, so that

$$u(t) = u_0 \sin(\omega_0 t + \theta) \tag{14.41}$$

for some constants u_0 and θ. Carrying out a partial fraction expansion of $y(s) = G(s)u(s)$ and inverse transforming the result, we can show that the response at the output contains a sinusoid at the same frequency. The amplitude of the output sinusoid is $|G(j\omega_0)|$ times that of the input sinusoid, and its phase is retarded by $\angle G(j\omega_0)$ relative to that of the input:

$$y(t) = |G(j\omega_0)|u_0 \sin(\omega_0 t + \theta + \angle G(j\omega_0)) + \cdots . \tag{14.42}$$

The terms that are represented by $\cdots$ correspond to exponentials at the natural frequencies of the system. For an asymptotically stable system, these terms die out, and only the sinusoidal response persists. We can therefore experimentally determine the frequency response on the basis of (14.42). The magnitude is determined by the amplification of the sinusoid between the input and the output, and the angle is determined by the phase retardation.

We can extend the preceding interpretation of the frequency response to the transfer function matrix of a multi-input, multi-output (MIMO, or multi-variable) system, by using it for each entry of the transfer matrix.

Control design for MIMO systems on the basis of frequency-response information is significantly more intricate than the corresponding classical approach to SISO systems, and is beyond the scope of this book. We thus focus on frequency responses associated with scalar transfer functions.

14.5 Piecewise LTI Models

With the expression (14.40) in hand for the general solution of a continuous-time LTI system, we are able to obtain useful analytical descriptions for the most common circuit models used in power electronics, namely, interconnections of LTI components and ideal switches. If the switch positions are frozen in any given configuration, the circuit model is LTI. Each change in switch condition (between open and closed) leads to a new configuration governed by a new LTI model. We have already encountered several instances of such *piecewise LTI* models.

For a piecewise LTI system, an LTI description can be associated with each switch configuration or topological state, so we can use expressions similar to (14.40) to solve for the behavior of the state variables in each of the circuit's successive configurations. We can then piece together these solutions from configuration to configuration by invoking the continuity of state variables. The final state in one configuration becomes the initial state for the next configuration. This is similar to what was done in Examples 13.8 and 13.9, and leads us naturally to discrete-time generalized state-space models.

Suppose that our piecewise LTI circuit is operated cyclically, taking a succession of N configurations in each cycle, with the kth cycle extending from time t_k to t_{k+1}. We number the configurations $0, 1, \ldots, N-1$ and let the state vector of the circuit be governed by the LTI description

$$\frac{dx}{dt} = A_\ell x(t) + B_\ell u(t) \tag{14.43}$$

in configuration $\ell(= 0, 1, \ldots, N-1)$. (It is sometimes useful or necessary to consider more general cases – for example, where the number and identity of the state variables can change from configuration to configuration or where the number of configurations can vary from cycle to cycle. The underlying ideas remain the same even in such cases.)

The constant matrices A_ℓ and B_ℓ in (14.43) are determined by the circuit topology and parameter values in configuration ℓ. As in the discussion preceding the sampled-data model (13.41), we assume that the input vector $u(t)$ in this configuration of the kth cycle is completely specified by a vector of determining variables, denoted by $p_\ell[k]$. Now applying (14.40) to the model (14.43), we can solve for the state at the end of configuration ℓ in terms of the state at the beginning of this configuration and the inputs acting during the intervening interval.

The transition times between configurations are of two types. Some transition times correspond to *directly controlled* changes in the conducting states of switches. Examples of this are transistors turning on and off or (forward-biased) thyristors turning on, in response exclusively to external control signals. These directly controlled transition times are determining variables, just as the $p_\ell[k]$ are. The remaining transition times correspond to *indirectly controlled*, state-dependent changes in switch status. Examples of this are diodes turning on when the reverse bias on them falls to zero, diodes and thyristors turning off when the current through them falls to zero, or controlled switches operating under the action of state-dependent feedback. These indirectly controlled transition times fall in the category of auxiliary variables, of the sort referred to in connection with the generalized state-space model (13.49).

We now combine the expressions obtained for the configurations from $\ell = 0$ to $\ell = N - 1$, equating the final state of one configuration to the initial state of the next, to obtain the desired cycle-to-cycle description. The result takes the form of the first equation in the generalized state-space description (13.49), if $x(t_k)$ is denoted by $x[k]$:

$$x[k+1] = \phi\left(x[k], p[k], w[k]\right) . \tag{14.44}$$

The auxiliary vector $w[k]$ comprises all the indirectly controlled transition times. We obtain the input vector $p[k]$ by stacking: (i) all the determining vectors $p_\ell[k]$; (ii) all the directly controlled transition times; and (iii) any remaining determining variables, such as those that specify the external reference signals applied to a feedback controller. (Determining variables in the third category actually do not appear in (14.44), but will appear in the constraint equations that determine the indirectly controlled transition times, so we include them in the definition of $p[k]$ to keep the notation manageable.)

We now identify the constraint equations that determine $w[k]$. For an indirectly controlled transition time, some state-dependent signal in the present configuration reaches a threshold level at some particular time. By equating the signal to the threshold, we obtain a constraint equation that involves the state at that transition time and the determining variables that govern that configuration.

Combining all such constraint equations for the indirectly controlled transition times, we obtain a set of constraints of the form given in connection with the generalized state-space description (13.49):

$$0 = \sigma\left(x[k], p[k], w[k]\right) . \tag{14.45}$$

We can also associate a set of output equations of the form (13.42) with the generalized state-space model (14.44) and (14.45). The variables of interest may be sampled values of components of the state vector, averages computed over the cycle, or harmonic components over the cycle. We can represent these and many other cases in the form

$$y[k] = \gamma\left(x[k], p[k], w[k]\right) . \tag{14.46}$$

It is beyond our scope here to discuss the issues involved in the numerical solution or simulation of piecewise LTI models. We simply note that such simulation involves two tasks: (i) simulating the LTI model that governs any given switch configuration; and (ii) recognizing the event that signals the transition from one configuration to the next.

14.6 Linearizing Discrete-Time Generalized State-Space Models

In Section 13.5, as well as Examples 13.8 and 13.9, we showed how nonlinear, time-invariant, sampled-data models in generalized state-space form arise when we describe the cycle-to-cycle behavior of piecewise LTI models for power electronic circuits. If the steady-state behavior of the power circuit is periodic, which is almost always the case, the steady state of the corresponding sampled-data model is constant. Once we know the constant steady state of the sampled-data model, we can find the periodic solution of the piecewise LTI model directly, if desired.

In this section we show that linearization of a nonlinear generalized state-space model yields a linear model in ordinary state-space form. Doing the linearization of a time-invariant sampled-data model at a constant steady state results in an LTI sampled-data model. Such an LTI discrete-time model is, explicitly or implicitly, the basis for most careful small-signal stability studies in power electronics. We discuss the analysis of stability and transfer function properties for discrete-time LTI models in Section 14.7.

Our starting point is the discrete-time generalized state-space description in (14.44), (14.45), and (14.46). We can easily specialize the results to ordinary discrete-time state-space descriptions by setting $w[k]$ and $\sigma(\cdot)$ to zero. The linearization procedure parallels that for continuous-time state-space systems in Section 14.3, except that now we also have the constraint equations associated with a generalized state-space description.

As in the continuous-time case, we use capital letters to denote the nominal values of the variables in (14.44)–(14.46). Thus, in the nominal operating condition, $x[k] = X[k], p[k] = P[k]$, $w[k] = W[k]$, and $y[k] = Y[k]$. The nominal values can be any values that together satisfy the equations, though the most important case for us is when the nominal solutions are constant. We now consider deviations from nominal and mark these quantities with a superscript $\widetilde{\ }$, so that away from the nominal solution we have

$$x[k] = X[k]+\widetilde{x}[k], \quad p[k] = P[k]+\widetilde{p}[k], \quad w[k] = W[k]+\widetilde{w}[k], \quad y[k] = Y[k]+\widetilde{y}[k]. \tag{14.47}$$

If the deviations are small, we deduce from (14.44)–(14.46) that, to a first-order approximation, the deviations satisfy

$$\begin{aligned} \widetilde{x}[k+1] &\approx \frac{\partial\phi}{\partial x}\,\widetilde{x}[k] + \frac{\partial\phi}{\partial p}\,\widetilde{p}[k] + \frac{\partial\phi}{\partial w}\,\widetilde{w}[k], \\ 0 &\approx \frac{\partial\sigma}{\partial x}\,\widetilde{x}[k] + \frac{\partial\sigma}{\partial p}\,\widetilde{p}[k] + \frac{\partial\sigma}{\partial w}\,\widetilde{w}[k], \\ \widetilde{y}[k] &\approx \frac{\partial\gamma}{\partial x}\,\widetilde{x}[k] + \frac{\partial\gamma}{\partial p}\,\widetilde{p}[k] + \frac{\partial\gamma}{\partial w}\,\widetilde{w}[k]. \end{aligned} \tag{14.48}$$

These expressions result from substituting (14.47) into (14.44)–(14.46), and then expanding $\phi(\cdot)$, $\sigma(\cdot)$, and $\gamma(\cdot)$ in multi-variable Taylor series, up to and including linear terms. The partial derivative symbols denote Jacobian matrices. In general, these partial derivatives are all functions of $x[k]$, $p[k]$, and $w[k]$, and must all be evaluated at the nominal solution, with $X[k]$, $P[k]$, and $W[k]$ respectively, when they are used in (14.48).

We can now explicitly solve the constraint expression in (14.48) for the auxiliary variables $\widetilde{w}[k]$, provided the matrix $\partial\sigma/\partial w$ is invertible (which it will be, if the classification of variables into state variables, input or determining variables, and auxiliary variables has been done correctly):

$$\widetilde{w}[k] \approx -\left[\frac{\partial \sigma}{\partial w}\right]^{-1}\left[\frac{\partial \sigma}{\partial x}\widetilde{x}[k] + \frac{\partial \sigma}{\partial p}\widetilde{p}[k]\right]. \tag{14.49}$$

Using (14.49) to eliminate the auxiliary variables in (14.48), we get a linear description in ordinary state-space form:

$$\begin{aligned}\widetilde{x}[k+1] &\approx \mathcal{A}[k]\widetilde{x}[k] + \mathcal{B}[k]\widetilde{p}[k],\\ \widetilde{y}[k] &\approx \mathcal{E}[k]\widetilde{x}[k] + \mathcal{F}[k]\widetilde{p}[k],\end{aligned} \tag{14.50}$$

where

$$\begin{aligned}\mathcal{A}[k] &= \frac{\partial \phi}{\partial x} - \frac{\partial \phi}{\partial w}\left[\frac{\partial \sigma}{\partial w}\right]^{-1}\frac{\partial \sigma}{\partial x}, \qquad & \mathcal{B}[k] &= \frac{\partial \phi}{\partial p} - \frac{\partial \phi}{\partial w}\left[\frac{\partial \sigma}{\partial w}\right]^{-1}\frac{\partial \sigma}{\partial p},\\ \mathcal{E}[k] &= \frac{\partial \gamma}{\partial x} - \frac{\partial \gamma}{\partial w}\left[\frac{\partial \sigma}{\partial w}\right]^{-1}\frac{\partial \sigma}{\partial x}, \qquad & \mathcal{F}[k] &= \frac{\partial \gamma}{\partial p} - \frac{\partial \gamma}{\partial w}\left[\frac{\partial \sigma}{\partial w}\right]^{-1}\frac{\partial \sigma}{\partial p}.\end{aligned} \tag{14.51}$$

If the original description was in ordinary state-space form rather than generalized form, the auxiliary variables and constraints would be absent, so $\mathcal{A}$, $\mathcal{B}$, $\mathcal{E}$, and $\mathcal{F}$ would simply be given by $\partial\phi/\partial x$, $\partial\phi/\partial p$, $\partial\gamma/\partial x$, and $\partial\gamma/\partial p$, respectively.

If the original nonlinear model is time invariant, so that $\phi(\cdot)$, $\sigma(\cdot)$, and $\gamma(\cdot)$ are not explicitly dependent on the time index k, and if in addition the nominal solution is constant, all the partial derivatives that define the linearized model will be constant. The linearized model is then LTI – the most important case in power electronics.

Computing a Constant Nominal Solution Suppose that the nominal solution of the sampled-data model is constant: $X[k] = X$, $P[k] = P$, and $W[k] = W$. To compute this constant nominal solution, we need to solve a time-invariant system of equations of the form (14.44) and (14.45), with these constant values substituted into them:

$$\begin{aligned}X &= \phi(X, P, W),\\ 0 &= \sigma(X, P, W).\end{aligned} \tag{14.52}$$

Here, P is known, as it is the specified vector of determining variables or inputs for the steady state, but X and W have to be found. The equations sometimes have a structure that permits a simple closed-form solution. More generally, however, the system (14.52) is nonlinear and has to be solved by iterative numerical methods. A standard Newton–Raphson approach assumes we have a good initial approximation of the constant steady-state solution and then uses the corresponding linearized model – specifically, the first two expressions in (14.48), with $\widetilde{p} = 0$ – to improve the approximation. The partial derivatives at each iteration are evaluated at the current estimate of the steady state. (Without a good initial guess, the iteration process may converge to a spurious solution of the nonlinear equations or may fail to converge.)

Example 14.6 Linearized Sampled-Data Models for Switching Converters under Duty-Ratio Control and Current-Mode Control

Let us take as our starting point the approximate sampled-data model obtained in (13.48) of Example 13.7:

$$x[k+1] = \phi\big(x[k], v_{\text{in}}[k], d_k\big). \tag{14.53}$$

Linearizing (14.53), we find

$$\tilde{x}[k+1] \approx \frac{\partial \phi}{\partial x}\tilde{x}[k] + \frac{\partial \phi}{\partial v_{\text{in}}}\tilde{v}_{\text{in}}[k] + \frac{\partial \phi}{\partial d}\tilde{d}_k. \tag{14.54}$$

The partial derivatives must be evaluated at the nominal operating condition. This nominal solution corresponds to the steady state obtained by fixing $d_k = D$, $v_{\text{in}} = V_{\text{in}}$, and having $x[k+1] = x[k] = X$ for all k. The resulting model is

$$x[k+1] \approx (I + TA_{d_k})x[k] + TB_{d_k}v_{\text{in}}[k], \tag{14.55}$$

where $A_{d_k} = d_k A_1 + d_k' A_0$ and $B_{d_k} = d_k B_1 + d_k' B_0$, as in (13.47). The corresponding steady state is governed by an equation in the form (14.52), which in this case can be solved explicitly to yield

$$X = \begin{bmatrix} I_L \\ V_C \end{bmatrix} = -A_D^{-1}B_D V_{\text{in}}. \tag{14.56}$$

This happens to be exactly the same equation that we obtained for the steady state of the averaged model in (14.11). Evaluating the derivatives in the linearized model (14.54) at the steady state, we find

$$\frac{\partial \phi}{\partial x} = I + TA_D, \qquad \frac{\partial \phi}{\partial v_{\text{in}}} = TB_D, \qquad \frac{\partial \phi}{\partial d} = T\left[(A_1 - A_0)X + (B_1 - B_0)V_{\text{in}}\right]. \tag{14.57}$$

Substituting these in (14.54), we obtain a linearized sampled-data model for duty-ratio control.

We showed in Example 13.8 that in current-mode control the duty ratio becomes an auxiliary variable, governed by a constraint of the form

$$0 = \sigma\big(x[k], v_{\text{in}}[k], i_P[k], d_k\big). \tag{14.58}$$

For the buck/boost converter, this constraint is

$$0 \approx i_L[k] + \frac{d_k T v_{\text{in}}[k]}{L} - i_P[k] + Sd_kT. \tag{14.59}$$

As noted in Example 13.8, we can solve for d_k from the constraint (14.59) and substitute it into (14.55) to obtain an ordinary state-space model. We can then linearize the resulting model if we so desire. However, we can obtain the same result with less messy computations by retaining the generalized state-space form and using (14.51). We have already done part of the work in obtaining (14.54). Linearizing the constraint, we get

$$0 \approx \frac{\partial \sigma}{\partial x}\tilde{x}[k] + \frac{\partial \sigma}{\partial v_{\text{in}}}\tilde{v}_{\text{in}}[k] + \frac{\partial \sigma}{\partial i_P}\tilde{i}_P[k] + \frac{\partial \sigma}{\partial d}\tilde{d}_k. \tag{14.60}$$

For the buck/boost converter, we find from (14.59) that

$$\frac{\partial \sigma}{\partial x} = [\ 1 \quad 0\ ], \qquad \frac{\partial \sigma}{\partial v_{\text{in}}} = \frac{DT}{L}, \qquad \frac{\partial \sigma}{\partial i_P} = -1, \qquad \frac{\partial \sigma}{\partial d} = T\left(S + \frac{V_{\text{in}}}{L}\right). \tag{14.61}$$

Using these expressions to solve (14.60) for $\tilde{d}_k$ and substituting into (14.55) results in the desired linearized sampled-data model for current-mode control. We defer the detailed expression to Example 14.8, where we use the linearized model to examine the stability of the nominal operating condition. Note that we have written the partial derivatives in (14.57) and (14.61) in terms of D for simplicity, but you can easily rewrite them in terms of the nominal value I_P of the commanded peak current.

A continuous-time parallel of the modeling results we have obtained for current-mode control can also be worked out, combining the averaged model in (14.10) of Example 14.3 with an appropriate constraint equation.

14.7 Analysis of Discrete-Time LTI Models

We have shown in the preceding section that discrete-time LTI models commonly arise in power electronics as descriptions of perturbations from cyclic steady state, when we examine samples of circuit variables taken once per cycle. Such models constitute the main route to analyzing the stability of cyclic nominal operation in power circuits. They are also the basis for designing discrete-time or digital control systems.

In this section we outline some of the basic tools and results needed to treat discrete-time LTI systems. The focus of our discussion is a model in the form of (14.50), but for the time-invariant case, hence

$$\begin{aligned} x[k+1] &= \mathcal{A}x[k] + \mathcal{B}p[k], \\ y[k] &= \mathcal{E}x[k] + \mathcal{F}p[k]. \end{aligned} \tag{14.62}$$

As we have seen, linearizing a nonlinear time-invariant discrete-time model around a constant operating point yields such a model.

14.7.1 Time-Domain Solution and Asymptotic Stability

Unlike the continuous-time case, the time-domain solution for discrete time is easy to describe. For the initial condition $x[0]$ and input $p[0]$, we can use (14.62) to determine $x[1]$ and $y[0]$. Knowing $x[1]$ and $p[1]$, we can similarly determine $x[2]$ and $y[1]$. Clearly, we can continue this iteration indefinitely, yielding

$$x[k] = \mathcal{A}^k x[0] + \sum_{i=0}^{k-1} \mathcal{A}^{k-1-i}\mathcal{B}p[i], \tag{14.63}$$

with a corresponding expression for the output. Compare this solution to the one in (14.39) for a continuous-time system.

The first term on the right in (14.63) constitutes the response to initial conditions alone, in the absence of external inputs, and is the natural response or free response or zero-input response. The second term is the forced response or zero-state response. Asymptotic stability of the natural response is evidently equivalent to having the entries of $\mathcal{A}^k$ decay asymptotically to zero as k goes to infinity.

As noted immediatedly following (14.33), the entries of the matrix exponential e^{At} comprise linear combinations of exponentials of the form $e^{\lambda_i t}$, where each λ_i is an eigenvalue of the matrix A, that is, a root of the characteristic polynomial $a(s) = \det(sI - A)$. Similarly, the entries of the matrix power A^k comprise linear combinations of geometric series (or "discrete-time exponentials") of the form λ_i^k, where each λ_i is an eigenvalue of the matrix A. For this discrete-time setting, the eigenvalues or natural frequencies may also be referred to as *natural ratios*.

In the continuous-time case, we arrived at insights into the matrix exponential via the Laplace transform. In the discrete-time case, a corresponding tool for understanding A^k is the $\mathcal{Z}$ transform. For a sequence $x[k]$, this is denoted by $x(z)$ or $\mathcal{Z}\{x[k]\}$ and defined by

$$x(z) = \mathcal{Z}\{x[k]\} = \sum_{k=0}^{\infty} x[k]z^{-k}. \tag{14.64}$$

For instance, the scalar geometric sequence $x[k] = a^k$, with a (possibly complex) geometric ratio a, has the transform $x(z) = 1 + az^{-1} + a^2z^{-2} + \cdots$, which can be condensed to the rational expression $x(z) = z/(z-a)$ for $|z| > |a|$, with a pole at the geometric ratio a. The definition of the transform applies to vector or matrix sequences too.

Asymptotic Stability From (14.63) and our observation relating the entries of A^k to the eigenvalues of A, it follows that the natural response of the system (i.e., the response to initial conditions, with the input identically at zero) will decay asymptotically to zero if and only if all the eigenvalues or natural ratios have magnitudes less than one, that is, lie strictly inside the unit circle in the complex plane. This condition is therefore necessary and sufficient for asymptotic stability of a discrete-time LTI system.

We can also show from (14.63) that the *forced* response will display discrete-time exponentials at the natural ratios, in addition to terms that reflect the properties of the input. As with continuous-time systems, asymptotic stability of the natural response turns out to be sufficient to guarantee a bounded response to bounded inputs.

Example 14.7 Stability of a Sampled-Data Model for a Buck/Boost Converter under Duty-Ratio Control

We start by computing the characteristic polynomial of the LTI sampled-data model that we obtained in (14.54) and (14.57) of Example 14.6 for the case of duty-ratio control of a buck/boost converter. We need to find the characteristic polynomial of the matrix $\partial\phi/\partial x = I + TA_D$ in (14.57). Using the expression for A_D from (14.10), we obtain for a buck/boost converter the characteristic polynomial

$$\det\begin{bmatrix} z-1 & -TD'/L \\ TD'/C & z-1+T/(RC) \end{bmatrix} = (z-1)\left(z-1+\frac{T}{RC}\right) + \frac{T^2D'^2}{LC}. \tag{14.65}$$

This characteristic polynomial has a simple relation to that of the averaged model in Example 14.4, namely, $\det(sI - A_D)$. Note that

$$\det(zI - I - TA_D) = T^2 \det\left(\frac{z-1}{T}I - A_D\right), \tag{14.66}$$

so substituting $(z-1)/T$ for s in the characteristic polynomial of the averaged model yields (apart from the constant factor T^2) the characteristic polynomial of the sampled-data model. Thus, λ_1 is a characteristic root of A_D if and only if $1 + T\lambda_1$ is a characteristic root of $I + TA_D$. While the natural response of the sampled-data model contains a geometric series of the form $(1 + T\lambda_1)^k$, that of the averaged model contains a continuous-time exponential of the form $e^{\lambda_1 t}$, whose values at the sampling instants are $(e^{\lambda_1 T})^k$. For small enough T, $e^{\lambda_1 T} \approx 1 + \lambda_1 T$, and the predictions of the two models are comparable. We noted in Example 14.4 that the characteristic roots of the averaged model for the parameter values in Example 12.1 constitute a lightly damped complex pair. You should verify that the corresponding characteristic roots of the sampled-data model form a complex conjugate pair within, and close to, the unit circle.

We can take our stability analysis one step further, treating the case of current-mode control in a buck/boost converter.

Example 14.8 Stability of a Sampled-Data Model for a Buck/Boost Converter under Current-Mode Control

Turning to the sampled-data model for current-mode control in Example 14.6, we need to find the characteristic polynomial of the matrix

$$\mathcal{A} = \frac{\partial \phi}{\partial x} - \frac{\partial \phi}{\partial d}\left[\frac{\partial \sigma}{\partial d}\right]^{-1}\frac{\partial \sigma}{\partial x} \tag{14.67}$$

in (14.51), where the partial derivatives are given by (14.57) and (14.61). If we denote $(LS + V_{\text{in}})^{-1}$ by μ to simplify the notation, we find that

$$\mathcal{A} = \begin{bmatrix} 1 - \mu(V_{\text{in}} - V_C) & TD'/L \\ -TD'/C - \mu(I_L L/C) & 1 - (T/RC) \end{bmatrix}, \tag{14.68}$$

from which finding $\det(zI - \mathcal{A})$ is straightforward.

If there is no stabilizing ramp, then $S = 0$ and $\mu = 1/V_{\text{in}}$. Using the steady-state relationships in (14.56), we find in this special case that $\mathcal{A}$ simplifies to

$$\mathcal{A} = \begin{bmatrix} -D/D' & TD'/L \\ -(TD'/C) - (DL/RCD'^2) & 1 - (T/RC) \end{bmatrix}. \tag{14.69}$$

The corresponding characteristic polynomial is

$$\det(zI - \mathcal{A}) = \left(z + \frac{D}{D'}\right)\left(z - 1 + \frac{T}{RC}\right) + \frac{TD}{RCD'} + \frac{T^2 D'^2}{LC}. \tag{14.70}$$

For small T, one root of the polynomial is close to (and slightly smaller than) $1 - (T/RC)$, and the other is close to (and slightly greater than) $-D/D'$. This result leads us to predict that the circuit will be unstable for $D > 0.5$, as one of the roots will then fall out of the unit circle, becoming more negative than -1.

We gain a more detailed insight into the dynamics of the circuit by examining $\mathcal{A}$ in (14.69) under the assumption that T is negligibly small. Recalling that this $\mathcal{A}$ governs the behavior of perturbations away from nominal in the absence of control, (14.69) shows that with $S = 0$ and $T \approx 0$ we have

$$\begin{aligned} \widetilde{i}_L[k+1] &\approx (-D/D')\widetilde{i}_L[k], \\ \widetilde{v}_C[k+1] &\approx \widetilde{v}_C[k] - (DL/RCD'^2)\widetilde{i}_L[k]. \end{aligned} \tag{14.71}$$

The first expression shows that the inductor current perturbations form a discrete-time exponential of ratio $-D/D'$, alternating in sign from one sample to the next, so the perturbations vary at one half the switching frequency. For $D < 0.5$, the magnitude of the inductor current perturbations decays exponentially. For $D > 0.5$, the magnitude of the perturbations grows, soon becoming large enough to fall beyond the purview of the linearized model (but limited in any case by the current threshold). This is an example of *ripple instability*. Simple calculations with the approximate inductor current waveforms in Fig. 14.7(a) and (b) confirm these predictions regarding stability of the perturbations.

The second expression in (14.71) shows that the capacitor voltage perturbations stay essentially constant, apart from a forcing term involving $\widetilde{i}_L[k]$. This term varies at one half the switching frequency and, for $D < 0.5$, has an exponentially decaying magnitude.

Evidently, steady-state operation with period T is impossible for $S = 0$ and $D > 0.5$, and the waveforms assume more complicated forms, corresponding either to periodic operation at some multiple of T (subharmonic operation) or to "chaotic" variation from cycle to cycle. These possibilities are represented in Fig. 14.7(c) and (d), respectively. Although the dc/dc conversion function is basically unaffected – the duty ratio on a longer time scale takes the necessary value – such operation is generally undesirable because of problems including filtering the switching ripple.

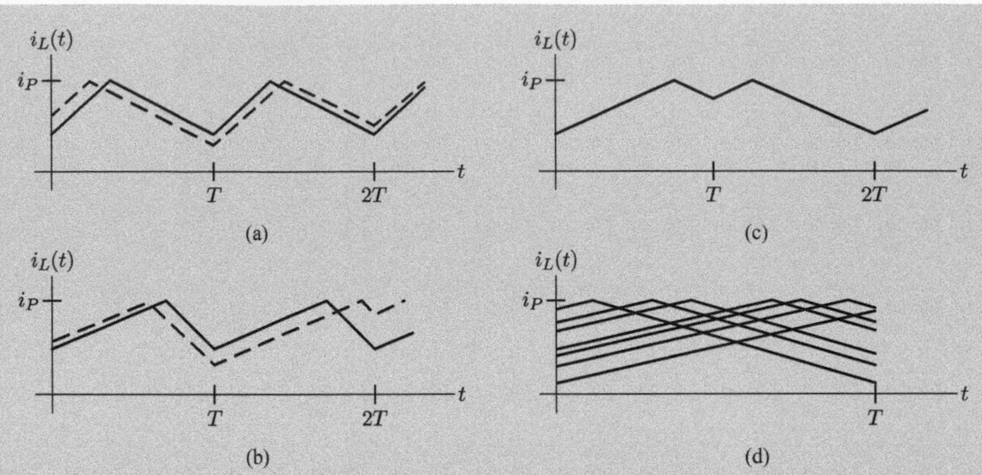

Figure 14.7 Inductor current waveforms with no stabilizing ramp. (a) $D < 0.5$. (b) $D > 0.5$. (c) Subharmonic operation for $D > 0.5$, period $2T$. (d) Several cycles superimposed to show "chaotic" operation for $D > 0.5$.

The stabilizing ramp in current-mode control is introduced precisely in order to overcome the stability limit on the allowable duty ratio. This sawtooth waveform *entrains* the converter waveforms, maintaining them at the switching frequency when they would otherwise display subharmonic or aperiodic behavior. We leave you to explore the effect of $S > 0$ on the characteristic roots of $\mathcal{A}$, and to obtain an approximate analysis based on simple calculations with the inductor current waveform in Fig. 13.10, Example 13.8.

14.7.2 Transfer Functions and Frequency Response

The transform-domain version of the expression (14.63) for the state of the LTI system in (14.62) is the expression

$$x(z) = (zI - \mathcal{A})^{-1} z x[0] + (zI - \mathcal{A})^{-1} \mathcal{B} p(z), \tag{14.72}$$

and the transform of the output is

$$y(z) = \mathcal{E} x(z) + \mathcal{F} p(z). \tag{14.73}$$

Transfer functions are computed by setting the initial conditions to 0 in the $\mathcal{Z}$ transform relationships. Using (14.72), we see that the transfer function matrix from p to x is $(zI - \mathcal{A})^{-1}\mathcal{B}$, and the transfer function matrix from p to y is

$$\mathcal{G}(z) = \mathcal{E}(zI - \mathcal{A})^{-1}\mathcal{B} + \mathcal{F}. \tag{14.74}$$

The transfer function is especially important for its role in defining the frequency response of a sampled-data system. The key result here is the following discrete-time version of (14.41) and (14.42).

If the input to a SISO discrete-time LTI system is

$$p[k] = p_0 \sin(\omega_0 kT + \theta), \tag{14.75}$$

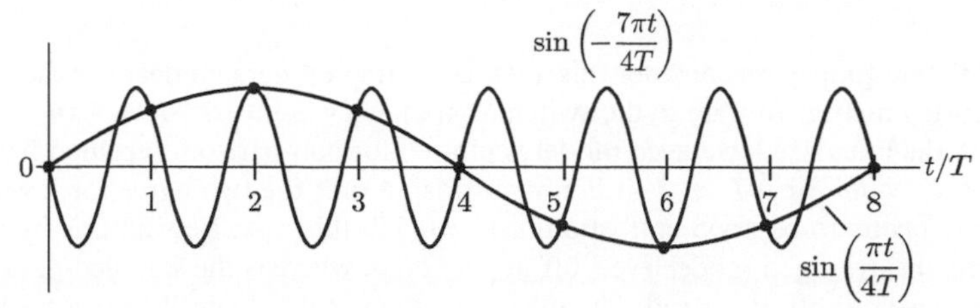

Figure 14.8 Aliasing of sampled sinusoids.

the output is

$$y[k] = \left|\mathcal{G}\left(e^{j\omega_0 T}\right)\right| p_0 \sin\left[\omega_0 kT + \theta + \angle\mathcal{G}(e^{j\omega_0 T})\right] + \cdots . \tag{14.76}$$

We term $\mathcal{G}(e^{j\omega T})$ the *frequency response* of the sampled-data system. The result in (14.76) shows its role in defining the response to samples of a sinusoid at any given frequency.

The frequency response of a sampled-data LTI system differs in an important way from that of a continuous-time system: it is a *periodic* function of ω, with period $2\pi/T$, because $e^{j\omega T} = e^{j(\omega+2\pi/T)T}$ for all ω. This difference reflects the fact that the samples of a sinusoid at the frequency ω, taken regularly at intervals of T, coincide with the samples of a sinusoid at a frequency displaced from ω by any integer multiple of $2\pi/T$; see Fig. 14.8. We refer to this phenomenon as *aliasing* and say that these sinusoids are *aliases* of each other. Hence, $y[k]$ in (14.76) is unchanged if ω_0 is increased by any integer multiple of $2\pi/T$.

Because $\mathcal{G}(z)$ has real coefficients, $\mathcal{G}(e^{-j\omega T})$ is the complex conjugate of $\mathcal{G}(e^{j\omega T})$. We therefore only need to know the frequency response over the interval $0 \le \omega \le \pi/T$. The response for $-\pi/T \le \omega \le 0$ can be obtained by conjugation, and the response for any other frequency can be obtained by invoking the periodicity of the frequency response. Hence, the highest frequency of interest when we describe or probe a sampled-data system is $\omega_{\max} = \pi/T$. The same limit is suggested by the fact that the fastest possible variation in the samples occurs when their signs alternate from cycle to cycle, corresponding to a period of $2T$, or a frequency of $2\pi(1/2T) = \pi/T$. This frequency is called the *Nyquist frequency*.

Example 14.9 Frequency Response of a Sampled-Data Model for a Buck/Boost Converter under Duty-Ratio Control

We return to the approximate linearized sampled-data model for a buck/boost converter under duty-ratio control obtained in (14.54) and (14.57) of Example 14.6. The transfer function from duty-ratio perturbations $\tilde{d}$ to perturbations $\tilde{v}_C$ in the sampled capacitor voltage is

$$\mathcal{G}(z) = \begin{bmatrix} 0 & 1 \end{bmatrix} \left(zI - \frac{\partial \phi}{\partial x}\right)^{-1} \frac{\partial \phi}{\partial d}, \tag{14.77}$$

where the partial derivatives are as given in (14.57). However, there is a shortcut to the answer. In Example 14.7 we related the characteristic polynomials of the linearized sampled-data model and the linearized averaged model in (14.12) of Example 14.4. Exploiting this again, we obtain the desired transfer function simply by replacing s with $(z-1)/T$ in the transfer function (14.28).

We find the frequency response for this linearized sampled-data model by replacing z by $e^{j\omega T}$ in $\mathcal{G}(z)$. The result is plotted in Fig. 14.9 up to the switching frequency $\omega_S = 2\pi/T = 3.14 \times 10^5$ rad/s. The frequency response of the linearized averaged model is plotted for comparison, repeated from Fig. 14.5(b). Because $(e^{j\omega T} - 1)/T \approx j\omega$ for $\omega T \ll 1$, it is not surprising that the two agree for low frequencies. In fact, the agreement happens to be good until about $\omega T = 0.5$ in this case. Beyond this frequency, and below $\omega_S/2$, the phase of the averaged model levels off at $-\pi/2$ rad, whereas the sampled-data model continues to add phase delay, leveling off at $-\pi$ rad. The magnitudes agree closely until they come nearer to $\omega_S/2$.

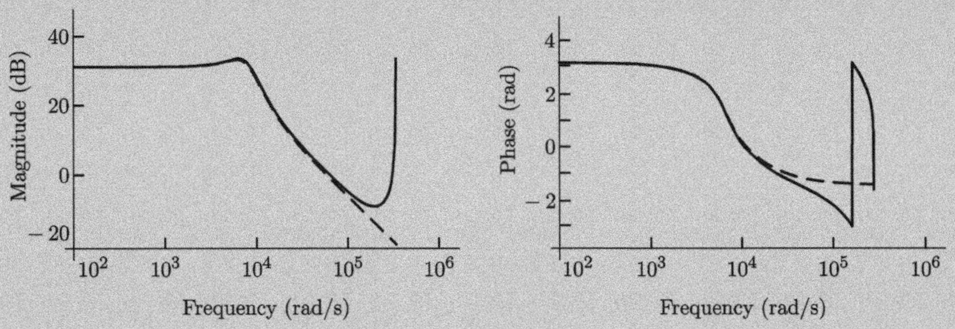

Figure 14.9 Frequency response of the linearized sampled-data model (solid line) and linearized averaged model (dashed line) of the buck/boost converter.

The response of the sampled-data model beyond $\omega_S/2$ is consistent with the symmetry and periodicity properties mentioned earlier. We could compute the response of the averaged model beyond $\omega_S/2$, but the response at these frequencies has no relevance to the actual behavior of the converter, for reasons discussed in Example 12.6 and immediately after.

Recall that the typical control for a switching converter involves a circuit similar to that of Fig. 12.9, Example 12.6, so what is really of interest is the response of the circuit to perturbations $\widetilde{m}(t)$ in the modulating signal $m(t)$. We have already argued that for *slow* variations in $m(t)$, $d(t) \approx m(t)/K$, so we can apply the averaged model. The sampled-data model is not restricted to slow variations in $m(t)$.

To see the effect of variations in $m(t)$ on the response of the linearized sampled-data model, let us suppose that

$$m(t) = K[D + \epsilon \sin(\omega t)], \tag{14.78}$$

where the amplitude ϵ is a small constant. Then $\widetilde{m}(t) = K\epsilon \sin(\omega t)$ and we can easily deduce from Fig. 12.9 that

$$\begin{aligned} \widetilde{d}_k &= \epsilon \sin[\omega(k + D + \widetilde{d}_k)T] \\ &\approx \epsilon \sin(\omega kT + \omega DT), \end{aligned} \tag{14.79}$$

provided $\widetilde{d}_k \ll D$. The phase offset ωDT or time offset DT is a result of the duty ratio in each cycle being determined by the value of $m(t)$ at the instant this signal crosses the sawtooth ramp, rather than at the beginning of the cycle. As the frequency response from $\widetilde{d}$ to $\widetilde{v}_C$ is given by $\mathcal{G}(e^{j\omega T})$, we see that

$$\widetilde{v}_C[k] \approx |\mathcal{G}(e^{j\omega T})|\epsilon \sin[\omega kT + \omega DT + \angle\mathcal{G}(e^{j\omega T})] + \cdots . \tag{14.80}$$

This expression holds even for $\omega > \pi/T$, so long as $\widetilde{d}_k$ is sufficiently small. However, if $\omega > \pi/T$, the sequence $\widetilde{d}_k$ in (14.79) can also be produced by an $\widetilde{m}(t)$ that varies at an alias frequency *below* π/T. The corresponding response mimics the response to the low-frequency alias. For example, varying $\widetilde{m}(t)$ at the switching frequency produces the same effect as a constant $\widetilde{m}(t)$. Hence, experimental frequency-response measurements using a network analyzer can produce misleading results above π/T – the network analyzer looks only at the input-frequency component of the output, but the dominant part of the output response will usually be at some lower frequency. We leave you to explore further the relationships among $m(t)$, d_k, $d(t)$, $\overline{v}_C(t)$, and $v_C[k]$ for variations in $m(t)$ at various frequencies.

The preceding example suggests how information provided by a sampled-data model can be used to describe the behavior of the underlying continuous-time model. You need to keep in mind the relation between the two classes of models whenever you use a sampled-data model.

14.8 Feedback Control Design

The subject of control design is vast. We can do no more in the rest of this chapter than suggest, through examples, how the dynamic models we have developed for power circuits can be mated with the systematic design methods available in the general control literature. Our focus is on continuous-time LTI models, though we shall illustrate the use of discrete-time LTI models as well.

The configuration and notation we assume here are those introduced in Section 12.5 and redrawn more simply in Fig. 14.10. We limit ourselves to the case where both $g(s)$ and $h(s)$ have their poles strictly in the left half-plane, so the loop transfer function $\ell(s) = h(s)g(s)$ is stable, and assume no hidden poles. We showed in Section 12.5 that certain important performance objectives could be translated into requirements on the nominal loop gain $\ell(j\omega) = h(j\omega)g(j\omega)$. These in turn impose requirements on the frequency response $h(j\omega)$ of the compensator for a given nominal plant frequency response $g(j\omega)$. We usually want a high loop-gain magnitude at low frequencies, where the reference signals and disturbances are significant, and a small loop-gain magnitude at high frequencies, where the measurement noise and modeling errors are large. For good transient performance, we want the crossover frequency ω_c (at which the magnitude of the loop gain is 1) to be as high as possible. The control design task in this setting is to choose the compensator so that these requirements are met, while maintaining stability of the closed-loop system.

To proceed further with this formulation, the stability of the closed-loop system must be related to the loop gain.

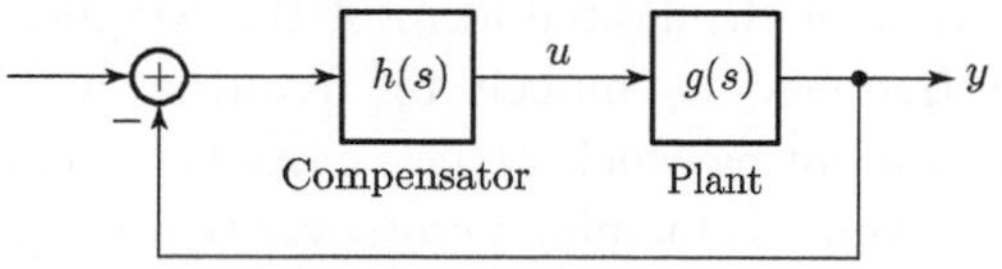

Figure 14.10 Configuration for classical control design.

14.8.1 Nyquist Stability Criterion

The key result here is the Nyquist stability criterion, whose general statement and proof are standard in undergraduate textbooks on control. We briefly state it here for our special case of a stable loop transfer function, but only to convey the flavor of the criterion. The criterion is stated in terms of the *Nyquist plot* of the loop transfer function, which is the trace of $\ell(j\omega)$ in the complex plane as ω ranges up from $-\infty$ to $+\infty$. The Nyquist criterion for this case of stable $\ell(s)$ then tells us that *the closed-loop system is stable if and only if the Nyquist plot has no net encirclements of the point* -1. It is also often useful to work with the *Nyquist half-plot*, which is the trace of $\ell(j\omega)$ as ω ranges up from 0 to $+\infty$. The full Nyquist plot is obtained by combining the half-plot with its reflection in the real axis of the complex plane.

The simplicity and generality of the Nyquist criterion make it an exceptionally powerful tool. An immediate consequence, for example, is the result on robust stability that we quoted in Section 12.5: If the loop gain of a stable closed-loop system is perturbed from $\ell(j\omega)$ to $\ell(j\omega)+\widetilde{\ell}(j\omega)$, while keeping the number of unstable open-loop poles unchanged, and if $|\widetilde{\ell}(j\omega)| < |1+\ell(j\omega)|$, the system remains closed-loop stable. The inequality ensures that the Nyquist plot is not perturbed enough to change the number of encirclements of -1, so stability is preserved.

14.8.2 A Design Approach

To extract some general guidelines for control design, we impose a further restriction: we choose the compensator to give a large and positive loop gain at low frequencies, so $\ell(0) \gg 1$. Note that for any physical system, $\ell(s)$ has more poles than zeros, so that as ω increases from 0 to ∞ the magnitude of the loop gain eventually decreases steadily.

For this case, the point on the Nyquist plot corresponding to $\omega = 0$ lies on the positive real axis of the complex plane. As ω increases from 0 to $+\infty$, the Nyquist half-plot generally has the form of a contracting clockwise spiral, approaching the origin asymptotically at an angle of $-\delta\pi/2$ rad, where δ is the number of poles plus the number of right-half-plane zeros minus the number of left-half-plane zeros. There may be local departures from this general form, with counterclockwise movements in the vicinity of frequencies determined by the locations of the left-half-plane zeros, and expansion rather than contraction in the vicinity of resonances.

A convenient and usually very successful strategy for satisfying the Nyquist criterion in this case is to ensure that, as ω increases from 0 to $+\infty$, the Nyquist half-plot enters the unit circle without encircling the point -1 and is then confined to the unit circle for higher frequencies. This is the case in Fig. 14.11(a). To obtain stability *robustness*, we should keep -1 *well away* from the region enclosed by the Nyquist plot, so that perturbations of $\ell(j\omega)$ caused by modeling errors do not shift the Nyquist plot so much that it encloses -1.

The magnified view of the desired form of the Nyquist plot shown in Fig. 14.11(a) marks the gain crossover frequency ω_c, namely, the frequency where the loop gain magnitude is unity, $|\ell(j\omega_c)| = 1$. The amount by which $\angle\ell(j\omega_c)$ exceeds $-\pi$ rad is referred to as the *phase margin*. Also marked in the figure is the phase crossover frequency ω_o, defined by $\angle\ell(j\omega_o) = -\pi$ rad. The factor $1/|\ell(j\omega_o)|$ by which the loop gain magnitude at ω_o is smaller than unity is termed the *gain margin*. (The figure actually shows $1 - |\ell(j\omega_o)|$.)

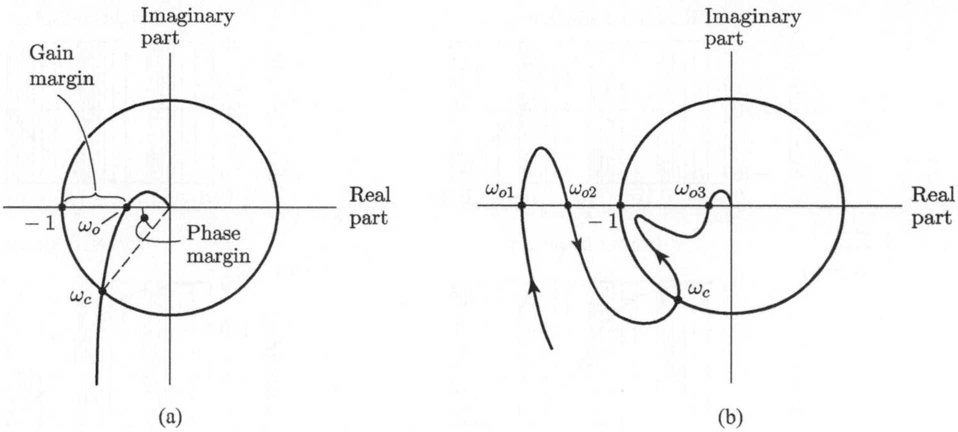

Figure 14.11 (a) Magnified view of the desired behavior of the Nyquist plot near the crossover points. (b) More complicated behavior near crossover.

The gain and phase margins serve as simple measures of the distance of the Nyquist plot from the point -1, and hence as measures of stability robustness. For the plot in Fig. 14.11(a), the phase margin is about $\pi/4$ rad (or 45°), and the gain margin is about 2 (or 6 dB). These margins allow us to tolerate an additional phase delay of $\pi/4$ rad in the loop gain at the gain crossover frequency, or a doubling of the loop gain magnitude at the phase crossover frequency, before the Nyquist plot encloses -1 and the closed-loop system goes unstable. We would consider such margins to be safe if we were fairly sure of our model in the vicinity of the crossover points.

However, the situation may be too complicated to convey with just a pair of crossover points. For example, the Nyquist plot shown in Fig. 14.11(b) has multiple phase crossover frequencies, each of which may be given an interpretation. The phase crossover at ω_{o2}, for instance, indicates that *reducing* the loop gain magnitude by the factor $|\ell(j\omega_{o2})|$ causes the system to go unstable, which shows that we have a *conditionally stable* system. (Transients that produce saturating signals often lead to behavior similar to that encountered by reducing the loop gain in a linear model. This leads us to avoid conditionally stable designs.) In this particular instance, however, even knowing the characteristics of all the crossover frequencies is not enough to indicate that the Nyquist plot goes disastrously close to -1 for a frequency between ω_c and ω_{o3}. Although we can often deduce the main features of a Nyquist plot by examining a few points on the plot, we cannot always do so.

14.8.3 Using Bode Plots

The information in the Nyquist half-plot can be displayed equivalently by a pair of plots showing the magnitude and phase of the loop gain as functions of frequency. As mentioned in Section 12.5.3, these plots are called *Bode plots* when we use log–log scales to display the magnitude and linear–log scales to display the phase. The log-magnitude is traditionally measured in decibels (dB), with the loop gain in dB being $20\log_{10}|\ell(j\omega)|$. The frequency axis is marked off in *decades*, with each decade representing a factor of 10 in frequency.

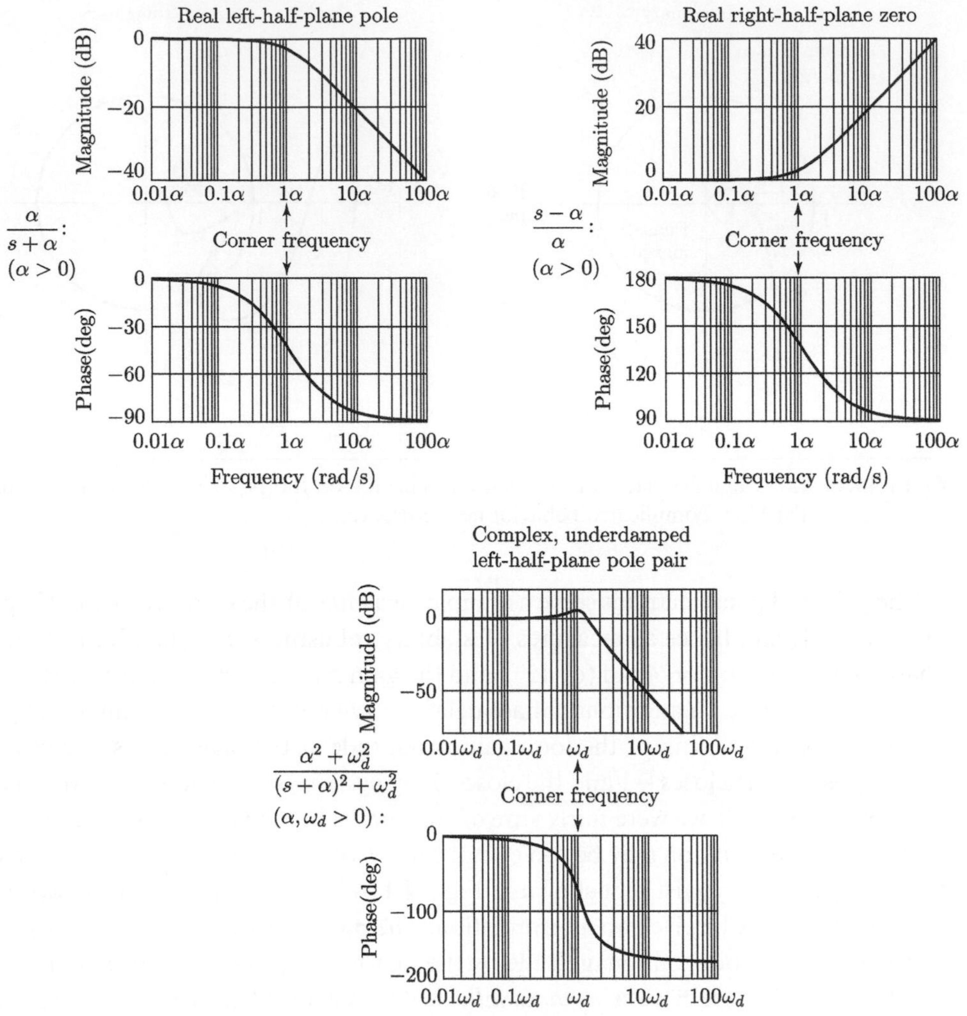

Figure 14.12 Bode plots for frequency responses of some pole and zero factors.

Note that the Bode plots of a *product* of two frequency responses are obtained by *adding* the respective Bode plots of the two factors, because the log-magnitudes and phases of the factors add. For example, as $\log|\ell(j\omega)| = \log|h(j\omega)| + \log|g(j\omega)|$ and $\angle\ell(j\omega) = \angle h(j\omega) + \angle g(j\omega)$, we can easily obtain the Bode plots of the loop gain $\ell(j\omega)$ by adding the corresponding plots of the frequency response $h(j\omega)$ of the compensator to those of the frequency response $g(j\omega)$ of the power circuit. The Bode plots of $h(j\omega)$ or $g(j\omega)$ in turn are the sums of the plots associated with their individual poles and zeros, which happen to be easy to sketch. These characteristics and the connection with the Nyquist plot make Bode plots very useful tools for design exploration.

Figure 14.12 shows the Bode plots for some factors associated with real and complex poles and zeros. The Bode plots of the *reciprocals* of these factors are simply derived from the given plots: they are just the *negatives* of the plots in Fig. 14.12.

Note that the "corner" frequency in each log-magnitude plot in Fig. 14.12 is directly determined by the associated pole or zero location. Except for a small range around the corner frequency, the slope at frequencies below the corner frequency is essentially 0, and above the corner frequency is essentially an integer multiple of 20 dB/decade. This multiple is -1 for each pole and $+1$ for each zero. Hence, for a complex pole or zero pair, the multiple is -2 or $+2$, respectively. The detailed behavior in the vicinity of the corner frequency for a complex pair is determined by the relative damping. The lighter the damping, the sharper the resonance at the corner frequency.

The phase plots also have a simple structure. The phase at low frequencies is essentially 0 rad for a real pole or zero in the left half-plane and π rad for one in the right half-plane. Complex pole or zero pairs have a phase of essentially 0 rad at low frequencies. The phase changes by an integer multiple of $\pi/2$ rad as ω increases past the corner frequency. This integer multiple is -1 for each left-half-plane pole or right-half-plane zero and $+1$ for each left-half-plane zero or right-half-plane pole. The change in phase happens mainly in the decade below and above the corner frequency. For a complex pair, the lighter the damping, the sharper the phase change around the corner frequency.

The Bode plots for a more complicated frequency response can be understood as the sum of the plots for its constituent poles and zeros. The log-magnitude plot has slopes that are essentially integer multiples of 20 dB/decade between the corner frequencies of these poles and zeros, and the phase plot is approximately an integer multiple of $\pi/2$ rad, except for transition regions around the corner frequencies. Of course, when two corner frequencies are less than two decades apart, the Bode plots in the frequency interval between them are dominated by features of the transition.

14.8.4 Designing the Bode Plots of the Loop Gain

The preceding discussion and the study of simple cases such as those in Fig. 14.12 show that, as long as we are away from the corner frequencies, we can approximate the phase of the loop gain fairly well from the slope of its log-magnitude Bode plot and some auxiliary information about right-half-plane zeros. For the case of stable $\ell(s)$ with $\ell(0) > 0$, which is what we are restricting ourselves to, the result is as described next.

Suppose the slope of the log-magnitude plot at ω is $20n(\omega)$ dB/decade, so $n(\omega)$ is approximately an integer in the regions between corner frequencies. Then, if $\ell(s)$ has no right-half-plane zeros with corner frequencies below ω, $\angle\ell(j\omega) \approx n(\omega)\pi/2$ rad; if the slope is approximately -20 dB/decade, the phase is approximately $-\pi/2$ rad, and so on. For the more general case where there are $r(\omega)$ right-half-plane zeros whose corner frequencies are below ω, the result is

$$\angle\ell(j\omega) \approx [n(\omega) - 2r(\omega)]\pi/2 \text{ rad}. \tag{14.81}$$

The approximate result in (14.81) has the virtue of being simple enough to yield simple design guidelines. A candidate compensator obtained by such approximations can be subjected to a more refined analysis on a second pass through the design.

Our design approach in Section 14.8.2 has two immediate consequences. As indicated in Fig. 14.11(a), we require the phase at the gain crossover frequency ω_c to be greater than $-\pi$ rad by an amount equal to the phase margin. Hence (14.81) implies the constraint $n(\omega_c) - 2r(\omega_c) > -2$. At crossover $n(\omega_c)$ must be negative, and if we limit ourselves to crossing over away from corner frequencies, it must be approximately an integer. We must therefore have $r(\omega_c) = 0$ and $n(\omega_c) = -1$. Our conclusions are:

- We must aim for crossover below the corner frequencies of all right-half-plane zeros. Thus, right-half-plane zeros impose a limit on the attainable closed-loop bandwidth. It turns out that the limitations imposed by right-half-plane zeros are fundamental, not a consequence of the particular design approach that we have followed.
- The slope of the log-magnitude of the loop gain at crossover must be approximately −20 dB/decade. This sets a limit on how fast we can roll off the loop gain magnitude from the desired large values at low frequencies to the required low values at high frequencies.

Before proceeding, we simplify our terminology. For the rest of this chapter, whenever we refer to the slope of a log-magnitude plot at ω, we mean $n(\omega)$ itself rather than the slope in dB/decade.

Example 14.10 Compensator Design for a Buck/Boost Converter under Duty-Ratio Control

We showed in Examples 12.1, 12.3, and 14.2 that some simple approaches to controlling a buck/boost converter led to unsatisfactory results. We tried open-loop control and proportional feedback control, with and without feedforward. We now follow the design approach just outlined.

Nominal Model We use the linearized model obtained in Examples 14.1, 14.3, and 14.4. In those examples we derived the nominal transfer function $g(s)$ between duty ratio perturbations and output voltage perturbations. The pole/zero diagram and Bode plots of the associated frequency response were shown in Fig. 14.5, using the parameter values given in Example 12.1. The Bode plots are repeated here in Fig. 14.13(a).

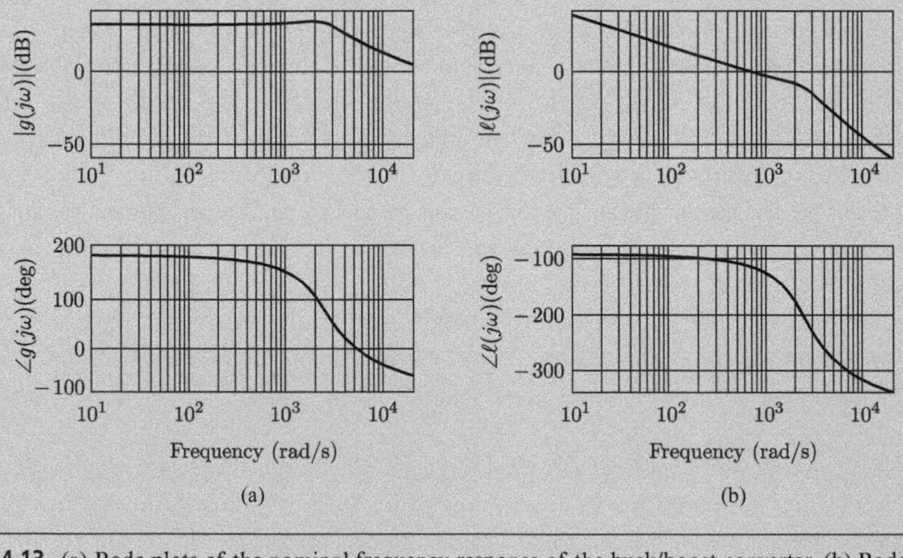

Figure 14.13 (a) Bode plots of the nominal frequency response of the buck/boost converter. (b) Bode plots of the loop gain with integral control.

The resonant peak in the log-magnitude plot marks the corner frequency associated with the complex pole pair. Beyond this, the log-magnitude starts to roll off with a slope of -2, but changes to a slope of -1 after the corner frequency associated with the right-half-plane zero. The phase angle correspondingly decreases by π rad in the vicinity of the resonance and a further $\pi/2$ in the vicinity of the corner frequency of the right-half-plane zero. Note that $\angle g(0) = \pi$ rad, that is, $g(0)$ is negative because of our choice of polarity for the definition of the output voltage v_o. Hence we require that the dc gain $h(0)$ of the compensator be negative in order to get $\ell(0) > 0$.

Compensator Design For the particular parameter values in our example, there is little separation between the corner frequencies of the pole pair and the zero, which makes designing for crossover in this region difficult. Moreover, such a design would be quite sensitive to the detailed features of the Bode plots in this region, and these features (such as the shape of the resonant peak) depend strongly on the component values and duty ratio. Let us therefore be less ambitious and aim for a loop crossover frequency that is below the resonant frequency.

Because the slope of $|g(j\omega)|$ below resonance is 0, the compensator's log-magnitude must roll off with slope -1 at crossover. A simple way to obtain a high loop gain at low frequencies and to roll off with a slope of -1 is to introduce a pole at (or near) the origin into the loop transfer function, so as to obtain (approximate) integral control action. Let us therefore pick $h(s) = -\beta/s$ (or $-\beta/(s+\epsilon)$ for some small $\epsilon > 0$), where β is a positive constant that we choose so as to obtain the desired crossover frequency. The negative sign in $h(s)$ is needed to obtain $h(0) < 0$.

The Bode plots of the resulting loop gain are shown in Fig. 14.13(b). Note that beyond crossover the loop gain magnitude continues to roll off, and its slope settles down to -2 beyond the corner frequency of the right-half-plane zero. To obtain a good gain margin and to limit the effects of modeling errors and switching ripple, we must sustain the roll-off beyond crossover. (A more refined converter model that included an equivalent series resistance or ESR of value R_C in series with the capacitor C would show the presence of a further zero at $-1/R_C C$. To counteract the effect of the ESR zero and maintain the roll-off beyond crossover, we could include an additional pole at approximately $-1/R_C C$ in the compensator.)

The integrator gain β fixes the crossover frequency ω_c. We need to pick ω_c small enough that the resonant peak in the loop gain magnitude is sufficiently below unity gain to give an adequate gain margin. The magnitude of the peak without the integrator is around 31 dB at 2000 rad/s. To obtain a gain margin of 9 dB, for example, we must pick β so that the integrator introduces an attenuation of 40 dB at 2000 rad/s. This requirement leads to the choice $\beta = 20$. The corresponding crossover frequency can be computed to be around 740 rad/s.

Evaluation We need to analyze and simulate several aspects of the performance of the preceding design in order to understand the trade-offs involved and to develop a basis for modifying or accepting the design. We describe some of the ingredients of such an evaluation here.

The block diagram in Fig. 14.14(a) represents the linearized model of the closed-loop converter under integral control. It shows the integral feedback from output voltage perturbations to duty-ratio perturbations, and also the dependence of the output perturbations on perturbations in the supply voltage. We have already derived the various transfer functions shown; see (13.5) and (14.3) in particular. An easy computation shows the closed-loop *audio susceptibility* transfer function to be

$$\frac{\widetilde{v}_o(s)}{\widetilde{v}_{\text{in}}(s)} = \frac{-(D'D/LC)s}{s[s^2 + (1/RC)s + (D'^2/LC)] - \beta(I_L/C)[s - (V_{\text{in}}/LI_L)]}. \tag{14.82}$$

This expression may be used to study the response of the closed-loop converter to small variations in the supply voltage. It generally also provides a reasonable approximation of the response to large variations. Note that – as expected with integral control – the dc value of this transfer function is 0, so the output is insensitive to constant offsets in the input voltage. The frequency response associated with this transfer function is shown in Fig. 14.14(b), and indicates poor rejection of input ripple at frequencies near 1700 rad/s. For example, the attenuation of input ripple at 120 Hz (740 rad/s) is less than 5 dB.

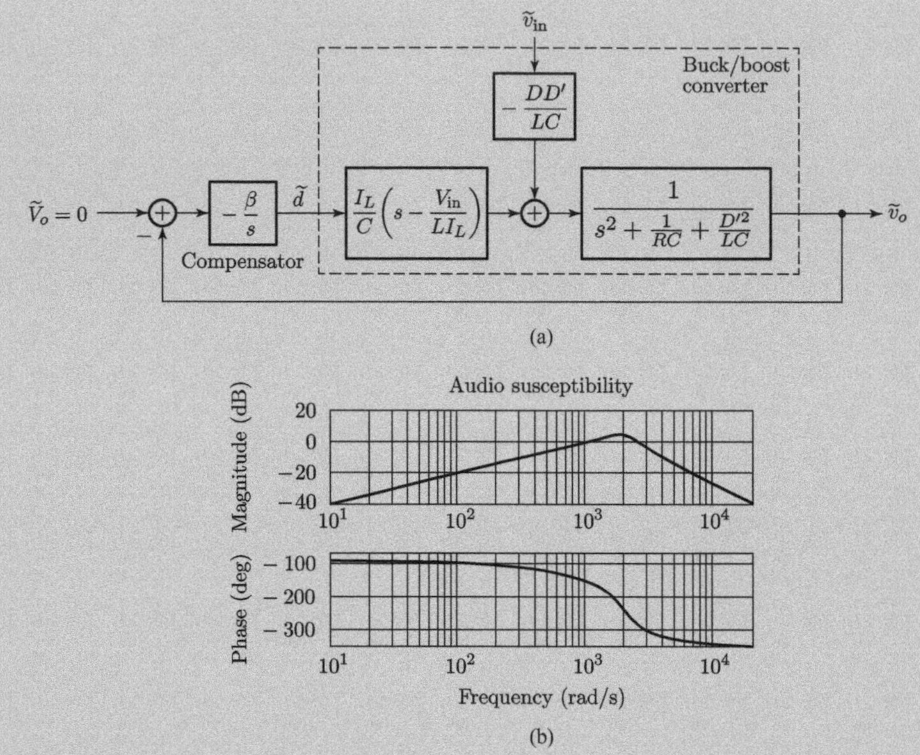

Figure 14.14 (a) Block diagram of the linearized model of the closed-loop converter with integral control. (b) Bode plots of the audio susceptibility frequency response, from input voltage perturbations to output voltage perturbations.

The denominator of the transfer function in (14.82) is the characteristic polynomial of the closed-loop system, and its roots yield the system poles. The locus of the closed-loop poles in our integral control scheme, as the compensator gain β varies from 0 to its design value of 20, is shown in Fig. 14.15(a). The complex poles of the closed-loop system are slightly closer to the imaginary axis than the poles of the open-loop converter alone. We therefore expect the time-domain response in closed loop to settle somewhat more slowly after a disturbance than in the open-loop case.

Figure 14.15(b) shows simulations of the response of an open-loop design with feedforward of the supply voltage, as in Fig. 12.3(b), Example 12.1, and the response of our closed-loop design here, without feedforward, when the source voltage steps from 12 V to 8 V. These simulations use a nonlinear averaged model that is less idealized than the model used for the simulations in Fig. 12.3. We model the diode and transistor as non-ideal switches, representing them by a 10 mΩ resistor in series with a 0.4 V dc source when they conduct. We also assume a capacitor ESR of 100 mΩ and an inductor ESR of 40 mΩ. As a result, the open-loop scheme, despite its feedforward, does not perform as well as it did in Example 12.1, and has significant steady-state error. Offsets in the nominal duty ratio, due to uncertainty about the exact value of V_{in} and implementation constraints, would cause additional error.

The initial response of the closed-loop system without feedforward resembles that of the open-loop system without feedforward, which we presented in Fig. 12.3(a). The output voltage magnitude changes by over 3 V before the feedback through the integral control becomes strong enough to pull it back toward nominal. Note that we manage to attain zero steady-state error despite the non-idealities in the circuit. The transient response of the closed-loop system would be far better if we incorporated feedforward.

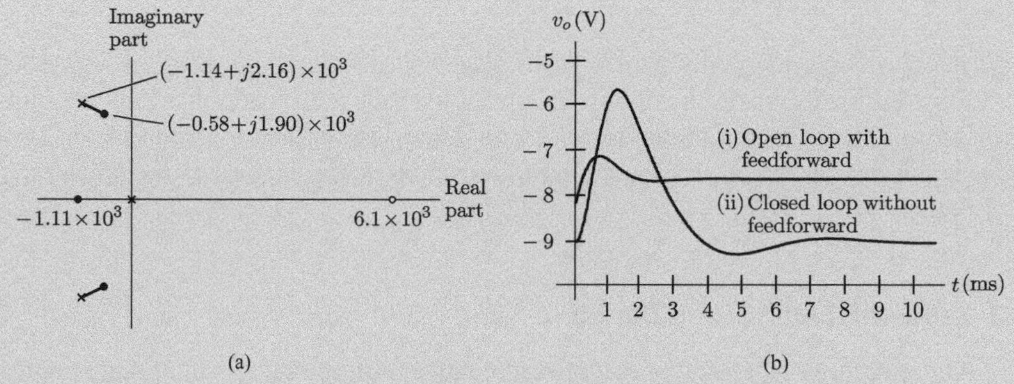

Figure 14.15 (a) Pole–zero diagram showing the open-loop poles (**x**) and zero (**o**) of the loop transfer function, and the closed-loop poles (•). (b) Response of the non-ideal buck/boost converter model to a step change in source voltage from 12 V to 8 V: (i) open loop with feedforward; (ii)‘closed loop without feedforward.

The integral control approach here is *not* robust to decreases in the load. For instance, if the load were dropped a factor of 10 by increasing the load resistor R from 2 Ω to 20 Ω, the resonant peak of the open-loop system would increase by about 20 dB, the distance of the open-loop poles to the imaginary axis would drop by a factor of 10, and the right-half-plane zero would move farther out by a factor of 10. If we retained the integral control designed for a load of 2 Ω, the closed-loop system would be unstable. We evidently need a different controller if such load variations are anticipated. The current-mode control design described later in this chapter is better behaved in this respect.

Implementation Figure 14.16 is a schematic representation of a possible implementation of our integral compensation, and worth comparing with Fig. 12.9. The output voltage $v_o(t)$ is first fed through an inverting amplifier of gain $-R_b/R_a$. The difference between this amplifier output and a reference voltage V_r is fed to an integrator, which produces the modulating signal $m(t)$. This signal is compared with the sawtooth waveform of amplitude K to generate the duty ratio.

The Zener diodes across the integrator capacitor C_1 in Fig. 14.16 limit $m(t)$ to an acceptable range between 0 and K, preventing *integrator windup*. Without these diodes, C_1 could charge up during extended transients to values well outside the range in which its voltage affects the duty ratio – and might then take a long time to come back within the range where the duty ratio responds to the feedback control signal. Usually, V_r is raised gradually to its nominal value when the circuit is first powered up. This so-called *soft start* limits the rate at which the duty ratio rises initially, and thereby reduces stresses in the circuit during start-up.

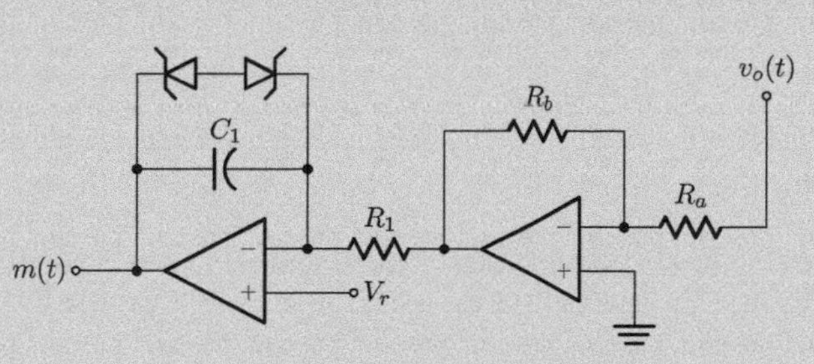

Figure 14.16 Schematic representation of a possible integral control implementation for the buck/boost converter.

14.9 Multi-Loop Control

The concepts and tools of classical control are aimed at designing feedback controllers for single feedback loops, where one measured signal is fed back to the controller, and one control signal is set by it. More commonly, however, several signals are available for feedback and several control signals need to be set, leading to multi-loop control design tasks of the type described in this section.

14.9.1 Sequential Multi-Loop Design

The usual way in which we use classical control to deal with this multi-loop control task is to *sequentially* design single-input, single-output loops that are obtained by pairing the outputs and inputs in some fashion. The specific features of the system and control problem often suggest a natural pairing of signals and a sequence for the control design, as illustrated in the next example.

Example 14.11 Speed Control of a DC Machine

Consider using a phase-controlled rectifier to control the rotor speed of a dc machine in the face of speed-dependent load torques. Our sole control signal is the firing angle α of the rectifier. The signals that can be fed back include the voltage v_d applied to the armature, the armature current i_d, and the rotor speed ω_r.

A block diagram of a typical multi-loop "cascade" design for speed control is shown in Fig. 14.17. The outer loop (or major loop or speed loop) compares the actual rotor speed with the commanded speed. The error indicates the electromagnetic torque that needs to be applied to the motor. As the average armature current $\overline{i}_d$ (averaged over the source period) determines the average electromagnetic torque, the speed-loop controller specifies a desired average armature current I_{ref}. (For simplicity, we have shown $\overline{i}_d$ rather than the instantaneous current i_d as being fed back – in effect assuming that the current-sensing circuitry delivers a good approximation to the local average current.)

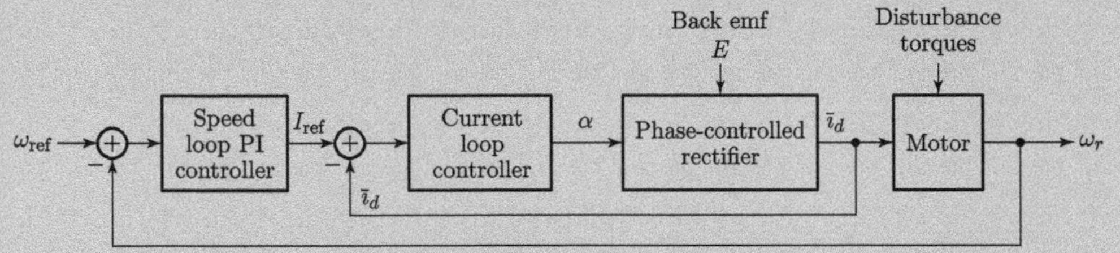

Figure 14.17 Multi-loop configuration for speed control of a dc motor using a phase-controlled rectifier.

The inner loop (or minor loop or current loop) is designed as outlined in Examples 12.2, 12.9, and 12.10, to make $\overline{i}_d$ follow I_{ref} independently of the speed, or equivalently the back-EMF. The purpose of this loop is thus to convert the phase-controlled rectifier into an approximate current source. As long as the variations in I_{ref} are considerably slower than the bandwidth of the current loop, we can assume that $\overline{i}_d \approx I_{\text{ref}}$. A more refined approximation would be

$$\overline{i}_d(s) \approx \frac{1}{sT_i + 1} I_{\text{ref}}(s), \tag{14.83}$$

where T_i is the dominant time constant of the current loop. These approximations of the inner current loop greatly simplify the analysis of the outer speed loop. However, the approximations are reasonable only if we take care to make the bandwidth of the outer loop significantly smaller than that of the inner loop.

With the preceding approximations, we can easily design a PI controller for the speed loop, using the design strategy described earlier. If the closed-loop system is stable and driven only by constant signals, the output of the PI controller is constant in the steady state. Its input must then be 0; that is, the speed error must be 0. The output of the PI controller in steady state is thus automatically set to the value of I_{ref} needed to generate the torque required by the load. We omit further consideration of the details. Once an initial design has been obtained, a more detailed and accurate model for the current loop can be used to assess overall system performance more accurately, using simulations and analysis. The initial design can then be refined.

The same cascade approach can also be extended to provide position control. For this, a position-control loop is constructed outside the speed loop, with the position error being the input to a position controller. A proportional control in this loop ensures that when the position error is 0, the speed command is 0.

14.9.2 Current-Mode Control

A similar sequential multi-loop approach in the case of high-frequency switching converters leads to the idea of current-mode control, which we mentioned in Examples 12.6, 13.8, 14.6, and 14.8. Rather than having the controller set the duty ratio in each cycle, we cause it to set the peak inductor current or peak switch current in each cycle. The most commonly used method involves turning on the switch every T seconds and turning it off when the inductor current i_L or the switch current reaches the specified peak value i_P or a threshold related to i_P. The duty ratio thus becomes an auxiliary variable, implicitly determined by i_L and the control variable i_P. In this section we discuss the case of converters operating in continuous conduction. The discontinuous conduction mode is actually easier to analyze, and we leave it to you to determine the requisite modifications, with Examples 12.5 and 12.18 as a starting point.

Figure 13.10 showed the key features of the inductor current waveform of a switched converter under current-mode control. The threshold signal for the inductor current in each cycle is obtained by subtracting a stabilizing ramp of slope S from the control signal i_P. As we noted in Example 14.8, the stabilizing ramp entrains the converter waveforms, maintaining them at the switching frequency. The ripple in the inductor current waveform in Fig. 13.10 is somewhat exaggerated, but needs to be large enough to permit robust operation of the threshold detection. If the ripple is too small, sensing the current variations reliably in the presence of switching and other noise is difficult.

Ideally, current-mode control results in an inner loop that regulates the inductor current i_L and its average $\overline{i}_L$ at approximately i_P. The controller in the outer ("current programming" or voltage regulation) loop specifies the value of i_P, and thereby $\overline{i}_L$, needed to regulate the average output voltage $\overline{v}_o$ at a desired value. We now examine the conditions under which we come close to this ideal operation.

If ST and $V_{\text{in}}T/L$ are small, the offset between i_P and $\overline{i}_L$ is small in the steady state. For the offset to be small during transients as well, we must ensure that the *slew rate* V_{in}/L of the inductor current is significantly higher than the rate of change of i_P. We can interpret this second condition as requiring the outer loop to have a much smaller bandwidth than the inner loop.

Under the preceding conditions, we can neglect the offset and dynamics of the inner loop – the dynamics attributable to the inductor – in the initial design of the outer loop controller. This approach allows us to obtain a simplified model for the outer loop and makes the design of its controller much easier. We can then assess and refine the controller through simulation and analysis with more detailed and accurate models that take into account the offset and dynamics of the inner loop. Of course, if the conditions for approximately ideal operation are not satisfied, we are forced to work with the detailed models from the beginning.

We have already shown how to obtain more detailed dynamic models for current-mode control. Examples 13.8 and 14.6 obtained a sampled-data model for the dynamics of the outer loop, with the inner loop closed. We used this model in Example 14.8 to analyze stability of the inner loop and bring out the role of the stabilizing ramp. It could further be used to design sampled-data controllers for the outer loop.

The fact that current-mode control approximately transforms the inductor into a controlled current source not only facilitates multi-loop control design but also underlies other properties and applications of current-mode control. For example, buck or buck-derived converters, which have no switch interposed between the inductor and the load, become relatively insensitive at the output to variations in the input voltage. Also, we can easily enforce power sharing when several paralleled converters under current-mode control feed a common load. Designing a compensator that can handle both continuous and discontinuous operation is usually simpler with current-mode control, because the inductor current dynamics does not affect either mode significantly. In boost converters, there is no switch between the inductor and the supply, so we can use current-mode control to draw a specified current waveform from the supply. This feature is commonly used in high-power-factor converters, as we shall see in Example 14.14.

Another advantage of current-mode control for many topologies is that we obtain direct control of the peak switch current in each cycle, and hence an automatic current limiting capability. This capability allows us to protect the converter from the effects of overloads or transformer saturation. (In implementing such a capability, we must guard against premature turnoff caused by current spikes generated at the switching instants by non-ideal behavior of the switches.)

Example 14.12 Compensator Design for a Buck/Boost Converter under Current-Mode Control

We discussed duty-ratio control of a buck/boost converter in Example 14.10. Current-mode control provides an illuminating contrast.

Nominal Model We begin by deriving a simplified averaged model for a buck/boost converter under current-mode control, using the approximation $\bar{\imath}_L \approx i_P$. Our starting point is the averaged model in Fig. 13.2(b) of Example 13.1, repeated here in Fig. 14.18, but with $\bar{\imath}_L$ replaced by i_P.

We need a description of the circuit that does not involve the duty ratio d, as d is only implicitly determined under current-mode control.

Applying (13.2) and (13.3) to eliminate d, or equivalently doing a "power balance" on the ideal transformer in Fig. 14.18, we get

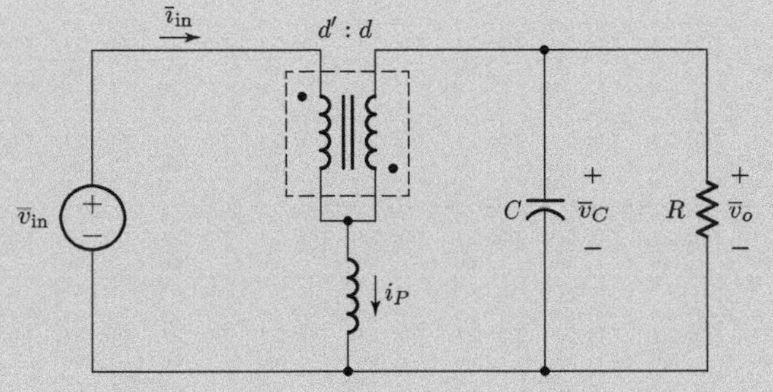

Figure 14.18 Averaged circuit of a buck/boost converter under current-mode control, assuming that $\overline{i}_L \approx i_P$.

$$\overline{v}_C C\frac{d\overline{v}_C}{dt} + \overline{v}_C\frac{\overline{v}_C}{R} + L\frac{di_P}{dt}i_P = \overline{v}_{\text{in}}\overline{i}_{\text{in}} = \overline{v}_{\text{in}}\left(i_P + C\frac{d\overline{v}_C}{dt} + \frac{\overline{v}_C}{R}\right). \tag{14.84}$$

We can now linearize this around a constant nominal operating condition and express the resulting LTI model in the transform domain (using our standard notational conventions), obtaining

$$\begin{aligned}\widetilde{v}_C(s) = \widetilde{v}_o(s) =& \frac{(V_{\text{in}} - sLI_P)R}{sRC(V_o - V_{\text{in}}) + (2V_o - V_{\text{in}})}\widetilde{i}_P(s) \\ &+ \frac{V_o + RI_P}{sRC(V_o - V_{\text{in}}) + (2V_o - V_{\text{in}})}\widetilde{v}_{\text{in}}(s).\end{aligned} \tag{14.85}$$

Using the steady-state relationships $V_o = -(D/D')V_{\text{in}}$ and $I_P D' = -V_o/R$ to simplify these expressions, we obtain

$$\widetilde{v}_o(s) = a_0\frac{1 - sT_z}{1 + sT_c}\widetilde{i}_P(s) + b_0\frac{1}{1 + sT_c}\widetilde{v}_{\text{in}}(s), \tag{14.86}$$

where

$$\begin{aligned} T_c &= \frac{1}{1+D}RC, \qquad T_z = \frac{LI_P}{V_{\text{in}}} = \frac{D}{D'^2}\frac{L}{R}, \\ a_0 &= -\frac{D'}{1+D}R, \qquad b_0 = -\frac{D^2}{D'(1+D)}. \end{aligned} \tag{14.87}$$

For the parameter values in Example 12.1, the control-to-output transfer function $g(s)$ in (14.86) has a pole at $-(1/T_c) = -3247$ rad/s, and a right-half-plane zero at $1/T_z = 6095$ rad/s. Compare these pole and zero locations with those in Fig. 14.5 for the case of duty-ratio control. The zero locations are identical. However, there is only one pole for current-mode control because we have neglected the dynamics of the inner loop. This pole is better damped than the poles that govern duty-ratio control.

The negative sign of $g(0)$ is an artifact of our sign convention for the output voltage. We cancel it by choosing our compensator transfer function $h(s)$ to also have a negative dc value, so that the dc loop gain $g(0)h(0)$ will be positive. To avoid having to account for these signs in the discussion that follows, we work in terms of $-g(s)$ and $-h(s)$. The Bode plots of the frequency response $-g(j\omega)$ are shown in Fig. 14.85(a). For our parameter values, the corner frequencies of the pole and zero are separated by less than a factor of two, so the corners are not strongly defined.

Compensator Design In order to have a good closed-loop bandwidth, we want a compensator that results in a crossover frequency not far below the frequency of the (right-half-plane) zero. Note that the frequency response of $-g(s)$ in Fig. 14.19(a) has a magnitude slope of about -1 in the region between

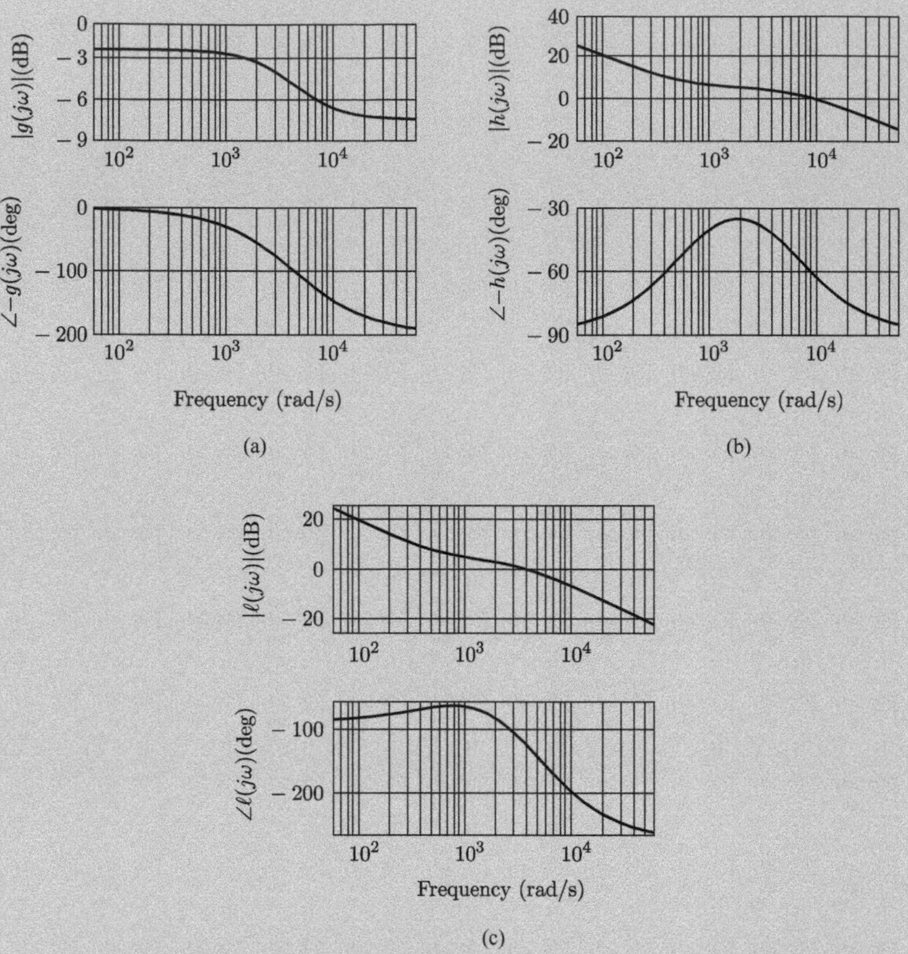

Figure 14.19 Bode plots of (a) the frequency response of $-g(s)$, where $g(s)$ is the control-to-output transfer function; (b) the frequency response of $-h(s)$, where $h(s)$ is the transfer function of the proposed compensator; and (c) the loop gain with this compensator.

the pole and zero frequencies, with an associated phase of $-\pi/2$ rad at a frequency of around 4000 rad/s. We therefore aim to have crossover occur at this frequency, with $-h(s)$ contributing little gain or phase at this frequency. However, we want the compensator to provide significant gain at lower frequencies and significant attenuation at higher frequencies. The Bode plots for the frequency response of a $-h(s)$ that has the required features are shown in Fig. 14.19(b), corresponding to the transfer function

$$h(s) = -h_0 \frac{1 + (s/\omega_1)}{s[1 + (s/\omega_2)]}, \tag{14.88}$$

with $h_0 > 0$. This transfer function is easily realized with an op-amp circuit.

For an initial attempt at picking compensator parameters, we choose the corner frequencies of the compensator to be $\omega_1 = 600$ rad/s and $\omega_2 = 6000$ rad/s. Then $|h(j\omega)|$ at $\omega = 4000$ rad/s is $1.402 \times 10^{-3} h_0$, while $|g(j\omega)| = 0.603$. To obtain crossover at this frequency we must have $h_0 = 1/(1.402 \times 10^{-3} \times 0.603) = 1183$. The loop gain corresponding to this choice is shown in Fig. 14.19(c). The crossover is at 4000 rad/s, as expected, and the phase margin is around 50°.

Evaluation We mention only a few of the many tests that should be made in evaluating the candidate design. It is important first to check that the conditions we used to justify setting $\bar{i}_L \approx i_P$ are indeed satisfied. We leave it to you to make the checks.

The response of the closed-loop converter to small input voltage variations $\tilde{v}_{\text{in}}$ can be studied using the linearized model shown in Fig. 14.20. To determine the response of the converter to an input step as large as -4 V, corresponding to the input voltage dropping from 12 V to 8 V, we may wish to use a more detailed nonlinear model. Figure 14.21 shows the response of both the linearized averaged model and a nonlinear switched model to a -4 V step in the input voltage. The nonlinear simulation assumes an ideal converter and uses $S = -V_o/L$ (the "optimum" slope, in the sense that it causes inductor current perturbations to disappear in essentially one cycle). Note the improvement over the case of duty-ratio control. Another difference from duty-ratio control is that increasing the load resistance from 2 Ω to 20 Ω actually improves the response to the input step. Again, we leave it to you to check this.

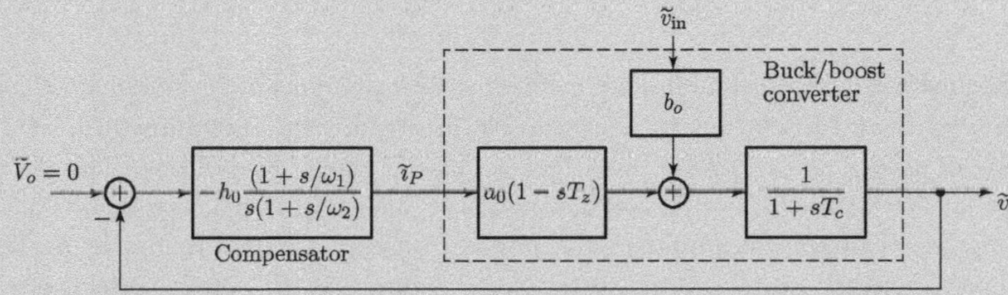

Figure 14.20 Block diagram of the linearized model of the closed-loop converter.

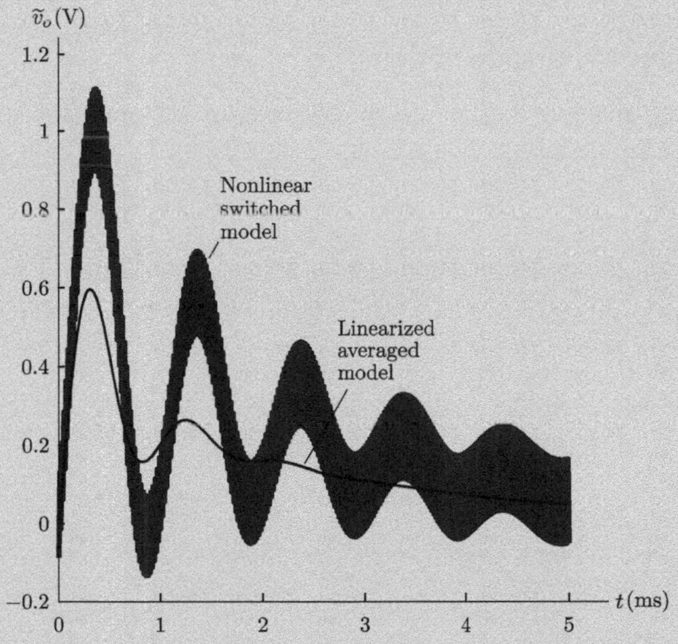

Figure 14.21 Response of the closed-loop converter to a step change in the input voltage from 12 V to 8 V, estimated using the linearized averaged model and a nonlinear switched model.

Further Issues Numerous other issues need to be dealt with before the design can be considered anywhere near complete. We may want to explore what happens with other compensator choices. For instance, a simple integral control can be used if we are satisfied with a crossover frequency sufficiently below the converter's pole frequency of 3247 rad/s. Also, either compensator will probably need additional roll-off beyond crossover, to cancel the effect of the zero contributed by capacitor ESR.

Various implementation issues need to be dealt with. We need to decide whether to sense the switch current or the inductor current, and how to sense it. The slope S of the stabilizing ramp has to be fixed on the basis of the largest duty ratio that we expect to encounter. Note that when the supply drops to 8 V in our example, we have $D > 0.5$, so without the stabilizing ramp the converter would not operate at the switching frequency. The control circuitry has to be arranged so that signal strengths are appropriate at each point, to permit reliable operation in the presence of various sources of noise, such as current spikes at the switching instants. More detailed consideration of such issues is beyond our scope here.

Sequential design of multi-loop control systems is not always possible or useful. In many cases, we have to consider various possible feedback paths simultaneously. This leads to the study of matrix generalizations of the classical LTI control results we have been using so far. We then have to work with plant, compensator, and loop transfer *matrices* that act on *vectors* of signals. The requisite matrix extensions of classical control theory to deal with issues of stability, performance, and robustness in multi-input, multi-output systems are now available and well supported by various computer tools, though they have yet to find significant use in power electronics.

14.9.3 Simultaneous Multi-Loop Design

The following example illustrates a somewhat more elaborate and non-sequential multi-loop feedback control design.

Example 14.13 Multi-Loop Control of a Buck/Boost Converter

In Example 14.10 we examined a switching converter control scheme in which the duty ratio is perturbed as a dynamically varying function of perturbations in the average output voltage. We now consider a modification of this approach, taking a cue from the current-mode control scheme discussed in Example 14.12. We now make the duty-ratio perturbations a function of perturbations in the average output voltage *and* the average inductor current. The particular control law that we have in mind sets

$$\widetilde{d}(t) = -h_I \widetilde{i}_L(t) - h_V \widetilde{v}_o(t) - h_N \int_0^t \widetilde{v}_o(\xi)\, d\xi, \tag{14.89}$$

where h_I, h_V, and h_N are *constant* feedback gains, which we choose so that the closed-loop system has specified natural frequencies.

The last term in (14.89) represents integral control action. If the closed-loop system is stable and driven only by constant signals, all variables must settle to constant values in the steady state. The integrand $\widetilde{v}_o$ in (14.89) must therefore settle to zero to prevent the integral from contributing a time-varying term to $\widetilde{d}(t)$ in (14.89). Hence, so long as we can find feedback gains that stabilize the system, we can guarantee zero steady-state error in the output voltage.

With (14.89) serving as a model for the feedback path, we need to introduce the transfer functions that characterize the forward path. In Examples 13.1, 14.1, 14.3, and 14.4, we have already shown how to derive the transfer functions in the following transform domain relationships:

$$\begin{aligned}\widetilde{\imath}_L(s) &= \frac{1}{a(s)}[b_1(s)\widetilde{d}(s) + b_2(s)\widetilde{v}_{\text{in}}(s)],\\ \widetilde{v}_o(s) &= \frac{1}{a(s)}[c_1(s)\widetilde{d}(s) + c_2(s)\widetilde{v}_{\text{in}}(s)].\end{aligned} \tag{14.90}$$

Our earlier results show that for the buck/boost converter (with no capacitor ESR),

$$a(s) = s^2 + \frac{1}{RC}s + \frac{D'^2}{LC}, \qquad c_1(s) = \frac{I_L}{C}\left(s - \frac{V_{\text{in}}}{LI_L}\right), \qquad c_2(s) = \frac{-D'D}{LC}. \tag{14.91}$$

Some straightforward computations similarly show that

$$b_1(s) = \frac{V_{\text{in}}}{LD'}\left(s + \frac{1+D}{RC}\right), \qquad b_2(s) = \frac{D}{L}\left(s + \frac{1}{RC}\right). \tag{14.92}$$

To get an equation that describes the closed loop, we now substitute (14.90) in the Laplace-transformed version of (14.89). Some rearrangement of the resulting equation then yields

$$\begin{aligned}&\big(s[a(s){+}h_I b_1(s) + h_V c_1(s)] + h_N c_1(s)\big)\widetilde{d}(s)\\ &\quad = -\big(s[h_I b_2(s) + h_V c_2(s)] + h_N c_2(s)\big)\widetilde{v}_{\text{in}}(s).\end{aligned} \tag{14.93}$$

Here, $\widetilde{v}_{\text{in}}$ represents an external drive, and $\widetilde{d}$ is now an internal variable of the closed feedback loop. This equation allows us to identify the characteristic polynomial of the closed-loop system as

$$a_H(s) = s[a(s) + h_I b_1(s) + h_V c_1(s)] + h_N c_1(s). \tag{14.94}$$

This is a third-order polynomial with three gains, and it turns out that we can actually obtain three *arbitrary* roots for this polynomial (provided any complex root is accompanied by its complex conjugate) by picking the gains appropriately.

To illustrate, we can use the parameter values of the buck/boost converter in Example 12.1 and pick the feedback gains to place the closed-loop eigenvalues at the locations shown in Fig. 14.22(a), namely, $-2477 \pm j4170$ and -4000. The corresponding closed-loop characteristic polynomial is $[(s + 2477)^2 + 4170^2](s + 4000)$. Equating it to the polynomial in (14.94) yields the gains $h_I = 0.119$, $h_V = -0.093$, and $h_N = -430$. Figure 14.22(b) shows the response of $\widetilde{v}_o, \widetilde{\imath}_L$, and the duty-ratio perturbations $\widetilde{d}$ in the resulting LTI model for a -4 V step in $\widetilde{v}_{\text{in}}$. This response provides a reasonable approximation to the response of the converter to a drop in supply from 12 V to 8 V. The response compares favorably with what we obtained using duty-ratio control and current-mode control.

Note that even though it is possible in principle to obtain a closed-loop characteristic polynomial with arbitrary roots (subject to complex roots being accompanied by their conjugates), we need to keep in mind the various constraints and assumptions that underlie our model. Most importantly, the closed-loop time constants should be significantly larger than twice the switching period, and the gains should be small enough for the duty-ratio perturbations to not leave the allowable range of $-D$ to D'. These considerations keep us from becoming too ambitious in our choice of closed-loop natural frequencies.

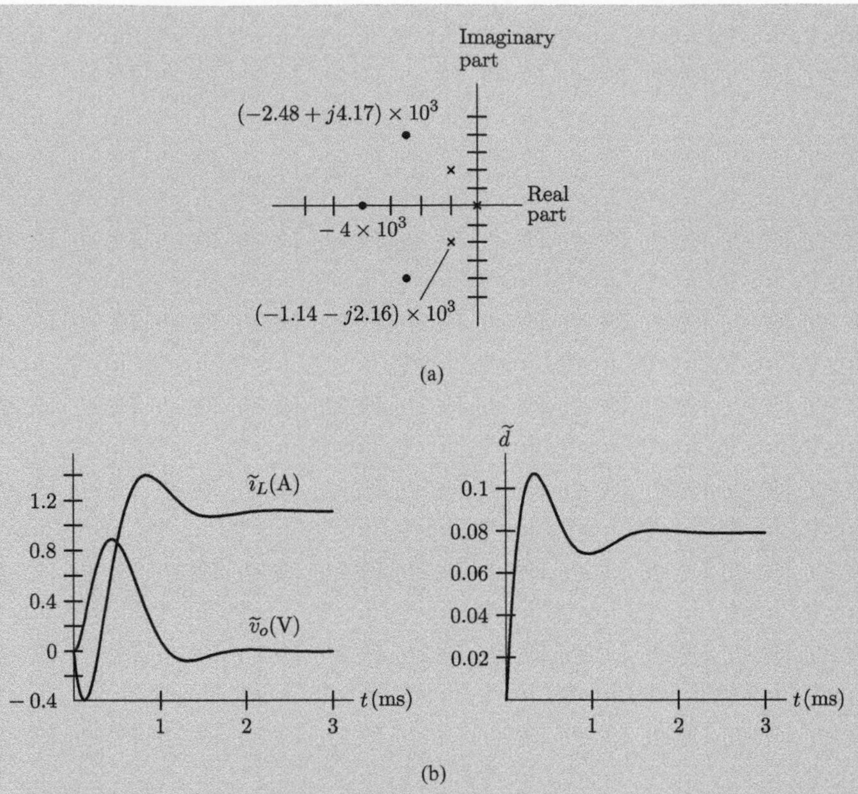

Figure 14.22 (a) Open-loop poles (**x**) and chosen closed-loop poles (•). (b) Closed-loop response of the linearized model to a −4 V step change in the input voltage.

In the following section, we show how to recognize the preceding simultaneous multi-loop design as an instance of state feedback in an LTI model.

14.10 State Feedback

The central idea of feedback control is to make the control inputs responsive to the state of the system. This immediately suggests that the variables that should ideally be measured and fed back to the controller are the state variables associated with some model of the system. If these variables cannot be measured directly, we can attempt to estimate them from available measurements, using so-called state *observers*. These simple ideas form the basis for a variety of linear and nonlinear (and usually multi-loop) control designs. Standard control textbooks treat linear state feedback for LTI systems, and also the design of state observers that provide asymptotically converging estimates of the state variables of such systems, when these variables cannot be directly measured. We only consider here the case where the state is available for measurement, and does not need to be estimated using an observer. Our treatment is for the continuous-time case, but the required changes for the discrete-time case should be clear.

14.10.1 Pole Placement by LTI State Feedback

Consider the LTI state-space model (13.29) or (14.14), repeated here for convenience:

$$\frac{dx}{dt} = Ax(t) + Bu(t), \tag{14.95}$$

where the state vector x has n components and the control input u has m components. Suppose that we choose the control vector to be

$$u(t) = Hx(t) + w(t), \tag{14.96}$$

where H is a constant state-feedback gain matrix of dimension $n \times m$, and $w(t)$ is some external input. This yields a closed-loop system described by the LTI state-space model

$$\frac{dx}{dt} = (A + BH)x(t) + Bw(t). \tag{14.97}$$

Its stability is determined by its natural frequencies, that is, the eigenvalues of $A + BH$, which are the roots of its characteristic polynomial,

$$a_H(s) = \det(sI - A - BH).$$

It turns out that under a "controllability" condition on the system in (14.95), H can be picked to get an $a_H(s)$ with arbitrary roots (again subject to complex roots being accompanied by their complex conjugates), but this is beyond our scope to pursue further here.

The multi-loop control law in Example 14.13 is an example of LTI state feedback. Without integral control, that is, when $h_N = 0$ in (14.89), the control is an LTI function of the state variables $\widetilde{i}_L$ and $\widetilde{v}_C$ of the converter and corresponds to LTI state feedback in the linearized averaged model (14.12). When integral control is present, the control law corresponds to LTI state feedback in an augmented LTI state-space model whose state variables are $\widetilde{i}_L$, $\widetilde{v}_C$, and the output of the integrator used for integral control.

We now briefly examine some notions of *nonlinear* state feedback.

14.10.2 Nonlinear State Feedback

Discussing nonlinear state feedback schemes in any generality is difficult, because they are designed to match or exploit the specific nonlinearities of a given system or model. However, the piecewise LTI nature of typical power converter models makes them amenable to certain design strategies. We mention only one approach, based on choosing switching surfaces in state space. Specific instances can be found in all categories of power converters.

Switching Surfaces Our control in power electronics is essentially limited to deciding when to turn switches on or off. In the intervals between switch operations, the state trajectories are usually well described by LTI models. As we noted in Section 14.5, the end point of the state trajectory for one switch configuration or topological state becomes the initial point for the state trajectory in the succeeding switch configuration. This allows us to describe entire state trajectories, including the effects of switching. A direct approach to control design is therefore to divide the state space into regions that are separated by *switching surfaces*, and to specify which switches are to be turned on or off when a state trajectory arrives at each of these surfaces.

Switching surfaces may be fixed, or varied as a function of external commands and feedback signals. The design of switching surfaces can be trivial for first-order systems and relatively simple for second-order systems, because the state trajectories and switching surfaces – or curves – are easy to visualize in the state plane. Higher-dimensional systems can sometimes be dealt with satisfactorily.

Different types of switching surfaces are encountered. A *refracting* surface introduces a discontinuity in trajectories that arrive at the surface but allows them to continue on past the surface. A switching surface is refracting when the components of the state velocity vectors normal to the surface point in the same direction on both sides of the surface. This situation is shown in Fig. 14.23(a).

Another type of switching surface, which the control literature calls a *sliding* surface, introduces a more severe discontinuity. The normal components of the state velocity vectors in this case both point toward the switching surface, as indicated in Fig. 14.23(b). Consequently, a trajectory that arrives at the surface remains on the surface, and the system is then said to be in a *sliding mode*. In practice, we impose a switching discipline that prevents the switching frequency (and associated switching losses) from becoming excessive, so the trajectory will actually "chatter" back and forth across the surface. However, we can imagine that in the limit of infinitely fast switching, the motion is more like a sliding than a chattering. The duty ratio during this fast switching must be such that the resulting average of the velocity vectors on the two sides lies along the surface. This average velocity along the surface determines the state equation during sliding. The sliding surface must be designed to attract trajectories and to induce an acceptable sliding motion once a trajectory hits the surface.

Several switching control laws in power electronics approach sliding-mode control in the limit of infinite-frequency switching. It is usually best to get an initial idea of the behavior of such systems by assuming ideal sliding, refining the analysis later by building on the insights obtained from the idealized analysis. We have already shown such an approach in the current-mode control design of Example 14.6. The switching surface there corresponded to the line $i_L = i_P$

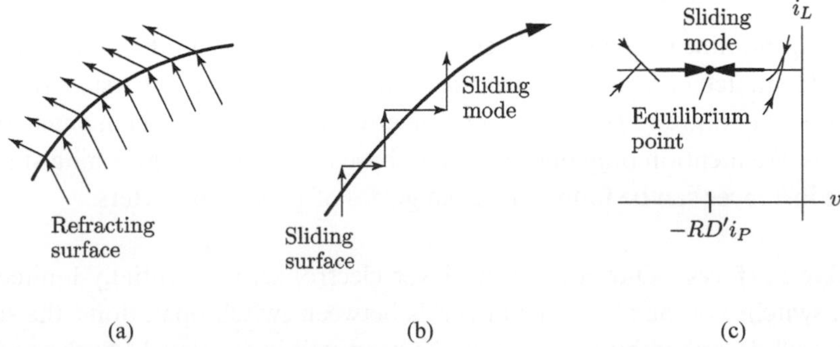

Figure 14.23 Switching surfaces: (a) refracting surface; (b) sliding surface; and (c) sliding surface in current-mode control.

in the state space, as shown in Fig. 14.23(c). We ignored offset and dynamics of the inner loop in our analysis, which corresponds to assuming ideal sliding. In that example, the switching line was actually varied slowly by the outer control loop, as a function of the deviations of the output voltage from the desired value. Hysteretic controllers provide another class of examples that are often best understood if we use sliding-mode ideas prior to more detailed analysis. (We mentioned a hysteretic controller for current-mode control in Example 13.8.)

14.11 Digital Control

We noted in Sections 13.4 and 13.5 that discrete-time models are well suited to describing power electronic circuits, because these circuits are usually operated cyclically, with controls that are specified by a finite number of variables in each cycle. The state variables in these discrete-time models comprise circuit quantities sampled each cycle. Sections 14.6 and 14.7 have made clear the role of discrete-time LTI models in describing the local dynamics and stability of a periodic operating mode. These models also form a natural starting point for the design of *discrete-time* or *digital controllers*.

Digital controllers process sequences of numbers (in contrast to analog controllers, which process continuous-time waveforms). In our applications, the numbers are generally obtained by sampling continuous waveforms using analog-to-digital converters. The discrete-time processing may involve only simple operations with latches, adders, registers, and so on, or more commonly may involve digital signal processing chips, field-programmable gate arrays, or microprocessors capable of more elaborate computations and equipped with significant memory. The numbers produced by the digital controller are translated into appropriate PWM pulses.

Digital controllers – and microprocessor-based controllers in particular – are valued for their flexibility, ability to carry out complicated computations and table look-ups, and insensitivity to noise and aging. With the high sampling and processing speeds that are now available, digital control has become much more important in a wide range of power electronics applications.

We have already mentioned that the results on continuous-time state feedback in Section 14.10 all have natural discrete-time parallels. Similarly, the classical control results in Sections 14.8 and 14.9 have their discrete-time parallels. Developing these parallels is beyond the scope of this book, however. We shall be content with one representative example.

Example 14.14 Digital Control of a High-Power-Factor AC/DC Switching Preregulator

Switching regulators are often operated off an ac supply by first rectifying the sinusoidal ac voltage and then filtering it with a large capacitor. A major difficulty with this approach, however, is that the source current, far from being sinusoidal, consists of a tall, narrow charging pulse in each cycle. This distorted current waveform has high harmonic content and leads to poor power factors, in the range of 0.5 to 0.7. Ideally, we want the source current to be sinusoidal, to minimize harmonic interference with other loads operating off the same source. We also want this sinusoid to be in phase with the source voltage, to maximize the power available for given constraints on the peak source current. (See Section 3.4 for detailed discussion of these requirements.)

One solution to this problem is to insert a high-frequency switching preregulator, such as that in Fig. 14.24, between the rectified ac supply and the switching-regulator load. (The figure incorporates both circuit-diagram and block-diagram conventions, but the description below should preclude confusion.) Let T_L denote the period of the (rectified) input voltage $v_{\text{in}} = V|\sin(\pi t/T_L)|$ and T_S the switching period of the preregulator, so that $T_L \gg T_S$. The circuit is basically a boost converter, and the input current equals the inductor current. We operate the circuit under current-mode control, but make the reference i_P for the inductor current proportional to v_{in}, so

$$i_P(t) = \mu v_{\text{in}}(t). \tag{14.98}$$

This approach leads to an input current that approximates a rectified sinusoid in phase with the input voltage. The proportionality factor μ in (14.98) is varied from one input cycle to the next, that is, every T_L seconds, to regulate the output voltage v_o around a desired nominal value V_o. With proper control, this configuration yields power factors in the range of 0.95 to 0.99, reduces the total harmonic distortion of the input current to as low as 3%, accommodates large variations in the input-voltage amplitude, and provides a better output voltage to drive the downstream regulators. We now develop and analyze one possible digital control scheme for this converter.

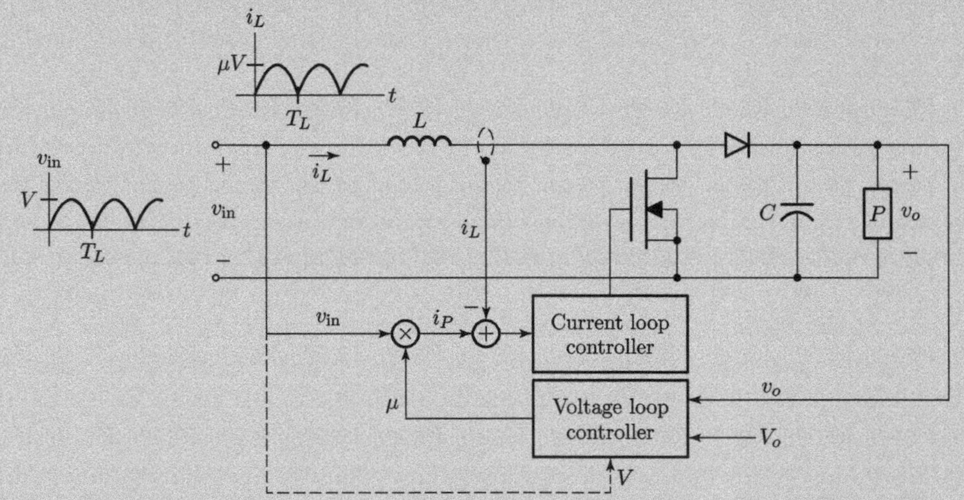

Figure 14.24 High-power-factor ac/dc switching preregulator.

We assume, as in Example 14.6, that the inner loop of the current-mode controller performs ideally and we therefore make the approximation $i_L \approx i_P$. This approximation turns out to yield tractable models and satisfactory controllers in many cases. We also model the load on the preregulator as a constant-power load that draws power P. This is a reasonable model for a load that comprises high-efficiency regulated switching converters. Other loads, such as resistive or current-source loads, can be incorporated into the following development if desired, with a slight increase in complexity.

Assuming lossless switching in the preregulator, and ignoring the switching ripple, we can now write the following power balance equation in the kth cycle, with μ in this cycle denoted by $\mu[k]$:

$$\begin{aligned} \frac{1}{2}C\frac{dv_o^2}{dt} &= v_{\text{in}}i_{\text{in}} - \frac{1}{2}L\frac{di_L^2}{dt} - P \\ &= \mu[k]v_{\text{in}}^2 - \mu^2[k]\frac{1}{2}L\frac{dv_{\text{in}}^2}{dt} - P, \end{aligned} \tag{14.99}$$

where we have substituted i_P for both i_L and i_{in}, and then made use of (14.98). This equation is useful in obtaining efficient simulations that ignore details at time scales smaller than T_S.

We simplify (14.99) still further to obtain the desired discrete-time model. Integrating the equation over the kth cycle, and denoting v_o^2 at the beginning of the kth cycle by $x[k]$, we find that

$$\frac{1}{2}C(x[k+1]-x[k]) = T_L\left(\mu[k]\frac{V^2}{2}-P\right),$$

or

$$x[k+1] = x[k] + \frac{T_L V^2}{C}\mu[k] - \frac{2T_L}{C}P. \qquad (14.100)$$

This is a first-order, LTI, discrete-time, state-space model, driven by the control input $\mu[k]$ and the (unknown) disturbance input P.

A natural way to regulate x to the desired value $X = V_o^2$ is to use a discrete-time version of PI control. The block diagram corresponding to this is shown in Fig. 14.25. We have isolated the factor $C/T_L V^2$ in the block diagram as a separate gain, in order to simplify the final design equations. Instead of the integrator that would be used in an analog PI scheme, we use an accumulator, whose output σ is governed by the equation

$$\sigma[k+1] = \sigma[k] + (x[k]-X). \qquad (14.101)$$

If the closed-loop system in Fig. 14.25 is stable and P is constant, the system reaches a constant steady state. In this steady state, the output of the accumulator is constant, so its input will be 0. In other words, if we can find gains h_1 and h_2 that stabilize the system, in the steady state $v_o^2 = x = X = V_o^2$, or $v_o = V_o$, as desired. (Actually, the value of v_o at the beginning of each input cycle is not quite the same as the average value of v_o over the cycle; the former, not the latter, will be regulated to V_o.)

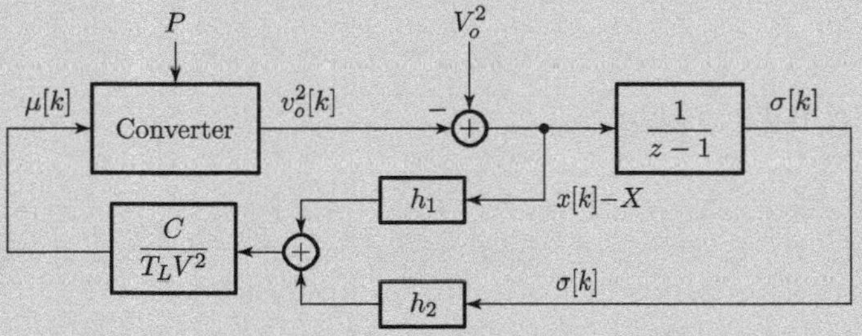

Figure 14.25 A discrete-time PI control design for the high-power-factor converter.

A state-space model for the closed-loop system is

$$\begin{bmatrix}\sigma[k+1]\\ x[k+1]\end{bmatrix} = \begin{bmatrix}1 & 1\\ h_2 & 1+h_1\end{bmatrix}\begin{bmatrix}\sigma[k]\\ x[k]\end{bmatrix} - \begin{bmatrix}1\\ h_1\end{bmatrix}X - \begin{bmatrix}0\\ 2T_L/C\end{bmatrix}P, \qquad (14.102)$$

for which the characteristic polynomial is $z^2-(h_1+2)z+1+h_1-h_2$.

The simulation in Fig. 14.26(a) shows the response of the circuit to a large perturbation in the required output voltage V_o, with the gains h_1 and h_2 picked to place both natural ratios at 0.5, corresponding to the characteristic polynomial $(z-0.5)^2$. The simulation in Fig. 14.26(b) displays the response (with these same gains) to a step increase in load power from 1100 W to 1650 W.

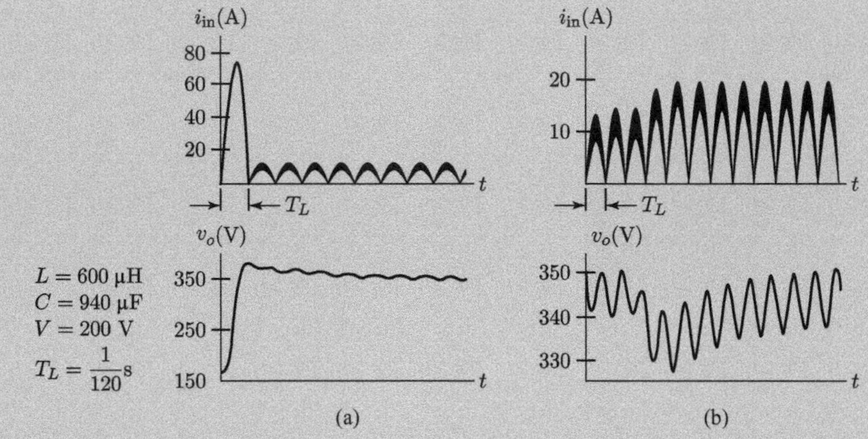

Figure 14.26 Response of the closed-loop system to (a) a large perturbation in the required output voltage V_o, and (b) a step in load power.

Approaches to dynamic modeling and control in power electronics are as varied as the circuits and systems that are encountered in power electronics practice. The applications of power electronics grow to embrace ever larger ranges of power and frequency, accompanied by commensurate innovations in materials, design, engineering, and operation. Nevertheless, the fundamental elements and principles of dynamic modeling and control for power electronic systems, as developed through Chapters 12–14, remain widely applicable.

Notes and Bibliography

Many of the references listed in Chapters 12 and 13 are relevant to this chapter as well. For linearization of state-space models, analysis of continuous-time and discrete-time LTI state-space models, and introductory aspects of feedback control in a state-space setting, see [1].

Our description of piecewise LTI models in Section 14.5 is based on [2], which contains references to related earlier work.

For more on analog and digital feedback control of power converters, see [3], [4], and [5].

The collection of articles in [6] conveys some sense of the range of nonlinear behaviors that power converters can exhibit.

1. A. V. Oppenheim and G. C. Verghese, *Signals, Systems and Inference*, London: Pearson, 2017.
2. G. C. Verghese, M. E. Elbuluk, and J. G. Kassakian, "A General Approach to Sampled-Data Modeling for Power Electronic Circuits," *IEEE Trans. Power Electronics*, vol. 1, pp. 76–89, Apr. 1986.
3. R. W. Erickson and D. Maksimović, *Fundamentals of Power Electronics*, 3rd ed., New York: Springer, 2020.
4. L. Corradini, D. Maksimović, P. Mattavelli, and R. Zane, *Digital Control of High-Frequency Switched-Mode Power Converters*, Chichester: Wiley-IEEE, 2015.

5. S. Buso and P. Mattavelli, *Digital Control in Power Electronics*, 2nd ed., San Rafael, CA: Morgan & Claypool, 2015.
6. S. Banerjee and G. C. Verghese (eds.), *Nonlinear Phenomena in Power Electronics: Attractors, Bifurcations, Chaos, and Nonlinear Control*, Chichester: Wiley-IEEE, 2001.

PROBLEMS

14.1 Verify that, if $f(x, u, t) = M(u, t)x$ for some matrix $M(u, t)$, then $\partial f/\partial x = M(u, t)$.

14.2 You will find it a worthwhile review to derive for yourself the various expressions presented in (i) Example 14.1; (ii) Example 14.2; (iii) Example 14.3; (iv) Example 14.4; (v) Example 14.5 for the buck/boost converter, and to then create your own versions of the plots shown in (vi) Fig. 14.3(b); (vii) Fig. 14.5(b); and (viii) Fig. 14.6 (and, for this latter figure, showing the relationship to the waveforms in Fig. 12.3). We leave you to pick some manageable subset of these tasks at each visit to this problem!

14.3 For the buck converter discussed in Example 5.5, develop analytical expressions and carry out numerical calculations and simulations similar to those done for the buck/boost converter in (i) Example 14.1; (ii) Example 14.2; (iii) Example 14.3; (iv) Example 14.4; (v) Example 14.5.

14.4 This problem deals with the buck converter discussed in Problem 5.22.

(a) Choosing $L = 3.6\,\mu\text{H}$ and $C = 5.6\,\mu\text{F}$, develop analytical expressions and carry out numerical calculations and simulations similar to those done for the buck/boost converter in (i) Example 14.1; (ii) Example 14.2; (iii) Example 14.3; (iv) Example 14.4; and (v) Example 14.5.

(b) Simulate and plot the output voltage transient when the load steps from 0.5 to $1\,\Omega$, and also when it steps in the other direction.

(c) Repeat part (b) for the case where $C = 7.5\,\mu\text{F}$ rather than $5.6\,\mu\text{F}$.

14.5 Examples 14.3 and 14.4 dealt with the steady state, linearization, and transfer functions of switched and averaged models of the buck/boost converter.

(a) Work out the detailed derivation of the linearized *switched* model, accounting for the fact that $\widetilde{q}$ is restricted to the values 0 and 1.

(b) Determine the transfer functions in the *averaged* model from duty-ratio perturbations and input voltage perturbations to inductor current perturbations. Compare to results obtained directly from the corresponding linearized averaged *circuit*.

(c) How do your analysis and results change if the switches, instead of being modeled as a short circuit when they conduct, are represented by a (small) resistance in series with a (small) dc voltage source?

14.6 Repeat Examples 14.3 and 14.4 for the case of a boost converter.

14.7 This problem deals with the state-space model (13.39) that we obtained in Example 13.6 to describe the ω_s-component of the response of a series-resonant *RLC* circuit. (See references [7] and [8] in Chapter 13 for related examples.)

(a) If we consider ω_s as a control input that can be varied with time rather being a constant, so it becomes $\omega_s(t)$, then this model is *nonlinear* rather than linear, and is still time *invariant*. Explain.

(b) Determine the steady state of this model if $\omega_s(t)$ is fixed at the constant value Ω_o, and $\widehat{v}_{\text{in}}(t)$ is fixed at the value given in (13.40).

(c) Linearize this model around the steady state you determined in (b), assuming $\omega_s(t) = \Omega_o + \widetilde{\omega}_s(t)$, where $\widetilde{\omega}_s(t)$ is a *small* perturbation that *varies slowly* relative to the nominal switching frequency Ω_o.

14.8 The general time-domain solution in (14.40) for the LTI model in (14.14) was derived using transform-domain arguments. For a direct check in the time domain, confirm that the expression in (14.40) satisfies (14.14) and takes the specified initial value $x(t_0)$ at t_0.

14.9 Show that the "rotation matrix" used to analyze the resonant converter in Example 13.9 is indeed the state-transition matrix of the circuit in each of its topological states.

14.10 Suppose the input $u(t)$ to the system in (14.14) is constant at the value u_k for t between t_k and t_{k+1}. Show that $x(t_{k+1}) = \mathcal{A}_k x(t_k) + \mathcal{B}_k u_k$ for matrices $\mathcal{A}_k$ and $\mathcal{B}_k$ that you should explicitly relate to A and B in (14.14).

14.11 We briefly described how the linearized model in (14.48) could be used in an iterative solution of (14.52) for computation of the steady state in a sampled-data model. Obtain a more detailed algorithm, and test it on the sampled-data model for a buck/boost converter under current-mode control, referring to Examples 13.8 and 14.6. Take the commanded peak current $i_P[k]$ to be constant at $I_P = 7\,\text{A}$, assume no stabilizing ramp ($S = 0$), and let all other parameter values be the same as those in Example 12.1.

14.12 Derive a continuous-time generalized state-space-averaged model for the behavior of a buck/ boost converter under current-mode control, following the route suggested in the paragraph after Example 14.6. Determine its steady state. Then linearize this model around the steady state and determine the transfer function from perturbations $\widetilde{i}_P$ in the commanded peak current to perturbation $\widetilde{v}_o$ in the output voltage. As a check, one pole of your transfer function should go to $-\infty$ when there is no stabilizing ramp ($S = 0$), leaving just one finite pole. Suppose we decide to allow perturbations $\widetilde{s}$ around a nominal stabilizing-ramp slope of S. What is the transfer function from $\widetilde{s}$ to $\widetilde{v}_o$? (For additional practice, repeat this for a boost converter.)

14.13 The stability of the sampled-data model for the buck/boost converter under current-mode control in Example 14.8 was studied by means of the matrix $\mathcal{A}$ in (14.68). Verify that the entries of this matrix are as claimed. Extend the exact and approximate analysis in that example to the case of a nonzero stabilizing ramp, $S \neq 0$, studying the locus of characteristic ratios as S varies. Verify your approximate analysis by appropriate calculations with the inductor-current waveform in Fig. 13.10. What would be a good choice for S?

14.14 For the linearized sampled-data model of the buck/boost converter under current-mode control, as given in Example 14.6, determine the transfer functions from perturbations $\widetilde{i}_P$ and $\widetilde{v}_{in}$ to $\widetilde{v}_o$. With the nominal value of i_P fixed at $I_P = 7$ A, and with all other parameter values the same as those in Example 12.1, plot the associated frequency responses for varying values of the stabilizing-ramp slope S.

14.15 Guided by the results in Example 14.9 on the frequency response of the sampled-data model of the buck/boost converter, explore the relationships among $m(t)$, d_k, $d(t)$, $\overline{v}_C(t)$, and $v_C[k]$ for variations in $m(t)$ at various frequencies.

14.16 Obtain linearized sampled-data models of operation in discontinuous conduction for the switched converters treated in Problems 13.8 and 13.9, for both duty-ratio control and current-mode control. Evaluate the stability of each model, compute relevant transfer functions, and compare your results with the results of Problems 12.3 and 12.4.

14.17 Find a linearized sampled-data model for the phase-controlled rectifier drive in Problem 13.10, relating perturbations $\widetilde{\alpha}$ in the firing angle to perturbations $\widetilde{i}_d$ in the armature current. Determine the associated transfer function.

14.18 Find a linearized sampled-data model for the resonant dc/dc converter in Example 13.9 and determine the associated characteristic polynomial. Using the notation in Example 13.3, let $L = 20\,\mu\text{H}$, $C = 10\,\text{nF}$, $V_1 = 14$ V, and $f_s = 400$ kHz. Plot the locus of natural ratios as V_2 varies from 0 V to 13 V.

14.19 This problem suggests one of many approaches to finding a state-space *realization* of a scalar, rational transfer function $h(s)$, that is, a state-space model whose transfer function is $h(s)$.

(a) Suppose $h(s) = h_0/(s + a)$. Find a first-order realization.

(b) If $h(s) = (e_1 s + e_0)/(s^2 + a_1 s + a_0)$, verify that a realization of the form (13.29) and (13.30) is obtained by choosing

$$A = \begin{bmatrix} 0 & 1 \\ -a_0 & -a_1 \end{bmatrix}, \quad B = \begin{bmatrix} 0 \\ 1 \end{bmatrix}, \quad E = [e_0 \quad e_1], \quad F = [0].$$

(c) An $h(s)$ whose numerator degree does not exceed its denominator degree can be expanded into a partial fraction expansion comprising a constant and terms that are as simple as those in (a) and (b). Show how state-space realizations of each of these terms can be combined to yield a state-space realization of $h(s)$.

14.20 Verify the Bode plots in Fig. 14.12, and use them to justify (14.81), which approximately relates the phase of the loop gain to the slope of its magnitude and the number of right-half-plane zeros. Also verify that the Nyquist plot for a stable $\ell(s)$ approaches the origin at an angle of $-\delta\pi/2$, where δ is defined in Section 14.8.2.

14.21 Check the results and claims regarding duty-ratio control of the buck/boost converter in Example 14.10. How does the closed-loop simulation in Fig. 14.15(b) compare with the response you would predict using the linearized model in Fig. 14.14(a)? Also compare with a simulation that combines the controller in this example with the averaged model in (13.36).

14.22 Check the results and claims regarding current-mode control of the buck/boost converter in Example 14.12. Are the conditions that justify neglecting the offset and dynamics of the inner loop satisfied for the parameter values in the example? Compute the locus of closed-loop poles for $R = 2\,\Omega$ and $R = 20\,\Omega$ as the gain h_0 in (14.88) is varied from 0 to 2000.

14.23 Check the results and claims regarding multi-loop control of the buck/boost converter in Example 14.13. With the voltage gain h_V and integral gain h_N fixed at the values in the example, determine and sketch the locus of closed-loop natural frequencies as the current gain h_I varies upward from 0. Obtain a simulation of the closed-loop system obtained by combining the control law in Example 14.13 with the averaged model in (13.36). How does its response to a step in the input voltage compare with the results in Fig. 14.15(b)?

14.24 Explicitly show that the multi-loop control design in Example 14.13 corresponds to state feedback in an augmented state-space model that comprises the buck/boost converter along with the integrator used for integral control.

14.25 Develop a multi-loop controller with integral control action for (i) a buck converter, and (ii) a boost converter, both in continuous conduction.

14.26 Generate a candidate controller for the boost converter operating in discontinuous conduction under current-mode control. For particular numerical values of the components, explore how well it handles transition to continuous conduction.

14.27 Consider the multi-loop configuration for motor speed control in Example 14.11. Suppose the transfer function that relates the motor speed ω_r to the average armature current $\bar{\imath}_d$ is $K/(sT_r + 1)$. Using the model in (14.83) for the current loop, show how to design the PI controller in the speed loop to obtain a crossover between $1/T_r$ and $1/T_i$. Also explore how the phase margin depends on the PI controller parameters.

14.28 Consider the digital controller for the high-power-factor ac/dc switching preregulator in Example 14.14. What gains h_1 and h_2 are needed to place both natural ratios at 0.5? What gains are needed to make the system "deadbeat," that is, to have it settle in a finite number of steps?

14.29 Suppose you want to implement state feedback in the discrete-time LTI system $x[k+1] = \mathcal{A}x[k] + \mathcal{B}p[k]$ but the time required at each sampling instant to compute the control is not negligible. You may then be forced to make the present control depend on the state at the *previous* instant. Suppose therefore that you use the control law $p[k] = \mathcal{H}_X x[k-1] + \mathcal{H}_P p[k-1]$.

(a) Write a state-space description for the augmented closed-loop system whose state vector at time k comprises $x[k]$ and $p[k]$ (the variables in $p[k]$ represent the memory of the controller and are natural state variables). Under an appropriate ("controllability") condition on the original system (beyond our scope to elaborate on here), you can place the natural ratios of the augmented system at arbitrary locations by properly choosing $\mathcal{H}_X$ and $\mathcal{H}_P$.

(b) Use the preceding results to modify the digital controller for the high-power-factor ac/dc switching preregulator in Example 14.14, assuming now that the controller requires one time step for its calculations.

14.30 Obtain a continuous-time averaged model for the high-power-factor ac/dc switching preregulator in Example 14.14 by taking the local average of (14.99) over an interval T_L. Show that under appropriate assumptions the result is a continuous-time LTI model for $\overline{v}_o^2$. How is the model modified if the load is a resistor R in parallel with the constant-power load P?

Part III

Components and Devices

15 Components and Devices: An Overview

A power circuit is typically composed of only a few kinds of components (other than its source and load): switches, and energy storage elements such as capacitors and inductors (or transformers). In its ideal form, each of these components is lossless and capable of operating at any frequency. In the ideal switch, for instance, the voltage across it is zero when it is on, the current flowing through it is zero when it is off, and the transition between these two states occurs infinitely fast. Although we did not explicitly discuss the means by which the actual switches are turned on and off (a topic covered in Chapter 23), an ideal switch responds infinitely fast to its drive signal and requires zero drive power. And, of course, there is no limit to the off-state voltage or the on-state current of the ideal switch. Similarly, the ideal capacitor and the ideal inductor are purely energy storage elements, and they each store energy in only one field type. The ideal capacitor contains no magnetic field energy, and the ideal inductor contains no electric field energy. Furthermore, neither ideal element dissipates any energy.

In reality, the components we use fall short of this ideal description, and so our power circuits are neither 100% efficient nor capable of operating at an infinitely high switching frequency. Improvements in power circuit behavior depend on the development of components with characteristics that are closer to ideal, and design techniques that compensate for the non-idealities. You thus need to understand the fabrication and material technologies on which power electronic components depend. Only through such an understanding can you properly apply power circuit components, as well as track and predict developments in the field.

As an example of the impact of advances in components, consider how the power supply for low-dc-voltage electronic loads has evolved. In the early 1950s, semiconductor power diodes became a practical replacement for vacuum tube (or selenium) rectifiers.† As the on-state voltage of a semiconductor diode (about 1 V) was much lower than that of a vacuum rectifier (about 30 V), the power supply efficiency increased significantly. By the mid 1960s, the switching speeds of bipolar power transistors had improved enough to make high-frequency dc/dc converters operating in the 10–20 kHz range possible. This improvement allowed replacement of linear-regulator technology with the lighter, smaller, and more efficient switching regulator. By the late 1970s, the power MOSFET became economically practical and permitted efficient operation in the 200 kHz to 1 MHz range, which further reduced the size of the required switching power supply. The more recent availability of semiconductor devices using wide-bandgap materials, such as SiC and GaN, and improved magnetic materials, have enabled switching frequencies

† Not many readers of this book will have experienced the joy of replacing the cat whisker and galena crystal with a germanium diode, and marveling at the increased volume of the local radio station.

in the 10s and even 100s of MHz. In each case, a component improvement permitted the "breakthrough" in power electronic technology.

15.1 Practical Semiconductor Switches

Typical current and voltage waveforms for a semiconductor switching device are shown in Fig. 15.1. When the switch is on, it exhibits a forward voltage drop V_{on}, which gives a conduction loss $p = V_{on}I$ as current I flows through it. Similarly, when the switch is off, it carries a small *leakage current* that produces an off-state loss. The leakage current is usually small enough that we can neglect this off-state loss. But when the switch is a Schottky diode, or when the power circuit tries to impose a voltage across a switch that is greater than the switch's *breakdown voltage*, the leakage current and consequent off-state loss can be significant.

A semiconductor switch also needs time to make the transition from one state to the other. During this *switching time*, $v_{sw}i_{sw} \neq 0$ and the device dissipates energy. The total energy lost in the turn-on and turn-off transitions, multiplied by the switching frequency f_{sw}, gives the net *switching loss* for the device.

15.2 Practical Energy Storage Elements

We can explain some of the dissipation that occurs in real energy storage elements by the resistances of the capacitor's leads and plates, and of the inductor's winding. However, calculating the loss is not simple because we must take into account the *skin effect*, which dictates where ac current flows in a conductor, and the *proximity effect*, which dictates how much circulating, or *eddy*, current is induced in nearby wires or other conductive material. In general, to reduce

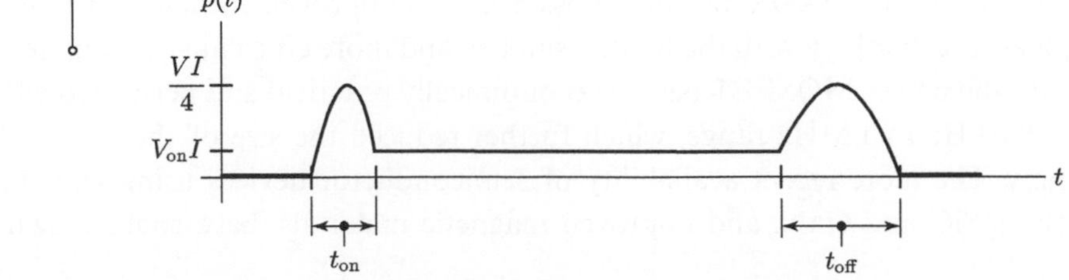

Figure 15.1 Voltage, current, and dissipation waveforms for a semiconductor switch.

this I^2R dissipation, we must make the energy storage element physically large to incorporate more conductor area. However, when the conductors are large enough for the skin and proximity effects to dominate the total loss, conductor fabrication becomes more complicated.

In addition to I^2R dissipation, the *change* of stored energy in the magnetic or dielectric materials also causes losses in inductors and capacitors. Generally, these losses are a nonlinear function of ac field strength, frequency, dc bias, and temperature. Both inductors and capacitors can be made from a variety of materials, which implies trade-offs. One material might have less loss at a given frequency than another, but have a lower permeability (or permittivity), a lower maximum field strength, or a higher cost.

Dissipation is not the only troublesome imperfection of inductors and capacitors. Because these elements can be used as filters, we also care about whether their impedances deviate from the anticipated $j\omega L$ or $1/j\omega C$. Capacitors have a parasitic series inductance resulting from the loop formed by their plates, their leads, and their connection to other components. Inductors have a parasitic parallel capacitance caused by the proximity of two windings at different potentials. We can neglect these parasitic components for both elements if the frequency is low enough. However, at very high frequencies the "capacitor" acts like an inductor and the "inductor" acts like a capacitor. As a result, a filter having these components does not behave as expected above a certain frequency.

15.3 Semiconductor Devices

In this section our purpose is to outline the main power semiconductor devices available today – diodes, transistors, and thyristors – and to discuss their terminal characteristics. In Chapter 16 we explain in more detail the physics underlying the behavior of these devices, and in Chapter 17 we discuss the unique characteristics of their design for power applications.

At this point, however, you should know that current in a semiconductor is carried by both positive charges (holes) and negative charges (electrons). When one of these charge carriers dominates the total concentration of carriers, we call it the *majority carrier*. We call the other type of carrier the *minority carrier*. Devices whose behavior is dominated by the distribution of majority carriers (*majority-carrier devices*) – the field-effect transistor, for example – can be switched much more quickly than can those devices that rely on the change of minority-carrier distributions (*minority-carrier devices*), such as the bipolar junction transistor. The carrier concentrations also determine the resistance of the different regions of a device.

15.3.1 Diodes

A diode is a switch whose state cannot be explicitly controlled, but is instead determined by the behavior of the circuit in which it is used. If the circuit tries to impose a positive voltage across the diode, it will turn on, and if the circuit tries to impose a negative current through the diode, it will turn off. Types of semiconductor diodes include the bipolar diode (based on a p–n semiconductor junction) and the Schottky diode (based on a metal–semiconductor junction). The schematic symbols and the i–v characteristics for these devices are shown in Fig. 15.2.

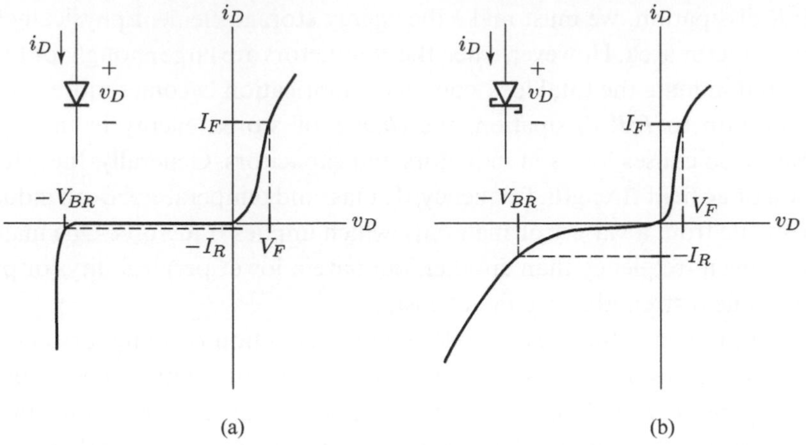

Figure 15.2 Symbols and i–v characteristics of power semiconductor diodes. (a) Bipolar junction diode. (b) Schottky diode.

At the rated current I_F, the forward voltage V_F is typically 1 V for a silicon bipolar power diode, and 0.6 V or less for a silicon Schottky diode. Diodes fabricated on wide-bandgap materials such as SiC and GaN have higher forward voltages. Note that above I_F the forward voltage increases at a faster rate with current than it does below I_F. The reason is that the resistive drop in the diode, which increases linearly with current, becomes comparable to, and then exceeds, the junction voltage, which increases logarithmically with current. The reverse voltage rating of a power diode, V_{BR}, sometimes called the *peak inverse voltage* (PIV), is determined by reverse leakage current. For a bipolar diode, V_{BR} can be in excess of 10 kV, and is usually specified at a leakage current I_R between 1 μA and 10s of milliamps, depending on the I_F rating of the diode. For reverse voltages greater than V_{BR}, the leakage current increases rapidly. For the same die size and voltage, the leakage current of a Si Schottky diode is much higher than that of a Si bipolar diode. For both diodes, leakage current increases exponentially with temperature.

Bipolar Diode In addition to the presence of a nonzero forward drop and leakage current, the bipolar diode departs from ideal in its switching behavior. Figure 15.3 shows typical voltage and current waveforms during turn-on and turn-off of a bipolar diode in a buck converter. Note that when the device first begins to carry current, its forward voltage rises to V_{fp}, known as the *forward recovery voltage*, a value higher than the static curve of Fig. 15.2(a) predicts. Typical values for V_{fp} range between 5 V and 20 V. This phenomenon, which is a function of the on-state current, I_o in this case, occurs because the excess charge requires time to build up to its final value in the device, and before it does, the resistance of the diode is very high. During the *forward recovery time* t_{fr}, the power dissipation in the diode is much higher than predicted from the diode's static characteristics. Also, this transient voltage is impressed across other devices in the power circuit (the transistor in this converter, for example), and this additional voltage stress must be considered when rating these other devices.

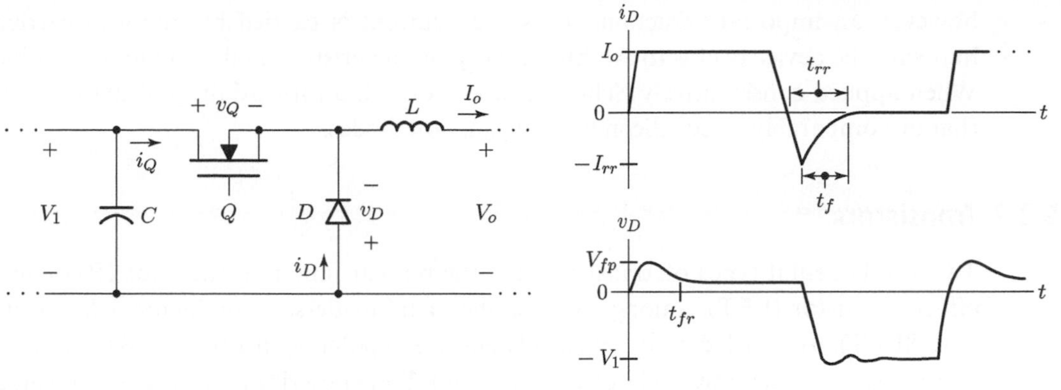

Figure 15.3 Switching waveforms for a bipolar diode used in a buck converter.

As the waveforms in Fig. 15.3 also show, when the bipolar diode turns off, its current first goes negative; then its reverse voltage begins to fall and the current returns to zero. This phenomenon is called *reverse recovery*, and it lasts for a time called the *reverse recovery time* t_{rr}. It occurs because the excess carriers in the device must be removed before the junction can withstand a reverse voltage. Another practical consequence of reverse recovery is that until the reverse current peaks at $-I_{rr}$ and starts to decrease to zero, the diode is still on. Therefore reverse recovery limits the maximum frequency at which we can use a diode.

Large values of $-I_{rr}$ increase the current stress on other semiconductor devices in the power circuit. During time t_f, which can be very short compared to t_{rr}, di/dt is high, and the parasitic inductance through which the reverse current flows imposes a high Ldi/dt voltage across these other devices. For this reason, we sometimes refer to diodes in terms of their "snappiness." A snappy diode is one whose current quickly drops to zero after I_{rr} is reached; in other words, the di/dt of these diodes is very high. In a non-snappy, or *soft-recovery*, diode, the recovery current takes a relatively long time to return to zero from $-I_{rr}$. That is, it exhibits a relatively small recovery di/dt.

In general, the higher the reverse voltage rating of a bipolar diode, the longer its forward and reverse recovery times. Reducing these recovery times (by reducing minority-carrier lifetimes) increases the forward voltage of the device. For this reason, diodes designed for low-frequency applications, such as 50 or 60 Hz rectifiers, have lower forward drops than diodes designed for high-frequency dc/dc or PWM converters. For instance, a typical 200 V *rectifier-grade* diode has a reverse recovery time of 50 μs and a forward drop of 1.2 V at its rated current. A typical 200 V, 30 A, *fast-recovery* diode has a reverse recovery time as low as 10s of nanoseconds, but a forward drop of 1.6 V.

Schottky Barrier Diode A Schottky diode does not generate significant excess carrier concentrations when conducting. Instead it carries current primarily by the drift of majority carriers. For this reason in the ideal case it does not have the forward or reverse recovery behavior exhibited by a bipolar diode, although its junction space-charge layer capacitance must still be charged and discharged. As the reverse voltage rating of a Si Schottky is raised above 100 V, however, an important fraction of its total current is carried by minority carriers. As this happens, the device begins to exhibit recovery characteristics similar to those of a bipolar diode. When applied conservatively, Schottky diodes exhibit a forward drop of about 0.4 V less than that of comparably rated silicon bipolar junction diodes.

15.3.2 Transistors

Two fundamental types of transistors are the bipolar junction transistor (BJT) and the field-effect transistor (FET), among which is the metal-oxide-semiconductor field-effect transistor (MOSFET). A third device, the insulated-gate bipolar transistor (IGBT), is an integrated combination of these two. The discussion that follows describes transistors fabricated in Si, but all three of these transistor types are also available in the wide-bandgap (WBG) materials SiC and GaN. The structure and characteristics of these transistors, particularly those made from GaN, differ in some respects from those for Si. These WBG devices are discussed in Chapter 17.

The schematic symbols and the *i–v* characteristics for the BJT and MOSFET are shown in Fig. 15.4.[‡] A significant distinction between the two devices is that the BJT is controlled by its base current, while the MOSFET by its gate voltage. The BJT is a minority-carrier device, so like the bipolar diode, excess charge must be supplied or removed to turn it on or off. As a majority-carrier device, the MOSFET does not have stored minority carriers, but it does have junction and oxide capacitances that must be charged or discharged to turn the device on or off. Therefore, neither the BJT nor the MOSFET can be switched instantaneously.

Note that both the BJT and MOSFET have a maximum permissible off-state voltage. They both also have a nonzero on-state voltage that depends on how much current is flowing and how hard the transistor is driven. Between the on and off states is a region of operation referred to as the *forward active region*. For a fixed base current (respectively, gate voltage), the collector (respectively, drain) current of both transistors remains roughly constant over the full range of the device voltage in this region. Because power transistors are used as switches that are either on or off, the forward active region of operation is of interest only during switching, when it must be traversed.

Bipolar Junction Transistor To be in the on state, the BJT requires the presence of base current i_B. The ratio of on-state collector current i_C to the base current i_B necessary to support this value of i_C is an important parameter called the *dc current gain*, h_{FE} (or β_F). Unlike signal-level transistors, for which $h_{FE} > 200$ is common, values for h_{FE} of 5 to 20 at their rated current are typical for power transistors.

Base current also affects the off-state characteristics of a BJT, as shown in Fig. 15.4(a). If the base current is sufficiently negative (resulting from reverse biasing the base–emitter junction during turn-off), the collector voltage can go as high as the collector–base junction breakdown voltage V_{CBO}.[§] If $i_B = 0$ in the off state, the collector voltage can go only as high as V_{CEO},

[‡] The linear and forward active regions of MOSFET operation are also referred to as the triode and saturation regions, respectively.

[§] This subscript notation defines the terminal conditions for the parameter. For example, V_{CBO} is the collector–base voltage with the emitter open.

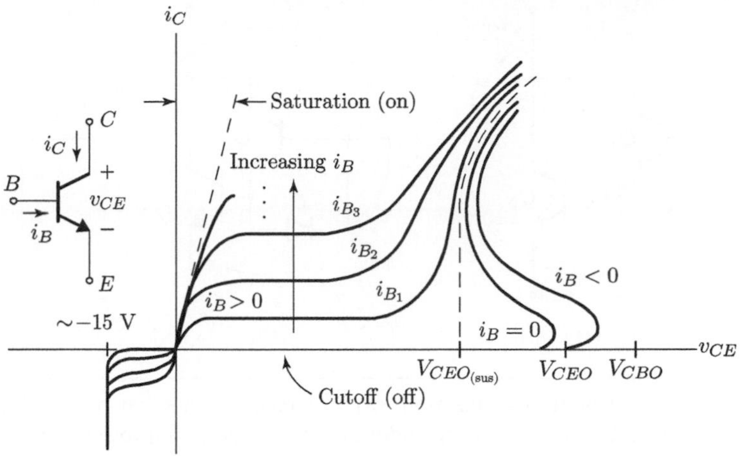

(a)

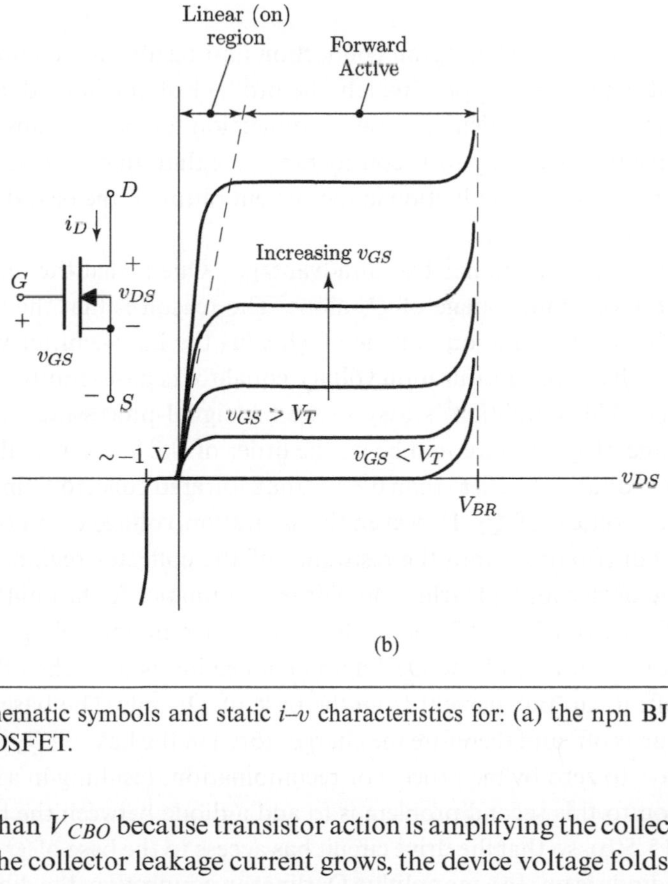

(b)

Figure 15.4 Schematic symbols and static *i–v* characteristics for: (a) the npn BJT; and (b) the n-channel MOSFET.

which is lower than V_{CBO} because transistor action is amplifying the collector leakage current. In either case, as the collector leakage current grows, the device voltage folds back to the *sustaining* voltage $V_{CEO(\text{sus})}$ – the voltage at which the leakage current amplified by the transistor action is in balance with, or sustains, the collector current.

Darlington Connection Because the current gain of a power BJT is so low, the connection of two BJTs as shown in Fig. 15.5(a) is often used to reduce the base drive requirement of the main

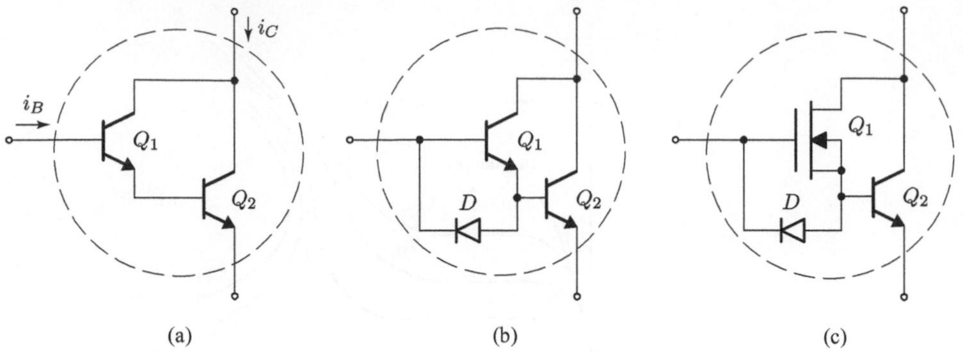

Figure 15.5 The Darlington connection. (a) The basic connection of two bipolar transistors. (b) The connection of (a) with the addition of a diode to improve its turn-off speed. (c) A MOSFET connected as the drive transistor Q_1.

device. This is known as a Darlington connection and results in a composite device having an effective current gain that is approximately the product of the individual gains, that is, $h_{FE} = I_C/I_B \approx h_{FE_1}h_{FE_2}$. It is a particularly useful connection for power transistors because the base current necessary to drive the pair is considerably less than that needed to drive Q_2 alone. The smaller base current allows a reduction in the current rating of the base drive circuit, simplifying its design.

The Darlington connection has two disadvantages. One is that the on-state drop of the pair is higher than the on-state voltage of Q_2 alone. The reason is that the lowest on-state voltage of the pair is the hard-saturation voltage of Q_1 plus the base–emitter voltage of Q_2, whereas Q_2, if used alone, has a hard-saturation voltage equal to its base–emitter voltage *minus* its base–collector voltage. This condition is easy to see for signal-processing transistors, in which the saturation voltage of Q_2 alone could be on the order of 0.2 V, whereas the Darlington pair can have an on-state voltage of no less than 0.8 V – the saturated collector–emitter voltage of Q_1 plus the base–emitter voltage of Q_2. However, the saturation voltage of a power device usually has an appreciable contribution from the resistance of the collector region, making the difference between a single device and a Darlington pair less dramatic. A Darlington pair exhibits an on-state voltage of between 2 and 5 V, depending on its current and voltage ratings.

The second disadvantage of the Darlington connection is that when the drive circuit tries to turn the pair off, it can pull current from the base of Q_1 only. The base current of Q_2 goes to zero when Q_1 turns off, and therefore the charge stored in the base of Q_2 is not forcibly removed. Instead, it decays to zero by the process of recombination, resulting in a long turn-off time. An effective solution to this second problem is to add a diode between the bases of Q_1 and Q_2, as shown in Fig. 15.5(b), so that the drive circuit has access to the base of Q_2 during turn-off, when $I_B < 0$ and the diode is on. In a monolithic Darlington connection, the diode is usually a separate device mounted in the same package and wire-bonded to the die containing the Darlington pair.

We can also use a MOSFET for Q_1, as shown in Fig. 15.5(c), resulting in a composite device with the MOSFET's ease of drive and the BJT's on-state drop. The larger the die area of the MOSFET, the smaller its on-state resistance, and the lower the on-state voltage of the pair.

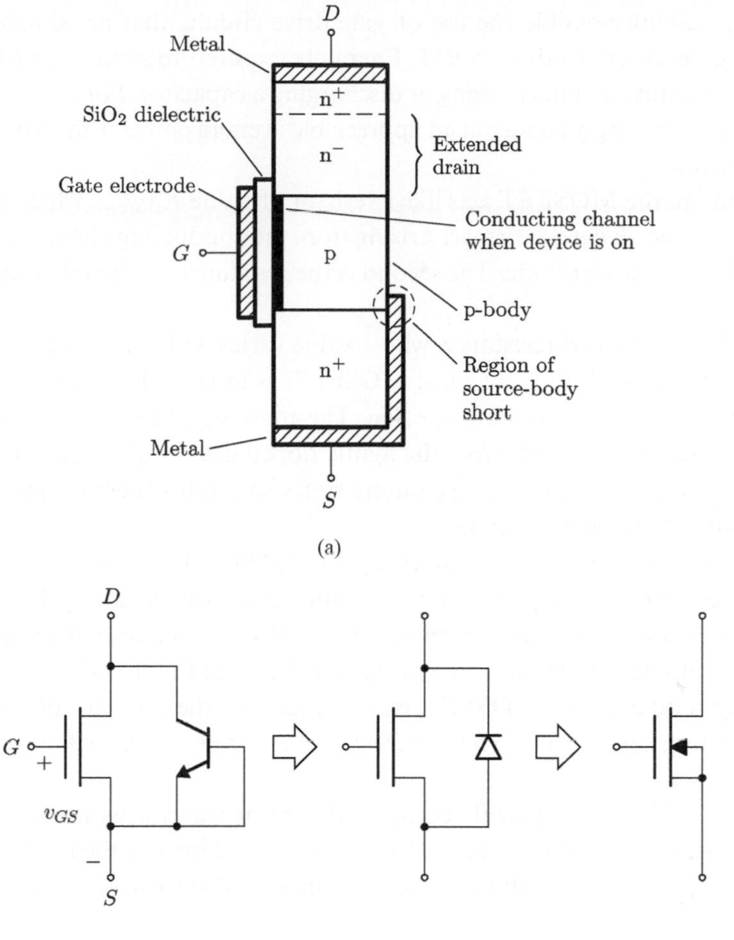

Figure 15.6 The n-channel power MOSFET. (a) Simplified structure. (b) The parasitic npn transistor with the source contact creating a base–emitter short, the parasitic body diode, and the MOSFET symbol showing the presence of the source–body short.

A monolithic derivative of this structure is the IGBT, which has largely displaced the Darlington connection.

MOSFET Figure 15.6(a) shows a simple representation of the *n-channel* vertical MOSFET structure typically used in discrete devices. "Vertical" refers to the direction of the drain current, which is orthogonal to the device substrate. When a positive voltage v_{GS} is applied from gate to source, a conducting n-type channel is formed beneath the gate electrode in the p-region, which is known as the *body* of the device. This channel connects the drain to the source, turning the device on.

The MOSFET turns on when v_{GS} exceeds the *threshold voltage* V_T, which is typically 2–5 V for Si devices. The gate acts like a capacitor, so when the MOSFET is either on or off, the gate power is zero, making possible the use of gate drive circuits that are simple and efficient compared to those necessary to drive a BJT. Energy is required to switch the MOSFET, however, as the process is equivalent to charging or discharging a capacitor. For this reason, the gate drive circuit is required to supply energy at an appreciable average power if the MOSFET is operated at high frequencies.

When on, the MOSFET acts like a resistor of value $R_{DS(\mathrm{on})}$, which is composed of two parts. The first is the *channel resistance* arising from the conducting channel created by the electric field beneath the gate electrode. The second is the resistance of the relatively long region comprising the drain.

Unlike the channel resistance, whose value varies with v_{GS}, the drain resistance is of constant value. Therefore $R_{DS(\mathrm{on})}$ for most MOSFETs is independent of v_{GS} for v_{GS} higher than a few volts above the threshold voltage V_T. The most significant part of $R_{DS(\mathrm{on})}$ for devices rated above about 100 V results from the lightly doped drain region, called the *extended drain*, which determines their maximum V_{DS} rating, so its length for these devices is strongly related to the transistor's breakdown voltage.

The three silicon layers comprising the MOSFET of Fig. 15.6(a) create a parasitic npn transistor structure. To prevent this parasitic device from affecting the behavior of the MOSFET, the p-body and n^+ source regions are shorted by the source contact, as shown. The result is the parasitic bipolar diode across the MOSFET shown in Fig. 15.6(b). This *body diode* is often useful in bridge circuits. The MOSFET symbol indicates the presence of the source–body short and the resulting body diode. The direction of the arrow indicates that the body is p-type and the channel is n-type.

Figure 15.7 shows typical voltage and current waveforms for the MOSFET as it turns on and off with a clamped inductive load. Such a load creates the most severe switching behavior, because the switched voltage cannot begin to fall (at turn-on) or rise (at turn-off) until the

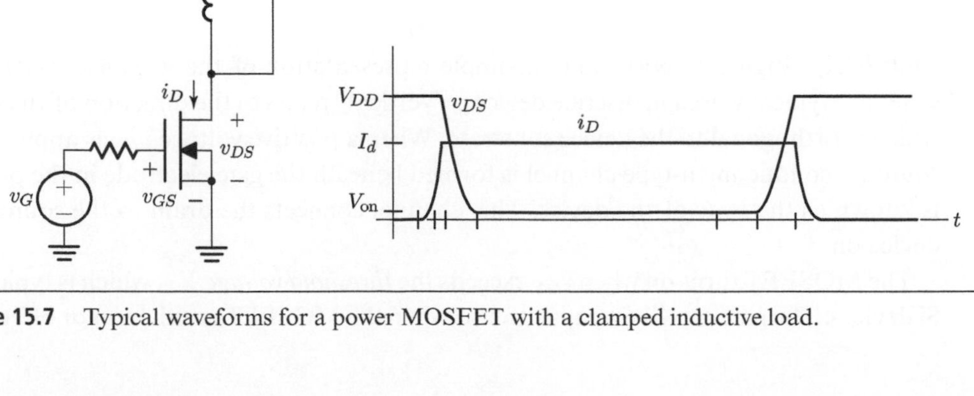

Figure 15.7 Typical waveforms for a power MOSFET with a clamped inductive load.

current has commuted from the diode to the transistor, or from the transistor to the diode. This is the condition used in device datasheets to specify turn-on and turn-off times. Note that the device does not begin to change state until the gate voltage equals V_T (typically 3–5 V). During each transition, as the drain voltage either falls or rises to its final value, the gate voltage stays constant at approximately the threshold voltage V_T. This plateau in the gate voltage waveform is called the *Miller effect*, which we explain in more detail in Chapter 23. The speed at which a MOSFET turns on or off is determined by the rate at which its parasitic capacitances can be charged or discharged, which is limited by the gate driver and the internal gate resistance.

Insulated Gate Bipolar Transistor The IGBT is a monolithically integrated Darlington-like connection of a MOSFET driving a BJT, as shown in Fig. 15.8(a). It has the benefits of the drive simplicity of a MOSFET and the high voltage capability of a BJT. Because of the way the two transistors are integrated, we can protect the MOSFET from the full off-state voltage of the IGBT by the space-charge layer at the collector junction of the BJT. The result is that a much smaller area on the device die is dedicated to the MOSFET. Note that because the two

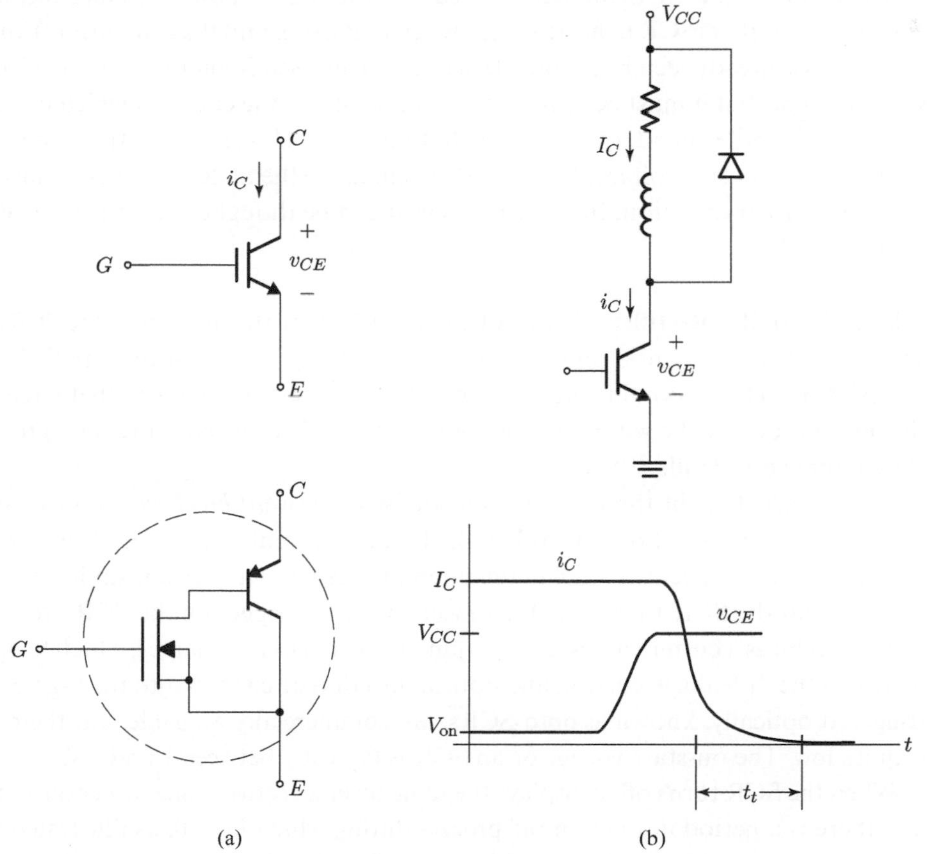

Figure 15.8 (a) The IGBT and its equivalent connection of MOSFET and BJT. (b) Waveforms at turn-off showing the slow "tailing" effect during t_t.

transistors are of opposite polarity (an n-channel MOSFET and a pnp BJT), the gate is driven with respect to the collector of the BJT.

The turn-on transition time of the IGBT is that of its BJT. The turn-off is much slower than would be possible with the BJT alone, however, because there is no way to aid the turn-off process by pulling negative current from the base of the BJT. Figure 15.8(b) shows how the collector current makes a quick change when the MOSFET first turns off, but then slowly tails to zero. The initial step corresponds to that fraction of the device current that had been flowing through the MOSFET. The tail during time t_t is the characteristic of a BJT turning off with an open-circuit base.

15.3.3 Thyristor

The name *thyristor* is generic for devices composed of four silicon layers (p–n–p–n) and exhibiting a regenerative internal mechanism that latches the device in the on state. As a family, thyristors are capable of switching voltages over 9 kV and currents over 8 kA (rms). (Such a device weighs over 5 kg.)

Members of the thyristor family used as switches in power circuits include the silicon-controlled rectifier (SCR), the triode ac switch (TRIAC), and the gate turn-off thyristor (GTO). As you have already seen in Chapter 4, the SCR is only semi-controllable; it can be turned on at a specified time, but it must be turned off by the action of the circuit in which it is connected. The GTO, on the other hand, can be turned both on and off from its control (gate) terminal, but it cannot block reverse voltage. The TRIAC is similar to the SCR, but can conduct and withstand voltage in either direction. In many respects, it can be thought of as a pair of SCRs connected in anti-parallel.

Silicon-Controlled Rectifier The most common form of the thyristor is the SCR. Figure 15.9(a) shows its schematic symbol and i–v characteristic. If $v_{AK} > 0$, a current applied to the gate turns the SCR on. This action initiates an internal regenerative mechanism that latches the device in the on state, even if the gate current is then removed. The gate power required to switch the SCR is therefore very small.

The voltage V_{BO} in the i–v characteristic is the *forward breakover voltage*. When subjected to a forward voltage above this value, the leakage current at junction J_1 acts similarly to gate current and is sufficient to initiate the internal regenerative mechanism in the absence of gate current, and the SCR turns on. Other mechanisms to trigger on the SCR are: a high dv_{AK}/dt, which produces a current across J_1 as its junction capacitance charges; a high temperature, which increases the J_1 leakage current; and optical injection of carriers into the p-gate region. Devices triggered optically, known as opto-SCRs, are commercially available, but their dv_{AK}/dt rating is quite low. The on-state voltage of an SCR is typically between 1 and 3 V.

When the SCR turns off, it displays the same reverse recovery phenomenon as a bipolar diode, and there is a period in the turn-off process during which $i_A < 0$, as illustrated in Fig. 15.9(b). Even after $i_A = 0$, however, the SCR is incapable of blocking a forward voltage until the charge remaining in the device has decayed below a critical value. The time required for this process is called the *circuit commutated turn-off time*, t_q, and ranges from about 5 to 100 μs. Reapplying a

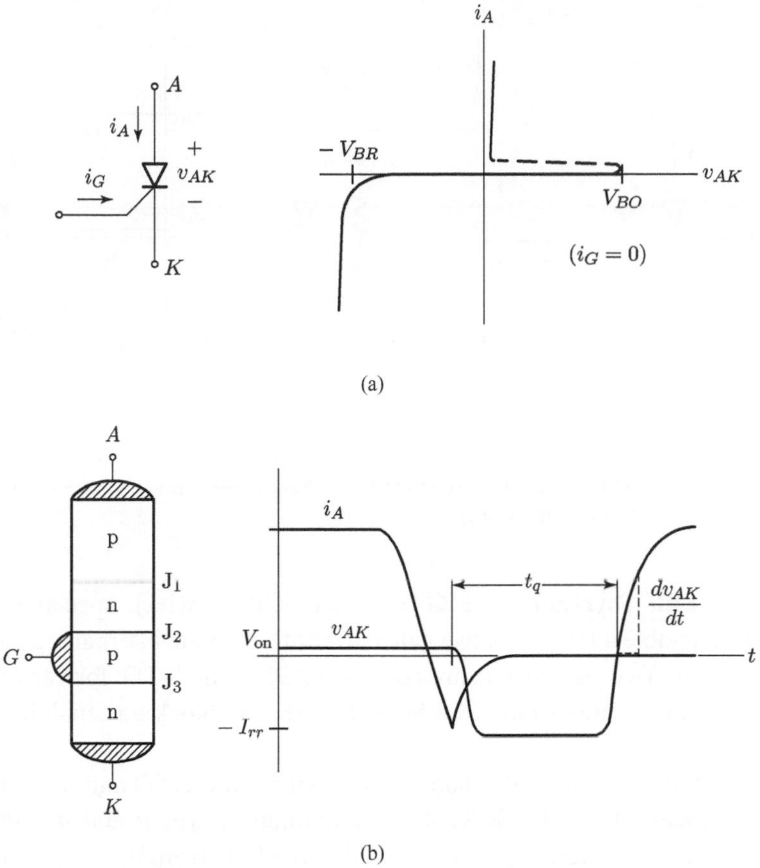

Figure 15.9 The SCR. (a) Schematic symbol and its static i–v characteristic. (b) Reverse recovery, showing the minimum off-state time t_q.

forward voltage to the SCR before t_q risks retriggering the regenerative process and turning the device on again.

There are ways to reduce t_q by adjusting the SCR's design, although an additional consequence is a higher on-state voltage. For this reason there are two standard grades of SCRs: *rectifier grade* and *inverter grade*. Rectifier-grade SCRs have a t_q in the 100 µs range and an on-state voltage between 1 and 2 V. Inverter-grade SCRs have a t_q of between 5 and 30 μs, and an on-state voltage between 2 and 3 V. While inverter-grade SCRs have historically been applied to motor drives or other PWM applications, and to high-frequency resonant converters, their use has been supplanted in most of these applications by the MOSFET or IGBT. Today the primary use of the SCR is in line-frequency rectifiers, as discussed in Chapters 4 and 9.

There is another special version of the SCR, called the *asymmetrical SCR* (ASCR), in which the reverse blocking capability of the device has been sacrificed for a much shorter t_q and a higher dv_{AK}/dt rating. The ASCR is useful in power circuits that have a diode in anti-parallel with each controllable switch, and therefore require no reverse blocking capability.

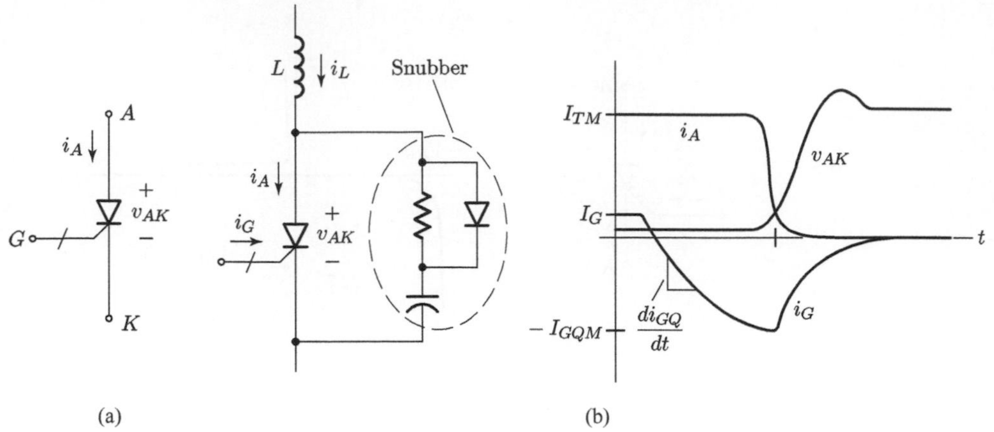

Figure 15.10 (a) The GTO symbol. (b) GTO switching waveforms when a snubber circuit is used to provide an alternate path for i_L.

Gate Turn-Off Thyristor The GTO is an SCR in which special provisions made during fabrication enable a large negative gate current to stop the internal regenerative process, turning the device off. Two parameters are compromised in the GTO: forward drop, which is typically 1 V more than a comparably rated SCR, and reverse blocking capability, which is typically less than 50 V.

Figure 15.10(a) shows the schematic symbol for the GTO and terminal variables during the turn-off process. Like the SCR, it has maximum dv_{AK}/dt and di_A/dt ratings. It also has a maximum anode current, I_{TM}, that can be turned off from the gate terminal. The *turn-off gain* of the GTO is defined as the ratio of I_{TM} to the peak negative gate current required to turn it off. Gains on the order of 3–4 are typical, so the gate drive must be capable of providing a large negative current I_{GQM} to turn off an anode current of I_{TM}. This negative current pulse has a minimum specified negative rate of rise, $-di_{GQ}dt$. During its on state the GTO requires a minimum gate current I_G to prevent the device from turning off at low anode currents.

The mechanism employed to turn off a GTO is to force the anode current into regions of the device area having low regenerative gain. Therefore, as this current-focusing process is taking place, the anode current must decrease in order to avoid localized regions of high current density and device failure. Any parasitic or load inductance in series with the GTO will prevent i_A from decreasing at the necessary rate. For this reason, GTOs are always employed with a snubber circuit, as shown in Fig. 15.10(b), to provide an alternative path for the current in the circuit inductance L. This snubber also determines the rate of rise of v_{AK} and its overshoot.

Triode AC Switch The TRIAC, whose symbol and i–v characteristic are shown in Fig. 15.11, is a thyristor capable of bilateral conduction. Because it is a slow device, having a dv/dt limit on the order of 100 V/μs, it is used almost exclusively for electric utility (50/60 Hz) applications such as light dimming. The TRIAC is bilateral, so we do not use the terms "anode" and "cathode" to identify its terminals. Instead, we identify the switched terminals as main terminals 1 and 2 (MT_1 and MT_2). We use MT_1 as the reference terminal for the gate and MT_2 voltages.

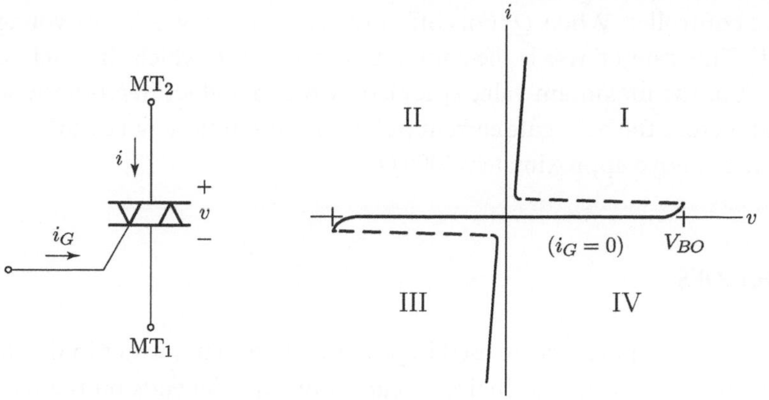

Figure 15.11 Symbol and i–v characteristic for the TRIAC.

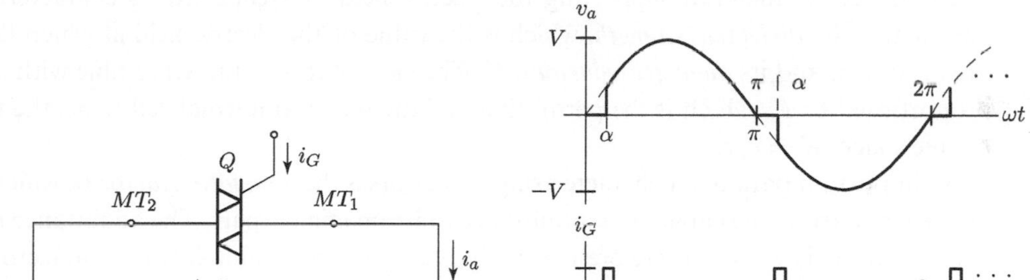

Figure 15.12 The TRIAC used as an ac controller to illustrate the commutating dv/dt to which it is subjected.

We can turn the TRIAC on with positive or negative gate current – and with MT_2 either positive or negative. The TRIAC conducts in quadrant I of its i–v plane if MT_2 is positive when gate current is applied, and in quadrant III if MT_2 is negative. The four possible triggering modes are therefore I+, I−, III+, and III−, with the "+" and "−" referring to the direction of the gate current. The triggering sensitivity of the TRIAC is different in each of the triggering modes, with I+ and III− being the most sensitive (requiring the least gate current) and therefore the most frequently used. In any of the modes, the gate characteristic, measured between G and MT_1, is similar to a forward-biased diode.

In practically all applications the TRIAC must block voltage immediately after turning off, so we replace the SCR parameter t_q with a parameter called the *commutating* dv/dt for the TRIAC. The meaning of this parameter is illustrated in Fig. 15.12, which shows TRIAC Q being used as an ac controller. When Q turns off at the zero-crossing of i_a, its voltage v_Q rises at an initial rate ωV. This rate of rise is the commutating dv/dt to which the TRIAC is subjected. It must be less than the maximum value specified by the manufacturer, otherwise the TRIAC will turn on again before the next gate current pulse. This limitation makes TRIACs unsuitable for use at frequencies above approximately 400 Hz.

15.4 Capacitors

Many types of capacitors are used in power circuits. They differ in the dielectric employed and in their physical form. The choice of capacitor type depends on the particular application. In addition to its capacitance value, a capacitor designed for power circuit applications usually has a maximum rms current specification because of internal dielectric and I^2R heating.

The dielectric material supporting the electric field in a capacitor is characterized by two parameters: its *dielectric strength*, which is the value of the electric field at which the dielectric breaks down; and its *dielectric constant*, K. The capacitance value achievable with a material is proportional to K, which is the permittivity of the material normalized to ϵ_o, the permittivity of free space: $K = \epsilon/\epsilon_o$.

An important parameter characterizing capacitors is the *dissipation factor D*, which is the ratio of the real part of the capacitor's admittance to its imaginary part. The admittance is frequency dependent, as is D, and therefore is usually presented in graphic form. A nonzero dissipation factor means that the capacitor exhibits losses. These losses are a function of construction method and material properties of the dielectric.

15.4.1 Film Capacitors

For applications that require large currents but relatively little capacitance, such as in snubbers or resonant tanks, a *film capacitor* is a common choice. It is simply two layers of metal foil separated by insulating film. This long dielectric–conductor sandwich is usually rolled into a cylinder. The two foil electrodes are offset so that only one sticks out beyond the dielectric at either end of the cylinder. Terminals are then connected to the edges of these electrodes.

The dielectric film is a thin, insulating sheet of plastic. Although many plastics can be used, the ones most commonly used in power electronic capacitors are polypropylene and polyester (Mylar). Polycarbonate is a very thermally stable dielectric with a very low dissipation factor, but is no longer generally available, as the major manufacturer of the film has stopped producing it. Polypropylene has more than twice the dielectric strength of polycarbonate (about 16 kV/mm for polypropylene versus 7.5 kV/mm for polycarbonate); polyester sits between (at about 11 kV/mm). The electrode can also be vacuum-deposited on the dielectric film to produce *metalized film*, yielding an electrode thickness of 0.01–0.05 μm; alternatively, a separate sheet of metal foil, 5–15 μm thick, can be used. Capacitors using the *metal foil* construction are capable of carrying considerably higher currents than those made with metalized film.

The dielectric properties of polypropylene are very stable when subjected to changes in temperature. The material's dielectric constant ($K \approx 2.3$) changes only a few percent when the temperature changes from 25 °C to 105 °C. Polycarbonate has a slightly higher dielectric constant ($K \approx 3.0$) and can be used at temperatures as high as 125 °C. The dielectric loss in both materials is very low, and is stable under temperature variations (a ΔD of approximately 0.02% for polypropylene and 0.1% for polycarbonate over a frequency range of 1–100 kHz is typical). The film capacitor's lead and electrode resistance are the dominant sources of loss.

The parasitic inductance of a film capacitor depends on the capacitor package's geometry and size. Typical resonant frequencies (where $\omega L = 1/\omega C$) range from 100 kHz for a 50 μF capacitor to 10 MHz for a 0.01 μF capacitor.

15.4.2 Electrolytic Capacitors

Electrolytic capacitors offer far greater capacitance per unit volume than do film capacitors. The trade-off is a much higher series resistance and an inability to withstand a reverse voltage of significant value. Because of their unipolar voltage limitation, they are most often used in dc filters in which a large capacitance is needed but the current (per unit of capacitance) is relatively small. There are two major types of electrolytic capacitors: aluminum and tantalum. Tantalum capacitors have more capacitance per volume, but they are more expensive because of the material's higher cost.

Aluminum Electrolytic Capacitors The first step in producing aluminum electrolytic capacitors is to chemically etch an aluminum foil, making its surface very rough. This process increases the surface area of the foil (which is connected to the positive terminal of the capacitor) by a factor of as much as 100. A very thin layer of aluminum oxide, Al_2O_3, is then formed on the foil's surface. This oxide, which has a dielectric constant of $K = 8.4$, is the dielectric in the capacitor. A liquid electrolyte forms the other "plate" of the capacitor as it comes in contact with the oxide. A second foil of aluminum makes electrical contact to this electrolyte, and is the negative terminal of the capacitor. As the electrolyte is not a very good conductor compared to a metal, the resistance of an aluminum electrolytic capacitor is relatively high.

The behavior of the electrolyte is the reason that the life of electrolytic capacitors is short compared to that of other components used in power circuits. The electrolyte slowly escapes through the seal of the package until enough has left to change the electric characteristics of the capacitor dramatically. The rate at which this occurs is temperature dependent; the higher the temperature, the shorter the life. The loss of electrolyte occurs during storage as well as during operation of the power circuit, although usually at a much slower rate because the capacitor is cooler. Before the electrolyte runs out, the most common cause of failure of an aluminum electrolytic capacitor is a short circuit through the oxide.

Tantalum Electrolytic Capacitors Tantalum oxide, Ta_2O_5, has a dielectric constant approximately three times that of Al_2O_3, giving tantalum capacitors a significant volumetric specific density advantage over aluminum capacitors. Three different methods are used to construct tantalum electrolytics. The *tantalum foil* construction is identical to that used for making aluminum electrolytics. The *wet slug* and *solid* constructions each start with a porous tantalum pellet made by sintering powdered tantalum. A wire, which forms the positive terminal, is attached to this pellet. In the wet slug construction, the negative electrode is a liquid electrolyte, which is absorbed by the porous pellet. The negative electrode of the solid tantalum electrolytic is manganese dioxide, created on the Ta_2O_5 surface by another chemical reaction.

The temperature dependency of tantalum capacitors is fairly linear over their operating range, varying from −5% at −55 °C to +5% at 125 °C. They are also stable with respect to applied voltage.

Electric Double-Layer Capacitors The electric double-layer capacitor (EDLC, DLC, or "super-capacitor") has an extremely high capacitance, on the order of 100s of farads, but very low voltage ratings, approximately 3 V. They are capable of very high rates of charge and discharge, and useful in pulsed power applications, such as absorbing regenerated braking energy in electric vehicles. While their volumetric specific energy density is relatively low compared to batteries, their volumetric specific power density is much higher. These capacitors can be used for energy storage, but not filtering applications, as their high-frequency performance is poor.

15.4.3 Multilayer Ceramic Capacitors

Multilayer ceramic capacitors are made by stacking many layers of thin, unfired ("green") ceramic sheets on which a conductor has been screen printed using conductive inks. This sandwich is then sintered to form a solid, monolithic capacitor.

There are two types of ceramic dielectrics, designated Class I and Class II. Class I ceramics have a dielectric constant (normalized permittivity) of between 10 and 500 that is very stable (typically ±3%) over wide ranges of voltage, time, and temperature. The dissipation factor of these materials is also low, typically 0.1% at 1 MHz. Because of their low permittivity, however, Class I ceramics have a low volumetric specific density compared to Class II ceramics. Capacitors designated NPO (or COG) are made of Class I ceramic.

Class II ceramics have much higher dielectric constants than Class I materials and come in two subgroups. The first has a dielectric constant that ranges from 500 to 3000 and is relatively stable over voltage, time, and temperature. The second has a relatively unstable dielectric constant ranging from 3000 to 20 000.

A three-symbol code is used to specify the performance of Class II ceramics. The first and second symbols specify the low and high limits of an operating temperature range. The third symbol specifies by how much the capacitance value at the temperature extremes will change from its value at 25 °C. Table 15.1 defines this code. Two common types are X7R (±15% from −55 °C to +125 °C) and Z5U (+22%, −56% from +10 °C to +85 °C).

The Class II code designation provides information only about the temperature stability of the capacitance. However, Class II ceramics have a strong voltage dependency about which the code is silent. This dependency can result in an 80% decrease in capacitance at rated voltage. Class II ceramics also exhibit a logarithmic degradation of permittivity over time, losing on the order of 10% of their capacitance at one year, and 20% at 10 years. Representative voltage and temperature dependencies for typical Class II ceramic capacitors are shown in Fig. 15.13.

Multilayer ceramic capacitors can have very low series resistance and inductance and, in addition to their use in high-frequency resonant circuits, are often used as high-frequency bypass

Table 15.1 Class II ceramic capacitor codes.

Low T (°C)		High T (°C)		Change from 25 °C value (%)	
X	−55	4	+65	A	±1.0
Y	−30	5	+85	B	±1.5
Z	+10	6	+105	C	±2.2
		7	+125	D	±3.3
		8	+150	E	±4.7
				F	±7.5
				P	±10
				R	±15
				S	±22
				T	+22, −33
				U	+22, −56
				V	+22, −82

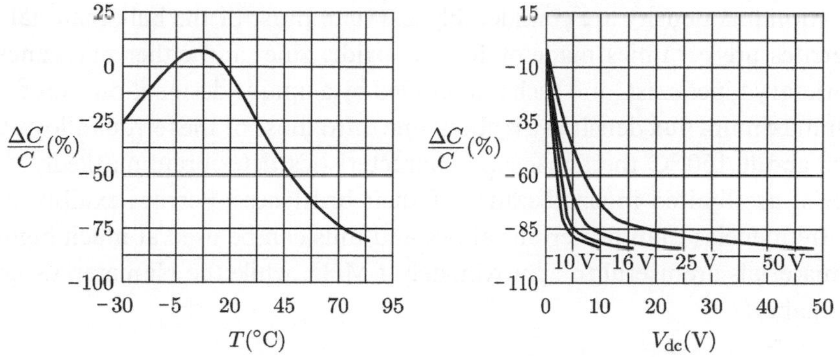

Figure 15.13 Representative temperature and voltage dependencies of Class II ceramic capacitors.

capacitors in parallel with electrolytics in filtering applications. In some cases they can provide energy densities comparable to electrolytic capacitors, but are relatively expensive.

15.5 Inductors and Transformers

Inductors and transformers used in power electronic circuits are often custom designed for the specific application. Seldom can these components be obtained from a manufacturer as a standard product. Even where off-the-shelf components are available, custom-designed components can often yield higher system performance. As a result, you must become familiar with the materials and manufacturing techniques used in producing magnetic components, subjects covered in Chapter 20. As a foundation for that presentation, in Chapter 2 we discussed the behavior of transformers and their circuit models, and here we provide an overview of loss mechanisms, materials, and the common physical forms of inductors and transformers.

15.5.1 Magnetic Materials

Most magnetic components used in power circuits contain a core of high-permeability material. Use of a core results in a smaller structure than otherwise, and the core prevents magnetic fields from extending great distances from the component. Such stray fields could cause interference or induction heating of other components or structures. Two broad classes of materials are used to make cores: most commonly ferrous alloys, and otherwise magnetic ceramics known as *ferrites*.

A great variety of ferrous alloys is available, and in several forms. These alloys differ in permeabilities, loss characteristics, and mechanical workability. For low-frequency applications (below 1 kHz), iron with a low silicon content, often known as *transformer steel*, is generally used. At higher frequencies, other elements are added to the iron to increase its permeability and lower its loss. These more exotic magnetic steels – containing various amounts of cobalt, nickel, molybdenum, and chromium – can be used to frequencies of about 20 kHz. They are very expensive relative to the silicon–iron materials. The ferrous alloys are available in either sheet or powdered form. When powdered, they can be used at frequencies higher than those cited, with nanocrystaline materials useful up to around 100 kHz. However, both their permeability and saturation flux density are considerably less than those of the bulk material.

Ferrites are ceramics made of ferrous oxide, zinc, and either manganese or nickel. Many proprietary types exist, and each is identified by a unique designation – for example, 4C4 or 3C6. Their maximum flux densities are about one third those of the ferrous alloys, and at temperatures above about 150 °C their magnetic characteristics deteriorate markedly. Cores made of this material are limited in size because of their brittleness. Ferrites exhibit much lower losses at high frequencies than do ferrous alloys and thus can be used at much higher frequencies. The Mn materials are useful to approximately 1 MHz, while the Ni materials are acceptable to 10s of megahertz.

15.5.2 Magnetic Core Geometries

Figure 15.14 shows some of the commonly used core shapes. The "C," "E," and "pot" cores are the easiest to utilize, because their windings can be made on an open bobbin. The bobbin is then slipped over the legs of the "C" or "E" cores or the center post of the "pot" core. A toroidal core, on the other hand, must be wound by threading the wire through the core, a process that can be automated, but which is generally more expensive than winding on a bobbin.

15.5.3 Losses in Magnetic Components

Loss in a magnetic component arises from two sources: the resistance of the copper windings, and hysteresis and eddy currents in the core. At high frequencies, the copper loss is aggravated by skin and proximity effects, but in general we can calculate it in a relatively straightforward manner. Core loss, however, is a nonlinear function of both frequency and flux level. For a particular material, core manufacturers generally provide data giving core loss in W/kg as a graph parametric in frequency and flux level, as shown in Fig. 15.15 for ordinary transformer steel. For ferrites, the loss data is normally provided in W/cm^3.

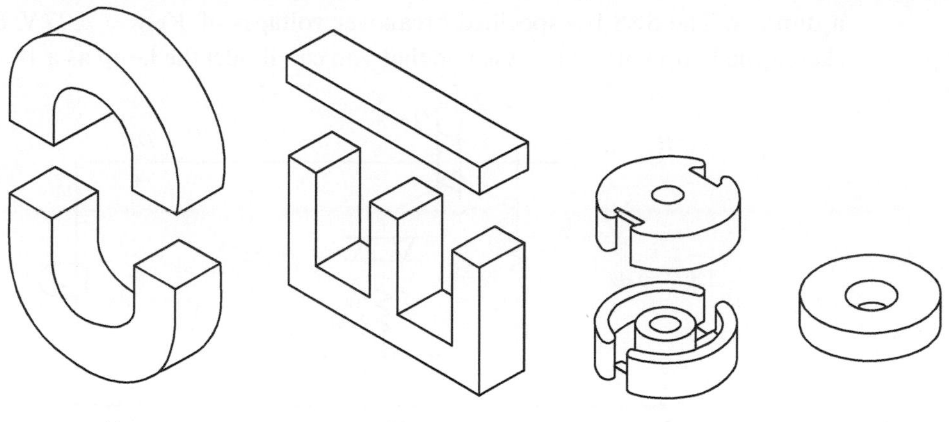

Figure 15.14 Core geometries used in power circuits. (a) Double "C" core. (b) "E" core with an "I" closing piece. (c) "Pot" core. (d) Toroidal core.

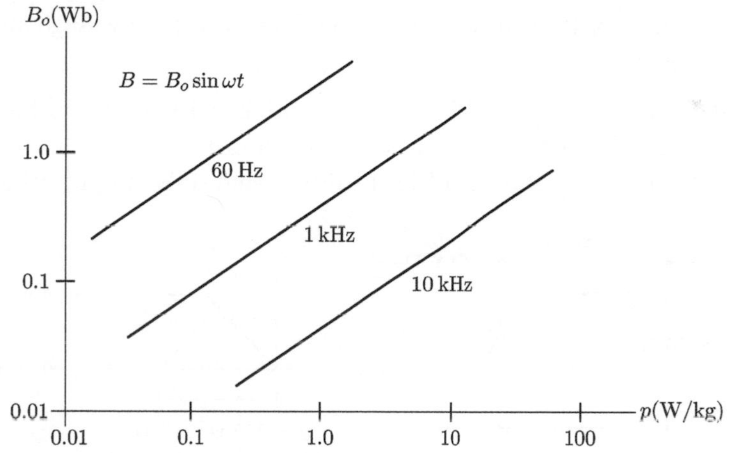

Figure 15.15 A core-loss graph for ordinary transformer steel.

PROBLEMS

15.1 A simple light dimmer circuit is shown in Fig. 15.16. When the TRIAC, Q_T, is off, C charges through R_1 and the lamp. The device Q_S is a *silicon bilateral switch* (SBS), one of a family of *trigger devices* designed to be used in thyristor gate-drive circuits. Think of it as a gateless TRIAC having symmetrical and low breakover voltages, $\pm V_{BO}$. If the magnitude of the voltage across Q_S exceeds V_{BO}, the SBS will turn on and thereafter behave as a TRIAC.

In this circuit, the dimming level is controlled by varying R_1. Note that the dimmer is connected between the hot (H) and dimmed hot (DH) nodes and does not require access to the neutral wire (N), which is often not available at a wall switch being retrofitted with a dimmer. The SBS has specified breakover voltages of $V_{BO} = \pm 27$ V. Determine and sketch the lamp voltage v_ℓ. Assume that you can model the lamp as a 100 Ω resistor.

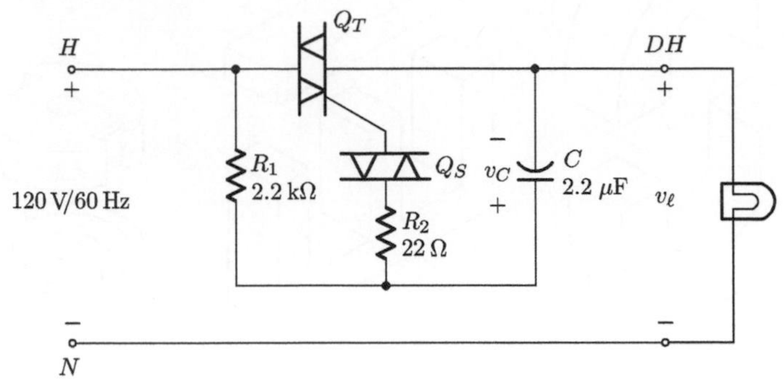

Figure 15.16 The light dimmer circuit of Problem 15.1. The TRIAC is fired by the SBS, Q_S.

15.2 Sketch the transistor voltage and current waveforms v_Q and i_Q for the down converter of Fig. 15.3. In what ways does the diode affect the voltage and current ratings of the transistor?

15.3 The converter of Fig. 15.17(a) is operating at 100 kHz, and the base drive signal to the transistor has a duty ratio of 50%. If the transistor voltage and current waveforms during switching are as shown in Fig. 15.17(b), what is the output voltage of the converter?

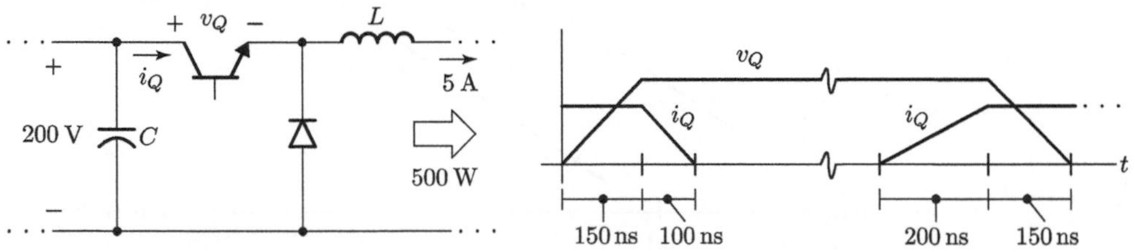

Figure 15.17 (a) The buck converter of Problems 15.3 and 15.4. (b) Voltage and current waveforms during switching for the converter of (a).

15.4 The buck converter of Fig. 15.17(a) is operating at a frequency of 20 kHz and delivering 500 W to the load. The diode is ideal, but the bipolar transistor has an on-state drop of 1.5 V, a leakage current of 2 μA, and nonzero switching times whose values are shown in Fig. 15.17(b). What is the efficiency of the converter? What is the efficiency if the switching frequency is increased to 50 kHz?

15.5 What is the maximum effective duty ratio that we can achieve in the 100 kHz converter of Problem 15.3 if we require the diode to turn on for an arbitrarily short time before the transistor turns on again?

15.6 In discussing the SCR, we said that the opto-SCR has a low dv_{AK}/dt rating. Explain why it is unsuitable for use as a bridge switch.

15.7 We stated that the current gain of a Darlington pair is *approximately* $h_{FE_1}h_{FE_2}$. If $h_{FE_1} = h_{FE_2} = 20$, what is the error in h_{FE} that results from using this approximation?

15.8 The diode in the buck converter of Fig. 15.18 is in series with 500 nH of parasitic lead inductance, and the diode current i_D is as shown. Calculate and plot the diode and transistor voltages v_D and v_Q and the transistor current i_Q. If the converter is operating at 200 kHz, what are the switching losses in the diode and transistor?

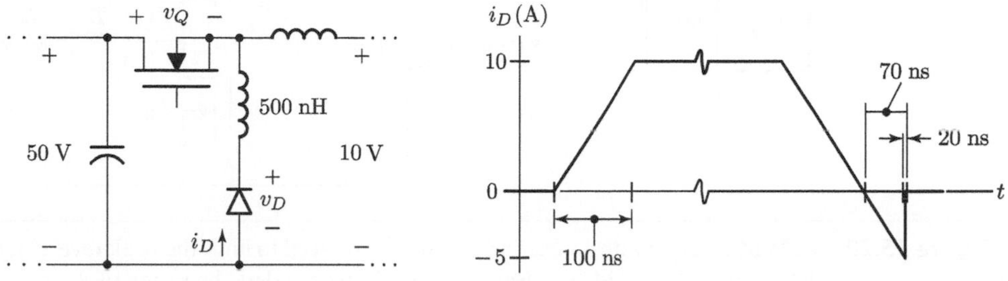

Figure 15.18 The buck converter of Problem 15.8, and its diode current waveform.

15.9 An inverter-grade SCR, Q, is used to excite a tank circuit, as shown in Fig. 15.19. Q has a reverse recovery time of $t_{rr} = 10\,\mu s$ and a recovery characteristic so snappy that you can assume $t_f \approx 0$. The diode D and the 20 Ω resistor form a snubber to maintain continuity of energy in L when Q turns off. The SCR turns on at $t = 0$. Calculate and sketch i_a and v_a.

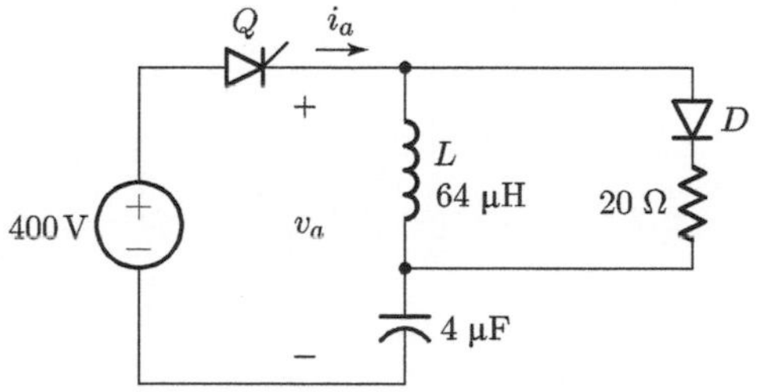

Figure 15.19 The circuit of Problem 15.9, in which v_a and i_a are calculated and sketched.

15.10 The circuit of Fig. 15.20 is a 5 kHz PWM inverter generating a sinusoidal short-term average voltage $\overline{v}_a$, so that $i_a = 200 \sin \omega_a t \approx \overline{v}_a/R$; that is, at the output frequency ω_a, the LR load appears resistive. The switching sequence for $\overline{v}_a > 0$ is as shown. The body diodes in Q_1 and Q_3 have reverse recovery times of $t_{rr} = 300$ ns, and t_f is a negligible part of this time. Because of this nonzero recovery time, an inductor, $L_s = 1\,\mu$H, is placed in series with the source V_d to limit the reverse recovery current when Q_1 and Q_3 toggle.

The diode D_s and resistor R_s dissipate the excess energy trapped in L_s at the end of the Q_1 or Q_3 recovery period. What is the required power rating of R_s?

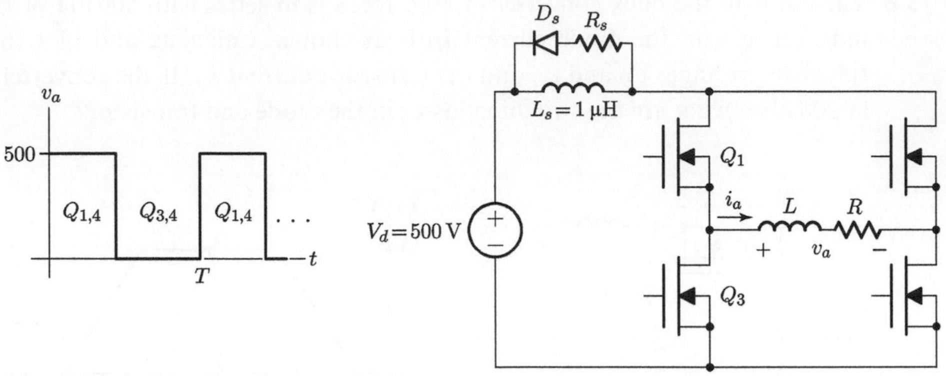

Figure 15.20 A PWM inverter with source inductance introduced to limit the peak reverse recovery currents of the Q_1 and Q_3 body diodes. Problem 15.10 considers the rating of R_s.

16 Review of Semiconductor Devices

A basic understanding of device physics, and how the physics relates to device behavior, are valuable assets for both interpreting a device's specification and its successful application.

In this chapter we review the fundamentals of semiconductor device behavior: the existence of holes and electrons, the methods by which charge transport occurs, and the physical details of a p–n junction in thermal equilibrium. We then derive the electric behavior of the bipolar diode, the bipolar junction transistor, and the metal-oxide-semiconductor field-effect transistor. Understanding how these three devices function, only a small further step is needed to understand the operation of the silicon-controlled rectifier and insulated-gate bipolar transistor discussed in Chapter 17.

We focus here on Si-based devices. We do not consider the effects of high voltage or high current density in these analyses, nor the behavior of devices fabricated with the *wide-bandgap* materials silicon carbide (SiC) and gallium nitride (GaN). We address these topics in Chapter 17.

16.1 Elementary Physics of Semiconductors

In this section our purpose is to review the basic physics of semiconductor devices. We first discuss the creation of holes and electrons, which are the positive and negative charge carriers in a semiconductor. We then describe the mechanisms of charge transport and analyze a p–n junction in thermal equilibrium.

16.1.1 Charge Carriers in a Semiconductor

The physical characteristic of semiconductors that distinguishes them from metals is the presence of both negative *and* positive mobile charge carriers. The negative charge carrier is the electron (as it is for a metal); the positive charge carrier is called a *hole*, which is the absence of an electron from the valence band of a semiconductor atom.

Intrinsic Semiconductor Figure 16.1(a) shows a simplified two-dimensional representation of a three-dimensional pure silicon lattice. Such a lattice is called *intrinsic* to distinguish it from an *extrinsic* lattice, or one that has a small percentage of atoms of an impurity present in the lattice. Each silicon atom forms a covalent bond with its four neighboring atoms, sharing its four outer-shell (valence) electrons so that all the atoms form completed shells of eight electrons.

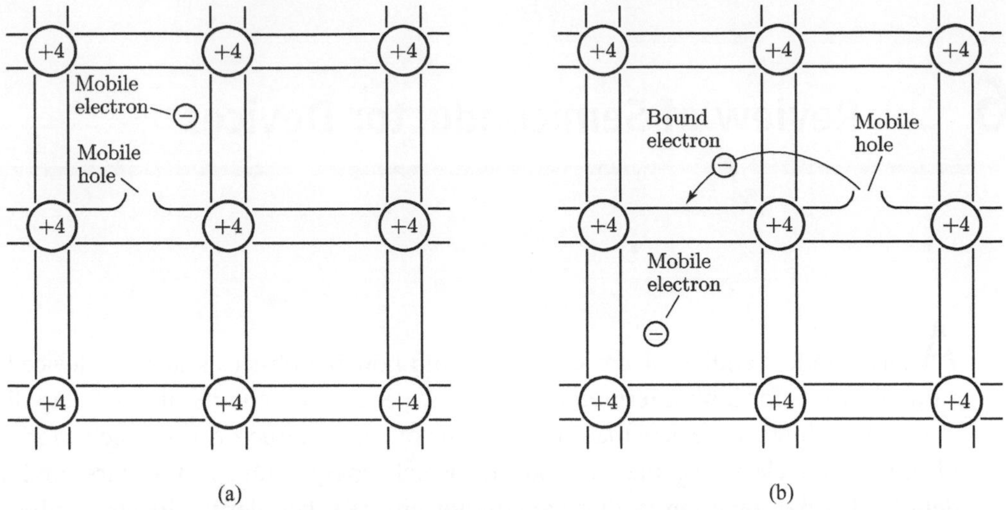

Figure 16.1 An intrinsic silicon lattice. (a) Creation of a mobile electron–hole pair by a broken covalent bond. (b) Hole movement caused by the "sliding" of bound electrons adjacent to the hole.

Although the silicon lattice is stable at room temperature, some electrons have sufficient thermal energy, E_g (known as the energy *bandgap*), to break their bonds, creating a sea of *mobile* (*free*) electrons. Besides creating a mobile electron, each broken bond becomes a positively charged location, called a hole in the lattice. In a metal, this positive charge remains tied to the atom and is therefore immobile. In a semiconductor, however, very little energy is required for a valence electron forming a bond with an adjacent atom to replace the electron whose escape caused this first broken bond, as shown in Fig. 16.1(b). In essence, the hole (which is the absence of a bound electron) has moved. Although the details of this movement are complicated, the hole in general acts like a mobile, positively charged particle with a positive mass.

In a sample of intrinsic semiconductor, one hole exists for every mobile electron (from here on we use the term "electron" to mean "mobile electron"). Owing to the dielectric relaxation phenomenon, which drives any net space charge to zero in a conductive medium, the holes and electrons reside in equal concentrations throughout the sample. We label these concentrations p and n, respectively, in units of number per cm^3. In intrinsic material, $n = p$, because the freeing of an electron always leaves behind a broken bond, creating a hole.

Because both electrons and holes can move, their individual concentrations can vary spatially, while at the same time maintaining spatial charge neutrality. This is the significant difference between a metal and a semiconductor. The absence of mobile positive charge carriers in a metal prevents the creation of non-uniform electron distributions. If a non-uniform distribution were present, it could not be neutralized by a corresponding distribution of positive charges, and the resulting electric field would force a uniform redistribution of electrons (dielectric relaxation).

Electrons and holes may *recombine* if they encounter one another. For low concentrations of mobile carriers, the recombination rate is approximately proportional to the product of their concentrations, or pn. In intrinsic material, with only the equilibrium thermal energy causing the creation (*generation*) of hole–electron pairs, the concentrations of holes and electrons are those necessary for the recombination and generation rates to be equal. This state is called *thermal equilibrium*, in which

$$n_o = p_o = n_i, \tag{16.1}$$

where the subscript "*o*" indicates thermal equilibrium, and n_i is the *intrinsic concentration*. The generation rate increases with temperature, so the value of n_i changes with temperature. Its specific temperature dependence is

$$n_i = CT^{3/2}e^{-E_g/2kT}, \tag{16.2}$$

where E_g is the energy bandgap in the semiconductor (1.1 eV for silicon), k is Boltzmann's constant (9×10^{-5} eV/K), T is the temperature in K, and C is a constant that is a function of the material. For silicon, $n_i \approx 1.4 \times 10^{10}/\text{cm}^3$ at 25 °C. Note the strong dependence of n_i on E_g. It is the differences in E_g among different semiconductors (e.g., Ge, SiC) that is principally responsible for the unique behaviors of devices based on different materials.

Extrinsic Semiconductor To change the ratio of hole to electron concentrations in thermal equilibrium, we can dope the silicon – which is in column IV of the periodic table – with elements from either columns V or III of the table.† Atoms of elements from column V, such as phosphorous or arsenic, have five valence electrons. When such an atom forms a covalent bond with four neighboring silicon atoms, as shown in Fig. 16.2(a), its fifth outer-shell electron is only loosely bound to the atom. Very little energy is required to free it, thus creating a free electron. Conversely, elements from column III, such as boron, have three valence electrons. When such an atom forms a bond with silicon atoms, as shown in Fig. 16.2(b), the shell is incomplete (less than eight electrons). A nearby valance electron can move to this shell with very little change in energy, indicating the creation of a hole.

At room temperature the thermal energy is high enough to free the fifth valence electron from most of the column V impurity atoms. These join the sea of mobile electrons. For this reason, we call column V elements *donor* impurities. The positively ionized atom left behind does not behave as a hole, because it takes too much energy for an adjacent valence bond electron to replace the lost electron. Instead, the positive charge remains fixed at the site of the donor atom. The concentration of these fixed, ionized donor sites depends on how the impurity was introduced into the lattice. Its value can vary spatially, and is designated $N_D^+(x)$, where x is a spatial coordinate.

Similarly, at room temperature most of the column III impurities have filled their valence shells by accepting an electron from a nearby silicon atom, creating a hole in the process. For this reason, we call column III elements *acceptor* impurities. The acceptor atom with a completed outer shell now has a net negative charge and remains fixed in the lattice with a spatially dependent concentration, designated $N_A^-(x)$.

Some of the electrons and holes in an extrinsic semiconductor are also created in pairs by thermal generation. However, over the operating temperature range of semiconductor

† This is also true for SiC, but doping GaN is done with elements from columns II and IV since Ga is a column III element.

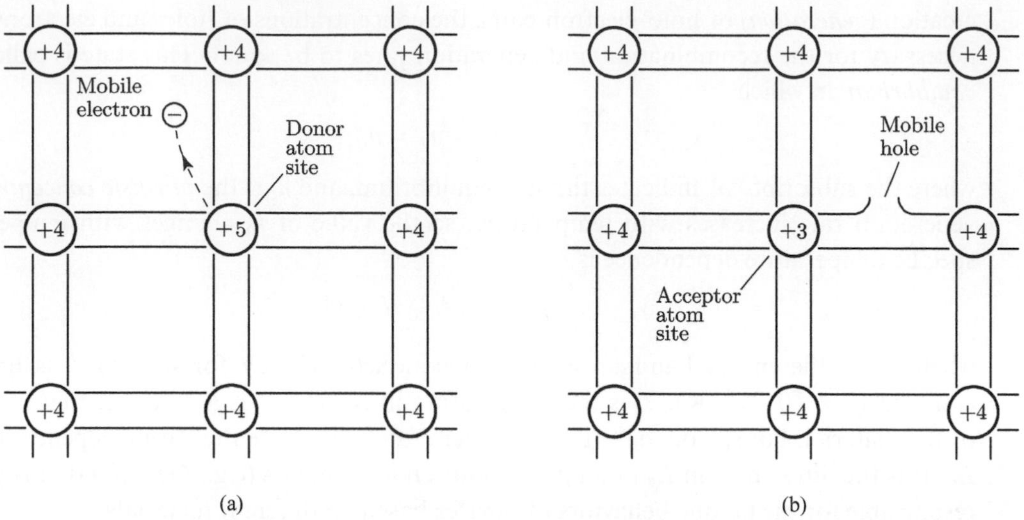

Figure 16.2 Extrinsic semiconductor lattices showing the creation of mobile holes and electrons. (a) A lattice doped with a column V (donor) impurity. (b) A lattice doped with a column III (acceptor) impurity.

devices, the concentration of the species contributed by the impurity dominates the ionization process. For example, the concentration of electrons in material doped with donor impurities is approximately equal to the concentration of the donors, or N_D. Holes, however, are still created only by the breaking of silicon atom bonds, and therefore the concentrations of holes and electrons are no longer equal. To determine the carrier concentrations, we must now combine the constraint of charge neutrality with the balance of thermal generation and recombination.

The charge neutrality constraint that results from the dielectric relaxation phenomenon, which forces any space charge density to zero throughout the sample, requires that

$$p_o - n_o + N_D^+ - N_A^- = 0. \tag{16.3}$$

At room temperature and above, the concentrations of ionized impurities are very close to the total impurity concentrations N_D and N_A. That is,

$$N_D^+ \approx N_D, \qquad N_A^- \approx N_A. \tag{16.4}$$

Because the impurity concentration is always very small compared to the concentration of silicon atoms, the thermal generation rate in extrinsic silicon is nearly the same as it is in intrinsic silicon. This generation is balanced by recombination in thermal equilibrium. The $n_o p_o$ product in thermal equilibrium, which is proportional to the recombination rate, is thus the same as for intrinsic material, so

$$n_o p_o = n_i^2, \tag{16.5}$$

but now $n_o \neq p_o$.

We can find the thermal equilibrium concentrations n_o and p_o for an extrinsic semiconductor from (16.3) and (16.5). The donor and acceptor concentrations usually differ by several orders of magnitude in an extrinsic semiconductor material. Therefore, the total impurity concentration is dominated by the concentration of one type of impurity. This dominant impurity must have a concentration much greater than n_i; otherwise the material will behave no differently from the intrinsic material. For example, if $N_D \gg N_A$,

$$n_o \approx N_D, \qquad p_o \approx \frac{n_i^2}{N_D}. \tag{16.6}$$

In this case there are many more electrons than holes, and the material is called *n-type*. Similarly, when $N_A \gg N_D$,

$$p_o \approx N_A, \qquad n_o \approx \frac{n_i^2}{N_A}, \tag{16.7}$$

and the material is called *p-type*.

16.1.2 Charge Transport in Semiconductors

Charge transport (current flow) in a semiconductor is created by two mechanisms: *drift*, which is charge flow forced by an electric field; and *diffusion*, which is flow produced by a concentration gradient.

Drift Current The presence of an electric field exerts a $q\vec{E}$ force on each particle of charge q, which accelerates the particle until it encounters an obstruction. The result is an average velocity proportional to the electric field,[‡] and a *drift current* density for electrons and holes,

$$\vec{J}_e^{\,\text{drift}} = qn\mu_e\vec{E} + qp\mu_h\vec{E}, \tag{16.8}$$

where n and p are the electron and hole concentrations, and the constants of proportionality $\mu_{e,h}$ are the *mobility* for electrons and holes, whose values vary with material, carrier type, and carrier concentration. The electric field $\vec{E}$ is called the *drift field*. Conduction in metals is by drift.

Diffusion Current Carrier gradients cause a net *diffusion* of carriers from regions of high concentration to regions of low concentration, resulting in a *diffusion current*. The average velocity of a carrier in diffusion is proportional to the size of the gradient, and its direction is opposite to that of the gradient (down the gradient). The diffusion current density for positively charged particles is

$$\vec{J}^{\,\text{diff}} = -qD\vec{\nabla}\mathcal{K}(x, y, z), \tag{16.9}$$

where $\mathcal{K}(x, y, z)$ is the spatially dependent concentration, and the constant of proportionality D between gradient and average carrier velocity is the *diffusion constant*. Its value, like that for mobility, depends on carrier type and material. For negatively charged particles, the diffusion current flows in the direction of the gradient.

‡ At high fields this proportionality no longer holds, and the velocity is approximately constant at the electron or hole *saturation velocity*.

Diffusion is a transport mechanism that does not occur in metals because dielectric relaxation prevents maintenance of a non-uniform charge distribution. Semiconductors, however, have two types of mobile charge carriers – holes and electrons – and an equal non-uniform distribution of both can exist without creating a net charge. The existence of diffusion as a charge transport mechanism in semiconductors is the essential difference between them and metals. This mechanism allows us to make electronic devices from semiconductor materials.

In one dimension, we can express the drift and diffusion current densities in terms of the physical and electric properties of the material:

$$J^{\text{drift}} = J_e^{\text{drift}} + J_h^{\text{drift}} = q\mu_e nE + q\mu_h pE, \tag{16.10}$$

$$J^{\text{diff}} = J_e^{\text{diff}} + J_h^{\text{diff}} = qD_e\frac{\partial n}{\partial x} - qD_h\frac{\partial p}{\partial x}. \tag{16.11}$$

The variables n, p, and E are functions of the position x. Therefore, even if the total current density $J = J^{\text{drift}} + J^{\text{diff}}$ is constant throughout the sample, the individual drift and diffusion current densities, as well as the individual hole and electron current densities, are generally functions of position.

The parameters μ_e and μ_h are the mobilities of the electrons and holes, respectively. They are complicated functions of temperature, carrier concentration, and electric field. However, for the purpose of first-order device analysis, you can consider them to be constants that relate the average velocity of a charged particle to the strength of the electric field.

The parameters D_e and D_h are the diffusion constants of the electrons and holes, respectively. They relate the average velocity of a charged particle to the spatial gradient of its concentration. The diffusion constant for each carrier type is related to the mobility of that carrier by the *Einstein relation*,

$$\frac{D_e}{\mu_e} = \frac{D_h}{\mu_h} = \frac{kT}{q}. \tag{16.12}$$

For a material in which the current density is proportional to the electric field, we define the constant of proportionality as the *conductivity* of the material, σ. From (16.10) you can see that, for a semiconductor,

$$\sigma = q\mu_e n + q\mu_h p. \tag{16.13}$$

Note that σ is a function of the carrier concentrations. In extrinsic material, the concentration of one carrier type is usually much higher than that of the other, so the conductivity is dominated by one of the terms in (16.13).

Table 16.1 gives the nominal values of important physical and electrical parameters for various semiconductors. They vary to differing degrees with temperature and carrier concentration. The energy bandgaps E_g for GaN and SiC are an order of magnitude larger than those for Ge or Si. For this reason GaN and SiC are known as wide-bandgap materials. The parameter E_c, the *critical field*, is the field at which the material breaks down due to an unconstrained increase in mobile carriers, resulting from carrier–lattice collisions. Its value is a function of doping concentration, and the value in the table is for intrinsic material. The high values of

Table 16.1 Nominal parameters of various semiconductor materials at 25 °C.

Parameter	Si	Ge	GaN	SiC	Units
Bandgap (E_g)	1.12	0.66	3.4	3.26	eV
Critical field (E_c)	3×10^5	10^5	3×10^6	3×10^6	V/cm
Intrinsic concentration (n_i)	1.4×10^{10}	3×10^{13}	1.6×10^{-10}	8.2×10^{-9}	/cm^3
Electron mobility (μ_e)	1360	3900	1250	900	cm^2/V-s
Hole mobility (μ_h)	490	1900	200	100	cm^2/V-s
Saturation drift velocity (v_{sat})	10^7	6×10^6	2.5×10^7	2.7×10^7	cm/s
Electron diffusion constant (D_e)	34	100	25	22	cm^2/s
Hole diffusion constant (D_h)	12	50	5	3	cm^2/s
Permittivity (ϵ)	11.8	16	8.9	9.7	ϵ_o(F/m)
Thermal conductivity (κ)	1.5	0.6	1.6	3.6	W/cm-K

E_c and thermal conductivity κ for SiC are the primary reasons it is favored as a material for power devices. Although its electron and hole mobilities are lower than those for Si, most power devices operate at high electric fields during conduction. Under this condition, the carrier velocities *saturate* at v_{sat} because of lattice and carrier–carrier collisions. As the table shows, SiC and GaN have a higher v_{sat} than Si. High-voltage MOSFETs, IGBTs, and Schottky barrier diodes are available in SiC. GaN is primarily used for very fast-switching devices in lower-power applications.

Example 16.1 A Conductivity Calculation

The main terminals of power devices are most often on opposite surfaces of the silicon die (the piece of silicon on which the device is made), in order to maximize ohmic contact area. The die is usually mounted so that these surfaces are in the horizontal plane, and current flowing between the terminals is then flowing vertically. Hence, these power devices are called *vertical* devices. One consequence of this design is that the device current must flow through the substrate of the die, that thick part of the die necessary for mechanical integrity, and this substrate has resistance. Figure 16.3 shows a square die of n-type silicon with a size and thickness representative of the substrate of a 100 A device, perhaps a diode. What is the voltage drop across the substrate when the device is carrying a rated current of 100 A?

We first calculate the resistance between the top and bottom surfaces of the die:

$$R = \frac{\ell}{A\sigma} \approx \frac{\ell}{A\left(q\mu_e n\right)} = \frac{125 \times 10^{-4}}{0.25\left(1.6 \times 10^{-19} \times 200 \times 10^{18}\right)} = 1.56\,\text{m}\Omega. \tag{16.14}$$

Note that the mobility value used in this calculation is 200 cm^2/V-s from a tabulation of mobility as a function of concentration, rather than the value 1360 cm^2/V-s of Table 16.1. The reason is that the value given in the table is for doping concentrations low enough to not cause the lattice structure to deviate much from that of intrinsic silicon. However, at a concentration of 10^{18}/cm^3, the impurity atoms in the lattice begin to "get in the way" of the mobile carriers, reducing their mobility.

With the device carrying 100 A, the voltage drop across the substrate is 156 mV. Compared to a power diode forward voltage of 1 V, this substrate drop is small but not negligible. The power dissipated in the substrate at 100 A is 15.6 W during conduction, which is again small but not negligible compared with a dissipation of 100 W at the junction.

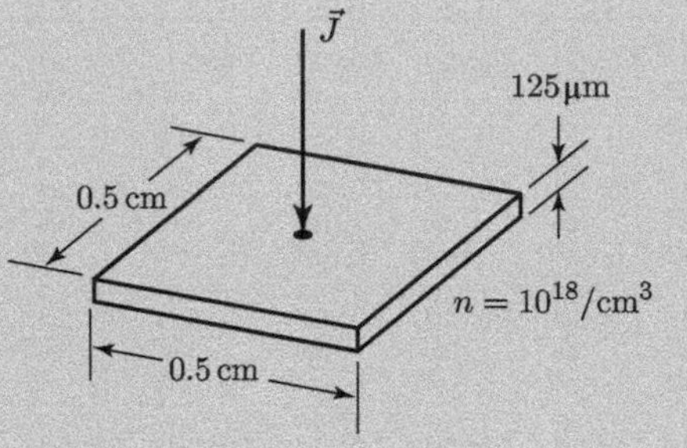

Figure 16.3 Silicon chip representing a power-device substrate.

16.1.3 The p–n Junction in Thermal Equilibrium

When n-type and p-type materials are "brought together" to form a junction, infinite gradients of both hole and electron concentrations initially appear at the point of contact (known as the metallurgical junction). The resulting diffusion currents sweep holes away from the p side, and electrons away from the n side, leaving behind the immobile charge of the ionized impurity atoms. This immobile *space charge*, which is positive on the n side and negative on the p side, creates an electric field pointing from the n side toward the p side of the junction.

The presence of this electric field now causes hole and electron drift currents to flow in directions opposite to their diffusion currents; holes drift back toward the p side and electrons drift back toward the n side. We are assuming that there is no external stimulation of the junction – and that it is in thermal balance with its surroundings – so a steady state is reached in which the net electron and hole currents are independently zero. This condition is called *thermal equilibrium*. Note that although the total current of each carrier type is zero, each has large, but equal and oppositely directed, drift and diffusion components.

The net charge density $\rho(x)$, electric field $E(x)$, and electrostatic potential $\psi(x)$ within the p–n junction in thermal equilibrium are shown in Fig. 16.4. The region of nonzero charge density in the vicinity of the junction is called the *space-charge layer* (SCL). The rest of the device is known as the *neutral region*.

The variation of the space-charge density is difficult to determine analytically, but near the p–n junction, where the material is almost completely depleted of mobile carriers, $\rho(x)$ is a constant proportional to the density of ionic impurity. Near the edge of the SCL, the charge density tapers back to zero over a distance with characteristic length known as a *Debye length*. As this distance is typically short compared to the length of the SCL, the charge distribution is approximated as stepping to zero at the edge of the SCL, as shown by the solid lines in Fig. 16.4. This is the *abrupt SCL model*.

We can find the electric field from Gauss's law. It is zero in the neutral regions: if it were not, a drift current would flow, with no equal and opposite diffusion current. Within the SCL, the electric field is not zero, but triangular in distribution. Using the abrupt SCL model, we obtain the peak electric field:

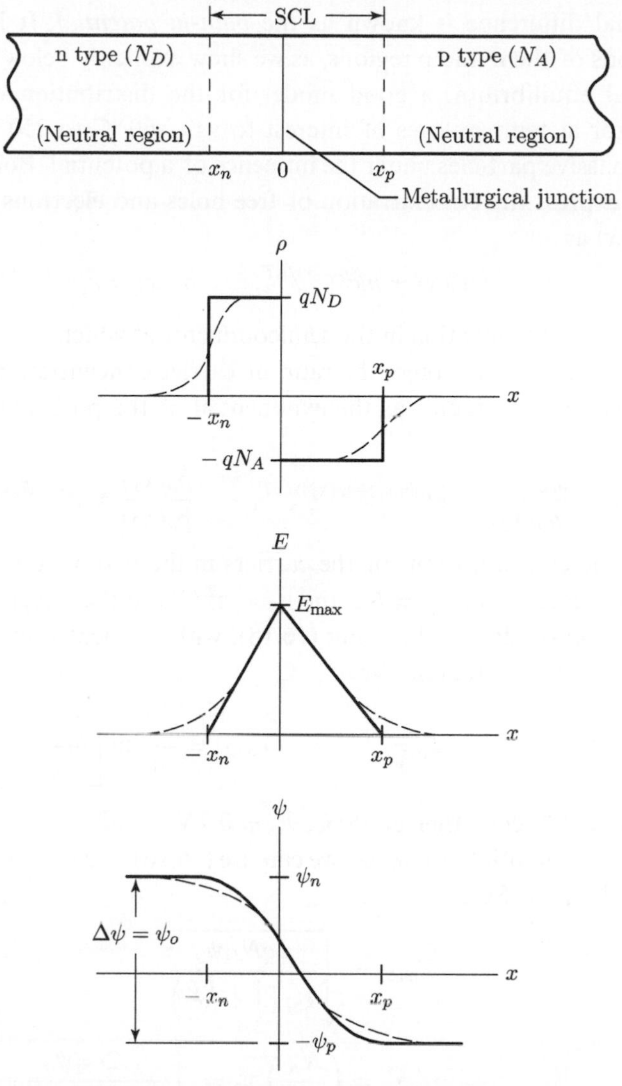

Figure 16.4 The space charge $\rho(x)$, electric field $E(x)$, and electrostatic potential $\psi(x)$ across a p–n junction in thermal equilibrium. The dashed lines show actual distributions; the solid lines are used for the abrupt SCL model.

$$E_{\max} = \frac{qN_D x_n}{\epsilon_{Si}} = \frac{qN_A x_p}{\epsilon_{Si}}, \tag{16.15}$$

where x_n and x_p are the SCL lengths in the n-type and p-type materials, respectively. We can obtain the potential difference $\Delta\psi$ between the two neutral regions by integrating the expression for the electric field from x_n to x_p:

$$\Delta\psi = E_{\max}\left(\frac{x_n + x_p}{2}\right) = \psi_o. \tag{16.16}$$

This potential difference is known as the *built-in potential*. It is a function of the doping concentrations of the n and p regions, as we show explicitly below.

In thermal equilibrium, a good model for the distribution of holes and electrons in a semiconductor at temperatures of interest to us (−60 °C to 250 °C) is that of a gas of non-interactive massive particles under the influence of a potential. Boltzmann statistics then apply, and we can express the concentration of free holes and electrons in terms of the electrostatic potential $\psi(x)$ as

$$n_o(x) = n_i e^{q\psi(x)/kT}, \qquad p_o(x) = n_i e^{-q\psi(x)/kT}, \tag{16.17}$$

where $\psi(x) = 0$ is the location in the semiconductor at which $n_o = p_o = n_i$.

Using (16.17), we can express the ratio of carrier concentrations at two different points, x_1 and x_2, in the semiconductor as the exponential of the potential difference between the two. That is,

$$\frac{n_o(x_1)}{n_o(x_2)} = e^{q(\psi(x_1)-\psi(x_2))/kT}, \qquad \frac{p_o(x_1)}{p_o(x_2)} = e^{-q(\psi(x_1)-\psi(x_2))/kT}. \tag{16.18}$$

We know the concentrations of the carriers in the two neutral regions ($n_n = N_D$ and $p_n = n_i^2/N_D$ in the n region and $p_p = N_A$ and $n_p = n_i^2/N_A$ in the p region). Hence, we can determine ψ_o as a function of doping by using (16.18), with x_1 located in the neutral n region, and x_2 located in the neutral p region:

$$\psi_o = \psi(x_1) - \psi(x_2) = \frac{kT}{q} \ln\left[\frac{N_A N_D}{n_i^2}\right]. \tag{16.19}$$

If $N_A = N_D = 10^{16}/\text{cm}^3$, then at 25 °C, $\psi_o = 0.7\,\text{V}$.

Knowing the potential difference, we can use (16.16) and (16.19) to find the peak electric field and the length of the SCL:

$$E_{\max} = \sqrt{\frac{2qN_D\psi_o}{\epsilon_{\text{Si}}\left(1+\frac{N_D}{N_A}\right)}}, \tag{16.20}$$

$$x_n = \left(\frac{N_A}{N_D}\right) x_p = \sqrt{\frac{2\epsilon_{\text{Si}}\psi_o}{qN_D\left(1+\frac{N_D}{N_A}\right)}}. \tag{16.21}$$

Example 16.2 SCL Width in an Asymmetrically Doped p–n Junction

Most p–n junctions have significantly different impurity concentrations on either side of the junction. Such a junction is called an *asymmetrical junction*. In this example we show that the SCL extends primarily into the more lightly doped side of the junction.

We assume that in the p–n diode of Fig. 16.5(a), $N_A = 10^{18}\,\text{cm}^{-3}$ on the p side and $N_D = 10^{15}\,\text{cm}^{-3}$ on the n side. The resulting built-in potential ψ_o is

$$\psi_o = \frac{kT}{q} \ln\left[\frac{N_A N_D}{n_i^2}\right] = 0.73\,\text{V}. \tag{16.22}$$

The distances that the SCL extends into the n and p regions in thermal equilibrium are therefore

$$x_n = \sqrt{\frac{2\epsilon_{\text{Si}}\psi_o}{qN_D\left(1+\dfrac{N_D}{N_A}\right)}} = 0.96\ \mu\text{m}, \qquad x_p = \frac{N_D}{N_A}x_n = 0.96\,\text{nm}. \tag{16.23}$$

Figure 16.5(b) shows the space-charge density, electric field, and electrostatic potential for this device. Note that, as x_p is so much smaller than x_n, nearly all the voltage drop appears across the n region side of the SCL.

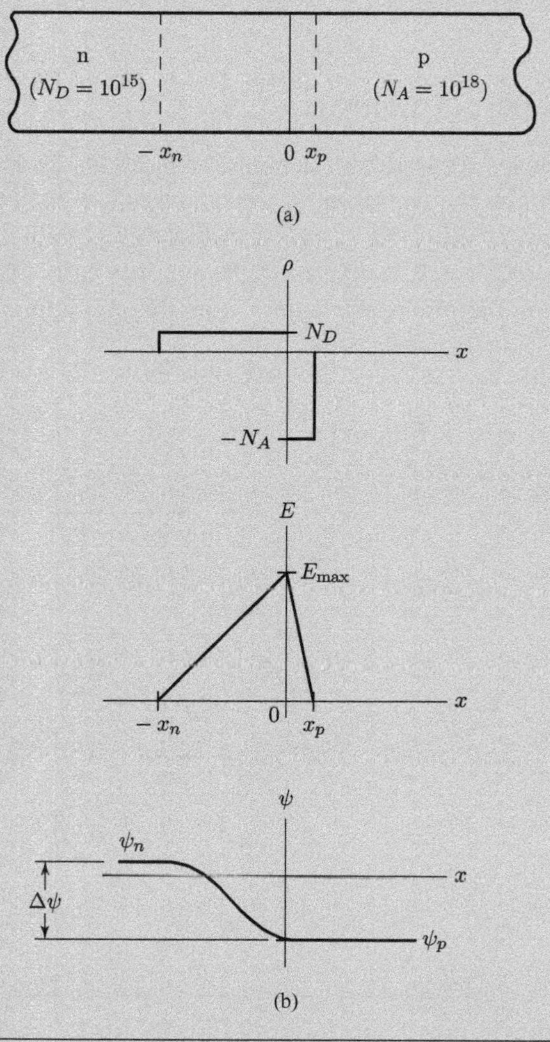

Figure 16.5 (a) An asymmetrically doped abrupt junction p–n diode. (b) Space charge, electric field, and potential distributions for the diode of (a).

16.2 Simple Analysis of a Diode

We now determine the relationship between an applied voltage and the resulting diode current. While we are focused on the diode in this analysis, all power semiconductor devices, except the Schottky barrier diode, contain junctions that behave similarly. We also illustrate the capacitive behavior of the SCL, which is often of critical importance in the application of semiconductor devices. The simple one-dimensional diode that we analyze in this section is shown in Fig. 16.6. We assume the abrupt SCL model, and indicate the edges of the SCL with dashed lines and their associated reference coordinates 0^- and 0^+. Voltage v_A is applied to the terminals of the diode as shown.

16.2.1 Dependence of the SCL Width on v_A

The voltage v_A must appear across the neutral regions of the diode and/or the SCL. Although we do not go through the derivation here, under *low-level injection* most of v_A appears across the SCL. By "low-level injection" we mean that throughout the diode the minority carrier concentrations remain small, compared to the majority carrier concentrations, even though the diode may not be in thermal equilibrium. Under this condition, the voltage across the SCL is

$$\Delta\psi = \psi_o - v_A. \tag{16.24}$$

We find the peak electric field and the n- and p-side widths of the SCL by substituting $\psi_o - v_A$ for ψ_o in (16.20) and (16.21):

$$E_{\max} = \sqrt{\frac{2qN_D(\psi_o - v_A)}{\epsilon_{\text{Si}}\left(1 + \frac{N_D}{N_A}\right)}}, \tag{16.25}$$

$$x_n = \sqrt{\frac{2\epsilon_{\text{Si}}(\psi_o - v_A)}{qN_D\left(1 + \frac{N_D}{N_A}\right)}}, \tag{16.26}$$

$$x_p = \sqrt{\frac{2\epsilon_{\text{Si}}(\psi_o - v_A)}{qN_A\left(1 + \frac{N_A}{N_D}\right)}}. \tag{16.27}$$

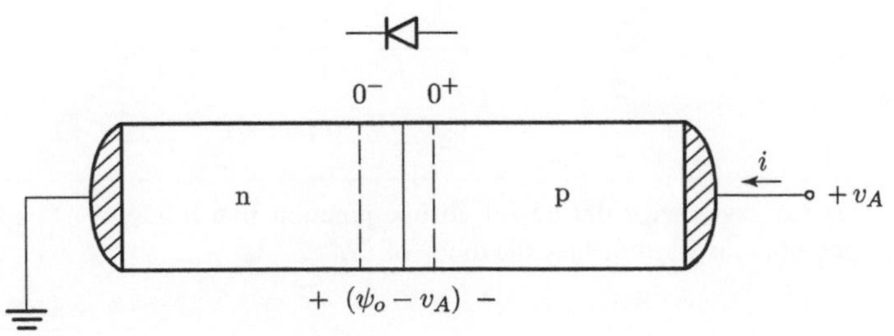

Figure 16.6 Simple one-dimensional diode with voltage v_A applied to its terminals.

Note that a positive v_A reduces the SCL width, whereas a negative v_A increases the width. Under forward bias, the applied voltage never gets so big that the SCL width reaches zero. Under reverse bias, the SCL can grow substantially if $v_A < 0$ and $|v_A| \gg \psi_o$. The electric field given by (16.25) is of particular importance because a given material will have a critical electric field, E_c, at which the material breaks down.

Remember that regions outside the SCL are neutral. Therefore only the SCL can contain the electric field producing the potential $\psi_0 - v_A$. For the junction to support a large reverse bias, the width of the SCL must increase considerably. All power devices except for the Schottky barrier diode rely on a p–n junction for their voltage blocking capability. As a consequence, providing room for the necessary growth of the SCL often dictates the size (generally the die thickness) of the device. This has additional consequences for high-voltage devices, such as increased on-state resistance.

16.2.2 Incremental Capacitance of the SCL

Changing the voltage across the SCL of a p–n diode also changes $\rho(x)$. This change of charge causes a current to flow through the diode, allowing us to model the SCL as a capacitor. Unlike the common parallel-plate capacitor, however, the charge in this junction capacitor is not directly proportional to the voltage across it. The reason is that the charge is stored in the SCL, and the dimensions of this depend nonlinearly on the applied voltage. The relationship between the stored SCL charge, Q, and v_A is

$$Q = qN_D x_n A = A\sqrt{2q\epsilon_s N_D(\psi_o - v_A)}, \tag{16.28}$$

where A is the cross-sectional area of the device.

We can find the incremental capacitance C_{inc} of the SCL for an applied voltage v_A. In reverse bias, where $v_A < 0$ and $|v_A| \gg \psi_o$, this capacitance is

$$C_{\text{inc}} = \frac{dQ}{d|v_A|} \approx A\frac{\sqrt{q\epsilon_s N_D/2}}{\sqrt{|v_A|}} \propto \frac{1}{\sqrt{|v_A|}}. \tag{16.29}$$

For power devices, C_{inc} is somewhat irrelevant, since v_A varies between essentially 0 (device on) and a large negative v_a (device off). The charge represented by (16.28) is supplied and extracted from the SCL as the device switches from on to off, and this current has both switching-speed and loss implications. The charge required by the SCL between turn-on and turn-off is therefore often provided by device data sheets.

16.2.3 Carrier Concentrations at the Edge of the SCL

Recall that (16.18) gives the relationship between the carrier concentrations at two points and the potential difference between them. However, (16.18) applies only in thermal equilibrium. Strictly speaking, therefore, we cannot use it to determine concentrations throughout the diode when an applied voltage, v_A, is present. However, because for reasonable values of v_A the SCL is disturbed only slightly from thermal equilibrium, we can use (16.18) for points within the SCL. Specifically, we use (16.18) to relate the carrier concentrations at the n- and p-region edges of the SCL:

$$\frac{n_p(0^+)}{n_n(0^-)} = e^{-q(\psi_o - v_A)/kT}, \qquad \frac{p_n(0^-)}{p_p(0^+)} = e^{-q(\psi_o - v_A)/kT}, \tag{16.30}$$

where $n_p(0^+)$ is the concentration of electrons at the p-region edge of the SCL, and so on. We can rewrite these equations as

$$\frac{n_p(0^+)}{n_n(0^-)} = \frac{n_{po}}{n_{no}} e^{qv_A/kT}, \qquad \frac{p_n(0^-)}{p_p(0^+)} = \frac{p_{no}}{p_{po}} e^{qv_A/kT}, \tag{16.31}$$

where n_{po} is the concentration of electrons in the neutral n region *at thermal equilibrium*, and so on. We also know that $n_{no} = N_D$, $p_{po} = N_A$, $n_{po} = n_i^2/N_A$, and $p_{no} = n_i^2/N_D$.

Although (16.31) tells us the ratio of carrier concentrations under any applied voltage, we still need one more relationship to determine the actual values. This relationship comes from our previous assumption that the diode is in low-level injection, which means that $n_n(0^-) \approx n_{no}$ and $p_n(0^+) \approx p_{po}$. Substituting these approximations into (16.31) yields

$$n_p(0^+) = n_{po} e^{qv_A/kT}, \qquad p_n(0^-) = p_{no} e^{qv_A/kT}. \tag{16.32}$$

Thus, (16.32) relates the carrier concentrations at the SCL edges to the applied voltage v_A. We can obtain the carrier concentration profiles $n(x)$ and $p(x)$ in the two neutral regions by applying these boundary conditions to the solution of one-dimensional *diffusion equations* for holes and electrons. In doing so we focus on the *excess* carrier concentration, which is the difference between the total and the equilibrium concentrations. We denote this excess concentration with a prime. For example, $n'_p(0^+) = n_p(0^+) - n_{po}$.

16.2.4 Determination of Diode Current

The total diode current density J, which is independent of x, is

$$J = J_e(x_i) + J_h(x_i), \tag{16.33}$$

where

$$\begin{aligned} J_e(x_i) &= q\mu_e n(x_i)E(x_i) + qD_e \left.\frac{\partial n(x)}{\partial x}\right|_{x_i}, \\ J_h(x_i) &= q\mu_h p(x_i)E(x_i) - qD_h \left.\frac{\partial p(x)}{\partial x}\right|_{x_i}. \end{aligned} \tag{16.34}$$

We now posit that in extrinsic material under low-level injection – if minority-carrier current is a significant part of the total current – then the drift component of minority-carrier current is negligible; that is, diffusion dominates the minority-carrier current flow. Our argument goes as follows.

If the drift of majority carriers (holes in the p region, for example) is important, the drift of minority carriers at the same place will be unimportant. The reason is that the minority concentration is many orders of magnitude less than the majority concentration because of our assumption of low-level injection. Moreover, both concentrations are subjected to the same electric field, so the drift components of their currents are orders of magnitude different.

Therefore, if the minority carrier current is important (that is, if it is a significant part of the total current), it must be so because of its diffusion component.

The realization of this important result makes bipolar device analysis simpler. For the minority-current density, we can now avoid the nonlinearity caused by the product of carrier concentration and electric field in (16.34), because this term is, by our assumption, negligible. As a result, the continuity equation for the minority carriers, which in general is

$$\frac{1}{q}\frac{\partial J_e}{\partial x} + G - R = \frac{\partial n}{\partial t} \tag{16.35}$$

for the p region, now simplifies to

$$D_e \frac{\partial^2 n}{\partial x^2} + G - R = \frac{\partial n}{\partial t}. \tag{16.36}$$

The term R in (16.36) is the recombination rate of carrier pairs per unit volume and time, and the term G is the generation rate of carrier pairs. (The latter includes not only those carriers generated thermally, but also any carriers generated by other means, such as light.) The net recombination rate $R - G$ is zero in thermal equilibrium but nonzero when the device is out of thermal equilibrium.

For our analysis we assume a very simple form for $R - G$, one in which the net recombination rate is related linearly to the *excess* concentration of minority carriers. The constant of proportionality is $1/\tau_e$ or $1/\tau_h$, where τ_e and τ_h are known as the *minority-carrier lifetime* for electrons or holes, respectively.

As we are assuming that the regions of the diode outside the SCL are neutral, $n' = p'$ and

$$\frac{\partial n'}{\partial x} = \frac{\partial n}{\partial x} = \frac{\partial p'}{\partial x} = \frac{\partial p}{\partial x}. \tag{16.37}$$

If we combine these relationships with our assumed recombination model, and assume dc steady-state conditions, (16.36) becomes

$$D_e \frac{\partial^2 n'(x)}{\partial x^2} - \frac{n'(x)}{\tau_e} = 0, \tag{16.38}$$

or the diffusion equation for electrons when they are in the minority. We can write a similar equation for holes. Knowing the boundary conditions at the edges of the SCL from (16.32), we can solve these diffusion equations for the excess carrier concentrations in the two neutral regions of the diode.

The solution to (16.38) is a pair of real exponentials (one growing, the other decaying) with characteristic lengths $\sqrt{D_e \tau_e}$. If we assume the bar to be infinitely long, we can discard the growing term for physical reasons, leaving

$$n'_p(x) = n'_p(0)e^{-x/L_e}, \qquad L_e = \sqrt{D_e \tau_e}, \tag{16.39}$$

where L_e is the *diffusion length* for electrons. It is the mean distance that an electron can be expected to diffuse before it is lost to recombination. We now see that "infinitely long" means long compared to a diffusion length.

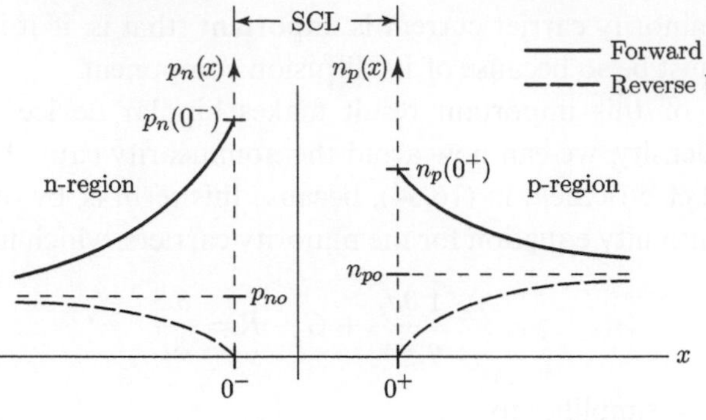

Figure 16.7 Minority-carrier distributions in a long diode under both forward and reverse bias.

The coefficient $n_p'(0)$ in (16.39) is related to the applied voltage through (16.32), so we can rewrite (16.39) in terms of v_A as

$$n_p'(x) = n_{po}\left(e^{qv_A/kT} - 1\right)e^{-x/L_e}. \tag{16.40}$$

A similar equation describes the profile of excess holes $p_n'(x)$ in the n region. Figure 16.7 illustrates these distributions for both forward and reverse bias.

Because the SCL is relatively narrow, we can assume that no appreciable generation or recombination occurs within it, even though the carrier concentrations are not at their equilibrium values. Because of this assumption, if we know the electron current density at any point within the SCL, we know it everywhere within the SCL. Similarly, if we know the hole current density at one point, we know it everywhere. Therefore, because we know the electron current density at the p-region edge of the SCL and the hole current density at the n-region edge of the SCL, we know these current densities everywhere within the SCL. Thus,

$$\begin{aligned} J_e(0^+) &= J_e^{\text{diff}}(0^+) = \frac{qD_e n_{po}}{L_e}\left(e^{qv_A/kT} - 1\right), \\ J_h(0^-) &= J_h^{\text{diff}}(0^-) = \frac{qD_h p_{no}}{L_h}\left(e^{qv_A/kT} - 1\right). \end{aligned} \tag{16.41}$$

We can now obtain the total diode current density by summing these two components:

$$\begin{aligned} J &= \left[\frac{qD_h p_{no}}{L_h} + \frac{qD_e n_{po}}{L_e}\right]\left(e^{qv_A/kT} - 1\right) \\ &= \left[\frac{qD_h n_i^2}{N_D L_h} + \frac{qD_e n_i^2}{N_A L_e}\right]\left(e^{qv_A/kT} - 1\right), \end{aligned} \tag{16.42}$$

which is the ideal semiconductor diode equation. The term in brackets is the *saturation current* density J_s. It is the magnitude of the *reverse leakage* current density of a diode. Multiplying (16.42) by the cross-sectional area of the diode yields the diode $v - i$ relation:

$$i = I_s\left(e^{qv_A/kT} - 1\right).$$

Example 16.3 Concentration Profiles and Currents in a Short Diode

In our analysis we assumed that the terminal contacts of the diode were infinitely far from the junction. As we found, "infinite" in this case means many diffusion lengths. In some diodes, however, the terminals are much closer together. The purpose of this example is to determine the carrier profiles and current densities in such diodes.

Figure 16.8 shows a p–n diode with p and n regions that are ℓ_p and ℓ_n long, respectively. We assume that these lengths are very short compared to the corresponding minority-carrier diffusion lengths, that is, L_e on the p side and L_h on the n side.

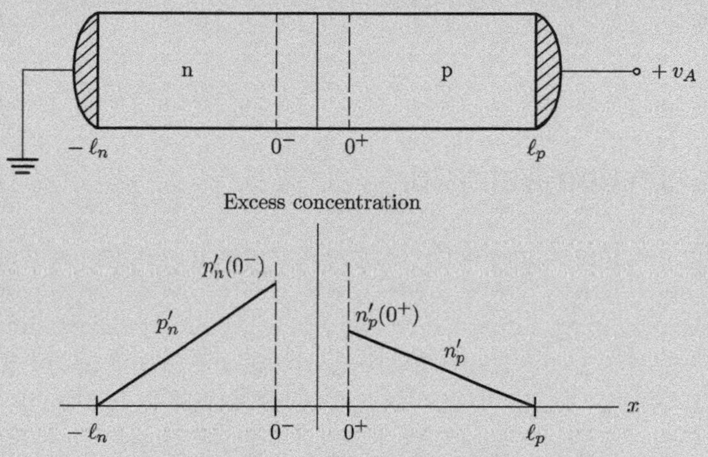

Figure 16.8 A short diode and its excess-carrier profiles when it is forward biased.

We must use both the growing and decaying solutions to the diffusion equations, (16.38), in order for the short diode to meet the boundary conditions at the SCL edge and at the ohmic contact (where the excess carrier concentrations are zero). The resulting solutions are of the form $\sinh(x/L_{e,h})$ and $\cosh(x/L_{e,h})$. The argument in these functions is always small in our diode, so the solutions are very close to straight lines. In other words, the dimensions of the diode are such that few carriers are generated in the n and p regions, and we can expect most of the excess carriers to diffuse from the junction to the ohmic contact before they are lost to recombination. Therefore, if we ignore the recombination and generation terms in (16.36), we see that the solutions to the diffusion equations are straight lines. In essence, we define a short diode as one in which recombination in the neutral regions can be ignored.

Figure 16.8 shows the excess carrier profiles based on our assumptions. The electron and hole current densities in the SCL are

$$\begin{aligned} J_e(0^+) &= J_e^{\text{diff}}(0^+) = \frac{qD_e n_{po}}{\ell_p}\left(e^{qv_A/kT}-1\right), \\ J_h(0^-) &= J_h^{\text{diff}}(0^-) = \frac{qD_h p_{no}}{\ell_n}\left(e^{qv_A/kT}-1\right), \end{aligned} \tag{16.43}$$

and the total diode current density is

$$J = \left[\frac{qD_h p_{no}}{\ell_n} + \frac{qD_e n_{po}}{\ell_p}\right]\left(e^{qv_A/kT}-1\right). \tag{16.44}$$

We have assumed that the SCL width is small compared to ℓ_n and ℓ_p, so the lengths of the neutral regions are not significantly reduced by the SCL. This condition is generally true for forward-bias conditions. Under reverse bias, however, the SCL width becomes much wider. As it does, the actual lengths that should be used in (16.44) get smaller and the reverse leakage current therefore gets larger. This increase in leakage current is not important until the SCL starts to reach the edge of the device, in which case the current begins to increase very rapidly. This condition is called *punch-through*. It is one breakdown mechanism that could limit the maximum reverse voltage, V_{BR}, that can be placed across a device. If the SCL punches through on the n side first, then

$$V_{BR} \approx \frac{qN_D\ell_n^2}{2\epsilon_s}\left(1 + \frac{N_D}{N_A}\right). \tag{16.45}$$

As most devices are designed to avoid this condition, the maximum reverse voltage is instead limited by *avalanche*, a phenomenon that we discuss in Chapter 17.

16.3 Bipolar Junction Transistor

The bipolar junction transistor (BJT) forms a critical part of the insulated gate bipolar transistor which, along with the MOSFET, has largely replaced the BJT in practice. The SCR is also often modeled as a latching connection of two BJTs. Therefore, understanding the physical behavior of a BJT is critical to understanding the behavior and limitations of these other devices.

The BJT consists of three layers of doped semiconductor, forming two junctions. The layers can be arranged as npn or pnp. Most power transistors are of the npn variety because of the superior performance of this arrangement. We first examine the behavior of the BJT using a very simple one-dimensional model.

16.3.1 Physical Model and Terminal Currents

Figure 16.9(a) is a one-dimensional model of an npn bipolar transistor. The doping concentrations in the emitter, base, and collector are N_{DE}, N_{AB}, and N_{DC}, respectively. The width of the base region is W, and the widths of the emitter and collector are long compared to the diffusion lengths in these regions.

You can apply everything you learned in the analysis of the p–n diode directly to the analysis of this transistor. We assume low-level injection throughout the device, so that applied terminal voltages appear only across the SCLs, and minority carriers flow only by diffusion. We therefore know the excess carrier concentrations at the edge of each SCL as a function of the voltage applied between the neutral regions on either side of the junction. Hence, we can solve the diffusion equations for the carrier concentrations in the neutral regions.

Figure 16.9(b) shows the minority carrier concentrations for forward active operation, the operating condition in which the base–emitter junction is forward biased and the base–collector junction is reverse biased. Also, we assume that the lifetime of minority carriers in the base is long, or, alternatively, that the diffusion length of minority carriers in the base L_B is long compared to the base width W.

From the minority carrier distributions of Fig. 16.9(b), we can calculate the minority currents at the edges of an SCL and then sum them to obtain the total current density at the junction.

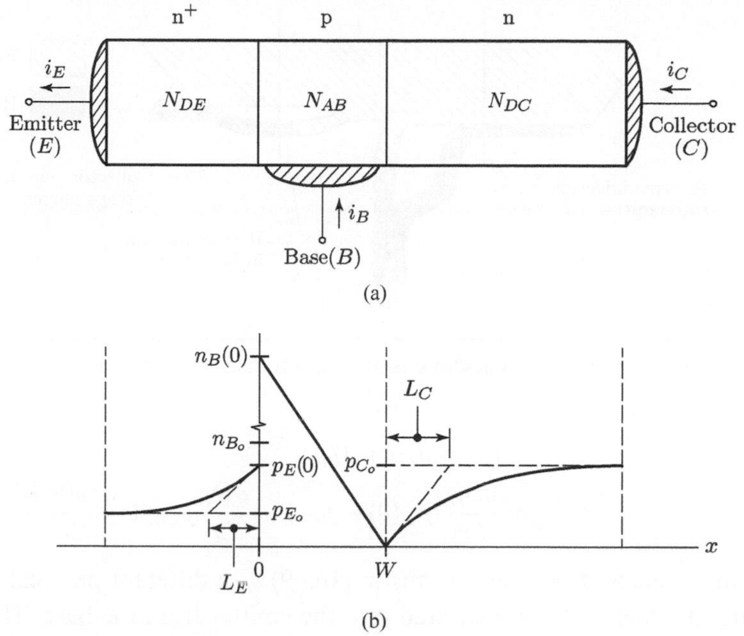

Figure 16.9 (a) One-dimensional model of an npn bipolar transistor. (b) Minority carrier concentrations for operation in the forward active gain region.

The total current flowing right to left at $x = 0$ is the emitter current, and that at $x = W$ is the collector current. The corresponding current densities are

$$J_E = J(0) = \frac{qD_E}{L_E}(p_E(0) - p_{Eo}) + \frac{qD_B n_B(0)}{W}, \tag{16.46}$$

$$J_C = J(W) = \frac{qD_C}{L_C}p_{Co} + \frac{qD_B n_B(0)}{W}, \tag{16.47}$$

where the directions of J_E and J_C are as defined in Fig. 16.9(a), and L_E and L_C are the minority carrier diffusion lengths in the emitter and collector.

In (16.46) and (16.47) we have ignored the difference between the emitter and collector currents resulting from the recombination of charge in the base. We assume that an electron injected at the emitter reaches the collector. Although these lost carriers are generally insignificant compared to the total number of electrons flowing from emitter to collector, the holes required to recombine with them may be an important part of the base current. In calculating the base current, we must therefore account for recombination in the base. If we again use a recombination model in which the rate is proportional to the total excess charge, the recombination current density supplied to the base J_{BR} is

$$J_{BR} = \frac{Q_B}{\tau_B} = \frac{q[n'_B(0) - n'_B(W)]W}{2\tau_B} \approx \frac{qn_B(0)W}{2\tau_B}, \tag{16.48}$$

since $n'_B(W) = -n_{Bo}$ and $n'_B(0) \gg n_{Bo}$. We can now specify the base current in terms of the emitter and collector currents, (16.46) and (16.47), and the additional recombination component, (16.48):

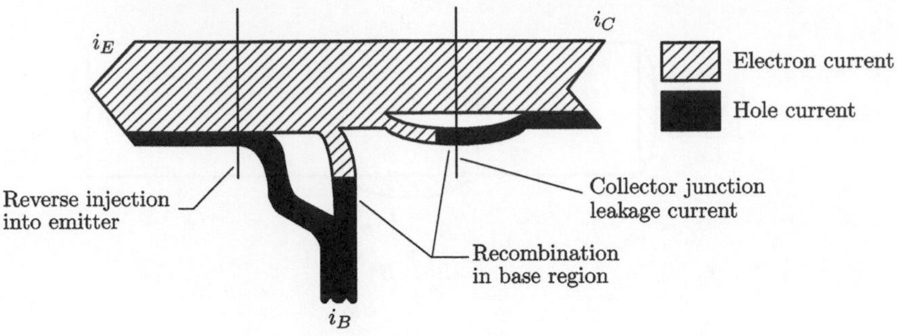

Figure 16.10 Current flow in a one-dimensional npn transistor, showing the origin of base current.

$$\begin{aligned} J_B &= (J_E - J_C) + J_{BR} \\ &= \frac{qD_E}{L_E}(p_E(0) - p_{Eo}) - \frac{qD_C}{L_C}p_{Co} + \frac{qn_B(0)W}{2\tau_B}. \end{aligned} \tag{16.49}$$

We can relate each of the three terms in (16.49) to a different physical process. The first term is the rate at which holes are injected into the emitter from the base. This process is known as *reverse injection* and usually accounts for the largest component of base current in a power BJT. The second term is the rate at which holes are injected into the base from the collector because of the reverse-biased collector junction. This component of the base current is usually negligible. The third term is the rate at which holes must be supplied to support recombination in the base. Figure 16.10 shows this relationship.

16.3.2 Device Parameters

Two related parameters characterize the current gain of a bipolar transistor. The ratio of collector current to base current is called the *short-circuit common-emitter current gain* β_F,

$$\beta_F = \frac{i_C}{i_B}. \tag{16.50}$$

The ratio of collector current to emitter current is called α_F, the *short-circuit common-base current gain*,

$$\alpha_F = \frac{i_C}{i_E}. \tag{16.51}$$

The subscript F denotes forward operation of the device, uppercase implying a ratio of large-signal quantities. As $i_B = (i_E - i_C)$, the relationship between these two current gains is

$$\beta_F = \frac{\alpha_F}{1 - \alpha_F}. \tag{16.52}$$

Because an ideal transistor requires no base current and $\beta_F = \infty$, we may view the two mechanisms giving rise to the base current shown in Fig. 16.10 as *defects* of two types. The first is the *recombination defect* δ_R and is the result of the impossibility of achieving a zero base width W or an infinite base lifetime τ_B. The second is the *emitter defect* δ_E, which reflects the impossibility of creating an emitter junction which, when forward biased, does not cause reverse injection. That is, it does not inject majority carriers (holes in this case) from the base into the emitter.

These defects are defined mathematically as the ratio of the corresponding component of the base current to the minority current (electrons in this case) injected into the base from the emitter. In terms of the physical parameters of the transistor, these ratios are

$$\delta_R = \frac{i_{BR}}{i_e(0)} = \frac{qn_B(0)W/2\tau_B}{qD_Bn_B(0)/W} = \frac{W^2}{2L_B^2}, \tag{16.53}$$

$$\delta_E = \frac{i_h(0)}{i_e(0)} = \frac{qD_E}{L_E}\frac{(p_E(0) - p_{Eo})}{qD_Bn_B(0)/W} \approx \frac{D_EWN_{AB}}{D_BL_EN_{DE}}, \tag{16.54}$$

where i_{BR} is the recombination component of the base current, and $i_h(0)$ – the hole current at $x = 0$ – is the component of the base current supplying the holes for reverse injection.

From (16.53) you can see the desirability of keeping the base very short compared to L_B in order to minimize δ_R. From (16.54) you can see that the emitter should be doped as heavily as possible, compared to the base, in order to keep δ_E small.

Two other parameters that are useful in relating the behavior of a transistor to its physical construction are the *base transport factor* and the *emitter efficiency*. The base transport factor α_T is the ratio of the electron current at the collector junction and the electron current injected at the emitter. It is related to the recombination defect by

$$\alpha_T = 1 - \delta_R. \tag{16.55}$$

The emitter efficiency γ_{E} is the ratio between the electron current injected into the base from the emitter and the total emitter current. It is related to the emitter defect by

$$\gamma_{\mathrm{E}} = \frac{1}{1 + \delta_E}. \tag{16.56}$$

From these definitions, we can show that

$$\frac{i_C}{i_B} = \beta_F = \frac{1 - \delta_R}{\delta_E + \delta_R} = \frac{\alpha_T\gamma_{\mathrm{E}}}{1 - \alpha_T\gamma_{\mathrm{E}}}, \tag{16.57}$$

$$\frac{i_C}{i_E} = \alpha_F = \frac{1 - \delta_R}{1 + \delta_E} = \alpha_T\gamma_{\mathrm{E}}. \tag{16.58}$$

Example 16.4 Calculating the Gain of a Transistor

For the transistor in Fig. 16.9, assume that $N_{DE} = 10^{17}\,\mathrm{cm}^{-3}$, $N_{AB} = 10^{15}\,\mathrm{cm}^{-3}$, and $N_{DC} = 10^{14}\,\mathrm{cm}^{-3}$. The minority-carrier lifetime in both the base and collector regions is 1 μs, but as the emitter is so heavily doped, its lifetime there is 10 ns. The width of the base region is $W = 10^{-3}$ cm, and the emitter and collector regions are long compared to their diffusion lengths.

We first calculate the diffusion lengths in the base and emitter. At the low levels of doping in the base and collector, the diffusion constant for electrons in the base is $D_{eB} = 40\,\mathrm{cm}^2/\mathrm{s}$. The diffusion length is therefore $L_B = 6.3 \times 10^{-3}$ cm. In the emitter, however, the doping concentration is so high that the mobilities, and therefore the diffusion constants, are reduced from their values for intrinsic material. We assume that $D_{hE} = 5\,\mathrm{cm}^2/\mathrm{s}$ for holes in the emitter, which gives $L_E = 2.2 \times 10^{-4}$ cm.

With these numbers, we can now calculate the defects for this transistor:

$$\delta_R = \frac{W^2}{2L_B^2} = 12.6 \times 10^{-3}, \tag{16.59}$$

$$\delta_E = \frac{D_E W N_{AB}}{D_B L_E N_{DE}} = 5.7 \times 10^{-3}. \tag{16.60}$$

These defect values result in a forward current gain of

$$\beta_F = \frac{1-\delta_R}{\delta_E + \delta_R} = 54. \tag{16.61}$$

16.3.3 Saturation

When we operate a transistor in its forward active region, the more base current we deliver, the higher the collector current becomes. If the load on the transistor causes the collector–emitter voltage to drop as the collector current rises – as it would in the common–emitter amplifier circuit of Fig. 16.11(a) – a point can be reached at which the base–collector junction becomes forward biased. This condition, in which both the collector and emitter junctions are forward biased, is called *saturation*. Figure 16.11(b) shows the excess carrier profiles in the transistor under this condition.

As a transistor enters saturation from the forward active region, its base current must be increased to support the additional requirements of increased base charge and minority carrier injection into the collector. The transistor's gain therefore appears to be decreasing. If the base current is increased further, a point is reached where, for a given collector current, the collector–emitter voltage cannot be reduced further because both diodes are strongly forward biased. This point is called *hard saturation*, and is identified in the typical collector *i*–*v* characteristics of an npn transistor of Fig. 16.11(c). This is the region in which a power device is "on."

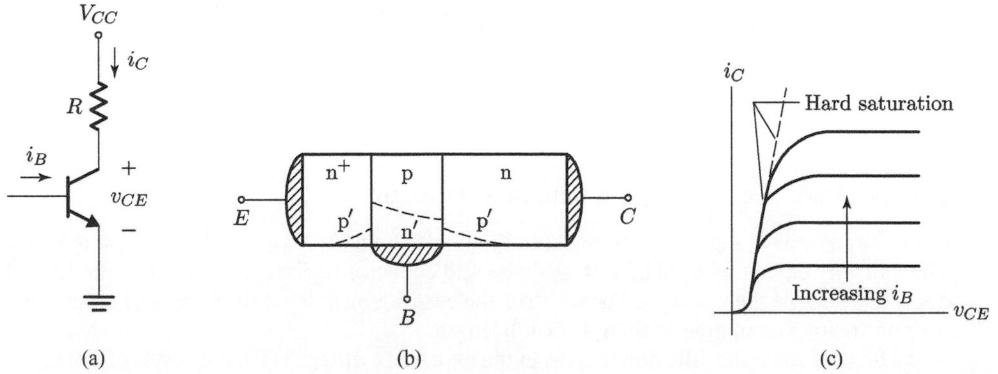

Figure 16.11 The saturation region of bipolar transistor operation. (a) Common-emitter amplifier circuit. (b) Excess carrier profiles in a saturated npn transistor. (c) The collector characteristics for an npn transistor.

16.4 MOSFET

Figure 16.12 shows a lateral MOSFET geometry. The term "lateral" (as opposed to "vertical") refers to the direction of current flow between the drain D and source S when the MOSFET is conducting. Both the drain and the source are made of heavily doped n-type silicon, while the *body*, or *substrate*, is made of relatively lightly doped p-type silicon. The gate G is a conductive plate (usually made of polysilicon, but sometimes made of metal) that is electrically isolated from the rest of the device by an oxide insulator called the *gate oxide*.

Looking just at the silicon structure of the device, we would conclude that no current could flow between the drain and source terminals because they are isolated by a pair of back-to-back p–n junctions ($n^+p\text{-}n^+$). This is, in fact, the case when the MOSFET is off. It will withstand a voltage in either direction, and the SCL supporting this voltage will grow mostly into the p region. One end of the gate oxide is therefore exposed to this off-state voltage, as well.

If the drain and source are both grounded, and we then apply a positive voltage to the gate, an electric field perpendicular to the plane of the gate appears in the oxide layer. This field originates on positive charge residing on the gate electrode and terminates on negative charge within the p region of the silicon. At low gate voltages, this negative charge comes from the immobile acceptor ions that are left unneutralized when holes flow away from the oxide–silicon interface. In other words, an SCL grows into the silicon from the interface.

Because of the electric field in the SCL, the potential in the silicon just under the oxide is higher than it is in the neutral p region. The silicon is in thermal equilibrium (no current is flowing), so the relationships between carrier densities and the potential given by (16.17) apply here. Therefore, although the concentration of holes decreases as the potential at the surface increases, the concentration of electrons is growing. When the potential becomes great enough, there are actually more electrons than holes, and a thin layer of effectively n-type silicon, called a *channel*, is created just beneath the oxide. When this condition occurs, there is no longer a p–n junction between the drain and source. Instead, there is an $n^+n\text{–}n^+$ structure that acts like a resistor and permits current to flow. It is a very large resistor, however, because the concentration of electrons under the oxide is very low.

As we continue to increase the gate voltage, the concentration of electrons in the n-type channel layer grows. Eventually the electron concentration equals that of the original hole concentration N_A, a condition called the *onset of inversion*. At this gate-source voltage, known as the *threshold voltage*, V_T, the resistance of the n-type channel, begins to fall rapidly as the

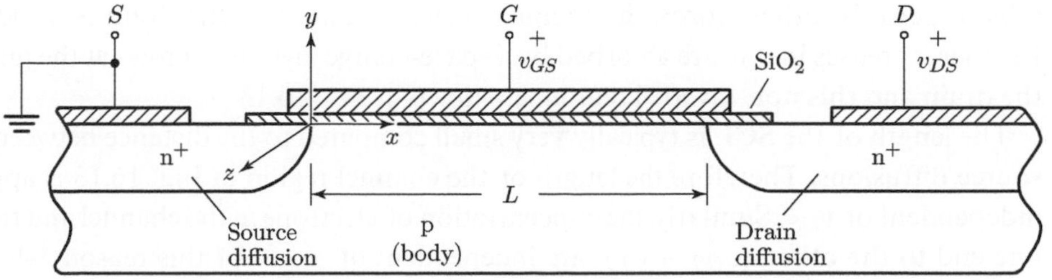

Figure 16.12 The lateral MOSFET.

gate voltage is raised. For $v_{GS} > V_T$, the concentration of electrons becomes very large, and the resistance of the channel becomes a strong function of the gate voltage.

The MOSFET of Fig. 16.12 is an *n-channel enhancement mode* lateral MOSFET. The term "enhancement" refers to the fact that the channel conductance is enhanced by the application of gate voltage. The device is off if $v_{GS} \leq V_T$. It is also possible to make p-channel MOSFETs by reversing the dopant types, in which case the gate voltage must be negative with respect to the source in order for the device to be turned on.

16.4.1 MOSFET Characteristics

In the preceding discussion we assumed a zero drain–source voltage, or at least one that was negligible compared to V_T. Such an assumption results in a negative charge distribution in the channel that is independent of x. If, however, there is current flowing from drain to source, the voltage at the drain edge of the channel will be higher than it is at the source end. The electric field in the oxide will therefore be weaker at the drain edge, and the number of electrons on which it terminates will be correspondingly fewer. As a result, the conductivity of the channel varies with x.

For $v_{DS} < (v_{GS} - V_T)$, the drain current as a function of v_{GS} and v_{DS} can be shown to be

$$i_D = \frac{\mu_e Z C_o}{L}\left[(v_{GS} - V_T)v_{DS} - \frac{v_{DS}^2}{2}\right], \tag{16.62}$$

where C_o is the oxide capacitance per unit area, L is the channel length, and Z is the z-axis dimension of the channel. If v_{DS} is small compared to $(v_{GS} - V_T)$, the quadratic term in (16.62) may be ignored and the drain current is directly proportional to the drain–source voltage. The ratio i_D/v_{DS} is a conductance of value

$$\frac{i_D}{v_{DS}} = g_{ch} = \frac{\mu_e Z C_o}{L}(v_{GS} - V_T). \tag{16.63}$$

Because the relationship between drain voltage and drain current is linear for any gate voltage, this region of operation is commonly called the *linear* region. Note that this name corresponds to the fully-on condition of the device, or the saturation condition in bipolar transistor terminology.

When v_{DS} reaches a value such that the voltage across the oxide at the drain end of the channel is reduced to V_T, that is, when $v_{DS} = v_{GS} - V_T$, the electron density at that end of the channel falls to zero. In other words, the region under the gate near the drain is no longer inverted. Further increases in v_{DS} are absorbed by a space-charge layer that grows at the junction between the drain and this non-inverted region, as shown in Fig. 16.13.

The length of the SCL is typically very small compared to the distance between the drain and source diffusions. Therefore the length of the channel region in Fig. 16.13 is approximately L, independent of v_{DS}. Similarly, the concentration of electrons in the channel and the voltage from one end to the other, $(v_{GS} - V_T)$, are independent of v_{DS}. For this reason, when v_{DS} exceeds $(v_{GS} - V_T)$, the current flowing through the channel no longer depends on the drain–source voltage, and the following relationship exists between the drain current and the gate voltage:

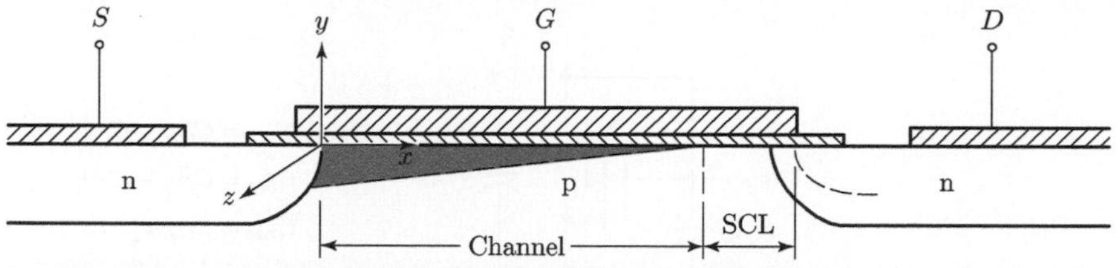

Figure 16.13 Inverted channel and SCL in a MOSFET with $v_{DS} > v_{GS} - V_T$.

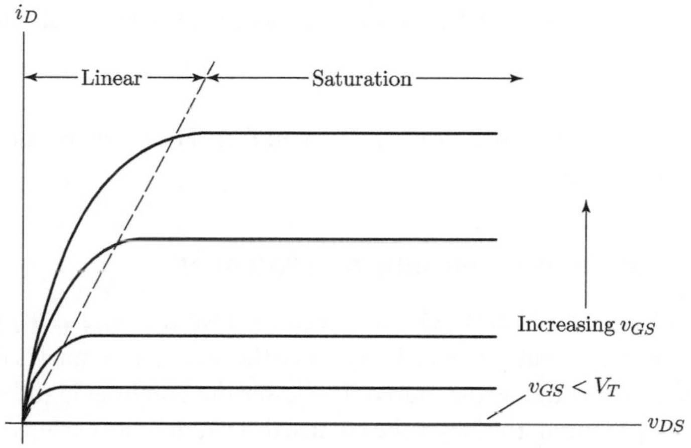

Figure 16.14 Common-source i–v characteristics for an n-channel MOSFET.

$$i_{D(\text{sat})} = \frac{\mu_e Z C_o}{2L}(v_{GS} - V_T)^2. \tag{16.64}$$

We call this region of operation *saturation*, although it corresponds to a condition we call forward active in bipolar transistor terminology.

The relationship between $i_{d\text{sat}}$ and v_{gS} in saturation is known as the *transconductance*, g_m,

$$g_m = \frac{di_{d(\text{sat})}}{dv_{gs}} = \frac{\mu_e Z C_o}{L}(v_{GS} - V_T), \tag{16.65}$$

and is typically on the order of $100\,\Omega^{-1}$.

Figure 16.14 shows the complete common-source (voltages referenced to the source) i–v characteristics for an n-channel MOSFET. Note that although these characteristics appear similar to those of the bipolar transistor, the value of the drain current depends on the gate voltage squared, rather than linearly on the base current.

MOSFET Capacitances All semiconductor devices exhibit capacitance characteristics between their terminals. These capacitances arise from junction SCL charging and discharging (as in the diode), or from metal-oxide-semiconductor layers present in the MOSFET. Figure 16.15 shows these capacitances for a MOSFET, illustrating their nonlinear nature and their relationship to

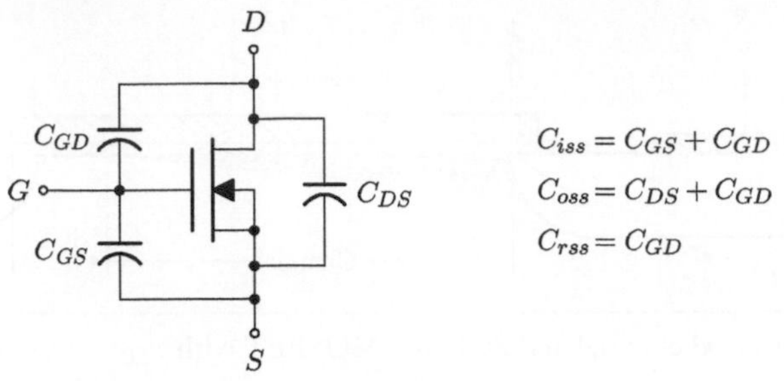

Figure 16.15 Intrinsic MOSFET capacitances as seen from the terminals and their relation to measured datasheet parameters.

measured datasheet parameters C_{iss}, C_{oss}, and C_{rss}. The origin of these capacitances is explored more fully in Chapter 17.

16.4.2 The Dependence of V_T on Substrate Potential

The value of v_{GS} at which the channel becomes inverted ($v_{GS} = V_T$) depends on the potential of the neutral p-type substrate, or body. Once the inverted channel is formed, it has (at its source end) the same potential as the source. The lower the potential in the neutral p region compared to the source potential, the larger the voltage drop across the depleted SCL region and the larger the electric field at the oxide interface. We can multiply this electric field, adjusted for the different dielectric constants of silicon and oxide, by the thickness of the oxide to find V_T.

Although we do not develop explicit relationships here, note that the lower the substrate potential, the higher the threshold voltage, and vice versa. For this reason, connecting the substrate to a known potential is important, so that the threshold voltage will have a fixed value. Because the substrate forms a p–n diode with both the drain and the source, connecting it to the lower of the drain and source potentials is necessary. Normally, the source is the lower potential, so a typical MOSFET has its substrate connected to its source, forming what is known as the *source–body short*. This creates a diode between the source and drain represented by the arrow in the middle of the MOSFET symbol.[§]

16.5 The Safe Operating Area

Except for voltage breakdown, the *i–v* characteristics shown in Fig. 15.4(a) and (b) do not indicate the operational limits of the device. Figure 16.16 shows a transistor's *i–v* characteristics with several other limits imposed. The area within these limits is called the *safe operating area*, or SOA, of the device.

[§] Although technically redundant, there is a common convention to include an external diode symbol between the source and drain in schematics. This only functions to clutter the diagram.

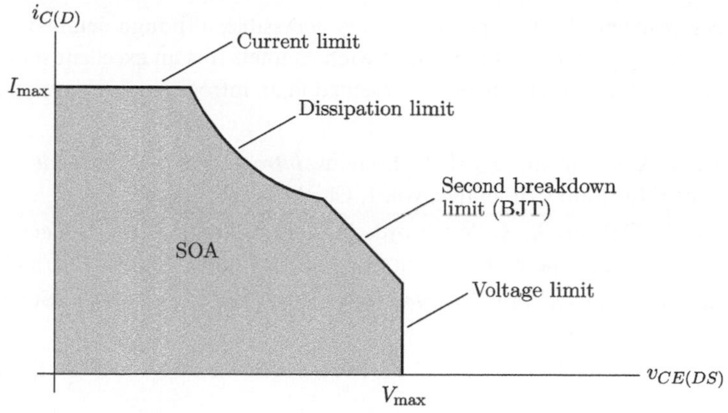

Figure 16.16 The safe operating area (SOA) of a BJT or MOSFET.

One boundary of the SOA is the maximum device current, specified as I_{max}. The current-carrying capabilities of the wire bonds connecting the die to its package, and of the device itself, are responsible for this limit. The limit is a function of how long the current lasts. A BJT's upper current limit for very short pulses is typically less than a factor of two above its dc limit, while a MOSFET's upper current limit is typically three to four times higher. The power (dissipation) rating of the device forms another boundary of the SOA. This maximum power increases as the pulse duration decreases for constant repetition rate.

The final boundary of the SOA for the BJT depends on a phenomenon called *second breakdown*. This is a problem that occurs only in BJTs, when both the voltage and current are high. It is a destructive breakdown that causes a self-sustaining collapse in the voltage in a localized region of the collector–base junction. All the current flowing through the device focuses on this spot, creating such a large dissipation in such a small area that the device fails. An ideal MOSFET does not experience this problem, although its parasitic bipolar transistor can. However, because the base of the parasitic bipolar is shorted to its emitter, second breakdown is seldom a problem in practical MOSFETs.

Notes and Bibliography

There are many textbooks and reference books on the fundamentals of semiconductor devices that take the student from the basic physics of carrier concentrations and charge transport through to the i–v characteristics of modern electronic devices. Volumes 1 and 2 of the Semiconductor Electronics Education Committee (SEEC) series [1, 2] are some of the oldest, yet still very informative and useful, presentations on semiconductor physics and bipolar devices. Volume 1 provides the physical background needed to understand all semiconductor devices, and Volume 2 applies this understanding to the bipolar diode and transistor.

The Modular Series on Solid State Devices [3, 4, 5, 6] is an excellent reference for the fundamentals of semiconductor devices. Volume 1 presents the fundamentals of semiconductor physics, and then Volumes 2, 3, and 4 give extensive coverage of the p–n diode, the bipolar junction transistor, and field-effect devices, respectively. The material is very accessible, although detailed, the figures are informative, and there are many problems at the end of each chapter. It is an excellent set of texts to read as a review and as an advancement of the material presented in an introductory course on semiconductor devices.

1. R. B. Adler, A. C. Smith, and R. L. Longini, *Introduction to Semiconductor Physics*, Semiconductor Electronics Education Committee, vol. 1, Chichester: Wiley, 1964.
2. P. E. Gray, D. DeWitt, A. R. Boothroyd, and J. F. Gibbons, *Physical Electronics and Circuit Models of Transistors*, Semiconductor Electronics Education Committee, vol. 2, Chichester: Wiley, 1964.
3. R. F. Pierret, *Semiconductor Fundamentals*, Modular Series on Solid State Devices, vol. 1, Reading, MA: Addison-Wesley, 1983.
4. G. W. Neudeck, *The pn Junction Diode*, Modular Series on Solid State Devices, vol. 2, Reading, MA: Addison-Wesley, 1983.
5. G. W. Neudeck, *The Bipolar Junction Transistor*, Modular Series on Solid State Devices, vol. 3, Reading, MA: Addison-Wesley, 1983.
6. R. F. Pierret, *Field-Effect Devices*, Modular Series on Solid State Devices, vol. 4, Reading, MA: Addison-Wesley, 1983.

PROBLEMS

16.1 The intrinsic carrier concentration in a semiconductor is a strong function of temperature. In silicon, it is $n_i^2 = C_1 T^3 e^{-12\,800/T}$, where T is in Kelvin. The electron and hole mobilities also vary with temperature. We can approximate their relationship to T as $\mu_e = C_2 T^{-2.2}$ and $\mu_h = C_3 T^{-2.2}$. From the values given in Table 16.1 for n_i, μ_e, and μ_h at 300 K, find the values of these parameters at $-50\,°\text{C}$ and $+150\,°\text{C}$.

16.2 An n-type silicon sample has a conductivity of $5 \times 10^{-2}\,\Omega^{-1}\,\text{cm}^{-1}$ at 300 K. Based on your results in Problem 16.1, find the conductivity of this sample at 200 K and 400 K. (*Note:* The sample may not be extrinsic at all three temperatures.)

16.3 Gallium arsenide (GaAs) is a compound semiconductor with a structure similar to silicon except that it consists of alternating Ga and As atoms. (That is, each Ga atom has four As atoms as nearest neighbors, and each As atom has four Ga atoms as nearest neighbors.) The ion core of Ga has a charge of $+3q$ and the ion core of As has a charge of $+5q$ (where q is the electronic charge).

(a) Draw a two-dimensional representation of what you expect to happen when Mg, Al, or Si is substituted for a Ga atom, or when Si, P, or S is substituted for an As atom (a total of six cases). For each case, state whether the impurity is a donor, an acceptor, or neither.

(b) Assume that Si atoms are added to GaAs with a concentration of $10^{15}\,\text{cm}^{-3}$, and that Si will substitute for As four times as often as it substitutes for Ga. Calculate the conductivity of the resulting sample at room temperature using the following parameters for GaAs:

$$n_i^2 = 80 \times 10^{12}\,\text{cm}^{-6}, \qquad \mu_e = 8800\,\text{cm}^2/\text{V-s}, \qquad \mu_h = 400\,\text{cm}^2/\text{V-s}.$$

16.4 Assume that you have a chamber that extends to $\pm\infty$ in the x direction and has a unit cross-sectional area in the y–z plane. Within this chamber particles are distributed uniformly in the y–z plane but have the following distribution in the x direction at $t = 0$: $C(x) = (1 + \cos kx)\,\text{cm}^{-3}$. These particles obey the following diffusion equation:

$$\frac{\partial C(x,t)}{\partial t} = D\frac{\partial^2 C(x,t)}{\partial x^2}.$$

Find an expression that describes $C(x,t)$ for $t > 0$. (*Hint:* Use complex notation and assume a solution of the form

$$C(x,t) = C_o + \text{Re}\{C_1\, e^{at} e^{bx}\}.$$

The resulting solution provides an estimate of the time required for an excess minority carrier profile in a bipolar device to flatten out after the excitation has been turned off.)

16.5 Calculate the built-in potential ψ_o for a p–n junction when both sides of the junction are doped at (a) $10^{14}\,\text{cm}^{-3}$, (b) $10^{16}\,\text{cm}^{-3}$, and (c) $10^{19}\,\text{cm}^{-3}$. For each case, calculate the reverse voltage at which the peak electric field is 3×10^5 V/cm. (This is approximately the voltage at which avalanche, a breakdown mechanism, occurs.) For each voltage, find the distance that the SCL extends into either the n or the p regions.

16.6 For each of the cases in Problem 16.5, find the applied forward bias voltages at which high-level injection is reached. High-level injection is defined as the point at which the maximum excess minority carrier concentration equals the dopant level.

16.7 The diode shown in Fig. 16.17 has a p region that is doped much more heavily than its n region. Assume that ℓ, the width of both regions, is very small compared to the characteristic diffusion length of the minority carrier.

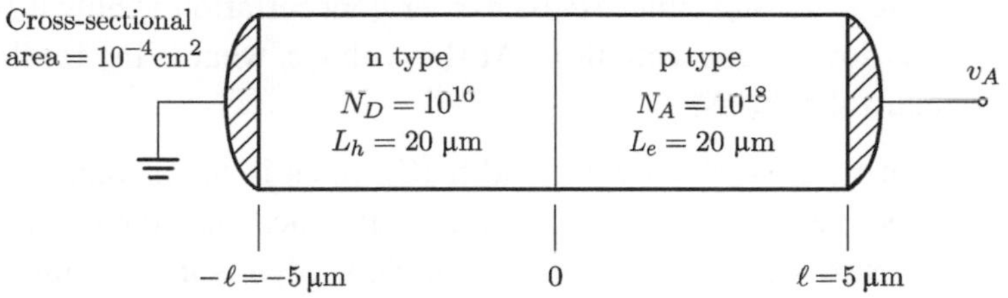

Figure 16.17 The diode structure analyzed in Problem 16.7.

(a) Find an expression for the total current flowing through the device when a voltage v_A is applied with the polarity shown. Assume that the SCL width is small compared to the length of the device. Evaluate this current when $v_A = 0.6$ V.

(b) Find an expression for x_n, the distance the SCL extends into the n region, as a function of the applied voltage v_A. Calculate this distance for $v_A = +0.5$, 0, and -100 V.

(c) When $v_A < 0$, x_n can become a significant fraction of the length of the n region. Derive an expression for the leakage current that flows under reverse bias that accounts for this extension of the SCL. Assume no recombination or generation in the SCL. Graph this current as a function of the applied reverse voltage. (You should observe a condition, called punch-through, where the leakage current gets very large.)

16.8 A forward-biased voltage, v_A, is applied across a diode with neutral regions that are the same length as the corresponding minority carrier diffusion length, as shown in Fig. 16.18. (Ignore the width of the SCL.)

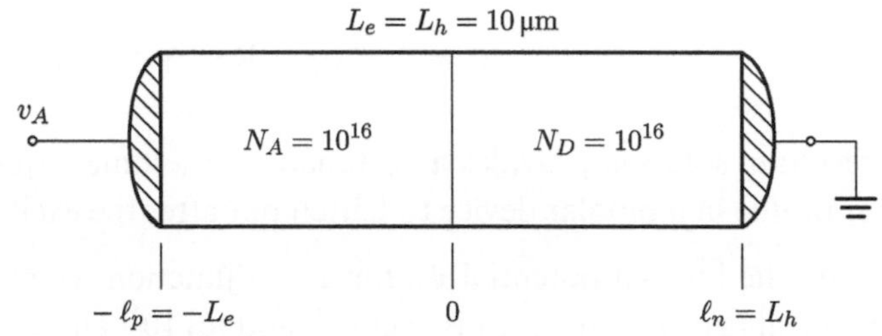

Figure 16.18 A diode whose neutral regions have lengths equal to their minority carrier diffusion lengths. This diode is analyzed in Problem 16.8.

(a) Find, sketch, and label p', n', J_e (drift and diffusion), and J_h (drift and diffusion) in the two neutral regions.

(b) Find, sketch, and label the electric field in the neutral regions. Integrate the expression for this field to determine the total voltage drop across the neutral regions.

(c) Assuming that all the applied voltage appears across the SCL, find the voltage at which the maximum excess carrier concentration is only a factor of five away from the dopant concentration. At this voltage, what is the total voltage drop across the neutral regions?

16.9 An npn transistor's base has width W, which is small compared to its diffusion length L_e. Assume that the applied voltages are such that the excess electron concentration at the base side of the base–emitter SCL edge is $n'(0^+)$, and that the excess electron concentration at the base side of the base–collector SCL edge is zero.

(a) Assuming that the minority carrier lifetime in the base is infinite, solve for the excess electron concentration in the base, and approximate the net recombination current density in the base as a function of the real lifetime, τ_e, and $n'(0^+)$.

(b) Find the ratio of this recombination current to the electron current injected by the emitter. (This is the approximate base defect of the transistor.)

(c) Instead of using the assumption in (a), solve for the exact excess electron concentration in the base for a finite minority carrier lifetime τ_e.

(d) Using your result in (c), find the difference between the electron current injected by the emitter and the electron current collected at the collector. Express this difference as a ratio of the electron current injected by the emitter. (This is the exact base defect of the transistor.)

(e) At what value of W/L_e will the approximate base defect differ by a factor of two from the exact base defect?

16.10 The npn transistor with doping levels and dimensions shown in Fig. 16.19 has an infinite minority carrier lifetime throughout.

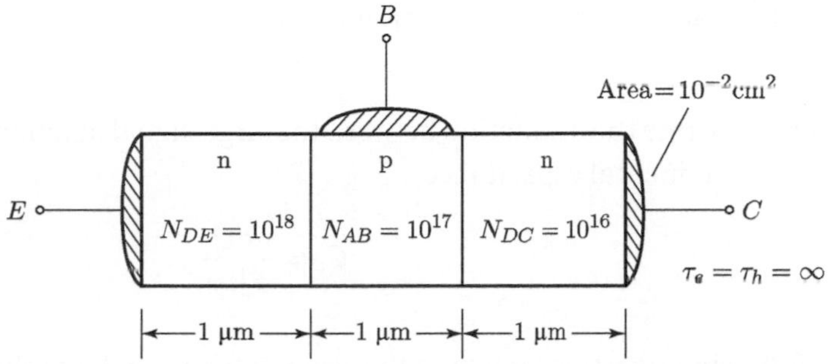

Figure 16.19 Dimensions and doping concentrations for the transistor analyzed in Problem 16.10.

(a) Find α_F, α_R, β_F, β_R, I_{ES}, and I_{CS} for this transistor.

(b) If the applied base–emitter voltage is $v_{BE} = 0.6$ V and $v_{BC} = 0$ V, find the current that flows into the base terminal.

(c) With $v_{BE} = 0.6$ V, at what base collector voltage v_{BC} will the base terminal current be twice that found in (b)? How much has the collector terminal current changed from its value under the conditions of (b)? (*Note:* This value could be considered the onset of saturation.)

(d) For the base–collector voltage found in (c), sketch the excess carrier profiles in the transistor.

16.11 The npn transistor shown Fig. 16.20 has a base whose width is small compared to the minority carrier diffusion length L_e, but emitter and collector lengths that are very long compared to their (equal) diffusion lengths L_h.

(a) For the forward mode condition ($v_{BE} > 0$ and $v_{BC} \approx 0$), sketch and label the excess minority carrier concentrations and the hole and electron currents (both drift and diffusion) throughout the transistor.

(b) Repeat (a) for the reverse mode condition ($v_{BC} > 0$ and $v_{BE} \approx 0$).

(c) What is the emitter defect of this device?

16.12 Sketch and label the SCL charge density as a function of v_A for $-1\text{ V} \le v_A \le \psi_o$ for a p–n junction with doping concentrations $N_D = N_A = 10^{16}\text{ cm}^{-3}$ and $A = 1\text{ cm}^2$. Using your

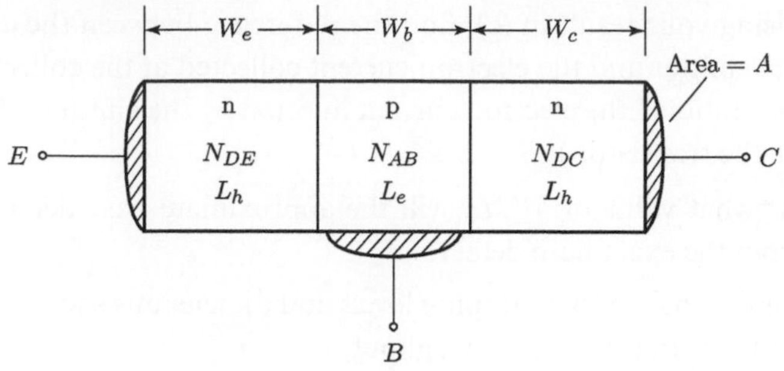

Figure 16.20 The transistor analyzed in Problem 16.11.

sketch, illustrate the distinction between the large-signal junction capacitance $C_o = Q/v_A$ and the incremental capacitance

$$C_i = \frac{dQ}{d\,|v_A|}.$$

16.13 The following questions relate to the asymmetrical pn diode shown in Fig. 16.21.

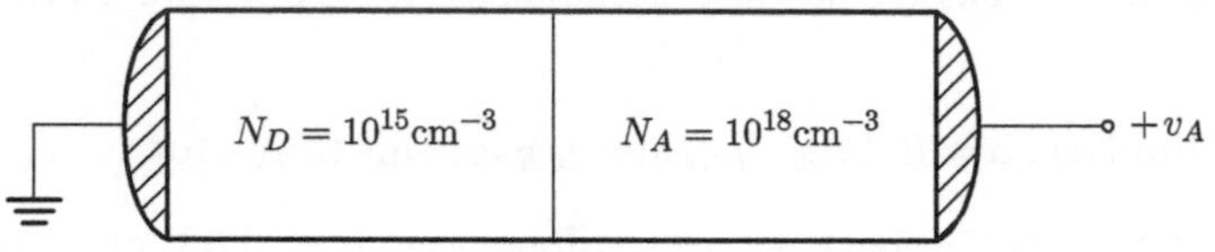

Figure 16.21 The p–n diode that is the subject of Problem 16.13.

(a) Find the charge in the SCL as a function of the applied voltage v_A.

(b) Find an expression for the large-signal junction capacitance $C_o = Q(v_A)/|v_A|$. Graph and label it as a function of v_A for $-10\,\text{V} \geq v_A \geq -200\,\text{V}$. (*Note:* For this and the following parts, assume that $|v_A|$ is very much larger than ψ_o, so that $(\psi_o - v_A) \approx -v_A$.)

(c) Find an expression for $C_i = d\,Q(v_A)/d\,|v_A|$. Graph and label it as a function of $-v_A$ for $-10\,\text{V} \geq v_A \geq -200\,\text{V}$.

(d) Find an expression for $C_{\text{energy}} = E_c(v_A)/(0.5\,v_A^2)$, the ratio between the stored electrostatic energy and one half the voltage squared. Graph and label it as a function of $-v_A$ for $-10\,\text{V} \geq v_A \geq -200\,\text{V}$. (*Note:* The proper way to calculate energy in a capacitor is

$$E_c(v_A) = \int_0^{Q(v_A)} v\,dQ = \int_0^{v_A} v\,\frac{dQ}{dv}\,dv = \int_0^{v_A} v\,C_i\,dv,$$

ignoring the effect of ψ_o near $v = 0$.)

16.14 An abrupt junction bipolar diode with ohmic contacts has a voltage, v_A, applied across its terminals, as shown in Fig. 16.22. Use the abrupt depletion region model for the SCL.

(a) If $N_A = N_D = 10^{16}\ \text{cm}^{-3}$, what is the junction's built-in potential ψ_o?

(b) For the doping concentrations of (a), at what applied voltage will the space-charge layer be (i) half as long as it is when $v_A = 0\ \text{V}$? (ii) 10 times as long as it is when $v_A = 0\ \text{V}$?

(c) If you want the maximum electric field strength in the SCL to be $E_{\text{max}} = 3 \times 10^5\ \text{V/cm}$ when $v_A = -200\ \text{V}$, what value of doping concentration should you use? (Assume a symmetrical junction where $N_A = N_D$.)

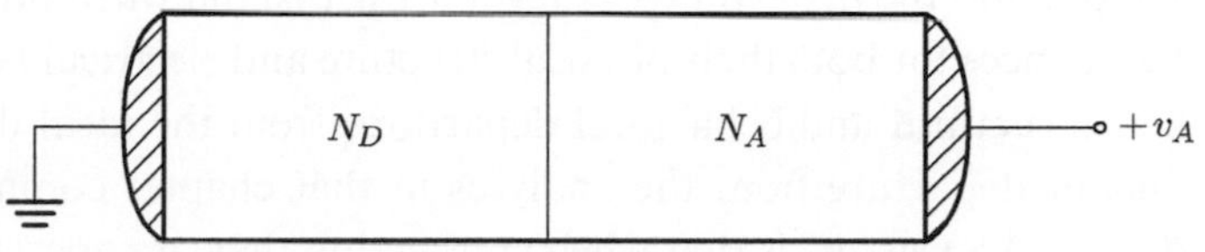

Figure 16.22 The abrupt junction bipolar diode analyzed in Problem 16.14.

17 Power Semiconductor Devices

Power semiconductor devices are distinguished by the high current densities at which they operate while on, and the high voltages they must withstand when off. These requirements have serious consequences for both their physical structure and electrical behavior. In this chapter we consider their structural and behavioral departures from the ideal devices studied in Chapter 16. A significant departure from the analyses in that chapter common to almost all bipolar power diodes and transistors is that when conducting they are operating in a regime known as *high-level injection*, where the excess minority concentration becomes comparable to or greater than the equilibrium majority concentration, whereas in Chapter 16 we considered only low-level injection (excess minority concentration $\ll$ equilibrium majority concentration). High-level injection is also referred to as *conductivity modulation* for reasons to be explained.

We first consider the bipolar power diode, which contains a wide space-charge layer to support high voltage, and exhibits current and voltage transients when switching between states. We also consider the Schottky diode, which, while it cannot support the high voltages of the bipolar diode, has a lower forward drop at comparable currents and much better switching characteristics. We then look at the power BJT, whose structure, which is required to support high voltage and high current simultaneously, reduces its usable switching frequency. Next we consider the power MOSFET, including the *superjunction* MOSFET, through which current travels vertically instead of laterally as discussed in Chapter 16. We then review the IGBT as a combination of the power BJT and MOSFET, which provides a very high voltage and current capability at the expense of switching speed. We also discuss the favorable characteristics of devices fabricated in the wide-bandgap materials SiC and GaN. Following this we consider the SCR as the regenerative connection of an npn and a pnp transistor, and the TRIAC. We conclude with a description of the device datasheet and its interpretation.

17.1 Bipolar Diode

The characteristics limiting the reverse voltage rating of a bipolar diode are leakage current and avalanche breakdown. Although we discussed both phenomena in Chapter 16, a number of simplifying assumptions are not consistent with the behavior of a device designed specifically for power applications. To block a high reverse voltage, a power bipolar diode requires a wide space-charge layer at the junction. This has implications for its behavior in both forward and reverse bias. Under reverse bias the leakage current of the device is substantially larger than that given by (16.44). And in forward bias the additional resistance caused by the increased device dimension necessitated by the wide SCL produces a resistive forward voltage drop in addition to the junction drop. This drop is mitigated by conductivity modulation of the region, but results in a forward voltage transient before the region if fully conductivity modulated. In addition, at high currents the excess charge in the neutral regions on either side of the junction is substantial and requires time (known as the *reverse recovery time*) to be removed when switching from conducting to blocking.

17.1.1 Leakage Current

In Chapter 16 we showed that a forward-biased diode whose neutral regions are in low-level injection – and are long compared to the minority carrier diffusion lengths – have the excess carrier profile shown in Fig. 17.1 and an i–v relationship of

$$J = \left[\frac{qD_h n_i^2}{N_D L_h} + \frac{qD_e n_i^2}{N_A L_e}\right]\left(e^{qv_A/kT} - 1\right) = J_s\left(e^{qv_A/kT} - 1\right). \tag{17.1}$$

In reverse bias, that is, $v_A \ll 0$, $J = -J_s$, where J_s is the leakage current density.

For Si at room temperature and under forward bias, J_s as defined by (17.1) is typically very small compared to J. For instance, if we assume that $v_A = 0.6\,\text{V}$, J is 2.6×10^{10} times greater than J_s. Such a small leakage current results in negligible reverse-bias power dissipation, even when the diode is blocking several thousand volts.

However, J_s is proportional to n_i^2, which is a strong function of temperature. That is,

$$n_i^2 \propto T^3 e^{-E_g/kT}. \tag{17.2}$$

At 25 °C, $n_i^2 \approx 2 \times 10^{20}/\text{cm}^6$, but at 175 °C (a typical maximum operating temperature for a power diode) $n_i^2 = 3.8 \times 10^{27}/\text{cm}^6$. This seven orders of magnitude increase gives a leakage current whose effects are no longer negligible.

Carrier Generation in the SCL and Diffusion Regions We know that not all the current in a reverse-biased diode is accounted for in (17.1). As discussed in Chapter 16, electron–hole

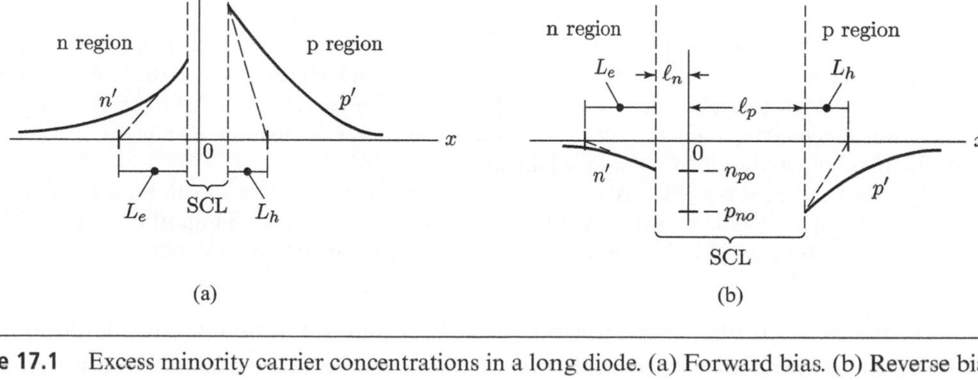

Figure 17.1 Excess minority carrier concentrations in a long diode. (a) Forward bias. (b) Reverse bias.

pairs are thermally generated throughout the device at a rate that depends on the intrinsic concentration n_i, which in turn depends on temperature. In thermal equilibrium, this generation is exactly canceled by recombination, which occurs at a rate proportional to the product of the carrier concentrations. The result is that $n_o p_o = n_i^2$ *everywhere* in the device. When the diode is reverse biased, however, it is no longer in thermal equilibrium. The excess carrier concentrations within the space-charge layer and the adjacent diffusion regions are substantially below their equilibrium values n_o and p_o. In these regions, then, the thermal generation rate, which is independent of n and p, exceeds the recombination rate. The carrier concentrations do not build up because the thermally generated carriers are swept out of the SCL by the intense electric field that exists there. Holes flow to the p region's contact and electrons flow to the n region's contact. These carrier flows give a component of the measured reverse-bias leakage current not accounted for by (17.1).

The generation rate in the depleted and diffusion regions is proportional to n_i, which is a function only of material and temperature. If we define a characteristic lifetime τ_{SCL} so that the generation rate is $G = n_i/\tau_{SCL}$, the total current density caused by thermal generation in these regions is

$$J_{SCL} \approx q(L_e + L_h + \ell_p + \ell_n)G = \frac{q(L_e + L_h + \ell_p + \ell_n)n_i}{\tau_{SCL}}, \tag{17.3}$$

where $(L_e + L_h + \ell_p + \ell_n)$ is the width of the SCL and adjacent diffusion regions.

Because the width of the SCL and the diffusion regions are so long in a power device, we make τ_{SCL} as large as possible. This lifetime is related to, but is usually longer than, the low-injection-level minority-carrier lifetimes τ_e and τ_h. In general, J_{SCL} is much larger than J_s at room temperature, but at high operating temperatures J_s is larger.

Example 17.1 Calculation of Leakage Current Components

For the diode described by (17.1), assume that $\tau_e = \tau_h = 0.2\,\mu s$, $\tau_{SCL} = 10\,\mu s$, and $N_A = N_D = 3 \times 10^{15}/cm^3$. These doping concentrations are such that for a reverse voltage of $v_A = -100$ V, the SCL of this symmetric diode extends approximately 4.6 μm into both the n and p regions, that is, $\ell_p + \ell_n = 9.2\,\mu m$.

We calculate J_s at 25 °C using (17.1) and parameters from Table 16.1, obtaining $2.4 \times 10^{-10}\,A/cm^2$. For this value of J_s, a forward-biased current density of $3\,A/cm^2$ requires $v_A = 0.58$ V. This current density is very low compared to the typical current density for a power diode but is about the level at which this diode will no longer be in low-level injection. Although we do not yet have an accurate model for the diode in high-level injection, we can expect it to carry a forward current density of $50\,A/cm^2$, resulting in $v_A \approx 1$ V. Operating at this current density, the diode's dissipation is $50\,W/cm^2$.

The minority carrier diffusion lengths ($\sqrt{D\tau}$) in this diode are $L_e = 26\,\mu m$ and $L_h = 15\,\mu m$. The total width of the SCL and diffusion regions is therefore approximately 50 μm at −100 V. At room temperature, this width yields a thermally generated leakage current $J_{SCL} = 1.12 \times 10^{-6}\,A/cm^2$. Although this leakage current is much larger than J_s, it causes a reverse-bias power dissipation of only $0.112\,mW/cm^2$, which is negligible compared to the forward-bias dissipation.

At 175 °C, $J_s = 4.6 \times 10^{-3}\,A/cm^2$ and $J_{SCL} = 5.4 \times 10^{-3}\,A/cm^2$. With these currents, the reverse-bias dissipation is $1\,W/cm^2$, which is still small but is quickly becoming important. At 200 °C, $J_s = 28 \times 10^{-3}\,A/cm^2$, $J_{SCL} = 13 \times 10^{-3}\,A/cm^2$, and the reverse dissipation is $30\,W/cm^2$.

A precise quantitative prediction of leakage current requires consideration of effects in addition to those discussed here. For instance, temperature affects lifetimes, and generation

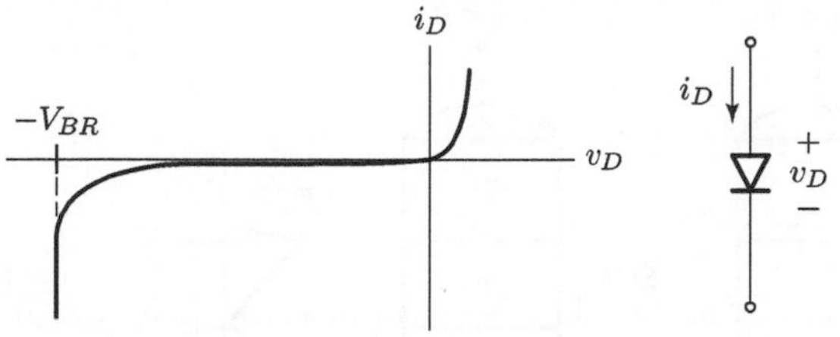

Figure 17.2 The i–v characteristic of a power diode, showing the fast rise in reverse leakage current when the avalanche breakdown voltage $-V_{BR}$ is reached.

within the SCL is not uniform. However, we can draw some important conclusions from the preceding analysis:

- Generation in the wide SCL and diffusion regions of a power device will yield substantially more leakage current at 25 °C than is predicted by the classic diffusion model, (17.1).
- To keep the leakage current acceptably small, the minority carrier lifetime in the SCL should be as long as possible.
- The component of leakage current resulting from thermal generation in the SCL, J_{SCL}, grows as n_i, which approximately doubles for every 11 °C increase in T between −50 °C and 200 °C. But J_s grows as n_i^2, so it eventually dominates at high temperatures. This increase in reverse leakage current limits the maximum operating junction temperature of a power device.

17.1.2 Avalanche Breakdown

We assumed in Chapter 16 that applying an electric field to a semiconductor causes the holes and electrons to flow with a velocity proportional to the electric field. Actually, the particles frequently collide with the lattice imperfections (impurity atom sites, physical defects, or phonons). What we really meant was that the *average* velocity of the particles is proportional to the electric field. If the electric field is strong enough, the velocity of a particle becomes large enough that, during a collision, energy is imparted to the lattice to create an electron–hole pair. This process is called *impact ionization*. It is a multiplicative process: the newly generated particles are accelerated by the field, collide with the lattice, and create additional electron–hole pairs, and so on.

As the reverse bias voltage across a diode increases, so does the peak electric field in its SCL. In Si, when the field reaches a magnitude of about 3×10^5 V/cm, the impact ionization process begins. This field is up to an order of magnitude larger for wide-bandgap materials like SiC or GaN. The number of electron–hole pairs generated in the SCL by impact ionization grows very quickly with increasing electric field above this value. Because these generated carriers add to the thermally generated ones to give more reverse leakage current, the diode's reverse bias characteristic looks like that shown in Fig. 17.2. When impact ionization reaches an infinite rate, we say that the diode is in *avalanche breakdown*.

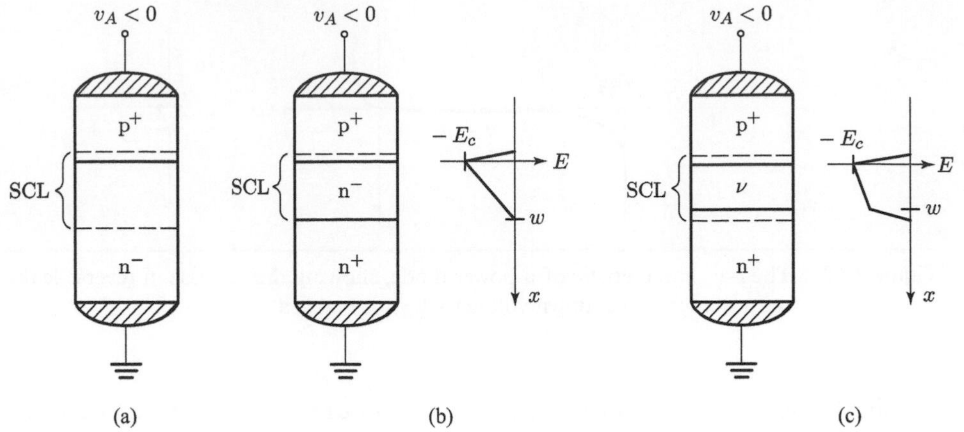

Figure 17.3 Evolution of the pin diode structure. (a) Conventional asymmetric diode with a large n^- substrate resistance. (b) The diode of (a) with the neutral part of the n^- region replaced by a low-resistance n^+ region. (c) The diode of (b) with the n^- region replaced by a very lightly n-doped ν region, producing a pin (or pνn) diode. The resulting field distributions for (b) and (c) are also shown.

The onset of avalanche breakdown occurs approximately when the peak electric field in the SCL reaches a critical value, E_c. This statement is not strictly accurate, because the avalanche process requires that impact ionization occur over some minimum distance so the multiplication process can sustain itself. Because the gradient of the electric field is much greater in the SCL of a heavily doped diode than in the SCL of a lightly doped diode, the distance over which ionization occurs is shorter. The peak electric field can therefore be greater in a heavily doped diode before avalanche begins. For a doping level of $10^{16}/\text{cm}^3$ (typical of a 40–50 V device), $E_c \approx 3.7 \times 10^5$ V/cm, but for a doping level of $10^{14}/\text{cm}^3$ (typical of a 800–1000 V device), $E_c \approx 2.1 \times 10^5$ V/cm.

17.1.3 pin Diode

A reverse-biased, asymmetrically doped diode supports nearly all the voltage across the SCL in the lightly doped n^- region. Making a vertical power diode by simply diffusing the p^+ region into an n^- substrate, as shown in Fig. 17.3(a), would result in a large n^- substrate resistance and a large resistive component of the forward drop. The reason is that the substrate must be about 500 μm thick to maintain mechanical integrity, which introduces a large series resistance. However, because the SCL has a width of between 10 and 200 μm, sandwiching a relatively thin n^- layer between p^+ and n^+ layers, as shown in Fig. 17.3(b), can lower the resistive component of the forward drop. The figure shows the field distribution when the voltage is such that the critical field, E_c, occurs at the junction.

Doping the n^- region so lightly that it is nearly intrinsic (a doping level well below n^- sometimes referred to as ν) further improves performance. The resulting electric field magnitude is shown in Fig. 17.3(c), with v_A such that the field is again E_c at the junction. Because the field distribution is almost rectangular, the i region width is half that of the n^- region of Fig. 17.3(b) for the same voltage rating. The diode of Fig. 17.3(c) is called a *pin diode*. In a practical device, the intrinsic region is really very slightly n-type (ν) or p-type (π), and the device is known as a pνn or pπn diode. Almost all power diodes have this type of structure, which we refer to generically as a pin diode. The highly doped n^+ region quickly terminates the field and is sometimes known as a *field stop* layer.

Example 17.2 Electric Field and Breakdown Voltage of a pin Diode

If the doping concentration in the n^- region of the diode shown in Fig. 17.3(b) is $N_D = 2 \times 10^{14}/\text{cm}^3$, what is the minimum width, w, of this region resulting in the maximum breakdown voltage? We assume that the SCL ends at the n^-–n^+ junction, so all the voltage is dropped across the n^- region. When the device just reaches avalanche breakdown, the field is $-E_c$ at the p^+–n^- junction and 0 at the n^-–n^+ junction, as shown in the figure. Using Gauss's law and $\epsilon_{Si} = 11.8\,\epsilon_o$, we determine $E_{n^-}(x)$, the field in this region:

$$\frac{dE_{n^-}}{dx} = +\frac{qN_D}{\epsilon_{Si}} = -3.08 \times 10^7\ \text{V/cm}^2, \tag{17.4}$$

$$E_{n^-}(x) = -E_c + 3.08 \times 10^7 w. \tag{17.5}$$

The width w necessary to create this condition is determined by setting (17.5) to zero. For a doping concentration of $N_D = 2 \times 10^{14}$, the critical field $E_c = 2 \times 10^5$ V/cm, giving $w = 65\,\mu\text{m}$. Since the E field is triangular, V_{BR} is easily calculated:

$$V_{BR} = \left(\frac{1}{2}\right) E_c w = \left(\frac{1}{2}\right)(2 \times 10^5\ \text{V/cm})(66\ \mu\text{m}) = 660\ \text{V}. \tag{17.6}$$

If we now reduce N_D to $10^{12}/\text{cm}^3$ to create the ν region shown in Fig. 17.3(c), what is the resulting width w required to support the same 660 V? The field in the ν region is now trapezoidal, and if we ignore the voltage contributions of the narrow SCLs in the p^+ and n^+ regions, the resulting field in the ν region is

$$E_\nu(x) = -E_c + 1.546 \times 10^5 x. \tag{17.7}$$

We calculate the required width by integrating E_ν and solving for $x = w$ at $V = 660$ V. The resulting width is $w = 33\,\mu\text{m}$, which is a substantial reduction.

17.1.4 Forward Bias

We based the analysis leading to the pn diode i–v characteristic, (17.1), on the assumption that both sides of the junction were in low-level injection. This assumption is not valid for the intrinsic region of a pin diode because in this region $n \approx n' \gg n_o$ and $p \approx p' \gg p_o$ at any reasonable injection level, and since neutrality requires $n' = p'$, $n \approx p$ in the neutral regions. When the total carrier concentration in a region is dominated by the excess concentration, as in this case, we say that the region is in *high-level injection*, also called conductivity modulation, because the conductivity is no longer determined by the majority doping concentration – it is now a function of injection level.

The diode we choose for analysis is shown in Fig. 17.4, along with its minority carrier concentrations. We assume that the width of the intrinsic region W_i is comparable to the characteristic diffusion length. The intrinsic region is in high-level injection everywhere, but the

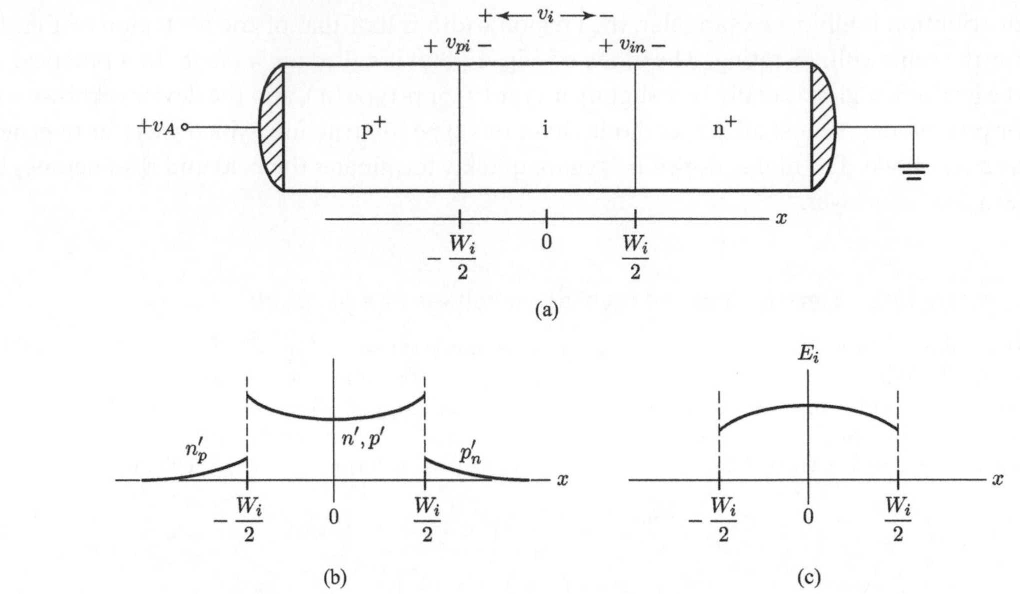

Figure 17.4 (a) A one-dimensional pin diode structure. (b) Carrier concentrations with the intrinsic region in high-level injection. (c) The electric field in the intrinsic region.

heavily doped p^+ and n^+ regions remain in low-level injection. Because the end regions are so heavily doped, the diffusion of minority carriers in these regions is negligible, as it was in the heavily doped side of the asymmetric diode analyzed in Chapter 16. Only hole current flowing by drift is significant in the p^+ region, and only electron current flowing by drift is significant in the n^+ region. These holes and electrons flow into the intrinsic region where they recombine. It is this recombination that determines the terminal current.

Because $n' = p'$, in high-level injection the recombination rate for holes and electrons is the same and is characterized by the *ambipolar lifetime*, τ_a. Furthermore, as the majority and minority concentrations are the same, we can no longer make the approximation that the minority current flows only by diffusion. The continuity equations in the intrinsic region must now contain the drift term that we eliminated in our analysis of the diode in low-level injection. Including the drift term, the continuity equations are

$$\begin{aligned} \frac{\partial n}{\partial t} &= \frac{1}{q}\frac{\partial J_e}{\partial x} - \frac{n'}{\tau_a} \\ &= \mu_e \left(n\,\frac{\partial E}{\partial x} + E\,\frac{\partial n}{\partial x} \right) + D_e\,\frac{\partial^2 n}{\partial x^2} - \frac{n'}{\tau_a}, \end{aligned} \tag{17.8}$$

$$\begin{aligned} \frac{\partial p}{\partial t} &= -\frac{1}{q}\frac{\partial J_h}{\partial x} - \frac{p'}{\tau_a} \\ &= -\mu_h \left(p\,\frac{\partial E}{\partial x} + E\,\frac{\partial p}{\partial x} \right) + D_h\,\frac{\partial^2 p}{\partial x^2} - \frac{p'}{\tau_a}. \end{aligned} \tag{17.9}$$

The concentrations are equal to each other and dominated by their excess in high-level injection, that is, $n \approx n' = p \approx p'$. Recognizing this, (17.8) and (17.9) can be rewritten as

$$\frac{\partial n'}{\partial t} = \mu_e \left(n' \frac{\partial E}{\partial x} + E \frac{\partial n'}{\partial x} \right) + D_e \frac{\partial^2 n'}{\partial x^2} - \frac{n'}{\tau_a}, \tag{17.10}$$

$$\frac{\partial n'}{\partial t} = -\mu_h \left(n' \frac{\partial E}{\partial x} + E \frac{\partial n'}{\partial x} \right) + D_h \frac{\partial^2 n'}{\partial x^2} - \frac{n'}{\tau_a}. \tag{17.11}$$

Multiplying (17.10) by D_h and (17.11) by D_e and adding the two equations gives

$$(D_e + D_h)\frac{\partial n'}{\partial t} = (\mu_e D_h - \mu_h D_e)\left(n' \frac{\partial E}{\partial x} + E \frac{\partial n'}{\partial x} \right) + 2D_e D_h \frac{\partial^2 n'}{\partial x^2} - (D_e + D_h)\frac{n'}{\tau_a}. \tag{17.12}$$

Substituting for μ_e and μ_h in (17.12) using the Einstein relation, $\mu = (q/kT)D$, and simplifying yields

$$(D_e + D_h)\frac{\partial n'}{\partial t} = 2D_e D_h \frac{\partial^2 n'}{\partial x^2} - (D_e + D_h)\frac{n'}{\tau_a}. \tag{17.13}$$

The continuity equations in high-level injection are

$$\frac{\partial n'}{\partial t} = D_a \frac{\partial^2 n'}{\partial x^2} - \frac{n'}{\tau_a}, \tag{17.14}$$

$$\frac{\partial p'}{\partial t} = D_a \frac{\partial^2 p'}{\partial x^2} - \frac{p'}{\tau_a}, \tag{17.15}$$

where D_a is the *ambipolar diffusion constant*,

$$D_a = \frac{2D_e D_h}{D_e + D_h}. \tag{17.16}$$

Because $n' = p'$, (17.14) and (17.15) are identical. We solve them to determine the spatial dependencies of the carrier concentrations and the diffusion components of the currents in the intrinsic region. As in our previous analyses, the concentrations vary exponentially with x. But the characteristic diffusion length, called the *ambipolar diffusion length*, $L_a = \sqrt{D_a \tau_a}$, is now the same for both holes and electrons.

There is also a drift component of current in the intrinsic region, and the drift field gives rise to a voltage drop, v_i, which adds to the on-state voltage of the diode. Using a straightforward calculation, we can obtain the value of this drift field, E_i:

$$J_e = q\mu_e \left(\frac{kT}{q} \frac{dn'}{dx} + n' E_i \right), \tag{17.17}$$

$$J_h = q\mu_h \left(-\frac{kT}{q} \frac{dp'}{dx} + p' E_i \right). \tag{17.18}$$

Because $J_e + J_h = J$ (the total diode current), which is constant throughout the device, and $n' = p'$,

$$E_i = -\frac{kT}{q} \left(\frac{\mu_e - \mu_h}{\mu_e + \mu_h} \right) \frac{1}{n'} \frac{dn'}{dx} + \frac{J}{q(\mu_e + \mu_h)n'}. \tag{17.19}$$

The first term in (17.19) is the field necessary to ensure quasi-neutrality when $\mu_e \neq \mu_h$ (and, therefore, $D_e \neq D_h$). The second term is simply the ohmic drop in the conductivity modulated intrinsic region. The conductivity of this region now depends on injection level, and is $\sigma_i = q(\mu_e + \mu_h)n'$.

An interesting consequence of high-level injection is that the electric field E_i is independent of injection level and therefore of current. We can show this condition as follows. The solution to the diffusion equation, (17.14), is a pair of exponentials with characteristic length L_a. Thus, the first term in (17.19), in which the spatial derivative of n' is divided by n', does not depend on n'. And as we assumed that all the current flowing through the diode results from recombination within the intrinsic region – and as this recombination is proportional to the excess carrier concentration – we know that $J \propto n'$. The second term in (17.19) is therefore also independent of n'. Thus, the electric field E_i, and therefore the voltage v_i, are dependent only on physical parameters and the length of the intrinsic region.

The i–v characteristic of the pin diode of Fig. 17.4(a) can be determined by assuming that $D_e = D_h = D_a$ and therefore $\mu_e = \mu_h$ in the intrinsic region. This is a reasonably accurate assumption in regions of very high-level injection. Because the p^+ and n^+ end regions of the diode are heavily doped, their conductivity is high and we can ignore the voltage drop across these regions. With these assumptions the terminal voltage v_A is the sum of the junction drop and the drop v_i across the wide intrinsic region. Although we do not give the details of the calculation here, the intrinsic region drop is

$$v_i = \int_{-W_i/2}^{W_i/2} E_i\, dx = \frac{2kT}{q} \sinh\left(\frac{W_i}{2L_a}\right) \tan^{-1}\left[\sinh\left(\frac{W_i}{2L_a}\right)\right]. \tag{17.20}$$

For $W_i/2 \leq L_a$,

$$v_i \approx \frac{2kT}{q}\left(\frac{W_i}{2L_a}\right)^2, \tag{17.21}$$

and for $W_i/2 \gg L_a$,

$$v_i \approx \frac{\pi kT}{2q} e^{(W_i/2L_a)}. \tag{17.22}$$

If we assume v_i is small enough to ignore, the v–i relationship of a pin diode is

$$J = \frac{2qL_a n_i}{\tau_a} \tanh\left(\frac{W_i}{2L_a}\right)\left(e^{qv_A/2kT} - 1\right). \tag{17.23}$$

Note that the current is now proportional to $e^{v_A/2}$. Using $L_a = \sqrt{D_a \tau_a}$ to substitute for τ_a results in a relationship similar to that for the short diode of Example 16.3, the major difference being that for a given current the pin diode produces twice the forward drop of the short diode. We have also ignored any ohmic drop in the intrinsic region, which, depending on temperature and the width of the region, would produce an additional drop not present in the short diode.

Example 17.3 Forward Drop in the Intrinsic Region

We consider a 600 V, 100 A pin diode with an intrinsic region width W_i of 240 μm and an ambipolar carrier lifetime of 1 μs. The ambipolar diffusion constant for Si at 300 K is approximately 15 cm^2/s, yielding a diffusion length $L_a = \sqrt{D_a \tau_a}$ of 40 μm. Since $W_i/2 > L_a$, we calculate the drop using (17.22), which gives a value of 0.79 V. This is a significant additional forward drop contributed by the field in the intrinsic region.

17.1.5 Additional Effects of High Current Densities

Two phenomena reduce the ambipolar diffusion length L_a at the high carrier concentrations reached when the diode's current density gets very large. One is *carrier–carrier scattering*, which reduces the mobility of the carriers and therefore the ambipolar diffusion constant D_a. The other is *Auger recombination*, which reduces the ambipolar lifetime τ_a. As the diode current increases, the corresponding reduction in L_a caused by these two effects makes the voltage across the intrinsic region v_i grow quickly, as expressed by (17.21) or (17.22).

The injection-level dependence of L_a, the increased significance of end-region diffusion currents, and the increase of the $n_o p_o$ product cause the i–v characteristic of a power diode, (17.23), to bend at high currents. Figure 17.5 shows the curve that results from these effects.

17.1.6 Temperature Dependence of *J*

We show in (17.23) that the diode current is proportional to n_i if the injection level is not too high. The temperature dependence of n_i is

$$n_i \propto T^{3/2} e^{-E_g/2kT}. \tag{17.24}$$

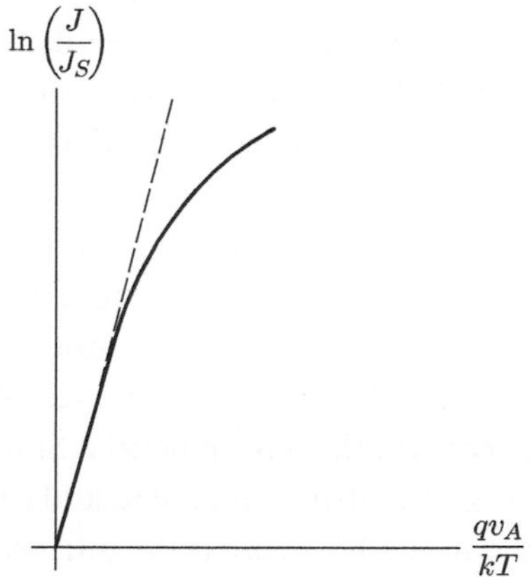

Figure 17.5 The i–v characteristic of a typical power diode, showing the curve bending at high currents because of high-level injection effects.

Because n_i increases more rapidly than $e^{qv_j/kT}$ decreases with increasing temperature, J increases with temperature for a given L_a and τ_a. Conversely, the forward drop decreases with temperature if the forward current is fixed. Although we do not discuss here how L_a and τ_a vary with temperature, the effect of these additional temperature dependencies is to increase the current flowing through a pin power diode with temperature for a given applied voltage.

17.2 Switch Transitions in a pin Diode

We now turn to the transitions of a diode between its static on and static off states. For simplicity, we ignore the capacitance of the SCL and focus on the dynamics of the excess stored charge in the neutral regions outside the SCL. We can easily add the effects of the SCL to those we describe here. We also ignore the change in the lifetime in the intrinsic region as the device goes from no excess carriers to high-level injection, and the effect of the doping concentration in the intrinsic region on the nature of the transitions.

The two salient phenomena during switching are the *forward recovery* voltage and the *reverse recovery current*. While all bipolar diodes exhibit these phenomena, they are greatly aggravated in power diodes because of their wide SCL.

17.2.1 Forward Recovery

We consider first the turn-on transition, in which the diode current steps instantly from zero to I_F. As Fig. 17.6(a) shows, the terminal voltage v_A first rises to a value V_{fp} that is much higher than the static forward drop. It then decays to the steady-state value v_F, determined by the current I_F. This process is called *forward recovery*, and V_{fp} is known as the *forward recovery voltage*. The duration of the forward recovery process is called the forward recovery time t_{fr}.

We can explain the overshoot in the diode voltage by considering how the carrier concentrations in the intrinsic region change during the turn-on transition. Initially, there are no excess carriers. However, as the forward current flows, holes drift through the p$^+$ region and are injected into the intrinsic region. Similarly, the electrons drift through the n$^+$ region and are injected into the intrinsic region. Over time, these injected carriers accumulate, as indicated by the profiles shown in Fig. 17.6(b).

Consider the carrier profiles at t_1, a short time after the forward current starts flowing. As Fig. 17.6(b) shows, the carriers build up near the junctions but not in the center of the intrinsic region. Some of the holes injected into the intrinsic region at the p$^+$–i junction add to the accumulation of carriers at this junction, and the rest flow across the intrinsic region to add to the accumulation of carriers at the i–n$^+$ junction. Similarly, some of the electrons injected at the i–n$^+$ junction flow across the intrinsic region to add to the accumulation at the p$^+$–i junction.

How do these injected carriers flow across the intrinsic region? Not by diffusion, because at this time there is no gradient to the carrier profiles in most of the i region. Therefore, flow in the middle of the i region must be by drift. The voltage that results from this drift is very large. The reason is that the i region is not in high-level injection, so its conductivity, $\sigma_i = q\mu_e n + q\mu_h p$, is

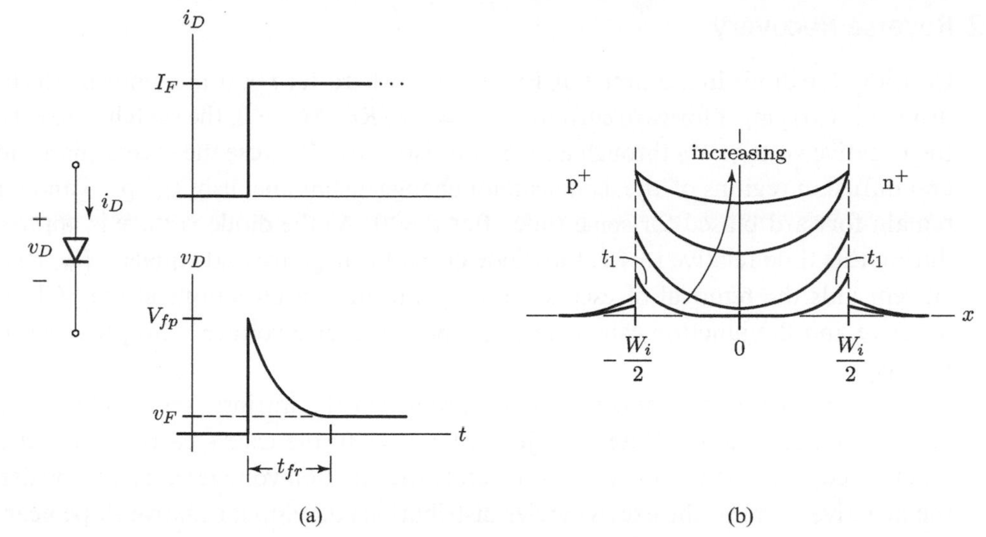

Figure 17.6 Forward recovery of a pin diode. (a) Typical diode voltage and current waveforms during the turn-on transition. (b) The time evolution of excess carrier concentrations in the neutral regions of the device.

very low and yet the region is carrying the full current I_F. Even if the region is not intrinsic, but is a lightly doped ν or π region, its resistance is high compared to its value upon reaching high-level injection. This resistance is the source of the peak transient voltage V_{fp} of the waveforms of Fig. 17.6(a).

As time passes, the growing carrier concentrations in the middle region modulate the region's conductivity and reduce its resistance. As a result, the voltage across the middle region drops. When the carrier concentrations reach their final value, so too does the diode's voltage. The charge in the middle region at steady state is proportional to the forward current ($I_F \propto Q/\tau_a$). Hence, the time required to supply this charge – in other words, the duration of the forward recovery transition – is approximately τ_a.

Example 17.4 Magnitude of the Forward Recovery Voltage

Assume that we have a pνn diode with the center region doped at $N_D = 10^{14}/\text{cm}^3$ and designed to withstand 400 V. This voltage requires a ν-region width of $W_\nu \approx 20\,\mu\text{m}$. The cross-sectional area of the device is $A = 0.5\,\text{cm}^2$, and its forward current steps from zero to 50 A at $t = 0$. What is V_{fp}?

Initially, the resistance of the ν region is

$$R_\nu = \frac{W_\nu}{\sigma A} = \frac{W_\nu}{q(\mu_e n + \mu_h p)A} = 185\,\text{m}\Omega, \tag{17.25}$$

where $n = N_D$, $p = N_A = n_i^2/N_D$ is negligible because it is several orders of magnitude less than N_D, and $\mu_e \approx 1350\,\text{cm}^2/\text{Vs}$ at this concentration. With $I_F = 50$ A, the drop across this region is 9.25 V, which is the diode's peak forward recovery voltage V_{fp}.

17.2.2 Reverse Recovery

Consider the diode in the circuit of Fig. 17.7(a), where, for $t < 0$ and ignoring the forward diode drop, it is carrying a forward current of $I_F = V_F/R_F$. At $t = 0$, the switch connects the diode to the negative voltage V_R through a second resistor, R_R. Because the excess charge in the i region and diffusion regions of the diode cannot change instantaneously, the p^+–i and i–n^+ junctions remain forward biased for some time after $t = 0$. As the diode voltage is approximately zero during this time relative to V_R, the diode current is negative and equals $-V_R/R_R$. This reverse current aids the removal of excess charge, until the concentrations at the SCL edges become negative and the junction can begin to support a reverse voltage. This process is called *reverse recovery*.

Figure 17.7(b) shows how the carrier profiles in the intrinsic region change as the reverse recovery current flows. Note that, just after $t = 0$, the excess carrier concentrations at the junction edges are still positive, and therefore the junction voltages are also positive. To support the negative current, the excess carrier distributions develop a negative slope near the junction edges. Holes diffuse out of the intrinsic region, are injected into the p^+ region, and drift to the left ohmic contact. Similarly, electrons leave the intrinsic region at the i–n^+ junction and drift to the right contact. So long as the excess concentrations at the SCL edges are greater than zero, and $|V_R| \gg v_F$, the total diode current equals $-V_R/R_R$. Continuing recovery current flow eliminates the excess carrier concentrations in the intrinsic region, as the profiles in Fig. 17.7(c) show.

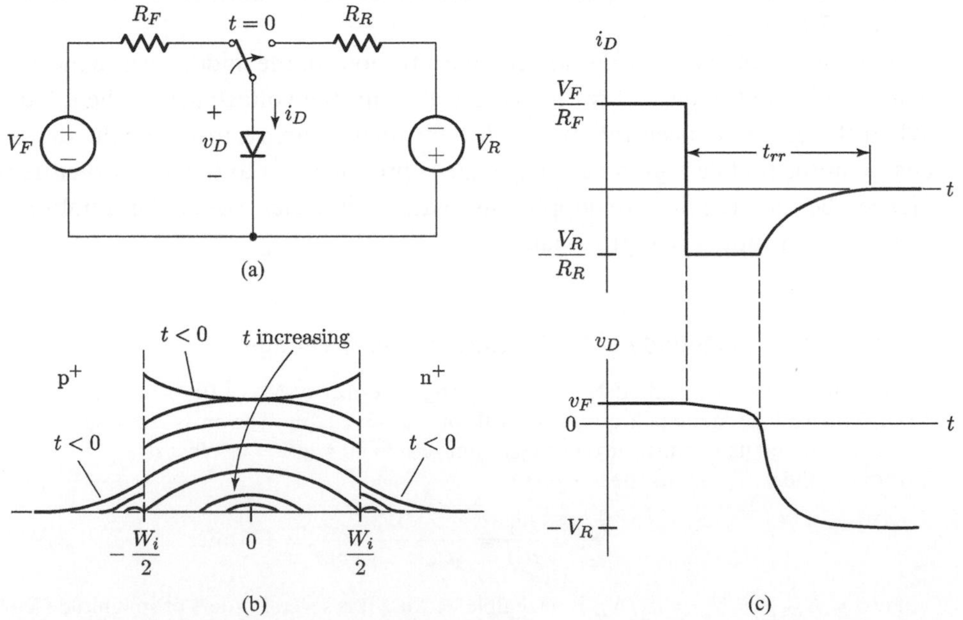

Figure 17.7 (a) A hypothetical circuit to illustrate pin diode reverse recovery. (b) Carrier profiles in the intrinsic region during the turn-off transition. (c) Waveforms showing reverse diode current during the reverse recovery time t_{rr}.

After the excess carrier concentrations at the junction edges reach zero, the concentration gradients decrease and the diode current can no longer be maintained at $i_D = -V_R/R_R$. As $|i_D|$ decreases, the drop across R_R also decreases. The difference between V_R and the voltage across R_R appears across the diode SCLs, which begin to widen. The dynamics of this process produce an approximately exponential rise of i_D to zero and fall of v_D to $-V_R$, as Fig. 17.7(c) shows. During the initial phase of the reverse recovery process, the diode voltage changes only slightly from its static forward value v_F (the small decrease results from the change in sign of v_i that occurs when the current changes direction).

The duration of reverse current flow in the diode is t_{rr}, the reverse recovery time. Based on our discussion so far, t_{rr} is directly proportional to the reverse current I_R and the charge Q that is initially stored in the intrinsic region. However, reverse current does not remove all of this stored charge; some of it recombines. How much depends on the magnitude of the negative current relative to the forward current. The initial stored charge is $Q = I_F\tau_a$. If $I_R \gg I_F$, almost all the stored charge is swept from the intrinsic region in a time much less than τ_a, and only a small amount will be removed by recombination. But if $I_R \leq I_F$, a substantial fraction of the charge will recombine. In the limit of $I_R = 0$, when all the charge is removed through recombination, the reverse recovery time is $t_{rr} \approx \tau_a$.

From the preceding qualitative discussion of the turn-on and turn-off transitions of a pin diode, we can conclude that the less charge the diode stores for a given forward current, the faster the diode will switch. As $Q = I_F\tau_a$, the way to store less charge is to shorten the lifetime. However, doing so reduces L_a, resulting in less conductivity modulation of the intrinsic region and a higher on-state drop, v_i, across it, as given by (17.21) and (17.22). For this reason, the forward drop of a fast bipolar diode is typically greater than that of a slower bipolar diode.

17.3 The Schottky Barrier Diode

When metal leads are attached to a semiconductor device, the interface between the metal and semiconductor is usually ohmic. An *ohmic contact* is one in which the resistances of the contacting materials govern current flow. Obtaining a good ohmic contact between a semiconductor and a metal is difficult. A great deal of research has been directed at developing an understanding of this interface and the metallurgy necessary to produce ohmic contacts. When we fail to pay attention to these processes, the result is a contact with a poor rectifying characteristic, which was the basis for early "cat whisker" crystal receivers and point-contact transistors. However, making a good, large-area, rectifying contact is even more difficult than making a good ohmic contact. A diode made with a metal–semiconductor junction is called a *Schottky barrier diode*, or SBD. The symbol for a Schottky diode is shown in Fig. 17.8.

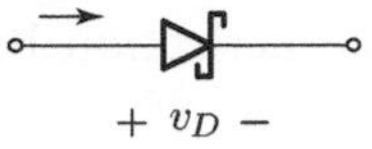

Figure 17.8 The circuit symbol for a Schottky barrier diode.

In principle, either n-type or p-type semiconductor material can be used to make a Schottky diode, although n-type material is used because in Si electron mobility is greater than that for holes. SBDs can be fabricated in Si, SiC, GaN, or GaAs, but we address here only Si and SiC devices. Silicon SBDs typically have low reverse blocking voltage ratings – with maximums on the order of 100 V – while SiC SBDs are available with ratings near 2 kV. As we demonstrate in this section, Schottky diodes offer two advantages over bipolar diodes. The first is that, for the same current density, the forward voltage drop of a Si SBD is several tenths of a volt less than that of a Si p–n diode. Although, because of its larger bandgap, the forward drop of a SiC p–n diode ($\sim 3-3.5$ V) is much higher than that of a Si p–n diode, the forward drop of a SiC SBD is similar to that of a Si p–n diode, making the SBD an attractive application of SiC. The second is that the current in a Schottky flows only by drift, so there is no need to accumulate or remove excess carriers, thereby eliminating forward and reverse recovery phenomena.

The disadvantage of Schottky diodes is that they have more leakage current when reverse biased than do bipolar diodes. This is especially true for Si Schottkys. Schottky diodes using SiC, however, can have leakage currents in the μA range at reverse blocking voltages in excess of 1 kV. The parameters of SiC Schottkys are also less temperature dependent than those for Si devices, especially the forward voltage drop. However, the forward drop for a SiC device is usually higher than for a Si device.

17.3.1 The Schottky Barrier

Current flow across a rectifying metal–semiconductor junction is the result of *thermionic emission* of electrons from the metal into the semiconductor and from the semiconductor into the metal. In thermal equilibrium, these two flows are equal and there is no net current. When a voltage is applied to the junction, one direction of electron emission dominates and a net electron current results.

Because free electrons in the semiconductor are more energetic than those in the appropriately chosen metal, more electrons initially flow into the metal. They leave behind a positive space charge layer of donor atoms and create a compensating negative charge on the metal's surface.

17.3.2 Forward Bias

We can now determine the effect of an applied junction voltage, v_j, on the flow of electrons across the Schottky barrier junction. As for a bipolar diode, v_j will change the SCL width. The junction is highly asymmetric, so nearly all of v_j appears across the semiconductor side of the SCL. If $v_j > 0$ (that is, the diode is forward biased), the electric field of SCL origin is reduced, reducing the potential barrier to electron emission from the semiconductor. However, the barrier to emission from the metal has been changed only slightly. The result is a forward current that is a strong function of v_j and consists of electrons flowing from the semiconductor to the metal.

Although we do not show the derivation here, the *i–v* characteristic of a *forward-biased* Schottky diode is

$$J_F = J_o\left(e^{qv_j/kT} - 1\right), \tag{17.26}$$

which is identical in form to the characteristic of a bipolar diode. However, the current coefficient J_o is of a substantially different form from J_s, the leakage current as defined in (17.1) for a bipolar diode. The current coefficient J_o has different values in forward and reverse bias, as we show later.

The current coefficient J_o for a forward-biased silicon Schottky diode is typically four to six orders of magnitude larger than J_s for a silicon bipolar diode. For this reason, its forward drop is 0.23–0.34 V lower than that of a bipolar diode operating at the same current density. This lower on-state voltage is a Schottky's major advantage.

Forward Voltage Drop The forward drop of a Schottky diode contains a component in addition to v_j that is equal to the voltage across the semiconductor. The resistance of this region has two parts: (i) the lightly doped material into which the SCL grows in reverse bias; and (ii) the heavily doped substrate. Figure 17.9 illustrates this structure. For low-voltage Schottkys (< 40 V), these two resistances are comparable because, even though the lightly doped region has a higher resistivity, the heavily doped region is longer (typically 500 μm versus 5 μm). The total resistance of a 1 cm^2 silicon die might be 0.5–1 mΩ, giving a drift region drop of 50–100 mV at a current of 100 A.

The resistance of the lightly doped region quickly dominates the resistance of the drift region as the Schottky voltage rating increases from approximately 40 V. The reason is that a higher voltage rating requires a longer lightly doped region and a lower doping concentration, both of which increase the resistance of this region. Unlike in the bipolar diode, conductivity modulation does not reduce this resistance when the junction is forward biased. The large resistance is one of the reasons that silicon Schottky diodes are not generally available at voltage ratings above approximately 100 V.

Temperature Dependence of the Forward Drop The forward voltage drop of a Schottky depends on temperature for two reasons: (i) the i–v characteristic, (17.26), has an explicit temperature dependence; and (ii) the resistance of the drift region in the semiconductor increases with temperature because electron mobility decreases as temperature increases. The mechanism responsible for this decrease is the increase in the amplitude of lattice vibrations. The first

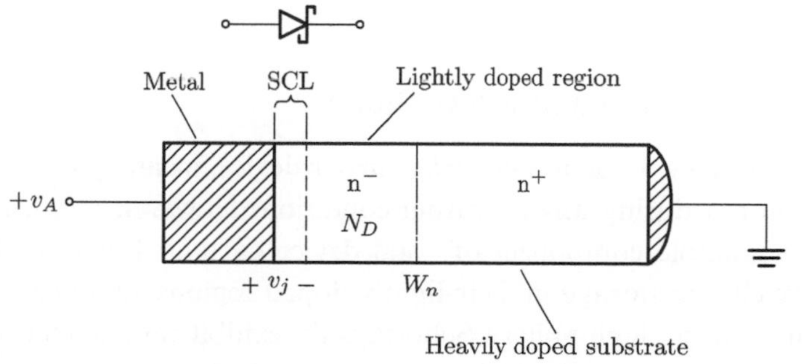

Figure 17.9 The Schottky diode structure showing the metal contact, the lightly doped region which supports the reverse voltage, and the heavily doped substrate.

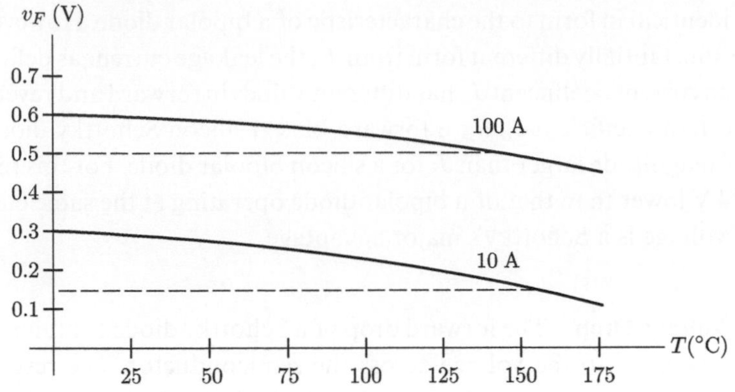

Figure 17.10 The forward voltage drop v_F as a function of temperature for a 50 V, 100 A Schottky diode operating at two different currents.

mechanism dominates the temperature dependence of the forward drop in almost all practical Schottky diodes.

The forward voltage as a function of T is shown in Fig. 17.10 for a 50 V, 100 A Schottky diode. The influence of the drift region resistance at high currents is apparent from the different slopes of the two curves in the figure.

17.3.3 Reverse Bias

Under reverse bias, the barrier to the flow of electrons from the semiconductor is greatly increased, virtually preventing electron flow from the semiconductor to the metal. Reverse current consists primarily of electrons flowing from the metal to the semiconductor. Because of the voltage dependence of the metal barrier height, this current is a strong function of reverse voltage, even though it is still much smaller than the forward current. This reverse voltage dependency makes the leakage current much greater than $-J_O$. The dissipation resulting from this leakage current is the principle determinant of the reverse voltage rating of a Schottky diode.

17.3.4 Switch Transitions in a Schottky Diode

The ideal Schottky diode is a *majority carrier* device, meaning that, generally, negligible excess charge is stored during any operating condition. However, minority carriers do constitute a normally negligible component of total device current. But in high-voltage Schottkys there is minority charge storage in their lightly doped regions, and because in high-voltage devices this region is wide, high-voltage Schottkys do exhibit reverse recovery behavior. In addition, the Schottky diode contains SCL capacitance, which must be charged and discharged during switching.

17.3.5 Selection of Barrier Metal

Because the potential of the Schottky barrier depends on the work function of the metal contact, we can change the forward and reverse characteristics of the diode by using different metals. Chrome, molybdenum, platinum, and tungsten are typical choices. Chrome gives the lowest junction drop but the highest reverse leakage current, so it is usually reserved for low-temperature applications. Tungsten is commonly used because, although it has the highest forward drop of the metals mentioned, it also exhibits the lowest leakage current and so can be used at high temperatures.

17.4 Power BJT

As we noted in Chapter 16, although the BJT has been largely displaced by the power MOSFET and IGBT, familiarity with its operation and characteristics is necessary for understanding the behavior of the SCR and IGBT. The power BJT differs from the BJT discussed in Chapter 16 primarily in its structure required to support high voltages, which results in corresponding behavioral differences.

17.4.1 Structure and Operating Regions of the Power BJT

Figure 17.11(a) shows a cutaway view of a typical npn bipolar power transistor. It is a vertical device, with the collector connection on the bottom (substrate) of the die and the base and emitter connections on the top. Note from the top view in Fig. 17.11(a) that the base and emitter metal traces are *interdigitated*. This pattern is used to keep the distance between the base contact and the centerline of the emitter (line A–A') short, the reasons for which we discuss later.

Looking at the profile of the doping concentration along line A–A', you can see the one-dimensional representation of the transistor shown in Fig. 17.11(b). Note that the collector of this transistor has two regions: a lightly doped ν region not present in a conventional (signal-level) transistor, and a heavily doped substrate. The purpose of the ν collector region, which is more lightly doped than the p-type base, is identical to that for the pin diode – to make the collector–base junction SCL grow into it, rather than into the base. The SCL can then be allowed to punch through to the substrate region, as shown in the plot of E in Fig. 17.11(b). As with the similar pin diode structure, we can nearly double the breakdown voltage for a given ν-region width by allowing this punch-through to occur.

When this transistor is in its on state, the entire ν region is in high-level injection so that its contribution to the on-state drop is small, and current flows by drift. For this reason, we sometimes refer to the ν region as the *drift region*. The operating condition during the drift region's transition to high-level injection results in four, rather than the usual three, distinct regions of operation.

The first is the off-state region, also known as *cutoff*. In this region both the collector–base and emitter–base junctions are reverse biased, the ν region is depleted, and the collector current

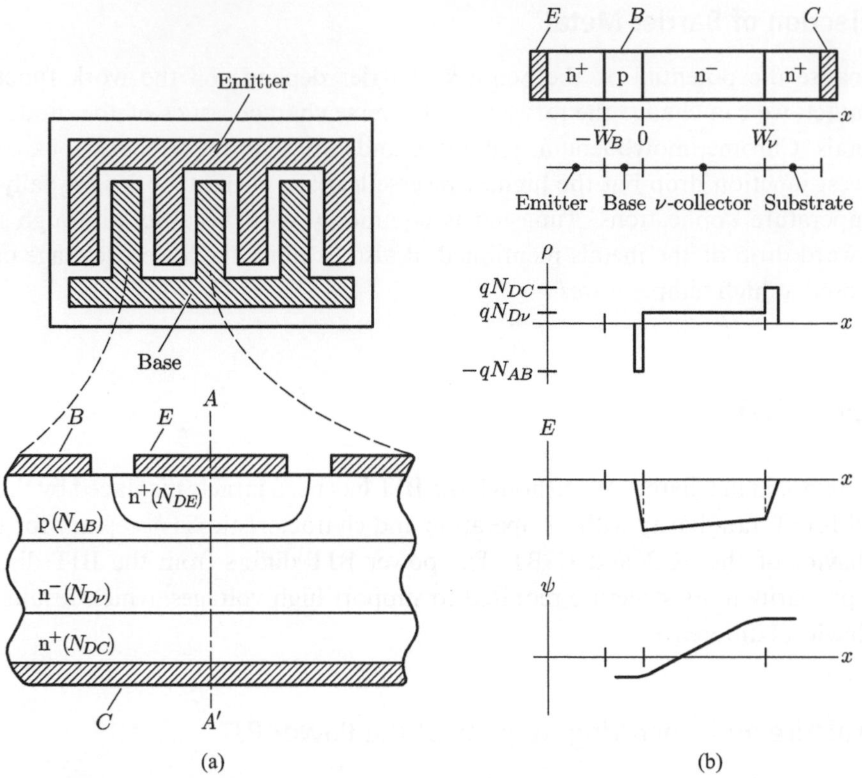

Figure 17.11 The npn power bipolar junction transistor. (a) Top and side-section views of the transistor. (b) One-dimensional representation of the power transistor doping profile along the line A–A' bisecting the emitter, and the charge, electric field, and potential profiles of the reverse-biased collector–base junction.

is approximately zero. The second is the *forward-active* region, where the collector–base junction is strongly reverse biased, while the emitter–base junction is forward biased. Operation in this region resembles that of a conventional transistor. At low values of v_{CB}, corresponding to low values of v_{CE}, when the ν region is not yet in high-level injection (conductivity modulated) the transistor enters the third region, called *quasi-saturation*, corresponding to a forward-biased p–ν junction that injects excess carriers into the ν region, so that *part* of this region is now in high-level injection. The fourth region of operation, *hard saturation*, occurs when the entire ν region is conductivity modulated, and v_{CE} assumes its lowest on-state value for a given current. The transistor characteristics in each of the four operating regions in the i_C–v_{CE} plane are shown in Fig. 17.12.

The Cutoff Region One might reasonably expect the voltage rating of the transistor in cutoff to be determined by avalanche at the collector base junction. However, this is only true if the emitter is open-circuited. In this case the breakdown voltage is V_{CBO}. But the emitter is infrequently open-circuited, resulting in transistor action which produces a multiple (β_F) of the

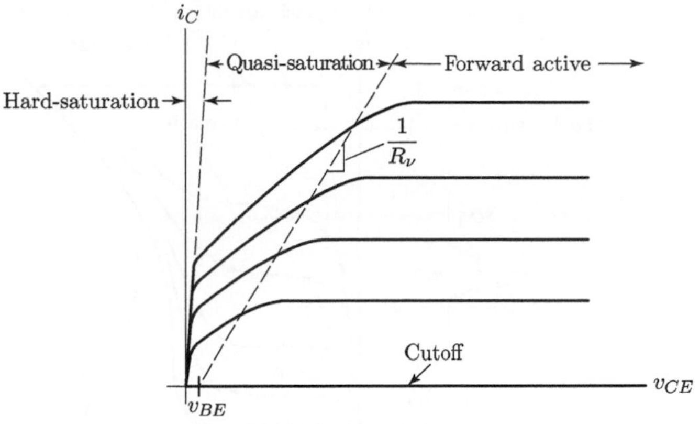

Figure 17.12 The i_C–v_{CE} curves for an npn power transistor with a ν collector region.

base current at the collector junction. For this reason the maximum collector–emitter voltage rating of a BJT is a function of the conditions existing at the base terminal.

If the base terminal is left open, the only source of majority carriers (holes) to support excess charge in the base is the leakage current of the collector junction. But this leakage current has the same effect as the base current in determining excess base charge and therefore the emitter current. Because of transistor action, this emitter current is larger than the "base" (collector leakage) current by a factor β_F, and as the base is open, this increased emitter current will increase the collector current by the same amount. The multiplicative avalanche process further increases the collector current as v_{CE} is increased, creating a positive feedback process, which, at some value of collector leakage current, becomes unstable, resulting in unconstrained collector current. The collector–emitter voltage at which this occurs is known as the *sustaining voltage*, $V_{CEO(sus)}$, which is lower than V_{CBO}. If the base terminal is not open, V_{CEX} (where X refers to the base terminal variable) is a function of i_B, decreasing for $i_B > 0$ and increasing for $i_B < 0$, as shown in Fig. 17.13.

The Forward-Active Region From its off state, the transistor enters the forward-active region when the emitter junction is forward biased and minority carriers are injected into the base region. In most of the forward-active region, the collector is strongly reverse biased and the entire ν collector is depleted. At small values of reverse bias, however, the ν collector is *not* depleted but presents a region of high resistance to the flow of collector current. The value of this resistance, R_ν, is

$$R_\nu = \frac{W_\nu}{q\mu_e N_{D\nu} A_\nu}, \tag{17.27}$$

where A_ν is the cross-sectional area of the ν region.

Figure 17.14(a) shows a simple model for the transistor when it is operating in the forward-active region but at low collector–base voltages where the p–ν junction SCL is very narrow. Figure 17.14(b) shows the corresponding minority carrier distributions. The p–ν junction SCL is not shown because it is so narrow at the low v_{CE}. The transistor will continue to operate in

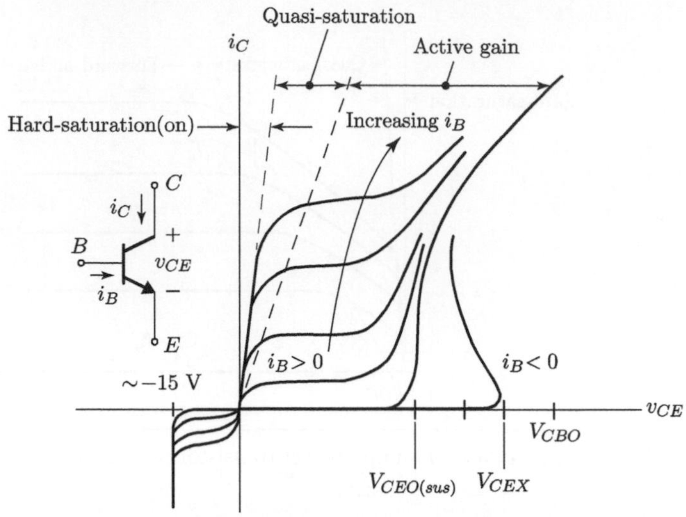

Figure 17.13 Typical i_C–v_{CE} curves for an npn power transistor showing the sustaining breakdown voltage as a function of i_B and its relationship to V_{CBO}.

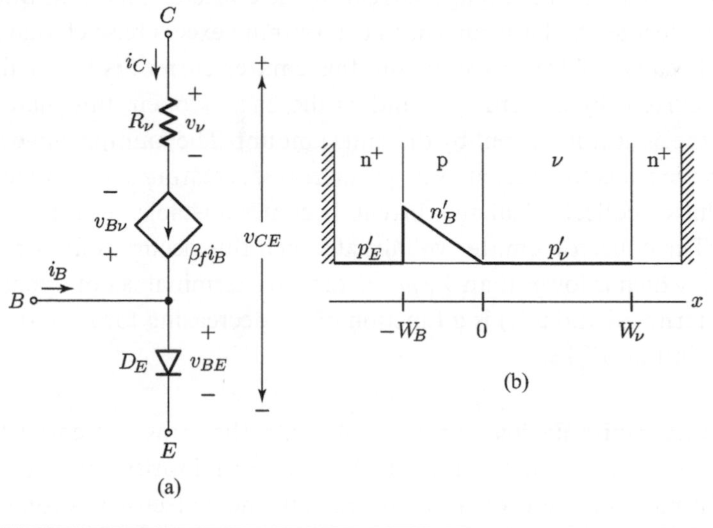

Figure 17.14 (a) A simple model for the power BJT operating in the forward-active region. (b) Excess minority carrier distributions in the power BJT operating in the forward-active region.

this region so long as the collector–base junction (p–ν) remains reverse biased. If we assume that $v_{BE} \approx 0.8$ V, this junction is reverse biased if $v_{CE} > i_C R_\nu + 0.8$ V. On the v_{CE}–i_C plane, this boundary between the forward-active and quasi-saturation regions is a straight line with a slope $1/R_\nu$, intersecting the v_{CE} axis at $v_{BE} = 0.8$ V, as shown in Fig. 17.12. The region between this boundary and the $+v_{CE}$ axis defines forward-active operation.

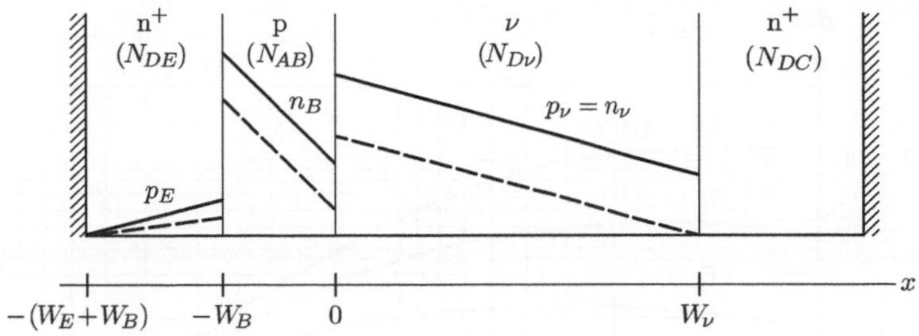

Figure 17.15 Carrier distributions for a transistor in hard saturation. The dashed profiles are those at the boundary of quasi- and hard saturation.

The Quasi-Saturation Region At the quasi-saturation region boundary with the forward-active region the p–ν junction voltage is zero and v_{CE} is the sum of $i_c R_\nu$ and v_{BE}. As v_{CE} continues to decrease, the p–ν junction becomes forward biased and the ν region begins to enter high-level injection and conductivity modulation, reducing R_ν. To match the decrease in v_{CE}, i_C must decrease, which is what we see in Fig 17.12.

The Hard-Saturation Region If, for a given collector current, we drive the transistor with a high enough base current i_B, we will push it into the hard-saturation region. Figure 17.15 shows the carrier profiles for a transistor in this condition. The slopes of the profiles in the base and ν regions are the same as they were at the onset of saturation, but the levels have shifted upwards.

Although the collector current has remained constant, the hole current injected into the emitter from the base increases by the same factor that the excess electron concentration at $x = -W_B$ has been increased. Furthermore, because the ν–n^+ junction is now forward biased, holes originating as base current will be injected into the n^+ collector (similar to reverse injection of holes into the emitter). For these reasons, the required base current quickly increases as we try to push the transistor further into the hard-saturation region, and the gain falls rapidly.

The common-emitter current gain $\beta_F = I_C/I_B$ is a very important parameter for users of power transistors (this gain is often labeled h_{FE} in specification sheets). But unlike users of signal-processing transistors, who only want to know β_F in the forward-active region, we want to know β_F at the onset of saturation in order to determine the minimum base current needed to ensure the on state. As noted above, β_F falls rapidly at high current levels as the transistor is pushed into hard saturation. While signal level BJTs have active region β_Fs in the 100s, power BJT's gains in saturation are in the low double or even single digits. But the gain also falls at very low current levels because of the base current needed to support recombination in the emitter–base SCL. This also corresponds to a decrease in the emitter efficiency, γ_E, the ratio of electron current to hole current at the emitter junction discussed in Section 16.3.2. We will see that this current dependence of γ_E is critical to the operation of the SCR. A typical specification of β_F as a function of collector current at constant v_{CE} is shown in Fig. 17.16. Note that the gain is specified at $v_{CE} = 4$ V, which is not very deep into hard saturation. Driving the transistor harder to reduce v_{CE} will further reduce the gain. Note that the gain is also temperature dependent.

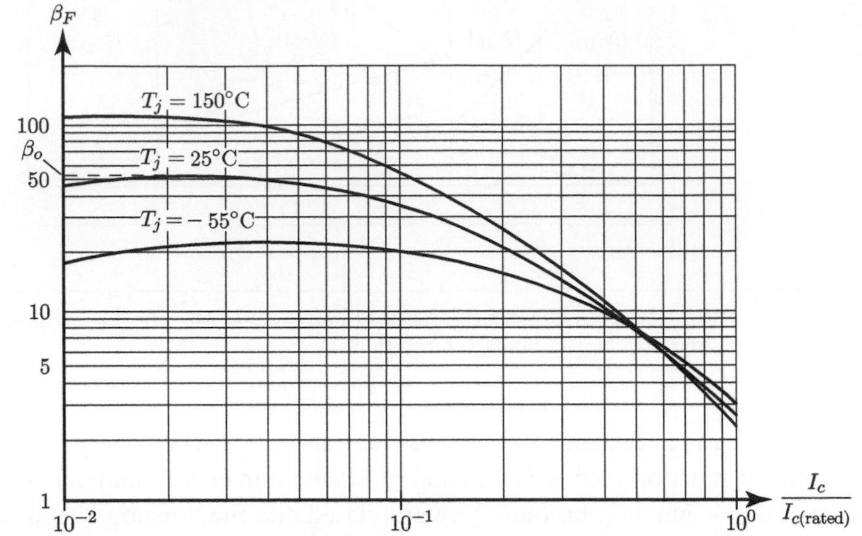

Figure 17.16 A typical specification of common-emitter current gain (β_F) as a function of collector current while holding $V_{CE} = 4$ V. The parameter β_o is the value of β_F specified in the datasheet.

Switch Transitions Figure 17.17 illustrates typical switching waveforms for a bipolar transistor. Consider the turn-on transition first. At $t = 0$ the device is cut off and the base current makes a step change from 0 to $+I_{B_1}$. The transistor must first pick up the load current, as it commutates from the diode, before v_{CE} can fall, since the diode cannot support reverse voltage while it is conducting forward current. The collector voltage falls quickly at first but then "tails off" for the last few volts, reaching its final value only after the *voltage tail time*, t_{vt}. This behavior is similar to the forward recovery of a bipolar diode, and is caused by the gradual drop in the resistance of the transistor's ν-collector region as it becomes conductivity modulated.

During turn-off, when the base current switches from $+I_{B_1}$ to $-I_{B_2}$, the excess charge in the base and ν regions begins to decrease, but their gradients remain unchanged so i_C remains constant. When the excess concentration at the collector ν–n junction reaches zero, we enter the quasi-saturation region and the fraction of the ν-collector that is conductivity modulated decreases, increasing its resistance and hence v_{CE}. This accounts for the slow rise of v_{CE} before i_C begins to decrease. The period over which this occurs is known as the *storage delay time*, t_s. As i_C begins to fall, the diode turns on and I_d begins to commutate from the transistor to the diode. The current i_C decreases as the base charge declines via the negative base current and recombination. The transistor enters cut-off when $i_C = 0$. During turn-off $-i_{B2}$ is pulling charge out of the base region, and until the excess concentration at the emitter–base junction reaches zero, the junction remains forward biased but conducting negative current. Once $i_C = 0$, v_{BE} can fall to the negative drive voltage $-V_x$.

During both the turn-on and turn-off transients, the transistor is simultaneously supporting nonzero i_C and v_{CE} and dissipating energy. This switching loss and the duration of the turn-on and turn-off processes determine the maximum switching frequency at which a BJT can operate efficiently.

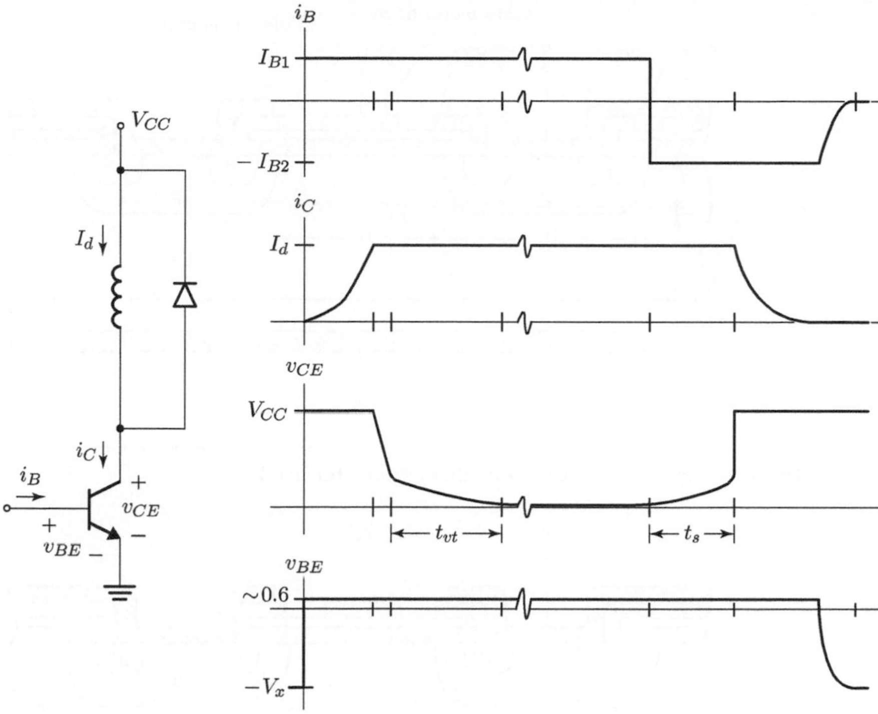

Figure 17.17 Typical switching waveforms exhibited by a power BJT.

17.5 Power MOSFET

Figure 17.18 shows the structure of an n-channel power MOSFET. Note that the drain contact is on the bottom of the die, rather than on the top as in a lateral signal-processing MOSFET, resulting in a device in which the current flows vertically. This structure provides for a thick lightly doped n-region, called the *extended drain*, to block the off-state voltage without compromising surface area. It also gives maximum area to both drain and source contacts in order to produce a low-resistance connection to the package terminals. The polysilicon (or metal) electrode gate is insulated from the source metal that covers it by a layer of SiO_2. The gate electrode is connected by metal fingers that make contact through windows etched in the SiO_2. The fabrication of power MOSFETs uses a double-diffusion process to create the n^+ and p regions, and the device is often referred to as a *vertical double-diffused* MOSFET (VDMOSFET or DMOS). The structure of Fig. 17.18 has a lateral channel. While lateral-channel designs currently dominate SiC MOSFETs, more modern Si power MOSFETs have a vertical channel which we discuss shortly. But we discuss the lateral-channel structure here because it is somewhat easier to visualize the various aspects of its operation.

Between the n^+ source and the n^- drain regions are *wells* of p-type material. These wells are known as the *body* region of the device. The channel of the MOSFET is formed on the surface of these p wells just beneath the gate oxide. Note that the p wells are shorted to the source by the source electrode. We described the reason for this source–body short in Section 16.4.2.

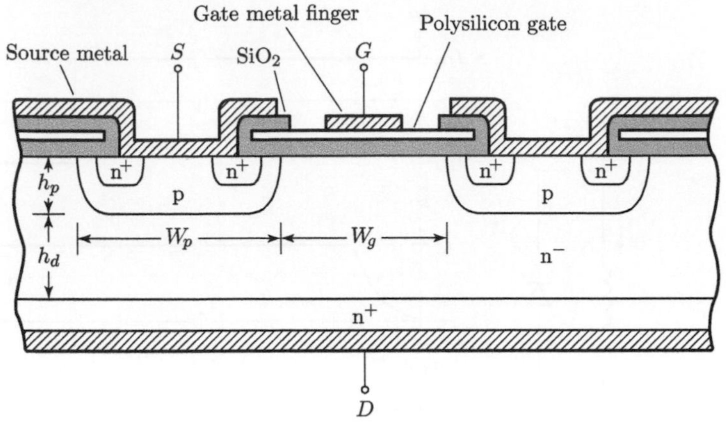

Figure 17.18 Structure of a vertical n-channel power MOSFET.

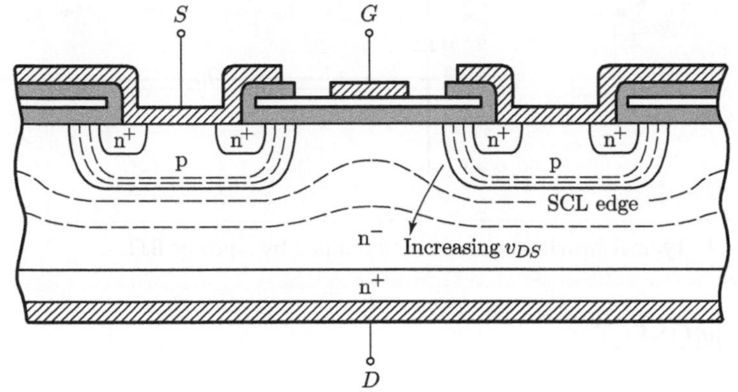

Figure 17.19 An illustration of how the SCL grows in a power MOSFET with increasing v_{DS}, pinching off the n^- region between the p wells.

The lightly doped n-type drain region is unique to power MOSFETs, and is provided to allow growth of a long SCL, permitting the device to block a high voltage when it is off. In this respect, the region functions like the ν region in a power BJT. This lightly doped drain region is frequently referred to as the *extended drain* or *drift region*. As the SCL grows, it *pinches off* (depletes) the region between the p wells, as shown in Fig. 17.19 (the gate electrode acts as a field plate to promote this pinch-off). This feature of a power MOSFET's structure is important because it keeps the gate oxide from being subjected to the full drain voltage. In fact, the voltage just underneath the gate oxide typically reaches only 5–10 V with respect to the gate electrode, even though the drain voltage may be 200–400 V. As a result, we can make the gate oxide relatively thin, which keeps both the gate threshold voltage V_T and the gate energy loss low.

A further distinction between the signal-level and power MOSFETs is the relationship between i_D and v_{GS}. That for the signal-level MOSFET was given in (16.64), which shows the quadratic relationship resulting in the non-uniform spacing of the curves in Fig. 16.14. At the high gate voltages characteristic of power MOSFETs, the carrier velocity saturates at v_{sat}, which is on the order of 10^7 cm/s for Si, and is no longer given by $\mu_e E$. Effectively, μ decreases as E (and V_{DS}) increases. An approximation to the i_D–v_{GS} relation better than (16.64) for the power MOSFET is

$$i_D = kv_{\text{sat}}(v_{GS} - V_T). \tag{17.28}$$

This is not particularly important for applying power MOSFETs, since they are seldom operated in their linear region. But it is important if one calculates transconductance as discussed with respect to the Miller effect in Chapter 16.

DMOS On-State Resistance Figure 17.20(a) shows how the drain current flows through the DMOS of Fig. 17.18 when it is on. As the current flows through the extended drain, it focuses in the area between the p wells, called the *neck* region. From there it must further focus in the thin entrance to the p-well channels on either side of the neck region. This latter focusing would result in a large resistance if the current had to flow through a progressively smaller cross-sectional area of the n^- region to get to the channel entrance. Fortunately, the region just below the gate oxide (and between the p wells) is *accumulated*, meaning that it has a very high concentration of mobile electrons, which makes it much more conductive than the bulk n^- region. The drain current between the p wells therefore tends to first flow directly upward to this layer (taking advantage of the full width between the p wells) and then flows horizontally (with little resistance) to the channel entrances.

Figure 17.20(b) shows how the MOSFET's total drain–source resistance, R_{DS}, can be broken down into four components: R_{xd}, R_{neck}, R_{accum}, and R_{ch}. For a DMOS designed for 400 V or above, R_{xd} is typically much larger than R_{neck}, R_{accum}, and R_{ch}, and thus dominates R_{DS}. For a DMOS designed for 100 V and below, however, $R_{\text{ch}} + R_{\text{neck}}$ is typically one third to one half of R_{DS}. Power MOSFETs are characterized by a normalized parameter called *specific on-state resistance*. It is defined as

$$R_{DS(\text{sp})} = (R_{DS})(A),$$

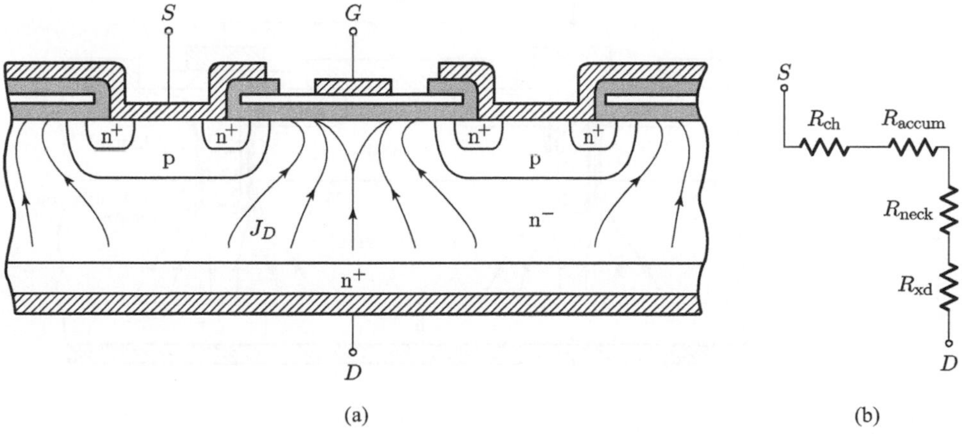

Figure 17.20 (a) Current flow in a power MOSFET. (b) A model of R_{DS}, showing its components.

where A is the active area of the device. This parameter is voltage specific, and illustrates what minimum R_{DS} is possible for a given device area and therefore current rating. For an ideal device when $R_{DS(\text{on})}$ is dominated by the extended drain, this relationship is

$$R_{DS(\text{on})} \propto V_{BR}^{2.5}/\text{Area}. \tag{17.29}$$

For example, a device designed to withstand 400 V would have $4\sqrt{2}$ times the resistance of a device designed to withstand only 200 V, if they both have the same die area. Actual devices obey this relationship only at higher voltage levels (above approximately 100 V). For devices rated below 100 V, $R_{DS(\text{on})}$ is higher than (17.29) predicts because the channel and neck resistances become significant and because less of the total die area is actually used.

Note that current in the MOSFET is carried by majority carriers, which is why it is called a majority carrier device. Unlike the pin diode or BJT, the extended drain cannot be conductivity modulated to reduce its on-state resistance.

17.5.1 Trench or UMOS

Figure 17.21 illustrates the structure of a vertical-trench power MOSFET, sometimes referred to as a UMOS. Its operation is similar to that of the DMOS of Fig. 17.18, except that the channel is now vertical at the oxide walls of the trench. The benefits of this structure are that there is no neck region when the device is on, and the cells can be fabricated at a higher density than those of the DMOS design, that is, there are a greater number of channels per unit chip area. In addition there exists an accumulation of carriers at the gate oxide where the n-drift region is penetrated by the gate trench, reducing the resistance in this region. The result is a lower $R_{DS(\text{on})}$ relative to the planar DMOS device, particularly for lower-voltage devices where the channel and accumulation region resistances are significant relative to that of the n-drift region.

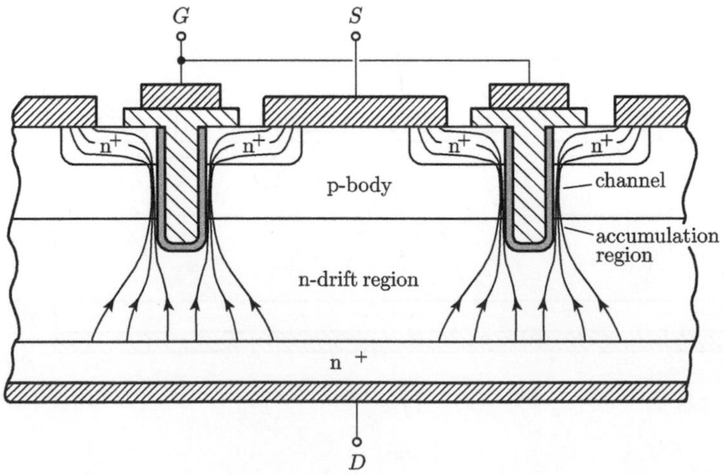

Figure 17.21 The trench or UMOS structure, also illustrating the drain current paths.

UMOS On-State Resistance Figure 17.21 shows the current paths in a trench-gate (UMOS) device. The principle contributions to $R_{DS(\text{on})}$ are the drift region and channel. In low-voltage devices the substrate (the n$^+$ drain) and the source metal contact also contribute. In high-voltage devices (above about 100 V), the resistance of the drift region dominates because of its required depth to support the high voltage.

17.5.2 Temperature Dependence of $R_{DS(\text{on})}$

Because the high-voltage vertical Si MOSFET's resistance is dominated by the extended drain, the temperature dependence of $R_{DS(\text{on})}$ is the result of the temperature dependence of μ_e. For the range of interest (0–200 °C), mobility decreases with temperature as $\mu_e \propto T^{-2.2}$, which means that $R_{DS(\text{on})}$ increases with temperature. A 100 °C rise produces an increase of approximately 90%.

17.5.3 DMOS Dynamic Behavior

Because the MOSFET is a majority carrier device, it does not have the buildup of excess carrier concentrations that controls the dynamic behavior of the BJT. The transient performance of the MOSFET is governed only by the oxide and SCL capacitances, and by the impedances that limit our ability to charge and discharge these capacitances. (Some of these impedances are internal to the device but most are external.) We discussed these capacitances briefly in Chapter 16, and described how they were characterized on datasheets. Figure 17.22 shows their physical origins in a DMOS. Comparable capacitances exist between the different elements of the UMOS structure.

Between the gate and the source are two parallel capacitances that comprise C_{GS}. One, C_{GS1}, is the result of source metalization actually covering the polysilicon gate, but being insulated from

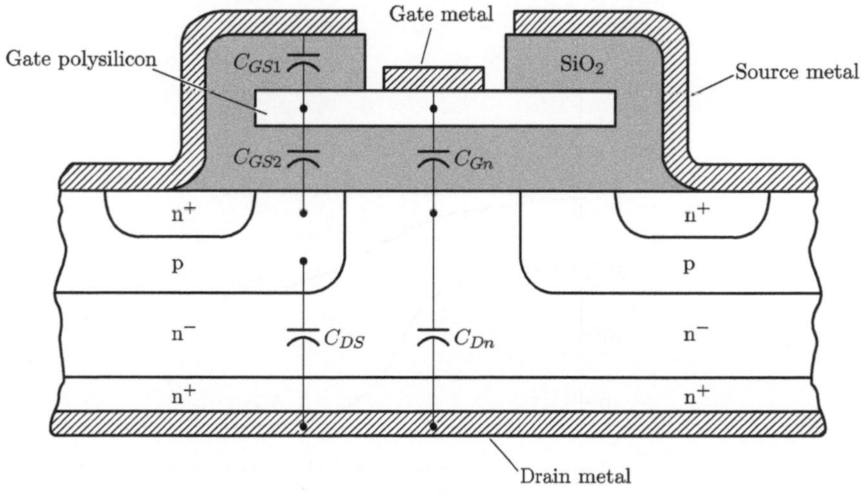

Figure 17.22 Origins of capacitances in a vertical MOSFET.

it by an oxide layer. Part of this capacitance is also the result of the overlap of the polysilicon and the n^+ source diffusion. Thus C_{GS1} is an oxide capacitance, which is not a function of the voltage across it. The second capacitance making up C_{GS} is the gate-to-p-well capacitance of the channel C_{GS2}. Although there is a resistance, R_p, between the p-well side of this capacitor and the source contact, it is small enough to ignore for typical rates of change of gate voltage. As C_{GS2} is the series combination of the oxide capacitance and the p-well's depletion capacitance, it is a strong function of the gate voltage when $v_{GS} < V_T$ (the capacitance gets smaller as the voltage increases). Once the gate voltage is above threshold, however, C_{GS2} is dominated by the oxide capacitance, which is constant.

Between the drain and the source is the capacitance of the drift region's space-charge layer, C_{DS}. Its value varies inversely with the square root of the drain–source voltage.

Finally, between the gate and drain is C_{GD}, a series combination of two capacitors. One, C_{Gn}, is the oxide capacitance, which is independent of voltage. The other, C_{Dn}, is the SCL capacitance between the drain and the neck–oxide interface. It gets smaller as the drain voltage rises. Note that these two capacitances are "connected" at the plane in which the accumulation layer grows.

At low values of v_{DS}, the SCL does not extend into the neck region, so $C_{DG} = C_{Gn}$. At higher voltages, the neck region becomes depleted, so C_{Dn} appears in series with C_{Gn} and reduces the net capacitance C_{DG}. As the drain voltage rises even higher, however, pinching-off of the neck region becomes complete, and very little incremental increase in drain voltage appears at the neck–oxide interface. Therefore C_{GD} becomes very small under this condition.

Note that as the channel and the accumulation layer form, the node connecting C_{Dn} and C_{Gn} is shorted to the source by the low resistances of these regions. Under this condition, C_{GD} becomes zero, C_{Gn} appears in parallel with the other gate–source capacitances, and C_{Dn} appears in parallel with C_{DS}.

The values of C_{GD}, C_{GS}, or C_{DS} are not provided in datasheets. Instead, they can be calculated from the measured incremental capacitances C_{iss}, C_{oss}, and C_{rss}, which are functions of V_{DS} and provided by the manufacturer in graphical form as shown in Fig. 17.23 for a typical DMOS. The relationship between these two sets of capacitances was illustrated in Fig. 16.15.

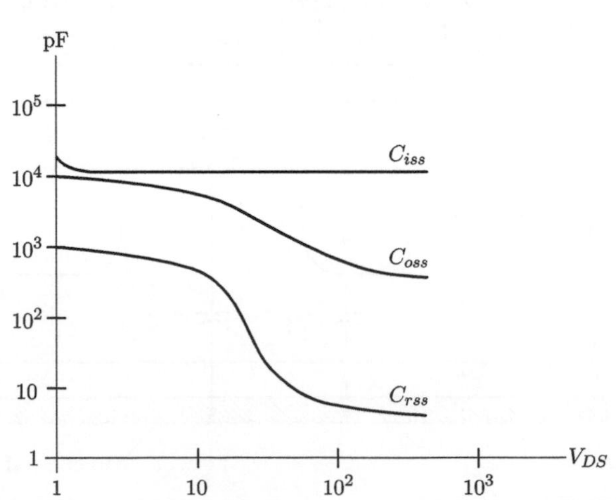

Figure 17.23 Dependence of DMOS datasheet capacitances C_{iss}, C_{oss}, and C_{rss} on V_{DS}.

17.6 Superjunction MOSFET

The blocking voltage, V_{BR}, of a conventional MOSFET is governed by the avalanche breakdown of the drift region, the extended drain for high-voltage devices. Increasing the blocking voltage is achieved by increasing the length of the extended drain and/or reducing its majority carrier concentration, simultaneously increasing the specific on-state resistance, which, in the simplest case of uniform current flow, is given by (17.29). The relationship between the minimum R_{drift} and V_{BR} is complex and constrained by what is known as the *silicon limit*, which for an abrupt diode is

$$R_{\text{drift(min)}} \propto V_{BR}^{2.5}. \tag{17.30}$$

Figure 17.24 shows the structure of a *superjunction* or SJFET. The p body is augmented with narrow columns of p^- regions extending vertically into the n^- drain region. The n^- drain is now also comprised of narrow columns. The result is two parallel pin diodes between the source and drain: n^+–π–p and n^+–ν–p. There exists a p^-–n^- junction between the columns which is reverse biased when the device is blocking (off). Since these columns are so lightly doped, they completely deplete at a low voltage. The resulting electric fields within the columns require a two-dimensional analysis, but insight into their behavior can be obtained by considering the stylized columns of Fig. 17.25.†

The equipotential lines shown in Fig. 17.25(a) are as they would be if the p–i–n columns were separated. However, since they share a boundary, the equipotentials as shown cannot remain since the potential must be continuous across a boundary unless there is an impulse of charge at the interface, which in this case does not exist. Consequently the equipotential

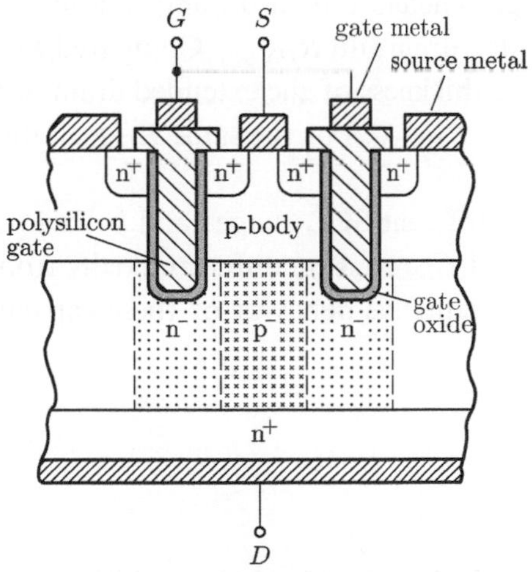

Figure 17.24 Basic structure of a superjunction MOSFET (SJFET).

† This representation is a derivative of that presented in R. Siemieniec, C. Braz, and O. Blank, "Design considerations for charge-compensated fast-switching power MOSFET in the medium-voltage range," *IET Power Electronics*, vol. 11, no. 4, pp. 638–645, 2018.

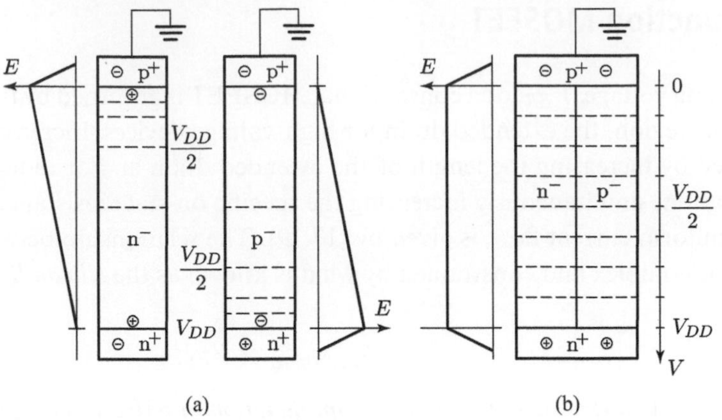

Figure 17.25 (a) A stylized representation of the n^- and p^- columns of the SJFET showing the E-field in each if they were separated. (b) The equipotential lines for an applied arbitrary voltage of V_{DD}.

lines must be distorted to achieve continuity. If there is a charge balance between the n^- and p^- columns at the junction between them, the result is equally spaced equipotentials (a result we will not prove here). If the columns are very narrow, the approximately equal spacing will penetrate the column widths, as shown in Fig. 17.25(b). Equal spacing of the equipotentials is the result of a constant E-field. Thus the combined columns behave as a single pin diode, with the near-intrinsic region forming the extended drain of the MOSFET and providing a higher breakdown voltage than a conventional device with the same extended drain thickness.

An additional benefit of the SJ structure is that because the columns are so narrow, they deplete at a low voltage. Therefore their doping concentrations can be increased, reducing the resistance of the extended drain and $R_{DS(\text{on})}$. Compared with a conventional MOSFET with the same voltage rating, the thickness of the extended drain of the SJFET can be thinner, resulting in a further reduction of $R_{DS(\text{on})}$. For the same $R_{DS(\text{on})}$, the die size can be reduced by a factor of 2–5.

Because of the very different SCL shapes and behaviors as a function of V_{DS}, the dependencies of C_{iss}, C_{oss}, and C_{rss} on V_{DS} differ markedly from those shown in Fig. 17.23 for the DMOS structure. Representative behavior of these capacitances for the SJFET are shown in Fig. 17.26.

17.7 IGBT

The circuit symbol, equivalent circuit, and physical structure of a lateral channel IGBT are shown in Fig. 17.27. Although the physical architecture of the device incorporates a pnp BJT, its symbol showing an npn BJT is designed to represent its application behavior. The structure is labeled consistent with the circuit symbol, not the terminals of the pnp BJT. An IGBT structure using a trench channel is shown in Fig. 17.28. It is the design most commonly used because of its smaller contribution to the on-state drop of the device relative to a lateral design, as discussed in Section 17.5.1. Note that the n–p–n–p IGBT structure is similar to a thyristor. However, the

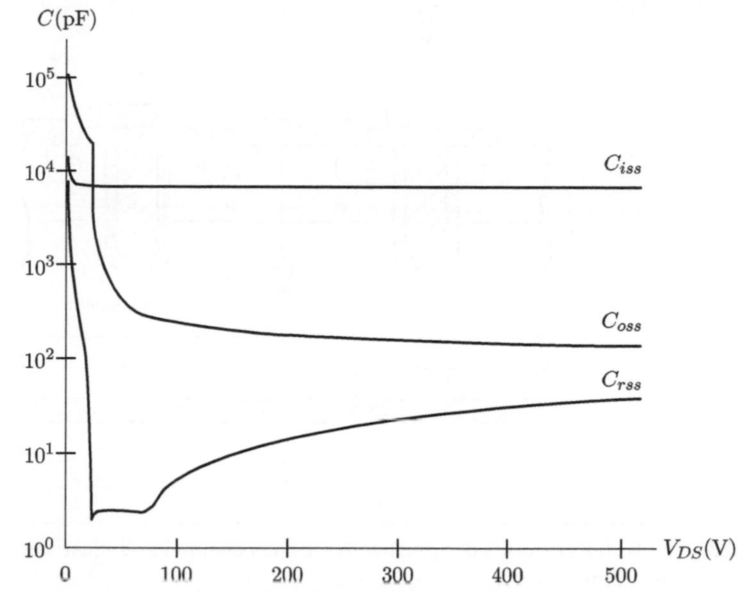

Figure 17.26 Capacitances vs. V_{DS} for a 600 V, 60 A SJMOSFET (at $T_C = 100\,°\text{C}$).

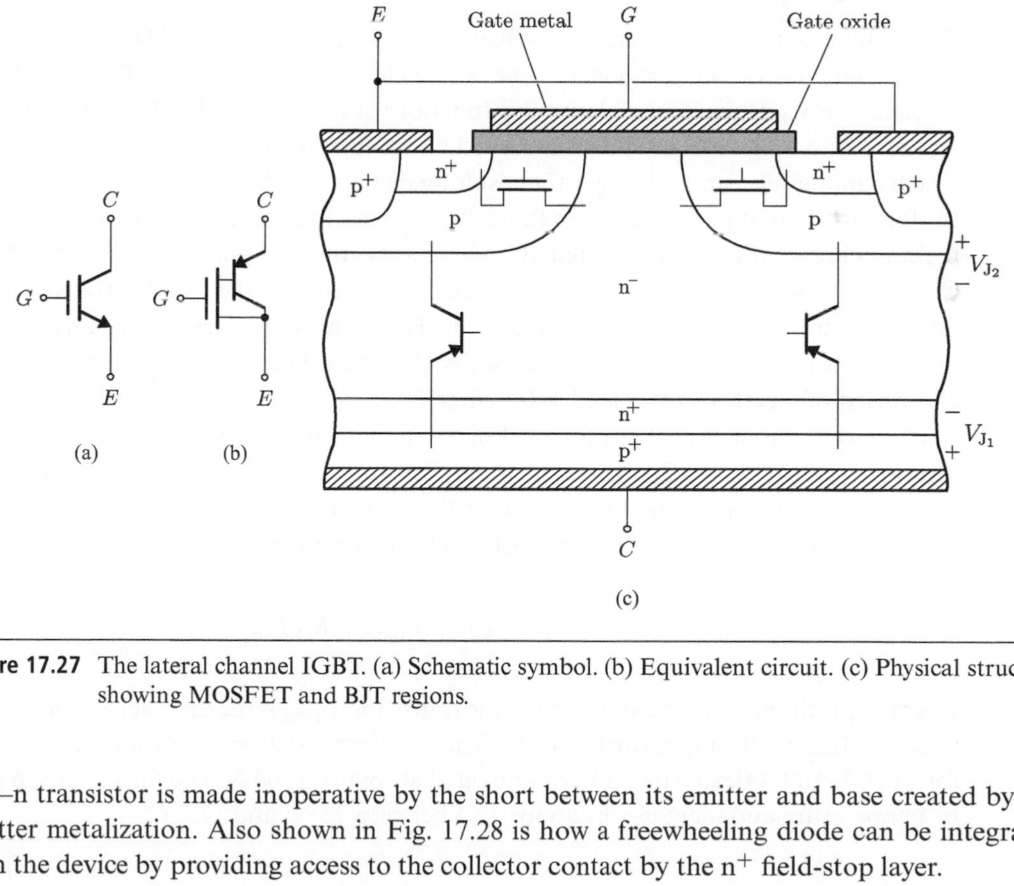

Figure 17.27 The lateral channel IGBT. (a) Schematic symbol. (b) Equivalent circuit. (c) Physical structure showing MOSFET and BJT regions.

n–p–n transistor is made inoperative by the short between its emitter and base created by the emitter metalization. Also shown in Fig. 17.28 is how a freewheeling diode can be integrated with the device by providing access to the collector contact by the n^+ field-stop layer.

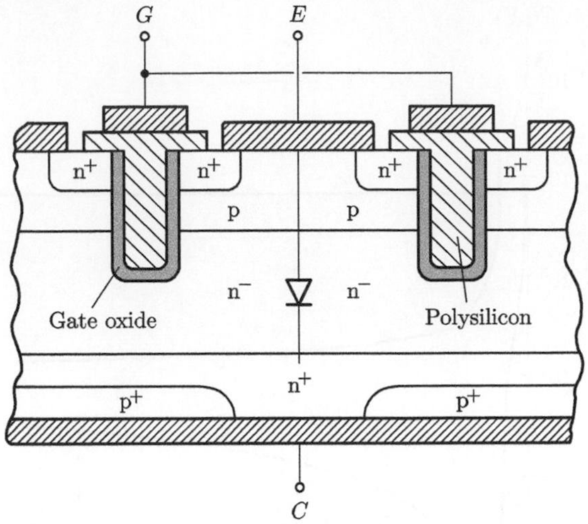

Figure 17.28 The IGBT structure designed with a trench channel and illustrating how a freewheeling diode can be integrated.

17.7.1 On State

With reference to Fig. 17.27(c), the n-channel MOSFET driving the BJT is comprised of the n^+ region as its source, the p region as its body (which is also the collector of the pnp BJT), and the n^- region as its drain. The n^- layer also functions as the base of the BJT, and the p^+ layer is its emitter. With $V_{CE} > 0$, the IGBT is turned on by applying a positive voltage to the gate (V_{GE}) which supplies electrons through the resulting channel to the n^- base region. Simultaneously, the forward-biased p^+–n^- junction floods the n^- base region, resulting in a large fraction of the region being conductivity modulated, reducing its contribution to the on-state drop $V_{CE(\text{on})}$. Contrary to what one might infer from the IGBT symbol, the bipolar transistor conducts a relatively small fraction of the device current. This is due to its very small gain as a consequence of its very wide base region and poor emitter efficiency. The majority of the collector current is comprised of electrons from the FET channel.

While one might expect the forward drop of the IGBT to be that of the pnp transistor in saturation, the MOSFET channel is in parallel with the p–n^- junction which is the collector junction of the pnp transistor. Therefore the p–n^- junction remains slightly reverse biased, preventing full saturation of the transistor. The forward drop is then

$$V_F = V_{J_1} + V_{\text{base}} + R_{\text{ch}} I_{\text{ch}},$$

where V_{J_1} is the emitter–base voltage of the transistor, V_{base} is the drop across the n^- base region, and I_{ch} is frequently approximated as I_C. The total forward drop is typically on the order of 1.5 V for a 600 V/100 A device to 3.5 V for one rated at 6500 V/1000 A. The forward drop is a function of temperature and increases by about 35% between 25 °C and 175 °C.

17.7.2 Off State

In the off state the channel is not inverted (that is, not conducting) and the p–n$^-$ junction J_2 is reverse biased and supports V_{DS}. The n$^+$ layer at the collector junction is a buffer, or *field-stop* layer that prevents the SCL from punching through to the collector. As in the pin diode, this permits the lightly doped n$^-$ region to be completely depleted and block a high V_{DS}. It is in this regard that the IGBT differs from a BJT in its voltage capability. If such a highly doped buffer layer were present at the emitter side of the base region (to which the BJT SCL spreads), the transistor's emitter efficiency would be very low, resulting in such a low gain that the device would be unusable.

In this state there is a collector–emitter leakage current across J_2 whose value at rated V_{CE} is the parameter I_{CES} known as the *collector cut-off current*, which ranges from 10s of μA to 10s of mA for devices with V_{CE} ratings of 100s of volts to 1000s of volts, respectively, at 25 °C, and increases with temperature.

17.7.3 Turn-on and Turn-off Dynamics

Like the MOSFET, the dynamic behavior of the IGBT is determined in large measure by its internal capacitances. But unlike the MOSFET, the IGBT is controlled by minority carrier as well as majority carrier dynamics. Minority carrier lifetime in the n$^-$ base region limits the turn-off time of the device. The datasheet specifications are shown in Fig. 17.29, along with their relationships to the capacitances seen from the device terminals. The measurements and relationships are similar to those for the DMOS of Fig. 17.23.

The IGBT turn-on and turn-off transients are shown in Fig. 17.30 assuming a clamped inductive load. That is, the inductor current is constant at I_L and circulating through the

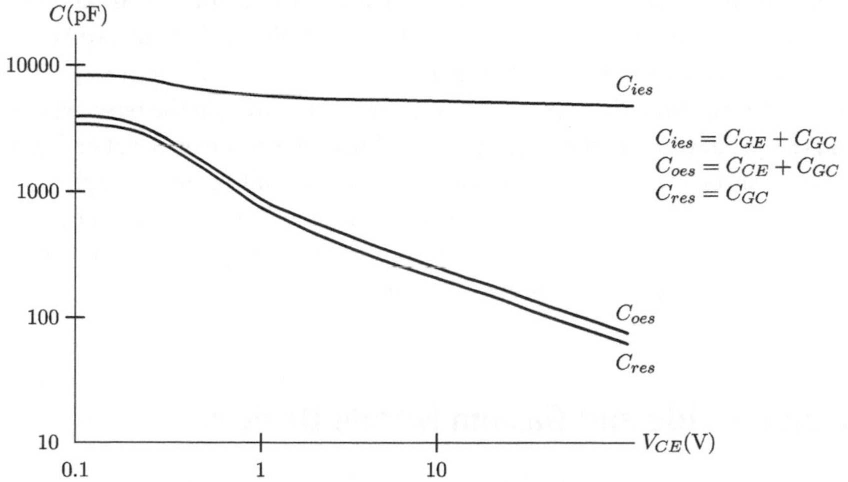

Figure 17.29 Typical variation of datasheet measurements of the input (C_{ies}), output (C_{oes}), and reverse transfer (C_{res}) capacitances with V_{CE} for a 600 V, 40 A trench gate IGBT.

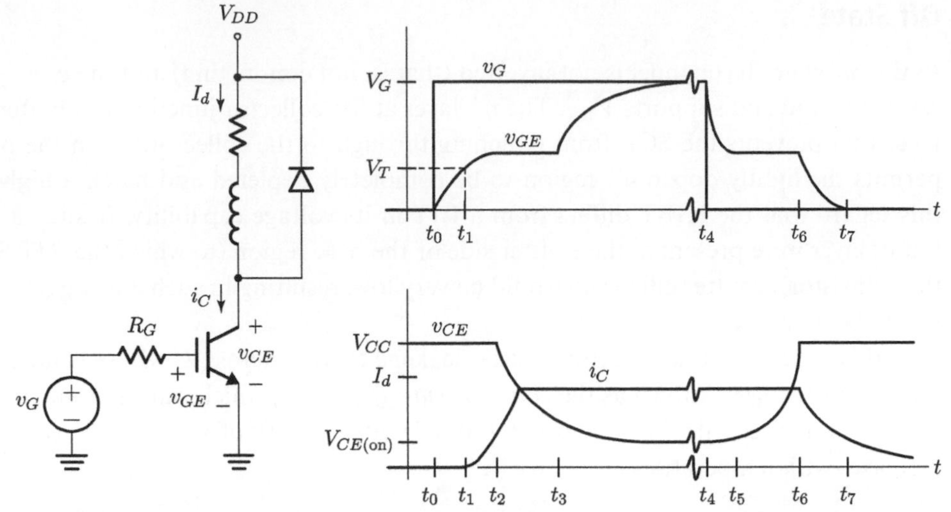

Figure 17.30 (a) The clamped inductive load circuit used to illustrate the IGBT switching behavior. (b) The IGBT turn-on and turn-off transients.

freewheeling diode and $v_{CE} = V_{CC}$. During turn-on between t_o and t_1, v_G rises exponentially to V_T at t_1 as the gate capacitance is charged. A further increase in v_G causes the collector current to rise as the diode current falls but v_{CE} remains at V_{CC}. When $i_C = I_L$ the diode turns off and v_{CE} falls as C_{CE} discharges. During the interval t_2–t_3, v_{GE} remains approximately constant resulting from the Miller effect as discussed in Chapter 23. At t_3 the Miller effect has diminished and v_{GE} continues its journey to V_G, the maximum gate voltage, while v_{CE} reaches $V_{CE(\text{on})}$.

Because there is no external access to the n$^-$ base region, the turn-off process cannot be accelerated by applying negative base current to extract the stored base charge. When $v_G = 0$ at t_4, v_{GE} begins to decrease and again exhibits a Miller-effect plateau between t_5 and t_6 as v_{CE} begins to rapidly rise. The diode remains off, and $i_C = I_L$ until $v_{CE} = V_{CC}$ and the diode turns on. At this time there is still excess charge remaining in the base, which slowly recombines. Because the lifetime in the base region is long, the result is a persisting *tail* in the collector current as shown in the waveform of i_C. Since this tail occurs at high v_{DS}, it can account for an appreciable fraction of the turn-off loss. The length of this tail is a function of the lifetime in the n$^-$ region, and for IGBTs designed for high frequency (50–100 kHz) the tail time is reduced by reducing lifetime at the expense of higher on-state losses.

17.8 Silicon Carbide and Gallium Nitride Devices

As mentioned in Chapter 16, the electronic and thermal properties of the wide-bandgap materials silicon carbide (SiC) and gallium nitride (GaN) provide improved power-device performance relative to Si. The primary reasons for this are the high values of their critical field and thermal conductivity. Although power devices are commercially available in both materials,

SiC dominates the high-voltage, high-current devices, while GaN is generally limited to voltages less than 650 V and moderate currents. The large bandgap for both devices is reflected in their large forward-junction drops, which are a disadvantage in bipolar but not field-effect devices. The wide bandgap would also permit these devices to operate at higher temperatures than Si, but packaging and bonding technologies at present limit maximum device temperature to less than 200 °C.

17.8.1 SiC Devices

Except for their increased threshold voltage (V_T), SiC MOSFETs are practically drop-in replacements for their Si counterparts. But the efficiency benefit of SiC is obtained when the on-state drop is dominated by the device drift region and the SiC drop is significantly lower than that for a Si device, which occurs around 900 V. The drift region, as illustrated in Figs. 17.20 and 17.21, must be thick enough to support the rated device voltage. Since SiC has a critical electric field E_B that is 10 times that of Si, the thickness of the drift region can be reduced significantly. Additionally, to support the same voltage in this thinner drift region, the E-field must be increased by increasing the doping density by the inverse of the square of the dimensional decrease. Combining these two changes, and accounting for the decreased electron mobility in SiC relative to Si, results in a (theoretical) reduction of the drift-region resistance by a factor of about 500 for devices of similar area. A further advantage of SiC MOSFETs is that the temperature dependence of $R_{DS(\text{on})}$ is substantially less than that for Si devices, as illustrated in the next section on the datasheet.

On a per-unit basis, the device capacitances of SiC and Si MOSFETs are comparable because the dielectric constants of the two materials are approximately equal. However, for devices of equal voltage rating, and taking advantage of a thinner SiC die to achieve reduced $R_{DS(\text{on})}$, for equal die area the SiC device will have a substantially higher C_{oss} due to the reduced width of its drain–source space-charge layer. Reducing the die area will reduce C_{oss} but increase $R_{DS(\text{on})}$. There is, therefore, a trade-off between $R_{DS(\text{on})}$ and C_{oss} depending upon the circuit design and whether switching or conduction dominates the device loss. But, for a given voltage rating, SiC devices have smaller $R_{DS(\text{on})}C_{oss}$ products than Si, and therefore their use will result in lower losses.

The combination of a thinner SiC device and its high thermal conductivity allows the device to operate in ambient temperatures considerably higher than for Si. But other system components, particularly capacitors, can still limit the maximum ambient temperature, while bonding and packaging considerations limit the die temperature.

Unlike Si SBDs, which are limited to relatively low voltages, SiC SBDs are available with voltage ratings on the order of 10 kV. The wide bandgap reduces the rate of thermally generated carriers, reducing considerably the leakage current which limits the voltage rating of Si SBDs.

17.8.2 GaN Devices

Unlike SiC devices, those fabricated in GaN exhibit important structural and behavioral differences from their Si counterparts. Early GaN MOSFETs were depletion-mode devices, meaning they were normally "on." But manufacturers have developed the technology to produce enhancement-mode MOSFETs which are now in common use. Most GaN devices are not fabricated on a native substrate, that is, the starting wafer is not pure GaN. A thin layer of GaN is epitaxially deposited on a wafer of Al_2O_3 (sapphire), SiC, or Si. All three exhibit a lattice mismatch to GaN, but SiC comes closest at approximately 3.5%. But because of the significant cost differential, Si is commonly used for power GaN devices. Due to the relatively low maturity and high cost of native GaN substrates, the majority of current commercial GaN devices are lateral instead of vertical. While this reduces their achievable voltage capability, it provides an opportunity for their monolithic integration. Their favorable high-frequency behavior is due in part to a threshold voltage of less than 2 V and a very small $R_G C_{iss}$ product, allowing fast transitions of the gate voltage. Device processing incorporates a layer of AlGaN on top of the GaN, resulting in what is known as a *heterojunction* between the two. For reasons not discussed here, this structure increases the mobility of electrons in the FET channel. The device is known as a *high-electron-mobility transistor* (HEMT).

A significant departure from the behavior of Si and SiC devices is a phenomenon known as *dynamic on-state resistance* exhibited by GaN MOSFETs. It manifests as an $R_{DS(\text{on})}$ at turn-on after blocking a high voltage that is multiples of the datasheet static $R_{DS(\text{on})}$. The increase is proportional to the magnitude of the blocking voltage. The cause is thought to be trapping of carriers in the buffer layer between the channel and oxide. The result is an on-state loss that is higher than anticipated based on datasheet parameters. One useful rule of thumb is to double the datasheet value of $R_{DS(\text{on})}$ for design purposes.

Since GaN devices are most often used in very high-frequency converters, loss in the output capacitance, C_{oss}, can be significant. This loss is due to hysteresis in the capacitor dielectric's D–H trajectory, which is a function of frequency.

17.9 Thyristor

As noted in Chapter 15, thyristor is the generic term for variants of a four-layer semiconductor device, the most common of which is the SCR. The structure is shown in Fig. 17.31(a). For the lower current ratings (10–100 A) the SCR is built on a small die of silicon, but for higher current ratings (100–6000 A), it is fabricated on an entire wafer. The heavily doped p region on the bottom of the device is the *anode*, and the heavily doped n region on the top of the device is the *cathode*. The p region under the cathode is the *gate*, and it is connected to the gate metalization on the top surface, as shown in Fig. 17.31(a). The interdigitation of the gate and cathode metalization is typically non-existent or sparse in an SCR. Figure 17.31(b) shows the top view of a device whose gate is restricted to the center of the wafer. Known as a *center-gate geometry*, it is used extensively for low-frequency SCRs. The gate pattern shown in Fig. 17.31(c) is called an *involute*, which has the important property of constant spacing between two gate fingers across the entire device.‡ A typical spacing is 3–5 mm for a wafer-sized device. Interdigitation using this or other patterns is utilized in high-frequency thyristors.

‡ Herb Storm of GE CR&D was the first to recognize the benefit of the involute structure and apply it as a thyristor gate geometry.

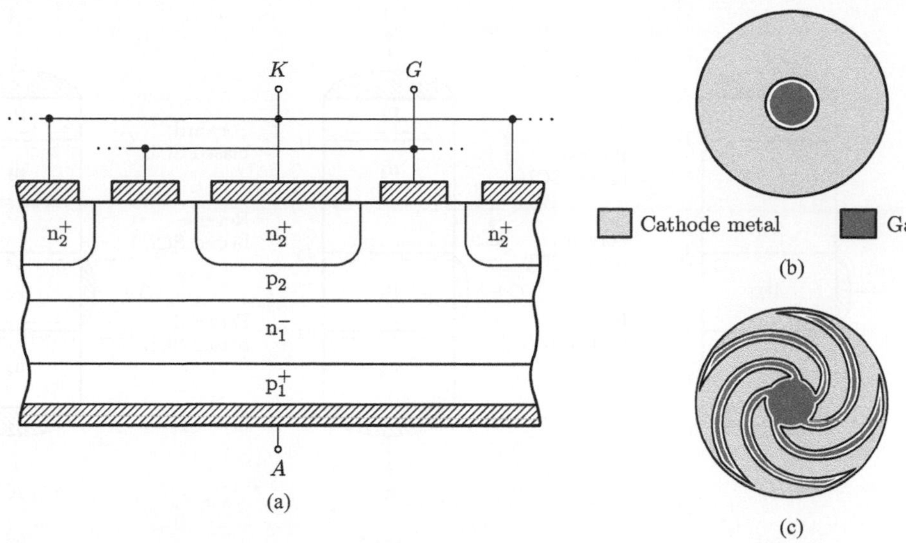

Figure 17.31 Structure of an SCR. (a) Cutaway view showing the four layers and terminal connections. (b) Top view of center-gate geometry. (c) Top view of involute gate geometry.

17.9.1 Overview of SCR Operation

The purpose of this section is to briefly describe the off-state characteristics, the regenerative turn-on process, and the on-state characteristics of the SCR. The sections that follow give greater detail on each aspect of the device's operation. Throughout these discussions, we use the one-dimensional model of the SCR shown in Fig. 17.32(a). (Note that the anode is now drawn on the top.) For future reference we have labeled each of the four layers (p_1, n_1, p_2, and n_2) and each of the three junctions (J_1, J_2, and J_3).

Off State When off, the SCR can block either a reverse or a forward anode–cathode voltage, v_{AK}. When v_{AK} is negative, junctions J_1 and J_3 are reverse biased (assuming that the gate terminal is left open), and J_2 is forward biased. The doping levels on each side of junction J_3 are very high, so this junction breaks down at a relatively low voltage. However, the n_1 region is both lightly doped and long. Therefore J_1 is able to block a large reverse anode–cathode voltage. As the n_1 region is much more lightly doped than the p_1 region, the SCL at J_1 grows mostly into this lightly doped n_1 region rather than into the more heavily doped p_1 region, as shown in Fig. 17.32(b).

When v_{AK} is positive, junctions J_1 and J_3 are forward biased and junction J_2, the only reverse-biased junction, withstands the voltage. As the n_1 region is also more lightly doped than the p_2 region, the SCL again grows mostly into the n_1 region, as shown in Fig. 17.32(c).

The n_1 region is therefore used to block both polarities of voltage when the SCR is off. Its doping level and length must be chosen to give the desired breakdown voltage. Note that breakdown can occur in two ways: *punch-through*, where the SCL extends far enough to reach the opposite edge of the n_1 region, and *avalanche*, where the peak electric field becomes high enough to cause impact ionization. Most SCRs are designed with the n_1 region long enough to cause avalanche to be the breakdown mechanism.

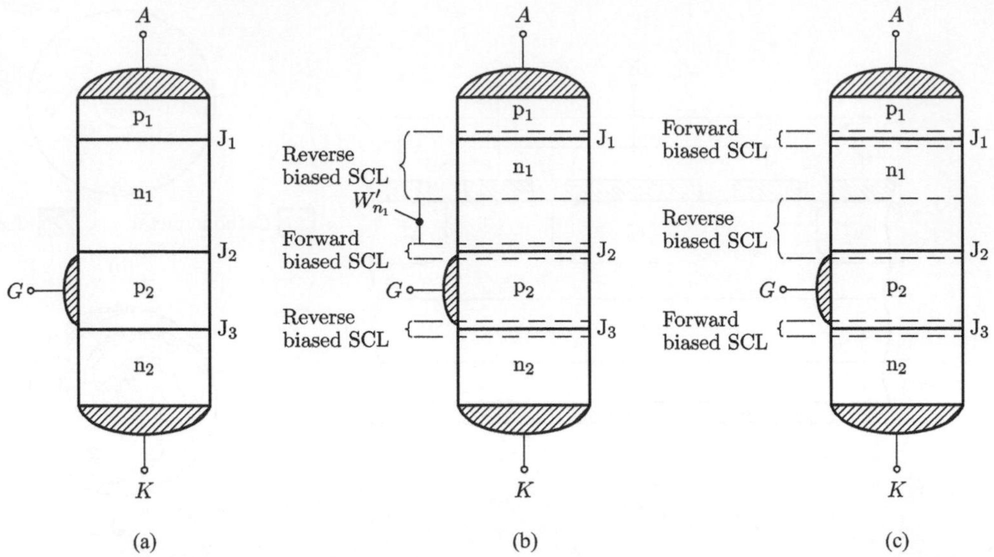

Figure 17.32 A one-dimensional model of the SCR. (a) The four layers and their junctions. (b) Space-charge layers when the SCR is blocking a reverse voltage. (c) Space-charge layers when the SCR is blocking a forward voltage.

Regenerative Turn-On Process To understand how the SCR can go from a forward-biased off state to its conducting on state, it is useful to view the device as a cross connection of two transistors. Figure 17.33(a) shows how the four-layer device can be considered as the interconnection of two three-layer devices – one a pnp and the other an npn transistor. The base of each transistor is connected to the collector of the other. Figure 17.33(b) illustrates this two-transistor SCR model.

Note that the base of each transistor is driven with a current that is β_F times its collector current. As long as the product $\beta_1\beta_2 > 1$, we can show that – once some current flows – the two transistors will drive each other harder and harder until they saturate. A short pulse of current into the gate terminal is one way to start this regenerative process. To desensitize the turn-on process to noise at the gate or leakage current at J_2 (which serves as base current for Q_2), Q_2 is designed to have a very low β_F at low emitter currents. The natural drop-off of β_F at low currents, illustrated in Fig. 17.16, is enhanced in Q_2 by fabrication techniques, one of which is to use the cathode metalization to provide a partial short between the gate and the cathode, as illustrated in Fig. 17.34. In simplest terms this can be modeled as a resistor between the base and emitter of Q_2, which shunts some of the gate current around J_3.

Note that the Q_1 collector leakage current functions as gate current and can turn the device on if it is too high. This is the reason that the SCR is a *silicon* device, and cannot be fabricated in germanium as the leakage current of a Ge transistor is orders of magnitude greater than that of Si.

The turn-on process is initiated by the injection of carriers into the n_2 region. A number of mechanisms can do this, some of which are more useful than others. Since collector leakage at

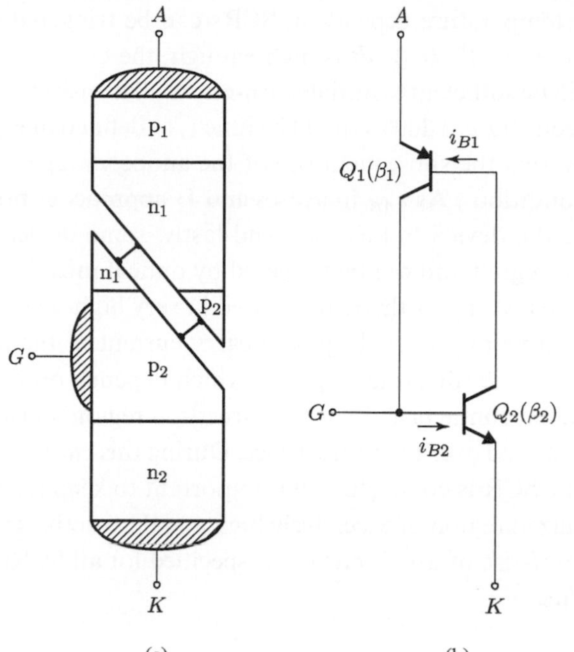

Figure 17.33 A two-transistor model of the SCR. (a) The four-layer device represented as the interconnection of two three-layer devices. (b) The two-transistor model of the interconnection of (a).

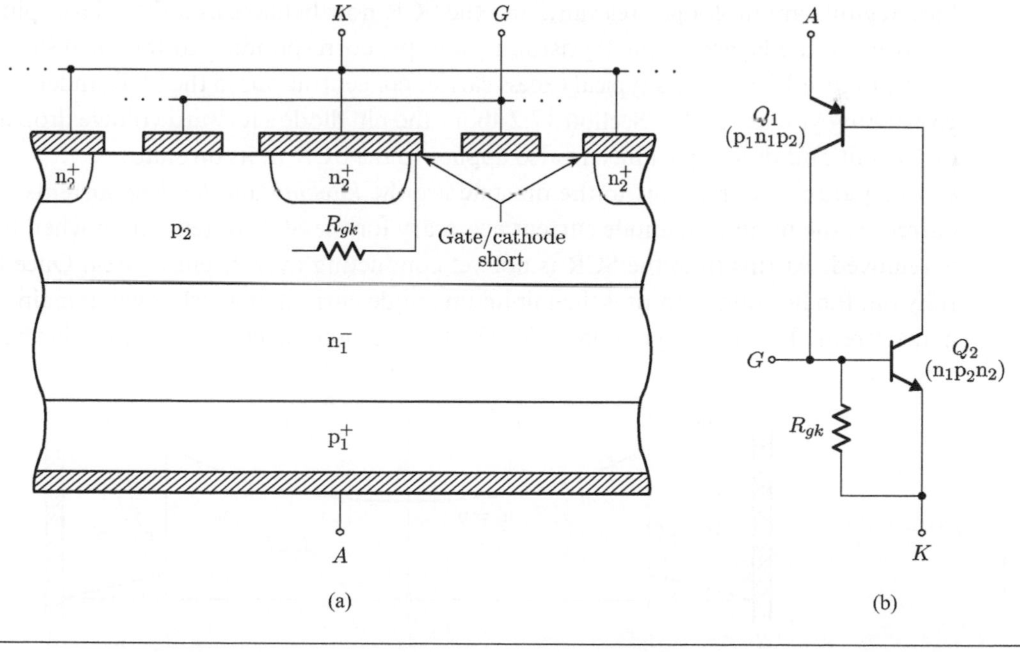

Figure 17.34 (a) A gate–cathode short produced on the die by making contact between the gate region and the cathode metalization. (b) The two-transistor SCR model showing the location of the gate–cathode short R_{gk}.

Q_1 is strongly temperature dependent, SCRs can be triggered if the device is operated above its temperature limit. If dv_{AK}/dt is high enough, the current at the Q_1 collector junction (J_2) capacitance will be sufficient to initiate turn-on, so all transistors have a maximum dv_{AK}/dt limit, typically between 200 and 2000 V/μs. (The time t_q as defined in Fig. 15.9 is usually specified under the assumption that the subsequent rise of the anode voltage is at this maximum rate, which is a worst-case condition.) As v_{AK} increases and J_2 approaches breakdown, its increasing current will also cause the device to turn on. And lastly, some devices are designed to allow optical access to the p_2 region and can be triggered by optical injection of carriers. This mechanism is very useful when a string of thyristors is used in very high-voltage applications, such as HVDC.

As an SCR turns on, it first begins to carry current in the regions of the die near the gate lead. Only after some time (the length of which depends on how finely interdigitated the gate and cathode metal contacts are) has the turned-on region spread sufficiently that the current is uniformly distributed over the entire device. During this *current-spreading* process, in which only a fraction of the SCR is conducting, it is important to keep the total anode current low enough to avoid a localized region of excessively high and destructive current density. For this reason, a maximum rate-of-rise of anode current is specified for all SCRs. Typically, di_A/dt must be kept below 1000 A/μs.

17.9.2 On State

As the two transistors drive each other into saturation, the excess carrier concentrations in their base regions reach high-level injection. At this point, the actual doping concentrations in the base regions are no longer relevant, and the SCR now behaves as a three-layer pin diode, with the two middle layers of the thyristor, n_1 and p_2, corresponding to the intrinsic region of the diode. Figure 17.35 shows typical excess carrier concentrations in the SCR under this condition. Everything you learned in Section 17.2 about the pin diode's forward voltage drop as a function of current and design parameters also applies to the SCR in its on state.

Two parameters relevant to the on state are the *latching* and *holding* currents. The latching current is the minimum anode current necessary for the SCR to remain on when the gate drive is removed. At this time the SCR is not yet conducting over its entire area. Once the device is fully on, the holding current is the minimum anode current at which it will remain on without a gate current. The holding current is lower than the latching current because during turn-on the

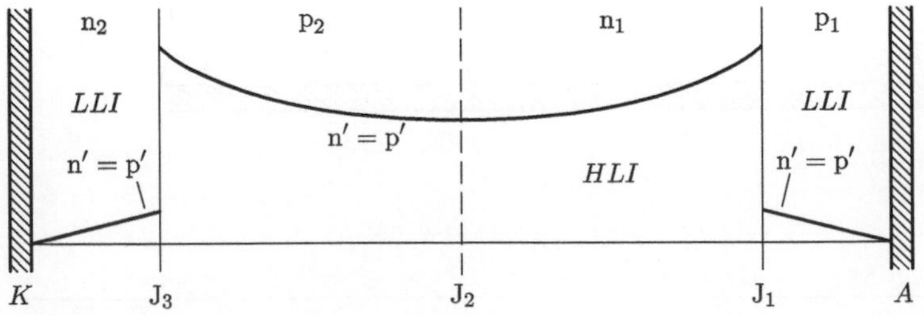

Figure 17.35 Excess carrier concentrations in an SCR in its on state. The two middle regions are in high-level injection, causing behavior like a pin diode.

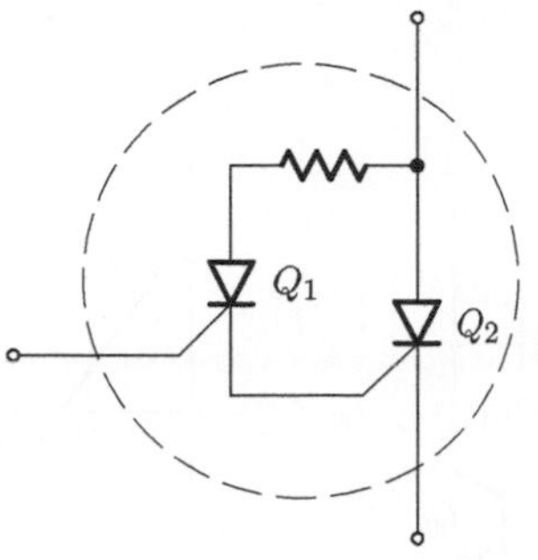

Figure 17.36 Darlington-type connection of two SCRs to produce an amplifying gate.

gate current aids the injection of electrons from the J_3 emitter as injection spreads from the J_3 region near the gate contact. When the device is fully on, the J_3 junction is uniformly forward biased (the device is conducting across its entire area) and providing adequate injection to the p_2 region to maintain the regenerative process without help from the gate.

We can connect two SCRs in a Darlington-like way, as shown in Fig. 17.36, in order to reduce the gate current requirement. In large devices, Q_1 is integrated on the same die as the main SCR, Q_2, and is known as an *amplifying gate*. A special version of this device uses a photosensitive SCR for Q_1. A photo-diode (LED) is then mounted in the same package and optically coupled to Q_1. The combination provides a means of electrically isolating the gate drive circuit from the SCR. However, these *opto-SCRs* often have a very low dv_{AK}/dt rating.

17.9.3 Turn-Off Process

Once the SCR is on, it behaves as a pin diode because the two center regions are well into high level injection for typical anode currents. When the power circuit tries to reverse the anode current, the SCR displays a reverse recovery characteristic similar to that of the pin diode. Current will flow in the negative direction until the excess charge stored in the center regions of the SCR is either removed or lost to recombination.

The same trade-off made between the speed of recovery and the on-state voltage of a pin diode can be made for an SCR. *Rectifier-grade* SCRs trade speed for a reduced forward drop. They generally have a simple center-gate electrode geometry, and are intended for line commutation applications of up to about 400 Hz. *Inverter-grade* SCRs are intended for higher-frequency applications. These devices are generally constructed with a highly interdigitated gate geometry. The lifetime for excess carriers in the base regions is made relatively short to give a fast reverse recovery at the expense of a higher forward voltage drop. Inverter-grade SCRs generally have a forward drop that is about 50% higher than a rectifier-grade device with comparable voltage and current ratings. As noted earlier, SCRs have largely been replaced by MOSFETs or IGBTs in high-frequency applications.

Turn-Off Time and Rate Effects When the SCR is off and blocking a forward voltage, it can be pushed into its regenerative turn-on mode if the current flowing through the device

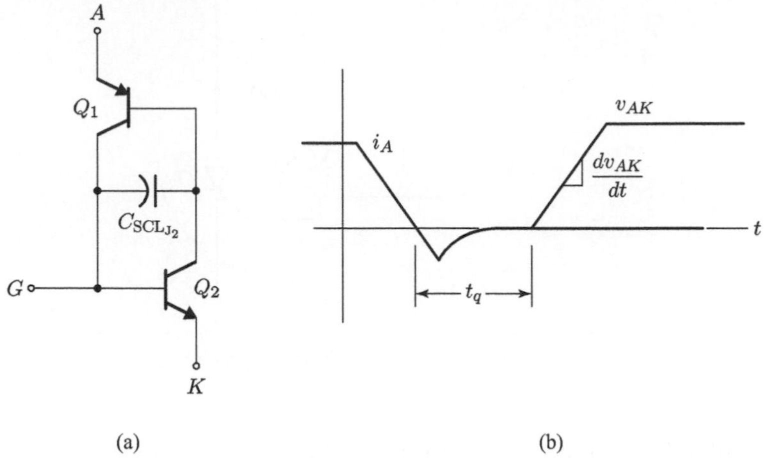

Figure 17.37 (a) Two-transistor model of an SCR showing the capacitance of the SCL at junction J_2. (b) Reapplied forward voltage at the end of the minimum turn-off time t_q.

exceeds a critical value. One way to generate this current (besides driving the gate terminal) is to apply a positive change in voltage across the SCL of junction J_2. As Fig. 17.37(a) shows, the junction can be modeled as having a parallel capacitance (because of the SCL), and the current flowing through this capacitance when dv_{AK}/dt is positive acts just like gate current. Therefore, if dv_{AK}/dt is large enough, the regenerative turn-on mechanism in the SCR can be triggered. For this reason, there is a maximum specified off-state dv_{AK}/dt above which the device is not guaranteed to remain off. Rates of 1000–2000 V/μs are typical for inverter-grade devices.

Note that the cathode short, which reduces the gain of Q_2, also helps to make the SCR less sensitive to dv_{AK}/dt. In addition, applying a negative voltage across the gate–cathode terminals when the SCR is off also helps reduce the chance of false triggering.

Because the J_2 SCL capacitance is nonlinear, the maximum allowable dv_{AK}/dt is specified at the largest value of this capacitance. This condition occurs when $v_{AK} = 0$. Moreover, the SCR is most sensitive to triggering immediately after it has been turned off, as excess charge still exists in the middle base regions. For these reasons, the maximum allowable dv_{AK}/dt is specified at a fixed time after the anode current reverses and the SCR begins its reverse recovery phase. This time is called the *circuit commutated turn-off time* and is designated by t_q. A forward voltage can safely be reapplied (meaning that retriggering will not take place) at the specified rate so long as it is delayed for a time t_q past the time that the anode current reverses. The relationships among i_A, v_{AK}, and t_q during the turn-off process are shown in Fig. 17.37(b). Typical values for t_q are 100–200 μs for rectifier-grade SCRs and 10–20 μs for inverter-grade SCRs.

The circuit commutated turn-off time t_q depends on whether the circuit in which the SCR is connected permits the flow of significant reverse recovery current. In a situation in which an anti-parallel diode is connected across the SCR (a common configuration), the switch current will reverse through the diode, but little reverse current will be carried by the SCR. The result is that elimination of the charge stored in the base regions of the SCR is not aided by reverse current. Instead, the charge will decay through recombination at a rate determined by the excess carrier lifetimes in the base regions. In this case, t_q will be longer than when an anti-parallel diode is absent. Datasheets generally give t_q under both conditions for inverter-grade SCRs, and their values differ by 50%–100%.

17.9.4 Gate Turn-Off Thyristor

So far we have assumed that a thyristor, once it is on, will not turn off until the power circuit drives its anode current to zero. But the two-transistor thyristor model of Fig. 17.33(b) implies that if we draw enough current out of the gate terminal of the device, we can reduce i_K to a value insufficient to maintain the regenerative process and the thyristor will turn off. We cannot do this in a conventional SCR, because the gate terminal affects only the region of the device in the immediate vicinity of the gate contact. The lateral voltage drop that would accompany the flow of current from regions remote from the gate contact effectively prevents the behavior of these regions being influenced by anything happening at the gate terminal. Therefore, while a negative gate current might stop the regenerative process in the immediate vicinity of the gate, the rest of the SCR would remain on. However, by using special design techniques, including dense gate interdigitation and reducing the gain of Q_1, we can design a thyristor to be turned off by a large negative gate current. Such a device is known as a *gate turn-off* (GTO) thyristor. One consequence of the structure necessary to create a GTO is that it loses its reverse blocking capability, now sustaining a reverse voltage on the order of only 50 V.

17.9.5 TRIAC

We described the operating characteristics of the TRIAC in Chapter 3. The device is a five-layer structure, as shown in Fig. 17.38. The MT_1 and MT_2 contacts short the n_2^+ region to the p_2

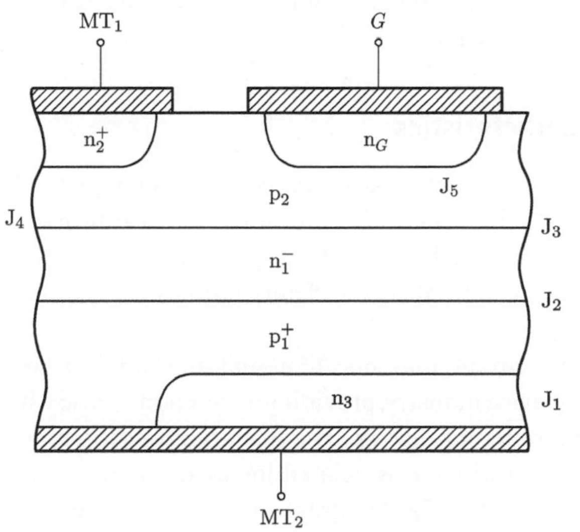

Figure 17.38 The TRIAC structure.

region and the n_3 region to the p_1^+ region. The result is the equivalent of two anti-parallel SCR structures: p_1^+–n_1^-–p_2–n_2^+ and n_3–p_1^+–n_1^+–p_2. The n_G region is the TRIAC gate, and it is shorted to the p_2 region by the gate contact.

The possible modes of operation of the TRIAC are defined in Section 15.3.3 as I+, I−, III+, and III−. In mode I+, the device is operating as a conventional SCR, with the anode being the p_1 region. The gate is the p_2 region. It is connected directly to the gate terminal, because the gate metal extends beyond the edge of the n_G diffusion, shorting the terminal to the p_2 region. Therefore, triggering in this mode is effected in the normal manner, and this is the mode with the most sensitive triggering characteristic.

III− is the next most sensitive mode. In this mode the n_3 region is the cathode and the p_2 region the anode. Triggering is accomplished by pulling the gate negative with respect to MT_1. This forward biases junction J_5, which functions as the base–emitter junction of the npn transistor composed of n_1, p_2, and n_G. The injected current (electrons) is collected at J_3, which is reverse biased in quadrant III. The resulting electron flow into the n_1 region is effectively base current for the $p_1n_1p_2$ transistor, and it begins to turn on, initiating regeneration and turning the TRIAC on.

17.10 Datasheet

All semiconductor devices are accompanied by a datasheet containing parameters that characterize the device. The meaning of some of the parameters is self-evident – $I_{\max}$, $V_{\max}$, T_J for example. Others are unique to the device type – V_T for the MOSFET or β_F for the BJT. Other parameters are more universal, but often complicated to interpret and apply. One such example is the safe operating area (SOA). Furthermore, many parameters are given at a specified temperature which is usually different from that in the application environment, requiring derating. And lastly, many parameters and operating characteristics are given in parametric graphs. In this section we discuss some of these relevant but somewhat confusing parameters, focusing on the MOSFET.

17.10.1 Switching Characteristics

Switching times and energy loss are based on the clamped inductive load circuit of Fig. 17.39, where the freewheeling diode D can be either an external diode or the body diode of the complementary MOSFET in a bridge leg. The inductor L is large enough for the device under test to be switched at a frequency and duty ratio which maintains the inductor current at the test value.

Although this circuit may not represent the application configuration, its use is nearly standard in the device industry, providing a common metric with which to compare the switching performance of different devices. The definition of switching times is shown in Fig. 17.39(b). These parameters and their associated losses E_{on} and E_{off} are given in the datasheet for one (typically a low) value of R_G, but graphs of these parameters are also given that are parametric in R_G. The junction temperatures at which the single data point and graphs are measured is also given, but may not be the same. For example, the single datasheet parameter may be at

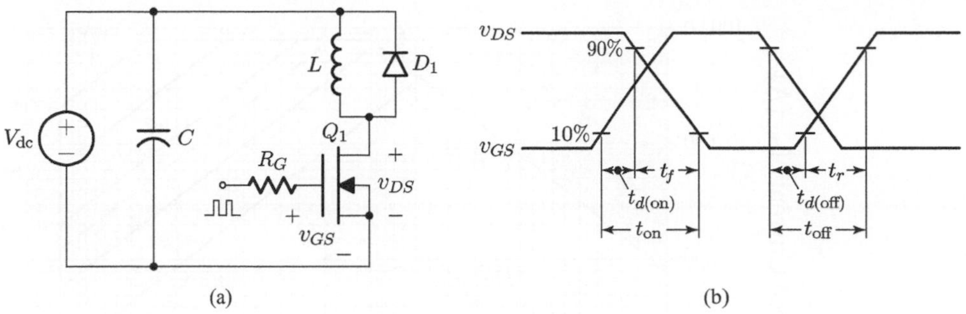

Figure 17.39 (a) Clamped inductive load test circuit used to determine MOSFET switching times and energy. (b) Waveforms of v_{GS} and VDS showing how the different switching parameters are defined.

$T_J = 175\,°C$ while the graphs give losses at $T_J = 25\,°C$. There may be a graph included which provides the temperature dependence of loss versus T_J for one value of R_G from which the temperature dependence at other values of R_G may be inferred.

17.10.2 Thermal Parameters

Keeping T_J below its maximum allowed temperature, which is generally given on the first page of the datasheet, is critical to maintaining the integrity of the device. The primary parameter for the circuit designer is $R_{\theta JC}$, the *thermal resistance* between the device junction and its case. Thermal design and the use of this parameter is covered in Chapter 25, but in summary it relates the dissipated power to the difference between case and junction temperatures. It is a single number with units of °C/W. Further thermal characterization is provided in a graph of *thermal impedance* versus pulse width with duty ratio as a parameter. The use of this graph is also covered in Chapter 26.

Another datasheet parameter related to the device's thermal behavior is maximum power dissipation, P_D. This number is generally surprisingly high because it is calculated based on $T_J = T_{J\max}$ and $T_C = 25\,°C$, conditions difficult to replicate in practice. A graph is always included which provide a power derating curve for $T_J = T_{J\max}$ and $T_{J\min} < T_C < T_{J\max}$.

17.10.3 Device Capacitances

The device capacitances C_{iss}, C_{oss}, and C_{rss} are given in two places on the datasheet: a single numerical value with test conditions noted, and, because these capacitances are nonlinear, a graph of their incremental values versus V_{DS}. Since C_{oss} figures prominently in switching behavior, two additional linear parameters are provided: the *energy-related* capacitance $C_{o(er)}$ and the *time-related* capacitance $C_{o(tr)}$. The former is the capacitance which stores the same energy as C_{oss} as V_{DS} rises from 0 to a specified value. The latter is the capacitance which, when charged at the device's specified continuous drain current I_D, results in the same charging time to the specified V_{DS} as C_{oss}.

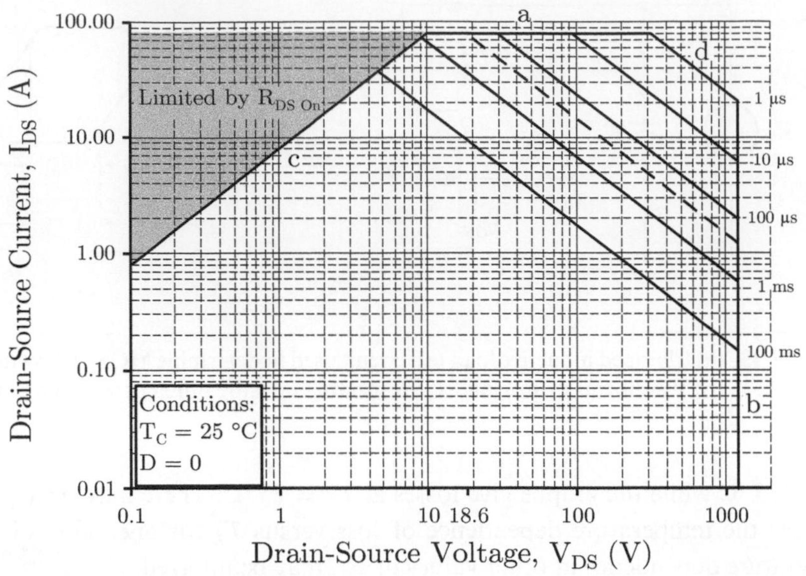

Figure 17.40 The safe operating area (SOA) graph for the Wolfspeed C3M0075120D SiC MOSFET. (Used with permission of Wolfspeed, Inc.)

17.10.4 Safe Operating Area

The SOA is presented in graphical form as shown in Fig. 17.40 and provides a region in the i_D–v_{DS} plane in which the device will operate safely at the specified T_C, which is usually 25 °C. It is most useful in applications where high I_D and V_{DS} occur simultaneously, such as non-switching (linear) audio amplifiers, or during the switching transient. Boundary (a) is defined by the datasheet parameter $I_{D(\text{pulse})}$, which is limited by the die bonding system and packaging constraints, and boundary (b) is constrained by the datasheet value of $V_{DS(\text{max})}$, sometimes identified as V_{DSS}. Boundary (c) shows the constraint on V_{DS} at a given I_d due to $R_{DS(\text{on})}$, that is, $V_{DS} = R_{DS(\text{on})}I_D$. The 45° lines (d) are the constant power loci for single pulses of different widths, each of which results in $T_J = T_{J(\text{max})}$. Since the test condition is $T_C = 25$ °C, for most applications the curves (d) have to be derated, reducing the SOA. The process of derating these curves is described in Chapter 25.

17.10.5 Diode Parameters

The body diode of the MOSFET is characterized in the datasheet with respect to forward voltage, continuous current, pulse current capability, and reverse parameters. The most useful of these is usually the reverse recovery time, t_{rr}.

17.10.6 Graphed Parameters

In addition to those graphed parameters already mentioned, other device parameters are also graphed as functions of T_C, T_J, V_{DS}, or V_{GS}. Their use is self-evident. The most useful of

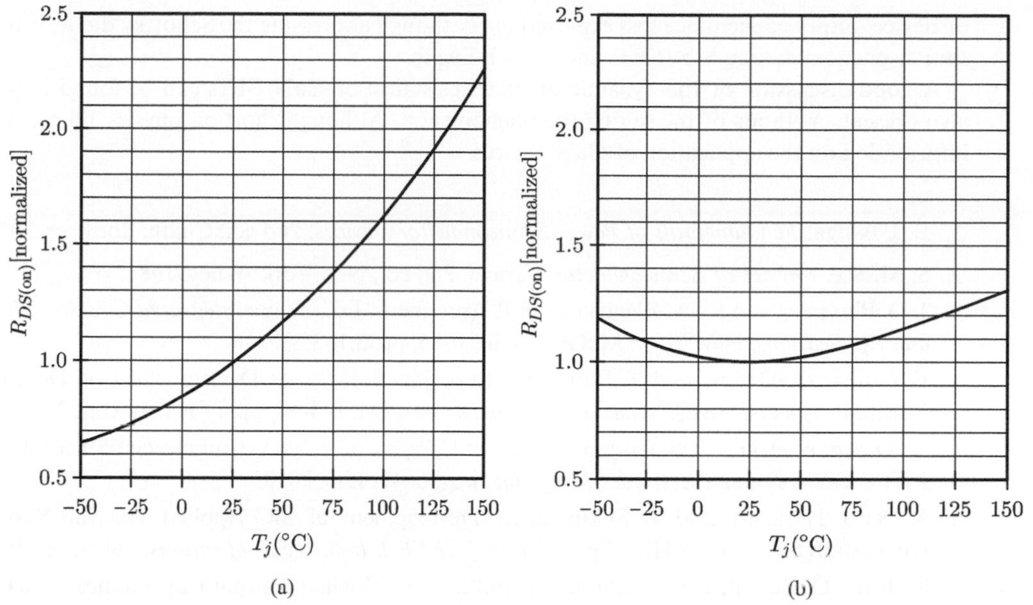

Figure 17.41 The temperature dependence of $R_{DS(on)}$ for (a) an IPW60R018CFD7 Si MOSFET, and (b) an IMW65R027M1H SiC MOSFET. (Used with permission of Infineon Technologies.)

these graphs are those which characterize $R_{DS(on)}$ as functions of T_J, I_D, and V_{GS}. Figure 17.41 illustrates the behavior of $R_{DS(on)}$ versus T for both a Si and SiC MOSFET which have similar voltage ratings. The datasheet generally gives a value of $R_{DS(on)}$ at $T_J = 25\,°\text{C}$, and the graph provides its temperature dependence normalized to this value. The Si MOSFET exhibits substantial variation with T_J, increasing by approximately 125% over its operating temperature range. However, because of different mechanisms affecting channel and drift region mobilities in the SiC device, its $R_{DS(on)}$ variation with T_J is substantially less, increasing by only 30% over the same temperature range. There are frequently additional graphs which illustrate how V_{GS} and I_D affect $R_{DS(on)}$.

Notes and Bibliography

The best general references on the material presented in this chapter are [1] and [2]. Baliga's book has extensive coverage of all the devices discussed in this chapter, but focuses more on device physics than phenomenological behavior. Sze, while not devoted to power devices, is another source of how and why device parameters vary with temperature, dopant concentration, etc. A comprehensive treatment of the Schottky diode and the physics of thermionic emission is also provided.

Chapter 3 of [3] by J. D. Plummer and R. D. Blanchard provides an excellent discussion of the power MOSFET development and addresses the difference in the i_D–v_{GS} relationship between signal- and power-level devices. Udrea et al. [4] provides an excellent discussion of superjunction technology, with an extensive reference list for those wishing to explore the technology further.

A good appreciation of the benefits of SiC can be found in [5] and [6]. The Shinkansen application in [6] is especially interesting for the system implications of changing from Si to SiC devices. The problem of device output capacitance loss explored in [7], though addressing the Schottky diode, is relevant to the similar phenomenon in Si SJFETs and GaN HEMTs.

A good discussion of the dynamic on-state resistance of GaN FETs can be found in [8]. The paper also presents methods of measuring the phenomenon. Although short on physics, [9] provides extensive information on the application of GaN devices.

1. B. J. Baliga, *Fundamentals of Power Semiconductor Devices*, 2nd ed., Cham: Springer, 2019.
2. S. M. Sze, *Physics of Semiconductor Devices*, 2nd ed., New York: Wiley, 1981.
3. J. D. Plummer and R. A. Blanchard, in P. Antognetti, Ed., *Power Integrated Circuits: Physics, Design, and Applications*, New York: McGraw-Hill, 1986, pp. 3.1–3.58.
4. F. Udrea, G. Deboy, and T. Fujihara, "Superjunction Power Devices, History, Development, and Future Prospects," *IEEE Trans. Electronic Devices*, vol. ED-64, pp. 713–727, Mar. 2017.
5. A. O. Adan, et al., "The Current Status and Trends of 1,200 V Commercial Silicon-Carbide MOSFETs," *IEEE Power Electronics Magazine*, pp. 36–47, June 2019.
6. K. Sato, H. Kato, and T. Fukushima, "Development of SiC Applied Traction System for Next Generation Shinkansen High-Speed Trains," *IEEE J. Industry Applications*, vol. 9, pp. 453–459, 2020.
7. Z. Tong, G. Zulauf, J. Xu, J. Plummer, and J. Rivas-Davila, "Output Capacitance Loss Characterization of Silicon Carbide Schottky Diodes," *IEEE J. Emerging and Selected Topics in Power Electronics*, vol. 7, pp. 865–878, June 2019.
8. R. Li, X. Wu, S. Yang, and K. Sheng, "Dynamic On-State Resistance Test and Evaluation of GaN Power Devices Under Hard- and Soft-Switching Conditions by Double and Multiple Pulses," *IEEE Trans. Power Electronics*, vol. 34, pp. 1044–1053, Feb. 2019.
9. A. Lidow, J. Strydom, M. de Rooij, and D. Reusch, *GaN Transistors for Efficient Power Conversion*, 2nd ed., Chichester: Wiley, 2015.

PROBLEMS

17.1 In determining the voltage v_i across the intrinsic region of the pin diode we stated that "$D_e = D_h = D_a$ and therefore $\mu_e = \mu_h$ in the intrinsic region." Justify the "therefore" in this statement.

17.2 Reconsider Example 17.2 but with a doping density in the n^- region of $10^{13}/\text{cm}^3$ and a calculated n^- region width of 125 μm, which now creates a punch-through pin diode. Sketch the electric field in the diode and calculate its breakdown voltage. At this doping density the critical field is approximately $E_c = 10^5$ V/cm.

17.3 In Section 17.4.1 we noted that the collector side of the collector junction is lightly doped to allow the SCL to expand into it. Why could we not instead create an intrinsic region in the base with a highly doped field-stop layer at the emitter junction to allow the SCL to grow into the base? (*Hint:* consider the issues discussed in Section 16.3.2.)

17.4 Figure 17.42(a) shows the vertical profile of a 77 mm-diameter fast-recovery rectifier rated at 3000 A_{rms}. Use the parameters as given in Fig. 17.42(a) for parts (a) and (b) of the problem.

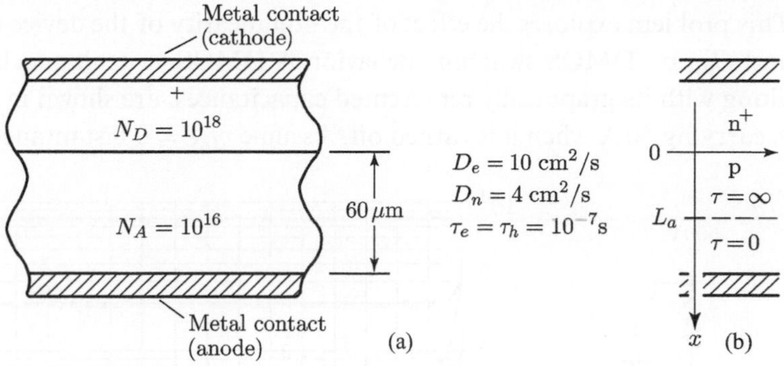

Figure 17.42 The 77 mm-diameter fast-recovery diode of Problem 17.4, and its lifetime profile.

(a) Determine the value of current density at which the device enters high-level injection on the p side of the SCL.

(b) Assuming no *bulk resistive drop*, that is, no drop outside the SCL, determine the terminal voltage of this diode at the rated current. Use the values for diffusion constants and lifetimes given in Fig. 17.42.

(c) What is the terminal voltage at the rated current if you do not ignore the drop across the p region outside the SCL? For the purposes of this calculation, assume that you can characterize the p region by an infinite lifetime within a diffusion length, L_a, of the junction and by a lifetime of zero throughout the rest of the p region, as shown in Fig. 17.42(b).

17.5 Figure 17.43 conceptually illustrates a MOSFET drain region whose physical dimensions and resistance you are to compare for fabrication in Si and SiC. Use the parameters for Si and SiC in Table 16.1 and assume that the drain in both cases will be required to block 1 kV.

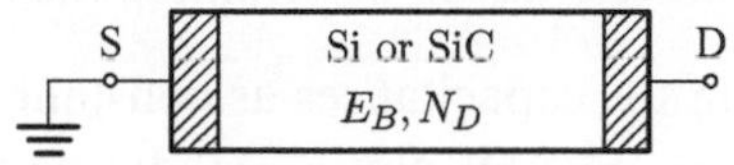

Figure 17.43 A stylized representation of the drain region of a MOSFET having length ℓ_d to be analyzed for fabrication in Si and SiC.

(a) Sketch the E-field and label its maximum value for both cases if the drain is fully depleted when the device is blocking 1 kV.

(b) Under the conditions of (a), determine the length ℓ_d for both materials.

(c) Determine the doping density, N_D, consistent with the answers to (a) and (b).

(d) Determine and compare $R_{DS(\mathrm{on})}$ for the Si and SiC drains.

17.6 This problem explores the effect of the nonlinearity of the device capacitances C_{iss}, C_{rss}, and C_{oss} on DMOS switching behavior. A DMOS switching a clamped inductive load, along with its graphically represented capacitances, are shown in Fig. 17.44. The device is carrying 50 A when it is turned off. Assume $v_{GS} = 0$ instantaneously at $t = 0$.

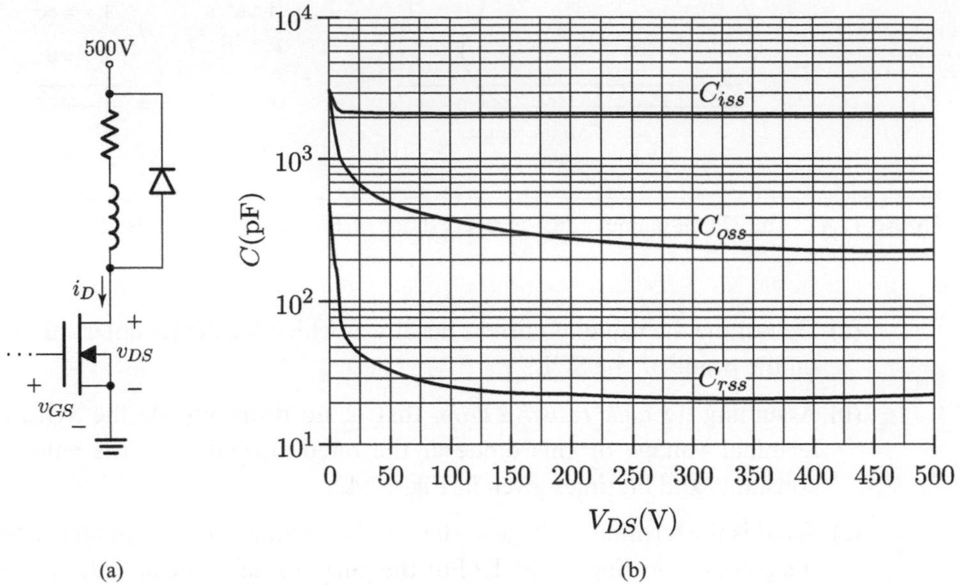

Figure 17.44 (a) A DMOS switching a clamped inductive load. (b) Capacitances as a function of V_{DS} for the DMOS of (a).

(a) Redraw Fig. 17.44(a) with its terminally viewed capacitances in terms of C_{iss}, C_{rss}, and C_{oss}. (You might want to refer to Fig. 16.15).

(b) Which capacitance(s) will dominate the turn-off transient?

(c) Draw the circuit relevant to determining the transient.

(d) Approximating the capacitances as constant at their average values during discrete intervals of $\Delta v_{DS} = 50$ V, plot $v_{DS}(t)$ during turn-off and estimate its rise time.

(e) The time-related output capacitance, $C_{o(tr)}$, for this device is 370 pF. Compare your estimated rise time to that calculated using $C_{o(tr)}$.

18 Introduction to Magnetics

Magnetic components such as inductors and transformers are present in the vast majority of power electronic circuits. Inductors store energy in the conversion process, filter switching ripple (as part of input and output filters, for example), create sinusoidal variations of voltage or current (paired with capacitors, as in resonant converters), limit the rate of change of current (as in snubber circuits), and limit transient current. Transformers provide isolation between two parts of a system, transform impedances, force current sharing (interphase transformer), provide phase shifting (three-phase delta–Y transformer), store and transfer energy (flyback transformer), and sense voltages and currents (potential and current transformers).

Unlike other components used in power electronic circuits, magnetic components are often not available in their required forms. Even when a predesigned magnetic component is available, the large number of parameters that characterize it – inductance(s), voltage, current, energy, frequency, turns ratio, leakage, dissipation, form factor – can often be adjusted to create a custom-designed component that provides much better performance. Moreover, customized magnetics provide circuit design opportunities that are simply not available with standard components, opening routes to higher efficiency and performance. Effective power electronics design thus requires a solid understanding of the principles governing the behavior of magnetic components.

In this chapter we provide an introduction to both the analysis and synthesis of magnetic components. The subsequent three chapters refine and extend the methods introduced here. Problems of analysis may be complex, but they are often straightforward and yield unique answers. Synthesis, however, is an engineering challenge because of the number of parameters involved. A design is never uniquely appropriate. Rather, a design is judged "best" according to criteria established by the application. For instance, if an inductor must not exceed a certain volume, several different designs using different materials and construction methods might be satisfactory. The "best" solution then becomes somewhat subjective and relies heavily on experience.

18.1 Inductor

From an electromagnetic fields point of view, an inductor is a circuit element that stores energy only in the form of magnetic fields. Its physical basis of operation is well described by the magnetostatic form of Maxwell's equations:

$$\nabla \times \vec{H} = \vec{J} \qquad \text{(Ampère's law)}, \tag{18.1}$$

$$\nabla \cdot \vec{B} = 0 \qquad \text{(divergence law)}. \tag{18.2}$$

The relationship between the magnetic field intensity H and the magnetic flux density B in a particular material is summarized by the *permeability*, μ. The permeability of a material is defined as the ratio

$$\mu = \frac{B}{H}.$$

The permeability of a material is specified as normalized to the permeability of free space, $\mu_0 = 4\pi \times 10^{-7}$ H/m. For magnetic materials, this normalized permeability may range from unity up to 100 000, and is not constant but varies as a nonlinear function of the flux density level B. We discuss this behavior in Section 18.2.

From (18.1) and (18.2) we can calculate the magnetic field created by a current distribution. However, in order to determine the voltage present at the terminals of the wire that carries this current, we need one more of Maxwell's equations, namely, Faraday's law:

$$\nabla \times \vec{E} = -\frac{\partial \vec{B}}{\partial t} \qquad \text{(Faraday's law)}. \tag{18.3}$$

Applying Stokes' theorem to (18.3) gives the familiar integral form of Faraday's law:

$$\oint_{\ell} \vec{E} \cdot d\vec{\ell} = -\frac{d}{dt} \iint_{S} \vec{B} \cdot d\vec{S}, \tag{18.4}$$

where S is any surface bounded by the closed contour ℓ. In an inductor, ℓ is the path formed by the winding and closed between the external terminals, and S is the cross-sectional surface area defined by this winding, or generally the number of turns times the area enclosed by each turn. If the winding is made with a good conductor, then $\vec{E} \approx 0$ within the wire itself so that only the electric field between the terminals of the inductor contributes to the line integral of $\vec{E}$ in (18.4). The left-hand side of (18.4) is then the negative of the terminal voltage v. Hence, we can rewrite (18.4) as

$$v = \frac{d}{dt} \iint_{S} \vec{B} \cdot d\vec{S} = \frac{d\lambda}{dt}, \tag{18.5}$$

where λ is a parameter called the *flux linkage*, defined as

$$\lambda = \iint_{S} \vec{B} \cdot d\vec{S}. \tag{18.6}$$

The difference between flux, Φ, and flux linkage, λ, can best be understood by considering a single-layer coil of N turns. The flux is the integral of B over the area enclosed by a single turn. The flux linkage is the integral of B over the surface of an area whose perimeter is the wire, a surface that looks something like a screw. In this case, $\lambda = N\Phi$. Thus we might think of flux linkage as the flux linking the helical surface defined by the wire, or as the average flux linking each turn, times the number of turns.

If the relationship between λ and the current that caused the magnetic field is linear – which it is if μ is constant, as can be seen from (18.1) and (18.2) – the constant of proportionality is the *inductance* L of the magnetic component, that is,

$$\lambda = Li. \tag{18.7}$$

18.1.1 Calculating Inductance

We do not discuss the details of how to solve the preceding equations to get an exact solution for the magnetic field. In most practical cases it is sufficiently accurate to assume that the magnetic field follows a defined path with a given cross-sectional area over which the field is uniform. Once you know the magnetic field intensity $\vec{H}$, you can calculate the inductance using the geometric parameters of the problem. The following steps summarize the method of solution.

1. Choose the closed-loop path ℓ along which the value of $\vec{H}$ is desired.
2. Invoke the divergence law, (18.2), which implies that the normal component of $\vec{B}$ is continuous across a boundary between two materials having different permeabilities, to relate $\vec{H}$ on the two sides of any discontinuity of permeability along ℓ.
3. Use Ampère's law, (18.1), to equate the line integral of $\vec{H}$ around the closed path to the surface integral of the current density. This results in an expression for $\vec{H}$ in terms of the current i, the number of turns N, and the length of the path ℓ.
4. Find the flux linked by the winding by integrating $\vec{B}$ over the surface formed by every turn. If all the turns are the same, the solution equals the number of turns times the integral of $\vec{B}$ over the surface defined by one turn. The result is the flux linkage of the winding, λ.
5. Divide λ by i to obtain the inductance, L.

The flux linkage λ, which has the units volt-seconds, specifies the state (i.e., the energy) of the inductor. Faraday's law, (18.3), relates the time derivative of this flux linkage to the voltage across the winding. Therefore, the change in flux linkage from one time to another is the integral of the voltage over this time:

$$\lambda(t) = \lambda(0) + \int_0^t v(\tau)\, d(\tau). \tag{18.8}$$

Example 18.1 Applying the Concept of Flux Linkage

Consider an inductor originally designed to support a maximum ac voltage of $V_0 \sin \omega t$. If we increase the number of turns by 50% and hold the amplitude of the voltage at V_0, how low can the frequency go before the peak flux density in the core exceeds that of the original design?

The number of turns N_2 on the new inductor is $1.5N_1$, where N_1 is the original number of turns. The flux linkage increases by the same amount if we hold the peak flux density constant at B_o. Thus

$$\lambda_2 \leq N_2 B_o = 1.5 N_1 B_o = 1.5\lambda_1. \tag{18.9}$$

So, if $\lambda_2 \leq 1.5\lambda_1$, applying (18.8) yields $\omega_2 \geq \omega_1/1.5$.

Only in the case of constant permeability can we strictly define inductance. For constant permeability, the value of inductance as defined by (18.7) will be directly proportional to the permeability of the material, the area of the winding, and the square of the number of turns, and inversely proportional to the length of the magnetic path. If these physical parameters are constant, the inductance will not vary with time, and we express its terminal behavior by the familiar relationship

$$v = L\frac{di}{dt}. \tag{18.10}$$

Example 18.2 An Inductance Calculation

Figure 18.1 shows an inductor constructed of windings on a doughnut-shaped, or toroidal, core. An inductor constructed in this way is known as a *toroidal inductor*. Let us calculate its inductance.

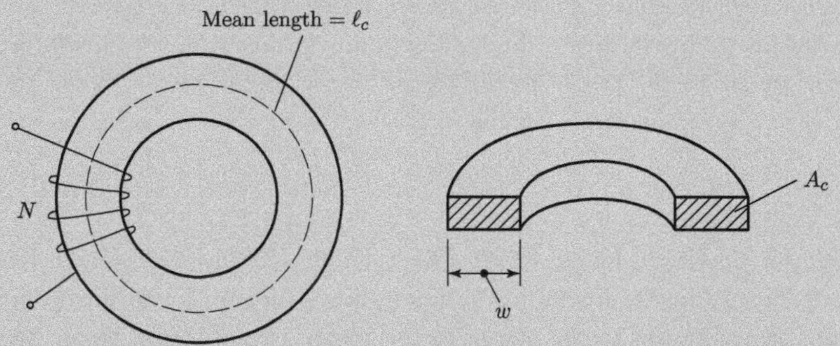

Figure 18.1 Toroidal inductor and cross-section of toroid core.

We immediately confront a problem when we apply Ampère's law to determine H: the value of H we obtain depends on whether we choose a contour at the inner or outer radius of the toroid or somewhere in between. We can usually obtain an answer that is accurate enough by assuming that H in the core is constant at the value that results from using a contour at the mean radius, with its circumference chosen as the core length ℓ_c in Ampère's law. This approximation is good enough if the inner radius is large compared with the core's radial dimension w. Applying Ampère's law with this approximation and substituting $B = \mu_c H$ yields

$$B = \frac{\mu_c N i}{\ell_c}, \tag{18.11}$$

where μ_c is the permeability of the core. We are assuming B to be uniform within the core, so the flux is

$$\Phi = BA_c = \frac{\mu_c A_c N i}{\ell_c}. \tag{18.12}$$

Because the area of each turn has the same value A_c, λ is simply

$$\lambda = NBA_c = \frac{\mu_c A_c N^2 i}{\ell_c}. \tag{18.13}$$

Dividing λ by the current yields the inductance,

$$L = \frac{\mu_c A_c N^2}{\ell_c}. \tag{18.14}$$

Substituting in some values helps to develop a feeling for the sort of inductance values achieved with particular inductor parameters. If $\mu_c = 200\,\mu_0$, $\ell_c = 15\,\text{cm}$, $A_c = 1\,\text{cm}^2$, and $N = 25$, then

$$L = \frac{(200 \times 4\pi \times 10^{-7})(10^{-4})(625)}{0.15} = 105\,\mu\text{H}$$

18.1.2 Electric Circuit Analogs

The result of any inductance calculation is an expression of the general form

$$L = N^2 \mathscr{P}. \tag{18.15}$$

The term $\mathscr{P}$, the *permeance* of the inductor, is determined by the geometry of the core and the magnetic properties of the core material.† The N^2 term is simply a scaling factor. In (18.15), $\mathscr{P} = \mu_c A_c/\ell_c$. As you will see, permeance and its inverse, *reluctance*, $\mathscr{R}$, are useful quantities in the analysis of complex magnetic structures.

If we assume a magnetic structure consisting of windings on a core made of linear magnetic material, the relationship between flux, Φ, and ampere-turns, Ni, as given by (18.12), is always of the form

$$\Phi = \frac{Ni}{\mathscr{R}} = \mathscr{P} Ni, \tag{18.16}$$

where $\mathscr{R}$ and $\mathscr{P}$ are the reluctance and permeance, respectively, of the magnetic field path used to calculate B from Ampère's law. From (18.12), the reluctance of the toroidal core is

$$\mathscr{R} = \frac{\ell_c}{\mu_c A_c}. \tag{18.17}$$

The divergence law, (18.2), states that magnetic field lines must close on themselves. But having the choice of a path to follow, the flux would follow the low-reluctance path of a highly permeable material – such as that represented by (18.17) – rather than the high-reluctance path of air. The mechanism is analogous to the way in which current flows through the wires comprising a circuit rather than through the surrounding air. This observation and the form of (18.16) suggest an analogy with Ohm's law, where $Ni \to v$, $\Phi \to i$, and $\mathscr{R} \to R$. Such an analog is a valuable tool for analyzing or designing complex magnetic structures, for it allows exploitation of a strong intuitive understanding of electric circuits. Because of this analog, we refer to paths of high permeability as *magnetic circuits*.

† Industry datasheets often refer to the permeance provided by a core set as the "A_L" value, in units such as $n\text{H}/turn^2$.

The expressions for the reluctance of a magnetic path and the resistance of a conductive path have similar forms. They depend directly on the length of the path and inversely on the cross-sectional area and permeability (respectively conductivity) of the material. The ampere-turns (Ni) is known as the *magnetomotive force* (MMF) and is analogous to the electromotive force (EMF) in an electric circuit. The magnetic flux is analogous to the electric current. The divergence law (18.2) is the basis for forming the equivalent to Kirchhoff's current law, KCL, in a magnetic circuit.

There is, however, a practical difference between the behaviors of electric and magnetic circuits. An electric circuit is defined by copper wires and other components whose conductivities are generally more than 12 orders of magnitude greater than that of the material surrounding the circuit (air or epoxy-glass board, for example). In contrast, a magnetic circuit is composed of materials whose permeances are only a few orders of magnitude greater than that of its surrounding medium (air, for example). In fact, air frequently forms part of the magnetic circuit, as you will see. Therefore, a measurable amount of flux "leaks out" of the magnetic path defined by the materials, and closes on itself through alternative paths in air. This flux is known as *leakage flux*, and we frequently assume it to be negligible, at least in a first approximation to a solution of the magnetic circuit.

Example 18.3 Using a Circuit Analog to Calculate Inductance

Figure 18.2 shows an inductor made of a three-legged core with a winding of N turns on the center leg. The two outer legs have cross-sectional areas different from that of the center leg. What is the inductance of the winding?

We first assume that the permeability of the core is much larger than that of free space, so all the flux is in the core. The core is the magnetic circuit, which may be broken into three branches: the center leg; the branch to the right, consisting of both horizontal and vertical core segments; and the branch to the left, which is symmetrical with that to the right. The length of each branch is approximately its mean path length. These lengths are shown in Fig. 18.2(a) as ℓ_1 and ℓ_2.

The permeance of each branch depends only on the geometry and material properties of the core, so we may calculate it independent of the winding details. The result is

$$\mathscr{P}_1 = \text{permeance of center leg} = \frac{\mu_c A_1}{\ell_1} = \frac{1}{\mathscr{R}_1}, \tag{18.18}$$

$$\mathscr{P}_2 = \text{permeance of outer legs} = \frac{\mu_c A_2}{\ell_1 + 2\ell_2} = \frac{1}{\mathscr{R}_2}. \tag{18.19}$$

The resulting analog electric circuit is shown in Fig. 18.2(b). The winding and its current are represented by a voltage source of value Ni. Polarity is determined by the right-hand rule and the definition of positive Φ in the analog circuit. We use branch reluctances rather than permeances because we can solve this circuit more easily in terms of resistances rather than conductances.

The flux linked by the winding is that in the center leg, Φ_1. We find the inductance by calculating this flux, multiplying by N to get λ, and dividing by i:

$$\Phi_1 = Ni\frac{1}{\mathscr{R}_1 + (\mathscr{R}_2/2)} = \frac{2A_1A_2\mu_c Ni}{2A_2\ell_1 + A_1(\ell_1 + 2\ell_2)}, \tag{18.20}$$

so

$$L = \frac{N\Phi_1}{i} = \frac{2A_1A_2\mu_c N^2}{2A_2\ell_1 + A_1(\ell_1 + 2\ell_2)}. \tag{18.21}$$

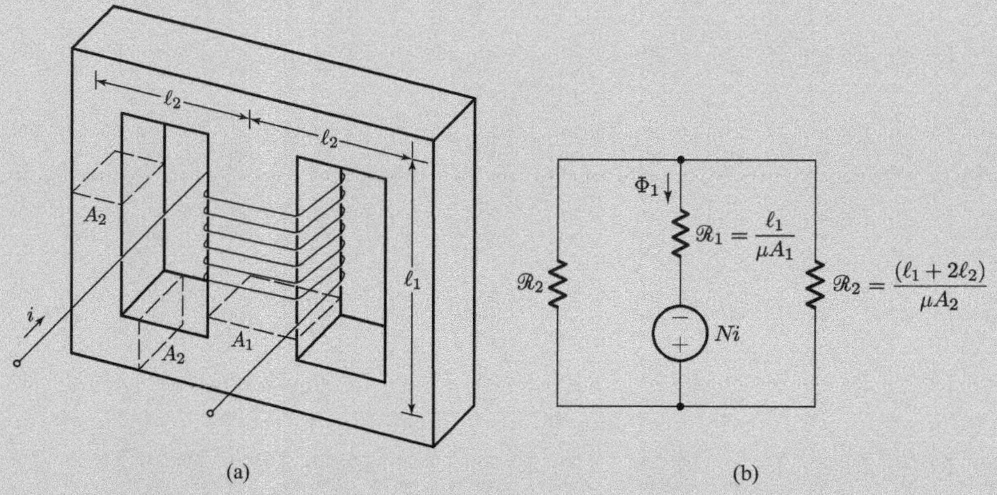

Figure 18.2 (a) Magnetic structure with one winding of N turns and three legs. The two outer legs have cross-sectional area A_2, and the inner leg has area A_1. The core material has permeability μ_c. (b) Circuit analog for the structure of (a).

The inductances we have calculated so far, (18.14) and (18.21), are linear functions of core permeability. This approach may be undesirable because – excepting certain low-permeability and "distributed-gap" magnetic materials such as powdered iron – the permeability of a magnetic material is not closely specified, and varies with flux level. To desensitize the value of inductance to variations in μ_c, we often place an air gap in the magnetic circuit.

Example 18.4 Inductor with an Air Gap in the Core

An air gap in a magnetic circuit makes the inductance independent of flux level (current) and increases the energy storage density of the structure. A core with an air gap in one leg is shown in Fig. 18.3(a). What is the inductance of this structure?

We can calculate the reluctances of the core and air gap by again using the mean path length approximation:

$$\mathscr{R}_c = \frac{\ell_c}{\mu_c A}, \qquad \mathscr{R}_g = \frac{g}{\mu_0 A}. \tag{18.22}$$

We now calculate the inductance, using reluctances as in Example 18.3:

$$\Phi = \frac{Ni}{\mathscr{R}_c + \mathscr{R}_g} = \frac{Ni}{(\ell_c/\mu_c A) + (g/\mu_0 A)}, \tag{18.23}$$

so

$$L = \frac{N\Phi}{i} = N^2\left(\frac{1}{\mathscr{R}_c + \mathscr{R}_g}\right) = \frac{\mu_0 A N^2}{(\mu_0/\mu_c)\ell_c + g}. \tag{18.24}$$

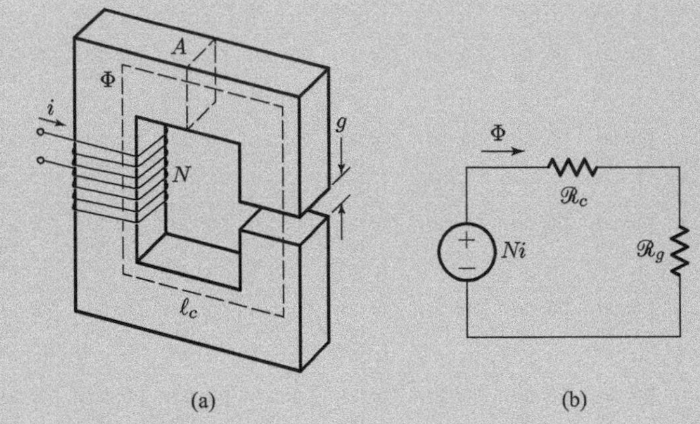

Figure 18.3 (a) An inductor made from a core with an air gap in one leg. The cross-sectional area of the core is uniform and equal to A. (b) Circuit analog for the inductor of (a).

Note that the value of the inductance becomes independent of the core's magnetic properties if

$$g \gg \frac{\ell_c \mu_0}{\mu_c}. \tag{18.25}$$

Because the magnetic properties of most core materials vary with temperature, flux level, sample, and manufacturer, a gap is frequently required to make the inductance value predictable and stable.

18.1.3 Second-Order Effects in Determining Inductance

So far, our calculations of inductance have been based on the assumption that all the flux linked by the winding is contained by the core. Because the permeabilities of the core and air differ by only a few orders of magnitude, an additional *leakage* component of linked flux is almost always present. Additionally, the flux across any gap contains a further component that bulges outside the core cross-section. The bulge is called a *fringing field*, and it increases the effective cross section of the gap. Both these effects increase the inductance from the value calculated on the basis of the assumptions that all flux is contained by the core and that gap flux is normal to the gap faces.

Figure 18.4(a) shows the same structure as that of Fig. 18.3(a), except that the leakage flux Φ_ℓ and the gap flux including fringing Φ_g are shown. The circuit analog for this inductor is shown in Fig. 18.4(b). The reluctance of the gap, $\mathcal{R}'_g$, is now smaller than before because in addition to the flux going straight across the gap, additional flux is present in the gap fringing field. The location of the leakage branch $\mathcal{R}_\ell$, and its representation as a single branch in shunt with the source, are approximations. The reason is that some of the leakage flux passes through some of the core (the leakage flux at the inside corners, for example) and is more accurately represented by a series of L-shaped networks, each containing part of the core (series branch) and part of the leakage flux (shunt branch). Seldom is this detailed representation of leakage flux necessary.

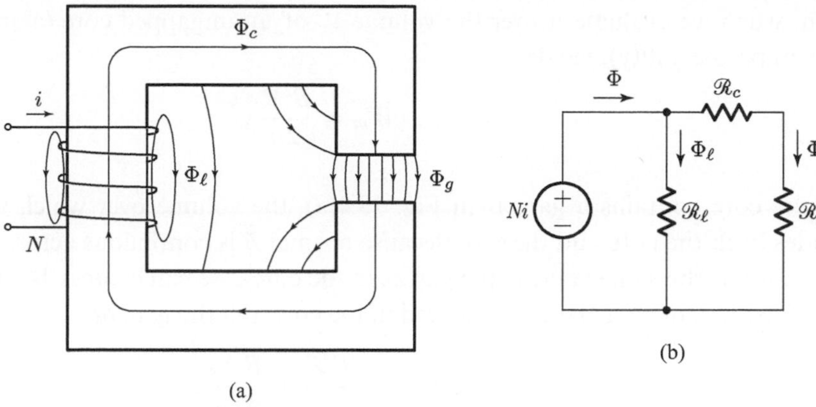

Figure 18.4 (a) Inductor showing leakage flux, Φ_ℓ, and fringing flux, Φ_g, at the edges of the gap. (b) Circuit analog for the magnetic structure of (a).

The inductance of the inductor of Fig. 18.3(a), calculated on the basis of the model of Fig. 18.4(b), is

$$L = \frac{N\Phi}{i} = N^2 \left(\frac{1}{\mathcal{R}_\ell} + \frac{1}{\mathcal{R}_c + \mathcal{R}'_g} \right). \tag{18.26}$$

As $\mathcal{R}_\ell > 0$ and $\mathcal{R}'_g < \mathcal{R}_g$, this calculation results in an inductance value greater than that given by (18.24).

The importance of these second-order effects depends on the aspect ratios, or relative dimensions, of the core and the gap. A gap whose length is small compared to the length of the sides of the core cross-section produces a much smaller second-order increase in λ than one whose length is long compared to these dimensions. The fringing flux is a much smaller fraction of the total gap flux in the former case than in the latter case. Tabulations of inductance correction factors as a function of gap aspect ratio are available. Also, a core with a relatively short mean length exhibits a larger second-order increase from corner leakage than does a long core.

This qualitative understanding of second-order effects usually permits adequate inductor design. A finite-element numerical analysis would yield a more accurate design.

18.1.4 Energy Storage

In circuit-variable terms, the magnetic stored energy in an inductor W_m is

$$W_m = \frac{1}{2} L i^2 = \frac{1}{2L} \lambda^2. \tag{18.27}$$

We can also express this energy in terms of fields as

$$W_m = \frac{1}{2} \iiint_{\text{volume}} \vec{B} \cdot \vec{H} \, d\mathcal{V}, \tag{18.28}$$

which, when we evaluate it over the volume $\mathcal{V}_c$ of an ungapped core (assuming constant and uniform permeability), yields

$$W_m = \frac{B^2 \mathcal{V}_c}{2\mu}. \tag{18.29}$$

If the core contains a gap, as in Fig. 18.4(a), the volume over which we integrate (18.28) includes both the core and the gap. Because normal $\vec{B}$ is continuous across discontinuities in μ (that is, it has the same value in the gap as in the core), we can express W_m in terms of B as the sum of expressions for the energy stored in the core and the gap, or

$$W_m = \frac{B^2 \mathcal{V}_c}{2\mu_c} + \frac{B^2 \mathcal{V}_g}{2\mu_0}. \tag{18.30}$$

Comparing the stored energy *densities* $w = B^2/2\mu$ in the core and the gap yields

$$\frac{w_g}{w_c} = \frac{\mu_c \mathcal{V}_g}{\mu_0 \mathcal{V}_c}. \tag{18.31}$$

If the cross-sectional areas of the gap and the core are equal (which may not always be true), we can rewrite (18.31) in terms of permeabilities and mean path lengths as

$$\frac{w_g}{w_c} = \frac{\mu_c g}{\mu_0 \ell_c}. \tag{18.32}$$

For an iron core, $\mu_c \approx 10^4 \mu_0$ and the ratio (18.32) is typically much greater than 1. The significance of this fact is that most of the energy stored in the inductor is stored in the fields in the gap rather than in the core. That is,

$$W_m \approx \frac{B^2 \mathcal{V}_g}{2\mu_0}. \tag{18.33}$$

Note that this condition is the same as the condition (18.25) for the inductance to be relatively independent of core properties. The volume of the gap times the energy density in the gap then gives a good approximation to the total stored energy in the inductor. If the core is iron with a saturation flux density of $B_{\max} = 1.6\,\text{T}$, the maximum energy density in the gap is

$$w_g = \frac{(1.6)^2}{2\mu_0} = 0.78\,\text{J/cm}^3. \tag{18.34}$$

However, if the core is ferrite, in which the saturation flux density is typically 0.3 T, the stored energy density is only 0.036 J/cm^3. These numbers are very valuable when you are designing an inductor, because specifying the values of inductance and current determines the total stored energy and thus the volume of the gap.

Energy storage considerations can drive the design of many inductors, such as filter or "choke" inductors. One design approach for such an inductor might proceed as follows:

1. Calculate the gap volume required to store the energy without causing saturation.
2. Assume that the length of the gap is small (perhaps by a factor of 10) compared to the length of the sides of the core cross-section. This assumption reduces the second-order effects of fringing.

3. Determine A_c and g so that steps 1 and 2 are satisfied.
4. Determine the number of turns needed to obtain the desired value of inductance.
5. Determine the window area (the area enclosed by the core in Fig. 18.3) required to hold this number of turns for the wire size necessary to keep the resistance below an acceptable level.
6. Check that (18.25) is satisfied for the now known length of the magnetic path.

Further considerations and approaches for designing magnetic components are introduced both below and in subsequent chapters.

18.2 Saturation, Hysteresis, and Residual Flux

In Example 18.4 we showed that one benefit of a gap in a magnetic core is that it makes the value of inductance less sensitive to variations of the magnetic properties of the core material. Another advantage of a gap is that it limits the detrimental effects of *remnant*, or *residual*, magnetization. Remnant magnetization is the flux remaining in the core when the magnetic field intensity H reaches zero. Figure 18.5 shows the relationship between B and H for a typical magnetic material. H is proportional to the current flowing in the winding, and B, through (18.13) and (18.8), is proportional to the integral of the voltage across the winding. This integral is the flux linkage λ.

Three important characteristics of the relationship are shown in Fig. 18.5. First, the path followed when the flux is increasing is not the same as that followed when the flux is decreasing. This behavior is called *hysteresis*. Second, note that neither path passes through the origin. When $H = 0$ the flux density is not zero but has a value $\pm B_r$ called the *remnant magnetization*, or *residual-flux density*. When $B = 0$, the magnetic field is not zero but is equal to $\pm H_c$, a parameter called the *coercive force* of the material. Third, the slope of the path, which is the incremental permeability μ_i, decreases rapidly with increasing B (ultimately reaching μ_0) after some maximum $B = B_s$ is reached. This state of low incremental permeability is called *saturation*, and B_s is the *saturation-flux density*.

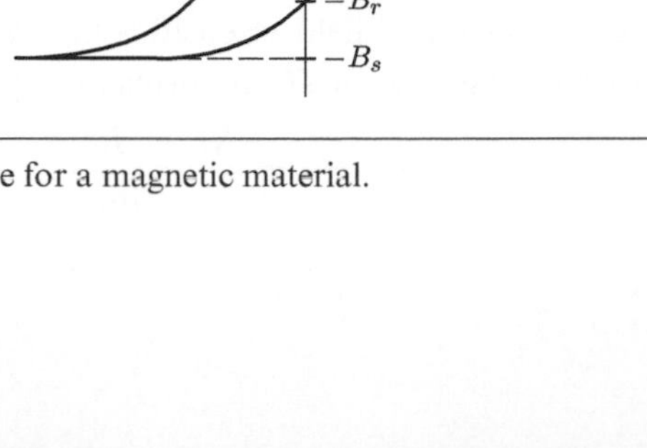

Figure 18.5 Typical B–H curve for a magnetic material.

18.2.1 Saturation

Although some applications exploit the saturation characteristics of magnetic cores (for example, magnetic amplifiers, fluorescent lamp ballasts, constant-voltage transformers, and self-oscillating transistor inverters), in most applications saturation is avoided. One reason is that in saturation the condition (18.25) necessary to make the value of L insensitive to material properties is often no longer satisfied, because μ_c has become very small. Another reason is that in saturation H increases very rapidly with B, resulting in large changes in current for relatively small increases in the volt–time product, λ. The final reason is that saturation reduces the coupling between windings in a transformer, because the permeance of the core is no longer much larger than that of the leakage paths.

Avoiding saturation is sometimes particularly difficult during transient circuit operation. A power circuit operating in the cyclic steady state sequentially imposes positive and negative volt-seconds of equal value across any inductors in the circuit. Stated another way, the integral of the inductor voltage over one cycle is zero, as it must be if λ, and hence the current, is not to increase without limit. When we know the value of volt-seconds to be applied, we can (loss permitting) design the inductor for cyclic steady-state operation so that its flux density varies from just under $+B_s$ to just over $-B_s$. However, during the first cycle, when the circuit is first turned on, saturation may occur because the flux may not be starting from $-B_s$ (assuming that the first half-cycle is to be positive), and therefore reaches $+B_s$ after fewer volt-seconds than anticipated. For instance, if $B(0) = 0$, only half the normal change in λ is actually available for the first cycle. This is known as the *start-up problem*.

We could solve the start-up problem by designing the inductor to support twice the normal λ. This solution, however, would require four times the maximum energy storage capability and therefore a much larger physical volume of the inductor. A better solution is to control the power circuit, limiting the volt-seconds to a lower than normal value during the first cycle.

18.2.2 Residual Flux Density

The start-up problem is aggravated when the core material is characterized by a B–H curve exhibiting a high value of residual flux density. Consider the curve of Fig. 18.5 and assume that the last time the power circuit was operated, B and H were at point A, at which time the circuit was turned off. When the current (and therefore H) in the inductor decays to zero, B decays to B_r. The result is that if the circuit were turned on again in a positive half-cycle, the volt-seconds available before saturation is reached would be much smaller than in the case of starting from $B = 0$.

We can reduce the start-up problem created by residual magnetism by introducing a gap in the core. Considering the structure of Fig. 18.3, we can write

$$H_g g + H_m \ell_c = Ni, \qquad B = \mu_0 H_g = \mu_c H_m,$$

where H_m and H_g are the magnetic field intensities in the magnetic material (core) and gap, respectively. Solving these equations for H_m in terms of B yields

$$H_m = \frac{Ni}{\ell_c} - \frac{Bg}{\mu_0 \ell_c}. \tag{18.35}$$

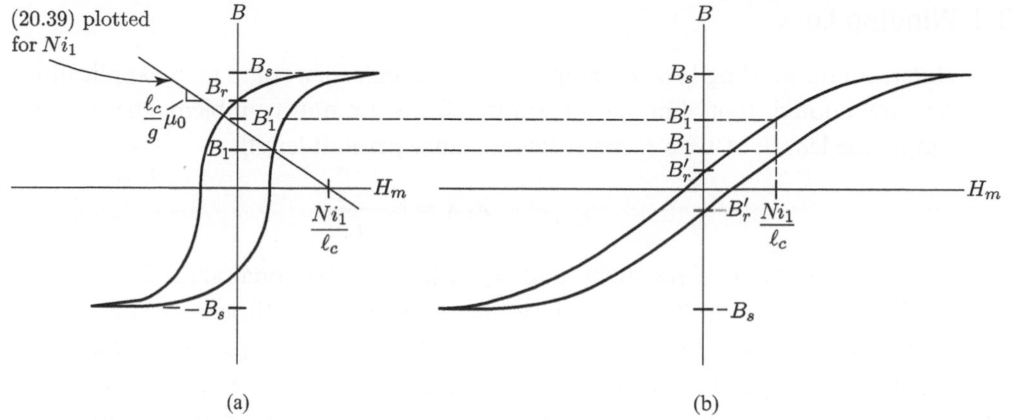

Figure 18.6 (a) Core hysteresis loop and air-gap load line for Ni_1. (b) Resulting locus of operation illustrating how the value of residual flux density is reduced from B_r to B_r' by the introduction of a gap in the core.

We must satisfy both (18.35) and the B–H_m relationship shown in Fig. 18.5. We can obtain the resulting relationship between ampere-turns (Ni) and B by graphically combining (18.35) and Fig. 18.5. We do so by putting the two functions on the same B versus H_m axes, and plotting their intersections as we vary Ni. Figure 18.6(a) shows the two functions on the common axes, with (18.35) plotted for an arbitrary value of Ni. The intersections of the two functions are plotted on the same set of axes as shown in Figure 18.6(b). The process can be visualized as plotting the B intersections while "sliding" the load line along the H_m axis. Note that the incremental inductance is smaller, B_s is unchanged because it is a property of the core material, and the residual flux at zero current is much less than before, reducing the start-up problem, as we desired.

The volt-seconds required to saturate a magnetic circuit (starting from $B = 0$) is $\lambda_s = NB_sA_c = Li_s$, which is independent of the gap length g. Changing the gap length reduces the inductance, but it also increases the current i_s at which saturation occurs. The start-up problem is most frequently associated with transformers, which are often constructed on cores with no gaps.

18.3 Losses In Magnetic Components

We have assumed that magnetic components are lossless, except when we considered loss in our discussion of the relationship between window area and wire size. The size of a magnetic element, however, is often directly related to its acceptable loss. Moreover, in many applications these losses are the largest contributor to total circuit loss.

Magnetic component losses can be grouped into three broad categories: in the windings, in the magnetic core, and in other conductive materials resulting from induced eddy currents. The last of these is often small, so we focus here on winding and core losses.

18.3.1 Winding Loss

Determining winding loss (or copper loss) is straightforward in many applications. At frequencies low enough that skin and proximity effects are not a problem, the winding resistance is simply the length of wire ℓ_w times its resistance per unit length,

$$R_{\text{Cu}} = \ell_w \frac{\rho_{\text{Cu}}}{A_w}, \tag{18.36}$$

where ρ_{Cu} is the wire resistivity and A_w is its cross-sectional area. The resistivity of copper is $1.7 \times 10^{-8}\ \Omega$-m at 20 °C. However, its temperature coefficient of resistivity is $0.0039\,K^{-1}$, so its resistivity rises by nearly 40% for a 100 °C rise in temperature. Thus, it is important to specify resistivity at the maximum temperature of interest when calculating losses. The criteria for selecting a wire size are functions of the application, and most often relate to acceptable losses or maximum temperature rise. Wire parameters are tabulated and easily obtainable. Typical design current densities range from 100 to 500 $\text{A}_{\text{rms}}/\text{cm}^2$. A good metric for back-of-the-envelope designs is the resistance of AWG 10 wire, which is 1 mΩ/ft (or 3.28 mΩ/m), and resistance increases by a factor of two for approximately every three size reductions in wire gauge.

Example 18.5 Current Density and Wire Size Selection

Wire size is often selected in terms of a maximum permissible current density. The permissible current density, however, itself is often a function of allowable loss or temperature rise. Consider inductors wound on an RM or "rectangular modular" core. The RM family of cores represents a standardized core type that comes in a range of sizes and is available in a variety of ferrite materials. These cores provide a nearly square footprint to facilitate dense packing of components on a printed circuit board. Each RM core size (from a small RM4 core up to a large RM14 core) allows a certain total dissipation for a given core "hot spot" temperature rise, assuming the core is cooled by natural convection. Figure 18.7 plots the total loss allowed for a 25 °C temperature rise in different RM core sizes versus the winding window areas of their bobbins, as determined from manufacturer data. Temperature rise is proportional to dissipation over a reasonable range, allowing dissipation for other temperature rises to be easily determined.

Power dissipation constraints such as those in Fig. 18.7 can be related to limits on winding loss and current density. For example, consider an inductor design where copper loss is dominant and in which the full winding window is used (as might be true in a filter inductor). Figure 18.8 plots the maximum rms current density that yields conduction loss within the power dissipation limit of Fig. 18.7 for different core sizes and associated winding window areas. It assumes only low-frequency conduction loss is significant. The presence of other losses (such as core loss or proximity loss) would reduce allowable current density, while higher permissible temperature rise would increase it. This plot assumes a winding packing factor (or window utilization factor) of $k_u = 0.64$, where k_u is defined as the ratio of copper area to total window area. The permissible current density varies significantly, decreasing for larger core sizes and winding areas.

The fact that allowable current density (and hence allowable power dissipation density) decreases with increasing size is to be expected from fundamental scaling considerations. The volume of a magnetic component increases as the cube of linear dimension, as does the total power dissipated for a given current density. However, the surface area through which the generated heat must be removed only increases as the square of linear dimension. Consequently, we should expect that at larger size scales (i.e., for larger inductors), we will have to reduce power dissipation density – and hence current density – to accommodate heat transfer limitations.

The rms current densities shown in Fig. 18.8 assume that only low-frequency conduction loss is significant; the presence of other losses (such as core loss, proximity effect loss, etc., as described below) would reduce allowable current density. For example, in a design where core loss equals copper loss, the

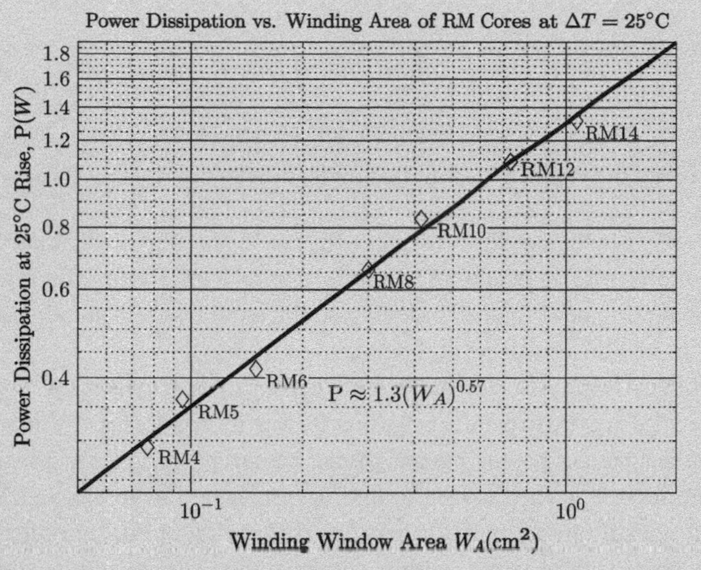

Figure 18.7 Allowable power dissipation for a 25 °C temperature rise versus the winding area provided by the bobbin for different RM core sizes.

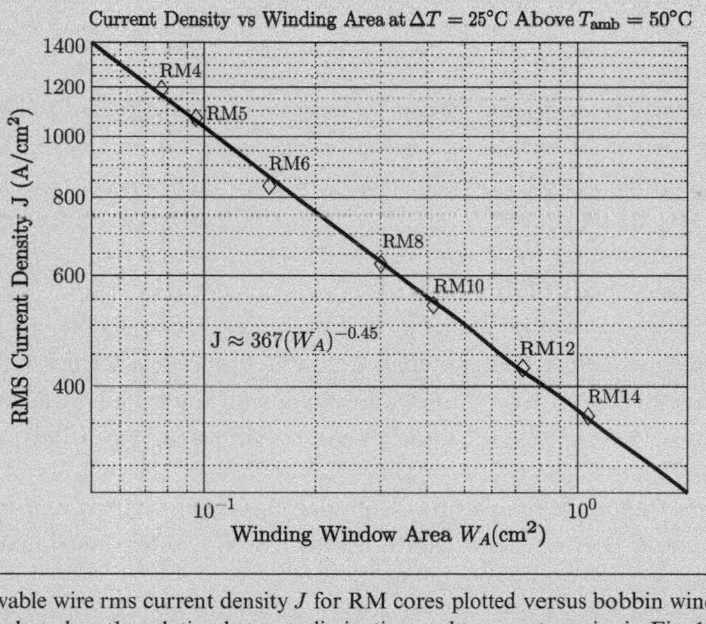

Figure 18.8 Allowable wire rms current density J for RM cores plotted versus bobbin window area W_A. This plot is based on the relation between dissipation and temperature rise in Fig. 18.7.

allowable rms current density would be expected to decrease by a factor of $1/\sqrt{2}$ from that shown. Higher ambient temperatures, lower allowed temperature rises, and worse cooling conditions would all likewise reduce allowable current density. Conversely, lower ambient temperatures, larger allowed temperature rises, and improved inductor cooling (e.g., via forced air cooling) would all increase allowable current density.

In choosing a current density (and wire size) for a given magnetics design, a designer often leverages some knowledge about power dissipation and temperature rise for a given core type, such as that embodied in Fig. 18.7. In the absence of specific knowledge, one can start with an educated guess at a current density limit based on size and operating conditions (perhaps informed by a plot such as Fig. 18.8). One can then refine the design using improved thermal models, finite element simulations, or measurements of a prototype. Chapter 21 will treat inductor design and sizing approaches in more detail.

The skin effect is the tendency for high-frequency ac current to be carried only near the surface of a conductor, and not penetrate to its interior regions. Essentially, eddy currents driven by the changing magnetic field of the ac current in the conductor itself cause the interior region of the conductor to be shielded from ac fields and currents.

Skin depth is a measure of the lateral depth of penetration of ac current in a conductor. Technically it is the exponential length constant with which current amplitude decreases with depth into the conductor, but one can often accurately compute loss by assuming that ac current is carried uniformly within a skin depth of the surface. The loss calculated by assuming uniform conduction within one skin depth matches that for the actual exponential current distribution in the deep skin-depth-limited case. A more detailed treatment of current distribution and loss is given in Chapter 21. We express skin depth δ as a function of frequency and the conductivity and permeability of the conductor, in this case copper:

$$\delta_{\mathrm{Cu}} = \sqrt{\frac{2}{\omega \mu_{\mathrm{Cu}} \sigma_{\mathrm{Cu}}}}. \tag{18.37}$$

The skin depth in copper is about 0.86 cm at 60 Hz and 25 °C. Thus, little 60 Hz current flows in the center of conductors having radii larger than a few centimeters. For this reason, aluminum conductors of power transmission lines can be reinforced with a steel core without compromising current capacity ($\delta_{\mathrm{Al}} \approx 1.25\,\delta_{\mathrm{Cu}}$). Skin depth decreases as $1/\sqrt{f}$, so is increasingly important as frequency increases. At the 10 MHz switching frequency of some power converter designs, skin depth is only ~21 μm.

Where necessary, the limitations of skin depth can be avoided by using stranded wire in which each strand is insulated. Commercially available wire of this kind is known as *litz wire*, a name derived from the German *Litzendraht*, meaning many-stranded. True litz wire is also *transposed*, a construction technique in which each strand occupies each possible position in the wire bundle for a length equal to the *transposition length* divided by the number of conductors in the wire. This ensures that each strand links the same flux, such that there are no circulating currents between strands. By using litz strands of radius on the order of a skin depth, the full conductor cross section can be used for carrying current.

The proximity effect is a further high-frequency conduction loss mechanism that occurs when ac fields from a first conductor (or portion of conductor) impinge on a second conductor portion, and induce eddy currents and associated losses in the second conductor portion. These losses manifest themselves as an increase in the apparent ac resistance of the first conductor. Important sources of proximity effect fields in magnetic components include fringing fields near the gap of an inductor (discussed in Chapter 21) and fields building up across multiple layers of winding in the window of an inductor or transformer. For this reason, high-frequency magnetic components often do not have conductors placed immediately adjacent to gaps, and use relatively few winding layers.

Because the proximity effect relates to the losses due to eddy currents, it is sometimes found that using *less* conductor in an inductor or transformer (thus providing less area in which eddy currents may flow) can reduce ac resistance. This counterintuitive result is not found at dc, or with skin effect only. The proximity effect can make it desirable to use litz wire with strands that are only a fraction of a skin depth in diameter.

18.3.2 Core Loss

The two basic types of materials used for inductor and transformer cores in power electronic equipment are ferrous alloys and the magnetic ceramics known as ferrites. Two mechanisms produce core loss in ferrous alloys: hysteresis and eddy currents. These effects are also present in ferrites, but additionally dielectric and domain-wall resonance loss mechanisms may be significant, depending on the frequency of operation. For practical purposes, we separate the loss into (i) eddy current losses and (ii) other core losses including hysteresis loss. We treat each of these in turn.

Eddy current loss is simply resistive loss caused by currents driven by voltages induced in the magnetic material by changes in the flux passing though it. If the material were non-conducting, the eddy current losses would be zero. All magnetic material has some degree of conductivity, with metallic materials having substantial conductivity, MnZn ferrites having lower conductivity, and NiZn ferrites having extremely low conductivity.

Because eddy currents are driven by a voltage induced in the material through Faraday's law, (18.3), they flow in a plane perpendicular to $\vec{B}$. The eddy current loss grows quickly with the cross-sectional area over which the induced driving voltage can build up and in which eddy currents can flow. Cores made of metallic material are thus often constructed of laminations parallel to $\vec{B}$ to reduce the driving voltage for a given segment of conductive magnetic material. These laminations are insulated from one another, usually by the oxide formed when the material is annealed. Figure 18.9 shows three laminations and the paths of the eddy currents flowing in them. The loss produced by these currents is proportional to v^2/R_c, where v is the induced driving voltage and R_c is the resistance of the path through which the current flows. As R_c is proportional to the length of the path but v is proportional to the area enclosed by the path, loss will increase rapidly with increasing lamination thickness.

The resistance R_c is linearly proportional to the material's resistivity. Various high-resistivity ferrous alloys have been developed to reduce the eddy current contribution to total core loss. For steel, the simplest way of increasing resistivity is to add a small amount (1%–3%) of silicon to it. More exotic low-loss ferrous alloys contain various amounts of molybdenum, chromium, and nickel.

Another reason to laminate a core, especially for high-frequency use, is that the magnitude of B in the core falls off exponentially from its value at the core surface, with a characteristic length equal to the skin depth of the material. The skin depth in 2.5% silicon steel at 10 kHz is approximately 0.05 mm. A solid core of this material would contain flux only in a shell about

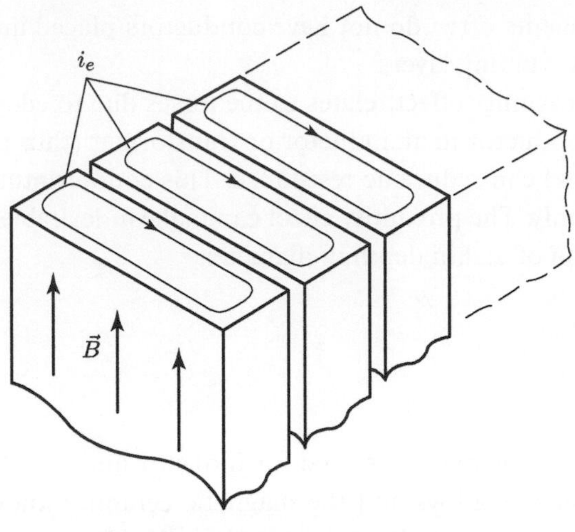

Figure 18.9 A section of a laminated core, showing the location of eddy currents i_e.

0.05 mm thick. That is, if the core cross-sectional dimensions were much larger than 0.05 mm, the center of the core would not contain any flux.

Stamped lamination thicknesses of about 0.025 mm are available, but laminating more thinly makes manufacturing the core expensive, and the insulation between the laminations begins to occupy an excessive fraction of the total core volume (approximately 17% for 0.025 mm laminations). Laminated steel cores can be used effectively to frequencies of about 20 kHz, but beyond this frequency, nanocrystalline cores (comprising extremely thin ribbons of material), powdered core materials (small particles of iron suspended in a binder), or ceramic (ferrite) cores are more commonly used.

The second class of losses, namely hysteresis loss, relates to effects on the scale of the magnetic particles and domains, and can thus be characterized in terms of loss per unit volume of the material. Hysteresis loss is the result of unrecoverable energy expended to rotate magnetic domains within a magnetic material. The energy lost per cycle per unit volume of material is

$$W = \oint dw_m = \oint \vec{H} \cdot d\vec{B}, \tag{18.38}$$

which is the area enclosed by the material's hysteresis loop. The total hysteresis loss is then the product of this area, the frequency, and the core volume. As a practical matter, however, the shape of the hysteresis loop can itself be a function of the rate at which the domains are rotated, and hence of the drive frequency.

The net effect of hysteresis and other core-loss mechanisms is a loss that varies with frequency and flux magnitude, with dependencies that are a function of the particular material. The generalized relationship for these dependencies, assuming a sinusoidal flux density, is embodied in the empirical Steinmetz equation:

$$P_\text{V} = kf^\alpha B^\beta. \tag{18.39}$$

P_V is the loss per unit volume, and the parameters k, α, and β are determined from manufacturers' data or from measured data showing loss as a function of frequency and flux density. Typical values of α may range from 1 to 2.6, and typical values of β may range from 1.6 to 3.5. However, it should be emphasized that this "Steinmetz" loss model is empirical in nature. Consequently, a given parameter set (k, α, β) is only useful over a limited frequency and temperature range, and for sinusoidal drive waveforms. The appropriate values to use may change with large changes in frequency, and the calculations are only loosely accurate for non-sinusoidal drive waveforms, or even for sinusoidal drives with dc offsets in flux. Nonetheless, core loss calculations based on this simple equation are sufficient to make at least approximate calculations of loss, even when these conditions are violated.

Example 18.6 Designing a Filter Inductor

Figure 18.10 shows the output filter of a 50 V dc/dc converter. We assume that the capacitor is large enough that we can consider its voltage to be constant. The average output current of the converter is 10 A, and the filter input voltage v_d is a square wave, switching between 0 and 100 V at a frequency of 500 kHz ($T = 2\,\mu$s), as shown. The 25 μH inductor is to be designed using a ferrite material with $B_{\max}$ ($< B_s$) $= 0.3$ T and $\mu_c > 10^3\,\mu_0$.

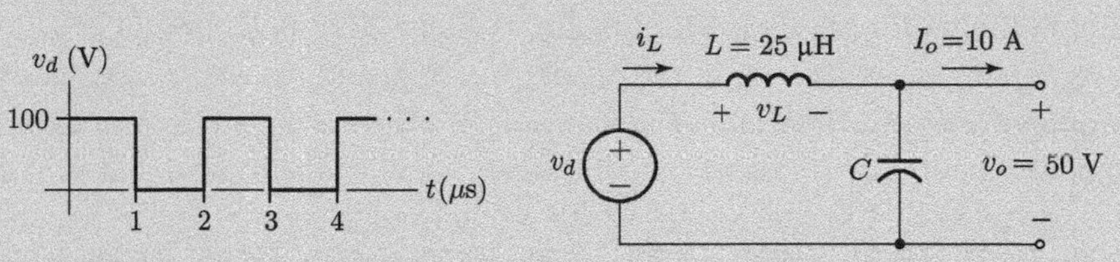

Figure 18.10 An LC filter at the output of a 50 V dc/dc converter. The effect of the converter switches is represented by the equivalent source v_d.

Under steady-state operation, the voltage v_L across the inductor is a square wave, symmetrical about zero, having an amplitude of 50 V. The peak inductor current is

$$I_p = i(0) + \frac{1}{L}\int_0^{T/2} v_L\,dt = i(0) + \frac{1}{25 \times 10^{-6}}\int_0^{1\times10^{-6}} 50\,dt = i(0) + 2\,\text{A}.$$

Since $\langle i_L \rangle = 10$ A, $i(0) = 9$ A and $i_p = 11$ A.

To design the inductor, we need to specify a core (defining the core cross-sectional area A_c and magnetic path length ℓ_c), the number of turns N, and the gap g. The relationships these parameters must satisfy are

$$\lambda_s = LI_p = 2.74 \times 10^{-4}\ \text{V-s} \le NA_cB_{\max}, \tag{18.40}$$

$$L = \frac{\mu_0 N^2 A_c}{g} = 25\,\mu\text{H}, \tag{18.41}$$

$$\frac{\ell_c}{\mu_c} \ll \frac{g}{\mu_0}. \tag{18.42}$$

One way to proceed is to make an educated guess at a core size, thereby determining A_c, use (18.40) to select the number of turns N, and use (18.41) to determine the resulting value for g. Finally, we must make sure that g is consistent with (18.42).

We will select a core from the RM family of cores mentioned above. Chapter 20 has a table providing data for this core family. Suppose we pick an RM10 core, which has a minimum core cross-sectional area $A_{c,\min} = 86.6\,\text{mm}^2$ (for determining saturation), an effective core cross-sectional area $A_c = 98.0\,\text{mm}^2$ (for determining inductance), an effective magnetic path length of $\ell_c = 44.0$ mm, an effective core volume

$V_{c,e} = 4.3\,\text{cm}^3$, a mean length of turn ℓ_t of 52 mm, and a thermal resistance (hot-spot temperature rise per W of dissipation) of 30 °C/W.

The constraint (18.40) gives us $N \geq 10.6$ turns to avoid saturation. Selecting $N = 11$, (18.41) gives us a desired gap $g = 0.6$ mm. This gap length is reasonable with respect to the core geometry, and easily satisfies (18.42). Note that the required magnetic energy storage is provided by the gap volume, as indicated in (18.33).

A last consideration is wire size, which must be large enough for the wire to carry the inductor current without getting too hot, while still fitting in the core window. In this case, litz wire comprising 650 paralleled AWG 44 strands (yielding a wire bundle having approximately 1.83 mm outer diameter) is probably a good choice. This winding will fit nicely within the 11.9 mm by 4.25 mm winding window of an RM10 bobbin, and will give sufficiently small winding loss to avoid undue temperature rise.

This filter can have a serious start-up problem, for if C is very large, its voltage at start-up rises slowly relative to the switching period. Then $v_L \approx v_d$ for many cycles, and i_L will increase well beyond the 11 A for which we designed the inductor. To avoid this, the duty ratio of the converter switches can be controlled to start from a small duty ratio and to slowly bring the output voltage up to 50 V. This is called a *soft start*.

18.3.3 A Simple Loss Model

A simple model for losses in an inductor of value L is illustrated in Fig. 18.11. The resistor R_{Cu} represents the resistance of the copper winding, and the energy dissipated in this resistor is the copper loss in the inductor. The value of R_{Cu} is determined by wire gauge, winding length, temperature, and any frequency dependencies, such as due to skin and proximity effects. Core loss is modeled by the resistor R_c, whose value is a function of both frequency and flux level. Because the inductor voltage is also a function of these variables, the most convenient location for R_c is in parallel with the inductor, as shown.

Example 18.7 Determining Parameters in a Loss Model

Here we determine the values of R_{Cu} and R_c in the model of Fig. 18.11 for the inductor designed in Example 18.6.

The inductor current, as calculated in Example 18.6, is a 500 kHz triangle wave having a peak–peak amplitude of 2 A added to a dc component of 10 A. The waveform is heavily dominated by its dc component: the rms value of the ac component is only 0.577 A as compared to the 10 A rms of the dc component, with an overall waveform rms value of 10.002 A. Considering that commercially available 650-strand/44 AWG litz wire has a dc resistance of 18.8 mΩ/m at 75 °C and the RM10 bobbin provides a mean length of turn

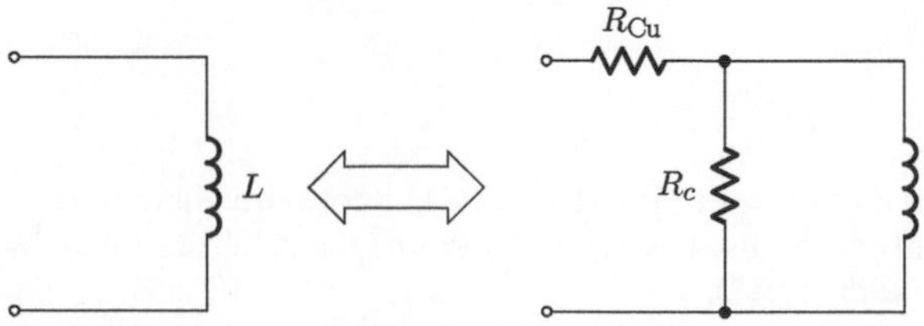

Figure 18.11 A simple circuit model for an inductor. The resistors R_{Cu} and R_c represent winding and core losses, respectively.

$\ell_t = 52$ mm, the dc resistance of the wire for 11 turns is predicted to be approximately 11 mΩ. The winding loss based on this dc resistance is approximately 1.1 W. Considering possible ac loss impacts, the skin depth in copper at 500 kHz and 75 °C is ~0.1 mm, larger than the diameter of the AWG 44 litz wire strands (~0.051 mm), indicating that skin effect will not be significant and suggesting that proximity effect loss will be considerably mitigated. This prediction is confirmed via numerical simulations that indicate the dc loss estimate is adequate. We thus use $R_{\mathrm{Cu}} = 11$ mΩ.

Determining the resistance R_c is more complicated. Although the peak flux in the core is 0.3 T, the amplitude of the ac component of this flux, B_{ac}, is much smaller:

$$B_{\mathrm{ac}} = \frac{1}{2}\frac{\lambda_{\mathrm{ac,pp}}}{NA_c} = \frac{1}{2}\frac{5.0 \times 10^{-5}}{(11)(98 \times 10^{-6})} = 0.023\ \mathrm{T}. \tag{18.43}$$

Furthermore, $B(t)$ is not sinusoidal as assumed in the Steinmetz equation; it has a substantial dc offset and a triangular ac component at a fundamental frequency of 500 kHz. Nonetheless, to estimate core loss we can crudely approximate $B(t)$ as a sinusoid with an amplitude of 0.023 T, understanding that there may be considerable error in the result. (Means for better estimating core loss in such non-sinusoidal situations are treated in Chapter 21.)

Suppose the core is made of ML95S ferrite material from Hitachi Metals. This material has a relative permeability greater than 1000 and a saturation flux density of over 0.4 T, making it suitable for the proposed design. At 100 °C, the Steinmetz core-loss parameters suitable for the 500 kHz–1 MHz range are $k = 8.09 \times 10^{-6}$, $\alpha = 2.011$, and $\beta = 3.536$ when the frequency f is expressed in hertz, ac flux density amplitude B is expressed in tesla, and loss density P_{v} is expressed in mW/cm^3. Substituting in, we calculate a loss density $P_{\mathrm{v}} = 3.76$ mW/cm^3. Multiplying by the effective core volume of 4.3 cm^3, we get a core loss of 0.016 W, far less than the winding loss. We can conclude that even a significant error in this core loss calculation will not much affect the overall loss or temperature rise.

Ignoring the drop across R_{Cu}, the voltage across R_c is a square wave having a 50 V amplitude. Thus the value of R_c that gives a loss of 0.016 W is

$$R_c = \frac{V_{\mathrm{rms}}^2}{P} = \frac{50^2}{0.016} = 156\ \mathrm{k\Omega}. \tag{18.44}$$

The total loss estimated for the inductor at the designed operating point is approximately 1.3 W. Using the nominal thermal resistance of the RM10 core (30 °C/W), this gives a reasonable predicted core "hot-spot" temperature rise of ~33 °C. The initial temperature estimate used to calculate winding resistance (75 °C) is thus conservative for ambient temperatures of up to ~42 °C. If the converter were to be operated in ambient temperatures substantially higher than this, one would want to revisit the winding resistance, loss, and temperature rise calculations, and refine the values used in the loss model.

18.3.4 Why Use a Core?

With the problem of saturation, the losses created by hysteresis and eddy currents, and the expense and weight of the magnetic material, you might well ask, "Why construct an inductor with a magnetic core?" After all, an air-core inductor would not present any of these problems. Although air-core inductors are sometimes used, there are several reasons for utilizing a core of magnetic material instead. One is that the magnetic fields are contained by the core. They do not extend great distances from the winding, as they would in an air-core inductor, reducing the problem of adverse effects on nearby equipment or circuits. Containing the magnetic field is also very important for a transformer if the coupling between two windings is to be good. Another reason is that, for a given value of inductance and a given overall volume, the resistance of the winding is much smaller if a core is used, since the number of turns, and thus wire length, is reduced.

18.4 Transformers

Transformers are paired inductors that are coupled through a shared magnetic circuit, that is, two or more windings that link common flux. Figure 18.12 shows a two-winding transformer. The core provides the low-reluctance magnetic circuit through which flows most of the flux generated by the windings. Transformers come in various forms and are used in power electronic circuits to perform a number of different functions. Low-frequency (50, 60, or 400 Hz) power transformers are usually made of laminated steel cores and function to step the line voltage up or down, to provide electrical isolation, and/or to provide phase shifting in systems using multiphase line supplies (12-pulse rectifiers, for instance). In high-frequency converter applications, the transformer provides isolation and voltage transformation, and also stores energy when used in circuits such as the flyback converter. Powdered magnetic alloys or ferrites are used in high-frequency transformers to control core losses. Additionally, transformers sometimes provide the coupling between gate-drive or base-drive circuits and the high-power devices being driven, or serve as voltage or current sensors in feedback control systems. Transformers are frequently the most costly components in power electronic equipment.

18.4.1 The Ideal Transformer

If the same flux is linked by both windings of a two-winding transformer – and this is the only flux they link – we say that the windings are perfectly coupled. As the flux linked by each turn is the same for both windings, the voltages induced in each turn are also identical. Moreover, the total voltage across a winding is directly proportional to its number of turns. The ratio of the two terminal voltages is therefore equal to the turns ratio, that is,

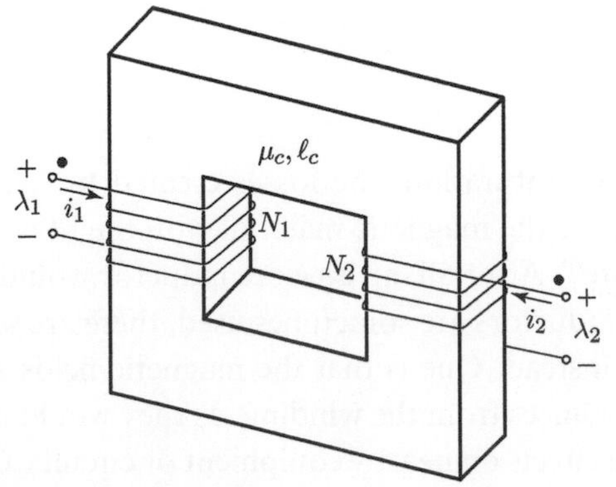

Figure 18.12 Structure of a two-winding transformer.

$$\frac{v_1}{v_2} = \frac{N_1}{N_2}. \tag{18.45}$$

The source of the magnetic field in a transformer is the algebraic sum of the Ni values of the windings. We use a *dot convention* to indicate winding polarity. Current into the dotted ends of windings produces aiding flux. An application of the right-hand rule to Fig. 18.12 shows that current into the dotted end of either winding produces clockwise flux. We can now use Ampère's law to calculate the magnetic field intensity:

$$H = \frac{N_1 i_1 + N_2 i_2}{\ell_c}. \tag{18.46}$$

If the reluctance of the magnetic path were zero, which would be the case if $\mu_c = \infty$ and there were no air gap, H would have to be zero to avoid infinite B. But H can be zero only if the Ni products for the two windings sum to zero. In this case, the ratio of the two currents is equal to the negative inverse of the ratio of the number of turns on the windings, or

$$\frac{i_1}{i_2} = -\frac{N_2}{N_1}. \tag{18.47}$$

The directions of the currents are opposite to each other – one flows into the dot while the other flows out.

If the terminal variables are sinusoidal, we can divide (18.45) by (18.47) to illustrate the impedance-transforming property of a transformer. If $\widehat{V}_1, \widehat{V}_2, \ldots$ are the complex amplitudes of the sinusoidal terminal variables, then

$$-\frac{\widehat{V}_2}{\widehat{I}_2} = Z_2 = \left(\frac{N_2}{N_1}\right)^2 \frac{\widehat{V}_1}{\widehat{I}_1} = N^2 Z_1, \tag{18.48}$$

where N is the turns ratio N_2/N_1.

Equations (18.45) and (18.47) describe the behavior of an *ideal transformer*, which has the circuit symbol shown in Fig. 1.3(a). A real transformer differs from the ideal in three ways. First, the voltages are not exactly related by (18.45), because, owing to leakage, not all the flux linked by one winding is linked by the other. Second, finite permeability prevents the currents from being related by (18.47). A nonzero difference between them is necessary to create flux in the core. This difference is called the *magnetizing current*. Third, the relationships (18.45) and (18.47) for the ideal transformer are not functions of frequency; that is, an ideal transformer works at dc, but a real transformer does not. In spite of these differences, the ideal transformer is extremely useful in modeling real transformers, in the same way that the ideal current source is useful in modeling transistors.

18.4.2 Magnetizing Inductance

For two windings to be magnetically coupled, a B-field must be created by one and linked by the other. As previously noted, only in the hypothetical case of infinite permeability can B exist without H, and therefore without a net Ni. The closest we can come to achieving this ideal condition is to use a highly permeable core with no gap. In this case what we get from one winding, if the other is open-circuited, is simply a very large – but finite – inductance called the

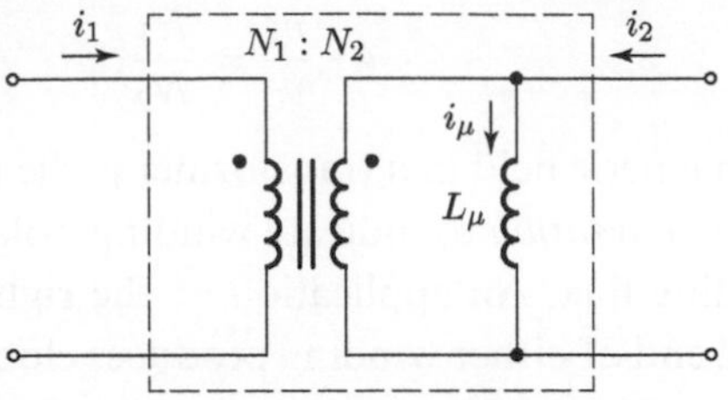

Figure 18.13 (a) A two-winding transformer with a magnetic core. (b) Model of the transformer of (a), assuming perfect coupling but finite magnetizing inductance L_μ.

magnetizing inductance. The magnetizing inductance will have one of two values, depending on which winding the inductance is being measured from, and these values are related by the turns ratio squared.

Figure 18.13 is a model for a two-winding transformer with perfect coupling but a finite magnetizing inductance. In this model we made use of the ideal transformer. The magnetizing inductance L_μ is in shunt with the ideal transformer. We can place L_μ on either side of the ideal transformer, so long as its value is correct for that side.

The magnetizing inductor current i_μ is known as the *magnetizing current*. It is the current that prevents the Ni products of the real transformer's windings from summing to zero. The presence of the magnetizing inductance is the reason that a real transformer will not work with dc. (The magnetizing current would be infinite.)

Example 18.8 Simple Use of the Transformer Model

Figure 18.14 shows the model of a transformer with a magnetizing inductance L_μ, having N_1 primary and N_2 secondary turns, connecting a sinusoidal voltage source of amplitude V_1 to a resistor of value R. The core has a cross-sectional area A_c and a saturation flux density B_s. We want to know the primary current i_1 and how low ω can go before the core saturates.

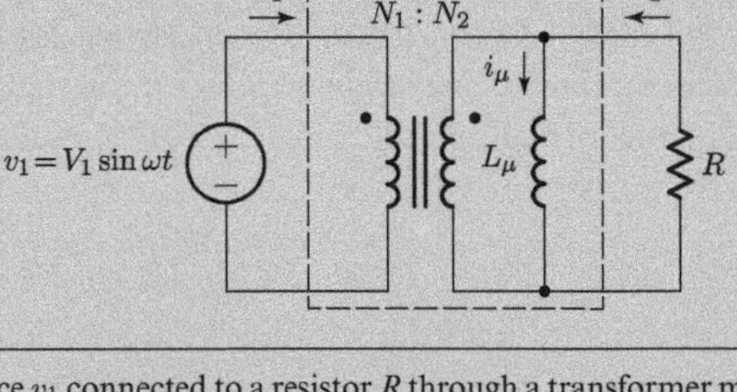

Figure 18.14 A voltage source v_1 connected to a resistor R through a transformer model including a magnetizing inductor L_μ.

The current i_1 is simply the source voltage divided by the secondary impedance transformed to the primary. That is,

$$i_1 = \mathrm{Im}\left[V_1 e^{j\omega t} N^2 \frac{R + j\omega L_\mu}{R(j\omega L_\mu)}\right] = \frac{N^2 V_1 \sqrt{R^2 + (\omega L_\mu)^2}}{R\omega L_\mu} \sin\left[\omega t - \tan^{-1}\left(\frac{R}{\omega L_\mu}\right)\right]. \tag{18.49}$$

We determine the lowest frequency at which this transformer can operate without saturating the core by equating the two expressions for λ_s in (18.13) and (18.8) and assuming that $\lambda(0) = -\lambda_s$. (This assumption ignores the start-up problem.) The result is

$$\lambda_s = N_X A_c B_s = -\lambda_s + \frac{1}{\omega_{\min}} \int_0^{\pi} V_X \sin \omega t \, d(\omega t) = \frac{V_X}{\omega_{\min}}, \tag{18.50}$$

$$\omega_{\min} = \frac{V_1}{N_1 A_c B_s}. \tag{18.51}$$

Note that we do not specify N_X as the primary or secondary turns, or V_X as the primary or secondary voltage, in (18.50). Only when we determine the side on which we will measure the winding voltage V_X do we specify N_X, and obtain (18.51).

18.4.3 Leakage Inductance

Not all the magnetic flux created by one winding of a two-winding transformer follows the magnetic circuit and links the other winding. Some flux escapes the core and returns through the air. Further, all the flux generated by a winding does not enter the core, but follows a path in the air space between the winding and core, or between layers of the winding. The result is that each winding produces some flux that is not linked by the other, causing imperfect coupling. This effect is modeled by series *leakage inductances*, which are shown as L_{ℓ_1} and L_{ℓ_2} in Fig. 18.15. The transformer winding voltages are no longer related simply by the turns ratio, because we must now subtract the drop across L_ℓ from the terminal voltages to get the ideal transformer winding voltages.

Leakage is more important in transformers than inductors. Leakage flux in an inductor links some of the winding, but does not contribute to flux in the core, resulting in an inductance greater than that calculated assuming all flux is constrained to the core. This is only a problem when a specific value of inductance is necessary – for a resonant circuit, perhaps. Whereas it interferes with the basic operation of a transformer. Understanding the origin of leakage allows you to design transformers more effectively, minimizing this parameter.

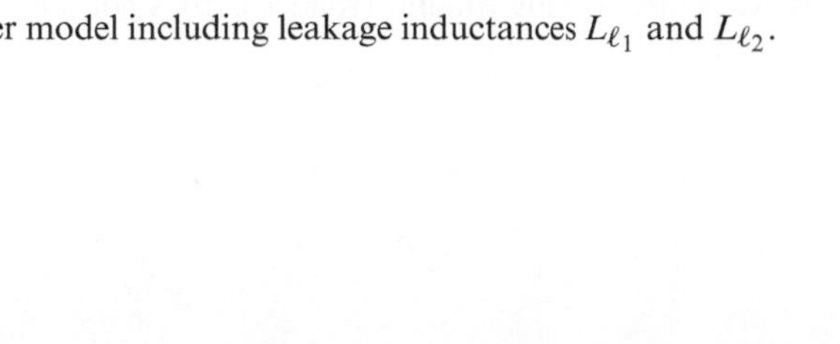

Figure 18.15 Transformer model including leakage inductances L_{ℓ_1} and L_{ℓ_2}.

There are several ways of designing a transformer to minimize leakage inductance. The first is to place one winding over the other. This configuration ensures that all the flux *in the core* generated by the inner winding is linked by the outer winding, eliminating leakage caused by the inner winding. However, some flux generated by current in the outer winding is not linked by the inner one, so the total leakage is not zero. The second approach recognizes that the leakage flux increases with the winding thickness, or *build-up*, suggesting that long windings with little build-up generally yield lower values of leakage than short, fat windings. A third technique is to use litz wire, or a conventionally twisted bundle of insulated wires, to make the winding and then connect the strands within the bundle in series or parallel to form the primary and secondary windings. This approach is theoretically no better than overlapping the windings, but practically can yield a somewhat higher copper packing factor. The use of thin, wide-foil conductors results in the lowest achievable value of leakage, because no space is wasted between conductors and the packing factor can be very high. As we mentioned in Section 18.3.1, the problem with this technique is that manufacturing the winding is more difficult than if it were made of wire.

Because the leakage flux in transformers and inductors occupies space external to the core, it can cause electromagnetic interference (EMI) in both its own circuit and external equipment. One technique used to limit the influence of these leakage fields is to place a shorted one-turn wide-foil winding, called a *shield* or *bucking* winding, around the entire magnetic circuit, including the windings, as shown in Fig. 18.16(a). Leakage flux induces currents in this winding, which in turn create bucking flux to reduce the fields outside the structure. The net magnetizing flux enclosed by this winding is approximately zero, so the winding has no effect on the magnetizing inductance or the coupling between windings. Another technique is to use what is known as a *pot core*. This is a shell-like core composed of two half-core pieces that completely encloses the winding, as shown in Fig. 18.16(b). The windings are placed around the centerpost (usually on a removable plastic bobbin) and the two core pieces are held in place by a screw or clamp.

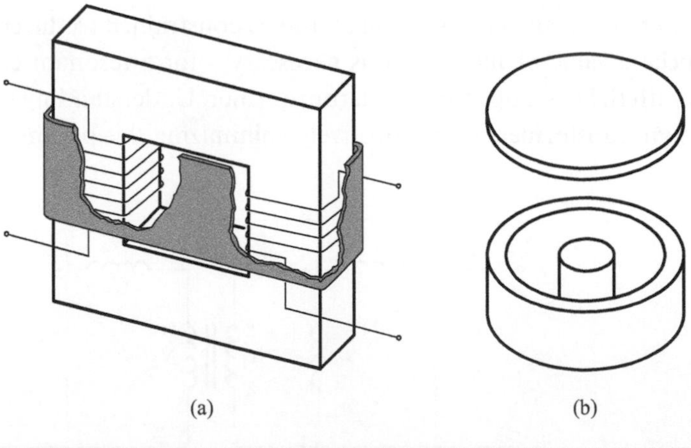

Figure 18.16 Two construction techniques for reducing EMI. (a) A transformer structure with a shorted one-turn leakage flux bucking winding (shaded). (b) A pot-core structure, in which the windings are placed around the centerpost.

18.4.4 Equations for a Non-Ideal Transformer

The equations describing the relationships among the terminal voltages and currents of the two-winding transformer model of Fig. 18.15 have the same form as those for any linear, time-independent two-port inductive circuit:

$$v_1 = L_1 \frac{di_1}{dt} + L_{12} \frac{di_2}{dt}, \tag{18.52}$$

$$v_2 = L_{21} \frac{di_1}{dt} + L_2 \frac{di_2}{dt}. \tag{18.53}$$

As the transformer is reciprocal, $L_{12} = L_{21} \equiv L_m$. By making the appropriate open- and short-circuit tests on the circuit of Fig. 18.15, we can determine the coefficients of these equations. The results are

$$L_m = L_\mu \left(\frac{N_2}{N_1}\right), \qquad L_1 = L_{\ell_1} + L_m \left(\frac{N_1}{N_2}\right), \qquad L_2 = L_{\ell_2} + L_m \left(\frac{N_2}{N_1}\right). \tag{18.54}$$

In these equations, L_μ is the magnetizing inductance referred to the N_1 side, L_m is the *mutual inductance* between terminal pairs 1 and 2, and L_1 and L_2 are the *self inductances* of terminal pairs 1 and 2, respectively. For a transformer constructed as in Fig. 18.12, and modeled with L_μ referred to the N_1 side, as in Fig. 18.15, the parameters L_μ and L_m are

$$L_\mu = \frac{N_1^2 A_c \mu_c}{\ell_c}, \qquad L_m = \frac{N_1 N_2 A_c \mu_c}{\ell_c}. \tag{18.55}$$

Example 18.9 Extracting Parameters from Transformer Measurements

The following measurements are made at the terminals of a transformer having a primary-to-secondary turns ratio of 5:

1. A 100 kHz voltage with an amplitude of 25 V is applied to the primary. The open-circuited secondary voltage is measured and found to be 4.59 V.
2. An impedance analyzer is used to measure the inductance of the primary at 100 kHz, with the secondary open-circuited. The inductance is recorded as 173.3 μH.
3. An impedance analyzer is used to measure the inductance of the secondary at 100 kHz, with the primary open-circuited. The inductance is recorded as 14.36 μH.

What are the parameters of this transformer if we use the model of Fig. 18.15?

To simplify our analysis, we reflect all secondary parameters and variables to the primary of the transformer, and identify them by using a prime (for example, $L'_{\ell_2} = 5^2 L_{\ell_2} = 25 L_{\ell_2}$). For the conditions of the first measurement, we find the voltage across L_μ, which is the primary voltage reduced by the reactive divider consisting of X_{ℓ_1} and X_μ (or L_{ℓ_1} and L_μ), so

$$V'_2 = \left(\frac{L_\mu}{L_{\ell_1} + L_\mu}\right) V_1, \tag{18.56}$$

and

$$\frac{L_\mu}{L_{\ell_1} + L_\mu} = \frac{5 \times 4.59}{25} = 0.918.$$

In this equation we have used the value of L_μ as seen from the primary. From measurement 2 we know the total inductance measured at the primary with the secondary open, namely $L_{\ell_1} + L_\mu$, so we infer that $L_\mu = 159$ μH and $L_{\ell_1} = 14.3$ μH.

We obtain another relationship among the parameters from measurement 3. Reflecting the magnetizing inductance L_μ to the secondary, we find a secondary-referred magnetizing inductance of 6.36 μH, making the secondary-side leakage inductance $L_{\ell_2} = 14.36 - 6.36 = 8.00\ \mu\text{H}$.

The final model, with these calculated parameters, is shown in Fig. 18.17.

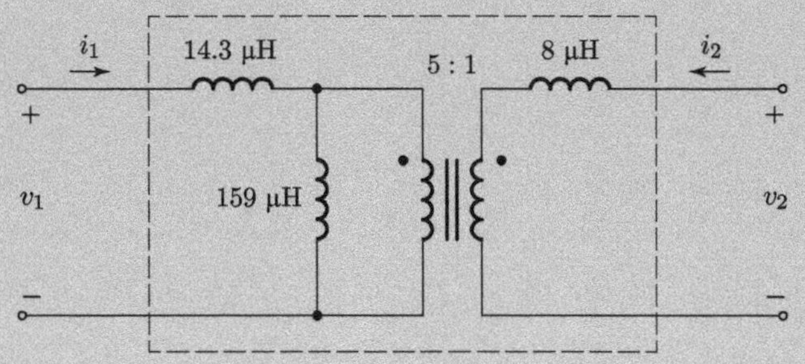

Figure 18.17 Transformer model with its calculated parameters.

18.4.5 Energy Storage Transformers

Most transformers are designed to maximize the magnetizing inductance, because this inductance seldom plays a role in the functioning of the circuit. In some types of power electronic circuits, however, the magnetizing inductance of the transformer does play an important role and is required to store appreciable energy. This is the case with the transformer-coupled flyback converter discussed in Chapter 7, and also with resonant converters such as the LLC converter in Chapter 10, in which the transformer magnetizing inductance is used as part of the resonating tank circuit.

When a transformer is used to store energy, the core must either be made of a low-permeability material or contain a gap. The reason is that the stored energy density in highly permeable material is very low (as discussed in Section 18.1.4).

18.4.6 Current and Potential Transformers

The *current transformer* (CT) and *potential transformer* (PT) are generally used in instrumentation to sense a current or voltage, respectively. In some cases their secondary signals also power the sensing circuit, which usually consists of low-current gates or operational amplifiers.

Ideally, the primary and secondary currents in a CT are related by the inverse of the turns ratio. Practically, they differ by the size of the magnetizing current, which is at a minimum when the magnetizing inductance is at a maximum. A CT is therefore usually made from an uncut core of high-permeability material, such as the Mo alloys. To minimize the size of the transformer, the secondary is usually terminated in a very low impedance, keeping the volt-seconds applied to the core as small as possible. The secondary termination is known as the CT *burden*.

The PT ideally relates the primary and secondary terminal voltages by the turns ratio. In practice, the leakage inductances cause drops that prevent the terminal voltage from being related exactly by the turns ratio, so the secondary of a PT should be terminated in a high impedance. Magnetizing current is generally of little consequence, unless it becomes large enough to affect circuit operation. The techniques used to reduce leakage (discussed in Section 18.4.3) can be effectively employed in the design of PTs.

Example 18.10 Analysis of a Current Transformer

A CT constructed of a tape-wound toroid core is designed to measure a sinusoidal current having a maximum amplitude of 5 A. The primary consists of a single turn, and the secondary is 500 turns of AWG 34. The circuit in which the transformer is used is shown in Fig. 18.18. The operational amplifier (assumed to be ideal) presents an incremental short-circuit to the transformer secondary. The relationship between the primary current i_1 and the output voltage v_o is

$$v_o = -1000/500 \cdot i_1 = -2i_1.$$

The parameters of the core are also shown in Fig. 18.18. We are interested in determining the minimum frequency at which the transformer will function without saturating.

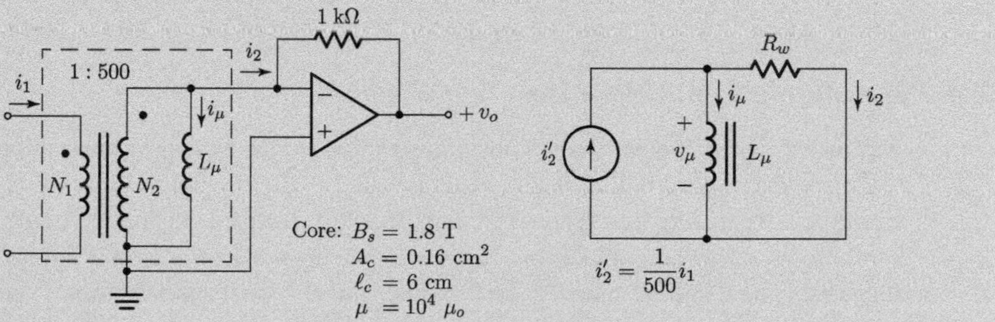

Figure 18.18 Current transformer, with an operational amplifier used as its burden.

Because the operational amplifier is effectively a short-circuit to the CT secondary, the voltage at the magnetizing inductance results only from the drop across the secondary impedance (winding resistance and secondary leakage). For simplicity, we assume that the secondary impedance is dominated by winding resistance, shown as R_w in Fig. 18.18. The resistance of AWG 34 wire is 8.56 mΩ/cm. Assuming a mean length/turn of 2 cm gives $R_w \approx 8.56\,\Omega$. We also assume that L_μ is large enough to make the magnetizing current negligible until saturation occurs.

The maximum permissible λ, calculated on the secondary side using (18.50), is

$$\lambda_{\max} = N_2 A_c B_s = 500(1.6 \times 10^{-5})(1.8) = 1.44 \times 10^{-2}\text{ V-s.} \tag{18.57}$$

The peak V_μ of the sinusoidal voltage v_μ across L_μ is

$$V_\mu = R_w I_2 = 8.56(0.01) = 0.086\text{ V.} \tag{18.58}$$

Equating the integral of v_μ to $\lambda_{\max}$ gives

$$\begin{aligned} \frac{1}{2}\int_0^{\pi/\omega} 0.086 \sin\omega t\, dt &\le 1.44 \times 10^{-2}\text{ V-s} \\ \omega &\ge 6\text{ rad/s} = 0.95\text{ Hz.} \end{aligned} \tag{18.59}$$

We now check our assumption that the magnetizing current is negligible. The parameters of the CT result in a magnetizing inductance of 0.84 H as measured from the 500-turn secondary. With a secondary current $i_2 = 10\,\text{mA}$, the peak voltage V_μ across the magnetizing inductance is 0.086 V. This voltage gives a magnetizing current I_μ at 1.9 Hz of

$$I_\mu = \frac{0.086}{(0.84)(11.9)} = 8.6\,\text{mA}.$$

This magnetizing current produces an error of almost 100% in the measured primary current – an unacceptably high error. If we desire an error of less than 3%, the magnetizing current must be less than 0.3 mA, requiring a minimum frequency (ω_{min}) such that

$$3 \times 10^{-4} \le \frac{0.086}{0.84\,\omega_{\text{min}}}. \tag{18.60}$$

This constraint results in

$$\omega_{\text{min}} \ge 341\,\text{rad/s} = 54.3\,\text{Hz}. \tag{18.61}$$

In this example, then, the error constraint is more severe than the constraint of avoiding saturation. Thus the minimum permissible frequency is approximately 54 Hz.

18.4.7 Transformers with More Than Two Windings

A transformer with more than two windings is needed for many power circuit topologies. While there are many possibilities in the general case, for the three-winding case there are two very common forms. The first, shown in Fig. 18.19(a), forms a magnetic circuit in which all the windings link the same flux. This is the *series approach*, as the circuit analog in Fig. 18.19(b) suggests. (The windings need not be separate, and may be wound on top of one another.) If the transformer is ideal, the winding voltages are related by the corresponding turns ratios, as for a two-winding transformer. However, the currents are not necessarily related by the inverses of the

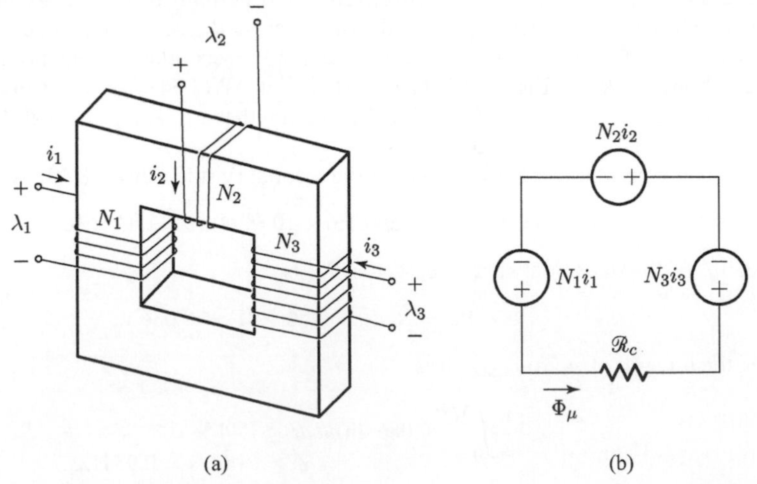

Figure 18.19 (a) Structure of a three-winding transformer with a series magnetic circuit. (b) Circuit analog for the transformer of (a).

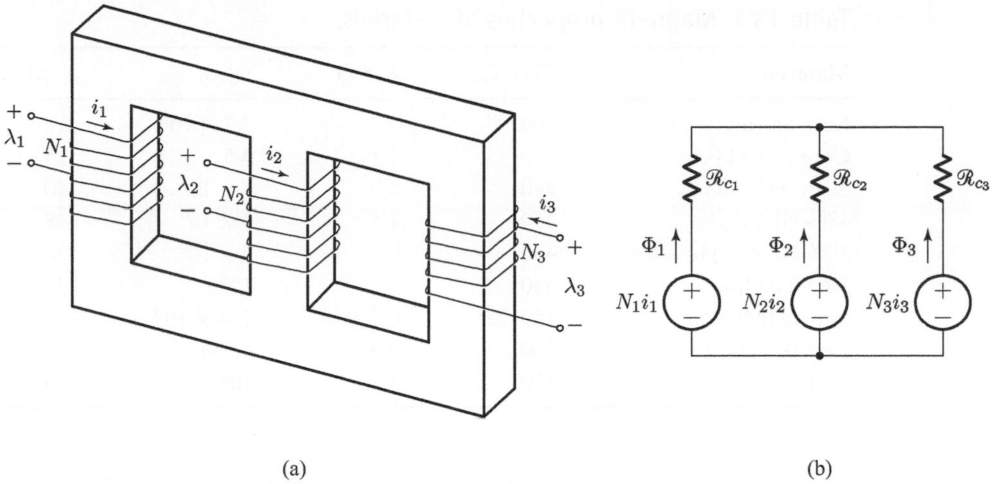

Figure 18.20 (a) Structure of a three-winding transformer with a parallel magnetic circuit. (b) Circuit analog for the transformer of (a).

turns ratios. The only constraint on the winding currents is that the Ni products of the windings sum to Φ_μ, which is approximately zero in a transformer not designed to store energy.

Figure 18.20 shows another possible structure for the magnetic circuit of a three-winding transformer. All the individual winding fluxes (and hence their flux linkages and voltages) can be different for this core and winding configuration, but they must sum to zero. This is called the *parallel approach*. If the transformer is ideal, the winding currents are related pairwise by the inverses of their corresponding turns ratios, and the v/N quotients of the windings sum to zero. The dot convention relating winding voltages, however, cannot be used for this structure. The dot convention holds only for windings which share the same flux path – the three windings in Fig. 18.19, for example.

18.5 Magnetic Material Properties

The Curie temperature T_C, saturation flux density B_s, relative permeability μ, and resistivity ρ of some common magnetic materials are listed in Table 18.1. The various alloys of iron are relatively standard. Different material manufacturers or distributors will give them different names, but in general every manufacturer of such materials has products that match those in the table.

The low-silicon irons (Si $\leq 1\%$) are relatively high-loss, high-saturation flux density materials, used primarily at line frequencies for power transformers, motors, and relays. The higher-silicon-content, generally grain-oriented, steels exhibit lower loss than low-silicon steels but are more expensive. They are used in high-performance line-frequency applications, such as high-efficiency motors or toroidal power transformers. Specialized nickel alloys have a saturation flux density approximately 25% lower than that of silicon alloys, but have substantially higher resistivities, which makes them well suited for high-frequency applications. The permeabilities

Table 18.1 Magnetic properties of materials.

Material	T_C (°C)	B_s (T)	μ/μ_0	ρ (μΩ-cm)
Low-Si iron (0.25%)	760	2.2	2.7×10^3	10
Core iron (1%)	810	2.1	4.5×10^3	25
Si steel (2.5%)	780	2.0	5×10^3	40
48% Ni alloy	450	1.5	4×10^4	48
80% Ni, 4% Mo alloy	460	0.8	5×10^4	58
50% Co alloy	950	2.3	10^4	35
Ferrite (Mn–Zn)	150–225	0.4–0.8	$1\text{–}4 \times 10^3$	—
Ferrite (Ni–Zn)	300	0.3	4–500	—
Metallic glass	370	1.6	10^4	125

of these materials are the highest of any soft magnetic material. Cobalt alloys have the highest saturation flux density of any available material. Cobalt and nickel alloys are very expensive, relative to silicon alloys. Ferrite is a ceramic made of various combinations of primarily ferrous oxide and manganese or nickel, and zinc. A large number of proprietary formulations exist, but in general the MnZn ferrites exhibit higher permeabilities and saturation flux densities than do the NiZn ferrites. The MnZn materials can be used in power applications up to a frequency of approximately 1 MHz. The NiZn materials can be used at frequencies to beyond 30 MHz, depending upon the details of their formulation. Amorphous metal, or metallic glass, material has excellent magnetic properties but is available only as ribbon having a thickness of about 0.05 mm. It is very brittle and difficult to work, but is available as a toroid or in a cut-core form (two U-shaped pieces).

Notes and Bibliography

One of the best and most comprehensive treatments of magnetic circuits is still [1]. It provides insightful representations of magnetic-circuit models and behavior, and contains a very thorough treatment of the design of inductors. Unfortunately this book is out of print, but most technical libraries have a copy. Be attentive to the units.

A wealth of practical information about the design of magnetic components is contained in [2] and [3]. McLyman's book is especially useful for the design of magnetics in high-frequency converters. It contains an array of extraordinarily useful tables, graphs, and charts illustrating material properties, design rules, units, conversion factors, and more.

The physics, processing, manufacturing, and properties of ferrites are well covered in [4], while [5] describes the evaluation and performance of a wide range of commercial NiZn ferrite materials in the HF frequency range of 3–30 MHz.

Methods for measuring core loss can be found in [5], [6], and [7]. Han et al. [7] address measurement techniques and material data for frequencies up to 100 MHz.

Sullivan [8] discusses the use of aluminum in place of copper for high-frequency transformer windings. Pollock et al. [9] describe the publicly available tool LitzOpt developed by Sullivan. The tool combines two-dimensional field simulations with an eddy current loss model to estimate winding loss, including proximity effects owing to the gap fringing fields, and field build-up across the core window. Confirmation of loss predictions in Example 18.7, including ac effects, were made using LitzOpt.

In [10], Steinmetz proposed the form of the Steinmetz equation, (18.39), but frequency was absent because he limited it to 200 Hz. Nevertheless, he is credited with the eponymous extended formula. An interesting history of the early studies of core loss may be found in [11].

1. MIT EE staff, *Magnetic Circuits and Transformers*, Cambridge, MA: MIT Press, 1943.
2. W. T. McLyman, *Transformer and Inductor Design Handbook*, 4th edn., Boca Raton, FL: CRC Press, 2011.
3. S. Smith, *Magnetic Components, Design, and Applications*, New York: Van Nostrand, 1985.
4. E. C. Snelling, *Soft Ferrites: Properties and Applications*, London: Iliffe Books, 1969.
5. A. J. Hanson, J. A. Belk, S. Lim, C. R. Sullivan, and D. J. Perreault, "Measurements and Performance Factor Comparisons of Magnetic Materials at High Frequency," *IEEE Trans. Power Electronics*, vol. 31, pp. 7909–7925, Nov. 2016.
6. "IEEE standard for test procedures for magnetic cores," IEEE Standard 393–1991, Mar. 1992.
7. Y. Han, A. Li, G. Cheung, C. R. Sullivan, and D. J. Perreault, "Evaluation of Magnetic Materials for Very High Frequency Power Applications," *IEEE Trans. Power Electronics*, vol. 27, pp. 425–435, Jan. 2012.
8. C. R. Sullivan, "Aluminum Windings and Other Strategies for High-Frequency Magnetics Design in an Era of High Copper and Energy Costs," *IEEE Applied Power Electronics Conference (APEC)*, pp. 78–84, Anaheim, 2007.
9. J. D. Pollock, T. Abdallah, and C. R. Sullivan, "Easy-to-Use CAD Tools for Litz-Wire Winding Optimization," *IEEE Applied Power Electronics Conference (APEC)*, pp. 1157–1163, Miami Beach, Feb. 2003.
10. C. P. Steinmetz, "On the Law of Hysteresis," *Trans. AIEE*, vol. 9, pp. 621–758, 1892.
11. R. R. Kline, *Steinmetz: Engineer and Socialist*, Baltimore, MD: Johns Hopkins University Press, 1992.

PROBLEMS

18.1 What is the inductance of the structure in Fig. 18.2 if the winding is placed on an outer leg?

18.2 Repeat Example 18.6 by expressing the peak stored energy in terms of the fields in the gap.

18.3 If the filter inductor is the only cause of loss in the converter referred to in Fig 18.10, what is the efficiency of the converter, based on the model derived in Example 18.7?

18.4 Redo the calculation of minimum frequency in Example 18.8 by using the secondary-side voltage, and show that the answer is the same as that obtained using the primary voltage.

18.5 Consider the circuit of Fig. 18.10 when the converter is first turned on. If the capacitor and load have values such that the peaks of i_L at the end of each switching cycle can be approximated as rising *linearly* to their final values in 10 switching cycles (2 ms), plot the flux in the inductor as a function of time, assuming that $B_r = 0$.

18.6 Figure 18.21(a) shows a 12-pulse rectifier made by connecting a pair of 6-pulse rectifiers in parallel through an *interphase transformer*. The three-phase ac sources feeding the six-pulse rectifiers are phase-shifted by 30° with respect to each other.

(a) Assume that the interphase transformer is ideal, and explain how it functions to maintain continuous current in the two six-pulse bridges.

(b) In terms of the peak ac line voltage V_s and ω, what maximum number of volt-seconds (λ) is the interphase transformer core required to support?

(c) In practice, the interphase transformer does not have an infinite magnetizing inductance. An appropriate model for the transformer with finite magnetizing inductance, L_μ, is shown in Fig. 18.21(b). Where does the magnetizing current i_μ come from, and how does it affect circuit operation?

(d) If the peak of the line–line, 60 Hz, ac voltage is 680 V, what minimum value of L_μ allows continuous conduction of the six-pulse bridges down to a dc load current of 10 A?

18.7 The following specifications are for an inductor needed to model the main field magnet of MIT's magnetically confined fusion research machine, Alcator-C: $L = 350$ H, $R = 133\ \Omega$, and $I_{max} = 1.39$ A. You are to design an inductor to meet these specifications.

(a) If a maximum flux density of 1.2 T is assumed, determine the required *volume* of the air gap?

(b) What is the 60 Hz Q of this inductor?

(c) Obtain a suitable design by assuming: (i) negligible fringing at the air gap; (ii) negligible leakage inductance; (iii) a winding packing factor, k_u, defined as the ratio of copper area to total window area, of 0.5; and (iv) a current density of no more than 2000 A/cm^2.

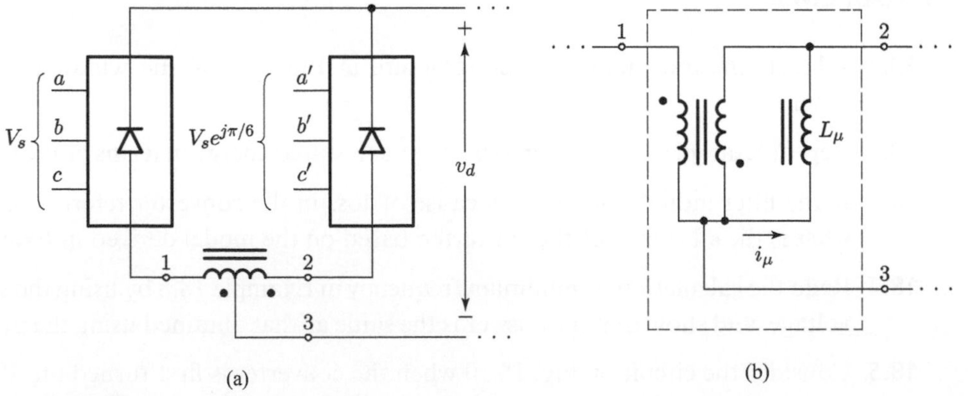

Figure 18.21 (a) The 12-pulse rectifier made by connecting two 6-pulse bridges in parallel through an interphase transformer. (b) The model of the interphase transformer with finite magnetizing inductance for Problem 18.6.

(d) What is the weight of your design? Use the internet to obtain the physical and electric characteristics of copper wire. You may specify any core size, but your design should minimize the total weight of copper and iron.

18.8 Design an inductor for a 100 W, 500 kHz boost converter operating in continuous conduction mode from a nominal input voltage of 12 V into an output voltage of 36 V. The desired inductance is 3 μH, and it is specified that the inductor be implemented with Ferroxcube 3F3 ferrite material in an RM core. Data for the RM10 core size is provided in Example 18.6. The RM12 core size has the following parameters: a minimum core cross-sectional area $A_{c,\min} = 121\,\text{mm}^2$ (for determining saturation), an effective core cross-sectional area $A_c = 140\,\text{mm}^2$ (for determining inductance), an effective magnetic path length of $\ell_c = 56.9\,\text{mm}$, an effective core volume $V_{c,e} = 7.97\,\text{cm}^3$, a mean length of turn ℓ_t of 61 mm, and a thermal resistance (hot-spot temperature rise per W) of $R_{th} = 23\,°\text{C/W}$. You may assume that the core can be custom gapped to a desired value. For this design you may assume a maximum allowable flux density of 0.3 T and a core material relative permeability μ_r of 2000. Assume a maximum allowable current density in the windings of 500 A/cm^2. (The specified windings must fit within the winding area of the core, and preferably on a winding bobbin, also known as a "coil former." The RM12 bobbin has a winding window area of $W_{A,B} = 73.0\,\text{mm}^2$.)

(a) Calculate the ripple ratio and peak inductor current for operation at full load (100 W).

(b) Design an inductor that meets the required specification. For simplicity, you may neglect inductor core and winding losses and inductor temperature rise. You should provide key information regarding your design (e.g., core, gap value, number of turns, wire gauge) along with calculations necessary for evaluating your design (e.g., peak core flux density, current density, etc.).

(c) Make an approximate calculation of inductor loss. To do so, compute the losses in the inductor as the sum of winding and core losses. Winding power loss may be (crudely) approximated as the dc winding resistance times the rms inductor current squared. (In more sophisticated calculations, skin effect may be considered in the windings. Proximity effect loss in the windings might also be included for the highest accuracy.) Core loss in a 3F3 material core can be computed by approximating the ac flux in the core as sinusoidal, and calculating the core loss as

$$P_c = 3.6 \times 10^{-9} \cdot f_{sw}^{2.4} \cdot B_{pk}^{2.25} \cdot V_c, \tag{18.62}$$

where P_c is the core loss in mW, f_{sw} is the switching frequency in Hz, B_{pk} is the peak ac flux swing in T, and V_c is the volume of the core in cm^3. One may estimate the expected "hot-spot" temperature rise of the core as the total dissipation times the core thermal resistance R_{th}.

18.9 A three-winding transformer is constructed on a three-legged core, as shown in Fig. 18.22(a). The cross-sectional area of the core is uniform, and the windings have N_1, N_2, and N_3 turns. Winding N_3 is terminated in a resistor of value R. If sinusoidal

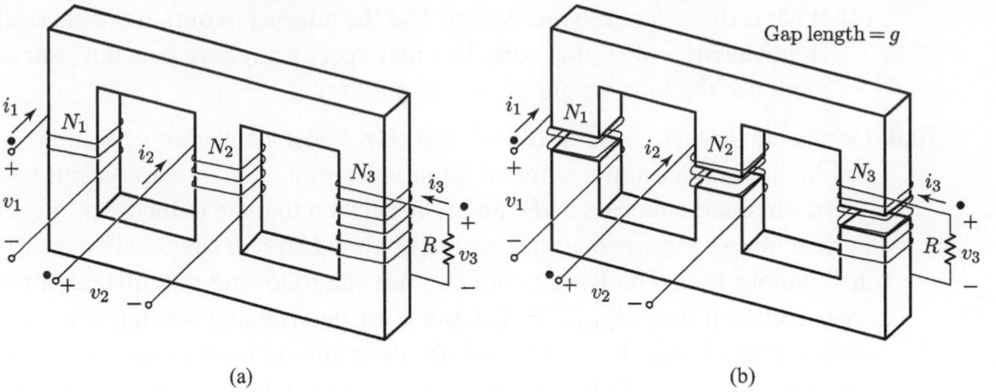

Figure 18.22 (a) The three-legged, three-winding transformer analyzed in Problem 18.9. The core has a uniform cross section. (b) The transformer of (a) with gaps of the same dimensions placed in each of the three legs.

voltages with amplitudes V_1 and V_2 are applied to windings N_1 and N_2, respectively, what is the amplitude I_3 of the current i_3 in winding N_3? Assume that no leakage occurs.

18.10 The three legs of the transformer of Problem 18.9 are now modified to have uniform gaps as shown in Fig. 18.22(b). Assuming that the permeability of the core is infinite, determine the amplitude of the currents i_1 and i_2 in terms of V_1, V_2, N_1, N_2, and ω. (*Hint*: Use a circuit analog and make sure that your answer does not depend on the core cross section.)

18.11 Relate the reluctances, $\mathscr{R}_{c_1}$, $\mathscr{R}_{c_2}$, and $\mathscr{R}_{c_3}$ to their physical origins in the three-winding transformer with parallel magnetic circuit of Fig. 18.20. That is, identify the segments of the core corresponding to each of the lumped reluctances.

18.12 The three-winding transformers of Figs. 18.19 and 18.20 are connected identically in the following manner. A sinusoidal voltage source of amplitude V_1 is connected to winding N_1, and resistors of value R_2 and R_3 are connected to windings N_2 and N_3, respectively. Determine and compare the resulting winding voltages and currents for each of these two core configurations.

18.13 Figure 18.23 shows the structure, idealized B–H curve, and reluctance model for a permanent magnet. Parameters of permanent magnets include the residual induction B_r, which represents the maximum flux density a magnet can produce under "magnetic short-circuit" conditions, and the coercive force H_c, which represents the applied external magnetic field amplitude at which no net flux density results (and is equivalent to the net internal MMF per length of the magnet). Assume a Nd–Fe–B magnet having H_c = 920 kA/m, B_r = 1.25 T and a linearly transitioning B–H characteristic as shown in Fig. 18.23. Consider the characteristics of such a magnet of 1 mm length l_m (and magnetized in this direction) and 9 mm^2 cross-sectional area.

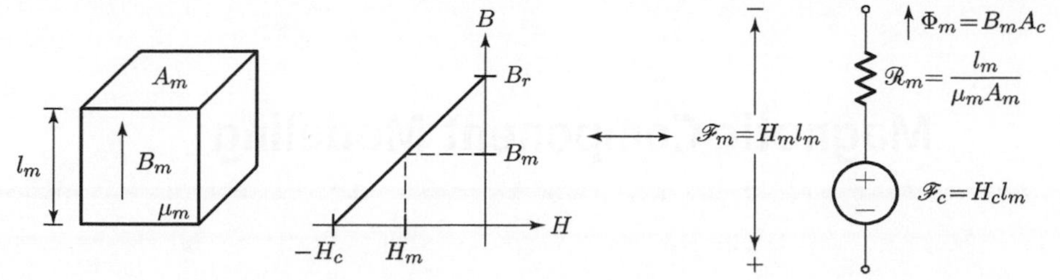

Figure 18.23 Magnet structure, idealized B–H curve, and reluctance model for a permanent magnet as considered in Problem 18.13.

(a) Find the reluctance model for this magnetic element, including the parameter values for MMF $\mathscr{F}_c$ and reluctance $\mathscr{R}_m$.

(b) The magnet will be placed in a single-loop magnetic circuit with a cross-sectional area of $9\,\text{mm}^2$. What should be the reluctance of the magnetic path provided by the rest of the magnetic circuit (i.e., that "loads" the permanent magnet) if one desires an equivalent flux through the circuit of 0.5 T?

18.14 Explain why the dot convention cannot be used to describe the relationship of the voltages in the magnetic structure of Fig. 18.20.

19 Magnetic Component Modeling

In this chapter we describe methods used to model magnetic components in both the electrical and magnetic domains. Magnetic components such as inductors, coupled inductors, and transformers lie at the heart of most power electronic circuits. Multi-port components especially can exhibit quite complex behavior, and advancing the performance of power electronic circuits often relies on understanding and leveraging this behavior. The various models developed here are valuable for expressing the behavior of these components, for designing them, and for representing them in circuit modeling and design. Since all figures in this chapter are of transformer models, we have dispensed with our convention, introduced in Chapter 1, of framing the transformer model with a dashed box to avoid ambiguity between the model and external elements.

19.1 Mathematical Representation: Inductance Matrix

Mathematically, a single-winding magnetic component – an inductor – can be represented by the relationship between the winding's flux linkage λ and its current i. For an ideal linear inductor this results in a simple relation $\lambda = Li$, where the inductance L is a positive constant. Differentiating this gives us the familiar relationship in terms of voltage, $v = L\,di/dt$. Introducing a second winding in proximity to the first expands this relationship to a pair of coupled equations:

$$\begin{bmatrix} \lambda_1 \\ \lambda_2 \end{bmatrix} = \begin{bmatrix} L_{11} & L_{12} \\ L_{21} & L_{22} \end{bmatrix} \begin{bmatrix} i_1 \\ i_2 \end{bmatrix}. \tag{19.1}$$

This is again $\boldsymbol{\lambda} = Li$, but now $\boldsymbol{\lambda}$ is a vector of flux linkages, i is a vector of currents, and L is an *inductance matrix*. Introducing P windings (and therefore P ports) expands the inductance matrix to $P \times P$ representing P coupled equations.

The inductance matrix description of the system captures the effects of coupling among windings. Owing to reciprocity, an inductance matrix must be symmetric. In the two-winding case, this means that $L_{12} = L_{21} = L_M$, where L_M is known as the mutual inductance of the two windings, and may be positive or negative, depending upon the coupling direction. Just as a single-winding inductance L must be positive to satisfy conservation of energy, L must be positive semidefinite. For the two-winding case this means that $|L_M| \leq \sqrt{L_{11}L_{22}}$. The coupling coefficient k is a number between 0 and 1, defined as $k = |L_M|/\sqrt{L_{11}L_{22}}$. A coupling coefficient of zero means that the windings link no common flux (so act as independent inductors), and a coupling coefficient of unity means that the same flux links both windings (so there is no leakage flux between windings). Because the inductance matrix must be symmetric, a P-winding device has $P(P+1)/2$ independent parameters: P parameters are the diagonal elements L_{ii} of the inductance matrix, known as self-inductances, and $P(P-1)/2$ parameters are the off-diagonal elements L_{ij}, known as mutual inductances. Thus, while a two-winding transformer has three independent parameters (L_{11}, L_{22}, L_M), a three-winding transformer has six, and a four-winding transformer has ten. This quadratic growth in complexity with the number of ports P can make modeling and analysis of multi-winding components challenging.

Example 19.1 Inductance Matrix and the Magnetic Circuit

The inductance matrix for a linear multi-winding magnetic structure can be determined directly from its magnetic circuit. Consider the transformer shown in Fig. 19.1(a), for example. Two windings having N_1 and N_2 turns respectively are magnetically coupled via a core having reluctance $\mathscr{R}_c$. In addition, each winding has a leakage flux path (e.g., partly through the air) that does not couple the other winding; we designate the reluctances of the two leakage paths as $\mathscr{R}_{\ell 1}$ and $\mathscr{R}_{\ell 2}$ respectively. The magnetic circuit for this system is shown in Fig. 19.1(b).

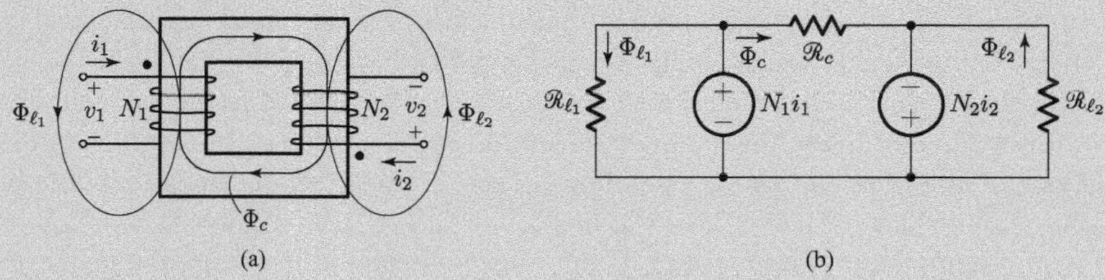

Figure 19.1 (a) A two-winding transformer showing magnetizing Φ_c and leakage $(\Phi_{\ell 1}, \Phi_{\ell 2})$ fluxes. (b) The corresponding magnetic circuit model.

The net fluxes through the MMF sources are most easily found by superposing the effects of the two sources separately. This readily shows that

$$\Phi_1 = N_1 i_1 \left(\frac{1}{R_c} + \frac{1}{R_{\ell 1}}\right) + \frac{N_2 i_2}{R_c}, \tag{19.2}$$

$$\Phi_2 = \frac{N_1 i_1}{R_c} + N_2 i_2 \left(\frac{1}{R_c} + \frac{1}{R_{\ell 2}}\right). \tag{19.3}$$

Recognizing that the flux linkage of a winding is the flux though the associated MMF source times its number of turns, we can find the flux linkages in terms of the winding currents as

$$\begin{bmatrix}\lambda_1\\ \lambda_2\end{bmatrix} = \begin{bmatrix} N_1^2\left(\frac{1}{\mathscr{R}_c} + \frac{1}{\mathscr{R}_{\ell 1}}\right) & \frac{N_1 N_2}{\mathscr{R}_c} \\ \frac{N_1 N_2}{\mathscr{R}_c} & N_2^2\left(\frac{1}{\mathscr{R}_c} + \frac{1}{\mathscr{R}_{\ell 2}}\right)\end{bmatrix}\begin{bmatrix} i_1 \\ i_2\end{bmatrix} = \begin{bmatrix} L_{11} & L_M \\ L_M & L_{22}\end{bmatrix}\begin{bmatrix} i_1 \\ i_2\end{bmatrix}. \tag{19.4}$$

As expected, the computed inductance matrix is symmetric, and its diagonal elements are positive. Its determinant is

$$N_1^2 N_2^2\left[\left(\frac{1}{\mathscr{R}_c} + \frac{1}{\mathscr{R}_{\ell 1}}\right)\left(\frac{1}{\mathscr{R}_c} + \frac{1}{\mathscr{R}_{\ell 2}}\right) - \frac{1}{\mathscr{R}_c^2}\right],$$

which is evidently nonzero as it must be, to ensure, for any physical non-negative values of N_1, N_2, $\mathscr{R}_c$, $\mathscr{R}_{\ell 1}$, and $\mathscr{R}_{\ell 2}$, that the inductance matrix is positive semidefinite. As can be seen, the magnetic circuit for a multi-winding magnetic component provides a means to calculate its inductance matrix. The elements of the inductance matrix for a magnetic component can also be determined from experiment or finite element simulation, based on

$$L_{11} = \left.\frac{\lambda_1}{i_1}\right|_{i_2=0}, \qquad L_{22} = \left.\frac{\lambda_2}{i_2}\right|_{i_1=0}, \qquad L_M = \left.\frac{\lambda_1}{i_2}\right|_{i_1=0} \text{ or } \left.\frac{\lambda_2}{i_1}\right|_{i_2=0}. \tag{19.5}$$

19.2 Circuit Representations for Two-Port Transformers

While the inductance matrix description is a useful mathematical representation of a magnetic component, it is not particularly convenient for circuit modeling and design. In this section we consider circuit models for magnetic components, focusing on the two-port case.

One circuit model for a two-winding transformer, shown here in Fig. 19.2, has already been introduced in the previous chapter. When the turns ratio $N_1 : N_2$ in the model is the *physical* turns ratio of the component, each of the inductances in the model has a physical meaning: the magnetizing inductance L_{μ_1} captures the energy stored in the flux common to the two windings; the primary-side leakage inductance L_{ℓ_1} captures energy stored in the flux coupling winding 1 but not winding 2; and the secondary-side leakage inductance L_{ℓ_2} captures energy stored in the flux coupling winding 2 but not winding 1. There is a further implicit assumption that all of the flux coupling any turn of a given winding couples *all* of the turns of that winding – that is, there is no "intra-winding" leakage flux. If this implicit assumption is violated, the above description becomes only approximate.

Considering the results of the previous chapter, and comparing with the magnetic circuit description in Example 19.1, we can find the model parameters as

$$L_{\mu_1} = \frac{N_1^2}{\mathscr{R}_c}, \qquad L_{\ell_1} = \frac{N_1^2}{\mathscr{R}_{\ell 1}}, \qquad L_{\ell_2} = \frac{N_2^2}{\mathscr{R}_{\ell_2}}. \tag{19.6}$$

The parameters of this model can also be directly related to the inductance matrix parameters as follows:

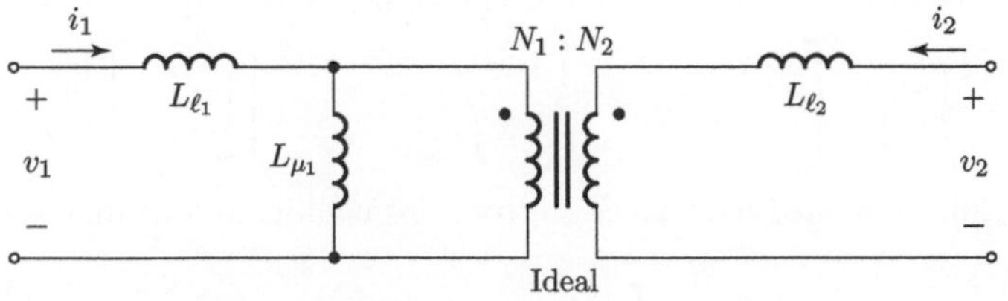

Figure 19.2 A circuit model for a two-winding transformer with magnetizing and leakage inductances.

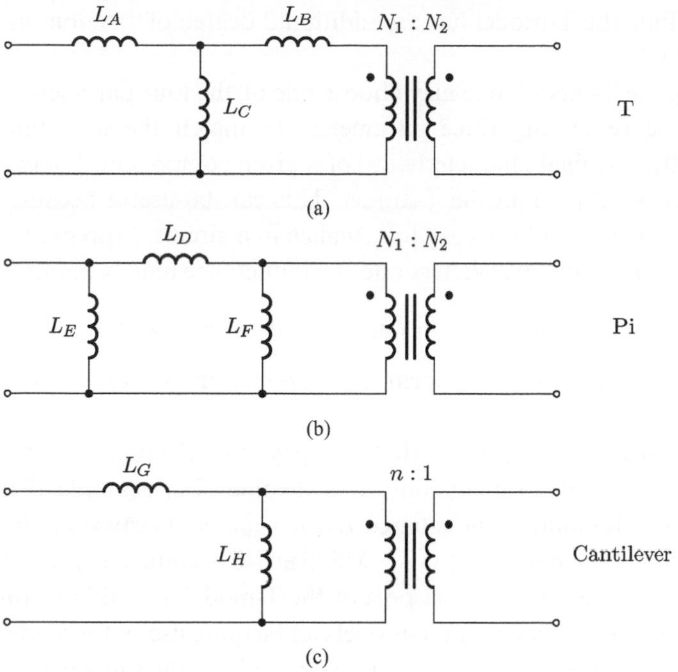

Figure 19.3 Circuit models for two-winding transformers: (a) T-model, (b) π-model, and (c) cantilever model. With appropriate parameters, these models provide terminal relations identical to those of the inductance matrix.

$$L_M = L_{\mu 1} \left(\frac{N_2}{N_1} \right), \tag{19.7}$$

$$L_{11} = L_{\ell_1} + L_{\mu 1}, \tag{19.8}$$

$$L_{22} = L_{\ell_2} + L_{\mu 1} \left(\frac{N_2}{N_1} \right)^2. \tag{19.9}$$

The circuit model of Fig. 19.2 is one version of the "T-model" of a two-winding transformer. The reason it is named the T-model becomes more clear when the secondary-side leakage is reflected to the primary side of the ideal transformer. This version of the T-model (with revised inductance labels) is shown in Fig. 19.3(a); the three inductances in the model form a T.

As introduced in the discussion of the inductance matrix, there are three parameters that define the terminal behavior of a (linear) two-winding magnetic component, namely L_{11}, L_{22}, and L_M. Within physical bounds these parameters may be thought of as three independent degrees of freedom in the possible behavior of a component. The T-model as shown in Fig. 19.3(a) actually has *four* free parameters that can be adjusted to match the characteristics of a given component: the three inductances (L_A, L_B, and L_C) and the turns ratio N_2/N_1. If the turns ratio N_2/N_1 in the T-model is selected as the physical turns ratio of the component, one has exactly the three remaining parameters necessary to match the three degrees of freedom in the inductance matrix. However, there is no requirement that the turns ratio in the T-model be set to the actual physical turns ratio, and in some cases the actual turns ratio may not be known. Thus, the T-model has an additional degree of freedom in modeling a given physical component.

In using the T-model, one may choose one of the four parameters for modeling convenience, and use the remaining three parameters to match the inductance matrix parameters (or, equivalently, terminal characteristics) of a given component. For example, one common choice is to select $N_2/N_1 = 1$ in the T-model. This can be useful because the action of a 1 : 1 ideal transformer is particularly easy to visualize in a circuit. Moreover, in this case the relationship between the T-model parameters and the inductance matrix parameters becomes quite simple:

$$L_A = L_{11} - L_M, \qquad L_B = L_{22} - L_M, \qquad L_C = L_M. \tag{19.10}$$

(Indeed, these relationships appear in the model simplification approach introduced in Section 19.4.4.)

It is important to emphasize that non-physical selection of one parameter can result in the other parameters also taking non-physical values. For example, if the turns ratio N_2/N_1 is set to unity and the mutual inductance L_M is negative because of the coupling direction of the windings, the T-model inductor L_C will likewise become negative (though the total inductance measured at any external circuit port of the T-model will still remain positive).[†] The additional degree of freedom in using the T-model can be quite useful for modeling and circuit design even though the parameters in the model no longer have the same meaning as when purely physical parameter selections are used.

While the T-model is the most widely used model for two-winding transformers, other highly useful models are also available. A second model – the π-model – is shown in Fig. 19.3(b). The π-model can be particularly attractive in cases where a shunt inductor across the input or output port of the transformer model is useful (e.g., where it simplifies analysis or can be absorbed as part of circuit operation.) The π-model can be directly synthesized from the T-model, or conversely, by using the classical Y–Δ transformation to convert the T (or Y) inductor network into a π (or Δ) network. This transformation has been known for more than a century in the context of three-phase power systems. As with the T-model, the π-model has four parameters, allowing one to be freely selected.

A third popular model – the cantilever model – is shown in Fig. 19.3(c). This model is so named because the arrangement of inductors in the model resembles a cantilever, though it could equally be called an L-model. Unlike the T- and π-models, the cantilever model provides exactly three parameters (L_G, L_H, and n), as needed to match the three degrees of freedom of the inductance matrix. The cantilever model can be understood as equivalent to the T-model with one inductance (L_B) set to zero, and equivalent to the π-model with one inductance (L_E) set to infinity. As described in a subsequent section, the cantilever model parameters can be directly related back to the parameters of the inductance matrix. This model is particularly suited for use in circuit design, as it only uses two inductors to model transformer behavior, making it easier to visualize circuit operation than using a model with three inductors. Nonetheless, since the parameters are all non-physical, it does not provide a good physical understanding for transformer design.

[†] While counterintuitive, some circuit designs take advantage of this "negative inductance" effect, see [2].

19.3 Determining Transformer Parameters

In Example 18.9 we outlined a process for finding the parameters of a two-winding transformer. In this section we provide an expanded treatment of transformer parameter determination, including discussion of some important measurement and computational issues to consider. As will be seen in subsequent sections, these issues can become even more critical when determining the parameters of many-winding transformers.

The parameters of a transformer (e.g., its magnetizing and leakage inductances, or its inductance matrix parameters) can be determined experimentally through measurements. To do so, we typically measure inductances, voltage gains, and other quantities at various ports of a device, with other ports under open- or short-circuit conditions (or with appropriate terminations), and calculate the model parameters from these measurements. We can also estimate transformer parameters from a detailed physical model of the transformer (e.g., a finite element model). For example, one can determine model parameters from the numerically simulated magnetic energy storage under different drive conditions.

Major challenges to accurately determining transformer parameter values include the required measurement accuracy, the numerical sensitivity of the required calculations, and – in some cases – the assumptions underlying the models themselves. Similarly, when using the T-model of Fig. 19.2 with a physical turns ratio, the presence of significant intra-winding leakage (in which the same flux does not link all turns of a winding) can cause apparently non-physical (e.g., negative) calculated leakage inductances. Numerical sensitivities become especially acute in cases where there are very large or very small magnetic couplings, which in some cases can yield non-physical parameter values. Numerical problems can also be especially challenging in the multi-winding case, as there are $P(P+1)/2$ parameters to be determined for a P-winding component. Lastly, factors such as transformer capacitances can sometimes adversely influence measurements, especially at high frequencies.

The principal approach to addressing the above issues is to select measurements that can be made with good accuracy, and calculations that – where possible – avoid the need to make sensitive calculations such as taking differences of large but almost equal numbers. Inductance measurement (e.g., using a vector impedance analyzer) and port-to-port voltage-gain measurements (made with calibrated probes) can usually be made with good accuracy, and are preferred. Current measurements at shorted ports are to be avoided if possible, especially at high frequencies, because current measurements are often insufficiently accurate, and imposing a sufficiently good short circuit is difficult.

A challenge in finding the parameters of highly coupled transformers is that the leakage inductances can be extremely small, compared to the magnetizing inductance. One approach to tackling this is extracting the T-model parameters of a 1 : 1 transformer (or of a transformer one chooses to model as having a 1 : 1 turns ratio) is to make an inductance measurement that avoids exciting the magnetizing inductance. Consider the model of Fig. 19.2 for the case of a 1 : 1 turns ratio. If we tie the negative terminal of the primary side to the negative terminal of the secondary side, and then measure the inductance at the port comprising the positive terminals of the two sides, we will bypass the magnetizing inductance to measure the total leakage inductance $L_{\ell_1} + L_{\ell_2}$. Knowing the sum of the leakage inductances is often useful, especially in the case of a symmetric structure in which the leakage inductances are equal. This same measurement is useful for finding the inductance matrix parameters, regardless of the turns ratio: it gives us $L_{11} + L_{22} - 2L_M$, as is clear from (19.10). Measuring the inductances of the primary and secondary ports with the other port open, L_{11} and L_{22} respectively, enables us to find the three parameters of the inductance matrix using only three measurements from an impedance analyzer.

Example 19.2 Transformer Parameter Calculation

In this example, which presents actual experimental measurements, we identify both inductance matrix parameters and circuit model parameters for a flyback transformer having a physical turns ratio of $N_1 : N_2 = 5 : 1$. The following measurements are made:

- An impedance analyzer is used to measure the inductance of the primary, with the secondary open-circuited, yielding $L_{11} = 1987\,\mu\text{H}$.
- An impedance analyzer is used to measure the inductance of the secondary, with the primary open-circuited, yielding $L_{22} = 79.98\,\mu\text{H}$.
- An ac voltage of arbitrary amplitude 1.047 V is applied to the primary, with the secondary loaded only with a high-impedance oscilloscope probe (secondary approximately open-circuited), resulting in a measured voltage amplitude of 0.2082 on the secondary.

The impedance analyzer measurements directly determine L_{11} and L_{22}. The mutual inductance may be calculated based on the relationship between v_1 and v_2 (or λ_1 and λ_2) in (19.1) when only winding 1 is driven ($i_2 = 0$):

$$L_M = L_{11}\frac{v_2}{v_1} = 395.1\,\mu\text{H}. \tag{19.11}$$

The inductance parameters in the model of Fig. 19.2 (with $N_1 : N_2 = 5 : 1$) may be found in three steps. First, using the primary-to-secondary voltage gain measurement, we can find the voltage division between the total primary inductance and the magnetizing inductance, and use this to compute the magnetizing inductance $L_{\mu 1}$ from the measured total inductance L_{11}. Knowing this, we can then compute the leakage inductances from the L_{11} and L_{22} measurements by subtracting out the magnetizing component reflected to the appropriate side:

$$L_{\mu 1} = L_{11}\left(\frac{N_1}{N_2}\frac{v_2}{v_1}\right) = 1976\,\mu\text{H}, \tag{19.12}$$

$$L_{\ell 1} = L_{11} - L_{\mu 1} = 11.39\,\mu\text{H}, \tag{19.13}$$

$$L_{\ell 2} = L_{22} - \frac{N_2^2}{N_1^2}L_{\mu 1} = 0.94\,\mu\text{H}. \tag{19.14}$$

19.4 Multi-Winding Model

As described previously, the terminal behavior of a P-winding component is determined by $P(P+1)/2$ parameters (self and mutual inductances). A circuit model that captures this degree of complexity thus requires at least that many free parameters. In this section we describe two related circuit models with numbers of parameters that scale precisely as needed to capture the P-winding case. The required model parameters can be determined directly through port measurements, as illustrated in the previous section, though the parameters do not all have direct physical meaning. We also introduce a method that can sometimes be used for simplifying coupled magnetic-circuit models.

19.4.1 Extended Cantilever Model

Most circuit models do not extend cleanly to capture the full behavioral complexity of multi-winding transformers. A widely used circuit model that does extend is a generalization of the cantilever model of Fig. 19.3. This model provides precisely the number of parameters to match the degrees of freedom of the P-winding case, and has parameters that can be directly determined from terminal measurements on a physical device or from the inductance matrix for a component.

Figure 19.4 shows the extended cantilever model for components having from one through four ports (or windings). The circuit model for the single-port case is simply an inductor across the port, with the single model parameter ℓ_{11} being the inductance. To extend from one port to two, we add a "leakage" inductor ℓ_{12} and an ideal transformer of turns ratio $1:n_2$, with the secondary (n_2-turn side) of the ideal transformer forming the second port of the circuit model. The negative terminal of the primary (1-turn side) of the transformer is connected directly to the negative terminal of the first port, and the positive terminal of the primary is connected to the positive terminal of the first port via the inductor ℓ_{12}. This two-port circuit model is the same as the cantilever model shown in Fig. 19.3, but with the parameters renamed. It has exactly the three independent parameters (ℓ_{11}, ℓ_{12}, and n_2) needed to model the port behavior of a two-winding transformer.

To add a third port, we introduce a further ideal transformer of turns ratio $1:n_3$, with its secondary forming the third port, as shown in the three-winding circuit of Fig. 19.4. We have thus added three independent parameters (n_3, ℓ_{13}, and ℓ_{23}) to the two-winding circuit, for the total of six parameters needed for a three-winding component.

A powerful feature of the extended cantilever model is that for any number of ports P, it provides exactly the $P(P+1)/2$ independent parameters necessary to capture the port relations. In particular, a P-port cantilever model has one "magnetizing" inductance (ℓ_{11}), $P-1$ transformers with associated turns ratios (n_2 to n_P) and an inner network having P vertices fully interconnected via $P(P-1)/2$ "leakage" inductors, for a total of $1+(P-1)+P(P-1)/2 = P(P+1)/2$ parameters.

Looked at another way, to extend a magnetic system from $P-1$ to P ports, the inductance matrix gains an additional row and column with P new parameters (L_{1P} to L_{PP}). Likewise, to expand a corresponding transformer circuit model from $P-1$ to P ports, one needs to add an additional port and circuit elements providing P new degrees of freedom. To expand an extended cantilever model from $P-1$ to P ports, one adds an ideal transformer of turns ratio $1:n_P$ whose secondary is the additional port, along with a set of $P-1$ inductors from the transformer primary to nodes associated with each of the other $P-1$ ports. One thus adds P new parameters

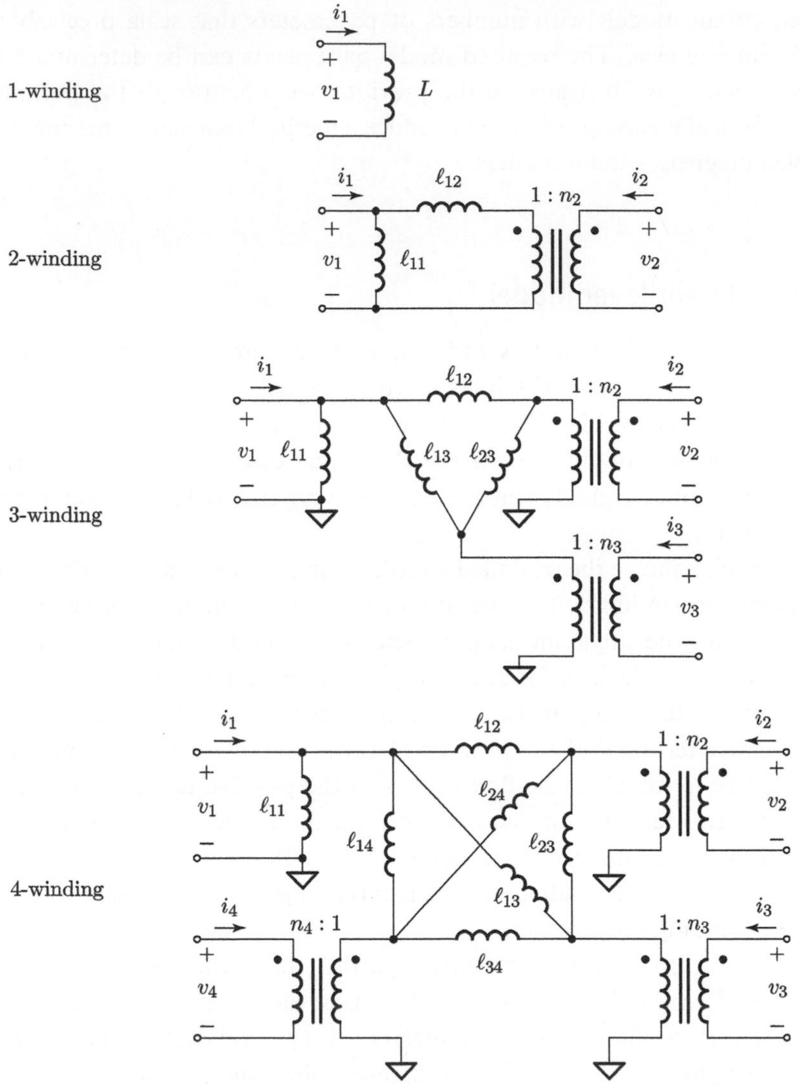

Figure 19.4 The extended cantilever model shown for one port (an inductor) up through four ports. The model is readily extensible to an arbitrary number of ports.

($P - 1$ inductors and one turns ratio), for a total of $P(P+1)/2$ parameters. This methodology can be seen in the expansion of the model from one through four ports in Fig. 19.4.

An important feature of the cantilever model is that its parameters can be extracted directly from experimental measurements on a device, without requiring sensitive numerical manipulations to determine them. The recommended procedure for determining the model parameters from port measurements on a device is as follows:

- Inductance ℓ_{11} (the "primary inductance") is found as the inductance at port 1 with the other ports open-circuited.

- Parameters n_1 to n_M (the "effective turns ratios") can be found by measuring the voltage gain from port 1 to each other port, with all these other ports open. The voltage gain from port 1 to port j is n_j. Any negative value of n_j can be made positive by changing the winding polarity at that port.
- Inductances ℓ_{ij} (the "effective leakage inductances") can be found by driving a voltage at port i with all other ports short-circuited, and measuring the current at port j. At (angular) frequency ω, ℓ_{ij} can be calculated as $\ell_{ij} = V_i/(2\pi\omega n_i n_j I_j)$, where we define $n_1 = 1$.

Of these measurements, the last set is the most fraught, owing to the difficulty of making accurate current measurements while maintaining low "short-circuit" impedances at the ports.

The parameters of the cantilever model can also be related directly to the inductance matrix parameters of a magnetic component, simplifying the task of creating a cantilever circuit model from an inductance matrix description, and vice versa. The inductance matrix representation $\boldsymbol{\lambda} = Li$ has the symmetric inductance matrix L with elements $\{L_{ij}\}$. For any such system, we can also find the inverse inductance matrix $B = L^{-1}$, which is also symmetric and has elements $\{B_{ij}\}$, giving us the alternative matrix description $i = B\boldsymbol{\lambda}$. By superposition of flux linkage drives (or equivalently voltage drives) at the individual ports of the cantilever model with the other ports shorted, it can be shown that the parameters of the inverse inductance matrix are

$$B_{jj} = \frac{1}{n_j^2}\sum_{k=1}^{n}\frac{1}{\ell_{jk}}, \qquad B_{jk} = \frac{-1}{n_j n_k \ell_{jk}} \quad (j \neq k), \tag{19.15}$$

where we define $n_1 = 1$, $\ell_{jj} = \infty$ for $j \neq 1$, and $\ell_{jk} = \ell_{kj}$. Finding B from the cantilever model parameters and inverting it gives us the inductance matrix.

Likewise, we can start with the inductance matrix and find the cantilever model parameters. From the above we already have the "leakage" inductances of the cantilever model in terms of the parameters of the inverse inductance matrix:

$$\ell_{jk} = \frac{-1}{n_j n_k B_{jk}} \quad (j \neq k). \tag{19.16}$$

Driving port 1 of the model with a current, with the other ports open-circuited, and comparing the voltages obtained with those in the inductance matrix description, we can find the other cantilever model parameters in terms of the inductance matrix parameters:

$$\ell_{11} = L_{11}, \qquad n_j = \frac{L_{1j}}{L_{11}}. \tag{19.17}$$

19.4.2 An Alternative "Necessary and Sufficient" Circuit Model

The extended cantilever model provides a sufficient number of free parameters to capture the behavior of an arbitrary P-winding magnetic component, yet only employs the minimum number of parameters necessary to do so. We might thus describe it as a "necessary and sufficient" circuit model. Perhaps the main shortcoming of the extended cantilever model is the need to measure short-circuit currents for direct experimental determination of the model parameters – measurements that are difficult and prone to error. It would be advantageous to

Table 19.1 Recommended measurements for determining the parameters in the "necessary and sufficient" circuit model of Fig. 19.5.

Measurement	Winding 1	Winding 2	Winding 3	Measure	Result
m_1	v_1	OC	OC	Z_1	$= j\omega L_m$
m_2	v_1	OC	OC	v_2/v_1	$= n_2$
m_3	v_1	OC	OC	v_3/v_1	$= n_3$
m_4	v_1	SC	OC	v_3/v_1	$= n_3 \cdot L_2/(L_1 + L_2)$
m_5	v_1	OC	SC	v_2/v_1	$= n_2 \cdot L_3/(L_1 + L_3)$
m_6	SC	v_2	OC	Z_2	$= n_2^2 \cdot j\omega(L_1 + L_2)$
m_7	SC	OC	v_3	Z_3	$= n_3^2 \cdot j\omega(L_1 + L_3)$
m_8	SC	v_2	OC	v_3/v_2	$= n_3/n_2 \cdot L_1/(L_1 + L_2)$
m_9	SC	OC	v_3	v_2/v_3	$= n_2/n_3 \cdot L_1/(L_1 + L_3)$

Table 19.2 Method for determining the parameters in the "necessary and sufficient" circuit model of Fig. 19.5 from the measurements in Table 19.1.

Parameter	Calculation
L_m	$m_1/(j\omega)$
n_2	m_2
n_3	m_3
L_1	$m_6 \cdot m_8/(j\omega \cdot m_2 \cdot m_3)$ or $m_7 \cdot m_9/(j\omega \cdot m_2 \cdot m_3)$
L_2	$m_4 \cdot m_6/(j\omega \cdot m_2^2 \cdot m_3)$
L_3	$m_5 \cdot m_7/(j\omega \cdot m_2 \cdot m_3^2)$

have a model that instead only requires measurement of one-port impedances and two-port voltage gains in order to determine the necessary model parameters.

Figure 19.5 shows a different "necessary and sufficient" circuit model for a three-port magnetic component that has this desirable property. This model replaces the Δ connection of "leakage" inductors in the three-port extended cantilever model with a Y connection of inductors (the topologically dual arrangement). This Δ–Y transformation of the model's internal network of inductors makes it easy to determine the inductor values using measurements of voltage gain instead of the voltage-to-current gain needed by the extended cantilever model. The 1 : 1 transformer shown is not strictly necessary, and can be eliminated by coupling the input port directly to the internal inductor network. Table 19.1 shows the recommended measurements for determining the parameters in this three-port model, while Table 19.2 shows the calculations recommended to determine the parameters from the measurements. In the table, OC means apply open circuit; SC means apply short circuit; Z_n means measure impedance at port n; and v_n/v_m means measure the voltage gain from port m to port n.

A key desirable trait of this model is that the parameters can be determined using only very simple and accurate measurements, with computations that are well conditioned. Versions of this alternative model can be synthesized for systems with more than three ports, but the benefits for parameter determination, compared to the extended cantilever model, become less clear.

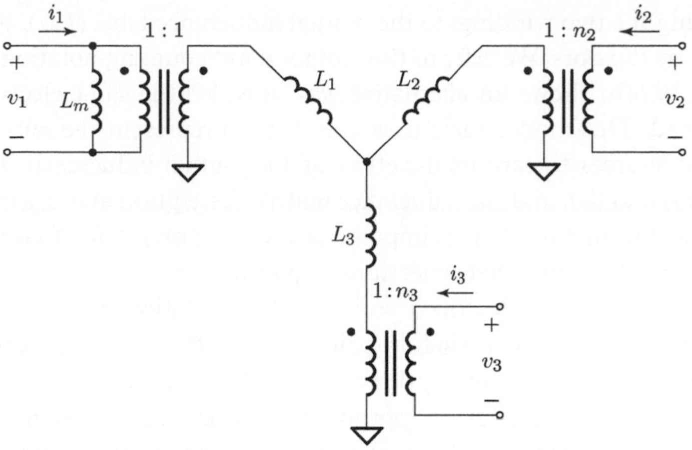

Figure 19.5 A three-port "necessary and sufficient" model with parameters that can be determined using only impedance and voltage gain measurements [5].

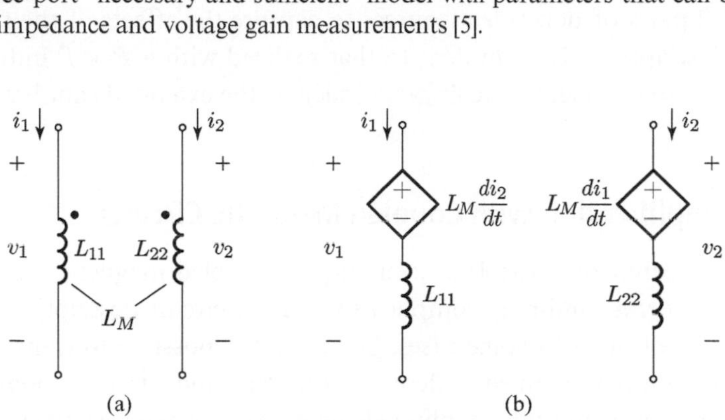

Figure 19.6 Equivalent models for a two-port network with self inductances L_{11} and L_{22} and mutual inductance L_M. (a) The network represented with coupling notation. (b) A model using inductors to represent the self inductances and a pair of dependent sources to represent the mutual inductance.

19.4.3 Further Representations of Transformer Winding Couplings

As we have seen, the relations among different ports in a magnetically coupled circuit can be represented with an inductance matrix description, or by using any of a variety of circuit models comprising inductors and ideal transformers. In this subsection we introduce a further way in which magnetic relationships are sometimes represented in circuits, and also a further way in which these relationships can be expressed in terms of conventional circuit elements.

Figure 19.6(a) shows an alternative circuit representation for a two-port magnetic system having self-inductances L_{11} and L_{22} and mutual inductance L_M, implementing the inductance matrix relationship (19.1) (e.g., as in the system of Example 19.1). The two dotted windings in Fig. 19.6(a) are magnetically coupled. The self-inductance of each winding is indicated next to the winding (L_{11} and L_{22}), while the mutual inductance between the two windings is indicated via lines linking the two windings to the mutual inductance label (L_M), with the coupling direction indicated by the dots. We refer to this notation as "coupling notation."

Figure 19.6(b) shows an alternative way in which the same electrical port relationship can be described. This model again uses inductors to represent the self-inductances, but a pair of dependent sources to capture the effect of the mutual inductance. The circuit descriptions of Fig. 19.6(a) and (b), and the inductance matrix description above, are all equivalent in terms of the electrical terminal relations imposed at the two ports (and of course also match the models of Fig. 19.3 with appropriate selections of parameters).

In general, a circuit described with coupling notation may have P windings with mutual inductances between each winding pair (yielding P self-inductances and $P(P-1)/2$ mutual inductances). Arbitrary coupling directions can be realized among all winding pairs by adjusting the signs of the mutual terms appropriately. (Changing the sign of a mutual inductance is equivalent to reversing the polarity of one dot for that coupling relationship.) One could likewise represent such a system with P inductors to capture the effects of the self-inductances, and $P(P-1)/2$ pairs of dependent sources to capture the effects of the mutual inductances. Each of these descriptions is equivalent to that realized with a $P \times P$ inductance matrix, or with a "necessary and sufficient" circuit model such as the extended cantilever model.

19.4.4 Model Simplification with Coupled Magnetic Circuits

The general behavior possible in a multi-port coupled magnetic system is necessarily complicated, with correspondingly complicated general circuit descriptions (e.g., as detailed in the last few subsections.) In some cases, however, it is possible to transform a circuit model with magnetic coupling into an equivalent "decoupled" model that includes inductances but requires no transformers or mutual coupling elements or dependent sources. When possible to realize, such a model can be useful for easily visualizing the behavior of a system. Here we introduce a technique for developing such a simplified model.

Simplified models comprising only inductors may only become possible when the various electrical ports are connected directly together (i.e., without electrical isolation). To illustrate this, Fig. 19.7(a) shows such a circuit with three terminals and three branches, in which two of the branches are magnetically coupled. This circuit diagram uses the coupling notation of the previous subsection. For comparison, Fig. 19.7(b) shows an alternative description for the same system, using inductors to represent self-inductance terms and dependent sources to represent the magnetic coupling term. In Fig. 19.7(c) we introduce a different arrangement of dependent voltage sources that has the same v–i relationships at its terminals as the circuits in Fig. 19.7(a) and (b). This arrangement is carefully selected such that each dependent voltage source depends only on the derivative of its current. We can thus think of each of those sources as implementing the constitutive relation of an appropriately valued inductor. Figure 19.7(d) shows a circuit model that has the same terminal relationships as those of Figs. 19.7(a)–(c) but uses only inductors. This is a "decoupled" circuit model for the system.

Note that the models of Fig. 19.7(a)–(d) can be used to represent a wide range of coupled subcircuits. For example, a system with the polarity of one of the dots reversed in Fig. 19.7(a) can be simply represented by replacing M_{12} by $-M_{12}$ in the models of Fig. 19.7(b) and (c).

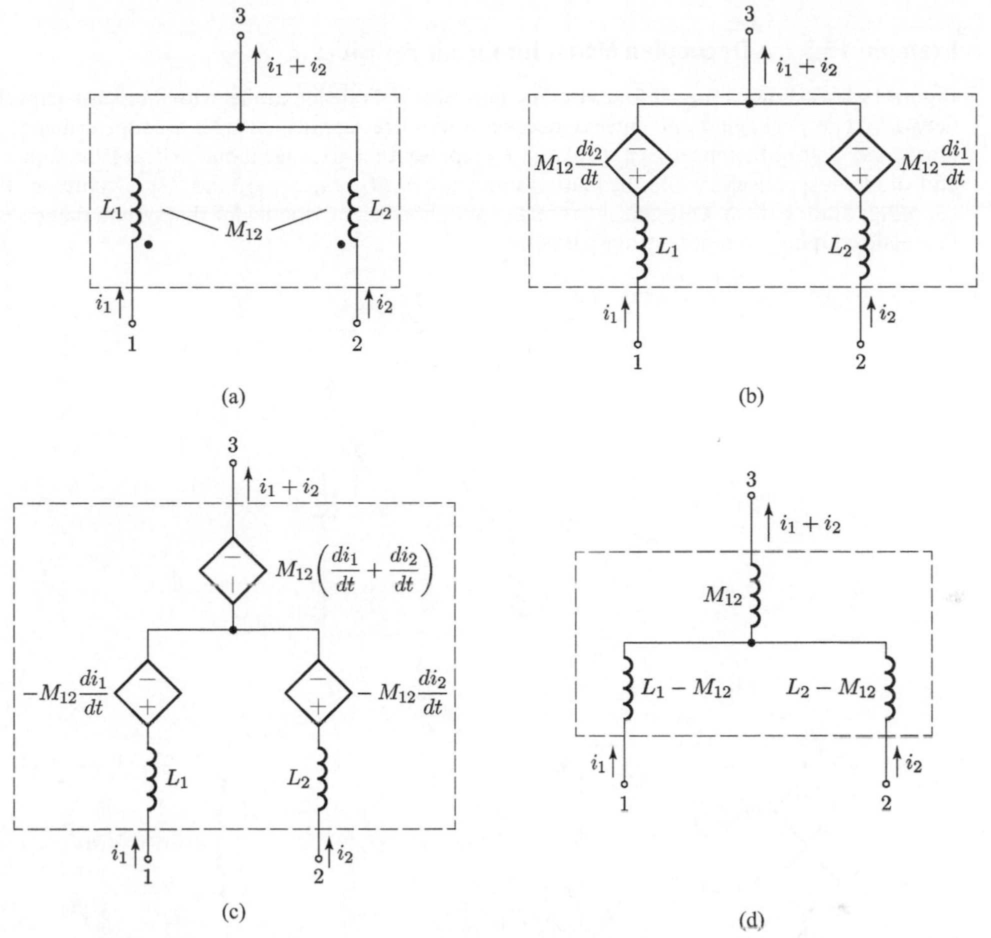

Figure 19.7 Equivalent models for three coupled inductors. (a) Using coupling notation. (b) Using inductors for self-inductance terms and dependent sources for the magnetic coupling term. (c) An alternative arrangement of dependent sources. (d) A decoupled model created by transforming the dependent sources in(c) into equivalent series inductors.

Likewise, a two-terminal subcircuit can be represented by leaving one of the terminals open-circuited in the models of Fig. 19.7(a)–(d).

The decoupled subcircuit model of Fig. 19.7(d) can be substituted for any coupled subcircuit having one of the forms of Fig. 19.7(a)–(c) without changing circuit terminal behavior. By repeatedly making such substitutions, we can often develop a simple decoupled model for an entire system.

Cases such as the particular one in Fig. 19.7 often occur in practice. For example, in addition to describing magnetic components such as transformers, circuit models with magnetic coupling are often used to represent the parasitic magnetic interactions of circuit elements and interconnections in a system. This can be especially useful in modeling filters, component packages, circuit boards, busbars, and similar items. In such cases there is often no electrical isolation, and the circuit imposes closely related currents and/or voltages at the ports of the magnetic model. These characteristics often make it possible to realize a simplified circuit model.

Example 19.3 A Decoupled Model for Circuit Parasitics

Figure 19.8(a) shows a model for a circuit comprising a diode in parallel with a resistor–capacitor snubber network. The packaging and interconnect parasitics are represented with a coupled magnetic circuit as illustrated. Self-inductances L_{1A}, L_{1B}, and L_2 represent the package inductances of the capacitor, resistor, and diode, respectively, while the mutual inductances M_{1A1B}, M_{1A2}, and M_{1B2} represent the magnetic couplings among them. Our goal is to create a simplified circuit model for this system that does not require magnetic coupling to represent the parasitics.

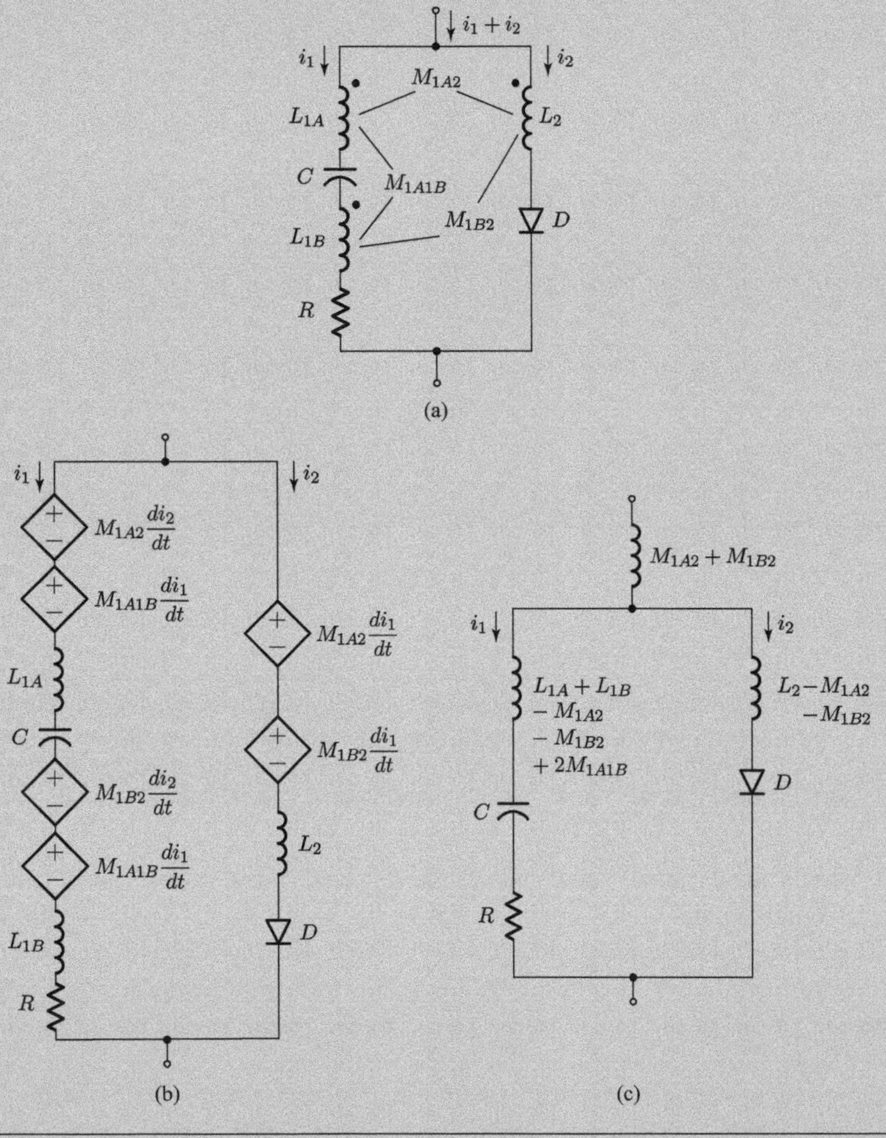

Figure 19.8 Models for a diode in parallel with an RC snubber network including interconnect and packaging parasitics. (a) Parasitics modeled as a coupled magnetic circuit using coupling notation. (b) Parasitics modeled using dependent sources to represent the mutual inductance terms. (c) Parasitics modeled using only inductances, to arrive at a decoupled model.

To develop a "decoupled" circuit model, a version of the circuit is drawn using pairs of dependent sources to represent the mutual inductance terms, as shown in Fig. 19.8(b). Subcircuits having the form of Fig. 19.7(b) are identified and replaced with decoupled versions having the form of Fig. 19.7(d), rearranging the order of elements appearing in series as needed to enable this step to be repeated. Inductors appearing in series are combined. The result of these steps is shown in Fig. 19.8(c), in which the net effect of the parasitics is more simply represented as a set of three inductances in a T network.

19.5 Extending the Magnetic Circuit Model Concept

The previous sections explored circuit models that represent the electrical behavior of magnetic systems. In this section we focus on magnetic circuits, extending the magnetic circuit concept described in the previous chapter. First, we address how loss in a magnetic system can be represented in a magnetic circuit model. Second, we explore how two magnetic systems that are linked electrically by a set of windings can be represented with a single magnetic circuit model. As will be seen, these extensions to the magnetic circuit analogy can be very useful in representing the behavior of magnetic systems.

19.5.1 Modeling Loss in a Magnetic Circuit

A time-varying magnetic flux following a given path (or flowing in a given magnetic circuit branch) may have associated losses owing to core eddy currents, shorted turns, and other conductive paths around a core. Such losses can be modeled with a winding N_{loss} on that flux path coupled to an electrical resistance R_{loss}, as illustrated in Fig. 19.9. A changing flux ϕ in the magnetic path induces a voltage $v_{\text{loss}} = N_{\text{loss}}\frac{d}{dt}\phi$ across the resistor, driving current $i_{\text{loss}} = -v_{\text{loss}}/R_{\text{loss}}$ and resulting in power dissipation $v_{\text{loss}}^2/R_{\text{loss}}$ in the resistor. Such a representation can also be used to approximate the impact of many other sources of loss.

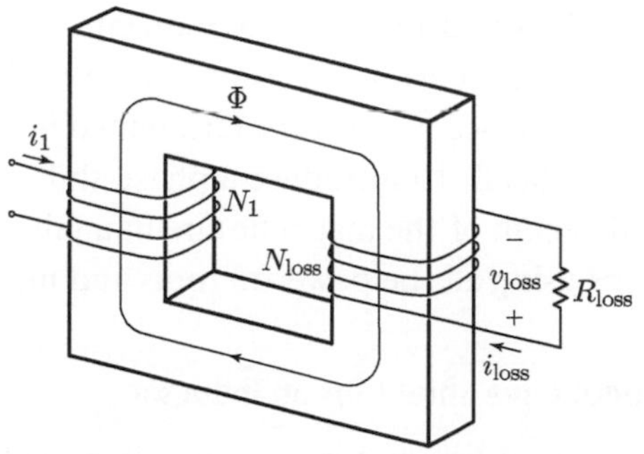

Figure 19.9 Loss induced by a time-varying magnetic flux in a given path is modeled using a winding N_{loss} around that path, connected to a resistor R_{loss}.

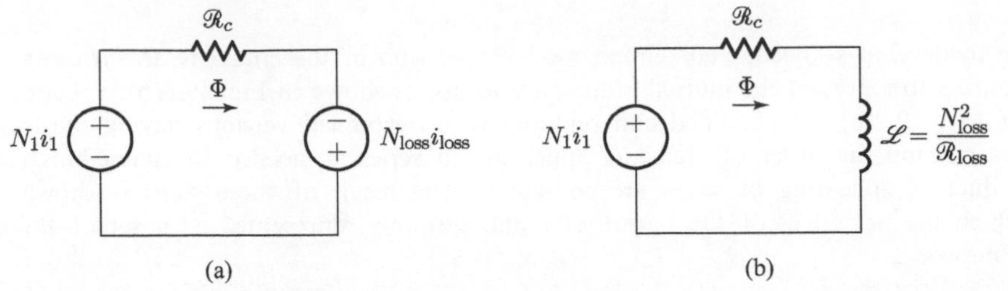

Figure 19.10 (a) A circuit model for the system of Fig. 19.9 in which the winding N_{loss} is represented by the MMF source $N_{\text{loss}} i_{\text{loss}}$. (b) An alternate model in which the winding N_{loss} coupled to resistor R_{loss} is represented by a transferance $\mathcal{L} = N_{\text{loss}}^2 / R_{\text{loss}}$.

Figure 19.10(a) shows a simple circuit model for the system of Fig. 19.9, in which $\mathcal{R}_c$ is the core reluctance and the "loss winding" is represented by an MMF source $N_{\text{loss}} i_{\text{loss}}$. From the KVL equation for this simple magnetic circuit we get

$$N_1 i_1 = \phi \cdot \mathcal{R}_c - N_{\text{loss}} i_{\text{loss}}. \tag{19.18}$$

From the constraint imposed by the resistor, we get

$$N_{\text{loss}} i_{\text{loss}} = -N_{\text{loss}} v_{\text{loss}} / R_{\text{loss}} = -\left(\frac{N_{\text{loss}}^2}{R_{\text{loss}}} \right) \cdot \frac{d}{dt} \phi. \tag{19.19}$$

Using this result, and defining a constant $\mathcal{L} = N_{\text{loss}}^2 / R_{\text{loss}}$, we can rewrite (19.18) as

$$N_1 i_1 = \phi \cdot \mathcal{R}_c + \mathcal{L} \frac{d}{dt}(\phi). \tag{19.20}$$

This equation directly shows the constraint imposed by the resistor-coupled winding on the magnetic circuit. Examining (19.20) and matching it to the magnetic circuit model of Fig. 19.10(b), we can see that the resistor-coupled winding acts exactly as the magnetic circuit analog of an inductor, which we denote as the *transferance* $\mathcal{L}$ (units of siemens).

A magnetic circuit model including transferance elements is governed by differential equations rather than simple algebraic ones, as seen in (19.20). Consequently, the responses in the circuit need no longer be instantaneous with the drive inputs, may be frequency dependent, and may include both forced and natural-response components. Indeed, loss is sometimes introduced in a system just to enable such behavior. For example, in the *shaded-pole* motor, a shorted auxiliary winding (which can be modeled as one turn terminated in a resistor) is placed around a portion of each motor pole specifically to introduce a phase-shifted component of magnetic flux in the system. A substantial benefit of the magnetic circuit analogy is that we can treat the behavior of such magnetic systems using all the powerful tools and methods developed for electric circuits.

Example 19.4 A Conductive Shield for an Inductor

Figure 19.11 shows the cross-section of a special pot-core inductor having a low-permeability centerpost and outer shell, and high-permeability end-caps. Also shown are the magnetic field lines of the fringing fields outside of the core. Because of the distributed gap provided by the centerpost and outer shell, these fringing fields can be significant. Inductor fringing fields at high frequencies can be a source of EMI as well as eddy current losses in surrounding elements. The fringing fields can be reduced by wrapping a shorted single-turn layer of copper foil around the outer circumference of the pot core (except where the inductor leads come out). In this example we develop a magnetic circuit model to estimate the effect of such a conductive shield on the fringing fields and inductor characteristics.

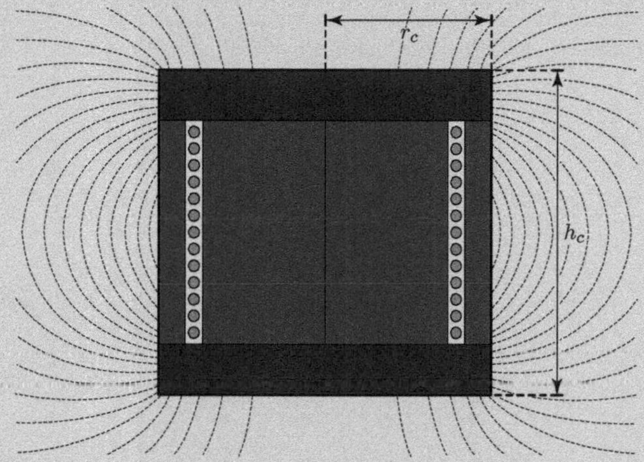

Figure 19.11 A pot-core inductor having a low-permeability centerpost and outer shell, and high-permeability end-caps. The pot core is cylindrical with a winding around the centerpost but inside the outer shell. Also shown are the fringing field lines outside of the core.

A magnetic circuit for the inductor of Fig. 19.11 without a conductive shield is shown in Fig. 19.12(a). The model includes the reluctances of the centerpost, $\mathscr{R}_{\text{cpost}}$, the outer shell, $\mathscr{R}_{\text{shell}}$, and the distributed fringing field path around the pot core, $\mathscr{R}_{\text{fringe}}$, but treats the high-permeability end-caps as having zero reluctance. The reluctances of the centerpost and outer shell are straightforward to compute based on their geometry, but finding a value for the fringing reluctance is more difficult. For dimensions where $h_c > (2/3)r_c$, the reluctance of the fringing field path may be approximated as

$$\mathscr{R}_{\text{fringe}} \approx \frac{0.9}{\mu_0 \pi r_c}. \tag{19.21}$$

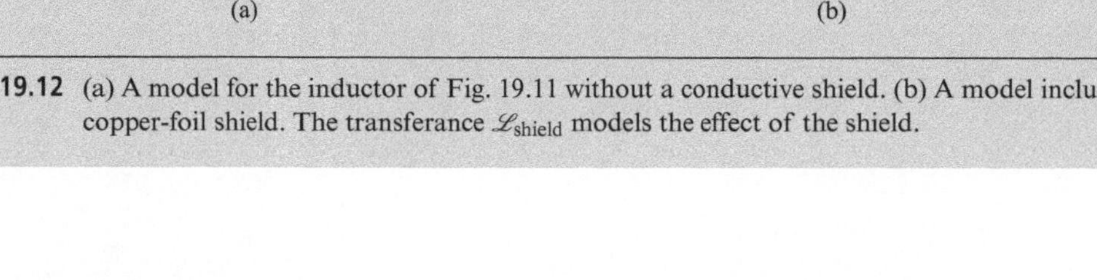

Figure 19.12 (a) A model for the inductor of Fig. 19.11 without a conductive shield. (b) A model including the copper-foil shield. The transferance $\mathscr{L}_{\text{shield}}$ models the effect of the shield.

A magnetic circuit model for the inductor of Fig. 19.11 including the conductive shield is shown in Fig. 19.12(b). The transferance $\mathscr{L}_{\text{shield}}$ models the effect of the conductive shield. Because the shorted shield winding links only the flux in the fringing field path, the transferance appears in the branch of the magnetic circuit carrying the fringing flux ϕ_{fringe}.

We reduce the high-frequency fringing flux by proper design of the shield. To find the appropriate value of $\mathscr{L}_{\text{shield}}$, we need to calculate the effective resistance of the shorted shield winding. We make the simplifying assumption that the shield winding encompasses the circumference of the pot core across the full core height h_c and has a thickness t_w, giving it a dc resistance of

$$R_{\text{dc}} = \frac{2\pi\, r_c\, \rho_{\text{Cu}}}{h_c\, t_w}, \tag{19.22}$$

where ρ_{Cu} is the resistivity of the copper shield conductor. The effective ac resistance R_{ac} of the shield matches this dc resistance for frequencies at which skin effect in the shield winding is not important. It can be shown that this is true for frequencies where

$$\omega \ll \frac{2\,\rho_{\text{Cu}}}{t_w^2\, \mu_{\text{Cu}}}, \tag{19.23}$$

where μ_{Cu} is the permeability of the copper shield conductor. While we do not derive it here, a general expression for the ac resistance valid at all frequencies (including skin effect, when applicable) can be found as:

$$R_{\text{ac}} = R_{\text{dc}} \cdot \Delta \cdot \frac{\sinh(2\Delta) + \sin(2\Delta)}{\cosh(2\Delta) - \cos(2\Delta)}, \tag{19.24}$$

where Δ is the ratio of the shield winding thickness to the skin depth δ,

$$\Delta = \frac{t_w}{\delta} = t_w \cdot \sqrt{\frac{\omega\, \mu_{\text{Cu}}}{2\,\rho_{\text{Cu}}}}. \tag{19.25}$$

We can find the value of the transferance associated with the conductive shield as $\mathscr{L}_{\text{shield}} = 1/R_{\text{ac}}$.

Considering the extended magnetic circuit model of Fig. 19.12(b), the shield winding has a significant impact in reducing the fringing flux from its low-frequency value above a cutoff frequency of

$$\omega_{\text{cutoff}} \approx \frac{(\mathscr{R}_{\text{cpost}} \parallel \mathscr{R}_{\text{shell}}) + \mathscr{R}_{\text{fringe}}}{\mathscr{L}_{\text{shield}}}. \tag{19.26}$$

At frequencies far below ω_{cutoff} the fringing fields are not greatly affected, while for frequencies far above ω_{cutoff} the fringing fields are suppressed.

It should be noted that with the addition of the shield winding, the effective inductance provided by the inductor becomes frequency dependent, owing to the frequency dependence of magnetic flux and associated energy storage. At frequencies far below ω_{cutoff}, the transferance $\mathscr{L}_{\text{shield}}$ has negligible impact and the inductance of the inductor is

$$L_{LF} \approx \frac{N^2}{\mathscr{R}_{\text{cpost}} + \mathscr{R}_{\text{shell}} \parallel \mathscr{R}_{\text{fringe}}}, \tag{19.27}$$

while for frequencies far above ω_{cutoff}, the transferance $\mathscr{L}_{shield}$ acts as an "open circuit" in the magnetic circuit, reducing the effective inductance to

$$L_{HF} \approx \frac{N^2}{\mathscr{R}_{\text{cpost}} + \mathscr{R}_{\text{shell}}}. \tag{19.28}$$

Note that this high-frequency inductance is what we would have predicted for all frequencies if we had not included the fringing path, and – depending upon the geometry of the system – it might or might not be substantially smaller than the low-frequency inductance.

19.5.2 Modeling Electrical Coupling of Magnetic Circuits

In an electric circuit, one can model magnetic coupling between two electrical subsystems with a transformer. Here we describe how in a *magnetic* circuit one can model *electrical* coupling between two magnetic subsystems, such as might occur when two separate magnetic circuits are linked by electrical connections of windings. We also show how to account for parasitic effects such as loss in the electrical coupling network.

Figure 19.13 shows a system with two separate magnetic cores linked by the electrical connection of a pair of windings. The two magnetic core structures share no common flux, but the electrical connection of winding N_X on the first core to the winding N_Y on the second core imposes electrical coupling between the two magnetic structures. This results in a pair of constraints on the magnetic circuit variables. First, because the electrical connection forces the voltages v_X and v_Y (and associated flux linkages λ_X and λ_Y) to be equal and opposite, there is a relationship imposed between the fluxes through the core paths associated with windings N_X and N_Y:

$$N_X \Phi_X = -N_Y \Phi_Y. \tag{19.29}$$

Second, the MMFs due to windings N_X and N_Y are related by the current i_W:

$$i_W = \frac{\mathscr{F}_X}{N_X} = \frac{\mathscr{F}_Y}{N_Y}. \tag{19.30}$$

The constraints of (19.29) and (19.30) are exactly those of the magnetic circuit analog of an ideal transformer (i.e., the ideal transformer relations, but for fluxes Φ and MMFs $\mathscr{F}$ instead of currents i and voltages v). Figure 19.14 shows a magnetic circuit for the system of Fig. 19.13 using this representation. The coupling between the two magnetic subsystems imposed by the electrical connection of windings N_X and N_Y is represented by the magnetic circuit analog of an ideal transformer. This analog thus allows us to represent the electrical couplings among different magnetic subsystems entirely in the magnetic circuit domain.

The magnetic circuit can also model non-idealities (e.g., loss) in the electrical connection. For example, suppose we include a series resistance R_W in the network electrically connecting windings N_X and N_Y, as shown in Fig. 19.15(a). This might represent an actual physical resistor,

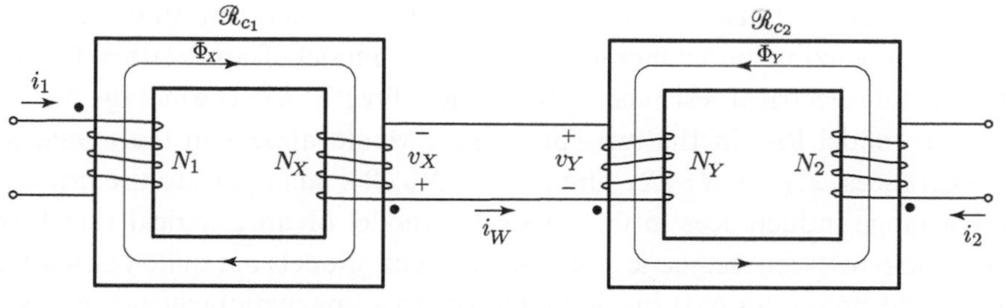

Figure 19.13 Two magnetic structures linked by an electrical connection. This connection imposes electrical coupling between the two magnetic systems.

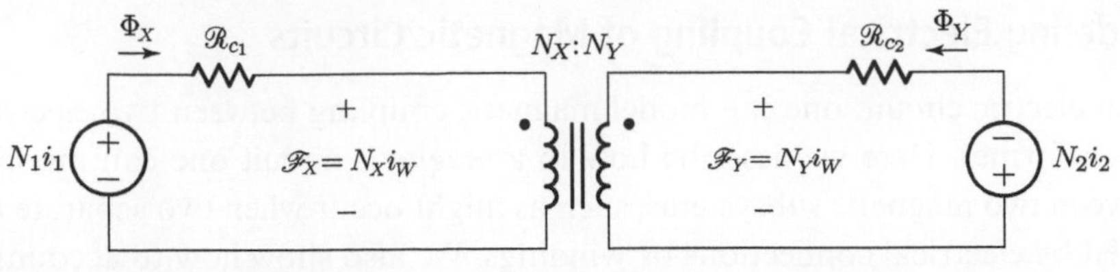

Figure 19.14 A model for the system of Fig. 19.13. The coupling between the two magnetic subsystems imposed by the electrical connection of windings N_X and N_Y is represented by the ideal $N_X : N_Y$ transformer.

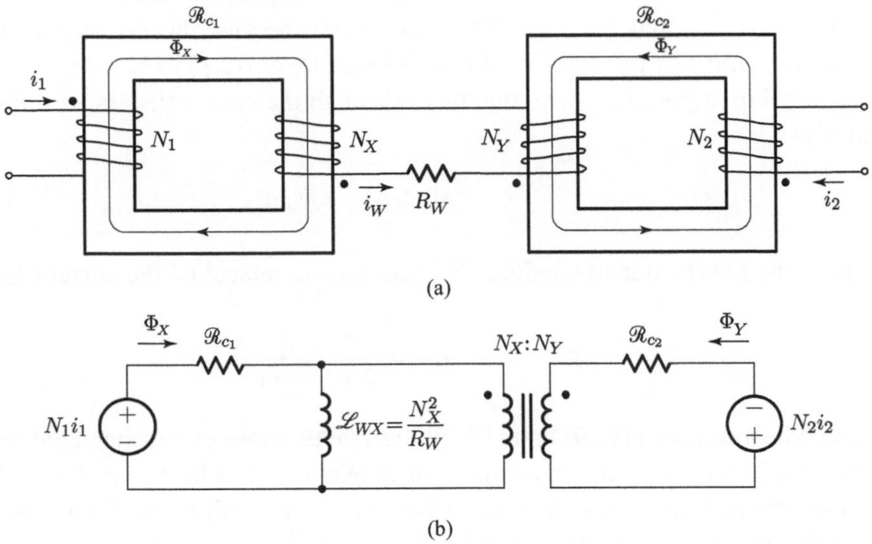

Figure 19.15 (a) Two magnetic structures linked by the electrical connection of a pair of windings via a series resistance R_W. (b) A magnetic circuit model for this system, in which the effect of resistance R_W is modeled as a shunt transferance $\mathscr{L}_{WX}$.

or the parasitic resistance associated with the two windings. In the magnetic circuit domain, this can be represented with transferance $\mathscr{L}_{WX}$ across the N_X side of the magnetic circuit "transformer," as illustrated in Fig. 19.15(b). Interestingly, this shunt transferance is analogous to the magnetizing inductance in the physical T-model of an electrical transformer. Were we to also include electrical resistances R_X and R_Y directly across windings N_X and N_Y respectively (e.g., to model loss in the two cores) these would appear in the magnetic circuit model as transferances $\mathscr{L}_{lX} = N_X^2/R_X$ and $\mathscr{L}_{lY} = N_Y^2/R_Y$ analogous to the primary- and secondary-side leakage inductances in the physical T-model of an electrical transformer. The parallels observable between magnetic and electric circuit models are quite fascinating. For example, in an electric circuit, an N-turn winding linked to a (magnetic) reluctance $\mathscr{R}$ yields an inductance $L = N^2/\mathscr{R}$, while in a magnetic circuit an N-turn winding linked to an (electric) resistance R yields a transferance $\mathscr{L} = N^2/R$.

19.6 Physical Electric Circuit Models for Magnetic Systems

In the previous sections, the focus has largely been on circuit models that can represent the terminal behavior of a magnetic component, without regard to whether or not they capture the underlying physical operation. An exception to this is the T-model with physical turns ratio, introduced in the previous chapter and repeated here in Fig. 19.2. The elements and structure of this electric circuit model can be directly related to the magnetic circuit model of Fig. 19.1(b), and capture the magnetic energy stored in its flux paths.

It would be desirable to have an electric circuit model that directly captures the physical behavior of a magnetic structure in the general case (e.g., for many windings and/or for other magnetic circuit structures). Such an electrical model can provide insight into how a magnetic component physically functions in the context of a circuit. Moreover, as will be shown, it can be useful for synthesizing a magnetic component structure that provides a desired electrical behavior. In this section, we introduce a general means to directly map between a magnetic circuit model and a corresponding electric circuit model that physically represents its structure and energy storage.

To develop a direct mapping between a magnetic circuit model and a corresponding electric circuit model, it is useful to first recognize that there is a form of topological duality between them. The *through*-variable in a magnetic circuit is flux ϕ, which is related to the *across*-variable $v = N\frac{d}{dt}\phi$ in a corresponding electric circuit. The across-variable in a magnetic circuit is magnetomotive force (MMF) $\mathscr{F}$, which is related to the through-variable $i = \mathscr{F}/N$ in a corresponding electric circuit. Thus, allowing for appropriate modifications (differentiation of variables, scalings by N), one can expect the structures or graphs of corresponding magnetic and electric circuits to be topological duals. This structural duality underpins the mapping between a magnetic circuit model and a corresponding electric circuit model.

The duality between the structures of corresponding magnetic and electric circuit models does not however, address the fact that there are significant differences between the representations involved. For example, magnetic circuit models represent energy storage with reluctances, while the corresponding electric circuit models represent it with inductances. These differences are reflected in the fact that transposition between across- and through-variable relationships in corresponding magnetic and electric circuit models also has differentiation and scaling involved. We can thus expect that manipulations beyond topological duality alone are necessary to map between a magnetic circuit model and a corresponding electric circuit model.

A general process for synthesizing an electric circuit model from a magnetic circuit model is shown in Fig. 19.16, using the magnetic circuit model of Fig. 19.1(b) as an example. As mentioned, a magnetic circuit model has MMF $\mathscr{F}$ as the across-variable (units of amperes) and flux ϕ as the through-variable (units of volt-seconds), such that the "resistors" in the model are reluctances $\mathscr{R}$ (units of henrys^{-1}). Turns N is unitless, though some developments explicitly incorporate turns in the units.

As a first step, we take the dual of the magnetic circuit, yielding a new circuit in which the through-variable is MMF, the across-variable is flux, and the "resistors" are permeances $\mathscr{P} = 1/\mathscr{R}$ (units of henrys). This provides the structural portion of the mapping.

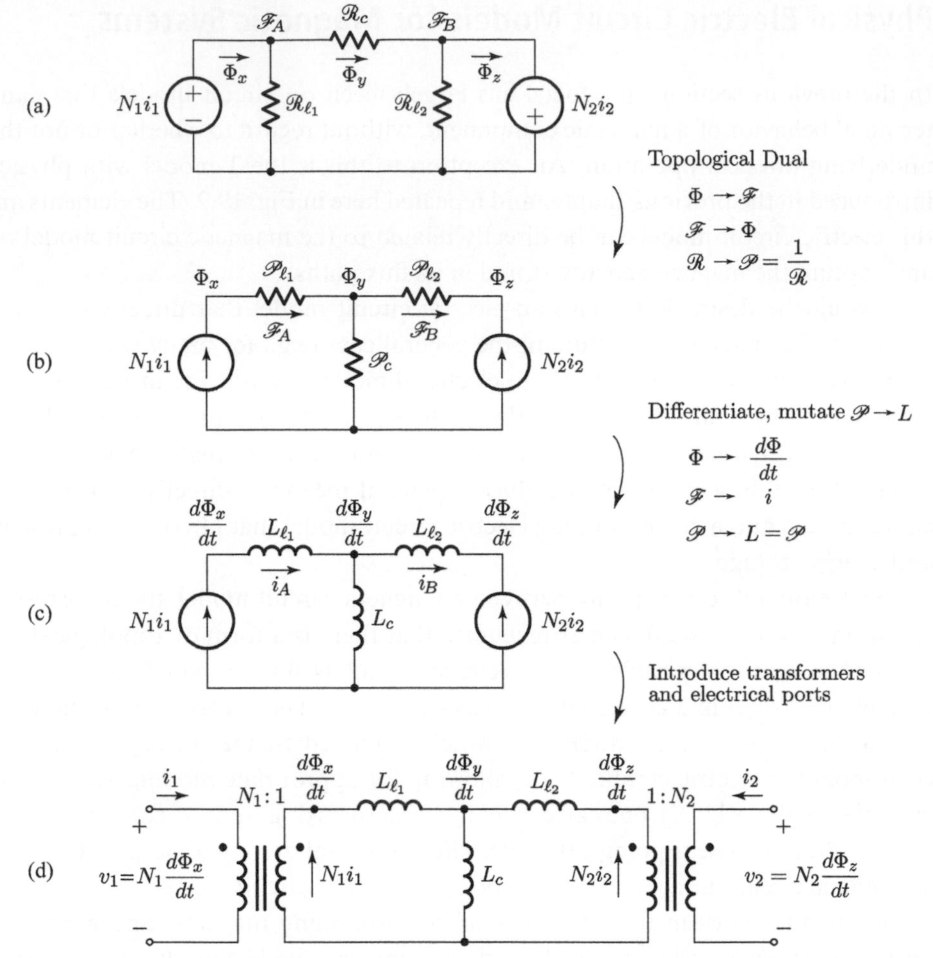

Figure 19.16 A process for transforming a magnetic circuit model into a corresponding electric circuit model that provides a physical representation of its structure and energy storage. This process can also be reversed to synthesize a magnetic circuit model from an electric circuit model.

As a second step, we change the across-variable from ϕ to $\frac{d}{dt}\phi$ (changing units from volt-seconds to volts), change the through-variables $\mathscr{F}$ to equal-valued currents i (maintaining units of amperes), and mutate the permeance "resistors" $\mathscr{P}$ into equal-valued inductors L (retaining their units of henrys). This step converts the circuit variables to voltages and currents while preserving the physical relations imposed by the circuit elements. That is, setting $\mathscr{F}_x = i_x$ and $\mathscr{P}_x = L_x$, the terminal relation $\phi_x = \mathscr{P}_x \cdot \mathscr{F}_x$ becomes $\phi_x = L_x \cdot i_x$, which, when differentiated, becomes the new terminal relation $\frac{d}{dt}\phi_x = L_x \cdot \frac{d}{dt}i_x$.

Third, to provide the equivalent electrical port currents and voltages, we replace the sources $N_x i_x$ with transformers having an $N_x : 1$ turns ratio, yielding electrical port input currents i_x and input voltages $v_x = N_x \frac{d}{dt}\phi = \frac{d}{dt}\lambda_x$. This results in an electric circuit model that corresponds to the original magnetic circuit model. A last, optional, step not shown in Fig. 19.16 is to further manipulate and simplify the circuit, such as by reflecting the "internal" inductors to other sides of the transformers and/or by combining transformers.

The correspondence of the electric circuit model of Fig. 19.16(d) to the original magnetic circuit model of Fig. 19.16(a) is apparent, with the π network in the magnetic circuit model becoming a T network in the electric circuit model (its topological dual) and each inductance in the electric circuit model representing a reluctance branch (flux path) in the magnetic circuit model. Simplifying this electric circuit model by repositioning and combining the two transformers yields the traditional physically based electric circuit model of Fig. 19.2 and the associated magnetizing and leakage parameters of (19.6). We thus obtain the physical result we would expect for the example structure of Fig. 19.1(a).

Importantly, the process of Fig. 19.16 can be used to generate a corresponding electric circuit model for *any* magnetic circuit model. The resulting electric circuit model may have more or fewer parameters than needed to represent the terminal behavior of a general N-port magnetic component, but will exactly reflect the behavior of the magnetic circuit model involved, with each flux path accounted for through the duality between the magnetic and electric representations. Likewise, the process can be used in reverse to synthesize a magnetic circuit structure having a desired electrical circuit representation, though it does not guarantee that the resulting magnetic circuit model is easy to realize in a practical magnetic component.

To further underscore the direct correspondence of these models, consider how magnetic energy storage is represented in a magnetic circuit as compared to in an electric circuit. Magnetic energy storage W_M of a magnetic flux ϕ_x in a given volume of length l_x, cross-sectional area A_x, and permeability μ_x can be written as

$$W_M = \frac{1}{2} \frac{l_x}{\mu_x A_x} \phi_x^2. \tag{19.31}$$

In a magnetic circuit model with a branch of reluctance $\mathscr{R}_x$ carrying flux ϕ_x, we find that

$$W_M = \frac{1}{2} \mathscr{R}_x \phi_x^2, \tag{19.32}$$

where

$$\mathscr{R}_x = \frac{l_x}{\mu_x A_x}. \tag{19.33}$$

Thus, the stored magnetic energy in a given flux path is related to the "power dissipation" in the branch reluctance of the associated magnetic circuit model. The relations imposed in the transformation of Fig. 19.16 give us $\mathscr{R}_x = 1/L_x$ and $\phi_x = L_x \cdot i_x$. Substituting these relations into the above expression for stored energy gives us $W_M = \frac{1}{2} L_x i_x^2$, the energy storage in a linear inductor L_x with current i_x. This demonstrates that the magnetic energy storage implied for each branch of the magnetic circuit model is exactly reflected in the associated inductor of the corresponding electric circuit model.

Example 19.5 Magnetic Circuit Synthesis

It is possible to use the process of Fig. 19.16 in reverse to synthesize a magnetic circuit model from an electrical circuit representation of a desired magnetic element. One can then proceed from the magnetic circuit model to a component design. While there is no guarantee that a component having such a synthesized magnetic circuit model can be easily realized in practice (e.g., without significant parasitic paths not captured in the model), it can nonetheless form a useful basis for developing a design.

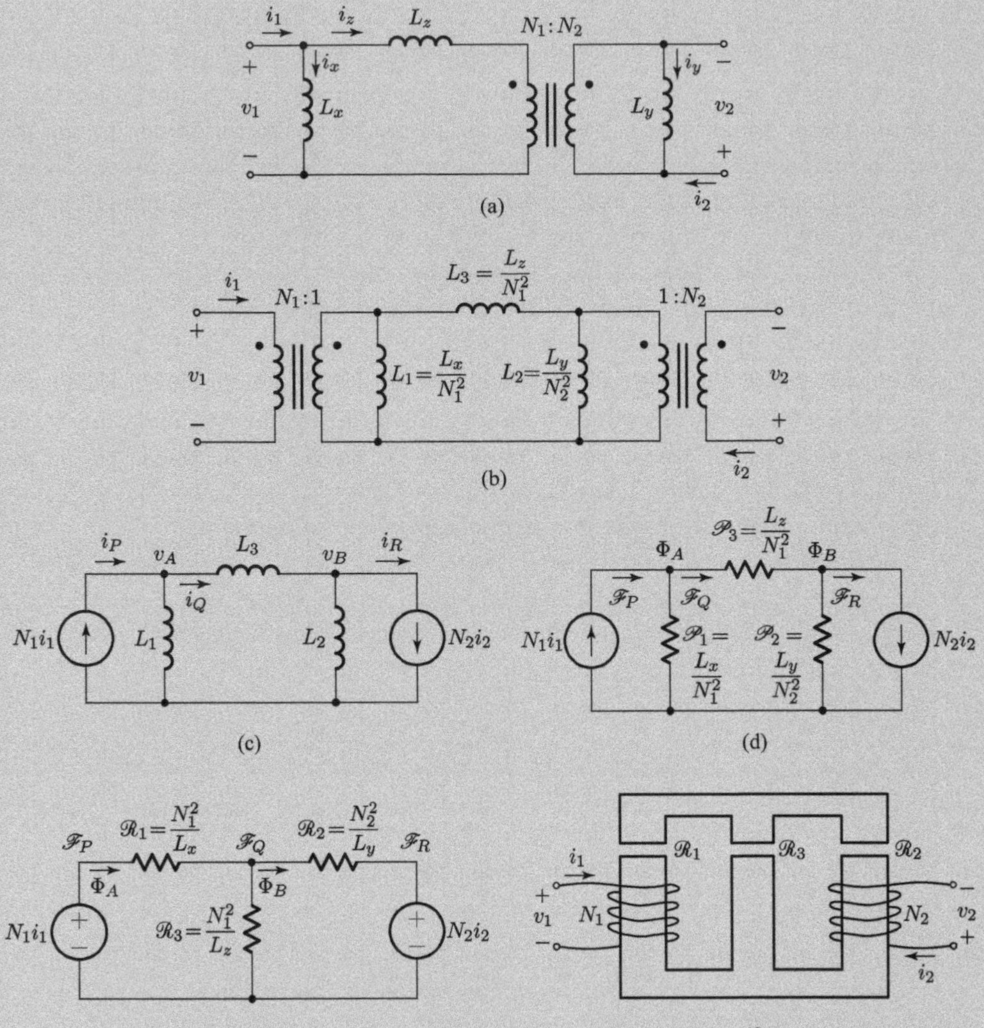

Figure 19.17 (a)–(e) The process of Fig. 19.16 used in reverse to synthesize a magnetic circuit model from an electric circuit model. (f) A possible realization of the magnetic circuit.

Figure 19.17 shows an example of such a process. Suppose that a certain circuit design requires a magnetic component with the electrical model shown in Figure 19.17(a), where the inductance and current of each inductor in the circuit model has been determined. As a first step, the circuit model is reconfigured to have the general structure shown in Fig. 19.16(d) in which each input is coupled by an $N_x : 1$ transformer to an inner network comprising only inductors (with an appropriate value of N_x for each input) as shown in Fig. 19.17(b). As a second step, the input transformers are replaced with equivalent current sources that drive appropriate currents (e.g., $N_x i_x$) into the inner network as shown in Fig. 19.17(c). Next, we change the across-variable to flux ϕ, the through-variable to MMF $\mathscr{F}$, and accordingly mutate the inductors to permeance "resistors" $\mathscr{P}$ as shown in Fig. 19.17(d). Finally, we take the topological dual of this circuit, yielding a magnetic circuit model Fig. 19.17(e) that corresponds to the electric circuit model in Fig. 19.17(a).

To create an appropriate magnetic component, we can seek a magnetic structure that provides this magnetic circuit model, such as that suggested in Fig. 19.17(f). Knowing the required currents and inductances gives us the energy storage requirement for each element in the original electric circuit model and the associated requirement for the corresponding reluctance in the magnetic circuit model. We can seek appropriate core parameters (e.g., cross-sectional areas(s), permeability, lengths/gaps, etc.) and absolute numbers of turns (maintaining turns ratios) to realize these requirements in the magnetic circuit model branches while meeting flux density and other constraints in the physical implementation.

Notes and Bibliography

The transformation between Y and Δ connections in electrical networks was first presented by Kennelly [1].

Neugebauer *et al.* [2] shows the use of a non-physical turns ratio in the T-model of a magnetic structure, and exploits the negative branch inductance that can result in the model.

The extended cantilever model for many-winding transformers is introduced in [3], and issues associated with determining its model parameters are explored in [4], which also presents techniques for making accurate current measurements while maintaining low "short-circuit" impedances at the ports. An overview of modeling techniques for transformers is presented in [5], which provides a model that requires measurement of only one-port impedances and two-port voltage gains in order to determine the necessary cantilever model parameters. This paper also includes a discussion of numerical sensitivity in parameter extraction, and first proposed the alternative to the cantilever model having better parameter measurement properties.

The extensions of magnetic circuit models to encompass loss and electrical coupling were first introduced in [6], which also presents interesting applications of these concepts. Cherry [7] first introduced the topological duality between electric circuit and magnetic circuit systems, forming the basis for direct physical circuit representations of magnetic systems. Variants and examples of this direct physical circuit representation approach can be found in [5] and [8]. The step of taking a topological dual strictly applies only to a planar network. However, the technique of [9] can be used to derive the dual even in cases where the magnetic circuit model is non-planar.

An example application of transferances in magnetic circuit analysis and design can be found in [10], which explores the design of "fractional-turn" transformers.

The inductor structure that serves as the basis for Example 19.4 was introduced in [11]; this approach enables implementation of very low-loss inductors at multi-megahertz frequencies.

1. A. E. Kennelly, "Equivalence of Triangles and Three-Pointed Stars in Conducting Networks," *Electrical World and Engineer*, vol. 34, pp. 413–414, 1899.
2. T. C. Neugebauer, J. W. Phinney, and D. J. Perreault, "Filters and Components with Inductance Cancellation," *IEEE Trans. Industry Applications*, vol. 40, pp. 483–490, Mar./Apr. 2004.
3. R. W. Erickson and D. Maksimovic, "A Multiple-Winding Magnetics Model Having Directly Measurable Parameters," *IEEE Power Electronics Specialists Conference (PESC)*, Fukuoka, Japan, pp. 1472–1478, 1998.
4. M. Shah and K. D. T. Ngo, "Parameter Extraction for the Extended Cantilever Model of Magnetic Component Windings," *IEEE Trans. Aerospace and Electronic Systems*, vol. 36, pp. 260–266, Jan. 2000.

5. A. J. Hanson and D. J. Perreault "Modeling the Magnetic Behavior of N-Winding Components," *IEEE Power Electronics Magazine*, pp. 35–45, Mar. 2020.
6. E. R. Laithwaite, "Magnetic Equivalent Circuits for Electrical Machines," *Proc. IEE*, vol. 114, pp. 1805–1809, 1967.
7. E. C. Cherry, "The Duality Between Interlinked Electric and Magnetic Circuits and the Formation of Transformer Equivalent Circuits," *Proc. Physical Society*, Section B, vol. 62, pp. 101–111, 1949.
8. R. Severns and G. Bloom, *Modern DC-to-DC Switchmode Power Converter Circuits*, New York: Van Nostrand, 1985.
9. A. Bloch, "On Methods for the Construction of Networks Dual to Non-Planar Networks," *Proc. Physical Society*, vol. 58, pp. 677–694, 1946.
10. M. K. Ranjram, I. Moon, and D. J. Perreault, "Variable-Inverter-Rectifier-Transformer: A Hybrid Electronic and Magnetic Structure Enabling Adjustable High Step-Down Conversion Ratios," *IEEE Trans. Power Electronics*, vol. 33, pp. 6509–6525, Aug. 2018.
11. R. Yang, A. J. Hanson, B. A. Reese, C. R. Sullivan, and D. J. Perreault, "A Low-Loss Inductor Structure and Design Guidelines for High-Frequency Applications," *IEEE Trans. Power Electronics*, vol. 34, pp. 9993–10005, Jan. 2019.

PROBLEMS

19.1 Consider the gapped transformer shown in Fig. 19.18. The winding orientations and transformer dimensions are as illustrated. The core has cross-sectional area A_c and permeability μ_c. For the purposes of this problem, you may neglect any leakage fields (and associated inductances) and any fringing fields at the gap.

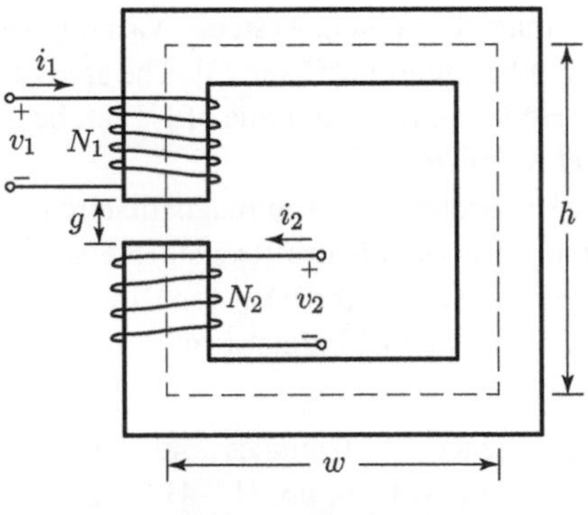

Figure 19.18 The gapped transformer considered in Problem 19.1. The core has cross-sectional area A_c and permeability μ_c. The dimensions shown are the effective magnetic path lengths for the structure.

(a) Draw a magnetic circuit model for this transformer, including expressions for the parameter values.

(b) What is the magnetizing inductance of this transformer as seen from winding N_1?

(c) For this and subsequent parts, you may assume that the core permeability μ_c is sufficiently high that the reluctance of the core is negligible compared to that of the gap. Assume that the core material has a saturation flux density limit of B_{SAT}. If winding N_2 is open-circuited, what is the maximum current winding N_1 can carry without saturating the core?

(d) If voltage v_1 is positive, will the voltage v_2 be positive or negative?

(e) Derive the inductance matrix description for this transformer, and find expressions for L_{11}, L_{12}, and L_{22}.

19.2 Find the cantilever model for the transformer of Example 19.2, including the values of all of its parameters.

19.3 The paper "Transformer Synthesis for VHF Converters" by A. Sagneri *et al.* (IEEE International Power Electronics Conference, June 2010) describes the design of an isolated dc/dc converter operating at 75 MHz along with methods for implementing its transformer. Because it operates at such a high frequency, the design can use a small, coreless printed-circuit-board transformer as its main energy storage element. However, the transformer must have closely controlled parameters to make the converter operate properly. The inductance matrix description of the transformer has $L_{11} = 10.1$ nH, $L_{22} = 44$ nH, and $L_{12} = 10.8$ nH.

(a) Draw the circuit and find the parameters of an extended cantilever model of the transformer. The model parameters needed are ℓ_{11}, ℓ_{12}, and n_2.

(b) Given known physical numbers of turns of one turn for the primary and three turns for the secondary, find a physically based electric circuit model for this transformer, including values for the primary (winding 1) leakage inductance L_{ℓ_1}, the magnetizing inductance referred to the primary L_{μ_1}, and the secondary (winding 2) leakage inductance L_{ℓ_2}.

19.4 Consider both the inductance matrix description of a three-winding transformer and the extended cantilever model for it as shown in Fig. 19.4.

(a) By superposition of flux linkage drives (or equivalently voltage drives) at the individual ports of the cantilever model with the other ports shorted, find the elements of the inverse inductance matrix, and show that they match the expressions of (19.15).

(b) Considering the response of the system at various ports to a current at port 1 only, derive the parameter value relationships in (19.17) for the three-winding case.

19.5 Consider the gapped transformer shown in Fig. 19.19. The winding orientations and transformer dimensions are as illustrated. The core has cross-sectional area A_c and permeability μ_c. For the purposes of this problem, you may treat the core permeability as infinite.

(a) Draw a magnetic circuit model for this transformer, including expressions for the parameter values.

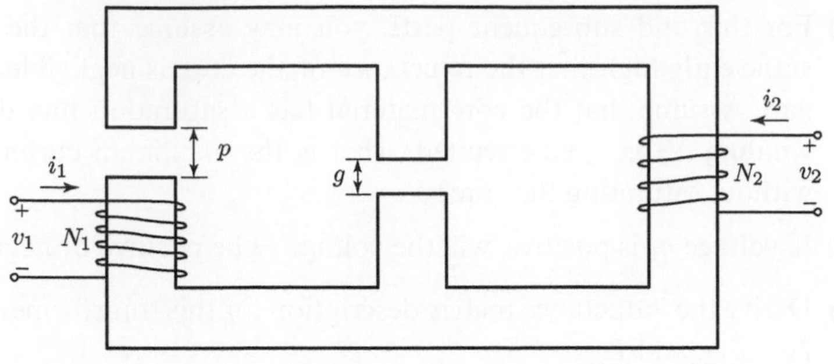

Figure 19.19 The gapped transformer considered in Problem 19.5 having cross-sectional area A_c and very high permeability μ_c.

(b) Derive an inductance matrix description for this magnetic circuit, and find expressions for L_{11}, L_{12}, and L_{22}.

(c) Draw the circuit and find the parameters of a cantilever model of the transformer. The model parameters needed are ℓ_{11}, ℓ_{12}, and n_2.

(d) Develop an electric circuit model for this transformer that provides a direct physical representation of its flux paths and energy storage.

19.6 Consider the gapped three-winding transformer shown in Fig. 19.20. The winding orientations and transformer dimensions are as illustrated. The core has cross-sectional area A_c and permeability μ_c. For the purposes of this problem, you may treat the core permeability as infinite and neglect any fringing fields at the gaps and any leakage fields.

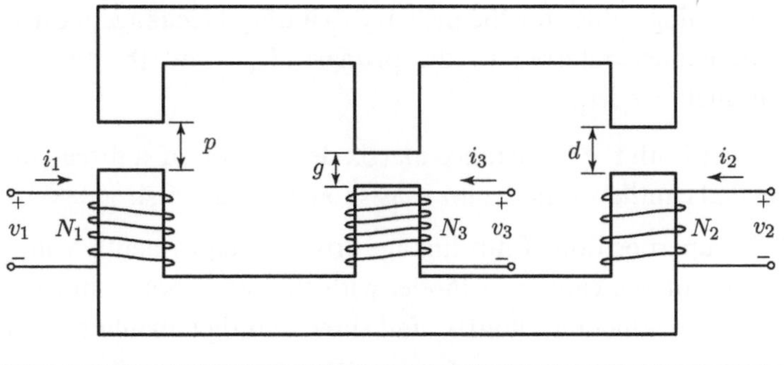

Figure 19.20 Structure of the three-winding transformer with multiple gaps considered in Problem 19.6. The core has cross-sectional area A_c and (very large) permeability μ_c.

(a) Draw a magnetic circuit model for this transformer, including expressions for the parameter values.

(b) Develop an electric circuit model for this transformer that provides a direct physical representation of its flux paths and energy storage.

(c) Develop either an extended cantilever model or the alternative "necessary and sufficient" model representation for this transformer.

19.7 Figure 19.9 shows a two-winding transformer with one winding terminated in a resistance R_{loss}. The transformer core has a cross-sectional area A_c, an effective core length ℓ_c, and a permeability μ_c. The number of primary turns is N_1 and the number of secondary turns is N_{loss}.

(a) Find a magnetic circuit model for this system, assuming that the primary winding is driven with a current $i_1(t)$; indicate all the parameter values.

(b) Use the above model to find the time response for the core flux $\phi(t)$ if the current driving the winding is $i_1(t) = Iu(t)$.

(c) For the same conditions as part (b), find the voltage response $v_{\text{loss}(t)}$ based on your magnetic circuit domain results above.

(d) Create an electric circuit model for the system and calculate the same response as part (c) based on this model. Do the responses predicted by the two models match?

20 Introduction to Magnetics Design

The previous chapters on magnetics provided the key concepts needed to analyze, model, and design magnetic components such as inductors and transformers. The purpose of this chapter is to refine and extend the methods introduced there, with a focus on techniques for magnetic component design. In particular, we introduce design methods and sizing considerations for efficiently converging on an appropriate design. This includes, for example, approximate methods for sizing the magnetic core for an inductor or transformer, and metrics for comparing magnetic materials. Fundamental trade-offs – such as how magnetic components scale with power or frequency – are also explored. The approaches introduced in this chapter can be employed using the simple models for magnetic component losses described in Chapter 18. Chapter 21 will introduce more detailed methods for calculating high-frequency losses (skin and proximity effects, and core losses) than introduced thus far; these methods may be used to augment the design techniques introduced here in cases where the high-frequency losses are a critical design factor.

Magnetic components serve a diverse range of functions in power electronics, which impose various requirements on their design. For example, a filter, or "choke," inductor might be designed for a nearly constant current, while that for a resonant converter might be designed for a high-frequency ac current. Whereas a filter inductor design might be constrained by core saturation concerns, a resonant inductor design might be dominated by ac loss considerations. The diversity of requirements and their impact on design become more pronounced in multi-winding components. Frequency and power levels likewise have a large impact on design, influencing the kinds of materials and fabrication techniques that can be used. Moreover, even for a given type of component, a design may be bounded by a variety of design goals and physical constraints (e.g., efficiency requirements, temperature/thermal limits, saturation constraints, isolation requirements, etc.)

A consequence of this diversity of requirements, constraints, and design options is that magnetics design is complex. There is no single-pass approach that achieves high performance in all circumstances (or even many circumstances). We must instead often work iteratively, exploring different possible options, checking constraints, and adjusting a design until our goals are achieved. The main focus of this chapter is to provide broad guidelines and identify good starting points, rather than establishing prescriptive rules for complete designs. The last section of this chapter provides a representative approach to inductor design, leveraging the techniques previously introduced.

20.1 Filter Inductor Design and Core Factor

Filter inductors are extremely common in power electronics applications. Characteristic of a filter inductor is that its current contains an ac ripple component that is relatively small compared to its dc component. In this section we introduce an approximate approach for sizing the core for a filter inductor. This approach centers on realizing a desired set of inductor specifications without saturating the magnetic core. This core sizing approach can also be used as a starting point for some other kinds of saturation-limited magnetic components, with the caveat that it is more likely to provide a lower bound on size than an accurate prediction in those cases.

Consider the basic structure of a cored inductor with a gap as shown in Fig. 20.1(a). This illustration represents the characteristics of a variety of different core shapes, including pot, RM, and "EI" (incorporating elements shaped like an E and an I) cores. The core itself may be characterized by a core area A_c, a magnetic path length ℓ_c, and a core window area (or winding window area) W_A. The core also has a gap g. The core may be further characterized by a mean length of turn ℓ_t (also sometimes referred to as MLT), which represents the average length of a turn for a winding filling the core window area. A winding of N turns filling the window will have a wire length of $N \cdot \ell_t$. The winding itself may be characterized by the number of turns, the wire resistivity ρ, and its cross-sectional area A_W. We further assume that the winding fills the window with a *packing factor* (or *window utilization factor*) k_u representing the fraction of the window area used by the winding. Core shape, insulation requirements, and so on influence k_u, but it can be estimated from a wire data table. The core material is characterized by a saturation flux density B_{sat} above which all the magnetic domains are aligned and its incremental permeability is μ_o. One can also specify a maximum flux density $B_{\max}$ below which the material maintains an acceptably high relative permeability.

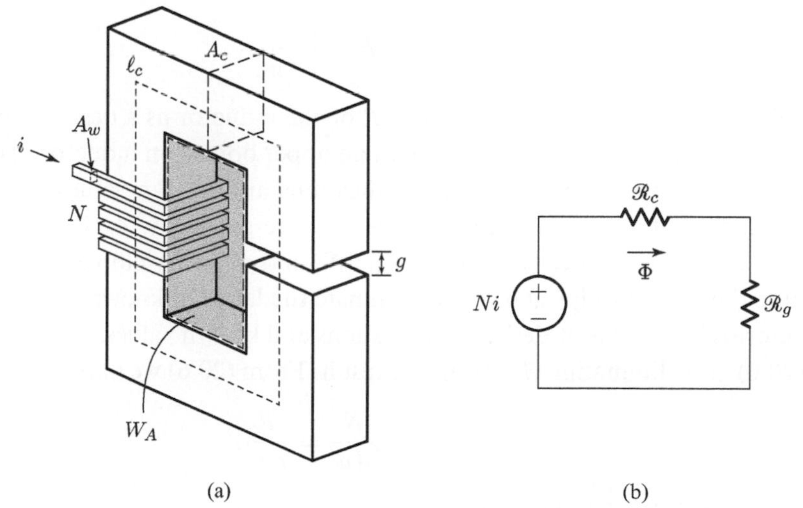

Figure 20.1 (a) A gapped cored inductor illustrating some of the important geometric features. (b) A magnetic circuit model for the inductor of (a).

To develop a metric for the required core size of a filter inductor, we first identify some of the basic requirements on the core. Using the magnetic circuit model of Fig. 20.1(b), we obtain the relation

$$Ni = \Phi \cdot (\mathscr{R}_c + \mathscr{R}_g) \approx \Phi \cdot \mathscr{R}_g, \tag{20.1}$$

where we assume that the core reluctance is small compared to the gap reluctance. To prevent the core flux density from exceeding a maximum allowed value $B_{\max}$ when the inductor current is at its maximum value $I_{\max}$, we require

$$NI_{\max} \approx \Phi \cdot \mathscr{R}_g \leq (B_{\max} A_c) \cdot \left(\frac{g}{\mu_0 A_c} \right), \tag{20.2}$$

where we have assumed that the effective gap area and core area are the same. Simplifying this expression, we get the constraint

$$NI_{\max} \leq \frac{B_{\max}\, g}{\mu_0}. \tag{20.3}$$

Under the same assumptions, we can express the inductance of the inductor as

$$L \approx \frac{N^2 \mu_0 A_c}{g}. \tag{20.4}$$

A further constraint on a practical inductor is that we must be able to fit the winding turns within the core window area. Using the definitions above, this constraint may be expressed as

$$NA_W \leq k_u\, W_A. \tag{20.5}$$

Assuming that the winding fully utilizes the core window at the specified packing factor k_u, and has a mean length of turn ℓ_t, the dc resistance for the inductor winding R can be found as

$$R = \frac{\rho\, \ell_t\, N}{A_W}. \tag{20.6}$$

We will treat the allowable dc resistance R of the inductor as a design specification; for purely dc waveforms this also indirectly provides an upper bound on inductor dissipation of $I_{\max}^2 R$.

Based the above, we have design specifications and known values L, R, $I_{\max}$, $B_{\max}$, μ_0, ρ, and k_u; some unknowns that characterize the core geometry, A_c, W_A, and ℓ_t; and some other unknowns that are needed for design, A_W, N, and g. To develop a core sizing metric, we will combine the relationships above and eliminate the latter unknowns to express the desired core characteristics in terms of design specifications and known values. We start by combining (20.5) and (20.6) and eliminating A_W from the result. From (20.6) we have

$$\frac{N}{A_W} = \frac{R}{\rho\, \ell_t}. \tag{20.7}$$

Multiplying this by (20.5) we get

$$N^2 \leq \frac{k_u W_A R}{\rho\, \ell_t}. \tag{20.8}$$

From (20.4) we have

$$N^2 = \frac{gL}{\mu_0 A_c}, \tag{20.9}$$

and by substituting (20.9) into (20.8) we can eliminate N, yielding

$$\frac{gL}{\mu_0 A_c} \leq \frac{k_u W_A R}{\rho \ell_t}. \tag{20.10}$$

To eliminate g from (20.10) we recognize from the maximum flux density limit (20.3) that

$$g \geq \frac{\mu_0 N I_{\mathrm{max}}}{B_{\mathrm{max}}}. \tag{20.11}$$

To eliminate N from this expression, we recognize that the limit on flux density in the core gives us

$$N \geq \frac{L I_{\mathrm{max}}}{B_{\mathrm{max}} A_c}, \tag{20.12}$$

which, when substituted into (20.11), gives us

$$g \geq \frac{\mu_0 L I_{\mathrm{max}}^2}{B_{\mathrm{max}}^2 A_c}. \tag{20.13}$$

Substituting (20.13) into (20.10), we obtain

$$\frac{L^2 I_{\mathrm{max}}^2}{B_{\mathrm{max}}^2 A_c^2} \leq \frac{k_u W_A R}{\rho \ell_t}, \tag{20.14}$$

which may be rearranged to provide the key result,

$$\frac{L^2 I_{\mathrm{max}}^2 \rho}{B_{\mathrm{max}}^2 R k_u} \leq \frac{A_c^2 W_A}{\ell_t}. \tag{20.15}$$

The right-hand side of 20.15 defines a quantity K_g called the *core factor*,

$$K_g = \frac{A_c^2 W_A}{\ell_t}. \tag{20.16}$$

The core factor, which has units of linear dimension to the fifth power (e.g., m^5), expresses a geometrical property of a magnetic core. Given a design requirement for a filter inductor as expressed by the left-hand side of (20.15), we know the minimum value of K_g that a core must have to implement the inductor. A designer may calculate K_g for a given core from datasheet information, or leverage pre-tabulated values of K_g, which are available for a variety of cores. For example, Table 20.1 shows a variety of data about the RM (rectangular modular) family of ferrite cores, including the core factor K_g. The physical structure of RM-type cores is illustrated in Fig. 20.2. The core halves are designed to clamp around a toroidal bobbin and winding such that the inductor presents an approximately square footprint. The cylindrical centerpost may be machined down on one or both core halves to provide a gap for energy storage.

Table 20.1 Core data for the RM (rectangular modular) family of ferrite cores.

	RM4	RM5	RM6	RM8	RM10	RM12	RM14
Effective magnetic path length ℓ_c (mm)	22.7	22.4	28.6	38.0	44.0	56.9	70.0
Effective core area $A_{c,e}$ (mm^2)	14.0	23.7	36.6	64.0	98.0	140	178
Minimum core area $A_{c,\min}$ (mm^2)	10.7	17.3	30.2	53.5	86.6	121	165
Core window area $W_{A,c}$ (mm^2)	15.6	18.2	26.0	48.9	69.5	110	155
Bobbin window area $W_{A,b}$ (mm^2)	7.7	9.5	15	30.0	41.5	73.0	107
Bobbin mean turn length ℓ_t (mm)	20	25	30	42	52	61	71.5
Effective core volume $V_{c,e}$ (mm^3)	318	530	1050	2430	4310	7970	12 500
Core set weight (g)	1.7	3.0	5.5	13	23	42	74
Effective surface area $A_{s,e}$ (mm^2)	586	787	1130	2020	2960	4460	6820
Thermal resistance R_{th} (°C/W)	86	69	60	38	30	23	19
Core area product $A_{c,e}W_{A,b}$ (mm^4)	1.1×10^2	2.3×10^2	5.5×10^2	1.92×10^3	4.07×10^3	1.02×10^4	1.90×10^4
Core factor K_g $A_{c,\min}^2 W_{A,b}/\ell_t$ (mm^5)	4.4×10^1	1.1×10^2	4.6×10^2	2.0×10^3	6.0×10^3	1.8×10^4	4.07×10^4

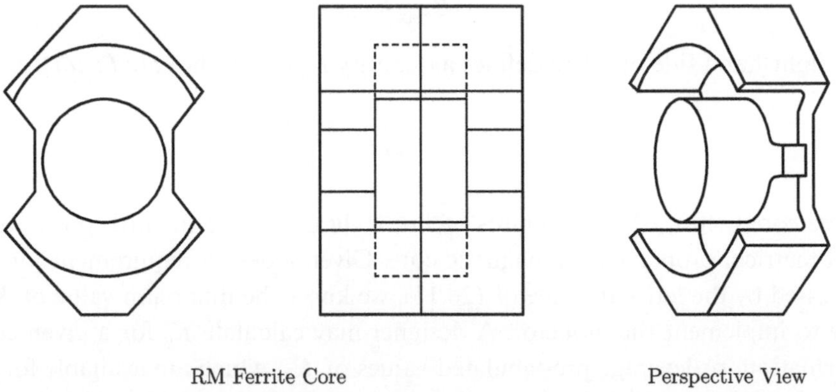

Figure 20.2 Representative core shape for the RM family of ferrite cores. The illustration shows the geometry of the RM8, RM10, and RM12 cores.

Example 20.1 Selecting a Core for a Filter Inductor

Example 18.5 describes the design of a filter inductor, but provides no basis for selection of the inductor core. Here we show how the K_g method can be used to select an appropriate core for the inductor. The 25 μH inductor is specified to carry a peak current $I_{\max}$ of 11 A, and is to be designed with a ferrite material having a maximum allowable flux density of $B_{\max} = 0.3$ T. We desire a dc winding resistance R of 10 mΩ, and assume a wire resistivity of $\rho = 2.09 \times 10^{-8}$ Ω-m (copper at 75 °C) and a winding packing factor $k_u = 0.5$. Based on this information, we require a minimum core factor of

$$K_g \geq \frac{L^2 I_{\max}^2 \rho}{B_{\max}^2 R k_u} = \frac{\left(25 \times 10^{-6}\right)^2 (11)^2 \left(2.09 \times 10^{-8}\right)}{(0.3)^2 \left(10^{-2}\right)(0.5)} = 3.5 \times 10^{-12}\ (\text{m}^5). \tag{20.17}$$

Converting the units to mm^5 and consulting Table 20.1, we see that the RM10 core (with $K_g = 6.0 \times 10^3$ mm^5) is the smallest RM-type core that meets the core factor requirement. This is the core size that is used in Example 18.6, yielding an appropriate inductor design for the application.

20.1.1 Core Factor Calculations and Published Data

As described above, published core factor data is available for a wide variety of cores, including in references cited in the bibliography. However, care is necessary when consulting such data. First, it should be noted that some developments of core factor include the packing factor k_u as part of the definition of K_g rather than as a design variable (i.e., in the numerator of K_g rather than the denominator of the design constraints). Such references typically include a nominal value of k_u as part of the tabulated core factor data, which should be accounted for when using such data.

In addition, when calculating core factors or leveraging published core factor data, careful attention should be paid to the specific values employed in (20.16). As illustrated in Table 20.1, a magnetic core can be characterized as having a minimum core cross-sectional area $A_{c,\min}$ (for determining saturation) and an effective core cross-sectional area $A_{c,e}$ (for determining inductance). While these two values would ideally be the same, they can be significantly different in some core geometries. Redeveloping K_g while considering the distinction between these values, it can be shown that one should employ the minimum core cross-sectional area $A_{c,\min}$ to calculate K_g rather than the effective cross-sectional area $A_{c,e}$. Nonetheless, tabulated values of K_g sometimes use the effective cross-sectional area, making the resulting K_g values optimistic in cases where $A_{c,\min}$ is significantly smaller than $A_{c,e}$. Likewise, the expression for K_g includes the winding window area W_A. For typical designs which use a bobbin or "coil former" on which the conductor is wound, K_g is more appropriately calculated using the bobbin window area $W_{A,b}$ than the window area of the core itself $W_{A,c}$. As can be seen in Table 20.1, the bobbin window area can be considerably smaller than the core window area, yielding smaller values of K_g when it is employed. Nonetheless, some tabulated values of K_g use $W_{A,c}$, and one should account for this if a bobbin is to be employed in the inductor design. In Table 20.1, K_g is calculated using $A_{c,\min}$ and $W_{A,b}$.

20.1.2 Limitations of K_g

The derivation of core factor considers core saturation constraints and dc winding resistance, but neglects ac conduction loss and core loss effects. It is thus mainly useful for providing an initial estimate for required core size in applications such as filter inductors where core loss and ac conduction loss components are expected to be small and the design is limited by core saturation constraints.

It is also worth noting that while the design constraint in (20.15) includes an allowed value of dc winding resistance R, there is no guarantee that an inductor design based on the resulting K_g will be thermally acceptable. That is, for a selected value of R, the loss may be sufficiently high to cause the inductor temperature rise to be unacceptable. Indeed, as R specified for a design is made larger, the (dc-related) dissipation gets larger and the core size (based on the resulting K_g) gets smaller, each of which tends to increase temperature rise. Thus, with the "K_g method" for selecting a core the designer must still separately check the thermal performance of a resulting design, and iterate the design if the temperature rise is too high (e.g., by starting with a smaller value of R and a larger resulting value of K_g). We address thermal considerations in magnetics design more directly in the next section.

20.2 Thermal Constraints in Magnetics Design

The core factor metric developed in the previous section is based upon realizing a desired energy storage within a specified flux density limit (i.e., without saturating the magnetic core). While saturation is an important constraint, magnetics designs are very often thermally limited. That is, the minimum size of a component is often dictated by the requirement that its power dissipation does not make it overheat. The next chapter addresses methods for calculating power dissipation in magnetics under a wide variety of conditions, while the chapter on thermal modeling and heat sinking treats heat transfer and temperature rise in detail. In this section, we provide a simplified treatment of the relationship between power dissipation and temperature rise in magnetic components, with the goal of providing basic guidance for magnetics design under thermal constraints.

Core manufacturers sometimes provide a thermal resistance R_{th} for a core, usually specified in °C/W. This number is intended to allow one to calculate the core "hot-spot" temperature rise ΔT (usually the core centerpost temperature rise) above ambient for a given total component dissipation P_{diss} (core plus winding dissipation) based on the equation

$$\Delta T = R_{th} \cdot P_{\text{diss}}. \tag{20.18}$$

Published thermal resistance data for RM-type cores are listed in Table 20.1, for example. When available, this information is extremely useful because it helps the designer estimate how much power dissipation is permissible for an acceptable temperature rise. (What temperature rise is acceptable is very application dependent, but allowed temperature rises in the range of 25–50 °C are not unusual.)

The model of (20.18) is grossly oversimplified, and cannot possibly represent well all situations of interest. First, the temperature rise of a component necessarily depends on the heat transfer properties of its surroundings (e.g., how heat is able to move away from the component). Second, the specific source and location of the dissipation can affect the hotspot location and temperature rise, as can the detailed thermal properties of the core material, the winding insulation and structure, the bobbin, and so on. However the data necessary for such models are not typically available, and making your own measurements is necessary to leverage this approach. The model of (20.18) is intended to represent "typical" conditions, though published R_{th} data usually comes without any specification of an exact use case to which the numbers apply. Such R_{th} numbers may be expected as most accurate for simply wound and insulated components with good conditions for passive cooling (e.g., a wire-wound inductor that is board mounted and cooled via conduction and natural convection). Given these limitations, it is not surprising that published R_{th} values for a given core can vary somewhat from manufacturer to manufacturer (and even sometimes in different publications from the same manufacturer!). Such variations may simply reflect differences in assumptions and operating conditions underpinning the data.

Despite these limitations, when applied conservatively, published R_{th} data as in Table 20.1 and in the references at the end of the chapter can be extremely useful, providing first-order estimates of temperature rise and allowed power dissipation for a magnetic component. These estimates can then be refined based on more detailed thermal modeling (e.g., using the methods shown in the thermal chapter, finite-element modeling, etc.) and/or based on measurements on experimental prototypes.

Often, no thermal resistance information is available, either because it is not provided for a particular core or because a custom core is being used. In this case, one can make initial thermal estimates based on available data from cores having similar properties. Here we plot data from RM-type ferrite cores for this purpose. Because heat transfer must take place through the surface of a component, it is useful to express thermal properties in terms of the surface area available for heat transfer. Figure 20.3 plots thermal resistance R_{th} versus effective surface area $A_{s,e}$ of RM-type ferrite cores along with a fitting function. A different view of this same information is shown in Fig. 20.4, which shows the effective heat transfer coefficient h_{tc} versus effective surface area $A_{s,e}$ for different RM type cores, along with a fitting function. The coefficient h_{tc} is defined as the heat transferred per surface area per degree of temperature rise above ambient of the core "hot-spot" temperature.† It gives an estimated power dissipation P_{diss} of a magnetic component having effective surface area A_{se} and temperature rise ΔT of

$$P_{\text{diss}} = h_{tc} A_{s,e} \Delta T. \tag{20.19}$$

One can use the data of Figs. 20.3 and 20.4 to provide very rough initial estimates for temperature rise and heat transfer for a variety of core types as a function of effective surface area. It is recognized that these plots do not resolve any of the limitations of the simplified model of (20.18) described above, and that the data need to be applied even more conservatively

† This is slightly different than a traditional heat transfer coefficient for convective cooling, which expresses heat transfer per area per degree of difference between the surface of the object and the fluid temperature.

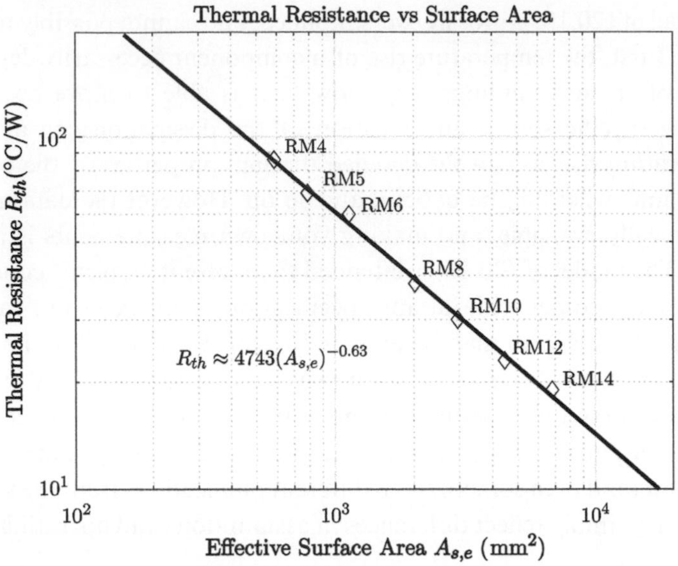

Figure 20.3 Thermal resistance R_{th} versus effective surface area $A_{s,e}$ for different RM core sizes. The effective surface area is approximated as the surface area of the smallest cube that would enclose a core with a full-window winding.

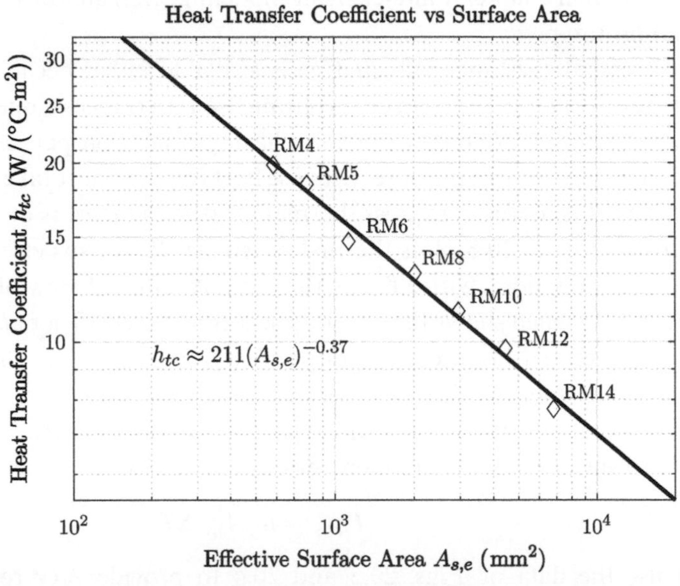

Figure 20.4 Effective heat transfer coefficient h_{tc} versus effective surface area $A_{s,e}$ for different RM core sizes.

when applied to magnetic components whose shapes, materials, and constructions vary widely from those with RM cores (such as by derating h_{tc}). Nonetheless, the results match other cases of passively cooled magnetics surprisingly well. While achievable heat transfer is often one of the greatest uncertainties in power magnetics design, this approach gives at least a reasonable starting point when other data are unavailable.

20.3 Inductor Energy Storage Limits

Given the difficulty of modeling temperature rise for magnetic components in the general case, it is hard to arrive at a core sizing metric for filter inductors that accounts for both thermal limits and flux density limits. Nonetheless, given specified limits on both power dissipation and flux density for a given magnetic core, it is possible to identify a limit on energy storage for a filter inductor built with that core.

Consider the design of an inductor on a specified core where the inductor current is nearly constant, such that dc winding loss is the only important loss mechanism. We assume the ability to select the core gap g and the number of turns N, with wire cross-sectional area A_W selected as large as possible based on the core window area W_A and packing factor k_u as per (20.5). Assuming the core permeability is sufficiently high that the energy is stored in the gap, and that the gap area is close to the core area A_c, the limit on energy storage imposed by the maximum flux density $B_{\max}$ may be expressed as the product of the maximum energy density and the gap volume,

$$W_M = \frac{1}{2}LI^2 \leq \frac{B_{\max}^2}{2\,\mu_0} \cdot gA_c. \tag{20.20}$$

This expression sets a *minimum* gap length g to achieve a specified energy storage W_M.

We can express the impact of a power dissipation limit $P_{d,\max}$ (e.g., as set by temperature rise constraints) on inductor current carrying capability as

$$P_d = I^2R = I^2\frac{\rho_{\mathrm{Cu}}N\ell_t}{A_W} \leq P_{d,\max}. \tag{20.21}$$

Combining this with the relationship (20.5) between turns, wire area, and window area, we get a power-dissipation-based constraint on the maximum ampere-turns for the inductor,

$$NI \leq \sqrt{\frac{P_{d,\max} \cdot k_u \cdot W_A}{\rho_{\mathrm{Cu}} \cdot \ell_t}}. \tag{20.22}$$

Using the expression for inductance in (20.4), we can express the energy storage of the inductor in terms of ampere-turns as

$$W_M = \frac{1}{2}LI^2 = \frac{1}{2} \cdot \frac{L}{N^2} \cdot (NI)^2 = \frac{1}{2} \cdot \frac{\mu_0 A_c}{g} \cdot (NI)^2. \tag{20.23}$$

Combining 20.22 with 20.23, we get

$$W_M = \frac{1}{2}LI^2 \leq \frac{1}{2} \cdot \frac{\mu_0 A_c}{g} \cdot \frac{P_{d,\max} \cdot k_u \cdot W_A}{\rho_{\mathrm{Cu}} \cdot \ell_t}, \tag{20.24}$$

which sets a *maximum* gap g that we can have to get a given energy storage within the power dissipation constraint.

By simultaneously applying (20.20), which sets a minimum gap length based on flux density, and (20.24), which sets a maximum gap length based on power dissipation, we can find an optimum gap length g_{opt} that provides the maximum energy that can be stored while meeting both of the constraints:

$$W_{M,\max} = \frac{1}{2} \cdot \frac{\mu_0 A_c}{g_{\text{opt}}} \cdot \frac{P_{d,\max} \cdot k_u \cdot W_A}{\rho_{\text{Cu}} \cdot \ell_t} = \frac{B_{\max}^2}{2\,\mu_0} \cdot g_{\text{opt}} A_c, \tag{20.25}$$

thus,

$$g_{\text{opt}} = \frac{\mu_0}{B_{\max}} \sqrt{\frac{P_{d,\max} \cdot k_u \cdot W_A}{\rho_{\text{Cu}} \cdot \ell_t}} \tag{20.26}$$

and

$$W_{M,\max} = \frac{1}{2} L I^2 \Big|_{\max} = \frac{1}{2} B_{\max} A_c \sqrt{\frac{P_{d,\max} \cdot k_u \cdot W_A}{\rho_{\text{Cu}} \cdot \ell_t}}. \tag{20.27}$$

Selecting the number of turns N appropriately allows us to achieve this maximum energy storage capability W_M across a range of inductances L and maximum current values $I_{\max}$, limited only by the requirement of an integer number of turns:

$$L = \frac{N^2 \mu_0 A_c}{g_{\text{opt}}}, \qquad I_{\max} = \frac{B_{\max} g_{\text{opt}}}{\mu_0 N}. \tag{20.28}$$

A refined version of the above derivation that accounts for a difference between the minimum core cross-sectional area $A_{c,\min}$ (related to maximum flux density) and the effective core cross-sectional area $A_{c,e}$ (for determining inductance and energy storage) yields modest changes to the above results. In particular, we find that the optimum gap includes a new factor, the ratio of $A_{c,e}$ to $A_{c,\min}$,

$$g_{\text{opt}} = \frac{\mu_0}{B_{\max}} \sqrt{\frac{P_{d,\max} \cdot k_u \cdot W_A}{\rho_{\text{Cu}} \cdot \ell_t}} \cdot \frac{A_{c,e}}{A_{c,\min}}, \tag{20.29}$$

such that the resulting maximum energy storage is determined by the minimum core cross-sectional area $A_{c,\min}$:

$$W_{M,\max} = \frac{1}{2} L I^2 \Big|_{\max} = \frac{1}{2} B_{\max} A_{c,\min} \sqrt{\frac{P_{d,\max} \cdot k_u \cdot W_A}{\rho_{\text{Cu}} \cdot \ell_t}}. \tag{20.30}$$

The inductance and allowed maximum current in terms of the optimal gap g_{opt} in (20.29) and the number of winding turns N then become

$$L = \frac{N^2 \mu_0 A_{c,e}}{g_{\text{opt}}}, \qquad I_{\max} = \frac{B_{\max} g_{\text{opt}}}{\mu_0 N} \cdot \frac{A_{c,\min}}{A_{c,e}}. \tag{20.31}$$

The above results allow us to identify the maximum energy we can continuously store using a given inductor core with constraints on allowable flux density and dissipation. Given a thermal model such as that of (20.18), we can also characterize maximum energy storage under constraints of allowable flux density and temperature rise. For example, Fig. 20.5 shows

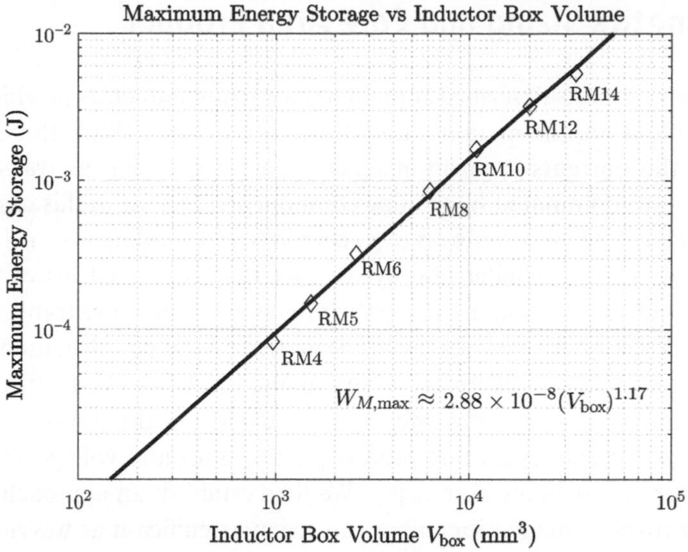

Figure 20.5 Maximum energy storage $W_{M,\max}$ versus inductor box volume V_{box} of inductors wound on RM cores for $B_{\max} = 0.3\,\text{T}$, $\Delta T_{\max} = 25\,°\text{C}$, $\rho_{\text{Cu}} = 2.09 \times 10^{-8}\,\Omega$-m and $k_u = 0.5$.

the maximum energy storage capability versus "box volume" for inductors based on RM-type cores with $B_{\max} = 0.3\,\text{T}$ and $\Delta T_{\max} = 25\,°\text{C}$ (with $\rho_{\text{Cu}} = 2.09 \times 10^{-8}\,\Omega$-m and $k_u = 0.5$.) Maximum energy storage is computed via (20.30) with power dissipation determined from the temperature rise limit using (20.18). Box volume is the volume of the smallest cubic box in which the wound inductor would fit. This plot is based on the data for RM-type cores in Table 20.1. As per (20.30), the achievable energy storage for a given core size is proportional to $B_{\max}$. Likewise, energy storage increases proportional to the square root of allowed power dissipation $P_{d,\max}$, or approximately as the square root of allowed temperature rise $\Delta T_{\max}$. (But because copper resistivity ρ_{Cu} increases with temperature, maximum energy storage goes up slightly more slowly than the square root of allowed temperature rise.) The plot and fitting function of Fig. 20.5 can be used to estimate how physically large an inductor must be (and what size inductor core is needed) to achieve a given energy storage under typical design constraints.

Two other interesting observations can be made from Fig. 20.5. First, the achievable energy storage (or energy density) that can be obtained with inductors using typical magnetic materials is not very high. For comparison, a typical 400 V aluminum electrolytic capacitor having a box volume of $10^4\,\text{mm}^3$ might store on the order of 10 J, providing four orders of magnitude higher energy storage density at that volume. Second, considering the fitting function provided in Fig. 20.5, it can be seen that inductor energy storage scales somewhat faster than proportionally with box volume, such that achievable energy density increases with volume. Conversely, achievable energy storage density *reduces* at small scales, which is not a desirable property for miniaturization. This trend is not an accident of the data, but instead reflects fundamental scaling properties of magnetic components.

20.4 AC Magnetics Sizing and Core Area Product

The previous sections described approaches for sizing inductors in which core saturation and/or dc winding loss are the main limiting design constraints. While this is often the case for filter inductors, the currents in many magnetic elements – such as transformers and "resonant" inductors – are dominated by their ac components. Design in this case is often constrained by loss associated with ac waveforms (e.g., core loss and winding loss including skin and proximity effects), and is made challenging by the fact that ac-related losses are often determined by detailed aspects of the design (e.g., how windings are layered in a transformer, or the proximity of windings to the gap in an inductor). While we defer detailed calculation of these ac-related losses and means for mitigating them to the next chapter, we can still establish some general approaches to the design of ac magnetics. We begin by exploring the power-handling characteristics of inductors and transformers that process purely sinusoidal voltages and currents, and use this as a basis for approximate core sizing. We then establish an approach for selecting the number of winding turns in magnetic components having significant ac waveform content.

20.4.1 Power Capability of an Inductor with Sinusoidal Waveforms

Consider an inductor having a purely sinusoidal voltage and current at a given operating frequency (e.g., as might occur for an inductor in a resonant converter), that is, $v_L(t) = V_L \cos(2\pi ft)$ and $i_L(t) = I_L \sin(2\pi ft)$. We might characterize the rating of such an inductor by its *apparent power*,

$$S = v_{L,\mathrm{rms}} i_{L,\mathrm{rms}} = \frac{1}{2} V_L I_L, \tag{20.32}$$

which has units of V-A, which for an inductor is the same as its reactive power Q.

Let us consider a simplified analysis in which we assume that the core flux density is limited to a peak value B_0 (for example, by saturation or, more typically, by core loss limits) and the winding current density is limited to a peak value J_0 (e.g., by winding loss limits). In fact, this is an oversimplification: permissible flux density and current density are typically dependent on heat transfer and, as we will see, can often be traded off against one another to optimize a design. Moreover, allowable current density is also a function of the details of the winding design. Nonetheless, we can still gain insight into ac-dominated magnetic components by temporarily treating B_0 and J_0 as design-independent constants.

Consider the inductor structure of Fig. 20.1 with a full-window winding. Accounting for the limited allowed winding current density J_0 and the maximum packing factor k_u of conductor in the window as per (20.5), we can express the maximum allowable winding current amplitude as $I_L = (1/N)J_0 k_u W_A$. Likewise, recognizing that $\lambda = NB_0A_c$ and $v_L(t) = d\lambda/dt$, we get $V_L = 2\pi fNB_0A_c$. Putting these together, we can re-express the apparent power in (20.32) as

$$S = \frac{1}{2}(2\pi fNB_0A_c)\left(\frac{1}{N}J_0 k_u W_A\right), \tag{20.33}$$

which simplifies to

$$S = \pi f B_0 J_0 k_u \cdot (A_c W_A). \tag{20.34}$$

According to (20.34), under the above assumptions we can express the power capability of an ac inductor simply in terms of its operating frequency (f), basic design limits (B_0, J_0, and k_u), and a geometric factor $A_c W_A$, which is known as the *core area product*. Stated another way, given an apparent power requirement S for an ac inductor, we can express a minimum core area product required to implement it as

$$A_c W_A \geq \frac{S}{\pi f B_0 J_0 k_u}. \tag{20.35}$$

The *core area product* $A_c W_A$, which has units of linear dimension to the fourth power, can thus be thought of as a metric for sizing the core for an ac inductor. As with metrics such as K_g, it is frequently tabulated for various cores to aid the designer in choosing an appropriate core size for a design. The core area product for the RM family of cores is provided in Table 20.1, where the effective core area $A_{c,e}$ and the bobbin window area $W_{A,b}$ are used to calculate core area product.

20.4.2 Power Capability of a Transformer with Sinusoidal Excitation

In considering ac magnetic components, one could likewise ask how the size of a transformer is related to its power handling capability. To analyze this, we consider a 1 : 1 transformer operating with purely sinusoidal voltages and currents and feeding a resistive load as illustrated in Fig. 20.6. (A 1 : 1 turns ratio is nice for conceptual simplicity, though the turns ratio does not matter to the results. Likewise, a resistive load is selected so that we can consider real power P, as opposed to apparent power S.) As in our previous analysis, we consider limits B_0 and J_0 on the peak ac core flux density and peak winding current density.

Figure 20.7 shows an example structure for such a transformer. We consider the case where the winding window is fully used (with a packing factor k_u) and where half of the window area is devoted to the primary winding and half the window area is devoted to the secondary winding. We further consider the case where the magnetizing current is small compared to the load current (such that $i_2 \approx i_1$) and neglect the effects of leakage inductance (such that $v_2 \approx v_1$). Under the above conditions we can express the transformer voltages as $v_1(t) = v_2(t) = V_T \sin(2\pi f t)$ and transformer currents as $i_1(t) = i_2(t) = I_T \sin(2\pi f t)$, where $I_T = V_T/R$. We can express the average (or real) power P being processed by the transformer as

$$P = \frac{1}{2} V_T I_T. \tag{20.36}$$

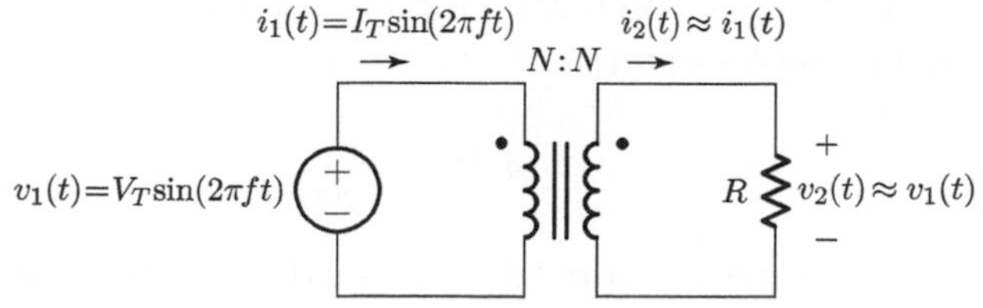

Figure 20.6 A 1 : 1 transformer operating with sinusoidal voltages and currents.

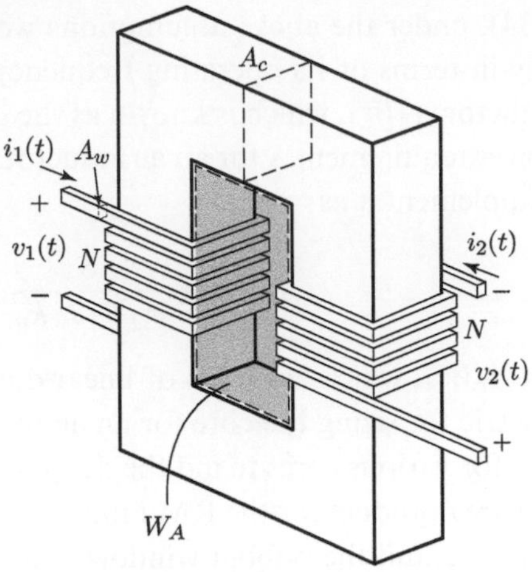

Figure 20.7 Structure of a 1 : 1 transformer illustrating the core area A_c and window area W_A.

Accounting for the limited allowed winding current density J_0 and the maximum packing factor k_u of conductor in the window, we can express the maximum allowable winding current amplitude as $I_T = (1/2N)\, J_0 k_u W_A$. Likewise, recognizing that the maximum allowable flux density B_0 yields maximum flux linkage values $\lambda_1 = \lambda_2 = \lambda = NB_0A_c$ and that $v_1(t) = v_2(t) = d\lambda/dt$, we get a maximum allowed transformer voltage amplitude $V_T = 2\pi fNB_0A_c$. Putting these together, we can re-express the average power P in (20.36) as

$$P = \frac{1}{2}(2\pi fNB_0A_c)\left(\frac{1}{2N}J_0k_uW_A\right), \tag{20.37}$$

which simplifies to

$$P = \frac{1}{2}\pi fB_0\, J_0\, k_u \cdot (A_c W_A). \tag{20.38}$$

According to (20.38), under the above assumptions we can express the power handling capability of a transformer simply in terms of its operating frequency, f; basic design limits B_0, J_0, and k_u; and its core area product, $A_c W_A$. For a reactive load instead of a resistive load, the expression in (20.38) would be the same, except with apparent power S replacing average (or real) power P. (This makes sense since we obtained the result based on permissible amplitude limits on transformer voltage and current, without any dependency on their relative phase; that is, we've really calculated a volt-ampere limit for the transformer.) Stated another way, given an apparent power handling requirement S for a transformer, we can express a minimum core area product required to implement it as

$$A_c W_A \geq \frac{2S}{\pi fB_0\, J_0\, k_u}. \tag{20.39}$$

Note that this limit on core area product is exactly twice that calculated for an ac inductor in (20.35); the factor of two arises from the fact that the transformer has two windings, transferring input energy to the output rather than storing it.

As with an ac inductor, we can see that the core area product $A_c W_A$ is closely tied to the volt-ampere capability of a transformer. It can likewise be used to as an approximate means of identifying required transformer "size," given design limits for B_0, J_0, and k_u. As we will see, just as with ac inductors, this is only a very crude measure since appropriate values for B_0 and J_0 in fact depend upon design details and can even be traded against one another to optimize a design. Moreover, selection of "size" based on $A_c W_A$ does not guarantee that the resulting design will meet thermal, efficiency, or other requirements of a design.

Example 20.2 Pre-Design Estimate for Size and Mass of a Transformer

Figure 20.8(a) show an isolated dc/dc converter called the double-ended bridge converter. As described in Chapter 7, this converter can provide galvanic isolation, transformation, and regulation of the output. However, for applications where regulation is not needed, this converter is sometimes operated in a "square-wave" operating mode, as illustrated in the waveforms of Fig. 20.8(b). In this operating mode the converter acts as a "dc transformer," providing galvanic isolation and – neglecting the effects of transformer parasitics and device drops – an approximately fixed transformation ratio:

$$\frac{V_{\text{out}}}{V_{\text{in}}} = \frac{N_S}{N_P}. \tag{20.40}$$

One benefit of this operating mode is that the input and output filters (including inductor L) can be quite small. Likewise, because the transformer operates with (approximately) square-wave voltages and currents at nearly unity power factor, it too can be relatively small and efficient.

In this example we show how core area product calculations can be used to develop a pre-design estimate of the required size and mass of the transformer, which is typically the largest and heaviest component of such a converter. One might be particularly interested in such an estimate at the outset of a design for a mass-sensitive aerospace application, for example. We treat a design in which $V_{\text{in}} = 400\,\text{V}$, $I_{\text{in}} = 1\,\text{A}$, and $f = 1/T = 200\,\text{kHz}$, and neglect the impact of transformer parasitics (e.g., magnetizing and leakage inductances) on our estimates.

As illustrated in Fig. 20.8(b), the flux linkage (shown for the primary winding) is triangular, with a peak value of

$$\lambda_{pk} = \frac{V_{\text{in}} T}{4}, \tag{20.41}$$

which lets us express the peak amplitude of the triangular core flux density B_{pk} as

$$B_{pk} = \frac{V_{\text{in}}}{4 f N_P A_c}, \tag{20.42}$$

where N_P is the number of primary transformer turns and A_c is the transformer core area. We can thus express the transformer core area in terms of the peak flux density as

$$A_c = \frac{V_{\text{in}}}{4 f N_P B_{pk}}. \tag{20.43}$$

Considering a fully utilized core window area W_A with half of the window allocated to the primary and half allocated to the secondary at a copper packing factor k_u, we can express the rms current density in the windings as

$$J_{\text{rms}} = \frac{2 N_P I_{\text{in}}}{k_u W_A}, \tag{20.44}$$

which gives us an expression for the required window area in terms of current density:

$$W_A = \frac{2 N_P I_{\text{in}}}{k_u J_{\text{rms}}}. \tag{20.45}$$

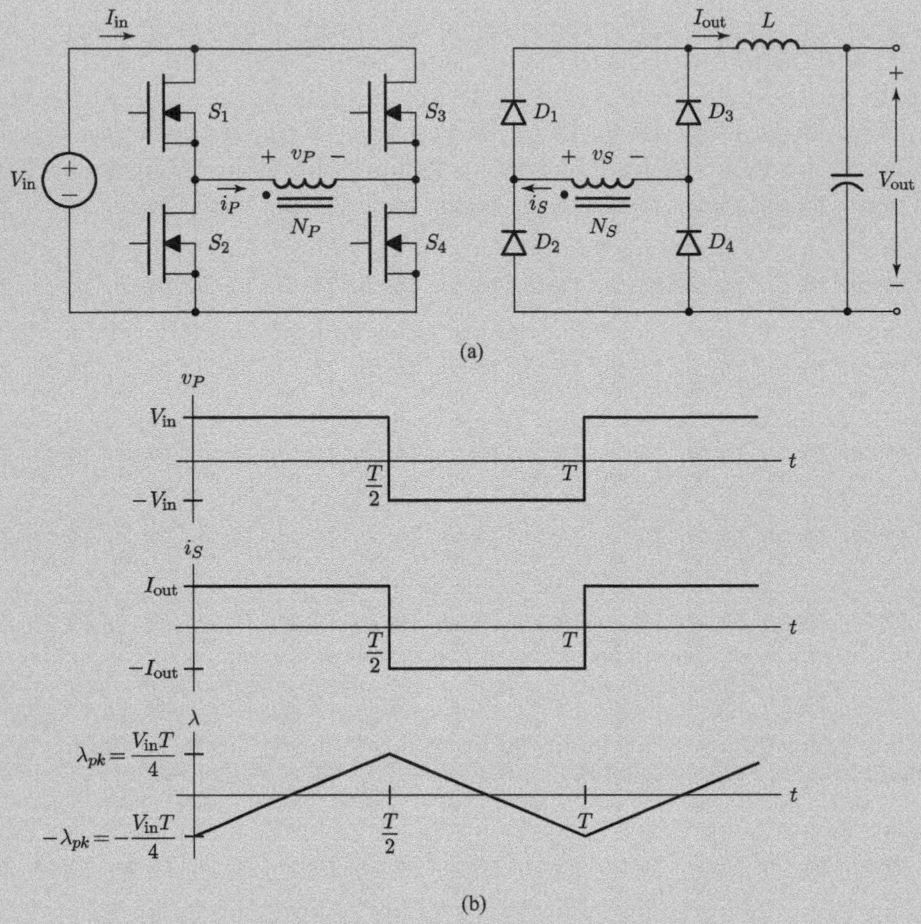

Figure 20.8 (a) A double-ended bridge converter. (b) Waveforms for the circuit of (a) when the converter is operated as a "dc transformer."

Taking the product of (20.43) and (20.45) and simplifying, we get a minimum required value for the core area product:

$$A_c W_A \geq \frac{V_{in} I_{in}}{2 k_u f B_{pk} J_{rms}}. \tag{20.46}$$

This expression closely resembles that of (20.39), but differs slightly as we are treating a transformer with non-sinusoidal waveforms and are expressing the result in terms of rms current density instead of peak current density.

To estimate the required transformer size and mass, we judiciously select parameter values to apply in (20.46). We select $B_{pk} = 0.1\,\text{T}$, as manufacturer's data give a reasonable core loss density of $P_V = 200\,\text{mW/cm}^3$ at 200 kHz and 100 °C for N87 core material at this flux density. (We treat the triangular flux density waveform as being "approximately sinusoidal" for the purposes of using manufacturer's loss data.) We further select $k_u = 0.35$ and $J_{rms} = 200\,\text{A/cm}^2$ as being achievable for many designs. Employing these values in (20.46) we get

$$A_c W_A \geq \frac{(400)(1)}{2\,(0.35)\,(2 \times 10^5)\,(0.1)\,(2 \times 10^6)} = 1.4 \times 10^{-8}\,\text{m}^4. \tag{20.47}$$

Considering the core area product data provided in Table 20.1, we find that an RM14 core meets this requirement.

It should be emphasized that this core area product calculation is only as good as the parameter estimates that are used, and in no way guarantees that a resulting transformer will be viable in terms of efficiency, temperature rise, transformer construction, transformer parasitics, and so on. For example, dissipation is only treated indirectly through selection of B_{pk} and J_{rms}. Overly aggressive values for these parameters will result in impractically small designs that will overheat, while overly conservative values for these parameters will yield designs that are larger than necessary.

We can partially check the impact of the parameter selections by calculating the potential losses and temperature rise that they imply for the selected core. For example, we can find the core loss P_c in the proposed RM14 core at the limiting flux density B_{pk} as the product of the loss density and core volume:

$$P_c = P_V V_{c,e} = (2.0 \times 10^5)(1.25 \times 10^{-5}) = 2.5\,\text{W}. \tag{20.48}$$

Similarly, using the core and material parameters defined previously, we can calculate winding loss at the specified rms current density J_{rms} as

$$P_w = J_{\text{rms}}^2\, k_u\, W_A\, \ell_t\, \rho_{\text{Cu}}\, F_R, \tag{20.49}$$

where $F_R = R_{\text{ac}}/R_{\text{dc}}$ is an "ac resistance factor" to account for the impact of skin and proximity effects on the effective winding resistances. Using $\rho_{\text{Cu}} = 2.25 \times 10^{-8}$ Ω-m (the resistivity of copper at 100 °C), $F_R = 2.0$ (a reasonable value for a litz wire design), and the parameters of an RM14 core, we get

$$P_w = (2 \times 10^6)^2(0.35)(1.07 \times 10^{-4})(0.0715)(2.25 \times 10^{-8})(2.0) = 0.5\,\text{W}. \tag{20.50}$$

Using the thermal model of (20.18) we get an estimated temperature rise for this total dissipation:

$$\Delta T = R_{th} \cdot (P_c + P_w) = (19)(3.0) = 57\,^\circ\text{C}, \tag{20.51}$$

which might be considered acceptable. Note that while these calculations do not give the loss and temperature rise of an optimized transformer design, they do suggest that the selected parameters and estimated RM14 core size are reasonable.

Given the above, we can make a first-order estimate of transformer weight. Assuming that a fraction k_u of the bobbin window is filled with copper (8.96 g/cm^3) for a mass of 23.8 g and the remaining fraction $1 - k_u$ is filled with insulation (e.g., Mylar at 1.39 g/cm^3) for a mass of 6.9 g, and adding in the weight of an RM14 core set (74 g), bobbin (5.5 g), and clips (3 g), we get a first-order estimate of 113 g for the transformer weight.

20.4.3 Selecting Turns to Minimize Loss

In this subsection we consider how one can optimally select the number of winding turns in an ac magnetic component to minimize overall loss. The design of inductors and transformers having purely ac waveforms is often limited by core loss and ac winding loss concerns. Core loss becomes a dominant consideration because the ac flux density amplitude allowed by core loss is usually less than the saturation flux density. Winding loss for ac currents requires more careful treatment than dc winding loss owing to the impact of skin and proximity effects, which increase loss. (We treat ac winding loss calculation in detail in the next chapter; here, we simply treat winding loss owing to an ac current as being some ac resistance factor F_R higher than one would obtain for a dc current having the same rms value.) As will be shown, by selecting the number of winding turns correctly, we can trade winding loss against core loss to achieve a minimum total loss.

Loss Minimization for AC Inductors Consider the design of an inductor of inductance L to carry a sinusoidal current of amplitude I_{ac} at a frequency f. We assume that the inductor is designed on a core having core volume V_c, core area A_c, window area W_A, and mean length of turn ℓ_t. Our goal is to select the optimum number of turns N that minimizes the total power dissipation in the design. It is assumed that in conjunction with selecting N we will choose a gap that provides the desired inductance,

$$g \approx \frac{N^2 \mu_0 A_c}{L}, \tag{20.52}$$

and that the number of turns selected will be sufficient to keep the inductor from saturating:

$$N \geq \frac{L I_{ac}}{B_{max} A_c}. \tag{20.53}$$

Of course, as the number of turns N is increased, the wire cross-sectional area A_W must be decreased to fit the winding in the winding window, and current density in the wire increases.

The Steinmetz core loss model expresses the core loss per unit volume as a power law with ac flux density amplitude B,

$$P_v = k f^\alpha B^\beta, \tag{20.54}$$

where P_v is the loss per unit volume, and the parameters k, α, and β are determined from manufacturer's or measured data. We can express the ac flux density in terms of the number of winding turns N as

$$B = \frac{L I_{ac}}{N A_c}. \tag{20.55}$$

Considering a core volume V_c, we can express the total core loss P_c in terms of the selected number of turns as

$$P_c = V_c k f^\alpha \left(\frac{L I_{ac}}{A_c}\right)^\beta \cdot N^{-\beta} = C_c \cdot N^{-\beta}, \tag{20.56}$$

where C_c is a design-specific constant for the purposes of optimizing N.

Assuming a fully utilized winding window, we can express the winding loss as

$$P_w = \frac{1}{2} I_{ac}^2 F_R \left(\frac{\rho_{Cu} \ell_t}{k_u W_A}\right) \cdot N^2 = C_w \cdot N^2, \tag{20.57}$$

where C_w is a design-specific constant for the purposes of optimizing N.

Together, (20.56) and (20.57) give us a total dissipation P_d in terms of the number of turns, N:

$$P_d = P_c + P_w = C_c \cdot N^{-\beta} + C_w \cdot N^2. \tag{20.58}$$

Notice that as we increase the number of turns N the core loss decreases while the winding loss increases. There should thus be some value of N that minimizes the sum of these losses. Ignoring the integer nature of N, we can find an approximate value for the optimal number of turns N_{opt} by differentiating (20.58) with respect to N and setting the result to zero:

$$\frac{dP_d}{dN} = -\beta\, C_c \cdot N^{-\beta-1} + 2\, C_w \cdot N = 0. \tag{20.59}$$

Solving this for $N = N_{\text{opt}}$, we get

$$N_{\text{opt}} = \left(\frac{\beta\, C_c}{2\, C_w}\right)^{1/(\beta+2)} \tag{20.60}$$

which gives us

$$P_{c,\text{opt}} = C_c \cdot \left(\frac{\beta\, C_c}{2\, C_w}\right)^{-\beta(\beta+2)}, \qquad P_{w,\text{opt}} = C_w \cdot \left(\frac{\beta\, C_c}{2\, C_w}\right)^{2/(\beta+2)}. \tag{20.61}$$

A useful insight can be drawn from examining the ratio of core loss to winding loss in an optimized design. Taking the ratio of the values in (20.61) and simplifying yields

$$\frac{P_{c,\text{opt}}}{P_{w,\text{opt}}} = \frac{2}{\beta}. \tag{20.62}$$

This reveals that the ratio of core loss to winding loss in an optimized design depends upon the core-loss characteristics of the magnetic material. Values for the Steinmetz parameter β typically fall somewhere between 2 and 3 for ferrite materials, such that an optimum design will have core loss close to but smaller than winding loss. For example, for a value of $\beta = 2.5$ (in the middle of the range for ferrites), core loss would ideally be 80% of winding loss for minimum total loss. One can thus optimize a design for minimum total loss by selecting a number of turns N such that the core loss to winding loss ratio of (20.62) is met. A common but incorrect rule of thumb for magnetics design is that the number of turns N should be specified to make core loss and winding loss equal. As (20.62) shows, this would only be optimal for $\beta = 2$. A correct rule of thumb is to increase the number of turns N until the incremental decrease in core loss due to an incremental change in N is equal to the incremental increase in winding loss due to that change.

In optimizing a design for minimum loss, it is recognized that in reality we must use an integer value of N close to the calculated value N_{opt}. Moreover, the selected number of turns must be sufficient to meet saturation limits as indicated in (20.53).

Loss Minimization for Transformers We can similarly optimize turn count to minimize loss in transformers. Consider the design of a transformer providing a turns ratio of $N_1 : N_2$. We assume an absolute number of primary turns $N_P = mN_1$ and a corresponding absolute number of secondary turns $N_S = mN_2$, such that by scaling m we adjust the absolute numbers of turns while preserving turns ratio. This scaling in absolute numbers of turns (via m) allows us to trade copper loss against core loss to minimize total loss. N_P and N_S must be whole numbers of turns in practice, but we neglect this constraint for analysis purposes.

Consider the design of a transformer on a core having volume V_c, core area A_c, window area W_A, and mean length of turn ℓ_t. We assume that the core is fully wound, with half the winding window dedicated to the primary and half of the window dedicated to the secondary, and with each winding implemented with the same packing factor k_u. We further neglect magnetizing current, such that the two winding currents are related by the turns ratio, and neglect leakage flux such that the two terminal voltages are related by the turns ratio. Under these assumptions, each winding has the same current density. We also make the simplifying assumption that the ac winding loss can be calculated just as for low-frequency waveforms but with an additional ac resistance factor F_R to account for skin and proximity effects.

We assume that the transformer is operated with a purely ac flux waveform, and is designed for a peak primary-referred flux linkage $\lambda_{P,\mathrm{pk}}$. For a sinusoidal primary voltage of amplitude V_s at frequency f, the peak flux linkage would be $\lambda_{P,\mathrm{pk}} = V_s/(2\pi f)$. We further assume that the transformer is designed for a primary-side rms current of $I_{P,\mathrm{rms}}$.

Based on the above assumptions and operating conditions we can find expressions for the power dissipation in the core and windings. We can calculate the peak flux density in the core from the primary flux linkage $\lambda_{P,\mathrm{pk}}$ as

$$B = \frac{\lambda_{P,\mathrm{pk}}}{N_P A_c}. \tag{20.63}$$

From this we can use the Steinmetz loss model to express the total core loss P_c in terms of the number of primary turns N_P as

$$P_c = V_c k f^{\alpha} \left(\frac{\lambda_{P,\mathrm{pk}}}{A_c}\right)^{\beta} \cdot N_P^{-\beta} = C_c \cdot N_P^{-\beta}, \tag{20.64}$$

which gives us the same general form for core loss as (20.56).

We can also write an expression for the current density in the windings in terms of the number of primary turns:

$$J_{\mathrm{rms}} = \frac{N_P I_{P(\mathrm{rms})}}{\frac{1}{2} W_A k_u}. \tag{20.65}$$

Note that while we have calculated this current density based on the primary winding, it applies to the secondary winding as well. Using this we can express the total winding loss in the core window as

$$P_w = J_{\mathrm{rms}}^2 k_u W_A \ell_t \rho_{\mathrm{Cu}} F_R = \frac{4 I_{P(\mathrm{rms})}^2 \ell_t \rho_{\mathrm{Cu}} F_R}{W_A k_u} \cdot N_P^2 = C_w \cdot N_P^2, \tag{20.66}$$

which gives us the same general form for core loss as (20.57).

Note that the total dissipation $P_d = P_c + P_w$ has the exact same form as in (20.58) but with N_P replacing N and with different values for the design-dependent constants C_c and C_w as determined in (20.64) and (20.66). Consequently, we can find the optimal number of primary turns N_P and optimal breakdown of core and copper loss using these values for C_c and C_w in (20.60)–(20.62).

We conclude that we achieve minimum loss at the same ratio of core loss to copper loss in a transformer as for an inductor. We achieve this minimum loss by scaling the absolute numbers of turns in the transformer (N_P and N_S together) while keeping the turns ratio $N_P : N_S$ at the desired value. It should be noted that in some transformer designs (e.g., providing a large step-down ratio to a low voltage) one cannot achieve the optimum numbers of turns for minimum loss because one hits the physical limit of a single turn for one of the windings. Such transformers can be significantly winding-loss dominated. Techniques for realizing fractional *effective* turns can be useful for reducing loss in such cases, as explored in [9].

Example 20.3 Transformer Turns Optimization

Here we explore selection of the numbers of turns for the transformer sized in Example 20.2. The waveforms for the dc/dc converter transformer in this design are shown in Fig. 20.8, with $V_{\text{in}} = 400$ V, $I_{\text{in}} = 1$ A, and $f = 1/T = 200$ kHz. The converter is to be designed for a transformation ratio of $N_P : N_S$ of 1 : 1 (i.e., so that it provides galvanic isolation without voltage transformation). We assume the same core size (RM14), magnetic material (N87), packing factor ($k_u = 0.35$), ac resistance factor ($F_R = 2.0$), and wire resistivity ($\rho_{\text{Cu}} = 2.25 \times 10^{-8}$ Ω-m) as considered previously.

From (20.41), the transformer sees a primary-side flux linkage of

$$\lambda_{P,\text{pk}} = \frac{V_{\text{in}}}{4f}. \tag{20.67}$$

At 200 kHz and 100 °C, N87 has the Steinmetz parameters $kf^{\alpha} = 1.26 \times 10^{-6}$ and $\beta = 2.59$ for loss density P_{v} in W/cm^3 and flux density B in mT. For an RM14 core, this gives us the following design coefficient C_c from (20.64) (in W):

$$\begin{aligned} C_c &= V_c kf^{\alpha} \left(\frac{V_{\text{in}}}{4fA_c} \right)^{\beta} \\ &= (12.5)(1.26 \times 10^{-6}) \left(\frac{400 \times 10^3}{4(2 \times 10^5)(1.78 \times 10^{-4})} \right)^{2.59} \\ &= 1.35 \times 10^4 \text{ W}. \end{aligned} \tag{20.68}$$

Likewise, from (20.66) we can find the design coefficient C_w (in W):

$$\begin{aligned} C_w &= \frac{4\, I_{P(\text{rms})}^2\, \ell_t\, \rho_{\text{Cu}}\, F_R}{W_A\, k_u} \\ &= \frac{4\,(1)^2\,(0.0715)\,(2.25 \times 10^{-8})(2.0)}{(1.07 \times 10^{-4})\,(0.35)} \\ &= 3.44 \times 10^{-4} \text{ W}. \end{aligned} \tag{20.69}$$

Solving this for $N_P = N_S = N_{\text{opt}}$ based on (20.60), we get

$$N_{\text{opt}} = \left(\frac{\beta\, C_c}{2\, C_w} \right)^{1/(\beta+2)} = \left(\frac{(2.59)(1.35 \times 10^4)}{(2)(3.44 \times 10^{-4})} \right)^{0.218} = 47.85, \tag{20.70}$$

which motivates us to choose $N_P = N_S \approx N_{\text{opt}} = 48$. For these numbers of turns we get, from (20.64) and (20.66),

$$P_c = C_c \cdot N_P^{-\beta} = (1.35 \times 10^4) \cdot (48)^{-2.59} = 0.60 \text{ W} \tag{20.71}$$

and

$$P_w = C_w \cdot N_P^2 = (3.44 \times 10^{-4})(48)^2 = 0.79 \text{ W}. \tag{20.72}$$

The ratio of core loss to winding loss in the optimum design is as derived in (20.62). The total power dissipation for the optimized design is $P_d = P_c + P_w = 1.4$ W.

The individual loss components and total loss for the transformer are plotted against the number of primary turns in Fig. 20.9. The total loss and loss distribution at the optimum number of turns are as expected from the calculations above. At least in this example, one can achieve close to optimum loss across a reasonable variation in the number of primary turns about the optimum. This can be valuable for addressing practical constraints in the transformer design, such as the requirement for integer numbers of turns, limitations in available wire sizes, and so forth.

The total power dissipation for the optimized design (1.4 W) is well below the 3.0 W "sum of losses" estimate in Example 20.2. The "sum of losses" estimate was simply based on assuming operation at both the flux density and current density limits selected for the $A_c W_A$ calculations. In the optimized design we do not prescribe maximum values for current density or flux density, but rather choose optimum values for them by trading winding loss against core loss to minimize total loss. For the optimum number of turns in this design, we obtain from (20.42) an ac flux density $B = 59\,\text{mT}$ and from (20.44) an rms current density $J_{rms} = 256\,\text{A/cm}^2$, below the assumed limit for flux density and above the assumed limit for current density in Example 20.2.

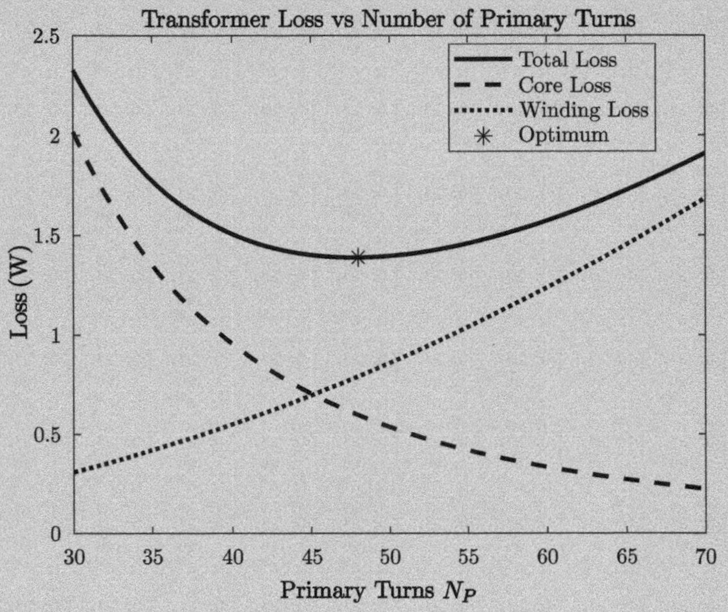

Figure 20.9 Total loss, P_d, core loss, P_c, and winding loss, P_w, versus number of primary turns N_P for the proposed transformer design. The total loss at the optimum number of turns for minimum loss ($N_{opt} = 48$) is indicated by the asterisk.

20.5 Performance Factor for Magnetic Materials

One aspect of designing a magnetic component is selecting a suitable magnetic material. There are many characteristics to consider in selecting a magnetic material, such as permeability, saturation flux density, core loss, temperature characteristics, and cost, among others. Which aspects are most important depends upon the application. For example, to minimize size in a filter inductor design, a material with a high saturation flux density might be preferred, while for an ac inductor one might instead focus on core loss characteristics in selecting a material. Likewise, in designing a current transformer one often prefers a very high-permeability material to reduce magnetizing current error, while in an inductor design there is no benefit in terms of size or efficiency to having a permeability above a certain minimum threshold.‡

Given the diverse requirements of different applications, and the multiplicity of requirements on a magnetic material even for a given application, no single metric can adequately define the "goodness" of a magnetic material. Nonetheless, in applications where core loss is a major consideration, a metric known as the *performance factor* can be useful for comparing different magnetic materials and their performance as a function of frequency.

To derive this metric, recall the expression in (20.34) for the power-handling capability (or apparent power S) of an inductor with sinusoidal waveforms, which we rewrite here with the factors expressed in a different order:

$$S = \pi A_c W_A J_0 k_u \cdot (B_0 f). \tag{20.73}$$

The power handling is proportional to factors involving the core geometry ($A_c W_A$), allowable winding current density and packing factor ($J_0 k_u$), and allowable flux density and operating frequency ($B_0 f$). Equation (20.38) shows that the power handling capability of a transformer is defined by these same factors. Of these factors, it is the last one ($B_0 f$) that directly relates to the performance of the magnetic material. Suppose we define the allowable flux density B_0 as the frequency-dependent value $\widehat{B}$ that results in a specified core loss density P_V in a material. That is, at any frequency f_o, $\widehat{B}(f_o)$ is defined as whatever flux density gives the specified loss density P_V (e.g., 200 mW/cm^3) in the material at f_o. Given this definition, the product $\widehat{B}f$ defines a material-dependent metric that is proportional to the power handling capability of an inductor or transformer using the material at the specified loss density P_V. $\widehat{B}f$ is known as the magnetic material performance factor, and is sometimes provided by manufacturers (or calculated by designers) to help select among materials.

Performance factor might be thought of as defining the power handling capability of a material when operated at a specified loss density. Various values of loss density are used in computing performance factor, with values in the range 100–500 mW/cm^3 being typical. The appropriate loss density to consider depends upon the cooling limitations anticipated for a given design, with higher values of loss density appropriate to smaller designs, larger allowed temperature rises, and higher-performance cooling methods.

Figure 20.10 plots the performance factor of three representative MnZn ferrite materials versus frequency for a loss density of $P_V = 200$ mW/cm^3. It can be seen that the different materials provide different levels of performance (in terms of core-loss-limited power handling capability). One would find qualitatively similar characteristics for other materials. Other considerations equal, the material providing the highest performance factor at the design frequency of interest would be preferred in an application where core loss is a dominant constraint.

Note that each material has some frequency range over which it gives its best performance. In practical terms, this means that if one were to design a set of loss-constrained ac inductors at a variety of frequencies on a given core size using the same magnetic material, one designed

‡ This perhaps surprising result is detailed in [12]. In fact, many high-frequency inductors are made with NiZn ferrites having relative permeabilities only in the tens.

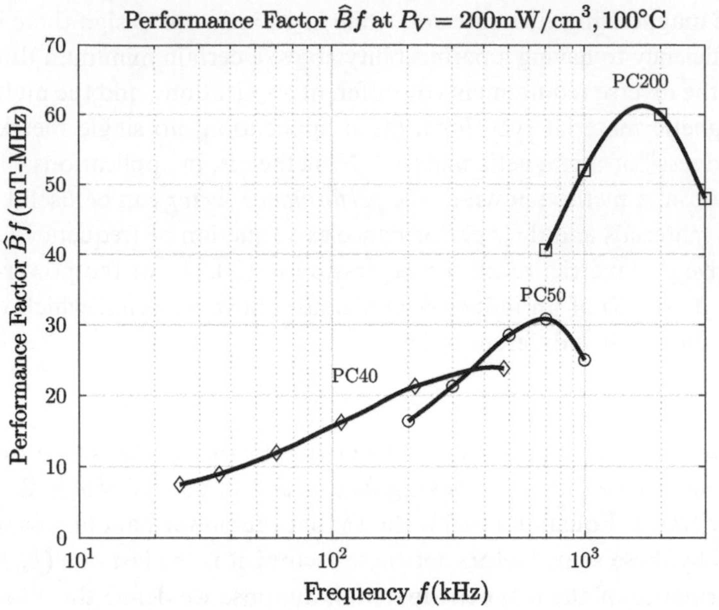

Figure 20.10 Performance factor $\widehat{B}f$ for three representative MnZn ferrite materials at a loss density $P_V = 200\,\text{mW/cm}^3$ and $100\,°\text{C}$. Symbols represent points for which core loss data is available.

at a frequency near where the performance factor peaks would tend to achieve the highest power-handling capability. (This strictly applies to frequencies (e.g., up to a few megahertz) where litz wire or other techniques can be employed to mitigate variations in winding loss with frequency.) Likewise, one can generally realize the smallest ac inductor for a given power handling requirement near the frequency where the material performance factor peaks.

While a given material has a specific frequency range over which it provides a high performance factor, the highest available performance factor considered across different ferrite materials tends to rise with frequency as suggested by Fig. 20.10. This trend of available performance factor increasing with frequency continues (with transition to low-permeability NiZn ferrite materials) out to at least 10 MHz, albeit with modest improvements versus frequency. This in turn suggests that – to the extent that winding loss can be addressed – there is potential for improvement in the power density of magnetic components to at least 10 MHz. At sufficiently high frequencies there are no adequately high-performing magnetic materials, and coreless or "air core" inductor designs become prevalent.

20.6 An Iterative Inductor Design Algorithm

As described in the introduction, there is no single-pass design approach that will yield an optimized magnetics design for a wide range of circumstances. Much of the material in this chapter has thus focused on topics such as identifying good design starting points and optimizing a design under a specific set of conditions. Nevertheless, one can take an *iterative* approach to realizing an optimized magnetics design. This section outlines one such approach, focusing on optimizing an inductor design based on a set of available magnetic core sizes.

The design algorithm may be described in terms of the following six steps:

1. Select an initial core and magnetic material.
2. Find the minimum number of turns $N_{\min}$ that will avoid saturation.
3. Find the number of turns $N_{\text{opt}} \geq N_{\min}$ that minimizes total power dissipation $P_d = P_c + P_w$.
4. Check if total power dissipation P_d is acceptable (in terms of efficiency, temperature rise, etc.). If not, select a larger core and/or a better magnetic material and return to step 2. Otherwise, proceed to step 5.
5. Check if a design based on the same material and next-smaller available core size has already failed at step 4. If not, and the present design has excess margin (in terms of dissipation, temperature rise, saturation, etc.), choose the next-smaller core size and return to step 2. Otherwise, proceed to step 6.
6. Select a gap to provide the specified inductance and complete the design.

In step 1 an initial core size and magnetic material are selected. Depending upon the inductor type (e.g., filter inductor, ac inductor, etc.), one might select an initial core size based on the K_g method (Section 20.1), on energy storage and dissipation limits (Section 20.3), or on $A_c W_A$ estimates (Section 20.4). Likewise, depending upon inductor type one might select an initial magnetic material based on its saturation flux density, performance factor (Section 20.5), temperature characteristics, and so on. It should be noted that while good initial selections will reduce the number of iterations to realize a design, poor initial selections will still yield the same ultimate result, such that "blind guessing" is a perfectly valid way to initiate a design.

In step 2, the minimum number of turns to avoid saturation is determined. Based on the constraint in (20.12), one way to express the minimum number of turns is

$$N_{\min} = \frac{L I_{\max}}{B_{\max} A_c}, \tag{20.74}$$

where $B_{\max}$ is chosen to be below the saturation flux density by some margin (e.g., 75% of B_{sat}).

In step 3, the optimum number of turns $N_{\min}$ to minimize loss is determined. In situations where winding loss is proportional to N^2, the results of Section 20.4.3 may be employed. Even in situations where those results do not apply, a similar approach can be employed. One may also simply calculate the loss for different numbers of turns $N \geq N_{\min}$ in a brute force fashion to find the minimum loss point. Methods for determining winding and core loss in complex cases are detailed in the next chapter. One can also apply numerical tools for loss calculation. Note that because power dissipation typically increases with temperature, the loss calculations should assume a temperature that is at or higher than the maximum acceptable temperature for the design.

In step 4, the computed power dissipation is evaluated. Power dissipation may be deemed too high based on its efficiency impact or because it will result in too high a temperature rise (e.g., as per a thermal model such as in Section 20.2, the prediction of a numerical tool, or experimental measurement of core thermal properties). The main decision in this step is whether or not to select a larger core size and return to step 2, though one may opt to select a better core material and return to step 2 if one is not already using the best-performing material available.

In step 5, one evaluates whether a smaller core might be viable. If there is significant margin in allowable power dissipation at the present core size, and the next smaller core size has not yet been evaluated, one can return to step 2 with the next-smaller core size to see if it will work. If there is not significant margin or if one is happy with the present core size, one instead proceeds to step 6.

In step 6 a design with the proposed core is finalized. A key element of this is selecting an appropriate gap to give the desired inductance. An approximate value for the required gap is

$$g \approx \frac{\mu_0 A_{c,e} N_{\text{opt}}^2}{L} - \frac{\mu_0}{\mu_c} \ell_c, \tag{20.75}$$

though this may need to be slightly adjusted in practice (e.g., to account for gap fringing fields, etc.).

Notes and Bibliography

References [1] and [2] provide nicely tabulated data for a wide variety of core types, including calculated values for core factor K_g. While these data are extremely useful, the reader is encouraged to take care with their application as described in Section 20.1.1.

The K_g method for selecting a core is most often used for the design of filter inductors. However, this approach can also be effectively applied to the design of multi-winding magnetic components that have a single magnetic flux path and modest winding current and core flux ripple, such as some coupled inductors. A detailed treatment of the multi-winding case is provided in [2], including a method to allocate the winding window among windings.

Further discussion of thermal modeling of inductors and transformers can be found in [1], [3], [4], and [5]. Chapter 3 of [3] includes an alternative treatment of the design of inductors for maximum energy storage under power dissipation constraints, and [4] expands (20.18) to separate core and winding loss and their respective effects on core and winding temperatures.

Core area product $A_C W_A$ is tabulated in [1] for a wide range of cores. It is employed in a variety of manners for transformer design and optimization, as illustrated in [1], [6], and [7].

Trading core loss against winding loss to minimize the total loss is illustrated in Chapter 4 of [3], in [7] for transformers, and in [8] for inductors. [9] introduces techniques for realizing fractional effective turns to achieve the desired loss optimum in large step-down transformers. Winding loss predictions in inductors and transformers, including ac effects, can be made using the numerical tool LitzOpt, developed by Prof. Charles Sullivan and described in [10]. LitzOpt combines two-dimensional field simulations with an eddy current loss model to estimate winding loss, including proximity effects owing to the gap fringing fields and field build-up across the core window.

The concept of magnetic material performance factor was established by Mulder [11] for comparing ferrite materials, though this reference is unfortunately difficult to obtain. Hurley [3] is more readily available and illustrates the calculation of performance factor and its application in transformer optimization. For cases where winding loss variations with frequency become critical, modified performance factor metrics can be used, as described in [12], which also provides measured performance factor data for a variety of materials in the HF frequency range (3–30 MHz).

The scaling properties of magnetic components have important implications for design and optimization of power electronics. The impact of frequency scaling on size is explored in [2], [6], [13], and [14]. The impact of dimensional scaling on performance is explored in [15] and [16].

The iterative design process described in Section 20.6 is loosely based on one described in [8].

1. W. T. McLyman, *Transformer and Inductor Design Handbook*, 4th edn., Boca Raton: CRC Press, 2011.
2. R. W. Erickson and D. Maksimović, *Fundamentals of Power Electronics*, 3rd edn., Cham: Springer, 2020.
3. W. G. Hurley and W. H. Wölfle, *Transformers and Inductors for Power Electronics*, Chichester: Wiley, 2013.
4. F. F. Judd and D. R. Kressler, "Design Optimization of Small Low-Frequency Power Transformers," *IEEE Trans. Magnetics*, vol. 13, pp. 1058–1069, July 1977.
5. P. A. Kyaw, M. Delhommais, J. Qiu, C. R. Sullivan, J.-L. Schannen, and C. Rigaud, "Thermal Modeling of Inductor and Transformer Windings Including Litz Wire," *IEEE Trans. Power Electronics*, vol. 35, pp. 867–881, Jan. 2020.
6. W.-J. Gu and R. Liu, "A Study of Volume and Weight vs. Frequency for High-Frequency Transformers," *1993 IEEE Power Electronics Specialists Conference*, pp. 1123–1129, June 1993.
7. W. G. Hurley, T. Merkin, and M. Duffy, "The Performance Factor for Magnetic Materials Revisited," *IEEE Power Electronics Magazine*, pp. 26–34, Sep. 2018.
8. C. R. Sullivan, "Basic Power Inductor Design," *TechRxiv*, https://doi.org/10.36227/techrxiv.14370581.v1, preprint (2021).
9. M. K. Ranjram, I. Moon, and D. J. Perreault, "Variable-Inverter-Rectifier-Transformer: A Hybrid Electronic and Magnetic Structure Enabling Adjustable High Step-Down Conversion Ratios," *IEEE Trans. Power Electronics,* vol. 33, pp. 6509–6525, Aug. 2018.
10. J. D. Pollock, T. Abdallah, and C. R. Sullivan, "Easy-to-Use CAD Tools for Litz-Wire Winding Optimization," *IEEE Applied Power Electronics Conference*, pp. 1157–1163, Feb. 2003.
11. S. A. Mulder, "Loss formulas for power ferrites and their use in transformer design," *Phillips Components*, pp. 1–16, Eindhoven: Philips, 1994.
12. A. J. Hanson, J. A. Belk, S. Lim, C. R. Sullivan, and D. J. Perreault, "Measurements and Performance Factor Comparisons of Magnetic Materials at High Frequency," *IEEE Trans. Power Electronics*, vol. 31, pp. 7909–7925, Nov. 2016.
13. D. J. Perreault, J. Hu, J. M. Rivas, Y. Han, O. Leitermann, R. C. N. Pilawa-Podgurski, A. Sagneri, and C. R. Sullivan, "Opportunities and Challenges in Very High Frequency Power Conversion," *2009 IEEE Applied Power Electronics Conference*, pp. 1–14, Feb. 2009.
14. W. Odendaal and J. Ferreira, "Effects of Scaling High-Frequency Transformer Parameters," *IEEE Trans. Industry Applications,* vol. 35, pp. 932–940, Jul./Aug. 1999.
15. A. Rand, "Inductor Size vs. Q: A Dimensional Analysis," *IEEE Trans. Components Parts*, pp. 31–35, Mar. 1963.
16. C. R. Sullivan, B. A. Reese, A. L. Stein, and P. A. Kyaw, "On Size and Magnetics: Why Small Efficient Power Inductors Are Rare," *2016 International Symposium on 3D Power Electronics Integration and Manufacturing*, June 2016.

PROBLEMS

20.1 This problem concerns the design of a 200 μH inductor for a buck converter operating at 250 kHz from a 200 V input to a 100 V output. The average output current of the converter is 4.5 A. To limit conduction losses the inductor is to be designed to provide a dc resistance of no more than 50 mΩ when the inductor winding temperature is 75 °C. (Assume the resistivity of copper to be $\rho = 2.08 \times 10^{-8}$ Ω-m at this temperature.)

The inductor is to be designed using an RM-type core in a ferrite material with $B_s >$ 0.4 T and $\mu_c > 10^3\,\mu_o$. It has been decided to limit the steady-state core flux density $B_{\max}$ (peak flux density under steady-state operation) to no more than 0.32 T. A conservative a priori estimate of winding packing factor $k_u = 0.5$ on a winding bobbin is assumed as a design starting point.

(a) Use the K_g method (core factor) to select the smallest-sized RM-type core that may be expected to meet the requirements of this design.

(b) Complete the design of the inductor using this core, specifying the gap g, the number of turns N, and the wire gauge. Constrain the winding design so the rms current density in the wire does not exceed 500 A/cm^2, and confirm that the required number of turns will fit in the bobbin window area. (You may consult a wire data table.) Estimate the dc resistance of your inductor, and compare it to the original design target.

20.2 As mentioned in Section 20.1, a magnetic core can be characterized as having a minimum core cross-sectional area $A_{c,\min}$ (for determining saturation) and an effective core cross-sectional area $A_{c,e}$ (for determining inductance). Re-derive the expression for K_g considering the use of these two core-area values, to yield the following expression for K_g:

$$K_g = \frac{A_{c,\min}^2\, W_A}{\ell_t}. \tag{20.76}$$

(*Hint:* Use $A_{c,\min}$ when expressing items such as allowable core flux, and $A_{c,e}$ when considering inductance and energy storage, considering the effective gap cross-sectional area to be $A_{c,e}$.)

20.3 This problem concerns the design of an input filter inductor for a dc/dc converter. The filter inductor is to have an inductance $L = 10\,\mu$H and be rated to carry a (nearly) dc current of up to $I_{\max} = 20$ A while having a maximum power dissipation $P_{d(\max)}$ of 1 W or less and a core temperature rise of less than 25 °C above an ambient temperature of 50 °C. The inductor is to be designed using an RM-type core with a ferrite magnetic material having relative permeability $\mu_r > 1000$ and with a maximum core flux density $B_{\max}$ of 0.3 T or below. Beyond these specifications, the goal is to have as small an inductor as possible.

(a) Identify the smallest-sized RM-type core to meet the requirements of this design. For initial sizing purposes you may assume an achievable packing factor of $k_u = 0.5$.

(b) Design a complete inductor meeting the requirements. Beyond selecting a final core, this includes selecting an appropriate gap, the number of winding turns, and wire size for the selected core. (You may wish to consult the internet for wire data, and might choose to implement the winding with a single wire or paralleled smaller wires. It may also be helpful to reference published data and examine the available dimensions of appropriate core bobbins.) Provide estimates for the maximum loss, temperature rise, and flux density for your design to confirm that it meets the requirements.

20.4 Example 20.2 uses the core area product $A_c W_A$ to estimate the required size of a transformer for a double-ended bridge converter operating at $f = 200\,\text{kHz}$. It has been determined – based on the performance of available semiconductor devices – that the converter can instead be designed for the same specifications at 500 kHz operation.

(a) Use core area product calculations to estimate the RM core size needed to implement the transformer for $f = 500\,\text{kHz}$. As with Example 20.2, assume an achievable winding packing factor $k_u = 0.35$, and a current density $J_{\text{rms}} = 200\,\text{A/cm}^2$. The new transformer is to use a core of N49 ferrite material, which allows a flux density of $B_{pk} = 72\,\text{mT}$ at 500 kHz and 100 °C for a loss $P_V = 200\,\text{mW/cm}^3$.

(b) Identify the optimal number of primary turns and resulting transformer power dissipation for your core selection of part (a). You may follow the general approach of Example 20.3 in doing so. At 500 kHz and 100 °C, the Steinmetz loss coefficients for N49 ferrite are $kf^\alpha = 2.15 \times 10^{-6}$ and $\beta = 2.675$ when P_V is in W/cm^3 and B is in mT.

(c) Using the thermal model of Section 20.2 and core data of Table 20.1, estimate the temperature rise of the transformer.

20.5 This problem concerns the design of an inductor for a resonant converter. The converter requires a 10 μH inductor that carries a 4 A (peak) sinusoidal current at 1 MHz. It has been decided to design the inductor with PC200 ferrite core material. At 1 MHz and 100 °C, the Steinmetz loss coefficients for PC200 ferrite are $kf^\alpha = 6.03 \times 10^{-3}$ and $\beta = 2.631$ when P_V is in mW/cm^3 and B is in mT.

(a) Use core area product calculations to estimate the required RM core size needed to implement the inductor. For your initial core area product calculations you may assume an achievable winding packing factor $k_u = 0.4$, an allowable peak current density $J_0 = 200\,\text{A/cm}^2$ (rms current density $J_{\text{rms}} = 141\,\text{A/cm}^2$), and an allowable core loss dissipation density of $P_V = 200\,\text{mW/cm}^3$.

(b) Identify the optimal number of turns and resulting inductor dissipation for your core selection of part (a). In making your winding loss calculations, you may assume an ac resistance factor $F_R = 2.0$, and wire resistivity $\rho_{\text{Cu}} = 2.25 \times 10^{-8}\,\Omega$-m (copper at 100 °C).

(c) Using the thermal model of Section 20.2 and core data of Table 20.1, estimate the temperature rise of the inductor designed with this core.

(d) Select an appropriate gap size to realize the desired inductance with the selected core and optimized number of turns.

(e) (optional) Select an appropriate litz wire for this design. You may use a numerical tool such as LitzOpt to do so. What is the ac resistance and winding dissipation? Is the proposed ac resistance factor achieved? If not, how might one re-optimize the design?

21 Magnetics Loss Analysis and Design

Power magnetics are often constrained by loss. Consequently, the ability to accurately predict the loss of a magnetic component is extremely valuable for design. The techniques for modeling magnetics loss introduced in the previous chapters are useful, but do not adequately cover all situations. In this chapter we introduce refined methods to predict winding and core losses in magnetic components, with particular emphasis on factors (such as proximity effect) that become dominant at high frequencies and on cases where the waveforms are not purely dc or sinusoidal. We also introduce design techniques and guidelines for reducing magnetic component loss. We first treat winding (or conductor) loss analysis, and then treat core loss.

21.1 Magnetic Diffusion

To predict winding losses in high-frequency magnetic components and model phenomena such as skin and proximity effects, it is important to understand how magnetic fields propagate within conductors and how ac currents are carried as a result. In this section we start from Maxwell's equations to develop the necessary framework for such an understanding. While the mathematics for this can appear somewhat hairy at first, the results are extremely useful and provide a great deal of insight into the behavior of practical devices.

21.1.1 Derivation of the Magnetic Diffusion Equation

We start with the magneto-quasistatic (MQS) form of Maxwell's equations (excepting Gauss's law, which is not needed here):

$$\nabla \times \vec{H} = \vec{J} \qquad \text{(Ampère's law)}, \tag{21.1}$$

$$\nabla \cdot \vec{B} = 0 \qquad \text{(divergence law)}, \tag{21.2}$$

$$\nabla \times \vec{E} = -\frac{\partial \vec{B}}{\partial t} \qquad \text{(Faraday's law)}. \tag{21.3}$$

We assume the linear relationship $\vec{B} = \mu \vec{H}$ between the magnetic field intensity $\vec{H}$ and the magnetic flux density $\vec{B}$, where μ is the permeability of the material. We further assume that conductive materials of interest follow Ohm's law,

$$\vec{J} = \sigma \vec{E} \qquad \text{(Ohm's law)}, \tag{21.4}$$

where σ is electrical conductivity.

Combining Faraday's law, Ohm's law and material properties we get

$$\nabla \times (\vec{J}/\sigma) = -\frac{\partial(\mu\vec{H})}{\partial t}, \tag{21.5}$$

which, combined with Ampère's law, gives us

$$\nabla \times \left(\frac{\nabla \times \vec{H}}{\sigma}\right) = -\frac{\partial(\mu\vec{H})}{\partial t}. \tag{21.6}$$

Assuming constant values of μ and σ, we can rearrange this as

$$\frac{1}{\mu\sigma}\nabla \times \nabla \times \vec{H} = -\frac{\partial \vec{H}}{\partial t}. \tag{21.7}$$

If we apply the identity $\nabla \times \nabla \times \vec{H} = \nabla(\nabla \cdot \vec{H}) - \nabla^2\vec{H}$ and recognize that for constant μ we get $\nabla \cdot \vec{H} = 0$ because of the divergence law (21.2), we get the following key result:

$$\frac{1}{\mu\sigma}\nabla^2\vec{H} = \frac{\partial \vec{H}}{\partial t}. \tag{21.8}$$

Equation (21.8) is the *magnetic diffusion equation*, a vector form of the diffusion equation. It holds in regions of constant conductivity and permeability, and has a diffusion coefficient (or diffusivity) of $1/\mu\sigma$. It governs how magnetic fields propagate within a conductor (under MQS conditions). In Cartesian coordinates, this means[†]

$$\begin{aligned}
\frac{1}{\mu\sigma}\left[\frac{\partial^2 H_x}{\partial x^2} + \frac{\partial^2 H_x}{\partial y^2} + \frac{\partial^2 H_x}{\partial z^2}\right] &= \frac{\partial H_x}{\partial t},\\
\frac{1}{\mu\sigma}\left[\frac{\partial^2 H_y}{\partial x^2} + \frac{\partial^2 H_y}{\partial y^2} + \frac{\partial^2 H_y}{\partial z^2}\right] &= \frac{\partial H_y}{\partial t},\\
\frac{1}{\mu\sigma}\left[\frac{\partial^2 H_z}{\partial x^2} + \frac{\partial^2 H_z}{\partial y^2} + \frac{\partial^2 H_z}{\partial z^2}\right] &= \frac{\partial H_z}{\partial t},
\end{aligned} \tag{21.9}$$

where the magnetic field $\vec{H} = H_x\vec{x} + H_y\vec{y} + H_z\vec{z}$ has components H_x, H_y, and H_z that may each be a function of x, y, z, and t.

21.1.2 The One-Dimensional Case

An understanding of how magnetic fields behave within conductors can be developed from the one-dimensional case. Consider the case where we have $\vec{H} = H_z(x,t)\vec{z}$ (z-directed fields corresponding to y-directed current densities). For this case we get the following version of the magnetic diffusion equation:

$$\frac{1}{\mu\sigma}\frac{\partial^2 H_z}{\partial x^2} = \frac{\partial H_z}{\partial t}. \tag{21.10}$$

[†] In Cartesian coordinates, $\nabla^2\vec{H} = (\nabla \cdot \nabla H_x)\vec{x} + (\nabla \cdot \nabla H_y)\vec{y} + (\nabla \cdot \nabla H_z)\vec{z}$

This one-dimensional diffusion equation says that the rate of change of H_z at any point x is proportional to the second derivative of H_z with respect to x. That is, the field $H_z(x,t)$ at a point x will increase if it is (spatially) concave up and decrease if it is concave down. Thus, regions of "field valleys" tend to fill in and regions of "field hills" tend to disperse towards field profiles that are flat though not necessarily level (i.e., having constant gradient).

Diffusion equations characterize systems in which a flow (of something) is proportional to its density gradient. In the case of (21.10), if we think of H_z as a density (e.g., of z-directed field lines), the flow of these field lines in the x direction is proportional to the gradient of H_z in the x direction (with proportionality constant $-1/\mu\sigma$). To see why this is the case, consider the following equation for an incremental region Δx centered at x (i.e., between $x - \Delta x/2$ and $x + \Delta x/2$):

$$-\frac{1}{\mu\sigma}\left[\left.\frac{\partial H_z}{\partial x}\right|_{(x-\Delta x/2)} - \left.\frac{\partial H_z}{\partial x}\right|_{(x+\Delta x/2)}\right] = \left.\frac{\partial H_z}{\partial t}\right|_x \Delta x. \tag{21.11}$$

The left-hand side of (21.11) may be thought of as the difference between the flow into the region at $x - \Delta x/2$ and out of the region at $x + \Delta x/2$ (where flow at a location is proportional to the density gradient at that location), while the right-hand side of (21.11) can be thought of as the time rate of build-up (e.g., of field lines) in the incremental region between $x - \Delta x/2$ and $x + \Delta x/2$. Dividing each side by Δx and taking the limit as $\Delta x \to 0$ gives us (21.10). We can thus conclude that magnetic fields literally diffuse within a conductive medium, with a diffusion coefficient that is a function of the permeability and conductivity of the medium.

21.1.3 Solution for the One-Dimensional Case

The diffusion equation (21.10) is both linear and separable. With sinusoidal drive currents and corresponding magnetic fields, we can express its sinusoidal steady-state solution as

$$H_z(x,t) = \mathrm{Re}\{\widehat{H}_z(x)e^{j\omega t}\}, \tag{21.12}$$

where

$$\widehat{H}_z = \widehat{c}_\pm e^{\mp\sqrt{j\omega\mu\sigma}x} = \widehat{c}_\pm e^{\mp(1+j)x/\delta}, \tag{21.13}$$

in which the length constant δ is the skin depth,

$$\delta = \sqrt{\frac{2}{\omega\,\mu\,\sigma}}. \tag{21.14}$$

We can thus write the sinusoidal steady-state solution for this case as

$$H_z(x,t) = \mathrm{Re}\left\{\left[\widehat{c}_+ e^{-(1+j)x/\delta} + \widehat{c}_- e^{(1+j)x/\delta}\right]e^{j\omega t}\right\}, \tag{21.15}$$

where the complex constants $\widehat{c}_+$ and $\widehat{c}_-$ are selected to meet the boundary conditions of the problem. The magnetic fields tend to grow and/or decay exponentially with distance in a conductor with a length constant of the skin depth δ. The first term in (21.15) can be expressed as

$$H_{z+}(x,t) = \mathrm{Re}\{\widehat{c}_+ e^{-(1+j)x/\delta}e^{j\omega t}\} = \mathrm{Re}\{\widehat{c}_+ e^{-x/\delta}e^{j(\omega t - x/\delta)}\}, \tag{21.16}$$

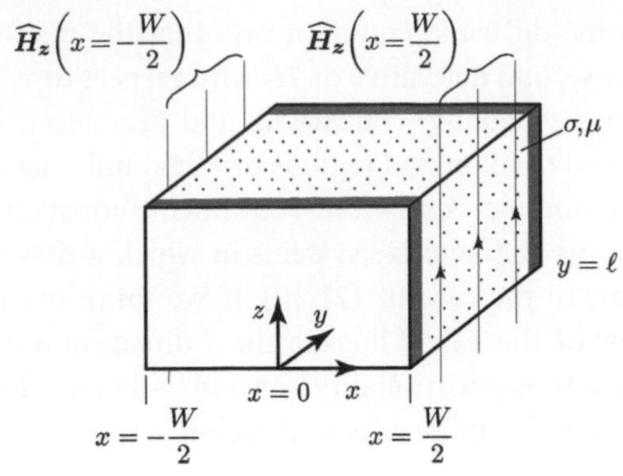

Figure 21.1 A one-dimensional system comprising a section of conductor having conductivity σ and permeability μ.

which is a "diffusion wave" which travels in the $+x$ direction and decays exponentially with a length constant δ. The second term in (21.15) is likewise a diffusion wave, except traveling (and decaying) in the $-x$ direction.

We are often interested in how currents are carried in a conductor; this can be determined from the field distribution by Ampère's law (21.1). For the present one-dimensional case this gives us

$$\vec{J} = -\frac{\partial H_z}{\partial x}\,\vec{y}, \tag{21.17}$$

resulting in a y-directed current density. From the sinusoidal steady-state field solution of (21.15) we get a current density in sinusoidal steady state of

$$J_y(x,t) = \mathrm{Re}\Big\{\Big[\widehat{c}_+ \tfrac{1+j}{\delta} e^{-(1+j)x/\delta} - \widehat{c}_- \tfrac{1+j}{\delta} e^{(1+j)x/\delta}\Big] e^{j\omega t}\Big\}. \tag{21.18}$$

Figure 21.1 shows an example one-dimensional system to which these equations apply. A section of conductor having width W, length ℓ, and infinite height is bounded at $x = -W/2$ by a z-directed sinusoidal field $\widehat{H}_z(-W/2)$ and at $x = +W/2$ by a z-directed sinusoidal field $\widehat{H}_z(W/2)$. The conductor has conductivity σ and permeability μ, and is terminated in perfectly conducting endplates at $y = 0$ and $y = \ell$. This system will have z-directed magnetic fields and y-directed current densities in the conductor as per (21.16) and (21.18), with the constants $\widehat{c}_+$ and $\widehat{c}_-$ determined from the field boundary conditions.

21.1.4 Skin Effect With Balanced Fields

Here we explore how currents are carried in a conductor as a function of skin depth (or, equivalently, as a function of frequency). Figure 21.2 shows a section of conductor like that in Fig. 21.1 driven with y-directed currents by a pair of distributed sinusoidal current sources, each having an amplitude of $K_d/2$ amperes per meter of height. The conductor is connected to the current sources via the perfectly conducting endplates. We will use the results above to determine the distribution of the z-directed magnetic field and y-directed current density in this

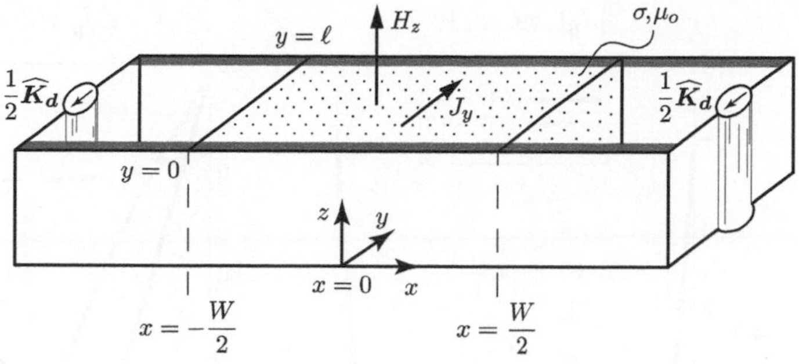

Figure 21.2 A one-dimensional system for illustrating the field and current distributions in sinusoidal steady state owing to magnetic diffusion. A balanced pair of distributed sinusoidal current sources drives y-directed current into the conductor of Fig. 21.1. The distributed sources have units of A/m, such that the total value of the y-directed current is $|\widehat{K}_d|$ times the height of the conductor in the z direction.

system. As will be seen later, this system can be used to approximate the behavior of foil and other magnetic windings.

We can represent the total distributed sinusoidal current in the conductor as $k_d(t) = \text{Re}\{\widehat{K}_d e^{j\omega t}\}$, where $\widehat{K}_d = K_d$ for a cosine wave of amplitude K_d, with the current sources on the two sides of the conductor each providing half of this current. With no additional impinging fields, this system is that of Fig. 21.1 with anti-symmetric boundary conditions on the fields at the conductor surfaces: $\widehat{H}_z(-W/2) = \widehat{K}_d/2$ and $\widehat{H}_z(W/2) = -\widehat{K}_d/2$. (That these are the correct boundary conditions is easily confirmed by taking Ampère's law around various contours in Fig. 21.2.) Applying these boundary conditions in (21.15), we can show that $\widehat{c}_+ = -\widehat{c}_-$, and we get (after a bit of algebra) the z-directed magnetic field:

$$\widehat{H}_z(x) = \frac{1}{2}\widehat{K}_d \frac{e^{-(1+j)x/\delta} - e^{(1+j)x/\delta}}{e^{(1+j)W/(2\delta)} - e^{-(1+j)W/(2\delta)}}. \tag{21.19}$$

From (21.17), we in turn get the y-directed current density:

$$\widehat{J}_y(x) = \frac{1}{2}\widehat{K}_d \frac{\frac{1+j}{\delta}e^{-(1+j)x/\delta} + \frac{1+j}{\delta}e^{(1+j)x/\delta}}{e^{(1+j)W/(2\delta)} - e^{-(1+j)W/(2\delta)}}. \tag{21.20}$$

Equations (21.19) and (21.20) are (literally) complex, but some facts are immediately apparent from their form: the magnetic field for this case in (21.19) is an odd function of x ($\widehat{H}_z(x) = -\widehat{H}_z(-x)$) such that it is anti-symmetric about the center of the conductor ($x = 0$). Similarly, the current density in (21.20) is an even function of y ($\widehat{J}_y(x) = \widehat{J}_y(-x)$) such that the current density is symmetric about the center of the conductor.

Further insight can be gained from magnitude and phase plots of the solution. Figure 21.3 shows plots of the magnitude and phase of the current density versus location in the conductor for $K_d = 1$ A/m at different values of skin depth δ relative to conductor width W. At low frequency, where the skin depth is on the order of the conductor width or larger, the current density is approximately uniform across the conductor, as it would be for dc currents. This corresponds to an approximately linear profile of magnetic field magnitude within the

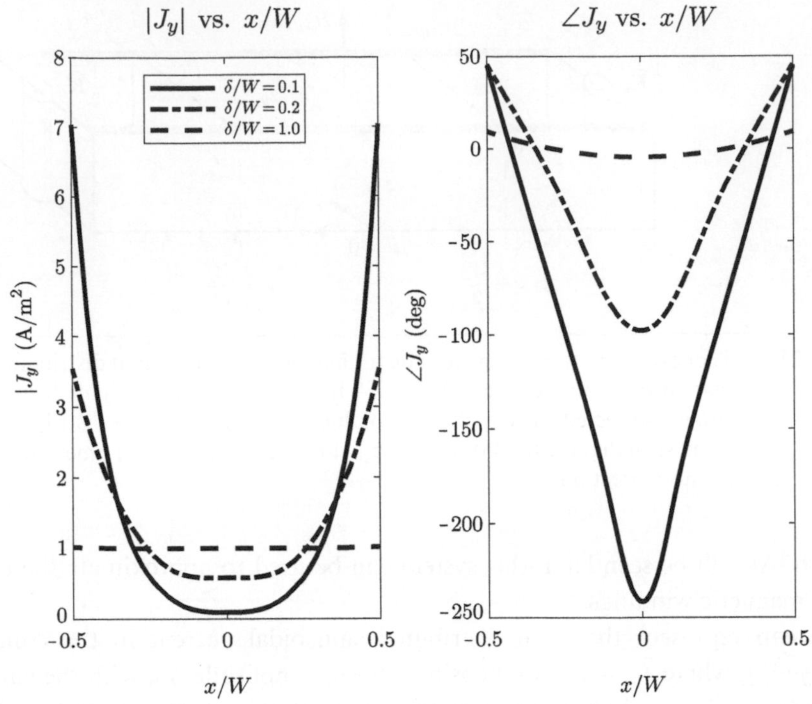

Figure 21.3 Plots of the magnitude and phase of the y-directed current density versus position in the conductor in the system of Fig. 21.2 for different values of the skin depth δ with respect to conductor width W. Each plot is for $K_d = 1$ A/m.

conductor. At high frequency, where the skin depth is much smaller than the conductor width, magnetic diffusion results in the current density falling exponentially within the conductor with length constant δ (corresponding to exponential fall-off of the magnetic field within the conductor as well). Essentially, at high frequency the changing magnetic fields penetrating the conductor induce eddy currents which tend to reinforce the currents at the surfaces of the conductor and suppress currents deeper within the conductor. For small skin depth relative to conductor width, one has the famous "skin effect" in which current is only carried near the surface of the conductor. Significant currents only exist within a few skin depths of the surface.‡

It is noteworthy that for skin depths on the order of the conductor width and below, the phases of the current density and H field each vary substantially with x. Thus, for the case of small skin depth, if we look at vertical "slices" of the conductor, each slice not only has a different current density amplitude, but a different phase as well. Consequently, the mean square current density (and loss) to carry a given net current increases as the skin depth becomes smaller. This trend is apparent in the magnitude plots of Fig. 21.3, as each plot is for the same net current.

‡ That ac currents should have a non-uniform distribution across a conductor's cross section was first noted by Gustav Kirchhoff in the 1850s. Detailed analysis of the decay of ac fields and currents in conductors was carried out by Charles Niven and Horace Lamb in the early 1880s, followed by a full description of the skin effect of current-carrying conductors by Oliver Heaviside in 1885.

The symmetry of the current density about $x = 0$ arises from the anti-symmetry of the fields at $x = -W/2$ and $x = W/2$. Were the boundary field on one side of the conductor stronger, the resulting current density would be weighted towards that surface. More generally, the surface(s) of a conductor that will carry high currents in the skin-depth-limited case are those at which high fields exist to drive magnetic diffusion.

21.1.5 Proximity Effect

Skin effect arises from the diffusion of magnetic fields due to currents carried by a conductor. However, it is also possible to have field and current density profiles in a conductor when there is no net current flow. This can arise from an externally generated magnetic field impinging on a conductor and inducing eddy currents within it. This phenomenon is commonly known as the "proximity effect" because the impinging field is typically due to currents in other conductors in proximity to the one in question.

Consider the conductor of Fig. 21.1 immersed in an external z-directed sinusoidal magnetic field having a magnitude of K_p A/m. The external field imposes symmetric boundary conditions on the fields at the conductor surfaces: $\widehat{H}_z(-W/2) = \widehat{K}_p$ and $\widehat{H}_z(W/2) = \widehat{K}_p$. Applying these boundary conditions in (21.15) we can show that $\widehat{c}_+ = \widehat{c}_-$, resulting in the z-directed magnetic field

$$\widehat{H}_z(x) = \widehat{K}_p \frac{e^{-(1+j)x/\delta} + e^{(1+j)x/\delta}}{e^{(1+j)W/(2\delta)} + e^{-(1+j)W/(2\delta)}}. \tag{21.21}$$

From (21.17), we in turn get the y-directed current density,

$$\widehat{J}_y(x) = \widehat{K}_p \frac{\frac{1+j}{\delta} e^{-(1+j)x/\delta} - \frac{1+j}{\delta} e^{(1+j)x/\delta}}{e^{(1+j)W/(2\delta)} + e^{-(1+j)W/(2\delta)}}. \tag{21.22}$$

In this case, we can see that the magnetic field within the conductor is an even function of x, while the current density is an odd function of x. The conductor thus carries no *net* current, but rather has only "circulating" currents (or eddy currents) within it, with the y-directed current density at some position $x = x_0$ being exactly the opposite of that at $x = -x_0$. We can also establish that there is no net current in the conductor directly from the field boundary conditions. Applying the integral form of Ampère's law,

$$\oint_{\ell} \vec{H} \cdot d\vec{\ell} = \iint_{S} \vec{J} \cdot d\vec{A}, \tag{21.23}$$

to a contour encircling the conductor, we get a net y-directed current per unit height in the conductor of $\widehat{H}_z(-W/2) - \widehat{H}_z(-W/2) = 0$.

Figure 21.4 shows the magnitude and phase of the induced current density in (21.22) for $K_p = 1$ A/m at different values of skin depth with respect to conductor width. For symmetric locations about $x = 0$ the current density magnitudes are the same and the current density phases are different by 180° (in keeping with an odd-symmetric current density profile), showing that current only circulates within the conductor. It can also be seen that the mean square of the circulating current density grows quickly as δ/W becomes small. Of course, while the current density in (21.22) imparts no net current to the conductor, it does induce loss. This eddy current loss (or "proximity" loss) can be quite significant in high-frequency designs, as will be shown later.

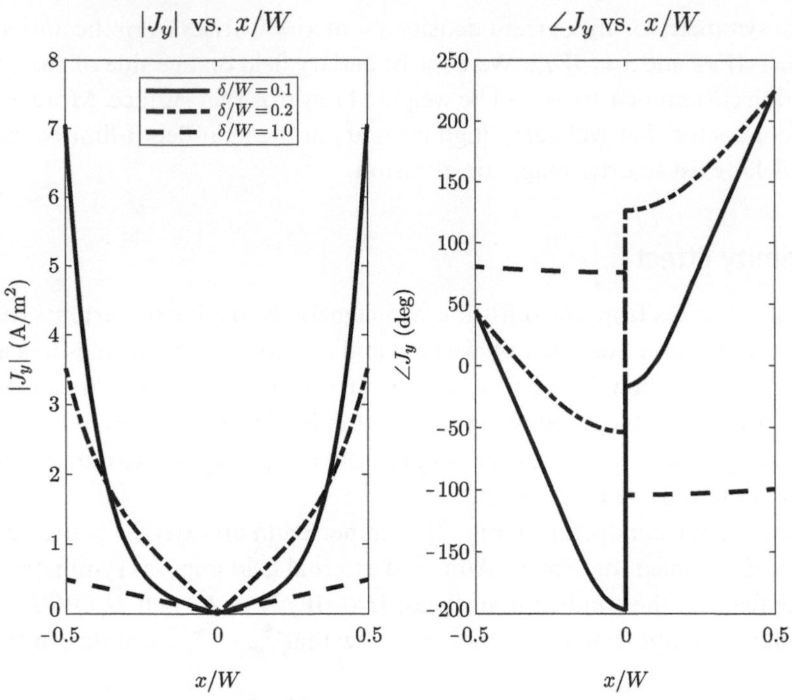

Figure 21.4 Plots of the magnitude and phase of the y-directed current density versus position in the conductor in the system of Fig. 21.1 for the case $\widehat{H}_z(-W/2) = \widehat{H}_z(W/2) = \widehat{K}_p$ at different values of the skin depth δ with respect to conductor width W. This is the case of an external z-directed field of amplitude K_p impinging on a conductor carrying no net current. The plots are each for $K_p = 1$ A/m.

21.1.6 Field and Current Decomposition

The previous two subsections illustrated solutions to the magnetic diffusion equation in the system of Fig. 21.1 for two special cases: (i) antisymmetric (odd) magnetic fields with symmetric (even) current densities, and (ii) symmetric (even) magnetic fields with antisymmetric (odd) current densities. Interestingly, we can represent any situation as a linear combination of these two special cases by even/odd decomposition. (Even/odd decomposition and some of its properties are reviewed in Section 8.2.1.) This is a useful decomposition because we can interpret the first case as representing current carried by the conductor in a balanced fashion (e.g., with balanced skin effect) and the second case as associated with proximity-effect currents circulating within the conductor but delivering no net current.

Again consider the system of Fig. 21.1 with boundary conditions at the conductor surfaces $\widehat{H}_z(-W/2)$ and $\widehat{H}_z(W/2)$. We can decompose the z-directed magnetic field $\widehat{H}_z(x)$ into an odd component $\widehat{H}_s(x)$ and an even component $\widehat{H}_p(x)$:

$$\widehat{H}_z(x) = \widehat{H}_s(x) + \widehat{H}_p(x), \tag{21.24}$$

where

$$\widehat{H}_s(x) = \frac{1}{2}\left(\widehat{H}_z(x) - \widehat{H}_z(-x)\right), \qquad \widehat{H}_p(x) = \frac{1}{2}\left(\widehat{H}_z(x) + \widehat{H}_z(-x)\right). \tag{21.25}$$

The y-directed current density $\widehat{J}_y(x)$ can be decomposed into an even component $\widehat{J}_s(x)$ related to $\widehat{H}_s(x)$ and an odd component $\widehat{J}_p(x)$ related to $\widehat{H}_p(x)$:

$$\widehat{J}_y(x) = \widehat{J}_s(x) + \widehat{J}_p(x), \tag{21.26}$$

where

$$\widehat{J}_s(x) = \frac{1}{2}\left(\widehat{J}_y(x) + \widehat{J}_y(-x)\right), \qquad \widehat{J}_p(x) = \frac{1}{2}\left(\widehat{J}_y(x) - \widehat{J}_y(-x)\right). \tag{21.27}$$

As described above, we might think of $\widehat{J}_s(x)$ as the "balanced skin effect" component of $\widehat{J}_y(x)$ and $\widehat{J}_p(x)$ as the "proximity effect" component of $\widehat{J}_y(x)$.

We can obtain the boundary conditions for the magnetic field components as

$$\widehat{H}_s(W/2) = -\widehat{H}_s(-W/2) = \frac{1}{2}\left(\widehat{H}_z(W/2) - \widehat{H}_z(-W/2)\right), \tag{21.28}$$

$$\widehat{H}_p(W/2) = \widehat{H}_p(-W/2) = \frac{1}{2}\left(\widehat{H}_z(W/2) + \widehat{H}_z(-W/2)\right). \tag{21.29}$$

Leveraging the results of the previous two subsections, we can formulate the decomposed solution for the current density. Adapting (21.20) with the boundary condition (21.28), we get the balanced skin effect component of current,

$$\widehat{J}_s(x) = \frac{\left(\widehat{H}_z(-W/2) - \widehat{H}_z(W/2)\right)}{2} \cdot \frac{\frac{1+j}{\delta}e^{-(1+j)x/\delta} + \frac{1+j}{\delta}e^{(1+j)x/\delta}}{e^{(1+j)W/(2\delta)} - e^{-(1+j)W/(2\delta)}}, \tag{21.30}$$

and adapting (21.22) with the boundary condition (21.29), we get the proximity effect component of current,

$$\widehat{J}_p(x) = \frac{\left(\widehat{H}_z(W/2) + \widehat{H}_z(-W/2)\right)}{2} \cdot \frac{\frac{1+j}{\delta}e^{-(1+j)x/\delta} - \frac{1+j}{\delta}e^{(1+j)x/\delta}}{e^{(1+j)W/(2\delta)} + e^{-(1+j)W/(2\delta)}}. \tag{21.31}$$

There are two main benefits of this decomposition. First, given the boundary fields, we can use (21.30) and (21.31) to immediately find the two current density components and the total current density without any further algebra. Second, it turns out that we can calculate loss owing to the balanced skin effect component $\widehat{J}_s(x)$ and loss owing to the proximity effect component $\widehat{J}_p(x)$ separately and add them to get the correct total loss in the conductor. This is possible because $\widehat{J}_s(x)$ is even and $\widehat{J}_p(x)$ is odd, such that they are spatially orthogonal (i.e., the integral of their product is zero). Since loss is proportional to the space integral of $\widehat{J}_y\widehat{J}_y^* = (\widehat{J}_s + \widehat{J}_p)(\widehat{J}_s^* + \widehat{J}_p^*)$, and cross terms between $\widehat{J}_s$ and $\widehat{J}_p$ die under integration, we can calculate the losses due to $\widehat{J}_s$ and $\widehat{J}_p$ separately and simply add them to get total loss. This is what is meant in the literature by the "orthogonality of skin effect and proximity effect."

Example 21.1 Current Distribution in a Conductor

Figure 21.5 shows a one-dimensional system in which a distributed sinusoidal current source $\widehat{K}_d$ drives current in a conductor from one side. This system matches that of Fig. 21.1 with field boundary conditions $\widehat{H}_s(-W/2) = \widehat{K}_d$ and $\widehat{H}_s(W/2) = 0$. We will compute the current distribution in the conductor directly and using the decomposition described above.

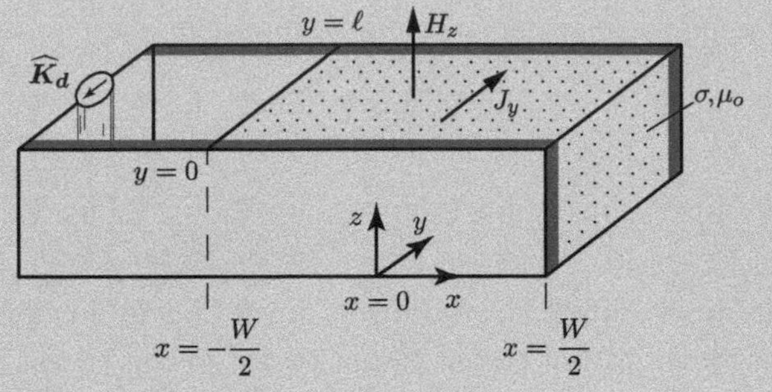

Figure 21.5 A one-dimensional system in which a distributed sinusoidal current source $\widehat{K}_d$ drives y-directed current into the conductor of Fig. 21.1.

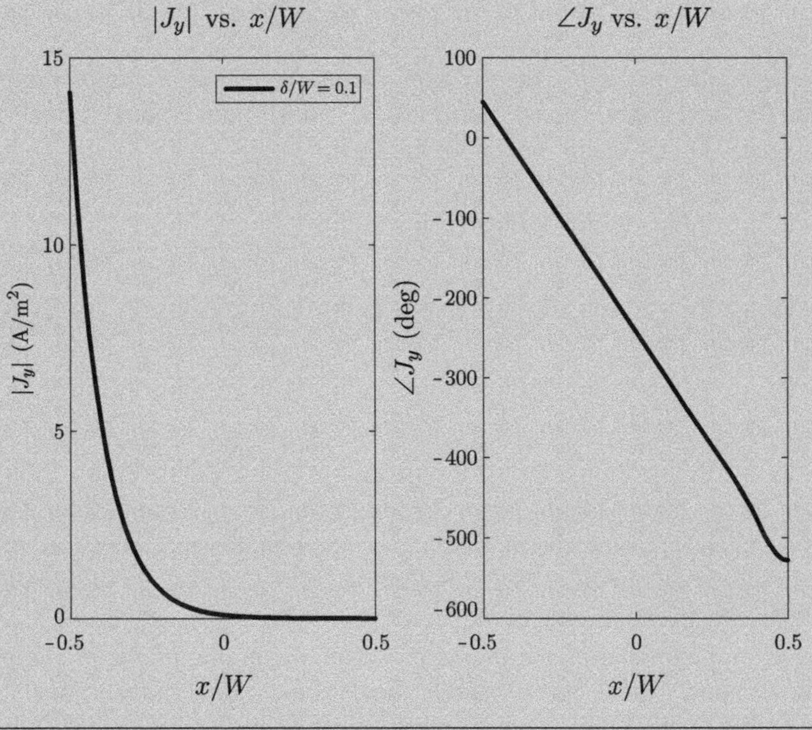

Figure 21.6 Plots of the magnitude and phase of the y-directed current density in the conductor versus position in the system of Fig. 21.5 for $\delta = 0.1\,W$ and $\widehat{K}_d = 1$ A/m.

We can directly calculate the current distribution in the conductor using the solution in Section 21.1.3 and finding the constants $\widehat{c}_+$ and $\widehat{c}_-$ based on the field boundary conditions. With this approach and a bit of algebra, we get the following expression for the y-directed current density:

$$\widehat{J}_y(x) = \widehat{K}_d \frac{\frac{1+j}{\delta} e^{(1+j)W/(2\delta)} e^{-(1+j)x/\delta} + \frac{1+j}{\delta} e^{-(1+j)W/(2\delta)} e^{(1+j)x/\delta}}{e^{(1+j)W/\delta} - e^{-(1+j)W/\delta}}. \tag{21.32}$$

This solution is plotted in Fig. 21.6 for skin depth $\delta = 0.1\,W$ and current drive $\widehat{K}_d = 1$ A/m. Because the conductor is much wider than the skin depth and is driven from the left side (yielding a high field boundary condition at $x = -W/2$ and zero field boundary condition at $x = W/2$), we observe skin-depth-limited conduction on the left side of the conductor. That is, the current density magnitude is highest at the left surface and decays exponentially (with length constant δ) further into the conductor. This is consistent with the observation in Section 21.1.4 that the conductor surface(s) that will carry current in the skin-depth limit are those with high fields to drive magnetic diffusion. We might consider this to be "single-sided" skin effect conduction, in comparison to the double-sided (or "balanced") skin effect conduction seen in Fig. 21.3 for the system of Fig. 21.2.

A schematic representation of this "single-sided" current profile is shown in in Fig. 21.7(a). Two current arrows, each representing a current of $\frac{1}{2}\widehat{K}_d$ per unit height, are placed adjacent to the left conductor surface, indicating that a net current $\widehat{K}_d$ per unit height is carried in a narrow region close to the left surface of the conductor.

The decomposed solution for the current (21.26) can be found by applying (21.30) and (21.31) with the field boundary conditions $\widehat{H}_s(-W/2) = \widehat{K}_d$ and $\widehat{H}_s(W/2) = 0$. Again considering a skin depth $\delta = 0.1W$ and current drive $\widehat{K}_d = 1$ A/m, the balanced skin effect component $\widehat{J}_s(x)$ is plotted in Fig. 21.8, while the proximity effect component $\widehat{J}_p(x)$ is plotted in Fig. 21.9.

Examining Figs. 21.8 and 21.9, one can appreciate that the balanced skin and proximity components of current density will reinforce near the left surface of the conductor and cancel near its right surface. Indeed, the plot of the sum of $\widehat{J}_s$ and $\widehat{J}_p$ is exactly the same as that shown in Fig. 21.6. A schematic representation of the net current profile and its components is shown in Fig. 21.7(b), where each current arrow represents a net current of $\frac{1}{2}\widehat{K}_d$ per unit height. The top two current arrows represent the current arising from the balanced skin effect component, with current concentrated equally near the left and right surfaces of the conductor. The bottom two current arrows represent the current arising from the proximity effect component, with current concentrated equally but opposite in direction at the left and right surfaces. Summing the balanced skin effect and proximity effect currents, the currents concentrated near the right surface of the conductor cancel while those near the left surface reinforce, leaving us with the net single-sided conduction pattern shown in Fig. 21.7(a).

The current distribution in the conductor for the system of Fig. 21.5 clearly exhibits single-sided skin effect conduction, with current concentrated near the left surface of the conductor as shown in Fig. 21.6. This arises because the drive current and associated fields are imposed at the left side of the conductor. At the same time, we can represent this single-sided current distribution as the sum of a balanced skin effect term and a proximity effect term. Indeed, we would get the same one-sided conduction with the superposition of a balanced current drive like that in Fig. 21.2 (generating balanced skin effect current) with an additional external field imposed upon the conductor (generating proximity effect current): the net fields and current distribution in the conductor would be identical. Thus, for the purposes of calculating current densities and losses in the conductor, we can feel free to describe the system in either way.

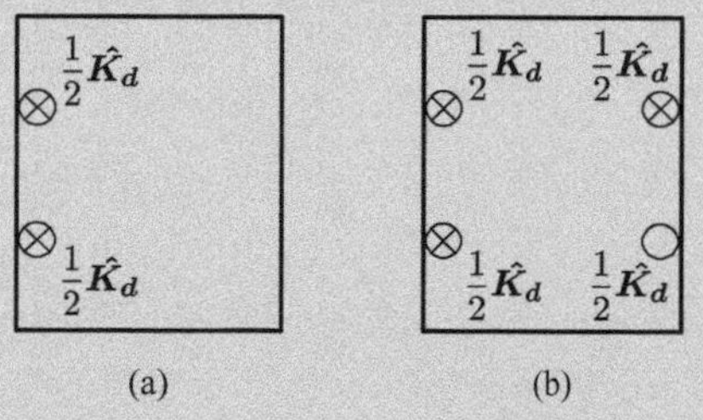

Figure 21.7 Schematic representations of the current profile illustrated in Fig. 21.6. Each current arrow represents a net current of $\frac{1}{2}\widehat{K}_d$ per unit height. (a) The current represented as single-sided skin effect. (b) The current represented as the superposition of a balanced skin effect current (top two current arrows) and a proximity effect current (bottom two current arrows).

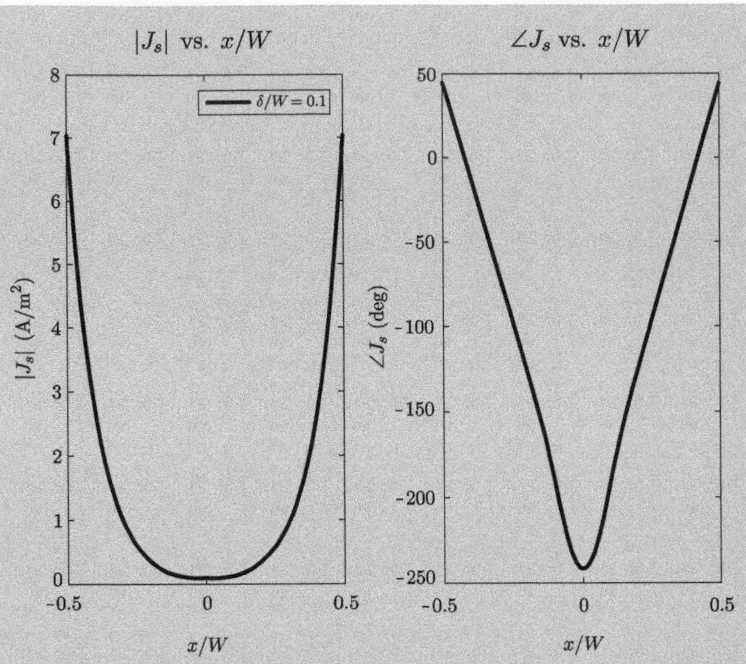

Figure 21.8 Plots of the magnitude and phase of the balanced skin effect component of current density versus position in the system of Fig. 21.5 for $\delta = 0.1W$ and $\widehat{K}_d = 1$ A/m.

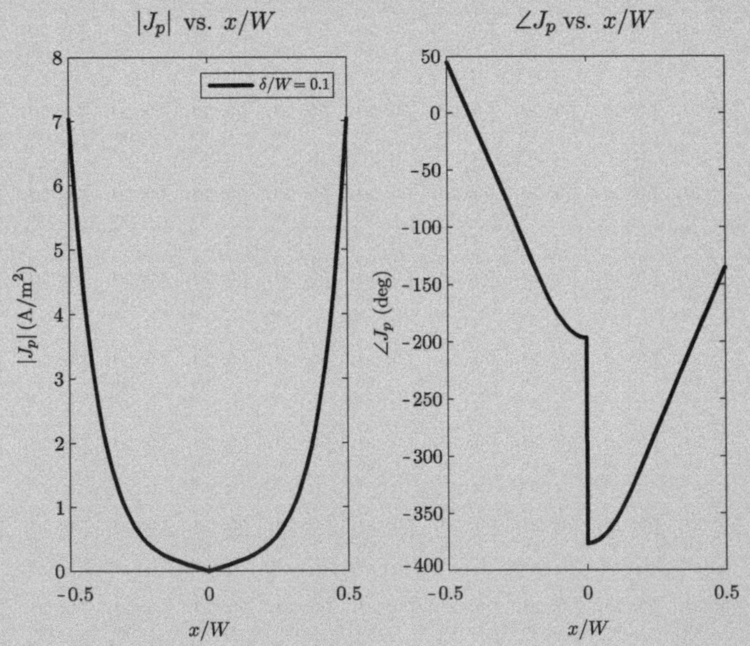

Figure 21.9 Plots of the magnitude and phase of the proximity effect component of current density versus position in the system of Fig. 21.5 for $\delta = 0.1W$ and $\widehat{K}_d = 1$ A/m.

21.2 Winding Loss Calculation

In this section we introduce methods for calculating winding losses in magnetic components, starting from some simple cases of interest and working towards more complex ones. We base much of our analysis on the solutions for linear segments of conductor as considered in the previous section. Losses for this case can accurately approximate the losses in rectangular conductors wrapped around a cylinder (e.g., as might occur with a foil winding wound around the centerpost of a pot core). When the width of the conductor is less than 10% of the radius of the cylinder on which it is wound, the error in approximating it as linear is on the order of 0.1% or less. We will also describe how these solutions can be used to approximate the loss in wire windings, and introduce further techniques for estimating loss in Litz and other stranded wire configurations.

21.2.1 Loss Calculation Basics

Consider a conductor of rectangular cross section with width W, length ℓ, height h, and conductivity σ, such as in Fig. 21.1 but with finite height. If the fields are purely z-directed (owing to the boundaries around the conductor) and the conductor carries a y-directed sinusoidal current density $\widehat{J}_y(x)$, we can find the time-average power dissipation density as

$$P_V = \frac{1}{2}\widehat{J}_y \cdot \widehat{E}_y^* = \frac{|\widehat{J}_y|^2}{2\sigma}, \tag{21.33}$$

where we have applied Ohm's law. $\widehat{J} = \sigma \widehat{E}$ to relate electric field to current density. The time-average dissipation in the whole conductor section is then

$$P_{\text{cond}} = \iiint_V \frac{|\widehat{J}_y|^2}{2\sigma}\, dx\, dy\, dz = \frac{\ell h}{2\sigma}\int_{-\frac{W}{2}}^{\frac{W}{2}} |\widehat{J}_y|^2\, dx. \tag{21.34}$$

Given the solution to the magnetic diffusion equation, we can calculate conduction loss for any sinusoidal waveform of interest.

Consider the case where our rectangular conductor is driven with a current from a single side, as shown in Fig. 21.5. If the conductor is of height h and carries a net current $\widehat{I}$, then we get $\widehat{K}_d = \widehat{I}/h$ in the current density solution (21.32).

At sufficiently low frequencies, $\delta \gg W$. In this case, the magnetic field and current diffuse fully, and current distributes uniformly across the conductor. Indeed, in the limit $\delta \gg W$ and with $\widehat{K}_d = \widehat{I}/h$, (21.32) becomes

$$\widehat{J}_y(x) \approx \frac{\widehat{K}_d}{W} = \frac{\widehat{I}}{hW}. \tag{21.35}$$

From (21.34), the power dissipated in the conductor in this case is

$$P_{\text{cond}} = \frac{1}{2}\frac{\ell}{\sigma h W}|\widehat{I}|^2 = \frac{1}{2}R_{\text{dc}}|\widehat{I}|^2, \tag{21.36}$$

where we recognize

$$R_{\rm dc} = \frac{\ell}{\sigma h W} \tag{21.37}$$

as the dc resistance of the conductor.

At high frequencies where $\delta \ll W$, we see a very different situation. In this case, current density decays exponentially from the left surface of the conductor as suggested by Fig. 21.6. In particular, for this case the current density magnitude in (21.32) is well approximated as

$$|\widehat{J}_y(x)| \approx |\widehat{K}_d|\frac{\sqrt{2}}{\delta}e^{-(x+\frac{W}{2})/\delta} = \frac{\sqrt{2}|\widehat{I}|}{\delta h}e^{-(x+\frac{W}{2})/\delta}. \tag{21.38}$$

Carrying out the integral in (21.34) for this case leads to an interesting result:

$$P_{\rm cond} = \frac{1}{2}\frac{\ell}{\sigma h\delta}|\widehat{I}|^2 = \frac{1}{2}\frac{W}{\delta}R_{\rm dc}|\widehat{I}|^2. \tag{21.39}$$

This equation indicates that at high frequencies where $\delta \ll W$, the loss is a factor W/δ higher than it would be for the same current at low frequency. Indeed, the loss is exactly the same as if the current $\widehat{I}$ were carried uniformly in a conductor of width δ. For this reason, designers sometimes think of skin-effect currents as being carried uniformly within a skin depth of the surface in the deep skin-depth limit, even though the current density is actually distributed exponentially versus depth and with substantial phase variation, as shown in Fig. 21.6.

In many instances it is useful to characterize the conduction loss at high frequency in terms of the loss one would get at low frequency where skin and proximity effects are not significant. One can express this loss increase as a multiplier on the effective resistance of the conductor over the dc resistance. We thus define the *ac resistance factor* F_R as the (frequency-dependent) ratio of ac resistance to dc resistance: $F_R = R_{\rm ac}/R_{\rm dc}$. Thus, we might express the conduction loss for a high-frequency ac waveform as

$$P_{\rm cond} = \frac{1}{2}F_R R_{\rm dc}|\widehat{I}|^2. \tag{21.40}$$

For single-sided conduction in the deep skin-depth limit, we get $F_R = W/\delta$ as per (21.39). It turns out that the ratio of conductor width to skin depth appears quite frequently in conduction loss calculations, so it is given its own symbol: $\Delta = W/\delta$. So, in the deep skin-depth limit with single-sided conduction we get $F_R = \Delta$. As we will see, the ac resistance factor for other circumstances can be represented as other functions of Δ.

21.2.2 Conductor Loss Calculation Using Boundary Conditions

To calculate losses under more general conditions, it is useful to have an expression for the losses in a conductor based on its field boundary conditions. Figure 21.10 shows a section of conductor with high-permeability regions above and below it (as might occur in many magnetic components). As with its infinitely high counterpart in Fig. 21.1, the conductor has y-directed currents (into the page) and z-directed magnetic fields around and within it. For simplicity we assume that the sinusoidal fields bounding the conductor on the left (H_{za}) and right (H_{zb}) are equal (or possibly opposite) in phase, as would be typical in inductors and single-phase

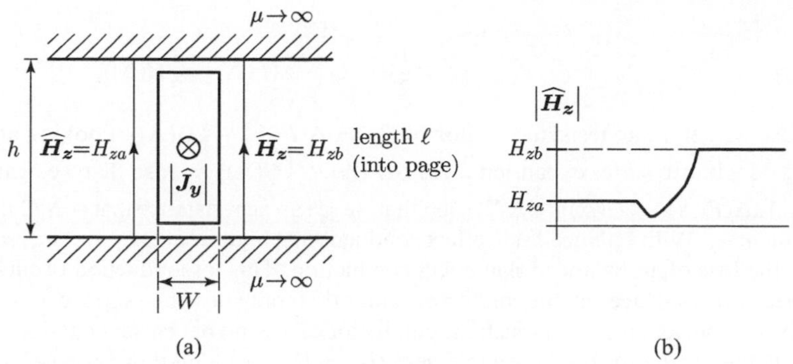

Figure 21.10 (a) A section of conductor between high-permeability regions, with z-directed magnetic fields and y-directed current. (b) The H field magnitude versus position.

transformers. For conductor widths W less than a skin depth ($\Delta < 1$) the field magnitude in the conductor transitions approximately linearly between H_{za} and H_{zb}. For conductor widths of a few skin depths the field is reduced near the middle of the conductor (as shown), and for conductor widths of many skin depths ($\Delta \gg 1$) the field will drop to zero in the middle of the conductor.

To calculate average power dissipation, we can start with a solution for $\widehat{J}_y(x)$ (e.g., the decomposed solution of (21.26), (21.30), and (21.31)) and apply (21.34). Doing so, after copious amounts of algebra, we get the following result:

$$P_{\text{cond}} = \frac{1}{2} h^2 R_{\text{dc}} \Delta \left[(H_{za} - H_{zb})^2 F(\Delta) + 2 H_{za} H_{zb} G(\Delta) \right], \tag{21.41}$$

where

$$F(\Delta) = \frac{\sinh(2\Delta) + \sin(2\Delta)}{\cosh(2\Delta) - \cos(2\Delta)}, \qquad G(\Delta) = \frac{\sinh(\Delta) - \sin(\Delta)}{\cosh(\Delta) + \cos(\Delta)}, \tag{21.42}$$

$$R_{\text{dc}} = \ell/(\sigma h W), \qquad \Delta = W/\delta.$$

Example 21.2 Three Loss Cases

Here we calculate the loss in the conductor of Fig. 21.10 for three cases of interest. First, consider the loss in the conductor when carrying a sinusoidal current of amplitude I with a single-sided field boundary such that $H_{za} = I/h$ and $H_{zb} = 0$. In this case we will have unbalanced skin effect at high frequency. Applying (21.41), we get

$$P_{\text{cond, us}} = \frac{1}{2} |I|^2 R_{\text{dc}} \cdot \Delta F(\Delta). \tag{21.43}$$

Thus, for this case we get an ac resistance factor of $F_R = \Delta F(\Delta)$. This is consistent with the limiting cases considered previously, since for $\Delta \ll 1$ a Maclaurin series expansion gives $\Delta F(\Delta) \approx 1$, such that we see the dc resistance at low frequency, while $\Delta F(\Delta) \approx \Delta$ when $\Delta \gg 1$, such that we see an ac resistance that is Δ times the dc resistance at high frequency.

Next, we consider the loss for the conductor carrying the same ac current I but with balanced boundary conditions $H_{za} = I/(2h)$ and $H_{zb} = -I/(2h)$. In this case, we expect to see balanced skin effect at high frequency. Applying (21.41) we get

$$P_{\text{cond,bs}} = \frac{1}{2}|I|^2 R_{\text{dc}} \cdot \Delta(F(\Delta) - \frac{1}{2}G(\Delta)). \tag{21.44}$$

In this case we get an ac resistance factor of $F_R = \Delta(F(\Delta) - \frac{1}{2}G(\Delta))$. Looking at low frequencies where $\Delta \ll 1$, a Maclaurin series expansion gives $\Delta(F(\Delta) - \frac{1}{2}G(\Delta)) \approx 1$, so that we again see the dc resistance. For $\Delta \gg 1$, $\Delta(F(\Delta) - \frac{1}{2}G(\Delta)) \approx \Delta/2$, such that we see an ac resistance that is $\Delta/2$ times the dc resistance at high frequencies. With balanced skin effect conduction at high frequency, we thus see half the ac resistance and half the loss of unbalanced skin effect conduction. This loss reduction occurs because the current is carried near both surfaces of the conductor rather than only near one surface.

Finally, we consider a case in which the conductor carries no net current but does carry circulating (eddy) currents. If the conductor is exposed to a field $H_{za} = H_{zb} = I/(2h)$ it has proximity effect loss owing to the eddy currents. Applying (21.41) in this case we get

$$P_{\text{cond,p}} = \frac{1}{2}|I|^2 R_{\text{dc}} \cdot \frac{1}{2}\Delta G(\Delta). \tag{21.45}$$

A Maclaurin series expansion gives $\Delta G(\Delta) \approx \Delta^4/6$, such that for $\Delta \ll 1$ the proximity effect loss goes as Δ^4 (and hence as ω^2). At such low frequencies, the proximity-effect loss is very modest as compared to the loss that would be induced if the conductor were to carry a net current I. For $\Delta \gg 1$, $\Delta G(\Delta) \approx \Delta$, and the proximity loss exactly equals that which would be induced if the conductor were to carry a current I with balanced skin effect. Moreover, the proximity loss at high frequency in this example is exactly the difference between the loss that would exist at high frequency ($\Delta \gg 1$) between carrying current with single-sided skin effect and balanced skin effect. This is consistent with the interpretation of Fig. 21.7 of single-sided skin effect being the superposition of a balanced skin effect component and an equal-amplitude (but spatially orthogonal) proximity-effect component.

More generally, the loss associated with the proximity-effect component of a magnetic field (e.g., as per (21.29)) is tied to the function $\Delta G(\Delta)$. In particular, proximity loss is proportional to the square of the amplitude of the proximity effect component of the field (here $I/(2h)$), the square of the height h, the conductor dc resistance R_{dc}, and the function $\Delta G(\Delta)$.

21.2.3 Multilayer Winding Loss Calculation

Armed with the results of the previous subsection, we can now calculate the winding loss of magnetic components having multiple layers of windings. Figure 21.11 shows a pot-core transformer with a three-turn primary winding and a three-turn secondary winding in which each winding comprises series-connected concentric turns of foil conductor. Each winding thus comprises three layers of conductor, with each layer carrying the same winding current. It is assumed that the primary and secondary windings carry sinusoidal currents of I amperes peak in opposite directions.

We begin by identifying the magnetic field distribution in the core window, and in particular at the boundaries of the winding conductors, as this can be used to compute winding loss. The top portion of Fig. 21.12 shows a closeup of the winding window in Fig. 21.11. By applying Ampère's law around contours like that of the dashed line, we can find the magnetic field between conductor layers in the winding window. Recognizing that each winding layer carries a net current I in amplitude, and assuming that the core permeability is sufficiently high that H fields in the core are negligible, we find that the amplitude of the z-directed H field to the left of the first primary layer is zero and is I/h between the first and second primary layers. We do not

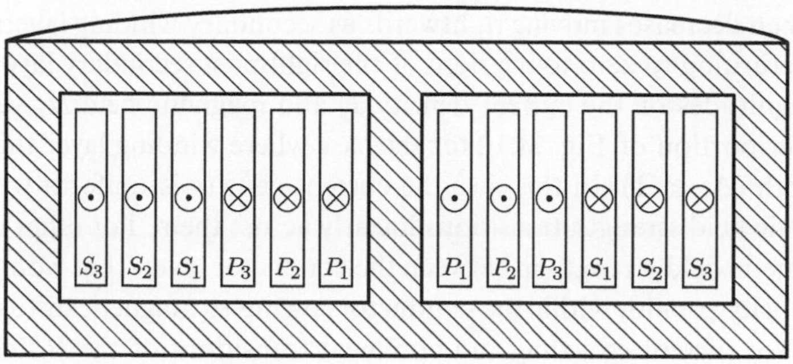

Figure 21.11 Cross section of a pot-core transformer. The primary (P) and secondary (S) windings each comprise three turns of foil conductor. The two windings carry oppositely directed sinusoidal currents of I A peak.

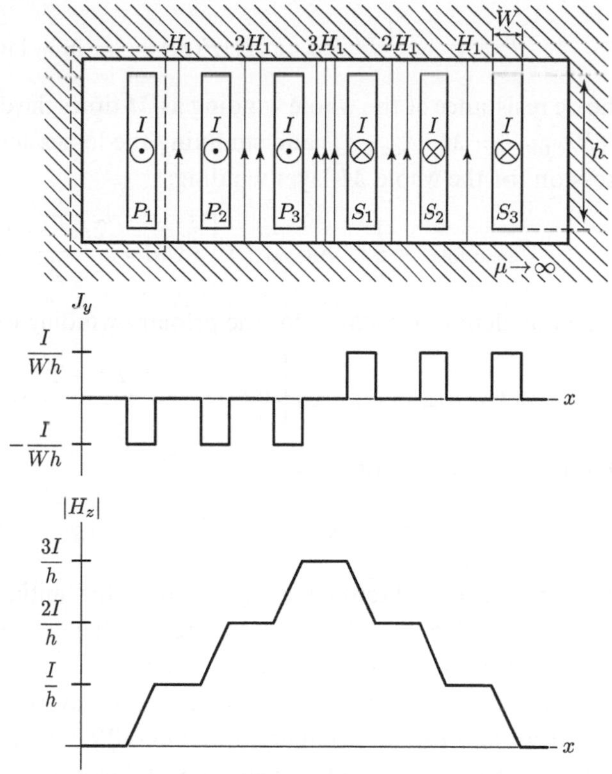

Figure 21.12 Close-up view of the core window in Fig. 21.11, with plots of current density J_y and H field magnitude $|H_z|$ versus position x across the window.

distinguish between the winding height and that of the window, defining both to be h. Continuing to expand our contour to encompass more layers, we find that the field amplitude is $2I/h$ between the second and third primary layers, and $3I/h$ to the right of the third primary layer. The field strength then decreases moving rightwards as secondary winding layers are encompassed in the contour, returning to zero to the right of the rightmost secondary layer.

The magnitudes of the current density J_y and magnetic field H_z versus position are shown in the lower portion of Fig. 21.12 for the case where winding layer thickness is much less than a skin depth ($\Delta \ll 1$). In this case the current density is uniform within the conductors and the magnetic field strength transitions linearly across them. For other cases, one would find the same magnetic field strength in between the conductor layers, but would have different field and current density profiles within the conductors owing to magnetic diffusion effects. In particular, we would see current crowding and nonlinear transitions of the H field within the conductors as illustrated in the previous section. The magnetic fields within the core window can be a substantial source of the leakage inductances of a transformer.

In general, for an M-layer primary winding in which each winding layer carries a net current I, we find a field mI/h to the right of the mth layer. We can use (21.41) to calculate the power dissipation in the mth layer as

$$P_{\text{cond,m}} = \frac{1}{2} I^2 R_{\text{dc, layer}} \Delta \left[F(\Delta) + 2m(m-1)G(\Delta) \right]. \tag{21.46}$$

Considering the dc resistance of the whole winding as M times the dc resistance of an individual winding layer ($R_{\text{dc,total}} = M \cdot R_{\text{dc,layer}}$) and summing the losses across the M layers, we get the following expression for the whole M-layer winding:

$$P_{\text{cond, total}} = \frac{1}{2} I^2 R_{\text{dc, total}} \Delta \left[F(\Delta) + \frac{2M^2-2}{3} G(\Delta) \right]. \tag{21.47}$$

This gives us an equivalent ac resistance for the primary winding of

$$R_{\text{ac, total}} = R_{\text{dc, total}} \left[\Delta F(\Delta) + \frac{2M^2-2}{3} \Delta G(\Delta) \right], \tag{21.48}$$

which corresponds to an ac resistance factor $F_R = R_{\text{ac, total}}/R_{\text{dc, total}}$ for the winding of

$$F_R = \Delta F(\Delta) + \frac{2M^2-2}{3} \Delta G(\Delta). \tag{21.49}$$

This equation is known as as Dowell's equation after the initial developer of this analysis. Considering the results of Example 21.2, we can interpret the first term in (21.49) as arising from single-sided skin effect in the primary winding, and the second term as arising from proximity effect owing to fields from some layers impinging on other layers.

We can arrive at an alternative formulation of Dowell's equation by separately considering the even and odd components of the fields about each layer. This analysis yields an equivalent ac resistance for the primary winding of

$$R_{\text{ac, total}} = R_{\text{dc, total}} \left[\Delta \left(F(\Delta) - \frac{1}{2} G(\Delta) \right) + \frac{4M^2-1}{6} \Delta G(\Delta) \right], \tag{21.50}$$

which corresponds to an ac resistance factor of

$$F_R = \Delta \left(F(\Delta) - \frac{1}{2} G(\Delta) \right) + \frac{4M^2-1}{6} \Delta G(\Delta). \tag{21.51}$$

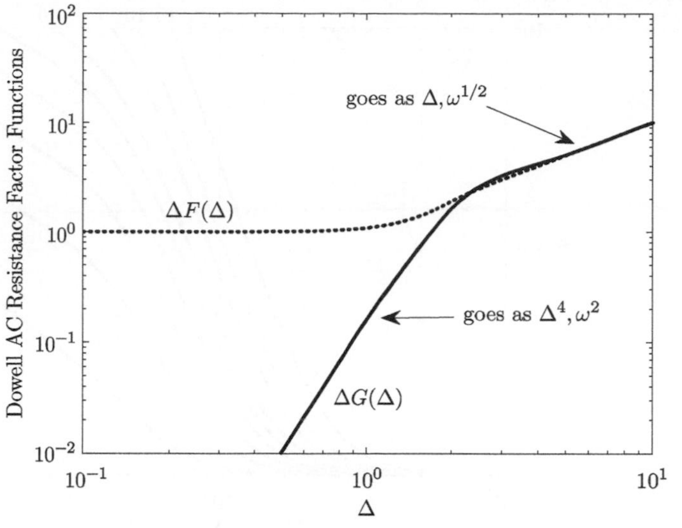

Figure 21.13 Log–log plots of $\Delta F(\Delta)$ and $\Delta G(\Delta)$ versus Δ.

Considering the results of Example 21.2, the first term in the ac resistance factor of (21.51) arises from balanced skin effect in the winding, while the second term arises from proximity effect associated with net fields in the window impinging on the winding layers. The total loss and F_R is of course the same in either formulation, with the difference in loss breakdown simply relating to whether single-sided skin effect is considered as purely skin effect or as the superposition of balanced skin effect and a proximity effect component (e.g, as illustrated in Fig. 21.7). In either of these interpretations, the contribution of proximity effect to total loss and ac resistance factor is associated with $\Delta G(\Delta)$, and grows quadratically with the number of layers M.

As described above, the two component functions $\Delta F(\Delta)$ and $\Delta G(\Delta)$ combine to determine the ac resistance factor F_R of a particular winding based on the number of winding layers. Figure 21.13 shows log–log plots of $\Delta F(\Delta)$ and $\Delta G(\Delta)$ versus Δ. Since Δ is proportional to $\sqrt{\omega}$, this plot also illustrates how these functions change with frequency. For $\Delta \gg 1$, $\Delta F(\Delta) \approx \Delta G(\Delta) \approx \Delta$. Of interest for many practical designs, however, are the behavior of these functions for values of Δ near and below 1. Series expansions of these functions give

$$\Delta F(\Delta) \approx 1 + \frac{4}{45}\Delta^4 - \frac{16}{4725}\Delta^8 + O(\Delta^{12}) \tag{21.52}$$

$$\Delta G(\Delta) \approx \frac{1}{6}\Delta^4 - \frac{17}{2520}\Delta^8 + O(\Delta^{12}). \tag{21.53}$$

If only terms up to the order of Δ^4 are used, for values of Δ up to 1.2 the error in approximating $\Delta F(\Delta)$ with its series expansion is less than 1.2% and the error in approximating $\Delta G(\Delta)$ with its series expansion is less than 8.4%.

Figure 21.14 shows log–log plots of winding ac resistance factor F_R against normalized winding thickness Δ, parameterized by the number of winding layers M. It can be seen that while $\Delta \approx 1$ is adequate to keep F_R small for a single-layer winding, F_R can grow extremely rapidly with winding layer thickness (as Δ^4) in multi-layer windings owing to proximity effect.

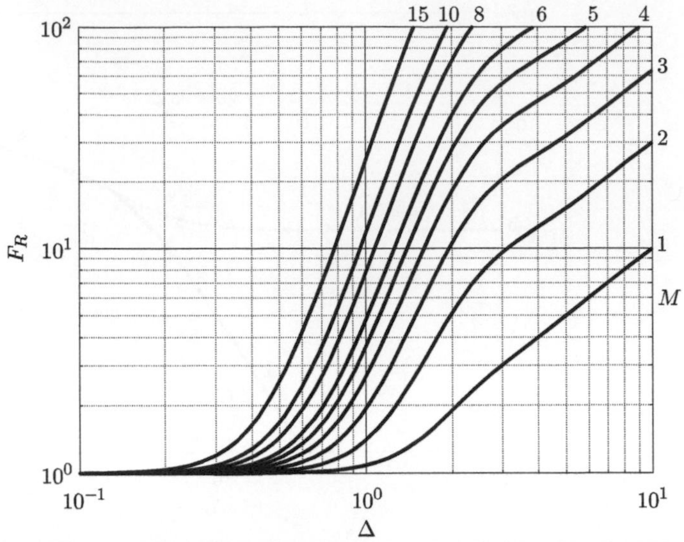

Figure 21.14 Log–log plots of ac resistance factor F_R versus normalized layer thickness Δ, parameterized in the number of winding layers M.

Indeed, for multi-layer windings, proximity effect is the dominant contributor to ac conduction loss effects.

For winding design, we are interested in minimizing loss. For sinusoidal waveforms this means minimizing the ac resistance of the winding at the frequency of interest, which we can do by appropriately selecting the winding thickness W, or equivalently by selecting Δ. (Here we consider sinusoidal currents at a specified frequency; we address current waveforms having content at multiple frequencies later.) The ac resistance is the dc resistance times F_R. We can express the total dc resistance of the winding as

$$R_{\text{dc, total}} = \frac{l_W}{\sigma h W} = \frac{l_W}{\sigma h \delta}\frac{1}{\Delta} = R_\delta \frac{1}{\Delta}, \tag{21.54}$$

where l_W is the length of the whole M-layer winding and R_δ is the dc resistance of a winding of width $W = \delta$. For a given geometry and operating frequency, R_δ is thus a known constant. Combining this with (21.48) the ac resistance of the winding can thus be expressed as

$$R_{\text{ac, total}} = R_\delta \left[F(\Delta) + \frac{2M^2 - 2}{3} G(\Delta) \right]. \tag{21.55}$$

For a single-layer winding ($M = 1$), $R_{\text{ac, total}} = R_\delta F(\Delta)$. From Fig. 21.13 we can infer that $F(\Delta)$ goes as $1/\Delta$ for small Δ, and becomes 1 for Δ far above 1. Thus, at small Δ the ac resistance of a single-layer winding will decrease with increasing Δ and will asymptote to R_δ at large Δ. This behavior is to be expected: for windings of small Δ, we can reduce the dc resistance (and hence the ac resistance) by increasing winding thickness. However, once the winding thickness becomes large with respect to a skin depth ($\Delta \gg 1$), additional conductor width does not contribute to carrying current or to further reducing ac resistance. Interestingly, for a single-layer winding

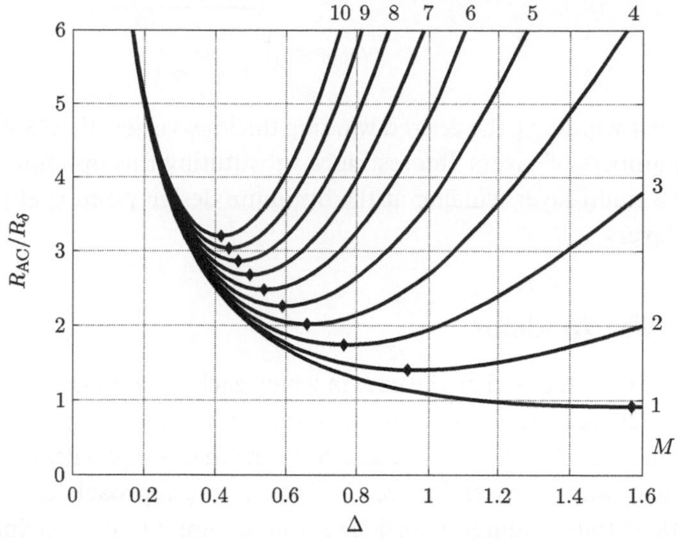

Figure 21.15 Plots of R_{ac}, normalized to the dc resistance of a winding one skin-depth thick, R_δ, versus normalized layer thickness Δ, parameterized in the number of winding layers M. The Δ providing the minimum R_{ac} is indicated by a diamond for each curve.

of large Δ, the loss and ac resistance are just as if the ac current were carried uniformly in a winding one skin-depth thick (and designers sometimes informally think of it this way). It is worth noting, however, that $F(\Delta)$ does not decrease monotonically to 1, but rather decreases to a minimum value of approximately 0.917 at $\Delta = \pi/2$ before transitioning to its asymptotic value of 1 at large Δ. This means that in principle one can achieve an ac resistance slightly lower than R_δ an an operating frequency of interest. In practice, designers often simply size the conductor width in a single-layer winding to be larger than a skin depth and treat it as if it were a skin depth thick for loss calculation purposes.

For multi-layer windings, ac loss is much more strongly sensitive to Δ, so greater care is necessary in selecting winding thickness. Figure 21.15 plots R_{ac}/R_δ vs. Δ for different numbers of layers M. The optimum value of Δ for minimum loss is also indicated for each curve. It can be seen that for multi-layer windings, the optimum Δ gives an ac resistance that is far below what would be achieved at larger or smaller values of Δ. As the number of layers M increases, the achievable value of R_{ac}/R_δ gets worse, and the required Δ to achieve this minimum gets smaller. These effects result from proximity effect becoming the dominant component of ac loss in multi-layer windings. Unlike in a single-layer winding in which over-sizing a conductor does not dramatically hurt performance, circulating currents in over-sized multi-layer windings can be a substantial source of loss. This often leads to the counter-intuitive result that increasing conductor cross section *increases* conduction loss. Sometimes, less (copper) is more (efficiency).

To find the optimum winding thickness for the multi-layer case, we again seek the minimum of (21.55). To simplify our calculations we leverage the power series expansions in (21.52) and (21.53), keeping only terms up to Δ^4. This yields an expression for the optimum normalized winding thickness:

$$\Delta_{\text{opt}} = \sqrt[4]{\frac{15}{5M^2 - 1}}. \tag{21.56}$$

For multi-layer windings, the desired winding thickness is *less* than a skin depth and gets smaller for higher numbers of layers. Interestingly, substituting this optimum back into (21.49) reveals that F_R for a multi-layer winding at the optimum design point is always 4/3, regardless of the number of layers.

21.2.4 Modeling Wire Windings

Up to now we have focused on the case in which each layer comprises a single conductor having a rectangular cross section, as might be found with foil windings. To model wire windings, we can approximate each layer of wire as a foil turn that is equivalent for purposes of calculating inter-layer fields and ac losses. Figure 21.16 shows one approach for transforming wire windings into equivalent foil windings. Consider a round-wire winding having conductor diameter D, conductivity σ_W, and carrying a current I_W. A set of N_ℓ turns wound as a helix forms a winding layer of height h. We first convert the round conductors in a layer to square conductors having the same dc resistance. The square conductors thus have a side length

$$W = \sqrt{\frac{\pi}{4}}\, D. \tag{21.57}$$

We then eliminate the insulation and spacing between the square conductors in the layer, forming a single "collapsed" foil conductor of width W and height $h' = N_\ell W$ carrying a current $I = N_\ell I_W$. We define the ratio of collapsed foil winding height h' to the original full height h as the *porosity factor* η,

$$\eta = \frac{N_\ell W}{h}. \tag{21.58}$$

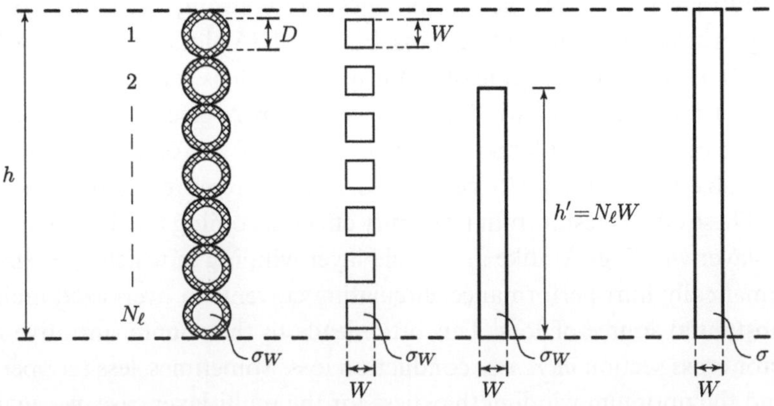

Figure 21.16 A transformation process for modeling a layer of a round-wire winding as an equivalent layer of foil. Each of the N_ℓ turns in the round-wire layer carries a current I_W; the equivalent foil layer carries a current $I = N_\ell I_W$.

To create a foil conductor of height h, we stretch the collapsed conductor in height by factor $1/\eta$ and scale its conductivity by a factor η such that the resulting full-height foil conductor has a conductivity $\sigma = \eta\sigma_W$, thus preserving dc resistance. We now have an equivalent full-height foil conductor to which we can apply the methods already developed.

This transformation is certainly an engineering approximation, and stretching the conductor and applying a porosity factor to its conductivity is particularly non-physical. One implication of this step is changing the skin depth from a value of δ_W in the wire to $\delta = \delta_W/\sqrt{\eta}$ in the equivalent foil conductor. Likewise, in finding the normalized thickness $\Delta = W/\delta$ we are using both a non-physical thickness W and a non-physical skin depth δ. Nonetheless, this transformation does preserve vertical field intensities and low-frequency conduction loss between the original wire winding and the equivalent foil winding, so might be viewed as a reasonable, if rough, approximation to reality. Applying a porosity factor and stretching a conductor can also be employed to make up the difference between winding height and actual window height (e.g., owing to the presence of a bobbin) which we have neglected thus far, and is also sometimes employed to handle partial-height winding layers.

A consequence of the approximate nature of the wire-to-foil transformation shown in Fig. 21.16 is that Dowell's equation tends to underestimate the loss of multi-layer round-wire windings, especially at frequencies for which the wire diameter is larger than a skin depth. An alternative approach to calculating the loss in round-wire windings is to directly estimate round-wire loss. Because the ac loss in a many-layer winding is highly dominated by proximity effect, we focus here on proximity loss.

Consider the loss in a cylindrical conductor of length ℓ, diameter D, and conductivity σ immersed in a uniform sinusoidal magnetic field of peak amplitude H oriented perpendicular to the axis of the conductor. For the case where the conductor diameter is small compared to a skin depth, the magnetic field inside the conductor approximately equals the imposed field, and the power dissipation in the conductor owing to eddy currents may be found as

$$P = \left[\frac{\pi}{32}\left(\frac{D}{\delta}\right)^4\right]\cdot\frac{|H|^2}{\sigma}\cdot\ell = \widehat{G}\cdot\frac{|H|^2}{\sigma}\cdot\ell. \tag{21.59}$$

A more general loss expression applicable to any wire diameter is

$$P = \left[-2\pi\gamma\frac{\mathrm{ber}_2(\gamma)\,\mathrm{ber}'(\gamma)+\mathrm{bei}_2(\gamma)\,\mathrm{ber}'(\gamma)}{\mathrm{ber}^2(\gamma)+\mathrm{bei}^2(\gamma)}\right]\cdot\frac{|H|^2}{\sigma}\cdot\ell = \widehat{G}\cdot\frac{|H|^2}{\sigma}\cdot\ell, \tag{21.60}$$

where

$$\gamma = \frac{D}{\sqrt{2}\delta}. \tag{21.61}$$

(The Kelvin functions ber and bei are available in many mathematical toolboxes.)

The above expressions enable us to express the proximity loss in a round-wire conductor in terms of its length, conductivity, the amplitude of the imposed sinusoidal magnetic field, and a function $\widehat{G}$ that depends on the wire diameter and skin depth. If we know the packing of round-wire conductors in a window, we can express the proximity loss per unit area in terms of the H field amplitude in the window. For example, for the square-packed winding illustrated in Fig. 21.17 we can express the proximity loss per unit area of the winding as

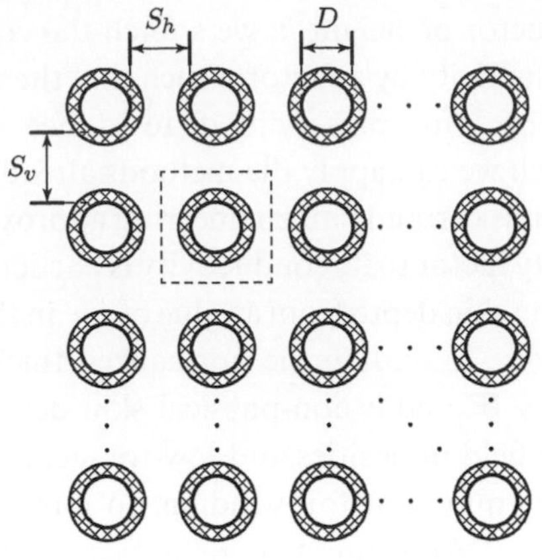

Figure 21.17 A square-packed array of round wires.

$$P_A = \frac{\widehat{G}}{\sigma} \cdot \frac{\ell}{(D+s_v)(D+s_h)} \cdot |H|^2. \tag{21.62}$$

Given knowledge of the magnetic field amplitude versus position (e.g., based on the known current pattern in the windings), we can find the overall proximity loss in a winding by integrating this expression over the extent of the winding.

The approach embodied in (21.62) is highly effective for calculating ac loss in many-layered windings. One minor limitation is that while the loss expression (21.60) is precisely correct for an isolated wire in a uniform field, it does not account for field interactions among closely packed wires. Consequently, it tends to somewhat overestimate loss in windings comprising wires of more than a couple of skin depths in diameter. However, optimized designs tend not to use such large wire diameters, so this limitation is mainly a concern in cases that are far from optimum.§

Example 21.3 AC Resistance Factor of a Generator Winding

Figure 21.18 shows the cross section of an armature winding slot for a high-speed generator. The goal is to find the ac resistance factor of the armature winding. The machine is wound with type 8 litz wire having a rectangular cross section and comprised of $n_s = 19$ strands of 24 AWG wire. The slot carries $N = 9$ two-layer bundles of series-connected turns of litz wire, resulting in a winding with 18 total layers. The litz wire bundles are separated from the slot walls and from each other by thin layers of insulation, and the nine bundles occupy a slot width $S_x = 12.8$ mm and a slot height of $S_z = 5.7$ mm. Each litz bundle carries a sinusoidal current of amplitude $I = 38$ A at a frequency of 2 kHz, such that each strand carries a net current of $I_s = 2$ A. It is assumed that the copper wire operates at $T = 100\,°$C with a wire conductivity of $\sigma = 4.4 \times 10^7$ S/m.

To find the ac resistance factor F_R we first calculate the winding loss. We calculate the total loss as arising from two orthogonal components: (i) conduction currents representing the net current carried by the winding (possibly impacted by balanced skin effect), yielding a loss $P_{\mathrm{cond,bs}}$; and (ii) circulating currents within the strands associated with proximity effect, yielding a loss $P_{\mathrm{cond,p}}$.

Since we have many layers, we can neglect any impact of balanced skin effect on current-carrying loss (i), since for any frequency at which balanced skin effect influences loss it will be greatly outweighed by the

§ For situations where it is of concern, [3] provides a numerically based expression for $\widehat{G}$ that addresses this limitation.

proximity effect contribution (ii). We can thus safely approximate the loss owing to net current carried in the winding simply using the dc resistance of the winding:

$$P_{\text{cond, bs}} \approx \frac{1}{2} I^2 R_{\text{dc}} = \frac{1}{2} I^2 \frac{4 N \ell}{\sigma\, n_s\, \pi\, D^2}, \tag{21.63}$$

where ℓ is the length of a single litz winding turn.

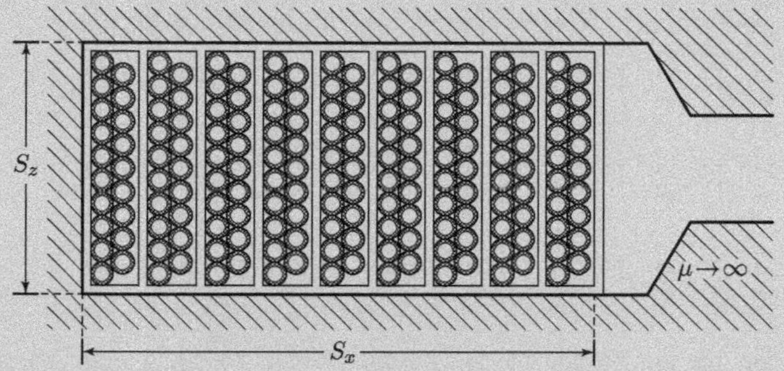

Figure 21.18 Cross section of an armature winding slot for a high-speed generator. The slot contains nine series-connected litz wire bundles, each having 19 strands of 24 AWG wire formed in two layers.

To calculate the proximity-effect contribution to total loss, we adopt the direct round-wire calculation approach described above. To simplify the analysis, we model the winding as having strands uniformly distributed across the region defined by S_x and S_z. We further model the magnetic field in the slot as being purely vertical (z-directed) with a magnitude rising linearly from zero at the left edge of the slot ($x = 0$) to a maximum value at the right edge of the winding region ($x = S_x$). Given these modeling assumptions we can express the amplitude of the magnetic field as

$$|H_z(x)| = \frac{N I}{S_z} \cdot \frac{x}{S_x} \tag{21.64}$$

for $0 < x < S_x$. Similar to (21.62), we can express the power dissipated per unit cross-sectional area of the winding region owing to proximity effect as

$$P_{A,\text{prox}} = \frac{\widehat{G}}{\sigma} \cdot \frac{\ell N n_s}{S_x S_z} \cdot |H_z|^2 \tag{21.65}$$

for $0 < x < S_x$.

From (21.14) we find a skin depth of $\delta = 1.69$ mm for the specified operating frequency and wire conductivity, significantly larger than the 24 AWG strand diameter of $D = 0.51$ mm. For the frequencies of interest we can thus use the simplified expression for $\widehat{G}$ in (21.59). Substituting this into (21.65), we obtain

$$P_{A,\text{prox}} = \frac{\pi D^4}{32\,\delta^4\,\sigma} \cdot \frac{\ell N n_s}{S_x S_z} \cdot |H_z|^2 \tag{21.66}$$

for $0 < x < S_x$. Substituting in our expression for magnetic field strength in (21.64) and integrating over the active slot area gives us the proximity loss from the cross-slot field:

$$P_{\text{cond, p}} = S_z \int_0^{S_x} P_{A,\text{prox}}\, dx = \frac{\pi N^3 n_s \ell \omega^2 \mu_0^2 \sigma D^4}{384\, S_z^2} I^2. \tag{21.67}$$

To find the ac resistance factor, we calculate the ratio of the total loss to that which we would have if there were only loss from carrying current through the dc resistance of the winding:

$$F_R = \frac{P_{\text{cond, bs}} + P_{\text{cond, p}}}{\frac{1}{2} I^2 R_{\text{dc}}} \approx 1 + \frac{\pi^2 N^2 n_s^2 \omega^2 \mu_0^2 \sigma^2 D^6}{768\, S_z^2}. \tag{21.68}$$

This expression holds for angular frequencies ω where the strand diameter D is small compared to a skin depth (as it is for the stated operating frequency of the generator). To address higher frequencies (e.g., perhaps for high-frequency PWM ripple current owing to a connected power converter), we could re-derive the result with an improved expression for $\widehat{G}$, such as that in (21.60).

Substituting the parameters for the generator into (21.68), we evaluate the ac resistance factor to be $F_R = 1.1$ at the specified operating frequency. By using litz wire we thus limit the impact of proximity effect in the winding to a 10% increase in loss over what we would obtain at low frequency.

21.2.5 Other Winding Configurations

In the transformer winding pattern shown in Fig. 21.12, the magnetic fields in the core window build up across the layers of each winding. This field build-up exacerbates proximity loss and can result in high ac resistance of a multilayer winding. To mitigate proximity loss, designers sometimes employ more complicated winding stack-ups. One such example is illustrated in Fig. 21.19, in which layers of primary winding and secondary winding are interleaved. As compared to the winding pattern of Fig. 21.12, this interleaved winding pattern yields lower fields in the core window, resulting in reduced ac loss and resistance in the windings and lower transformer leakage inductances. However, realizing such an interleaved pattern can be much more complex from a winding perspective. Moreover, the larger number of interfaces between the primary and secondary windings results in significantly higher inter-winding capacitance and poses greater challenges for achieving inter-winding insulation.¶ The capacitive parasitics and losses owing to interleaved construction can sometimes overwhelm the benefits of reducing proximity effect. To balance the benefits and drawbacks of interleaving, partially interleaved designs are often realized (e.g., by stacking a partial primary winding, then a complete secondary winding, and then the remainder of the primary). Careful attention to the number of winding layers used and to the way those layers are arranged is often important to achieving high efficiency at high frequency.

Another approach to reducing proximity loss is to use litz wire as illustrated in Example 21.3. With litz wire one can take advantage of a small wire strand diameter D (reducing proximity effect), while still achieving an overall large conductor cross-sectional area through the use of many (n_s) strands in parallel. The expression in (21.68) in Example 21.3 is often used to help guide the design of litz wire windings (e.g, for selecting D and n_s to limit proximity loss). Given an allowed wire bundle cross-sectional area, one can jointly select D and n_s to minimize loss. Having more strands of smaller diameter tends to reduce the ac resistance factor F_R of a winding, as illustrated in (21.68). However, as one goes to more strands of finer wire, strand insulation and

¶ Safety regulations often impose substantial insulation requirements between primary and secondary windings that can have a substantial impact on transformer construction.

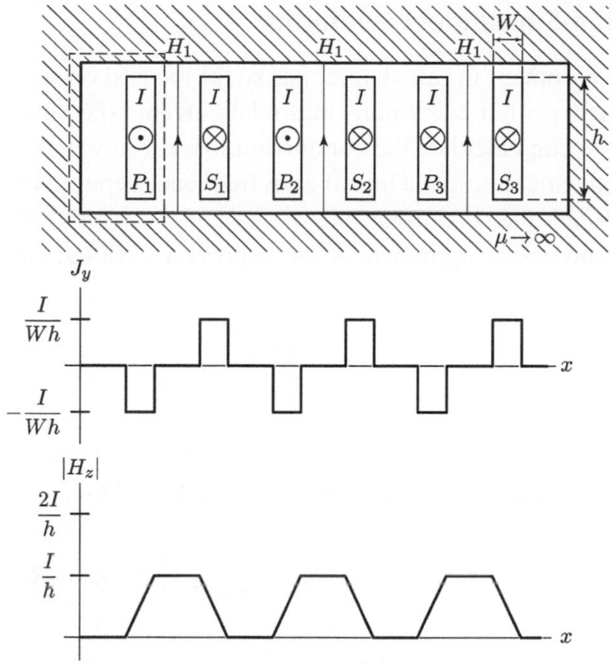

Figure 21.19 Close-up view of the window of a pot-core transformer having fully interleaved windings. Plots of current density J_y and H field magnitude $|H_z|$ versus position x across the window are also shown for the case where $\Delta \ll 1$.

strand packing reduce the conductor fraction in the overall wire bundle cross section, increasing the dc resistance associated with a given wire bundle cross-sectional area. Moreover, at the time of writing, litz wire with strand diameters smaller than 50 AWG is expensive and difficult to obtain. Thus, in selecting a litz wire there is often an optimum combination of n_s and D based on practical wire construction concerns and availability.

Dowell's equation applies to windings in which the layers of each winding are connected in series such that they necessarily carry the same net current. In some cases, however, it is desired to have different layers connected in parallel (e.g., to achieve high current-carrying capability). While the general techniques described here are suitable for modeling such windings, paralleled layers or turns are harder to model than series-connected ones because the net current distribution among the paralleled windings is not known a priori. Instead of carrying the same net current (as series-connnected conductors must), paralleled conductors instead have the same voltage across them. Any difference in flux linked by two paralleled conductors drives (by Faraday's law) a circulating current in the two conductors (governed by Ohm's law) that results in a net current imbalance between them. To achieve good current sharing between paralleled conductors, one must ensure that they link the same flux, with careful attention to minimizing any differential flux linkage between them. Litz wire accomplishes equal flux linkage of the strands through symmetry by transposing the position of the individual conductors over the length of the winding. In cases where equal flux linkage is not achieved, careful calculation is needed to identify the current distribution among the paralleled conductors.

21.2.6 Loss with Non-Sinusoidal Waveforms

While our development in this chapter has so far focused on sinusoidal waveforms, power magnetic components often carry non-sinusoidal currents (i.e., comprising multiple frequencies). Fortunately, we can calculate the conduction loss for any periodic waveform by using Fourier decomposition, calculating the loss at each frequency separately and adding the individual loss components to get the total loss.

Following the above approach, if we express a periodic current waveform $i(t)$ of period $T = 2\pi/\omega_1$ as

$$i(t) = \sum_{n=0}^{\infty} I_n \cos(n\omega_1 t + \phi_n), \tag{21.69}$$

where I_n is the amplitude of the nth harmonic current (with I_0 being the dc current component I_{dc} and $\phi_0 = 0$), we can express the total winding loss P as

$$P = I_0^2 R_{\text{dc}} + \sum_{n=1}^{\infty} \frac{1}{2} I_n^2 F_R(n\omega_1) R_{\text{dc}}, \tag{21.70}$$

where $F_R(\omega)$ is the ac resistance factor of the winding at angular frequency ω. This may be conveniently represented in terms of rms amplitudes of the different frequency components, where we recognize that $I_{0(rms)} = I_0 = I_{\text{dc}}$ and $I_{n(\text{rms})} = I_n/\sqrt{2}$ for $n \geq 1$:

$$P = \sum_{n=0}^{\infty} I_{n(\text{rms})}^2 F_R(n\omega_1) R_{\text{dc}}. \tag{21.71}$$

One challenge that arises in optimizing a winding design for non-sinusoidal waveforms is how to size the conductors (e.g., foil layer thickness, wire diameter, or litz strand diameter) considering ac loss effects at different frequencies. This can be done in a brute force fashion using the approach above to compute total loss for different conductor sizes and then selecting the conductor size providing the lowest loss. However, it would be nice if we could identify an "effective" ac frequency ω_{eff} that properly incorporates information about the frequency content of a waveform to enable us to treat loss calculation and optimization just as with sinusoids, including leveraging expressions such as (21.51) and (21.68). That is, we would like to express the loss of a winding with non-sinusoidal current simply as

$$P = I_{\text{rms, tot}}^2 F_R(\omega_{\text{eff}}) R_{\text{dc}}, \tag{21.72}$$

where $I_{\text{rms, tot}}$ is the rms of the current and ω_{eff} is an effective frequency such that $F_R(\omega_{\text{eff}})$ gives us the ac loss increase. We could then optimize the winding design just as if we had a sinusoidal current at angular frequency ω_{eff}. It turns out that we can do so, at least under limited conditions.

Consider that for a multi-layer winding we can approximate the ac resistance factor versus frequency as one plus a constant times ω^2, at least over a range of frequencies where the conductor width is similar to or smaller than a skin depth (e.g., see Figs. 21.13 and 21.14, and (21.68)). Considering the form of (21.68), we might express ac resistance factor F_R at a frequency

ω in this range in terms of ac resistance factor at some fundamental frequency ω_1:

$$F_R(\omega) = 1 + (F_R(\omega_1) - 1)\left(\frac{\omega^2}{\omega_1^2}\right). \tag{21.73}$$

Equating (21.71) and (21.72), and expressing $F_R(n\omega_1)$ as in (21.73), we can solve for $F_R(\omega_{\text{eff}})$ as

$$F_R(\omega_{\text{eff}}) = 1 + (F_R(\omega_1) - 1)\frac{\sum_{n=0}^{\infty}(n\, I_{n(\text{rms})})^2}{I^2_{\text{rms, tot}}}. \tag{21.74}$$

Comparing the form of this expression to that of (21.73), we can solve for ω_{eff}:

$$\omega_{\text{eff}} = \omega_1\sqrt{\frac{\sum_{n=0}^{\infty}(n\, I_{n(\text{rms})})^2}{I^2_{\text{rms, tot}}}}. \tag{21.75}$$

The above expression allows us to find an effective frequency ω_{eff} for a non-sinusoidal waveform that allows us to use all the tools we have developed for predicting loss and optimizing windings in the sinusoidal case. This expression applies to waveforms including both dc and harmonic content, and will give accurate results so long as the conductor dimensions are on the order of a skin depth or smaller for the highest frequency of importance in the waveform. Note that while (21.75) expresses ω_{eff} in terms of Fourier components, it can also be expressed in terms of the time-domain properties of the non-sinusoidal waveform $i(t)$. In particular, because

$$\sum_{n=0}^{\infty}\left(n\,\omega_1\, I_{n(\text{rms})}\right)^2 = \left[\left(\frac{d}{dt}i(t)\right)_{\text{rms}}\right]^2, \tag{21.76}$$

we can express ω_{eff} as

$$\omega_{\text{eff}} = \frac{\left(\frac{d}{dt}i(t)\right)_{\text{rms}}}{I_{\text{rms, tot}}}. \tag{21.77}$$

This expression is particularly convenient in that it can be readily calculated from time-domain simulations (e.g., in SPICE) without having to explicitly compute Fourier coefficients.

21.2.7 Electrical Representation of Winding Loss

Thus far we have largely explored how to calculate winding loss in magnetic components without worrying too much about how to represent it in terms of electrical port variables. For a single-winding component driven by a sinusoidal current $i_1(t) = \text{Re}\{\widehat{I}_1\, e^{j\omega t}\}$ we can express the dissipation in terms of its ac winding resistance R,

$$P = \frac{1}{2}\widehat{I}_1\,\widehat{I}_1^*\,R = \frac{1}{2}\,|\widehat{I}_1|^2\,R, \tag{21.78}$$

where R may be frequency dependent owing to skin and proximity effects (e.g., $R = R_{\text{dc}}F_R(\omega)$).

What may be more surprising is that for a multi-winding component, we cannot always represent loss with a single resistance for each winding as our intuition might suggest. Rather, to handle the general case we need to include *mutual* resistance between each pair of windings, even if those windings are electrically isolated. Considering sinusoidal voltages and currents

$i_x(t) = \text{Re}\{\widehat{I}_x e^{j\omega t}\}$ and $v_x(t) = \text{Re}\{\widehat{V}_x e^{j\omega t}\}$, respectively, we can represent a two-port magnetic component with winding loss as

$$\begin{bmatrix} \widehat{V}_1 \\ \widehat{V}_2 \end{bmatrix} = \begin{bmatrix} j\omega L_{11} + R_{11} & j\omega L_M + R_M \\ j\omega L_M + R_M & j\omega L_{22} + R_{22} \end{bmatrix} \begin{bmatrix} \widehat{I}_1 \\ \widehat{I}_2 \end{bmatrix}, \tag{21.79}$$

where we have included winding self resistances R_{11} and R_{22} and mutual resistance R_M to model loss. (In a general two-port system, we can have different resistances R_{12}, R_{21}. However, because of reciprocity in the kinds of magnetic systems we consider, $R_{12} = R_{21} = R_M$. This arises for the same reason that a two-winding magnetic component has a single mutual inductance L_M.) Calculating the cycle-average power entering port x as $P_x = \frac{1}{2}\text{Re}\{\widehat{V}_x \widehat{I}_x^*\}$, it can be shown that the total average power dissipation due to winding loss (i.e., associated with R_{11}, R_{22}, and R_M) is

$$P = \frac{1}{2}\widehat{I}^{\text{T}} R \widehat{I}^* = \frac{1}{2}\begin{bmatrix} \widehat{I}_1 \\ \widehat{I}_2 \end{bmatrix}^{\text{T}} \begin{bmatrix} R_{11} & R_M \\ R_M & R_{22} \end{bmatrix} \begin{bmatrix} \widehat{I}_1 \\ \widehat{I}_2 \end{bmatrix}^*. \tag{21.80}$$

This may also be written as

$$P = \frac{1}{2}|\widehat{I}_1|^2 R_{11} + |\widehat{I}_1||\widehat{I}_2|\cos(\angle\widehat{I}_1 - \angle\widehat{I}_2)\, R_M + \frac{1}{2}|\widehat{I}_2|^2 R_{22}, \tag{21.81}$$

where the power dissipation associated with the first and second ports is

$$P_1 = \frac{1}{2}|\widehat{I}_1|^2 R_{11} + \frac{1}{2}|\widehat{I}_1||\widehat{I}_2|\cos(\angle\widehat{I}_1 - \angle\widehat{I}_2)\, R_M, \tag{21.82}$$

$$P_2 = \frac{1}{2}|\widehat{I}_2|^2 R_{22} + \frac{1}{2}|\widehat{I}_1||\widehat{I}_2|\cos(\angle\widehat{I}_1 - \angle\widehat{I}_2)\, R_M. \tag{21.83}$$

Thus, loss depends not only on I^2R for each winding but on the product of the two current magnitudes, the cosine of their phase difference, and the mutual resistance R_M.

Example 21.4 Self- and Mutual Resistance Calculation

To illustrate why mutual resistance is sometimes needed to represent winding loss in magnetic components, consider the system of Fig. 21.20, which shows a "pot-core" magnetic component having three single-layer foil windings and gaps in both the centerpost and outer shell. We assume that windings 1 and 2 carry net sinusoidal currents $\widehat{I}_1$ and $\widehat{I}_2$, but that winding 3 is open-circuited (i.e., $\widehat{I}_3 = 0$); this might be the case in an energy storage transformer in which winding 1 is the primary, winding 2 the secondary, and winding 3 an electrostatic shield placed between them. As before, we "unwrap" the conductors into a Cartesian coordinate system with y-directed currents and z-directed fields as illustrated in the right portion of Fig. 21.20. The height of the windings (and the window) are each h, the lengths of the windings are each ℓ, and the widths of windings 1–3 are W_1–W_3.

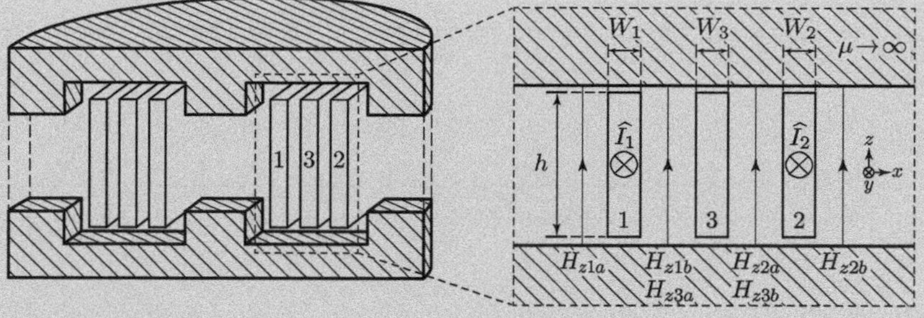

Figure 21.20 Cross-sectional views of a gapped pot-core magnetic component with three foil winding layers. Current and field designations for the winding layers (e.g., as in Fig. 21.10) are also shown.

We consider the winding loss for four cases (each with $\hat{I}_3 = 0$):

- Case 1: $\hat{I}_1 = I, \hat{I}_2 = 0$
- Case 2: $\hat{I}_1 = 0, \hat{I}_2 = I$
- Case 3: $\hat{I}_1 = I, \hat{I}_2 = -I$
- Case 4: $\hat{I}_1 = I, \hat{I}_2 = I$

In each case we first identify the (z-directed) H field boundary conditions at the surfaces of all the winding layers (e.g., as illustrated in Fig. 21.10), then find the associated current distribution and losses for those conditions. We could calculate loss for the general case using the formulation in (21.41). However, for simplicity we consider the situation in which the widths of the windings (W_1–W_3) are all large compared to a skin depth, such that the fields and current densities in the middle of the windings are zero. For this situation we can calculate loss just as if the winding currents were carried uniformly within a skin depth of the surface.

Figure 21.21 shows the fields and currents for each case. In case 1, only winding 1 carries net current. The resulting fields and winding currents are shown in Fig. 21.21(a). The boundary fields for this case (as defined in Fig. 21.20) are $H_{z1a} = I/2h$, $H_{z1b} = H_{z3a} = -I/2h$, $H_{z3b} = H_{z2a} = -I/2h$, and $H_{z2b} = -I/2h$. With this field distribution, winding 1 carries current I with balanced skin effect (i.e., with the current divided equally between its two surfaces), while windings 2 and'3 each exhibit proximity-effect currents owing to the field from the current in winding 1. We can calculate the total loss owing to all of these currents, and ascribe it to the self-resistance R_{11} of winding 1:

$$P_{C1} = \sum_j \frac{1}{2} R_j (I_j)^2 = 6 \cdot \frac{1}{2} \frac{\ell}{\sigma h \delta} \left(\frac{I}{2}\right)^2 = \frac{1}{2} I^2 R_{11}. \tag{21.84}$$

From this we can find the high-frequency ac self-resistance of winding 1:

$$R_{11} = \frac{3}{2} \frac{\ell}{\sigma h \delta}.$$

In case 2, shown in Fig. 21.21(b), we have the complementary situation in which only winding 2 carries net current. The boundary fields for this case are $H_{z1a} = I/2h$, $H_{z1b} = H_{z3a} = I/2h$, $H_{z3b} = H_{z2a} = I/2h$, and $H_{z2b} = -I/2h$. Winding 2 carries current I with balanced skin effect, while windings 1 and 3 each exhibit proximity effect currents owing to the fields from the current in winding 2. Repeating the same loss calculation for this situation, we get the high-frequency ac self resistance of winding 2:

$$R_{22} = \frac{3}{2} \frac{\ell}{\sigma h \delta}.$$

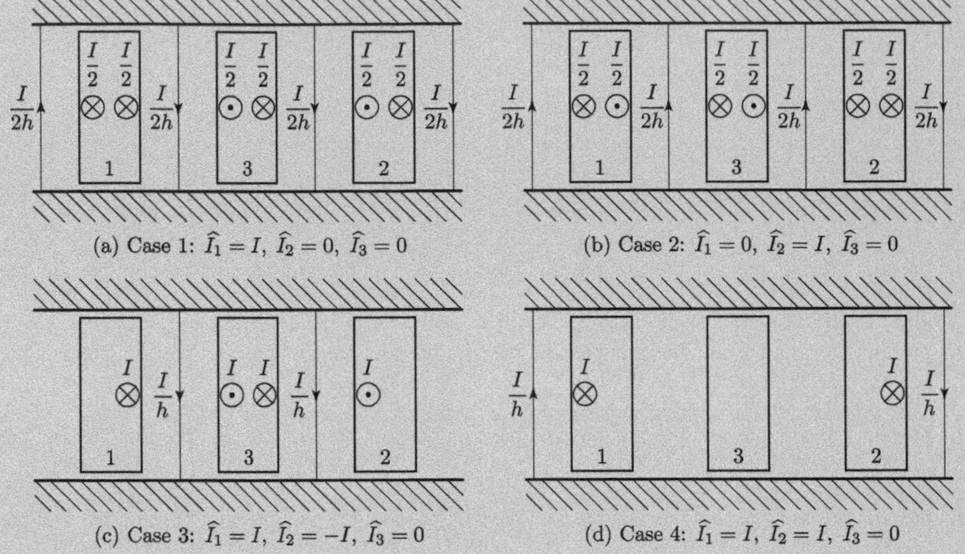

Figure 21.21 Field and current patterns for the system of Fig. 21.20 for four cases. (a) Case 1: $\widehat{I}_1 = I, \widehat{I}_2 = 0, \widehat{I}_3 = 0$. (b) Case 2: $\widehat{I}_1 = 0, \widehat{I}_2 = I, \widehat{I}_3 = 0$. (c) Case 3: $\widehat{I}_1 = I, \widehat{I}_2 = -I, \widehat{I}_3 = 0$. (d) Case 4: $\widehat{I}_1 = I, \widehat{I}_2 = I, \widehat{I}_3 = 0$.

In cases 3 and 4, both windings 1 and 2 carry net current. In case 3, shown in Fig. 21.21(c), winding 1 carries net current I and winding 2 carries net current $-I$. The boundary fields for this case are $H_{z1a} = 0$, $H_{z1b} = H_{z3a} = -I/h$, $H_{z3b} = H_{z2a} = -I/2h$, and $H_{z2b} = 0$, and the resulting power dissipation is

$$P_{C3} = \sum_j \frac{1}{2} R_j (I_j)^2 = 4 \cdot \frac{1}{2} \frac{\ell}{\sigma h \delta} (I)^2 = 2 \frac{\ell}{\sigma h \delta} (I)^2. \tag{21.85}$$

In case 4, shown in Fig. 21.21(d), windings 1 and 2 each carry net current I. The boundary fields for this case are $H_{z1a} = I/h$, $H_{z1b} = H_{z3a} = 0$, $H_{z3b} = H_{z2a} = 0$, and $H_{z2b} = -I/h$, and the resulting power dissipation is

$$P_{C4} = \sum_j \frac{1}{2} R_j (I_j)^2 = 2 \cdot \frac{1}{2} \frac{\ell}{\sigma h \delta} (I)^2 = \frac{\ell}{\sigma h \delta} (I)^2. \tag{21.86}$$

Notice that the loss is quite different for cases 3 and 4 despite the fact that both windings 1 and 2 have the same current amplitudes and carry current similarly (i.e., with single-sided conduction) in each case. The loss is different because of the difference in proximity effect from fields impinging on winding 3. The proximity-effect loss in winding 3 depends upon both the magnitudes and relative phases of the currents in windings 1 and 2. This underscores the need for a mutual resistance term to model loss, as provided in (21.80)–(21.83).

To find the mutual resistance for this example, we can take the difference in loss between cases 3 and 4, and match that to the difference predicted by (21.81) for the two cases:

$$P_{C3} - P_{C4} = \frac{\ell}{\sigma h \delta} (I)^2 = -2 R_M I^2, \qquad R_M = -\frac{1}{2} \frac{\ell}{\sigma h \delta}. \tag{21.87}$$

The results above are for the limiting case in which the windings are wide compared to a skin depth. The techniques described earlier in this section can be used to analyze the general case, yielding the following values for the self- and mutual ac resistances of the system:

$$R_{11} = \frac{\ell}{\sigma h \delta}\left[F\left(\frac{W_1}{\delta}\right) - \frac{1}{2}G\left(\frac{W_1}{\delta}\right) + \frac{1}{2}G\left(\frac{W_2}{\delta}\right) + \frac{1}{2}G\left(\frac{W_3}{\delta}\right)\right],$$
$$R_{22} = \frac{\ell}{\sigma h \delta}\left[F\left(\frac{W_2}{\delta}\right) - \frac{1}{2}G\left(\frac{W_2}{\delta}\right) + \frac{1}{2}G\left(\frac{W_1}{\delta}\right) + \frac{1}{2}G\left(\frac{W_3}{\delta}\right)\right], \quad (21.88)$$
$$R_M = -\frac{1}{2}\frac{\ell}{\sigma h \delta}G\left(\frac{W_3}{\delta}\right).$$

From this more general result, we can see that the mutual resistance approaches zero at low frequency and in other situations where the eddy currents in winding 3 are negligible (i.e., for $W_3/\delta \ll 1$). In this circumstance the mutual resistance becomes negligible and we can accurately model winding loss well using only self-resistances.

We can conclude from this example that mutual resistance can be necessary to accurately model loss in magnetic components, at least in situations where loss is determined by how fields from different windings interact.

21.3 Core Loss

As compared to the great detail with which winding loss can be analyzed, modeling core loss is often a very “rough and ready” affair. In part this arises from the fact that the underlying physical behaviors leading to core loss are complex, with details that vary considerably among materials and operating conditions. Compounding this challenge is the frequent dearth of data for magnetic materials measured under the conditions in which they will be used (e.g., with dc bias and/or non-sinusoidal ac flux waveforms). Consequently, the designer is often faced with predicting core loss based on inadequate information, sometimes forcing the adoption of significant “safety margins” for core loss estimates. Nonetheless, knowing what factors to pay attention to and how to best leverage available data can help to mitigate these challenges. In this section we describe some of the factors to consider in using magnetic materials, and describe methods for predicting core loss in power electronics applications based on widely available data.

21.3.1 Eddy Current Loss and Core Lamination

We have seen that magnetic diffusion and eddy currents can have a significant impact on field distribution and loss in windings. They can also be important in magnetic cores, especially in large cores and/or with core materials having high conductivity. Eddy currents driven by the ac flux carried in a core are one contributor to core loss. Moreover, when core cross-sectional dimensions become larger than a skin depth of the core material, magnetic diffusion effects can cause undesired concentrations of flux near the core surface and under-utilization of the center of the core.

Consider the system of Fig. 21.22, which shows a cylindrical core section having radius R and length ℓ which carries an axially directed flux density $B(t)$ (e.g., from a winding on the core). A shell forming part of the core material is also shown; it has radius r and incremental thickness dr.

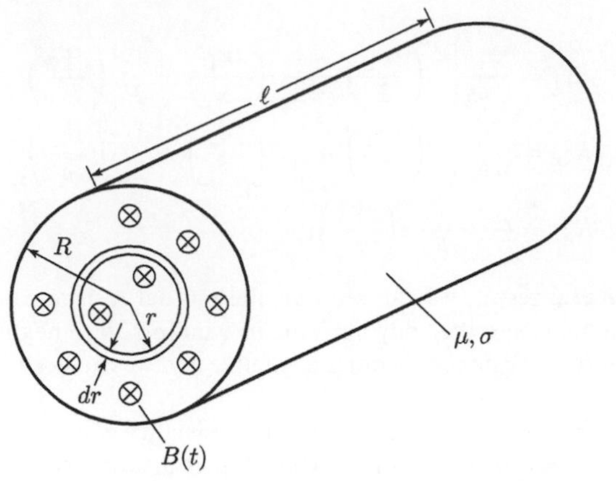

Figure 21.22 A cylindrical section of magnetic core having radius R, length ℓ, permeability μ, and conductivity σ. The core section carries a substantially uniform flux density $B(t)$ in the axial direction.

Eddy currents will flow azimuthally in the shell in response to the flux passing within its radius. We assume that $B(t)$ is substantially uniform across the core area, implying that the core diameter is on the order of a skin depth or below, such that eddy currents do not greatly affect the field distribution.

To calculate the core loss associated with the eddy currents, consider a toroidal "shell" of incremental thickness dr and inner radius $r < R$ through which azimuthal eddy currents can flow in response to the magnetic flux passing through its inner radius. Given the azimuthal resistance of the incremental shell,

$$R_{\text{shell}}(r) = \frac{2\pi r}{\sigma \ell (dr)}, \tag{21.89}$$

and a voltage driving eddy current around the shell,

$$V_{\text{shell}}(r,t) = \frac{d}{dt}\int B \cdot dA = \pi r^2 \left(\frac{dB}{dt}\right), \tag{21.90}$$

we get an instantaneous power dissipation in the shell of

$$p_{\text{shell}}(r,t) = \frac{V_{\text{shell}}^2}{R_{\text{shell}}} = \frac{1}{2}\pi r^3 \sigma \ell \left(\frac{dB}{dt}\right)^2 dr. \tag{21.91}$$

Integrating this over the radius of the core cylinder we get a total instantaneous power dissipation in the cylindrical core section of

$$p_{\text{cyl}}(t) = \int_0^R p_{\text{shell}}(r,t) = \frac{1}{8}\pi R^4 \sigma \ell \left(\frac{dB}{dt}\right)^2. \tag{21.92}$$

Considering that the cylinder cross-sectional area is $A = \pi R^2$ and its total volume is $A\ell$, we can calculate the instantaneous power dissipation per unit volume of the core as a function of time:

$$p_{\text{v,eddy}}(t) = \frac{\sigma}{8\pi} A \left(\frac{dB}{dt}\right)^2. \tag{21.93}$$

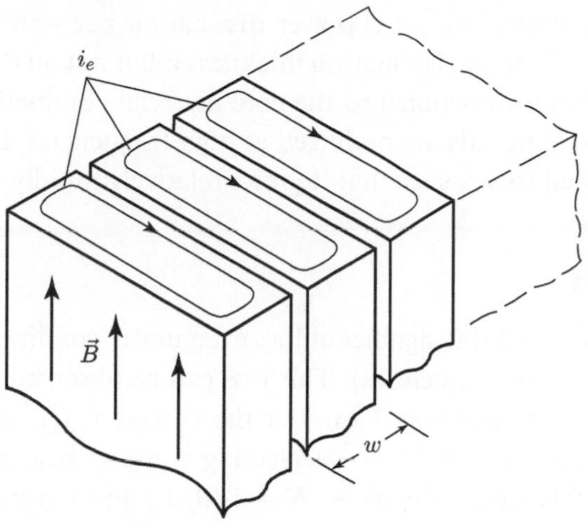

Figure 21.23 A section of a laminated core, showing the location of eddy currents, i_e, in the laminations.

For a sinusoidal drive $B(t) = |B|\cos(\omega t + \phi)$, this gives an average power dissipation per unit volume of

$$P_{\mathrm{v,eddy}} = \frac{\sigma\omega^2}{16\pi} A\,|B|^2. \tag{21.94}$$

What is notable about this result is that the power dissipation per unit volume is a function of the core cross-sectional area A, growing larger with core area. That is, larger-cross-section cores will not only have larger power dissipation, but a larger *density* of power dissipation than smaller cores. We thus cannot simply treat core loss due to eddy currents on a per-unit-volume basis independent of the core dimensions. This behavior arises because a larger core cross-sectional area increases the region through which eddy currents can circulate, resulting in higher loss density.

Core laminations are often used to address both the growth in eddy current loss with core area and to prevent flux crowding owing to magnetic diffusion effects. As illustrated in Fig. 21.23, laminations are placed such that the thin dimension w is perpendicular to the direction of flux in the core. This confines the induced eddy currents to circulate within a narrow region defined by the laminations. Selecting w to be a skin depth or less also enables full utilization of the core area even when core cross-sectional dimensions are large compared to a skin depth. An analysis similar to that above shows that with such laminations the power dissipation per unit volume owing to eddy currents is

$$p_{\mathrm{v,eddy}}(t) = \frac{\sigma w^2}{12}\left(\frac{dB}{dt}\right)^2. \tag{21.95}$$

For a sinusoidal flux density, this gives an average power dissipation per unit volume of

$$P_{\mathrm{v,eddy}} = \frac{\sigma w^2\omega^2}{24}\,|B|^2. \tag{21.96}$$

Thus, with laminations we get a power dissipation per unit volume owing to eddy currents $P_{\mathrm{v,eddy}}$ that depends on the lamination thickness w but not on the overall core dimensions.[‖] Eddy current loss is also proportional to the core material conductivity σ, which is one reason that low-conductivity materials are preferred at high frequencies. Lastly, we note that eddy current loss is proportional to ω^2, such that it grows relatively rapidly with frequency.

21.3.2 Hysteresis Loss

Magnetic cores can exhibit significant loss even under conditions where eddy current losses are negligible (e.g., at low frequencies). This loss can be observed directly in the v–i measurements of inductors and transformers. Consider the voltage $v_L(t)$ and current $i_L(t)$ of the ungapped inductor illustrated in Fig. 21.24. Neglecting winding resistance, we can relate the inductor voltage to core flux density $v_L(t) = N A_c (dB/dt)$ and the inductor current to core magnetic field strength $i_L(t) = (1/N) H \ell_c$. With a sinusoidal drive, the energy into the inductor over a cycle is

$$\begin{aligned} W_{\text{in}} &= \int_0^T v_L(t)\, i_L(t)\, dt \\ &= \int_0^T \left(N A_c \frac{dB}{dt}\right)\left(\frac{1}{N} H \ell_c\right) dt \\ &= \ell_c A_c \oint H \cdot dB. \end{aligned} \tag{21.97}$$

This gives an average energy loss per unit core volume of

$$W_{\text{v}} = \oint H \cdot dB. \tag{21.98}$$

Given an ideal linear relation $B = \mu H$, there would be no power dissipation in the core. However, real core materials exhibit hysteresis in their B–H relationships, as illustrated in the right part of Fig. 21.24 and described in Chapter 18. Equation (21.98) tells us that the energy dissipated per unit core volume per cycle is the area encompassed within the hysteresis loop. The hysteresis loop can be observed by measuring inductor voltage $v_L(t)$ and current $i_L(t)$. In practice, hysteresis loops and core loss are often determined by measuring current in the drive winding (to determine H) and measuring the voltage induced on a separate open-circuited sense winding (to determine dB/dt) so that winding losses are not captured in the calculation.

A static hysteresis loop (or static B–H loop) is one that is measured at sufficiently low excitation frequency that core eddy currents and other effects are not important contributors to loss. At these frequencies, the loop shape is not a significant function of frequency, though it does depend on the amplitude of applied ac flux density B. The area of such a loop, A_{hys}, represents the energy per volume needed to rotate the magnetic domains of the material. Given A_{hys} for an ac flux density of interest, we can thus estimate the low-frequency core loss density as

‖ For sufficiently thin lamination insulation and sufficiently high frequency, capacitive displacement currents can pass across insulation layers, reducing the lamination effectiveness. However, this is not a significant factor in most instances.

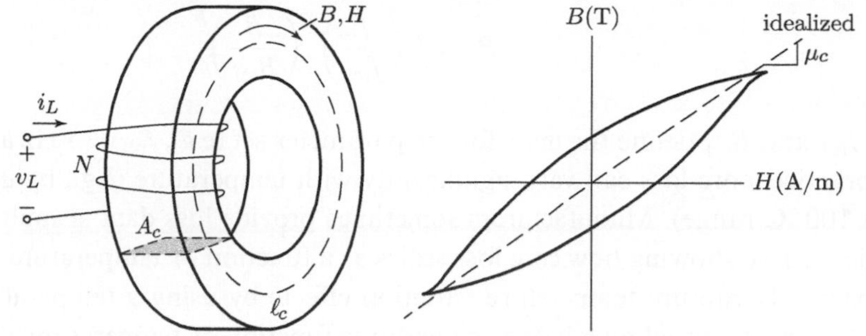

Figure 21.24 A toroidal inductor and a B–H plot showing hysteresis in the inductor core material. The hysteresis loop can be observed by measuring inductor voltage $v_L(t)$ and current $i_L(t)$.

$$P_{\text{v, hys}} = \frac{1}{T} A_{\text{hys}} = \frac{\omega}{2\pi} A_{\text{hys}}. \tag{21.99}$$

The calculation in (21.99) suggests that core loss density is proportional to frequency. This behavior is referred to as hysteresis loss, and is a useful way to model low-frequency core loss. Of course, the calculation in (21.97) includes *all* core loss. For excitation frequencies where core eddy currents and other loss effects become important, the shape of the hysteresis loop changes, becoming larger with higher frequency. Consequently, we cannot think of core loss as being proportional to frequency except at low frequencies where A_{hys} is constant.

Some core loss models try to capture loss behavior across frequency by decomposing loss as the sum of a static hysteresis loss term (proportional to ω), an eddy-current loss term (proportional to ω^2), and one or more further terms with other frequency dependencies (sometimes described as "excess loss" or "anomalous loss" components). However, this kind of loss book-keeping is not particularly helpful for most high-frequency power magnetics designs. Rather, it is more effective to employ empirical models for loss such as the Steinmetz equation.

21.3.3 The Steinmetz Equation

In practice, core loss data is often provided (or measured) as loss per unit volume versus ac flux density for sinusoidal excitation at various frequencies of interest. This representation assumes that the core cross-sectional dimensions are small with respect to a skin depth or that the core material is laminated such that a "loss per volume" description is meaningful. These data fit the famed Steinmetz equation well, at least over a limited range of frequencies f and flux densities B:

$$P_{\text{v}} = k f^{\alpha} B^{\beta}. \tag{21.100}$$

Here, P_{v} is the power loss per unit volume, and the parameters k, α, and β are fitting parameters used to match the data. Typical values of α range from 1 to 2.6, and typical values of β range from 1.6 to 3.5. The Steinmetz model is purely empirical in nature, and a given parameter set (k, α, β) may only be expected to give accurate loss predictions for sinusoidal waveforms over a limited range of f and B. When applying the Steinmetz equation with provided parameters, it is important to be careful about the units that are specified. While (21.100) is the conventional form of the Steinmetz equation, a more useful form from a units perspective is

$$P_v = k\left(\frac{f}{f_{ref}}\right)^{\alpha}\left(\frac{B}{B_{ref}}\right)^{\beta}, \tag{21.101}$$

where f_{ref} and B_{ref} define the units for the parameter set (e.g., $f_{ref} = 1$ Hz and $B_{ref} = 1$ T).

In practice, core loss can vary significantly with temperature (e.g., by up to a factor of two over a 100 °C range). Manufacturers sometimes provide loss data at multiple temperatures or provide a curve showing how core loss scales as a function of temperature. In this case, one can approximately capture temperature variation effects by using a temperature-dependent value for k. A curve fit based on a linear or quadratic function of temperature T is often adequate to capture the variation, e.g. $k(T) = k_0 + k_1T + k_2T^2$ with fit parameters (k_0, k_1, k_2). Alternatively, one can typically make a conservative loss estimate by using the core loss data at the highest temperature provided.

Core loss can also be a function of the dc H field in the core, typically increasing with the magnitude of the bias field. Thus, the core loss of the inductor in a buck converter might vary with load current even though the ac component of the flux waveform remains unchanged. Loss increases exceeding a factor of three have been observed with a dc bias field in some materials. As with temperature, the effect of the dc bias H_{dc} on core loss can be partially captured by making one or more of the Steinmetz parameters (k, α, β) functions of H_{dc}. Unfortunately, the information necessary to do so is only rarely available without making one's own measurements. In the absence of data, designers are sometimes simply conservative in estimating core loss for designs with dc bias fields.

21.3.4 Loss Calculation for Non-Sinusoidal Waveforms

Many power magnetic components operate with non-sinusoidal voltages and flux densities. For example, the inductor in a PWM buck converter experiences a square-wave voltage resulting in a triangular flux density. Predicting core loss in non-sinusoidal cases is challenging because (i) core loss data is typically only provided by manufacturers for sinusoidal flux density waveforms, and (ii) one cannot generally calculate core loss as the sum of losses owing to the individual frequency components of a waveform.

In cases where a flux density waveform is not too highly non-sinusoidal, one might approximate it as a sinusoid for loss calculation purposes, and provide a generous margin to account for the approximation. Alternatively, one can measure the loss of candidate core materials with waveforms approximating those in an intended application, but this is extremely labor intensive. It would thus be desirable to be able to *approximately* estimate core loss for non-sinusoidal waveforms based on only manufacturer-provided data (e.g., Steinmetz parameters k, α, β valid for a frequency range of interest). Here we introduce one approach for doing so known as the "improved generalized Steinmetz equation," or iGSE.

The iGSE is an empirical loss formula that gives the same results as the conventional Steinmetz equation for sinusoidal waveforms, provides reasonable predictions for limiting cases, and is physically plausible. Despite its simplicity, the iGSE often gives very good loss estimates for non-sinusoidal waveforms using only conventional Steinmetz model parameters. We start by introducing the iGSE for the special case of waveforms having no minor B–H loops, that is, in which there is a single section of the waveform where flux is constant or rising, and a single section where flux is constant or falling. This is equivalent to an applied voltage waveform having a single section of positive or zero voltage, and a single section of negative or zero voltage. We then treat its general application.

Recognizing that the time-averaged core loss ought to depend on both the peak-to-peak flux density swing ΔB and the instantaneous rate of change of flux density dB/dt, the iGSE estimates the power loss per unit volume P_V as

$$P_V = k_i\,(\Delta B)^{\beta-\alpha}\frac{1}{T}\int_0^T \left|\frac{dB}{dt}\right|^\alpha dt, \qquad (21.102)$$

where α and β are conventional Steinmetz parameters. The constant k_i is determined based on the Steinmetz parameter k such that the formulation of (21.102) gives the same result as the conventional Steinmetz equation for a sinusoidal waveform. Matching the above result with a sinusoidal flux density to that of the Steinmetz equation yields

$$k_i = \frac{k}{\pi^{\alpha-1}\,2^{\beta-1}\int_0^{2\pi}|\cos(\theta)|^\alpha\, d\theta}. \qquad (21.103)$$

For values of α between 0.5 and 3, this expression can be approximated with very high accuracy as

$$k_i = \frac{k}{\pi^{\alpha-1}\,2^{\beta-1}(1.1044 + (6.8244/(\alpha + 1.354)))}. \qquad (21.104)$$

Given (21.102) and (21.104) and a suitable set of Steinmetz parameters (k, α, β) we can estimate the loss of a non-sinusoidal waveform having no minor loops. The Steinmetz parameter set used should well represent the core loss over a frequency range encompassing the content of $B(t)$. For a waveform with widely separated frequency components, finding such a parameter set may be a challenge; this is one limitation of the iGSE. Other considerations and limitations in selecting Steinmetz parameters (e.g., with respect to temperature and H field bias) pertain to the iGSE just as for the conventional Steinmetz equation.

Many power magnetic components experience piecewise linear (PWL) flux density waveforms (i.e., owing to applied voltage waveforms that may be represented as a sequence of constant voltages). Consider a PWL flux density waveform having no minor loops that is generated by a sequence of fixed voltages V_j having respective durations Δt_j. In this case we can rewrite (21.102) as

$$P_V = k_i\,(\Delta B)^{\beta-\alpha}\frac{1}{T}\sum_j \left|\frac{V_j}{NA_c}\right|^\alpha (\Delta t_j), \qquad (21.105)$$

where N is the number of winding turns and A_c is the core cross-sectional area. We can thus very easily calculate core loss for this special case.

Example 21.5 Transformer Loss Calculation with the iGSE

Here we apply the iGSE to calculate the core loss of a transformer for an active-clamp forward converter. As described in Chapter 7 and illustrated in Fig. 21.25, an active-clamp forward converter imposes a rectangle-wave voltage across the transformer primary, resulting in a triangle-wave core flux density. We consider a converter design operating from an input voltage of $V_1 = 100\,\text{V}$ at a duty ratio $D = 0.3$ and a switching frequency of $f = 500\,\text{kHz}$.

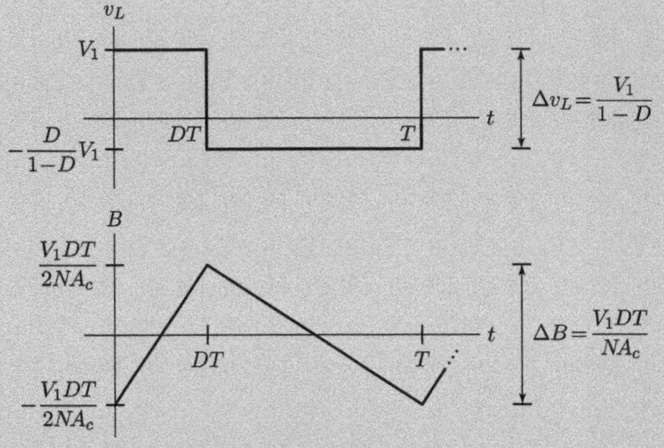

Figure 21.25 Transformer primary voltage and core flux density waveforms for an active-clamp forward converter.

The tranformer is wound with $N = 5$ primary turns on an RM10 core ($A_c = 98.0\,\text{mm}^2$, $V_c = 4.31\,\text{cm}^3$). The transformer core material is ML95S ferrite from Hitachi Metals. For the specified duty ratio, the core flux density waveform is dominated by its fundamental and second harmonic components. At 100 °C, the Steinmetz core-loss parameters suitable for the 500 kHz to 1 MHz range are $k = 8.09 \times 10^{-6}$, $\alpha = 2.011$, and $\beta = 3.536$ when the frequency f is expressed in hertz, the ac flux density amplitude B is expressed in tesla, and the loss density P_V is expressed in mW/cm^3.

To estimate the core loss owing to the triangular flux density waveform we use the iGSE formulation in (21.105). We first calculate k_i using (21.104), which yields $k_i = 1.40 \times 10^{-7}$ mW/cm^3. Applying (21.105) gives a loss density of

$$P_\text{V} = k_i \left(\frac{V_1 DT}{NA_c}\right)^{\beta-\alpha} \left\{ D\left[\frac{V_1}{NA_c}\right]^\alpha + (1-D)\left[\frac{DV_1}{(1-D)NA_c}\right]^\alpha \right\}, \tag{21.106}$$

which yields $P_\text{V} = 116\,\text{mW/cm}^3$, resulting in an estimated core loss of 0.499 W.

Up to now, we have considered loss calculation with the iGSE for flux density profiles without minor loops (i.e., in which the core flux density rises or is constant in the first portion of the cycle and falls or is constant in the remaining portion of the cycle). Some waveforms, however, exhibit minor loops, as illustrated in Fig. 21.26. Minor-loop behavior occurs when there are multiple alternating sections of positive and negative applied voltage (and hence rising and falling flux density) within a period, as often occurs with PWM waveforms for example. It has been found that accurately modeling core loss in such instances requires explicitly addressing minor-loop behavior. The iGSE has been specifically formulated to do so as described below.

To calculate core loss for waveforms with minor loops using the iGSE, the waveform is first disaggregated (in time) into multiple waveform patterns representing the major loop and individual minor loops. This process is illustrated conceptually in Fig. 21.27. Essentially, sections of the flux density waveform associated with minor loops are abstracted from the overall

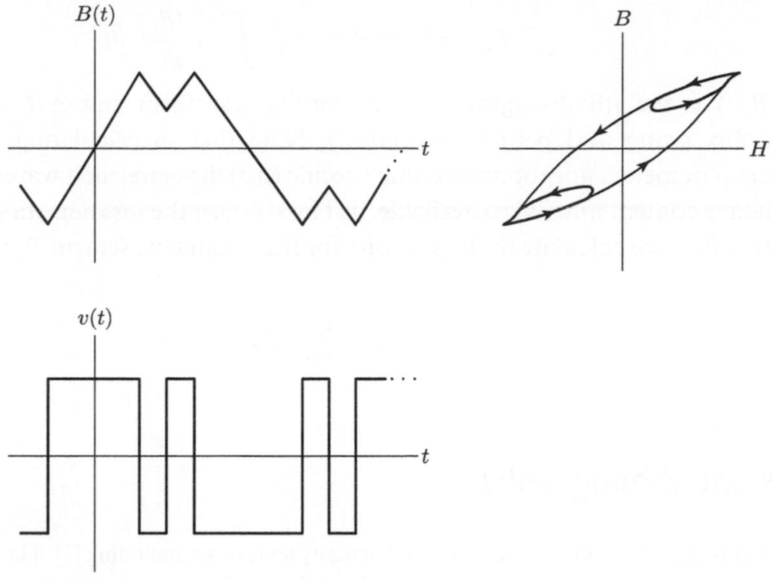

Figure 21.26 A voltage waveform, core flux density waveform, and associated B–H loop illustrating minor-loop behavior.

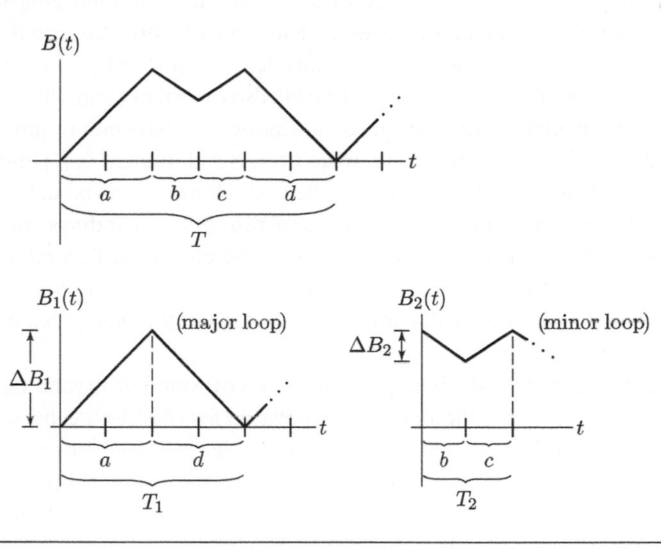

Figure 21.27 Separation of a flux density waveform with a minor loop into one representing the major loop component and a second representing the minor loop component. Each separate waveform has its own period and peak-to-peak flux swing.

waveform and defined as separate individual waveforms, each with their own period, leaving behind a major loop waveform with its own period.

The loss density is calculated for each of the disaggregated waveforms individually according to the iGSE formulation. Thus, the jth disaggregated (minor or major loop) waveform has a loss density of

$$P_{V,j} = k_i\,(\Delta B_j)^{\beta-\alpha}\frac{1}{T_j}\int_0^{T_j}\left|\frac{dB_j}{dt}\right|^{\alpha} dt, \qquad (21.107)$$

where $B_j(t)$ is the jth disaggregated flux density waveform and ΔB_j and T_j are its peak-to-peak flux swing and period, respectively. Note that in calculating $P_{V,j}$ one should use Steinmetz parameters appropriate to that specific (jth) disaggregated waveform (i.e., suitable for its frequency content and, when available, dc bias). Given the loss densities of the disaggregated waveforms $P_{V,j}$, we calculate the loss density for the original waveform P_V as their time-weighted average:

$$P_V = \sum_j \frac{T_j}{T} P_{V,j}. \qquad (21.108)$$

Notes and Bibliography

Magnetic diffusion and skin effect are treated in many textbooks, including [1]. The decomposition of fields and current densities into orthogonal skin and proximity effect components are described and exploited in [2] and [3]. For cases where it is of interest, the contribution of balanced skin effect to ac loss of a round wire may be found in [2].

The classic paper by Dowell [4] introduces techniques for analyzing loss and ac resistance factor in transformer windings, including Dowell's equation (21.49). Note that the results can be expressed in a variety of different (but numerically equivalent) forms; see [5] for another common form of this development, for example. It is noteworthy that [4] also describes means for computing transformer leakage inductance based on energy stored in the core window. Transformer optimization using these techniques, including with non-sinusoidal operating waveforms, is well treated in [6] and [7].

The results of Dowell are based on simplified one-dimensional models that are well applicable to foil conductors and can approximate loss in layers of round-wire windings when the wire diameter is on the order of a skin depth and below. For larger wire diameters, Dowell's results tend to under-predict loss. As shown in [8], a variety of other methods can also be used to accurately estimate loss in round-wire windings when the winding diameter is small compared to a skin depth. One such method is used in [8] towards the optimization of litz wire designs.

The approach of directly calculating proximity loss of round-wire windings embodied in (21.59)–(21.62) was first proposed in [2] using the analytical expression for round-wire loss in (21.60). While this analytical solution is exact for isolated round-wire conductors, it tends to over-predict loss for packed round-wire windings having wire diameters of more than a couple of skin depths because it does not account for the interactions among wires. Closer matching for large wire diameters can be achieved using the loss function $\widehat{G}$ provided in [3], which explicitly includes the effects of wire spacing.

Analytical calculations can be augmented with numerical tools for loss calculation, such as finite element tools. Sullivan [8] treats the optimization of litz wire windings in detail, including modeling insulation and other effects, and applying a generalized version of (21.68). The online numerical tool LitzOpt described in [9] is also useful for optimizing designs with litz wire.

Finite element simulations can provide very good accuracy and can handle complex geometries. However, they can often require significant setup time, demand careful validation of results, and do not readily yield the design insights provided by the analytical techniques detailed here.

The use of an "effective" frequency in optimizing the design of a winding for non-sinusoidal waveforms was introduced in [8] in the context of litz wire selection, and further developed in [6]. Additional examples of its application in winding design can be found in [7]. The electrical representation of winding loss in magnetic components including mutual resistance terms is explored in [10]. Methods for calculating current distributions in paralleled conductors are treated in [11] and [12].

Methods for measuring core loss may be found in [13] and [14]. In addition to commercial datasheets, core loss data for some high-frequency materials may be found in [14] and [15]. The impact of dc H-field bias on core loss is nicely described in [16], which also provides Steinmetz "premagnetization" curves for a few materials indicating how k, α, and β vary with H_{dc} for several common materials. It is worth noting, however, that there can even be variations in core loss that depend on the history of the bias field [15].

There is a rich literature on modeling of core loss. Unfortunately, there is no approach based on fundamental principles that works well across the diversity of magnetic materials and operating conditions relevant to power electronic systems. Consequently, empirical loss models based on measured data are the designer's best tools at present. As has been described, loss data are typically only available for sinusoidal waveforms. While there have been a variety of approaches developed to leverage such data for non-sinusoidal waveforms, the iGSE approach developed in [17] as an extension of the generalized Steinmetz equation [18] has the benefit of being relatively simple and very effective. Further examples of applying the iGSE may be found in [7]. Limitations of the iGSE (including for waveforms with "constant flux" segments) are explored in [19], which also proposes a method to improve core loss estimates using additional material parameters.

There are many other considerations in determining the loss of magnetic components that are not covered here. For example, how windings are terminated can be important to loss, including in planar magnetics [20], in which the windings are implemented directly as part of a printed circuit board (e.g., see [21]). Likewise, the capacitance of magnetic components can have a significant impact on their operation and loss (e.g., [22, 23]), as can how the magnetic materials themselves are handled and processed.

Oliver Heaviside's 1885 description of the skin effect in current-carrying conductors is nicely described in [24].

1. H. A. Haus and J. R. Melcher, *Electromagnetic Fields and Energy*, Englewood Cliffs, NJ: Prentice-Hall, 1989.
2. J. A. Ferreira, "Improved Analytical Modeling of Conductive Losses in Magnetic Components," *IEEE Trans. Power Electronics*, vol. 9, pp. 127–131, Jan. 1994.
3. X. Nan and C. R. Sullivan, "Simplified High-Accuracy Calculation of Eddy-Current Loses in Round Wire Windings," *IEEE Power Electronics Specialists Conference*, pp. 873–897, 2004.
4. P. L. Dowell, "Effects of Eddy Currents in Transformer Windings, *Proc. IEE*, vol. 13, pp. 1387–1394, Aug. 1966.
5. R. W. Erickson and D. Maksimovic, *Fundamentals of Power Electronics*, 3rd edn., Cham: Springer, 2020.
6. W. G. Hurley, E. Gath, and J. G. Breslin, "Optimizing the AC Resistance of Multilayer Transformer Windings with Arbitrary Current Waveforms," *IEEE Trans. Power Electronics*, vol. 15, pp. 369–376, Mar. 2000.
7. W. G. Hurley and W. H. Wölfle, *Transformers and Inductors for Power Electronics*, Chichester: Wiley, 2013.
8. C. R. Sullivan, "Optimal Choice for Number of Strands in a Litz-Wire Transformer Winding,: *IEEE Trans. Power Electronics*, vol. 14, pp. 283–291, Mar. 1999.
9. J. D. Pollock, T. Abdallah, and C. R. Sullivan, "Easy-to-Use CAD Tools for Litz-Wire Winding Optimization," 2003 *IEEE Applied Power Electronics Conference* (APEC), pp. 1157–1163, Feb. 2003.
10. J. H. Spreen, "Electrical Terminal Representation of Conductor Loss in Transformers," *IEEE Trans. Power Electronics*, vol. 5, pp. 424–429, Oct. 1990.

11. M. Chen, M. Araghchini, K. K. Afridi, J. H. Lang, and D. J. Perreault, "A Systematic Approach to Modeling Impedances and Current Distribution in Planar Magnetics," *IEEE Trans. Power Electronics*, vol. 31, no. 1, pp. 560–580, Jan. 2016.
12. M. Solomentsev and A. J. Hanson, "Modeling Current Distribution Within Conductors and Between Parallel Conductors in High Frequency Transformers," *2021 IEEE Applied Power Electronics Conference*, (APEC), pp. 1701–1708 2021.
13. IEEE, "IEEE Standard for Test Procedures for Magnetic Cores," IEEE Standard 393-1991, Mar. 1992.
14. Y. Han, A. Li, G. Cheung, C. R. Sullivan, and D. J. Perreault, "Evaluation of Magnetic Materials for Very High Frequency Power Applications," *IEEE Trans. Power Electronics*, vol. 27, pp. 425–435, Jan. 2012.
15. A. J. Hanson, J. A. Belk, S. Lim, C. R. Sullivan, and D. J. Perreault, "Measurements and Performance Factor Comparisons of Magnetic Materials at High Frequency," *IEEE Trans. Power Electronics*, vol. 31, pp. 7909–7925, Nov. 2016.
16. J. Mühlethaler, J. Biela, J. W. Kolar, and A. Ecklebe, "Core Losses Under the DC Bias Condition Based on Steinmetz Parameters,"*IEEE Trans. Power Electronics*, vol. 27, pp. 953–963, Feb. 2012.
17. K. Venkatachalam, C. R. Sullivan, T. Abdallah, and H. Tacca, "Accurate Prediction of Ferrite Core Loss with Nonsinusoidal Waveforms using only Steinmetz Parameters," *IEEE Workshop on Computers in Power Electronics*, pp. 36–41, 2002.
18. J. Li, T. Abdallah, and C. R. Sullivan, "Improved Calculation of Core Loss with Nonsinusoidal Waveforms," *IAS Annual Meeting*, pp. 2203–2210, 2001.
19. J. Mühlethaler, J. Biela, J. W. Kolar, and A. Ecklebe, "Improved Core-Loss Calculation for Magnetic Components Employed in Power Electronic Systems," *IEEE Trans. Power Electronics*, vol. 27, pp. 964–973, Feb. 2012.
20. Z. Ouyang and M. A. E. Andersen, "Overview of Planar Magnetic Technology: Fundamental Properties," *IEEE Trans. Power Electronics*, vol. 29, pp. 4888–4900, Sep. 2014.
21. M. K. Ranjram, P. Acosta, and D. J. Perreault, "Design Considerations for Planar Magnetic Terminations," *20th Workshop on Control and Modeling for Power Electronics*, pp. 1–8, June 2019.
22. L. Casey, A. Goldberg, and M. F. Schlecht, "Issues Regarding the Capacitance of 1–10 MHz Transformers," *IEEE Applied Power Electronics Conference* (APEC), pp. 352–359, 1988.
23. J. Biela and J. W. Kolar, "Using Transformer Parasitics for Resonant Converters: A Review of the Calculation of the Stray Capacitance of Transformers," *IEEE Trans. Industry Applications*, vol. 44, pp. 223–233, 2008.
24. Paul Nahin, *Oliver Heaviside: The Life, Work, and Times of an Electrical Genius of the Victorian Age*, Baltimore, MD: Johns Hopkins University Press, 2002.

PROBLEMS

21.1 Consider the system of Fig. 21.2. Apply Ampère's law around various contours to confirm that the fields at the conductor surfaces are $\widehat{H}_z(-W/2) = \widehat{K}_d/2$ and $\widehat{H}_z(W/2) = -\widehat{K}_d/2$ as described in Section 21.1.4. You may assume that there is no external impinging field, that is, that the H field is zero to the left of the left-side distributed current source and to the right of the right-side distributed current source.

21.2 Derive the proximity loss expression (21.59) for a round wire of diameter D and length ℓ immersed in a sinusoidal magnetic field of magnitude H oriented perpendicular to the conductor. You may assume that the field inside the conductor is uniform, which is consistent with the wire diameter being small compared to the skin depth δ. (*Hint:* Apply Faraday's law to determine the eddy current density in the wire. You may assume that the length of the wire is much larger than its diameter. To do this you may want to consider loops formed by incremental slices of the conductor as suggested in Fig. 21.28.

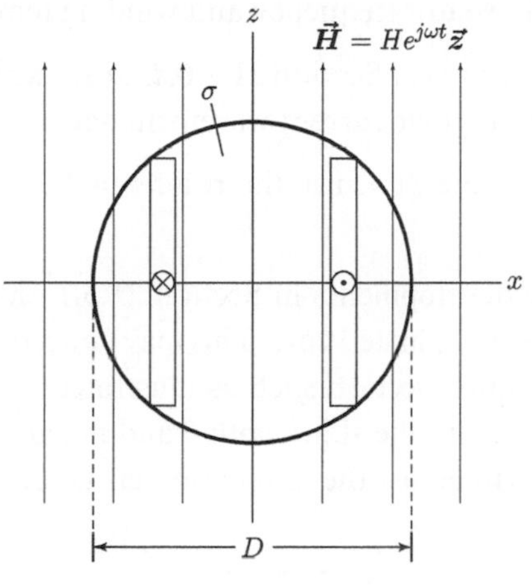

Figure 21.28 Cross section of a wire having diameter D and conductivity σ immersed in a sinusoidal magnetic field of angular frequency ω and amplitude H. A representative loop comprising a pair of incremental slices of the conductor is also shown. Proximity effect current flows in this loop owing to the time-varying flux linking it.

21.3 A pot-core transformer for a high-frequency converter has a primary winding comprising five turns of copper foil of thickness d_1 (arranged as five layers of one turn each). Over this is wound a secondary winding comprising one turn of copper foil of thickness d_2.

(a) If the transformer is designed for sinusoidal currents at 400 kHz, find the optimum thicknesses of the windings (d_1 and d_2) based on skin and proximity effects. You may assume that the inner radius of the winding window is more than a factor of 10 larger than the foil thicknesses.

(b) Assuming the transformer carries a primary current of magnitude I, roughly sketch how the magnitude of the H field varies across the winding layers.

(c) How might you wind the transformer differently to reduce proximity effect?

21.4 This problem regards estimating the winding losses of an automotive alternator that is being reverse engineered. *Pay attention to the conditions for which various parameters are given.* You may assume that the alternator currents are sinusoidal, and that the geometry of the alternator is such that the one-dimensional Dowell model adequately describes the system.

The alternator is wound with wire having a (measured) diameter that is 0.2 times the skin depth of copper at the maximum alternator frequency and 25 °C. A preliminary calculation of winding loss at this frequency and 175 °C (max winding temperature) was based on the (known) rms winding current, the dc resistance of the winding measured at 25 °C, and a scaling factor for the resistance increase over temperature. However, subsequent measurements have revealed that due to proximity effect, the ac resistance of the winding at maximum frequency and 25 °C is actually four times the dc resistance at 25 °C. The question is: by what factor should we increase our estimate of power dissipation for maximum alternator frequency and winding temperature?

21.5 Considering the results of Section 21.2.6, find an expression for the effective frequency ω_{eff} for a balanced triangular current waveform having no dc value and a period $T = 2\pi/\omega_1$.

21.6 Considering Example 21.4, use the results of Section 21.2.2 to derive the value of R_{11} given in (21.88).

21.7 Considering the developments in Section 21.3.1, derive (21.96) for the eddy current loss density $P_{v,\text{eddy}}$ in a laminated core. This may be accomplished by treating the eddy currents in a single lamination section such as illustrated in Fig. 21.29. Assume that the width w is less than or equal to the skin depth δ and much less than the lamination height h, and that $B(t)$ is uniform across the lamination cross section.

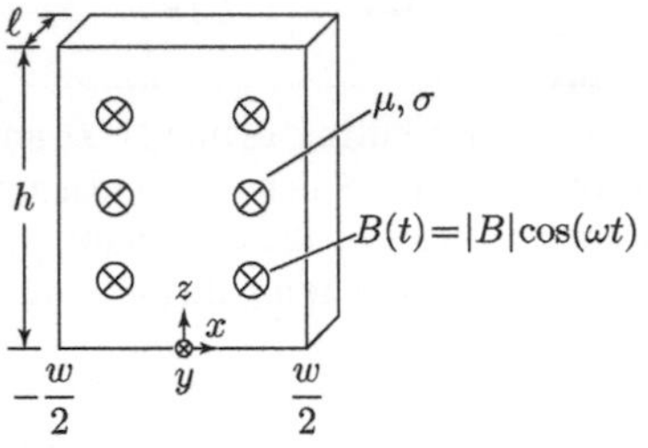

Figure 21.29 Cross section of a lamination having width w and conductivity σ carrying a sinusoidal flux density $B(t) = |B|\cos(\omega t)$.

21.8 Consider the double-ended bridge converter of Fig. 21.30(a). The details of this converter are described in Chapter 7. The transformer primary waveform v_P is shown in Fig. 21.30(b). (Note that we define the duty ratio D in terms of the period T of the full-wave rectified voltage v_d instead of the switching period of the primary-side switches.) Assume that the transformer is designed with N primary turns wound on a core with cross-sectional area A_c and core volume V_c, and is constructed of a material having Steinmetz parameters k, α, and β.

Apply the iGSE to write an expression for the core loss in the transformer as a function of k, α, β, D, T, N, A_c, V_c, and V_1. (Note, however, that the iGSE can sometimes underestimate core loss for this type of waveform [19].)

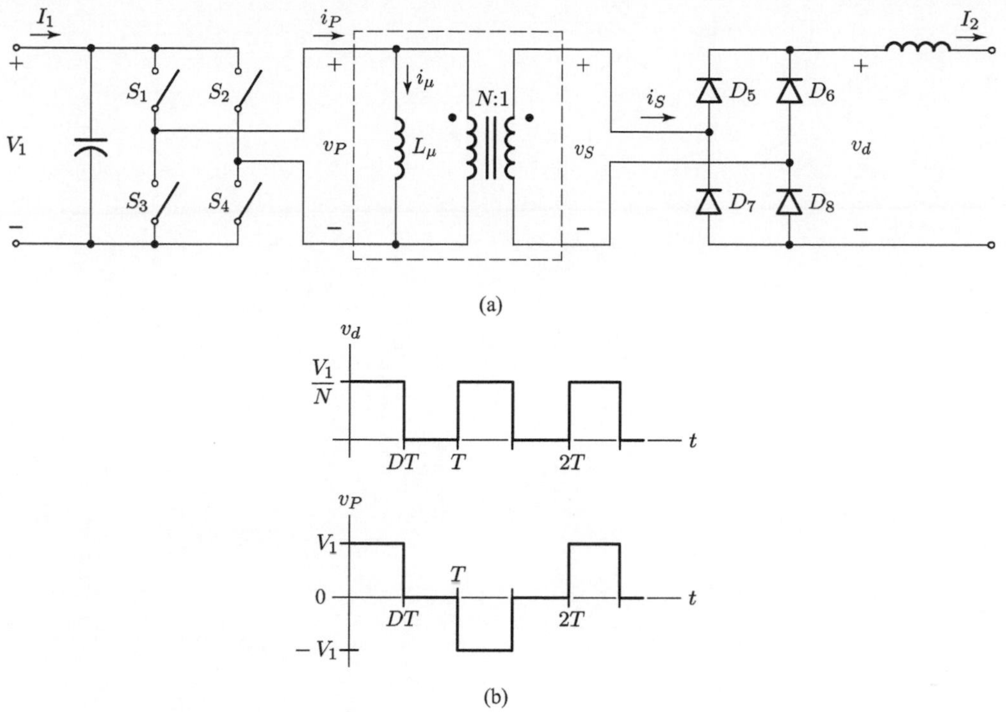

Figure 21.30 (a) Double-ended bridge converter. (b) Output voltage v_d and transformer primary voltage v_P.

Part IV

Practical Considerations

22 Practical Considerations: An Overview

The topics we have addressed so far – power circuits, control, and components – do not cover all the issues encountered in designing a power electronic system. Among those we have deferred are: (i) providing gate and base drives to the power semiconductor switches; (ii) using forced commutation to turn off SCRs; (iii) controlling the transient voltages and currents that accompany switching in practical circuits; (iv) contending with EMI created by fast switching waveforms; and (v) providing a thermal environment that allows system components to operate within their temperature ratings. We address these five topics in Part IV.

22.1 Gate and Base Drives

The proper design and operation of the circuits that drive the gates of thyristors and the bases or gates of transistors are crucial to the integrity of a power electronic system. These circuits function not only to turn the switches on and off in a prescribed manner, but often also to control the switches in response to sensed signals that predict impending failure – an excessive voltage or current, for instance.

The gate or base drive requirements depend upon the device being controlled. The BJT requires a continuous base current. For an npn BJT, a positive base current is required to turn the device on, and a negative current to turn it off quickly. The MOSFET and IGBT require a positive current pulse to charge the gate capacitance above V_T, and a negative pulse to discharge C_{GS} and turn the device off. Since the use of MOSFETs dominates that of BJTs, and the drive issues are not significantly different (the most significant being that the BJT requires continuous base current, whereas the MOSFET or IGBT requires only an impulse of gate-charging current), we concern ourselves here with MOSFET drive issues.

The thyristor is triggered by a positive gate current pulse, but cannot be turned off via the gate unless it is specifically designed for such control (e.g., the GTO). A conventional thyristor is turned off by load circuit behavior causing the anode current to reverse (line commutation), or by using an auxiliary circuit to force the current to reverse (forced commutation). Because the IGBT has replaced the thyristor in many applications, forced commutation is now seldom employed. We provide an example of a forced commutation circuit in this overview but do not explore the technique further.

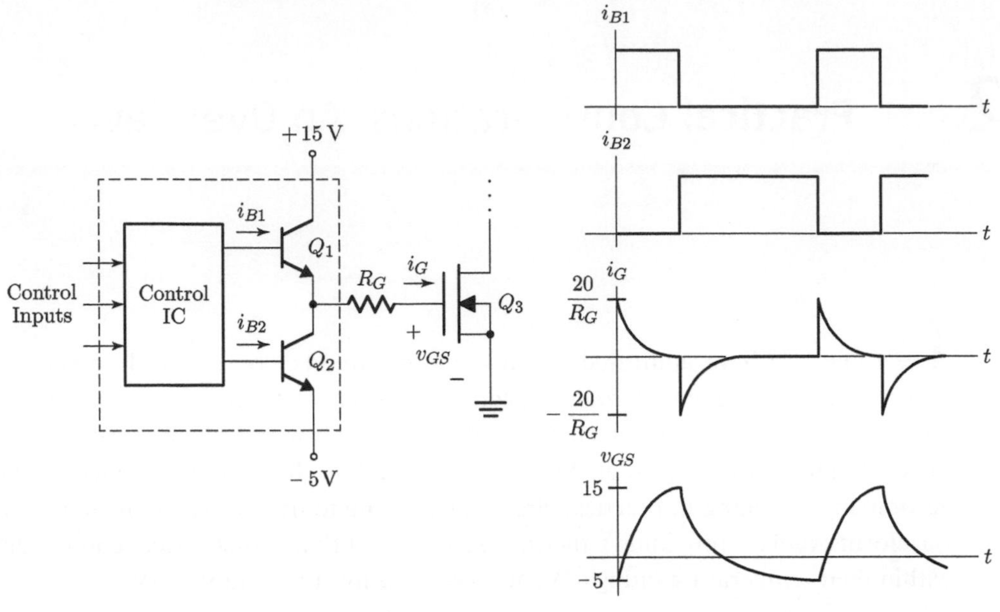

Figure 22.1 A commonly employed circuit to drive the gate of a MOSFET. Except for R_G, this drive circuit is available as an IC.

22.1.1 A Basic Gate or Base Drive Circuit

Figure 22.1 illustrates the essential elements of a gate drive circuit. The control IC takes inputs from a feedback loop or an external source to generate the timing and duration of the current pulses applied to the gates of Q_1 and Q_2. Turning Q_1 on provides a positive current pulse to Q_3's gate, turning it on, after which v_{GS} is maintained at 15 V. To turn Q_3 off, Q_1 is turned off while Q_2 is turned on, providing a pulse of reverse current to discharge C_{GS}, and bringing v_{GS} down to -5 V. The step in i_G may excite ringing of C_{GS} with wiring and package inductance, and this is damped by R_G. The circuit within the dashed box is available from a number of manufacturers as an IC. When the MOSFET is on, $v_{GS} = 15$ V, which is well above V_T, assuring Q_3 is driven deep into its triode region. The negative gate voltage applied at turn-off provides faster removal of charge from C_{GS} than if the emitter of Q_2 were simply grounded.

Driving the transistors in a bridge leg provides a challenge, because the upper device's source terminal is at the bus voltage when the transistor is on. The drive circuit for this device must therefore be at a voltage that is *level-shifted* from ground. The solution to this problem is discussed later.

22.1.2 Thyristor Gate Drives

Thyristor and MOSFET gate drives must provide a short pulse of current to the device. However, whereas a negative current is required to turn off the MOSFET, a negative gate current does little to enhance the turn-off process of most thyristors. The notable exception is the gate-turn-off

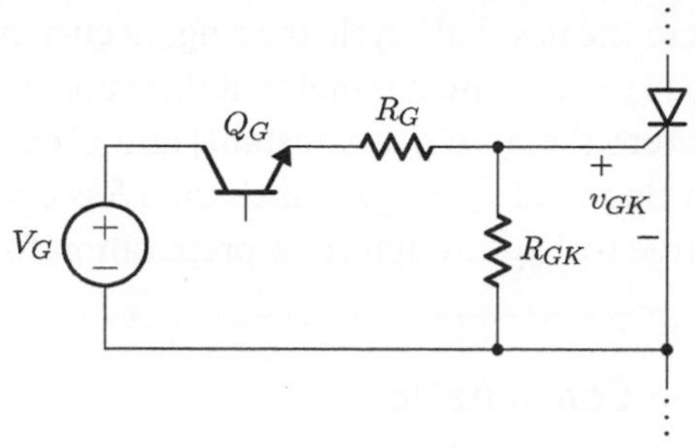

Figure 22.2 A simple thyristor gate-drive circuit. The resistor R_{GK} desensitizes the device to noise and rate-induced triggering.

thyristor (GTO). Thus, a circuit as simple as that of Fig. 22.2 is often satisfactory for triggering a thyristor. The resistor R_{GK} desensitizes the turn-on process to noise and dv/dt triggering.

The values of V_G and R_G in Fig. 22.2 are typically 20 V and 20 Ω for devices rated for currents below approximately 100 A. Because the thyristor can exhibit a peak value of v_{GK} during turn-on that can exceed 10 V, V_G must be large enough to ensure that the gate current does not reverse during this interval, as discussed in Section 23.3.1.

22.2 Thyristor Commutation Circuits

Except for the specially constructed GTO, all members of the thyristor family of switching devices are turned off by the behavior of the power circuit in which they are connected. In phase-controlled applications, such as those discussed in Chapter 4, and the cycloconverter applications in Chapter 11, the SCR is turned off by the reversal of the ac line voltage. Before the advent of high-current and -voltage MOSFETs and the IGBT, thyristors were used as switches in dc and ac motor drives. In a dc drive, the dc voltage was controlled by a dc/dc converter (known at the time as a *chopper*) using an SCR as the switch. In ac drives, six SCRs in a three-phase bridge produced a variable frequency and voltage waveform. In both of these applications, the thyristor needs to be turned off by the action of an auxiliary *commutation circuit* designed to force the switch current to zero. The process of turning off the thyristor with a commutation circuit is called *forced commutation*.

22.2.1 A Simple Commutation Circuit

For historical and cultural purposes, a simple commutation circuit is presented here. The basic issues of forced commutation can be illustrated by examining the behavior of the circuit of Fig. 22.3. Assume that Q is on and C is charged to $v_C = V_{dc}$ through R. The switch S is closed at $t = 0$ to initiate the turn-off of Q.

When we ignore the effect of R, the only purpose of which is to recharge C, the resulting waveforms of the circuit variables are shown in the figure. During the first half-cycle of the L–C

ring, i_L adds to i_Q, but in the next half-cycle the ringing current reduces i_Q until $i_Q = 0$ at t_1 and Q turns off. At this time $i_d = I_{dc}$, and it remains at this value until $v_d = 0$ at t_3. Between t_1 and t_3 the voltage across L is zero (because i_L is constant) and C charges linearly. When D turns on at t_3, L and C once again ring until $i_L = 0$, at which time S is opened. The commutation capacitor C now begins to recharge to V_{dc} through R, in preparation for the next commutation.

22.2.2 Limitations of Forced Commutation

The waveforms of Fig. 22.3 illustrate several limiting features of forced commutation. First, Q does not turn off at $t = 0$, when the commutation process is started. A delay dependent on both the dc current I_{dc} and the resonant frequency of the commutation circuit must be considered, if switch timing is critical. Some commutation circuits do not exhibit this delay, but it is present in many practical circuits. Second, the SCR current i_Q is increased from its load value to I_{Qp} during commutation. This increase affects the required rating of Q. Third, and most important, is the time t_{off} during which the SCR must recover its forward blocking capability. In other words, the commutation circuit must ensure that $t_{off} > t_q$, where t_q is the manufacturer's specified circuit commutated turn-off time for Q. The time t_{off} is usually a function of L, C, and I_{dc}.

In addition to these three limitations, all of which can be inferred from the waveforms of Fig. 22.3, the commutation circuit dissipates energy, and thus represents both a cause of reduced efficiency and a source of heat. The principal cause of this energy loss in the circuit of Fig. 22.3 is the resistor R. Although we have ignored it in our description of circuit operation, this resistor

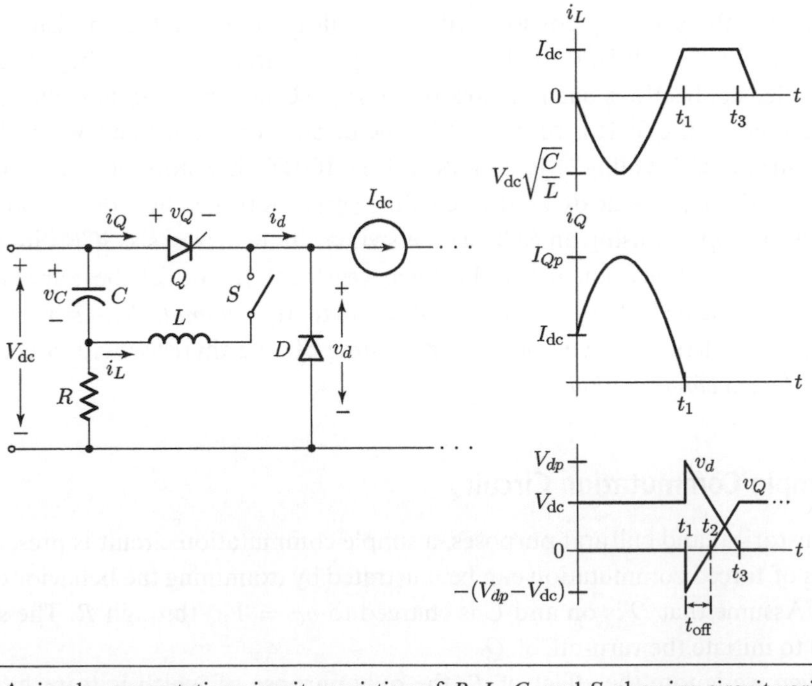

Figure 22.3 A simple commutation circuit consisting of R, L, C, and S, showing circuit variable behavior during commutation.

has a voltage across it during the entire commutation period. It therefore dissipates energy during this entire time, not just when C is charging (or discharging) to V_{dc}, to establish the proper circuit conditions for the next commutation event.

22.3 Snubbers, Clamps, and Soft Switching

The term *snubbing* generally refers to control of any unwanted overcurrent or overvoltage transients, or rates-of-rise of current or voltage, that occur during switching. It also refers to the control of the switching trajectory locus in the switch's i–v plane. Such *switching locus control* is sometimes necessary to prevent the simultaneous presence of a high switch voltage and a high switch current.

When circuit layouts create inductive loops around the switching device, producing an additive $L\,di(t)/dt$ voltage at turn-off, a clamp can be employed to limit the peak transient voltage. Unlike a snubber, the clamp does not control the rate-of-rise of current or voltage, nor switching trajectories.

Soft-switched power converters achieve both gentle switching transitions and low switching loss. There are various techniques by which soft switching can be achieved, but they all operate to reduce the overlap of switch voltage and current during transitions. Constraining the voltage to a low value while the current rises or falls is known as *zero-voltage switching* (ZVS). Constraining the current in similar fashion is *zero-current switching* (ZCS).

22.3.1 Snubbers

Figure 22.4 illustrates both a condition requiring a snubber and the basic operation of a simple snubber circuit. The inductance L_p represents the parasitic inductance in series with the transistor. When the transistor is turned off, we assume that its current falls linearly to zero in time t_f, creating a voltage $L_p I_{\text{dc}}/t_f$ across L_p. The resulting transistor voltage and current waveforms are shown in Fig. 22.4(a), along with the transistor's switching locus in the i_Q–v_Q plane. When Q begins to turn off, its voltage jumps immediately to

$$V_{p1} = V_{\text{dc}} + L_p \frac{I_{\text{dc}}}{t_f}. \tag{22.1}$$

This transition is represented by the line from point a to point b in the i_Q–v_Q plane. The transistor current then falls from I_{dc} at point b to 0 at point c in time t_f. During this time of high dissipation, both i_Q and v_Q have large values. When $i_Q = 0$, v_Q drops abruptly to V_{dc} at point d.

A snubber circuit designed to reduce both the large voltage transient in v_Q and the dissipation in the transistor during turn-off is shown in Fig. 22.4(b). The shunt circuit consisting of R_s and C_s provides an alternative path for the inductor current when Q turns off. The resulting waveforms of v_Q and i_Q and the switching locus are also shown. The resistor R_s is necessary not only to damp the oscillation between L_p and C_s but also to reduce the peak value of i_Q when Q turns on again and C_s discharges. Although this snubber has reduced the energy dissipated in the transistor, the net energy dissipated in the circuit has been increased from that in the snubberless

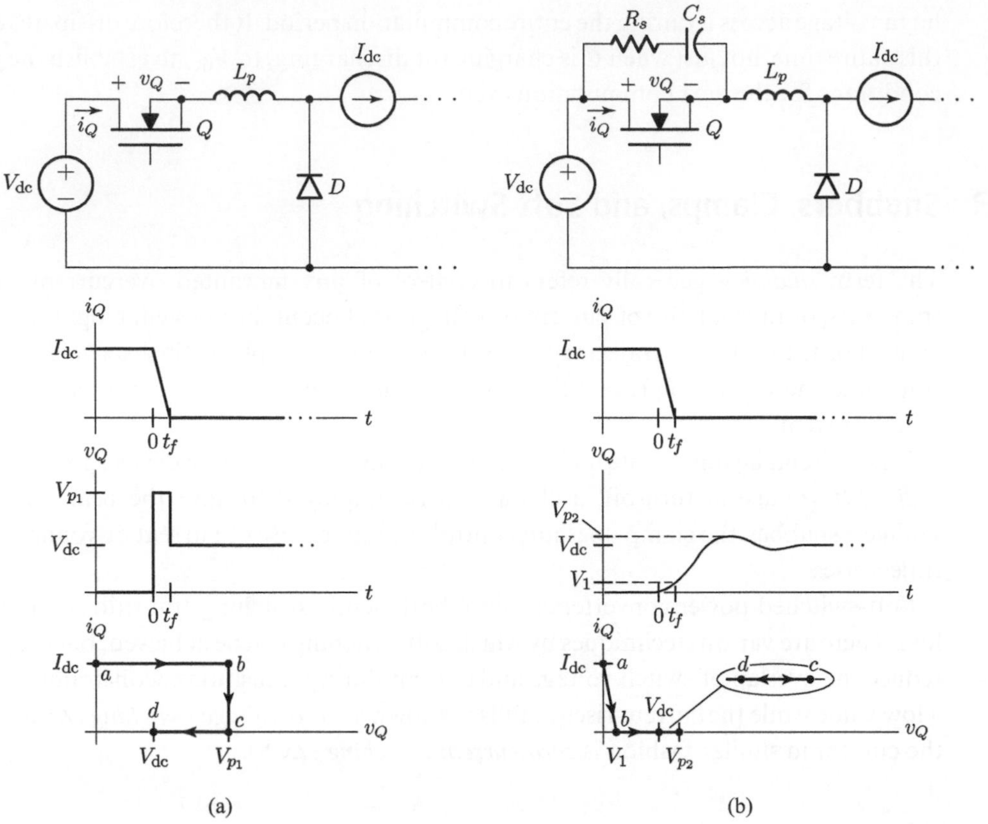

Figure 22.4 (a) A circuit containing parasitic inductance L_p that causes a transient overvoltage when Q is turned off. (b) A simple snubber circuit placed across Q to reduce the switching overvoltage and improve the switching locus.

circuit of Fig. 22.4(a). Snubbers are designed to reduce the voltage, current, and thermal stresses on switches, but they do not increase circuit efficiency unless a means is provided for recovering the energy trapped in the snubber inductors and capacitors.

An extension of snubbing called *soft switching*, discussed in Chapter 25, uses a resonant circuit during switching to create conditions of zero voltage and/or zero current to reduce switching loss.

22.3.2 Voltage Clamps

An easily visualized clamp circuit is shown in Fig. 22.5, in which the transistor's transient voltage during switching is limited to V_C, which is regulated by a separate circuit. Clamps can also consist of Zener diodes or metal oxide varistors placed directly across the clamped element, but the range of clamped voltages is limited to the available devices break-over voltages.

22.3.3 Soft Switching

Figure 22.6 illustrates ZVS and ZCS transitions for each of switch turn-off and turn-on. In a ZVS turn-off transition (top left of Fig. 22.6), the voltage across the switch is kept low as

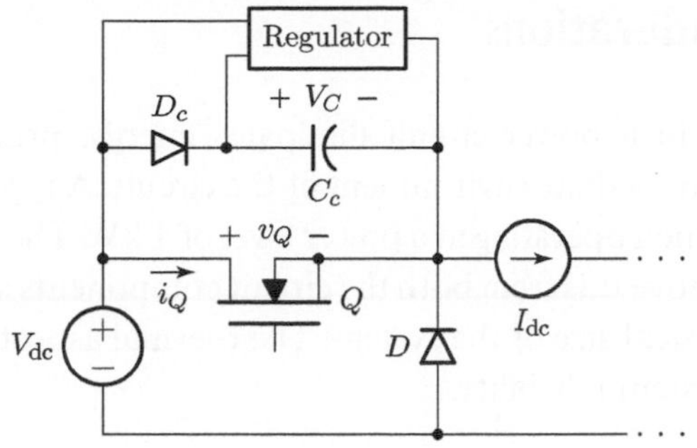

Figure 22.5 A clamp circuit designed to limit the peak transistor voltage to V_C.

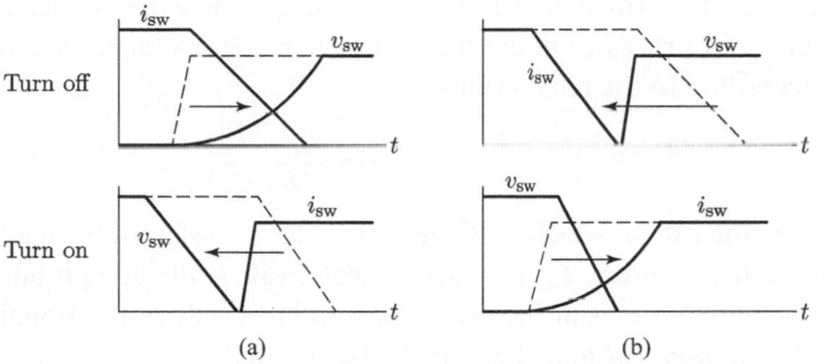

Figure 22.6 Switching waveforms for turn-off and turn-on transitions with zero-voltage and zero-current switching. Their hard-switched counterparts are shown in dashed lines.

the switch current falls. This is often accomplished by placing a capacitor across the switch during turn-off (e.g., as in the simple snubber of Fig. 22.4(b)). For soft switching, however, some means must be provided to losslessly recover the energy stored on the capacitor. This is often accomplished by returning the switch voltage to zero prior to turn-on (through some means not shown in Fig. 22.4(b)), enabling ZVS turn-on of the switch as illustrated in the bottom left of Fig. 22.6.

Zero-current switching is the dual of zero-voltage switching. In a ZCS turn-on transition (bottom right of Fig. 22.6), some mechanism is used to keep the switch current low as the switch voltage falls. One means of doing this is to place an inductor in series with the switch during turn-on. For soft switching, the energy stored in the inductor at turn-on should be losslessly recovered. This is often accomplished through some means of returning the switch current to zero prior to switch turn-off, enabling ZCS turn-off of the switch, as illustrated in the top right of Fig. 22.6).

Soft switching can be used to achieve higher circuit efficiency by reducing switching loss, and/or to enable the use of higher switching frequencies than would otherwise be possible. It can also help to reduce the electromagnetic interference associated with high-frequency switching, without incurring high snubber losses. It is for these reasons that soft switching is often employed in high-performance designs.

22.4 Thermal Considerations

Every component in a power circuit dissipates energy, primarily as heat, and this must be removed from the immediate environment of the circuit. A typical power circuit might exhibit an efficiency of 95% when operating at a power level of 1 kW. The 50 W of loss is in the form of heat, and the need to remove this from both the circuit components and the equipment enclosure often determines the physical size of the system. The thermal aspects of a power electronic system are major factors in system reliability.

22.4.1 Thermal Models

Thermal modeling is critical to the design of satisfactory heat transfer systems. These models most often take the form of electric network analogs, where the circuit elements represent various components and processes in the thermal system. For example, steady-state heat conduction behaves according to the relationship

$$Q_{12} = \frac{T_1 - T_2}{R_{\theta 12}}, \tag{22.2}$$

where Q_{12} is the rate at which heat energy is being transferred from a body at temperature T_1 to a body at temperature T_2. This heat transfer rate is directly proportional to the difference between the temperatures of the two bodies, and inversely proportional to the parameter $R_{\theta 12}$, which is the *thermal resistance* between bodies 1 and 2.

The relationship in (22.2) is of the same form as Ohm's law. Therefore we can represent it as an electric circuit, as shown in Fig. 22.7. In this analog, T_1 and T_2 are the voltages of terminals 1 and 2, respectively, measured with respect to a common reference. The current through the resistor of value $R_{\theta 12}$ is the electric analog of the rate Q_{12} at which heat is being transferred from body 1 to body 2.

Figure 22.7 and the relationship in (22.2) model a steady-state heat conduction process. In many situations the thermal behavior of a power electronic system is dynamic. For instance, in a thyristor used in a pulsed application, where the anode current is high but brief and infrequent, the generation of the thermal energy is equally brief. This energy is initially stored in the thermal mass of the device, and the internal device temperature rises rapidly. The stored heat energy

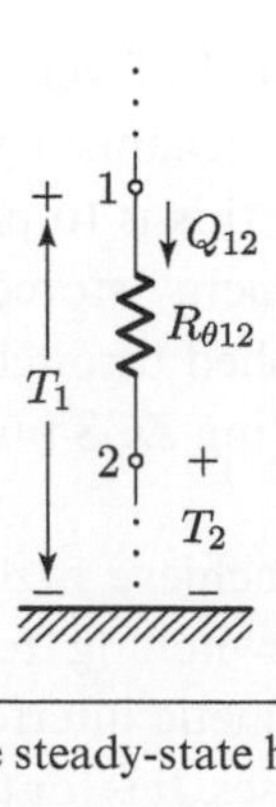

Figure 22.7 The electric circuit analog for the steady-state heat transfer relationship, (22.2).

is then conducted through the case and its interfaces to the external environment, where it is ultimately dissipated. In this situation, the thermal mass of the device plays an important role: it stores energy. The appropriate model is one that includes energy storage, such as a capacitor, and therefore exhibits dynamic behavior. This type of model is known as a *transient thermal model.*

22.4.2 Heat Sinking

A ubiquitous feature of power electronic equipment is the *heatsink*, often an extruded aluminum structure with fins. Heat from components mounted on the heatsink is transferred to the sink, from which it is then transferred to the environment. Although heatsinks are most frequently applied to cooling semiconductor devices, mounting other components – such as resistors, capacitors, and even inductors – on a heatsink to reduce their temperatures is not unusual.

Because of their complex shapes, modeling heatsinks in detail is difficult. However, for design purposes, characterizing the performance of a specific heatsink geometry by a single thermal resistance between the sink and the ambient environment, $R_{\theta SA}$, is generally sufficient. This number, which is provided by the manufacturer, is an approximation, in that the sink is assumed to be an isotherm, and the environmental conditions in which it will be used are unknown. But if the heatsink is not too large relative to the size of the heat-generating components mounted on it, the isothermal approximation is good. For example, a heatsink with a mounting surface area of 100 cm^2 would not be too far from isothermal if a device in a TO-247 package (approximate surface area of 2 cm^2) were mounted on it. Manufacturers assume that environmental conditions permit air to flow unobstructed through the fins. This flow results from the buoyancy created by the thermal gradient of the air between the (vertical) fins, or is forced by a fan. The former is known as *natural convection* and the latter as *forced convection*. The specified thermal resistance of extruded finned heatsinks is conditional on the sink being mounted vertically, which is seldom the case in practice, so additional care must be taken in the choice of sink.

PROBLEMS

22.1 Figure 22.8 presents a MOSFET drive circuit as an alternative to that of Fig. 22.1. The base current I_B is large enough that $\beta_F I_B R_C > 15$ V.

(a) Sketch the MOSFET gate current i_G and v_{CE}, the collector–emitter voltage of the drive transistor.

(b) What are the disadvantages of using this simpler drive circuit?

22.2 The circuit of Fig. 22.9 shows a *totem-pole* gate drive circuit, consisting of Q_1 and Q_2, driving a MOSFET, Q_3, whose gate is modeled as a capacitor of constant value equal to 100 nF. The drive transistors Q_1 and Q_2 are assumed to switch in zero time. The inverter in the gate circuit of Q_2 creates a complementary drive for the totem pole: when Q_1 is on,

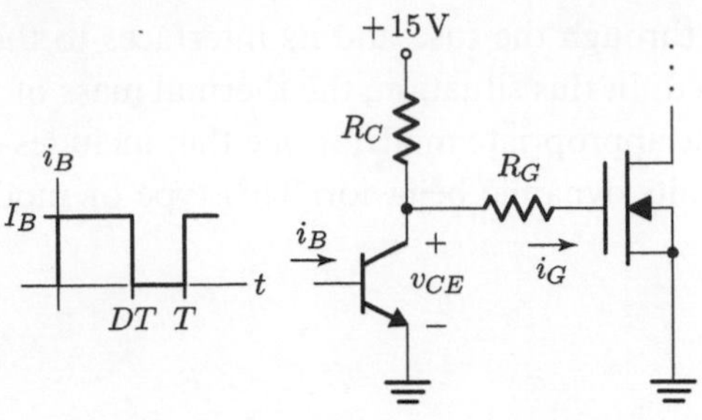

Figure 22.8 The simplified MOSFET gate drive circuit considered in Problem 22.1.

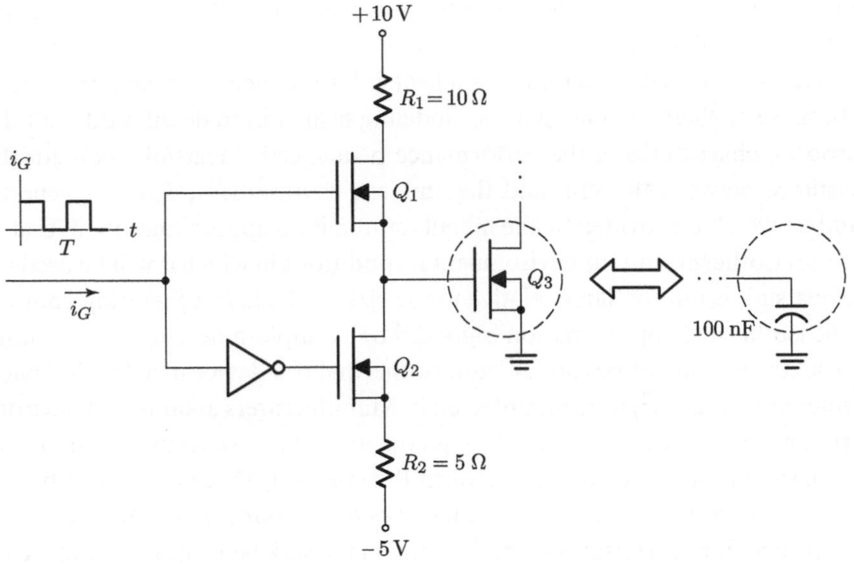

Figure 22.9 The totem-pole gate drive circuit analyzed in Problem 22.2.

Q_2 is off, and vice versa. If the switching frequency of Q_3 is 100 kHz, what are the powers dissipated in R_1 and R_2?

22.3 Consider the waveforms in the simple forced commutation circuit of Fig. 22.3.

(a) Determine I_{Qp}, V_{dp}, and t_{off} in terms of the circuit parameters.

(b) Determine and sketch v_C.

(c) If the switching frequency of Q were f, determine an expression for the average power dissipated in R.

22.4 If $R_{\theta 12}$ in Fig. 22.7 represents the thermal resistance between the case of a transistor and the ambient environment, what does Q_{12} correspond to? What are the units of $R_{\theta 12}$?

22.5 Figure 22.10 illustrates the application of a turn-off snubber. The elements R_s and C_s act to prevent v_Q from becoming excessively high when Q turns off. If the snubber were not

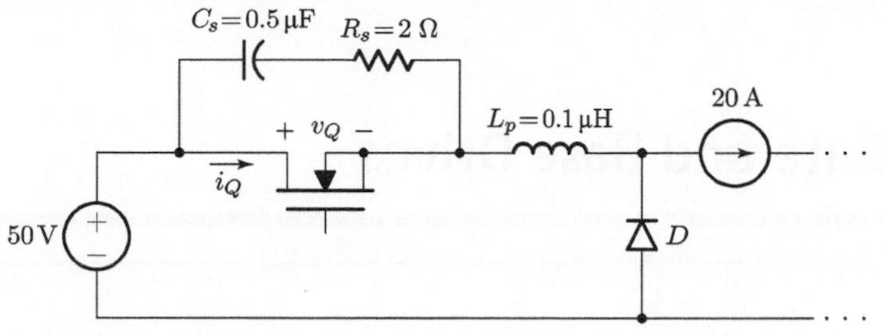

Figure 22.10 A circuit illustrating the application and performance of a turn-off snubber.

present, the abrupt interruption of current in the parasitic inductance L_p could cause v_Q to exceed the rating of Q.

(a) Sketch $v_Q(t)$ if the snubber, consisting of R_s and C_s, were not present and i_Q fell linearly with a fall time of 50 ns.

(b) Now assume that Q turns off instantaneously and that the snubber is in place. Draw the equivalent circuit you would use to determine $v_Q(t)$ and specify initial conditions.

(c) Determine and sketch $v_Q(t)$ for the conditions of (b).

23 Gate and Base Drives

The three-terminal power devices introduced and discussed in Part III – the bipolar junction transistor, the power MOSFET, the insulated gate bipolar transistor, and the thyristor – each requires the application of a control terminal current or voltage to cause the device to switch. So far we have assumed the necessary gate- or base-drive waveforms are present. In this chapter we address the detailed drive requirements of these devices and how the necessary drive waveforms are created.

Each family of controlled power devices has unique drive requirements. The bipolar transistor needs a continuous drive current to supply the base defect. The MOSFET or IGBT gate looks like a capacitor and therefore needs a gate circuit capable of driving a capacitive load. That is, it must be capable of delivering an initial pulse of current and then acting like a voltage source. The thyristor gate must be provided with enough current to initiate the regenerative process that turns the device on. A lower value of continuous gate current may be required, depending on the anode circuit. The thyristor gate current must be synchronized with the correct anode–cathode voltage condition ($v_{AK} > 0$). In the case of the gate turn-off thyristor, the turn-off drive generally requires a separate, more powerful circuit than the turn-on drive. Thyristor gate circuits also influence the susceptibility of the device to rate effects.

A gate or base drive actually consists of two parts. The first is a signal processing circuit that senses circuit or system variables, some of which may come from feedback loops, and creates the desired timing and wave shapes. It usually consists of a microprocessor. The second part of the drive, the output stage, takes the low-level signals from the first part and transforms them into the levels of voltage and current required by the main power device. These gate or base variables can be tens of volts and tens of amperes, so this output stage is really a power circuit itself. It sometimes contains circuitry to give the gate or base waveforms a special shape. We do not address the design of these circuits because their functions are available in a variety of commercial integrated or packaged multi-chip circuits. These components often have provisions to sense switch voltage and current to detect and respond to overload conditions.

A frequently encountered complication in the design of gate or base drives is that power devices within a circuit do not share a common emitter, cathode, or source connection. In this case, the drives must be isolated from each other, which often requires either a separate power supply for each drive circuit, or a bootstrap arrangement to shift the reference level of the higher-voltage switch. Examples of this problem are the drive to the upper switches in a buck converter or full-bridge circuit. Electrically isolating the control components from the power device they are driving may also be necessary.

In this chapter we discuss the requirements for driving the four device types, and their differentiating characteristics. Although, as noted earlier, MOSFETs and IGBTs have largely supplanted the application of BJTs, we include a brief discussion of BJT drive requirements as they are still used in some applications.

23.1 MOSFET and IGBT Gate Drives

One of the principal advantages of the MOSFET is that it requires no dc gate-drive current. It does, however, require the transfer of charge to and from the gate electrode in order to turn the channel on and off. This transfer must be rapid to obtain fast switching speeds, and the result is that peak gate currents can be very high.

A concern for the effects of parasitic impedances distinguishes MOSFET gate drive circuit design from BJT base drive circuit design. MOSFETs are capable of switching with rise and fall times of less than 10 ns, but practically achievable times are often limited to more than this by the presence of resistance and inductance in the drive circuit. Because of the high switching speeds of MOSFETs, parasitic impedances can also aggravate the problem of EMI.

23.1.1 The Miller Effect

We can now explain the plateau, a consequence of the Miller effect,† in the gate voltage of the MOSFET switching waveform shown in Fig. 15.7, reproduced here in Fig. 23.1. Figure 23.2 shows the saturation-region MOSFET model based on (16.64) embedded in a simple circuit that is in the process of switching between its off and on states. The device capacitances C_{DG} and C_{GS} are included in the model (C_{DS} is not included as it is irrelevant to the discussion).

The conventional explanation of the Miller effect identifies the cause to be the magnification of the input impedance through feedback in a linear amplifier. A similar argument can be made for the effect in a MOSFET, where the additional gate current required to charge or discharge C_{DG}, which is a function of dv_{DS}/dt, results in an apparent increase in C_{GS}. But here we present an alternate explanation by employing the relationship between i_D and v_{GS}.

At t_o, v_G steps to V_G and the gate capacitance begins to charge. While there is a slight increase of i_D before t_1, i_D does not change appreciably until t_1, when $v_{GS} = V_T$. When $i_D = I_d$ at t_2, the diode current has completely commutated to the transistor, and v_{DS} rapidly decreases while the transistor traverses the saturation region. During this traversal between t_2 and t_3, $i_D = i_{D(\text{sat})} = I_d$. Since $i_{D(\text{sat})} \propto (v_{GS} - V_T)^2$, and V_T is constant, for $i_{D(\text{sat})}$ to be constant at I_d, v_{GS} must also be constant. Hence the plateau in v_{DS} until t_3, when the transistor leaves saturation and enters its linear region as v_{GS} increases to v_G. During turn-off the process simply reverses. During turn-on, the gate drive must source the current required to discharge C_{DG} from V_{DD} to $-V_G$ (resulting in $V_{DS} = 0$) and charge C_{GS}, while at turn-off the gate drive must sink the required charging currents.

† Named after John Milton Miller, who published an analysis of the effect in 1920 in the context of vacuum tube amplifiers.

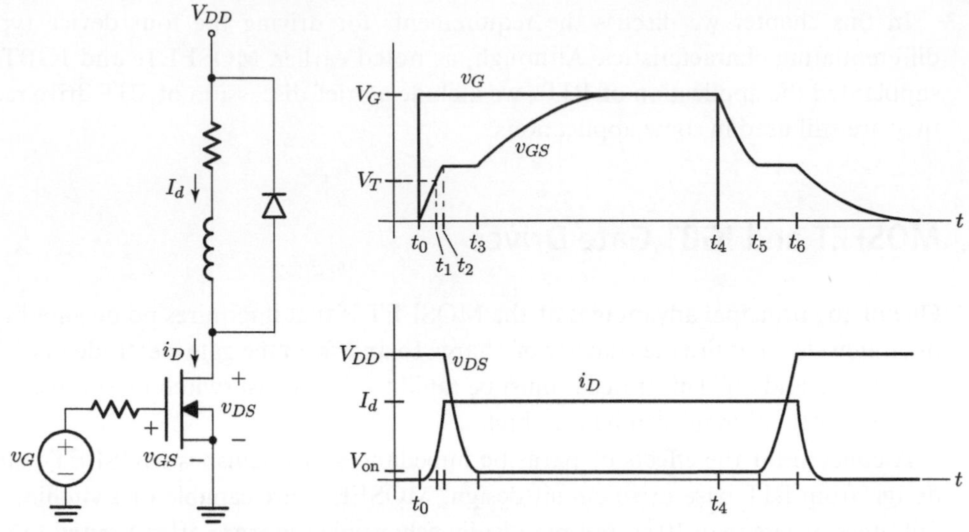

Figure 23.1 MOSFET switching transients, showing the Miller effect.

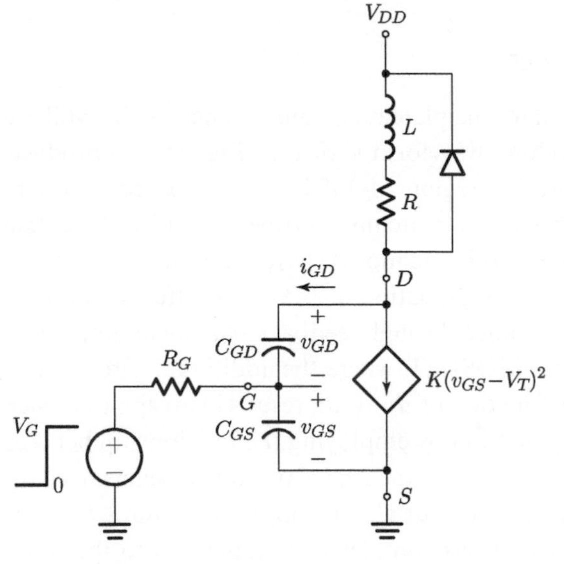

Figure 23.2 The MOSFET saturation region model including device capacitances used to illustrate the Miller effect.

23.1.2 Gate Drive Requirements

A MOSFET or IGBT gate is essentially a nonlinear capacitor which is charged and discharged by the gate drive circuit to turn the device on and off, respectively. The drive must switch the gate voltage between some value below the gate threshold V_T, where the device is off, and some value above V_T, where the device is on. Power MOSFETs have a V_T that is typically between 3 V and 6 V. The available gate current is also a critical gate drive parameter, because it and resistance

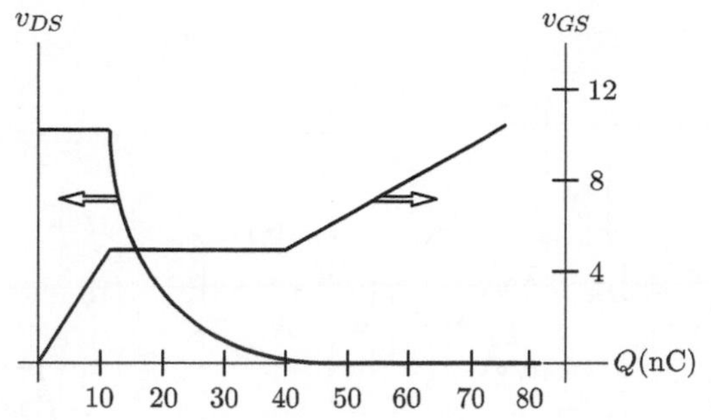

Figure 23.3 Typical gate charge versus gate source voltage characteristic for a power MOSFET driving an inductive load.

in the gate network determine how quickly the gate capacitance can be charged or discharged. Since the requirements for driving MOSFETs and IGBTs are similar, we discuss gate drives in the context of the MOSFET.

The MOSFET gate characteristics are most conveniently represented in the form of a gate charge versus gate source voltage graph, such as the one in Fig. 23.3. From this graph we can determine the influence of the gate drive circuit on the rise and fall times of the drain current or voltage, as well as the gate circuit power requirements. For instance, if the gate of the device represented by Fig. 23.3 were driven by a circuit capable of sourcing 1 A, the fall time of v_{DS} is approximately

$$t_f = \frac{30\,\text{nC}}{1\,\text{A}} = 30\,\text{ns},$$

where 30 nC is approximately the change in gate charge during the plateau in v_{GS} as v_{DS} falls. This calculation assumes that the driver current rise time is negligible, which is often not true.

Example 23.1 Use of the MOSFET Gate Charge Characteristic

Figure 23.4 shows a simplified schematic of the output stage of the UCC27511 single-channel low-side gate drive IC. It is capable of sourcing 4 A through Q_1 and sinking 8 A through Q_2. The resistors R_1 and R_2 provide a means of controlling the switching slew rates. For this example we will assume $R_1 = 0$ and the drive supplies a turn-on current step of 4 A. Also shown is the gate charge characteristic of the MOSFET in this application. We want to determine the drain voltage fall time t_{vf} and the power dissipated in the drive circuit if the switching frequency is 500 kHz.

The fall time is approximately equal to the length of the plateau region in the gate charge characteristic, because during this time the falling drain voltage causes the gate voltage to be constant due to the Miller effect. Thus,

$$t_{vf} = \frac{50\,\text{nC}}{4.0\,\text{A}} = 12.5\,\text{ns}. \tag{23.1}$$

The required drive power is the switching frequency times the energy required to deliver the charge from the 15 V driver supply to bring the gate voltage to 15 V. Assuming a linear Q versus V relationship shown by the dotted line in Fig. 23.4(b), the required gate power is

$$P = \int_0^Q v\,dq = \frac{1}{2}QVf = \frac{1}{2}(140 \times 10^{-9})(15)(5 \times 10^5) \approx 0.5\,\text{W}. \tag{23.2}$$

This power does not include all the losses in the UCC27511.

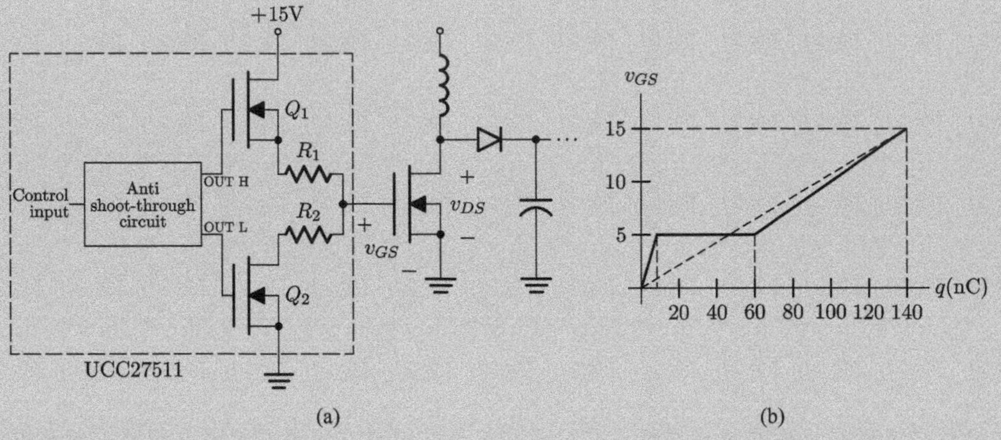

Figure 23.4 (a) A MOSFET in a direct converter being driven by a UCC27511 gate drive IC, whose simplified output stage is shown in the dashed-line box. (b) The gate charge characteristic for the MOSFET with its linear approximation.

23.1.3 The Kelvin Connection for SiC MOSFETs

Since SiC devices are often employed at high switching speeds, some are mounted in a package that provides direct access to the source pad on the die through what is known as a *Kelvin* connection. The gate drive is applied between the gate terminal and the Kelvin connection. This eliminates the resistance, inductance, and voltage drop associated with the source-to-package bonding wire. The result is a considerable improvement in rise and fall times. For example, the IMW65R027M1H SiC device whose $R_{DS(\mathrm{on})}$ is shown in Fig. 17.41(b) has datasheet values of 13.6 ns and 14.2 ns, for t_r and t_f, respectively. This device also comes in a package with a Kelvin connection (IMZA65027M1H). Its corresponding rise and fall times are 4.2 ns and 8.4 ns, representing significant reductions from the device without the Kelvin connection.

23.1.4 Gate Drive Loss and Shoot-Through

The mechanism giving rise to loss associated with driving the MOSFET gate can be understood by assuming that C_{GS} is linear. We know that when charging a linear capacitor from a voltage source (as in the drive circuit of Fig. 23.4) the stored energy is only 50% of that extracted from the source. The other half is dissipated in resistance between the source and capacitor, and is independent of the resistance. When the capacitor is discharged, its stored energy is dissipated, so the total energy lost is CV^2, or QV. Using this linear analysis for Fig. 23.4(b), we obtain an energy loss per cycle of $15\,\mathrm{V} \times 140\,\mathrm{nC} = 2.1\,\mu\mathrm{J}$. At a switching frequency of 1 MHz this loss represents 2.1 W of power. Most of this loss accrues to the driver, but MOSFET gate resistance is responsible for some of it. This loss is unavoidable unless the gate is driven by a resonant circuit, which is seldom the case.

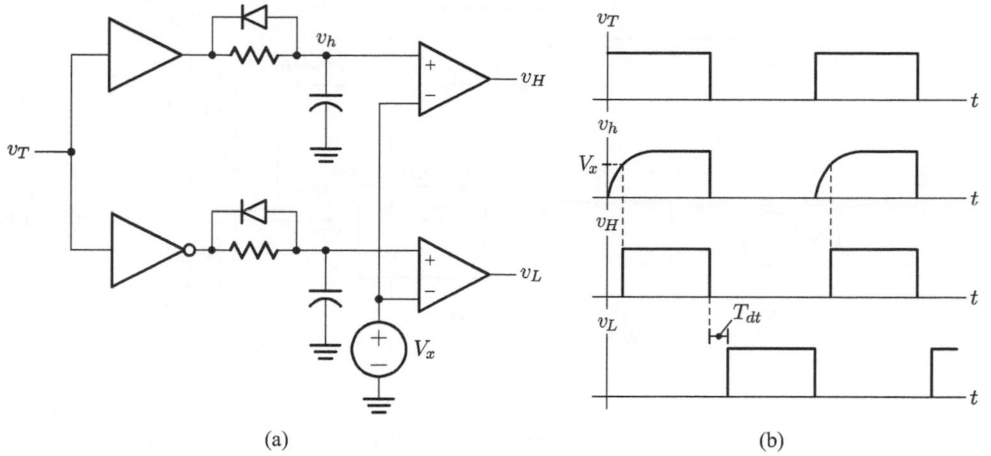

Figure 23.5 (a) A circuit used to provide shoot-through protection to a bridge leg. (b) The waveforms of the variables in (a) showing the dead time T_{dt}.

A further source of gate drive loss is due to shoot-through in the output stage of some driver ICs. Shoot-through occurs during switching if the timing of the pull-up and pull-down transistors does not provide sufficient time to avoid both devices conducting simultaneously during the transient. Modern gate drive ICs usually provide internal measures to limit this occurrence.

The gate drive signals for a half-bridge (as illustrated in Fig. 23.6 in the discussion of gate drive design) must be timed to avoid shoot-through in the bridge legs of the power stage. This is done by providing a *dead time* in the drive to the two bridge leg devices. During the dead time the gate drive holds both gates low, assuring the device turning off is completely off before the complementary device begins to turn on. The dead time is thus a function of the switching speed of the devices. During the dead time load current commutates from the switch turning off to the complementary switch's diode.

A design to illustrate how we can provide shoot-through protection is shown in Fig. 23.5(a). The logic-level timing waveform v_T is applied through complementary logic gates to op-amp comparators whose logic outputs provide the low- and high-side signals to the high- and low-side gate driver output stages. The resulting waveforms are shown in Fig. 23.5(b). The RC network provides the required dead time T_{dt}. The diodes shunting the resistor allow the capacitors to be rapidly discharged, creating a level low output with no delay.

23.1.5 Level Shifting and Gate Drive Designs

As noted earlier, there are a number of commercial ICs that are designed for, or are adaptable to, driving MOSFETs or IGBTs. We consider here the driver requirements for a SiC MOSFET bridge leg, which is more complicated than driving a single device because the high-side device, Q_1, requires a voltage level shift from ground. The driver for Q_1 must be floating relative to system ground because Q_1's gate control voltage is relative to its flying source terminal, and it is at a potential higher than V_{DD} when on.

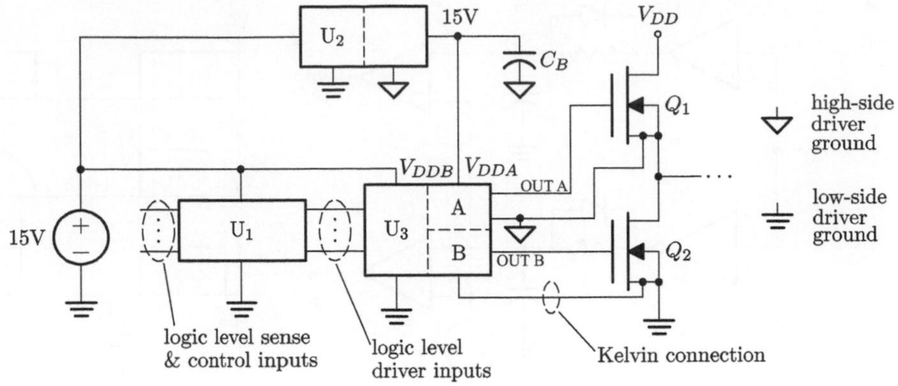

Figure 23.6 A typical functional configuration for driving the two transistors in a bridge leg. (U_1 signal processor; U_2 MGJ2D151SJ isolated dc/dc converter; U_3 UCC21530 isolated dual channel gate driver.)

A block diagram of a typical driver configuration, including commercially available ICs for each block, is shown in Fig. 23.6. The controller receives inputs from relevant circuit variables (e.g., device current, load voltage, or temperature), and provides at its output the appropriately timed gate signals for Q_1 and Q_2. These signals are processed by the driver IC to generate the required gate currents. The driver IC specified in this design provides both +4 A and −6 A outputs and isolation between the gate driver's outputs and inputs, as well as isolation between the two outputs. The isolated dc/dc converter is capable of an output current of 80 mA and provides a floating supply voltage to the isolated part of the driver IC driving Q_1, the high-side transistor. The low-side transistor is driven by the same IC, since the output channels are isolated from one another. Since the low-side transistor's gate remains near system ground, no isolated supply is required to power this second channel. The separate grounds connected to the sources of the MOSFET symbols are a representation of the Kelvin connection described in Section 23.1.3. The purpose of the capacitor C_B is the subject of Problem 23.1. The isolation coupling can be capacitive, magnetic, or optical, depending on the device chosen. For applications with lower drive current requirements there are ICs, such as the HIP 8081, which incorporate many of these functions.

An alternative to the isolated dc/dc converter is a circuit known as a *bootstrap*, shown in Fig. 23.7. When Q_2 is "on," the capacitor charges to $V_C = V_{\text{driver}}$. When Q_2 turns off, D becomes reverse biased, preventing C from discharging to $V_{DD} - V_{\text{driver}}$ when Q_1 turns on, and C becomes a floating supply for the gate driver. The capacitor must be large enough to maintain a fairly constant voltage, and Q_2 must be "on" for a time sufficient to charge C, thus limiting the maximum duty ratio of Q_1. The diode must be fast to avoid its reverse recovery current from discharging C. These constraints are absent when using an isolated supply.

Common-Mode Transient Immunity U_2 and U_3 in Fig. 23.6 each incorporate an isolation barrier between their inputs and outputs. The differential voltage between the grounds of their input and output voltage supplies switches between V_{DD} (when Q_1 is conducting) and 0 (when

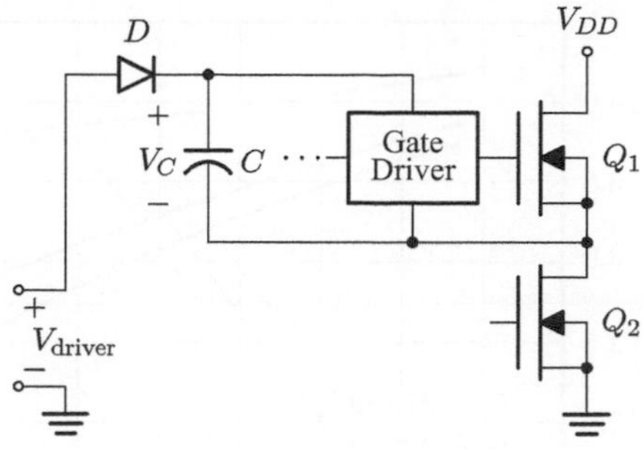

Figure 23.7 A bootstrap circuit creating a floating supply for high-side switch drivers.

Q_2 is conducting). Depending on the rate-of-rise or -fall during the switching transient, and the corresponding dv/dt of the voltage between grounds, noise can be conducted through the isolation barrier, causing failure of the isolating device to function properly. For this reason isolation devices, such as U_2 and U_3, have a parameter known as the *common-mode transient immunity* (CMTI) specified in V/ns; this specification for U_2 is $>$ 200 V/ns, and $>$ 100 V/ns for U_3. The voltage of concern is *common mode* because it is common to both the ground and positive output of the isolating ICs.

The CMTI has become an important parameter with the advent of SiC and GaN devices as their switching transients can be very fast. Some newer Si superjunction MOSFETs can also be applied at switching frequencies where attention must be paid to the CMTI of isolation devices in their gate drive circuits.

23.1.6 Parallel Operation of MOSFETs

Although multi-chip MOSFET modules are available for high-current applications, paralleling discrete devices sometimes results in improved performance. Discrete devices spread the heat over a larger area and provide greater flexibility in the thermal design, and, unlike chips in a module, they can be geometrically arranged to reduce asymmetry in their layout and circuit interconnections.

Because they are basically resistive devices having a positive temperature coefficient of resistance, MOSFETs, particularly Si MOSFETs, do not exhibit the current-hogging behavior of paralleled bipolar transistors. However, the negative temperature dependence of V_T, as illustrated in Fig. 23.8, works in opposition to the moderating effect of the $R_{DS(\text{on})}$ behavior with temperature. The result can lead to current imbalance, particularly during switching (*dynamic current balance*) when the lower V_T of the hotter device causes it to conduct earlier and stop conducting later than the cooler device. A further complication is created by the manufacturer-specified tolerance of V_T, which varies by device and manufacturer but a value of $\pm 25\%$ around the datasheet's "typical" value is representative. This means there can be variations of V_T among devices "straight out of the box." Therefore, to obtain the best performance of paralleled devices their thermal systems should be closely coupled and the devices selected for their V_T match.

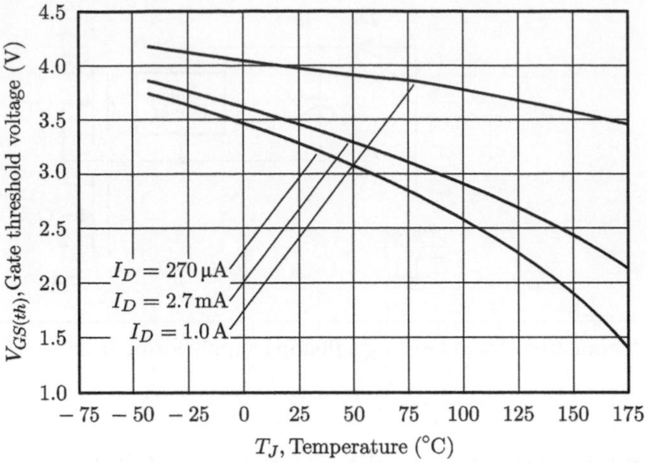

Figure 23.8 Typical normalized temperature dependence of the IRFP4868PbF MOSFET's threshold voltage, $V_{GS(th)}$, which we call V_T in the text. (Courtesy of Infineon)

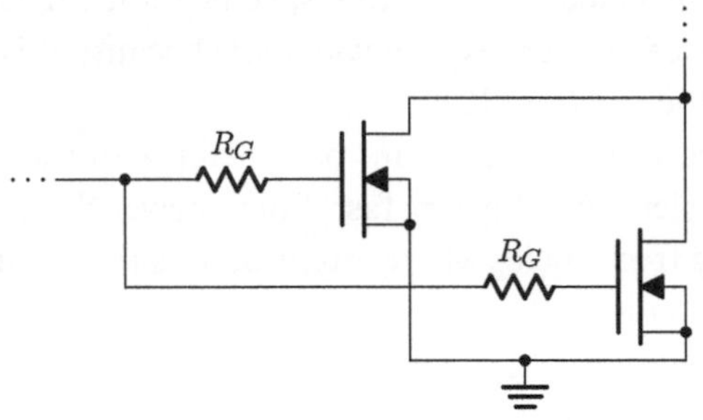

Figure 23.9 Paralleled MOSFETs illustrating the incorporation of resistors R_G to eliminate parasitic oscillations.

An additional challenge when paralleling MOSFETs is that they exhibit a tendency to oscillate when paralleled, particularly in older device designs, and measures must be taken to suppress these oscillations to avoid breakdown of the gate dielectric.

The oscillations are generally at frequencies in excess of 100 MHz and occur only during the switching transient when the devices are in their active gain region. For this reason they are difficult to observe. In early devices, where multiple chips were packaged together to achieve high-current capability, the problem was particularly acute. The oscillations can be damped sufficiently by placing small resistors in series with the gate lead, as shown in Fig. 23.9. A value of $R_G = 5\,\Omega$ is typically adequate and interferes little with the switching times of the devices. In many cases, the resistance of the polysilicon gate electrode is sufficient to eliminate the oscillation problem, but prudent design would include R_G anyway, because process changes made by device manufacturers cannot be anticipated.

Because R_{DS}(on) for SiC devices has a much weaker temperature dependence than for Si devices, additional care must be taken to avoid temperature deviations among paralleled SiC MOSFETs. Additionally, if the Kelvin source connection is used, there is a possibility that some

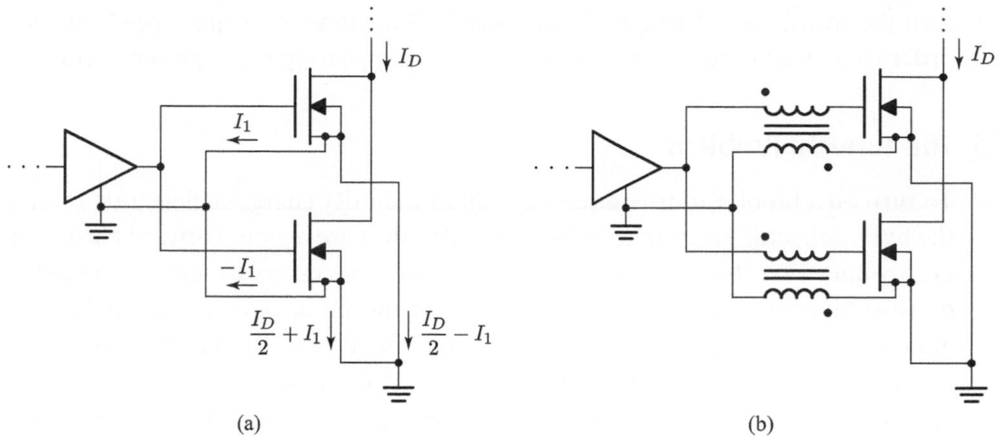

Figure 23.10 Driving paralleled SiC MOSFETs using the Kelvin source connection. (a) A fraction of the source currents circulating through the Kelvin leads. (b) Common-mode chokes to eliminate the spurious Kelvin lead currents.

fraction of the transistor source current could flow through the Kelvin leads, as illustrated in Fig. 23.10(a). This situation poses a potential for oscillations at the gate, and the possibility of fusing the internal Kelvin source connection as it is not designed to carry significant current. The use of common-mode chokes in series with the individual gate and source connections, as illustrated in Fig. 23.10(b), will present a low impedance to the gate current, but a high impedance to the circulating Kelvin source currents.

When paralleling a large number of MOSFETs, which may be the case for a large motor drive, a single gate drive IC usually cannot supply the necessary aggregate gate current. In this case the IC should drive emitter or source followers which drive the individual gates.

23.1.7 The EMI Problem

Because MOSFETs can switch in times on the order of nanoseconds, their switching transients can excite ringing oscillations in drain currents or voltages, which in turn can create serious EMI problems. The ringing originates with the oscillatory circuits formed by the parasitic circuit and device inductances and the parasitic circuit and device capacitances. Circuit layout that minimizes parasitic inductances and capacitances, particularly in the gate drive circuit, is necessary to reduce this problem. Control of the switching speed of the device is also an approach to reducing EMI. The turn-on and turn-off switching transients create different levels of EMI, so from an efficiency standpoint t_{on} and t_{off} may be controlled to different values.

23.2 Bipolar Transistor Base Drives

The base drive design for a bipolar transistor needs to achieve several goals. First, it must turn the transistor on in the shortest possible time to minimize turn-on losses. Second, it must provide sufficient steady-state base current to keep the device saturated. Third, it is generally required to turn the transistor off as quickly as possible. Sometimes switching speeds are compromised in order to prevent excessive $L\,di/dt$ voltages from appearing in the power circuit.

23.2.1 The Turn-On Problem

To turn on a bipolar transistor, an amount of majority charge sufficient to at least forward bias the base–collector junction must be supplied to the base region. Forward biasing of this junction occurs only after the base charge reaches that value necessary to support the collector current at quasi-saturation, $i_{C(\text{sat})}$. If the base current is constant at $i_B = i_{C(\text{sat})}/\beta$ (where β is the current gain at the edge of quasi-saturation), the collector current will rise asymptotically to $i_{C(\text{sat})}$ with a time constant that depends on the base defect. This rise time can be unacceptably long. We can reduce it considerably by first applying a large pulse of base current to quickly charge the base region and then reducing the drive to the value necessary to support the steady-state collector current. A base drive waveform of this type is shown in Fig. 23.11(a) and is frequently called a *pedestal and porch* waveform.

The product $I_{B_1} t_1$ of Fig. 23.11(a) is approximately the charge Q_F needed to bring the excess base concentration up to the level necessary to support $i_{C(\text{sat})}$. The larger we make I_{B_1}, the smaller will be t_1, the transistor's turn-on time. The maximum value of I_{B_1} is constrained either by base drive circuit limits or by the safe current limits of base leads and connections inside the transistor package.

We can create a practical approximation to the waveform of Fig. 23.11(a) with the circuit of Fig. 23.11(b). If we assume that $v_{BE} = 0.8$ V, the currents I_{B_1} and I_{B_2} are

$$I_{B_1} = \frac{V_B - 0.8}{R_1}, \qquad I_{B_2} = \frac{V_B - 0.8}{R_1 + R_2}. \tag{23.3}$$

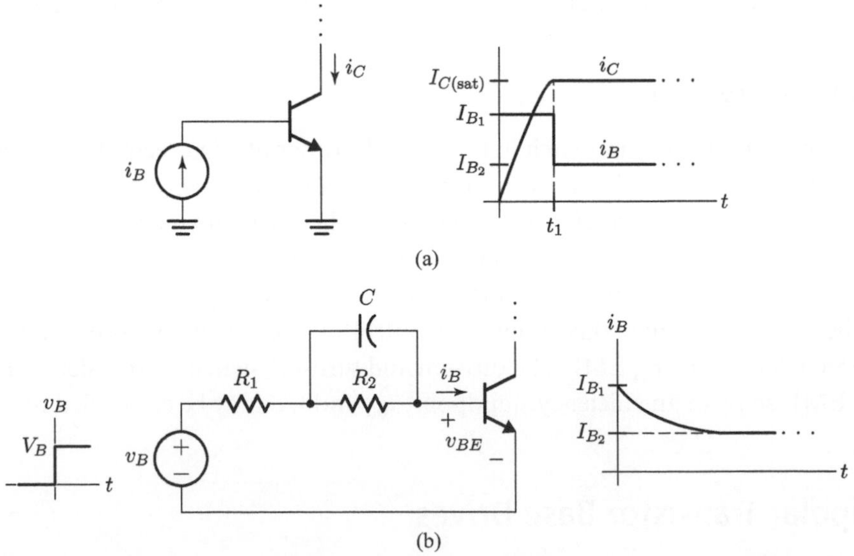

Figure 23.11 (a) Bipolar transistor base current waveform (pedestal and porch) required for fast turn-on. (b) Standard approach to synthesizing the waveform.

As β is not known precisely, and even less is usually known about Q_F, the values of R_1, R_2, and C have to be estimated or determined experimentally. Given V_B, one procedure is to use $\beta_{(min)}$ from the device datasheets and $I_{C(max)}$ for the circuit to determine the value of $R_1 + R_2$. This resistance is divided between R_1 and R_2 so that I_{B_1} is limited as discussed. The value of C is then determined experimentally. As shown in Section 23.2.2, however, these minimum values of R_1 and R_2 increase the transistor's turn-off time if its gain is larger than the minimum value $\beta_{(min)}$.

23.2.2 The Turn-Off Problem

To turn off a bipolar transistor, the base charge must be brought to zero. This condition can occur if the base circuit is simply opened, but the process can be considerably accelerated by reversing the base current and forcibly removing base charge. The removal of charge, in either case, occurs in two distinct stages. During the first stage, stored quasi-saturation charge, Q_S, is removed. This is the charge in excess of the charge Q_F needed to bring the transistor to the edge of quasi-saturation. The transistor's collector–emitter voltage remains low (but not zero) while Q_S is being removed, resulting in a delay between when the turn-off process is initiated at the base and when it becomes apparent at the collector when viewed on the scale of the full off-state voltage or on-state current. This time is known as the *storage delay time*, and is shown as t_s in Fig. 23.12(a). The second stage manifests itself as the fall in collector current as the charge concentration and its gradient at the collector junction both approach zero, as illustrated in Fig. 23.12(b).

Although the collector current fall time may be small, the storage delay time can cause control problems, such as the requested duty cycle being smaller than that realized or a phase shift occurring between control input and converter output. The problem is aggravated at higher frequencies. The obvious solution is to avoid operating deep in quasi-saturation. (Power BJTs

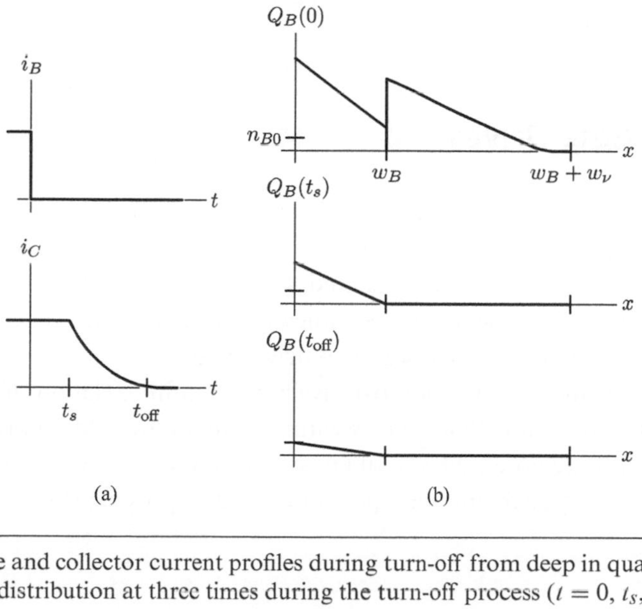

Figure 23.12 (a) Base and collector current profiles during turn-off from deep in quasi-saturation. (b) Base charge distribution at three times during the turn-off process ($t = 0$, t_s, and t_{off}).

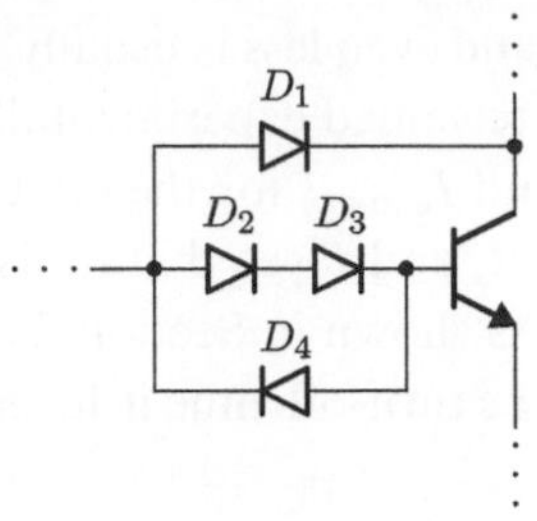

Figure 23.13 Anti-saturation clamp circuit.

can seldom be driven hard enough to enter the hard saturation region.) We could do so, using the circuit of Fig. 23.11(b), if we selected the value $R_1 + R_2$ with knowledge of the precise value of β. However, we cannot know β precisely, because it is a function of too many variables (temperature and current, for instance). A more practical solution is to use an *anti-saturation clamp*, or a base drive circuit that senses v_{CE} and controls the base current to keep this voltage above some minimum, that is, not too deep into quasi-saturation. The price we pay for either of these solutions is a slightly higher on-state voltage.

Figure 23.13 shows an anti-saturation clamp circuit. The clamp prevents the collector–base junction from being forward biased by more than the forward drop of the diode, or about 0.4 V, thereby keeping the transistor at the edge of saturation. Base current beyond that necessary to maintain this collector voltage is bypassed through the diode to the collector circuit. A problem with clamping a power device is that, because of ohmic drops, $v_{CB(\mathrm{sat})} > 0$ (for an npn device), even though the junction itself is forward biased. The diode must clamp the collector to a voltage (measured with respect to the base) that is higher than V_{CB} to prevent a hard forward bias on the junction. Therefore the anode of the clamp is elevated by placing additional diodes in series with the base – for instance, D_2 and D_3 in Fig. 23.13. Diode D_4 permits the flow of negative base current during turn-off.

23.3 Thyristor Gate Drives

Thyristor gates pose a unique problem during turn-on. Because the gate–cathode drop contains a component resulting from the product of anode current and the bulk resistance of the cathode region, the gate characteristic presented as a load to the drive circuit is time dependent. In particular, the gate voltage increases radically during the turn-on process, which, among other problems, can destroy a poorly designed drive circuit.

Converters using thyristors almost invariably require isolation of one or more of the thyristors' gates. The reason is that devices within a circuit are seldom arranged with common cathodes, or the cathode potential is frequently flying between ground and a high-voltage rail. Either opto-couplers or transformers are generally used to provide isolation. Opto-couplers require a power supply and amplifying circuitry on the thyristor side of the coupler. Transformers can couple enough energy to drive gates directly but require provisions for resetting the flux in cores.

23.3.1 Gate Drive Requirements

In our discussion of the physics of thyristor operation, we showed that the turn-on process requires that sufficient charge be supplied to the p base region to increase α_2 to the point of regeneration ($\alpha_1 + \alpha_2 = 1$). Once regeneration commences, the charge is maintained by the flow of anode current. Thus the gate current must be a pulse with an area large enough to charge the base to the device's regenerative threshold. For the SCR, the gate plays only a minor role in the turn-off process. For the GTO thyristor, however, the turn-off gate requirements are very strict. We discuss the problem of GTO turn-off drives separately.

Manufacturers generally use two parameters to characterize the gates of thyristors: V_{GT} and I_{GT}. The first is the minimum gate voltage guaranteed to trigger a thyristor. The second is the minimum gate current at which a thyristor is guaranteed to trigger. Note that V_{GT} is not v_G at $i_G = I_{GT}$. Instead, these parameters describe the extremes of the triggering thresholds of the gate variables, as measured for a large sample of the device. Furthermore, these values are measured *before* anode current flows.

Two straightforward gate drive circuits are shown in Fig. 23.14. They are similar to the bipolar base drives of Fig. 23.3, but the element values and the operation of the circuits are different. The circuit of Fig. 23.14(a) provides gate current during the time when Q_1 is off. The value of this current, $i_G \approx V_G/R_1$, should be larger than I_{GT} for the particular thyristor type being used. When Q_1 is on, it holds the gate of Q_T close to the cathode potential, providing a degree of immunity to gate noise and off-state dv/dt for Q_T. But this circuit again has the drawback that, if the base drive to Q_1 fails, Q_T turns on.

The alternative circuit shown in Fig. 23.14(b) has the dual advantages that (i) it is fail safe with respect to loss of base drive; and (ii) the negative pulse of gate current at turn-off assists the thyristor in supporting the highest possible reapplied dv/dt. The pedestal and porch turn-on waveform generated by this circuit has the following desirable effects on the performance of the main device.

- The pedestal gives the gate an initial charge, enabling the thyristor to support the maximum di/dt at turn-on.
- If the load is inductive and the anode current has not reached its latching value by the time the pedestal has expired, the porch provides gate current until the latching current is reached, keeping the device from turning off prematurely.
- If the thyristor is required to remain on for low anode currents, the porch permits the anode current to fall below the holding current.
- If Q_T were a GTO thyristor, the porch would be necessary because manufacturers generally specify a low level of continuous drive to offset the high holding current for these devices.

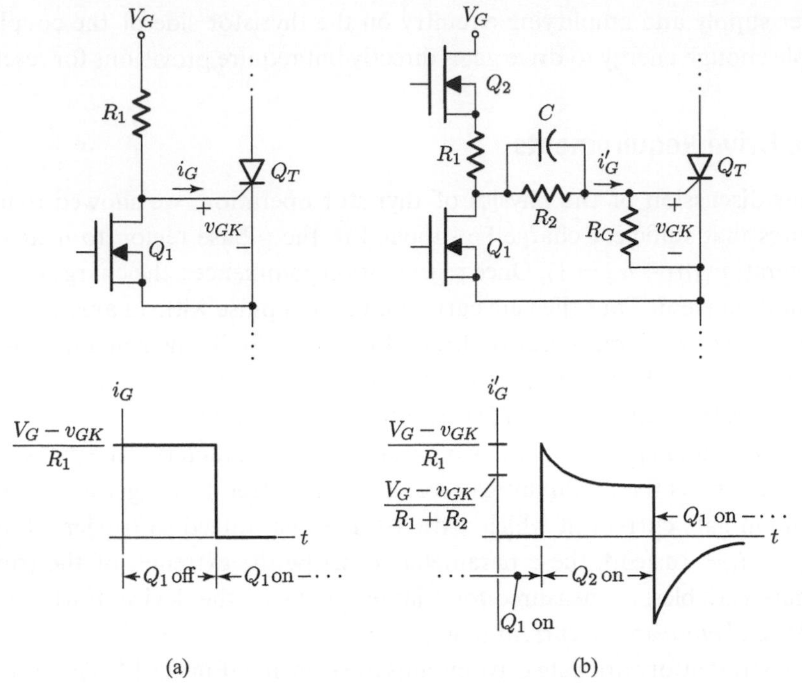

Figure 23.14 (a) Simple thyristor turn-on drive circuit; when Q_1 is on, Q_T is off. (b) Pedestal and porch drive and turn-off assist to maximize reapplied dv/dt.

The element values in the circuits of Fig. 23.14 are not particularly critical. The resistor R_G in Fig. 23.14(b) desensitizes the device to noise and dv/dt turn-on. Its value is usually suggested by the manufacturer and ranges from 10 Ω to 1 kΩ, depending on device size. A capacitor is sometimes placed in parallel with R_G to further enhance noise and dv/dt performance. Except for considerations of gate dissipation and efficiency of the drive circuit, there is little to discourage over-driving the thyristor. This situation contrasts with a bipolar drive, where excess base charge results in increased switching times. The only critical requirement is to ensure that the gate current remains positive during the turn-on process.

As mentioned previously, when the device is turning on with a high rate of rise of anode current di/dt, the initially restricted turned-on region presents a high resistance to anode current, resulting in a large voltage at the gate terminal. If V_G is smaller than this voltage, the gate current will actually reverse for a short time while the anode current is rising. This condition reduces the speed of the turn-on process and increases dissipation. To ensure positive gate current at all times, the manufacturer generally provides a gate drive specification of the form "20 V behind 20 Ω" for a gate current of 1 A. Figure 23.15 is a photograph of the gate current of a thyristor subjected to a high di/dt at turn on with an improperly designed gate drive, showing a substantial negative excursion. As we have suggested, high turn-on di/dt operation of thyristors requires special care in the design of the gate drive. The design rule is to get as much charge into the gate as quickly as possible. The di/dt specification in datasheets is generally predicated on a gate current rise time shorter than some specified maximum.

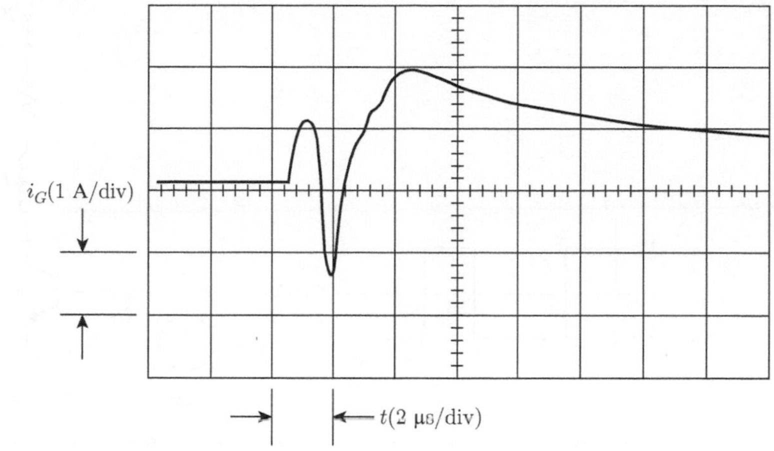

Figure 23.15 Gate current of an improperly driven thyristor with a high di/dt at turn-on.

Notes and Bibliography

A good general reference for power MOSFETs, including the design of gate drives, is [1].

The unique aspects of paralleling SiC MOSFETs are discussed in [2] and [3]. In particular, [2] discusses the challenge of static and dynamic current sharing, while [3] addresses the potential problem associated with using the Kelvin source connection of paralleled devices and presents design solutions.

An analysis of oscillations in paralleled MOSFETs and a discussion of corrective measures are presented in [4].

1. D. A. Grant and J. Gower, *Power MOSFETs: Theory and Applications*, New York: Wiley, 1989.
2. Z. Zheng, A. Lenze, D. Levett, K. Mainka, and M. Zhang, "A Practical Example of Hard Paralleling SiC MOSFET Modules," *PCIM Asia*, pp. 108–114, June 2019.
3. A. Lenze, D. Levett, Z. Zheng, and K. Mainka, *Hard Paralleling of SiC MOSFET Based Power Modules*, Bodo's Power Systems, pp. 26–28, Sept. 2020. (Also available through Infineon website.)
4. J. G. Kassakian and D. M. Lau, "An Analysis and Experimental Verification of Oscillations in Paralleled Power MOSFETs," *IEEE Trans. Electron. Devices*, vol. 31, pp. 959–963, July 1984.

PROBLEMS

23.1 What is the purpose of the capacitor C_B in the circuit of Fig. 23.6? (*Hint:* Consider that the dc/dc converter U_2 is capable of supplying only 80 mA.)

(a) Explain why C_B is necessary and how it functions in the circuit.

(b) If the gate driver of Fig. 23.4(a) were used to drive Q_1 in Fig. 23.6, and Q_1 had the gate charge characteristics of Fig. 23.4(b), what value of C_B would you specify to limit Δv_{C_B} to 10%?

(c) How long will it take to recharge C_B to 15 V?

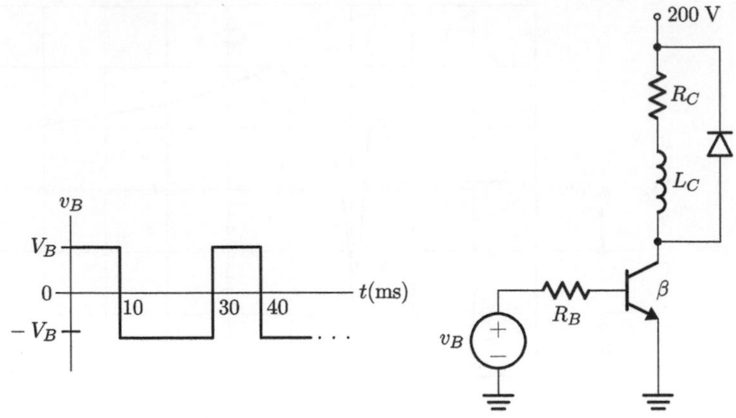

Figure 23.16 The circuit whose base drive parameters are determined in Problem 23.2.

23.2 The circuit shown in Fig. 23.16 has the following parameters:

$$L_C = 0.2\,\text{mH}, \qquad 5\,\Omega < R_C < 25\,\Omega, \qquad 12 < \beta < 20.$$

(a) Specify values for V_B and R_B that guarantee saturation for the specified ranges of R_C and β.

(b) In practice, the base drive voltage v_B does not have zero rise and fall times. Comment on the minimum values you would try to achieve if you were designing the complete drive circuit.

23.3 The circuit of Fig. 23.17(a) exhibits the base and collector current waveforms shown. In order to avoid the storage delay time of 0.1 μs, the base circuit is changed to that shown in Fig. 23.17(b). The presence of C_B still allows the base region to be charged swiftly, and R_B can be sized to avoid driving the transistor deep into quasi-saturation. This problem is concerned with determining values for R_B and C_B.

(a) Base current is principally composed of carriers supplying recombination in the base and ν regions and reverse injection into the emitter. Because both processes are proportional to the total stored charge in the base, the base current is also proportional to the total stored charge. This stored charge is composed of two parts: the first, Q_F, is that necessary to bring the transistor to the edge of quasi-saturation; the second, Q_S, is the charge in excess of that value. Removal of this excess stored charge causes storage delay time. Assuming that the lifetime in the base is long compared to the total switching time (0.5 μs), determine the values of Q_F and Q_S for the transistor in Fig. 23.17(a).

(b) To avoid storage delay time, we do not want to supply more base charge than necessary to keep the transistor at the edge of quasi-saturation. However, we would like to supply this charge as quickly as possible. Determine values for R_B and C_B that achieve these two goals. Sketch and dimension the resulting base current at both turn-on and

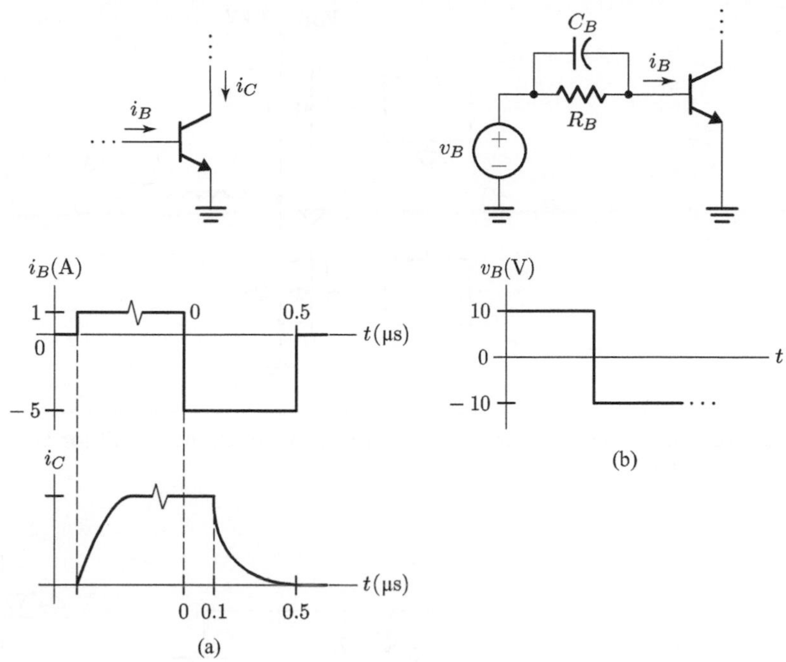

Figure 23.17 The base circuits discussed in Problem 23.3.

turn-off. (*Hint:* Assume some small resistance in series with the base so that an *RC* time constant can be defined and then let this resistor approach zero. Make suitable approximations.)

23.4 The paralleled MOSFETs shown in Fig. 23.9 each have the gate charge characteristic shown in Fig. 23.4(b). If they are driven at a common point by the current-limited output stage of the UCC27511 shown in Fig. 23.4(a), what will the fall time, t_{vf}, be?

23.5 Figure 23.18 shows one pole of a bridge circuit using two SiC MOSFETs. In this problem we determine the minimum CMTI required of the high-side driver, U_1. We consider the case of Q_2 turning off and I_o *eventually* commutating to the body diode of Q_1.

(a) Why does I_o *eventually* commutate to Q_1's body diode instead of instantaneously? In the MOSFET model of Fig. 15.6(b), assume the MOSFET transistor element turns off instantaneously.

The SiC MOSFETs are rated at 1200 V and 30 A (e.g., a CREE C3M0075120J). Although C_{oss} varies with V_{DS}, for a high-voltage SiC device it is nearly constant above a few hundred volts. Assume for this problem a constant value of 60 nF.

(b) Draw the incremental circuit relevant to calculating the voltage of the isolated high-side ground as a function of time during the switching period.

(c) Calculate the minimum CMTI required of the high-side driver U_1. (*Hint:* Make sure the relationship between C_{oss} for Q_1 and Q_2 in your circuit of part (b) is correct.)

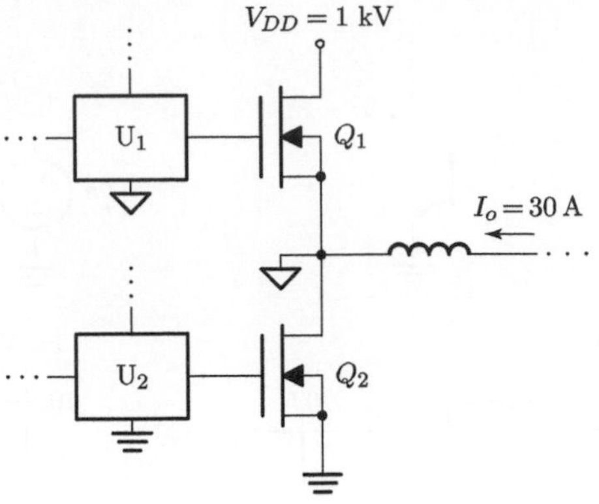

Figure 23.18 The bridge leg using SiC MOSFETs for which the CMTI for U_1 is determined in Problem 23.5.

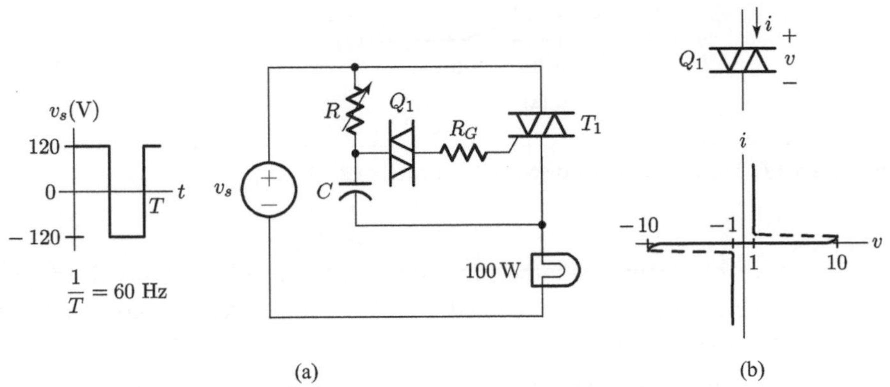

Figure 23.19 The incandescent light dimmer of Problem 23.6. (a) The circuit, utilizing an SBS (Q_1) trigger device. (b) The i–v characteristic of Q_1.

23.6 Figure 23.19(a) is a simple incandescent light dimmer circuit utilizing a TRIAC, T_1, and a *silicon bilateral switch* (SBS), Q_1. The SBS is one of a class of *trigger*, or *breakover*, devices developed for simplifying TRIAC gate drive circuits. The i–v characteristic of the SBS is shown in Fig. 23.19(b). The device is an open circuit until its voltage reaches ± 10 V, at which point the voltage *breaks back* to ± 1 V.

To simplify the analysis, assume that the light dimmer is operating from a square-wave source instead of the sinusoidal line voltage. What is the necessary range of the variable resistor R if the *power* to the 100 W lamp is to be varied between 10% and 95%? Assume that the lamp, which is rated at 120 V_{rms}, can be modeled as a simple resistor.

23.7 When the transistor Q in the circuit of Fig. 23.20 is on, we want it to exhibit a voltage of $v_{CE(sat)} = 2.4$ V with $v_{BE} = 1.2$ V. An anti-saturation clamp (ASC) is used in the circuit to prevent Q from going too deeply into quasi-saturation.

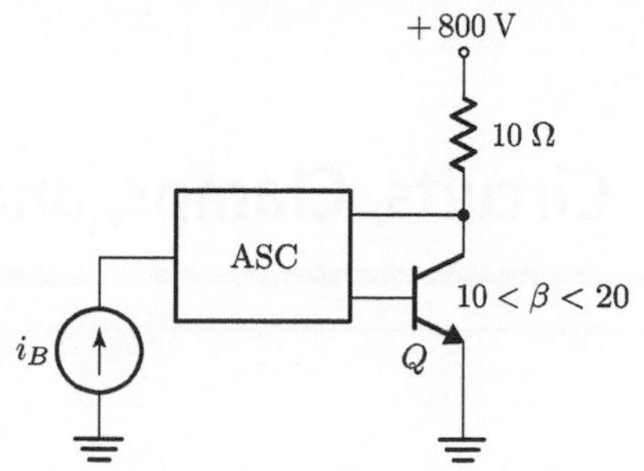

Figure 23.20 A transistor being driven through an anti-saturation clamp (ASC). Design of the clamp is the subject of Problem 23.7.

(a) Specify the network inside the box labeled ASC in Fig. 23.20.

(b) What is the minimum value that i_B can have and still maintain Q at the edge of saturation over the specified range of β?

23.8 Redo the calculation of the gate power required to turn on the MOSFET in Example 23.1 using the exact v_{GS} versus Q characteristic instead of the linear approximation. What was the error introduced by the approximation? Considering only the two MOSFET drive transistors, and both the turn-on and turn-off events, what is the total loss in the drive circuit? (N.B. You do not need to know the values of R_1, R_2, or $R_{DS(\text{on})}$.)

24 Snubber Circuits, Clamps, and Soft Switching

When a power semiconductor switch is either on or off, its power dissipation is relatively small. However, the transition from one state to the other is often not so benign, and can impose simultaneous high voltage and high current on the switching device. Special efforts are often required to ensure that the device will survive this most stressful part of its operating cycle. We have already seen in Fig. 22.4(b) how a snubber circuit consisting of C_S and R_S can moderate the effects of parasitic inductance. In this chapter we explore more fully the use of clamps and snubbers to protect switching devices, and illustrate how the trajectories of voltage and current during switching can be controlled to minimize device dissipation.

There are several reasons why we might need to control, or limit, the voltage and current in a switch when it makes a transition between states. One is the requirement that the voltage and current remain within the safe operating area (SOA) of the device. The SOA for a bipolar transistor is shown in Fig. 24.1. This area is bounded by constraints on peak current, peak voltage, peak power, and a region of voltage and current that could result in second breakdown. The SOA constraints of some common power transistors are detailed in Chapters 16 and 17. Of course, the switching of a controllable switch is often accompanied by a change in state of a diode. Making a transition slow enough to keep the diode's forward recovery voltage and reverse recovery current from excessively stressing the controllable switch is often important, and diodes can have their own SOA constraints that must be respected.

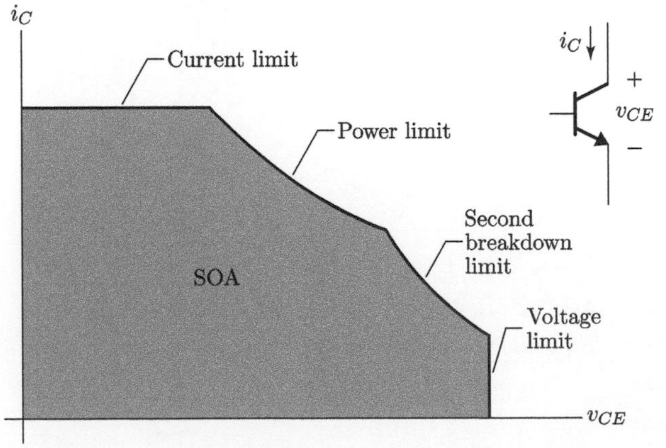

Figure 24.1 A typical safe operating area (SOA) for a bipolar junction transistor.

A second reason to control switching transitions is to keep the rates of change of the device voltage and current low enough to provide correct and reliable operation. The false triggering of SCRs caused by an excessive dv/dt, and the destructive results of current crowding in large-area devices when they are exposed to an excessive di/dt, are examples of why this limitation is important.

A third reason for paying attention to switch transitions relates to EMI. High dv/dt at a circuit node can increase capacitive noise injection into other circuit nodes, including sensitive nodes such as in control circuits. Likewise, high di/dt in a circuit loop can increase the noise magnetically coupled into other loops. The degree to which parasitic circuit resonances (e.g., owing to parasitic inductances and capacitances) are excited can also depend on the dv/dt and di/dt of switching transitions. Controlling the transitions is a key tool for mitigating such EMI effects.

A final reason to control switching transitions is to limit the dissipation that occurs in a device during switching. This switching loss, defined as the energy dissipated in each transition times the number of transitions that occur each second, is converted to heat that must be removed from the device to prevent the junction temperature from exceeding a specified limit. For a given thermal design, the less energy dissipated per transition, the higher the switching frequency can be.

Satisfying these concerns is an important practical aspect of power circuit design. One approach is to add special components to the power circuit to limit the voltage and current in a switch during switching transitions. These components form what is called a *snubber circuit*. There are two basic kinds of snubbers: one to control the turn-off transition, and the other to control the turn-on transition. The first controls the rate of rise of the switch voltage. The second controls the rate of rise of switch current. Both use small energy storage elements – a capacitor and an inductor, respectively. The initial portion of the chapter discusses the design and operation of these snubber circuits.

Another approach to addressing switching transitions is to design the power circuit itself to inherently provide desirable transitions. *Soft-switched* circuits use zero-voltage switching and/or zero-current switching to reduce switching loss, enabling higher switching frequencies than otherwise achievable. Soft switching can also help address other concerns, such as SOA limits and EMI generation, but these advantages often come at the expense of system complexity and/or control constraints. The final portion of the chapter provides an introduction to soft-switching techniques.

24.1 Turn-Off Snubber

To understand the need, design, and operation of a snubber circuit, consider the buck converter of Fig. 24.2(a). This is typical of many power circuits, in that an inductor current commutates between a transistor and a diode. The turn-off snubber is designed to modify the voltage and current waveforms during the transistor turn-off, in order to reduce the power the transistor dissipates and present a more acceptable transition between devices. We illustrate snubber design for an example converter with a bipolar transistor, because bipolar transistors often require a snubber for acceptable performance.

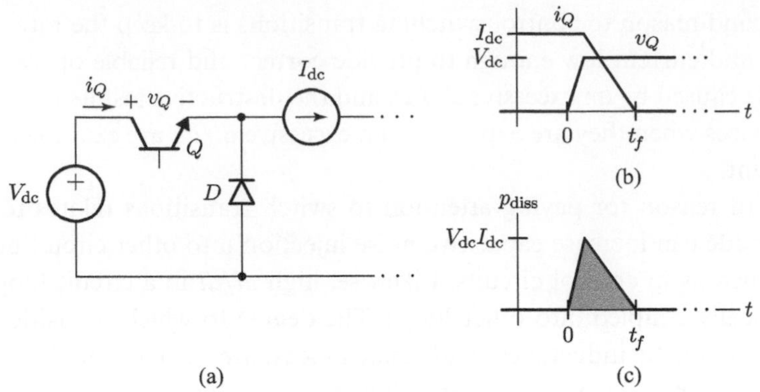

Figure 24.2 (a) A buck converter used in snubber examples. (b) Voltage, v_Q, and current, i_Q, during turn-off. (c) Dissipated power versus time during turn-off.

24.1.1 Turn-Off Dissipation and the SOA

In the following calculations we assume that all switch waveforms make linear transitions from their starting value to their final value. We also neglect the effect of device capacitances and assume that diodes operate ideally. Although these assumptions are not exactly correct, they are accurate enough for our present purposes.

When the transistor is turned off, its voltage and current waveforms are typically as shown in Fig. 24.2(b). Note that the current in the transistor cannot change until the diode turns on, which cannot occur until the voltage across the transistor reaches V_{dc}. The turn-off process lasts for a fall-time t_f, which is determined by the parameters of the transistor and its drive circuit. (Transistor switching times are often indicated in datasheets for nominal drive conditions.)

Figure 24.2(c) shows the power dissipated in the transistor, p_{diss}, as a function of time during the turn-off transition. Its peak value is the product of V_{dc} and I_{dc}. If we consider an example application in which $I_{dc} = 50$ A and $V_{dc} = 400$ V, we get a peak instantaneous transistor dissipation of 20 kW! The energy lost per transition is

$$W_{diss} = \int_0^{t_f} v_Q i_Q \, dt = \frac{V_{dc} I_{dc} t_f}{2}. \tag{24.1}$$

For a switching time t_f of 0.5 μs (a typical value for a power bipolar transistor) this gives $W_{diss} = 5$ mJ. If the switching frequency were 20 kHz (quite low by modern standards, but typical of converters at this rating built with bipolar transistors), 100 W would be dissipated in the transistor because of the turn-off transition alone. The widespread adoption of Si MOSFETs and devices constructed from wide-bandgap materials has been in part driven by their much shorter switching times, which reduce switching loss and enable higher switching frequencies to be achieved. At the same time, reducing t_f tends to increase the dv/dt and di/dt during the transition, which poses its own challenges.

Figure 24.3 is a superposition of the v_Q–i_Q switching locus on the transistor's SOA. It shows that if the transistor is operated near its rating, the SOA can easily be exceeded, with respect to

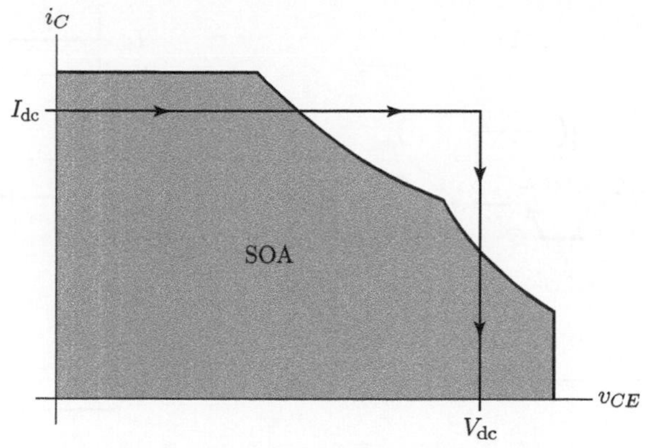

Figure 24.3 Switching locus of the turn-off transition for the waveforms of Fig. 24.2(b).

both peak power and second-breakdown limits, when the switch waveforms are of the form and relationship shown in Fig. 24.2(b).

A properly designed turn-off snubber both reduces the switch dissipation during turn-off and keeps the switching locus within the SOA of the switch (and, when necessary, of the diode). We now address the design of such a snubber.

24.1.2 A Basic Turn-Off Snubber

If either the peak or the average power dissipation caused by the waveforms of Fig. 24.2(b) is too high for the transistor, or if – as shown in Fig. 24.3 – the switching locus exceeds the SOA, a turn-off snubber can be used to change these waveforms. The capacitor C_s in the circuit of Fig. 24.4(a) is a simple snubber. Now when the transistor is turned off, its voltage and current waveforms look like those shown in Fig. 24.4(b).

Note that because C_s supplies a third path through which I_{dc} can flow, it is no longer necessary for v_Q to rise to equal V_{dc} before i_Q can begin to fall. Instead, any difference between I_{dc} and i_Q flows through C_s until the diode picks up the load current. The charging of this capacitor, rather than the switching time of the transistor, governs the rate at which v_Q rises from zero to V_{dc}. Because the voltage across C_s increases at a rate proportional to $1/C_s$, we can choose a large enough capacitor to keep the rise in voltage to only a small fraction, γ_v, of V_{dc} by the time i_Q reaches zero. The consequence of including C_s is that the peak power and total dissipation in the transistor at turn-off are much smaller than they were before the inclusion of C_s, as illustrated by Fig. 24.4(c).

In addition to reducing the dissipation in the transistor, the snubber capacitor also maintains the switching locus in the SOA of the transistor, and limits the peak dv/dt at the switching node during the transition. Figure 24.5 shows the turn-off switching locus as modified by the presence of C_s.

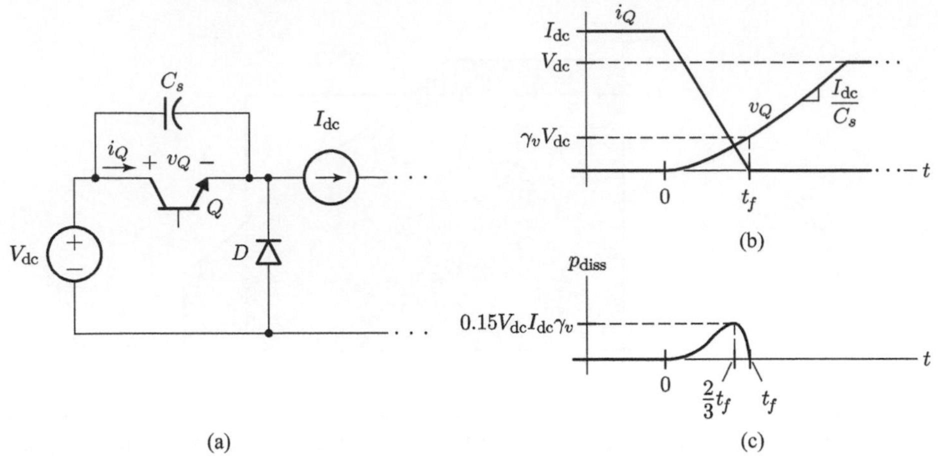

Figure 24.4 (a) A basic turn-off snubber consisting of capacitor C_s. (b) Transistor voltage and current waveforms during turn-off. (c) Dissipation in the transistor during turn-off.

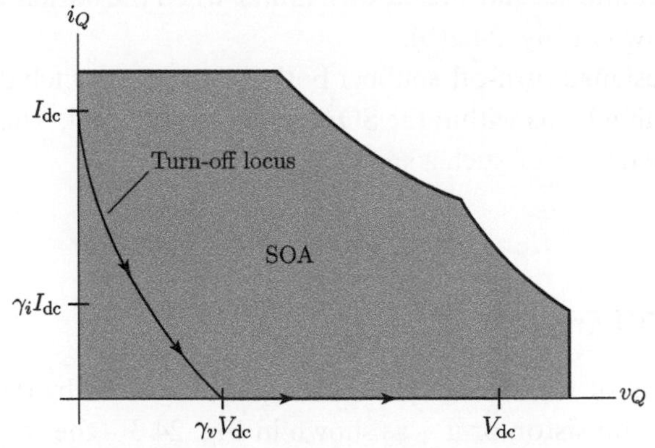

Figure 24.5 The turn-off switching locus of the transistor in the circuit of Fig. 24.4 using a turn-off snubber, forcing the locus to remain within the SOA.

An analysis of the transistor loss at turn-off is revealing. If we assume that the transistor current falls linearly to zero in a time t_f, the transistor current is

$$i_Q = I_{dc}\left(1 - \frac{t}{t_f}\right) \tag{24.2}$$

for $0 < t < t_f$. Consequently, the snubber capacitor current during this time is

$$i_C = I_{dc}\left(\frac{t}{t_f}\right) \tag{24.3}$$

and the switch voltage is

$$v_Q = \frac{1}{C_s}\int_0^t i_C\, dt = \frac{I_{dc}\, t^2}{2\, t_f\, C_s} \tag{24.4}$$

for $0 < t < t_f$. From these expressions we can find the energy dissipated in the switch during its turn-off transition:

$$W_{sw} = \int_0^{t_f} v_Q \, i_Q \, dt = \int_0^{t_f} \frac{I_{dc}\, t^2}{2\, t_f\, C_s} I_{dc} \left(1 - \frac{t}{t_f}\right) dt = \frac{(I_{dc})^2 (t_f)^2}{24 C_s}. \tag{24.5}$$

This equation indicates that the turn-off loss in the transistor is proportional to the square of both the turn-off time t_f and the current being switched, and is inversely proportional to the capacitance C_s. The peak dv/dt at turn-off is likewise inversely proportional to C_s.

Example 24.1 Choosing a Value For C_s

When the current flowing through the transistor in the circuit of Fig. 24.4(a) reaches zero at $t = t_f$, the voltage across the transistor is $v_Q = \gamma_v V_{dc}$. By evaluating (24.4) at $t = t_f$ we can relate C_s to $\gamma_v V_{dc}$:

$$C_s = \frac{I_{dc} t_f}{2 \gamma_v V_{dc}}. \tag{24.6}$$

Substituting this in to (24.5), we can obtain an alternative expression for the turn-off loss in the switch:

$$W_{sw} = \left(\frac{1}{12}\right) \gamma_v V_{dc} I_{dc} t_f. \tag{24.7}$$

If we choose C_s so that $\gamma_v = 0.2$ for $V_{dc} = 400$ V, $I_{dc} = 50$ A, and if we assume $t_f = 0.33$ μs, then we require $C_s = 0.1$ μF and get $W_{diss} = 0.12$ mJ. This capacitor selection yields a very modest peak transistor dv/dt at turn-off of

$$\frac{dv}{dt} = \frac{I_{dc}}{C_s} = 0.5\,\text{V/ns}. \tag{24.8}$$

24.1.3 A More Practical Snubber

A capacitor by itself is not a sufficient snubber, because when the transistor is turned on, C_s will discharge through Q. The resulting transistor current can be very large and lead to failure. To avoid this problem, we add a resistor and a diode to C_s, as shown in Fig. 24.6. The purpose of the resistor R_s is to limit the discharge current and absorb the capacitor energy when the transistor is turned on, and the purpose of the diode D_s is to allow the charging current to bypass the resistor during the turn-off transition. The resulting circuit is sometimes called an RCD snubber.

A trade-off is necessary in determining the value of R_s. It must be small enough to ensure that the capacitor is fully discharged during the shortest time that Q might be on, but large enough to prevent the discharge current from exceeding the transistor rating. R_s typically dissipates the vast majority of the energy W_C stored in C_s at turn-off, so must have an appropriate power dissipation rating. Given a stored energy of $W_C = 0.5 C_s V_{dc}^2$, R_s should be rated for a power dissipation of W_C times the switching frequency f_{sw}. We consider the impact of this snubber loss on efficiency in detail below, but it is typically preferable to use as small a capacitor as possible while meeting other design goals.

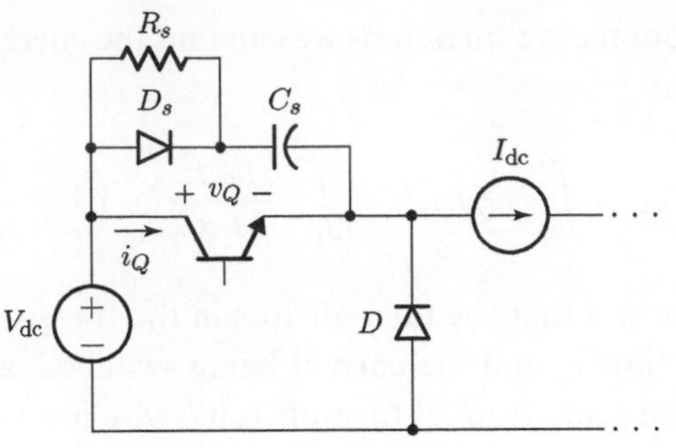

Figure 24.6 A practical RCD turn-off snubber circuit consisting of C_S, R_S, and D_S.

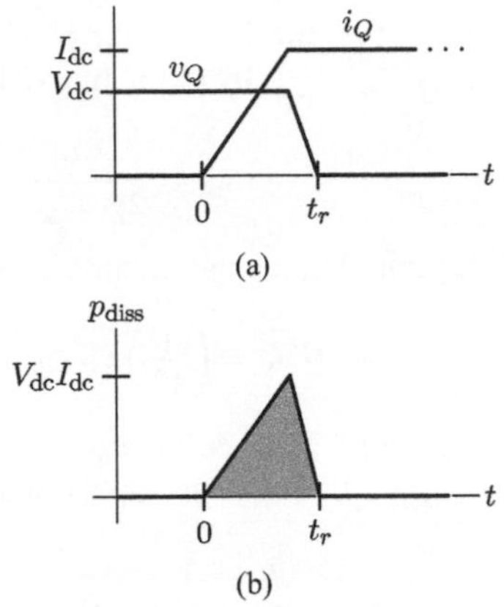

Figure 24.7 (a) Waveforms during turn-on of the transistor in the converter of Fig. 24.2(a). (b) Power dissipated in the transistor as a function of time.

24.2 Turn-On Snubber

The transistor voltage and current waveforms during turn-on in the converter of Fig. 24.2(a) are shown in Fig. 24.7(a). Initially i_Q rises, but until this current equals I_{dc}, v_Q must remain at V_{dc} because D must be on to carry the difference between I_{dc} and i_Q. When the load current has completely commutated from the diode to the transistor, D turns off and v_Q falls to zero. As during turn-off, the transistor voltage and current during turn-on are simultaneously nonzero, leading to switching loss and the possibility of leaving the SOA. A turn-on snubber is designed to modify the switching waveforms to reduce this loss and maintain operation within the SOA.

24.2.1 Turn-On Dissipation and the SOA

Device dissipation during turn-on can be calculated using (24.1), with t_f replaced by the rise time t_r. A typical value of t_r for a 400 V, 50 A Si bipolar transistor is 0.5 μs, though modern devices can be orders of magnitude faster. This value for t_r results in a peak power of 20 kW and an energy loss of 5 mJ during each turn-on transition. With a switching frequency of 20 kHz (suitable for a Si bipolar transistor of this type), turn-on causes another 100 W of dissipation in the transistor. One can see why the faster turn-on times of modern devices are important to reducing switching loss and achieving higher switching frequencies.

The switching locus in the i_Q–v_Q plane for this transition is identical to the one shown in Fig. 24.3 for the turn-off transition, except the path is taken in the reverse direction. Once again, the transistor's peak power and second-breakdown limits are exceeded.

24.2.2 A Basic Turn-On Snubber

The solution to the turn-on problem is the dual of the solution to the turn-off problem. A snubber inductor, L_s, is placed in series with the transistor, as shown in Fig. 24.8(a), and the waveforms of Fig. 24.8(b) result. Notice that with L_s in place, it is no longer necessary at turn-on for the transistor to support V_{dc} until the diode current reaches zero. Instead, v_Q can decrease to zero while the diode remains on. Whatever fraction of V_{dc} that does not appear across the transistor appears across L_s. The snubber inductor's voltage then determines the rate at which i_Q increases. When the current reaches its final value of I_{dc}, the diode turns off, and the voltage across the inductor drops to zero.

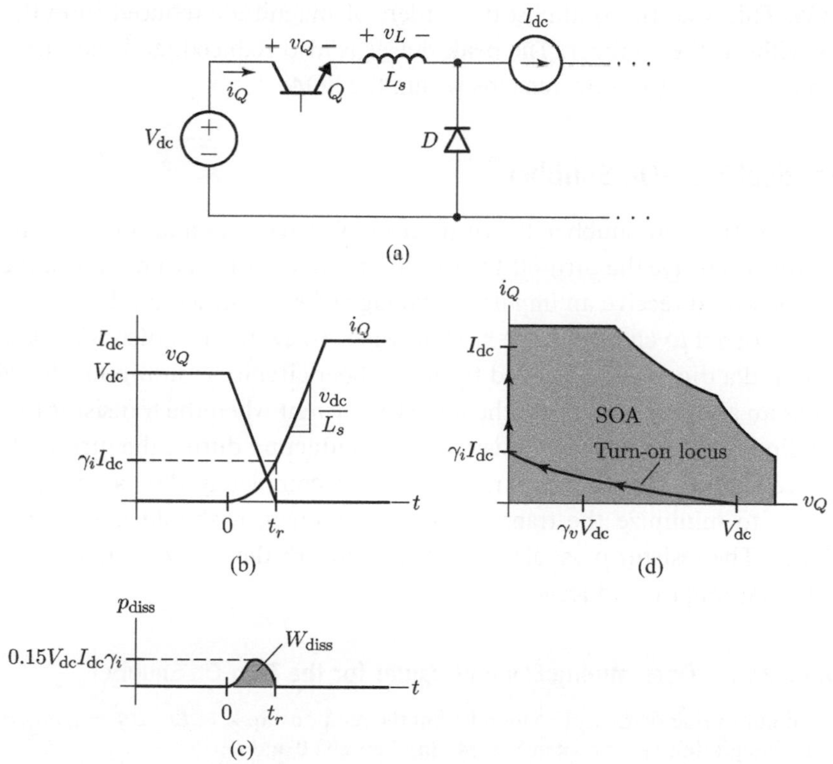

Figure 24.8 (a) A basic turn-on snubber consisting of the inductor L_s. (b) Collector voltage and current waveforms. (c) Device dissipation during turn-on. (d) The turn-on switching locus with a snubber.

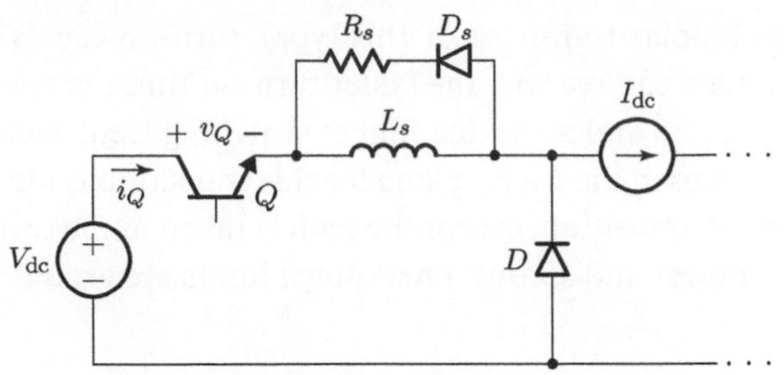

Figure 24.9 A practical *RLD* turn-on snubber circuit.

Because the rate at which the snubber inductor current rises is proportional to $1/L_s$, we can choose a value for L_s to limit the transistor current to a fraction γ_i of I_{dc} by the time v_Q reaches zero. Assuming a linear transition in transistor voltage with $t_r = 0.1$ μs and a snubber inductance giving $\gamma_i = 0.33$, we can calculate the resulting dissipation as we did in (24.7):

$$W_{diss} = \left(\frac{1}{12}\right)\gamma_i V_{dc} I_{dc} t_r = 0.06\,\text{mJ}. \tag{24.9}$$

At the switching frequency of 20 kHz, the resulting average turn-on dissipation *in the transistor* is 1.2 W. This quantity is almost two orders of magnitude reduced from the transistor turn-on losses without the snubber. The peak power is also reduced, and the turn-on switching locus, shown in Fig. 24.8(d), now remains within the SOA.

24.2.3 A Practical Turn-On Snubber

As with the turn-off snubber, the turn-on snubber requires additional components to allow the transistor to survive the turn-off transition. In this case, if nothing were added to the circuit, the transistor would receive an impulse of voltage when it turned off, because the inductor current would be forced to change in a very short time. To avoid this potentially destructive transient, a resistor and a diode can be added to the snubber circuit, as shown in Fig. 24.9. The resistor R_s provides an alternative path for the inductor current when the transistor turns off. The purpose of the diode D_s is to keep the resistor from conducting during the turn-on transition.

The resistance R_s must be large enough to completely discharge L_s during the off state. However, to minimize the transistor's voltage stress, it should not be made any larger than necessary. The resistor must also be rated to absorb the energy (and average power) associated with discharging the inductor.

Example 24.2 Determining Element Values for the Turn-On Snubber

The minimum value of L_s can be found from the relationship $\lambda = Li$. At $t = t_r$, $i_Q = \gamma_i I_{dc}$ and λ can be found by integrating v_L, shown in Fig. 24.8(a), between 0 and t_r:

$$\begin{aligned} \lambda &= \int_0^{t_r} v_L\,dt = \frac{V_{dc} t_r}{2} = L_s \gamma_i I_{dc}, \\ L_s &= \frac{V_{dc} t_r}{2\gamma_i I_{dc}} = 1.2\,\mu\text{H}. \end{aligned} \tag{24.10}$$

The resulting value of 1.2 μH is based on our previous parameter values of $V_{dc} = 400$ V, $I_{dc} = 50$ A, $\gamma_i = 0.33$, and $t_r = 0.1$ μs.

If the shortest time that we expect the transistor to be off is 2.5 μs, or 5% of a 20 kHz cycle, we need an L_s/R_s time constant of about 0.5 μs to ensure that L_s is completely discharged. Meeting this condition requires a resistor of value $R_s \approx 2.5\,\Omega$.

24.3 Combined Turn-On/Turn-Off Snubber

When both a turn-on and a turn-off snubber are used in a power circuit, we do not have to include a separate diode–resistor pair for each. Instead, one snubber can be used to correct the problem that required a resistor and diode in the other snubber. Two ways of doing so are described below.

Figure 24.10(a) shows the first approach, in which a standard turn-off snubber, composed of C_s, R_s, and D_s, is used in conjunction with a turn-on snubber inductor, L_s. The behavior of the circuit at turn-on is identical to that shown in Fig. 24.8, except that the discharge current of C_s must be added to i_Q. During turn-off, however, the circuit behavior is quite different from that illustrated in Fig. 24.4. They are similar until t_1, when C_s is charged to V_{dc}. But at t_1, when $v_C = V_{dc}$ and D turns on, L_s is still carrying a current $i_L = I_{dc}$, and therefore L_s and C_s will ring for a quarter of a cycle until $i_L = 0$. At this time, D_s turns off and R_s damps the transient.

Figure 24.10(b) shows the waveforms of the combined snubber at turn-off, and Fig. 24.10(c) shows the equivalent circuits during different time periods. When analyzed using the specified initial conditions, these circuits yield the waveforms of Fig. 24.10(b). The incremental ringing voltage across Q is

$$\Delta v_Q = I_{dc}\sqrt{\frac{L_s}{C_s}}. \qquad (24.11)$$

This incremental voltage is added to V_{dc} to give the peak transistor voltage, which is equal to the peak voltage across C_s.

Example 24.3 Transient Voltage Produced by a Combined Snubber

In Examples 24.1 and 24.2, we found values for the snubber capacitor and inductor of $C_s = 0.066$ μF and $L_s = 1.2$ μH. If we use these values for L_s and C_s in the combined snubber of Fig. 24.10(a), we can calculate the peak transistor voltage V_Q, using (24.11):

$$V_Q = V_{dc} + \Delta v_Q = V_{dc} + I_{dc}\sqrt{\frac{L_s}{C_s}} = 400 + 50\sqrt{18} = 613\,\text{V}. \qquad (24.12)$$

This value is large relative to V_{dc} and illustrates the need to carefully consider the values of elements used in the design of a combined snubber.

A second way in which a turn-on and a turn-off snubber can be combined is shown in Fig. 24.11(a). The behavior of this circuit is essentially the dual of the behavior exhibited by

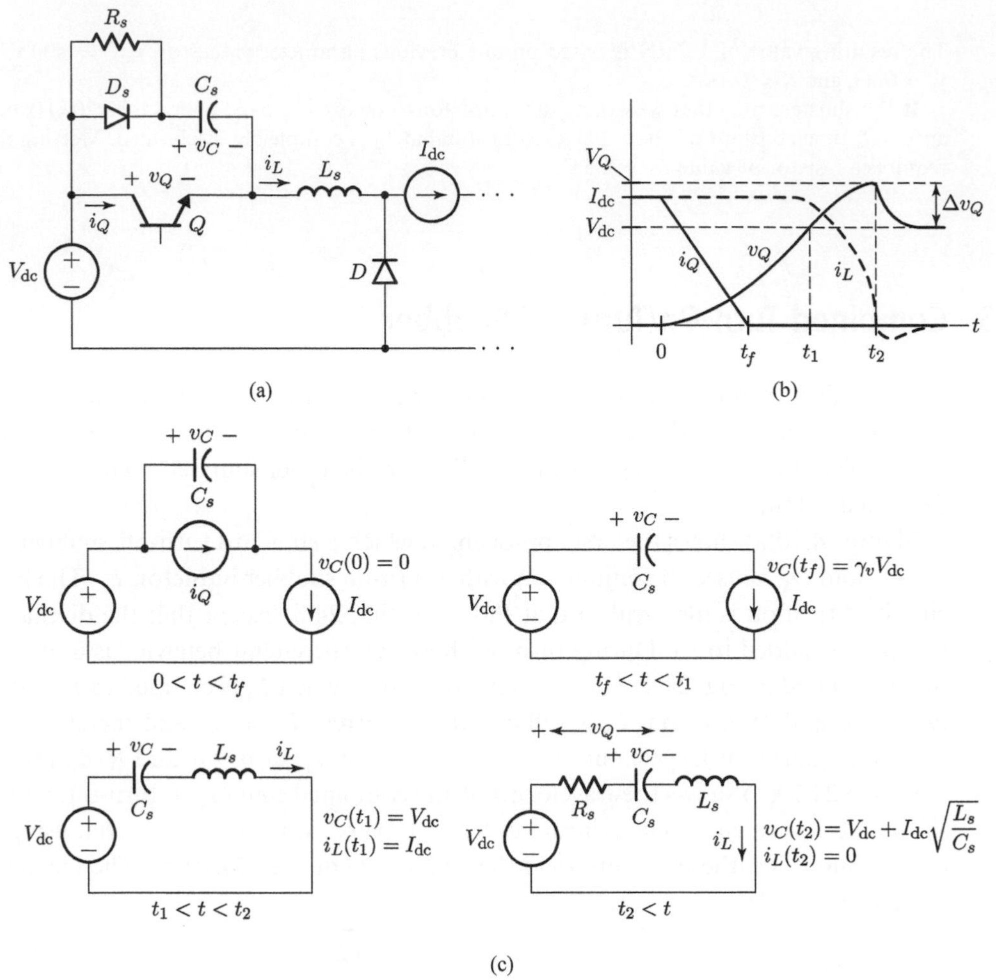

Figure 24.10 (a) A combined turn-on/turn-off snubber circuit with the resistor placed to discharge C_s. (b) Waveforms at turn-off. (c) Equivalent circuits and initial conditions for different time periods during the turn-off transient.

the combined snubber of Fig. 24.10. During turn-off, the waveforms of v_Q and i_Q are identical to those shown in Fig. 24.4, except that the discharge voltage across R_s contributes to v_Q. But at turn-on the circuit exhibits a period of second-order behavior. After $v_Q = 0$ at t_r, L_s charges linearly to I_{dc} at t_1, at which time D turns off. During this period, D_s is reverse biased and $v_C = V_{dc}$. After D turns off at t_1, L_s and C_s ring until $v_C = 0$ and D_s turns on at t_2. With R_s now in the circuit, i_Q settles to I_{dc} in an overdamped manner. The peak current in the transistor occurs at t_2 and is equal to

$$I_Q = I_{dc} + \Delta i_Q = I_{dc} + V_{dc}\sqrt{\frac{C_s}{L_s}}. \tag{24.13}$$

The waveforms of i_Q, v_Q, and v_C during turn-on are shown in Fig. 24.11(b).

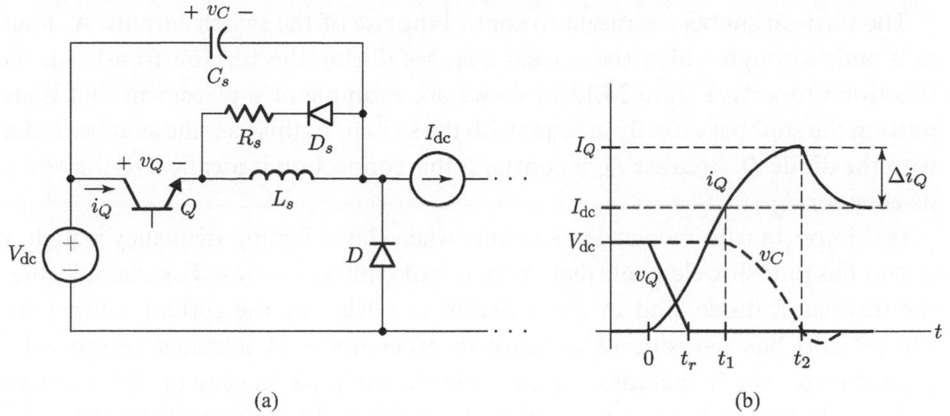

Figure 24.11 (a) A combined turn-on/turn-off snubber circuit with the resistor placed to discharge L_S. (b) Voltage and current waveforms at turn on.

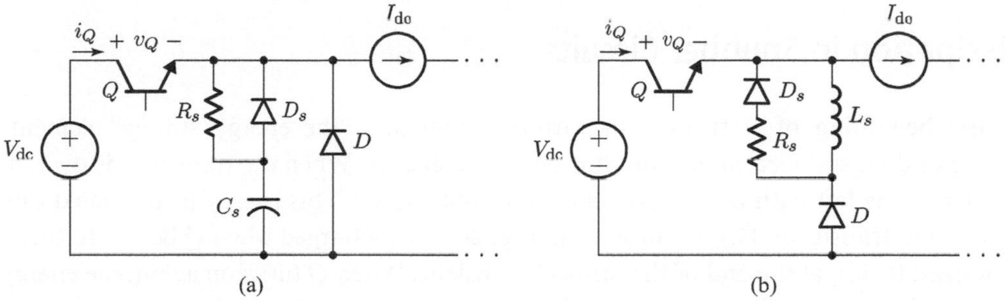

Figure 24.12 Alternative positions for the turn-on and turn-off snubber circuits. (a) Turn-off snubber. (b) Turn-on snubber.

24.4 Alternative Placements of the Snubber Circuit

In our discussions so far, we have placed the turn-off snubber in parallel with the transistor, and the turn-on snubber in series with the transistor. We now explore alternative placements for these snubber circuits.

The job of the turn-off snubber is to control the rise of v_Q. Because the collector of the transistor is ideally connected to a stiff voltage source V_{dc}, we also control v_Q in principle if we control the voltage between the emitter of Q and ground (or any node at incremental ground). For example, Fig. 24.12(a) shows an alternative placement where C_S is connected to ground.

For the turn-off snubber to work properly when connected in positions different from directly across the switch, it is essential that the terminal of the switch to which the snubber is not connected – the collector in the case of Fig. 24.12(a) – be at incremental ground. If there is any appreciable impedance between this switch terminal and incremental ground, such as lead inductance, the alternative placements suggested here will not protect the transistor to the same extent that a snubber connected directly across the switch would.

The turn-on snubber is meant to control the rise of the switch current. As long as it is placed in a path through which the current changes during the turn-on transition, the snubber will function properly. Figure 24.12(b) shows one example of a placement that is an alternative to putting the snubber directly in series with the switch. In this case the snubber inductor is in series with the diode D. Because I_{dc} is constant, this connection is identical to the series connection in its effect on i_Q.

It is important to recognize, especially when the switching frequency is high, that the power circuit has parasitic elements that provide some snubber action. For example, the node to which the transistor, diode, and output inductor (modeled by the current source) are connected in Fig. 24.2(a) has parasitic capacitance to other nodes at incremental ground. This parasitic capacitance, which includes the intrinsic device capacitances of the transistor and diode, collectively functions as a turn-off snubber. Similarly, the loop formed by the input source, transistor, and diode has a parasitic inductance that provides some turn-on snubbing action. In such cases there may be a need for only minimal additional snubber elements.

24.5 Dissipation in Snubber Circuits

At the beginning of a turn-on or turn-off transition, the energy storage element in the corresponding snubber circuit contains no stored energy. When the transition is over, however, the element is left with a nonzero value of stored energy. This energy is dissipated during the next switch transition. For instance, C_s in Fig. 24.6 is uncharged when Q begins to turn off and is charged to V_{dc} at the end of the turn-off transient. When Q turns on again, the energy stored in C_s is dissipated in R_s. Our purpose in this section is to evaluate the amount of energy lost in the process of resetting the energies stored in L_s and C_s to zero, as it is this loss that sets the required power rating of R_s. Note that this loss is in addition to that dissipated in the device during switching.

24.5.1 Separate Turn-Off and Turn-On Snubbers

We first consider the case of the turn-off snubber of Fig. 24.6. The energy stored in C_s at the end of the turn-off transient is

$$W_C = \frac{1}{2} C_s V_{dc}^2. \tag{24.14}$$

Using the expression for C_s in (24.6), we find the stored energy is

$$W_C = \frac{t_f I_{dc} V_{dc}}{4\gamma_v} = 8.25\,\text{mJ}. \tag{24.15}$$

For the parameters in Example 24.1, this evaluates to 8.25 mJ. Note that this loss in the snubber is $t_f/(2t_f\gamma_v)$ times the energy dissipated in the transistor without a snubber capacitor, as in (24.1). For small values of γ_v the snubber loss is larger than the loss in the transistor without a snubber, and increasing the snubbing action by making γ_v smaller increases the snubber loss.

The snubber loss for a turn-on snubber is determined through a calculation similar to that done above for the turn-off snubber. The energy stored in the snubber inductor just before turn-off is

$$W_L = \frac{1}{2} L_s I_{\text{dc}}^2. \tag{24.16}$$

Using (24.10), we find

$$W_L = \frac{t_r I_{\text{dc}} V_{\text{dc}}}{4\gamma_i}. \tag{24.17}$$

For the parameter values in Example 24.2, this evaluates to 1.5 mJ. Again, this energy is larger than that dissipated in the transistor without a turn-on snubber.

24.5.2 Combined Snubber

We have shown that, when the turn-on and turn-off snubbers are combined as in Fig. 24.10(a), the capacitor is charged to a voltage that exceeds V_{dc} by Δv_Q. The peak energy stored in C_s is thus

$$W_C = \frac{1}{2} C_s (V_{\text{dc}} + \Delta v_Q)^2. \tag{24.18}$$

Using (24.11), we can rewrite (24.18) as

$$W_C = \frac{1}{2}(C_s V_{\text{dc}}^2 + L_s I_{\text{dc}}^2) + V_{\text{dc}} I_{\text{dc}} \sqrt{L_s C_s}. \tag{24.19}$$

The first component of the right-hand side of (24.19) is exactly the amount of energy that would be lost if the two snubbers were not combined. The last term is a further, and not negligible, contribution to the energy stored in C_s. However, this additional energy is not dissipated. Instead, it is returned to the input voltage source by the current that discharges the capacitor. Therefore the actual energy lost in the combined snubber is equal to the energy that would be lost if the snubbers were not combined:

$$W_{\text{diss}} = \frac{1}{2}(C_s V_{\text{dc}}^2 + L_s I_{\text{dc}}^2). \tag{24.20}$$

24.5.3 Increased Circuit Dissipation Caused by Snubbers

We have seen that the introduction of a turn-off or turn-on snubber reduces the transistor loss of the respective transition, but introduces loss in the snubber resistor owing to the stored energy left in the snubber capacitor or inductor. In most cases, the introduction of a snubber increases total loss. For example, consider the total turn-off loss W_{off} with a turn-off snubber, comprising the loss in the switch W_{sw} plus the loss in the snubber resistor, which is equal to the energy W_C stored in the capacitor at turn-off. Using (24.5), this can be expressed in terms of the snubber capacitance as

$$W_{\text{off}} = W_{\text{sw}} + W_C = \frac{(I_{\text{dc}})^2 (t_f)^2}{24 C_s} + \frac{1}{2} C_s V_{\text{dc}}^2, \tag{24.21}$$

and by applying (24.6) it can be expressed in terms of γ_v as

$$W_{\text{off}} = V_{\text{dc}} I_{\text{dc}} t_f \left[\frac{1}{4\gamma_v} + \frac{\gamma_v}{12} \right]. \tag{24.22}$$

For values of $\gamma_v < 1$, W_{off} decreases monotonically with γ_v, or equivalently increases monotonically with C_s. From this the total turn-off loss W_{off} exceeds that without a snubber for $\gamma_v < 3 - \sqrt{6} \approx 0.55$, and if $\gamma_v < 0.5$ the snubber resistor loss alone exceeds the turn-off loss without a snubber. We can thus expect that introduction of a snubber will typically increase total loss and reduce efficiency owing to the energy dissipated in the resistor. Of course, loss in a snubber resistor may be more acceptable than in the switching device, making an increase in loss acceptable.

It is natural to wonder if the energy trapped in the snubber capacitor and inductor at the switching transitions might be somehow recovered instead of dissipated in a snubber resistor. This goal has led to the concept of energy recovery snubber circuits and ultimately to the development of soft-switched converters, which possess both gentle switching transitions and low switching loss. We take up the topic of soft switching in Section 24.6.

24.5.4 Voltage Clamps

The turn-off snubber, by limiting the rate of rise of v_Q, protects the switch from simultaneously high voltage and current. There are situations, however, in which a protection circuit is needed that simply limits the maximum voltage across the switch. This special circuit, shown in Fig. 24.13, is referred to as a *voltage clamp*. In its simplest form it consists of a capacitor, C_c, which is charged to the maximum allowable (clamp) voltage V_C, and some regulation circuitry that maintains this voltage. There is also a diode, D_c, that directs the load current I_{dc} into C_c when v_Q reaches V_C. If C_c is large enough, it absorbs this current for the rest of the switch transition without a significant change in its voltage.

A voltage clamp is useful in the circuit of Fig. 24.13 when the loop formed by V_{dc}, Q, and D has parasitic inductance. In the absence of this inductance, the voltage across the transistor would never exceed V_{dc}. When loop inductance is present, however, v_Q rises above V_{dc}, as was shown in connection with combined snubbers in Fig. 24.10(b). Clamps can be especially important in

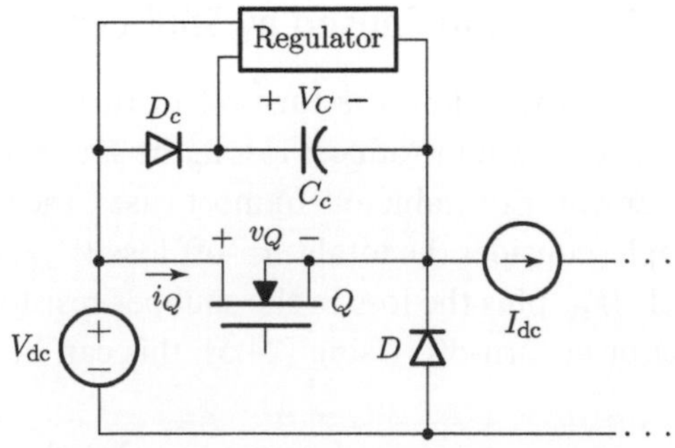

Figure 24.13 A clamp circuit designed to limit the peak transistor voltage to V_C.

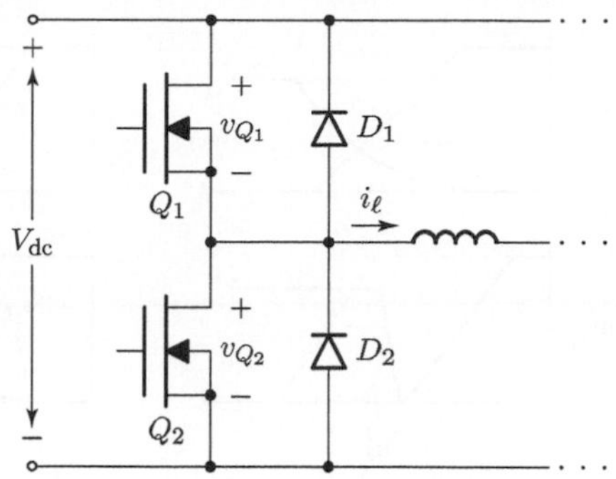

Figure 24.14 Part of a PWM bridge inverter in which the operation of D_1 and D_2 can be viewed as clamping v_{Q_2} and v_{Q_1}, respectively. D_1 and D_2 might be the intrinsic body diodes of the power MOSFETs or might be discrete diodes.

circuits in which the transistor switches current in a transformer winding – such as in a flyback converter – owing to the presence of transformer leakage inductance.

We can also limit the peak voltage across the transistor by choosing an appropriate turn-off snubber capacitance. Equation (24.11) shows that the larger the capacitor, the smaller the amount by which v_Q exceeds V_{dc}. The disadvantage of making C_s large is that the dissipation, as given by (24.20), also increases. If only the peak transistor voltage needs to be limited, a clamp circuit is often preferable to a snubber because the loss incurred is lower.

Although the clamp of Fig. 24.13 consists of a specially constructed voltage source – C_c, D_c, and the regulator – many power circuits have voltages available to which v_Q can be clamped. For instance, in the half of a PWM bridge inverter shown in Fig. 24.14, D_1 can be thought of as clamping v_{Q_2} to V_{dc} when Q_2 turns off. However, in order to function properly, the circuit construction must be such that the parasitic inductance in that part of the circuit consisting of the transistors, diodes, and V_{dc} is small enough to permit the load current i_ℓ to commutate from Q_2 to D_1 without creating a voltage transient that would manifest itself as an unacceptable increase in v_{Q_2}. It is also worth noting that some power devices can act as their own clamp. Devices such as power MOSFETs are often specified with a repetitive avalanche energy rating that indicates how much energy a device can safely absorb in voltage breakdown at turn-off (e.g., owing to parasitic inductance).

24.6 Soft Switching

Soft-switched power converters achieve both gentle switching transitions and low switching loss. There are various techniques by which soft switching can be achieved, but they all operate to reduce the overlap of switch voltage and current during transitions. A soft-switched transition might be characterized as providing ZVS when the switch voltage is kept small during the transition, and/or as providing ZCS when the switch current is kept small during the transition.

Figure 24.15 illustrates ZVS and ZCS transitions for each of switch turn-off and turn-on. In a ZVS turn-off transition (top of Fig. 24.15(a)), the voltage across the switch is kept low

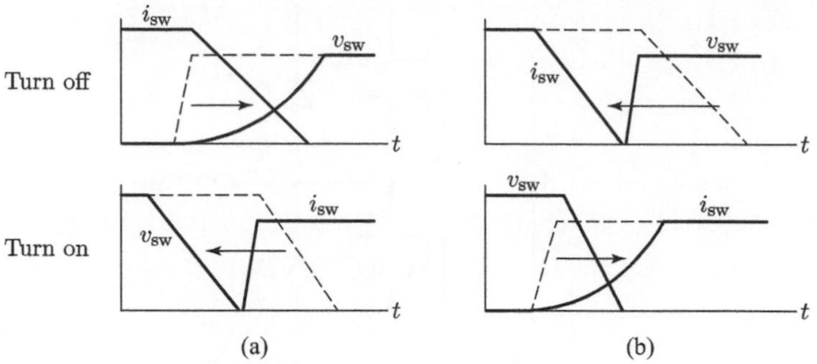

Figure 24.15 Switching waveforms for turn-off and turn-on transitions with (a) ZVS and (b) ZCS. The dashed lines show hard-switched waveforms.

as the switch current falls. This is often accomplished by placing a capacitor across the switch during turn-off (e.g., as in the simple snubber of Fig. 24.4). For soft switching, however, some means must be provided to losslessly recover the energy stored in the capacitor. This is often accomplished by returning the switch voltage to zero prior to turn-on (through some means not shown in Fig. 24.4), enabling ZVS turn-on of the switch, as illustrated at the bottom of Fig. 24.15(a).

Zero-current switching is the dual of zero-voltage switching. In a ZCS turn-on transition, illustrated at the bottom of Fig. 24.15(b), some mechanism is used to keep the switch current low as the switch voltage falls. One means of doing this is to place an inductor in series with the switch during turn-on, as in Fig. 24.8. For soft switching, the energy stored in the inductor at turn-on is recovered losslessly by using some means to return the switch current to zero prior to switch turn-off, as illustrated at the top of Fig. 24.15(b).

Soft switching can be used to achieve higher circuit efficiency by reducing switching loss and/or to enable the use of higher switching frequencies than otherwise possible. It can also help to reduce EMI associated with high-frequency switching without incurring high snubber losses. It is for these reasons that soft switching is often employed in high-performance designs.

24.6.1 Implementation of Soft Switching

As suggested by Fig. 24.15, soft switching is accomplished either by delaying the rise of switch voltage or current at a transition, or by arranging the switching event to take place at a time when voltage or current will naturally be zero. In some converters (e.g., resonant converters, as described in Chapter 10) soft-switched operation is readily achieved, while in other cases it may require additional circuit or control complexity, impose additional conduction losses and/or peak device stresses, and introduce significant design or operating constraints. Nonetheless, the benefits often make these trade-offs worthwhile.

While many taxonomies for soft-switched converters have been proposed, none is entirely satisfactory. We might loosely categorize designs into those that realize soft switching though the design and control of the main power stage itself, and those that leverage additional circuitry (e.g., operational near the switching transitions) to achieve soft switching. In the former, the power stage design and operation are governed by requirements for achieving soft switching, and circuit control is likewise constrained (e.g., requiring variable-switching-frequency techniques). In the latter, the main power stage may be designed and controlled flexibly (e.g., similar to a conventional PWM converter), but potentially complicated auxiliary circuits are needed to provide soft-switched transitions. In both categories there can be significant limitations in load or voltage ranges over which soft switching is achievable, further complicating design and control.

Example 24.4 ZVS Resonant-Transition Buck Converter

Figure 24.16 shows a zero-voltage-switched resonant-transition buck converter and its operating waveforms. The basic circuit topology is exactly that of a conventional buck converter with a capacitor placed across the switch. The capacitor placement suggests that this converter uses ZVS turn-off and turn-on transitions. The capacitor, which could instead be placed across the diode, includes, and may comprise, switch and diode capacitances. The topology of Fig. 24.16(a) is associated with a general class of converters known as quasi-square-wave converters, but to fully define a soft-switched converter one must also specify its design (e.g., including component value selections) and control.

Figure 24.16(b) shows the converter's operating sequence. At the beginning of a cycle, the switch voltage v_{DS} is ideally zero and the switch turns on with ZVS, beginning phase 1 of the operating cycle. The switch is held on for a time t_{on} until the inductor current reaches a value $i_{L,pk}$. The inductor current i_L ramps up linearly during phase 1, delivering energy from the input to the output and storing energy in the inductor. Once the condition $i_L = i_{L,pk}$ is reached, the switch is turned off and phase 1 ends. Capacitor C_s enables ZVS turn-off of the switch as illustrated in the top of Fig. 24.15(a). During phase 2, inductor L resonates with capacitor C_s, charging it until the diode turns on, beginning phase 3. During phase 3 the inductor current ramps down linearly, delivering the inductor's accumulated energy to the output. Phase 3 ends when the inductor current has ramped down to zero and the diode turns off. During phase 4, inductor L again resonates with capacitor C_s for up to a half of a resonant cycle. Phase 4 persists until either (i) switch voltage v_{DS} reaches zero, at which point the switch body diode turns on, or (ii) v_{DS} rings down for half a resonant cycle to a minimum voltage $V_{\text{in}} - 2V_{\text{out}}$. In case (i), which occurs for voltages $V_{\text{out}} \geq 0.5V_{\text{in}}$, one can turn on the switch with zero-voltage switching, as at the bottom of Fig. 24.15(a). In case (ii), perfect ZVS is lost but the switch can be turned on when v_{DS} is at a minimum (minimizing capacitor discharge loss). A new cycle begins with switch turn-on.

Observing the inductor current in Fig. 24.16(b), we can see that the converter operates similarly to a buck converter at the edge of discontinuous conduction (i.e., with a triangular inductor current swinging between zero and a peak value), but with the slope changes at the peak and valley of the inductor current waveform softened by the resonant transitions. As with a DCM converter, the resonant-transition buck converter requires a relatively small-valued inductor operating at a high ripple current.

The converter may be controlled using either on-time control (by directly setting the switch on-time t_{on}) or current-mode control (by setting an inductor current switching threshold $i_{L,pk}$). Modulating $i_{L,pk}$ provides direct control over the inductor current of the converter, with a cycle-average current of $i_{L,pk}/2$ delivered through the inductor to the output capacitor and load. Current control in this fashion is appealing from a control perspective. Modulating t_{on} achieves approximate control over inductor current and does not require current sensing, which is advantageous for high-frequency operation.

Unlike many PWM converters, the resonant-transition buck converter operates with variable switching frequency. Based on the inductor waveform of Fig. 24.16(b) we can approximate the switching period of this converter as

$$
\begin{aligned}
T &\approx \frac{Li_{L,pk}}{V_{\text{in}} - V_{\text{out}}} + \frac{C_s V_{\text{in}}}{i_{L,pk}} + \frac{Li_{L,pk}}{V_{\text{out}}} + \pi\sqrt{LC_s} \\
&\approx \frac{Li_{L,pk}}{V_{\text{in}} - V_{\text{out}}} + \frac{Li_{L,pk}}{V_{\text{out}}} + \pi\sqrt{LC_s}.
\end{aligned}
\tag{24.23}
$$

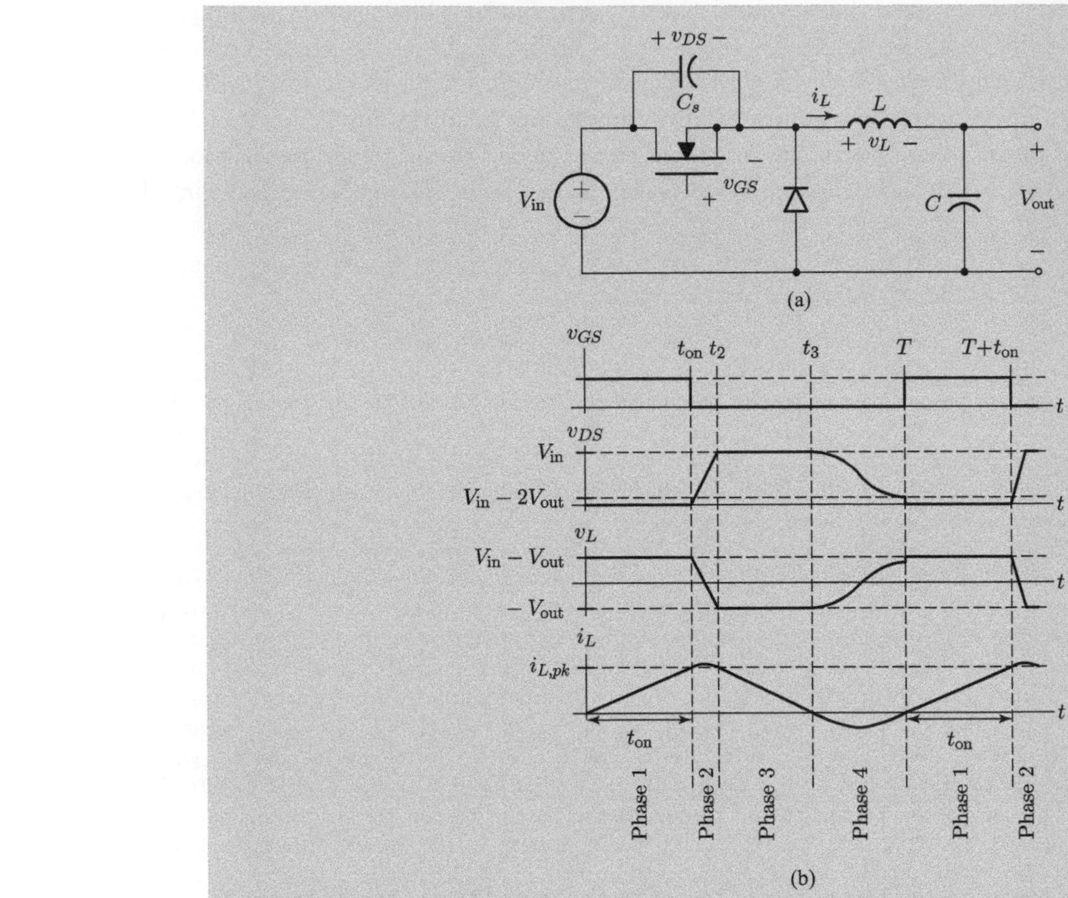

Figure 24.16 (a) A ZVS resonant-transition buck converter. (b) Its switching waveforms.

The switching period is determined by the operating waveforms and depends on both the component values and the current delivered to the output.

This converter offers true zero-voltage switching for a limited operating range $V_{in} \leq 2V_{out}$, but can often be operated with reasonable efficiency somewhat past this range (e.g., up to $V_{in} \approx 3V_{out}$). ZVS turn-on of the switch is often achieved by using a circuit based on a high-speed comparator, to detect the ring-down of the switch drain–source voltage, though response delays through the detector circuit can constrain the achievable operating frequency. Note that – depending on V_{out} – there can also be a minimum allowable current $i_{L,pk}$ to ensure that the voltage fully transitions during phase 2. Consequently, this converter cannot operate with ZVS down to zero output current.

Benefits of this converter include simple structure and control, requiring only a very small-valued inductor, and achieving high switching frequencies at excellent efficiency. Depending on the application, drawbacks include its constrained high-efficiency operating range, its variable-frequency operation, and the need for unusual control circuitry.

The resonant-transition converter of Example 24.4 achieves soft switching by ensuring that the inductor current has the proper polarity and sufficient amplitude at device turn-off to provide resonant transitions in the switch voltage.

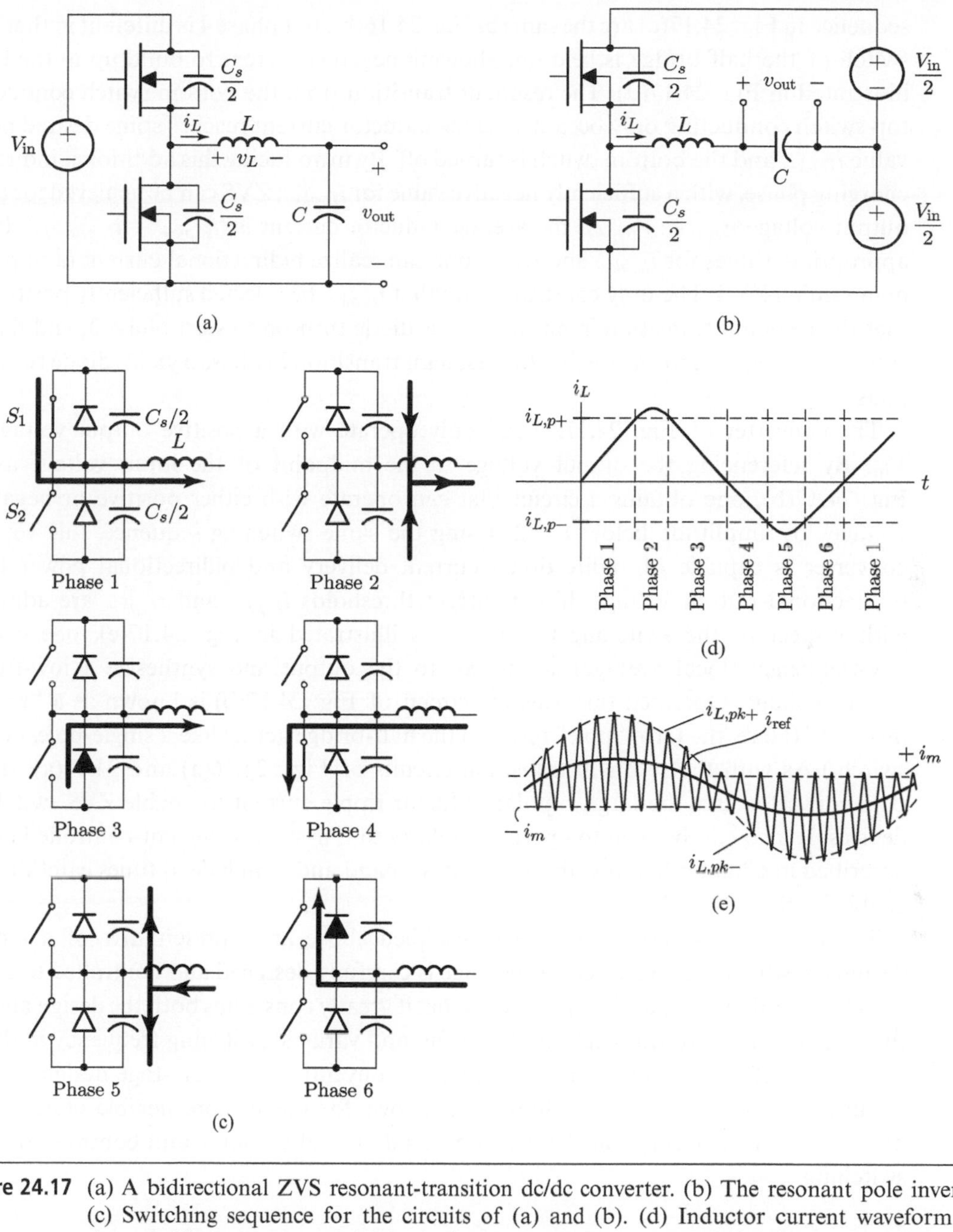

Figure 24.17 (a) A bidirectional ZVS resonant-transition dc/dc converter. (b) The resonant pole inverter. (c) Switching sequence for the circuits of (a) and (b). (d) Inductor current waveform for the switching sequence of (c) for both circuits. (e) Many cycles of inductor current for the inverter of (b).

Variants of this basic resonant transition idea can offer greater flexibility and additional functionality. For example, consider the converter circuit of Fig. 24.17(a) and the associated switching sequence in Fig. 24.17(c). This is similar to the resonant-transition buck converter of Fig. 24.16(a) except that it uses an active device in place of a diode. Phases 1 to 3 of the switching sequence in Fig. 24.17(c) are the same as Fig. 24.16(b), but phase 4 is different in that the bottom switch of the half-bridge is held on, allowing negative current to build up in the inductor, as illustrated in Fig. 24.17(d). The resonant transition from the bottom switch conducting to the top switch conducting only occurs once the inductor current reaches some desired non-positive value $i_{L,pk-}$ and the bottom switch is turned off. By introducing this additional inductor current charging phase, with a sufficiently negative value for $i_{L,pk-}$, ZVS can be achieved for any positive output voltage $v_{\text{out}} < V_{\text{in}}$. As the average inductor current is $(i_{L,pk+} + i_{L,pk-})/2$, by choosing appropriate values for $i_{L,pk+}$ and $i_{L,pk-}$ one can realize bidirectional current (and power) while maintaining ZVS. The only constraints are that $i_{L,pk+}$ be selected sufficiently positive to ensure that the resonant transition in phase 2 yields diode turn-on to start phase 3, and that $i_{L,pk-}$ be sufficiently negative to ensure that the resonant transition in phase 5 yields diode turn-on to start phase 6.

The converter of Fig. 24.17(a) can only operate with a positive output voltage less than V_{in}. By referencing the output voltage to the midpoint of the input voltage as shown in Fig. 24.17(b), one obtains a circuit that can operate with either positive or negative output voltages of amplitude below $V_{\text{in}}/2$. Using the same switching sequence, this version of the converter is capable of bidirectional current delivery and bidirectional power flow into a bidirectional output voltage. If the current thresholds $i_{L,pk+}$ and $i_{L,pk-}$ are adjusted slowly with respect to the switching frequency, as illustrated in Fig. 24.17(e), one can deliver a low-frequency (local average) ac current to the output and synthesize a low-frequency ac output voltage. Operated this way, the circuit of Fig. 24.17(b) is known as a "resonant pole inverter." (Here, the term "pole" refers to the half-bridge acting like a single-pole, double-throw switch.) As with the resonant-transition circuits of Figs. 24.16(a) and 24.17(a), the resonant pole converter uses the large, bipolar inductor ripple current to enable ZVS switching of the devices. It also can be seen to operate similarly to a hysteretic current-controlled converter, as described in Chapter 8, but with the hysteresis band and switch dead times carefully selected to enable ZVS.

The resonant-transition converters described above are representative of a large class of designs in which the converter power stage is carefully designed and controlled to achieve soft-switching. A downside to this approach is that it greatly constrains both the design and control of the system (e.g., requiring high ripple current and variable switching frequency in the examples above.) A different approach is to augment a conventional power stage design with auxiliary circuitry that enables soft switching. This allows for much more flexible design and control of the main power stage, at the expense of additional circuitry and controls to achieve soft switching.

Example 24.5 ZVS Resonant-Transition Boost Converter

Boost converters supplying high output voltages often suffer from large switching loss, limiting achievable efficiency and switching frequency. Significant losses often arise from reverse recovery of the diode and from discharging/charging the switch and diode capacitances when the switch turns on. One approach to address these losses is to use a zero-voltage-transition (ZVT) boost converter. ZVT converters incorporate an auxiliary circuit that enables the main power stage (e.g., a boost converter) to operate with zero-voltage switching.

Figure 24.18 shows a ZVT boost converter. It can be seen as a conventional PWM boost converter (comprising switch Q, diode D, inductor L, and capacitor C) augmented with an auxiliary circuit (comprising snubber capacitor C_s, switch Q_s, diode D_s, and inductor L_s) that is sometimes humorously called a "baby boost" circuit. The auxiliary circuit acts to allow the main power stage to operate with ZVS, while introducing only small additional size and loss. ZVS turn-off of transistor Q is enabled by capacitor C_s placed across it. To enable ZVS turn-on of transistor Q and soft turn-off of diode D, the auxiliary circuit is triggered to operate shortly before the main switch turns on, and it ceases to operate just after the main switch is on.

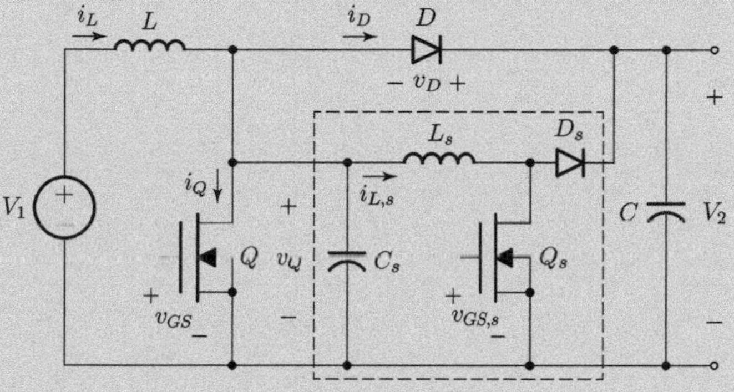

Figure 24.18 The zero-voltage-transition (ZVT) boost converter.

Figure 24.19 shows the operating sequence of the ZVT boost converter. For simplicity, we assume that the main inductor L and main capacitor C are large, so i_L and V_2 are constant over a switching cycle. During phase 1 of operation, diode D carries i_L and the auxiliary circuit is off (with Q_s and D_s off and $i_{L,s} = 0$). To switch the converter state (i.e., to transfer current from D to Q), the auxiliary circuit is triggered by turning on Q_s, beginning phase 2. Q_s is turned on with ZCS as L_s has no current at the start of phase 2. During phase 2, $i_{L,s}$ ramps up linearly while i_D ramps down linearly. When i_D reaches zero, diode D turns off (softly, with low dv/dt), initiating phase 3. During phase 3, $i_{L,s}$ continues to increase as L_s resonates with C_s, resonantly discharging v_Q down to zero. Once v_Q reaches zero, the body diode of Q turns on and clamps the transistor voltage to zero, beginning phase 4. At any point after this, phase 5 can be initiated by turning on Q and turning off Q_s. Thanks to the resonant discharge of C_s by the auxiliary circuit, switch Q turns on with ZVS. By contrast, the turn-off of Q_s is "hard," but as Q_s is only required to operate transiently, it can be small compared to the main device, and switched with low loss. During phase 5, diode D_s conducts and $i_{L,s}$ decreases while switch current i_Q increases. D_s turns off once $i_{L,s}$ reaches zero, beginning phase 6, during which the full inductor current i_L is carried by Q. Phases 2 to 5 thus serve to enable i_L to be commutated from D to Q with soft switching. Once the switch Q has been held on for its desired on-time, it may be turned off with ZVS owing to the presence of C_s. This initiates phase 7, during which C_s is charged up by i_L until diode D turns on, returning the converter to phase 1.

The ZVT boost converter has some important merits: it enables conventional design and PWM control of the converter power stage, while providing for ZVS soft switching of the main devices. Moreover, this is accomplished using only a simple auxiliary circuit that introduces modest additional power stage stress. Nonetheless, while the auxiliary circuit has small-valued components and a very low average power rating, it has high peak voltage and current stresses. Also, while the auxiliary switch is operated with ZCS turn-on, it has a hard turn-off, introducing some amount of switching loss. The ZVT boost converter thus partially addresses boost converter switching loss using only modest additional circuitry.

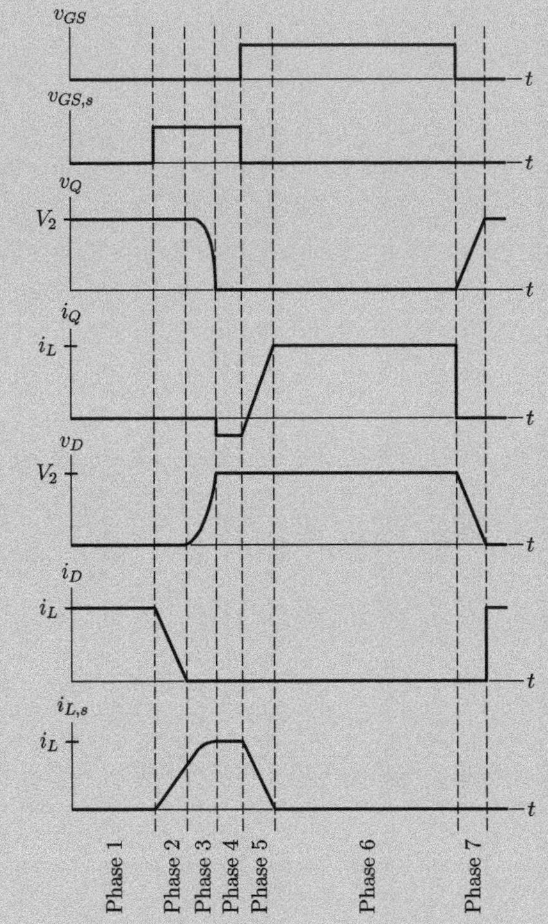

Figure 24.19 Operating sequence of the ZVT boost converter of Fig. 24.18.

The ZVT boost converter of Example 24.5 is representative of a broad class of converters in which auxiliary circuitry is employed to enable soft switching, while preserving desirable design and control aspects of the power stage. Many such approaches exist, differing in the soft switching strategies employed, the degrees of loss reduction achievable, the size and stress of the auxiliary circuitry, complexity, and operating constraints.

This approach is not limited to dc/dc converters, but can be employed in many other converters. For example, Fig. 24.20 shows the circuit of an *auxiliary resonant commutated pole* (ARCP) converter comprising a half-bridge power stage connected to an inductively filtered source or load (represented by inductor L_2) and augmented with an auxiliary circuit for soft switching. With appropriate but somewhat complex control this circuit enables ZVS switching between Q_1 and Q_2 for any polarity and amplitude of current i_2, while providing ZCS switching of the auxiliary devices $Q_{s,A}$ and $Q_{s,B}$. The ARCP thus represents a flexible building block for soft-switched PWM inverters and rectifiers.

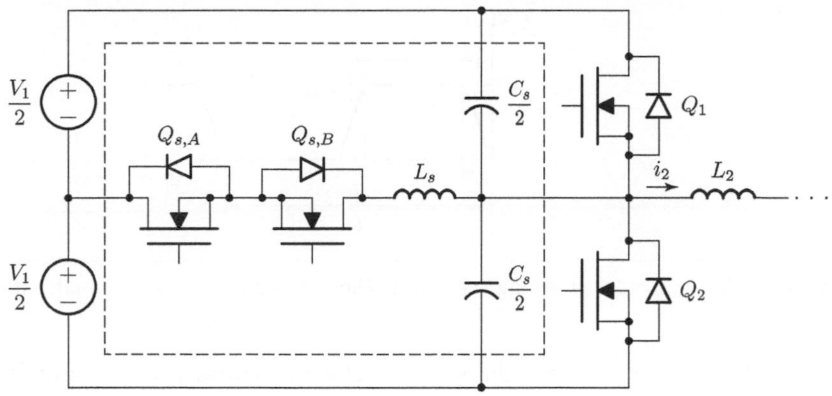

Figure 24.20 The auxiliary resonant commutated pole (ARCP) converter.

As suggested above, there are innumerable ways soft switching can be implemented, with the selected approach closely tied to circuit design and control. To select a suitable approach, one should consider both the application requirements and the type(s) of semiconductor devices being used. Broadly speaking, ZVS (and especially ZVS turn-on) is strongly preferred for circuits operating at high voltages and/or at very high frequencies. This is because ZCS switching cannot eliminate the loss owing to the discharge of intrinsic switch capacitance when a switch is turned on under voltage. ZCS switching (and especially ZCS turn-off) is particulary helpful in high-current designs, in cases where there is large parasitic inductance in series with a device (e.g., owing to packaging), and with device types (such as thyristors) which only turn off at zero current.

The real-world benefits of soft switching are closely linked to semiconductor device characteristics. For example, some devices (e.g., some thyristors) will not switch on properly under zero voltage and thus do not benefit from ZVS turn-on. Moreover, our simplified discussion of ZVS has ignored device output capacitance. In practice, the intrinsic device capacitance will augment, or in some cases replace, C_s. However, device capacitances often come with significant equivalent series resistance (ESR), so even resonant-charging and -discharging them incurs a degree of loss. Worse still, the output capacitances of some devices can exhibit rate-dependent hysteretic charge and discharge losses that are even larger than may be accounted for by ESR alone, especially when operating at extremely high frequencies (e.g., tens of megahertz and above).

24.6.2 EMI Reduction from Snubbers and Soft Switching

One of the benefits of snubbers and soft switching is reduction in EMI arising from switching transitions. The practical amount of EMI mitigation achieved through these methods depends heavily on details of circuit layout, construction, and parasitics, and is often determined experimentally. Nonetheless, it is useful to understand the impact of slowing down voltage and/or current transitions on the frequency content of converter waveforms, as the switching transitions are the source of EMI.

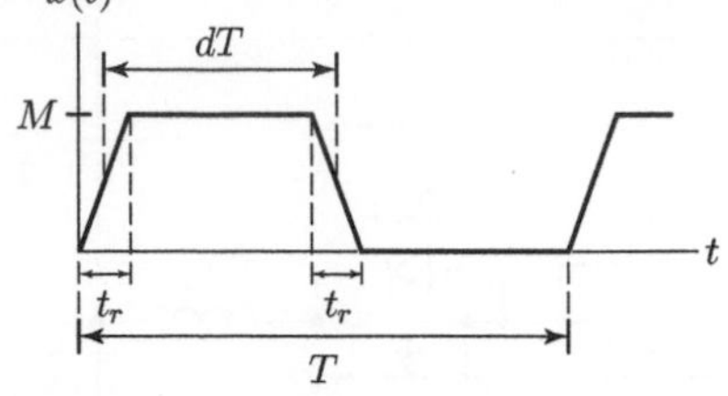

Figure 24.21 A trapezoidal switching waveform. The rise and fall times are equal, with value t_r.

Figure 24.21 shows a trapezoidal waveform having period $T = 1/f_s$ and duty ratio d, with rise and fall times t_r. This waveform can be used to crudely approximate the voltage- and current-switching waveforms of many PWM converters in the absence of parasitics.

A single period of this waveform can be thought of as the result of *convolving* a rectangular pulse of width dT with one of width t_r. The Fourier transform of a single rectangular pulse of width W in time is a sinc function in frequency, with a magnitude that falls off as $1/f$, and with its first null at $1/W$. The transform of the convolution of two such pulses is the *product* of the individual transforms, and has a magnitude that falls off as $1/f^2$ in frequency.†

It should then not be surprising to know that for the periodic trapezoidal waveform in Fig. 24.21, the magnitude of the nth Fourier harmonic is proportional to

$$\left(\frac{\sin(n\pi d)}{n\pi d}\right) \cdot \left(\frac{\sin(nf_s\pi t_r)}{nf_s\pi t_r}\right). \tag{24.24}$$

For $nf_s t_r \ll 1$, the second term is approximately 1 (i.e., the value of a sinc function near the origin), and the envelope of the first term falls off as $1/n$, or 20 dB/decade in frequency. As $nf_s t_r$ exceeds 1, for frequencies $f > 1/t_r$, the envelope of the second term also starts to fall off as $1/n$, contributing an additional 20 dB/decade of roll-off in frequency. More generally, smoother transitions in time lead to faster decay of spectral content at high frequencies.

Notes and Bibliography

The snubber literature is extensive, and the references listed here represent only a cross section of papers addressing important issues. McMurray's paper [1] is the classic snubber reference. Criteria and detailed analytic techniques for designing conventional *RLC* snubbers are presented in [1], [2], and [3]. Only circuits resulting in second-order behavior are considered in [1] and [2], but [3] considers the effects of parasitic capacitances, such as those associated with junctions and transformer windings, and presents a technique for analyzing the resulting third-order network. An in-depth treatment of snubbers, including their history, categorization, design, and numerous examples, can be found in the book by Severns [4] (free download). Severns also provides a deep bibliography on the topic.

† More generally, if all time-derivatives of the waveform up to order k are well-behaved non-impulsive functions, then the magnitude falls off as $1/f^{k+1}$; see W. M. Siebert, *Circuits, Signals, and Systems*, MIT Press, 1985.

The literature on soft-switching techniques is likewise extensive. More detail about the example approaches shown in this chapter can be found in [5]–[9], including for resonant-transition converters, [5] and [6], the resonant pole inverter, [7], zero-voltage-transition converters, [8], and the auxiliary resonant commutated pole inverter, [9]. Numerous additional approaches to soft switching are overviewed in [10] and [11]. Resonant converters are also designed to implement soft switching, as are inverters for radio-frequency applications; these classes of converter are examined in Chapter 10.

Device considerations for soft-switched converters and treatment of device capacitance losses may be found in [12] and [13], and references therein. The impact of finite rise/fall times on waveform spectral content is detailed in [14].

1. W. McMurray, "Optimum Snubbers for Power Semiconductors," *IEEE Trans. Industry Applications*, vol. 8, pp. 593–600, Sept./Oct. 1972.
2. W. McMurray, "Selection of Snubbers and Clamps to Optimize the Design of Transistor Switching Converters," *IEEE Trans. Industry Applications*, vol. 16 pp. 513–523, July/Aug. 1980.
3. K. Harada, T. Ninomiya, and M. Kohno, "Optimum Design of RC Snubbers for Switching Regulators," *IEEE Trans. Aerospace and Electronic Systems*, vol. 15, pp. 209–218, Mar. 1979.
4. R. P. Severns *Snubber Circuits for Power Electronics*, Rudy Severns, https://rudys.typepad.com/files/snubber-e-book-complete.pdf, 2008.
5. C. P. Henze, H. C. Martin, and D. W. Parsley, "Zero-Voltage Switching in High Frequency Power Converters Using Pulse Width Modulation," *IEEE Power Electronics Specialists Conference (PESC)*, pp. 33–40, Kyoto, Apr. 1988.
6. V. Vorperian, "Quasi-Square-Wave Converters: Topologies and Analysis," *IEEE Trans. Power Electronics*, vol. 3, pp. 183–191, Apr. 1988.
7. D. Divan and G. Skibinski "Zero Switching Loss Inverters for High Power Applications," *IEEE Industry Applications Society Annual Meeting*, pp. 627–634, Atlanta, Oct. 1987.
8. G. Hua, C.-S. Leu, Y. Jiang, and F. C. Lee, "Novel Zero-Voltage-Transition PWM Converters," *IEEE Trans. Power Electronics*, vol. 9, pp. 213–219, Mar. 1994.
9. R. W. De Doncker and J. P Lyons, "The Auxiliary Resonant Commutated Pole Inverter," in *IEEE Power Electronics Specialists Conference (PESC)*, pp. 248–253, Boston, June 1991.
10. M. D. Bellar, T.-S. Wu, A. Tchamdjou, J. Mahdavi, and M. Ehsani, "A Review of Soft-Switched DC–AC Converters," *IEEE Trans. Industry Applications*, vol. 34, pp. 847–860, July/Aug. 1998.
11. S. A. Q. Mohammed and J.-W. Jung, "A State-of-the-Art Review on Soft-Switching Techniques for DC–DC, DC–AC, AC–DC, and AC–AC Power Converters," *IEEE Trans. Industrial Informatics*, vol. 17, pp. 6569–6582, Oct. 2021.
12. G. Zulauf, S. Park, W. Liang, K. N. Surakitbovorn, and J. M. Rivas-Davila, "C_{OSS} Losses in 600 V GaN Power Semiconductors in Soft-Switched, High- and Very-High-Frequency Power Converters," *IEEE Trans. Power Electronics*, vol. 33, pp. 10748–10763, Dec. 2018.
13. G. Zulauf, Z. Tong, J. D. Plummer, and J. M. Rivas-Davila, "Active Power Device Selection in High- and Very-High-Frequency Power Converters," *IEEE Trans. Power Electronics*, vol. 34, pp. 6818–6833, July 2019.
14. H. W. Ott, *Electromagnetic Compatibility Engineering*, Chichester: Wiley, 2009.

PROBLEMS

24.1 In this chapter we assumed that, during a switch transition, the voltage and current waveforms of a transistor without a snubber changed linearly from their starting values to their final values. For this problem, assume that these waveforms change quadratically, that is, the *rate* at which the waveforms change grows linearly with time. If the total time required to complete the transition (t_f or t_r) is the same as in the linear case, what happens to the amount of energy dissipated in the transistor when these alternative shapes are assumed?

24.2 In some power circuits, the load does not have the large series inductor (modeled as the current source I_{dc}) shown in Fig. 24.2. In such a case the current drawn by the load changes during the switch transitions. Draw the switch waveforms for both the turn-on and turn-off transitions when the load is purely resistive and no snubbers are used. How much energy is dissipated in the transistor during each transition?

24.3 We added the resistor R_s in Fig. 24.6 to the basic turn-off snubber of Fig. 24.4(a) to limit the discharge current when Q turns on. We then added D_s to bypass this resistor during turn-off, giving the complete turn-off snubber of Fig. 24.6. The purpose of this problem is to illustrate the importance of D_s. Assume the circuit parameters of Example 24.1, with $R_s = 5\,\Omega$, and determine and sketch v_Q during turn-off if D_s is not present. Compare this waveform to that of v_Q in Fig. 24.4(b), which is the same as v_Q at turn-off in the circuit containing D_s, in Fig. 24.6.

24.4 Consider the buck converter with a turn-off snubber shown in Fig. 24.6. Assume the following parameters: $V_{dc} = 400\,\text{V}$, $I_{dc} = 50\,\text{A}$, $C_s = 0.25\,\mu\text{F}$, and $R_s = 5\,\Omega$. If the transistor is turned on for only $2\,\mu\text{s}$, sketch and dimension v_Q during the turn-off transition. Also sketch the locus of the turn-off transition on the i_Q–v_Q plane.

24.5 Determine and sketch the voltage v_Q at turn-off for the circuit of Fig. 24.9, assuming that the element values are those determined in Example 24.2.

24.6 For the combined turn-on/turn-off snubber shown in Fig. 24.10, we stated that of the total energy stored in the snubber capacitor, given by (24.19), part was dissipated in the resistor and part was returned to the input voltage source. Show that this is true so long as the transistor does not turn on before the capacitor has had a chance to discharge to V_{dc}.

24.7 Consider the buck converter with only a turn-on snubber inductor shown in Fig. 24.8. Assume that $V_{dc} = 400\,\text{V}$, $I_{dc} = 50\,\text{A}$, and $L_s = 16\,\mu\text{H}$. Without an alternative path for the snubber inductor current to flow, we stated that when the transistor turns off, an impulse of voltage appears across the inductor and, correspondingly, the transistor. Actually, the transistor voltage quickly reaches its avalanche breakdown level, at which point the snubber inductor current can continue to flow through the transistor. Assume that this transistor breakdown voltage is 500 V. Sketch as a function of time the snubber inductor's voltage and the transistor's voltage and current waveforms. When does the transistor come out of avalanche? How much energy is dissipated in the transistor while it is avalanching?

24.8 Determine and draw, as was done in Fig. 24.10(c), the equivalent circuits for the combined snubber of Fig. 24.11(a).

24.9 The combined snubber of Fig. 24.10(a) is constructed using the values for V_{dc}, I_{dc}, L_s, and C_s from Example 24.3.

(a) What is the minimum value of R_s that gives the desired overdamped response after t_2?

(b) Redraw the waveforms of Fig. 24.10(b) to scale. That is, calculate and identify the times t_1 and t_2, and also calculate and identify the time at which v_Q first passes through V_{dc} after t_2.

(c) Using duality, repeat (a) and (b) for the combined snubber of Fig. 24.11(a).

24.10 In high-frequency converters, the nonlinear junction capacitance of diodes and transistors often forms part of a turn-off snubber. Consider the buck converter of Fig. 24.4(a) in which the capacitor C_s is nonlinear. Specifically, its incremental capacitance is

$$C_{sl} = \frac{dQ}{dv} = C_o\sqrt{\frac{V_{dc}}{v_C}}, \tag{24.25}$$

where Q is the charge in the capacitor and C_o is the value of C_{si} when the voltage across the capacitor v_C equals V_{dc}. Assume that $V_{dc} = 400$ V, $I_{dc} = 50$ A, and $C_o = 0.1$ μF. In addition, assume that, when the transistor turns off at $t = 0$, its current drops linearly to zero in a time $t_f = 0.25$ μs.

(a) Sketch the waveform of v_Q during turn-off.

(b) What is $\gamma_v = v_C(t_f)/V_{dc}$?

(c) Calculate the energy stored in C_s when $v_C = V_{dc}$.

(d) If the snubber capacitor were linear, what value of capacitance would give the same γ_v? Compare the energy stored in this linear capacitor when $v_C = V_{dc}$ to your answer for (c).

24.11 Consider a converter with RCD turn-off snubber and having switch turn-off waveforms as shown in Fig. 24.5. While inclusion of the snubber reduces turn-off loss in the device, it incurs loss in the resistor due to dissipation of the energy stored in C_s.

(a) Show that for snubber capacitances where $\gamma_v < 1$, total loss (in the switch and in R_s combined) increases monotonically with C_s, or equivalently decreases monotonically with γ_v.

(b) Show that the total switching loss exceeds that without a snubber for $\gamma_v < 3 - \sqrt{6} \approx 0.55$.

24.12 The resonant transition buck converter of Fig. 24.16(a) is only one of a family of ZVS resonant-transition dc/dc converters.

(a) Formulate the structure of a resonant-transition boost converter having a single active switch and a diode.

(b) Sketch the operating waveforms for the resonant-transition boost converter following the same format as shown for the resonant-transition buck converter in Fig. 24.16(b).

(c) For what output voltage range can this converter achieve ZVS turn-on of the active switch?

24.13 Consider the resonant pole inverter of Fig. 24.17(b), its operating sequence as illustrated in Fig. 24.17(c), and its inductor current as shown in Fig. 24.17(d). The top switch is turned on until inductor current i_L rises to a positive value $i_{L,pk+}$. The top switch is then turned off (under ZVS conditions) and the resonant capacitors ring with the inductor until the bottom diode conducts, at which point the bottom switch can be turned on (under ZVS conditions). The other half of the cycle is essentially the reverse of the first half.

(a) Derive the minimum current $i_{L,pk+}$ that will enable zero-voltage turn-on of the bottom device. Express this minimum current as a function of C_s, L, V_{in}, and v_{out}. You may assume that v_{out} does not change during a switching cycle.

(b) Calculate the turn-off loss of the top switch, assuming that at turn-off the current in the switch falls linearly to zero in a time t_f. As this converter is designed to operate with zero-voltage switching, you may assume the switch current reaches zero before the switch voltage rises to the bus voltage V_{in}.

24.14 The ZVT boost converter of Fig. 24.18 uses a "baby boost" auxiliary circuit to enable soft switching of the main power stage devices. Figure 24.22 shows a corresponding ZVT buck converter incorporating a "baby buck" auxiliary circuit to create zero-voltage switching of the main power devices. Devise an operating sequence for the switches in Fig. 24.22 to attain zero-voltage turn-on and turn-off of buck switch Q and zero-current turn-on of auxiliary switch Q_s. Plot the waveforms of this ZVT buck converter in a manner similar to that done for the ZVT boost converter in Fig. 24.19.

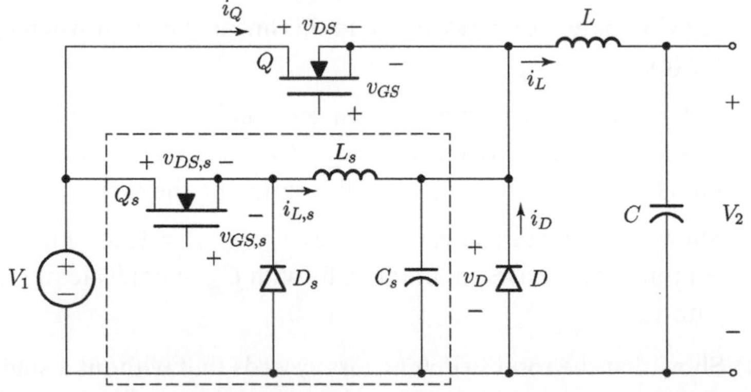

Figure 24.22 A zero-voltage-transition (ZVT) buck converter.

25 Thermal Modeling and Heat Sinking

An unfortunate consequence of our preoccupation with things electrical is that the problems of heat sinking and thermal management are frequently ignored until forced on us by sound, sight, or smell. The insatiable need to make things smaller – and the possibility of doing so by using higher frequencies and new components and materials – aggravates the problem of heat transfer, because such improvements in power densities are seldom accompanied by corresponding improvements in efficiency. Thus we are stuck with the task of getting the same heat out of a smaller volume while disallowing any increase in temperature.

The diversity of heat sources within power electronic apparatus produces a challenging cooling problem. Unlike signal processing circuits – where most heat-generating components come in a common, small, and low-profile rectangular package – energy processing circuits contain components of odd shapes and orientations. Even those of the same type come in many different forms and packages. Inductors, for instance, can be small or large, round or rectangular, and with loss dominated by core or copper. Each possesses special requirements and presents a unique thermal problem. The task of integrating these parts into a reliable piece of equipment becomes as much a thermo-mechanical challenge as the circuit design was an electrical challenge.

Heat transfer occurs through three mechanisms: conduction, convection, and radiation. In conduction the heat transfer medium is stationary, and heat is transferred by the vibratory motion of atoms or molecules. Convective heat transfer occurs through mass movement – the flow of a fluid (gas or liquid) past the heat-generating apparatus. In *natural convection*, the buoyancy created by temperature gradients causes the fluid to move; in a *forced-convection* system, the mass flow is created by pumps or fans.

Heat transfer by radiation turns the heat energy into electromagnetic radiation, which is absorbed by other elements in the environment. Radiation as a mechanism of heat transfer is important for space applications but less so for terrestrial power electronic systems. Heat transferred through radiation is a function of the temperatures, T_S and T_R respectively, of the radiating element's surface and the receiving surface (which may be a surface in the surrounding environment at a remove from the hot component). Specifically, $Q_{\text{rad}} \propto T_S^4 - T_R^4$. Radiation may be important in equipment where this difference is large. However, the strong nonlinearity of this relationship and the relatively low incidence of its importance do not justify the complexity of considering radiation in detail here. Therefore, in this chapter we focus on conduction and convection.

If only these two mechanisms are considered, the design will be conservative, as any heat transferred through radiation will reduce the temperature of the apparatus below the design temperature. The exception is in enclosures, where radiation from hot components may be absorbed by those at lower temperatures, causing these latter components to operate at higher temperatures than anticipated. In such cases, radiation shields – shiny metal partitions – can be employed to isolate the offending or affected components.

The material in this chapter will not give you novel ideas for designing heat transfer systems. The problem is too application specific to permit such a discussion to be of value. Instead, we first describe the parameters governing the performance of any such system. Then we consider the modeling of both steady-state and transient thermal behavior, as applicable to power electronic systems. Some straightforward examples of specific designs will be presented to illustrate the discussions.

25.1 Static Thermal Models

Circuit theory is the *lingua franca* of engineering for good reason. The elegance and simplicity of its canonical formulations (for example, KCL and KVL) permit complex problems to be approached in an organized way, and the insights gained through such formulations are extremely valuable in predicting system behavior. Therefore, many engineering problems in contexts other than electrical engineering – particularly in the setting of "flows" driven by "gradients" – are cast in terms of circuit models before being analyzed. One of these contexts is heat transfer.

25.1.1 Analog Relations for the DC Case

The rate at which heat energy is transferred by conduction from a body at temperature T_1 to another at temperature T_2 is denoted by Q_{12}. It is well modeled as linearly proportional to the temperature difference between the two bodies, $T_1 - T_2$, and inversely proportional to a physical parameter called the *thermal resistance* between them, R_θ:

$$Q_{12} = \frac{T_1 - T_2}{R_\theta} = \frac{\Delta T}{R_\theta}. \tag{25.1}$$

The analogy with Ohm's law is evident, and we can make the following assignment of variables:

$$T_{1,2} \Longleftrightarrow v_{1,2}, \qquad Q_{12} \Longleftrightarrow i, \qquad R_\theta \Longleftrightarrow R. \tag{25.2}$$

Note that the analog of thermal power is i, not vi. If heat leaves body 1 only through the interface characterized by R_θ, then i is not only analogous to Q_{12}, but, because we are considering only steady-state conditions, it also represents the rate at which energy is being converted to heat in body 1. In the context we are interested in, body 1 would be a packaged electrical network, and Q_{12} would represent the rate at which electrical energy is being converted to heat (dissipated) in the package, that is, $p_{\text{diss}} \Longleftrightarrow i$. The thermal management problem is to design a heat transfer system (that is, R_θ) that constrains ΔT to the value dictated by component ratings and ambient conditions.

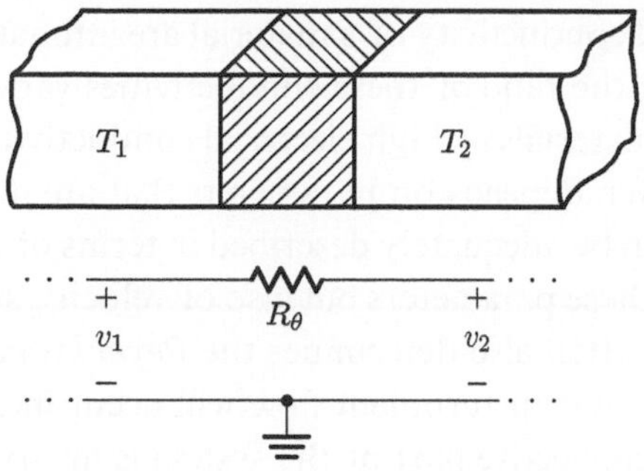

Figure 25.1 Electrical analog of a simple two-body thermal system. The cross-hatched region is the thermal interface characterized by the longitudinal thermal resistance R_θ.

Figure 25.1 illustrates the electrical analog for the simple two-body system just discussed. The bodies are at temperatures T_1 and T_2 and are connected thermally through the cross-hatched interface, which can be characterized by a thermal resistance of value R_θ. If the units of T are °C and the units of Q are watts (W), then thermal resistance has the units °C/W. As with electric circuits, where parallel resistances can be combined into a single equivalent resistance, parallel thermal paths can be characterized by thermal resistances and combined into an equivalent single thermal resistance.

Convection is the mechanical transport of heat by a moving fluid. The fluid (air, for instance) can move because of gravitational forces caused by density gradients, in which case the process is called natural convection, or the fluid can be driven (perhaps by a fan), resulting in what is called forced convection. Convection is a somewhat more complex process than conduction and can be described by the relation

$$Q_{12} = h(\Delta T, v)A(T_1 - T_2), \tag{25.3}$$

where v is the fluid velocity. The parameter $h(\Delta T, v)$ is termed the *film coefficient of heat transfer*; it depends on fluid velocity and the difference between the inlet and outlet temperatures. The cross-sectional area of the interface is A. Over the range of temperature differentials of interest in our application, h is fairly constant. With respect to fluid velocity, significant changes in h occur when the flow regime changes from laminar to turbulent. Within each regime, however, h improves only slowly with increased velocity. For forced convection, h is independent of ΔT. Within these limits, the product hA may be modeled as constant, giving to (25.3) the same form as (25.1), with $R_\theta = 1/hA$. Thus the electrical analog shown in Fig. 25.1 is appropriate for representing convective as well as conductive heat transfer.

25.1.2 Thermal Resistance

As we have just shown, a thermal resistance can be used to model both conductive and convective heat transfer. The physics governing thermal conduction is much like that for electrical conduction, and the thermal resistance or conductance can be described in terms of parameters abstracted from the physics of the process (for example, conductivity) and geometry. In fact, thermal and electrical conductivity of a material are intimately related by the *Wiedemann–Franz law*, which states that the ratio of these conductivities varies linearly with temperature T (so is fixed at a given T) – materials of high electrical conductivity are also good thermal conductors.

Convection, however, depends on parameters that are not so easily abstracted. For instance, while conductivity can be adequately described in terms of material type and temperature, h is a function not only of these parameters but also of velocity, surface characteristics, and geometry. The latter parameter often also determines the *Reynolds number*, a dimensionless number that indicates whether laminar or turbulent flow will occur in a particular situation. Furthermore, the geometry of the convective part of the system is invariably complex, consisting quite often of a finned aluminum extrusion. Thus the equivalent thermal resistance model for convective transfer from a specific heatsink type in a variety of environments (for example, forced or natural convection) is tabulated by the manufacturer. For this reason, the following discussion is directed at determining the equivalent thermal resistance for those parts of the system where heat transfer is by conduction.

The analog of electrical resistivity (Ω-m, or more commonly, Ω-cm) is thermal resistivity, ρ_θ, in units of °C-cm/W (or K-cm/W, as Kelvin is often used instead of Celsius). In terms of ρ_θ and physical dimensions, the longitudinal thermal resistance of a piece of material of cross-sectional area A and length l is

$$R_\theta = \frac{\rho_\theta l}{A} \ {}^\circ\text{C/W}. \tag{25.4}$$

An alternate definition of thermal resistance, used for sheet material, incorporates the sheet thickness. The conductance through a sheet of unit area and specified thickness is given by the parameter h_c having units of W/K-cm^2. Therefore the resistance of an area A of the sheet is

$$R_\theta = \frac{1}{h_c A} \ {}^\circ\text{C/W}. \tag{25.5}$$

The thermal resistivities of various materials used in heat transfer paths in electronic equipment are shown in Table 25.1. Mylar, and less commonly mica, is used to provide electrical isolation between electrically hot components (for example, the semiconductor device package and the heatsink). Mica has a much higher dielectric strength, is more impervious to mechanical puncture, and can be cleaved to produce thinner sheets than Mylar, but is more expensive. Beryllia (BeO) and alumina (Al_2O_3), and recently aluminum nitride (AlN), are also used to provide electrical isolation, most frequently within device packages.

Silicone grease impregnated with metallic oxides, such as ZnO_2, is used to fill imperfections such as scratches on mating surfaces in a heat transfer path – between the bottom of a device package and the top of a heatsink, for instance. The need to fill these voids with something other than air is apparent from the table. Filled silicone grease is also referred to as *thermal grease* or *thermal compound* (or "goop," for reasons that become clear when you use it). Anodizing is frequently used to create an attractive or black surface on aluminum components. Since the resulting oxide is very thin, it contributes little to the thermal resistance of a path. Although the oxide is a good insulator, it is unwise to rely on it in lieu of a dielectric material for providing electrical isolation between surfaces.

Table 25.1 Thermal resistivities of materials used in electronic equipment.

Material	Thermal resistivity (°C-m/W)
Still air	30.50
Mylar	6.35
Silicone grease	5.20
Mica	1.50
Filled silicone grease	1.30
Filled silicone rubber	1.0
Alumina (Al_2O_3)	0.06
Silicon	0.012
Beryllia (BeO)	0.01
Aluminum nitride (AlN)	0.0064
Aluminum	0.0042
Copper	0.0025

Example 25.1 A Calculation Using An Electrical Analog

Figure 25.2(a) shows a resistor embedded in the center of a 10 cm-long block of aluminum whose ends are at temperature T_A. What is the temperature of the resistor if it is dissipating 50 W and the ambient temperature is $T_A = 75\,°C$?

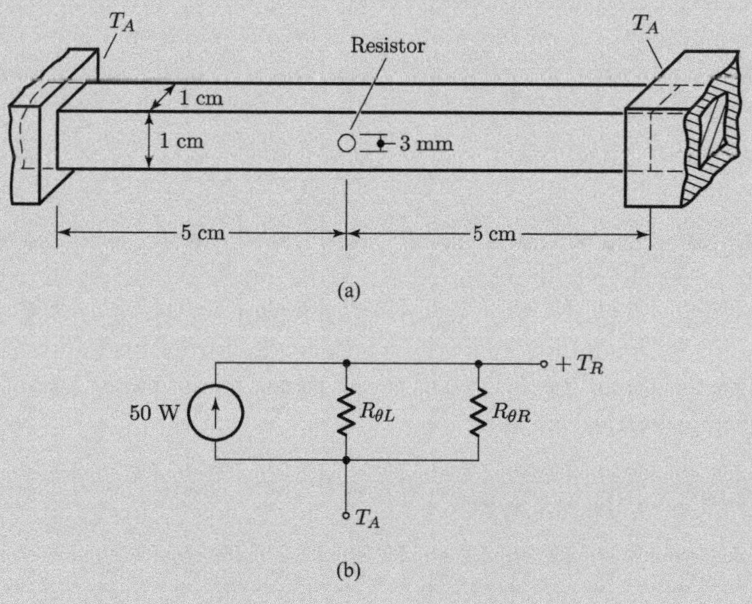

Figure 25.2 (a) A thermal system consisting of a resistor embedded in the center of an aluminum block. (b) The electric circuit analog for the thermal system of (a).

As the length of the block (10 cm) is much longer than the radius of the resistor (3 mm), we can assume that the detailed pattern of heat flow in the vicinity of the resistor is unimportant. The resulting analog circuit model is shown in Fig. 25.2(b), where $R_{\theta L}$ and $R_{\theta R}$ are the thermal resistances of the aluminum bar to the left and right of center. The value of these resistances is

$$R_{\theta L} = R_{\theta R} = \frac{(0.42)(5)}{1} = 2.1\,°\text{C/W}. \tag{25.6}$$

At a dissipated power of 50 W, the temperature of the resistor, T_R, is

$$T_R = 75 + 50\left(\frac{2.1}{2}\right) = 127.5\,°\text{C}.$$

25.2 Thermal Interfaces

A critical part of any heat transfer system is the interface between mechanical components in the thermal path. Some issues related to these interfaces in the context of our application were raised in the previous section. The geometry of most interfaces can be modeled as two parallel planes with material of a specific thermal resistivity between. If the material is of thickness δ and of area A, the thermal resistance of the interface between the planes is

$$R_{\theta i} = \frac{\rho_\theta \delta}{A}. \tag{25.7}$$

Consider, for example, a device in a TO-220 package mounted on a heatsink. Between the device and heatsink is a 1.6 mm alumina pad because electrical isolation between the device and heatsink is required. The mating surface area of a TO-220 package is approximately 0.95 cm^2, giving a thermal resistance between the case and the sink of

$$R_{\theta CS} = \frac{(6)(0.16)}{0.95} = 1.01\,°\text{C/W}. \tag{25.8}$$

Thus, the difference between the case and sink temperatures increases by 1.01 °C for each watt of thermal power being transported across the interface. A dissipation of 15 W is not unusual for a device in a TO-220 package, giving a temperature rise of 15.2 °C across just the alumina interface. This amount, which does not include the thermal resistance of the interfaces between the alumina pad and the TO-220 case or heatsink, is significant, and illustrates the price paid for requiring electrical isolation.

Example 25.2 A Thermal System

The physical structure depicted in Fig. 25.3 is typical of the thermal system that results from mounting a semiconductor die. The device itself is bonded to the header using solder or epoxy; the header is made part of a package that is mounted to a heatsink (perhaps with some intervening insulating material), and the heatsink is thermally connected to the ambient environment, generally through free or forced (fan) convection. A highly detailed model for this system is shown in Fig. 25.4(a), where each part of the system

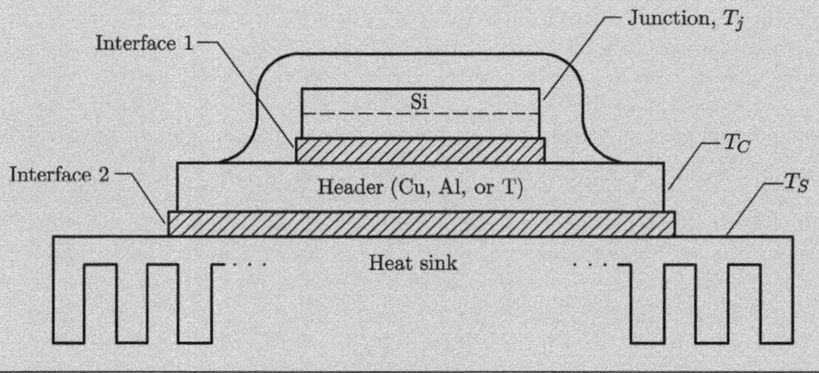

Figure 25.3 Typical mechanical structure used for mounting semiconductor die.

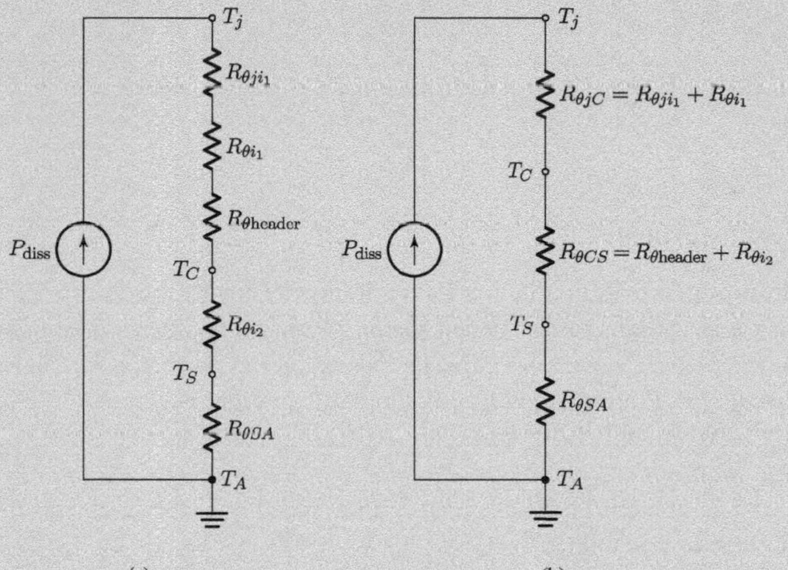

Figure 25.4 (a) A static thermal model for Fig. 25.3. (b) A simplified model of the circuit in (a).

is explicitly represented by its equivalent thermal resistance. The model also shows the relationship between certain node voltages and the temperatures they represent. The current source represents the rate at which electrical energy is dissipated in the device, P_{diss}. The physical location within the device of this dissipation is the node to which the source is connected, the junction in this case.

Some of the thermal resistances shown in Fig. 25.4(a) are so small relative to others that they can be neglected. Because the header is made of copper or aluminum, its vertical thermal resistance is negligible, as is that of the thermal grease (assuming that it is applied properly, which it often is not!). Others of the identified resistances are frequently lumped together, such as those for the silicon and bonding material, which are generally inaccessible to the circuit designer. Implicit in the element $R_{\theta SA}$ are the thermal resistance of the sink extrusion between the region on which the package is mounted and the surfaces from which heat is being removed by convection, and the thermal resistance representing the convection process. Figure 25.4(b) shows the simplified model.

The variable of interest in the models of Fig. 25.4 is T_j, the "junction" temperature of the device. The term "junction" is used rather loosely to represent in lumped form the source of heat in the device. In reality, this source seldom exists as a simple plane. In the MOSFET it is not a junction at all. Nevertheless, the term persists, and manufacturers determine $R_{\theta jC}$ empirically, which takes into account the actual geometry of the heat-producing region. To determine T_j, we need to know P_{diss} as well as the thermal resistances between the junction and the ambient environment.

The dissipation in the device is a function of its electrical environment (for example, its current, voltage, and switching loci). For the purposes of this example, we assume that these calculations have been made for our device, and that the result is $P_{\text{diss}} = 25$ W. The physical configuration is a TO-247 package mounted on an extruded, finned, free convection-cooled sink without any insulating interface but with thermal grease. Typical values of the thermal resistances are $R_{\theta jC} = 1.1$, $R_{\theta CS} = 0.12$, and $R_{\theta SA} = 1.8$, all in units of °C/W. The last parameter needed is the ambient temperature, which is not "room temperature," but that of the air in the vicinity of the sink. We take it to be 40 °C. We determine the temperature drop between nodes in the model by using Ohm's law, obtaining

$$T_S = 40 + (25)(1.8) = 85\,^\circ\text{C}, \qquad \text{T}_\text{C} = 85 + (25)(0.12) = 88\,^\circ\text{C}, \qquad \text{T}_\text{j} = 88 + (25)(1.1) = 115.5\,^\circ\text{C}.$$

Is this an adequate design? The answer depends on the type of device being cooled. If it were a Si MOSFET, the design gives a good margin between predicted junction temperature and typical maximum limits of 150 °C. For a thyristor the design is marginal.

25.2.1 Practical Interfaces

Mechanical interfaces are not parallel in practice. They contain surface imperfections, such as scratches, and a characteristic called *run-out*. Run-out is the maximum deviation from flatness that a surface exhibits over a specified lateral distance. It is measured in (linear dimension)/(linear dimension), for example, cm/cm. A standard aluminum extrusion exhibits run-out that is typically 0.001 cm/cm. Both scratches and run-out degrade the thermal performance of an interface. Run-out is generally not under the design engineer's control. Therefore we must measure or estimate it and make proper allowance for it. However, run-out is seldom an issue with modern commercial heatsinks.

The use of thermal grease has already been mentioned. It is designed to reduce the degrading effects of surface scratches and other small imperfections, but is not designed to remedy the effects of run-out. Although our primary focus is on basic principles, thermal grease is so frequently misused that a brief departure from "principle" to "practice" is justified. The problem arises from a belief that if a little is good, a lot is better. However, the art of applying thermal grease is much like that of watering plants – too much and it's dead. Silicone grease is highly viscous and refuses to "squish out" when squeezed between header and sink by mounting hardware. In such cases, a thin layer of grease can remain in the interface, giving rise to a significant thermal resistance that was not anticipated in the thermal design. The grease should be applied sparingly, and then wiped off, removing almost all traces. Thermal grease oozing from under device packages is a sign of poor construction and potential thermal problems.

When electrical isolation between device and heatsink is required, a pad made of silicone (or other conformable material) can be used as an interface material. These "squishy" materials fill scratches and other surface imperfections when mounting pressure is applied and are available

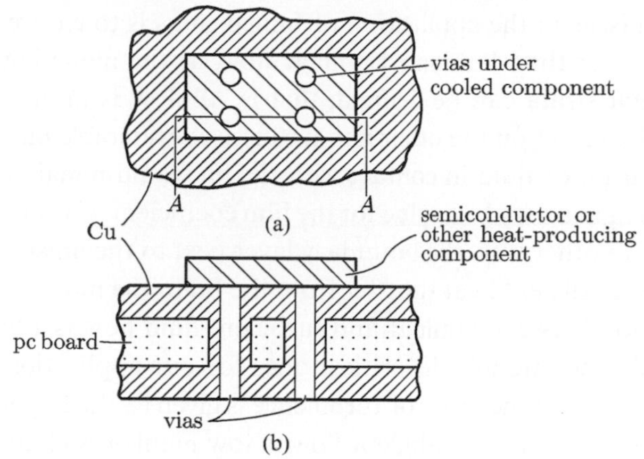

Figure 25.5 An illustration of the use of vias as heatsinks for components mounted directly to a PC board. (a) Top view of device showing the location of four vias. (b) Cross section A–A illustrating the copper plating in the vias.

as sheets, or in shapes conforming to most device package geometries. A unique consideration when using these pads is that the resulting thermal resistance between the two surfaces is a function of mounting pressure. The relationship among $1/h_c$, thickness, and pressure is usually provided by the manufacturer.†

The printed circuit board (PCB) presents unique thermal design problems. One can mount on the board heatsinks to which are attached the semiconductor devices; however, this occupies valuable board real estate. If the thermal requirements are not too severe, a common approach is to mount the device directly on the board over vias‡ that thermally connect the upper and lower layers of copper foil. The device is thermally connected to the upper foil layer, which spreads and dissipates the heat as well as transferring heat through the vias, where it is spread on the lower foil layer. This heat sinking method for PC boards is illustrated in Fig. 25.5.

25.2.2 Convective Interface

Even though we showed that both conduction and convection processes could be modeled by similar electrical analogs, our discussion so far has focused on conduction interfaces. However, all conduction actually leads to a convective interface. Heat is removed from a conventional finned sink by air flowing over the fins. A more sophisticated system incorporating a heat exchanger probably uses a liquid to move the heat from one place to another. These are convective processes. As mentioned earlier, the physics governing these processes is beyond our scope here. However, a short discussion of the application of finned sinks is helpful.

† Manufacturers of sheet material often use the term *thermal impedance* instead of resistance to denote $1/h_c$.

‡ A via is a hole through the board, the interior wall of which has been plated with copper. It may or may not connect different layers of copper foil. Vias may optionally be filled with copper or another material to reduce thermal resistance.

The critical issue in the application of finned sinks is to ensure that air flow through the fins is turbulent rather than laminar. Laminar flow, as the name implies, is the flow of a fluid in such a way that strata can be defined; that is, all flow is in one direction, with no mixing of strata. Turbulent flow, on the other hand, causes considerable mixing. Without such mixing, the particular stratum of fluid in contact with the fin would remain in contact with it for its entire length, resulting in a very low value for the film coefficient of heat transfer h, discussed in Section 25.1.1. Stated another way, the boundary layer next to the fin surface remains intact in laminar flow, preventing efficient heat transfer from the fin to the moving stream of air (or other fluid).

The transition between laminar and turbulent fluid flow is a function of many variables; fin geometry and flow rate are the critical ones for our application. The relationship among fin spacing, flow rate, and the onset of turbulence is given by the Reynolds number. A high Reynolds number is characteristic of turbulent flow; a low number is characteristic of laminar flow. The Reynolds number Re for a fluid flowing at velocity v through a channel of width w is

$$Re = \frac{\rho v w}{\eta},$$

where ρ is the fluid density, and η is its coefficient of viscosity. This expression shows that fluid flowing in a wider channel will enter the turbulent flow regime at a lower velocity than that through a narrower channel. The point here is that, like thermal compound, more is not necessarily better. Because of reduced turbulence and flow rate, many closely spaced fins and a large surface area could result in poorer thermal performance than fewer but more widely spaced fins and a smaller surface area.

Although we have been using the context of semiconductor heatsinks for this discussion, it is equally appropriate to the cooling geometry associated with other components. Closely spaced parts impede proper convective flow for the same reasons that too closely spaced fins do.

25.3 Transient Thermal Models

So far our discussion and models have been limited to systems in which both the energy being dissipated and the temperatures within the system are constant. Our models do not represent the thermal processes associated with start-up, where dissipation may be constant, but temperatures are climbing – or pulsed operation, where temperatures may be constant, but dissipation is not. The latter situation is the more important, for under such conditions the permissible instantaneous dissipation can be much higher than predicted by static thermal models. Essentially, the heat capacity of components or their constituent parts creates a low-pass filter, which in the limit of small bandwidth only responds to the dc in $p_{\text{diss}}(t)$.

Heat capacity is a measure of the energy required to raise the temperature of a mass by a specific amount. In SI units, it is specifically the energy in joules required to raise one kilogram of the material by one degree centigrade, and has the units /°C-kg. Water has one of the largest thermal capacities of any fluid at room temperature: 4.2×10^3 J/°C-kg. Masses in a thermal system, then, constitute thermal energy storage devices, and thermal systems containing mass will exhibit dynamic behavior.

25.3.1 Lumped Models and Transient Thermal Impedance

Since thermal power is Q, the analog of thermal capacitance is electrical capacitance in the circuit model for heat transfer. In its simplest form, then, the dynamic model for a mass being supplied with heat energy is an RC circuit, as in Fig. 25.6. If the heat source is constant, the final temperature is analogous to the final capacitor voltage, that is, $R_\theta Q$. Previously we have dealt with the steady-state solutions to such thermal systems. Now we are concerned with the transients leading to these steady states.

The temperature curve of Fig. 25.6(d) predicts the temperature T_1 as a function of time for a step P_o in thermal power. If this curve is normalized by the step amplitude (P_o in this case), the resulting vertical scale has the units of thermal impedance, that is, °C/ W. An experimentally or theoretically determined normalized curve of this kind is useful for predicting temperatures during thermal transients. The normalized quantity is a function of time and is called the *transient thermal impedance*, denoted by $Z_\theta(t)$:

$$Z_\theta(t) = \frac{T(t)}{P_0}. \tag{25.9}$$

It is important to note that in order to properly represent the physics of the situation, thermal capacitances in a system model should always be connected to "ground," that is, a reference temperature.

Distributed Models Energy in a mass is stored in a continuum. However, as is done for a transmission line, this continuum system may be modeled by an interconnection of lumped electrical elements. Consider, for instance, the mass of Fig. 25.6 with a source of heat energy applied at one end. The mass can be broken into an arbitrary number of sections, each assumed to be at a uniform temperature. Each such section is characterized by a heat capacity, and the sections are interconnected through a thermal resistance. This thermal resistance is that of the mass section between the interfaces with adjoining sections. This multi-lump model of the mass of Fig. 25.6 is shown in Fig. 25.7. The number of "lumps," five, was chosen arbitrarily.

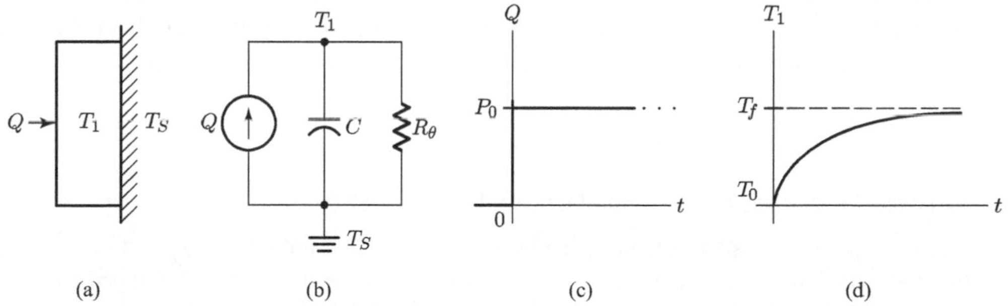

Figure 25.6 (a) A simple thermal system consisting of a mass at temperature T_1 being supplied heat Q and in contact with a sink at temperature T_S. (b) A single-lump dynamic model for the system shown in (a). (c) A step in thermal power exciting the thermal system of (a). (d) The temperature response of node T_1 to the excitation of (c).

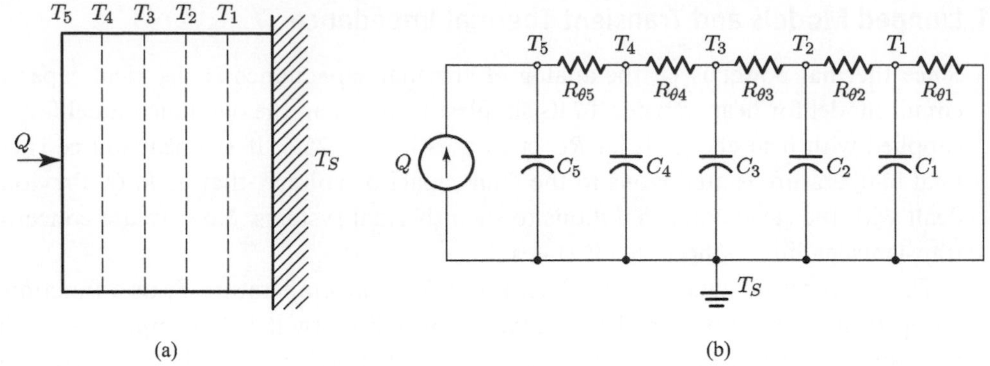

Figure 25.7 (a) The thermal system of Fig. 25.6(a) divided into five "lumps." (b) The lumped electrical analog model for the thermal system of (a).

When a continuous system is modeled by lumps, each lump displays the aggregate behavior of the physical piece of the system it represents. The model of Fig. 25.7 has been constructed so that the node voltages represent the section temperature aggregated at the interface. The number of lumps that should be chosen to represent a system depends not only on the spatial resolution of interest but on the bandwidth of the behavior being modeled. For instance, if Q is constant, no dynamics are excited, the bandwidth of the behavior is small, and a one-lump static model is adequate. However, if Q varies with time at a rate much greater than $(R_\theta C)^{-1}$ for the segments of Fig. 25.7, more lumps would be needed to accurately model the behavior of the system.

The device and package structure of Fig. 25.3 contains several thermal masses that contribute dynamics to its thermal behavior. These dynamics are important when the device is forced to dissipate high levels of power for short periods of time. "Short" is relative to the $R_\theta C$ time constant of the structure's electrical model. For very short pulses, the mass of the silicon is most important in determining the excursion of the junction temperature T_j. As the pulse gets longer, the mass of the header and then the heatsink become important. Manufacturers usually provide transient thermal impedance curves as functions of duty ratio and pulse width in the specification sheets for their devices. An illustrative set of curves for a SiC MOSFET in a TO-247 package is shown in Fig. 25.8. Only the bottom curve in this family is $Z_\theta(t)$ as defined by (25.9). The other curves are parametric in duty ratio for a series of pulses having a pulse width given on the x-axis.

Example 25.3 Transient Thermal Design for a MOSFET

The Wolfspeed C3M0075120D SiC MOSFET rated at $V_D = 1200\,\text{V}$ and $T_j = 175\,°\text{C}$, whose transient thermal impedance characteristics are shown in Fig. 25.8, is used in an 800 V clamped inductive switching application. We consider an example in which it is subjected to repetitive 35 A current pulses with a duty ratio of 0.1 at a frequency of 10 kHz. The gate is driven between −4 V and +15 V. It has already been determined that the device will remain within its safe operating area. We want to determine the maximum allowable heatsink thermal resistance, $R_{\theta SA}$, to maintain the junction at a conservative temperature of 125 °C for an ambient temperature of 40 °C.

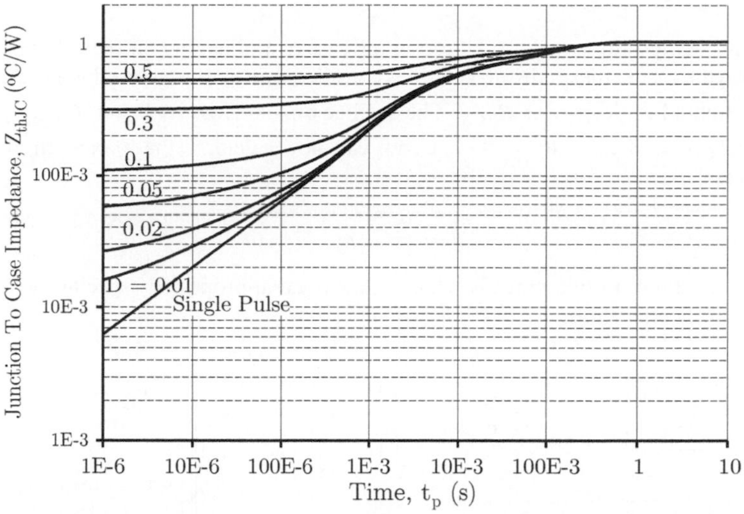

Figure 25.8 Transient thermal impedance, $Z_\theta(t)$, parametric in duty ratio and functions of pulse width t_p, for a 1200 V, 32 A Wolfspeed C3M0075120D SiC MOSFET in a TO-247 package. (Used with permission of Wolfspeed, Inc.)

The device dissipation has two parts: on-state and switching losses. Since they occur at different times during the pulse and are each short compared to a thermal time constant, they can be treated independently and their results added.

Energy is lost during the on state at a power of $R_{DS(\mathrm{on})}I_{DS}^2$. But $R_{DS(\mathrm{on})}$ is a function of both junction temperature and drain current, so we must consult Fig. 25.9(a) which is taken from the device datasheet. The figure shows $R_{DS(\mathrm{on})}$ at $I_{DS} = 35$ A at $T_j = 175\,°$C and 25 °C. We interpolate $R_{DS(\mathrm{on})}$ to a value of 121 mΩ at $T_j = 125\,°$C. The on-state power during a pulse is then

$$P_{\mathrm{on}} = (0.121)(35)^2 = 148\,\mathrm{W}. \tag{25.10}$$

The switching loss is determined from the loss versus I_{DS} curves using the datasheet graphs shown in Fig. 25.9(b). Switching loss is not a strong function of temperature, so the measurement condition of $T_j =$ 25 °C instead of 125 °C is relatively immaterial. The E_{total} curve at $I_{DS} = 35$ A gives $E_{\mathrm{total}} \approx 2$ mJ/cycle. Since the switching times are on the order of tens of nanoseconds for this device, the instantaneous switching power is very high, though the average power loss associated with switching is not.

The transient thermal impedance Z_θ presented to the 10 µs, 0.1 duty ratio current pulses is given by Fig. 25.8 as approximately 0.12 °C/W. But Z_θ for the very short (tens of nanoseconds) pulses of power during switching is not available from Fig. 25.8. The timescale of these switching power pulses is extremely short compared to both the available time constants of the system and of the on-state power pulses. We can estimate temperature rise by including the switching energy with the longer timescale of the on-state power pulses (as the on-state pulses are still very short compared to the known system time constants). Distributed over the $t_{\mathrm{on}} = 10$ µs duration of the conduction period, the switching energy provides an additional equivalent on-state power of

$$P_{\mathrm{sw,equiv}} = E_{\mathrm{total}}/t_{\mathrm{on}} = (2 \times 10^{-3})/(10 \times 10^{-6}) = 200\,\mathrm{W}. \tag{25.11}$$

Therefore, we use $Z_{\theta jc}$ and $P_{\mathrm{on}}+P_{\mathrm{sw,equiv}}$ to determine the maximum T_C allowed to maintain $T_j < 125\,°$C:

$$\Delta T_{jc} = Z_\theta(P_{\mathrm{on}} + P_{\mathrm{sw,equiv}}) = 0.12(148 + 200) = 41.8\,°\mathrm{C},$$
$$T_C \leq 125 - 41.8 = 83.2\,°\mathrm{C}.$$

The thermal power transferred through the case to ambient includes both the conduction and switching loss. Since the case is thermally massive, it is considered an isotherm; we therefore use the average total power to be dissipated to determine $R_{\theta CA}$. The average on-state loss is $\left\langle P_{\text{on}} + P_{\text{sw,equiv}} \right\rangle = 0.1 \times 348 = 34.8$ W. Using $\Delta T_{CA} = 83.2 - 40 = 43.2\,°\text{C}$, we can now calculate the maximum allowable heatsink thermal resistance $R_{\theta CA}$:

$$R_{\theta CA} \leq \frac{\Delta T_{CA}}{\langle P \rangle} = \frac{43.2}{34.8} = 1.24\,°\text{C/W}. \tag{25.12}$$

Thermal resistance in this range is achievable with an appropriately specified extruded aluminum heatsink using natural convection.

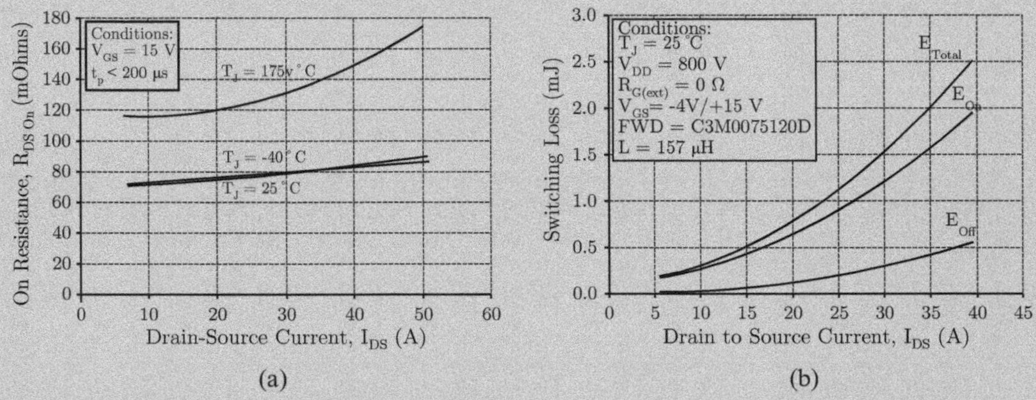

(a)

(b)

Figure 25.9 Specifications for the C3M0075120D SiC MOSFET. (a) $R_{DS(\text{on})}$ versus temperature and I_{DS}. (b) Switching loss versus I_{DS}. (Used with permission of Wolfspeed, Inc.)

Example 25.4 Derating the Safe Operating Area

A device's maximum allowed average power dissipation, $P_{D(\max)}$, is given in its datasheet but is usually specified at a case temperature $T_C = 25\,°\text{C}$, accompanied by a graph derating $P_{D(\max)}$ for higher values of T_C, as shown in Fig. 25.10(a). The SOA graph provided in datasheets is derived from single-pulse measurements (duty ratio $D=0$) and also at $T_C = 25\,°\text{C}$ with curves parametric in pulse width, as illustrated by Fig. 25.10(b). A case temperature of 25 °C seldom conforms to the application, where the case temperature is generally much higher. So we need to modify the datasheet SOA to reflect our application, derating the device and producing a smaller SOA. The process requires the use of the P_D derating and transient thermal impedance curves from the datasheet, shown in Figs. 25.10(a) and 25.8, respectively.

The maximum voltage and current constraints of the SOA are unchanged, as is the line constrained by $R_{DS(\text{on})}$. We need to determine new coordinates for the constant power constraints for the various pulse widths, at our application $T_C = T_{Ca}$, using the transient thermal impedance curves. The coordinates are scaled by δ_P, the ratio of $P(T_{Ca})$, the maximum permissible average dissipation at T_{Ca}, to $P(25°)$, the allowable dissipation at $T_C = 25\,°\text{C}$, numbers obtained from Fig. 25.10(a).

We calculate the maximum allowable dissipation for our pulse if $T_j = 175\,°\text{C}$ and $T_C = 25\,°\text{C}$, and scale it by the ratio δ_P to obtain $P_p(T_{Ca})$, the maximum pulsed power. This allowable dissipation at T_{Ca} then allows us to calculate new $V_{DS} - I_d$ coordinates on the SOA graph for our application case temperature and pulse width.

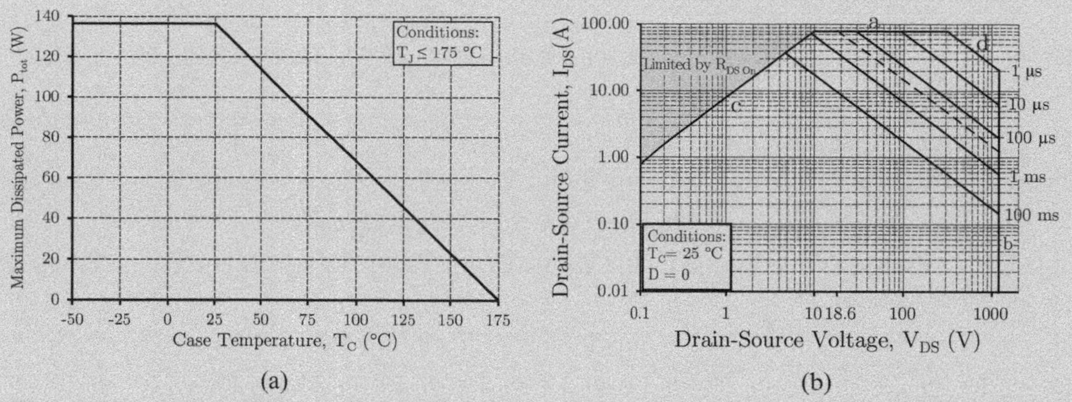

Figure 25.10 SiC MOSFET C3M0075120D specifications. (a) Maximum power dissipation derating curve. (b) The safe operating area, where the dashed line defines the boundary for a 100 µs pulse at $T_C = 100\,°\text{C}$. (Used with permission of Wolfspeed, Inc.)

We illustrate the process by derating the iso-power line in Fig. 25.10(b) for a 100 µs pulse at $T_{ca} = 100\,°\text{C}$. Figure 25.10(a) gives us $P(25°) = 136$ W, $P(100°) = 68$ W, and $\delta_p = 0.5$. From Fig. 25.8, for a single 100 µs pulse we estimate $Z_\theta = 0.06\,°\text{C/W}$, which we use to calculate $P_p(25°)$, the 100 µs pulse power if $T_C = 25\,°\text{C}$ and $T_j = 175\,°\text{C}$, which we scale by δ_p to give us $P_p(100°)$:

$$P_p(25°) = \frac{175 - 25}{0.06} = 2917\,\text{W}, \tag{25.13}$$

$$P_p(100°) = P_p(25°) \times \delta_p = 1488\,\text{W}. \tag{25.14}$$

We can now calculate a pair of coordinates on the iso-power limit line for a 100 µs pulse with $T_C = 100\,°\text{C}$. Choosing $I_D = 80$ A (the maximum specified pulse current),

$$V_{DS} = \frac{P_p(100°)}{I_D} = \frac{1488}{80} = 18.6\,\text{V}. \tag{25.15}$$

We now have one point on the line, but the line is iso-power with a slope of −1 so we can draw the new constraint on the SOA graph, as indicated by the dashed line in Fig. 25.10(b).

Notes and Bibliography

The volume of work published in the general area of heat transfer is massive. The references selected here are representative of those that are accessible to the non-specialist. An undergraduate text covering most topics of interest to the designer of electronic equipment, although not in this context, is [1]. The book is liberally illustrated and contains numerous examples. Lienhard and Lienhard [2] is a very comprehensive text with numerous examples and problems addressed to juniors through graduate students. Among its unique inclusions are photographs of Ludwig Prandtl, Osborne Reynolds, and Ernst Kraft Wilhelm Nusselt, whose namesakes are the Prandtl, Reynolds, and Nusselt numbers, important parameters in heat transfer. It is inexpensive and available as an e-book.

Lee [3] is focused on specific heat exchange technologies. The book includes extensive analyses of the different devices used for heat transfer. Steinberg [4] is short on theory but long on practical applications.

The numerous examples reflect the author's own experience in the military/avionics area. A lot of practical data is presented, and there is a good discussion of fluid-based heat transfer systems, including heat pipes.

A concise discussion and mathematical statement of the Wiedman–Franz law can be found on p. 150 of Kittel [5].

The TI application note [6] provides an extensive discussion of using thermal vias for heat sinking on printed circuit boards.

1. F. M. White, *Heat Transfer*, Boston, MA: Addison-Wesley, 1984.
2. J. H. Lienhard IV and J. H. Lienhard V, *A Heat Transfer Textbook*, 5th edn., Mineola, NY: Dover Publications, 2019.
3. H. Lee, *Thermal Design: Heat Sinks, Thermoelectrics, Heat Pipes, Compact Heat Exchangers, and Solar Cells*, Chichester: Wiley, 2010.
4. D. S. Steinberg, *Cooling Techniques for Electronic Equipment*, 2nd edn., Chichester: Wiley Interscience, 1991.
5. C. Kittel, *Introduction to Solid State Physics*, 6th edn., New York: Wiley, 1986.
6. Texas Instruments, *ANM-2020 Thermal Design By Insight, Not Hindsight*, Application Report SNVA419C-April 2010-, revised April 2013.

PROBLEMS

25.1 A double-insulated window is made of panes of glass 4 mm thick spaced 1 cm apart. Window glass has approximately the same thermal resistivity as SiO_2, 100 °C-cm/W. If the interior temperature of the building is 25 °C and the outside temperature is 0 °C, what is the rate of heat lost by conduction in kW/m^2?

25.2 The CRC *Handbook of Chemistry and Physics* (35th edn.) defines thermal conductivity of materials as "the quantity of heat in calories which is transmitted per second through a plate 1 cm thick across an area of 1 cm^2 when the temperature difference is 1 °C." The value for dry compact snow is 0.000 51. What is the thermal resistivity of dry compact snow in units of °C-cm /W?

25.3 An isolating interface of alumina having a thickness of 1 mm is placed between the device package and the heatsink in Fig. 25.3 (Example 25.2). What is the junction temperature T_j, if other parameters of the example remain unchanged?

25.4 Figure 25.11 shows two identical devices, Q_1 and Q_2, mounted on a common heatsink. The devices are in TO-220 packages and have a thermal resistance from junction to case of $R_{\theta jC} = 1.2\,°\mathrm{C/W}$. The interface between the case and sink has a thermal resistance of $R_{\theta CS} = 0.20\,°\mathrm{C/W}$, and the thermal resistance between the sink and ambient is $R_{\theta SA} = 0.8\,°\mathrm{C/W}$.

(a) Draw the static thermal model for the thermal system of Fig. 25.11.

(b) If the devices are dissipating the same power, and $T_A = 40\,°\mathrm{C}$, what is the maximum total power that can be dissipated if $T_{j(\max)} = 150\,°\mathrm{C}$?

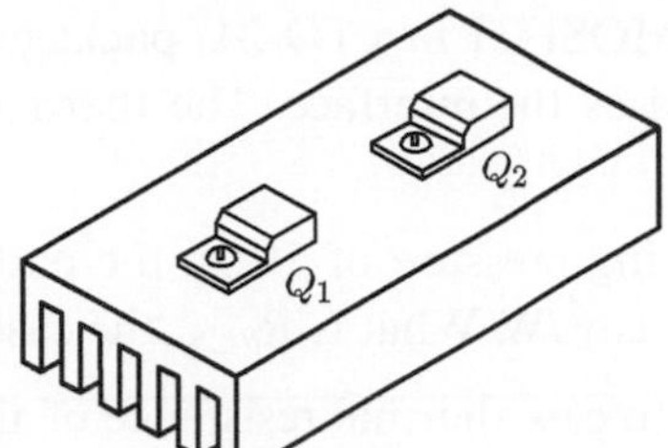

Figure 25.11 Two devices mounted on a common heatsink analyzed in Problem 25.4.

(c) What is the maximum possible power dissipated if only one of the devices is operating?

25.5 A transistor in a TO-3 case has a junction-to-case thermal resistance of 1 °C/W and is to be used in an environment having an ambient temperature of 60 °C. The transistor is to be isolated from its heatsink by a Mylar spacer having a thickness of 0.1 mm, and the available heatsink has a specified value of sink-to-ambient thermal resistance of $R_{\theta SA} = 2\,°\text{C/W}$.

(a) Determine and draw the static thermal model for this system.

(b) What is the maximum power that can be dissipated by the device if its junction temperature must be less than 150 °C?

25.6 Figure 25.12 shows the internal structure and dimensions of a power diode mounted in an axial lead package. The diode is cooled by conduction through its leads, which are soldered to terminals that are assumed to be at temperature T_A. Heat is generated at the junction of the diode, which is planar and centered between the two surfaces.

(a) Draw the analog circuit model for the thermal system of Fig. 25.12.

(b) If the maximum permissible junction temperature of the diode is $T_j = 225\,°\text{C}$, what is the maximum permissible dissipation for $T_A = 75\,°\text{C}$?

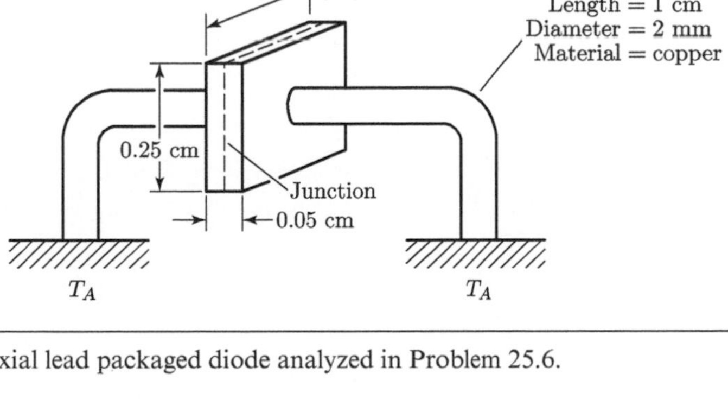

Figure 25.12 The axial lead packaged diode analyzed in Problem 25.6.

25.7 A superjunction MOSFET in a TO-247 package is mounted to a heatsink with a 0.5 mm-thick silicone pad as the interface. The thermal contact area of a TO-247 package is 2.5 cm^2.

(a) At the mounting pressure of 10 psi the pad has a thermal resistance-area product, Z_{th}, of 0.6 °C-cm²/W. What is $R_{\theta CS}$, the case to sink thermal resistance?

(b) The junction to case thermal resistance of the MOSFET is $R_{\theta jC} = 0.3$ °C/W, and at a junction temperature T_j of 150 °C its on-state resistance is $R_{DS(\text{on})} = 40$ m. If the heatsink temperature can be maintained at 50 °C, what is the maximum continuous current that the device can conduct?

25.8 What is T_f in Fig. 25.6(d)?

25.9 The SiC MOSFET characterized by the transient thermal impedance curves of Fig. 25.8 is subjected to an overload condition that is cleared by a protection circuit in 3 μs. The MOSFET had been operating at a junction temperature of $T_j = 150$ °C. How much energy can the device be allowed to dissipate during the fault to maintain $T_j \leq 200$ °C?

25.10 Consider the "single pulse" thermal response of a system (e.g., as illustrated in Fig. 25.8). This response $Z_\theta(t)$ is in fact the thermal step response of the system. That is, $Z_\theta(t)$ represents the temperature rise response over time to a unit step in input power at $t = 0$.

(a) Show that if one can treat the dynamic thermal system as a linear, time-invariant (LTI) system (e.g., the circuit elements in the model of Fig. 25.6(b) are LTI), then we can write the temperature response to a short pulse in power of amplitude P starting at $t = 0$ and having duration t_1 as

$$\Delta T_{jC} = P[Z_\theta(t) - Z_\theta(t - t_1)].$$

(b) For the same LTI system assumption, what would be the temperature rise response to a sequence of two pulses of amplitude P, each of duration t_1, one starting at $t = 0$ and the second starting at $t = t_2 (> t_1)$?

25.11 Using the CREE SiC SOA of Fig. 25.10(b), determine the derated limiting boundary for a 1 ms pulse if the case temperature is 125 °C. What is the maximum allowable I_D?

26 Electromagnetic Interference and Filtering

The rapid switching transitions of a power converter are potential sources of electromagnetic interference (EMI), both for the converter itself and for the systems to which it is connected. Adequate filtering at the input and output of the converter is important, both to obtain acceptable performance and to prevent interference with other equipment. In this chapter, we consider the sources of EMI in a converter, how EMI is measured and modeled, and how it can be mitigated, with a focus on conducted (rather than radiated) EMI.

We begin by considering basic measures of switching noise and EMI in converters, and how they are measured according to EMI standards issued by regulating organizations – industrial, national, or international. We then address the basics of filter design and damping, and outline the impact of common parasitics on filter performance. We demonstrate how to decompose EMI noise and other waveforms into common- and differential-mode components, and show how each of these components can be addressed by appropriate filter design. Lastly, we describe aspects of filter components and layout that can affect filter performance, and consider means for addressing the challenges that can arise.

26.1 EMI Specifications and Measurement

Input and output filters are important in practical converters, and can often take up an appreciable fraction of overall system volume. In addition to meeting ripple specifications imposed by a given load or source, filters can also be important to prevent radio-frequency interference with nearby equipment. Interference can be conducted via the input or output of the converter, or can be radiated, often with the input and/or output interconnections of the converter serving as antennas. Input and output filters can thus serve to reduce both conducted and radiated emissions.

In simple cases, converter filtering requirements may be specified in terms of time-domain measurements (e.g., peak or rms voltage or current ripple). However, filtering requirements to limit conducted emissions (that is, noise conducted through power leads) are frequently specified in the frequency domain, and almost universally so for converters connected to a power system, such as the ac grid or a vehicle power bus. Conducted EMI specifications are often made in terms of the noise spectrum in the power leads, detected within a defined bandwidth (e.g., 9 kHz) as measured across a defined frequency range (often 150 kHz to 30 MHz). Still-higher frequencies are governed by radiated EMI specifications; emissions at these frequencies are detected using antenna-based measurements, and are less related to converter filter design.

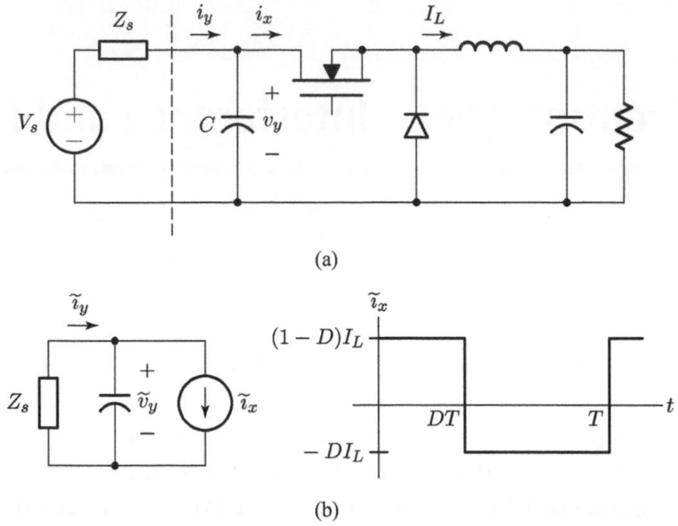

Figure 26.1 (a) A buck converter with an input filter capacitor C, fed from a source with impedance Z_s. (b) An input-side ac ripple model for the system of (a).

The standards for measuring conducted EMI are highly detailed, with carefully specified measurement conditions. This is necessary to make such measurements repeatable – important for demonstrating compliance with EMI standards. For this reason, EMI compliance testing is often carried out in a dedicated commercial facility. Pre-compliance testing is often carried out using the same basic approach, though without adhering to all of the details. This is often adequate for filter development, especially if a significant safety margin is included in the filter design. In this section we give an introduction to the basic approach used for such measurements. More detailed treatments are provided in references at the end of the chapter.

To appreciate the challenges in making conducted EMI measurements, consider the simple buck converter of Fig. 26.1(a). Capacitor C serves as a simple input-side filter, and the converter is connected to a source voltage V_s having a source impedance Z_s. Figure 26.1(b) shows an input-side ac ripple model for this system. Both the input ripple voltage $\widetilde{v}_y$ and input ripple current $\widetilde{i}_y$ are strong functions of the source impedance. For example, as $|Z_s| \to 0, \widetilde{i}_y \to \widetilde{i}_x$ and $\widetilde{v}_y \to 0$, while as $|Z_s| \to \infty, \widetilde{i}_y \to 0$ and $\widetilde{v}_y \to \widetilde{i}_x/(j\omega C)$. Depending on the composition of the source impedance, there may also be resonances between Z_s and C that cause radical variations in $\widetilde{v}_y$ and $\widetilde{i}_y$ across frequency. The measured responses accordingly depend upon the details of the source impedance at ripple frequencies.

Figure 26.2 shows an EMI test setup configured to address this challenge. A device known as a line impedance stabilization network, or LISN, is placed between the equipment under test (here, a buck converter and its input filter) and the input source. A spectrum analyzer having a known input impedance (e.g., $R_{50} = 50\,\Omega$) is connected to the measurement port of the LISN, and the spectral content of the voltage at that port (v_{LISN}) is used as a measure of emissions. At low (source) frequencies, L_{LISN} acts as a short and C_{LISN} as an open, and power is provided from the source to the equipment under test as in normal operation. At high (EMI)

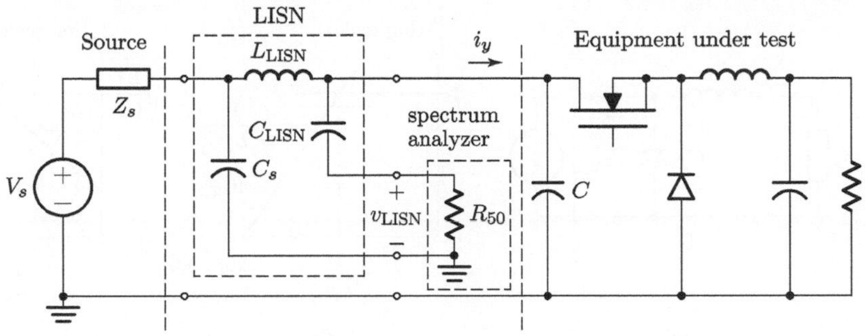

Figure 26.2 An EMI test setup. A LISN provides a controlled source impedance for the equipment under test. A spectrum analyzer measures the conducted EMI.

frequencies, L_{LISN} acts as an open and C_{LISN} as a short, and the converter looks back into the well-controlled impedance of the spectrum analyzer, the ripple current from the converter determining v_{LISN}. This approach provides consistency, enabling validation against a set of specifications, independent of the high-frequency impedance of the source.

EMI specifications define the LISN component values to be used; typical values are $C_{\text{LISN}} = 0.1\,\mu\text{F}$, $L_{\text{LISN}} = 50\,\mu\text{H}$, and $C_s = 1\,\mu\text{F}$. Likewise, the settings of the spectrum analyzer and many other aspects of the test are defined in the specifications, with the spectral content of v_{LISN} being compared against allowable limits to determine compliance. Unlike the setup shown in Fig. 26.2, many EMI specifications require that a LISN be used for *all* power leads between the source and equipment under test; the reason for this will be addressed in Section 26.3. The LISN is only present during EMI testing, and the converter and filter must perform acceptably in its absence with whatever source impedance is found in practice.

26.2 Filter Design

EMI specifications often constrain noise to extremely low values (e.g., ripple currents at the LISN input in the tens of μA or less). This is to avoid interference with radios and other electronic systems that are sensitive to RF noise. A simple capacitor filter as illustrated in Fig. 26.1 is often inadequate for achieving the necessary attenuation, so one must employ more sophisticated filters. For such designs we focus on the converter current ($\tilde{\imath}_x$) to source current ($\tilde{\imath}_y$) transfer function, $H_i(j\omega)$.

26.2.1 The *LC* Filter Stage

One basic filter building block is an *LC* filter, as illustrated in Fig. 26.3(a). Figure 26.3(b) illustrates the magnitude of the ripple current $\tilde{\imath}_x$ to filtered current $\tilde{\imath}_y$ transfer function $H_i(j\omega) = \tilde{\imath}_y/\tilde{\imath}_x$, while Fig. 26.3(c) shows the magnitude of the filter impedance $Z_x(j\omega) = \tilde{v}_x/\tilde{\imath}_x$. In each case, the behavior is shown for $Z_s = 0$, and we ignore behavior near the filter cutoff frequency $\omega_f = 1/\sqrt{L_f C_f}$, which we treat later. The current transfer function for the filter is

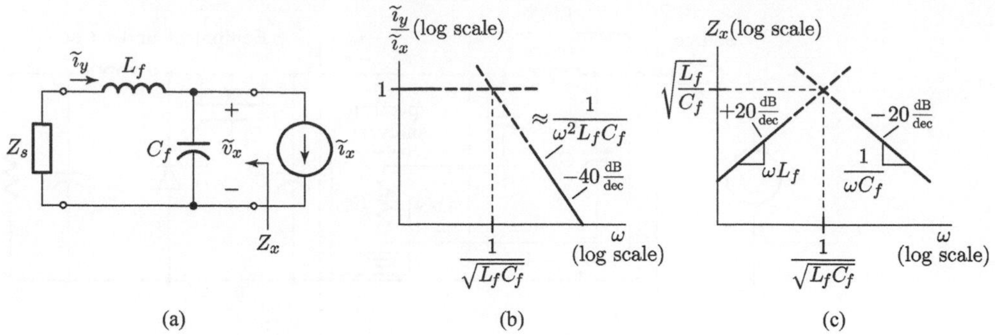

Figure 26.3 (a) An LC filter driven by a ripple current $\tilde{i}_x$ and loaded by impedance Z_s. (b) Asymptotes of the ripple current $\tilde{i}_x$ current to filtered current $\tilde{i}_y$ transfer function magnitude $|H_i|$. (c) Asymptotes of the impedance magnitude $|Z_x|$. (Both (b) and (c) for $Z_s = 0$.)

$$H_i(j\omega) = \frac{\tilde{i}_y}{\tilde{i}_x} = \frac{1}{1 - \omega^2 L_f C_f}. \tag{26.1}$$

For low frequencies ($\omega \ll \omega_f$), ripple current $\tilde{i}_x$ is passed to $\tilde{i}_y$ without attenuation. At high frequencies ($\omega \gg \omega_f$), the filter transfer function magnitude rolls off as $1/\left(\omega^2 L_f C_f\right)$. We can use this relationship to select a cutoff frequency providing the desired degree of attenuation in $\tilde{i}_y$ at the frequencies of interest. The attenuation provided by the filter is determined by current division between the series (L_f) and shunt (C_f) paths. Because the series impedance increases as ω, and the shunt impedance as $1/\omega$, the transfer function magnitude far beyond cutoff behaves as $1/\omega^2$, or -40 dB/decade.

A second aspect of interest is the filter output impedance $Z_x(j\omega)$, which is

$$Z_x(j\omega) = \frac{\tilde{v}_x}{\tilde{i}_x} = \frac{j\omega L_f}{1 - \omega^2 L_f C_f} \tag{26.2}$$

for $Z_s = 0$. Far below cutoff, the filter impedance is dominated by the inductor L_f, while far above cutoff, the filter output impedance is dominated by the capacitor C_f ($Z_x \approx 1/(j\omega C_f)$). Providing a sufficiently low filter impedance can be one basis for choosing the filter characteristic impedance $Z_f = \sqrt{L_f/C_f}$.

26.2.2 *LC* Filter Damping

One concern with the simple LC filter shown in Fig. 26.3(a) is its behavior near its cutoff frequency, where L_f and C_f resonate. As seen in (26.1) and (26.2), both the current transfer function H_i and the filter impedance Z_x have unbounded magnitude as ω approaches ω_f. While some damping would always exist in practice, large peaking in either Z_x or H_i causes undesirable effects on converter operation. First, any converter current components near the filter resonance (e.g., owing to load variations on the converter, noise, etc.) could cause unacceptable excursions in $\tilde{v}_x$ and $\tilde{i}_y$. This is compounded by the fact that undamped resonances in the input filter tend to make the converter more susceptible to transients and noise in the source voltage. Lastly, as we will see in the next subsection, converters have a tendency to provide "negative damping" of their input filters, which can lead to oscillations if the filter is not damped enough.

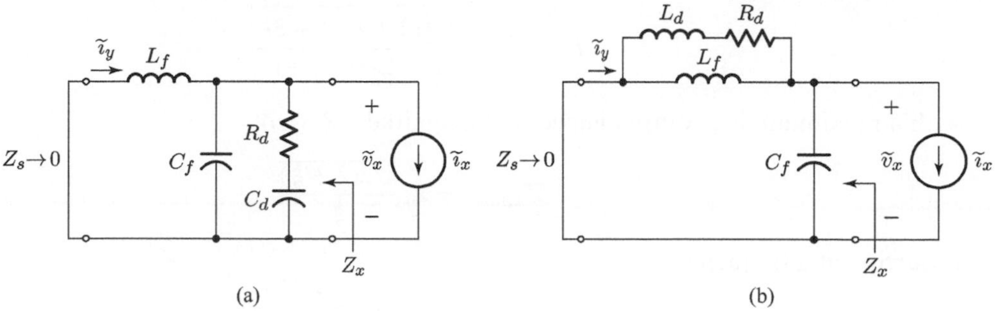

Figure 26.4 Damped LC filters with driving ripple current $\tilde{\imath}_x$ and filtered current $\tilde{\imath}_y$ into a source impedance $Z_s \to 0$. (a) An $R_d C_d$ damping leg. (b) An $L_d R_d$ damping leg.

A solution to this issue is to add explicit damping to the filter (e.g., a damping resistor R_d) that acts to limit the peaking in H_i and Z_x near the cutoff frequency. We would like to do this in a manner that avoids imposing large power-frequency currents or voltages on the damping resistor (to preserve efficiency) and does not greatly hurt the high-frequency performance of the filter. Two common methods for achieving this are shown in Fig. 26.4.

Consider the filter of Fig. 26.4(a), which includes an $R_d C_d$ damping leg. If C_d is significantly larger than the filter capacitor C_f, then we can consider it as approximately a "short" near $\omega_f = 1/\sqrt{L_f C_f}$, so R_d can act approximately as a parallel damping resistor for the $L_f C_f$ resonance. At the same time, C_d blocks dc, so R_d sees no dc dissipation. The current gain of this filter is

$$H_i(s) = \frac{\tilde{\imath}_y}{\tilde{\imath}_x} = \frac{1 + R_d C_d s}{1 + R_d C_d s + L_f(C_d + C_f)s^2 + L_f R_d C_d s^3}, \tag{26.3}$$

while the filter impedance is

$$Z_x(s) = \frac{L_f s(1 + R_d C_d s)}{1 + R_d C_d s + L_f(C_d + C_f)s^2 + L_f R_d C_d s^3}, \tag{26.4}$$

which we evaluate at $s = j\omega$ to obtain the filter frequency responses.

Selection of a damping resistor value is a "Goldilocks" endeavor. If one selects too small a value of R_d (e.g., $R_d \to 0$), then capacitor C_d is effectively in parallel with C_f and one gets large resonant peaks in H_i and Z_x near $\omega_{df} = 1/\sqrt{L_f(C_d + C_f)}$. If one picks too large a value of R_d (e.g., $R_d \to \infty$), then R_d is effectively an open, and does not damp the resonance at $\omega_f = 1/\sqrt{L_f C_f}$. There is, however, a value of R_d that is "just right" in the sense of minimizing the peak value of Z_x across frequency.

The optimum value of R_d to minimize the peaking in filter impedance Z_x depends upon the relative values of the damping capacitor and the filter capacitor (that is, on their ratio $\xi = C_d/C_f$), with higher values of C_d (and ξ) leading to smaller maxima of $|Z_x|$. In particular, it can be shown that, for a given undamped natural frequency $\omega_f = 1/\sqrt{L_f C_f}$ and characteristic impedance $Z_f = \sqrt{L_f/C_f}$, the optimum value is

$$R_{d,\text{opt}} = Z_f \sqrt{\frac{(2+\xi)(4+3\xi)}{2\xi^2(4+\xi)}}, \tag{26.5}$$

with a maximum filter output impedance magnitude $|Z_x(j\omega)|$

$$Z_{x,\max} = Z_f \frac{\sqrt{2(2+\xi)}}{\xi} \tag{26.6}$$

occurring at a frequency

$$\omega_{\max} = \omega_f \sqrt{\frac{2}{2+\xi}}. \tag{26.7}$$

Figure 26.5 shows normalized plots of output impedance and current gain for the filter of Fig. 26.4(a) optimized at different values of ξ (and hence with different values of C_d and R_d), where each design has the same low- and high-frequency asymptotes. While increasing ξ has a modest impact on filter current gain peaking, it has a major impact on the maximum filter impedance $Z_{x,\max}$. This is valuable, because, as detailed below, we often seek to keep maximum filter output impedance low.

The damping scheme of Fig. 26.4(a) is quite popular; its main drawback is the relatively large value and size of damping-leg capacitor C_d. While C_f needs to have excellent high-frequency performance, C_d is mainly operative at low frequencies (near cutoff), and significant parasitics (ESR and ESL) are often tolerable. For this reason, aluminum electrolytic capacitors are often

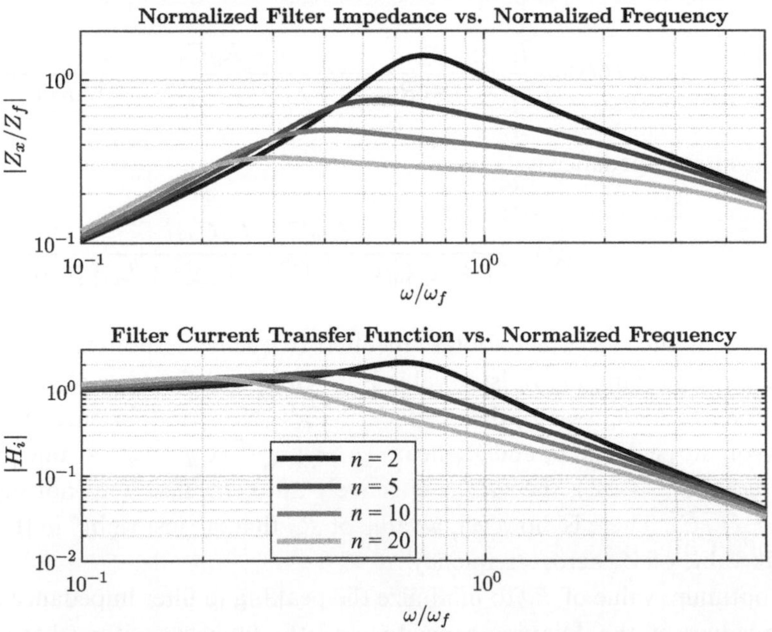

Figure 26.5 Normalized filter output impedance and current gain magnitudes $|Z_x/Z_f|$ and $|H_i|$ versus normalized frequency ω/ω_f for optimally designed filters of the type in Fig. 26.4(a) at different capacitor ratios $\xi = C_d/C_f$.

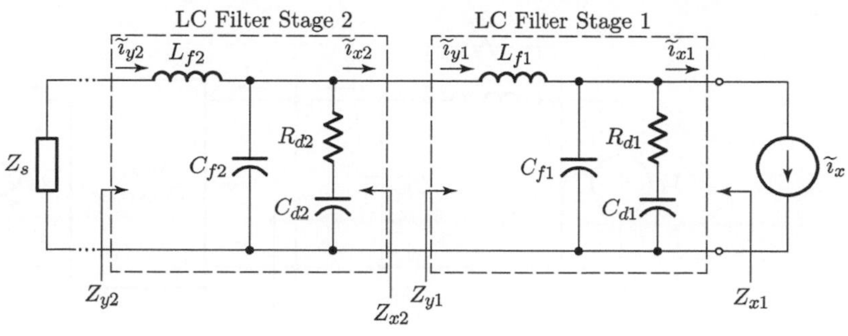

Figure 26.6 A multi-stage *LC* filter.

used for C_d, and sometimes for R_d by utilizing their ESR, though the variation in ESR with time and temperature can be problematic.

Another popular damping scheme is shown in Fig. 26.4(b). In this approach, the damping inductor L_d is *smaller* than L_f by some factor, such that it acts as an approximate short near ω_f, and R_d appears approximately in parallel with L_f for damping. Similar design equations to those above exist for selecting optimal damping for this design. (Alternatively, simulation experiments can readily reveal the best damping resistor R_d for a given ratio L_d/L_f.) A key benefit of the damping scheme of Fig. 26.4(b) is that L_d carries no dc current, and hence can be quite small. The largest drawback of this approach is that the damping leg hurts the high-frequency filtering performance, with a high-frequency current gain that goes as $1/\left(\omega^2(L_f||L_d)C_f\right)$ instead of $1/\left(\omega^2 L_f C_f\right)$.

When a large amount of filter attenuation is needed, it is often preferable to implement a multi-stage filter, as illustrated in Fig. 26.6. Though this requires more components, they can have lower values and thus be physically smaller, leading to a more compact design. Each stage is often designed with its own damping, with stage k loaded by the filter impedance of stage $k+1$. While design and optimization of multi-stage filters is beyond our scope, it is worth noting that it is generally best *not* to make the filter cutoff frequencies of the different stages coincide. Rather, one can get fewer inter-stage interactions (and lower achievable impedance $Z_{x1,\max}$) by staggering the cutoffs of the filter stages, with the first stage – that nearest the converter – having the lowest cutoff frequency and subsequent stages having increasingly higher cutoffs and lower impedances Z_{xn}. A consequence of this is that the attenuation provided by each stage will not be identical.

26.2.3 Converter and Input Filter Interactions

Up to now we have mainly considered the interactions between a converter and its input filter in terms of ripple attenuation. However, it is also important to address the potential for low-frequency dynamic interactions between them. A converter is usually designed assuming that it is fed from an ideal source. However, if the impedance Z_x presented by a filter to a converter is not sufficiently small, deviations from expected behavior can occur. In some cases the resulting

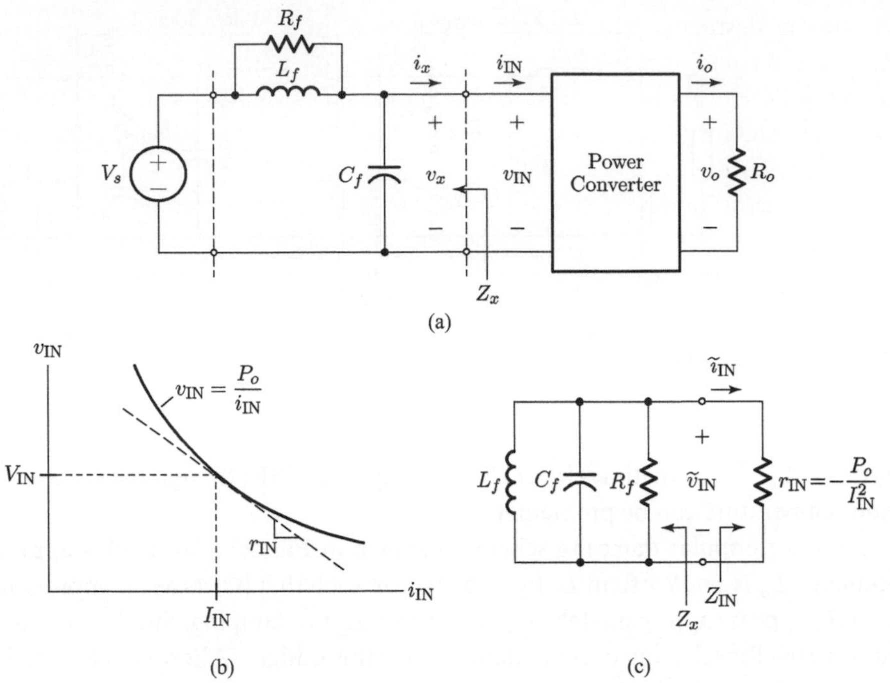

Figure 26.7 (a) A converter with a simple input filter. (b) The v–i relation of the converter; the slope represents the converter's input resistance at a given operating point. (c) An incremental model for the converter and filter.

system can even become unstable, resulting in sustained oscillations between the converter and its input filter.

To appreciate how this can arise, consider the system of Fig. 26.7(a) comprising a dc/dc converter and a simple input filter. Assume the converter is ideal, in that it is 100% efficient and has perfect control, instantly transferring power from its input to maintain its output voltage at some reference value $v_{o,\text{ref}}$, despite any input voltage variations. Thus, for some fixed output power P_o, the input voltage and current satisfy $i_{IN} = P_o/v_{IN}$. Figure 26.7(b) shows the v–i characteristic observed at the input of this ideal converter for an output power $P_o = v_{o,\text{ref}}^2/R_o$. At a fixed operating point $v_{IN} = V_{IN}$, the converter would draw a fixed current $i_{IN} = I_{IN}$. If v_{IN} varies incrementally, however, (i.e., $v_{IN} = V_{IN} + \tilde{v}_{in}$ for some small variation $\tilde{v}_{in}$), then we should expect that the current drawn by the converter will also vary incrementally (i.e., $i_{IN} = I_{IN} + \tilde{i}_{in}$ with some small variation $\tilde{i}_{in}$). The relationship among these incremental variations is the incremental (or linearized) converter input impedance at the specified operating point:

$$Z_{in} = \left.\frac{\tilde{v}_{in}}{\tilde{i}_{in}}\right|_{V_{IN}, I_{IN}} = \left.\frac{\partial v_{IN}}{\partial i_{IN}}\right|_{V_{IN}, I_{IN}} = -\frac{P_o}{I_{IN}^2} = -\frac{V_{IN}}{I_{IN}} = r_{IN}. \tag{26.8}$$

Because i_{IN} decreases when v_{IN} increases and vice versa, the converter looks incrementally like a *negative resistance*. This negative resistance characteristic means that the converter can potentially destabilize the overall system. The filter must thus be designed with sufficient damping to counteract the destabilizing effect of the converter's negative input resistance.

Figure 26.7(c) shows a linearized model for the system of Fig. 26.7(a) at a given operating point. The stability of the overall system depends upon there being net positive resistance loading the LC network, that is, upon having $R_f || r_{IN} > 0$. This in turn requires that $R_f < |r_{IN}|$. Given (26.8), this condition is the hardest to meet at low input voltage and high input current. In cases where this condition is not met, we can expect sustained oscillations (or worse) in the converter input voltage and source current at the resonant frequency of the input filter.

We have illustrated the destabilizing effect of a converter's negative input impedance for a specific case. More generally, if the converter's control bandwidth exceeds the cutoff frequency of the filter (such that it acts like our idealized example), we must ensure that the maximum magnitude of the filter output impedance $Z_{x,\max}$ is smaller than $|Z_{IN}|$ of the converter to ensure stability of the overall system (and much smaller to guarantee good damping). It is in part for this reason that a lot of attention is paid to limiting filter output impedance, as described in the previous subsection.

Beyond ensuring stability, it is also highly preferable that the presence of the input filter does not affect the converter control dynamics. This is important because the converter and its controller are usually designed without regard to the presence of an input filter. The details of how this is best accomplished are beyond the scope of this text, but a sufficient set of conditions to achieve this is: (i) the maximum filter output impedance magnitude $Z_{x,\max}$ should be much smaller than $|Z_{IN}|$ with idealized control, as above; and (ii) at every frequency, the filter output impedance magnitude $|Z_x(j\omega)|$ should be much smaller than the small-signal input impedance magnitude of the converter $|Z_{IN}(j\omega)|$ for the case where the converter control is fixed (e.g., operating at a constant duty cycle). Meeting this extended set of conditions again calls for filter designs providing low output impedance Z_x.

Example 26.1 A Preliminary Input Filter Design for a Buck Converter

Consider the preliminary design of an input filter for a synchronous buck converter as illustrated in Fig. 26.8(a). The converter is rated to operate from an input voltage V_1 in the range 30–40 V and provide an output voltage $V_2 = 15$ V at an output power P_o in the range 50–150 W. The details of the buck converter power stage are not yet defined, but it is known that the converter will operate deep in continuous conduction mode at a switching frequency of $f_s = 500$ kHz. The input filter should limit the source current i_y to have no more than 30 mA peak-to-peak ac ripple, assuming the source impedance is negligible.

Figure 26.8(b) shows the current i_x driving the filter, where I_L is the average of inductor current i_L. For simplicity, we approximate current i_x as a square wave of duty ratio D pulsing between 0 and I_L, where $D = V_2/V_1$ varies between 0.375 and 0.5, while $I_L = P_o/V_2$ varies between 3.33 and 10 A. The peak amplitude of the nth harmonic of such a square wave is

$$I_{x,n} = \frac{I_L}{n\pi}\sqrt{2 - 2\cos(2\pi nD)}. \tag{26.9}$$

For the range of duty ratios involved, the ripple in i_x is dominated by its fundamental component. Moreover, given the steep roll-off of an LC filter, the ripple in i_y is heavily dominated by its fundamental. We thus design the filter to attenuate the fundamental, and neglect higher harmonics for simplicity. We approximate the drive waveform with its fundamental: a sinusoid at $\omega_s = 3.14 \times 10^6$ rad/s having amplitude $I_{x,1} = (I_L/\pi)\sqrt{2 - 2\cos(2\pi D)}$. The worst case is for $D = 0.5$ and $I_L = 10$ A, yielding a fundamental amplitude

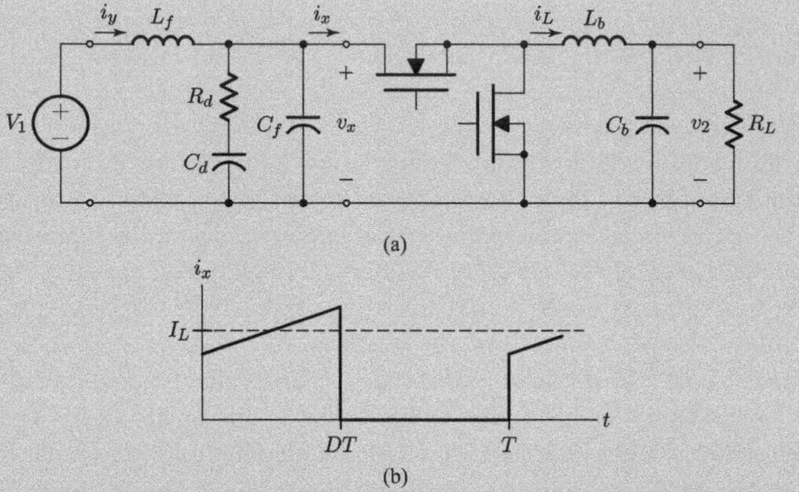

Figure 26.8 A synchronous buck converter with an LC input filter.

$I_{x,1} = 6.36\,\text{A}$. If the ripple in i_y were purely from its fundamental, a peak-to-peak ripple of 30 mA would correspond to a fundamental amplitude $I_{y,1} = 1.5 \times 10^{-2}\,\text{A}$. To account for the simplifications made and the effects of component variations and unmodeled parasitics, and to provide some design margin, we design the filter to achieve ripple that is a factor of three smaller: $I_{y,1} = 5 \times 10^{-3}\,\text{A}$.

We can set the cutoff frequency of the filter $\omega_f = 1/\sqrt{L_f C_f}$ based on the high-frequency asymptote of the current transfer function H_i:

$$\frac{|I_{y,1}|}{|I_{x,1}|} = \frac{1}{\omega_s^2 L_f C_f} = \frac{\omega_f^2}{\omega_s^2}. \tag{26.10}$$

This gives us the desired cutoff frequency

$$\omega_f = \omega_s \sqrt{\frac{|I_{y,1}|}{|I_{x,1}|}} = 8.8 \times 10^4\,\text{rad/s}. \tag{26.11}$$

We must also design the maximum output impedance of the filter to be much smaller than the magnitude of the negative incremental resistance (26.8) posed by the converter. This limit is worst at the lowest input voltage and highest power, resulting in $r_{\text{IN}} = -6\,\Omega$. To facilitate meeting this, we can select a low filter characteristic impedance, e.g. $Z_f = \sqrt{L_f/C_f} = 0.2\,\Omega$.

Together, the target values ω_f and Z_f let us select appropriate component values. Given $C_f = 1/(\omega_f Z_f) = 56.81\,\mu\text{F}$, a filter capacitor $C_f = 60\,\mu\text{F}$ is selected. This may be constructed from a bank of ceramic or film capacitors, for example, which have low parasitics and high current capability. Likewise, given $L_f = Z_f/\omega_f = 2.27\,\mu\text{H}$, a custom inductor having this value may be designed or an off-the-shelf 2.2 μH inductor selected. To damp the filter we may select $\xi = C_d/C_f = 3$, which gives $C_d = 180\,\mu\text{F}$. From (26.5) and (26.6), this selection yields a damping resistor $R_d = 0.14\,\Omega$ and a maximum filter impedance across frequency of $Z_{x(\text{max})} = 0.21\,\Omega$.

While this preliminary design is a good baseline, once the power stage is fully specified, we will need to validate that the filter meets the filtering requirements and satisfies the additional impedance requirements described above. Careful attention to component selection, parasitics, and layout will also be needed to ensure that the actual filter will provide the necessary performance.

26.3 Common-Mode and Differential-Mode Variables

The currents at the input terminals of a converter can be decomposed into two components – differential-mode and common-mode currents – which will be further defined in Section 26.3.1. The importance of this decomposition is that it guides an effective structuring of the EMI filter. We now look at the origin of the differential- and common-mode currents.

Up to now our discussions have included a subtle assumption: that our circuits are referenced to a common potential (e.g., ground) that acts as a single node. We have not concerned ourselves with cases that do not fit this model, or with what might happen with a practical conductor serving as a common connection. In this section we introduce ways of treating systems that do not fit the earlier assumption.

As an introduction, consider the system of Fig. 26.9(a), which shows a boost converter fed from a grounded source via a pair of cables. This system might represent a load in an automobile fed from the battery via a boost converter, for example, with ground in the diagram representing the vehicle chassis (to which the negative terminal of the battery is connected). The converter is fed through two cables (connected to the two battery terminals) that are intended to carry the positive and return current to the converter and load. The intended current path feeding the system is shown, looping from the source, out the positive cable to the converter and load,

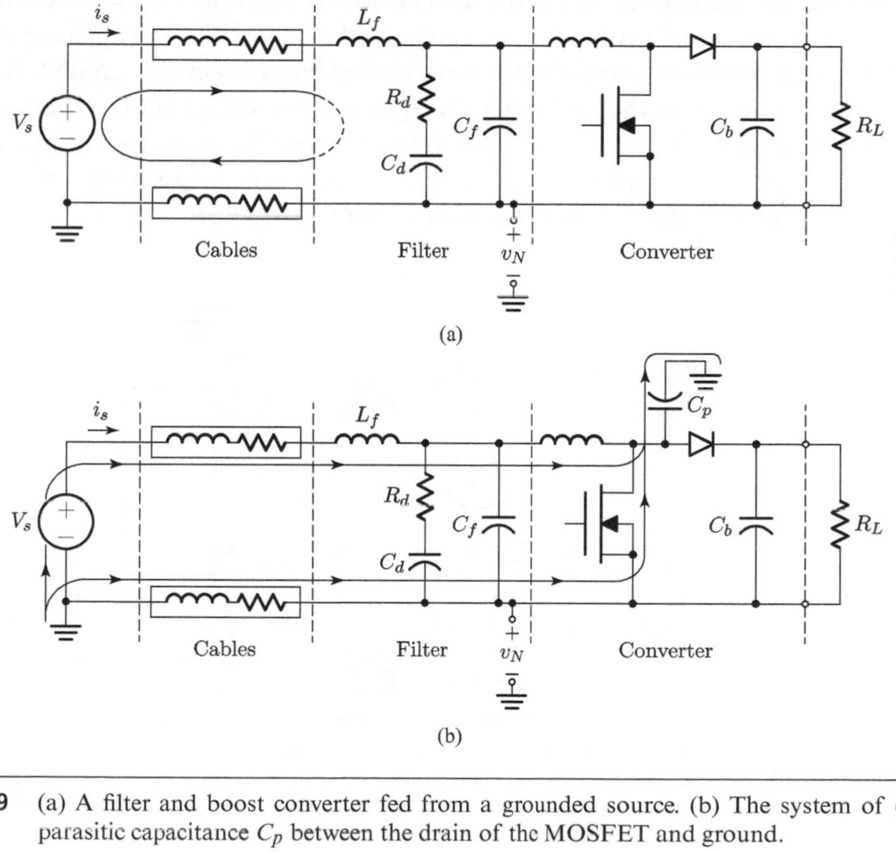

Figure 26.9 (a) A filter and boost converter fed from a grounded source. (b) The system of (a) with a parasitic capacitance C_p between the drain of the MOSFET and ground.

and back through the return cable. This is known as the differential-mode current. Given that the cables have some nonzero impedance (e.g., owing to inductance and resistance), there will be a nonzero voltage v_N between the negative terminal of the load and the chassis ground. This voltage thus deviates from what we would expect if the impedances of practical conductors were negligible.

Consider now the system of Fig. 26.9(b), which is identical to that of Fig. 26.9(a) except for the addition of a parasitic capacitance C_p between the drain of the MOSFET and ground. Such a parasitic capacitance might exist, for example, because the power MOSFET is heatsunk through a thin insulation pad to a case connected to the vehicle chassis. Importantly, capacitance C_p provides a path for high-frequency ac currents (e.g., caused by the switching of the MOSFET) to flow to ground. Figure 26.9(b) also shows an ac current path that loops from ground, through the return cable and converter, and back to ground through capacitor C_p, bypassing the filter. High-frequency currents through such a path, known as a common-mode path, can represent a substantial source of EMI, and must be addressed through appropriate filtering. Indeed, the possibility of such currents and noise is why many EMI standards require a LISN and EMI measurements on *both* power leads of a converter, not just a single lead.

While the above example is cast in terms of a dc automotive system, similar issues arise in many power electronic systems, including those connected to the ac grid. Figure 26.10 shows the typical setup of a device connected to a single-phase ac grid (e.g., in a home). Power is delivered from the ac grid (source v_{ac}) to the load device via a pair of cables: "hot" and "neutral." The neutral is tied to earth ground at some point (within the service-entry panel in a home), such that the voltage on the neutral is close to earth ground but is not exactly ground. A ground wire may also be connected to the load (typically to its case for safety) but is not intended to carry any current except during a fault (e.g., a short to the case). This ac system is similar to the dc system of Fig. 26.9(b), where parasitic capacitance to earth (e.g., between the device and its grounded case) can allow undesired common-mode current i_{CM} to flow.

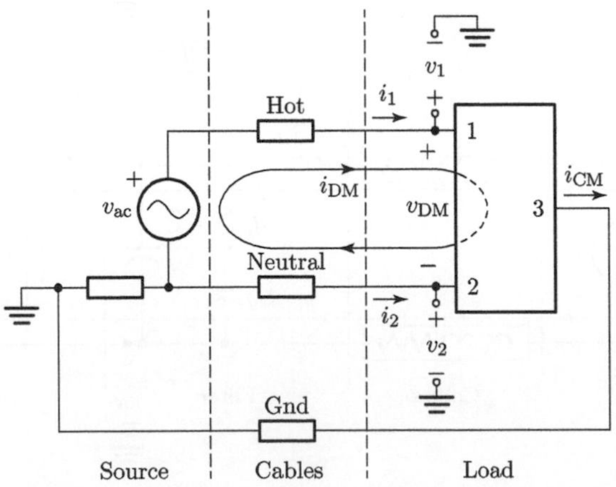

Figure 26.10 A device connected to the single-phase ac grid between "hot" and "neutral," and a ground wire connecting the case to ground for safety. Capacitance between device and ground forms a path for undesired common-mode current i_{CM} to flow.

The common- and differential-mode currents can be measured directly. If we clamp both the hot and neutral wires of Fig. 26.10 together in a current probe, we measure the common-mode current, $i_{CM} = i_1 + i_2$. Clamping both wires together in the probe again but reversing the direction the neutral wire passes through the probe, we now measure twice the differential-mode current, $2i_{DM} = i_1 - i_2$.

In the remainder of this section we describe means for representing and filtering the voltages and currents that can arise in systems like those in Figs. 26.9 and 26.10.

26.3.1 Common-Mode and Differential-Mode Definitions

We introduce a convenient means of representing the voltages and currents in systems like those of Figs. 26.9 and 26.10. The current i_1 flows from the ac source via the "hot" cable into terminal 1 of the device, and the current i_2 flows into terminal 2 through the "neutral" cable.

In the absence of an alternate return path for the current, we would have a current $i_1 = -i_2 = I$ circulating through the source and device, but this need not be the case in a situation such as that shown in Fig. 26.10. The alternate path in this case must support a current $i_1 + i_2$, which is referred to as the *common-mode* current i_{CM}:

$$i_{\mathrm{CM}} = i_1 + i_2. \tag{26.12}$$

It is helpful to think of the situation as still involving an underlying circulating current I that is, however, perturbed by an amount $i_1 - I$ to produce the actual current i_1 entering terminal 1 of the device, and is perturbed by an amount $-i_2 - I$ to produce the actual current $-i_2$ leaving terminal 2 of the device. The choice of I that has the least aggregate perturbation is the average value of i_1 and $-i_2$. This particular choice of circulating current I is referred to as the *differential-mode* current i_{DM},

$$i_{\mathrm{DM}} = \frac{i_1 - i_2}{2}. \tag{26.13}$$

With these definitions, the currents into terminal 1 and out of terminal 2 of the device can be rewritten in terms of their differential-mode and common-mode components:

$$i_1 = i_{\mathrm{DM}} + \frac{i_{\mathrm{CM}}}{2}, \qquad -i_2 = i_{\mathrm{DM}} - \frac{i_{\mathrm{CM}}}{2}. \tag{26.14}$$

As we shall see in the next section, it is possible – and often convenient – to design filtering operations that are separately aimed at the differential-mode and common-mode components; this is a very effective means of meeting EMI specifications.

We may also define common-mode and differential-mode components of the device terminal *voltages*. In the ideal case where $i_{\mathrm{CM}} = 0$, the circulating current $i_1 = -i_2 = i_{\mathrm{DM}}$ delivers power into the device via the voltage $v_1 - v_2$, and this is referred to as the differential-mode voltage, v_{DM},

$$v_{\mathrm{DM}} = v_1 - v_2. \tag{26.15}$$

With this definition, the power delivered to the device by the source in the usual case when i_{CM} is small is approximately $v_{\mathrm{DM}} i_{\mathrm{DM}}$. The difference between this and the actual delivered power is

$$v_1 i_1 + v_2 i_2 - v_{\mathrm{DM}} i_{\mathrm{DM}} = v_1 i_1 + v_2 i_2 - (v_1 - v_2)(i_1 - i_2)/2,$$

which can be rewritten as $v_{CM}i_{CM}$ if we define the common-mode voltage v_{CM} as the average of the voltages at the input terminals:

$$v_{CM} = (v_1 + v_2)/2. \tag{26.16}$$

26.3.2 Common-Mode and Differential-Mode Filtering

A key reason for expressing input waveforms in terms of their common- and differential-mode components is that one often uses different methods to filter these different components. Consider the design of a filter for an ac-grid-connected device like that in Fig. 26.10, which might be a power-factor-corrected power supply and its load, for example.

Figure 26.11 shows filters for such a system, in which the device is simply represented as a differential-mode current $i_{x,DM}$ from one device input terminal to the other and a common-

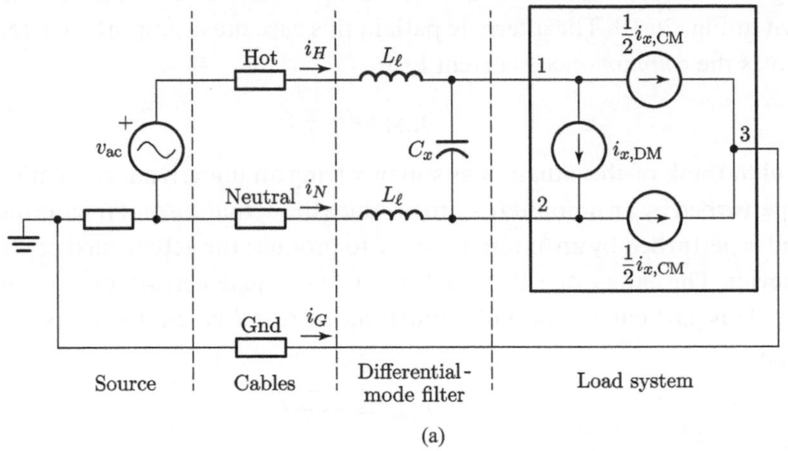

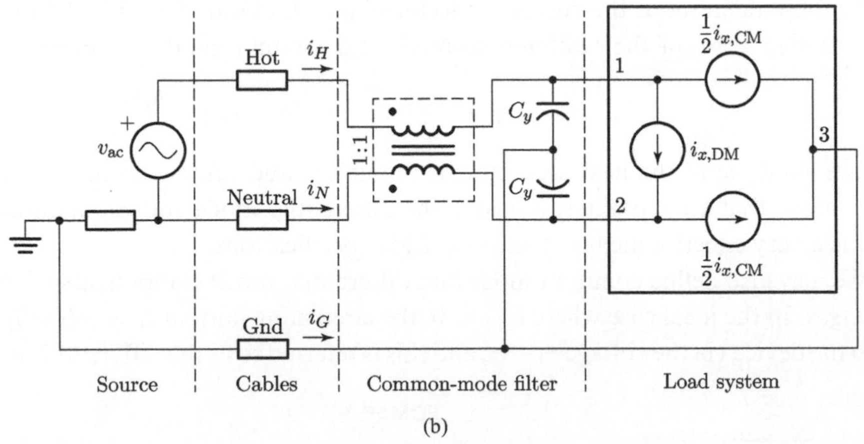

Figure 26.11 A load, modeled as common- and differential-mode current sources, connected through filters to the ac grid. (a) The connection using a differential-mode filter. (b) The connection using a common-mode filter including a common-mode choke.

mode current $i_{x,\mathrm{CM}}$ split equally between the input terminals. Figure 26.11(a) shows a system with an LC filter comprising a filter capacitor C_x connected across the device input terminals, and an inductor L_ℓ in series with each power lead. This filter is suitable for attenuating the high-frequency component of the differential-mode current $i_{x,\mathrm{DM}}$, while passing its power-frequency component to the source. While this structure is suitable as a differential-mode filter, it provides no attenuation of the common-mode component of the terminal currents ($\frac{1}{2}i_{x,\mathrm{CM}}$ into each terminal). This is because capacitor C_x acts as a "virtual open" to common-mode current.

Figure 26.11(b) shows a system with a filter suitable for attenuating the common-mode component of the terminal currents. This structure comprises a capacitor C_y connected between each of the device inputs and ground, and a 1 : 1 transformer connected with its windings in series with the two power leads, to act as a "common-mode choke." The filter capacitors C_y shunt the high-frequency common-mode terminal currents $\frac{1}{2}i_{x,\mathrm{CM}}$ to local ground (returning locally through the device). At the same time, to the extent that the transformer acts ideally, it will prevent any common-mode current component from passing through the power leads, while freely passing differential-mode current. The fact that it inhibits the flow of common-mode current but not differential-mode current is why a 1 : 1 transformer connected this way is referred to as a common-mode choke.

Figure 26.12(a) shows a filter including the components of both Figs. 26.11(a) and (b). It is suitable for attenuating both common- and differential-mode noise. The magnetizing inductance L_μ and leakage inductances L_ℓ of the common-mode choke are shown as explicit circuit elements. In reality, the ability of the common-mode choke to inhibit common-mode current is limited by its finite magnetizing inductance L_μ, which carries the common-mode component of the line current $i_H + i_N$. Because the large differential-mode component of line current $\frac{1}{2}(i_H - i_N)$ ideally does not magnetize the core of the common-mode choke, one can use an ungapped core and achieve a large impedance to common-mode current flow with a very small magnetic component.

In practice, a common-mode choke is typically wound with its two windings widely separated on the core, such that it exhibits significant leakage inductances L_ℓ. These leakage inductances are used to help provide differential-mode filtering. Discrete inductors may also be included in series with the hot and neutral leads to increase the effective value of L_ℓ, but if so, must be sized to carry the line current.

Figure 26.12(b) shows a common-mode model for the filter in Fig. 26.12(a), where we represent the source side of the filter as an ideal common-mode voltage $v_{y,\mathrm{CM}}$ expressed in terms of voltages v_H and v_N. This model may be derived by setting differential components of voltages and currents to zero, treating the ideal transformer as an open circuit to common-mode currents, and recognizing that the top and bottom portions of the resulting circuit are symmetric about ground and may be combined.

As seen in Fig. 26.12(b), common-mode filtering is provided by the large magnetizing inductance L_μ (augmented by a leakage component $L_\ell/2$) and net capacitance $2C_y$. Having the large inductance L_μ filtering common-mode current is useful because the allowable value of capacitors C_y connected to ground in ac systems is typically quite small (e.g., $< 10\,\mathrm{nF}$). This is because it is not permissible to allow significant power-frequency current (e.g., 60 Hz current) to flow from the ac source into ground via the capacitors C_y.

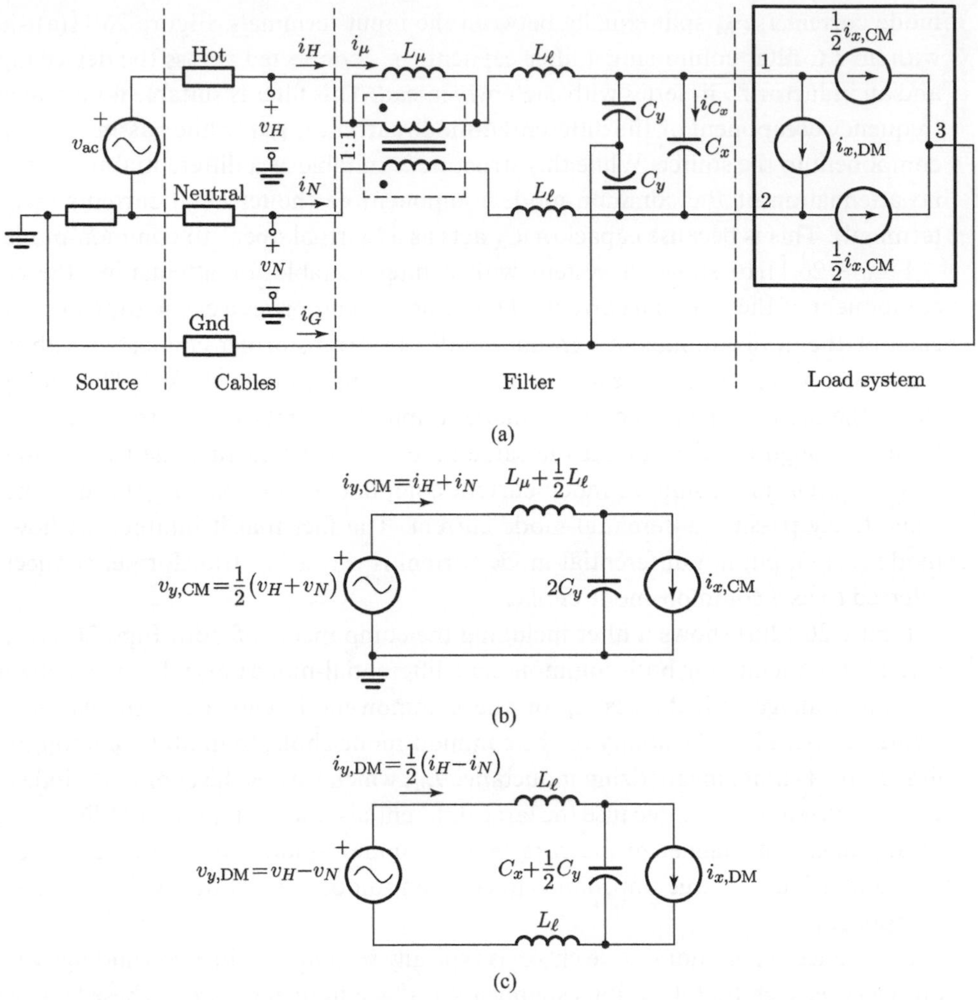

Figure 26.12 (a) A load and filter connected to the ac grid. The load is modeled with common- and differential-mode current sources. (b) A model representing common-mode filtering in (a). (c) A model representing differential-mode filtering in (a).

Figure 26.12(c) shows a differential-mode model for the filter in Fig. 26.12(a), where we represent the source side of the filter as an ideal differential-mode voltage $v_{y,\text{DM}}$ expressed in terms of voltages v_H and v_N. This model may be derived by setting common-mode components of voltages and currents to zero, recognizing that the ideal transformer acts as a short to differential-mode current, and observing that ground becomes a virtual open by symmetry.

As seen in Fig. 26.12(c), the inductance for differential-mode filtering is provided by the leakage inductances L_ℓ, which appear in series to differential-mode currents for a total inductance of $2L_\ell$. The capacitance for differential-mode filtering is provided by capacitor C_x augmented in parallel by the effective series connection of capacitors C_y, for a total capacitance $C_x + C_y/2$.

By appropriate selection of components in the filter of Fig. 26.12(a), we can thus filter both common-mode and differential-mode noise. Higher-order filters can be similarly constructed.

As a final practical note regarding EMI filters for ac line applications, the capacitor names C_x and C_y were chosen for a specific reason: x-type capacitors are a class of capacitors that are specifically designed (and safety rated) for use across the ac line, while y-type capacitors are rated for placement between the line and ground. One would thus use x- and y-type capacitors for C_x and C_y, respectively, in an ac-line filter.

26.4 Parasitics and Circuit Layout

A challenge in designing power electronic equipment is the impact of component and layout parasitics on circuit performance. Filters are particularly sensitive to parasitics, especially when seeking high levels of attenuation. In this section we describe how some common component and circuit parasitics affect the performance of filters. We also describe some approaches towards reducing the impact of these parasitics.

Component parasitics – such as the parasitic resistance and inductance of capacitors – are an important consideration in filter design. Figure 26.13(a) shows a model for a practical capacitor, including its ESR R_c and equivalent series inductance (ESL) L_c, which model the conduction loss and parasitic magnetic energy storage in the capacitor, respectively. A practical capacitor might thus be thought of as a series *RLC* network for the purposes of filter design. Figure 26.13(b) shows the asymptotes of the impedance magnitude $|Z_c|$ of a practical capacitor, for the case where $R_c > \sqrt{L_c/C}$ (which is typical of electrolytic capacitors, for example). At low frequencies, the capacitor behaves nearly ideally, with its impedance magnitude falling as $1/\omega C$. At intermediate frequencies the capacitor impedance is dominated by its ESR R_c (so looks resistive), while at high frequencies the capacitor impedance is dominated by its ESL L_c, with impedance magnitude rising as ωL_c. A capacitor for which $R_c < \sqrt{L_c/C}$ (as is typical for a ceramic or film capacitor) will have $|Z_c|$ with the same low- and high-frequency asymptotes, but with a sharp resonant drop to its minimum value R_c at $\omega = 1/\sqrt{L_c C}$. For either case, the frequencies where these parasitic effects become important depend upon component size and value, occurring at lower frequencies for larger-valued and physically larger capacitors.

The basic structure of an *LC* filter and its ideal current gain are shown in Fig. 26.3(a) and (b), respectively. Figure 26.14(a) shows a model for an *LC* filter in the presence of capacitor parasitics R_c and L_c, while Fig. 26.14(b) shows its current gain. Unlike the ideal filter in which current gain above cutoff falls as $|H_i(j\omega)| \approx 1/(\omega_s^2 L_f C_f)$ (-40 dB/dec) indefinitely, the current gain with capacitor parasitics rolls off only as $|H_i(j\omega)| \approx R_c/(\omega L_f)$ (-20 dB/dec) for frequencies where ESR dominates the capacitor's impedance, and becomes constant at $|H_i(j\omega)| = L_c/(L_f + L_c) \approx L_c/L_f$ for frequencies where ESL dominates the capacitor's impedance. Where an ideal *LC* filter might provide a tremendous amount of attenuation, a practical *LC* filter limited by capacitor parasitics exhibits much worse performance at high frequencies. One must thus pay careful attention to the capacitor parasitics when estimating the achievable attenuation of a filter.

In some cases, to achieve low shunt impedance over a wide frequency range, capacitors of different types, values, and physical sizes might be placed in parallel in a filter, with attention

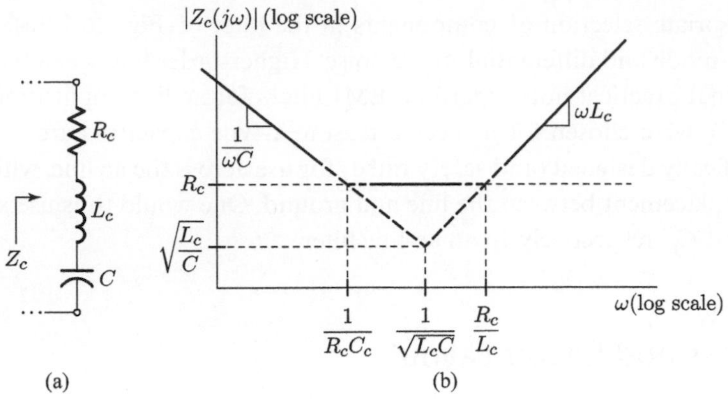

Figure 26.13 (a) A model for a practical capacitor with ESR (R_C) and ESL (L_C). (b) Asymptotes for its impedance magnitude for the case where $R_C > \sqrt{L_C/C}$.

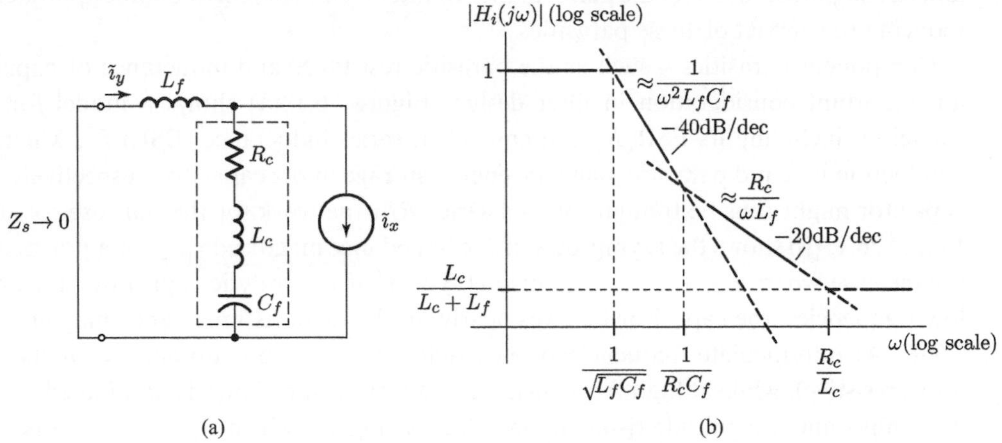

Figure 26.14 (a) A ripple current source $\tilde{\imath}_X$ driving an LC filter (damping not shown), in which the capacitor C_f has ESR R_c and ESL L_c. (b) Asymptotes of the filter input-to-output current-transfer function magnitude $|H_i|$ for the case where $R_c > \sqrt{L_c/C_f}$.

paid to possible resonances between the paralleled capacitors. Alternatively, or additionally, one might select components of a first filter stage to provide good attenuation over a first frequency range, and augment it with a second filter stage that provides good attenuation over a second (e.g., higher) frequency range.

Just as capacitors have parasitics, a practical inductor can exhibit an "equivalent parallel capacitance" (EPC) C_l that spoils its performance at high frequencies (i.e., reduces its high-frequency impedance). For a well-designed filter inductor, this parasitic often becomes important at frequencies higher than those at which capacitor parasitics become important. Nonetheless, at frequencies for which C_l limits the impedance of an inductor, it can cause significant degradation in achievable filter attenuation.

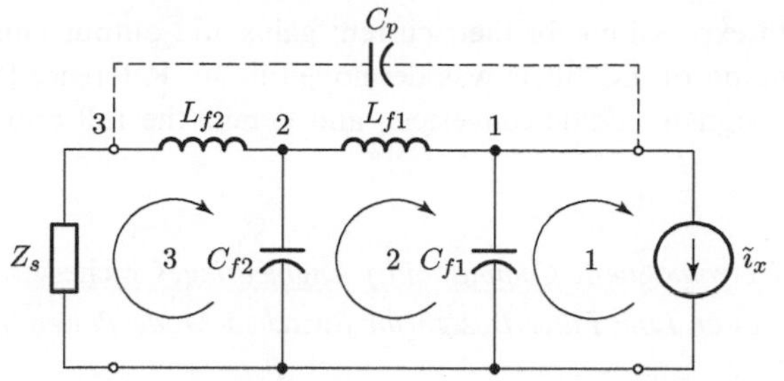

Figure 26.15 A multi-stage filter with a parasitic capacitance C_p bridging across the filter.

In addition to component parasitics, there can be substantial degradation in filter performance owing to layout parasitics and component-to-component coupling. Consider the multi-stage filter of Fig. 26.15 (shown without damping), which has a parasitic capacitance C_p bridging across the filter. Such a capacitance can easily arise due to layout (e.g., from circuit-board traces of the input and output connections of the filter passing near one another). Given the high attenuation theoretically achievable by such a filter, even a very small (e.g., pF-scale) capacitance C_p can cause enough noise from the filter input (node 1) to be injected at the filter output (node 3) to dramatically reduce filter performance. It is for this reason that designers often seek to keep the "quiet" of the filter physically far and/or shielded from both the filter "noisy" port and the converter power stage. (Lamentably, this is not always possible to do.) Likewise, it is also desirable to minimize the conductor area of "noisy" circuit nodes to reduce capacitance to sensitive ones.

Filter performance can also be harmed by magnetic coupling. For example, mutual inductance between the input (loop 1) and output (loop 3) of the filter in Fig. 26.15 can inject noise at the filter output. Reducing mutual inductance between "noisy" and "quiet" loops through spacing, shielding, and minimizing loop size is thus desirable. Similarly, inadvertent mutual coupling between inductors (such as L_{f1} and L_{f2}) from fringing fields can increase noise transmission, as can magnetic coupling between inductors and sensitive circuit loops. Careful design, shielding, and positioning of magnetic components and circuit loops are thus critical to achieving high filter attenuation.

Parasitic coupling such as described above is often the source of seemingly anomalous behavior that can be frustrating to address. To reduce coupling, commercial EMI filters sometimes place each filter stage inside its own shielded enclosure (with pass-throughs between stages), but this is impractical to do in many power electronics designs. Instead, the designer must often simply be clever about layout and component design to minimize parasitic coupling.

Notes and Bibliography

The excellent book by Ott [1] is a treasure trove of information about EMI measurements and noise reduction techniques. While older, the book by Nave [2] also has a wealth of useful information and focuses specifically on filters for power electronics. A variety of power filter topologies can be found in [3], along with expressions for their current gains and output impedances. The technique described for optimal damping of LC filters was developed in [4]. Reference [5] provides an extensive treatment of input filter design for dc/dc converters, and details the full criteria to avoid input filter/converter interactions.

1. H. W. Ott, *Electromagnetic Compatability Engineering*, Chichester: Wiley, 2009.
2. M. J. Nave, *Power Line Filter Design for Switched-Mode Power Supplies*, New York: Van Nostrand Reinhold, 1991.
3. T. K. Phelps and W. S. Tate, "Optimizing Passive Input Filter Design," *Proceedings of POWERCON 6*, pp. G1.1–G1.10, 1979.
4. R. W. Erickson, "Optimal Single Resistor Damping of Input Filters," *1999 IEEE Applied Power Electronics Conference*, pp. 1073–1079, 1999.
5. R. W. Erickson and D. Maksimović, *Fundamentals of Power Electronics*, 3rd edn., Cham: Springer, 2020.

PROBLEMS

26.1 Create a preliminary filter design similar to that in Example 26.1, but for a converter that has an input voltage V_1 of 20–25 V and an output voltage V_2 of 10 V at a power P_o of 50–100 W.

26.2 Considering the system of Fig. 26.10, show that we can express the total instantaneous power flowing into terminals 1 and 2 of the device as $P = v_1 i_1 + v_2 i_2 = v_{\text{DM}} i_{\text{DM}} + v_{\text{CM}} i_{\text{CM}}$.

26.3 Starting from the filter in Fig. 26.12(a), derive the common-mode and differential-mode models of Figs. 26.12(b) and (c).

26.4 Figure 26.16 shows an "EMI noise separator." This circuit is designed to measure the common- and differential-mode voltages of the two input terminals (at the left of the circuit), appearing across the two terminating resistors as voltages v_{CM} and v_{DM}, while maintaining the input impedances seen at ports 1 and 2 (between each of the left-hand input terminals and ground) at 50 Ω.

Difficulties constructing such a circuit include managing the parasitic inductances and capacitances of the transformers. (This is addressed in part in the referenced paper through the use of "transmission-line transformers" and careful modeling and design.) For the purposes of this problem, consider the two transformers to be ideal.

(a) Show that the voltage developed across the 25 Ω terminating resistor is the common-mode component of the voltages v_1 and v_2.

(b) Show that the voltage developed across the 100 Ω terminating resistor is the differential-mode component of the voltages v_1 and v_2.

(c) Show that the input impedance seen looking into port 1 (the port associated with v_1, that is, between the top left terminal and ground) is 50 Ω regardless of the voltage v_2. By symmetry, both ports 1 and 2 have 50 Ω terminations to ground.

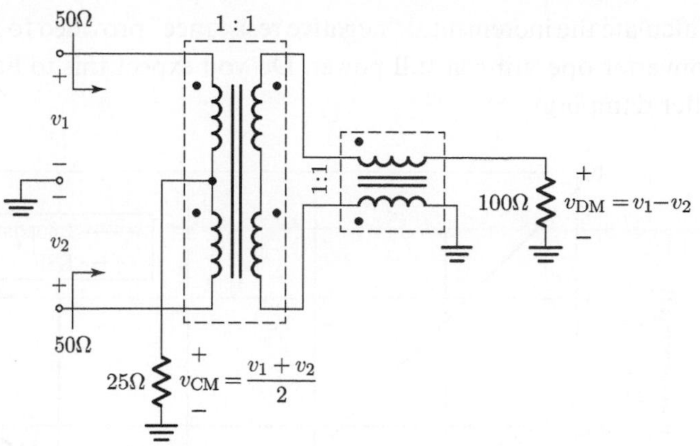

Figure 26.16 An EMI noise separator explored in Wang et al., "Characterization, Evaluation and Design of Noise Separator for Conducted EMI Noise Diagnosis," *IEEE Trans. Power Electronics*, vol. 20, pp. 974–982, July 2005.

26.5 A filter with capacitor parasitics is shown in Fig. 26.14(a).

(a) Derive the asymptotes for the current gain magnitude $|H_i(j\omega)|$ shown in Fig. 26.14(b) for this non-ideal filter.

(b) Given a filter inductor $L_f = 10\,\mu\text{H}$ and a filter capacitor with a parasitic inductance of $L_c = 10\,\text{nH}$, what current gain magnitude does this filter asymptote to at high frequencies, expressed in dB?

(c) If the capacitor has a very small ESR ($R_c \ll \sqrt{L_c/C_f}$), above what frequency will we start to approach this asymptote?

26.6 Consider a buck converter and LC input filter like that in Fig. 26.8. The 25 W buck converter operates in heavy continuous conduction mode at a switching frequency of 200 kHz, and generates a 5 V output from a 9 V input. The filter capacitor C_f is a 220 μF Sanyo OSCON capacitor with an rms current rating of 3.7 A. Its impedance characteristic is shown in Fig. 26.17. The filter inductor L_f is 220 μH, and is considered ideal for the purposes of this problem. The input source supplying the buck converter has negligible output impedance.

(a) Estimate the ESR and ESL of the filter capacitor C_f. (You may treat the ESR as constant at its minimum value, neglecting variations with frequency.)

(b) Select numerical values for the damping components C_d and R_d such that there is less than 10 dB of peaking in the transfer function from filter current input to filter current output $H_i(j\omega)$.

(c) Plot $|H_i(j\omega)|$, including the effect of filter capacitor parasitics. How does this compare to a filter with an ideal capacitor?

(d) Calculate the incremental "negative resistance" provided to the filter by the closed-loop converter operating at full power. Do you expect this to have a significant impact on filter damping?

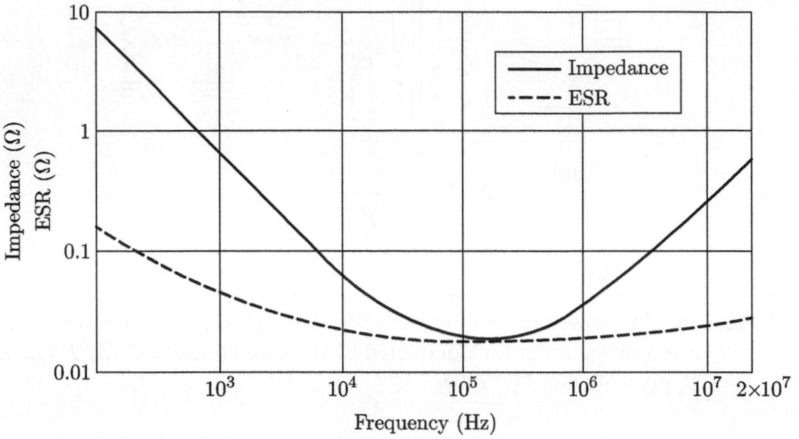

Figure 26.17 Impedance characteristics of a 220 μF OSCON capacitor.

Index

α_T, *see* base transport factor
β_F (current gain), *see* BJT
γ, *see* margin angle
γ_E, *see* emitter efficiency
ω_s-component, 314, *see* control
ψ_o, 470
τ_a, *see* ambipolar lifetime

ac/ac converter, 12, 20–22, 289
  ac controller, 21
  solid-state transformer, 290
  with dc link, 21, 289
    energy storage in, 290
  with dc-link, 22
    sizing bus capacitor, 293
  See also cycloconverter
ac/dc converter, 13–18
  control of, 425–428
  filter, 16
  See also rectifier
active resonant-commutated pole, *see* ARCP
$A_C W_A$ (core area product), *see* inductor
Alcator-C, 578
ambipolar diffusion constant, 501
ambipolar diffusion length, 501
ambipolar lifetime, 500
Ampère's law, 547
ampere-turns, 550
angle of retard, *see* SCR
apparent power, 38, 624
  See also power factor
ARCP, 746
ASCR, *see* SCR
assumed states, 30
audio susceptibility, 349
average value analysis, 81
averaged circuit, *see* control
averaging
  generalized circuit, 354

base drive, 693–695, 704
  See also BJT
battery charger, 110
beat frequency, 352
BIBO, *see* control
bipolar junction transistor, *see* BJT
BJT, 2, 4, 478–482, 511–516
  ν region, 511
  base current, 480
  base drive, 704–716
    anti-saturation clamp, 716
    pedestal and porch, 714
    storage delay time, 715
    turn-off requirements, 715
    turn-on requirement, 714
  base transport factor, 481
  collector–base breakdown voltage, 442
  current gain, 442
    calculation of, 481
    common-base (α_F), 480
    common-emitter (β_F), 480, 515
    common-emitter (h_{FE}), 442
  cut-off, 512
  Darlington connection, 443
  drift region, 511
  emitter defect, 480
  emitter efficiency, 481, 515
  high-level injection, 511
  operating regions, 511
    forward active, 513
    hard-saturation, 515
    quasi-saturation, 514
    saturation, 482
  parameters, 480
  physical model, 478
  punch-through, 511
  recombination defect, 480
  reverse injection, 480
  safe operating area, 486, 487, 724
  second breakdown, 487
  sustaining voltage, 513
  switching dynamics, 515
  terminal currents, 478
  v–i characteristics, 442
Bode plot, 338, 407–413
Boltzmann's constant, 463
boost converter, 84, 106
  conversion ratio, 88
  switch implementation, 86
  with ZVS, 744–746
bootstrap circuit, 710
bridge circuit, 14
buck converter, 84
  as part of dc/ac bridge, 194
  as power amplifier, 180
  element values, 99
  EMI input filter, 779–780
  EMI measurements, 772
  linearizing, 331
  switch implementation, 84–86
  with bilateral power, 87
  with ZVS, 741–742
buck/boost converter, 90, 91
  as dc/ac converter, 196
  averaged circuit for, 326, 348
  control, 420–422
    compensator design, 410–414, 416–420
    current-mode, 369
    feedback, 320, 381
    feedforward, 317
    frequency response, 403
    in continuous conduction, 380
    open-loop, 317
    stability, 381, 401
  discontinuous conduction, 326, 332
  element values, 102
  linearized averaged model, 384, 388–390
  non-inverting, 94, 95
    See also SEPIC converter
  sampled-data model, 367
  state-space model, 362–364
    generalized, 369
  step response, 349

canonical cell, 82, 89, 92
capacitor, 452–455
as clamp, 139
ceramic, 454–455
age-dependent permittivity, 454
Class II codes, 454
classes, 454
structure, 454
use at high frequencies, 455
voltage-dependent permittivity, 454
dielectric parameters, 452
dissipation factor, 452
electric double-layer (DLC), 454
electrolytic, 453
aluminum, 453, 777
tantalum, 453
film
construction, 452
dielectric material, 452
metalized, 452
parasitic inductance, 453
polycarbonate, 452
polypropropylene, 452
losses in, 439
metal-foil, 452
x- and y-type, 787
cat whisker, 437
characteristic polynomial, 337, 387
roots, 337
charge multipliers, *see* switched-capacitor converter
chopper, *see* dc/dc converter
circuit commutated turn-off time, *see* SCR
circuit replacement by equivalent source, 33, 79
clamp circuit, 697, 698, 738–739
clamped inductive load, 446, 538
Clarke transform, 231–234
power invariance, 232
Clarke, Edith, 232
Cockcroft–Walton voltage multiplier, 131
coercive force, 555
common mode, 783
(*see also* EMI)
common-mode
choke, 785
transient immunity, 710
conductivity, 466
calculation of, 467
conductivity modulation, 494
See also high-level iinjection
control
ω_s-component, 314
averaging to, 350–354
local, 350
model for *RLC* circuit, 364
state-space model for, 364
bandwidth, 329
BIBO stability, 393
circuit averaging, 324–327
KVL and KCL, 324
LTI component, 324
of buck/boost converter, 326
of controlled rectifier, 325
switching function, 327
compensator design, 410–414, 416–420
constant frequency, 84
current-mode, 329, 369, 397, 415–420
digital, 425–428
of ac/dc preregulator, 425–428
duty-ratio, 397
dynamic behavior, 314–322
feedback, 318–322, 332–342, 381, 422–425
classical, 333
design, 405–413
nonlinear state, 423–425
pole placement, 423
feedforward, 316–318, 321, 329
frequency response, 393, 402–405
linearizing a circuit, 331–332
local ω_s-component, 316
local average, 315
loop gain, 338–340, 409
loop transfer function, 335, 339
MIMO, 393
models
ω_s-averaged circuit, 351
averaged-circuit, 323–329, 346–350
averaging a switch, 347–350
continuous-time LTI, 385–394
continuous-time state-space, 354–365
delay elements, 367
discrete-time, 399–405
dynamic, 322–323
for ω_s-component, 353
for controller, 372
for interconnected system, 372
generalized state-space, 368–372
linearization, 330
linearized, 329–332
linearizing, 378–382, 385, 395–398
LTI, 323
periodically varying, 357
piecewise LTI, 362, 394–395
sampled-data, 327, 365–368, 403
small-signal, 378
state-space, 323, 346, 359, 361–365
multi-loop, 414–422
design, 414
of buck/boost converter, 420–422
nonlinearities, 342
of dc machine, 414
open-loop, 316–318
phase crossover frequency, 341
phase margin, 341
PI controller, 320
PID controller, 320
robustness, 336, 340–342
sensitivity function, 335
stability, 330, 336, 381, 401–402
asymptotic, 400
state-space averaging, 363
transfer function
of plant, 334
system, 336
unity-gain crossover frequency, 338
controlled rectifier, 57
averaged-circuit model for, 325
feedback control of, 334
loop transfer function, 339
See also phase-controlled converter
converter
ac/ac, *see* ac/ac converter
ac/dc, *see* ac/dc converter
boost, *see* boost converter
buck, *see* buck converter
buck/boost, *see* buck/boost converter
Ćuk, *see* Ćuk converter
DAB, *see* dual active-bridge
dc/ac, *see* dc/ac converter
See also rectifier, inverter
dc/dc, *see* dc/dc converter
direct, *see* direct converter
dual active-bridge, *see* DAB
converter, 150
as SST, 303
flyback, *see* flyback converter
flyforward, *see* flyforward converter
forward, *see* forward converter
indirect, *see* indirect converter
matrix, *see* matrix converter
phase-controlled, *see* phase-controlled converter
resonant, *see* resonant converter

SEPIC, *see* SEPIC converter
switched-capacitor, *see* switched-capacitor converter
transformer-isolated, 192–194
up/down, *see* buck/boost converter
variable-frequency, 167
Zeta, *see* Zeta converter
core factor, *see* inductor
core loss, 561–684
data, 565
eddy current, 561, 675–678
ferrrous alloys, 561
hysteresis, 562, 678–679
laminations, 561
Steinmetz equation, 562, 679
improved, 680
with non-sinusoidal flux, 680–684
core volume $\mathcal{V}_c$, 554
crossover distortion, 181
reduction of, 181
Ćuk converter, 92, 94, 154
element values, 102
switch stress, 103
Curie temperature T_C, 575
current commutation, 35
current-source inverter
source inductor, 171
cycloconverter, 20, 21, 289
high-frequency-link, 299
naturally commutated, 294–297
switch inplementation, 300
with polyphase output, 297

D_a, *see* ambipolur diffusion constant
damped natural frequency, 349
damping factor, 352
datasheet, 538
dc machine
speed control of, 414
dc/ac converter, 167–201
as power amplifier, 181
control of, 168
harmonic cancellation, 177–179
harmonic elimination, 175–177
harmonic reduction, 172–175
See also waveform properties
high-frequency bridge, 182
inverter current control, 187–189
multi-level, 190–192
using flying capacitor, 190–192
notched waveforms, 175–177
power balance between ports, 196
pulse-width modulated, 179–187
pulse-width modulation
generating $d(t)$, 184
THD calculation, 183
unfolding bridge, 180
waveshaping, 180
stepped waveform, 179
compared to notched, 179
using buck/boost converters, 196
variable-frequency bridge, 167
with ac-voltage-source load, 170
control of, 170
dc/dc converter, 12, 18–20
discontinuous conduction, 106–108
filter element values, 95
isolated, 136–158
active clamp, 143
clamp energy recovery, 139
clamp voltage, 139
control of, 146
core resetting, 138
Ćuk converter, 154
double-ended, 145–150, 152
dual active-bridge, 150
flyback, 136, 153–154
flyforward, 141
forward, 136
hybrid bridge, 141
leakage inductance effects, 155–157
magnetizing current, 138
push–pull, 150
SEPIC, 154
transformer-coupled clamp, 140
voltage clamping, 138
with multiple outputs, 158
Zeta converter, 155
multi-level
using flying capacitor, 104–106
pulse-width modulated, 76
switch implementation, 23, 84–88
switch stress, 142
See also direct converter, indirect converter
deadbeat, 432
delay angle, *see* SCR
describing function, 264
device circuit symbols, 2
dielectric relaxation
in a metal, 462
in a semiconductor, 462
differential mode, 783
(*see also* EMI)
diffusion length, 475
diode, 3, 439–442, 494–511
analysis of, 472–478
bipolar, 440, 494–507
avalanche breakdown, 497
conductivity modulation, 495
fast-recovery, 441, 542
forward bias, 499
forward drop, 440, 441
forward recovery, 440, 504
high-current effects, 503
leakage current, 440, 495–497
pin diode, 498–507
reverse recovery, 441, 495, 506
reverse recovery time, 441
slow recovery, 441
switching dynmaics, 504
temperature dependence, 503
freewheeling, 32
peak inverse voltage, 440
punch-through, 478
reverse leakage current, 476
saturation current, 476
Schottky, 438, 494, 507–511
forward drop, 440, 441
leakage current, 440
recovery characteristics, 441
SiC, 529
snappy, 441
soft-recovery, 441
voltage rating, 440
Zener
as clamp, 139
direct converter, 82–88
ripple-frequency model, 96
switch stress, 103
with bilateral power, 86
discrete-time, *see* sampled-data
displacement factor, *see* power factor
distortion
measures and effects of, 38–44
distortion factor, *see* power factor
DLC, *see* electric double-layer
Dowell's equation, 660
down converter, *see* buck converter
dual converter, 72
duty ratio, 19, 77, 88
dynamic phasor, 351, 373

eddy current, 438
Edison, Thomas, 27, 49
eigenvalue, 31, 387
Einstein relation, 501
electric double-layer, *see* capacitor

electromagnetic interference, *see* EMI, 771
electron
  mobile, 462
EMI, 570, 599, 725, 747, 771–789
  LC filter, 773–777
    damping, 774–777
    negative damping, 775
    output impedance, 776
  circuit layout, 788–789
    magnetic coupling, 789
    parasitic capacitance, 789
  common- and differential-mode filter, 785
  common-mode, 783–784
    choke, 785
    filter, 785
    model, 785
  conducted, 771–773
  converter–filter interactions, 777–780
    negative resistance, 778
    stability, 778
  differential-mode, 783–784
    filter, 785
    model, 785
  element parasitics, 787–788
  filter design, 773–780
  input filter
    control dynamics, 779
  LISN, 772
  measurement, 771–773
  multi-stage filter, 777
  specifications, 773
equivalent series resistance, *see* ESR
ESR, 110, 350

FHA, *see* fundamental harmonic approximation
filter, 16
  effectiveness, 79
  for direct converter, 98
  higher-order, 81
  low-pass, 16, 76–78
  second-order, 16
  See also EMI
firing angle, *see* SCR
flux linkage λ, 547
flyback converter, 136
  in dc/ac application, 195
flyforward converter, 141
flying capacitor, 104–106, 190–192
forward converter, 136
  isolated single-ended, 136–144
forward Euler method, 359
Fourier series, 28, 43, 172–175, 314
fundamental harmonic approximation, 256

gallium nitride, *see* GaN
GaN, 494, 497
  dynamic on-state resistance, 530
  HEMT, 530
  MOSFET, 528–530
  See also MOSFET
gate drive, 693–695, 704
  See also MOSFET, SCR
gate turn-off thyristor, *see* GTO
Greinacher voltage multiplier, 131
GTO, 4, 450, 537
  snubber requirement, 450
  turn-off gain, 450
  turn-off requirements, 450

harmonic cancellation, 177–179
harmonic elimination, 175–177
heat sinking, 701, 753
heat transfer, 753
  conduction, 753
  convection, 753
    film coefficient of, 755
    flow boundary, 762
    forced, 755
    natural, 755
    Reynolds number, 756, 762
  junction temperature, 760
  radiation, 753, 754
  thermal interface, 758, 760
    materials, 756
    thermal grease, 756
  thermal system, 758
  through PC board, 761
heatsink, 701
  finned, 701, 761
h_{FE} (current gain), *see* BJT
high-level injection, 494, 499, 500, 502
  in BJT, 511
homogeneous response, 31
hysteresis, 555

IGBT, 2, 4, 524–528
  capacitances, 527
  collector current tail, 528
  collector cut-off current, 527
  dynamic behavior, 527
  equivalent circuit, 447
  forward drop, 526
  gate drive, 704–713
    requirements, 706
  off state, 527
  physical structure, 524
  switching dynamics, 448
indirect converter, 82, 89–91, 95
  conversion ratio, 89
  ripple-frequency model, 97
  switch implementation, 90
  variations of, 91
inductance
  calculation of, 547–549, 551
  definition of, 547
  flux linkage λ, 546
  second-order effects, 552–553
inductance matrix, *see* transformer
inductor, 455, 546–555
  ac
    loss minimization, 630–631
    power rating, 624–625
  core area product ($A_C W_A$), 625
  core factor, 615, 617–618
  core selection example, 617
  design example, 636–638
  energy storage, 553–555
  energy storage limit, 621–623
  filter, 618
  filter inductor, 613
    design of, 563
  loss model, 564
  losses in, 439
  mean length of turn, 613
  packing factor, 558
  shield winding, 598–600
    cutoff frequency, 600
    design of, 600
  toroidal, 548
  winding resistance (dc), 614
  with air gap, 551, 613
  with distributed gap, 551
insulated-gate bipolar transistor, *see* IGBT
inverter
  current-source inverter, 167, 171
    δ control of, 171
    θ control of, 171
    average power calculation, 171
    direction of power, 171
    need for bidirectional switches, 172
  definition, 14

multi-level, 237–240
neutral-point clamped, 238–240
See also NPC
phase-controlled, 57
power factor, 168
three-phase, 221–228
PWM of, 227–228
switching states, 224–227
voltage-source
non-unity-power-factor load, 169
voltage-source inverter
non-unity-power-factor load, 169

k_θ, *see* displacement factor
k_d, *see* distortion factor
Kelvin connection, 708, 710, 713
k_p, *see* power factor

L_a, *see* ambipolar diffusion length
leakage inductance, 155
level shifting, 709
light dimmer, 457
line-impedance stabilization network (LISN), *see* EMI
linear time-invariant (LTI), 322, 357
linearity, 356
LISN, *see* EMI
litz wire, 560
Litzendraht, 560
load regulation, *see* rectifier
local ω_s-component, 316
local average, 315, 323
local averaging, 346
local fundamental, 350
low-level injection, 478, 494

magnetic circuit, 550, 583
transference, 598
magnetic circuit model
for component design, 605
stored energy, 605
synthesizingfrom electric circuit model, 603
transference, 600
magnetic components, 455
core geometries, 456
losses in, 566
magnetic diffusion, 643–655
diffusion equation, 643–644
one-dimensional, 644–646
diffusion wave, 646
skin depth, 645
magnetic materials, 456, 575–576
ferrites, 456, 561, 576
ferrous alloys, 456, 561
magnetics, 545
ac, 624–634
ampere-turns, 549
core geometries, 615
design of, 612–638
electric circuit analog, 549
example, 550
permeance, 549
reluctance, 549
losses in, 456, 557–684
modeling, 597
scaling with volume, 558
materials
performance factor, 634–636
packing factor k_u, 613
residual flux, 555–557
RM core, 558
data, 615
saturation, 556
thermal constraints, 618–621
winding loss, *see* winding loss
window utilization factor k_u, 613
magnetizing current
resetting, 138
magnetomotive force, *see* MMF
matrix converter, 305
Maxwell's equations, 546, 546
McMurray, William, 302
mean length of turn, *see* inductor
metal-oxide field-effect transistor, *see* MOSFET
metal-oxide varistor
as clamp, 139
method of assumed states, 46
MMF, 550
MOSFET, 2, 4, 483–524
body diode, 446
calculating fall time, 707
capacitances, 485
carrier saturation velocity, 519
channel, 483
inversion of, 483
resistance, 446
characteristics, 484
datasheet, 538–541
DMOS, 517
drain resistance, 446
dynamic behavior, 521
enhancement mode, 484
extended drain, 517
GaN, 528–530
gate capacitance
charge vs. voltage, 707
gate drive, 446, 704–713
common-mode transient immunity, 710
dead-time, 709
EMI problem, 713
level shifting, 709
loss, 708
requirements, 706
shoot-through, 708
lateral, 483
linear region, 484
Miller effect, 705
on-state resistance, 519
parallel operation, 711
oscillation in, 711
parasitic transistor, 446
physical structure, 445
$R_{DS(\text{on})}$, 446, 521
regions of operation, 442
safe operating area, 486, 487
saturation region, 485
SiC, 529
common-mode choke, 713
Kelvin connection, 708, 710, 713
source–body short, 446
specific on-state resistance, 519
superjunction, 523–524
switching waveforms
with clamped inductive load, 446
threshold voltage, 445, 483, 486, 706
temperature dependence, 711
to replace diode, 122
transconductance, 485
trench, 520
UMOS, 520
v–i characteristics, 442
multi-level converter, 190
diode-clamped, 191
mutual inductance
coupling coefficient, 583

natural frequencies, 31, 336, 387
natural response, 389
neutral-point clamped inverter, *see* NPC
n_i, 463
nomenclature, 7
nominal operation, 313

NPC
  single-phase, 192
  three phase, 238–240
Nyquist criterion, 338
Nyquist plot, 406
Nyquist stability, 406–407

packing factor, 558
parasitics
  modeling of, 596
Park transform, 234
particular response, 31
peak inverse voltage, *see* diode
periodic operation, 314
  deviations from, 315
periodic steady state, 15, 33
permeability
  definition of, 546
  normalized to free space μ_0, 546
permeance, 549
PFC, 197–201
  control of, 201
  storage capacitor sizing, 200
phase-controlled converter, 16, 57–71
  as battery charger, 66
  commutation failure, 67–71
  control characteristic, 59
    linearizing, 59
  in dc-link converter, 291
  inversion limits, 67
  inversion region, 62
  magnet discharge example, 62, 72
    with commutating reactance, 70
  margin angle, 69, 71
  power factor, 63
  quadrants of operation, 61
  reactance factor, 70
  regulation curves, 65
  single-phase, 58
  three-phase bridge, 215
  with commutating reactance, 64
  See also controlled rectifier
phasor, 233
PIV, *see* peak inverse voltage
polyphase sources, 207–213
  balanced, 207
  creating rotating fields, 208
  phasor diagram, 208
  phasor representation, 207
  three-phase, 209
    Δ-connected, 210
    balanced load, 211
    calculating power, 212
    cancellation of triple-*n* harmonics, 212
    line voltages, 210–211
    neutral-point connection, 210
    phase-shifting connections, 211
    Y-connected, 210
power
  in electrical networks, 27
power factor, 39–44
  angle, 40
  apparent power, 38
  compensation of, 41
  displacement factor, 41
  distortion factor, 41
  in polyphase rectifier, 217
  of ac controller, 39
  of distorted waveforms, 41
  of single-phase bridge, 46
  of sinusoidal waveforms, 40
  reactive power, 40
power factor correction, *see* PFC
proximity effect, 438, 560
proximity loss, 665
pulse-width modulation, *see* PWM, 179–187
punch-through, 478
push–pull converter, 148
push–pull inverter, 278
PWM, 76, 84, 172
  modulation index, 180, 185
  naturally sampled, 184
  sine–triangle intercept, 184
  See also dc/ac converter

Q, *see* quality factor
quality factor, 253, 352

reactive power, 40, 187, 289, 293, 624
rectifier, 14
  12-pulse, 213
  6-pulse, 213
  commutating inductance, 35
  current commutation, 35
    equivalent circuit, 36
  load regulation, 31
  PFC, 197–201
  polyphase, 213–221
    12-pulse, 215
    6-pulse, 217
    commutation in, 216–221
    commutation modes, 217–221
    full-wave bridge, 213
    power factor, 217
  single-phase, 27–56
    bridge circuit, 44, 47
    discontinuous conduction, 31
    half-wave, 29
    load regulation, 47
    with capacitive filter, 34
    with center-tapped source, 46
    with freewheeling diode, 32
    with inductive load, 30
    with resistive load, 29
  switched-mode, 197
  synchronous, 87
reluctance, 549
residual-flux density, 555
resonant converter, 17, 249–283
  analysis, 256–260
  control
    adaptive-power tracking, 280
    burst-mode, 283
    drain modulation, 280
    envelope tracking, 280
    out-phasing, 280–283
    switching frequency, 267
  dc/dc, 264–271
    design, 267–268
    rectifier analysis, 264–266
    state-space model, 370
  modeling, 266–268
  radio-frequency, 271–283
    class D, 272–274
    class E, 274–278
    dc-voltage control, 280
    multi-switch, 278–280
    output control, 280–283
    push–pull, 278–280
  series-resonant, 256
    design, 259
    element values, 263
    output control, 257–259
  soft switching, 260–264
    above-resonance operation, 261
    below-resonance operation, 261
    circuit model, 261
  switch stress, 249
resonant dc/dc converter
  state trajectories, 359
resonant-transition converter, 741–747
Reynolds number, 756, 762
ripple instability, 401
ripple ratio, 99
root locus, 337, 344, 382

safe operating area, *see* SOA
sampled data, 365
saturation, 556
saturation-flux density, 555
SCC, *see* switched-capacitor converter
SCL, *see* space-charge layer
SCR, 2, 4, 448, 494, 530–537
  angle of retard, 59
  asymmetrical SCR (ASCR), 449
  characteristics, 57
  circuit commutated turn-off time, 448, 536
  delay angle, 59
  di/dt limit, 534
  dv/dt limit, 534, 536
  firing angle, 59
  forced commutation, 695–697
    limitations of, 696
  forward breakover voltage, 448
  gate drive, 694, 704, 716–718
    isolation, 716
    parameters, 717
    pedestal and porch, 717
    rate effects, 718
    requirements, 717
  holding current, 534
  inverter-grade, 449
  latching current, 534
  margin angle, 69
  opto-SCR, 535
  rectifier-grade, 449
  regnerative turn-on, 532
  reverse recovery, 448
  trigger mechanisms, 448
  two-transistor model, 532
second-order systems, 250–254
  frequency-domain response, 252–254
  quality factor, 254–256
  time-domain response, 250–252
semiconductor
  acceptor impurities, 463
  bandgap, 462
  carrier concentration, 464
    at SCL edge, 473
    calculation of, 464
    excess notation, 474
    in short diode, 477
    in thermal equilibrium, 463
    intrinsic, 463
    thermal generation of, 464
  carrier recombination, 462
    rate of, 462
  charge carriers, 461
  charge neutrality, 464
  charge transport, 465
    by diffusion, 465
    by drift, 465
  conductivity of, 466
  current densities
    calculation of, 466
  diffusion constants, 466
  donor impurities, 463
  doping, 463
  extrinsic, 461, 463
  hole, 462
  intrinsic, 461
  majority carrier, 439
  majority-carrier device, 439
  materials
    parameters of, 466
  minority carrier, 439
  minority-carrier device, 439
  p–n junction
    abrupt SCL model, 468
    built-in potential, 470
    in thermal equilibrium, 468
    space charge layer, 468
    space-charge layer, 472
  physics of, 461
semiconducutor
  GaN, 440
  SiC, 440
SEPIC converter, 94–95, 154
series-resonant converter
  switching-frequency dynamics, 351
shaded-pole motor, 598
shield winding, 598–600
SiC, 497, 528
  MOSFET, 529
  Schottky diode, 529
silicon carbide, *see* SiC
silicon controlled rectifier, *see* SCR
silicone grease, *see* thermal grease
skin depth, 560, 645
skin effect, 438, 560
sliding mode, 424
snubber, 50, 697
  alternative placements, 735
  basic operation of, 697
  combined turn-on/off, 733–734
    transient voltage, 733
  dissipation, 736
    in combined circuit, 737
    in turn-off circuit, 736
    in turn-on circuit, 736
    net circuit increase, 737
  EMI reduction, 747
  turn-off, 725–729
    basic circuit, 727
    loss, 727
    RCD circuit, 729
  turn-on, 730–733
    basic circuit, 731
    dissipation, 730
    RLD circuit, 732–733
SOA, 486, 540, 724, 726, 764–767
soft switching, 697–699, 725, 739–747
  EMI reduction, 747
  implementation, 740
  limitations, 747
solid-state transformer, 290, 302
space vector, 229–234
space-charge layer, 468, 472
  calculation of width, 470, 472
  incremental capacitance of, 473
space vector modulation, 229, 234–237
SST, *see* solid-state transformer
start-up problem, 556
state variables, 355, 361
  continuity property, 359
  flux and charge as, 363
  state evolution, 356
  state plane, 360
  state property, 358–361
  state trajectories, 359, 370
state-space averaging, *see* control
state-space models, 323
  for electrical circuits, 361–363
  numerical solution of, 359
state-transition matrix, 372
Steinmetz equation, 679
  improved, 680
switch stress, 103, 108, 142, 249
switch-stress factor, 103
switch-stress parameter, 103
switched-capacitor converter, 116–131
  characteristics, 116
  charge multipliers, 128
  charge transfer
    calculation of, 126
  Cockcroft–Walton voltage multiplier, 131
  fast-switching limit, 119, 129
    parasitic resistance, 120
    Thévenin equivalent, 120
    voltage ratio, 120, 121
  Greinacher voltage multiplier, 131
  hybrid converter, 132
  ladder network, 124

charge calculations, 127
slow-switching limit, 118, 127
Thévenin equivalent, 119
switch implementation, 121
two-state, 118
switching function, 327, 348
generation of, 328
modulating signal, 329
switching loss, 438, 725, 726
switching surface, 423
synchronous rectifier, 87, 122

Tellegen's theorem, 128
THD, 38, 42
calculation of, 43, 183
thermal grease, 756
thermal impedance, 761
thermal interface, 758, 760
convective, 761
silicone pad, 760
thermal models, 700, 753
circuit analog, 754, 757
static, 754
thermal power, 754
thermal resistance, 754, 755
thermal resistivity
of sheet, 756
transient, 762
design example, 764
distributed, 763
dynamic behavior, 764
heat capacity, 762
multi-lump, 763
thermal impedance, 763
thermal resistivity, 756
of materials, 756
units of, 756
three-phase systems, 229–234
(α, β, γ) coordinates, 230–234
common-mode component, 230
Clarke transform, 231–234
d-q-0 coordinates, 234
Park transform, 234
space vectors, 229–234
threshold voltage, *see* MOSFET
thyristor, 2, 4, 57, 530–538
family members, 448
See also SCR, TRIAC, GTO
time invariance, 356
topological duality, 603
total harmonic distortion, *see* THD
t_q, *see* circuit commutated turn-off time
transfer function, 335, 402
hidden poles, 336
transference, 598, 600
transform domain, 386
transformer, 5, 455, 566
π-model, 586
cantilever model, 586
extended, 589–591, 594
parameter extraction, 590
shortcoming, 591
current, 572–574
burden, 572
decoupled model, 594
design example, 627–629
dot convention, 5, 567, 575
electric circuit models, 602, 603
energy storage, 572
extracting parameters, 587–588
flux resetting, 138
flyback, 154
foil windings, 570
ideal, 5, 566–567
inductance matrix, 582
example, 583
reciprocity of, 582
symmetry of, 582
inter-winding leakage, 587
L-model, 586
leakage inductance, 6, 569–570, 584–588
loss minimization, 631–634
magnetic circuit
electrical coupling of, 601
magnetizing current, 5
magnetizing inductance, 5, 567–569
models, 566–575, 582–607
multi-winding, 588–597
negative inductance in, 586
of two-port, 584–586
with dependent source, 594
mutual inductance, 583
parameter extraction, 571–572
potential, 572
power rating, 625–629
schematic representation, 6
shield winding, 570
T-model, 585
terminal equations, 571–572
three-winding, 574
winding build-up, 570
tri-state inverter, 168, 170, 175
TRIAC, 58, 537
commutating dv/dt, 451
gate current requirements, 450
quadrants of operation, 451
triode ac switch, *see* TRIAC
triple-n harmonics, 175

uninterruptible power supply, *see* UPS
up converter, *see* boost converter
up/down converter, *see* buck/boost converter
UPS, 22

voltage-source inverter, 167
V_T, *see* threshold voltage

waveform properties, 172–175
even/odd decomposition, 173
half-wave repeating, 173
half-wave symmetric, 173, 175
orthogonality, 174
symmetry, 173–175
See also Fourier series
Westinghouse, George, 27, 49
Wiedemann–Franz law, 756
winding loss, 558–561, 624, 655–675
calculation
examples, 657
using boundary conditions, 656–658
calculation basics, 655–656
Dowell's equation, 660
electrical representation, 671–675
in multilayer winding, 658–664
effect of skin depth, 663
interleaving, 668
mutual resistanace, 672
porosity factor, 664
proximity effects, 665
resistance factor, 661, 666
self-resistance, 672
wire size, 558
with non-sinusoidal current, 670–671
winding models, 664–669
window utilization factor, 558

Z_θ, *see* transient thermal impedance
ZCS, 697, 739–747
zero-current switching, *see* ZCS
zero-voltage switching, *see* ZVS
Zeta converter, 155
ZVS, 697, 739–747